The Engineering of
Foundations, Slopes and
Retaining Structures

The Engineering of Foundations, Slopes and Retaining Structures

Second Edition

Rodrigo Salgado

CRC Press
Taylor & Francis Group
Boca Raton London New York

CRC Press is an imprint of the
Taylor & Francis Group, an **informa** business

Second edition published 2022
by CRC Press
6000 Broken Sound Parkway NW, Suite 300, Boca Raton, FL 33487-2742

and by CRC Press
2 Park Square, Milton Park, Abingdon, Oxon, OX14 4RN

CRC Press is an imprint of Taylor & Francis Group, LLC

© 2022 Taylor & Francis Group, LLC

First edition published by McGraw-Hill 2008

Library of Congress Cataloging-in-Publication Data
Names: Salgado, Rodrigo, author.
Title: The engineering of foundations, slopes and retaining structures /
Rodrigo Salgado.
Description: Second edition. | Boca Raton, FL : CRC Press, 2022. | Includes
bibliographical references and index.
Identifiers: LCCN 2021033593 (print) | LCCN 2021033594 (ebook) |
ISBN 9781138197633 (hbk) | ISBN 9781138197640 (pbk) |
ISBN 9781315213361 (ebk)
Subjects: LCSH: Foundations.
Classification: LCC TA775 .S2374 2022 (print) | LCC TA775 (ebook) |
DDC 624.1/5—dc23
LC record available at https://lccn.loc.gov/2021033593
LC ebook record available at https://lccn.loc.gov/2021033594

ISBN: 978-1-138-19763-3 (hbk)
ISBN: 978-1-138-19764-0 (pbk)
ISBN: 978-1-3152-1336-1 (ebk)

DOI: 10.1201/b22079

Typeset in Sabon
by codeMantra

To my students, present and past, partners in the search for knowledge and understanding...

Contents

8 Shallow foundations in soils: types of shallow foundations and construction techniques

9 Shallow foundation settlement

15   Pile groups and piled rafts   791

16   Retaining structures   819

Preface to second edition

... with all thy getting[,] get understanding.

Proverbs 4:7, King James Bible

I keep six honest serving-men

(They taught me all I knew);

Their names are What and Why and When

And How and Where and Who.

Rudyard Kipling

Geotechnical engineering is a fascinating subject. It deals with the safe and economical transfer to the ground, through foundations, of loads from structures of various types (such as buildings, bridges, and industrial facilities), functions, and sizes. It also addresses the stability of slopes, and, in its absence, the development of landslides. And it deals with the reconfiguration of a site's topography, using retaining structures, to accommodate space needs (and in the creation of underground space) or facilitate construction (as of bridges and overpasses). These three broad classes of problems are the focus of this text.

Geotechnical engineers draw on their knowledge of mechanics, of how to measure or estimate various soil and rock properties, of how structures respond to foundation movements, and of basic economic concepts to determine the best, most economical foundations for structures and the best solutions to problems involving slopes or retaining structures. Because of the need for knowledge from so many different fields, and because of the natural complexity of soils and rocks, many years of study and practice are required to master geotechnical engineering.

Albert Einstein once suggested that things should be made simple, but no simpler. This certainly applies to geotechnical engineering and to a text on the subject. Due to the large scope of the subject, simplicity is desired so that all of the information, analyses, and techniques can be understood. The essence of knowledge, however, is to have it so that we can apply it. Oversimplification leads to spotty knowledge, which is hard to put to use in realistic practical problems. In this text, I attempt to present the subject of geotechnical engineering in a logical framework, in a natural sequence, and in as simple a presentation as possible. I emphasize conceptual understanding and avoid an oversimplistic treatment of soil. An example of this philosophy is how I treated one of the keys to the successful practice of geotechnical engineering: the estimation of soil parameters for use in calculations. In practice, these parameters are not given to us so that all we have to do is to plug them into an equation or two and arrive at an answer. Much of the work in geotechnical design resides in determining or estimating these parameters, and I have devoted significant attention to this.

It has been almost 15 years since the first edition of TEOF was first published. In this second edition, I made significant updates to the soil mechanics chapters and the shallow and deep foundation design chapters. I have made improvements and updates and added examples and problems throughout the text.

My guiding principles remained the same. They closely relate to those key interrogative pronouns in Kipling's wonderful childrens' poem. First and foremost, I have emphasized conceptual understanding, aiming for the "simple but no simpler" explanations and discussions of the material. I am convinced that oversimplification can lead to students not learning what they need . Given that they may not have as many opportunities to revisit what they learned while in school during their professional careers, I have strived to present the correct concepts and theory in this text. I also believe that today's engineering education is the engineering practice of the future, so building a strong conceptual base in the fundamental geotechnical engineering courses is important to the entire profession.

A second guiding principle has been to provide methods, equations, and theory always accompanied by complete derivations whenever possible. This offers the reader and the student the best opportunity to fully understand the material.

A third guiding principle has been to provide methods based on realistic models that have been validated. Geotechnical engineering is not an abstract discipline. It exists to allow engineers to design real structures. I have provided theories with parameters with a clear meaning and the related guidance on how to obtain the values of these parameters required to perform realistic analyses and effectively design geotechnical structures.

Finally, I wrote this second edition with the same focus as the first: the reader. I wrote the text so that it is clear and easy to read, even when the material becomes somewhat challenging, and so that it will remain useful beyond the classroom.

The book is organized in four parts. The first (chapters 1 and 2) deals with an introduction to the discipline of geotechnical engineering. The second (chapters 3-7) covers soil mechanics in a fairly complete and modern manner and site investigations. The third (chapters 8-15) covers foundations. And the fourth (chapters 16 and 17) covers slopes and retaining structures. Three types of problems are proposed at the end of most chapters: conceptual, quantitative, and design. Conceptual problems are intended to test one's understanding of the topics. Quantitative problems are intended as relatively straightforward application of analyses and equations. Design problems simulate the types of decisions and calculations done in real problems, emphasizing how to start from information typically available in practice to arrive at intermediate or final results that are needed in design. Many of the examples and problems have been designed in a way that requires the solution to start at the same point as problems found in practice start, with the information typically collected from borings, penetration tests, and so on, from which stiffness or strength parameters of the soil or rock must be estimated. More challenging problems are labeled with an asterisk (*). The list of references is separated into cited references, references that are not cited but whose reading may benefit those with a deeper interest in a certain topic, and ASTM standards and procedures that are pertinent to each chapter. Throughout the book, sections containing more challenging material are identified with an asterisk (*) so that both readers and instructors know they can skip them if more basic coverage of the subject is desired.

The book can be used as a text in both undergraduate and graduate instruction. I have used it in both Geotechnical Engineering I and II – undergraduate-level courses, the first covering soil mechanics and the second foundations, slopes and retaining structures – at Purdue University and in graduate foundations classes. The students have appreciated the clarity of the exposition. I hope that you too will find it helpful.

I have been fortunate to count on several colleagues for reviews of chapters of this second edition. Prof. Monica Prezzi of Purdue University has read and commented on Chapters

13–17. I am very thankful for her tremendous attention to detail and for her insightful suggestions and comments. Prof. Fei Han, of the University of New Hampshire, reviewed Chapters 13–15 and made numerous valuable suggestions and comments. Prof. Dimitrios Loukidis of the University of Cyprus proofread Chapters 10 and 11. Additionally, Dr. Vibhav Bisht of Itasca Consulting Group, Inc., proofread Chapters 7–11, Christopher Henderson proofread Chapters 3 and 13, and Ms. Jeehee Lim of Purdue University proofread Chapters 4–6 and 10. Major Rameez Raja, a current graduate student at Purdue University, and Mr. Abhishek Sakleshpur, of Purdue University, proofread the entire text. I am very thankful to all for their helpful comments, which resulted in an improved text, and for their generosity with their time. My gratefulness to those who contributed with comments on the first edition endures.

My work was made much easier by the dedicated assistance of Rameez Raja, who helped me navigate the administrative and logistical hurdles involved in getting a book ready for submission to the publisher. Fei Han, Rameez Raja, and Jeehee Lim also helped draft the figures that are new to this edition.

I thank Tony Moore for his guidance in the preparation of this second edition and Frazer Merritt and Vijay Shanker for helping turn the manuscript into this final product.

Finally, we, as professionals and people, are the product of our interaction with family, friends, and colleagues throughout our lives. So there is a little bit of very many people in this book, and, even if I did not name them, the gratefulness is there.

Rodrigo Salgado
West Lafayette, October 3, 2021

Author

Rodrigo Salgado is the Charles Pankow Professor in Civil Engineering at Purdue University, Editor-in-Chief of the ASCE *Journal of Geotechnical and Geoenvironmental Engineering*, and author of over 150 journal publications, 100 conference publications, and 40 reports. He has supervised 32 Ph.D. students and was the recipient of prestigious awards, including the Sloan Best Paper Award, the ICE Geotechnical Research Medal, the Prakash Research Award, the ASCE Huber Research Prize, the ASCE Arthur Casagrande Award, and the IACMAG Excellent Contributions Award.

Chapter 1

The world of foundation engineering

> Technology moves so rapidly that the challenges and problems of tomorrow are certain to be different than those of today, and it is probable that most of tomorrow's problems will come as surprises, since our ability to predict the future is very poor. I believe that the best way to adapt is to have a thorough mastery of first principles, as these never change and can always be applied to new problems.
>
> James K. Mitchell

A profession can be broadly understood as a type of work done using specialized knowledge or skills acquired by attending and getting a degree from a higher education institution and by lifelong learning. The "first principles" that Prof. Mitchell refers to, in an engineering context, are most often scientific principles – science being one of the cornerstones of engineering. Additionally, every profession has its own language and its own set of rules, which the effective professional knows well. In this chapter, we will examine the current practice in geotechnical and foundation engineering. We will start by defining what is meant by geotechnical and foundation engineering and then go on to discuss the education and licensing processes a geotechnical engineer must go through to practice the profession in its plenitude. We follow that with a brief discussion of the geotechnical and foundation engineering industries. The discussion is pursued in the context of the "global economy," which to a considerable extent has become a reality. In the paradigm of the global economy, barriers to trade, including engineering services, weaken considerably, forcing the engineer to deal with technical, cultural, legal, and language issues that could be safely ignored in the past. This chapter concludes with a discussion of units, which are needed in any engineering calculation.

1.1 THE GEOTECHNICAL ENGINEERING INDUSTRY

1.1.1 Geotechnical engineering, foundation engineering, and geotechnical and foundation engineering problems

Engineering is the creative use of experience, empirical methods, and engineering science to solve engineering problems. Engineering involves the search for solutions to problems that require significant technical and scientific knowledge and that change the environment in some way or make new things possible. For example, a civil engineer may be interested in connecting one bank of a river to the other, while a chemical engineer may be interested in the manufacturing of a certain chemical in industrial scale.

An engineering problem is subjected to a number of constraints arising mainly due to economic and environmental reasons. More formally, an engineering problem may be defined in a general way as the need or wish to change something from a current state A to a new

DOI: 10.1201/b22079-1

state B in the most economical fashion, under safe conditions, with minimum or no harm to the environment. Additional requirements arising out of legal and aesthetic issues also need to be fulfilled. Fortunately, there are also a number of tools available to the engineer that facilitate achieving an optimal design solution: engineering science, empirical rules, and experimental techniques.

Geotechnical engineering is the engineering of problems involving the ground. Geotechnical engineers work on the analysis, design, and construction of dams, soil and rock slopes, tunnels through soil or rock, and foundations for various types of structures. Foundation engineers are called upon to determine the best way of transferring to the ground the loads from structures (such as buildings, warehouses, and bridges), machines, highway signs, and a variety of other sources. Therefore, we may define foundation engineering as the body of knowledge that enables solution of problems involving the safe and economical transfer of structural loads to the ground. A foundation engineer would design the foundation elements themselves, but, in the United States, structural engineers typically do this, using the information or data provided to them by geotechnical engineers.

Foundations may take many different configurations, being most generally classified as shallow (covered in Chapters 8–11) or deep (covered in Chapters 12–15). The most common types of shallow foundations are concrete footings, typically built of reinforced concrete in shallow excavations with plan areas most often in the $1-10\,m^2$ range. Piles are the most common type of deep foundations. These are slender structural elements made of wood, steel, or concrete that, in onshore applications, have diameters ranging from $0.1\,m$ to in excess of one meter and lengths ranging from a few to many tens of meters.[1] Foundation engineers are also called upon to safely and economically design and build retaining structures of all kinds. Construction dewatering and excavations are also often part of foundation works. Reinforced concrete design of foundations and retaining structures is often done by a structural engineer; in such cases, the best results are obtained when the structural and foundation engineers work in close cooperation.

The term geotechnical engineering has a more general meaning than foundation engineering, as the subject encompasses a broader spectrum of soil or rock problems, including geoenvironmental problems such as landfill design, waste containment, and groundwater cleanup.

As discussed in the previous paragraph, the design and analysis of slopes and retaining structures are an integral part of foundation engineering, but are seen more generally as well. In urban areas, where space is restricted, retaining structures allow the construction of underground facilities, such as parking garages, underground shopping malls, and subway stations. Although more incidental to foundation works, slope instability failures must also be prevented. In foundation works, temporary excavations are sometimes planned in such a way as to leave temporary soil slopes to support the excavation walls. Another situation where slope stability becomes an issue is when the building or structure is located in a hilly or mountainous area, in which a well-designed foundation will serve no purpose if the structure is founded on a potential sliding soil or rock mass. More broadly, slopes occur as part of road embankments, soil or rock cuts for various projects, and in dam projects.

1.1.2 Geotechnical engineering as a profession

It is useful to identify what distinguishes a profession from other occupations. The following excerpt from a New York Court of Appeals judgment (*In re Estate of Julius Freeman,*

---

[1] In offshore applications, piles can have diameters of several meters and lengths in excess of 100 m.

355 N.Y. S.2d 336, 1974) provides an excellent summary of the distinctions[2]: "A profession is not a business. It is distinguished by the requirements of extensive formal training and learning, admission to practice by a qualifying licensure examination, a code of ethics imposing standards qualitatively and extensively beyond those that prevail or are tolerated in the marketplace, a system for discipline of its members for violation of the code of ethics, a duty to subordinate financial reward to social responsibility, and, notably, an obligation on its members, even in nonprofessional matters, to conduct themselves as members of a learned, disciplined, and honorable occupation."

1.1.3 Education and professional licensing

Geotechnical and foundation engineers are, with some exceptions, first and foremost civil engineers. In the United States, civil engineering programs normally last 4 years. In other countries, programs may last 5 or even 6 years and typically require a thesis or project report at the end of the program. Students interested in geotechnical engineering or foundation engineering, particularly if graduating from a 4-year program, would benefit tremendously from obtaining a master's degree in geotechnical engineering after completion of the undergraduate program.

In many countries, a graduate of a civil engineering program is automatically licensed to practice engineering, sign plans, and supervise the work of other engineers. However, this is not true in the United States, where engineering programs vary in size, scope, and duration, and a system of licensure has been in place for a long time to ensure the quality of engineering work. Licenses to practice engineering are granted by professional engineering state boards. The licensing is a four-step process. The first requires obtaining a civil engineering degree (preferably from an institution accredited by the Accreditation Board for Engineering and Technology [ABET]). The second step is the taking of a fundamental engineering exam focusing on engineering science. Anyone who passes the exam (formerly known as the EIT exam, now known as the Fundamentals of Engineering or FE exam) becomes an engineer intern (EI) or, equivalently, and engineer in training (EIT). The EI then completes a number of years of actual professional experience. This number varies from state to state and is longer for engineers whose degrees are not from an ABET-accredited institution. Years spent on graduate studies count toward the required number of years of experience up to a limit. After satisfying the years of experience requirement, the engineer is entitled to take the professional engineer exam. Those who pass this exam become professional engineers in the state in which they take the exam. Professional engineers are held legally accountable for the work they do or supervise, as well as for any plans or documents to which they affix their seal. Professional engineers should only practice in their areas of expertise.

1.1.4 Professional standard of care

In the practice of a profession, there is a standard of care that professionals must meet. This standard of care requires the professional to practice with the skill and knowledge commonly possessed by fellow professionals working in that discipline. The standard of care is a function of the specific type of activity and of the gravity of the risk involved. For example, the standard of care for work related to a large dam may be different from that required for the foundations of a single-story building. Obvious ways in which this appears are the level and detail of site investigation done, the level of sophistication of analyses done for design, and the level of checking of the design. In large projects, consulting boards, constituted by

---

[2] This appeared in a column by Arthur Schwartz in *Engineering Times*, Vol. 25, No. 6, June 2003.

experienced engineers, are charged with reviewing designs. In geotechnical engineering, the standard of care is also linked to the geologic and geotechnical settings of the area in which the community developed. A local code (we discuss codes in Section 1.2) sets forth some parameters for the standard of care. Violation of a statute would most likely be viewed as a failure to meet due care standards.

Another aspect of due care is that the professional must stay current with engineering and scientific advances in his or her area of practice. It is important to distinguish staying current with the technological and scientific progress in the discipline from doing what everyone else is doing. This is not necessarily a defense against negligence. As a court stated: "Evidence of custom in the trade may be admitted on the issue of the standard of care, but is not conclusive." *Coburn v. Lenox Holmes, Inc.*, 186 Conn. 370, 381, 441 A.2d 620, 626 (1982). In the same vein, the famous judge Learned Hand stated in a very famous case: "... a whole calling may be unduly lagged in the adoption of new and available devices."[3]

The advantages and cost savings associated with new solutions will keep driving innovation. When an innovative solution or technique is used, there must be sufficient analyses, testing, or experience showing that it is sound and effective and that due care was exercised. A scientific basis for a method used in design will likely be compelling as a defense in legal action for negligence. Indeed, new technology may become so "imperative that even their universal disregard will not excuse their omission."[4]

Negligence is the failure to meet the applicable standard of care. In litigation, if the professional has not met the standard of care, the professional will likely be found liable for any damages resulting from his or her negligence. There is an exception to the negligence doctrine. Some applications are considered by courts to be so dangerous that courts apply a different doctrine to issue opinions: strict liability. In that case, even the absence of proof of negligence will not protect the engineer from liability. In geotechnical engineering, two areas in which strict liability applies, at least under some circumstances, are dam engineering and blasting. Strict liability may also apply when no significant uncertainties that would justify an error on the part of an engineer are present, such as furnishing "plans and specifications for a contractor to follow in a construction job," for which the court in *Broyles v. Brown Engineering Co.,*[5] found that the engineer "impliedly warrants their sufficiency for the purpose in view." Lastly, while engineers are virtually always held to a negligence standard when doing engineering work, contractors – such as mass producers of homes – are more often held liable based on the strict liability doctrine. When strict liability applies, so long as it can be shown that the engineering works actually caused the damage that a plaintiff is alleging and the damage is itself proved, the defendant will be found liable.

1.1.5 Professional ethics

As engineers progress in their careers, it is important that they conduct themselves in a proper, ethical way. Both the American Society of Civil Engineers (ASCE) and the National Society of Professional Engineers (NSPE) have ethics codes that should be read by young engineers and periodically reread by more experienced engineers.

---

[3] *The T.J. Hooper,* 60 F.2d 737 (2d Cir. 1932).
[4] *T.J. Hooper.* SUPRA.
[5] *Broyles v. Brown Engineering Co.,* 151 So. 2d 767 (Ala. 1963)

1.1.6 The players: owner, architect, developer, general contractor, consultant, specialty contractor, and regulatory agencies

Construction, whether of buildings, bridges, roadway, or any infrastructure project, usually involves a number of different entities. The relationships between these different entities vary significantly internationally and even within the same country, but it is possible to describe them in a general way (Figure 1.1). There is usually an owner or developer supplying the funds that make construction possible. The owner/developer often arranges for financing of the project with a financial institution, by issuing bonds or by direct financing. The owner/developer may hire an architect. The architect prepares the bidding documents (plans and specifications). The successful bidder becomes the general contractor for the project and is responsible for satisfactory completion of the project. If an architect is indeed hired by the owner, the general contractor will deal mostly with the architect.

The general contractor hires specialty contractors to complete specific parts of the project. For example, a specialized contractor will do all of the electric wiring of a building. Likewise, a foundations contractor will build or install the foundations if special equipment or skills are required for the foundations work or the general contractor does not have the expertise to do it. The owner/developer, the general contractor, or the specialty contractor may at any time hire consultants to carry out design, check design done by others, inspect installation/construction for conformance with design specifications, or serve as expert witnesses in litigation. The variations in contracting arrangements is however significant, and foundation/geotechnical engineers may end up as subcontractors to the architect, to the structural engineer, or to the general contractor. It is also possible that consultants and contractors may join to form a design–build team, which would carry out the project from the concept stage to conclusion. The main advantage of this approach is the undivided responsibility, given that all of the work is essentially done by a single entity.

Finally, regulatory agencies, such as environmental (e.g., the Environmental Protection Agency, EPA), work safety (the Occupational Safety and Health Administration, OSHA), or local (city or county building officials) agencies, will play a role at one or more stages of the process. It is usually necessary to get authorizations or permits to start a construction project. Inspections during and after construction to ensure compliance with codes and a

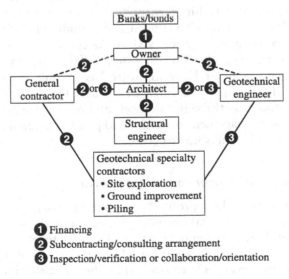

Figure 1.1 The players: relationships of those involved in a construction project.

minimum quality of work are also common. In the case of buildings, for example, a certificate of occupancy issued by a local agency is required at the end of the project.

The geotechnical or foundation engineer usually works for consulting companies, specialty contractors, regulatory agencies, or owners. Design, analysis, and inspection of foundation works are the usual activities in consulting companies. Design and management of field work are activities the geotechnical/foundation engineer would carry out when working for a specialty contractor.

1.1.7 Business and financial aspects of the geotechnical consulting and specialty contractor industries

1.1.7.1 Legal structures of firms

Engineering businesses, whether consulting companies or specialty contractors, are rarely sole proprietorships, which require no special action from the owner for their creation and are taxed through the individual's income tax (on Schedule C). The legal structures most commonly adopted by engineering businesses are the corporation, the partnership, the S corporation, and the limited liability company (LLC).

Corporations are legal entities separate from the individuals or other corporations who own it. In the Unites States, incorporation is done in one state, but if the corporation does business in any other state, it must register with that state as well. The equity of corporations is divided in a number of shares. These shares may be privately held by a relatively small number of individuals (these corporations are called private corporations) or may be widely held and traded in exchanges (in which case they are referred to as public corporations). The liability of corporations is limited to the equity or total net worth of the company. This is a desirable characteristic of a legal structure, as it means that no individual can be personally sued for holding equity in that corporation or caused to lose more than his or her equity in the company. An undesirable characteristic of corporations as separate legal entities is double taxation: taxes are paid both on the net income of corporations and on any dividends distributed to equity holders.

Partnerships avoid the double taxation associated with the corporation structure, but certain partnership forms have a serious drawback: partners are personally liable beyond any equity they hold in the partnership. S corporations were created mainly to offer small businesses protection from both double taxation and unlimited liability. In S corporations, the net income of the corporation is not taxed at the corporation level, passing through directly to the income tax of individuals holding shares of the corporation. Recently, a more advantageous, increasingly popular legal structure has become available in most states, namely the limited liability company or LLC. LLCs share the main features of S corporations (limited liability and pass-through taxation) and have some additional advantages, including flexibility in the management structure and profit allocation among members and the absence of a share-based ownership structure.

1.1.7.2 Metrics of the geotechnical consulting industry

Businesses exist primarily to make a profit. An engineering company is no exception. It is useful for foundation and geotechnical engineers to understand how revenues are obtained, classified, and recorded. It is also important to realize that these revenues must be sufficient to offset both direct and indirect costs for a business to exist as a going concern. This knowledge is helpful to engineers, as it allows them to better contribute to their companies' ultimate economic goal: maximization of profits.

Income Statement		
Gross revenues	$99,070	
Direct cost	$19,971	
Net revenues	$79,099	100%
Direct labor	$61,697	78%
Gross project income	$17,402	22%
Sales, general, and administrative expenses (SGA)	$7,910	10%
Income from operations	$9,492	12%
Interest income	$791	1%
Income before taxes	$10,283	13%
Income tax	$3,955	5%
Net income	$6,328	8%
All amounts in thousands of dollars		

Figure 1.2 Income statement of a fictitious geotechnical engineering consulting company.

Much of the information that is required to ascertain the financial performance of a company is available in an accounting report called the income statement. Figure 1.2 shows the income statement of a geotechnical engineering consulting company. The first item in the income statement is gross revenues, which comprise of engineering fees and other project revenues. Engineering fees represent the amount charged to clients for the services rendered by the company. Fees may be assessed on the basis of hours, weeks, or months worked or on a lump-sum basis. Additional revenues include those from laboratory testing done at the company or subcontracted to an external laboratory, equipment rental, and other works subcontracted to a third party. Work done by third parties is usually marked up, as the consulting company needs to identify the subcontractor and needs to manage and supervise the work. All subcontracting costs and other nonlabor direct costs (costs attributable to a project) are subtracted from gross revenues to obtain net revenues.

Direct labor is the amount of the company's total expenditures on labor that is charged to projects. The portion of the labor expenditures related to the general management of the company (administration, marketing, training, and other such activities) is not charged to projects and is referred to as indirect labor. An important metric in the consulting industry is chargeability, also known as direct labor utilization rate, defined as the ratio of direct labor to total labor. The higher this number is, the higher is the percentage of dollars spent on labor in direct support of projects, the lower is the overhead costs, and the better utilized are the company's personnel. Employees in most companies get paid whether they are working on a specific project or not. It is to the company's advantage to have them working on projects as often as possible so that their time can be directly charged to the project (that is, to the client) and not to the company.

The labor multiplier is the ratio of engineering fees to direct labor. The net multiplier, also known as effective multiplier, is the ratio of net revenues to direct labor. As indicated earlier, net revenues are engineering fees less project-related expenses. Net revenues are what is left after subtracting out costs of consultants and other project-specific expenses incurred to complete projects. The higher the net multiplier, the easier it is for the company to support its administrative structure and to have enough profits left to compensate its shareholders.

There is another accounting report, also produced every accounting period (typically, every quarter), called the balance sheet. The balance sheet lists all assets and liabilities of the firm. Assets are all that the firm owns that has value. Liabilities are in essence the firm's debt. Both assets and liabilities are classified as current (or short term) or long term. Current assets are assets that we expect will naturally convert to cash within 1 year. Current liabilities are debts that must be paid within 1 year. Subtraction of liabilities from assets gives us the firm's net worth, equity, or book value.

For the firm's management, there are two key ratios that can be obtained from the balance sheet: the financial leverage (debt-to-equity ratio) and the current ratio. A firm with a large ratio of debt to equity is said to be highly leveraged and may be vulnerable to sudden changes in business or economic conditions. Typically, a ratio of no more than two is believed to be acceptable. The current ratio, the ratio of current assets to current liabilities, indicates the firm's solvency or capacity to honor its debts in the short term. It is a measure of liquidity. Again, a ratio of at least two is deemed adequate.

1.1.7.3 Metrics of the specialty contractor industry

Specialty contractors generally enter into two types of contracts. Most commonly, they are paid for the amount of work actually done. As an illustration, a drilled shaft contractor is typically paid by the volume of soil excavated,[6] with adjustment for soil conditions (drilling in sand with a high water table will be priced differently from drilling in clay, for example). If micropiles are installed, pricing is typically by the length of micropiles installed, with adjustments for the diameter of the micropile and for the type of soil or rock the micropiles are installed in.

In the second type of contract, a contractor is paid a lump sum to complete the work. In this type of arrangement, it is in the contractor's interest to use the most economical solution, in terms of volume of excavation; volumes of concrete, grout, or cement; and weight of reinforcing steel as possible. This arrangement is most common in countries where contractors have significant in-house design capabilities, but it is also possible that the contractor will hire or partner with a consultant to help with the design. Since a crew working on a particular job is paid by the hour worked, it is obvious that it is in the contractor's interest to get the job finished as quickly as possible, without sacrificing quality.

An arrangement that avoids this last shortcoming is payment by unit time. This creates an indirect incentive for the specialty contractor to do the best possible job by eliminating the incentive to finish the work in a hurry.

1.1.7.4 Trends in the geotechnical and foundation engineering industry

Since the late 1980s, the market for environmental and geoenvironmental work (such as cleaning up contaminated soil sites and designing and building waste landfills) strengthened considerably. This type of work became a bigger source of firms' revenues when compared with foundation and traditional geotechnical engineering. As a result, geotechnical companies started focusing more on geoenvironmental work, and it became difficult for mid- to large-sized companies focusing exclusively on traditional work to remain independent and be able to compete effectively (although there are many small companies with a focus on

---

[6] Pricing is typically by depth excavated, varying with the diameter of the shaft, the general intent obviously being to have price track approximately the cost of excavation.

traditional geotechnical work). A series of mergers and acquisitions ensued, including take-over of some specialty contractors and consulting companies by large conglomerates dealing with waste management and related activities and even by large, diversified conglomerates and foreign corporations. Several of these companies are public corporations.

The trends in the industry remain the same in 2021: economies of scale and the marriage of more traditional work with work that is more in demand in related fields seem to make sense for the more efficient financial management of companies. This means the options for recent graduates interested in traditional geotechnical and foundation engineering are mostly either to work in a small, more focused company or to work for large corporations with interests in several different fields. Naturally, these two types of companies are man-aged in different ways and tend to work in projects of different magnitudes and scope.

After many decades of research on analytical and numerical models for modeling geo-technical problems, user-friendly software that take advantage of research results are now becoming more available. Increasing use of computer analysis in geotechnical and founda-tion design, with corresponding decline in reliance on rough rules of thumb and approxi-mate methods, is likely to continue.

1.2 FOUNDATION ENGINEERING TOOLS

1.2.1 Soil and rock mechanics: the underlying sciences

Engineers make use of a number of tools available to them. Empirical rules continue to play a role, but engineering science has become increasingly important in foundation and geotechni-cal engineering. Satisfactory analyses that can be used in design are available today in the fields of soil and rock mechanics. The tools necessary to characterize the subsoil or rock masses and estimate the quantities needed to analyze a variety of problems are available as well.

Soil mechanics is simply the application of the principles of continuum mechanics to soils. Continuum mechanics is applicable to soils in most cases because the scales of most problems are much larger than the sizes of soil particles. For example, there is a difference of at least three orders of magnitude between the size of a footing and the size of a sand particle. This remains true in the treatment of water flow through soil (viewed in this case as a porous medium), which relies on hydraulics or fluid mechanics principles. Because of the overwhelming importance of fractures in rocks, continuum mechanics is not as applicable as for soils, but remains an important part of the body of knowledge in rock mechanics. Both soil and rock mechanics are used in foundation engineering with two purposes: to compute the stability of soil structures and to estimate displacements and settlements of foundations and other soil and rock structures.

1.2.2 Codes and standards

Codes of practice and standards are frequently used tools by geotechnical and foundation engineers. In the context of civil engineering, a *code* of practice is a document that recom-mends practices and procedures for the design, manufacture, installation, maintenance, or utilization of structures or facilities. In the context of foundation engineering, a code of practice is a document that has the purpose of ensuring a minimum level of quality in the design, construction, or maintenance of the foundations of buildings and other structures, of retaining structures, and of soil and rock slopes. Since codes may have legal bearing, the engineer should as a rule adhere to them, for if the applicable code is not followed and the engineered structure does not perform as required, there is liability risk. However, it is not uncommon for codes to be outdated in their geotechnical requirements. In such cases, departures from the code may be necessary and do indeed happen.

Most local codes in the United States refer to the International Building Code (IBC), which was developed by the International Code Council (ICC), but often have amendments with respect to geotechnical design and construction. Another example of a code that imposes certain constraints on foundation design is American Association of State Highway and Transportation Officials (AASHTO) *LRFD Bridge Design Specifications,* which impacts the design of bridge foundations.

A *standard* is a document, established by consensus within an engineering organization, that sets rules and guidelines for certain activities, with the purpose of guaranteeing that they will be performed approximately in the same way regardless of the person or persons performing them. In the United States, the American Society for Testing and Materials (ASTM) is usually responsible for publishing standards for geotechnical tests, for test materials (for example, a "standard" sand), and for some products used in the practice of geotechnical engineering. Standards play an important role, as they allow design methodologies and analyses to be developed with basis on certain quantities that are meaningful only because engineers know that they are determined in a reproducible way.

1.2.3 The role of experience and empiricism

The role of experience in geotechnical engineering should not be understated. Fortunately, the experience of others is more easily recorded today in the scientific framework offered by soil mechanics and rock mechanics. Experience, therefore, does not necessarily mean personal experience and is not necessarily related to the age of the engineer; it is the collective experience of the profession, to which a young engineer has access through formal education, books, journal and conference papers, technical reports, and, perhaps most importantly, more experienced colleagues. Engineering experience is often organized by means of correlations and other means without a theoretical or scientific basis. These constitute an empirical knowledge basis that engineers have relied on for decades and even centuries. Gradually, scientific approaches are displacing purely empirical rules, but they continue to play a role in geotechnical and foundation engineering.

1.2.4 The role of publications: where to go for help

Publications play an important role in a professional field. In foundation and geotechnical engineering, scientific journals are the first source to which students and engineers may go for information and guidance when facing a problem containing questions that are not part of the engineer's routine. Journals publish a paper only after it has undergone peer review. These reviews are an important filter that minimizes (but does not eliminate) the chance that papers containing incorrect or inadequate material will appear in print. Some of the major journals that publish papers exclusively in English that may be of interest to foundation and geotechnical engineers include the *ASCE Journal of Geotechnical and Geoenvironmental Engineering, Geotechnique, Computers and Geotechnics, the Canadian Geotechnical Journal, Acta Geotechnica, the International Journal of Geomechanics, Soils and Foundations, Geomechanics and Geoengineering,* and the *Geotechnical Testing Journal.*

1.2.5 The role of conferences and short courses

As of 2001, four American states had passed laws requiring every licensed professional engineer (PE) to obtain a number of continuing education courses to renew their licenses.

In 2021, most, if not all states, require continuing education hours for license renewal. Professional development hours (PDHs) can usually be obtained by attending short courses and conferences or writing technical articles. This has spurred an increasing number of short course offerings from universities, firms, and professional organizations. Attendance of courses or conferences offers an opportunity to keep abreast of technical developments and to make invaluable contacts with professional colleagues.

1.2.6 The role of computers

Computers play an increasingly important role in geotechnical engineering. The research done on analytical and numerical methods in geotechnical engineering is finally bearing fruit, and more programs that can be useful to solve foundation engineering problems are appearing in the market.

1.3 SYSTEMS OF UNITS

Grams and pounds killed orbiter in NASA's Martian metric mix up. Confusion over metric and English measurements likely lost the $125 million Mars Climate Orbiter as it began orbiting Mars last week, NASA said. Two teams – in California and Colorado – each used different measurement systems. The probe probably burned in the Martian atmosphere.

Investors Business Daily (October 1, 1999)

1.3.1 Units

Engineers are destined to work with units: they must measure and predict sizes, lengths, stresses, deformations, and numerous other quantities. Quantities may be measured in different units, which developed in different parts of the world in different cultural contexts. While SI units are used throughout the world, a system much like the original English system of units continues to be used in the United States. Even if SI units are not fully adopted in the United States in the coming years, familiarity with the SI unit system is imperative for the US-based engineer to operate in these days of the global economy, where American companies are often involved in work in other countries. Even within the United States, confusion with units can be rather costly, as evidenced by the Mars orbiter incident related in the news clip in the box.

A system of units is the set of units necessary to express all quantities of a field of science. A unit system is built upon a set of fundamental or basic units, which are mutually independent units from which all other units in a system can be obtained. For example, units of mass, length, and time are usually taken as the fundamental units in mechanics. In this case, force, acceleration, and velocity units, for example, can all be derived from them. If problems in thermodynamics are to be solved, then units for temperature must be introduced, and electricity problems require the introduction of electric current units.

Newton's second law of motion, expressed as

$$F = \frac{ma}{g_c} \tag{1.1}$$

turns out to be very useful in understanding the different systems of units that exist. In Equation (1.1), F is force, m is mass, a is acceleration, and g_c is a constant used for unit conversion that depends on the system of units being used. Don't confuse g_c with the acceleration of gravity g.

There are three types of systems of units:

1. absolute or length–mass–time (LMT): mass is a fundamental quantity, and force is derived from mass through Newton's second law (Table 1.1);
2. gravitational or length–force–time (LFT): force is a fundamental quantity, and mass is derived from force through Newton's second law (Table 1.2); and
3. technical or engineering system: both mass and force are fundamental quantities (Table 1.3).

In the United States, many practicing engineers continue to use the so-called US Customary Units, largely based on the English units no longer in use in the UK and British Commonwealth. US units are based on the foot (ft) for length and pound for force. The greatest source of confusion is the use of the pound, a unit usually associated with mass, as a force unit. The only way that pound can be used as a unit of force and mass at the same time is to clearly distinguish a pound of mass (lb) and a pound of force (lbf) and to use $g_c = 32.173$ (ft/s²)(lb/lbf) in Newton's second law. This is the approach followed in the engineering systems. Alternatively, a separate unit of mass, the slug, must be defined based on the force unit (the pound). This is what is done in the US units version of the gravitational system.

It is interesting to note that Tables 1.1–1.3 contain examples of the absolute, gravitational, and engineering systems for both metric and US-based systems of units. The SI[7] is an absolute system in which the basic units are the meter, the kilogram, and the second. It is incorrect to refer to the SI as the metric system, as there are metric versions of the engineering and gravitational systems of units as well. The CGS system (with centimeter, gram, and seconds as its basic units) is no longer used anywhere. Table 1.4 gives the dimensions of the quantities used in mechanics.

Table 1.1 Absolute or LMT system

Quantity	US units	SI	CGS
L (length)	ft	m	cm
M (mass)	lb	kg	g
T (time)	s	s	s
F (force)	poundal	Newton (N)	dyne
g_c	1 (ft/s²) (lb/poundal)	1 (m/s²) (kg/N)	1 (cm/s²) (g/dyne)

Table 1.2 Gravitational or LFT system

Quantity	US units	Metric units
L (length)	ft	m
F (force)	lb	kg
T (time)	s	s
M (mass)	slug	MTU (mass technical unit)
g_c	1 (ft/s²) (slug/lbf)	1 (m/s²) (MTU/kgf)

---

[7] Short for Système International d'Unités, or International System of Units.

Table 1.3 Engineering system

Quantity	English system	Metric
L (length)	ft	m
M (mass)	lb	kg
F (force)	lbf	kgf
T (time)	s	s
g_c	32.174 (ft/s^2) (lb/lbf)	9.806 (m/s^2) (kg/kgf)

Table 1.4 Dimensions of the quantities of mechanics

Quantity	LMT	LFT
Length	L	L
Mass	M	$F L^{-1} T^2$
Time	T	T
Temperature	Q	Q
Force	$M L T^{-2}$	F
Mass density	$M L^{-3}$	$F L^{-4} T^2$
Unit weight	$M L^{2} T^{2}$	$F L^{3}$
Stress	$M L^{-1} T^{-2}$	$F L^{-2}$
Velocity	$L T^{-1}$	$L T^{-1}$
Acceleration	$L T^{-2}$	$L T^{-2}$
Volumetric flow rate	$L^3 T^{-1}$	$L^3 T^{-1}$
Angle	Dimensionless	Dimensionless
Angular velocity	T^{-1}	T^{-1}
Angular acceleration	T^{-2}	T^{-2}
Work, energy	$M L^2 T^{-2}$	$F L$
Power	$M L^2 T^{-3}$	$F L T^{-1}$
Moment of force	$M L^2 T^{-2}$	$F L$
Dynamic viscosity	$M L^{-1} T^{-1}$	$F L^{-2} T$
Kinematic viscosity	$L^2 T^{-1}$	$L^2 T^{-1}$
Surface tension	$M T^{-2}$	$F L^{-1}$

Force is often expressed in foundation engineering practice in terms of one thousand pounds or kilo-pounds (kips). There is still a residual use of US tons (2000 pounds) and, in countries using the SI system, of metric tons (the force associated with 1000 kg). An imperial ton (used, at least historically, in some British Commonwealth countries) is 2240 lb, which is practically the same as a metric ton. The abbreviations psf (pounds per square foot), tsf (tons per square foot), and pcf (pounds per cubic foot) are also quite common for stress and unit weight.

Appendix A contains a detailed list of quantities of interest in geotechnical and foundation engineering and the corresponding conversions between different systems of units.

1.3.2 Measurements and calculations

In the measurement and calculation of quantities, the concept of significant figures is important. Significant figures for a given quantity are all the zero and nonzero figures measured using an analog device and the first approximated digit, or, alternatively, all of the figures obtained using a digital instrument. In a foundation load test, for example, if an analog read-out unit is used, and the minimum subdivision is 1 kN, then loads can be given with a maximum accuracy of one-tenth of a kN, where the decimal is an approximate figure. A measured load of 10.3 kN has three, and 112.1 kN has four significant figures. Some additional rules may be helpful in identifying the number of significant figures of a measured or calculated quantity. These are as follows:

1. All nonzero numbers are significant (e.g., 2765 has four significant figures).
2. Zeros before a decimal are not significant (e.g., 0.55 has two significant figures).
3. Zeros between significant digits are significant (403 has three significant figures).
4. Zeros used to indicate precision are significant (e.g., 14.00 has four and 0.880 has three significant figures).
5. Placeholder zeros are not significant (e.g., 0.005 has one significant figure; 1500 has two significant figures, unless measurements were made to the degree of precision of the first or second zero, or the number 1500 resulted from mathematical operations on numbers with three or four significant figures).
6. All digits in the mantissa of a number expressed in scientific notation are significant (e.g., 2.60×10^{-4} has three significant figures).
7. Numbers that are exact or that are not measured (such as many conversion factors) have an infinite number of significant figures. For example, the conversion factor 100 cm per 1 m has an infinite number of significant figures.
8. The logarithm of a number has the same number of significant figures as the number (e.g., $-\log 0.0120 = 1.91$).
9. In multiplication, division, powers, and roots of numbers, the number of significant figures of the result is equal to the number of significant figures of the number with the least number of significant figures (e.g., $0.0267 \times 3.1 = 0.083$).
10. In addition and subtraction, the precision of the result (that is, the number of decimal places) must be the same as the least precise number used in the calculations (e.g., $0.664 - 0.65 = 0.01$, $27.6 + 31 = 58$, and $3.5 + 8.3 = 11.8$).
11. In averaging, an average must be expressed with the same precision (that is, with the same number of decimal places) as that of the values being averaged (e.g., $(36.2 + 36.7 + 36.4)/3 = 36.4$).

In the text, we at times present calculation results to more significant figures than would or should be used in practice if it makes it easier for readers to reproduce example calculations and feel comfortable about a match of what is typed in a calculator and what the calculator produces.

Example 1.1

In foundation design using US customary units, the capacity of foundation elements is often expressed in tons. One ton is a force equivalent to 2000 lb. If a foundation element has a capacity of 98.0 tons, what is its capacity in kN (1000 N)?

Solution:

Referring to Appendix A, one pound is equivalent to 4.45 N. So:

$$98.0 \, \text{tons} \approx (98.0 \, \text{tons})(2000 \, \text{lb} / \text{tons})(4.45 \, \text{N} / \text{lb})(10^{-3} \, \text{kN} / \text{N}) = 872 \, \text{kN}$$

The product of the numbers above is actually 872.2, but the number of significant figures in this case is only three. You know this by examining all the numbers involved: the force has three significant figures (the 9, 8, and 0 in 98.0); the conversion factor, 4.45, has three significant figures; and the conversion factor, 2000, has four. The least number is three, and so the result only has three significant figures. Note that Rule 7 does not apply, as the conversion factors in this case are not exact.

1.4 DIMENSIONLESS EQUATIONS AND DIMENSIONAL ANALYSIS

It is convenient to write engineering equations in dimensionless form. One of the main advantages of doing so is that equations look exactly the same in any system of units. In a dimensionless equation, there is one or more variables in the equation, called reference variables, that assume different values depending on the units in use. Note that reference variables are not measured quantities and thus fall under Rule 7 for significant figures, having an infinite number of significant figures.

Example 1.2

The shear modulus G_0 of sand at small strains at a point in a soil deposit is usually expressed as

$$\frac{G_0}{p_A} = C_g \frac{(e_g - e)^2}{1 + e} \left(\frac{\sigma'_m}{p_A} \right)^{n_g}$$

where C_g, e_g, and n_g = dimensionless constants, σ'_m = mean effective stress, e = void ratio at the point where G_0 is being calculated, and p_A = reference stress = 0.1 MPa = 100 kPa ≈ 1 tsf = 2000 psf. Calculate the shear modulus, in MPa, of a sand with a void ratio of 0.900 subjected to a stress σ'_m of 100 kPa if $C_g = 650$, $e_g = 2.90$, and $n_g = 0.50$.

Solution:

Since G_0 is desired in MPa, the value of p_A dividing it should be 0.1 MPa so that G_0/p_A has no dimensions. Notice that a different, independent value of p_A (100 kPa) can be used to normalize σ'_m, since σ'_m is given in kPa. Entering the equation with the given values of the other quantities, a value of $G_0 = 136.8$ MPa results. If the void ratio is known to three significant figures (the nine and the two 0's after the decimal point), as is usually the case, then this result should be rounded up to 137 MPa. If the void ratio is 0.90, then $G_0 = 140$ MPa. Note that p_A has an infinite number of significant figures.

Dimensionless equations are used throughout the text. This avoids the awkward alternative of using equations specifically derived for a certain set of units. When unit-dependent equations are used and results are desired in a different system of units, the quantities in the equation need first to be converted to the desired units, which is often a source of confusion and waste of time.

All physical equations express relationships between certain quantities that should be fully independent of the way we decide to measure them. In order for this to be true, the expression relating the quantities must be dimensionally compatible or homogeneous: the dimensions on both sides of an equation must be the same, in terms of the fundamental dimensions. Dimensional analysis can be helpful in (1) assessing whether a physical law can in fact exist, (2) determining the values of constants in expressions, and (3) predicting the form of physical laws.

Dimensional analysis is not as useful in soil and rock mechanics as it is in fields such as fluid dynamics. The difficulties reside in the scaling of soil and in the dependence of soil behavior on effective confining stress, which we discuss in detail in Chapters 4–6. Soil has an intrinsic scale, given by the size of its particles. A geotechnical problem has a characteristic scale. For example, in the calculation of the bearing capacity of a foundation element, the size of the element would define the characteristic scale of the problem. In laboratory experiments, if an experiment is performed inside a container, the size of the container must also be considered.

Dimensional analysis is sometimes used in a special soil modeling test called the centrifuge test. The centrifuge allows simulation of field-scale prototype problems using a small-scale model. In a centrifuge test, a relatively large box of soil is rotated at a fixed radius and at a pre-selected angular velocity. This applies a centripetal acceleration to the soil within the box that is many times the acceleration of gravity. While the stress due to soil self-weight at the surface of the soil within the box remains the same (zero) as it is at 1 g, the stress at any other depth within the soil is much higher than the stress at the same depth at 1 g. Although the test is not perfect (one of the difficulties is scaling down the size of soil particles), it does offer some insight to problems that are otherwise difficult to analyze. Geotechnical centrifuges are available in a number of educational and research institutions in the United States and the world. Although in the past they were used mostly in research projects, they are now equally as often used in practice (typically in large-budget projects dealing with more complex problems).

1.5 CHAPTER SUMMARY

Foundation and geotechnical engineers make use of **engineering science, empirical rules,** and **experience** to analyze and design foundations, slopes, and retaining structures. Geotechnical engineers develop their experience, knowledge, and engineering judgment from formal training at universities, consultation of publications, attendance of short courses and conferences, and, most importantly, from practicing and interacting with colleagues. After a number of years of education, practice under supervision of professional engineers, and the passing of a specific exam, an engineer in the United States becomes a PE, which in essence allows full practice of the profession in one of the states.

A foundation/geotechnical engineer usually works for a consulting company, a specialty contractor, or a government organization. Consultants are involved in the planning and design of foundations and other geotechnical structures, and inspection of their installation or construction. Contractors and specialty contractors do the actual installation/construction. Engineers working for specialty contractors do some design in-house, but would spend much of their time choosing and adapting equipment and methods of construction to each project, and making sure construction proceeds smoothly, quickly, and safely. Engineers in government organizations are often involved in managing or supervising/checking design and construction or doing research.

A sound background in soil and rock mechanics is essential for the successful practice of geotechnical and foundation engineering, as is the successful use and manipulation of units. In a global economy, familiarity with other countries' practices and, most importantly, with **SI units** is also essential. Engineering calculations always involve units. The **US Customary Units** are still used to considerable extent in the United States, while SI units are used everywhere else. US-based engineers must be capable of performing calculations with both systems of units to be effective. The use of dimensionless equations, in which quantities appear normalized with respect to a reference value of that same quantity, offers great advantages to engineers, which need not get confused or waste time converting between units.

1.6 WEBSITES OF INTEREST

1.6.1 Codes

www.intlcode.org
  International Code Council
http://www.aci-int.org/
  American Concrete Institute
www.aisc.org
  American Society of Steel Construction
http://www.cenorm.be/sectors/construction/eurocode.htm
  Eurocodes

1.6.2 Standards

www.astm.org
  American Society for Testing and Materials
www.iso.org
  International Standards Organisation
https://www.bsigroup.com/en-US/Standards/
  British Standards Institute

1.6.3 Journals

https://ascelibrary.org/journal/jggefk
  Journal of Geotechnical and Geoenvironmental Engineering
https://cdnsciencepub.com/loi/cgj
  Canadian Geotechnical Journal
https://www.icevirtuallibrary.com/toc/jgeot/current
  Geotechnique
https://ascelibrary.org/journal/ijgnai
  International Journal of Geomechanics
https://www.astm.org/geotechnical-testing-journal.html
  Geotechnical Testing Journal
https://www.tandfonline.com/loi/tgeo20
  Geomechanics and Geoengineering: An International Journal
https://www.journals.elsevier.com/computers-and-geotechnics
  Computers and Geotechnics

1.6.4 Professional organizations

www.asce.org
  American Society of Civil Engineers
www.nspe.org
  National Society of Professional Engineers
www.issmge.org
  ISSMGE – International Society for Soil Mechanics and Foundation Engineering
www.eeri.org
  Earthquake Engineering Research Institute
www.usucger.org
  United States Universities Council on Geotechnical Education and Research

1.7 PROBLEMS

1.7.1 Conceptual problems

Problem 1.1 Define all the terms in bold contained in the chapter summary.

Problem 1.2 Obtain a code (either a local code or a broader code if a local code is not available) and the ASTM standard for the cone penetration test (CPT). Can you distinguish between the goals of each document? Explain what the goals are and how they are achieved by the way the document is organized and written.

Problem 1.3 What are the engineering licensing requirements for your state or country?

Problem 1.4* Select a firm working in the foundation engineering industry. The company may be active in your area, you may have worked as an intern for it, or someone you know may have worked for this firm. Research the company's operations. Which niche(s) of the market is the firm in? Is it a private or public company? How is it organized? What positions do engineers occupy in this firm? Do you have access to the financial reports of the company? Can you identify any of the financial ratios we discussed? If so, would you be able to gauge how well this firm is doing?
*More challenging or demanding problems

Problem 1.5 Can you identify a major conference that has taken place this year with focus on foundation or geotechnical engineering? Which organization was mostly responsible for organizing the conference? Provide examples of topics that were discussed in the conference.

Problem 1.6 Visit the library and get acquainted with five journals in the field of geotechnical and foundation engineering. List them here.

Problem 1.7 Define system of units.

Problem 1.8 What is a fundamental quantity? A derived quantity?

Problem 1.9 What are significant figures?

1.7.2 Quantitative problems

Problem 1.10 Find the value of a pressure of 180 kPa in MPa, kgf/cm^2, tons per square foot (tsf), kilo-pounds per square foot (ksf), and pounds per square foot (psf).

Problem 1.11 Water has unit weight of 62.4 pcf. Starting from this number, obtain the unit weight of water in kN/m^3.

Problem 1.12 A clay has unit weight of 15 kN/m^3. What is its unit weight in pcf?

Problem 1.13 A load of 300 kN is applied on a square foundation element with side $B=2$ m. Assuming a construction tolerance of 5 cm for the sides of the foundation element, what is the range of the average pressure acting on the base of the element?

Problem 1.14 The small-strain shear modulus of a certain sand is given by

$$G_0 = 600 \frac{(29-e)^2}{1+e} \sigma_m'^{0.5}$$

for both G_0 and σ_m' in tsf. Find the equivalent dimensionless equation.

Problem 1.15 Calculate G_0 with the correct number of significant figures using the equation obtained in Problem 1.14 for a soil with $e=0.59$ and $\sigma_m'=350$ kPa.

Chapter 2

Foundation design

Build me straight, O worthy Master!

Stanch and strong, a goodly vessel,

That shall laugh at all disaster,

And with wave and whirlwind wrestle!

Henry Wadsworth Longfellow

The elements of the art of war are first, measurement of space; second, estimation of quantities; third, calculations; fourth, comparisons; and fifth, chances of victory.

Sun Tzu: The Art of War

The foundation design process requires organization and analysis of all the available information about the structure to be supported and the subsurface conditions. The optimal foundation solution transfers the structural loads to the ground in a way that minimizes costs over the life of the structure without sacrificing safety or performance. This requires engineers to consider all things that could go wrong, so that they can prevent them from happening. Although humankind has been building foundations for thousands of years, the sciences of soil mechanics and rock mechanics did not exist before the 20th century. Most foundations were built without what we would consider today to be a proper design. Foundation type and dimensions were determined based on empirical rules and arbitrary judgment calls. We now have the tools to build much more rational and economical foundations. This requires following a logical, orderly process of information analysis.

This chapter covers foundation design by referring to the *limit states design* (LSD) framework, in which design engineers must clearly identify everything that could go wrong with a foundation system or the structure that it supports and then take measures to make sure the probability that it will happen is sufficiently small. In particular, any state leading to structural damage or collapse can be checked using one of three frameworks: working stress design (WSD), load and resistance factor design (LRFD), and reliability-based design (RBD). This chapter discusses in detail these three approaches. It also covers in detail the estimation of foundation movements that could be problematic even if not leading to structural damage.

Although the general concepts presented in this chapter also apply to the design of slopes and retaining structures, specific discussion of the design of these geotechnical structures is reserved for subsequent chapters in which they are covered in detail.

DOI: 10.1201/b22079-2

2.1 THE DESIGN PROCESS

2.1.1 What constitutes foundation design

The foundations of a structure consist of one or more foundation elements. A foundation element is the transition element between the soil or rock and the structure or a component of the structure. Foundation design is the decision-making process of selecting the type of foundation elements to adopt, deciding where to place them in the ground, choosing their dimensions, and specifying how to build them. The foundation engineer also inspects or supervises the construction or installation of foundations and, in some cases, monitors their performance under actual structural loads. There is a natural sequence in the solution to a foundation engineering problem, which we will now discuss briefly.

2.1.2 The sequence in the solution to a foundation problem

2.1.2.1 Determination of the design loads

The structural loads are usually the product of the structural design, with the possible exception of extremely simple structures, for which the foundation engineer might need to estimate the loads to be supported by the foundations. Loads may be due to the self-weight of the structure (dead loads), the use or function of the structure (live loads), and other sources (water and earth pressures, wind, snow, wave action, and seismic loads). Dead loads are due to the self-weight of the different materials that constitute the structure or building in a permanent way, while live loads include the weight of people and objects that occupy the structure some, most, or all the time, but are not permanently attached to it. It is important to have all the following information for each load to be applied on a foundation element: (1) magnitude, (2) load direction, (3) point of application (centered or eccentric with respect to the foundation element), and (4) nature of loads (such as dead, live, wind, snow, and seismic).

Structural engineers calculate how these loads propagate throughout the structure, eventually resulting in the loads that must be carried by the foundations. These loads are provided to the geotechnical or foundation engineer by the structural engineer. Depending on the country, on the local practice, on the problem, and even on the engineers involved, varying degrees of interaction then takes place between the geotechnical and the structural engineer to converge on a final foundation design.

Table 2.1 shows typical loads for residential buildings with reinforced concrete (RC) structures and brick wall fillings. Using the numbers in the table, if we take a moderately high-rise residential building with 30 stories, a typical maximum column load for such a building would be 9000 kN (equivalent to 918 metric tons and 1010 tons). This load is lower if internal walls are built using the dry wall method, which are hollow walls essentially made of a light, gypsum-based plaster.

Table 2.1 Typical vertical loads for residential buildings with reinforced concrete frame and brick walls

Load type	Load per floor
Distributed load	12 kN/m²
Minimum column load	100 kN
Average column load	200 kN
Maximum column load	300 kN

Example 2.1

For a 15-story RC building with a frame structure, estimate the maximum column load. What is the load in tons?

Solution:

Using Table 2.1, we see that the maximum column load is 300 kN/floor. Having 15 floors, the result is:

$$300 \frac{kN}{floor} \times 15 \, floors = 4500 \, kN$$

$$4500 \, kN \frac{224.8 \, lb}{kN} \frac{ton}{2000 \, lb} = 505.8 \, tons \approx 500 \, tons$$

2.1.2.2 Subsurface investigation

In addition to having the values of the structural loads to be supported, the engineer must also assess the load-bearing capacity of the soil or rock. For soil sites, this is usually done in either of two ways: direct correlation of foundation capacity to the results of certain *in situ* tests or estimation of soil properties from laboratory or *in situ* tests that can then be used in stability or deformation analyses for the assessment of foundation capacity.

The *in situ* tests that are most often used in foundation design are the standard penetration test (SPT) and the cone penetration test (CPT), both of which are discussed in Chapter 7. *In situ* tests are performed at the site where construction will take place by loading the soil in some manner and simultaneously making related measurements that are a function of the initial soil properties.

Another way of estimating soil properties is by collecting both disturbed and undisturbed soil samples and testing them in the laboratory. The term undisturbed refers to the desirability of having the samples reflect the true state of the soil *in situ*. It is obviously impossible to obtain perfectly undisturbed samples, but it is possible in certain soils to obtain samples that are reasonably undisturbed. This requires selecting a suitable soil sample recovery technique and cautiously applying this technique. Additional information usually collected in the course of a site investigation program includes data on the general geology of the site and any related problems, as well as data on the groundwater pattern at the site. Chapter 7 covers subsurface investigation.

2.1.2.3 Selection of suitable types of foundation

Based on the information collected at the subsurface investigation stage, possible types of foundations are selected. This selection is based on considerations of constructability, cost, and performance. The optimal foundation type is the one that can be constructed or installed with the least difficulty and cost.

In the process of selecting possible foundation types, we rely on a number of different resources. Soil and rock mechanics help us both interpret site investigation results and make preliminary foundation capacity calculations. Constructability is of course closely linked with site conditions. As an example, shallow foundations may be too expensive or outright impractical if the groundwater table is very near the ground surface, as the walls of the excavations needed to construct them would likely cave in. As an additional example of a constructability problem, driving or jacking piles (discussed in detail in Chapter 12) into soil containing large boulders would not be possible. Some piles would reach their design depth, but many would be either blocked or damaged.

Codes of practice can be helpful in establishing the range of foundation load capacity to be expected for different types of terrain. Previous experience, whether personal experience or experience reported in the literature, is essential to a cost-effective and timely solution to foundation problems. Sometimes, experience is codified in empirical rules, which are often useful at this stage of the foundation design process. Lack of experience with a particular type of foundation may prevent its adoption in a situation where it would be ideal. The foundation engineer therefore needs to keep informed of developments of new equipment and new types of foundations.

Local factors are quite important. Local practice reflects, to some extent, local economics. For example, if an area is located near a steel pile producer, steel piles may be relatively inexpensive, as transportation costs are low. Otherwise, steel piles may not be cost-effective. Additionally, special equipment is often needed to install piles. If equipment is not locally available to install a certain type of pile, it would be impractical and expensive to specify this pile in design. However, economics does not always explain the dominance of certain foundation practices in a given market, and it is valid for the engineer to question established practices if they don't seem to be the best solution for a given project.

Project-specific factors are sometimes important. For example, there may be a tight deadline to be met, and this precludes any type of foundation that would be too slow to construct. It may not be possible to obtain a permit to lower the groundwater table, which may preclude the use of shallow foundations. Access to the site may also be a factor. It may not be possible for large equipment to be taken into a site, in which case certain types of piles may not be used. If the site is very muddy, then the mobility of any equipment necessary in the construction of the foundations becomes an issue. Lastly, foundations of any neighboring structures must not be compromised. When structures immediately next to the site are founded on shallow foundations, the option of choosing shallow foundations very near existing structures may not be available, unless measures are taken to make sure the neighboring foundations are not damaged during construction.

2.1.2.4 Final selection, placement, and proportioning of foundation elements

The core of the foundation design process is the selection, placement, and proportioning of foundation elements. If a single type of foundation has not yet been chosen, calculations are made at this point to assess which type is likely to be most economical. Traditionally, one of the first design decisions has been to select either shallow or deep foundations.[1] In the last three decades, another possibility has become available: piled rafts (or piled mats). A piled raft is a combination of the traditional mat foundation (a shallow foundation) with piles.

The first step in foundation load-carrying capacity calculation is to choose the depth at which to place the bases of the foundation elements. This choice is based to a large extent on the requirement that the load-carrying capacity be maximized, but a few other considerations are also needed, as discussed in Chapter 8. Once the depth of placement of the foundations is known, the foundation elements are proportioned for the given *design loads*. For a footing, this means establishing the plan dimensions based on available soil base resistance and the slab thickness based on the structural capacity of the footing. For a pile, it means choosing an appropriate cross section. For pre-fabricated piles, this means selecting

---

[1] Shallow foundations are structural elements that are installed to no more than about 2 m or 6 ft and that derive their resistance mostly from normal stresses that develop between their base and the soil. Deep foundations and piles in particular are installed to greater (sometimes much greater) depths, relying on base resistance and/or side resistance (resulting from shear stress that develops between the sides of the structural element and the soil).

the cross-sectional dimensions; for cast-in-place piles, there is also the need to specify the material properties and any reinforcement of the pile. Cast-*in situ* piles are most often constructed with concrete, but grout is also used.

The next step is to consolidate all this information in an organized way into a set of specifications. The specifications are later used by the contractor in the construction of the foundations and by the foundation inspector in ascertaining that construction is indeed done according to specifications. There is a trend in place to emphasize more performance when writing specifications, meaning that the specific way in which the work is to be done is not spelled out in detail, giving contractors more flexibility while holding them to performance goals.

2.1.2.5 *Construction*

In the construction stage, the ideas contained in the specifications are converted into reality. The general contractor can often handle the construction of shallow foundations, but the installation of deep foundations sometimes requires the hiring of a specialty contractor.

As far as construction inspection is concerned, in addition to simple observation of construction activities, tests of different types are often carried out. When cast-in-place concrete foundations are used, it is usual for a concrete supplier to determine the compressive strength of the concrete by testing a number of concrete cylinders (or cubes, as specified by British standards). The workability of the concrete required for successful construction varies according to foundation type and is also checked (usually by performing slump tests) whenever appropriate.

Foundation load tests provide the only way to assess directly the performance of foundation elements. In the most common form of load test, load is applied to a foundation element and a corresponding displacement measured. Most commonly, the element is loaded vertically and the vertical displacement (called the settlement) at the top of the foundation element is measured. The result of this test is a plot of load versus settlement, the slope of which represents the stiffness of the soil–foundation element system. So long as the element can take the design load without undergoing an excessively large settlement, the element is considered acceptable. In certain soils, where settlement occurs over a long period of time, this settlement is in general not reflected in load test results to any significant extent. This occurs because there is not enough time for any significant consolidation to develop, even in load tests in which the loads are applied relatively slowly (and take days, not hours, to complete).

Another type of test that has become common in recent years, particularly for deep foundations, is the integrity test. An integrity test is usually done after pile installation by hitting the pile head (or pile top) with a hammer (thus creating a wave that will travel down the pile) and analyzing the returned reflection of the pulse. The reflected signal can sometimes indicate the existence and even the location of any defects that may be present. These tests are covered in Chapter 14.

After construction of the structure is concluded, excessive movements may, on occasion, be observed. Sometimes there is no direct evidence of foundation movement, but rather a suspicion of it caused by the observation of cracking in structural elements. While foundations are the usual suspects, they are by no means always the cause of structural damage,[2] and careful assessment of any such situation is needed. If foundation movements are indeed larger than designed for, either the design or construction of the foundations was defective

---

[2] Structural damage is cracking of structural elements to the extent that their function (of carrying and transmitting loads down the structure all the way to the foundations) may be compromised.

(or both). It is customary in these cases to control foundation and structural movements carefully using appropriate surveying techniques. Ongoing (or accelerating) foundation movement would be very serious and require immediate action, commonly in the form of underpinning of the existing foundations, which consists of constructing new foundation elements to which the structure or foundations are connected for additional support. Fortunately, these cases are the exception and not the rule. Unfortunately, when they happen, they are not widely publicized, making it harder for the profession to learn from its mistakes. This chapter's case study summarizes the history of settlement of the Leaning Tower of Pisa and the corrective measures taken in the 1990s to contain the accelerating inclination of the tower.

2.2 LIMIT STATE DESIGN AND WORKING STRESS DESIGN

The modern view of the geotechnical design process is founded on the concept of *limit state*. A limit state exists on the boundary between acceptable and unacceptable states. They offer a solid conceptual basis for the definition of *failure*, which then becomes the achievement of an undesirable state as defined by the limit states. The *probability of failure* then becomes nothing more than the probability of achievement of an undesirable state.

There are two types of limit states: serviceability and ultimate limit states. A structure is said to be serviceable if it performs as intended without requiring significant unexpected repair or maintenance. A serviceability limit state (SLS), therefore, is a state such that even small undesirable changes to any variables defining it will cause the structure to cease to perform its intended function. Serviceability may be better understood with the aid of specific examples. A residential building, for example, is serviceable if it does not settle excessively. Excessive settlement, even if uniform, might lead to shearing of utility lines or access problems. Uneven settlement would distort parts of the structure, leading to cracking of panel walls, window or door jamming, and other architectural damage.[3] Serviceability can in many cases be restored by incurring additional costs related to repair or maintenance of the structure. Referring again to the case of a residential building, the repair of utility lines, the correction of access problems, and the sealing of cracks may be sufficient to restore serviceability.

Very large, uneven settlements[4] could also lead to an ultimate limit state (ULS). A ULS is a state at which the structure is marginally unsafe: any undesirable changes in pertinent variables would lead to a structure in a dangerous condition. So ULSs are associated with the concept of danger (or lack of safety) and usually result from some type of serious structural damage that might lead to partial or full structural collapse, usually preceded by the formation of extensive, open cracks. In foundation engineering, it is usually (but not always) true that prevention of serviceability limit states precludes ultimate limit states as well. Design practice, however, calls for the engineer to check each possible limit state independently, showing that none will occur under the proposed design.

The *working stress design* (WSD) or *allowable stress design* (ASD) method, as it is also known, has been in use in geotechnical engineering for more than a century. It is based on the concept of the *factor of safety* (FS). With respect to foundation design, the method calls for determining a load that, if applied to the foundation element, would lead to a ULS. This load, which we will refer to as the *ultimate load* Q_{ult} throughout this text, is then divided by the FS. The resulting load is called the *allowable load* Q_a:

---

[3] Architectural damage is the cracking of components of a building that don't have a structural function or superficial, slight cracking of structural members.

[4] Such settlements may develop if the soil can only carry the imposed loads after significant reduction in volume (as we will see in later chapters, soils become stiffer and stronger, and thus able to sustain loads more effectively, when denser).

Table 2.2 Factors of safety (*FS*)

			Soil exploration	
Category	Typical structures	Observations	Thorough	Limited
A	• Railway bridges • Warehouses • Blast furnaces • Retaining walls • Silos	• Maximum design load likely to occur often • ULSs with disastrous consequences	3	4
B	• Highway bridges • Light industrial and public buildings	• Maximum design load may occur occasionally • ULSs with serious consequences	2.5	3.5
C	• Apartment buildings • Office buildings	• Maximum design load unlikely to occur	2	3

Source: Modified after Vesic (1975).

$$Q_a = \frac{Q_{ult}}{FS} \tag{2.1}$$

The WSD method is based on the inequality:

$$Q_d \le Q_a \tag{2.2}$$

where Q_d = design load (or working load). Design loads are calculated by structural engineers as specified in the applicable code, usually as load combinations. All load combinations typically include dead load (the load associated with everything that is thought of as permanently attached to the structure) plus various combinations of live loads and other temporary loads (such as wind, fluid, earth pressure, and seismic loads).

Typical values of *FS* used in foundation design range from 2 to 4, depending on a variety of factors, such as type and importance of the supported structure, type of foundation, construction method and conditions, quality and quantity of subsurface investigation, and method of analysis (see Table 2.2). Table 2.2 should be used as a general reference for appropriate values of the safety factor. Since the *FS* values vary by foundation type, we will provide more specific guidance as to appropriate values of *FS* in later chapters, when we discuss the design of the various types of shallow and deep foundations.

Independent checking for excessive foundation settlement has long been recognized as necessary, although the reference to the term SLS is relatively recent. Tolerable settlements for a variety of structures are discussed later in this chapter. Another method that has often been used in practice to prevent excessive settlements is to increase the value of *FS* beyond what would be required based strictly on safety considerations. This use of the factor of safety is not consistent with the concept of a factor of safety and should, as a rule, be avoided.

2.3 RELIABILITY-BASED DESIGN (RBD) AND LOAD AND RESISTANCE FACTOR DESIGN (LRFD)

2.3.1 The design problem framed as a reliability problem

WSD is somewhat limited as a general design approach. It lumps all the uncertainty regarding a design problem in one number: the factor of safety. Conceptually, it is applicable only to

ULS checks. A more general approach would be desirable, one that can handle limit states of any type and that can separately treat the uncertainties associated with loads and resistance, and, in addition, consider the uncertainties associated with each source of load and each factor controlling resistance. *Reliability-based design* (RBD) and *load and resistance factor design* (LRFD), which may be viewed as simplified RBD, offer this measure of generality.

In the extract of Longfellow's poem, *The Building of the Ship*, epigraph to this chapter, a ship builder is ordered to build a ship "that shall laugh at **all** disaster." Nature has a way of reminding us that there is no such thing as a ship (recall the unsinkable Titanic) or any other thing designed and built by us that will "laugh at all disaster." That does not mean that we shall not continue to try, but we do so under the constraints imposed by economics! *Risk*, whether of disaster or something less, is the key concept in RBD. Risk for us refers to the likelihood (or probability) that a geotechnical system will fail to perform adequately. Another view of the same problem emphasizes not the *negative* (the failure to perform) but the *affirmative* (satisfactory performance). Instead of risk, we then speak of the reliability of the foundation or geotechnical system.

Figure 2.1 shows a hypothetical design problem involving two variables – resistance and load – both of which are random variables. Load and resistance are random variables because there are uncertainties in the characterization of the soil supporting the foundation, in the soil models and analyses we use for calculating settlements and catastrophic failure loads, in the geometry of the foundation elements, and in the characteristics and behavior of the superstructure. We will discuss how to specifically define load and resistance later in this chapter and in chapters dealing with specific geotechnical systems. For now, the fundamental concept with which we need to concern ourselves is that load and resistance may be

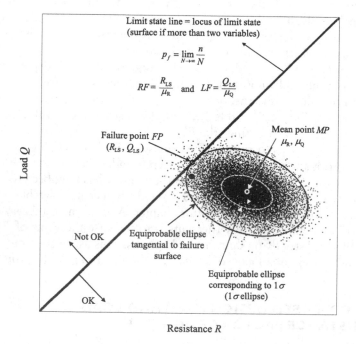

Figure 2.1 Design problem in a probabilistic framework, with load and resistance being random variables and a line separating acceptable from unacceptable outcomes. (modified after Salgado and Kim 2014; with permission from ASCE.)

defined differently depending on the limit state and that, however they are defined, we need the resistance to exceed the load for the design to be technically viable.

The line in Figure 2.1 separates acceptable pairings ($Q < R$) of load Q on the footing and resistance R of the footing from those that would lead to failure. The line is the locus of limit states, and states above the line are failure states. There would be a plot like this for each limit state that should be considered in design. The cloud of points in the figure represents pairings of load Q and resistance R around the mean point (μ_R, μ_Q) of these two variables. Each of these points is a simultaneous realization of random variables Q and R. Since realizations of Q and R are more likely to be close to their respective mean, there is a heavier concentration of points near (μ_R, μ_Q); however, if the load and resistance in a hypothetical problem deviate from their means by sufficiently large amounts, Q may end up exceeding R, which would, by definition, constitute failure.

The movie "Groundhog Day," in which a man is forced to relive the same day time after time until he finally gets to make the right decisions and do the right things, offers a way to understand the meaning of Figure 2.1. In a thought experiment, we could imagine N different versions of the world in which the same project is built at the same location. However, in each different version of the world, each pertinent variable (essentially the loads, geometrical configuration of the building and foundations, and the soil properties) takes a different value according to its inherent variability. This would happen despite attempts by builders and designers to adhere to the same plan. Consequently, the load Q and resistance R for a specific foundation element helping support the building would be different. If the N different pairs of Q and R in each of these different versions of the world were plotted in the same figure, we would get Figure 2.1. Abandoning our thought experiment, we could also obtain the cloud of points in the figure through theoretical simulations if the probability distributions of Q and R were known. This can most effectively be done using Monte Carlo simulations. The *probability of failure* can be estimated from the ratio of the number n of points above the limit state line shown in the figure to the total number N of points. Mathematically:

$$p_f = \lim_{N \to \infty} \frac{n}{N} \tag{2.3}$$

which means that N must be large enough for the n/N ratio to converge to the probability of failure p_f.

One of the key questions in the use of RBD in engineering design has been: what is an acceptable probability of failure? The answer to this question is necessarily subjective, as it requires the engineering profession or the design engineer to make judgments as to what level of risk would be tolerable. This relates back to our discussion in Chapter 1 of the appropriate standard of care for a given engineering problem. A very first answer can be arrived at by counting the known cases of failure within a certain period of time, calculating a probability of failure based on this count, and assessing whether society as a whole has accepted this level of risk as satisfactory. A target probability of failure can then be established, and a target reliability index can be calculated from it. Who should assess what an acceptable probability of failure for a given project is? It depends. In large projects, such as large dams or nuclear power plants, the owner would probably decide that, but with oversight by a consultancy board. For more routine work, reliability analysis would not typically be directly applicable, but design codes or guidelines or specifications may be based on reliability analyses. For example, for slopes involved in highway engineering, the owner (the department of transportation or an equivalent agency) would also make the decision (Salgado et al. 2013). In the design of foundations for residential or industrial facilities, a code of practice would have guidance in this regard, and this means that decisions would be made by committees of experts. When failure represents consequences that are unrelated to safety (consequences that may affect the functionality or serviceability of a structure, but not its structural integrity), higher probabilities of failure are acceptable.

2.3.2 Load and resistance factor design

The load and resistance factor design is an approach to design based on the following inequality:

$$(RF) R_n \geq \sum (LF_i) Q_{i,n} \tag{2.4}$$

where R_n is the *nominal resistance*, RF is the corresponding *resistance factor*, $Q_{i,n}$ are the *nominal loads* (with dead loads, live loads, and so on, identified each by a different value of i), and LF_i are the corresponding *load factors*. This inequality is viewed as superior to that corresponding to WSD, given in Equation (2.2), in part because it allows uncertainties related to load and resistance to be considered separately, rather than lumping them in a single factor of safety.

Also shown in Figure 2.1 are dispersion ellipses. A dispersion ellipse represents the locus of the resistance–load pairings with the same level of deviation from the mean point. There is one dispersion ellipse that is tangent to the failure line at a point known as the most likely "failure point" (FP), also known as the "design point." A quantity that is sometimes used in reliability analysis is the reliability index β, a relative measure of how far the mean point is from the design point. This distance is normalized by the overall variability of the design problem, as represented by the standard deviation of problem variables. The reliability index β may be related to the probability of failure (the higher the β is, the lower is the probability of failure), but the concept of a reliability index is not well defined for certain conditions, and it is therefore easier to work with the probability of failure.

From the point of view of development of material factors or resistance factors for LRFD, if the probability of failure calculated from the Monte Carlo simulations is the value that should be targeted in design, the relationship between the design point FP and the mean point MP yields directly the factors that would be used in design:

$$RF = \frac{R_{LS}}{\mu_R} \quad \text{and} \quad LF = \frac{Q_{LS}}{\mu_Q} \tag{2.5}$$

where R_{LS} and Q_{LS} are the resistance and load, respectively, at the design point FP, and μ_R and μ_Q are the means of resistance and load, respectively.

For different reasons, code designers may set load factors LF_i^* that must apply to a range of design settings. If a resistance factor is developed using reliability analysis, thus calculated from an equation like Equation (2.5), this resistance factor is consistent with the load factor calculated also using Equation (2.5). If it is to be used with a different load factor specified in a code, it must be adjusted. If we refer to the adjusted resistance factor as RF^*, it may be computed by requiring that inequality Equation (2.5) apply equally whether RF^* and LF_i^* or RF and LF_i are used, which leads to:

$$RF^* = RF \frac{\sum (LF_i^*) Q_{i,n}}{\sum (LF_i) Q_{i,n}} \tag{2.6}$$

The LRFD framework can be used for any limit state, whether a ULS or an SLS. Resistance factors have been developed for the design of foundations (Basu and Salgado 2012; Foye et al. 2006; Han et al. 2015; Kim et al. 2011), retaining structures (Kim and Salgado 2012a,b), and slopes (Salgado and Kim 2014) using reliability analysis. This, provided that the analyses are realistic and that probability distributions are representative and well defined, is by far the best way to determine resistance factors. An alternative is rough calibration to experience or limited tests, an approach that has been more common than would be desirable. We will discuss specific values of resistance and load factors for use in design in later chapters. The method is not well developed for serviceability limit states, so we will tend to focus on ultimate limit states.

2.4 LOAD AND RESISTANCE FACTOR DESIGN (LRFD) FOR ULTIMATE LIMIT STATES

Over the past four decades, the LRFD method has been brought into practice in the United States with the adoption of the American Concrete Institute (ACI) Building Design Code in 1963 (Goble 1999). In structural design practice, the LRFD is currently accepted worldwide along with WSD. AASHTO and FHWA have been the drivers of the use of LRFD for geotechnical design, with a mandate in the 1990s to use it in all transportation infrastructure designs. To understand how the inequality in Equation (2.4) works in geotechnical design, let us consider the example of a footing. The loads $Q_{i,n}$ are the total loads acting on the footing due to the dead loads, live loads, and the like, originating in the superstructure. The resistance R_n is the load-carrying capacity (also referred to as the bearing capacity) of the footing, which is obviously closely related to the properties of the soil in ways we discuss in Chapter 10. The word "nominal" means that both the loads and resistances are determined following certain guidelines. The word "characteristic" (as in characteristic resistance) is used interchangeably with "nominal" to represent the same thing.

An important distinction between LRFD and WSD is the specific assignment of uncertainty in LRFD to either loads or resistances, whereas in WSD they are blended in the concept of the safety factor. In LRFD, uncertainties related to loads are accounted for in the values of the load factors, while those associated with all the steps of the determination of soil resistances are accounted for in the values of the resistance factors.

With the trend toward the increased use of LRFD, codes in North America (AASHTO 1994, 1998; API 1993; MOT 1992; and NRC 1995) have implemented recommendations for the use of LRFD for geotechnical design over the past several years. The AASHTO (1994, 1998) code proposes the use of the same loads, load factors, and load combinations for foundation design as those used in structural design, which is a sound principle. Scott et al. (2003) did a comprehensive review of the major codes used for foundation design, with a focus on loads and load factors. When comparing the bridge and offshore codes to the building codes, there are many differences in the types of limit states considered for design and in the load types and load combinations defined for each limit state. Usually, a greater number of limit states and load types apply to the design of special structures such as bridges and offshore structures. However, certain types of loads appear in most design situations for all types of structures. These are dead loads, live loads, wind loads, and earthquake loads.

Table 2.3 shows the ranges of values of load factors for ULSs included in the following LRFD codes: "AASHTO LRFD Bridge Design Specifications (AASHTO 2020)," "Building Code Requirements for Structural Concrete (ACI 2019)," "LRFD Specification for Structural Steel Buildings (AISC 1994)," "Recommended Practice for Planning, Designing, and Constructing Fixed Offshore Platforms-LRFD (API 1993)," "Ontario Highway Bridge

Table 2.3 Load factors

Loads	AASHTO (2020)	ACI (2019)	AISC (1994)	API (1993)	MOT (1992)	NRC (1995)
Dead	1.25–1.95 (0.65–1.0)	1.4 (0.9)	1.2–1.4 (0.9)	1.1–1.3 (0.9)	1.1–1.5 (0.65–0.95)	1.25 (0.85)
Live	1.35–1.75	1.6 (1.0)	1.6	1.1–1.5 (0.8)	1.15–1.4	1.5
Wind	1.0	1.0	1.3	1.2–1.35	1.3	1.5
Seismic	1.0	1.0	1.0	0.9	1.3	1.0

Note: Values in parentheses apply when the load effects tend to resist failure for a given load combination, that is, when the loads have a beneficial effect.

Design Code (MOT 1992)," and "National Building Code of Canada (NRC 1995)." In general, for the bridge codes (AASHTO 2020 and MOT 1992) and the offshore foundation code (API 1993), the range of load factor values is rather wide compared with that for building or onshore foundation codes. For example, the range of values of load factors for dead loads in AASHTO and MOT extends from 1.25 to 1.95 and from 1.1 to 1.5, respectively, whereas the range for the building codes is from 1.2 to 1.4. The values of live load factors in the bridge and offshore foundation codes lie between 1.1 and 1.75. The values of live load factors for the two building codes referenced in Table 2.3 are 1.5 (NRC) and 1.6 (ACI).

Many different dead load types are considered in AASHTO (2020) and MOT (1992). These include the weight of the structural members, the weight of wearing surfaces such as asphalt, and earth pressure loads. A different value of load factor is applied to each of these load types. For example, in AASHTO (2020), while the value of the load factor for structural components is 1.25, the load factor values for the weight of wearing surfaces (pavement surfaces) and the vertical earth pressure applied to flexible buried structures are 1.5 and 1.95, respectively. The relatively high values of the load factors for the wearing surface weight and the earth pressure applied to buried structures reflect high variability in estimating the magnitude of the corresponding loads. On the other hand, the dead loads in the building codes such as ACI (2019) and NRC (1995) consist mostly of the weight of structural components, partitions, and all other materials incorporated into the building to be supported permanently by the structural components. The same load factor is used for all these loads because they are all treated simply as dead loads. The rather wide ranges for the dead load factors in the bridge codes, therefore, are associated with the various types of dead loads accounted for in the design of bridges.

When a live load is used together with other transient loads (that is, live, wind, or earthquake loads), the simultaneous occurrence of their maximum values is not likely, and some loads may counteract other loads when they occur together. To account for this, most codes, except for the bridge codes (AASHTO and MOT), apply a load combination factor <1.0 when more than two different transient loads are used in a load combination. As an example, NRC (1995) proposes a value of 0.7 for the load combination factor when both a live and a wind load are present. In that case, therefore, 70% of each factored load effect for both the live and the wind loads are considered in design:

$$Q = (LF)_D Q_D + 0.7\left[(LF)_L Q_L + (LF)_W Q_W\right]$$

(2.7)

where the various loads are denoted by the letter Q and a suitable subscript (D = dead, L = live, and W = wind).

The load combination factor usually varies with the number of transient loads that are present. That is, in the case where only one transient load applies, the value of the load combination factor is unity, but it is less than one otherwise. A different approach is used in the bridge codes (AASHTO and MOT), in which different values of the load factors are defined in different load combinations, instead of multiplying the proposed load factors for each load by the load combination factor. As an example, AASHTO defines one load combination when live load, but not wind load, is present:

$$Q = 1.25Q_D + 1.75Q_L \tag{2.8}$$

but defines another load combination when both live load and wind load are present:

$$Q = 1.25Q_D + 1.35Q_L + 0.4Q_W \tag{2.9}$$

A cautionary remark is in order: the advantages of LRFD come with a price. Every factor in a design problem, if changed, will necessarily change the value of the resistance factor. The reason for this is very simple: the probability distribution associated with that factor changes, and the resistance factor results from the consideration of the probability distributions of all variables or equations involved in design calculations. As a result, the values of the resistance factors vary with the type of foundation, methods of subsurface investigation, and methods of analysis used in calculating bearing capacities. One must not use a resistance factor developed for a given set of conditions for conditions that do not match them. This is not widely recognized and has created some confusion in early application of the method. With increasing familiarity with LRFD and with the fact that it is closely tied to the implicit variability observed in specific soils, in specific design methods, and in construction, the method should be applied correctly and consistently. The discussion of LRFD will continue in later chapters, as design methods for different types of foundations and geotechnical systems are introduced.

2.5 TOLERABLE FOUNDATION MOVEMENTS

2.5.1 Consideration of foundation settlement in design

As we saw earlier, we are concerned mostly with two goals in foundation design: (1) avoid danger (which is associated with ULSs) and (2) preserve serviceability of the structure (a serviceability loss is a SLS). Settlements can do both: disrupt the functionality of the structure and be so large and uneven that parts of the structure become so overstressed that they may fail. In an extreme, such failures may lead to partial or complete collapse of a structure. We will discuss how to estimate foundation settlements in other chapters. In this chapter, we are concerned with assessing when settlements are likely to lead to either SLSs or ULSs.

2.5.2 Settlement patterns

Structures settle because soil can be compressed and distorted upon loading. If a soil mass is uniform and the structure transfers loads to it uniformly, the structure will settle uniformly. This is not the usual settlement pattern, but it does on occasion happen. Figure 2.2, for example, shows what used to be a window, now partly underground, of a building that has experienced large uniform settlements. The picture shows clearly that uniform settlements have not led to cracking of the building.

Tilt may be a problem with tall structures. In this case, one side of the structure either is more heavily loaded, or is supported by more compressible materials, or both. As a result,

Figure 2.2 The large uniform settlement of a building in Ravenna, Italy, shown in this photo is noticeable by the arch of what used to be a window (or door), now partly underground.

more settlement takes place on one side than on the other, and the structure tilts. Tilt, even if not threatening to stability, leads to awkward if not outright unserviceable conditions. In Santos, Brazil, there are inhabited buildings where doors close on their own and a ball released from one end of an apartment will roll to the other end (clearly undesirable conditions!). The Millennium Tower in San Francisco, California, is experiencing at the time of this writing very significant leaning. While the building is still deemed inhabitable, there was a US$200 million claim filed by the condominium association against multiple parties, including engineering firms, that was ultimately successfully mediated. Still, settlement has continued unabated, even after early interventions, as of 2021.

Excessive tilt may lead to instability and collapse, as occurred to the Pavia Tower in Italy in the late 1980s. The most famous case of foundation tilt is the Leaning Tower of Pisa (Figure 2.3), which has been undergoing settlement for centuries. It was closed to tourist visitation in the early 1990s due to concerns about its stability, but was reopened in 2001 after lengthy stabilization work. An interesting aspect of the Tower is that it has the shape of a banana, a result of the fact that it was built in several stages, each stage spaced by many decades and even centuries of construction inactivity. Each time construction restarted, an attempt was made to correct the noticeable lean of the Tower, resulting in the banana shape. Even though the Tower of Pisa has been a success as a tourist attraction, no foundation engineer aspires to leaving behind a crooked building as his or her legacy. The Leaning Tower of Pisa is the focus of the case history at the end of this chapter.

An uneven settlement pattern, in which differential settlements develop between different points of the foundations, is potentially very damaging to the structure. It may also cause serviceability problems, such as door and window jamming and uneven floors. Because of its costly consequences, it is necessary to quantify damaging differential settlements, at least approximately. We accordingly reserve a substantial portion of the remainder of this chapter to this effort.

2.5.3 Crack formation

Cracking in a structure is closely related to both ultimate and serviceability limit states. It is therefore useful to understand how cracks form in response to foundation movement and how these cracks differ from those that develop due to shrinkage (of plaster, for example, upon drying) and other such local effects.

Figure 2.3 The Leaning Tower of Pisa: (a) a view of the tower; (b) and (c) the settlement difference between the two opposite sides of the tower. ((a) Courtesy of Michele Jamiolkowski.)

The key to understanding cracking due to foundation movement is to understand that the strains caused throughout the structure by foundation movement are not local in nature, so they vary little from a point to another in a wall or from one side of a wall to another. Let us say, for example, that all the cracking is due to the settlement of one footing supporting a specific column of a frame structure. The direction and intensity of the strains induced at locations a few inches or even a meter or two apart will be very similar given

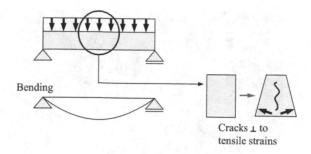

Figure 2.4 Cracking of beam at midspan due to bending: the tensile stress there is maximum and horizontal; as a result, vertical cracking develops.

that these locations are approximately at the same distance from the source of the deformations. This understanding allows us to identify certain characteristics that cracks caused by foundation movement have:

1. These cracks may have any direction, and this direction at a certain location in an element (such as a wall or beam) is determined by the prevailing direction of the strain field created there by the foundation movement.[5]
2. If multiple cracks exist at a specific location in a wall, beam, or some other element, these cracks have approximately the same direction, because the strains responsible for the cracking do not vary much across the relatively small distances between cracks. This means cracks with a web pattern (typical of shrinkage cracks) are never due to foundation movement.
3. The cracks appear on both sides of walls, beams, columns, or other elements because the strains due to foundation movement at a location in such a structural element are the same on either side of it.
4. If there are cracks caused by foundation movement in one floor of a building or structure, then there will also be similar cracks in other floors. The intensity of the cracking decreases as we move toward higher floors in frame structures because the structure absorbs and redistributes the internal loadings induced by the foundation movement, so that, if we are sufficiently far from the foundations, the effects tend to be small. However, for masonry structures with load-bearing walls, crack intensity does not reduce markedly with height because these structures do not redistribute internal loads as effectively.

The types of crack we deal with in buildings are tension cracks. These cracks form in brittle materials because their resistance to tension is small. The direction of a tension crack is perpendicular to the tensile stress that causes it. The easiest way to visualize this is to consider the simple case of a simply supported beam acted upon by a uniform load (Figure 2.4). The figure shows the resulting bending moment diagram, which has a peak exactly at midspan (at which the shear force is zero). Consequently, the tensile stress there will be maximum. If we increase the uniform load far enough, that tensile stress will exceed the strength of the material and a crack forms. The figure also shows how an element of the beam located exactly at midspan is shortened at the top and is lengthened at the bottom.

---

[5] Foundation movement imposes a pattern of deformation throughout the structure, which causes individual members to elongate (or compress) and distort in certain ways. Cracks will form normal to the direction of maximum tensile strains (direction of maximum elongation).

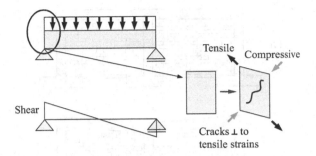

Figure 2.5 Cracking of a beam at the supports due to shearing: the maximum tensile stress associated with shearing makes an angle of 45° with the horizontal, and thus so does any cracking that develops.

Because the tensile stress at the bottom of the element is horizontal, the crack will be vertical. Bending cracks will not form elsewhere until cracking develops at midspan because that is where the largest tensile stress acts.

If we now examine the shear force diagram for the same beam (Figure 2.5), it is clear that such force is maximum at the supports, where the bending moment is zero. So that is where we would expect shear cracks to form first. The bending moment there is zero, so a condition of pure shear exists. Pure shear generates tensile and compressive principal stresses of equal magnitude with directions making 45° angles with the direction of shearing (which is vertical in our case). This means that we should expect diagonal cracks (cracks at 45° with the horizontal) to develop.[6] This is shown in the figure by the distorted element, which has one diagonal stretched, whereas the other diagonal is shortened. The crack will form with a direction normal to the stretched diagonal. Cross sections located between the supports and midspan will experience both shear and bending, so we can expect any crack that forms in one of these cross sections to have an angle with the horizontal between 45° and 90°, with the angle increasing as we move from the support to the center.

What happens to beams when distorted by bending or shearing also happens to buildings. Frame buildings, which are more flexible in shearing than in bending (because slabs and beams resist tensile forces very well, but parallel rows of columns can easily move with respect to each other in the vertical direction), tend to develop diagonal cracks. This is illustrated in Figure 2.6. The other extreme is that of buildings, namely unreinforced brick buildings, that resist shearing better than direct application of tensile stresses (that is, bending). This case is illustrated in Figure 2.7. It is also noticeable in Figure 2.7 that cracks tend to start from the corner of openings such as windows and doors. This also happens when shearing is the critical mode of loading (see Figure 2.8). The reason for that is the stress concentration that develops around openings as the stress field[7] reorganizes around the opening, which leaves a reduced cross-sectional area for load transfer, leading to larger stresses.

From the cracking pattern, it is typically possible to visualize the corresponding foundation movement. For example, in the case of Figure 2.9, the load-bearing wall is built on soil that is not uniform across the length of the wall. Near the right end of the wall, the soil is much more compressible, leaving the wall "hanging," creating tensile stresses that are nearly vertical and cracks that are nearly horizontal. As the distance from this weak support zone increases, the cracks transition toward the more vertical direction we would expect from bending-induced cracking.

---

[6] In Chapter 4, after we discuss the Mohr circle, we propose as a problem to use the Mohr circle to show that shear cracks are indeed diagonal cracks.

[7] A stress field is defined by the values and directions of stress at every point of a certain space or domain.

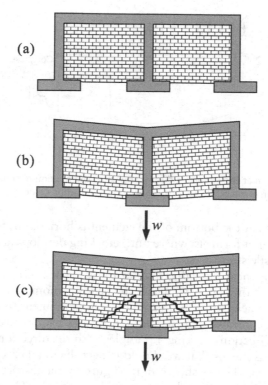

Figure 2.6 Cracking in a reinforced concrete frame building: (a) frame before movement; (b) movement causing distortion of the frame and panel walls; and (c) cracks that form perpendicular to the stretched diagonal.

Figure 2.7 Masonry building with an unreinforced load-bearing wall showing vertical cracking near midspan.

A final point to make is the following: foundation/geotechnical engineers are sometimes called to the site to assess cracks that are either extensive or very open (see Table 2.4 to correlate crack width with serviceability and safety). It is very important to keep in mind that cracking is not always caused by foundation movement (although this is often the case). A failing structural member, for example, can also cause excessive movements in parts of the

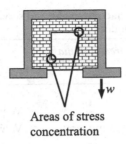

Areas of stress
concentration

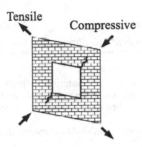

Tensile   Compressive

Figure 2.8 Stress concentration around openings causes cracks to form at the location of windows, doors, and other openings before they form elsewhere.

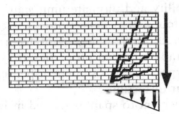

Figure 2.9 Cracking due to uneven soil support for a load-bearing wall.

structure that the failing member helps support. So a rather global view of the structure and foundations is required for us to fully understand cracking patterns and what they imply.

2.5.4 Quantification of tolerable settlements

2.5.4.1 Differential settlement and angular distortion

As discussed earlier, SLSs in geotechnical design are most often caused by excessive settlements. Excessive settlements can also lead to ULSs. To determine whether a foundation has settled or moved excessively, it is essential to first quantify the amount of foundation deformation using a suitable parameter. The value of this parameter could then be compared with the maximum tolerable value that the parameter can assume before a limit state is reached. We will examine three parameters: *differential settlement*, *angular distortion*, and *relative deflection*.

Referring to Figure 2.10a, the differential settlement Δw for any two columns of a frame building is the difference between the total settlements w_1 and w_2 of the two columns. If L is the span or distance between the two columns, the angular distortion α is defined as

$$\alpha = \frac{\Delta w}{L} \qquad (2.10)$$

Table 2.4 Cracking width and the associated damage and serviceability/safety issues for residential, commercial, and industrial buildings

Crack width (mm)	Degree of damage			Serviceability or safety issues
	Residential	*Commercial*	*Industrial*	
<0.1	None	None	None	None
0.1–1	Slight	Slight	Very slight	Cracks may be visible
1–2	Slight to moderate	Slight to moderate	Very slight	Possible penetration of humidity
2–3	Moderate	Moderate	Slight	Serviceability may be compromised
3–15	Moderate to severe	Moderate to severe	Moderate	ULSs may be reached
>15	Severe to dangerous	Moderate to dangerous	Severe to dangerous	Risk of collapse

Source: Modified after Thorburn (1985).

Very slight: visible on close inspection; correctable with interior design/decoration tools.
Slight: external cracks may need to be filled for watertightness; doors and windows may jam slightly.
Moderate: replacement of small amount of brickwork needed; service pipes may be severed; jamming doors/windows.
Severe: replacement of portions of walls needed; window/door frames distorted; uneven floors; service pipes severed; leaning or bulging walls.
Dangerous: beams lose bearing; walls require shoring; windows broken by distortion; danger of instability.

Figure 2.11 shows that the differential settlement alone is not sufficient for assessing tolerable deflections. Serviceability and ultimate limit states result from the distortion of the superstructure caused by foundation deflections. It is clear from the figure that the same differential settlement causes different degrees of distortion for different spans. If two structural frames with comparable material properties are subjected to the same angular distortion, however, we can expect to see a similar response; that is, they should crack at approximately the same angular distortion.

The ratio of differential settlement to span expressed in Equation (2.10) is actually the tangent of angle α, which in most cases is sufficiently small that it is approximately equal to the angle itself. Figure 2.10b and c illustrate the damage that large angular distortions can do to a building and its structure.

Figure 2.12 illustrates the concept of relative deflection. Relative deflection w_r is the vertical deflection (settlement) measured from a reference line. This reference line, in the case of Figure 2.12a, is obtained by joining the points of the foundations that are above all others (in the case of this figure, these are the two extreme points, which have coincidentally moved by the exact same amount). In more complex cases, when there are inflection points in the deformed configuration of the foundations, with parts of the foundations concave upward and other parts concave downward, more than one reference line exists. This is illustrated in Figure 2.12b, where the left side of the foundations is in a sagging configuration and has its own reference line, while the right side is in a hogging configuration and also has its own reference line. The reference line can be generally defined as the line joining two points of the foundation in such a way that the line lies always below (for hogging) or above (for sagging) all other foundation points. Hogging is not common in cases where only load-induced foundation settlement is involved. It tends to appear when excavation or tunneling activities are underway near the building and can therefore cause greater foundation movements near the edges of the buildings. Soil swelling can also generate hogging, but swelling tends to be uneven and is difficult to predict with any accuracy.

The maximum relative deflection $w_{r,max}$ is the maximum value of the relative vertical deflection w_r, that is, the maximum vertical distance between the deformed foundations

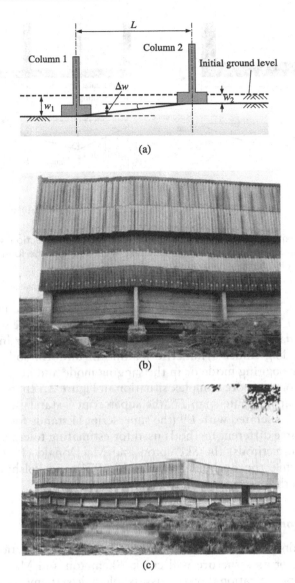

(a)

(b)

(c)

Figure 2.10 Angular distortion due to differential settlement: (a) definition; (b) large angular distortion leading to severe damage to a warehouse; and (c) panoramic view of the same warehouse with widespread and variable differential settlements.

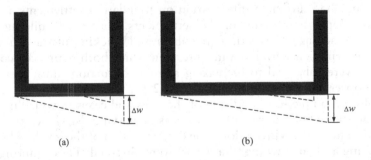

(a)   (b)

Figure 2.11 The same differential settlement causes more distortion in case (a) than in case (b) because of the shorter span in case (a).

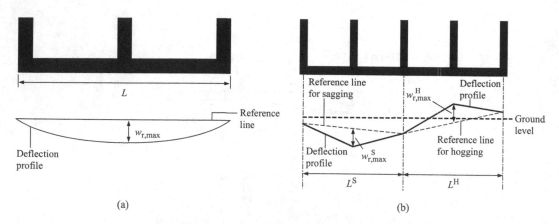

Figure 2.12 Foundation configurations and resulting maximum relative deflection $w_{r,max}$: (a) simple case of symmetric sagging and (b) more complex case where part of the foundation is in the sagging mode and part in the hogging mode.

and the reference line. The maximum relative deflection ratio is the ratio of $w_{r,max}$ to the horizontal projection L of the reference line, which coincides with the width of the building when the whole foundation system is either in the hogging mode or in the sagging mode. For a symmetric building under symmetric loading in uniform soil, the entire building would be either in the hogging mode or in the sagging mode and $w_{r,max}$ would be observed exactly at midspan. For the more complex situation in Figure 2.12b, the $w_{r,max}^S$ for sagging is associated with an appropriate span L^S (the superscript S stands for sagging), while the $w_{r,max}^H$ for hogging is associated with L^H (the superscript H stands for hogging).

We will examine three different methods used for estimating tolerable foundation movements. Of these three methods, the Skempton and MacDonald (1956) method relies on angular distortion, while the Burland and Wroth (1974) and Polshin and Tokar (1957) methods are based on the concept of relative deflection.

2.5.4.2 The Skempton and MacDonald (1956) study

Because cracking results from distortion, the larger the angular distortion α, the more likely it is that the building or its structure will crack. Skempton and MacDonald (1956) were among the first to propose a rational way to assess tolerable settlements for frame buildings. They did so by using the concept of the angular distortion.

Skempton and MacDonald (1956) developed a database of 98 buildings, both frame buildings and masonry buildings with load-bearing walls in the area of London, England. For each building, they had sufficient information on foundation settlements and on whether any cracking developed as a result of the settlements. For frame buildings, they distinguished between cracking of structural members and cracking of the panel walls. They combined that information with laboratory tests in which both reinforced concrete frames and brick walls were subjected to increasing angular distortions until they cracked. The results of these observations are shown in Figure 2.13, where cases in which cracking was observed are plotted on the left-hand side of the axes and cases in which no cracking was visible appear on the right-hand side of the axes as open symbols. The vertical axes have an angular distortion (α) scale, which allows the determination of the α values below which no structural cracking and no cracking of any sort were observed. These limiting values are as follows:

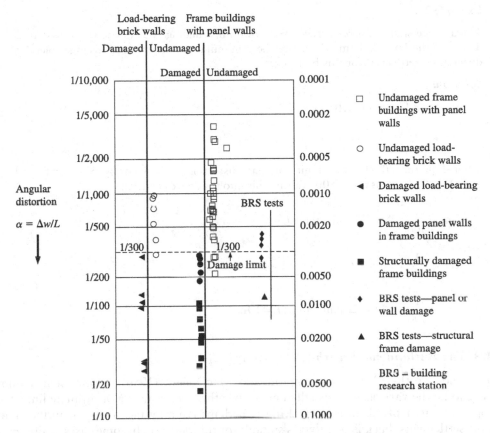

Figure 2.13 Cracking or the absence thereof plotted as a function of angular distortion for buildings in the London area and for laboratory tests on frames and brick walls. (Skempton and MacDonald 1956; courtesy of the Institution of Civil Engineers and Thomas Telford Limited.)

$$\alpha = \begin{cases} \dfrac{1}{170} & \text{(ULS: frame cracking)} \\[2mm] \dfrac{1}{300} & \text{(SLS: wall craking)} \\[2mm] \dfrac{1}{500} & \text{(unlikely to lead to either SLS or ULS)} \end{cases} \qquad (2.11)$$

where ULS = ultimate limit state and SLS = serviceability limit state.

According to these results, as long as the angular distortion is capped at 1/500, foundation settlement will not lead to either structural or architectural damage. This value already includes a reduction factor to account for uncertainties in both the observations of Skempton and MacDonald (1956) and the calculation of settlements; therefore, the design loads need not be magnified by load factors for settlement estimation. It is interesting to point out that construction methods have evolved over the past decades, and other methods of constructing walls in reinforced concrete buildings (such as the use of dry walls) are today more common than the use of masonry walls. Dry walls are more easily repaired, so the 1/500 criterion can still be used (and may be slightly conservative).

Example 2.2

The average span for a reinforced concrete frame building is 5 m. The shortest span is 3 m, and the largest, 10 m. Calculate the minimum, maximum, and average tolerable differential settlement for this building.

Solution:

According to Equation (2.10),

$$\Delta w = \alpha L$$

From Equation (2.11), the limit angular distortion prescribed by Skempton and MacDonald (1956) is $\alpha = 1/500$. Taking this into the above equation:

$$\Delta w = \frac{1}{500} \times 3\,\text{m} = 6\,\text{mm} \quad \text{for} \quad L = 3\,\text{m}$$

$$\Delta w = \frac{1}{500} \times 5\,\text{m} = 10\,\text{mm} \quad \text{for} \quad L = 5\,\text{m}$$

$$\Delta w = \frac{1}{500} \times 10\,\text{m} = 20\,\text{mm} \quad \text{for} \quad L = 10\,\text{m}$$

2.5.4.3　The Burland and Wroth (1974) study

Burland and Wroth (1974), as Skempton and MacDonald (1956) before them, assumed cracking to be the variable that usually defines whether serviceability or ultimate limit states develop. The onset of substantial cracking is perhaps the only means for quantification of tolerable settlements, but it is not always suitable for this purpose. In some cases, visible cracking is not a limit state; in others, it must be accepted and dealt with if economic foundations are desired. A measure of judgment is therefore needed in assessing the adequacy of tolerable settlements resulting from criteria based on the onset of cracking. Notwithstanding all this, avoiding substantial cracking (in terms of either crack intensity or crack width) is still a key design criterion as well as a yardstick for whether serviceability or safety has been impaired.

Table 2.4 summarizes the general relationship between crack width and limit states for three classes of buildings. In addition to crack width, the following are all important factors that we should consider in assessing serviceability and safety: the way in which a structure is supported (frame versus shear wall), the state of balance of the structure, and the stabilization with time of foundation movements. The table is not particularly useful in design, but it may be useful in judging the seriousness of any cracking that may develop in a finished building.

Burland and Wroth (1974) framed the discussion of tolerable settlements by using a beam analogy, in which a beam with span L, height H, and elastic properties G (shear modulus) and E (Young's modulus) is distorted and loaded in different ways until cracking starts in some part of the beam. Cracking starts when the tensile strain at some point reaches a critical value, which is generally in the -0.05% to -0.10% range (tensile strains being negative). Based on this simple analysis, by varying the relative flexibility of buildings in shear and bending (accomplished by changing the values of either E/G or L/H), they showed that buildings responded differently to differential settlements of their foundations depending on the ratio of their stiffness in shear to their stiffness in bending. Buildings that are stiffer in shear than in bending, such as those with load-bearing masonry walls, tend to develop vertical or subvertical midspan cracks (Figure 2.7). Buildings that are flexible in shear but stiff

(a)

(b)

Figure 2.14 Diagonal cracking in (a) a reinforced concrete frame building in Porto Alegre, Brazil, and (b) a reinforced masonry wall building in Santa Monica, California. ((b) Courtesy of EERC, University of California, Berkeley; photographer: Stojadinovic, Bozidar.)

in bending, such as frame buildings and buildings with reinforced load-bearing masonry walls, tend to develop diagonal cracks most often near the ends of the spans (Figure 2.14).

The beam analogy also allowed the development of a criterion for the hogging mode of deformation, in which the central part of the building either settles less than or moves up more than the edges. As shown in Figure 2.15, this mode of deformation can be quite damaging to masonry buildings. Figure 2.15c shows a gas station in Plymouth, Indiana, that underwent substantial and uneven foundation movement when tanks were extracted from the ground near its foundations. Note how the lack of restraint at the top of the structure, which is characteristic of unreinforced masonry structures, allows the cracks resulting from the foundation distortion to open wide. This point is illustrated in Figure 2.15b, which shows that, for the masonry wall without restraint at the top, the hogging mode is more severe than the sagging mode shown in Figure 2.15a. Which one (a or b) develops is largely a function of construction details, such as quality of the mortar. The construction of

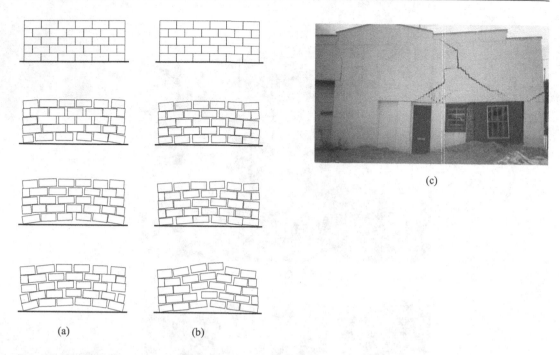

Figure 2.15 Hogging mode of foundation movement: (a) consequences of hogging for a structure with good-quality mortar, (b) consequences of hogging for a structure with poor-quality mortar, and (c) large, open cracks extending to roof of gas station in Plymouth, Indiana. (Courtesy of Aaron Humphrey.)

underground spaces and tunnels is another common cause of the hogging mode of deformations in structure foundations.

In their analysis of the hogging mode of foundation movement, Burland and Wroth (1974) assumed the neutral axis of the beam analog to be coincident with the bases of the footings, as the foundations are attached to the soil rather firmly, preventing the building from either elongating or contracting at that level. In practice, the hogging mode could result from placing the foundations on expansive soils. It is difficult to predict the movement of foundations on expansive soils, so the quantification of tolerable foundation movement in the hogging mode is not as useful as it is in the case of settlements.

The Burland and Wroth (1974) results are better expressed in terms of the maximum relative deflection ratio $w_{r,max}/L$ of the building. As seen previously (refer to Figure 2.12), two values of $w_{r,max}$ (one for sagging and one for hogging) may be defined for the same building. However, because at the design stage it is impossible to predict with any accuracy the final configuration of the building at the foundation level, we usually assume that the building foundations settle in a manner that leads to sagging. A single value of $w_{r,max}/L$ is then defined, where L in this case is the width of the building.

The Burland and Wroth (1974) criterion for frame buildings and load-bearing masonry wall buildings (in both the sagging and hogging modes) are plotted together with the Skempton and MacDonald (1956) criterion and the Polshin and Tokar (1957) criterion in Figure 2.16. The Polshin and Tokar (1957) criterion was developed in the former Soviet Union exclusively for load-bearing wall buildings. It was expressed in terms of maximum tolerable relative deflection ratios $w_{r,max}/L$. For plotting the Skempton and MacDonald (1956) criterion in Figure 2.16, Burland and Wroth found the equation relating the angular distortion α calculated for the end and center of a beam to w_r/L at midspan; they then

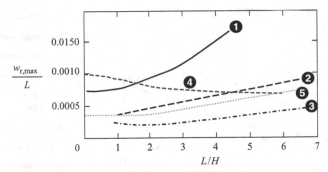

① Burland and Wroth (1974): frame structures; reinforced masonry buildings
② Burland and Wroth (1974): unreinforced masonry buildings (sagging)
③ Burland and Wroth (1974): unreinforced masonry buildings (hogging)
④ Skempton and MacDonald (1956)
⑤ Polshin and Tokar (1957)

Figure 2.16 Maximum tolerable settlement criteria. (Burland and Wroth 1974; courtesy of the Institution of Civil Engineers and Thomas Telford Limited.)

plugged in a value of $\alpha = 1/500$ (the maximum tolerable value of α) into this equation to obtain the maximum tolerable value of relative deflection, $w_{r,max}/L$. Note from Figure 2.16 that $w_{r,max}/L$ varies with the height-to-width ratio of the load-bearing wall.

In Figure 2.17, the three criteria are compared with case histories where no damage, slight damage, and substantial damage were observed. It is apparent from this figure that

- the Skempton and MacDonald criterion works well only for frame buildings;
- the Polshin and Tokar criterion works well only for masonry buildings; and
- the Burland and Wroth criterion works reasonably well for both frame and masonry buildings; it is also the only criterion that is applicable to the hogging mode.

Example 2.3

You need to design the foundations for a brick building with height of 6 m and width of 15 m. The building will be founded on clay. For a building on clay, sagging is expected. Determine the tolerable maximum relative deflection for this building.

Solution:

For sagging, considering that $H = 6$ m and $L = 15$ m, Figure 2.16 gives:

$$\frac{w_{r,max}}{L} = 0.5 \times 10^{-3} \text{ according to the Burland and Wroth (1974) criterion}$$

and

$$\frac{w_{r,max}}{L} = 0.4 \times 10^{-3} \text{ according to the Polshin and Tokar (1957) criterion.}$$

We will use the slightly less conservative Burland and Wroth (1974) equation:

$$w_{r,max} = 0.5 \times 10^{-3} \times 15 \text{ m} = 7.5 \text{ mm}$$

Note that this is not the tolerable maximum settlement of this building, which would be larger than this, as discussed in the next section.

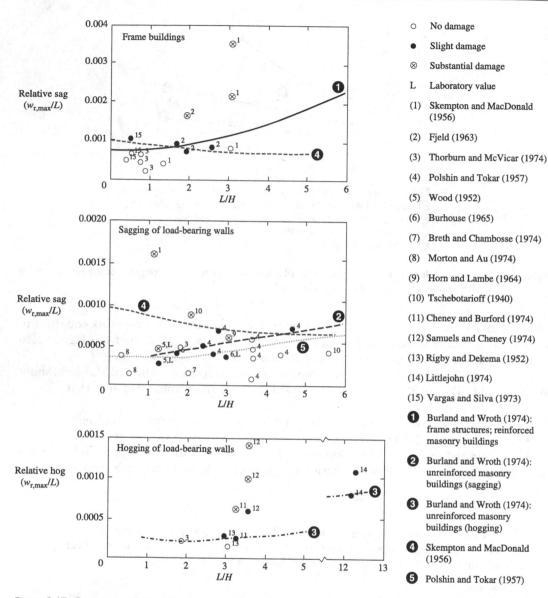

Figure 2.17 Comparison of maximum tolerable settlement criteria with settlements both causing damage and not for buildings and experiments. (Burland and Wroth 1974; courtesy of the Institution of Civil Engineers and Thomas Telford Limited.)

2.5.4.4 Tolerable total settlement of buildings

Settlements are calculated with basis on the loads on the foundations, the dimensions of the foundations, and the design soil profile. We have seen that differential settlements are undesirable because of all the problems they may lead to. Let us assume that our analyses for settlement calculation produce accurate results (we will discuss them in Chapter 9). A natural question that would follow at this point is: why not proportion the foundations in such a way as to have zero differential settlement everywhere? The answer to this question is that, unfortunately, differential settlements cannot be prevented at the design stage. Even if we proportion the foundations aiming to have each element settle the same, (1) the lack of

Table 2.5 Ratio of maximum total settlement w_{max} to maximum angular distortion α_{max}

Soil type	Isolated foundations	Mat foundations
Sand/sandy fill	$15L_R$	$20L_R$
Clay	$25L_R$	$30L_R$

Source: Modified after Skempton and MacDonald (1956).

L_R = reference length = $1\,m \approx 40\,in$.

soil profile uniformity across the site, (2) the different values of the ratios of actual load to design load and dead load to live load for different columns, and (3) the difference between as-built and design dimensions of the foundations will cause our total settlements to be different. However, we can limit the differential settlement by limiting total settlement because there is a direct relationship between the differential settlement between two foundation elements and the maximum of the two settlements.

Differential and total settlements are correlated because one needs the other to exist. There can be no differential settlement in the complete absence of total settlement. As the loads on two foundation elements are gradually increased, the three factors we listed (differences in the soil profile below each foundation, differences in dead-to-live load ratio between the two foundations, and differences in as-built dimensions) will have an opportunity to have an effect on the relative motion of the two footings. Thus, the greater the loads applied on the two foundations, the greater the total settlements for both and the greater the differential settlement between them.

The correlation between differential and total settlements is particularly strong for sands, where one foundation element might stand above a denser pocket and the other above a looser pocket of sand; in this case, it is clear that the greater the total movement of both foundations, the greater the difference in their settlements as well. Because clay deposits are more uniform than sand deposits, clays experience less differential settlement than sands for a given total settlement. Additionally, the correlation between differential and total settlements is not as strong for clays as it is for sands, particularly in the case of long-term (consolidation) settlement.[8] Terzaghi (1956), in a discussion of the Skempton and MacDonald (1956) paper, was critical of proposing correlations between maximum and differential settlements for deep clay layers, in which consolidation settlement predominates. If we accept this criticism, then, for clays, we would not attempt to correlate differential to total settlements but instead would estimate each separately and then design the foundations to prevent excessive values of both.

An additional factor in the correlation between total and differential settlements is the type of foundation. The stiffer the foundation (that is, the more the foundation operates as a unit, as opposed to as a collection of isolated elements), the lower the ratio of differential to total settlement. So a raft (mat) foundation (which in essence is a single reinforced concrete slab supporting all the columns of the structure) would show less differential settlement than isolated foundation elements (each supporting a single column) for the same maximum total settlement. Table 2.5 shows the values of this correlation as a function of soil and foundation type.

---

[8] Consolidation settlement is settlement that takes place over a long period of time in low-permeability soils (clays) as water flows out of the soil pores slowly to allow the soil skeleton to densify, become stiffer, and thus better support applied loads.

Example 2.4

Determine the maximum tolerable settlement for a building supported on isolated footings bearing on sand.

Solution:

From Table 2.5,

$$\frac{w_{max}}{\alpha_{max}} = 15 L_R = 15 \times 1\,\text{m} = 15\,\text{m}$$

Using $\alpha_{max} = 1/500$ from Equation (2.11):

$$w_{max} = 15 / 500 = 0.03\,\text{m} \quad \text{or} \quad 30\,\text{mm}.$$

This value is only slightly larger than 1 in., an approximate number that has traditionally been used for tolerable settlement of footings in sand.

Example 2.5

Assuming an average span of 8 m, what is the ratio of differential to total settlement implied by Table 2.5 for isolated footings on sand?

Solution:

Using Equation (2.10), we see that

$$\Delta w = \alpha L = \frac{1}{500} \times 8000\,\text{mm} = 16\,\text{mm}$$

From Table 2.5:

$$w_{max} = 15 L_R \alpha_{max} = 15 \times 1000\,\text{mm} \times \left(\frac{1}{500}\right) = 30\,\text{mm}$$

So the ratio of differential to total settlement is 16/30 mm = 0.53% or 53%.

Example 2.6

Consider the building of Example 2.3. What would the tolerable total immediate settlement be for this building?

Solution:

Considering that the load-bearing walls will be placed on strip footings, which are continuous and behave more like mat foundations than isolated foundations, the ratio of total settlement to angular distortion is obtained from Table 2.5 as

$$\frac{w_{max}}{\alpha} = 30 L_R = 30\,\text{m}$$

The angular distortion can be calculated approximately when we consider that the maximum differential settlement Δw is equal to $w_{r,max}$ and develops over half the width of the wall (see E-Figure 2.1):

$$\alpha = \frac{w_{r,max}}{\dfrac{L}{2}} = 2\frac{7.5 \times 10^{-3}\,\text{m}}{15\,\text{m}} = 1 \times 10^{-3}$$

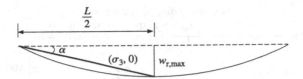

E-Figure 2.1 Relationship between maximum relative deflection and angular distortion for Example 2.6.

This gives:

$$w_{max} = 1 \times 10^{-3} \times 30\,m = 30\,mm$$

Note that this does not include the consolidation settlement, which would need to be computed separately and would have an effect on serviceability considerations.

2.5.4.5 Tolerable movements of bridge foundations

So far we have dealt with masonry and frame structures, focusing on buildings. Let us now examine another important type of structure: bridges. The concept of serviceability in bridges is related largely to the riding conditions. A wavy or cracked pavement will lead to uncomfortable and even dangerous riding/driving conditions, and either settlements or horizontal displacements of the foundations causing such conditions are clearly not tolerable. Another dimension to the problem is whether the owner of the bridge would find it more advantageous to save on the foundations (an initial cost) even if it means more frequent repair and maintenance of pavements (delayed costs).[9] Typically, for reasons related to how projects are funded, the preference is for more robust foundations and less frequent maintenance. In connection with this reasoning, defects caused by vertical movements are easier to correct than those caused by lateral movements, which may lead to an undesirable closing of expansion joints. Figure 2.18 shows a compilation made by Bozozuk (1978) of acceptable and unacceptable movements. Notice how bridges can tolerate as much as 100 mm of vertical settlement without major serviceability problems, but as little as 50 mm of horizontal movement can already cause problems. The accepted values of tolerable angular distortion for single-span and multiple-span bridges are 0.008 and 0.004, respectively (AASHTO 2005; Barker et al. 1991; Moulton et al. 1985).

2.5.4.6 Tolerable foundation movements of other types of structures

Structures with special features will be susceptible to movements in ways that are different from the ways in which bridges and buildings are affected. As an example, towers, silos, and very tall buildings must not rotate so much that the rotation becomes visible to the unaided eye, even if this much rotation does not lead to a stability problem or to any of the serviceability problems we have discussed. Taking the differential settlement between the two sides of the tall structure and dividing it by the width of the structure, the number above which rotation becomes visible is roughly 1/250. Table 2.6 contains tolerable values of α for some other structures.

---

[9] Although this concept is also applicable to building foundations, building maintenance and repair can be a lot more complicated and costly; additionally, there is often more than one owner in a building. So saving on the foundations while accepting more frequent repair is not an attractive option for building foundations.

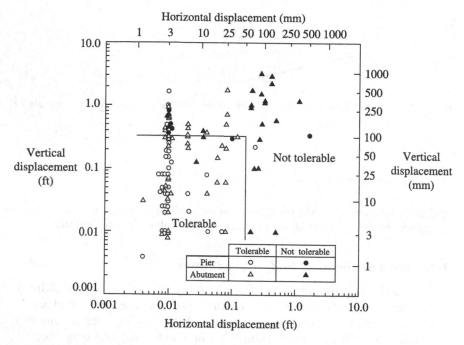

Figure 2.18 Tolerable movements for bridge foundations. (Bozozuk 1978, Figure 1, page 18. Reprinted with permission from TRB.)

Table 2.6 Tolerable angular distortion for special-purpose structures

Structure	$\alpha = \Delta w/L$
Machines	1/750
Frames with diagonals	1/600
Overhead cranes	1/300
Tilt of tall buildings/structures	1/250
Steel tanks	1/25

2.5.4.7 *Load factors for settlement computations*

As indicated previously, there is no need to use load factors to magnify load effects in foundation movement calculations because the values of tolerable movements found in the literature and in this text generally contain an adjustment for uncertainty (for example, the use of 1/500 instead of 1/300 for angular distortion). However, AASHTO, in their bridge design code, does use load factors different from 1 in some specific cases. For example, AASHTO recommends a value of 0.3 for the wind load factor. The use of a value less than 1 is based on the reasoning that time-dependent loads, such as wind loads, are not likely to remain at their maximum value for significant periods. Furthermore, live loads considered in bridge design are traffic loads that may be highly dependent on time compared with live loads in buildings that are mostly occupancy loads. However, the use of a load factor value of 1 may be more appropriate for SLS checks for foundations on nonplastic silts, sands, and gravels because the settlement of these soils is immediate and therefore would reflect the maximum wind load applied to the structure.

2.6 CASE STUDY: THE LEANING TOWER OF PISA (PART I)[10]

2.6.1 Brief history of the Tower of Pisa

The Tower of Pisa (Figure 2.3) located in the city of Pisa, in Tuscany, Italy, is a major tourist attraction. The city is located on the Arno River, northwest of Rome, and 10 km away from the coast. The tower stands 54 m tall and weighs 142,000 kN. It is composed of a tall ground story, six loggias (open galleries), and the belfry at the top. The foundation is a spread foundation in the form of a hollow cylinder with an outer diameter equal to 19.58 m and an inner diameter equal to 4.47 m (Mitchell et al. 1977). The hollow space appears to have been filled with rubble and mortar at the time of construction (Burland et al. 2003). The foundation is embedded ~3 m in the soil.

Construction of the tower to its present height was done in several stages over the course of centuries. The first stage extended from 1173 to 1178, when the ground story and the first two loggias were built. No construction activity took place during the next century. Construction was restarted in 1272, lasting until 1278. During this time, four more loggias were added. By 1272, it was evident that the tower had started to lean, and masons attempted to correct for the leaning by placing stones on plumb (along a vertical alignment), not according to the tower alignment. Because of this, the tower is curved, much like a banana. Another century passed, and, in 1370, the belfry was built, and construction was completed, after another decade of work. It is estimated that the lean of the tower at that time was 3.5°, corresponding to an angular distortion α equal to 0.061 or ~1/16.

2.6.2 Why the settlement?

The tower is located on top of 300 m of sediments deposited both by the Arno River and by the sea, at the time when the city was located in a coastal lagoon (many centuries ago, the Tyrrhenian Sea on the west coast of Italy reached the city of Pisa). Focusing on the layers nearer to the ground surface, the tower rests on about 9 m of a silty/sandy soil deposited by the river underlain by ~30 m of marine clay. Because the silt layer was more compressible on the south side of the tower, the settlement developed faster there than on the north side, resulting in the tower's present inclination (for more on this, see Part III of the case history in Chapter 9).

It is interesting to note the reason why no bearing capacity failure ever occurred. The century-long waiting periods between the constructions of the three stages of the Tower allowed the silts and clays to compress and strengthen (because denser soils are stronger), such that the soil was able to sustain the loads associated with subsequent construction. In Chapter 6, we discuss in detail why clays require long periods of time to compress under load. By 1838, the Tower had settled in excess of 3 m, and the base of the tower had completely disappeared into the ground. An architect named Gherardesca did not like the fact that people could no longer see the base of the Tower and had a walkway excavated around the Tower. This decision was certainly not a good idea from an engineering standpoint, as the removal of ground support only accelerated the Tower's inclination. By 1911, the inclination had reached 5.4° (corresponding to $\alpha = 0.094$); by 1990, it was 5.5° ($\alpha = 0.096$) with no signs of stabilization. Figure 2.19 shows the situation of the Tower in May 1993. Note the overhang of over 4 m.

---

[10] All references for this case history are at the end of Part IV in Chapter 11.

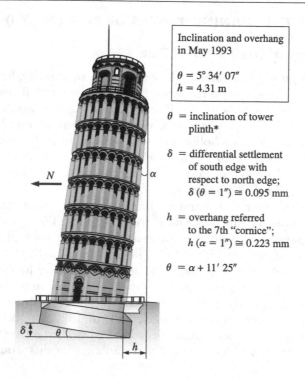

Inclination and overhang
in May 1993

$\theta = 5°\ 34'\ 07''$
$h = 4.31$ m

θ = inclination of tower
plinth*

δ = differential settlement
of south edge with
respect to north edge;
$\delta\ (\theta = 1'') \cong 0.095$ mm

h = overhang referred
to the 7th "cornice";
$h\ (\alpha = 1'') \cong 0.223$ mm

$\theta = \alpha + 11'\ 25''$

Figure 2.19 Situation of the Tower of Pisa in May 1993. (After Jamiolkowski 2006.)

2.6.3 Stabilization of the tower

Altogether, there have been 17 commissions set up to assess the stability of the Tower of Pisa over the 19th and 20th centuries. The process was always quite political and usually resulted in no measures being implemented. The 17th commission was different. It was set up in 1990, with Professor Michele Jamiolkowski of the Technical University of Turin as the chair. The commission was created and given complete autonomy by the Italian Prime Minister. Installation of the commission happened at the time when memories of the collapse of the Tower of Pavia, a city located just North of Milan, Italy, were fresh. This was certainly helpful in overcoming the political resistance that had been a problem for previous commissions.

One of the first moves of the new commission was to reinforce the lowest story of the tower using prestressed steel wires. This was done because long delays related to the politics surrounding the work of the commission were expected, and there was concern that the masonry composing the southern wall of the tower was severely overstressed. Additionally, Professor John Burland of Imperial College, a member of the commission, observed that the north side of the tower was actually moving up, while the south side continued to settle. In order to temporarily stabilize this rotation of the tower, a post-tensioned concrete ring was built around the base of the tower and 6,000 kN of lead ingots were stacked on it on the north side (see Figure 2.20). The lead ingots did stabilize and even reverse the lean slightly.

The lead ingots were not intended as a permanent solution, as the intent was always to reopen the tower to visitation by tourists, and the ingots were considered a visually unattractive solution. After an attempt to install ground anchors as a replacement for the ingots in 1995, as a result of which the tower lean increased in a single day the equivalent of a whole year's worth, the commission finally decided to proceed with soil extraction from

Figure 2.20 Placement of lead ingots on the north side of the tower. (Courtesy of Michele Jamiolkowski.)

under the north side of the tower as a definitive solution for its stabilization. This technique had been successfully used in the stabilization of the Mexico City Cathedral in the 1980s.

In late 1996, pilot tests of the underexcavation technique were done. Figure 2.21 shows the underexcavation technique in progress. The idea is simple (see Figure 2.22): to carefully and gradually remove soil from underneath the north side of the tower so that it will settle, therefore reducing the lean. However, some members of the commission had reservations about the technique. A concern expressed by one member of the commission was that soil underexcavation might actually accelerate the leaning and even lead to collapse of the tower

Figure 2.21 Underexcavation in progress: the small-diameter boreholes are used to remove soil from under the side of the Tower that has settled the least. (Courtesy of Michele Jamiolkowski.)

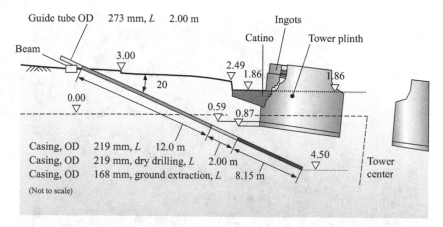

Figure 2.22 Schematic cross section of intervention by underexcavation (elevation in meters). (Courtesy of Michele Jamiolkowski.)

by removing support (load-carrying capacity) from the tower foundations, further stressing the already overloaded south side. Engineers must do that: seek every possible situation that may lead to failure to prevent it. A more natural expectation, however, would be that careful, slow extraction of soil would allow overlying soil to move down to occupy the newly created space, moving the tower down with it. This is indeed what happened, for the tilt decreased sharply with the start of drilling at the end of 1999, extending throughout the year 2000, and finally stabilizing in 2001, as shown in Figure 2.23. As an "insurance policy," in the words of Prof. Michele Jamiolkowski, steel cables (shown in Figure 2.24) were attached to the tower to keep it in place in case something went wrong.

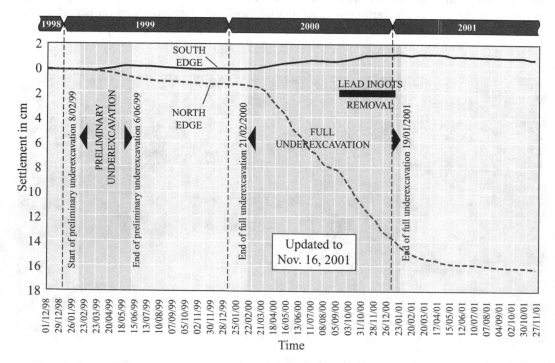

Figure 2.23 Evolution of tilt of the Tower of Pisa. (After Jamiolkowski 2006.)

Figure 2.24 Steel cables used to hold the Tower of Pisa in place in case intervention by underexcavation had produced unintended consequences. (Courtesy of Michele Jamiolkowski.)

A total of 41 small drill holes were extended roughly 5 m below the south side of the tower. The Tower, now with the same inclination it had in 1800, was reopened to tourists in a ceremony on June 17, 2001, with performances by Andrea Bocelli, the famous Italian tenor, and the Toscanini Orchestra of Parma. "Cavalleria rusticana," "Va pensiero," "Il trovatore," and other classical Italian music and songs were in the program.

End of Part I. In continuation of the Tower of Pisa's case history at the end of Chapter 6, we will explore in more detail the properties of the soil profile and, especially, the clay layer below the tower.

2.7 CHAPTER SUMMARY

2.7.1 Main concepts and equations

The foundation design and construction process typically follows the sequence: (1) determination or estimation of the loads on the foundations; (2) subsurface investigation; (3) choice of type of foundations to use; (4) determination of materials, dimensions, and locations of foundation elements; and (5) construction.

It is important to design foundations with reference to an appropriate and logical design framework. The **limit state framework** helps engineers to identify clearly whether they are trying to prevent **serviceability limit states** or **ultimate limit states**. The two most common approaches to check for ULSs are **working stress design (WSD)** and **load and resistance factor design (LRFD)**.

The WSD method is based on the inequality:

$$Q_d \leq Q_a \tag{2.2}$$

where Q_d is the **design load** and Q_a is the **allowable load,** which is given by

$$Q_a = \frac{Q_{ult}}{FS} \tag{2.1}$$

where Q_{ult} is the foundation load leading to an ULS and FS is the **factor of safety** (typically, a value between 2 and 3). The ultimate load is taken in the vast majority of cases as the load-carrying capacity of the foundation (other terms used for it are load capacity, resistance, or bearing capacity).

Geotechnical engineers sometimes use LRFD for foundations, but it is still much less used than WSD. The LRFD design inequality is written as

$$(RF)\,R_n \geq \sum (LF_i)\,Q_{i,n} \tag{2.4}$$

where LF, Q_n, RF, and R_n are the **load factor, nominal load, resistance factor,** and **nominal resistance,** respectively; the subscript i refers to the load type (**dead load, live load,** and so on). Equation (2.4) states that the factored load effects must not exceed the factored resistance for pre-defined possible limit states.

Most SLSs are related to the occurrence of excessive **settlements,** whether **total** or **differential.** The **angular distortion** α, which is used to gauge the likelihood of attainment of a limit state in frame structures, is expressed as the ratio of the differential settlement between two adjacent columns and the span between the columns:

$$\alpha = \frac{\Delta w}{L} \tag{2.10}$$

Structural damage (an ULS) is likely when α exceeds 1/170; damage to panel walls and architectural finishings (an SLS) is likely when it exceeds 1/300. A conservative tolerable value of α is often taken as 1/500 for frame structures.

The tolerable tilt of towers and other tall structures can be expressed in terms of the angular distortion α as well. In this case, the structure moves almost as a rigid body, and α is just a convenient way of quantifying the amount of tilt. Tilts corresponding to more than 1/250 would be noticeable to the unaided eye (an SLS).

For load-bearing masonry walls, tolerable criteria are better expressed in terms of the **relative deflection** ratio $w_{r,max}/L$. The tolerable values of $w_{r,max}/L$ (which are found in Figure 2.16) tend to increase slightly with L/H, where L = width or length of building or wall and H = height of building or wall.

Tolerable foundation movement for bridges may be expressed in terms of the absolute lateral and vertical deflections. Bridges can tolerate vertical deflections better than horizontal deflections (as seen in Figure 2.18). Maximum values of tolerable angular distortion tend to be smaller (of the order of 0.008) for single-span bridges than for multiple-span bridges (for which a value of 0.004 has been proposed).

2.7.2 Symbols and notations

Symbol	Quantity represented	US units	SI units
FS	Safety factor	Unitless	Unitless
L	Length or span	ft	m
L_R	Reference length	in.	m
LF	Load factor in LRFD	Unitless	Unitless
MP	Mean point	Unitless	Unitless
Q	Load on footing	tons	kN
Q_a	Allowable load	tons	kN

(Continued)

Symbol	Quantity represented	US units	SI units
q_b	Unit load across a footing	tsf	kPa
Q_d	Design load	tons	kN
Q_n	Nominal load in LRFD	tons	kN
Q_{ult}	Ultimate load	tons	kN
R	Resistance of footing	tons	kN
R_n	Nominal resistance in LRFD	tons	kN
RF	Resistance factor in LRFD	Unitless	Unitless
w_{max}	Maximum total settlement	in.	mm
w_r	Relative deflection	in.	mm
α	Angular distortion	Unitless	Unitless
α_{max}	Maximum angular distortion	Unitless	Unitless
β	Reliability index	Unitless	Unitless
Δw	Differential settlement	in.	mm

2.8 PROBLEMS

2.8.1 Conceptual problems

Problem 2.1 Define all the terms in bold contained in the chapter summary.

Problem 2.2 Classify each of the following as an ultimate limit state (ULS) or serviceability limit state (SLS):

a. The door of your new house jams when you attempt to close it.
b. A reinforced concrete beam on the third floor of a residential building with the structure completed has a crack with thickness of ~4 mm.
c. The pavement of a bridge is severely cracked.
d. The 20-story building you live in is noticeably tilted, but there is no evident cracking anywhere in the building.
e. A machine in an auto parts factory causes very severe vibration when powered up.

Problem 2.3 Discuss whether the following two statements are true or false, and what the final conclusion should be with respect to whether sands or clays allow larger settlements without achievement of a limit state. Take into consideration additional factors if appropriate.

Statement 1: Tolerable settlements in clays tend to be larger than in sands because loading due to settlement is distributed throughout the structure more slowly and efficiently than in sands.

Statement 2: Tolerable settlements in sands tend to be larger than in clays because most of the settlement takes place during construction, before the more sensitive finishings are put in.

2.8.2 Quantitative problems

Problem 2.4 For a 30-story reinforced concrete building with a frame structure, estimate the range of column loads to be expected.

Problem 2.5 Two columns are supported each by a separate footing. The differential settlement between the two columns is 25 mm, and the distance between them is 4 m. Compute the angular distortion. Would this differential settlement be acceptable?

Problem 2.6 Considering a pair of footings with a span of 30 ft and a ratio of differential to total settlement of three-fourths, what is the maximum tolerable total settlement of each footing?

Problem 2.7 The loads to be supported by a foundation element are 1900 kN (dead load) and 1000 kN (live load). Write the ULS design equations according to both WSD and LRFD using the ACI load factors.

2.8.3 Design problems

Problem 2.8 You have been asked to design the foundations for an interstate highway bridge. There is a wealth of information about the site of the bridge that was collected as part of the initial site investigation. You know the soil types and the strength parameters of the soils. Based on this information, what value of factor of safety would you use for the design of the foundations using working stress design (WSD)? Justify your answer.

Problem 2.9 You are to design the foundations for a residential building with load-bearing brick walls resting on sand. The wall has a width of 12 m and a height of 3 m. What is the maximum tolerable settlement you will use in your verification of serviceability? Show how you arrived at your answer.

Problem 2.10 A building with an RC frame will be founded on isolated foundations resting on sand. Considering that the spans range from 6 to 15 m: (a) What is the range you would expect the tolerable differential settlement to be in? (b) If you were to use a single differential tolerable settlement value for the entire building, what would you use to avoid any problems related to serviceability? (c) What is the maximum total settlement of any foundation element if serviceability problems related to settlement are to be avoided?

REFERENCES

References cited

AASHTO. (1994). *AASHTO LRFD Bridge Design Specifications: SI Units*. 2nd Edition, American Association of State Highway and Transportation Officials, Washington, DC.

AASHTO. (1998). *AASHTO LRFD Bridge Design Specifications*. 2nd Edition, American Association of State Highway and Transportation Officials, Washington, DC.

AASHTO. (2005). *AASHTO LRFD Bridge Design Specifications*. 3rd Edition, American Association of State Highway and Transportation Officials, Washington, DC.

AASHTO. (2020). *AASHTO LRFD Bridge Design Specifications*. 9th Edition, American Association of State Highway and Transportation Officials, Washington, DC.

ACI. (2019). *Building Code Requirements for Structural Concrete (318-19) and Commentary (318R-19)*. American Concrete Institute, Farmington Hills, MI.

AISC. (1994). *Load and Resistance Factor Design Specification for Structural Steel Buildings*. 2nd Edition, American Institute of Steel Construction, Chicago, IL.

API. (1993). *Recommended Practice for Planning, Designing, and Constructing Fixed Offshore Platforms-Working Stress Design*. API RP-2A, 20th Edition, American Petroleum Institute, Washington, DC.

Barker, R. M., Duncan, J. M., Rojiani, K. B., Ooi, P. S. K., Tan, C. K., and Kim, S. G. (1991). *Manuals for the Design of Bridge Foundations*. NCHRP Report No. 343. Transportation Research Board, National Research Council, Washington, DC.

Basu, D. and Salgado, R. (2012). "Load and resistance factor design of drilled shafts in sand." *Journal of Geotechnical and Geoenvironmental Engineering*, 138(12), 1455–1469.

Bozozuk, M. (1978). "Bridge foundations move." Transportation Research Record No. 678, Transportation Research Board, Washington, DC, 17–21.

Burland, J. B., and Wroth, C. P. (1974). "Settlement of buildings and associated damage." Settlement of Structures, Cambridge, UK, 611–654.

Foye, K. C., Salgado, R., and Scott, B. (2006). "Resistance factors for use in shallow foundation LRFD." *Journal of Geotechnical and Geoenvironmental Engineering*, 132(9), 1208–1219.

Goble, G. (1999). "Geotechnical related development and implementation of load and resistance factor design (LRFD) methods". NCHRP Synthesis of Highway Practice 276, Transportation Research Board, Washington, DC.

Han, F., Lim, J., Salgado, R., Prezzi, M., and Zaheer, M. (2015). "Load and resistance factor design of bridge foundations accounting for pile group-soil interaction." Joint Transportation Research Program Publication No. FHWA/IN/JTRP-2015/24, Purdue University, West Lafayette, IN.

Kim, D., and Salgado, R. (2012a). "Load and resistance factors for external stability checks of mechanically stabilized earth walls." *Journal of Geotechnical and Geoenvironmental Engineering*, 138(3), 241–251.

Kim, D., and Salgado, R. (2012b). "Load and resistance factors for internal stability checks of mechanically stabilized earth walls." *Journal of Geotechnical and Geoenvironmental Engineering*, 138(8), 910–921.

Kim, D., Chung, M., and Kwak, K. (2011). "Resistance factor calculations for LRFD of axially loaded driven piles in sands." *KSCE Journal of Civil Engineering*, 15(7), 1185–1196.

MOT. (1992). *Ontario Highway Bridge Design Code*. Ontario Ministry of Transportation, Downsview, Ontario, Canada.

Moulton, K. L., GangaRao, H. V. S., and Halvorsen, G. (1985). "Tolerable movement criteria for highway bridges." Final Report No. FHWA/RD-85/107, U.S. Department of Transportation, Washington, DC.

NRC. (1995). National building code of Canada. National Research Council of Canada, Ottawa, Canada.

Polshin, D. E. and Tokar, R. A. (1957). "Maximum allowable non-uniform settlement of structures." *Proceedings of 4th International Conference on Soil Mechanics and Foundation Engineering*, London, UK, 1, 102–105.

Salgado, R. and Kim, D. (2014). "Reliability analysis of load and resistance factor design of slopes." *Journal of Geotechnical and Geoenvironmental Engineering*, 140(1), 57–73.

Salgado, R., Woo, S. I., Tehrani, F. S., Zhang, Y., and Prezzi, M. (2013). "Implementation of limit states and load resistance design of slopes." Joint Transportation Research Program Publication No. FHWA/IN/JTRP-2013/23, Purdue University, West Lafayette, IN.

Scott, B., Kim, B. J., and Salgado, R. (2003). "Assessment of current load factors for use in geotechnical load and resistance factor design." *Journal of Geotechincal and Geoenvironment Engineering*, 129(4), 287–295.

Skempton, A. W., and MacDonald, D. H. (1956). "The allowable settlements of buildings." *Proceedings of Institution of Civil Engineers, Part III*, 5, 727–768.

Terzaghi, K. (1956). "Discussion on: A.W. Skempton and D.H. MacDonald, allowable settlements of buildings." *Proceedings of Institution of Civil Engineers, Part III*, 5, 775–777.

Thorburn, S. (1985). "The philosophy of underpinning." In *Underpinning*. Thorburn, S. and Hutchinson, J. F. (eds.), Surrey University Press, Surrey, UK, 1–39.

Vesic, A. S. (1975). "Bearing capacity of shallow foundations." In *Foundation Engineering Handbook*. 1st Edition, Chapter 3, Winterkorn, H. F. and Fang, H. Y. (eds.), Van Nostrand Reinhold, New York.

Additional references

Becker, D. E. (1997). "Eighteenth Canadian geotechnical colloquium: Limit states design for foundations. Part I. an overview of the foundation design process." *Canadian Geotechnical Journal*, 33(6), 956–983.

Becker, D. E. (1997). "Eighteenth Canadian geotechnical colloquium: Limit states design for foundations. Part II. development for the national building code of Canada." *Canadian Geotechnical Journal*, 33(6), 984–1007.

Grant, R., Christian, J. T., and VanMarcke, E. H. (1974). "Differential settlement of buildings." *Journal of the Geotechnical Engineering Division*, 100(GT9), 973–991.

Wahls, H. E. (1981). "Tolerable settlement of buildings." *Journal of the Geotechnical Engineering Division*, 107(GT11), 1489–1504.

Wahls, H. E. (1994). "Tolerable movement of buildings." *Proceedings of Settlement 94: Vertical and Horizontal Deformations of Foundations and Embankment*, College Station, TX, 2, 1039–1057.

Chapter 3

Soils, rocks, and groundwater

I met a traveller from an antique land,

Who said—'Two vast and trunkless legs of stone

Stand in the desert ... Near them, on the sand,

Half sunk a shattered visage lies, whose frown,

And wrinkled lip, and sneer of cold command,

Tell that its sculptor well those passions read

Which yet survive, stamped on these lifeless things,

The hand that mocked them, and the heart that fed;

And on the pedestal, these words appear:

My name is Ozymandias, King of Kings;

Look on my Works, ye Mighty, and despair!

Nothing beside remains. Round the decay

Of that colossal Wreck, boundless and bare

The lone and level sands stretch far away.'

Percy Bysshe Shelley

Foundation and geotechnical engineers must always deal with either soil or rock, and often both. So it is necessary for us to understand these two materials and their occurrence as well as possible. In this chapter, we cover the genesis of soils and rocks, their composition, and their main features. This chapter is not intended to provide either a detailed account of any aspect of the geology of soils and rocks or an exhaustive coverage of the subject. It aims to provide the basic geological knowledge required for solving routine engineering problems that engineers can build on by additional reading and the continuing practice of geotechnical engineering.

3.1 SOIL AND THE PRINCIPLE OF EFFECTIVE STRESS

3.1.1 What is soil?

Much of the work of foundation and geotechnical engineers involves soil. It is imperative therefore that we are able to define it. A simple definition of soil is that it is a particulate medium, with particles resulting from a variety of geologic processes, and is composed of a

DOI: 10.1201/b22079-3

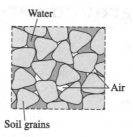

Figure 3.1 Soil element showing the soil particles and the void space filled with air and water.

variety of minerals, with particle sizes ranging from the order of 1 μm to that of 1 m.[1] The space between soil particles, referred to as the pore space, is filled with one or more of the following: air, some other gas, water, or some other liquid (Figure 3.1).

Beyond the amazing range of soil particle sizes, soil is complex in other ways. Soil particles may be arranged in densely packed states or very loose ones. Soils at depth exist under large stresses; soils close to the surface exist under small stresses. Particle arrangements can vary not only in terms of how dense they are but also in the way the particles are in contact with each other. The pore fluid chemistry can vary. In some soils, particles interact physiochemically with the pore fluid, and changes in pore fluid chemistry can affect soil behavior.

As a result of all these complexities, the mechanics and engineering of problems involving soils is challenging. In the past, this has led to perhaps an excessive use of arbitrary judgment in the solution of geotechnical engineering problems. At the present time, soil mechanics is developed to a degree such that the tools needed for rationally analyzing and solving foundation and geotechnical engineering problems are available.

3.1.2 Particle size

Soils are referred to by different names, depending on particle size. A common nomenclature is that of Table 3.1. A classification like that of Table 3.1 is important because soil particle size is a strong determinant of soil behavior. Of particular importance is the transition from sand to silt and clay size. Clay-sized particles are roughly the size needed to produce colloidal mixtures with water. Colloidal mixtures, in which more than one phase[2] is clearly distinguished (in contrast with solutions, where a single phase is observed), have special properties that, in part, help explain some of the behavior of clays.

The separation of soils in their different size fractions is done using sieves with different mesh sizes down to the upper boundary for silt size, corresponding to the No. 200 sieve (0.075 mm; see ASTM D6913 (2017c)).[3] For smaller particles, which pass the No. 200 sieve, an apparatus called a hydrometer is often used. This apparatus consists of a glass tube in which a mixture of soil and water is placed. The soil particles settle over time, during which

---

[1] A more general definition would include certain materials resulting from industrial processes, such as fly and bottom ash, that have been increasingly used in the construction of embankments, mechanically stabilized earth (MSE) walls, and other structures.

[2] Phase is a distinct state of matter in a system; that is, it is matter that is identical in chemical composition and physical state and separated from other materials by the phase boundary. In connection with saturated soils, one phase will be solid (the soil particles) and one, liquid.

[3] Material passing the No. 200 sieve is referred to as fines. So, clays and silts are fines. When they are present in a soil also containing sand, the percentage of fines, together with the plasticity of the fines, is a useful number in understanding the soil response. Plasticity in this context refers to the ability of the soil to deform without crumbling when mixed with water. The plasticity of the fines is assessed using Atterberg limit tests, to be discussed later in the chapter.

Table 3.1 Soil nomenclature based on particle size

Soil name	Particle size range (mm)	Passes sieve no.
Boulders	>300	—
Cobbles	75–300	—
Gravels	4.75–75	—
Sand	0.075–4.75	4
Silt	0.005–0.075	200
Clay	<0.005	—

time the soil concentrations are measured at various depths in the tube. According to Stokes' law, which can be used to analyze the interplay of gravitational, buoyancy, and drag forces on a sphere moving through a liquid, the smaller the particle, the more slowly it settles in water. This means that, at any given time, the proportion of soil particles remaining in suspension will be greater for smaller than for larger particles. The hydrometer test correlates the evolution of soil particle concentration in the tube as a function of depth and time with the particle sizes. ASTM D7928 (ASTM 2021) covers the test procedure in detail. Laser techniques (e.g., Mastersizer) have been increasingly used in place of the hydrometer. But most laboratories don't yet have the required equipment.

Particle size distribution curves are obtained based on the results of sieve or hydrometer analysis. Figure 3.2 shows particle size distribution curves obtained using both methods. Each point on a particle size distribution curve corresponds to a particle size (plotted in the horizontal axis in logarithmic scale) and the percentage of particles by weight with size less than that size (plotted in the vertical axis). It is usual to have percentages increase from 0 to 100 in the upward (positive) direction of the vertical axis, while particle size decreases from left to right along the horizontal axis. Instead of always examining the particle size distribution of soils, engineers often rely on one or more parameters defined based on the particle size distribution curve. Most commonly, these parameters are the particle sizes D_{10}, D_{50}, and D_{60} such that exactly 10%, 50%, and 60% of the soil (by weight) are composed of particles smaller than these sizes, respectively. The coefficient of uniformity C_U is given by

$$C_U = \frac{D_{60}}{D_{10}} \tag{3.1}$$

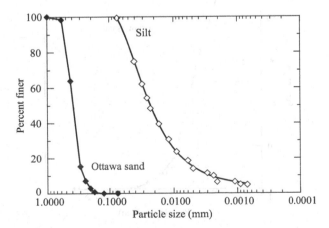

Figure 3.2 Particle size distribution of a sand obtained from sieve analysis and of a silt obtained from hydrometer analysis.

The D_{50} parameter is used to give a general indication of coarseness of the soil, while the coefficient of uniformity indicates whether the soil is well graded (poorly sorted) or poorly graded (well sorted). Soils with relatively uniform particle size have coefficients of uniformity with values <2. Well-graded soils have C_U values in excess of 4. Whether soils are relatively uniform, well graded, or somewhere between has an impact on various properties of the soil, such as hydraulic conductivity, stiffness, and shear strength.

A less useful parameter, the coefficient of curvature C_C, is defined as

$$C_C = \frac{(D_{30})^2}{(D_{10})(D_{60})} \tag{3.2}$$

Well-graded soils are expected to have C_C in the 1–3 range with, at the same time, a high value of C_U.

Example 3.1

Find the D_{10}, D_{50}, and D_{60} of Ottawa Sand from Figure 3.2, and calculate its coefficient of uniformity.

Solution:

E-Figure 3.1 shows the Ottawa Sand particle size distribution in Figure 3.2 replotted in a more convenient scale. The values of D_{10}, D_{50}, and D_{60} read from this figure are as follows:

$D_{10} = 0.265\,\text{mm}$
$D_{50} = 0.385\,\text{mm}$
$D_{60} = 0.415\,\text{mm}$

The coefficient of uniformity is calculated as

$$C_U = \frac{D_{60}}{D_{10}} = \frac{0.415}{0.265} = 1.57$$

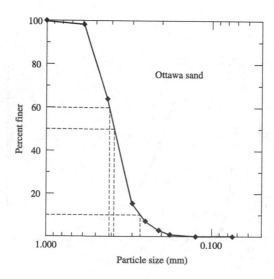

E-Figure 3.1 Particle size distribution of Ottawa Sand.

To a sand with this value of coefficient of uniformity, we would typically attach the adjective uniform. So Ottawa Sand with the particle gradation shown in E-Figure 3.1 is a uniform sand.

3.1.3 Unified Soil Classification System

Some practitioners find it helpful to classify soils according to their grain size distribution and plasticity characteristics using the Unified Soil Classification System (USCS). This system originated from the work of Arthur Casagrande in the first half of the 20th century concerning pavements for air strips and airports. It is now part of ASTM D2487 (ASTM 2017d) and is summarized in Table 3.2.

A significant criticism of the USCS is the 50% threshold used for percent fines in the soil. Research has shown (e.g., Salgado et al. 2000) that much smaller percentages of fines can lead to a soil in which larger particles, such as sand, are completely separated by fines, which therefore controls the soil response.

3.1.4 Composition of soil particles

A soil particle may contain one or more minerals. A mineral is a naturally occurring element or inorganic chemical compound that in most cases has a very well-ordered atomic arrangement (a crystalline structure). The minerals most commonly found in soils are silicates (quartz, feldspars, micas, and clay minerals), carbonates, oxides (iron and aluminum oxides), sulfate, and gypsum. Of these, carbonates are common in residual soils of carbonate rocks (such as limestone and dolomite) and in marine environments. Iron and aluminum oxides are present in tropical soils, which are often reddish in color due to the presence of iron oxides. Sulfate and gypsum occur in arid regions, such as parts of the western United States. Organic soils contain particles of organic origin (decomposed remains of living organisms).

3.2 GEOLOGY AND THE GENESIS OF SOILS AND ROCKS

Knowledge of engineering geology is very important to foundation and geotechnical engineers. This is particularly true for foundations in rock or in residual soils. It is desirable to discuss some of the basic aspects of soil and rock genesis and how they are important in foundation engineering problems. For more comprehensive reading on this topic, see Goodman (1992) or references listed in the Additional references section at the end of the chapter.

Earth is a giant recycling machine (Figure 3.3), constantly transforming soil into rock, solid rock into molten rock (magma), molten rock into rock again, and rock into soil. Additionally, materials get transported from one location to another. We will now discuss each of the materials and processes shown in Figure 3.3 in more detail.

3.2.1 Igneous rocks

The solid part of the earth, on top of which engineering activity virtually always occurs, is composed of plates. These plates, which have thicknesses ranging from tens to hundreds of kilometers, meet at three different types of boundaries: convergent, divergent, and transform. At convergent (subduction) boundaries, one plate dips beneath another. When that happens, the friction between the two plates generates heat. This is magnified by

Table 3.2 Unified Soil Classification System (USCS)

Major divisions			Group symbol	Typical names
Coarse-grained soils More than 50% retained on the 0.075 mm (No. 200) sieve	**Gravels** 50% or more of coarse fraction retained on the 4.75 mm (No. 4) sieve	Clean gravels	GW	Well-graded gravels and gravel–sand mixtures, little or no fines
			GP	Poorly graded gravels and gravel–sand mixtures, little or no fines
		Gravels with fines	GM	Silty gravels, gravel–sand–silt mixtures
			GC	Clayey gravels, gravel–sand–clay mixtures
	Sands 50% or more of coarse fraction passes the 4.75 mm (No. 4) sieve	Clean sands	SW	Well-graded sands and gravelly sands, little or no fines
			SP	Poorly graded sands and gravelly sands, little or no fines
		Sands with fines	SM	Silty sands, sand–silt mixtures
			SC	Clayey sands, sand–clay mixtures
Fine-grained soils More than 50% passes the 0.075 mm (No. 200) sieve	**Silts and clays** Liquid limit 50% or less		ML	Inorganic silts, very fine sands, rock flour, silty or clayey fine sands
			CL	Inorganic clays of low to medium plasticity, gravelly/ sandy/silty/lean clays
			OL	Organic silts and organic silty clays of low plasticity
	Silts and clays Liquid limit >50%		MH	Inorganic silts, micaceous or diatomaceous fine sands or silts, elastic silts
			CH	Inorganic clays or high plasticity, fat clays
			OH	Organic clays of medium to high plasticity
Highly organic soils			PT	Peat, muck, and other highly organic soils

Source: Modified after ASTM D2487 (ASTM 2017d).

Prefix: G = gravel, S = sand, M = silt, C = clay, O = organic.
Suffix: W = well graded, P = poorly graded, M = silty, L = $LL < 50\%$, H = $LL > 50\%$.

the high temperatures and pressures at great depths, causing the rock to melt into a fluid material, which has the same chemical composition as rock, called magma. The magma tends to be lighter than solid rock and to migrate upward. If it finds its way to the surface, volcanoes are created. Volcanoes allow magma to reach the surface quickly and to cool very fast, solidifying and generating magmatic rocks known as extrusive igneous rocks. It is also possible that the magma migrates upward but ends up never reaching the surface, solidifying very slowly within a mass of cooler rock. Such rocks are called intrusive igneous rocks.

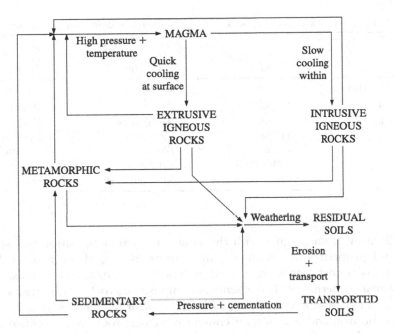

Figure 3.3 The geologic "recycling machine."

The rate of cooling turns out to be quite important in determining the appearance and properties of the resulting rock. Intrusive rocks are formed slowly, and their minerals have time to group during hardening, forming large grains of clearly distinguishable minerals. For example, we can easily identify, in a piece of granite, the minerals mica, feldspar, and quartz (Figure 3.4a). The mica shows as small dark plates if it is biotite mica, and as small grayish plates if it is muscovite mica. The feldspar crystals can be identified by their rectangular shape and shiny faces in either pink or gray color; the quartz, as glassy, rounded crystals that are most often clear to gray in color.

Minerals tend to group because chemical bonds between units of the same substance have a lower potential energy and are more stable than chemical bonds between units of different minerals. However, if the cooling process is too fast, extensive mineral coalescence is not possible. That is the case with extrusive rocks, which have a fairly homogeneous appearance with no clearly identifiable grains. The mineral groupings are still there, but they are very small. The classical example of an extrusive rock is basalt (Figure 3.4b).

(a) (b)

Figure 3.4 Samples of (a) granite and (b) basalt.

Table 3.3 Main igneous rock classification according to chemical composition and texture

	Rock type		
	Granitic	*Andesitic*	*Basaltic*
Coarse-grained texture	Granite	Diorite	Gabbro
Fine-grained texture	Rhyolite	Andesite	Basalt
Mineral composition	Quartz	Amphibole	Feldspar
	Feldspar	Feldspar	Pyroxene
	Micas	Biotite	Olivine
	Amphibole	Pyroxene	Amphibole
Rock color	Light	Medium dark	Dark gray to black

Rocks originating from magma with the same chemical composition will still differ in appearance and properties depending on the rate of cooling of the magma during rock formation. Extrusive rocks, as we have seen, will be fine-grained, while intrusive rocks will be coarse-grained. In terms of their chemical composition, rocks with large silica content and low iron and magnesium content are called granitic or felsic rocks. As the silica content decreases and the iron and magnesium content increases, rocks are classified as andesitic and then, for even lower silica contents and higher iron and magnesium contents, as basaltic. Table 3.3 provides the main igneous rocks and their most important features.

Intrusive rocks form at great depths. After long-lasting erosive action, intrusive rock masses are eventually exposed at the surface. While originally subjected to extremely large vertical stresses, they now find themselves under almost zero vertical stress. This subjects the rock mass to tensile strains that generate subhorizontal and/or subvertical fracture planes. This type of fracturing, common in granitic rock masses, is sometimes referred to as sheet jointing, as large sheets of granite are formed. Support for the stability of these sheets may be removed by excavations or voids created by erosion over geologic time, potentially leading to sliding (Terzaghi 1962). Another pattern of interest resulting from sheet jointing is the formation of large parallelepipeds of granite, which over time weather down to large rounded boulders floating in a soil matrix.

Basalt and other extrusive rocks tend to exist at the surface and immediately flow across large areas. This gives origin to what is referred to as basalt flows. The magma cools very fast, contracting in a way that generates predominantly vertical fractures, leading to relatively thin (of the order of 5–20 cm) columns of rock. The joint pattern is accordingly called columnar jointing.

3.2.2 Sedimentary rocks

Sedimentary rocks form as a result of the increasing compression and cementation of soil particles. This happens as soil deposits grow in depth, subjecting the deeper layers of material to increasingly greater stresses. Additionally, pore water containing a variety of cementing agents (ranging from weak clay-like materials and calcite to extremely strong oxides) in solution or colloidal suspension, flow through the soil pores, depositing these cements between soil particles. This process of rock formation is known in geology as diagenesis or lithification. It is not easy to define clearly what would be considered an aged, cemented soil and what would be considered a weak sedimentary rock. From a practical standpoint, it is not important whether we call a very strong and stiff material soil or rock, so long as we properly account for its strength and stiffness and how they might evolve as the material is loaded.

Sedimentary rocks are named after the predominant soils from which they originate. Sands give way to sandstones; clays to claystones; silts to siltstones; clays and silts to mudstones or shales; and gravels to conglomerates or breccias. Some sandstones are so strongly cemented by silica or quartz and are so strong that they are referred to as quartzites, a term also used for metamorphic rocks originating from sandstones. The range of shear strength and stiffness of sedimentary rocks is rather ample, depending mostly on the mineral composition of the original soil, the degree of compaction of the resulting rock, and the content and strength of the cement.

Carbonaceous rocks, such as limestone and dolomite, are sedimentary rocks resulting most commonly from the diagenesis of calcareous soils and sometimes from precipitation of calcite or dolomite from solution. These rocks are extremely soluble rocks, which frequently lead to the formation of karst. Karst is a term originally used to describe the landscape characteristic of the Karst region of Yugoslavia (Goodman 1992). In the United States, karst is found, for example, in southern Indiana, Kentucky, and parts of Florida. Karstic terrain presents challenges to the foundation engineer, particularly when it is in its advanced stage, when it usually contains underground cavities that can sometimes collapse, forming sinkholes. The process of site exploration in such terrains is very important, as it is desirable to learn as much as possible about the likelihood of existence of caverns and cavities and to assess the likelihood that their roof will not collapse. Another feature of karstic terrain is that the residual soil overlying rock can present very marked variations in elevation, with pinnacles of rock alternating with cavities. Figure 3.5 shows a typical mature karst terrain, where a cavity has collapsed.

3.2.3 Metamorphic rocks

Metamorphic rocks result from sedimentary, igneous, or other metamorphic rocks that undergo processes that subject them to extremely large pressures and high temperatures. The high temperatures cause the rocks to acquire a certain degree of fluidity. If pressures are then applied, minerals tend to realign themselves along the direction of the applied pressure. Granite, subjected to very high pressures and temperatures, becomes gneiss. Gneiss retains to a great extent the same colors and mineral composition as granite, but the mineral groupings cease to be distributed at random orientations, acquiring a predominant direction. In extreme cases of metamorphism, it is possible that bands of mica, feldspar, and quartz will

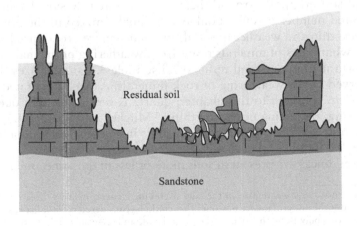

Figure 3.5 Illustration of mature karst terrain.

Table 3.4 Metamorphic rocks and likely parent rock

Metamorphic rock	Parent rock	Observations
Gneiss	Granite, schist	—
Quartzite	Sandstone	Extremely hard, strong rock
Slate	Shale	—
Schist	Shale	Differs from slate in that new minerals have formed
Marble	Limestone	—

form; the resulting rock is termed banded gneiss. Table 3.4 lists a few of the most important metamorphic rocks and their more common parent rocks.

Table 3.4 presents a simplified view of the complex progression of metamorphism in some cases. For example, shales are first metamorphosed into slates, but that can progress further to phyllite, schist, and then gneiss. A general understanding of the process is typically sufficient for foundation and geotechnical engineers, for if truly detailed geological knowledge is required, a geologist will be consulted.

3.2.4 Soil genesis: residual soils

Soil is created from the *weathering* of rocks. Weathering is the physical breakdown of rocks and/or the chemical transformation of rock minerals into other minerals. Chemical weathering is enhanced by relatively high temperatures and humidity, such as those that exist in tropical and subtropical areas. In arid regions, physical weathering predominates. Physical weathering usually precedes and facilitates chemical weathering, as it opens up fractures and other pathways for water to penetrate the rock and facilitate mineral degradation. For example, plant roots may penetrate some of the vertical fractures existing in granite due to the stress relief caused by erosion and in basalt from the fast cooling, opening them up for penetration of large amounts of water. The water is usually slightly acidic, reacting with the rock minerals to produce other minerals that are more chemically stable.

Soils that remain at the location where they were created by the weathering process are called *residual soils*. Figure 3.6 shows a typical granite residual soil profile. The degree of weathering increases as the distance from the bedrock increases, which really means weathering typically starts at the surface and makes its way down the rock, hence the more advanced state of weathering near the surface. Figure 3.7 shows granite boulders in place after the residual soil originally around them was excavated. Residual soils of basalt may have a more limited number of small boulders or cobbles mixed in the soil matrix, as the basalt columns are thin and weather into soils in a much more pronounced way.

Knowledge of what types of minerals result from weathering of the main minerals found in rocks is important to understand residual soils. Table 3.5 summarizes the main transformations observed in the weathering of rock minerals. As well illustrated by the beautiful poem used as an epigraph to this chapter, quartz is very stable and does not undergo chemical weathering, outlasting nearly anything.[4] Mitchell and Soga (2005) point out that quartz makes up only 12% of the weight of minerals in igneous rocks, compared with 59% for feldspars and 4% for micas, but that it is by a large margin the most common mineral in soils. Feldspars and micas are present in soils more or less in the same amounts. This speaks

---

[4] The poem is of course not about weathering or the sand stretching far away in every direction from Ozymandias' statue, but rather about the impermanence of values that could be considered materialistic, such as power, glory and possessions. We can only hope that art, as the poem itself, and science, as that covered in this book, fare better.

Figure 3.6 Typical residual soil of granite.

Figure 3.7 Granite boulders as part of a residual soil deposit.

to the chemical stability of quartz and the chemical vulnerability of other minerals, particularly feldspars. To illustrate how this knowledge can be used to understand the formation and behavior of residual soil, consider granite. Near the bedrock, where weathering is in its initial stages, there will be some boulders in a matrix composed predominantly of sand,

Table 3.5 Chemical weathering

Original mineral	Weathering product
Quartz	Quartz – quartz is very stable and is not broken down by chemical weathering
Feldspar Mica Ferromagnesian minerals	Clay minerals – the type of clay mineral formed depends on parent material and ambient conditions

with a silt and clay content of the order of 10%. In the initial stages of weathering, the clay results mostly from the chemical weathering of amphibole. Some of the feldspar may still exist as sand- or silt-sized particles. The sand fraction of the soil will also contain quartz and may contain a comparatively small number of sand-sized mica particles. Higher up in the soil profile, in a zone that has experienced weathering for a longer time, the boulders will have either disappeared or become much smaller, and the feldspars will have been converted into clay, as will much of the micas. In general, the further away from bedrock, the higher the clay content and the lower the sand content. Similar reasoning can be applied to residual soils of other igneous, metamorphic, and sedimentary rocks.

3.2.5 Transported soils

Erosion of residual soils and subsequent transportation to other locations by gravity, wind, or water leads to soil deposits that no longer have a close relationship with the parent rock. They are called *transported soils*. Their grain size distribution and fabric (the way in which particles are arranged) depend strongly on the transporting agent and the form of deposition. Among transported soils (Table 3.6), aeolian soils, alluvial soils, and colluvial soils deserve special mention.

Aeolian soils are soils transported and deposited by wind. The carrying capacity of wind, closely dependent on its velocity, is such that it transports and deposits particles of approximately the same size. Dune sand, for example, is an aeolian soil; it is characterized by a remarkably uniform particle size.

Alluvial soils are deposits of interlayered gravel, sand, silt, and clay. They form as rivers carry and deposit particles of varying sizes, depending on the water velocity, which

Table 3.6 Most common transported soils

Transporting agent	Soil designation	Comments
Wind	Aeolian (e.g., dune sand and loess)	Uniform particle sizes
Gravity (particles roll down slopes and cliffs)	Colluvial	Wide range of particle sizes
Water (streams and rivers)	Alluvial	Intermixed clay, silt, sand, and gravel layers
Water (lakes)	Lacustrine (e.g., varved clay)	Clay, silt, possibly interlayered due to seasonal factors
Water (oceans)	Marine (e.g., marine clay)	Possibly quick clay if pore fluid changes occur after deposition
Ice (glaciers advance and retreat in geologic time)	Glacial soils (e.g., till)	Wide range of particle sizes; overconsolidated

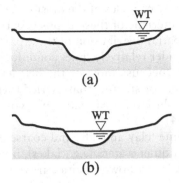

Figure 3.8 River cross section during (a) rainy periods and (b) dry periods.

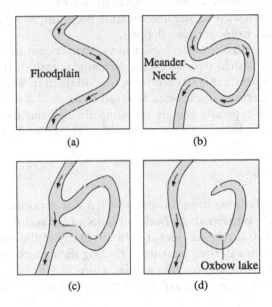

Figure 3.9 River bend: (a) original configuration; (b) sharpened by deposit on inside bank and removal of material from outside bank; (c) and (d) abandoned oxbow "lake."

changes across time and space. Locally, at a river cross section, water velocity drops with distance from the center of the river. It also varies seasonally, as the river expands laterally to occupy its flood plain during rainy seasons and shrinks during drier periods (Figure 3.8). As a result, the river tends to deposit soils on the inside of a bend and to remove soil from the outside. This, over time, increases the sharpness of the bend (Figure 3.9b); this process may go so far that eventually it becomes easier for the river to cut across the soil, leaving the bend behind as an abandoned oxbow lake (Figure 3.9c). The water velocity also varies along the axis of the river. Closer to its source, it tends to be steeper, and water velocities, higher. As the river approaches its delta, velocities drop substantially, and the river deposits larger volumes of soil (which tends to be made up of finer particles).

As a result of all these processes, alluvial deposits tend to have complex profiles with separate layers or lenses of gravel, sand, silt, and clay. When we know we are dealing with an alluvial deposit, we can usually take advantage of this knowledge for foundation engineering purposes by looking for layers of sand that are sufficiently thick and stiff on which to place foundations.

Colluvial deposits are formed by particles of the most varied sizes that roll down hills and slopes. The most striking characteristic of these deposits is the wide range of particle sizes and the rather heterogeneous spatial distribution of these particles.

Lacustrine soils are formed under relatively quiet conditions at the bottom of lakes, tending to be silty or clayey in size (because most of the heavier particles, sands and gravels, deposit onto the beds of the rivers or streams that carried them). Marine soils form as they sediment out of suspension or solution in sea water. We will discuss two marine soils with unusual behaviors in Section 3.4.

The environments in which fine (clay and silt) and coarse (sand and gravel) particles are deposited are quite different. In quiet waters (e.g., lakes), fine particles settle vertically. As the size of the particles increase, larger flow velocities are required to carry them. Thus, they are usually deposited in agitated waters (e.g., rivers), where they are dropped at points of reduced water flow velocity. An important consequence of this distinction is that clay and silt deposits tend to be much more homogeneous in the lateral direction than sand and gravel deposits. This has implications for a site investigation program, as we need far fewer borings or soundings in a laterally homogeneous deposit.

Glacial deposits have formed over geologic time as glaciers advanced and retreated many times, in synchrony with world climate cycles. When glaciers advanced, they would drag pieces of rock and soils with them and eventually unload them in some location. This action led to soils with varied particle sizes, often referred to as glacial tills. The soil underneath the glaciers would also be typically heavily overconsolidated[5] due to the large weight of the glaciers. The properties of glacial soils naturally reflect these two facts.

3.3 "CLASSIC" SOILS

Clays and sands, particularly as found in transported soil deposits, are the "classic" soils that have been extensively written about both in books and in technical articles. As a result, they are well understood. Gravels and silts are also generally well understood; there are substantial amounts of test data in the literature regarding their mechanical behavior. Gravels and nonplastic silts generally behave like sands, while plastic silts generally behave like clays. We will now discuss some basic aspects of the behavior of both sands and clays.

3.3.1 Silica sand

Silica sand, with particles made up of silicates, mostly quartz, is by far the most common type of sand (see Figure 3.10). The strong three-dimensional atomic arrangement of quartz has the important consequence that silica sand particles tend to be very strong, a fact that significantly affects its mechanical properties. Sands can be fine, medium-coarse, or coarse, depending on whether they are closer to the lower or upper bound of the sand range outlined in Table 3.1. With respect to size, they can also be classified as well graded if a variety of particle sizes are represented or poorly graded (uniform) if the particles are more or less of the same size (Figure 3.11).

Particle morphology (whether particles are more equidimensional or more elongated, the so-called sphericity of the particle, and whether the particle surface has a gentle curvature or sharp angles and "corners," quantified by the so-called roundness of the particle) will obviously affect the mechanical response of the soil. In general, angular, equidimensional

---

[5] An overconsolidated soil is a soil that has been subjected to a vertical effective stress greater than that experienced at present.

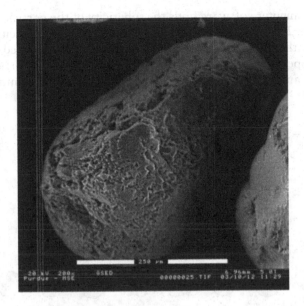

Figure 3.10 Particle of Ottawa Sand observed using a scanning electron microscope.

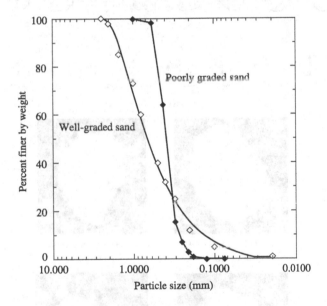

Figure 3.11 Examples of well-graded and poorly graded sand.

particles tend to lead to stronger, stiffer soil. In recent years, software tools have become available that allow a more precise, quantitative determination of these two quantities.

For calculations of particle morphology measures based on images of the particle, roundness is defined as the ratio of the average of the local minima of the radius of curvature within convex segments of the edge of the 2D particle projection to the radius of the largest circle inscribed in the projection of the particle (Wadell 1932), and sphericity is defined as the ratio of the width d_1 to the length d_2 of the particle:

$$S = \frac{d_1}{d_2}$$

(3.3)

Special-purpose computer software finds the sphericity of the particle by fitting an ellipse to the outline of the 2D projection of the particle. Although sphericity has been defined in different ways in the literature, Zheng and Hryciw (2015) showed that the definition in Equation (3.3) appears to best match the traditional, standard particle silhouettes proposed by Krumbein and Sloss (1951). Figure 3.12 illustrates the meaning of sphericity and roundness.

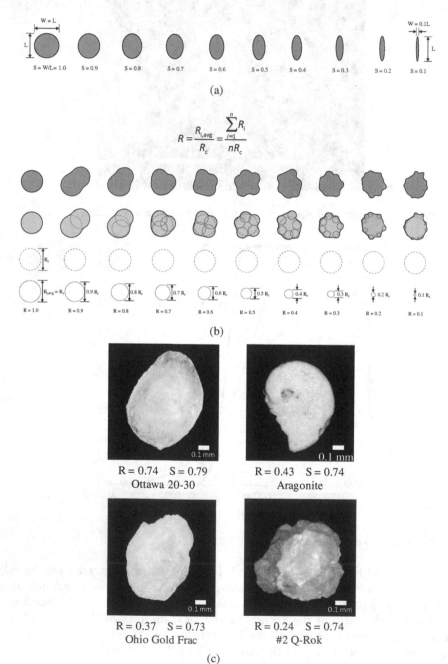

$$R = \frac{R_{i,avg}}{R_c} = \frac{\sum_{i=1}^{n} R_i}{nR_c}$$

Figure 3.12 Soil particle shape descriptors: (a) sphericity; (b) roundness; and (c) real particles with corresponding roundness and sphericity values.

Figure 3.13 Particles of nonplastic silt observed using a scanning electron microscope.

Sands appear in natural environments of many different types, including alluvial deposits, aeolian deposits, lacustrine deposits, and as residual soils of sandstone, granite, and other rocks. The characteristics of the sand particles, as well as the relative density and fabric of the soil, vary depending on the geologic setting in which it formed, as discussed briefly earlier. This should be recognized in planning subsoil investigations and developing a model of the subsoil for design.

Nonplastic silts tend to behave like sands, except that they drain more slowly, and in some cases substantially more slowly. Silt, as sand, may have particles that vary in angularity Figure 3.13 shows particles of a nonplastic silt.

3.3.2 Clays

3.3.2.1 Composition of clays

Clays are soils made up of small particles usually constituted of clay minerals. Different soil classification systems may define the size below which particles are considered to be of clay size differently. An upper limit of $5\,\mu m$ is most common. Clay minerals form most commonly from the chemical weathering of other minerals, mainly feldspars, micas, and ferromagnesian minerals. Clay minerals are quite resistant to any further weathering; accordingly, they are, for practical purposes, the end products of the weathering of different rock minerals. Clay minerals exist in the form of very small particles that typically contain a net negative charge and behave plastically when mixed with water (Mitchell and Soga (2005)) (Figure 3.14).

The fundamental building blocks of clay minerals (and indeed of many other minerals) are the silicon tetrahedron and the aluminum/magnesium octahedron. These are fundamental crystalline units where a central cation (typically, Si^{4+}, Al^{3+}, or Mg^{2+}) is surrounded by either oxygens or hydroxyls, with the oxygens and hydroxyls occupying the vertices of the tetrahedron or octahedron (Figure 3.15). A large number of silicon tetrahedra can be arranged as shown in Figure 3.16a to form a sheet of silica. The silica sheet is represented schematically by the geometric form also shown in Figure 3.16a. A similar arrangement of

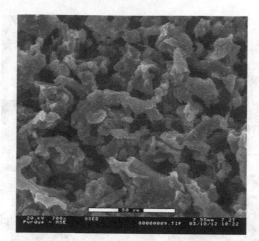

Figure 3.14 Particles of kaolinite clay observed using a scanning electron microscope.

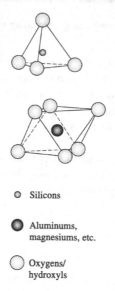

○ Silicons

● Aluminums,
    magnesiums, etc.

○ Oxygens/
    hydroxyls

Figure 3.15 The tetrahedral and octahedral units that, repeated in various ways, make up clay minerals.

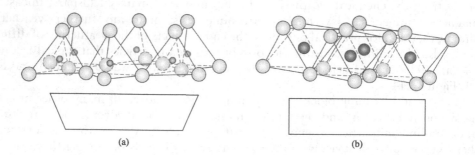

(a)   (b)

Figure 3.16 Silica and octahedral sheets: (a) tetrahedral arrangement in a silica sheet and schematic
           representation of the silica sheet; and (b) octahedral arrangement in an octahedral sheet and
           schematic representation of the octahedral sheet.

octahedra, shown in Figure 3.16b, leads to an octahedral sheet, represented schematically by a rectangle. If the central positions of the octahedra are filled by aluminum ions, the sheet is called a gibbsite sheet; if they are filled by magnesium ions, it is called brucite.

Figure 3.17 shows in schematic form how clay minerals result from the combinations of atoms into octahedral or tetrahedral units, which in turn are organized into silica, gibbsite, or brucite sheets. These sheets combine to form 1:1 or 2:1 silica–gibbsite or silica–brucite layers, which, in turn, are bonded together in various ways, forming the different clay minerals. The 1:1 and 2:1 layers are shown in Figures 3.18 and 3.19, respectively. These layers bond together because they are almost always not electrically neutral, having a net negative charge. This negative charge is due in large part to isomorphous substitution, by which either the silicon in the tetrahedron or the aluminum in the octahedron is replaced during

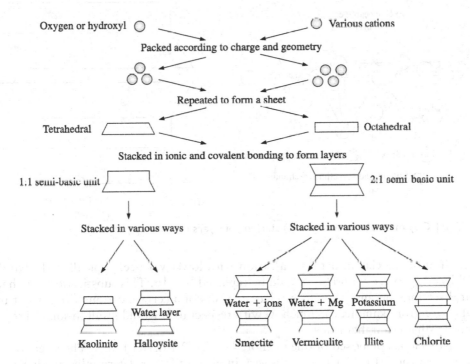

Figure 3.17 Clay minerals and their basic building blocks. (Mitchell and Soga (2005); Copyright © 2005; Reprinted with permission from John Wiley & Sons, Inc.)

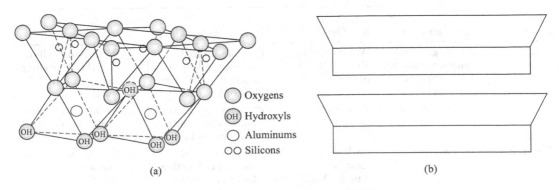

Figure 3.18 1:1 Clay mineral: (a) layer and (b) stacking of layers to make up particle.

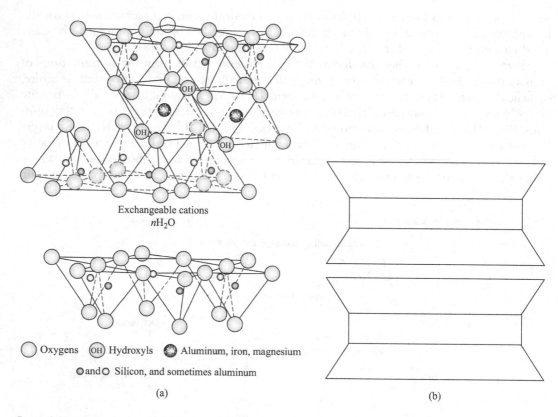

Exchangeable cations
$n\mathrm{H_2O}$

○ Oxygens   ⊙ Hydroxyls   ⊛ Aluminum, iron, magnesium

⊙ and ○ Silicon, and sometimes aluminum

(a)   (b)

Figure 3.19 2:1 Clay mineral: (a) layer and (b) stacking of layers to make up particle.

the formation of the clay mineral by a cation with lower valence. Typically, Al^{3+} would be replaced by Mg^{2+} or Fe^{2+}, and Si^{4+} would be replaced by Al^{3+}. This substitution may happen in natural settings because there is a relative excess of a certain cation with respect to the original cation (for example, too much Al with respect to Si would result in some Si replaced by Al in the silicon tetrahedra).

A clay particle is composed of one or more bonded 2:1 or 1:1 layers. The strength of the bonds between the 2:1 or 1:1 layers is a very important factor determining the behavior of

Table 3.7 Most common clay minerals

Clay mineral	Layer structure	Interlayer bonding	Particle dimensions	Interlayer swelling
Smectites	2:1 (w/gibbsite): montmorillonite 2:1 (w/brucite): saponite	Water+cations Van der Waals (weak)	Particles up to 2 microns in length and thickness = 10 Å to 1% of length	Weak bonds allow interlayer swelling by intrusion of water
Illite	2:1 (w/gibbsite)	Potassium (strong)	Particles up to several microns in length and thickness as low as 30 Å	Strong bonds allow little interlayer swelling
Kaolinite	1:1 (w/gibbsite)	Hydrogen bonds Van der Waals (very strong)	Particles up to 4 microns in length and thickness = 0.05–2 microns	Very strong bonds allow no interlayer swelling

the clay mineral. Table 3.7 lists the most common clay minerals (kaolinite, illite, and the smectites) and their most important properties. The two extremes in behavior are kaolinite and the smectites (notably, montmorillonite). Kaolinite typically results from the weathering of granite and rocks like it; smectites, from the weathering of basalt and similar rocks. A smectite that is well known in practice is bentonite, which took this name after the Benton Formation (formerly Fort Benton formation, in Wyoming, the USA). The interlayer bonds in kaolinite are of the very strong hydrogen bond (H-bond) type. The H-bonds develop between the oxygens of the silica sheet and the hydroxyls of the octahedral sheet of adjacent layers. In contrast, smectite has very weak interlayer bonding and, consequently, very thin particles. Due to the weak interlayer bonding, water can intrude reasonably freely, so smectites can swell significantly. Expansive soils tend to be rich in smectites. Kaolinite particles tend to be thicker than smectite particles, and water cannot break into the space between layers.

3.3.3 Clay–water systems

3.3.3.1 Double layer

As seen earlier, clay particles are constituted of one to several layers of one of the possible 1:1 or 2:1 units. Since the chemical bonds are much stronger within the layers than between layers, these particles tend to be platy, that is, wider and longer rather than thicker. Obviously, plates cannot be too large, for, at some ratio of thickness to length/width, they become vulnerable to breakage by bending. For many applications, however, they can be idealized as very extensive plates with a net negative charge near the surface of the particle. Except for very arid conditions, clays have considerable water contents. The water typically contains electrolytes (ions) dissolved in it. The interaction of a clay particle with surrounding water can be idealized as in Figure 3.20. Because of the net negative charge on the surface of the particle, cations tend to be attracted to the particle surface, and anions tend to be repelled. Therefore, the concentration of cations rises at an increasing rate as we approach the particle surface, and the concentration of anions mirrors that, decreasing sharply near the particle surface.

In essence, the clay particle generates a layer of "structured" water near it. The clay particle–water system is referred to as the double-layer system. When two particles approach each other, their respective double layers tend to resist this approach. So the double-layer system can be seen as an "enlarged" clay particle. To quantify the reach of a clay particle, we may define the double-layer thickness, $1/K$, as the distance from the centroid of the distribution of cation concentration as a function of distance from the particle surface. Mitchell and Soga (2005) show how $1/K$ can be calculated by solving a boundary-value problem, resulting in the following equation:

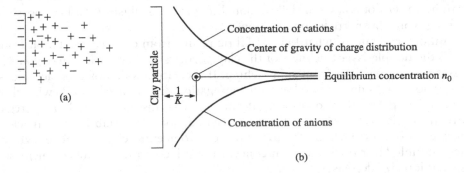

Figure 3.20 Clay–water system: (a) a platy clay particle, the water next to it, and the diffused ion cloud; and (b) concentration of cations and anions as a function of distance from the surface of the particle.

Table 3.8 Effects of changes in various factors on double-layer thickness

Change	Double layer	Explanation
Electrolyte concentration increases	Contracts	Smaller volume contains the number of cations required to balance the net negative charge of the particle
Cation valence increases	Contracts	Smaller number of cations is required to balance the net negative charge of the particle
Cation size increases	Expands	Number of cations required to balance net negative charge of the particle occupies larger volume

$$\frac{1}{K} = \sqrt{\frac{\varepsilon k_{\mathrm{B}} T}{2 n_0 e^2 v^2}} \qquad\qquad (3.4)$$

where ε = permittivity of the medium (typically, water), k_{B} = Boltzmann constant, T = absolute temperature in Kelvin, n_0 = electrolyte concentration, v = valence of the cations, and e = electronic charge. Another factor not appearing explicitly in Equation (3.4) is the size of the cations, which are assumed to be dimensionless particles (points) in the derivation of this equation. In reality, they do occupy space, and the larger they are, the larger the double-layer thickness $1/K$ is.

Table 3.8 summarizes the effects of changes in these three factors on double-layer thickness as well as the reasons for these effects.

3.3.3.2 Sedimentation of clay in water

Knowledge of the effects of double-layer thickness on clay sedimentation in water is useful to understand the different types of fabric resulting from how clay particles sediment in water. This is important because a nonnegligible percentage of the clay particles carried by streams and rivers are unloaded into lakes or oceans, where they find quieter waters and settle out of suspension. The first fact to recognize is that clay particles may or may not be aggregated. Particle aggregates are face-to-face combinations of particles (the combinations resulting from attraction forces that develop between particles when the distance between them becomes very small). When particles exist as part of aggregates, we say that the soil fabric is aggregated; otherwise, the fabric is referred to as dispersed. A fabric in which edge-to-face or edge-to-edge particle associations are prevalent is said to be flocculated; otherwise, we speak of a deflocculated fabric. Aggregation and flocculation are independent processes. A flocculated fabric can be either aggregated or dispersed. Likewise, a dispersed fabric can be either flocculated or deflocculated. Figure 3.21 illustrates all the possible fabrics resulting from clay particle combinations.

The main determinant of what kind of fabric results from clay sedimentation from suspensions is the double-layer thickness of the clay particles as they settle. Large double-layer thicknesses prevent particles from approaching. If the particles do not approach, the short-range attraction forces do not develop, and aggregation cannot develop. Likewise, flocculation also requires particles to come close (although not as close as required for aggregation). Therefore, if the clay is to deposit with an aggregated, flocculated fabric, the particles must be subjected to one or more of the following environmental changes (all of which lead to reduced double-layer thickness): ion concentration increases, ion valence increases, pH decreases, or ion size decreases.

Some of these conditions, particularly the first one listed, are likely to manifest in a marine deposition environment, for example. So, marine clays tend to be aggregated and

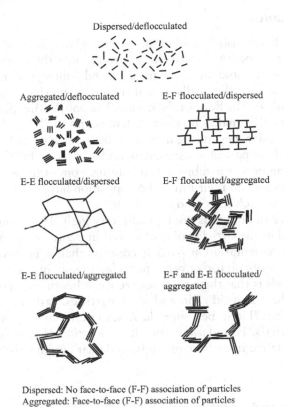

Dispersed/deflocculated

Aggregated/deflocculated   E-F flocculated/dispersed

E-E flocculated/dispersed   E-F flocculated/aggregated

E-E flocculated/aggregated   E-F and E-E flocculated/
                                  aggregated

Dispersed: No face-to-face (F-F) association of particles
Aggregated: Face-to-face (F-F) association of particles
Flocculated: Edge-to-edge (E-E) or edge-to-face
            (E-F) associations of aggregates
Deflocculated: No association between aggregates

Figure 3.21 Possible clay fabrics. (Van Olphen (1977); Copyright © 1977; Reprinted with permission from John Wiley & Sons, Inc.)

flocculated. This is not necessarily true in lacustrine environments, for example. Clay properties, notably hydraulic conductivity, are dependent on fabric. A very interesting situation develops when a clay deposits under a given set of conditions, which then change over geologic time. Quick clays result from this type of process, as discussed in the next section.

In general, for all other conditions being the same (and, in particular, for the same void ratio, a measure of how dense the soil is that we define more carefully in Section 3.5), it would be preferable for a foundation to bear on flocculated rather than on deflocculated fabric. In addition to the inherently stiffer structure of flocculated fabrics, the particles in a soil with deflocculated fabric will be relatively far apart for the void ratio to be the same as in the same soil with a flocculated fabric. This means the soil with deflocculated fabric will likely compress more under a given load.

3.4 "NONCLASSIC" SOILS*

The understanding of the mechanical response of transported soils of all particle sizes (gravels, sands, silts, and clays) is very good after about 100 years of testing and modeling. Soils not falling within this group are typically less understood. We will discuss a few of these "nonclassic" soils and their peculiarities next.

3.4.1 Carbonate sands

Carbonate sands exist both onshore and offshore throughout the world. They are plentiful in the Caribbean, off the coast of Brazil, in Australia, near the coast of the southeastern region of the United States, and in other tropical and subtropical regions of the world. Carbonate sand particles are the shells and skeletons of marine organisms. When these organisms die, they settle on the floor of the ocean. The only lasting part of these organisms is the calcium carbonate that makes up their skeletons and shells. Carbonate sand particles are hollow or highly angular, depending in part on how much particle breakage and crushing occur. Over time, these particles accumulate and eventually break under the weight of the particles above them or when subjected to loading from other sources. The angularity of carbonate sand particles and their mineral composition have the important consequence that the friction angles of such soils tend to be high.

The void ratios of carbonate sands are typically very high. This is due in part to the high particle angularity and intra-particle voids observed in many particles, but it may also be the result of very light cementation at particle contacts that, it is speculated, may develop soon after a particle deposits on top of other particles already in place. An additional peculiarity of carbonate sands is that their particles are weak because carbonates are weak, soft minerals. Because of the high void ratio and weak particles, carbonate sands are very contractive and compressible. These behaviors have important consequences for foundations installed in these materials. Foundations for oil exploration platforms and other offshore structures are often installed in carbonate sands, and their design must reflect the particular behaviors of these soils.

3.4.2 Marine clays and quick clays

Marine clays are clays that deposit in salt-water conditions, under conditions of high electrolyte concentration. As a result, these clays form with a relatively small double-layer thickness and tend to have flocculated fabric. Quick clays form when the salt water is leached out of the clay and replaced by fresh water. The leaching, which occurs over a long period of time, may happen for a number of reasons, notably land uplift due to the unloading upon glacial retreat but also possibly other mechanisms, including the development of an artesian condition sometime after the formation of the clay deposit. The replacement of water with high electrolyte concentration by water with low electrolyte concentration causes the double layer to expand. This leads to an unstable structure because the soil fabric was formed under conditions consistent with a smaller double-layer thickness. As a result, when disturbed, even if only slightly, particles will push each other apart, reducing shear strength significantly. This is illustrated quite dramatically by the Rissa quick clay slide of 1978 (the case study at the end of this chapter).

3.4.3 Expansive soils

Expansive soils are commonly soils containing large amounts of smectites. When these soils are exposed to water (because of rain, the breakage of water lines, or even just from watering lawns), water enters between the 2:1 layers because the interlayer bonding in smectites is very weak. Entry of water in the interlayer and interparticle spaces decreases the electrolyte concentration by dilution, which in turn leads to an increase in the double-layer thickness. This double-layer expansion and the water infiltration between the 2:1 layers are responsible for the expansion.

One of the serious problems with expansive soils is that the expansions that occur due to wetting of these soils are rarely uniform. So, a house on shallow foundations supported on these soils, for example, will typically get substantially distorted, which may lead to a serviceability limit state or worse. These soils are abundant in the United States, notably in the south and in the west. Average yearly damages due to expansive soils are quoted as greater than those caused by earthquakes, tornadoes, and other natural disasters combined. These damages could be greatly reduced if expansive soils were recognized more often as being present at construction sites. If they are identified in the course of the site investigation, suitable foundations can then be chosen; alternatively, appropriate measures could be taken to minimize the exposure of the soil to water, or additives could be added to the soil to minimize expansion.

3.4.4 Loess

Loess is a wind-blown silt deposit that covers about 11% of the area of the world, including parts of the midwestern and western United States. There is usually some light interparticle cementation in loess, due to either calcite or clay. The clay is usually smectite. Because of the way the particles are deposited and the presence of light cementation, loess exists at high void ratios, corresponding to unit weights as low as 11–12 kN/m³ (70–76 pcf). Another peculiarity of loess deposits is the presence of rootholes. When the deposit forms over tall grass, and the grass later dies and disappears, vertical rootholes develop. As a consequence, loess presents vertical cleavage. This tendency to cleave vertically combined with the light cementation leads to the formation of vertical cuts or slopes.

Upon saturation, the clay or calcite at the particle contacts will weaken or even dissolve, leading to large volume contractions and to sliding of the vertical slopes in loess deposits. If water cannot be prevented from reaching loess deposits, it is advisable to cause a forced collapse of the deposit and to flatten slopes before any construction takes place.

3.4.5 Organic soils

Organic soils result from the gradual decomposition of plants and roots. They are typically very soft, weak, compressible soils. They also are subject to pronounced secondary compression or creep. Peat is an organic soil. Even small loads applied on top of deposits of this material will lead to significant compression. The material is also weak, so slides or collapse ultimate limit states easily develop in organic soils.

3.4.6 Mixtures of sand, silt, and clay

Natural soil deposits often contain particles of various sizes. There are two ways in which the presence of smaller particle sizes affects the response of the soil: (1) by reducing the hydraulic conductivity of the material and thus making drainage slower and (2) by interfering with the interaction between the larger particles. The classical case is that of sands containing fines (either silt or clay). If the fines content is small, they may simply occupy the voids between the sand particles, not getting in between them. A fabric of this type is referred to as a nonfloating fabric (Salgado et al. 2000). The soil still behaves as a sand but with properties changed by the presence of the fines. If the fines content is large, the sand particles may be completely separated by fine particles, so that they are now floating in a silt or clay matrix. This soil then no longer behaves as a sand, but as a silt or clay. We will discuss specific values of limiting fines content as we discuss in later chapters the shear strength and stiffness of these soils.

3.5 SOIL INDICES AND PHASE RELATIONSHIPS

Soil indices are quantities that describe the relationships between weights and volumes of the soil particles, voids, and pore water. It would be impossible to quantify soil behavior without having the means to quantify how dense, how plastic, or how saturated soils are. Soil indices serve this purpose. Table 3.9 contains a fairly complete list of soil indices. The *void ratio e* is among the most important. The void ratio at a point is defined as the ratio of the volume of voids to the volume of solids (constituting the soil particles) prevalent at that point (a "point" here really means a small characteristic soil volume called the representative elementary volume, discussed later in this chapter) (Figure 3.22):

$$e = \frac{V_v}{V_s} \tag{3.5}$$

In sands, it is possible to define minimum and maximum void ratios. The minimum void ratio e_{min} corresponds to the densest state in which a sample of sand can form without significant particle crushing. The maximum void ratio e_{max} is the void ratio corresponding to the loosest possible arrangement for the sand that is not inherently unstable. An inherently unstable void ratio is one corresponding to a soil fabric (particle arrangement) that would collapse when subjected to small disturbances. An inherently unstable fabric is more likely in sands containing fines; at very large void ratios, fines separate sand particles, and it may not require much to dislodge these fines from between the sand particles, leading to collapse. ASTM has proposed standards [ASTM D4253 (ASTM 2016c) and ASTM D4254 (ASTM 2016d)] for the determination of e_{min} and e_{max} that produce consistent results, even when different operators perform the tests.

Table 3.9 Main soil indices

Soil index	Symbol	Definition
Specific gravity of solids	G_s	Ratio of unit wt. of solids to unit wt. of water, γ_s/γ_w
Porosity	n	Ratio of volume of voids to total volume of soil, V_v/V
Void ratio	e	Ratio of volume of voids to volume of solids, V_v/V_s
Specific volume	v	Ratio of total volume to volume of solids, V/V_s
Degree of saturation	S	Ratio of volume of water to volume of voids, V_w/V_v
Mass density of soil	ρ_m	Ratio of soil mass to total volume, m/V
Unit weight of soil	γ_m	Ratio of soil weight to total volume, W/V
Dry unit weight	γ_d	Ratio of weight of solids to total volume, W_s/V
Saturated unit weight	γ_{sat}	Value of γ_m when the soil is saturated
Buoyant unit weight of soil	γ_b	$\gamma_{sat} - \gamma_w$
Water content	wc	Ratio of weight of water to weight of solids, W_w/W_s
Minimum void ratio	e_{min}	Void ratio of soil in its densest state
Maximum void ratio	e_{max}	Void ratio of soil in its loosest state
Relative density	D_R	$(e_{max} - e)/(e_{max} - e_{min})$
Liquid limit	LL	Lowest water content at which soil exhibits viscous flow
Plastic limit	PL	Lowest water content at which soil exhibits plastic flow
Plasticity index	PI	$LL - PL$
Liquidity index	LI	$(wc - PL)/PI$

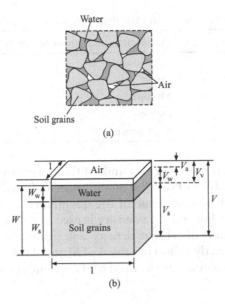

Figure 3.22 Soil and its three phases (solid particles, water, and air): (a) soil element and (b) phase diagram where the distinct phases have been conveniently combined into three separate layers.

ASTM has three vibrational procedures for determining e_{min}. If followed closely, these three methods produce similar results. Modified compaction test methods that yield smaller e_{min} values due to breaking and crushing of particles and consequent change in gradation must not be used for e_{min} determination.

The loosest possible stable condition that a soil can reach without risk of collapse is the condition under which e_{max} should be determined. There are several ways to form soil in approximately this condition: wet pluviation, in which sand particles settle through water (not to be used if fines are present, for they settle at different rates and will segregate from the larger soil particles); dry pluviation, in which particles are rained in through air and which also leads to segregation; and dry tipping. In the dry tipping method, sand is pluviated into a graduated cylinder using a funnel. The aim is to obtain sand in a loose state. The graduated cylinder is then inclined gently as close to the horizontal as possible, to obtain a state with practically no confining stress. By taking a reading of the average height of soil in the graduated cylinder, the volume is calculated and the unit weight calculated from the known mass. In any of the three cases, the procedure is repeated several times and the loosest condition is selected as the representative condition. ASTM D4254 (ASTM 2016d) provides three other methods with detailed guidance on obtaining e_{max}.

For clean silica sands, that is, sands containing no silt- or clay-sized particles, the values of e_{min} range from roughly 0.37 to 0.78, whereas e_{max} ranges from 0.65 to 1.24 (Bolton 1986; Cho et al. 2006; Han et al. 2018; Salgado and Prezzi 2007; Youd 1973). For sands with silt contents up to 20% by weight, both e_{min} and e_{max} decrease with the addition of fines. For sands with clay contents up to 10%, e_{min} decreases, but e_{max} may increase with the addition of fines. For more details on e_{max} and e_{min} of sands with fines, see Salgado et al. (2000) and Carraro et al. (2003).

The *relative density* D_R of a sand is defined as

$$D_R = \frac{e_{max} - e}{e_{max} - e_{min}} \tag{3.6}$$

Table 3.10 Relative density and the degree of compactness of sandy soils

D_R (%)	Qualitative assessment of degree of compactness
0–15	Very loose
15–35	Loose
35–65	Medium
65–85	Dense
85–100	Very dense

which is a number between 0 and 1 (or 0% and 100% if expressed as a percentage).

The relative density is useful in both the quantitative and qualitative characterization of soil. We will delay the discussion of some of the quantitative uses of D_R to later chapters, but refer to Table 3.10 to see how relative density is used to give a general idea of how dense sandy soils are. The usefulness of relative density will increase significantly as more attention is paid in practice to strictly following the ASTM standards.

The *water content, wc,* is defined as the ratio of the weight of water to the weight of solids for a given soil volume:

$$wc = \frac{W_w}{W_s} \tag{3.7}$$

The *degree of saturation* is the ratio of the volume of water contained in the soil pores to the total volume of pores:

$$S = \frac{V_w}{V_v} \tag{3.8}$$

The specific gravity of the solids G_s is the ratio of the unit weight of the solids γ_s to the unit weight of water γ_w. It is of the order of 2.6–2.7 for most soils. A useful relationship between void ratio, degree of saturation, and water content can be derived from Equations (3.7) and (3.8):

$$e = \frac{(wc)G_s}{S} \tag{3.9}$$

This relationship is particularly useful if the soil is known to be saturated because, in this case, void ratios can be determined directly from water contents, which in turn can be determined from disturbed soil samples, so long as their humidity is preserved immediately after sample recovery. Water contents are determined by weighing soil samples before and after oven-drying.

Atterberg (1911) proposed water content ranges for behaviors codified as "viscous" and "plastic." In the context of *Atterberg limits,* soil is said to exhibit viscous behavior if it does not retain its shape. There is a special ASTM procedure for the determination of the *liquid limit* (LL) of the soil, the lowest water content at which this viscous behavior is observed [ASTM D4318 (ASTM 2017a)]. The procedure consists of mixing soil with water and placing the resulting paste in a dish. A small groove is cut into the soil using a tool with a triangular shape. The dish is then dropped from a height of 1 cm (10 mm) as many times as needed for the groove to close. The procedure is repeated for various water contents; the water content at which the groove closes after 25 drops of the dish is the liquid limit. Because we are unlikely to prepare the soil at exactly the liquid limit, a linear fit to blow count versus water content data is used to determine *LL*.

Soil exhibiting plastic behavior can be molded into different shapes [such as thin cylinders, as in ASTM D4318 (ASTM 2017a)] without crumbling or cracking. The *plastic limit (PL)* is the lowest water content at which the soil behaves plastically. The test consists of preparing soil at various water contents and rolling the soil into 1/8-inch-thick threads. The water content for which the threads crumble is the plastic limit. The *plasticity index (PI)* is simply the difference between the *liquid limit* and the *plastic limit*. The *liquidity index (LI)* is defined as

$$LI = \frac{wc - PL}{PI} \tag{3.10}$$

The activity of a clay is given as (Skempton 1953):

$$A = \frac{PI}{\text{Clay fraction (\%)}} \tag{3.11}$$

Typical values of activity A are as follows: 0.4–0.5 (kaolinite clays), 0.75–1.25 (illite clays), 0.95 (London Clay), and values in excess of 2.0 for montmorillonite clays.

The usefulness of indices such as LL, PI, and A lies in the fact that there are correlations for shear strength and stiffness in terms of them, and some of these correlations may at times be useful. Atterberg limits are particularly useful for clays or clayey soils; they cannot be determined for sand and nonplastic silts but may be determined for plastic silts.

Other relationships (for derivations, see the chapter problems in Section 3.12) between soil indices that may be of interest are as follows:

$$e = \frac{n}{1-n} \tag{3.12}$$

$$n = \frac{e}{1+e} \tag{3.13}$$

$$\gamma_d = \frac{G_s \gamma_w}{1+e} \tag{3.14}$$

$$\gamma_m = (1+wc)\gamma_d \tag{3.15}$$

$$D_R = \frac{\gamma_{d,max}}{\gamma_d} \frac{\gamma_d - \gamma_{d,min}}{\gamma_{d,max} - \gamma_{d,min}} \tag{3.16}$$

$$\gamma_m = \gamma_w \left[G_s(1-n) + nS \right] = \gamma_w \left[\frac{G_s + eS}{1+e} \right] \tag{3.17}$$

Table 3.11 shows typical ranges for the main soil indices. Soil properties vary widely within these ranges, so the table should be used with caution in the estimation of soil properties. With respect to the Atterberg limits, the plasticity chart shown in Figure 3.23, developed by Arthur Casagrande, allows a quick way of placing a soil in broad categories that are then associated with behavior that will be more or less plastic. In later chapters, we will see that some correlations exist between the Atterberg limits and soil properties that can be used in design.

Table 3.11 Typical ranges for the main soil indices

Symbol	Sand	Silty sand	Clay
G_s	2.63–2.67	2.63–2.67	2.60–2.80
n	0.28–0.50	0.23–0.47	0.45–0.71
e	0.40–1.0	0.30–0.90	0.80–2.5
γ_{sat} (kN/m³)	18–21	18.5–22.5	15–18
wc for $S = 1$	0.15–0.38	0.10–0.35	0.28–1
e_{min}	0.4–0.5	0.3–0.4	NA
e_{max}	0.75–1	0.65–0.9	NA
LL (%)	NA	NA	30–200[a]
PL (%)	NA	NA	20–140
PI (%)	NA	NA	10–60

[a] The LL of smectites can be much higher than 200, but most clays would fall within this range.

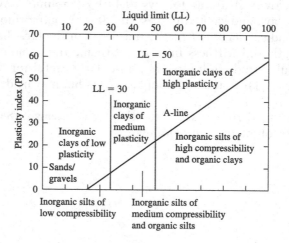

Figure 3.23 Plasticity chart. (After Casagrande (1932).)

Example 3.2

A 505-gram sample of soil was collected in the field from a certain depth in a saturated soil deposit. After oven-drying the sample, the dry mass was found to be 447 g. The specific gravity of the solids was measured as 2.64. Find the following:

1. *in situ* void ratio;
2. moist unit weight of the soil *in situ*; and
3. dry unit weight of the soil *in situ*.

Solution:

Refer to Equations (3.12)–(3.17) in the following.

1. Void ratio:

$$wc = \frac{505 - 447}{447} = 0.130$$

$$e = (wc)G_s = 0.130 \times 2.64 = 0.343$$

2. Moist unit weight:

$$n = \frac{e}{1+e} = \frac{0.343}{1.343} = 0.255$$

$S = 1$ in Equation (3.17) leads to

$$\gamma_m = \gamma_w \left[G_s(1-n) + nS \right] = 9.81 \left[0.255 + 0.745 \times 2.64 \right] = 21.8 \text{ kN/m}^3$$

3. Dry unit weight:

$$\gamma_d = \frac{\gamma_m}{1+wc} = \frac{21.8}{1+0.13} = 19.3 \text{ kN/m}^3$$

Example 3.3

The dry unit weight of a soil with porosity 0.313 is 18.2 kN/m³. Find the void ratio, the specific gravity of the soil solids, and the unit weight of solids.

Solution:

Void ratio:

$$e = \frac{n}{1-n} = \frac{0.313}{1-0.313} = 0.456$$

Using the equation:

$$\gamma_d = \frac{G_s \gamma_w}{1+e} = \frac{9.81 G_s}{1+0.456} = 18.2$$

$$G_s = 18.2(1+0.456)/9.81 = 2.7$$

and

$$\gamma_s = G_s \gamma_w = 2.7 \times 9.81 = 26.5 \text{ kN/m}^3$$

Appendix B has typical values of soil indices and properties pertaining to this and other book chapters.

3.6 EFFECTIVE STRESS, SHEAR STRENGTH, AND STIFFNESS

3.6.1 Interaction between soil particles and the effective stress principle

As we can see from the previous discussion, soils are extremely complex materials, presenting challenges that engineers do not have to face when dealing with other civil engineering materials, such as steel and concrete. In this text, we are mostly interested in the mechanical properties of soil, namely stiffness and strength. Soil stiffness and strength derive from the way particles interact in the presence or absence of pore fluid. In particular, the following interactions are of interest:

1. elastic deformation at contacts between soil particles;
2. bending of platy, elongated particles;

3. particle sliding/rolling;
4. particle crushing at contacts between soil particles; and
5. particle breakage.

Interactions (1) and (2) lead to elastic (recoverable) strains, while (3)–(5) lead to inelastic (irrecoverable) strains. Particle breakage or crushing occurs only in problems in which stresses become large. Recoverable shear strains in soils are strains less than ~10^{-5}. The small-strain shear and Young's moduli are denoted by G_0 and E_0, and the small-strain Poisson's ratio by v_0. The subscript zero is used here to indicate initial values of the soil parameters, that is, their values at very small strains.

3.6.2 The principle of effective stress

Naturally, since what happens at the contacts between soil particles is so important in determining the response of soil to load, any variable that affects the behavior of those contacts is an important variable in the quantification of soil behavior. The most important such variable is the stress actually experienced at the contacts between particles. Contact areas are very small, a tiny percentage of the area, A, of any planar surface crossing a soil mass, and impossible to quantify in the calculations done in practice. In the early part of the 20th century, Terzaghi introduced the concept of *effective stress* to represent the fraction of the total stress that is carried by saturated soil at the particle contacts. The effective stress principle states that the total force acting on an area, A, of soil is partly supported by the soil skeleton (through contacts between soil particles) and partly by the pore fluid. This is stated mathematically as

Total force on area A = Fraction of force carried by soil particles

+ Fraction of force carried by pore fluid

Dividing both sides of the equation by the area A, we get:

Total stress = Stress carried by the soil skeleton + pore pressure

or:

$$\sigma = \sigma' + u \tag{3.18}$$

where σ = total normal stress, σ' = effective normal stress, and u = pore pressure. Equation (3.18) has become known as the effective stress principle. The effective stress is not the true stress experienced at the contacts between soil particles, but rather an average stress, with respect to the area A, that can be attributed to the load-carrying capacity of the soil skeleton.

The stiffness and shear strength of soil depend on interactions (1)–(5) mentioned earlier between soil particles. Under low to moderate stresses, resistance of particles to sliding controls the soil response. This sliding resistance depends on the extent of particle contact areas and the forces across these contact areas. More compact particle arrangements under the action of larger confining stresses will be stiffer and stronger. So a volume decrease, which typically results from increasing stress, leads to stronger, stiffer soil. We can now relate that to Equation (3.18). If we apply an increment $d\sigma$ of stress under either isotropic or one-dimensional conditions to a fully saturated soil initially in equilibrium without the possibility of drainage, no change in volume is possible, and so the soil skeleton is not allowed to contract to develop additional stiffness. Without gaining stiffness, the soil skeleton cannot carry the additional stress; as a result, the additional stress must be carried by the pore water; in other words, $d\sigma = du$. For the soil to carry the additional stress, that is, for the effective stress to increase, water must be allowed to flow out of the pores so that the soil skeleton can contract. If full drainage is allowed, $d\sigma = d\sigma'$.

Most concepts and quantities in soil mechanics (such as stress, unit weight, or shear strength) refer to a single point within a soil mass. Each point within a soil mass may have a slightly different unit weight or shear strength, for example, because the soil particles would be packed slightly more densely (or loosely) at one point than at a nearby point. However, the concept that each soil property has well-defined values at each point within the soil implies continuity, while soil is clearly not a continuum. How do we reconcile the particulate nature of soil with the concept of continuity? The answer lies in what is meant by a "point." For our purposes, a "point" is a small three-dimensional element large enough to contain a sufficiently large number of particles (of the order of 1000 particles or more) to fully characterize in a macroscopic sense the soil (or the soil "behavior") within that element. This element is called in mechanics the representative elementary volume (REV). If the REV is small in comparison with the characteristic dimension of the problem to be analyzed, the hypothesis of soil as a continuum is perfectly acceptable. For example, a 1-meter footing is a thousand times greater than an average-sized sand particle, and as much as one million times larger than a clay particle. Clearly, soil can be modeled as a continuum for the analysis of footings supported in either clay or sand. On the other hand, if a thin needle is pushed through sand, the size of the needle is of the same order as the size of the sand particles, and it would be unreasonable to assume soil to be a continuum in this case. Thus, for the continuum hypothesis to apply, the characteristic dimension of the problem (the width of a footing, the diameter of a pile, or the height of a retaining wall or of a slope) must be a large multiple of the soil particle size (at least of the order of 30 times the size of the REV, and preferably higher).

3.6.3 Groundwater and the water table

We have seen throughout this chapter how soils and rocks have voids; so, water from rain and snow, and from streams, lakes, and oceans continuously enters and leaves the pores (and other voids) of soils and rocks. Groundwater may exist in soils under essentially hydrostatic conditions, or it may flow through the soil pores. If the soil is not capped at the top by an impervious layer, a water table exists. The water table is the surface at which the pore pressures are equal to zero. Under hydrostatic conditions, it is a level surface; otherwise, it is a sloping surface. The determination of the location of the water table or the establishment of any hydrodynamic conditions is important in the solution of any geotechnical problem.

Water is important in geotechnical problems in different ways. We examined in Section 3.1 how the existence of pore pressures reduces the effective stress actually carried by the soil skeleton. We will now see, through an example, how to calculate effective stresses under hydrostatic conditions.

Example 3.4

E-Figure 3.2 shows a sand deposit with a dry unit weight γ_d equal to 16 kN/m³ and a saturated unit weight γ_{sat} equal to 22 kN/m³ and four different hydrostatic conditions. Calculate the effective stress at a depth of 8 m for each of the four conditions depicted in the figure. For simplicity, assume the soil above the water table to be dry.

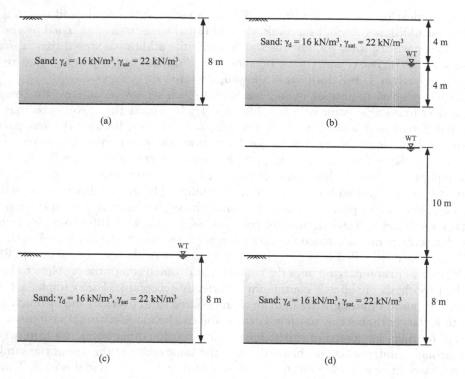

E-Figure 3.2 Soil profile with (a) deep water table (not shown), (b) water table 4 m below the ground surface, (c) water table at the ground surface, and (d) water table 10 m above the ground surface.

Solution:

The effective stress at a depth of 8 m is calculated using the principle of effective stresses
for the four cases as follows.
For deep water table:

$$\sigma' = \sigma - u = (16 \times 8) - 0 = 128\,\text{kPa}$$

For water table 4 m below the ground surface:

$$\sigma' = \sigma - u = (16 \times 4) + (22 \times 4) - (9.81 \times 4) = 113\,\text{kPa}$$

For water table at the ground surface:

$$\sigma' = \sigma - u = (22 \times 8) - (9.81 \times 8) = 98\,\text{kPa}$$

For water table 10 m above the ground surface:

$$\sigma' = \sigma - u = (9.81 \times 10) + (22 \times 8) - (9.81 \times 18) = 98\,\text{kPa}$$

3.6.4 Unsaturated soils*

Bishop (1959) generalized the principle of effective stress, applicable only to saturated soils,
to unsaturated soils:

$$\sigma' = (\sigma - u_a) + \chi(u_a - u) \tag{3.19}$$

where σ' is the effective stress, σ is the total stress, u_a is the pore air pressure (zero under 1 atmosphere), u is the pore water pressure, and the effective stress parameter χ represents the fraction of the matric suction that contributes to the effective stress; χ has been considered to be a function of the degree of saturation in such a way that $\chi = 0$ if $S = 0$ and $\chi = 1$ if $S = 1$. In the laboratory, u_a can be controlled, but, in real applications in the field, $u_a = 0$, so Equation (3.19) reduces to Equation (3.18) at either of these extremes.

3.7 GROUNDWATER FLOW

3.7.1 Effects of groundwater flow

Contrary to the popular saying, water does not simply "flow downhill." Water flows between two points when the energy per unit mass of water at one point is higher than at the other. This energy has elevation as a component, but it also has two others: deformation or pressure energy and kinetic energy. Groundwater flow may imply different values of pore pressure at different points within the soil. Water flow can therefore change the effective stress state within a soil, and that may even lead to ultimate limit states for a structure. It is therefore crucial to understand groundwater flow from the point of view of not only quantifying and understanding the flow itself, but also its effects on the soil through which the flow is happening.

We will now review the basics of groundwater flow in soils, which will be needed when we discuss clay consolidation in Chapter 6 and when we examine the stability of retaining structures in Chapter 16 and slopes in Chapter 17.

3.7.2 Elevation, kinetic, and pressure heads

In studying water flow in hydraulics, we define the *hydraulic head* as the amount of energy per unit weight of water. If we consider a water drop at an elevation z (measured with respect to a reference plane called the *datum*), some of the energy contained in this water drop will be potential energy, which is a result of the elevation z; some will be a pressure-related energy (related to the water pressure u); and some will be kinetic energy (related to the water flow velocity). Mathematically:

$$E = mgz + uV + \frac{1}{2}mv_t^2 \tag{3.20}$$

where E = total energy associated with the drop of water with mass m, volume V, and velocity v_t located at an elevation z with respect to the reference plane (the datum). The energy per unit weight of the water, referred to as the hydraulic head h of the water, can be calculated by dividing Equation (3.20) by the weight $W = \gamma_w V = mg$ of the water drop:

$$h = z + \frac{u}{\gamma_w} + \frac{v_t^2}{2g} \tag{3.21}$$

Equation (3.21) is known as the Bernoulli equation. In soils, the velocity v_t of water flow is comparatively quite small, and the third term of Equation (3.21) becomes negligible compared with the first two. So the hydraulic head, which has units of length, can be expressed with sufficient accuracy by

$$h = z + \frac{u}{\gamma_w} \tag{3.22}$$

We can rewrite Equation (3.22) in more compact form as

$$h = h_e + h_p \tag{3.23}$$

where h_e = elevation head and h_p = pressure head.

3.7.3 Darcy's law

The unit flow rate (specific discharge) v through the soil can be expressed through Darcy's law (Darcy 1856):

$$v = \frac{Q}{A} = -Ki \tag{3.24}$$

where v is the *specific discharge*, which has units of velocity [LT⁻¹]; Q is the *flow rate* [L³T⁻¹]; A is the cross-sectional area of the soil through which the water flows; K is the *hydraulic conductivity* [LT⁻¹] of the soil; and i is the *hydraulic gradient* (which is unitless). The minus sign means the flow occurs from points with higher to points with lower hydraulic head or potential.

The hydraulic conductivity is a function of both the soil and the fluid flowing through its pores. It is expressed in terms of fundamental soil and pore fluid properties as (Muskat 1937):

$$K = k \frac{\gamma_{fl}}{\mu} \tag{3.25}$$

where k = intrinsic permeability of the soil, γ_{fl} = unit weight of the flowing fluid, and μ = dynamic viscosity of the fluid. The unit in which k is expressed is called a darcy (one darcy equals $10^{-12}\,\text{m}^2$) in honor of the proponent of Darcy's law. Permeability has been shown to increase approximately with the square of particle size. Clays, having very small particle size, have very low hydraulic conductivity; they allow water to flow through their pores at very low rates. Obviously, the hydraulic conductivity of very viscous oil, for example, would be lower than that of water. Our focus in this text is on water. Figure 3.24 shows typical ranges for the values of both k and K for various soils and rocks; the K values are for groundwater flow.

The hydraulic gradient is the ratio of the hydraulic head differential dh between two points to the length dL of the flow path between the two points (Figure 3.25). Mathematically, for two points very near each other, we can write:

$$i = \frac{dh}{dL} \tag{3.26}$$

Equations (3.24) and (3.26) show that the hydraulic gradient drives the flow and determines the rate of flow through the soil. The hydraulic gradient increases with increasing difference in head between two points and with decreasing distance between the two points. Considering, in addition, that Equation (3.23) tells us that hydraulic head is composed of both an elevation and a pressure component, we can see that water flows from points with higher water pressure to points with lower water pressure for points at the same elevation, and that water flows from higher to lower elevations for points at the same water pressure. For other conditions, both pressure and elevation heads must be considered together, and it

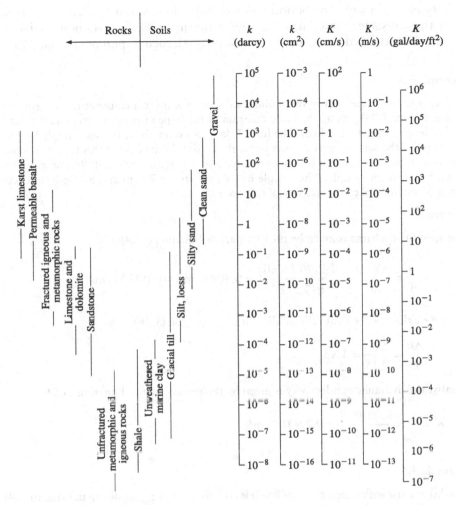

Figure 3.24 Typical values of permeability as well as hydraulic conductivity for water. (Reproduced from Freeze and Cherry (1979).)

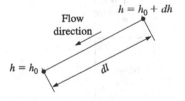

Figure 3.25 Hydraulic gradient between two points.

is perfectly possible for water to flow upward, for example, so long as the pressure at lower points is sufficiently high.

The hydraulic conductivity of laboratory samples can be measured using either a constant-head or a falling-head permeameter. An illustration of the measurement of K using a constant-head permeameter is given in Example 3.5. The interpretation of falling-head permeameter tests is discussed in Appendix C. Given the strong dependence of hydraulic conductivity on grain size distribution and soil state (density and fabric, in particular), the validity of these tests for materials occurring naturally in the field is questionable. In such cases, K may be estimated using an *in situ* test method, such as pumping tests.

Example 3.5

Consider the measurement of hydraulic conductivity K using a constant-head permeameter. E-Figure 3.3 shows a schematic diagram of this type of permeameter. As the name of the device suggests, the hydraulic head is kept constant above the soil sample by adding water at the same rate as it flows through the soil. In the case of this test, a constant differential head of 500 mm is maintained for 2 minutes, after which 1489 g of water has flown through the sample. The sample has a diameter of 7.2 cm and a length of 15 cm. What is the hydraulic conductivity of this soil?

Solution:

The specific discharge is given by the first part of Equation (3.24):

$$v = \frac{Q}{A} = \frac{\Delta V / \Delta t}{\pi r^2} = \frac{1489/(2\times60)}{\pi (7.2/2)^2} = 0.3048\,\text{cm/s} = 0.003048\,\text{m/s}$$

Now we calculate the hydraulic gradient using Equation (3.26):

$$i = \frac{\Delta h}{L} = \frac{0.5}{0.15} = 3.33$$

Finally, the hydraulic conductivity is given by the second part of Equation (3.24):

$$K = \frac{v}{i} = \frac{0.003048}{3.33} = 9.2\times10^{-4}\ \text{m/s}$$

Example 3.6

Two lakes exist with different water level elevations (z_{L1} and z_{L2}, as shown in E-Figure 3.4). These lakes are a distance L apart. The geology of the site is such that the lakes are completely isolated by impervious material, except at a specific location, where a horizontal

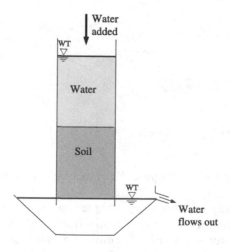

E-Figure 3.3 Schematic diagram of a constant-head permeameter.

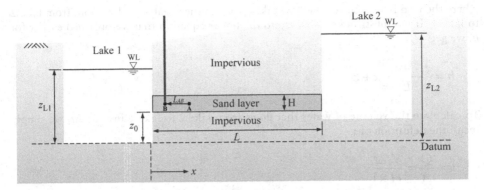

E-Figure 3.4 A sand layer of uniform thickness joining two lakes with different water elevations.

sand layer of thickness H joins the two lakes. (1) Find how the hydraulic head and pore pressure vary along the sand layer. (2) If the hydraulic conductivity of the sand layer is K, what is the specific discharge? (3) How much water flows from one lake to the other during a given time period Δt?

Solution:

Because the sand layer is horizontal, the elevation head at its two extremes is the same (z_0 for the base of the layer, as shown in E-Figure 3.4). The pressure heads are $z_{L1}-z_0$ and $z_{L2}-z_0$. So the hydraulic head is z_{L1} on the left boundary of the layer and z_{L2} on the right boundary of the layer. Note that this will be true of any elevation within the layer (that is, whether we consider the base or top of the layer), showing that, in this problem, h depends only on the horizontal coordinate x, not on y.

According to Equation (3.24):

$$v = -Ki = -K\frac{\partial h}{\partial x} = -K\frac{dh}{dx}$$

where the transformation to an ordinary differential equation is because h depends only on x. If we integrate this equation, we obtain:

$$h = -\frac{v}{K}x + C$$

where C is an integration constant. We can now use the boundary condition that $h = z_{L1}$ at $x = 0$ to find that:

$$C = z_{L1}$$

Taking that back into the equation for h:

$$h = z_{L1} - \frac{v}{K}x$$

which gives us the variation of h with x, which is obviously linear. We can now calculate v by simply plugging the other boundary condition ($h = z_{L2}$ at $x = L$) into the preceding equation:

$$v = -K\frac{z_{L2} - z_{L1}}{L}$$

where the minus sign means the flow takes place in the negative x direction, from lake 2 to lake 1. If we substitute this back into the linear equation that we obtained earlier for h, we get:

$$h = \frac{z_{L2} - z_{L1}}{L} x + z_{L1}$$

To calculate the volume of water that flows from lake 2 to lake 1 in time Δt, we simply recall the definition of v:

$$v = \frac{Q}{A} = \frac{V/\Delta t}{H \times 1}$$

So the volume that flows from lake 2 to lake 1 is equal to:

$$V = vH\Delta t$$

in m^3/m or ft^3/ft.

The specific discharge v is not the actual velocity of the water through the soil. To simplify the discussion, consider linear flow. The true velocity v^* of water will be the flow rate divided by the cross-sectional area actually filled by water, that is, the voids area $A_v = nA$. This means that the true velocity of a water drop is:

$$v^* = \frac{vA}{A_v} = \frac{v}{n} \tag{3.27}$$

One of the reasons we might want to know how long it takes for a water drop to travel from one point to another is in the study of contaminant transport (refer to Problem 3.23). The time required for a contaminant that finds its way into the ground to arrive at another location depends on v^*, not v.

3.7.4 Two-dimensional water flow through soil

Equation (3.24) tells us how fast water flows through soil at a point depending on what the hydraulic gradient is at the point. Where a water drop initially goes depends on the direction of maximum hydraulic head change per unit length; how fast the drop moves depends on the magnitude of the change in hydraulic head per unit length. This is very much analogous to the case of a ball released on a slope (Figure 3.26). It will roll down the slope in the direction of steepest drop in elevation head (because it is not subject to a pressure head, in contrast with our water drop, which is subject to both elevation and pressure heads), which is the same thing as saying that it maximizes the rate of change of potential energy. How fast the ball goes depends on how steep the slope is. When we developed Equation (3.24), we did not concern ourselves with the direction of flow, but we must now do so. If we are to use Equation (3.24) in two-dimensional or three-dimensional groundwater flow simulations, we must formulate our analysis independently of knowing the direction of flow and let that be a result of the analysis.

For two-dimensional flow, we can quantify the position occupied by a point or by a water drop by establishing a Cartesian reference system with axes x_1 and x_2. If we don't know *a priori* the direction of flow at a point in the x_1–x_2 space and therefore cannot write Equation (3.24) directly along the flow path, we can still write that the components of the

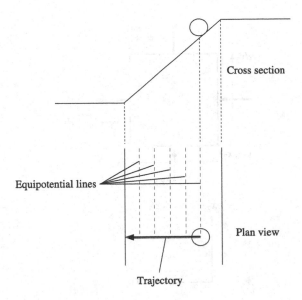

Figure 3.26 A ball released on a slope seeks the direction of maximum change of the potential energy due to gravity, that is, the steepest trajectory. The steepest trajectory is also perpendicular to the lines of equal potential energy (equipotential lines), shown as dashed lines in the plan view of the slope.

specific discharge in the directions x_1 and x_2 must be proportional to the components of the hydraulic gradient in those directions. Therefore, they must satisfy Equation (3.24) on their own:

$$v_1 = K_1 i_1 = -K_1 \frac{\partial h(x_1, x_2)}{\partial x_1} \tag{3.28}$$

and

$$v_2 = -K_2 i_2 = -K_2 \frac{\partial h(x_1, x_2)}{\partial x_2} \tag{3.29}$$

where the hydraulic conductivity has been assumed to be potentially different in the horizontal (K_1) and vertical (K_2) directions. We often assume the soil to be isotropic,[6] so that $K_1 = K_2 = K$. Dropping the notation $h(x_1, x_2)$ in favor of simply h (but remembering that the hydraulic head is a function of position in the x_1–x_2 space), we may more simply write:

$$v_1 = -K \frac{\partial h}{\partial x_1} \tag{3.30}$$

and

$$v_2 = -K \frac{\partial h}{\partial x_2} \tag{3.31}$$

---

[6] A soil is isotropic at a point if it has the same value of some property (in this case, K) in all directions at that point. If the soil is, in addition, homogeneous, the value of the property is the same for every point of the soil mass.

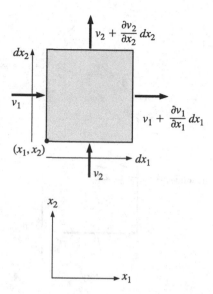

Figure 3.27 Elemental (control) volume through which the net flow of water must be zero to conserve mass.

Water flow must also satisfy the physical law of conservation of mass.[7] Given that water can be taken as incompressible for our purposes, we can satisfy this law by enforcing the requirement that the amount of volume that flows into an element is exactly matched by that flowing out of the element. This is shown in Figure 3.27, in which the volume of water flowing into the element (with cross-sectional area normal to the x_1 direction equal to $dx_2 \times 1$ and, conversely, cross-sectional area normal to the x_2 direction equal to $dx_1 \times 1$) is equal to:

$$dV = v_1 dx_2 dt + v_2 dx_1 dt$$

The volume of water exiting the element must be:

$$dV = \left(v_1 + \frac{\partial v_1}{\partial x_1} dx_1 \right) dx_2 dt + \left(v_2 + \frac{\partial v_2}{\partial x_2} dx_2 \right) dx_1 dt$$

For these two volumes to be the same, we must have:

$$\frac{\partial v_1}{\partial x_1} + \frac{\partial v_2}{\partial x_2} = 0 \tag{3.32}$$

If we now substitute Equations (3.30) and (3.31) into Equation (3.32), we obtain the following partial differential equation on h:

$$\frac{\partial^2 h}{\partial x_1^2} + \frac{\partial^2 h}{\partial x_2^2} = 0 \tag{3.33}$$

---

[7] In civil engineering, we deal only with Newtonian mechanics and have the luxury of not having to worry about the transformation of mass into energy according to the famous (and simplified) $E = mc^2$ equation attributable to Einstein.

The partial derivatives of Equation (3.33) may be rewritten as a linear operator operating on h as follows:

$$\left(\frac{\partial^2}{\partial x_1^{\,2}}+\frac{\partial^2}{\partial x_2^{\,2}}\right)h=0$$

We recognize $\dfrac{\partial^2}{\partial x_1^2}+\dfrac{\partial^2}{\partial x_2^2}$ as the two-dimensional Laplacian operator, which, considering all three dimensions, would be written as

$$\nabla^2=\frac{\partial^2}{\partial x_1^{\,2}}+\frac{\partial^2}{\partial x_2^{\,2}}+\frac{\partial^2}{\partial x_3^{\,2}} \tag{3.34}$$

In our case, the derivative with respect to x_3 is zero, as we are analyzing two-dimensional flow and therefore nothing varies in the x_3 direction. Our equation now becomes:

$$\nabla^2 h=0 \tag{3.35}$$

which is Laplace's equation, one of the classical partial differential equations.[8]

To solve a partial differential equation and obtain a solution that is meaningful in the context of engineering, we need boundary conditions (and, sometimes, although not here, initial conditions). A differential equation considered together with boundary and initial conditions is known as a boundary-value problem. Engineers find (sometimes) and use (almost all the time) solutions to boundary-value problems. If the boundary conditions are simple, analytical solutions can be found. Let's reconsider the case of Example 3.6.

Example 3.7

Solve Example 3.6 using Laplace's equation.

Solution:

As discussed in Example 3.6, h depends on x only. Laplace's equation reduces to:

$$\frac{d^2 h}{dx^2}=0$$

which, integrated, gives:

$$h=Cx+D$$

We have two integration constants, for which we have two boundary conditions ($h=z_{L1}$ at $x=0$ and $h=z_{L2}$ at $x=L$). Substituting these, one at a time, into the preceding equation, gives $D=z_{L1}$ and $C=\dfrac{z_{L2}-z_{L1}}{L}$, so that:

$$h=\frac{z_{L2}-z_{L1}}{L}x+z_{L1}$$

---

[8] We will study and use two others: the consolidation (heat) equation in Chapter 6 and the wave equation in Chapter 14.

which is the same equation we obtained in Example 3.6. From this point on, the solutions are the same. The difference in how we solved the same problem is that, in using the Laplace equation, we first obtained how h varies in space, and then everything else follows from that.

Two-dimensional flow is usually not as simple to analyze because the boundary conditions in most cases of interest in practice are not conducive to analytical solutions. When analytical solutions are not possible or are not easily obtained, we resort to numerical methods. The finite difference method, in which derivatives are replaced by ratios of finite differences, and the finite element method, in which the differential equation at a point in space is rewritten as an integral equation over a small element with small but not zero volume approximating the point, are used. It is beyond the scope of this text to describe this subject in detail.

In the past, when neither the finite difference nor the finite element method was available or practical, given the absence of computers, engineers resorted to drawing flow nets. The basic ideas behind flow nets (for a homogeneous and isotropic soil mass) are the following:

1. Flow trajectories (flow lines) are perpendicular to lines of equal hydraulic head (equipotential lines).
2. If we "guess" the shapes of the flow lines and equipotential lines correctly, and use the right number of each type of lines, we should be able to obtain a flow net that will be composed of approximately square units. A unit here means the space defined by two adjacent flow lines and two adjacent equipotential lines.

Figure 3.28 shows a flow net. Note how the boundary conditions here correspond to the hydraulic head being set at specific values at the boundaries AB and CD. So AB and CD are themselves equipotential lines. Between AB and CD, there is a head loss of Δh. Since there are 17 equipotential lines, there are $17-1 = 16$ incremental head losses between AB and CD. This allows computation of pore pressures at any point in the soil, as illustrated in the next example.

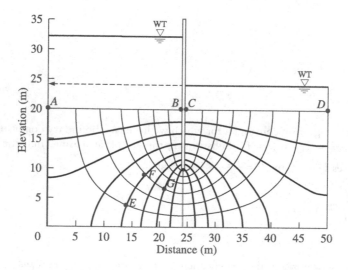

Figure 3.28 A correctly drawn flow net.

Example 3.8

For the flow net in Figure 3.28, calculate the pore pressures at points E, F, and G.

Solution:

We start by taking note of the elevations of points E, F, and G (with respect to our datum, for which $z = 0$). This gives:

$$z_E = 3.8\,\text{m}$$

$$z_F = 9\,\text{m}$$

$$z_G = 6.8\,\text{m}$$

Then we note that points E and F lie on the fifth equipotential line and that point G lies on the seventh equipotential line. We know further that the total hydraulic head is 32 m (12 m of pressure head plus 20 m of elevation head) at AB and 24 m at CD. This means we have a total head loss of 8 m. This means the head loss between each pair of equipotential lines is $8/16 = 0.5$ m. We can now calculate the head at E and F:

$$h_E = h_F = 32 - 4 \times 0.5 = 30\,\text{m}$$

and at G:

$$h_G = 32 - 6 \times 0.5 = 29\,\text{m}$$

The pore pressures can now be computed by using:

$$u = \gamma_w h_p = \gamma_w (h - z)$$

So we get, in sequence,

$$u_E = \gamma_w h_{pE} = \gamma_w (h_E - z_E) = 9.81(30 - 3.8) = 257\,\text{kPa}$$

$$u_F = \gamma_w h_{pF} = \gamma_w (h_F - z_F) = 9.81(30 - 9) = 206\,\text{kPa}$$

$$u_G = \gamma_w h_{pG} = \gamma_w (h_G - z_G) = 9.81(29 - 6.8) = 218\,\text{kPa}$$

3.7.5 Seepage forces

Recall from our discussion of Equations (3.20) and (3.21) that h is the energy per unit of weight of water. So, when water flows along a trajectory L with a head drop dh for a travel length dL, the energy available to one unit of weight of water has dropped by dh. If we multiply dh by γ_w, we obtain the energy loss per unit of volume of water. Where has that energy gone?

For the water to flow through soil, it has to overcome drag forces. Energy is required to keep the water flowing. This energy comes from the head loss. If the soil opposes water flow, in effect exerting a force on the water against the direction of flow, this means, by Newton's third law, that the water exerts a force on the soil skeleton as well. This force is called seepage force, and it may be calculated by realizing that the energy loss during water flow is "converted" (in effect, dissipated) into the work that must be done to overcome the resistance to flow by the soil skeleton. The work dW done by the *seepage force* f_s per unit volume of water is simply the force times the volume of water times the displacement dL along trajectory L at the point under consideration:

$$dW = V_w f_s dL$$

The energy loss during the flow along dL is:

$$dE = dh\gamma_w V_w$$

By requiring that $dE = dW$:

$$f_s = \gamma_w \frac{dh}{dL} = i\gamma_w \qquad (3.36)$$

Seepage forces become important in the following types of problems: (1) slope stability analysis (if water is flowing down the slope, it exerts a driving or destabilizing force on the slope (Chapter 17)) and (2) structures or situations where an upward flow of water may float the soil particles, eliminating effective stress and thus soil strength and stiffness (Problem 3.21).

Example 3.9

Calculate the seepage force between points G and F in the flow net in Figure 3.28.

Solution:

Our calculation is made easier by the fact that these two points lie exactly on the same flow line. We can take the distance between them as ~4.8 m. There are two incremental head losses of 0.5 m between G and F. This means the hydraulic gradient is equal to:

$$i_{GF} = \frac{\Delta h_{GF}}{L_{GF}} = \frac{2 \times 0.5}{4.8} = 0.208 \approx 0.21$$

The seepage force can now be calculated as

$$f_s = i_{GF}\gamma_w = 0.2 \times 9.81 \approx 2 \text{ kN/m}^3$$

3.8 CASE STUDY: THE RISSA, NORWAY (1978), QUICK CLAY SLIDES

In 1978, as a result of a landslide near Rissa, Norway, an area of roughly 0.329 km², which included seven farms, disappeared. The slide contained about 5.3–6 million m³ of debris – the biggest slide in Norway in the 20th century. Of the 40 people caught in the slide area, 1 died. The cause of this landslide was later determined to be the complete loss of strength of quick clay that was triggered by the excavation and stockpiling of 690 m³ of soil placed by the shore of Lake Botnen. The stockpiled soil was generated by excavation work from the construction of a new wing being added to an existing barn. The slide, which occurred over a period of 6 minutes, started at the lake shoreline and developed retrogressively landward (a common feature of quick clay slides). About 70–90 m of shore slid into the lake.[9]

---

[9] For more details on the Rissa landslide, several resources are available: The Norwegian Geotechnical Institute (NGI; http://www.ngi.no) offers both a video and an article by O. Gregersen (NGI publication No. 135). A brief summary is given in *Failures in Civil Engineering: Structural Foundation and Geoenvironmental Case Studies*, Shepherd, R. and Frost, J.D. (eds), ASCE, 92 pp.

3.9 CHAPTER SUMMARY

3.9.1 Main concepts and equations

Soil is a particulate medium composed of particles that may vary in many ways, including size, chemical composition, and shape. **Soil indices** (such as those listed in Table 3.9) help characterize soils by quantifying such things as void space, mass, weight, and pore space saturation. Classifications based on particle size lead us to the main types of soils we deal with, namely **clays, silts, sands,** and **gravels.**

Soils carry **stresses** through the contacts between adjacent soil particles. **Effective stresses** represent the share of any load applied on the soil that is carried by the **soil skeleton.** If the soil is saturated, the total stress at a point in the soil mass is carried by both the soil skeleton and the pore water. This is expressed by the **effective stress principle:**

$$\sigma = \sigma' + u \tag{3.18}$$

The stresses applied to the soil are converted into **effective stresses,** which in turn lead to **deformations** if drainage is allowed. These deformations may be recoverable or irrecoverable. **Recoverable deformation** is associated with elastic deformation of the soil particles as a result of forces transmitted between particles at their contact points. **Irrecoverable deformation** occurs when particles are displaced from their original positions, that is, when they slide with respect to neighboring particles or when particles break. Upon load removal, these particles will not return to their original positions.

Soils originate from the **weathering** of rocks. The rocks that give origin to soil are of three types: **igneous, sedimentary,** and **metamorphic.** If the soils remain in place, they are referred to as **residual soils,** otherwise, as **transported soils. Groundwater** often exists within the soil pores. **Darcy's law** is key to the quantification of groundwater flow in soils. According to this law, the **specific discharge** v is directly proportional to the **hydraulic gradient** i:

$$v = \frac{Q}{A} = -Ki \tag{3.24}$$

The constant of proportionality K in Equation (3.24) is known as the **hydraulic conductivity** of the soil. The hydraulic gradient i is simply the ratio of the difference in **hydraulic heads** between two points to the distance between the points. The hydraulic head is composed of an elevation component and a pressure component. Water flows from higher elevations to lower elevations, all other things being the same. Likewise, it flows from zones with higher pressures to zones with lower pressures, all other things being the same.

Groundwater flow must, in addition to taking place from points of higher to points of lower hydraulic head, also satisfy the **law of mass conservation,** which leads to the requirement that h must satisfy Laplace's equation:

$$\nabla^2 h = 0$$

where the Laplacian, in a Cartesian x_1–x_2–x_3 space, is defined as

$$\nabla^2 = \frac{\partial^2}{\partial x_1^2} + \frac{\partial^2}{\partial x_2^2} + \frac{\partial^2}{\partial x_3^2}$$

When flowing through soil, water exerts on it a **seepage force** f_s per unit volume of water. This force is given by

$$f_s = \gamma_w \frac{dh}{dL} = i\gamma_w$$

3.9.2 Symbols and notations

Symbol	Quantity represented	US unit	SI unit
A	Area	in.^{2}	mm^2
C_U	Coefficient of uniformity	Unitless	Unitless
D_{10}	Granulometric parameter	in.	mm
D_{50}	Granulometric parameter	in.	mm
D_{60}	Granulometric parameter	in.	mm
D_R	Relative density	%	%
e_{max}	Maximum void ratio	Unitless	Unitless
e_{min}	Minimum void ratio	Unitless	Unitless
G_s	Specific gravity of solids	Unitless	Unitless
i	Hydraulic gradient	Unitless	Unitless
K	Hydraulic conductivity	ft/s	m/s
k	Intrinsic permeability	in.^{2}	mm^2
LL	Liquid limit	Unitless or %	Unitless or %
n	Porosity	Unitless or %	Unitless or %
PI	Plasticity index	Unitless or %	Unitless or %
PL	Plastic limit	Unitless or %	Unitless or %
Q	Flow rate	ft^3/s	m^3/s
S	Degree of saturation	Unitless or %	Unitless or %
u	Pore pressure	psf	kPa
v	Specific discharge	ft/s	m/s
v	Specific volume	Unitless	Unitless
V_s	Volume of solids	ft^3	m^3
V_v	Volume of voids	ft^3	m^3
wc	Water content	Unitless or %	Unitless or %
W_s	Weight of solids (soil)	lb	kN
W_w	Weight of water	lb	kN
γ_d	Dry unit weight of soil	pcf	kN/m^3
γ_{fl}	Unit weight of fluid	pcf	kN/m^3
γ_m	Unit weight of soil	pcf	kN/m^3
γ_s	Unit weight of solids	pcf	kN/m^3
γ_w	Unit weight of water	pcf	kN/m^3
μ	Dynamic viscosity of a fluid	lb·s/ft^2	kN.s/m^2
ρ_m	Mass density of soil	pcf	kg/m^3
σ	Total stress	psf	kPa
σ'	Effective stress	psf	kPa

3.10 WEBSITES OF INTEREST

https://www.clays.org/
The Clay Minerals Society

3.11 PROBLEMS

3.11.1 Conceptual problems

Problem 3.1 Define all the terms in bold contained in the chapter summary.

Problem 3.2 Define and explain as needed:
   a. Diffuse double layer
   b. Flocculation and the conditions that cause it
   c. 2:1 Clay mineral structure; name the components
   d. Isomorphous substitution
   e. Fabric versus structure

Problem 3.3 Four different soils were formed of the same material and with the same void ratio, but with the following different fabrics: dispersed and deflocculated, aggregated and deflocculated, dispersed and flocculated, and aggregated and flocculated. Discuss the relative ease with which each would be compressed.

Problem 3.4 Starting from the definitions of the main soil indices, derive Equation (3.9).

Problem 3.5 Starting from the definitions of the main soil indices, derive Equation (3.12).

Problem 3.6 Starting from the definitions of the main soil indices, derive Equation (3.13).

Problem 3.7 Starting from the definitions of the main soil indices, derive Equation (3.14).

Problem 3.8 Starting from the definitions of the main soil indices, derive Equation (3.15).

Problem 3.9 Starting from the definitions of the main soil indices, derive Equation (3.16).

Problem 3.10 Starting from the definitions of the main soil indices, derive Equation (3.17).

Problem 3.11 The gradation curve for a sandy clay soil is given in P-Table 3.1. What are the weight percentages of sand-, silt-, and clay-sized material?

Problem 3.12 Discuss, at the level of the interaction between neighboring soil particles, how to discern between inelastic and elastic soil deformations.

Problem 3.13 Explain how the relationship between total and buoyant unit weights relates to Archimedes' principle.

P-Table 3.1 Particle size distribution of soil of Problem 3.11

Size (mm)	Percent passing by weight
1.5	100
1	92
0.5	81
0.1	62
0.07	58
0.05	54
0.01	42
0.005	37
0.001	24
0.0005	18
0.0002	6

3.11.2 Quantitative problems

Problem 3.14 A $9.61\,\text{cm}^3$ sample of soil weighs $19\,\text{g}$. After being dried in an oven, its weight is $17.5\,\text{g}$. If $G_s = 2.67$, find:
  a. Unit weight, γ_m
  b. Dry unit weight, γ_d
  c. Water content, wc
  d. Void ratio, e
  e. Porosity, n
  f. Degree of saturation, S

Problem 3.15 Find the dry unit weight, void ratio, porosity, and degree of saturation of a soil with a unit weight equal to $20\,\text{kN/m}^3$ and a water content of 15%. The specific gravity of the soil is 2.68.

Problem 3.16 A site consists predominantly of clay, and the water table is at the ground surface. Laboratory tests were performed by a geotechnical laboratory on clay samples extracted from a depth of $5\,\text{m}$. The following values were reported for four quantities:
Saturated unit weight $\gamma_{sat} = 14.9\,\text{kN/m}^3$
Initial void ratio $e_0 = 1.3$
Unit weight of solids $\gamma_s = 26.52\,\text{kN/m}^3$
Water content $wc = 48.1\%$
Which of the values for the four indices above are incorrect and what is the correct value for it? Consider the unit weight of water as $9.81\,\text{kN/m}^3$ in your calculations.

Problem 3.17 The saturated and dry masses of a sample of clay were determined to be equal to 1880 and $1450\,\text{g}$, respectively. Knowing that the dry unit weight of this clay is equal to $15\,\text{kN/m}^3$, calculate the wet unit weight, the void ratio, and the specific gravity of the clay. What would its wet unit weight be if the degree of saturation were equal to 60%?

Problem 3.18 Plot the total stress, effective stress, and pore pressure for the soil profile of P-Figure 3.1a. The unit weight of the sand is $21\,\text{kN/m}^3$ when saturated and $17\,\text{kN/m}^3$ when dry. The unit weight of the clay when saturated is $17\,\text{kN/m}^3$.

Problem 3.19 Redo Problem 3.18 with the water table now flush with the top of the sand layer (P-Figure 3.1b). Take into consideration capillary rise to compute the total stresses, effective stresses, and pore pressures in the clay layer.

Problem 3.20 Redo Problem 3.18 for two situations: (a) the water table is $5\,\text{m}$ above the ground surface (P-Figure 3.1c) and (b) the water table is $10\,\text{m}$ above the ground surface (P-Figure 3.1d).

Problem 3.21 Show that the hydraulic gradient needed for the seepage forces associated with vertical, upward flow to completely neutralize the soil self-weight (known as the critical gradient) is approximately equal to one.

Problem 3.22 Referring to E-Figure 3.4, calculate the hydraulic gradient, the specific discharge, and the volume of water that flows in 1 year between the two lakes if the water levels don't change. Take $H = 5\,\text{m}$, $L = 10,000\,\text{m}$, $z_{L1} = 100\,\text{m}$, $z_{L2} = 109\,\text{m}$, and $K = 10^{-4}\,\text{m/s}$.

Problem 3.23 Referring to Problem 3.22, how long would it take for a contaminant accidentally released during drilling operations at point A to reach the drinking well with tip at point B? The distance L_{AB} between points A and B is $500\,\text{m}$.

Problem 3.24 For the flow net shown in P-Figure 3.2, calculate the pore pressures at points A, B, and C and the seepage force between points B and C.

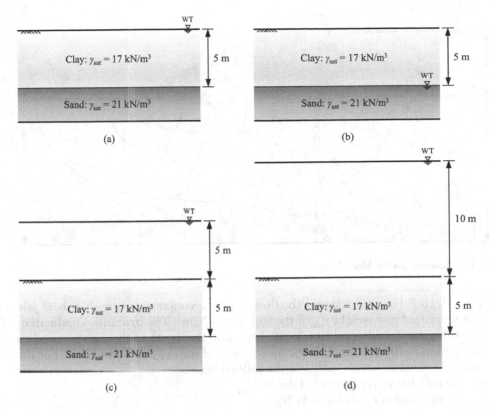

P-Figure 3.1 Soil profile for Problems 3.18–3.20.

Problem 3.25 Consider the measurement of hydraulic conductivity K using the falling-head test (see Appendix C). The diameter d and length L of the soil sample are 3.5 and 7 cm, respectively. The diameter d_p of the standpipe is 2 cm. During the test, the hydraulic head dropped from 30 to 18.5 cm in 2 minutes. Calculate the hydraulic conductivity of the soil.

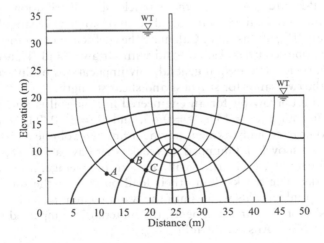

P-Figure 3.2 Flow net for Problem 3.24.

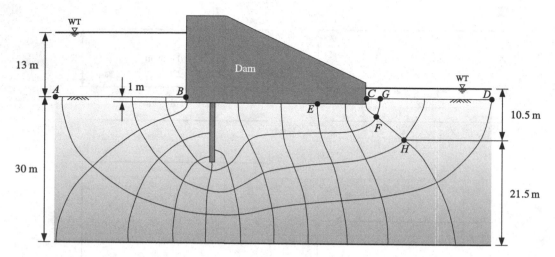

P-Figure 3.3 Flow net for Problem 3.26.

Problem 3.26 P-Figure 3.3 shows the flow net for a concrete dam with a sheet pile wall. The saturated unit weight γ_{sat} of the soil is 21 kN/m³. The hydraulic conductivity K of the soil is uniformly equal to 1.3×10^{-5} m/s, and the length of the dam perpendicular to the section is 100 m. Consider the unit weight of water γ_w as 9.81 kN/m³. For this dam and the foundation soil, calculate the following:

a. the number of equipotential drops N_d;
b. the number of flow channels N_f;
c. the pore water pressure u acting on the base of the dam at point E;
d. the seepage force f_s between points F and G, which are 2 m apart;
e. the factor of safety against piping approximately halfway between points F and G;
f. the flow rate Q beneath the dam; and
g. the vertical effective stress σ'_v at point H.

3.11.3 Design problems

Problem 3.27* P-Figure 3.4 shows the particle size distribution of an extensively researched Italian sand called Ticino Sand. Ticino Sand has subangular particles. Find the D_{10}, D_{50}, and D_{60} of this sand. Calculate the coefficient of uniformity. On the basis of this information, contrast Ticino Sand with Ottawa Sand (Figure 3.2), which has subrounded particles. Discuss, in general, any implications of differences or similarities between the two sands for stiffness and shear strength.

Problem 3.28* In a borrow pit for an engineered fill, the soil is found to have the following characteristics: $wc = 15\%$, $e = 0.60$, and $G_s = 2.70$. The soil will be used to construct a compacted embankment having a finished volume of 50,000 m³. The soil is excavated by a shovel and dumped on trucks that have a 5 m³ capacity each. When loaded to full capacity, each truck will contain, on average, a net mass of soil plus water of 4520 kg. The trucks dump their load on the fill, the soil is broken up and spread, and a sprinkler adds water until the water content wc is 18%. After thorough mixing by disks (or similar equipment), the material is compacted until the dry unit weight is 17.3 kN/m³. Answer the following:

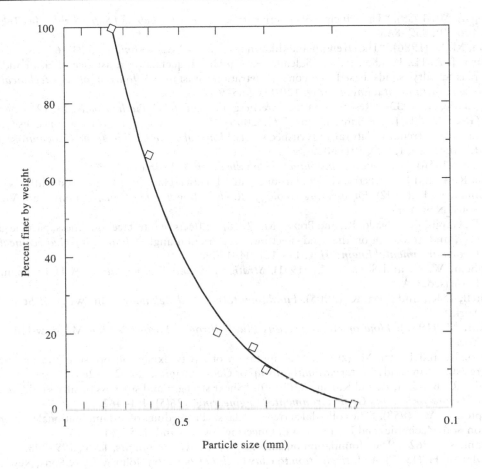

P-Figure 3.4 Particle size distribution of Ticino Sand.

a. How many full-capacity truck loads are required to construct this fill?
b. After completion, what is the expected volume of the pit from which the soil was removed?
c. How many gallons of water need to be added per truck load (assuming no loss by evaporation during hauling and handling)?
d. If the fill later becomes saturated (and does not change volume), what will be its saturation water content?
e. What will be the saturation water content of the fill if the soil swells to increase the original volume by 15%?

REFERENCES

References cited

Atterberg, A. (1911). "Uber die physikalische bodenuntersuchung und uber die Plastizitat der Tone." *Internationale Mitteilungen fur Bodenkunde*, 1, 10–43.

Bishop, A. W. (1959). "The principle of effective stress." *Teknisk Ukeblad I Samarbeide Med Teknikk*, 106(39), 859–863.

Bolton, M. D. (1986). "The strength and dilatancy of sands." *Géotechnique*, 36(1), 65–78.

Carraro, J. A. H., Bandini, P., and Salgado, R. (2003). "Liquefaction resistance of clean and non-plastic silty sands based on cone penetration resistance." *Journal of Geotechnical and Geoenvironmental Engineering*, 129(11), 965–976.

Casagrande, A. (1932). "Research on the Atterberg limits of soils." *Public Roads*, 13, 121–136.

Cho, G.-C., Dodds, J., and Santamarina, J. C. (2006). "Particle shape effects on packing density, stiffness, and strength: Natural and crushed sands." *Journal of Geotechnical and Geoenvironmental Engineering*, 132(5), 591–602.

Darcy, H. (1856). *Les fontaines publiques de la ville de dijon*. Dalmont, Paris.

Freeze, R. A. and J. A. Cherry. (1979). Groundwater. Prentice-Hall, Inc., Englewood Cliffs, NJ.

Goodman, R. E. (1992). *Engineering Geology: Rock in Engineering Construction*. John Wiley & Sons, New York.

Han, F., Ganju, E., Salgado, R., and Prezzi, M. (2018). "Effects of interface roughness, particle geometry, and gradation on the sand–steel interface friction angle." *Journal of Geotechnical and Geoenvironmental Engineering*, 144(12), 04018096.

Krumbein, W. C. and Sloss, L. L. (1951). *Stratigraphy and Sedimentation*. W.H Freeman, San Francisco, CA.

Mitchell, J. K., and Soga, K. (2005). *Fundamentals of Soil Behavior*. John Wiley & Sons, New York.

Muskat, M. (1937). *Flow of Homogeneous Fluids through Porous Media*. McGraw-Hill, New York.

Salgado, R. and Prezzi, M. (2007). "Computation of cavity expansion pressure and penetration resistance in sands." *International Journal of Geomechanics*, 7(4), 251–265.

Salgado, R., Bandini, P., and Karim, A. (2000). "Shear strength and stiffness of silty sand." *Journal of Geotechnical and Geoenvironmental Engineering*, 126(5), 451–462.

Skempton, A. W. (1953). "The colloidal activity of clays." Proceedings of 3rd International Conference on Soil Mechanics and Foundation Engineering, Switzerland, 1, 57–61.

Terzaghi, K. (1962). "Dam foundation on sheeted granite." *Géotechnique*, 12(3), 199–208.

Van Olphen, H. (1977). *An Introduction to Clay Colloid Chemistry*. John Wiley & Sons, New York.

Wadell, H. (1932). "Volume, shape, and roundness of rock particles." *The Journal of Geology*, 40(5), 443–451.

Youd, T. (1973). "Factors controlling maximum and minimum densities of sands." *In Evaluation of Relative Density and its Role in Geotechnical Projects Involving Cohesionless Soils*, ASTM International STP, Philadelphia, PA, 98–112.

Zheng, J. and Hryciw, R. D. (2015). "Traditional soil particle sphericity, roundness and surface roughness by computational geometry." *Géotechnique*, 65(6), 494–506.

Additional references

Fetter, C. W. (2001). *Applied Hydrogeology*. 4th Edition, Prentice Hall, Upper Saddle River, NJ.

Holtz, R. D. and Kovacs, W. D. (1981). *An Introduction to Geotechnical Engineering*. Prentice Hall, Upper Saddle River, NJ.

Szechy, K. (1966). *The Art of Tunneling*. Akademlai Klado Budapest, Budapest, Hungary.

Tarbuck, E. J., Lutgens, F. K., and Dennis, T. (2002). *The Earth: An Introduction to Physical Geology*. Prentice-Hall, Upper Saddle River, NJ.

West, T. (1995). *Geology Applied to Engineering*. Prentice-Hall, Upper Saddle River, NJ.

Wyllie, D. C. (1992). *Foundations on Rock*. E and FN Spon, London, UK.

Yeung, A. and Sadek, S. (2005). "Apparatus induced error in hydraulic conductivity measurement." *Geotechnical Testing Journal*, 28(5), 472–479.

Relevant ASTM standards

ASTM. (2013a). "Standard test method for permeability of rocks by flowing air." ASTM D4525, American Society for Testing and Materials, West Conshohocken, PA.

ASTM. (2013b). "Standard test methods for measurement of thermal expansion of rock using dilatometer." ASTM D4535, American Society for Testing and Materials, West Conshohocken, PA.

ASTM. (2014a). "Standard test methods for compressive strength and elastic moduli of intact rock core specimens under varying states of stress and temperatures." ASTM D7012, American Society for Testing and Materials, West Conshohocken, PA.

ASTM. (2014b). "Standard test methods for specific gravity of soil solids by water pycnometer." ASTM D854, American Society for Testing and Materials, West Conshohocken, PA.

ASTM. (2015). "Standard practice for preparation of rock slabs for durability testing." ASTM D5121, American Society for Testing and Materials, West Conshohocken, PA.

ASTM. (2016a). "Standard test method for splitting tensile strength of intact rock core specimens." ASTM D3967, American Society for Testing and Materials, West Conshohocken, PA.

ASTM. (2016b). "Standard test methods for creep of rock core under constant stress and temperature." ASTM D7070, American Society for Testing and Materials, West Conshohocken, PA.

ASTM. (2016c). "Standard test methods for maximum index density and unit weight of soils using a vibratory table." ASTM D4253, American Society for Testing and Materials, West Conshohocken, PA.

ASTM. (2016d). "Standard test methods for minimum index density and unit weight of soils and calculation of relative density." ASTM D4254, American Society for Testing and Materials, West Conshohocken, PA.

ASTM. (2016e). "Standard test method for calculating thermal diffusivity of rock and soil." ASTM D4612, American Society for Testing and Materials, West Conshohocken, PA.

ASTM. (2016f). "Standard test method for determining in situ modulus of deformation of rock using diametrically loaded 76-mm (3-in.) borehole jack." ASTM D4971, American Society for Testing and Materials, West Conshohocken, PA.

ASTM. (2017a). "Standard test methods for liquid limit, plastic limit, and plasticity index of soils." ASTM D4318, American Society for Testing and Materials, West Conshohocken, PA.

ASTM. (2017b). "Standard test methods for determining the amount of material finer than 75-μm (no. 200) sieve in soils by washing." ASTM D1140, American Society for Testing and Materials, West Conshohocken, PA.

ASTM. (2017c). "Standard test methods for particle-size distribution (gradation) of soils using sieve analysis." ASTM D6913, American Society for Testing and Materials, West Conshohocken, PA.

ASTM. (2017d). "Standard practice for classification of soils for engineering purposes (unified soil classification system)." ASTM D2487, American Society for Testing and Materials, West Conshohocken, PA.

ASTM. (2017e). "Standard practice for description and identification of soils (visual-manual procedures)." ASTM D2488, American Society for Testing and Materials, West Conshohocken, PA.

ASTM. (2018). "Standard test method for shrinkage factors of cohesive soils by the water submersion method." ASTM D4943, American Society for Testing and Materials, West Conshohocken, PA.

ASTM. (2019a). "Standard test methods for laboratory determination of water (moisture) content of soil and rock by mass." ASTM D2216, American Society for Testing and Materials, West Conshohocken, PA.

ASTM. (2019b). "Standard practices for preparing rock core as cylindrical test specimens and verifying conformance to dimensional and shape tolerances." ASTM D4543, American Society for Testing and Materials, West Conshohocken, PA.

ASTM. (2020a). "Standard test method for direct tensile strength of intact rock core specimens." ASTM D2936, American Society for Testing and Materials, West Conshohocken, PA.

ASTM. (2020b). "Standard test methods for determining the water (moisture) content, ash content, and organic material of peat and other organic soils." ASTM D2974, American Society for Testing and Materials, West Conshohocken, PA.

ASTM. (2021). "Standard test methods for particle-size distribution (gradation) of fine-grained soils using sedimentation (hydrometer) analysis." ASTM D7928, American Society for Testing and Materials, West Conshohocken, PA.

Chapter 4

Stress analysis, strain analysis, and shearing of soils

[W]hen you can measure what you are speaking about, and express it in numbers, you know something about it, but when you cannot express it in numbers, your knowledge is of a meagre and unsatisfactory kind...

Lord Kelvin

Ut tensio sic vis (strains and stresses are related linearly).

Robert Hooke

So I think we really have to, first, make some new kind of theories in which we take regard to the fact that there is no linearity condition between stresses and strains for soils.

J. Brinch Hansen

Foundation design requires that we analyze how structural loads are transferred to the ground, and whether the soil will be able to support these loads safely and without excessive deformation. Similarly, other geotechnical structures, such as slopes or retaining structures, rely on soil strength or stiffness to perform adequately. In our analyses and design of these systems, we treat soil deposits as continuous masses subjected to their own self-weight and to loads on their boundaries; as a result of these loads, stresses and strains appear in the soil. This chapter covers the basic concepts of the mechanics of soils needed for understanding stresses, strains, and their consequences. At the soil element level, coverage includes stress analysis, strain analysis, shearing and the formation of slip surfaces, and the laboratory tests used to study the stress–strain response and shearing of soil elements. At the level of boundary-value problems, coverage includes a relatively simple problem in soil plasticity, the development of Rankine states, with focus on the stresses generated inside a semi-infinite soil mass that has reached a Rankine state.

4.1 STRESS ANALYSIS

4.1.1 Elements (points) in a soil mass and boundary-value problems

In this chapter, soil is modeled as a continuum, that is, a material in which material points are infinitesimally near each other, leaving no gaps; however, it is obvious that soil is not in reality a continuum, being rather a collection of particles, with interparticle voids filled by gas or liquid. This paradox is solved by redefining what is meant by a point in the continuum. Instead of thinking of a point, we instead think of an element of soil, a small volume,

DOI: 10.1201/b22079-4

that is so small that we can think of it as being a point, yet large enough to contain a sufficient number of particles to be representative of how the soil will behave.

In this chapter, we are concerned with (1) how soil behaves at the element level and (2) how this behavior, appropriately described by suitable equations and ascribed to every element (that is, every point) of a soil mass subjected to certain boundary conditions, in combination with analyses from elasticity or plasticity theory, allows us to determine how boundary loads or imposed displacements result in stresses and strains throughout the soil mass (and, in extreme cases, in strains so localized and large that collapse of a part of the soil mass and all that it supports happens). At the element level, we must define precisely what the element is and how large it must be in order to be representative of soil behavior. We must also have equations that describe the relationship between stresses and strains for the element and that describe the combination of stresses that would lead to extremely large strains.[1] At the level of the boundary-value problem, we must define the maximum size an element may have with respect to characteristic lengths of the problem (for example, the width of a foundation) and still be treated as a point. This is important because we use concepts and analyses from continuum mechanics to solve soil mechanics problems, and so our elements must be points that are part of a continuous mass. We must then have analyses that take into account how elements interact with each other and with specified stresses or displacements at the boundaries of the soil mass to produce values of stress and strain everywhere in the soil mass. Further, we must also be able to analyze cases in which the stresses are such that large strains develop in localized zones of the soil mass. Concepts from plasticity theory are used for that.

It is clear from the preceding discussion that the mechanics of the continuum is a very integral part of soil mechanics and geotechnical and foundation engineering. We will introduce the concepts that are necessary gradually and naturally and with a mathematical treatment that is kept as simple as possible.

4.1.2 Stress

A significant amount of the work we do in geotechnical engineering is based on the concept of stress. *Stress* is a concept from the mechanics of continuous bodies. Because soil is not a continuous medium, it is useful to discuss the meaning of stress in soils. Consider a small planar area A passing through point P located within a soil mass (Figure 4.1). A normal force F_N and a tangential force F_T are applied on A (these forces result, as discussed previously, from boundary loads and the soil self-weight or other body forces, which propagate through the soil mass until they reach the small area A). If soil were a continuum, the *normal stress* σ acting normal to A at point P would be defined as the limit of F_N/A as A tends to the point P (that is, tends to zero, centered around P). The *shear stress* is defined similarly. Mathematically:

$$\sigma = \lim_{A \to 0} \frac{F_N}{A} \qquad (4.1)$$

$$\tau = \lim_{A \to 0} \frac{F_T}{A} \qquad (4.2)$$

---

[1] Extremely large strains, particularly extremely large shear strains, are closely tied to concepts of rupture, yield, and failure. These three terms do not completely define the range of problems we deal with, so we will introduce appropriate terms throughout this chapter and the remainder of the text.

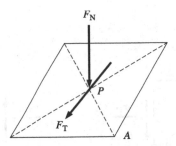

Figure 4.1 Definition of stress in soils: as the area A is allowed to shrink down to a very small value, the ratios F_N/A and F_T/A approach values σ and τ, the normal and shear stresses at P, respectively.

Because soil is not a true continuum, we must modify this definition. A point within a soil mass is defined as a volume V_0 that is still very small compared with the dimensions of the foundations, slopes, or retaining structures we analyze, but is sufficiently large to contain a large number of particles and thus be representative of the soil.[2] With this *representative elementary volume V_0*, often referred to as the REV, we associate a representative area A_0 (also very small, of a size related to that of V_0). So we modify Equations (4.1) and (4.2) by changing the limit approached by the area A from zero to A_0:

$$\sigma = \lim_{A \to A0} \frac{F_N}{A} \tag{4.3}$$

$$\tau = \lim_{A \to A0} \frac{F_T}{A} \tag{4.4}$$

The preceding discussion brings out one difference between soil mechanics and the mechanics of metals, for example. In metals, the REV is very small. The REV for a given metal is indeed so small that, in ordinary practice or introductory courses, we tend to think of it as being a point, forgetting that metals are also made up of atoms arranged in particular ways, so that they too have REVs with nonzero volume, although much smaller than those we must use in soil mechanics. In soil mechanics, our REVs must include enough particles that, statistically, this group of particles will behave in a way that is representative of the way larger volumes of the soil would behave.

4.1.3 Two-dimensional stress analysis

Stress analysis allows us to obtain the normal and shear stresses in any plane passing through a point,[3] given the normal and shear stresses acting on any two (in 2D) or three (in 3D) mutually perpendicular planes passing through the point.[4] We will see in Section 4.6 some examples of how these stresses can be calculated at a point inside a soil mass from a

---

[2] Mechanicians like to use the term "representative elementary volume" to describe the smallest volume of a given material that captures its mechanical properties.

[3] A point in the soil is indistinguishable, for our purposes, from a representative soil element, which has a very small volume (and so is a point for practical purposes), but is sufficiently large to be representative of the soil in its mechanical behavior.

[4] A more proper definition of stress analysis for advanced readers would be that stress analysis aims to allow calculations of the traction (which has normal and shear components) on a plane, given the stress tensor at the point.

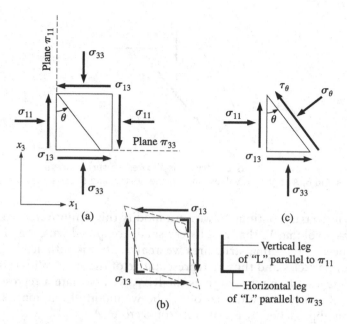

Figure 4.2 Representation of state of stress at a point: (a) Elemental representation of two-dimensional stress state at a point; (b) illustration of how the element would distort when acted upon by a positive σ_{13} with zero normal stresses σ_{11} and σ_{33}; (c) sectioned triangular prismatic element, where σ_θ and τ_θ depend on the angle θ.

variety of boundary loadings common in geotechnical engineering. In this subsection, we will discuss two-dimensional stress analysis, which is an easier introduction to the subject, and then follow up with three-dimensional stress analysis, required for working with more modern concepts and rigorous analyses in soil mechanics.

4.1.3.1 Stress state at a point

Figure 4.2a shows a small prismatic element of soil representing a "point" within the soil. The faces of the element are aligned with the directions of the reference axes x_1 and x_3. The soil element is acted upon by the normal stresses σ_{11} acting in the x_1 direction and σ_{33} acting in the x_3 direction, and the shear stresses σ_{13} and σ_{31}, which, due to the requirement of moment equilibrium, are equal in magnitude.[5] The first subscript of a stress component represents the direction normal to the plane on which it acts; the second, the direction of the stress component itself. A stress component with subscripts taking different values is a shear stress; one with subscripts taking identical values, a normal stress. For example, σ_{11} is the stress acting on the plane normal to x_1 in the x_1 direction; that is, it is a normal stress, while σ_{13} is the stress that acts on the plane normal to x_1 in the x_3 direction (and is thus a shear stress). It is simpler (although not required) to solve problems if we adopt the practice of choosing σ_{11} and σ_{33} such that $\sigma_{11} \geq \sigma_{33}$. So if the normal stresses are 1000 kPa and 300 kPa, then $\sigma_{11} = 1000$ kPa and $\sigma_{33} = 300$ kPa. Likewise, if the normal stresses are 100 kPa and –500 kPa, then $\sigma_{11} = 100$ kPa and $\sigma_{33} = -500$ kPa.

The plane where σ_{11} acts is denoted by π_{11}; likewise, π_{33} is the plane where σ_{33} acts. The stresses σ_{11}, σ_{33}, and σ_{13} are all represented in their positive directions in Figure 4.2a. This means normal stresses are positive in compression, and the angle θ is positive

---

[5] Note that many engineering texts use the notation τ_{ij} instead of σ_{ij} when $i \neq j$ to represent shear stress. We have retained here the traditional mechanics notation, in which σ is used for both normal and shear stresses.

counterclockwise from π_{11}. With respect to the shear stress σ_{13}, note that the prism shown in the figure has four sides, two representing plane π_{11} and two representing plane π_{33}. Looking at the prism from the left side, we may visualize the plane π_{11} as the vertical leg of an uppercase letter "L" and π_{33} as the horizontal leg of the uppercase letter "L". We then see that a positive shear stress σ_{13} acts in such a way as to open up (that is, increase) the right angle of the "L" formed by planes π_{11} and π_{33}. Figure 4.2b shows the deformed shape that would result for the element under the action of positive σ_{13} only.

4.1.3.2 Stress analysis: determination of normal and shear stresses in arbitrary plane

If we section the element of Figure 4.2a along a plane making an angle θ with the plane π_{11}, as shown in Figure 4.2c, a normal stress σ_θ and a shear stress τ_θ must be applied to this plane to account for the effects of the part of the element that is removed if we want the element to remain in equilibrium. A separate sign convention is useful in two-dimensional analysis to represent the normal and shear stresses at a plane. While the sign of the normal stress σ_θ is unambiguous (positive in compression), the shear stress on the sloping plane has two possible directions: up or down the plane. So we must decide which of these two directions is associated with a positive shear stress. The positive direction of the shear stress actually follows from the sign convention already discussed (that $\sigma_{13} > 0$ when its effect would be to increase the right angle of the uppercase "L" made up by π_{11} as its vertical and π_{33} as its horizontal leg). It turns out the shear stress τ_θ is positive as drawn in the figure, when it is rotating around the prismatic element in the counterclockwise direction. We will show why this is so later, when we introduce the *Mohr circle*.

Our problem now is to determine the normal stress σ_θ and the shear stress τ_θ acting on the plane making an angle θ with π_{11}. This can be done by considering the equilibrium in the vertical and horizontal directions and solving for σ_θ and τ_θ. The following equations result:

$$\sigma_\theta = \frac{1}{2}(\sigma_{11} + \sigma_{33}) + \frac{1}{2}(\sigma_{11} - \sigma_{33})\cos 2\theta + \sigma_{13}\sin 2\theta \tag{4.5}$$

$$\tau_\theta = \frac{1}{2}(\sigma_{11} - \sigma_{33})\sin 2\theta - \sigma_{13}\cos 2\theta \tag{4.6}$$

where the signs of σ_{11}, σ_{33}, and σ_{13} are as discussed earlier (positive in compression for the normal stresses and determined by the "L" rule in the case of σ_{13}).

4.1.3.3 Principal stresses and principal planes

Equations (4.5) and (4.6) tell us that σ_θ and τ_θ vary with θ. That means that the normal and shear stresses on each plane through a given point are a unique pair. There will be two planes out of the infinite number of planes through the point that are normal to the x_1–x_3 plane for which the normal stress will be a minimum and a maximum. These are called *principal stresses*. They are obtained by maximizing and minimizing σ_θ by differentiating Equation (4.5) with respect to θ and making the resulting expression for the derivative equal to zero. The largest principal stress is known as the *major principal stress*; it is denoted as σ_1. The smallest principal stress is the *minor principal stress*, denoted as σ_3.[6] The planes

---

[6] We are assuming here that the other principal stress, denoted σ_2, which appears in three-dimensional stress analysis, is no less than σ_3. If this assumption holds, then σ_2 is referred to as the intermediate principal stress.

where they act are referred to as the major and minor *principal planes*, denoted by π_1 and π_3, respectively. When we differentiate Equation (4.5) with respect to θ and make the resulting expression equal to zero, we obtain the same expression we obtain when we make τ_θ, given by Equation (4.6), equal to zero. This means that the shear stresses acting in the principal planes are equal to zero.

An easy way to find the angles θ_{p1} and θ_{p3} that the principal planes π_1 and π_3 make with π_{11} (measured counterclockwise from π_{11}) is then to make $\tau_\theta=0$ in Equation (4.6), which leads to:

$$\tan(2\theta_p) = \frac{2\sigma_{13}}{\sigma_{11}-\sigma_{33}} \tag{4.7}$$

When θ_p is substituted for θ back into Equation (4.5), we obtain the principal stresses σ_1 and σ_3, which are the two normal stresses acting on planes where $\tau_\theta=0$ and are also the maximum and minimum normal stresses for the point under consideration, given by

$$\sigma_1 = \frac{1}{2}(\sigma_{11}+\sigma_{33}) + \sqrt{\frac{1}{4}(\sigma_{11}-\sigma_{33})^2 + \sigma_{13}^2} \tag{4.8}$$

$$\sigma_3 = \frac{1}{2}(\sigma_{11}+\sigma_{33}) - \sqrt{\frac{1}{4}(\sigma_{11}-\sigma_{33})^2 + \sigma_{13}^2} \tag{4.9}$$

Given the definition of the tangent of an angle, there are an infinite number of values of θ_p that satisfy Equation (4.7). Starting with any value of θ_p satisfying Equation (4.7), we obtain additional values that are also solutions to Equation (4.7) by repeatedly either adding or subtracting 90°. Values of θ_p differing by 180° refer to the same material plane, so our solution to Equation (4.7) is a number between −90° and +90°. If $\sigma_{11}>\sigma_{33}$, we expect the major principal stress σ_1 to be closer in direction to σ_{11} (the larger stress) than to σ_{33} (the smaller stress); so, if the absolute value of the calculated value of the angle θ_p is less than 45°, $\theta_p=\theta_{p1}$; otherwise, $\theta_p=\theta_{p3}$. Once the angle θ_p for one of the principle planes is known, the direction of the other plane can be calculated easily by either adding or subtracting 90° to θ_p to obtain an angle with absolute value less than 90°. For example, if θ_{p1} is calculated as +25°, then θ_{p3} is equal to −65°. Alternatively, if θ_{p1} is found to be −25°, then θ_{p3} is calculated as −25°+90°=65°.

4.1.3.4 Mohr's circle

Moving ½ $(\sigma_{11}+\sigma_{33})$ to the left-hand side of Equation (4.5), taking the square of both sides of the resulting equation, and adding it to Equation (4.6) (with both sides also squared), we obtain:

$$\left[\sigma_\theta - \frac{1}{2}(\sigma_{11}+\sigma_{33})\right]^2 + \tau_\theta^2 = \frac{1}{4}(\sigma_{11}-\sigma_{33})^2 + \sigma_{13}^2 \tag{4.10}$$

Recalling the equation of a circle in Cartesian coordinates, $(x-a)^2+(y-b)^2=R^2$, where (a, b) are the coordinates of the center of the circle and R is its radius, Equation (4.10) is clearly the equation of a circle with center $C[(\sigma_{11}+\sigma_{33})/2, 0]$ and radius $R=\sqrt{\frac{1}{4}(\sigma_{11}-\sigma_{33})^2 + \sigma_{13}^2}$ in $\sigma-\tau$ space. Each point of this circle, which is referred to as the Mohr circle, is defined by two

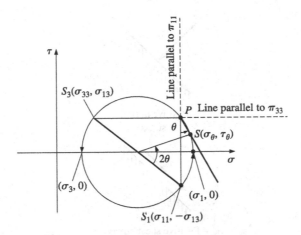

Figure 4.3 Mohr's circle corresponding to the state of stress shown in Figure 4.2.

coordinates: the first is a normal stress (σ), and the second is a shear stress (τ). Figure 4.3 shows the Mohr circle for the stress state ($\sigma_{11}, \sigma_{33}, \sigma_{13}$) of Figure 4.2a. According to Equations (4.5) and (4.6), the stresses σ_θ and τ_θ on the plane making an angle θ measured counterclockwise from plane π_{11} are the coordinates of a point on the Mohr circle rotated 2θ counterclockwise from point S_1 representing the stresses on π_{11}. Note that the central angle of the Mohr circle separating S_1 from S_3, which is 180°, is indeed twice the 90° angle separating the corresponding planes, π_{11} and π_{33}.

In Figure 4.3, the stress state ($\sigma_{11}, \sigma_{33}, \sigma_{13}$) plots as two diametrically opposed points: $S_1(\sigma_{11}, -\sigma_{13})$ and $S_3(\sigma_{33}, \sigma_{13})$. The figure illustrates the case of $\sigma_{11} > \sigma_{33}$. To understand why the shear stress σ_{13} plots as a positive number with σ_{33} and as a negative number with σ_{11}, refer to Figure 4.2b. With the positive values of σ_{11}, σ_{33}, and σ_{13} shown in Figure 4.2a, the prismatic element deforms as shown in Figure 4.2b, with the direction of maximum compression associated with the major principal stress σ_1. It is clear that the major principal plane, which is normal to the direction of maximum compression, is obtained by a rotation of some angle $\theta_{p1} < 90°$ counterclockwise with respect to π_{11}. This means that, in the Mohr circle, we must have a counterclockwise rotation $2\theta_{p1}$ from the point $S_1(\sigma_{11}, -\sigma_{13})$ associated with π_{11} to reach the point (σ_1, 0) of the circle. This implies that S_1 must indeed be located as shown in Figure 4.3, for if we had $S_1(\sigma_{11}, \sigma_{13})$ instead of $S_1(\sigma_{11}, -\sigma_{13})$, a counterclockwise rotation less than 90° would not take us to (σ_1, 0). So this means σ_{13} is plotted as negative if spinning clockwise around the element, as it does for plane π_{11}, and as positive if spinning counterclockwise, as it does for plane π_{33}. This is the basis for the convention we will use for plotting points in the Mohr circle: shear stresses rotating around the element in a counterclockwise direction are positive; they are negative otherwise (Figure 4.4). This sign convention is not independent from, but actually follows directly from the shear stress sign convention we adopted for the stress σ_{13} appearing in Equations (4.5) and (4.6).

4.1.3.5 Pole method

Mohr's circles have interesting geometric properties. One very useful property that every circle has is that the central angle of the circle corresponding to a certain arc is twice as large as an inscribed angle corresponding to the same arc (Figure 4.5). Applying this property to the Mohr circle shown in Figure 4.3, the angle made by two straight lines drawn from any point of the circle to point $S(\sigma_\theta, \tau_\theta)$ representing the stresses on the plane of interest and to point $S_1(\sigma_{11}, -\sigma_{13})$ is equal to θ. In particular, there is one and only one point P on the circle

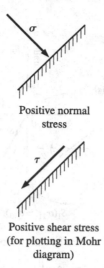

Positive normal
stress

Positive shear stress
(for plotting in Mohr
diagram)

Figure 4.4 Stress sign conventions.

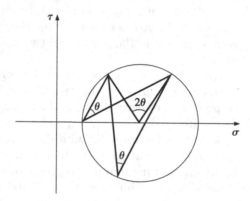

Figure 4.5 Geometric property of circles that a central angle 2θ produces the same arc as an inscribed angle θ.

with the property that the line joining P to point $S(\sigma_\theta, \tau_\theta)$ is parallel to the plane on which σ_θ and τ_θ act. The point P with this property is known as the *pole* of the Mohr circle. Based on the preceding discussion, the pole can be defined as the point such that, if we draw a line through the pole parallel to the plane where stresses σ_θ and τ_θ act, this line intersects the Mohr circle at a point whose coordinates are σ_θ, τ_θ.

To determine the pole, we need to know the orientation of at least one plane where the stresses are known. We can then use the known stresses (σ, τ) to find the pole by drawing a line through (σ, τ) parallel to the plane acted upon by these stresses. This line intersects the circle at a point: this point is the pole. Once we know the pole P, we can determine the stresses on any plane drawing a straight line through the pole parallel to the plane where the stresses are desired. This line intersects the circle at a point whose coordinates are the desired stresses.

We can use Figures 4.2a and 4.3 to illustrate the concept of the pole. Consider the element of Figure 4.2a. By plotting the Mohr circle for this state of stress, we obtain the expected diametrically opposed points S_1 and S_3 shown in Figure 4.3. If we look at point S_1 on the circle and consider the corresponding stresses shown in Figure 4.2a, we can easily determine

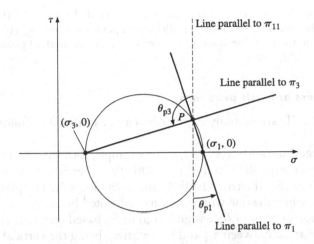

Figure 4.6 Illustration of the relationship of the principal directions and their representation in a Mohr's diagram.

the location of the pole. If we construct a line through S_1 that is parallel to the plane π_{11} where $(\sigma_{11}, -\sigma_{13})$ acts, we will determine the pole as the point where this line intersects the Mohr circle. In this case, since we are dealing with $\theta=0°$, the line is vertical, and the pole (point P in Figure 4.3) lies directly above S_1 (also shown in Figure 4.3). Likewise, if we look at point S_3 and draw a line through S_3 parallel to plane π_{33} on which $(\sigma_{33}, \sigma_{13})$ acts, we can also determine the pole as the point of intersection of this horizontal line with the Mohr circle. As expected, the pole is found to be directly to the right of S_3 and to coincide with the point determined previously by examining point S_1. This shows clearly that the pole is unique for a given stress state. Figure 4.6 illustrates for the same case how the principal directions and principal stresses would be determined once the pole is known.

Example 4.1

A state of stress is represented by the block in E-Figure 4.1a. Determine the location of the pole, and give its coordinates in the σ–τ system.

Solution:

First, we plot the Mohr circle (E-Figure 4.1b). Next, we determine the location of the pole. Using point (200, 100), we draw a line parallel to the plane on which σ_{11} (200 kPa)

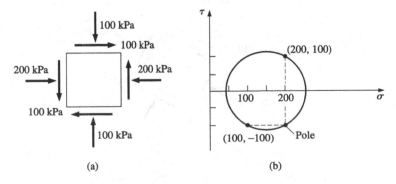

E-Figure 4.1 (a) Stress state for Example 4.1 and (b) the corresponding Mohr circle.

acts. In this case, it is a vertical line since π_{11} is vertical. Likewise, if the other point (100, −100) is chosen, we draw a line parallel to the plane on which σ_{33} (100 kPa) acts. In this case, it is a horizontal line since π_{33} is horizontal. Either method produces the location of the pole: (200, −100).

4.1.3.6 Solving stress analysis problems

The steps in solving a 2D stress analysis problem can be outlined as follows:

1. Choose the largest normal stress as σ_{11}. For example, if the normal stresses are 300 kPa and 100 kPa, then $\sigma_{11} = 300$ kPa and $\sigma_{33} = 100$ kPa. Likewise, if the normal stresses are 100 kPa and −300 kPa, then $\sigma_{11} = 100$ kPa and $\sigma_{33} = -300$ kPa. The plane where σ_{11} acts is denoted by π_{11}, while the one where σ_{33} acts is denoted by π_{33}.
2. If using Equation (4.5) or (4.6), assign a sign to σ_{13} based on whether it acts to increase or decrease the angle between π_{11} and π_{33} (with π_{11} being the vertical leg of the "L"; see Figure 4.2). The stress σ_{13} is positive if it acts in a way that would tend to increase the angle.
3. Recognize that the reference plane for angle measurements is π_{11} and that angles are positive counterclockwise.
4. Reason physically to help check your answers. For example, since $\sigma_{11} > \sigma_{33}$, the direction of σ_1 will be closer to that of σ_{11} than to that of σ_{33}. Whether σ_1 points up or down with respect to σ_{11} now depends on the sign of σ_{13}, for it tells us about the directions in which the element tends to be most compressed or extended. Naturally, σ_1 acts in the direction in which the element tends to be compressed the most.

Example 4.2

The state of stress at a point is represented in E-Figure 4.2. Find (a) the principal planes; (b) the principal stresses; (c) the stresses on planes making angles ± 15° with the horizontal. Solve both analytically and graphically.

Solution:

Analytical solution

Take $\sigma_{11} = 200$ kPa and $\sigma_{33} = 50$ kPa. So π_{11} makes an angle equal to +30° with the horizontal, and π_{33} makes an angle equal to −60° with the horizontal. To assign a sign to σ_{13}, we need to consider the right angle made by π_{11} and π_{33}; we must look at this angle as if

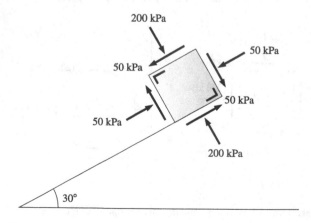

E-Figure 4.2 State of stress at a point (Example 4.2).

it were an uppercase letter "L", such that π_{11} is the "vertical" and π_{33} is the "horizontal" leg of the letter "L". Physically rotating the page until we see the "L" may be helpful in this visualization. The effect of the shear stress on the right angle made by π_{11} and π_{33} looked at in this manner is to reduce it; accordingly, $\sigma_{13} = -50\,\text{kPa}$. We are now prepared to solve the problem.

a. **Principal planes**
Substituting the values of σ_{11}, σ_{33}, and σ_{13} into Equation (4.7):

$$\tan\left(2\theta_p\right) = \frac{2(-50)}{200 - 50} = -\frac{2}{3}$$

from which:

$$\theta_p = \frac{1}{2}\arctan\left(-\frac{2}{3}\right) = -16.8°$$

The absolute value of $-16.8°$ is less than $45°$, so:

$$\theta_{p1} = -16.8°$$

$$\theta_{p3} = -16.8° + 90° = 73.2°$$

Graphically, θ_{p1} would show as an angle of $16.8°$ clockwise from π_{11} because the calculated angle is negative. If the answer is desired in terms of the angles that π_1 and π_3 make with the horizontal, then we need to add $30°$ (the angle that π_{11} makes with the horizontal) to these two results: π_1 is at an angle $13.2°$ and π_3 at $103.2°$ (or $-76.8°$) to the horizontal.

b. **Principal stresses**
The principal stresses can be calculated using either Equation (4.5) with $\theta = -16.8°$ and $73.2°$ or Equations (4.8) and (4.9). Using Equation (4.5):

$$\sigma_1 = \frac{1}{2}(200 + 50) + \frac{1}{2}(200 - 50)\cos\left[2(-16.8°)\right] + (-50)\sin\left[2(-16.8°)\right] = 215.1\,\text{kPa}$$

$$\sigma_3 = \frac{1}{2}(200 + 50) + \frac{1}{2}(200 - 50)\cos\left[2(73.2°)\right] + (-50)\sin\left[2(73.2°)\right] = 34.9\,\text{kPa}$$

Now using Equations (4.8) and (4.9):

$$\sigma_1 = \frac{1}{2}(200 + 50) + \sqrt{\frac{1}{4}(200 - 50)^2 + (-50)^2} = 215.1\,\text{kPa}$$

$$\sigma_3 = \frac{1}{2}(200 + 50) - \sqrt{\frac{1}{4}(200 - 50)^2 + (-50)^2} = 34.9\,\text{kPa}$$

c. **Stresses on plane making angles ±15° with horizontal**
These planes make angles $-15°$ and $-45°$ with π_{11}, respectively. Plugging these values ($-15°$ and $-45°$) into Equations (4.5) and (4.6):

$$\sigma_\theta = 125 + 75\cos\left[2(-15°)\right] - 50\sin\left[2(-15°)\right] = 214.9\,\text{kPa}$$

$$\tau_\theta = 75\sin\left[2(-15°)\right] + 50\cos\left[2(-15°)\right] = 5.8\,\text{kPa}$$

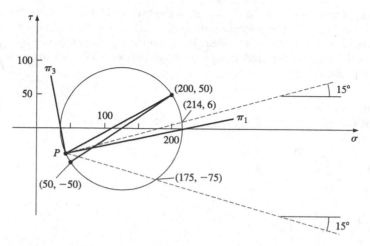

E-Figure 4.3 Mohr's circle and graphical solution of Example 4.2.

Not surprisingly, the σ_θ and τ_θ values calculated above are very close to the values for the major principal plane (215.1 and 0). You should verify that $\sigma_\theta = 175$ kPa and $\tau_\theta = -75$ kPa for the other plane (which makes an angle of $-45°$ with the horizontal).

<u>Graphical solution using the pole method</u>

The solution can be seen in E-Figure 4.3. The normal and shear stresses on plane π_{11} are 200 and 50 kPa, respectively; in plane π_{33}, they are 50 and -50 kPa.[7] These two points are diametrically opposite each other on the Mohr circle. If we plot these two points in $\sigma-\tau$ space and join them by a straight line, this line crosses the σ axis at the center of the circle. We can then easily draw the Mohr circle using a compass. The principal stresses are now easily read as the abscissas of the two points with $\tau = 0$.

If we now draw a line parallel to π_{11} through the point (200, 50), we obtain the pole P as the intersection of this line with the circle. If we draw a line parallel to π_{33} through (50, -50), we obtain the same result. The directions of π_1 and π_3 are obtained by drawing lines through P to the points $(\sigma_1, 0)$ and $(\sigma_3, 0)$ of the Mohr circle, respectively. The stresses at $\pm15°$ with the horizontal are found by drawing lines through the pole P making $\pm15°$ with the horizontal. These lines intersect the circle at two points with coordinates (214, 6) and (175, -75), respectively.

4.1.3.7 Total and effective stresses

When we plot Mohr's circles, we are representing the state of stress at a point in the soil mass. If there is a nonzero pore pressure u at this point, it is the same in every direction and thus does not affect the equilibrium of the point. We should remember that water cannot sustain shear stresses, so the presence of a pore pressure affects only normal stresses in the soil. It is useful to examine what happens if we plot both a total stress and an effective stress Mohr's circle in a reference system in which the horizontal axis is normal stress, irrespective of whether we are plotting total or effective stress. If we use Equation (4.10) first for total stresses and then substitute Equation (3.18) into Equation (4.10), we can see that the effective stress Mohr circle would be displaced along the σ axis with respect to the Mohr circle of total stresses by an amount equal to the pore pressure u (Figure 4.7).

---

[7] Note that, for Mohr's circle construction, counterclockwise shear stresses are plotted as positive, while clockwise shear stresses are plotted as negative.

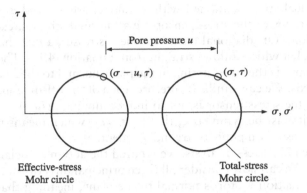

Figure 4.7 Principle of effective stresses: Illustration of the difference between the total and the effective stress states as represented using the concept of the Mohr circle.

4.1.4 Three-dimensional stress analysis*

Stress is a physical quantity described by a mathematical entity called *tensor*. In the case of stress, this tensor has rank two, which means that it has nine components. A vector in three-dimensional physical space, which has three independent components, is also a tensor, but of rank 1. Scalars are tensors of rank zero.

The *stress tensor* is defined at a material point P in space and is denoted by σ_{ij}, in which i and j can take the values 1, 2, or 3, representing the spatial orientations x_1, x_2, and x_3. So we can think of each component of a second-rank *tensor* as associated with two of the three coordinate axes: it corresponds to a force per unit area acting on a plane normal to one of the reference axes in the direction of one of the three axes. If we consider a plane through P, a unit force vector t_i, referred to in mechanics terms as traction, can be calculated on this plane from the stress tensor and the unit vector n_j normal to and pointing out of the plane using the equation:

$$t_i = -\sigma_{ij}n_j \tag{4.11}$$

where the repetition of the index j implies a summation over $j = 1, 2, 3$. So the equation above represents:

$$t_i = -\left(\sigma_{i1}n_1 + \sigma_{i2}n_2 + \sigma_{i3}n_3\right) \tag{4.12}$$

The minus sign in this equation results from the fact that the stress tensor is defined in such a way as to have the normal stresses on the plane pointing into the plane be positive (that is, compression is positive). If we now wish to calculate the normal stress on the plane, we can do so by taking the dot product of t_i by $-n_i$ (that is, we project t_i in the direction of $-n_i$):

$$\sigma = -t_i n_i = \sigma_{ij}n_i n_j \tag{4.13}$$

If the unit vector is aligned with one of the reference axes, then its components will be (1, 0, 0) if aligned with x_1, and (0, 1, 0) and (0, 0, 1) if aligned with the other two axes. If we calculate the normal stress acting in the direction of x_1, represented by the unit vector (1, 0, 0), we get:

$$\sigma = \sigma_{ij}n_i n_j = \sigma_{11}$$

Similarly, the normal stresses aligned with x_2 and x_3 are σ_{22} and σ_{33}. What this means is that the diagonal terms of the stress tensor are all normal stresses, each acting in one of the reference directions. Off-diagonal terms are shear stresses, a fact that we leave without explicit justification, but which follows straight from Equation (4.11). The stress tensor component σ_{12}, for example, is the stress acting in the plane normal to the x_1 axis in the negative direction of the x_2 axis. We can similarly interpret each of the off-diagonal terms, which are all shear stresses. If the stress tensor is not to induce any moment at the material point at which it is defined, it must be symmetric, that is, $\sigma_{ij} = \sigma_{ji}$, which means that only six of the possible nine stress tensor components are independent.

When we discussed 2D stress analysis, we ignored the action associated with one of the principal directions. We can now consider all three principal directions. Principal planes are those for which the traction vector is normal to the plane, meaning that it has no components tangential to the plane, so the shear stress is zero in it. This can be formulated mathematically in the form of the *eigenvalue* problem:

$$t_i = -\sigma_{ij} n_j = -\sigma_I n_j \tag{4.14}$$

where σ_I is a scalar. What Equation (4.14) states is simply that the traction computed from multiplying the unit vector on the plane by the stress tensor is aligned with the unit vector and has the magnitude of the normal stress σ_I on the plane. We will see that there will be three values of σ_I (σ_1, σ_2, and σ_3) that will be physically meaningful. Equation (4.14) can be rewritten in the form of a matrix equation:

$$\left(\sigma_{ij} - \sigma_I \delta_{ij}\right) n_j = 0 \tag{4.15}$$

Remembering that j is a dummy index, implying summation, the equation above represents three algebraic equations, one for each value of i (1, 2, and 3). If this system of three equations is possible and determinate, which is true if the determinant of $\left(\sigma_{ij} - \sigma_I \delta_{ij}\right)$ is different from zero, the solution to this equation is the null vector, which is of no interest. So the determinant of $\left(\sigma_{ij} - \sigma_I \delta_{ij}\right)$ must be equal to zero so that we can obtain a physically meaningful solution:

$$\det\left(\sigma_{ij} - \sigma_I \delta_{ij}\right) = 0 \tag{4.16}$$

This equation, fully expanded, is:

$$-\sigma_I^3 + I_1 \sigma_I^2 + I_2 \sigma_I + I_3 = 0 \tag{4.17}$$

where:

$$I_1 = 3\sigma_m = \sigma_{kk} = \sigma_{11} + \sigma_{22} + \sigma_{33} \tag{4.18}$$

$$I_2 = \frac{1}{2}\left(\sigma_{ij}\sigma_{ij} - \sigma_{kk}\sigma_{mm}\right) \tag{4.19}$$

$$I_3 = e_{ijk}e_{lmn}\sigma_{il}\sigma_{jm}\sigma_{kn} = e_{ijk}\sigma_{1i}\sigma_{2j}\sigma_{3k} = \det\left(\sigma_{ij}\right) \tag{4.20}$$

and σ_m is known as the mean or hydrostatic stress.

There are three solutions to this algebraic equation. It can be shown (e.g., Lubliner 2008) that these three solutions are real numbers, the three principal stresses. To determine the principal direction or plane corresponding to each of these real numbers, we substitute them, in turn, back into Equation (4.15). For each principal stress, this substitution leads to one system of equations, with two independent equations plus the equation corresponding to the requirement that n_i be a unit vector, that is, that $n_i n_i = 1$; the three equations can then be solved to find the unit vector representing the principal direction corresponding to each principal stress σ_1, $I = 1, 2, 3$.

Example 4.3

A stress state with respect to reference axes x_1, x_2, and x_3 is given by

$$\sigma = \begin{bmatrix} 0.1 & 0.6 & 0.0 \\ 0.6 & 1.2 & 0.0 \\ 0.0 & 0.0 & 0.3 \end{bmatrix} \text{MPa}$$

Determine the three stress invariants I_1, I_2, and I_3, the principal stress directions and the principal stresses corresponding to this stress tensor.

Solution:

The three principal invariants follow straight from Equations (4.18) through (4.20):

$$I_1 = \mathrm{tr}(\sigma) = \sigma_{kk} = 0.1 + 1.2 + 0.3 = 1.6 \text{ MPa}$$

$$I_2 = \frac{1}{2}\left(\sigma_{ij}\sigma_{ij} - \sigma_{ii}\sigma_{kk}\right) = \frac{1}{2}\left[2.26 - 1.6^2\right] = -0.15$$

$$I_3 = \det(\sigma) = 0.1(0.36) - 0.6(0.18) = -0.072$$

Equation (4.17) becomes:

$$-\sigma^3 + I_1\sigma^2 + I_2\sigma + I_3 = 0$$

$$\sigma^3 + 1.6\sigma^2 - 0.15\sigma - 0.072 = 0$$

which leads to the principal stresses:

$$\sigma_1 = 1.464 \text{ MPa}, \sigma_2 = 0.3 \text{ MPa and } \sigma_3 = -0.164 \text{ MPa}$$

To determine the major principal direction, we substitute 1.464 for σ_1 in Equation (4.15) to get:

$$\begin{bmatrix} 0.1-1.464 & 0.6 & 0 \\ 0.6 & 1.2-1.464 & 0 \\ 0 & 0 & 0.3-1.464 \end{bmatrix} \begin{Bmatrix} n_1 \\ n_2 \\ n_3 \end{Bmatrix} = \begin{Bmatrix} 0 \\ 0 \\ 0 \end{Bmatrix}$$

$$\begin{bmatrix} -1.364 & 0.6 & 0 \\ 0.6 & -0.264 & 0 \\ 0 & 0 & -1.164 \end{bmatrix} \begin{Bmatrix} n_1 \\ n_2 \\ n_3 \end{Bmatrix} = \begin{Bmatrix} 0 \\ 0 \\ 0 \end{Bmatrix}$$

$$-1.364n_1 + 0.6n_2 = 0$$

$$0.6n_1 - 0.264n_2 = 0$$

$$-1.164n_3 = 0$$

$$\begin{Bmatrix} n_1 \\ n_2 \\ n_3 \end{Bmatrix} = \begin{Bmatrix} 0.44 \\ 1 \\ 0 \end{Bmatrix} n_2$$

If we insist, as is customary, to have the principal direction represented by a unit vector (a vector with length equal to 1), then:

$$n_1^2 + n_2^2 + n_3^2 = 1 \Rightarrow n_2 = \pm \frac{1}{\sqrt{0.44^2 + 0 + 1}} = \pm 0.915$$

and the principal direction for $\sigma_1 = 1.464\,\text{MPa}$ is

$$\mathbf{n} = 0.403\mathbf{e}_1 + 0.915\mathbf{e}_2$$

where $\mathbf{e}_1$ and $\mathbf{e}_2$ are the unit vectors in the direction of x_1 and x_2, respectively. Substituting, one at a time, $0.3\,\text{MPa}$ and -0.164 for σ_1 in Equation (4.15), we obtain the other two directions:

$$\mathbf{n} = \mathbf{e}_3 \quad \text{for} \quad \sigma_2 = 0.3\,\text{MPa}$$

$$\mathbf{n} = -0.915\mathbf{e}_1 + 0.403\mathbf{e}_2 \quad \text{for} \quad \sigma_3 = -0.164\,\text{MPa}$$

where $\mathbf{e}_3$ is the unit vector in the direction of x_3.

If we subtract from the stress tensor its normal stress or confining stress content, we are left with a stress tensor that represents only the shear stress content of the stress tensor. This stress tensor is called the *deviatoric stress tensor*; it is expressed as

$$s_{ij} = \sigma_{ij} - \frac{1}{3}\delta_{ij}\sigma_{kk} \tag{4.21}$$

What this equation represents is the subtraction of the mean stress from each of the diagonal (normal stress) terms of the stress tensor. The tensor that is left, s_{ij}, can be thought of as an expression of the "shear stress content" in each reference plane at a point. Like the stress tensor σ_{ij}, s_{ij} also has three invariants, one of which is identically zero:

$$J_1 = s_{kk} = 0$$

The second invariant, which is a scalar indicator of the shear stress content of the stress tensor, is given by

$$J_2 = \frac{1}{2}\left(s_{ij}s_{ij} - s_{kk}s_{mm}\right) = \frac{1}{2}s_{ij}s_{ij} = \frac{1}{2}\sigma_{ij}\sigma_{ij} - \frac{1}{6}\sigma_{kk}\sigma_{mm} \tag{4.22}$$

The third invariant J_3 of the deviatoric stress tensor is its determinant. The three principal stresses and all invariants discussed above, such as I_1, I_2, and I_3 appearing in Equation (4.17), are invariant under a change in the coordinate system. It is common to use J_2 instead of I_2, and another quantity called the Lode's angle θ_L instead of I_3 or J_3. These three numbers (I_1, J_2, and θ_L) are sufficient to describe the physical effect of the stress tensor at a material point. In soil mechanics, the mean stress p, the Cambridge octahedral shear stress q, and Lode's angle θ_L are the most common choice for the three stress invariants. They are calculated as

$$p = \sigma_m = \frac{1}{3}I_1 = \frac{1}{3}(\sigma_1 + \sigma_2 + \sigma_3) \tag{4.23}$$

$$q = \sqrt{3J_2} = \frac{3}{\sqrt{2}}\tau_{oct} = \frac{1}{\sqrt{2}}\sqrt{(\sigma_1-\sigma_3)^2 + (\sigma_1-\sigma_2)^2 + (\sigma_2-\sigma_3)^2} \qquad (4.24)$$

and

$$\cos\theta_L = \frac{2\sigma_3 - \sigma_1 - \sigma_2}{2\sqrt{3J_2}} \qquad (4.25)$$

or, alternatively,

$$\cos 3\theta_L = \frac{3\sqrt{3}}{2}\frac{J_3}{(J_2)^{3/2}} \qquad (4.26)$$

These invariants have specific physical meanings that are helpful in understanding the effects of the stress tensor at a material point: p is a mean stress, an indicator of the confinement (mean normal stress in the three reference directions) of the stress tensor, q is an indicator of the shear stress content of the stress tensor, and θ_L, which conveys the same information as the third invariant of the stress deviator tensor, indicates how that shear stress is achieved. For example, the same value of q may be obtained by having $\sigma_1 = \sigma_2 > \sigma_3$ (we will see later that this is referred to as triaxial extension) or $\sigma_1 > \sigma_2 = \sigma_3$ (triaxial compression), but the response of the material may be different under triaxial extension and compression; it is possible to express this difference through θ_L. With these three numbers – p, q, and θ_L – all stress tensor effects can be properly captured and quantified. The three variables are used in the most successful theoretical framework for soil mechanics, called *critical-state soil mechanics*.

Example 4.4

Calculate the stress invariants p, q, and θ_L for the stress tensor of Example 4.3.

Solution:

The invariants J_2 and J_3 of the deviatoric stress tensor can be calculated from the values already calculated in Example 4.3:

$$J_2 = \frac{3}{2}\tau_{oct}^2 = \frac{3}{2}\times 0.469 = 0.703$$

$$J_3 = \det(s) = -0.433(-0.155) - 0.6(-0.14) = 0.151$$

$$p = \frac{I_1}{3} = \frac{1.6}{3} = 0.533\,\text{MPa}$$

$$q = \sqrt{3J_2} = \sqrt{3\times 0.703} = 1.452\,\text{MPa}$$

$$\cos 3\theta_L = \frac{3\sqrt{3}}{2}\frac{J_3}{(J_2)^{3/2}} = \frac{3\sqrt{3}}{2}\frac{0.151}{(0.703)^{3/2}} = 0.666$$

$$\theta_L = \frac{1}{3}\cos^{-1}(0.666) = 0.281\,\text{radians} = 16.1°$$

4.2 STRAINS*

4.2.1 Definitions of normal and shear strains

Analysis of geotechnical problems cannot be done only in terms of stresses. These stresses induce deformations, which are represented by strains. Strains can be normal strains or shear strains. At a given material point, a normal strain in a given direction quantifies the change in length (contraction or elongation) of an infinitesimal linear element (a very small straight line) aligned with that direction. The so-called *engineering shear strain γ* is a measure, at a given point, of the distortion (change in shape)[8] of a square element with sides aligned with respect to two reference axes.

Just as is true for stress, the state of strain at a point is described by a tensor: the *strain tensor*. There are three independent shear strain tensor components, one for each pair of reference axes. The component $\varepsilon_{13} = \frac{1}{2}\gamma_{13}$ expresses half the increase in the initial 90° angle formed by two perpendicular infinitesimal linear elements aligned with reference axes x_1 and x_3. The engineering shear strain is twice the corresponding strain tensor component:

$$\gamma_{ij} = 2\varepsilon_{ij} \tag{4.27}$$

Both normal and shear strains ε_{ij} can be expressed through

$$\varepsilon_{ij} = -\frac{1}{2}\left(\frac{\partial u_i}{\partial x_j} + \frac{\partial u_j}{\partial x_i}\right) \tag{4.28}$$

where $i, j = 1, 2, 3 =$ reference directions, $x_i =$ coordinate in the i direction, and $u_i =$ displacement in the x_i direction. The subscripts indicate the directions of linear differential elements and the directions in which displacements of the end points of the differential elements are considered. When $i = j$, the strain is a normal strain; it is a shear strain otherwise. There are theoretically nine numbers resulting from Equation (4.28), which together constitute the strain tensor. Of these nine, only six are independent. In this and in any mathematical aspect, the strain tensor is like the stress tensor discussed earlier.

Example 4.5

Derive, in a simple way, the expression for the normal strain at a point in the direction of reference axis x_1.

Solution:

Let's consider the case of E-Figure 4.4. For an undeformed soil mass, we have a differential element dx_1 aligned with the x_1 direction (E-Figure 4.4). We have labeled the initial point of the segment A and the end point B. In drafting this figure, we have corrected for rigid

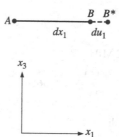

E-Figure 4.4 Normal strain ε_{11}: infinitesimal element dx_1 shown after correction for rigid body motion both before deformation (AB) and after deformation (AB*).

---

[8] We have avoided the duplication of symbols as much as possible, but there is no good alternative to using the traditional notation for engineering shear strain, which of course is the same as used for unit weight. The reader should observe the context in which symbols are used to avoid any confusion.

body translation in the x_1 direction. In other words, we are plotting the deformed element as if the displacement of A were zero for easier comparison with the original, undeformed element. This way, every displacement in the figure is relative to the displacement of A. If, after the soil mass is deformed, point B moves more in the positive x_1 direction than point A (that is, if the displacement u_1 of B is greater than that of A), as shown in E-Figure 4.4, then the element has clearly elongated (this elongation is seen in the figure as $B^* - B$).

Taking some liberty with mathematical notation, the unit elongation of dx_1 is the difference in displacement between points A and B ($u_{1B} - u_{1A} = B^* - B = \partial u_1$) divided by the initial length of the element (∂x_1), or $\partial u_1 / \partial x_1$. It remains to determine whether elongation is a positive or negative normal strain. To be consistent with the sign convention for stresses, according to which tensile stresses are negative, the normal strain ε_{11} in the x_1 direction is defined as

$$\varepsilon_{11} = -\frac{\partial u_1}{\partial x_1}$$

Note that this, indeed, is the expression that results directly from Equation (4.28) when we make $i=1$ and $j=1$.

Example 4.6

Derive, in a simple way, the expression of the shear strain at a point in the x_1–x_3 plane.

Solution:

Consider two differential linear elements, dx_1 and dx_3, aligned with the x_1 and x_3 axes, respectively, at a point within an undeformed soil mass (E-Figure 4.5). Now consider that the soil mass is deformed, and, as a result, points B and C (the end points of elements dx_1 and dx_3, respectively) move as shown (to new positions labeled B^* and C^*) with respect to point A (note that, as for Example 4.5, we are not representing rigid body translation in the x_1 and x_3 directions in the figure). We can see that both point B and point C have displacements that have components in both the x_1 and x_3 directions. Here we are interested in just the distortion of the square element made up of dx_1 and dx_3, not in the elongation or shortening of dx_1 and dx_3 individually. The distortion of the element clearly results from the difference in the displacement u_3 in the x_3 direction between A and B and in the displacement u_1 in the x_1 direction between A and C.

Taking again some liberties with mathematical notation, we can state that the differences in the displacements of points B and A (in the x_3 direction) and C and A (in the x_1 direction) can be denoted as ∂u_3 and ∂u_1, respectively. Since the deformations we are dealing with are small, the angle by which the differential element dx_1 rotates counterclockwise is approximately equal to ∂u_3 divided by the length of the element itself, or $\partial u_3 / \partial x_1$; similarly, the angle by which dx_3 rotates clockwise is $\partial u_1 / \partial x_3$. These two rotations create a reduction in the angle between the elements dx_1 and dx_3, which was originally 90°, characterizing a measure of distortion of the element. If we add them together, we obtain the absolute value of what has become known as the engineering shear strain γ_{13}; one half the sum gives us the absolute value of ε_{13}. It remains to determine whether this is a positive or negative distortion. Our sign convention for shear strains must be consistent with our shear stress sign convention. Recall from earlier discussion in this chapter that a positive shear stress was one that acted in a way that would tend to open up the 90° angle of the corner of our square.

Therefore, we will need a negative sign in front of our sum to obtain a negative shear strain for the reduction in the 90° angle we found to take place for the element in E-Figure 4.5:

$$\varepsilon_{13} = -\frac{1}{2}\left(\frac{\partial u_1}{\partial x_3} + \frac{\partial u_3}{\partial x_1}\right)$$

Note that this equation results directly from Equation (4.28) when we make $i=1$ and $j=3$ (or vice versa).

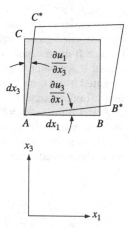

E-Figure 4.5 Shear strain ε_{13}: infinitesimal square element shown after correction for rigid body translation both before deformation (defined by $dx_1 = AB$ and $dx_3 = AC$) and after deformation (defined by $dx_1^* = AB^*$ and $dx_3^* = AC^*$).

Strains as expressed by Equation (4.28) are small numbers associated with small deformations and small rigid body rotations; there are other ways of defining strain that are more appropriate when elongations, contractions, distortions, and rotations become large, but these are outside the scope of this text. Equation (4.28) may also be used for increments of strains. It is appropriate in that case to use a "d" (the symbol for "differential") before the strain symbol (as in $d\varepsilon$ and $d\gamma$) to indicate that we refer to a strain increment.

As is true for stresses, there are also principal strains ε_1, ε_2, and ε_3 (and principal strain increments $d\varepsilon_1$, $d\varepsilon_2$, and $d\varepsilon_3$). These are strains (or strain increments) in the directions that remain perpendicular after deformation, that is, the directions in which there is no distortion (or no incremental distortion). Distortion happens any time the shape of an element changes. It is important to understand that, for $d\varepsilon_2 = 0$, a point in the soil experiences distortion as long as $d\varepsilon_1 > d\varepsilon_3$. To illustrate this, Figure 4.8b and c shows two alternative elements representing a point P. The larger outer element is aligned with the principal directions, and the deformed shape of the element (shown in Figure 4.8a) does not immediately convey the notion of distortion. However, if we instead focus on the smaller inner element, whose edges are not aligned with the principal directions, we clearly see the distortion that takes place as the element deforms (Figure 4.8b and c).

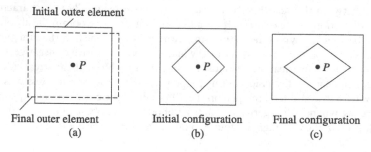

Figure 4.8 Alternative representations for the state of strain at a point: (a) an element aligned with the principal strain directions (vertical contraction and horizontal elongation); (b) the same element before deformation with an element inside it with sides oriented at 45° to the principal strain directions; (c) the same element after deformation, showing the distortion of the element with sides not aligned with the principal strain directions.

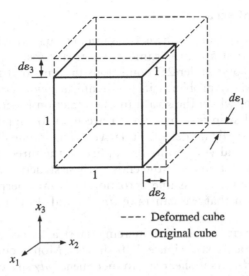

Figure 4.9 Calculation of volumetric strain.

The volumetric strain increment $d\varepsilon_v$, defined as minus the change in volume divided by the original volume (the negative sign being required to make contraction positive), can be easily determined in terms of a cubic element with sides with length initially equal to 1 and aligned with the principal directions (which means x_1, x_2, and x_3 are principal directions). The element is then allowed to expand as a result of elongations equal to du_1, du_2, and du_3 in the three reference directions. As the cube is aligned with the principal directions, there will be no distortion in the planes $x_1 x_2$, $x_1 x_3$, or $x_2 x_3$. It is apparent from Figure 4.9 that

$$dV = (1 + du_1)(1 + du_2)(1 + du_3) - 1$$

Referring back to our definition of normal strain and considering that the initial length of the sides of the cube are of unit length and the initial volume of the cube is also equal to 1, we can write the following for the volumetric strain increment:

$$d\varepsilon_v = -\frac{dV}{1} = 1 - (1 - d\varepsilon_1)(1 - d\varepsilon_2)(1 - d\varepsilon_3)$$

which, given that the strain increments are very small (and that second- and third-order terms would be extremely small and thus negligible), reduces to

$$d\varepsilon_v = d\varepsilon_1 + d\varepsilon_2 + d\varepsilon_3 \qquad (4.29)$$

The volume change at a point is clearly independent of the reference system and of any distortion, so the following equation would also apply even if x_1, x_2, and x_3 were not the principal directions:

$$d\varepsilon_v = d\varepsilon_{kk} = d\varepsilon_{11} + d\varepsilon_{22} + d\varepsilon_{33} \qquad (4.30)$$

where $d\varepsilon_{11}$, $d\varepsilon_{22}$, and $d\varepsilon_{33}$ = normal strains in the arbitrary directions x_1, x_2, and x_3.

4.2.2 Mohr's circle of strains*

The mathematics of stresses and strains is the same: normal strains play the same role as normal stresses, and shear strains, the same as shear stresses. So, just as it is possible to express the stresses at a point under 2D conditions in terms of a Mohr's circle, the same is possible for strains. Many problems in geotechnical engineering can be idealized as plane-strain problems, for which the strain in one direction is zero. For example, slopes, retaining structures, and strip footings, which are used to support lines of columns or load-bearing walls, are usually modeled as relatively long in one direction, with the same cross section throughout and with no loads applied in the direction normal to the cross sections. Except for cross sections near the ends of these structures, it is reasonable, based on symmetry considerations, to assume zero normal strain perpendicular to the cross section. The result is then that we can take $d\varepsilon_2=0$ and do our strain analysis in two dimensions.

In the case of the *Mohr circle of strains*, incremental strains, not total strains, are plotted in the horizontal and vertical axes. Figure 4.10 shows a Mohr's circle of strains plotted in normal strain increment de versus shear strain increment $\frac{1}{2}d\gamma$ space. As is true for stresses, each point of the Mohr circle represents one plane through the point in the soil mass for which the Mohr circle represents the strain state. The points of greatest interest in the Mohr circle of strains are:

1. the leftmost and rightmost points, $(d\varepsilon_3, 0)$ and $(d\varepsilon_1, 0)$, corresponding to the minor and major principal incremental strain directions;
2. the highest and lowest points, $(\frac{1}{2}de_v, \pm\frac{1}{2}d\gamma_{max})$, corresponding to the directions of largest shear strain ($\frac{1}{2}de_v=\frac{1}{2}(de_1+de_3)$ is the de coordinate of the center, and $\frac{1}{2}d\gamma_{max}=\frac{1}{2}(de_1-de_3)$ is the radius of the Mohr circle); and
3. the points where the circle intersects the shear strain axis, $(0, \pm\frac{1}{2}d\gamma^z)$.

The two points with zero normal strain increment correspond to the two directions along which $de=0$, that is, the directions along which there is neither incremental extension nor contraction. It is possible to define a separate reference system for each of these two directions such that x_1 in each system is aligned with the direction of zero normal strain. To clearly indicate that x_1 is a direction of zero normal strain, we can use a superscript z, as in x_1^z. This will be useful in our discussion of the dilatancy angle, which follows.

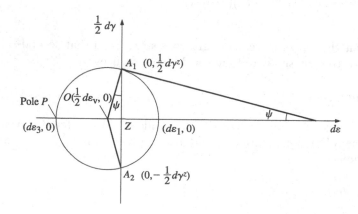

Figure 4.10 Mohr's circle of strains.

4.2.3 Dilatancy angle

The angle ψ shown in Figure 4.10, known as the *dilatancy angle,* is quite useful in understanding soil behavior. There are two ways of expressing the dilatancy angle based on the geometry of the Mohr circle of strains:

$$\sin\psi = \frac{OZ}{|OA_1|} = -\frac{\frac{1}{2}(d\varepsilon_1 + d\varepsilon_3)}{\frac{1}{2}(d\varepsilon_1 - d\varepsilon_3)} = -\frac{d\varepsilon_1 + d\varepsilon_3}{d\varepsilon_1 - d\varepsilon_3} = -\frac{d\varepsilon_v}{|d\gamma_{max}|} \tag{4.31}$$

$$\tan\psi = \frac{OZ}{|ZA_1|} = -\frac{\frac{1}{2}d\varepsilon_v}{\frac{1}{2}d\gamma'} = -\frac{d\varepsilon_v}{|d\gamma^z|} \tag{4.32}$$

where γ^z = shear strain in the $x_1^z - x_3^{zp}$ plane (Figure 4.11) and x_1^z is the direction of zero normal strain. The direction normal to x_1^z, in Figure 4.11 is represented by x_3^{zp}, where the superscript zp means that the direction x_3^{zp} is perpendicular to the direction of zero normal strain. There are in fact two distinct directions of zero normal strain, as will be shown later.

The dilatancy angle is related to the volumetric strain resulting from a unit increase in shear strain.[9] By definition, the dilatancy angle ψ is positive when there is dilation (volume expansion). This is apparent from Equations (4.31) and (4.32), for the dilatancy angle clearly results positive when volume expands, that is, when $d\varepsilon_v < 0$. Note that the denominators of Equations (4.31) and (4.32) are always positive (hence the absolute values taken), for the dilatancy angle is related to the volumetric strain increment resulting from a unit increment in shear strain, regardless of the orientation of the shear strain. In other words, the dilatancy angle would still be the same positive value if the element shown in Figure 4.11 were sheared to the left and not the right as shown.

The Mohr circle of Figure 4.10 corresponds to a state of dilation (expansion), as $d\varepsilon_v < 0$. Examining the state of deformation in the $x_1^z - x_3^{zp}$ reference system, expansion implies that the normal strain in the x_3^{zp} direction, normal to x_1^z, is negative (that is, that elongation takes place in the x_3^{zp} direction, as clearly shown in Figure 4.11). If we locate the pole in Figure 4.10, then draw two lines (one through A_1 and one through A_2); the two lines perpendicular to these two lines are the directions of zero normal strain, as we will discuss in detail later.

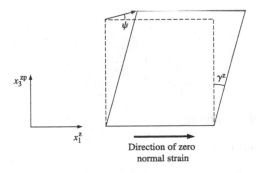

Figure 4.11 State of strain visualized for an element with one side aligned with the direction of zero normal strain.

---

[9] Technically, both the shear and volumetric strain increments in the definition of the dilatancy angle are plastic strain increments, a distinction that for our present purposes is not necessary to make.

Example 4.7

A soil element is subjected to the following incremental strains: $d\varepsilon_1 = 0.03\%$, $d\varepsilon_3 = -0.05\%$. Knowing that plane-strain conditions are in force (that is, the strain in the x_2 direction is zero), calculate the dilatancy angle.

Solution:

Because we know the principal strain increments, we can immediately calculate the dilatancy angle as

$$\sin\psi = -\frac{d\varepsilon_v}{d\gamma_{\max}} = -\frac{d\varepsilon_1 + d\varepsilon_3}{d\varepsilon_1 - d\varepsilon_3} = -\frac{0.03 - 0.05}{0.03 - (-0.05)} = 0.25$$

from which

$$\psi = 14.5°$$

In Problem 4.18, you are asked to continue this by plotting the Mohr circle, finding the pole for the case when the major principal strain increment is vertical, and determining the directions of the potential slip planes through this element (which is the subject of the subsequent section).

For a triaxial strain state in which $\varepsilon_2 = \varepsilon_3$, we may attempt to redefine the dilatancy angle by working with $d\varepsilon_s$, defined as

$$d\varepsilon_s = d\varepsilon_1 - 2d\varepsilon_3 \tag{4.33}$$

The volumetric strain in the triaxial case follows from Equation (4.29):

$$d\varepsilon_v = d\varepsilon_1 + 2d\varepsilon_3$$

Thus, the dilatancy angle for triaxial conditions is written as

$$\sin\psi = -\frac{d\varepsilon_1 + 2d\varepsilon_3}{d\varepsilon_1 - 2d\varepsilon_3} \tag{4.34}$$

A single expression for it, which applies to both plane-strain and triaxial conditions, is:

$$\sin\psi = -\frac{d\varepsilon_1 + kd\varepsilon_3}{d\varepsilon_1 - kd\varepsilon_3} = -\frac{d\varepsilon_v}{d\varepsilon_1 - kd\varepsilon_3} = -\frac{\dfrac{d\varepsilon_v}{d\varepsilon_1}}{2 - \dfrac{d\varepsilon_v}{d\varepsilon_1}} \tag{4.35}$$

where:

$$k = \begin{cases} 1 \text{ for plane-strain conditions} \\ 2 \text{ for triaxial conditions} \end{cases} \tag{4.36}$$

4.2.4 Strain variables used in critical-state soil mechanics*

Just as the stress tensor can be decomposed into two parts – one with the elements in its diagonal being the hydrostatic mean stress and the other having zero first invariant – the same can be done for the strain tensor:

$$e_{ij} = \varepsilon_{ij} - \frac{1}{3}\delta_{ij}\varepsilon_{kk} \tag{4.37}$$

where e_{ij} is the *deviatoric strain tensor*, a tensor with zero first invariant and thus one that expresses the shear strain "content" of the strain tensor.

For an incremental strain tensor, there are invariants just like there are invariants for the stress tensor, as seen earlier. For stress, p was chosen as representative of an isotropic or hydrostatic stress. The corresponding invariant for incremental strain is precisely $d\varepsilon_v = d\varepsilon_{kk} = d\varepsilon_{11} + d\varepsilon_{22} + d\varepsilon_{33}$. An isotropic strain increment is one in which the same elongation happens in every direction; for such a strain increment, there is no distortion, and $d\varepsilon_v$ describes this completely.

The strain increment variable corresponding to q is ε_q, which is given by

$$\varepsilon_q = \frac{\sqrt{2}}{3}\sqrt{3\varepsilon_{ij}\varepsilon_{ij} - \varepsilon_v^2} = \frac{1}{\sqrt{2}}\gamma_{oct} = \frac{2}{\sqrt{3}}\sqrt{L_2} \tag{4.38}$$

where γ_{oct} is the *octahedral shear strain* and L_2 is the second invariant of the deviatoric strain tensor $\mathbf{e}$ given by Equation (4.37). L_2 is given by

$$L_2 = \frac{1}{2}\left(e_{ij}e_{ij} - e_{kk}e_{mn}\right) = \frac{1}{2}e_{ij}e_{ij} = \frac{1}{2}\varepsilon_{ij}\varepsilon_{ij} - \frac{1}{6}\varepsilon_{kk}\varepsilon_{mm}$$

With this correspondence established, we can, for example, calculate the work done on a soil element due to an increment of strain as

$$dW = p\,d\varepsilon_v + q\,d\varepsilon_q \tag{4.39}$$

In addition to the volumetric strain and ε_q, there is also another strain invariant – Lode's angle for strain – that is used in advanced soil stress response simulations; it is given by

$$\cos\chi_L = \frac{2\varepsilon_3 - \varepsilon_1 - \varepsilon_2}{2\sqrt{3L_2}} \tag{4.40}$$

4.3 PLASTIC FAILURE CRITERIA, DEFORMATIONS, AND SLIP SURFACES

4.3.1 Mohr–Coulomb strength criterion

Soils are not elastic. Referring to the top stress–strain plot in Figure 4.12, if we apply repeatedly the same increment of shear strain to an element of soil (a process referred to as strain-controlled loading), the increment of stress that the soil element is able to sustain decreases continuously (a process that is sometimes referred to as *modulus degradation*, but is truly a reflection of inelastic processes leading to deformation that is not reversible upon load removal), until a state (represented by point F in the figure) is reached at which the stress increment is zero. At this point, if we continue to increase the strain, the stress will stay the same. The other stress–strain plot shown in the figure illustrates another possible response, whereby the stress peaks at point F. This second response (referred to as strain softening) is common in soils. The limiting or peak stress associated with point F in each case is usually what is meant by the shear strength of the soil, as it is the maximum stress the soil can take.

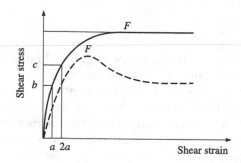

Figure 4.12 Nonlinearity of stress–strain relationship for soils and failure (the onset of very large deformations at some value of stress, represented by point *F*). Note that the first strain increment *a*
generates a stress increment *b*, but that the second strain increment generates a stress increment $c - b < b$.

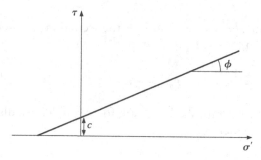

Figure 4.13 Mohr–Coulomb failure envelope.

Alternatively, we could have started loading the soil element by applying stress increments
to the element (which is referred to as stress-controlled loading). In this case, when the point
F at which the peak stress was observed during strain-controlled loading is reached, the
soil element will not be able to take any additional stress, and our attempt to apply another
stress increment to the soil element will result instead in an uncontrolled deformation. It is
not possible in the stress control case to plot the stress–strain relationship after point *F* is
reached.

The Mohr–Coulomb yield criterion (which we will also refer to as failure or strength criterion) has traditionally been used in soil mechanics to represent the shear strength of soil.[10]
It expresses the notion that shear strength of soil increases with increasing normal effective
stress acting on the potential shearing plane. Whereas in stress analysis we do not have to
concern ourselves with whether we are dealing with effective or total stresses, because the
analysis applies to both, we must now make a clear distinction. Soils "feel" only effective
stresses (that is, any deformation of the soil skeleton happens only in response to effective
stresses); thus, the response of soil to loading and the shear strength of soil depends only on
the effective stresses. Based on this consideration, we represent the Mohr–Coulomb criterion in $\sigma'-\tau$ space as two straight lines making angles $\pm\phi$ with the horizontal and intercept-
ing the τ axis at distances c and $-c$ from the σ' axis. Figure 4.13 shows only the line lying

---

[10]   The question of whether it should be used or, indeed, should ever have been used to represent the shear
strength response of soil is addressed in Chapter 5, when we discuss the mechanics of sands. Regardless, it is
a convenient means to introduce the notion of significant deformation when certain stress combinations are
imposed on a material, hence its use in this chapter.

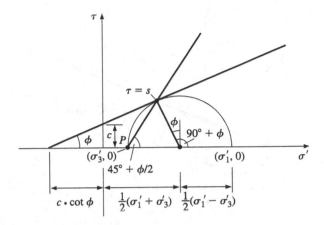

Figure 4.14 Mohr's circle at "failure."

above the σ' axis, since the diagram is symmetric about the σ' axis. The distance c is usually referred to as the cohesive intercept. Mathematically, the Mohr–Coulomb criterion may be represented in a simple way by

$$s = c + \sigma' \tan\phi \tag{4.41}$$

where s is shear strength of the soil and ϕ is its friction angle. For a given normal effective stress σ' on a plane, if the shear stress on the plane is $\tau=s$ as given by Equation (4.41), shearing or what is commonly referred to in engineering practice as "failure" of the material occurs. Failure here means the occurrence of very large shear strains in the direction of that plane and may best be referred to as plastic or shear failure.[11] This means that the soil cannot sustain shear stresses greater than the value given by Equation (4.41). Equation (4.41) is a straight line in σ'–τ space (as shown in Figure 4.13) that is referred to as the Mohr–Coulomb strength envelope. There can be no combination of σ' and τ that would lie above the Mohr–Coulomb strength envelope.

Figure 4.14 shows the Mohr circle for a soil element (or point) within a soil mass, at plastic failure, where σ' and τ act on the two planes corresponding to the two tangency points between the circle and the envelope (again, only the half of the diagram lying above the σ' axis is shown, as the part below the σ' axis is symmetric). All other points, representing all the planes where (σ', τ) do not satisfy Equation (4.41), lie below the Mohr–Coulomb envelope. If we know the directions of the principal stresses, we can determine the pole for this circle. Assuming a vertical major principal stress and a horizontal minor principal stress, the pole P lies at point $(\sigma'_3, 0)$. By simple geometry, the central angle corresponding to the arc extending from the pole to the point of tangency is $90° - \phi$. This means that the planes corresponding to the points of tangency lie at $\pm (45° + \phi/2)$ with the horizontal. One method of estimating the direction of a real shear (slip) surface passing through the soil element is to assume that it can be approximated by one of these two possible directions. Note that this refers to the direction of the slip surface at the point under consideration, and that at other points of the soil mass the slip surface direction may be different because the principal directions at those points may be different. This means that a slip surface through a soil mass is

---

[11] The use of just the word "failure" or to state that a soil has "failed" is not sufficiently specific and is the cause for considerable confusion among not only students, but professionals as well.

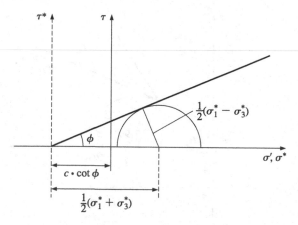

Figure 4.15 Caquot's principle: transformed normal stresses.

a surface tangent at every point to the direction estimated as described above, a direction making an angle of $\pm(45° + \phi/2)$ with the direction of the major principal plane at the point. In a subsequent subsection, we discuss slip surfaces in more detail, including their nature and the fact that, realistically, they are not truly surfaces (with zero thickness), but bands (with a small but nonzero thickness).

It is possible to find the relationship between the principal stresses σ_1' and σ_3' at failure from the geometry in Figure 4.14. It is easier to proceed if we define new, transformed normal stresses σ^* through:

$$\sigma^* = \sigma' + c \cot \phi \tag{4.42}$$

Taking Equation (4.42) into Equation (4.41) leads to:

$$s = \sigma^* \tan \phi \tag{4.43}$$

This transformation (sometimes referred to as Caquot's principle after the researcher who first made use of it) is represented graphically in Figure 4.15. A Mohr's circle tangent to the strength envelope, so at plastic failure, is also represented. We can write $\sin \phi$ in terms of the ratio of the radius of the Mohr circle to the distance from the center of the Mohr circle to the origin of the transformed system of stress coordinates (that is, where the σ^* and τ^* axes cross):

$$\sin \phi = \frac{\sigma_1^* - \sigma_3^*}{\sigma_1^* + \sigma_3^*}$$

The principal stress ratio σ_1^*/σ_3^* easily follows:

$$\frac{\sigma_1^*}{\sigma_3^*} = N \tag{4.44}$$

where N is known as the flow number, given by

$$N = \frac{1 + \sin \phi}{1 - \sin \phi} \tag{4.45}$$

If the relationship between the principal stresses in its original form is needed, we just need to use Equation (4.42) to rewrite σ_1^* and σ_3^* in Equation (4.44):

$$\frac{\sigma_1' + c\cot\phi}{\sigma_3' + c\cot\phi} = \frac{1 + \sin\phi}{1 - \sin\phi} = N$$

which can be rewritten as

$$\sigma_1' = N\sigma_3' + (N - 1)c\cot\phi \qquad (4.46)$$

This expression can be further rewritten as

$$\sigma_1' = N\sigma_3' + 2c\sqrt{N} \qquad (4.47)$$

by recognizing that

$$\cot\phi = \sqrt{\frac{1 - \sin^2\phi}{\sin^2\phi}}$$

Note that, for $c = 0$, Equation (4.46) reduces to:

$$\sigma_1' = N\sigma_3' \qquad (4.48)$$

In modern soil mechanics, instead of working with the ratio N of σ_1' to σ_3' for a plastic stress state, we work with the ratio M of q to p' also for a plastic state. M can be related to the friction angle (although the relationship changes depending on the loading path):

$$M = \frac{6\sin\phi}{3 - \sin\phi} \qquad (4.49)$$

for triaxial compression and

$$M = \frac{6\sin\phi}{3 + \sin\phi} \qquad (4.50)$$

for triaxial extension.

4.3.2 Slip surfaces*

Figure 4.16a shows a *slip surface* and an element of it, which is expanded in Figure 4.16b. A slip surface is also often referred to in the literature as a failure surface or the more technical *shear band*. This very thin band separates two soil masses moving with respect to each other. Within this very thin zone, shear strains are highly localized, achieving very high values, hence the representation of a slip surface element in Figure 4.16 as having nonzero thickness. The soil masses separated by the slip surface are often assumed in simplified analyses to be rigid (although in reality there can be some plastic deformation in the proximity of slip surfaces), hence the slip surface element shown in Figure 4.16b being bounded by two rigid blocks. One of the two soil masses separated by the slip surface is often stationary.

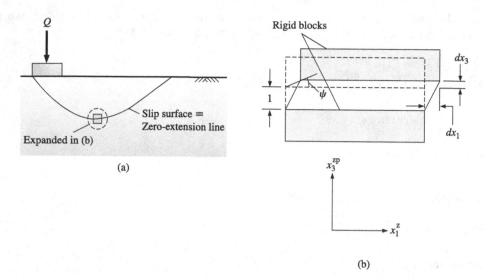

Figure 4.16 Details of slip surface element.

A slip surface develops when the shear stresses at every point of it exceed the corresponding shear strength. On the onset of failure, the rigid blocks on either side of the slip surface are connected to the slip surface and, being rigid, prevent any contraction or elongation in the direction of the slip surface. This means the normal strain in the direction x_1^z of the slip surface is equal to zero. The slip surface is, accordingly, a zero-extension line with direction given by an axis x_1^z. The state of strain at a point of the slip surface, referred to the axis x_1^z parallel to the slip surface and another x_3^{zp} normal to it, is represented in Figure 4.11. Taking the thickness of the slip surface as being equal to 1, the state of incremental infinitesimal strain for the slip surface element in Figure 4.16 can be expressed as follows:

$$d\varepsilon_{11}^z = 0$$

$$d\varepsilon_{33}^{zp} = -\frac{dx_3}{1} = -dx_3$$

$$\left| d\gamma^z \right| = \frac{dx_1}{1} = dx_1$$

$$d\varepsilon_v = d\varepsilon_{11}^z + d\varepsilon_{33}^{zp} = -dx_3$$

Recalling the definition of the dilatancy angle ψ, given in Equation (4.32), we can write:

$$\tan\psi = -\frac{d\varepsilon_v}{\left| d\gamma^z \right|} = \frac{dx_3}{dx_1} \tag{4.51}$$

Equation (4.51) shows that the dilatancy angle represents the angle that the motion of the top of the element makes with the horizontal. Equation (4.51) is instrumental in understanding the geometry of slip surfaces. Take, for example, the soil slope in Figure 4.17. It is common to see such slopes failing along curved slip surfaces. The shape of a potential slip surface can be determined if we take a center of rotation such that every point of the surface is at a variable distance r from the center of rotation. Once a relationship between r and the angle of rotation θ around the center of rotation is defined, the shape of the slip surface becomes known.

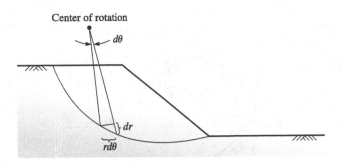

Figure 4.17 Slip surface geometry.

An infinitesimal rotation $d\theta$ with respect to the center of rotation corresponds to a tangent displacement $r d\theta$ along the slip surface and to a possible small increase in the radius (distance to the center of rotation) by an amount dr. So dr is analogous to dx_3, and $r d\theta$ is analogous to the dx_1 in Equation (4.51). Taking these into Equation (4.51) gives:

$$\tan \psi = \frac{dr}{r d\theta} \tag{4.52}$$

With the knowledge of one point of the slip surface, defined by r_0 and θ_0, integration of Equation (4.52) leads to

$$r = r_0 e^{(\theta - \theta_0) \tan \psi} \tag{4.53}$$

where r_0 is the radius corresponding to a reference angle θ_0.

Equation (4.53) is a geometrical shape referred to as a logarithmic spiral (or log spiral, for short). The implication is that a homogeneous soil mass, with the same dilatancy angle throughout, is expected to fail along a log spiral. We often approximate curved surfaces by planes or cylinders (straight lines or circles in cross section). Both are strictly applicable only if $\psi = 0$, for then Equation (4.53) reduces to that of a circle, and a straight line is nothing more than a circle with infinite radius. The approximation is sometimes justified when the maximum value $(\theta_{max} - \theta_0)$ of $\theta - \theta_0$ in Equation (4.53) is small. When we analyze a problem involving shearing under undrained conditions using total stresses and undrained shear strength, $\psi = 0$. We often do that for clays. The use of a slip surface with circular cross section then follows directly from Equation (4.53).

4.3.3 Slip surface direction

The Mohr circle can be used to determine the two directions of the two potential slip surfaces at a point within a soil mass. The direction of the slip surface at a point is the direction of its tangent at the point. The pole method can be used to do this graphically.

Assume that the major principal strain increment is vertical, therefore normal to the horizontal plane. In such case, the pole P is located at the leftmost point of the circle, as shown in Figure 4.18. We start by drawing lines PA_1 and PA_2 from P through the points A_1 and A_2 corresponding to zero normal strain. These lines are parallel to the planes that are perpendicular to the directions of zero normal strain (that is, the slip surface directions). So lines PB_1 and PB_2, normal to PA_1 and PA_2, respectively, are the directions of the two potential slip surfaces through the point under consideration.

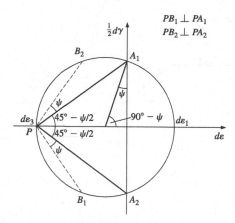

Figure 4.18 Determination of the direction of slip planes for a soil element.

From the geometry of the Mohr circle in Figure 4.18, it can be shown that the slip surfaces make angles equal to $45° + \psi/2$ and $-(45° + \psi/2)$ with the direction of the minor principal strain (which is horizontal in this case). If $\psi = \phi$ (a common assumption, even if tacitly made, in many geotechnical analyses), then the slip surface directions correspond to the directions of the planes where σ and τ satisfy the Mohr–Coulomb shear strength criterion, which are at $\pm(45° + \phi/2)$ to the horizontal. Even when ψ differs from ϕ, so long as the difference is not large, the direction of the slip surfaces can still be approximated as being $45° + \phi/2$ with respect to the horizontal. For dense sands, the directions of the two potential slip surfaces must be estimated from the dilatancy angle if accuracy is important. In dilative sands (sands that exist in a dense state and/or in a state of low effective confining stresses), the deviations from reality from assuming $\psi = \phi$ in analyses may be substantial and should not be ignored. This point will be illustrated in Chapter 10 in the context of bearing capacity analyses of footings.

4.3.4 The Hoek–Brown failure criterion for rocks

Although the Mohr–Coulomb plastic failure criterion is also used for rocks, Hoek and Brown (1980;1988) criterion is more practical, particularly for fractured rock, as its parameters have been related to quantities usually measured in site investigations in rock. The criterion is written as

$$\sigma_1' = \sigma_3' + q_u \sqrt{m \frac{\sigma_3'}{q_u} + s} \qquad (4.54)$$

where q_u = uniaxial, unconfined compressive strength of the rock; m and s = model parameters that depend on the degree of fracturing of the rock. The parameters m and s can be expressed in terms of the type of rock, degree of weathering, and frequency of discontinuities. This is covered in Chapter 7.

4.4 AT-REST AND ACTIVE AND PASSIVE RANKINE STATES

4.4.1 At-rest state

Let us consider a semi-infinite mass of homogeneous, isotropic soil (Figure 4.19a). A soil mass is homogeneous with respect to some physical property (e.g., unit weight, hydraulic conductivity, shear modulus, or shear strength) if the property has the same value at every

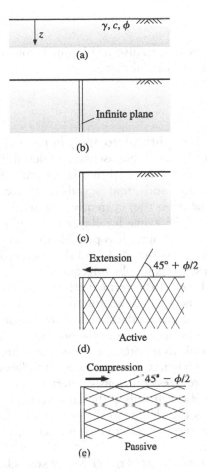

Figure 4.19 (a) A semi-infinite soil mass; (b) a semi-infinite vertical plate separating two halves of the semi-infinite soil mass; (c) vertical plate with soil completely removed from the left side; (d) soil in the active Rankine state, as illustrated by drawing the two families of possible slip surfaces at angles of $\pm(45° + \phi/2)$ with the horizontal; (e) soil in the passive Rankine state, as illustrated by drawing the two families of possible slip surfaces at angles of $\pm(45° - \phi/2)$ with the horizontal.

point of the soil mass. The soil is isotropic at a point with respect to some property if the value of the property at the point is the same in every direction. The soil deposit in the figure has a free surface that is level (horizontal); the soil deposit extends to infinity downward from the free surface and in the direction parallel to the free surface. Vertical and lateral stresses within this deposit are principal stresses because the shear stresses in horizontal and vertical planes are zero.

The soil deposit idealization of Figure 4.19a allows us to obtain very important results. If we assume that the soil deposit has not been disturbed after formation, the soil is said to be in a state of rest. The ratio $K_0 = \sigma'_h / \sigma'_v$ of the lateral to the vertical effective stress at any point in a soil mass in a state of rest is referred to as the *coefficient of lateral earth pressure at rest*. It is a very important quantity that appears often in geotechnical design. For a purely frictional soil, with strength parameters $c=0$ and nonzero ϕ, Jaky (1944) proposed an empirical relationship between K_0 and ϕ, given as[12]:

---

[12] For clays, some prefer to use 0.95 instead of 1 in Equation (4.55).

$$K_0 = 1 - \sin\phi \tag{4.55}$$

An alternative expression for K_0 can also be obtained from elasticity[13], in which case K_0 is related to the Poisson's ratio of the soil through:

$$K_0 = \frac{v}{1-v} \tag{4.56}$$

Equations (4.55) and (4.56) are difficult to apply in practice because soil does not have a single friction angle, a point that will be discussed in detail in Chapters 5 and 6, and the value of Poisson's ratio to use for a material that is not truly elastic is even more difficult to select. In normally consolidated soils, that is, soils that have never experienced a greater vertical effective stress than the stress they currently experience, K_0 is roughly in the 0.4–0.5 range for sandy soils and 0.5–0.75 range for clayey soils, with denser, stiffer soils having the values at the low end of these ranges. It is possible that an extremely dense sand, as an example, might have K_0 slightly less than 0.4, and slight exceedance of the high end of the range is also possible, but these cases will be infrequent.

In a normally consolidated soil, if the vertical effective stress is increased by an amount $d\sigma'_v$, the lateral effective stress increases by an amount $d\sigma'_h = K_0 d\sigma'_v$ to keep the ratio of σ'_h to σ'_v constant and equal to K_0. To consider the effects of stress history, assume that the vertical effective stress is then reduced by $d\sigma'_v$ down to the original value; if that is done, the lateral effective stress does not go back to its previous value, retaining a considerable fraction of the increase $d\sigma'_h$ it experienced when σ'_v was increased. It follows that soils that have experienced a greater vertical effective stress previously, referred to as overconsolidated soils, have higher K_0 than normally consolidated soils at the same vertical stress. Physically, the reason lateral stresses get locked in is the change in the density and fabric of the soil required to accommodate the increase in vertical effective stress, which is to some extent inelastic in nature and thus irrecoverable.

Brooker and Ireland (1965) investigated the effects of stress history on K_0, arriving at the following equation:

$$K_0 = K_{0,\mathrm{NC}}\sqrt{\mathrm{OCR}} \tag{4.57}$$

where OCR is the *overconsolidation ratio*, defined as

$$\mathrm{OCR} = \frac{\sigma'_{vp}}{\sigma'_v} \tag{4.58}$$

where σ'_{vp}=preconsolidation pressure, which is the maximum vertical effective stress ever experienced by the soil element and σ'_v is simply the current vertical effective stress.

4.4.2 Rankine states

4.4.2.1 Level ground

To investigate the behavior of soil deposits when subjected to relatively large strains, it will help us now to go through an imaginary exercise. Let us consider that we could insert a smooth, infinite, infinitesimally thin plane vertically into the soil deposit without disturbing

---

[13] To derive this equation, we can write the equation for the lateral strain in terms of its coaxial normal stress (that is, the lateral effective stress) and the two transverse normal stresses (the vertical and again the lateral effective stresses, but now in the other lateral direction) and make the lateral strain equal to zero.

the soil (Figure 4.19b) and thus keeping vertical shear stresses equal to zero. Let us consider further that we could remove all of the soil from one side of the plane (Figure 4.19c) so that we could now either push or pull on the plane in the horizontal direction, thereby causing the soil on the other side of the plane to either shorten in length or stretch in the horizontal direction.

When we pull the vertical plane horizontally away from the soil, it allows the soil to expand in the horizontal direction, which leads to a drop in the horizontal effective stress σ_h'. If we continue to pull on the plane, σ_h' continues to drop, while σ_v' remains unchanged. This process is illustrated by the Mohr circles in Figure 4.20, all with the same major principal stress $\sigma_1' = \sigma_v'$, but a decreasing minor principal stress $\sigma_3' = \sigma_h'$. This process, by which σ_h' decreases while σ_v' remains unchanged, cannot go on indefinitely; it in fact comes to an end when σ_v' / σ_h' becomes equal to the *flow number* $N = \sigma_1' / \sigma_3' = (1 + \sin\phi)/(1 - \sin\phi)$, at which point the Mohr circle touches the Mohr–Coulomb strength envelope (assumed here with $c = 0$), the plastic failure criterion is satisfied, and slip surfaces can potentially form anywhere in the soil mass. This state at which the whole soil mass is in a state of incipient collapse is known as an *active Rankine state*: "active" because the self-weight of the soil contributes to or is active in bringing it about; "Rankine" because it was first identified by Lord Rankine in the 19th century (Cook 1951; Rankine 1857).

The coefficient K_A of active earth pressure is the ratio of the lateral to the vertical effective stress in a soil mass in an active state, given by

$$K_A = \frac{\sigma_{hA}'}{\sigma_v'} = \frac{\sigma_3'}{\sigma_1'} = \frac{1}{N} = \frac{1 - \sin\phi}{1 + \sin\phi} \tag{4.59}$$

The vertical effective stress is the major principal effective stress in the active case, which means the leftmost point of the circle is the pole P_A (Figure 4.20). The direction of potential slip surfaces as estimated from the Mohr circle of stresses is simply the direction of the line joining the pole (the leftmost point of the circle for the active state, point P_A in Figure 4.20) and the point of tangency of the Mohr circle to the strength envelope. As shown in Figure 4.20, this direction makes an angle of $\pm(45° + \phi/2)$ with the horizontal. Figure 4.19d shows the two families of potential slip surfaces associated with the active Rankine state.

While the active state provides a lower bound to the value of the lateral earth pressure coefficient K, the passive Rankine state is on the other extreme, capping all possible values of K. The passive state is obtained by pushing the vertical plane in Figure 4.19c toward the soil mass, which increases σ_h' until σ_h' / σ_v' becomes equal to N. The progression from the at-rest to the *passive Rankine state* can be visualized through Mohr's circles as shown in Figure 4.20. The coefficient K_P of passive earth pressure is the ratio of the lateral to the vertical effective stress in a soil mass in a passive state, given by

$$K_P = \frac{\sigma_{hP}'}{\sigma_v'} = \frac{\sigma_1'}{\sigma_3'} = N = \frac{1 + \sin\phi}{1 - \sin\phi} \tag{4.60}$$

Recognizing that the lateral effective stress is the major principal effective stress in this case, we can find the pole P_P (see Figure 4.20). Connecting the pole to the points of tangency of the Mohr circle for the passive state and the strength envelope, we find that the directions of the potential slip surfaces in the passive Rankine state make angles of $\pm(45° - \phi/2)$ with the horizontal (Figure 4.20). Figure 4.19e shows the directions of the potential slip surfaces in the soil mass. Figure 4.21 shows the Mohr circles for the three states we have examined and how the corresponding lateral effective stresses relate to the vertical effective stress through the corresponding coefficient of lateral earth pressure. Note that K_0, K_P, and K_A are ratios of effective, not total, stresses.

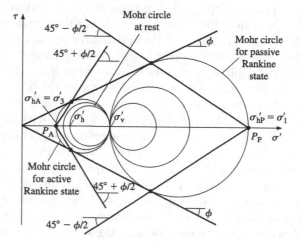

Figure 4.20 Sequence of Mohr's circles illustrating the horizontal unloading of a soil mass until the active Rankine state is reached and the horizontal compression of a soil mass until the passive Rankine state is reached.

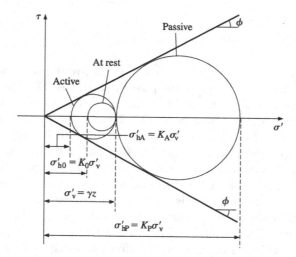

Figure 4.21 Mohr's circles representing the at-rest state and the active and passive Rankine states.

Active and passive earth pressure analyses can be easily extended to $c-\phi$ materials (a good reference for that is Terzaghi (1943)), but its usefulness in practice is rather limited. We will therefore not discuss this extension in this text.

Example 4.8

For a normally consolidated cohesionless soil with $\phi = 30°$, calculate the at-rest, active, and passive values of the earth pressure coefficient K.

Solution:

The coefficient K_0 of lateral earth pressure at rest is calculated using Equation (4.55):

$$K_0 = 1 - \sin\phi = 1 - \sin 30° = 0.5$$

The active and passive lateral earth pressure coefficients are calculated using Equations (4.59) and (4.60):

$$K_A = \frac{1 - \sin 30°}{1 + \sin 30°} = \frac{1}{3}$$

$$K_P = \frac{1 + \sin 30°}{1 - \sin 30°} = 3$$

Note how the ratio of passive to active pressures is 9 for a 30° friction angle.

4.4.2.2 Sloping ground*

Consider a soil mass sloping at an angle α_g with respect to the horizontal (Figure 4.22a). For simplicity, consider the soil mass to be free of water. Focusing on the prism of unit width shown in Figure 4.22a, we see that the weight of the prism per unit length into the plane of the figure is equal to γz. The weight of the prism is the only reason its base is subjected to a normal and a shear stress. By projecting the weight normally and tangentially to the base of the prism and dividing the component forces by the area of the base (which is $1/\cos\alpha_g$ for

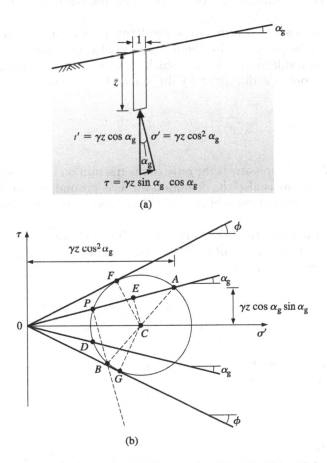

(a)

(b)

Figure 4.22 Active Rankine state in sloping ground: (a) a prism with unit width with base at depth z in a soil deposit with sloping surface; (b) Mohr's circle corresponding to active state at base of prism.

a prism of unit thickness), we obtain the following expressions for the normal and shear stresses at the base of the prism:

$$\sigma' = \gamma z \cos^2 \alpha_g \tag{4.61}$$

$$\tau = \gamma z \sin \alpha_g \cos \alpha_g \tag{4.62}$$

Note that the effective traction t' (the resultant of the normal and shear effective stresses) on the base of the prism of Figure 4.22a is vertical and given by

$$t' = \gamma z \cos \alpha_g \tag{4.63}$$

It is important to understand that t' is vertical, but is not σ'_v; that is, it is not the vertical effective stress and does not act on the horizontal plane.

We can plot the state of stress defined by Equations (4.61) and (4.62) in a Mohr's diagram as point A, as illustrated in Figure 4.22b. The distance OA from the origin of the $\sigma'-\tau$ space to point A is equal to $\gamma z \cos \alpha_g$, that is, t'. When OA is projected onto the σ and τ axes, we get Equations (4.61) and (4.62). The Mohr circle corresponding to the active Rankine state is drawn in the figure going through A. Because point A defines the stresses on a plane making an angle α_g with the horizontal, the pole P lies on the intersection of the Mohr circle with the OA line.

Now let us say we wish to determine the resultant stress on a vertical plane; we draw a vertical line through the pole, thereby obtaining point D as the intersection of this line with the Mohr circle. The resultant traction t'_A (which, note, is not the horizontal effective stress and is not even horizontal) is thus given by the line OD. As a result:

$$t' = OA$$

$$t'_A = OD$$

The coefficient of earth pressure is the ratio of the traction on the vertical plane to that on the horizontal plane. This is also the definition for level ground, but in that case, the traction is identical to the normal stress because the shear stresses on the horizontal and vertical planes are zero.

Let us now find the magnitude of the ratio of OD to OA using the Mohr diagram. We do that by analyzing the geometry of the Mohr diagram in Figure 4.22b. We first note that, because CE is perpendicular to OA,

$$EP = EA$$

We can now write:

$$\frac{t'_A}{t'} = \frac{OD}{OA} = \frac{OP}{OA} = \frac{OE - EP}{OE + EP} \tag{4.64}$$

But OE is simply:

$$OE = OC \cos \alpha_g \tag{4.65}$$

To find EP, we note that AEC is a right triangle and that $CA = CF$. We can now write:

$$EC = OC \sin \alpha_g$$

$$CA = CF = OC \sin \phi$$

So we have:

$$CA^2 = CE^2 + EA^2$$

which leads to

$$EA^2 = CA^2 - CE^2 - OC^2 \sin^2 \phi - OC^2 \sin^2 \alpha_g$$

and that, in turn, leads to

$$EA = OC\sqrt{\sin^2 \phi - \sin^2 \alpha_g} = OC\sqrt{\cos^2 \alpha_g - \cos^2 \phi} = EP \tag{4.66}$$

Taking Equations (4.65) and (4.66) into Equation (4.64), we obtain the active Rankine earth pressure coefficient for sloping ground:

$$K_A = \frac{t'_A}{t'} = \frac{\cos \alpha_g - \sqrt{\cos^2 \alpha_g - \cos^2 \phi}}{\cos \alpha_g + \sqrt{\cos^2 \alpha_g - \cos^2 \phi}} \tag{4.67}$$

The horizontal stress is obtained by projecting the resultant stress t'_A acting on the vertical plane onto the horizontal direction:

$$\sigma'_{hA} = K_A t' \cos \alpha_g = K_A \gamma z \cos^2 \alpha_g \tag{4.68}$$

while the shear stress is obtained by projecting t'_A onto the vertical direction:

$$\tau'_{vA} = K_A t' \sin \alpha_g = K_A \gamma z \sin \alpha_g \cos \alpha_g \tag{4.69}$$

The directions of slip planes can be determined in much the same way as we did that for level ground, with lines drawn from P through F and G yielding these directions.

4.5 MAIN TYPES OF SOIL LABORATORY TESTS FOR STRENGTH AND STIFFNESS DETERMINATION

4.5.1 Role of stiffness and shear strength determination

In the preceding sections, we examined stresses, strains, and the relationship between strains and stresses in soils, particularly shearing (plastic failure). We did so because they are an integral part of the calculations we do in the analysis of foundations, slope, retaining structures, and other geotechnical systems. In this section, we examine the important issue of how to measure the stress–strain properties of soils properly. This issue is fundamental both for work done in practice and for research on soil load response.

4.5.2 Stress (loading) paths

The mechanical response of soil is rigorously expressed in terms of the variables p, q, and θ_L defined in Equations (4.23) through (4.25):

$$p = \frac{1}{3}(\sigma_1 + \sigma_2 + \sigma_3) \tag{4.70}$$

$$q = \frac{1}{\sqrt{2}}\sqrt{(\sigma_1 - \sigma_3)^2 + (\sigma_1 - \sigma_2)^2 + (\sigma_2 - \sigma_3)^2} \tag{4.71}$$

$$\cot\theta_L = \frac{2\sigma_3 - \sigma_1 - \sigma_2}{\sqrt{3}(\sigma_1 - \sigma_2)} \tag{4.72}$$

Effective stress versions p' and q' of p and q can also be defined:

$$p' = p - u \tag{4.73}$$

$$q' = q \tag{4.74}$$

4.5.3 Loading paths and main laboratory tests

In the laboratory, soil samples may be subjected to loading that leads to different progressions in the values of the three stress invariants used to quantify the stress tensor in the soil. This progression is referred to as loading path, for it can be plotted in a three-dimensional space defined by three axes, one for each of the three stress invariants. The two stress spaces most often used are shown in Figure 4.23.

The loading paths of greatest interest in the laboratory are:

- isotropic compression and isotropic unloading,
- one-dimensional compression,
- triaxial compression,

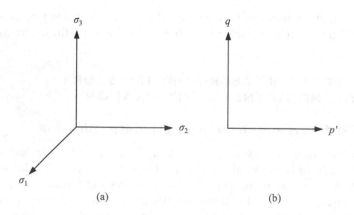

(a)   (b)

Figure 4.23 Most common reference systems for the representation of stress state: (a) system with principal stresses in the three reference axes (principal stress space); (b) system of reference with the Cambridge stress variables p' and q in two of the axes representing the isotropic stress and deviatoric stress components.

- triaxial extension, and
- simple shear.

The laboratory soil tests of greatest importance for what we cover in this text are the *one-dimensional compression* test (more routinely called the consolidation test), the direct shear test, the triaxial compression test, and the unconfined compression test (see Figures 4.24–4.26).

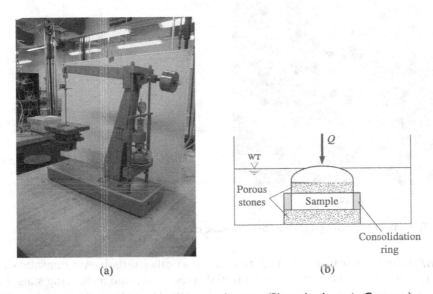

(a)   (b)

Figure 4.24 Consolidometer: (a) photo; (b) schematic diagram. (Photo by Antonio Carraro.)

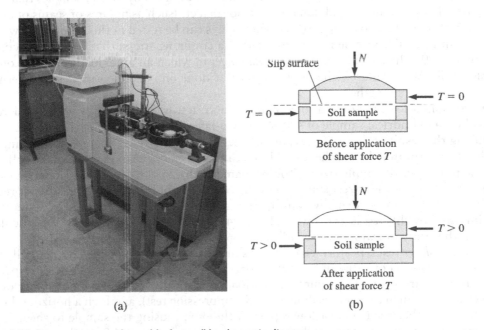

(a)   (b)

Figure 4.25 Direct shear machine: (a) photo; (b) schematic diagram.

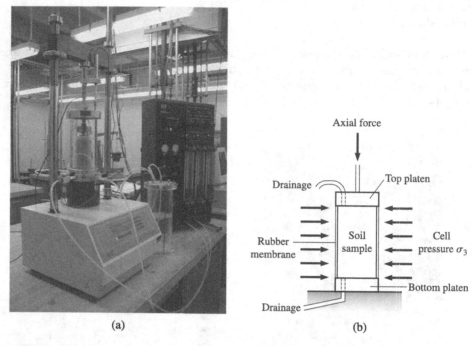

Figure 4.26 Triaxial test system: (a) photo; (b) schematic diagram.

The *consolidation* test is a one-stage test, whereas other tests have a consolidation stage (in which the sample is taken to a desired initial stress state) and a shearing stage (in which the shear strain is increased within the sample, often to values of the order of 20% or greater, with generally increasing shear stress as a result).

The *unconfined compression* test is a test that can be done only on soil samples that don't fall apart when unconfined and don't have major defects (such as fissures or seams of different materials cutting across the sample). Many clays can be tested in this manner, as we will see in Chapter 6. The sample is not subjected to a confining stress; that is, the total confining stress $\sigma_3 = 0$. The test has only a shearing stage, in which an axial load is applied on the sample until plastic failure. This is the usual test done on concrete samples to determine their compressive strength.

The consolidation (one-dimensional compression) test is performed by applying a vertical load on a cylindrical sample of soil restrained laterally by a steel ring (Figure 4.24) and measuring the resulting vertical deflection. When performed on clay samples, the sample is usually kept saturated during the test. Although there is more than one size of consolidometer, the most typical sample size is 70 mm (diameter) by 25 mm (thickness). Referring back to our discussion of the at-rest state and of the coefficient of lateral earth pressure at rest as a ratio of σ'_h to σ'_v for a normally consolidated soil, it is easy to understand why the stress path imposed by this test is as shown in Figure 4.27b. The consolidation test is particularly useful in the study of clays, and we will therefore rediscuss it in Chapter 6.

The *direct shear* test is performed on a soil sample that is also restrained horizontally by a split box constituted of a top and a bottom part that can move horizontally with respect to each other (Figure 4.25). The sample is first loaded vertically without the possibility of lateral expansion (much as in a one-dimensional compression test), and then a horizontal force is applied on either the upper or lower part of the box, causing the sample to shear in the horizontal direction. The stress path for this test is shown in Figure 4.27c. The consolidation

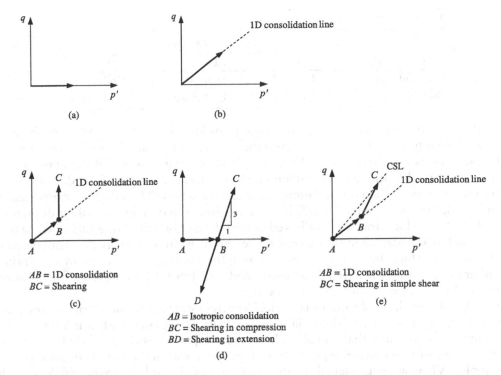

Figure 4.27 Stress paths for (a) isotropic compression, (b) 1D consolidation, (c) direct shear test, (d) drained triaxial compression and extension tests, and (e) simple shear test.

portion (AB) of the stress path is just like the stress path of a one-dimensional compression test. The shearing stage loading path is a vertical line, indicating that the shear stress added to the sample horizontally does not change p', just q. In truth, p' may increase slightly because of one of the shortcomings of the direct shear test, which is that there is some passive pressures that develop as some of the push imposed in the upper part of the sample ends up in part as a push against the lower part of the box on the other side of the sample; in other words, what, in concept, would be application of a pure shear force turns out to be largely a shear force, but one that also generates some normal stress in the soil. This also means that the shear strength of the soil is not the only resistance to failure; this passive resistance also contributes to the resistance to shearing of the sample. Another shortcoming of this test is that it forces shearing to develop horizontally instead of allowing it to develop naturally along an optimal shearing plane. Nonetheless, the direct shear test is used in practice often and, if well used, provides useful information regarding shear strength of the soil. It is particularly useful in the study of interface response when soil is in contact with a structural material such as concrete or steel. This will be discussed in more detail in Chapter 5.

In a *triaxial* (TX) test, a cylindrical sample, typically either 70 mm or 35 mm in diameter and at least twice that in height, is wrapped in a close-fitting, cylindrical, impervious membrane and placed between two platens. The bottom platen is attached to the base of the triaxial chamber and is stationary. The top platen is attached to a piston to which an axial force may be applied. During the test, the sample, platens, and piston are located within a plexiglass or aluminum cylinder that is sealed at the top and bottom. So, in the course of a test, the sample is first subjected to an all-around stress by increasing the air or water pressure around the sample (depending on the design of the system, the sample may be confined by air or water), and then to a change in the axial stress. The first stage is the consolidation

Table 4.1 Types of triaxial tests

	TX tests		
Drainage during	CD	CU	UU
Consolidation	Drained	Drained	Undrained
Shearing	Drained	Undrained	Undrained

stage (although the soil may or may not undergo volume change, depending on whether water is allowed to drain from or come into the sample); the second is the shearing stage.

The impervious membrane around the sample allows transmission of the applied pressure in the cell to the sample as a total stress. Drainage may be either allowed or prevented during either or both the consolidation and shearing stages. The effective stress state at the end of the consolidation stage will be different from the initial state only if drainage is allowed. Tests in which drainage is allowed during the consolidation stage are referred to as consolidated or C tests; when drainage is prevented, they are known as unconsolidated or U tests. Tests in which drainage is allowed during the shearing stage are known as drained (D); otherwise, tests are referred to as undrained (U). Table 4.1 summarizes the drainage conditions for the three main types of triaxial tests.

Table 4.2 shows the evolution of the effective stress state acting on a triaxial test sample tested under drained conditions. Samples are usually saturated with water. One of the advantages of doing so is that sample volume change can be easily measured by measuring the volume of water either expelled from pore space within the sample or sucked in by the sample. When water is sucked in, the void ratio of the soil increases, weakening the soil. When water is expelled, the void ratio drops and the sample gets stronger. There is no generation of pore pressures due to consolidation or shearing during drained tests, as the drainage lines are open during both consolidation and shearing. However, before the consolidation stage, we usually apply a pore pressure u_b (called back pressure) within the sample to dissolve air bubbles into the pore water, ensuring sample saturation. The back-pressure is increased gradually, at the same time as the cell pressure, so that there is no net change in effective stress in the sample. This pore pressure remains in place during the test (continues to be experienced by the sample) during consolidation or drained shearing. So the total stresses acting on the sample are larger than the effective stresses in Table 4.2 by a magnitude u_b. CD tests are commonly used for sands.

$$\sigma_c' = \sigma_c - u_b; \sigma_c = \text{cell pressure}; u_b = \text{back pressure};$$

$$\sigma_d = \text{applied axial force} / \text{sample cross-sectional area}$$

In CU tests, it is possible and desirable to measure the pore pressures during shearing. The effective stresses within the sample can then be calculated throughout shearing, and effective stress paths in terms of p' and q can be plotted. CU tests are commonly used for clays. Sands are sometimes tested this way as well to either enhance the understanding of the behavior of the soil or quantify its response to loads that are applied very fast, as during

Table 4.2 Evolution of applied effective stresses in consolidated drained triaxial (TXCD) tests

Stress	Initial	After consolidation	During shearing
$\sigma_3' =$	Zero	σ_c'	σ_c'
$\sigma_1' =$	Zero	σ_c'	$\sigma_c' + \sigma_d'$

earthquakes, when the rate of loading is much faster than the rate of dissipation of excess pore pressure generated by the loading.

In a triaxial test, $\sigma_2' = \sigma_3'$; it follows that p, q, p', and q' take the following definitions:

$$p = \sigma_m = \frac{1}{3}(\sigma_1 + 2\sigma_3) \tag{4.75}$$

$$p' = \sigma_m' = \frac{1}{3}(\sigma_1' + 2\sigma_3') \tag{4.76}$$

$$q = q' = \sigma_1 - \sigma_3 \tag{4.77}$$

The final element test of interest is the simple shear test. This test is nowadays mostly performed using the hollow cylinder device, whose discussion falls outside the scope of this text.

4.6 STRESSES RESULTING FROM THE MOST COMMON BOUNDARY-VALUE PROBLEMS

4.6.1 Elastic stress–strain relationship and elastic boundary-value problems

Stresses appear within a soil mass as a result of loads applied on its boundaries or loads applied within it, including the soil's own self-weight. Applied loads can be point loads, line loads, or distributed loads. The stresses decrease in magnitude with distance from the applied loads. Solutions from elasticity theory are available to calculate the stress components in two perpendicular planes (for two-dimensional problems) or three mutually perpendicular planes (for three-dimensional problems) at a point within a soil mass. For level-ground soil deposits, these directions are usually the horizontal and vertical directions.

Elasticity solutions are applicable so long as the deformations within the soil mass can be assumed to be very small (which means the soil mass is far from a state of collapse or sliding). The linear elastic stress–strain relationship of materials is determined by a pair of constants, referred to as the elastic pair. Solutions of interest in practice are usually formulated in terms of any two of the following four elastic constants: Young's modulus E, shear modulus G, Poisson's ratio ν, and bulk modulus K.

Young's modulus and Poisson's ratio are best understood by considering a stress state in which the only nonzero component is a normal stress acting in some direction (this is the case of an axially loaded bar, for example, which forms the basis for many common tests, including uniaxial extension tests of steel bars and the unconfined compression tests of concrete or soil). Young's modulus E is the ratio of this normal stress to the collinear normal strain. Poisson's ratio ν is the ratio of the strain appearing in a direction normal to the direction of the nonzero stress to the normal strain in the direction of the stress. The shear modulus G is the ratio of a shear stress to the corresponding shear strain. The bulk modulus K is the ratio of the mean stress to the volumetric strain. Table 4.3 shows the relationships between E, ν, G, K, and M.

An approach used to solve elastic boundary-value problems is usually the determination of strains from stresses, and integration of the strains, subject to compatibility, to yield the displacements. There are other approaches based on the concept of stress or strain potentials. Some of the solutions can be obtained by direct integration of the displacements from the Boussinesq solution (discussed later in this section). Whatever the approach, the solutions are typically somewhat involved and are outside the scope of this text; they can be found in

Table 4.3 Relationship between the five most common elastic constants

Elastic constant	Elastic pair	
	E, ν	K, G
Young's modulus $E =$	E	$\dfrac{9KG}{3K+G}$
Poisson's ratio $\nu =$	ν	$\dfrac{3K-2G}{6K+2G}$
Shear modulus $G =$	$\dfrac{E}{2(1+\nu)}$	G
Bulk modulus $K =$	$\dfrac{E}{3(1-2\nu)}$	K
Constrained modulus $M =$	$\dfrac{E(1-\nu)}{(1+\nu)(1-2\nu)}$	$\dfrac{3K+4G}{3}$

elasticity texts such as Timoshenko and Goodier (1970), Fung (1993), or Saada (2009). We will present only the results that will be of use to engineers solving foundation problems.

Elasticity solutions are given in terms of stresses, strains, and displacements. A few displacement solutions are useful in foundation engineering because they form the basis for the calculation of the settlement of shallow foundations; they are presented in Chapter 9. We restrict ourselves in this chapter to stresses created within soil masses by boundary loads.

A point anywhere in the soil can be expressed with respect to the origin by using Cartesian coordinates (x_1, x_2, and x_3), cylindrical coordinates (radius r, depth z, and angle θ), or spherical coordinates (radius ρ and angles θ and φ). The notation x, y, and z can be used interchangeably with x_1, x_2, and x_3, as convenient. The choice of which coordinate system to use depends, for any problem, on which system leads to an easier solution to the problem or the most compact equations. We will use σ_v for vertical stress whenever possible, instead of the σ_z that would typically appear in elasticity texts, to maintain consistency with normal soil mechanics practice and the other chapters. The same is not done for vertical strain, which is denoted by ε_z, because ε_v is reserved for volumetric strain.[14]

Loadings can be applied to the surface of an elastic soil mass in two ways: by application of point or distributed loads or by application (or imposition) of displacements. When loads are applied, that implies ideal flexibility. There is no internal rigidity to the loading. When displacements are imposed, there is rigidity implied. For example, we might push a rigid cylinder vertically down into a soil mass. This means every point of the cylinder in contact with the soil will settle by the same amount. This is what happens, as an example, when a rigid footing resting on soil is loaded. We will now discuss the stresses that are generated inside a soil mass by various types of loads.

4.6.2 Vertical point load on the boundary of a semi-infinite, elastic soil mass (Boussinesq's problem)

Figure 4.28 shows a vertical point load Q applied at a point on the boundary of a semi-infinite mass with Young's modulus E and Poisson's ratio ν. The point of application of Q

---

[14]Notation can be problematic in a discipline that deals with so many concepts and lies at the crossroads of many different branches of science. We will keep notation as simple, logical, and consistent as possible, but keeping track of it will still require some effort on the reader's part.

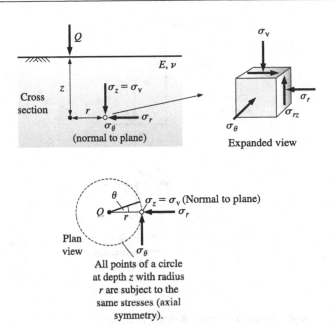

Figure 4.28 Vertical point load on the boundary of a soil mass (Boussinesq's problem).

is taken as the origin of the system of coordinates. The nonvanishing stress components are the normal stresses σ_r, σ_θ, and σ_v and the shear stress σ_{rz}; the other shear stresses are zero as a result of the symmetry of the problem. The equations are as follows (Boussinesq 1885):

$$\sigma_r = \frac{Q}{2\pi\left(r^2 + z^2\right)}\left[\frac{3r^2 z}{\left(r^2 + z^2\right)^{\frac{3}{2}}} - \frac{(1-2v)\left(r^2 + z^2\right)^{\frac{1}{2}}}{\left(r^2 + z^2\right)^{\frac{1}{2}} + z}\right] \tag{4.78}$$

$$\sigma_\theta = \frac{(1-2v)Q}{2\pi\left(r^2 + z^2\right)}\left[\frac{\left(r^2 + z^2\right)^{\frac{1}{2}}}{\left(r^2 + z^2\right)^{\frac{1}{2}} + z} - \frac{z}{\left(r^2 + z^2\right)^{\frac{1}{2}}}\right] \tag{4.79}$$

$$\sigma_v = \frac{3Qz^3}{2\pi\left(r^2 + z^2\right)^{\frac{5}{2}}} \tag{4.80}$$

$$\sigma_{rz} = \frac{3Qrz^2}{2\pi\left(r^2 + z^2\right)^{\frac{5}{2}}} \tag{4.81}$$

4.6.3 Vertical point load within a semi-infinite, elastic soil mass (Kelvin's Problem)

Figure 4.29 shows a vertical point load Q applied at a point within a semi-infinite mass with Young's modulus E and Poisson's ratio v. The point of application of Q is taken as the origin of the system of coordinates (radius r, depth z, and angle θ in cylindrical representation

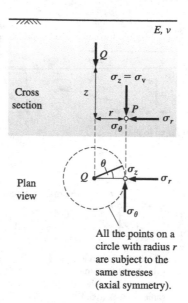

Figure 4.29 Vertical point load within a soil mass (Kelvin's problem).

or x_1, x_2, and $x_3=z$ in Cartesian coordinates). The nonvanishing stress components are the normal stresses σ_r, σ_θ, and σ_v and the shear stress σ_{rz}; the other shear stresses are zero as a result of the symmetry of the problem. The equations are as follows:

$$\sigma_r = \frac{Q}{8\pi(1-v)}\left[\frac{3r^2z}{\left(r^2+z^2\right)^{\frac{5}{2}}} - \frac{(1-2v)z}{\left(r^2+z^2\right)^{\frac{3}{2}}}\right] \tag{4.82}$$

$$\sigma_\theta = -\frac{Q}{8\pi(1-v)}\left[\frac{(1-2v)z}{\left(r^2+z^2\right)^{\frac{3}{2}}}\right] \tag{4.83}$$

$$\sigma_v = \frac{Q}{8\pi(1-v)}\left[\frac{(1-2v)z}{\left(r^2+z^2\right)^{\frac{3}{2}}} + \frac{3z^3}{\left(r^2+z^2\right)^{\frac{5}{2}}}\right] \tag{4.84}$$

$$\sigma_{rz} = \frac{Q}{8\pi(1-v)}\left[\frac{(1-2v)r}{\left(r^2+z^2\right)^{\frac{3}{2}}} + \frac{3rz^2}{\left(r^2+z^2\right)^{\frac{5}{2}}}\right] \tag{4.85}$$

4.6.4 Uniform pressure distributed over a circular area on the boundary of a semi-infinite, elastic soil mass

Consider the circular loaded area of diameter B and radius b as shown in Figure 4.30. If the distributed load is expressed as q_b and the center of the circle is taken as the origin of

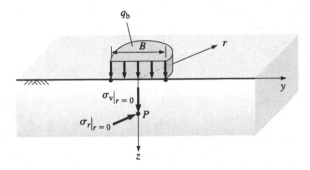

Figure 4.30 Circular loaded area on the boundary of a semi-infinite soil mass.

the system of coordinates, the nonvanishing stress components along the axis of the loaded area are as follows:

$$\sigma_v\big|_{r=0} = q_b \left[1 - \frac{z^3}{\left(b^2 + z^2\right)^{\frac{3}{2}}} \right] \tag{4.86}$$

$$\sigma_r\big|_{r=0} = \sigma_\theta\big|_{r=0} = \frac{1}{2} q_b \left\{ 1 + 2v + \left[\frac{z}{\left(b^2 + z^2\right)^{\frac{1}{2}}} \right]^3 - \frac{2(1+v)z}{\left(b^2 + z^2\right)^{\frac{1}{2}}} \right\} \tag{4.87}$$

4.6.5 Uniform pressure distributed over a rectangular area on the boundary of a semi-infinite, elastic soil mass

Figure 4.31 shows a rectangular loaded area, with plan dimensions B and L. If the magnitude of the distributed load is q_b, the vertical stress $\sigma_{33} = \sigma_v$ at depth $x_3 = z$ under any corner of the rectangle is given as (Newmark 1935):

$$\sigma_{33}\big|_{corner} = \sigma_v\big|_{corner} = \frac{q_b}{4\pi} \left[\frac{2mn\sqrt{C_1}}{C_1 + C_2} \frac{1 + C_1}{C_1} + \tan^{-1}\left(\frac{2mn\sqrt{C_1}}{C_1 - C_2} \right) \right] \tag{4.88}$$

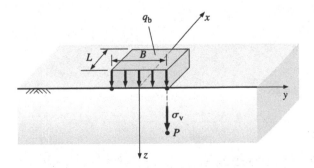

Figure 4.31 Rectangular loaded area on the boundary of a semi-infinite soil mass.

where:

$$m = \frac{B}{z} \tag{4.89}$$

$$n = \frac{L}{z} \tag{4.90}$$

$$C_1 = 1 + m^2 + n^2 \tag{4.91}$$

$$C_2 = (mn)^2 \tag{4.92}$$

The arctangent term in Equation (4.88) must be a positive angle in radians; thus, when $C_2 > C_1$ and a calculator gives a negative angle, we must add π to that angle.

Using Equation (4.88), we can calculate the vertical stress at any point with depth z by using the principle of superposition. For example, the vertical stress at a depth z below the center of a rectangular loaded area would be the sum of the stresses under the common corner of the four rectangular loaded areas with half the width and length of the original area that, considered together, constitute the loaded area. By using a combination of positive and negative loaded areas, it is possible to calculate the stress induced by a rectangle at a point not lying directly below the loaded area, as shown in Example 4.9.

Example 4.9

The load of a heavy building with plan dimensions 10 and 15 m is assumed to be uniformly distributed and equal to 150 kPa. Calculate the vertical stress induced by this load at a point P that is at a distance of 5 m from the smaller side and a distance of 5 m from the larger side of the building as shown in E-Figure 4.6 at a depth of 5 m.

Solution:

E-Figure 4.6 shows that we can do this calculation by considering the uniformly distributed load q_b applied over a large rectangle 1 ($ACGP$) with the corner lying directly above the point where the stress is desired. This generates an excess stress that is now compensated for by assuming $-q_b$ applied across rectangles 2 ($ABHP$) and 3 ($FDGP$). We are not done yet, because we have now removed the load across the intersection of rectangles 2 and 3 twice. We must add this area, rectangle 4 ($FEHP$), back in by considering a new load q_b applied over it. We can use Equation (4.88) directly.
For the large rectangle $ACGP$:

$$m = \frac{B}{z} = \frac{15}{5} = 3$$

$$n = \frac{L}{z} = \frac{20}{5} = 4$$

$$C_1 = 3^2 + 4^2 + 1 = 26$$

$$C_2 = (3 \times 4)^2 = 144$$

$$\sigma_v \big|_{corner} = \frac{q_b}{4\pi} \left[\frac{2 \times 3 \times 4\sqrt{26}}{26 + 144} \frac{1 + 26}{26} + \tan^{-1} \left(\frac{2 \times 3 \times 4\sqrt{26}}{26 - 144} \right) \right]$$

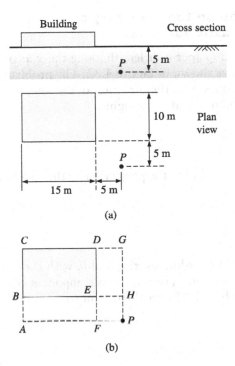

E-Figure 4.6 (a) Rectangular loaded area for Example 4.9; (b) composition of rectangles used in calculation.

$$\sigma_v\big|_{corner} = \frac{150}{4\pi}\left[\frac{2\times3\times4\sqrt{26}}{26+144}\frac{1+26}{26} + \tan^{-1}\left(\frac{2\times3\times4\sqrt{26}}{26-144}\right)\right] = \frac{150}{4\pi}\left[0.748 + \tan^{-1}(-1.037)\right]$$

We see that the arctangent argument is a negative number. If we take the arctangent of −1.037 in a calculator, we get −0.804, a negative angle in radians, not the positive angle producing the same arctangent, which is the one we need. We calculate that number by adding π to −0.804, obtaining 2.338. This gives us:

$$\sigma_v\big|_{corner} = \frac{q_b}{4\pi}\left[0.748 + 2.338\right] = 0.246q_b = 0.246\times150 = 36.9 \text{ kPa}$$

Calculations for the other three rectangles and the solution to the whole problem are shown in E-Table 4.1. We find that the vertical stress increment at point P due to the building is only 2 kPa. The small stress we calculated suggests that vertical stress dissipates very quickly as we move away from the building. In a problem at the end of the chapter, the reader is asked to calculate the stress at the same depth as in this example under the center of the building.

E-Table 4.1 Results for Example 4.9

Rectangle	1 (ACGP)	2 (ABHP)	3 (FDGP)	4 (FEHP)	Final
m	3	1	1	1	—
n	4	4	3	1	—
C_1	26	18	11	3	—
C_2	144	16	9	1	—
σ_v (kPa)	36.9	−30.6	−30.5	26.3	2.1

4.6.6 Vertical line load on the boundary of a semi-infinite, elastic soil mass

Consider a vertical line load Q (per unit length along the x axis) applied on the boundary of a semi-infinite soil mass as shown in Figure 4.32. The problem is a plane-strain problem (the normal strain in the x direction, the direction of the line, is zero everywhere in the soil mass). The depth z is related to r and θ through:

$$z = r\cos\theta \tag{4.93}$$

The distance y from the line load to the projection on the soil boundary of the point where the stresses are calculated is

$$y = r\sin\theta \tag{4.94}$$

It is best to use cylindrical coordinates x, r, and θ, with the line load aligned with the x axis. The equations for the nonzero, normal stress components that may be useful in typical foundation engineering problems, obtained from integration of the Boussinesq solution for a point load, are:

$$\sigma_r = \frac{2Q}{\pi}\frac{\cos\theta}{r} = \frac{2Q}{\pi}\frac{z}{r^2} \tag{4.95}$$

$$\sigma_x = \frac{2Qv}{\pi}\frac{\cos\theta}{r} = \frac{2Qv}{\pi}\frac{z}{r^2} \tag{4.96}$$

$$\sigma_v = \frac{2Q}{\pi}\frac{\cos^3\theta}{r} = \frac{2Q}{\pi}\frac{z^3}{r^4} \tag{4.97}$$

Given that the shear stresses $\sigma_{r\theta}$ and σ_{rx} are zero, σ_r is a principal stress.

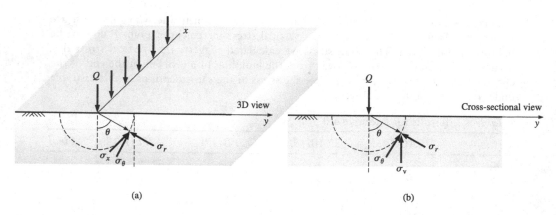

Figure 4.32 Vertical line load on the boundary of a semi-infinite soil mass.

4.6.7 Uniform pressure distributed over an infinitely long strip on the boundary of a semi-infinite, elastic soil mass

Figure 4.33 shows a strip load q_b applied on the boundary of a semi-infinite soil mass. The problem is a plane-strain problem (the normal strain in the direction of the strip is zero). This problem can be solved by integrating the solution for the vertical line load. In this case, the most compact representation of a point P is in terms of the angles θ_1 and θ_2 and the two vertical lines that pass through the edges of the loaded area. The nonvanishing stress components[15] are:

$$\sigma_{11} = \frac{q_b}{2\pi} \left[2(\theta_2 - \theta_1) - (\sin 2\theta_2 - \sin 2\theta_1) \right] \tag{4.98}$$

$$\sigma_{22} = \frac{2vq_b}{\pi}(\theta_2 - \theta_1) = \frac{2vq_b\alpha}{\pi} \tag{4.99}$$

$$\sigma_{33} = \frac{q_b}{2\pi} \left[2(\theta_2 - \theta_1) + (\sin 2\theta_2 - \sin 2\theta_1) \right] \tag{4.100}$$

$$\sigma_{13} = \frac{q_b}{2\pi} \left[(\cos 2\theta_1 - \cos 2\theta_2) \right] \tag{4.101}$$

where α is the angle formed by PP_1 and PP_2 (shown in Figure 4.33), which is given by

$$\alpha = \theta_2 - \theta_1 \tag{4.102}$$

The strip load is one of the most important boundary loadings in soil mechanics because the analysis of strip foundations serves as a basis for the core of shallow foundation design. Recall from our earlier discussion that when a load (as opposed to a displacement) is applied, as in this case, the results are strictly applicable to a flexible foundation. We will explore some additional facts about strip loads later in this chapter and in the Conceptual Problems section for this chapter.

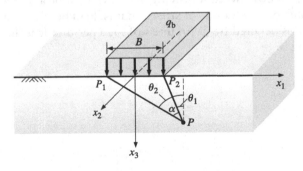

Figure 4.33 Uniform pressure over infinite strip on the boundary of a semi-infinite soil mass.

---

[15] A reminder of this type of stress notation, discussed earlier in this chapter: σ_{ij} is the stress component in the direction j acting on the plane normal to direction x_i; when $i = j$, σ_{ij} is a normal stress in that direction; when $i \neq j$, σ_{ij} is a shear stress, for it acts in a given plane in a direction tangential to it.

4.6.8 Rigid strip and rigid cylinder on the boundary of a semi-infinite, elastic soil mass

A rigid strip with width B, loaded by a central load Q (Figure 4.34), expressed as units of load per length settles uniformly when resting on a semi-infinite, elastic soil mass. The stress distribution q_b at the base of the strip is far from uniform, as larger stresses near the edges of the load are needed to keep the settlement uniform, given the proximity to the free boundary. If we orient the x axis along the boundary and normal to the strip, with $x=0$ at the center of the strip, we obtain the following equation for q_b from elasticity theory:

$$q_b = \frac{Q}{\pi\sqrt{\frac{1}{4}B^2 - x^2}} \tag{4.103}$$

Note that q_b is equal to

$$q_b = \frac{2Q}{\pi B} \tag{4.104}$$

at the center ($x=0$) and to infinity at the edges ($x=B/2$). The average unit load (stress) $q_{b,avg}$ is obviously equal to Q/B.

The solutions for a rigid cylinder (a cylindrical footing) with diameter B are very similar to those for the rigid strip (a strip footing). The contact stress (distributed load q_b), expressed in terms of cylindrical coordinates with origin at the center of the circular cross section of the cylinder, is given by

$$q_b = \frac{2Q}{\pi B\sqrt{B^2 - 4r^2}} \tag{4.105}$$

4.6.9 Approximate stress distribution based on 2:1 vertical stress dissipation

The vertical stress σ_v under a strip or rectangular load can be approximated by assuming that the stresses dissipate downward following a 2:1 ($V:H$) slope as shown in Figure 4.35. If the strip load is denoted by q_b (in units of stress, that is, load per unit length per unit length) and the resultant load is denoted by Q (in units of load per unit length), the resulting equation for a strip footing is:

$$\sigma_v = \frac{Q}{B+z} = \frac{q_b}{1+\dfrac{z}{B}} \tag{4.106}$$

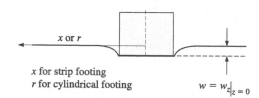

x or r

x for strip footing
r for cylindrical footing

$w = w_z|_{z=0}$

Figure 4.34 Cross section of either a rigid strip or a rigid cylindrical footing on the boundary of a semi-infinite, elastic soil mass.

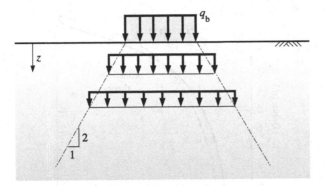

Figure 4.35 2:1 stress dissipation with depth.

For a rectangular footing, we have a load Q (units of load) distributed over an area BL, leading to q_b (units of stress). The equation for the stress at depth z is:

$$\sigma_v = \frac{Q}{(B+z)(L+z)} = \frac{q_b}{\left(1+\dfrac{z}{B}\right)\left(1+\dfrac{z}{L}\right)}$$

(4.107)

Example 4.10

Consider a 2-meter-wide strip load with a uniform magnitude of 150 kPa placed directly on the surface of the elastic half-space. Calculate and plot the distribution of the vertical stress with depth z right below the centerline ($y=0$ m) of the strip load using both elasticity theory and the 2:1 vertical stress dissipation method.

Solution:

The vertical stress from elasticity theory is given by Equation (4.100), where $\sigma_{33}=\sigma_v$. For points lying on the plane of symmetry of the load, $-\theta_2=\theta_1=\theta$, and θ is given as a function of depth $(z-x_1)$ by

$$\theta = \tan^{-1}\left(\frac{B/2}{z}\right)$$

Thus,

$$\sigma_v = \frac{q_b}{2\pi}\left(4\theta + 2\sin 2\theta\right)$$

$$\sigma_v = \frac{150}{6.28}\left(4\tan^{-1}\left(\frac{2/2}{z}\right) + 2\sin\left(2\tan^{-1}\left(\frac{2/2}{z}\right)\right)\right)$$

$$= 23.9\left(2\tan^{-1}\left(\frac{1}{z}\right) + \sin\left(2\tan^{-1}\left(\frac{1}{z}\right)\right)\right)$$

For the 2:1 vertical stress dissipation method, we have

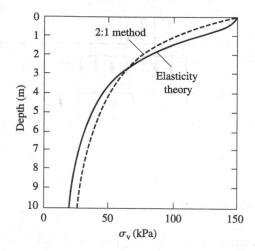

E-Figure 4.7 Distribution of vertical stress with depth along the centerline of the strip footing of Example 4.10 from both the elasticity theory and the 2:1 method.

$$\sigma_v = \frac{Q}{B+z} = \frac{q_b}{1+\dfrac{z}{B}} = \frac{150}{1+\dfrac{z}{2}}$$

E-Figure 4.7 shows results from the application of the preceding equations for the 0–10 m depth range. We observe that the approximate 2:1 method underestimates the value of the vertical stress within the depth range between 0 and $1.5B$ vertically below the load and overestimates it for larger depths.

4.6.10 Saint-Venant's principle

Saint-Venant's principle (Barré de Saint-Venant 1855) is occasionally useful in design calculations. It states that, if a point where we may want to calculate the stresses (or strains) is sufficiently removed from the loading causing these stresses (or strains), any statically equivalent load generates approximately the same stresses (or strains). Statically equivalent loads are loads with the same resultant and same moment with respect to an arbitrary point. What this principle states is that the details (that is, the geometry) of the loading have a negligible effect on the calculated stresses so long as the distance from the point where the stresses are desired to the load is much greater than the scale of the load.

Example 4.11

A uniform 2-meter-wide square 100 kPa load is placed directly on the surface of a semi-infinite, elastic soil mass. For this load and for a statically equivalent point load, calculate the vertical stress at points A^* and B^* located 2 m below points A and B, respectively (refer to E-Figure 4.8). Point A is located at the center of the square load; point B is located 3 m away from both sides of the square load (a distance 4.25 m from the closest corner of the footing along its diagonal). Discuss the accuracy/applicability of the approximation using a statically equivalent point load.

Solution:

To calculate the vertical stress at point A^*, we discretize the applied load as four squares, each with sides 1 m long, sharing point A at one corner. This way, the induced stress will be four times the stress calculated using Equation (4.88):

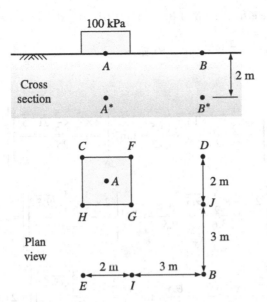

E-Figure 4.8 The square load and the points where the vertical stresses are to be calculated in Example 4.11.

$$\sigma_v\Big|_{corner} - \frac{q_b}{4\pi}\left[\frac{2mn\sqrt{C_1}}{C_1+C_2}\frac{1+C_1}{C_1} + \tan^{-1}\left(\frac{2mn\sqrt{C_1}}{C_1-C_2}\right)\right]$$

$$\sigma_v\Big|_A = 4\sigma_v\Big|_{corner} = \frac{100}{\pi}\left[\frac{2\times\frac{1}{2}\times\frac{1}{2}\sqrt{1.5}}{1.5\mid\frac{1}{16}}\frac{1+1.5}{1.5} + \tan^{-1}\left(\frac{2\times\frac{1}{2}\times\frac{1}{2}\sqrt{1.5}}{1.5-\frac{1}{16}}\right)\right]$$

$$= 31.83\left[0.653 + \tan^{-1}(0.426)\right] = 31.83\left[0.653 + 0.403\right] = 33.6 \text{ kPa} \approx 34 \text{ kPa}$$

Now we calculate the stress increase due to an equivalent load Q, obtained by multiplying the stress q_b by the area on which it acts:

$$Q = q_b \times \text{area} = 100 \times 4 = 400 \text{ kN}$$

We next use Equation (4.80):

$$\sigma_v = \frac{3Qz^3}{2\pi\left(r^2+z^2\right)^{\frac{5}{2}}} = \frac{3\times400\times2^3}{2\pi\left(0^2+2^2\right)^{\frac{5}{2}}} = 47.7 \text{ kPa} \approx 48 \text{ kPa}$$

This stress is much greater than the 34 kPa calculated earlier using the exact solution, indicating that the use of Equation (4.80) here is not appropriate because the point at which the stress is to be calculated is too close to the load.

To calculate the stress at point $B*$, we use the same two approaches. Using Equation (4.88) for rectangles $BECD$, $BDFI$, $BEHJ$, and $BIGJ$, the desired stress is expressed, by the superposition principle of elasticity, as

$$\sigma_v\Big|_{B,GHCF} = \sigma_v\Big|_{B,BECD} - \sigma_v\Big|_{B,BDFI} - \sigma_v\Big|_{B,BEHJ} + \sigma_v\Big|_{B,BIGJ}$$

Considering that the effects on B of $BDFI$ and $BEHJ$ are the same:

$$\sigma_v|_{B,GHCF} = \sigma_v|_{B,BECD} - 2\sigma_v|_{B,BDFI} + \sigma_v|_{B,BIGJ}$$

with:

$$\sigma_v|_{B,BECD} = \frac{100}{4\pi}\left[\frac{2 \times \frac{5}{2} \times \frac{5}{2}\sqrt{13.5}}{52.6}\frac{14.5}{13.5} + \tan^{-1}\left(\frac{2 \times \frac{5}{2} \times \frac{5}{2}\sqrt{13.5}}{-25.6}\right)\right] = 24 \text{ kPa}$$

$$\sigma_v|_{B,BDFI} = \frac{100}{4\pi}\left[\frac{2 \times \frac{3}{2} \times \frac{5}{2}\sqrt{9.5}}{23.6}\frac{10.5}{9.5} + \tan^{-1}\left(\frac{2 \times \frac{3}{2} \times \frac{5}{2}\sqrt{9.5}}{-4.6}\right)\right] = 22.7 \text{ kPa}$$

$$\sigma_v|_{B,BIGJ} = \frac{100}{4\pi}\left[\frac{2 \times \frac{3}{2} \times \frac{3}{2}\sqrt{5.5}}{10.6}\frac{6.5}{5.5} + \tan^{-1}\left(\frac{2 \times \frac{3}{2} \times \frac{3}{2}\sqrt{5.5}}{0.4}\right)\right] = 21.6 \text{ kPa}$$

This gives:

$$\sigma_v|_{B,GHCF} = 24 - 2\,(22.7) + 21.6 = 0.2 \text{ kPa}$$

or, for practical purposes, no stress increase.
Using a statically equivalent point load to approximate the rectangular load:

$$\sigma_v = \frac{3Qz^3}{2\pi\left(r^2 + z^2\right)^{\frac{5}{2}}} = \frac{3 \times 400 \times 2^3}{2\pi\left(32 + 2^2\right)^{\frac{5}{2}}} = 0.2 \text{ kPa}$$

Clearly, for point B^*, both methods produce the same result because point B^* is sufficiently removed from the load that its details (such as its geometry, that is, whether it is a point load or a distributed load with some shape) become unimportant, and any statically equivalent load will produce acceptable results.

4.7 TOTAL AND EFFECTIVE STRESS ANALYSES

As seen in both Chapter 3 and this chapter, when loads are applied on the boundaries of soil masses, total stress increments appear at every point (or element) of the soil mass. The magnitude and direction of the total stress increments appearing in an element of the soil mass depend on the magnitude and direction of the loads applied on the boundary of the soil mass, the geometry of the soil mass, and additional boundary conditions and the distances and the location of the element.

Each total stress increment is balanced by a pore pressure increment and an effective stress increment. The magnitudes of these two components depend on the hydraulic conductivity of the soil: in free-draining soils, pore pressures do not build up and the total stress increments are instantaneously and completely converted into effective stress increments. In soils with very small hydraulic conductivity, the stress increments are initially completely balanced by pore pressure increments; over time, the pore pressures dissipate as water flows from points with large to points with small hydraulic heads, and effective stresses increase by the same amounts as the pore pressures decrease until the pore pressures return to their initial values and the effective stress increments become equal to the total stress increments.

A soil mechanics problem can be analyzed, at least in simple applications, in terms of either total or effective stresses. If an effective stress analysis is used, all the quantities must be expressed in terms of effective stresses. For example, in stability problems, for which a slip surface will form in the case of loss of stability, the normal stress of interest on the slip surface is an effective normal stress and the shear strength parameters (c and ϕ) must be those consistent with effective stresses. Laboratory tests used to determine these parameters must then be performed sufficiently slowly for pore pressures not to appear (not an economical or practical proposition with slow-draining soils) or, alternatively, must be performed with equipment that can measure pore pressures that develop within the sample so that effective stresses can be obtained. On the other hand, if total stress analysis is used, then the total stresses resulting from the loading of the soil mass must be used in conjunction with a total stress-based, undrained shear strength.

In sands, effective stress analysis is almost always used because sands are free-draining, and induced pore pressures are equal to zero (the exception are seismic loads and other types of loads applied very quickly). In clays, in contrast, total stress analysis is more frequently used in routine practice. The main reason is that the determination of pore pressures, except in simple problems, is often difficult, requiring relatively sophisticated soil constitutive models and analysis. A point often missed is that effective stress analyses relate to pore pressures present during loading, not just the initial pore pressures in the soil. We will discuss effective and total stress analyses in more detail in the contexts of the load-carrying capacity of shallow and deep foundations (Chapters 10 and 13) and the stability of retaining structures (Chapter 16) and slopes (Chapter 17).

4.8 CHAPTER SUMMARY

4.8.1 Main concepts and equations

Stress analysis allows the determination of the **normal stress** and **shear stress** acting on any plane through a point. In two-dimensional stress analysis, the usual situation is that the stresses σ_{11}, σ_{33}, and σ_{13} acting on the perpendicular planes π_{11} and π_{33} passing through a point in the soil are known. The normal stresses σ_{11} and σ_{33} are positive in compression. For determining the sign of σ_{13}, we look at the planes π_{11} and π_{33} as making up an uppercase letter "L", with π_{11} being the vertical leg of the "L" and π_{33}, the horizontal leg. If the shear stress σ_{13} acts in directions that would tend to open up the right angle of the "L", then σ_{13} is positive. The stresses σ_θ and τ_θ acting on a plane making an angle θ with π_{11} (measured counterclockwise from π_{11}) are given by

$$\sigma_\theta = \frac{1}{2}(\sigma_{11}+\sigma_{33})+\frac{1}{2}(\sigma_{11}-\sigma_{33})\cos 2\theta + \sigma_{13}\sin 2\theta \tag{4.5}$$

$$\tau_\theta = \frac{1}{2}(\sigma_{11}-\sigma_{33})\sin 2\theta - \sigma_{13}\cos 2\theta \tag{4.6}$$

The **principal planes** are the planes where the shear stresses are equal to zero. The angles θ_p that the principal planes make with the reference plane π_{11} are determined from the equation:

$$\tan\left(2\theta_p\right) = \frac{2\sigma_{13}}{\sigma_{11} - \sigma_{33}} \tag{4.7}$$

The normal stresses on these planes are the principal stresses σ_1 and σ_3, given by

$$\sigma_1 = \frac{1}{2}\left(\sigma_{11} + \sigma_{33}\right) + \sqrt{\frac{1}{4}\left(\sigma_{11} - \sigma_{33}\right)^2 + \sigma_{13}^2} \tag{4.8}$$

$$\sigma_3 = \frac{1}{2}\left(\sigma_{11} + \sigma_{33}\right) - \sqrt{\frac{1}{4}\left(\sigma_{11} - \sigma_{33}\right)^2 + \sigma_{13}^2} \tag{4.9}$$

Stress analysis problems can also be solved graphically using the **Mohr circle** and the pole method; the **pole** is the point such that any line drawn through it parallel to a plane where we wish to determine the stresses intersects the Mohr circle at a point with coordinates equal to the stresses on that plane. A related sign convention exists for plotting shear stresses in a Mohr's diagram. Shear stresses that are applied on the face of an element in the counterclockwise orientation are positive for that purpose.

Soils are said to have attained a plastic state of "failure" (or to have sheared) when the principal stress ratio σ_1^*/σ_3^* satisfies the Mohr–Coulomb strength criterion, expressed mathematically as

$$\frac{\sigma_1^*}{\sigma_3^*} = N \tag{4.44}$$

where N is known as the flow number, given by

$$N = \frac{1 + \sin\phi}{1 - \sin\phi} \tag{4.45}$$

The Mohr circles can also be plotted for **strain increments**. The space used for plotting the Mohr circle of strain increments is defined by a horizontal axis where incremental **normal strains** $d\varepsilon$ are plotted and a vertical axis where incremental **shear strains** $\frac{1}{2}d\gamma$ are plotted. The pole method applies to the Mohr circle of strains as well. The **dilatancy angle** ψ can be visualized through the Mohr circle. The sine of this angle expresses the rate of increase in **volumetric strain** for a unit increase in maximum shear strain:

$$\sin\psi = -\frac{d\varepsilon_v}{\left|d\gamma_{max}\right|} \tag{4.31}$$

The dilatancy angle can be used to show that slip surfaces in soils with nonzero ψ are log-spirals. In a soil with $\psi = 0$, the slip surface is circular. The dilatancy angle is useful in other ways, as we will see in Chapter 5.

If undisturbed, soil exists in a **state of rest**, for which the ratio of lateral to vertical effective stress is the **coefficient of earth pressure at rest**, K_0. If allowed to stretch in the horizontal direction, a level soil mass reaches an **active Rankine state**, in which the whole soil is in a state of incipient plastic failure, with potential slip surfaces making angles of $\pm(45° + \phi/2)$ with the horizontal. The ratio of lateral to vertical effective stress in this case is K_A. If compressed in the horizontal direction, the soil reaches a **passive Rankine state,** in which the whole soil is in a state of incipient plastic failure, with potential slip surfaces making angles of $\pm(45° - \phi/2)$ with the horizontal. The ratio of lateral to vertical effective stress in this case is K_P. The values of K_0, K_A, and K_P are calculated using:

$$K_0 = K_{0,NC}\sqrt{OCR} \tag{4.57}$$

where $K_{0,NC}$ is roughly in the 0.4–0.5 range for sands and 0.5–0.75 range for clays, with higher values corresponding to looser, softer materials;

$$K_A = \frac{1}{N} = \frac{1 - \sin\phi}{1 + \sin\phi} \tag{4.59}$$

$$K_P = N = \frac{1 + \sin\phi}{1 - \sin\phi} \tag{4.60}$$

The stress–strain response and shear strength of soils are studied and measured in the laboratory using mostly the **consolidation, direct shear, triaxial,** and **unconfined compression tests. Stress paths** are a useful way of visualizing the loading of the samples in these tests. Stress paths are the evolution of p and q (total stress paths) or p' and q' (effective stress paths) during loading of a soil element or test sample.

For a triaxial test:

$$p = \sigma_m = \frac{1}{3}(\sigma_1 + 2\sigma_3) \tag{4.75}$$

$$p' = \sigma'_m = \frac{1}{3}(\sigma'_1 + 2\sigma'_3) \tag{4.76}$$

$$q = q' = \sigma_1 - \sigma_3 \tag{4.77}$$

Applications of loads on the boundary of a soil mass generate stress increments at every point within the soil mass that have magnitude and direction that depend on the magnitudes and directions of the loads, the geometry of the soil mass, and the distance from the point to the various loads. These stress increments have a pore pressure and an effective stress component, the magnitudes of which depend on the draining properties of the soil and drainage boundary conditions. An analysis in terms of total stresses is called **total stress analysis.** Likewise, an analysis in terms of effective stresses is called **effective stress analysis.**

4.8.2 Symbols and notations

Symbol	Quantity represented	US units	SI units
B, L	Dimensions	ft	m
c	Cohesive intercept	psf	kPa
e	Deviatoric strain tensor	in./in.	mm/mm
E	Young's modulus	psf	kPa

(Continued)

Symbol	Quantity represented	US units	SI units
G	Shear modulus	psf	kPa
K	Bulk modulus	psf	kPa
M	Constrained modulus	psf	kPa
N	Normal force	lb	kN
N	Flow number	unitless	unitless
Q	Vertical load	lb or lb/ft	kN or kN/m
q_b	Stress distribution on top of half-space (either as an imposed load or as reaction to imposed displacements by a rigid body)	psf	kPa
r	Radius of slip surface	ft	m
r, z	Cylindrical coordinates	ft	m
$\mathbf{s}$	Deviatoric stress tensor	psf	kPa
s	Shear strength of soil	psf	kPa
T	Tangential force	lb	kN
u_b	Back pressure	psf	kPa
x, y, z	Cartesian coordinates	ft	m
x_1, x_2, x_3	Cartesian coordinates	ft	m
γ_{ij}	Engineering shear strain corresponding to directions x_i and x_j, with $i \neq j$	in./in.	mm/mm
$\boldsymbol{\varepsilon}$	Strain tensor	in./in.	mm/mm
ε_1	Major principal strain	in./in.	mm/mm
ε_3	Minor principal strain	in./in.	mm/mm
ε_{11}	Normal strain in x_1 direction	in./in.	mm/mm
ε_{22}	Normal strain in x_2 direction	in./in.	mm/mm
ε_{33}	Normal strain in x_3 direction	in./in.	mm/mm
ε_{ij} (with $i=j$)	Normal strain	in./in.	mm/mm
ε_{ij} (with $i \neq j$)	Shear strain	in./in.	mm/mm
ε_v	Volumetric strain	in.^{3}/in.^{3}	mm^3/mm^3
θ	Cylindrical/spherical coordinate	unitless	unitless
θ_L	Lode's angle	unitless	unitless
ν	Poisson's ratio	unitless	unitless
π_{11}	Plane on which σ_{11} acts	unitless	unitless
π_{33}	Plane on which σ_{33} acts	unitless	unitless
ρ	Spherical coordinate	ft	m
$\boldsymbol{\sigma}$	Stress tensor	psf	kPa
σ	Normal stress	psf	kPa
σ^*	Transformed normal stress	psf	kPa
σ_1	Major principal stress	psf	kPa
σ_3	Minor principal stress	psf	kPa
σ_{11}	Normal stress acting on π_{11}	psf	kPa
σ_{13}	Shear stress acting on π_{33} in x_1 direction (numerically equal to σ_{31})	psf	kPa

(*Continued*)

Symbol	Quantity represented	US units	SI units
σ_{31}	Shear stress acting on π_{11} in x_3 direction (numerically equal to σ_{13})	psf	kPa
σ_{33}	Normal stress acting on π_{33}	psf	kPa
σ_c	Cell pressure	psf	kPa
σ_d	Applied axial stress (deviatoric stress)	psf	kPa
σ_θ	Normal stress acting on plane making angle θ with π_{11}, where θ is measured counterclockwise from π_{11}	psf	kPa
τ	Shear stress	psf	kPa
τ_θ	Shear stress acting on plane making angle θ with π_{11}, where θ is measured counterclockwise from π_{11}	psf	kPa
ϕ	Friction angle	deg	deg
ψ	Dilatancy angle	deg	deg

4.9 PROBLEMS

4.9.1 Conceptual problems

Problem 4.1 Define all the terms in bold contained in the chapter summary.

Problem 4.2* The concept of "stress" is crucial for geotechnical analyses. By considering soils with different particle sizes, give an example of a problem in which it is perfectly valid to use stresses in calculations and another in which it is not.

Problem 4.3 Consider Figure 4.8. Demonstrate visually that there is no distortion in the soil if $d\varepsilon_1 = d\varepsilon_3$ by replotting the figure for this condition.

Problem 4.4* What is a state of plastic equilibrium?

Problem 4.5* Why does a Poisson's ratio of 0.5 imply incompressibility?

Problem 4.6* Given that foundation loading in saturated clays is applied under undrained conditions, what can we state about the volumetric change undergone by the soil after load application? What value of Poisson's ratio would be used in settlement equations for saturated clays derived using elasticity assumptions? What does that imply about the mechanism by which immediate settlements develop in saturated clays?

Problem 4.7 When is a soil said to be in a state of rest?

Problem 4.8 What is the range of values that the coefficient of lateral earth pressure at rest can assume for typical sands and clays?

Problem 4.9 What is a Rankine state? What is the difference between an active and a passive Rankine state?

Problem 4.10 What is the pole method? If the stress state at a point in the soil mass is known, how can the pole be determined in the corresponding Mohr circle?

Problem 4.11 How can the direction of potential slip surfaces through a point in a soil mass be determined?

Problem 4.12* Consider drained tests (a triaxial compression and a triaxial extension test) on two soil samples. Is there a difference between the total stress path and effective stress path? Plot their stress paths in the p versus q space. During the shearing phase of the test, what is the slope of the stress path in p versus q space?

4.9.2 Quantitative problems

Problem 4.13 P-Figure 4.1 shows the state of stress at a point within a soil deposit. Using both an analytical approach and the pole method, determine: (a) the direction of the principal planes; (b) the magnitude of the principal stresses; and (c) the normal and shear stresses in a plane at 45° with the horizontal.

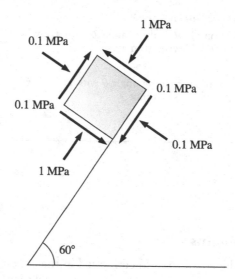

P-Figure 4.1 State of stress at a point for Problem 4.13.

Problem 4.14 For the state of stress defined in P-Figure 4.2: (a) draw the Mohr circle; (b) find the pole; identify it in the Mohr circle; and write its coordinates below; (c) find the principal planes and stresses (both graphically and analytically); (d) find the stresses (σ and τ) on the planes making angles of ±45° with the horizontal; indicate which is which; and solve both graphically and analytically.

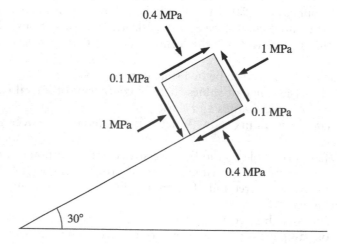

P-Figure 4.2 State of stress for Problem 4.14.

Problem 4.15 Repeat Problem 4.14 for the state of stress of P-Figure 4.3.

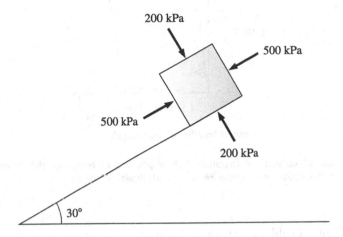

P-Figure 4.3 State of stress for Problem 4.15.

Problem 4.16 Consider the stress state of Problem 4.13. Redo the problem for shear stresses of 0, 0.2, and 0.3 MPa. What effect do you observe on (a) the magnitude and (b) the direction of the principal stresses as the applied shear stresses increase?

Problem 4.17* The unit block *abcd* is acted upon by loadings as shown in P-Figure 4.4. (a) Draw the Mohr circle of stress. (b) Locate the points on the Mohr circle indicating the stresses on face "*ad*" and "*dc*". (c) Find the pole. (d) Find the directions of principal planes and the magnitudes of the major and minor principal stresses. (e) Draw a smaller element inside *abcd* to represent the principal planes and stresses acting on them. (f) Solve the problem analytically and compare your results.

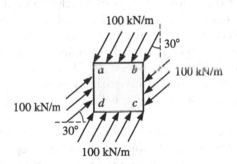

P-Figure 4.4 Element corresponding to Problem 4.17.

Problem 4.18 For the soil element of Example 4.7, consider the major principal strain increment to be vertical and (a) find the pole and (b) find the directions of the zero-extension lines (slip lines) through the element.

Problem 4.19* For the state of strain shown in P-Figure 4.5, (a) draw the Mohr circle, (b) find the pole, (c) find the principal directions and the values of the principal strain increments, and (d) find the directions of the zero-extension lines.

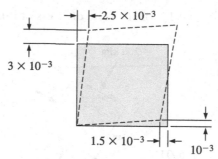

$$\text{Initial lengths of element sides} = 1$$

P-Figure 4.5 Incremental strains for Problem 4.19 expressed in terms of the original and deformed shapes of a square element with sides with length equal to 1.

Problem 4.20* Solve Problem 4.19 analytically.

Problem 4.21 A dry sand deposit has a friction angle of 35° and a unit weight of 20 kN/m³. What is the lateral effective stress and the coefficient of lateral earth pressure for (a) at-rest, (b) active, and (c) passive conditions at a depth of 10 m?

Problem 4.22 Consider the soil mass of Problem 4.19(c), retained in place by a semi-infinite, dimensionless plate. Consider the soil to reach a Rankine state with a dilatancy angle of 8°. If the plate moves in the direction away from the soil, resulting in an extension corresponding to a lateral strain increment of −0.5%, what is the vertical strain in the retained soil mass? What is the direction of the slip plane through the soil mass?

Problem 4.23 If the soil of Problem 4.22 has a friction angle of 34.5°, what is the error in the estimate of the slip plane direction based on friction angle?

Problem 4.24* Consider the following ways in which you could shear a triaxial sample: (a) reduce the radial stress while keeping the axial stress unchanged; (b) increase the radial stress while keeping the axial stress unchanged; (c) increase the axial stress while keeping the radial stress unchanged; (d) reduce the axial stress while keeping the radial stress unchanged. Plot (total stress) p–q diagrams for all four cases. Be specific about the directions of the total stress paths. Calculate also the following ratio for each case: $(\sigma_2 - \sigma_3)/(\sigma_1 - \sigma_3)$. What do you believe this ratio tells you about what you are doing to the sample?

Problem 4.25 Three vertical point loads (800, 200, and 250 kN) are applied to the surface of a soil mass along a straight line. The loads are separated by distances of 2 m. Using the Boussinesq equation, compute the vertical stress increase due to these loads at a depth of 2 m along their vertical lines of action.

Problem 4.26 A 100 kN point load is applied at a point on the surface of a soil mass. Calculate the vertical stress increase due to this load at a point with distances 1.5, 2, and 1 m with respect to the point of application of the load in the x, y, and z directions, respectively.

Problem 4.27 Two buildings, shown in P-Figure 4.6, are separated by 7 m. The uniformly distributed load of building 1, with dimensions 10 m × 15 m, is equal to 300 kPa; that of building 2, with dimensions 7 m × 25 m, is 500 kPa. Compute the vertical stress increase due to both buildings at point P, which is 8 m below the center of building 1.

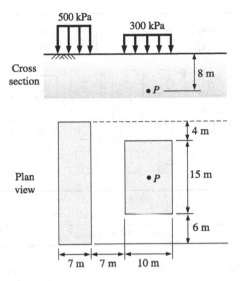

P-Figure 4.6 Two buildings of Problem 4.27.

Problem 4.28 A building is to be built on top of the soil profile in P-Figure 4.7. A solution based on footings[16] bearing on a layer of compacted soil is proposed. The areas of the footings are not yet available. The average unit load to be carried by the foundations will be approximately 100 kPa, but it is expected that it will be distributed among three classes of columns: A, B, and C. Columns C will carry 2 times the load carried by columns of class B, which, in turn, will carry 1.25 times as much load as columns of class A. Estimate the vertical stress increase due to the building loading for depths ranging from 2 to 13 m: (a) below point D (center of the corner column), (b) below point E, and (c) below the center of the building. Consider first the average unit load of 100 kPa applied over the entire plan area of the building and then the statically equivalent column (point) loads consistent with the 2 and 1.25 ratios discussed earlier.

Problem 4.29* In Section 4.6, we studied the strip loading (Figure 4.33). (a) Find the principal stresses σ_1 and σ_3 acting on principal planes π_1 and π_3 at point P with coordinates θ_1 and θ_2. Express the results for σ_1 and σ_3 in terms of α. (b) Show that the points belonging to a circle passing through P and through the two edges P_1 and P_2 of the strip load have the same state of stress as P. Write the equation for the maximum shear stress for this circle. (c) Sketch the trajectories of σ_1 and σ_3. (d) Using the equation for the maximum shear stress you obtained in (b), find the maximum shear stress for a given loading and the circle corresponding to it (that is, the value of α corresponding to maximum shear stress). The equations derived in Section 4.6 for σ_{11}, σ_{33}, and σ_{13} are repeated below for convenience:

$$\sigma_{33} = \frac{q_b}{2\pi}\left[2\left(\theta_2 - \theta_1\right) + \left(\sin 2\theta_2 - \sin 2\theta_1\right)\right]$$

$$\sigma_{11} = \frac{q_b}{2\pi}\left[2\left(\theta_2 - \theta_1\right) - \left(\sin 2\theta_2 - \sin 2\theta_1\right)\right]$$

---

[16] Footings are discussed in detail in Chapters 8–11.

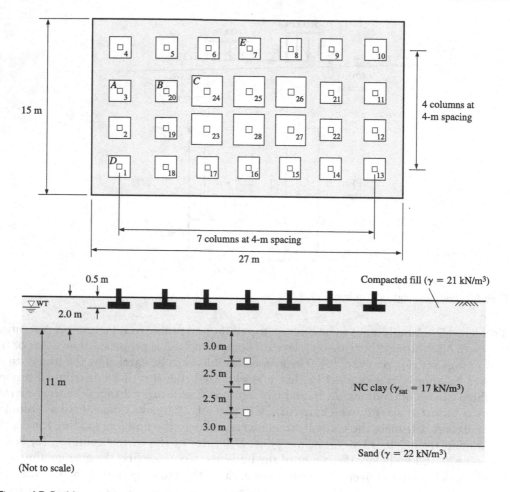

P-Figure 4.7 Building and its foundations for analysis in Problem 4.28.

$$\sigma_{22} = \frac{2vq_b}{\pi}(\theta_2 - \theta_1) = \frac{2vq_b\alpha}{\pi}$$

$$\sigma_{13} = \frac{q_b}{2\pi}\left[(\cos 2\theta_1 - \cos 2\theta_2)\right]$$

$$\alpha = \theta_2 - \theta_1$$

REFERENCES

References cited

Barré de Saint-Venant, A.-J.-C. (1855). "*Memoire sur la torsion des prismes: avec des considérations sur leur fléxion ainsi que sur l'équilibre intérieur des solides élastiques en général, et des formules pratiques pour le calcul de leur résistance à divers efforts s'exerçant simultanément.*" Mem. Divers Savants, 14, 233–560.

Boussinesq, M. J. (1885). *Application des potentiels à l'étude de l'équilibre et du mouvement des solides élastiques, principalement au calcul des deformations et des pressions que produisent, dans ces solides, des efforts quelconques exercés sur und petite partie de leur surface.* Gauthier-Villars, Paris, France.

Brooker, E. W., and Ireland, H. O. (1965). "Earth pressures at rest related to stress history." *Canadian Geotechnical Journal*, 2(1), 1–15.

Cook, G. (1951). "Rankine and the theory of earth pressure." *Géotechnique*, 2(4), 271–279.

Fung, Y. C. (1993). *A First Course in Continuum Mechanics.* 3rd Edition, Prentice-Hall, Upper Saddle River, NJ.

Hoek, E., and Brown, E. T. (1980). "Empirical strength criterion for rock masses." *Journal of the Geotechnical Engineering Division*, 106(GT9), 1013–1035.

Hoek, E., and Brown, E. T. (1988). "Hoek-Brown failure criterion–a 1988 update." *Proceedings of 15th Canadian Rock Mechanics Symposium*, Civil Engineering Department, University of Toronto, Toronto, Canada, 31–38.

Jaky, J. (1944). "The coefficient of earth pressure at rest." *Journal of Society of Hungarian Architects and Engineers*, 355–358.

Lubliner, J. (2008). *Plasticity Theory.* Dover Publications, New York.

Newmark, N. M. (1935). *Simplified computation of vertical pressures in elastic foundations. Circular No.24*, University of Illinois, Urbana, Illinois, USA.

Rankine, W. J. M. (1857). "On the stability of loose earth." *Philosophical Transactions of the Royal Society of London*, 147(1), 9–27.

Saada, A. S. (2009). *Elasticity Theory and Applications.* 2nd Edition, J. Ross Publishing, Plantation, FL.

Terzaghi, K. (1943). *Theoretical Soil Mechanics.* Wiley, New York.

Timoshenko, S. P., and Goodier, J. N. (1970). *Theory of Elasticity.* 3rd Edition, McGraw-Hill, New York.

Additional references

Duncan, J.M. and Chang, C.Y. (1970) "Nonlinear analysis of stress-strain in soils." *Journal of Soil Mechanics and Foundation Engineering Division*, 96(SM5), 1629–1653.

Fahey, M. and Carter, J.P. (1993). "A finite element study of the pressuremeter test in sand using a non-linear elastic plastic model." *Canadian Geotechnical Journal*, 30(2), 348–361.

Kondner, R. L. (1963). "Hyperbolic stress-strain response: cohesive soil." *Journal of Soil Mechanics and Foundation Engineering Division*, 89(SM1), 115–143.

Lee, J. H. and Salgado, R. (1999). "Determination of pile base resistance in sands." *Journal of Geotechnical and Geoenvironmental Engineering*, 125(8), 673–683.

Lee, J. H. and Salgado, R. (2000). "Analysis of calibration chamber plate load tests." *Canadian Geotechnical Journal*, 37(1), 14–25.

Michalowski, R. L., (2005). "Coefficient of earth pressure at rest." *Journal of Geotechnical and Geoenvironmental Engineering*, 131(11), 1429–1433.

Mindlin, R. D. (1936). "Force at a point in the interior of a semi-infinite solid." *Physics*, 7(5), 195–202.

Parry, R.H.G. (1995). *Mohr Circles, Stress Paths and Geotechnics.* E and FN Spon, London.

Poulos, H.G. and Davis, E.H. (1974). *Elastic Solutions for Soil and Rock Mechanics.* Wiley, New York.

Chapter 5

Shear strength and stiffness of sands

> There is in all things a pattern that is part of our universe. It has symmetry, elegance, and grace – those qualities you find always in that which the true artist captures. You can find it in the turning of the seasons, in the way the sand trails along a ridge....
>
> Frank Herbert, in "Dune": "The Collected Sayings of Muad'Dib" by the Princess Irulan.

The "way the sand trails along a ridge," the steepness of a sand mound, how able the sand is to sustain the weight of our bodies when we walk on it on the beach all depend on the mechanical properties of the sand, primarily on its internal shear strength and stiffness. Soils must be sufficiently stiff for foundations not to settle or move excessively when loaded. Additionally, the shear stresses induced by foundation loads must not exceed the shear strength of the soil, because this would lead to the formation of slip surfaces that might lead to the collapse of the foundation and the structure that it supports. Retaining structures and slopes also rely on shear strength and stiffness for satisfactory performance (not falling down or failing catastrophically and not moving excessively!). In this chapter, we discuss the shear strength and stiffness of sandy soils and how they can be quantified.

5.1 STRESS–STRAIN BEHAVIOR, VOLUME CHANGE, AND SHEARING OF SANDS

5.1.1 Stress ratio, dilatancy, and the critical state

Triaxial tests[1] are commonly used to study the behavior of sand and to determine its shear strength. Figure 5.1 illustrates idealized stress–strain curves for drained triaxial compression tests performed on three identical samples, each consolidated isotropically[2] to a different stress. The first sample is consolidated to $\sigma_3' = \sigma_{c0}'$; the second, to twice σ_{c0}'; and the last, to four times σ_{c0}'. The stress σ_c is the cell pressure (that is, the pressure applied on the boundaries of the sample from all directions), and σ_c' is the effective consolidation stress (the cell pressure minus the back pressure).

The plots in Figure 5.1a are in terms of $q = \sigma_1' - \sigma_3'$ versus the axial strain ε_a (a measure of the level of shear strain in the sample). We first note that the peak value of q (that is, the peak shear strength) increases by less than the factors of 2 and 4 by which σ_3' is multiplied for the second and third samples. This implies that there is a reduction in the *peak friction*

---

[1] See Chapter 4 for the different types of triaxial test and how each is performed.
[2] Isotropic consolidation is an increase in the all-around stress acting on the sample. In anisotropic consolidation, in contrast, stresses in different directions change by different amounts.

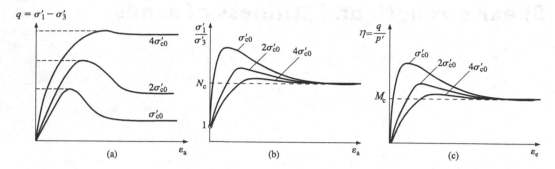

Figure 5.1 Stress–strain curves for drained triaxial compression tests on samples subjected to different confining stresses: (a) $q=\sigma'_1-\sigma'_3$ versus axial strain ε_a, (b) principal effective stress ratio σ'_1/σ'_3 versus axial strain ε_a, and (c) stress ratio $\eta=q/p'$ versus deviatoric strain ε_q.

angle with increasing confining stress[3]; this happens because peak shear strength depends not only on frictional strength, but also on the soil *dilatancy*, which will be discussed later in this section. After the peak, the shear stress levels off for all three samples and approaches a plateau, with the shear strength remaining practically unchanged with increasing axial strain. When this happens, the samples are said to have reached a *critical state*. A soil is said to be in a critical state if its void ratio is consistent with the applied stresses, including any shear stresses, so that any additional shearing will take place without further changes in volume or effective stress under either drained or undrained conditions. We will define the critical state with greater mathematical rigor in the next section.

As seen in Figure 5.1a, the critical-state shear strength increases by about the same factors 2 and 4 as the confining stress. This is so because the *critical-state friction angle* ϕ_c depends only on the coefficient of friction at the contacts between particles, and this remains the same for all three stress levels. In fact, *purely frictional state* would have been a better term for the critical state, but usage compels us to use the established term.

The plots in Figure 5.1b, in terms of the principal effective stress ratio σ'_1/σ'_3 (also a measure of the mobilized frictional strength in the sample) versus the axial strain ε_a (a measure of the level of shear strain in the sample), show the same observation in a different way. Using Equations (4.48) and (4.45), we can write:

$$\frac{\sigma'_1}{\sigma'_3} = N = \frac{1+\sin\phi}{1-\sin\phi} = \tan^2\left(45° + \frac{\phi}{2}\right)$$

(5.1)

which can be used to calculate a mobilized friction angle ϕ for each value of σ'_1/σ'_3. Note that the three curves shown in Figure 5.1b end at the same value of principal effective stress ratio $(\sigma'_1/\sigma'_3)_c=N_c=\tan^2(45°+\phi_c/2)$, where the subscript c indicates the value of the quantity at critical state, implying that the critical-state friction angle is the same regardless of the σ'_c value. It is also apparent in Figure 5.1b that the peak value of the principal effective stress ratio $\sigma'_{1p}/\sigma'_{3p} (= N_p)$, with the subscript p indicating the value of the quantity at peak strength, decreases with increasing consolidation stress.

Figure 5.1a also shows that the peak shear strength is reached at different values of axial strain ε_a depending on the level of consolidation stress $\sigma'_c = \sigma'_3$. Doubling of σ'_3 causes the strain required to reach both the peak and critical states to increase by about 20%–40%. For clean sands, these axial strains are typically in the 1%–7% range (but potentially can be

---

[3] Confining stress is a generic term that is usually associated with the mean effective stress $\sigma'_m = (\sigma'_1 + \sigma'_2 + \sigma'_3)/3$; in a triaxial test, after consolidation but before shearing, σ'_m is equal to the consolidation stress σ'_c.

as large as 10%) for peak strength and in excess of 25% (and possibly significantly greater than 25%) for the critical state in triaxial compression. It is apparent that increases in confining stress make the sand less brittle.

Another observation we can make regarding Figure 5.1a is that the sample subjected to the greatest confining stress $(4\sigma'_{c0})$ is stiffer than the one subjected to $2\sigma'_{c0}$, which in turn is stiffer than the one under σ'_{c0}. This is simply a reflection of the fact that sands are frictional materials and that larger confining stresses press particles closer together, which makes it harder to get them to move with respect to each other (thus making the soil stiffer).

Let us now consider volume change during the triaxial tests. We consider two consolidated drained triaxial compression (TXCD) tests on samples of the same sand. Both samples are tested at the same cell pressure σ'_c, but their relative densities are different: one is extremely loose; the other, very dense. Figures 5.2 and 5.3 illustrate the evolution of shear stress, volumetric strain, and dilatancy angle with axial strain for the very loose and very dense sand samples, respectively. The very loose sample contracts throughout the test, while the very dense sample is contractive initially, for very small strains, and dilative afterward.[4] Figures 5.4 and 5.5 represent the same process as Figures 5.2 and 5.3, but using the more general representations of stress and dilatancy and a more rigorous expression of the shear strain in the sample.

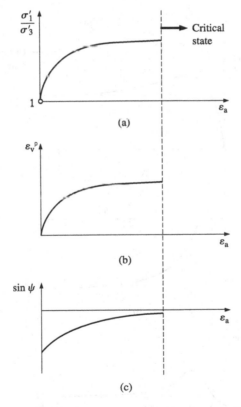

Figure 5.2 Triaxial test results for a contractive sand sample: (a) principal effective stress ratio σ'_1 / σ'_3, (b) volumetric strain ε_v, and (c) sine of the dilatancy angle ψ versus axial strain ε_a.

[4] Recall from Chapter 4 that contractive strains are positive, dilative strains are negative, and dilatancy angle is positive if the volume of the soil expands during shearing (see Equations (4.35) and (4.36)).

Figure 5.6 aids in explaining this behavior. It shows both the loosest and densest possible configurations for the sand particles. Observe that a dense sample, unless it is in the most compact state possible, will initially become denser as its particles occupy slightly more stable positions. It soon reaches a particle arrangement, as illustrated in Figure 5.7a, such that, in order for shearing to continue, particles must then climb over one another (Figure 5.7b), and expansion occurs until a configuration (the critical-state configuration) in equilibrium with the imposed stresses is reached. In the case of an ideally loose sample, shearing causes densification of the sand throughout the test because particles can always find more stable positions to fall into. It is important to understand that, in a real soil, there are particles of different sizes and shapes, so that the illustrations provided in Figures 5.6 and 5.7 refer to an ideal soil.

There is a clear relationship in Figures 5.2 and 5.3 between the principal effective stress ratio σ'_1 / σ'_3 and the dilatancy angle ψ or in Figures 5.4 and 5.5 between the stress ratio q/p' and the dilatancy D. Observe how, for the very dense sample, as illustrated in Figure 5.3, σ'_1 / σ'_3 is initially less than its value at critical state when the soil is contracting, but then becomes equal to that value exactly when the dilatancy angle becomes equal to zero, and then exceeds it when the dilatancy angle becomes positive and peaks when the dilatancy angle peaks. A similar link between σ'_1 / σ'_3 and the dilatancy angle exists for the contractive sand, and a similar response exists for q/p' and D.

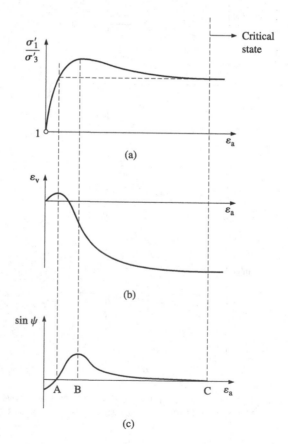

Figure 5.3 Triaxial test results for a dilative sand sample: (a) principal effective stress ratio σ'_1 / σ'_3, (b) volumetric strain ε_v, and (c) sine of the dilatancy angle ψ versus axial strain ε_a.

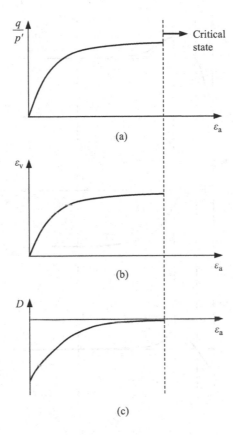

Figure 5.4 Triaxial test results for a contractive sand sample: (a) effective stress ratio q/p', (b) volumetric strain ε_v, and (c) dilatancy D versus axial strain ε_a.

5.1.2 Friction and dilatancy

In most conditions of interest in practice, the shearing of a sandy soil requires that energy be supplied to the soil so that: (1) particles overcome friction at the particle contacts and kinematic impediments generated by particle morphology (as quantified through roundness and sphericity) so that relative motion can occur and (2) particles climb over each other so they can move past adjoining particles along the slip plane. The climbing action associated with (2), if not compensated for by particles falling into the voids between other particles, results in an increase in volume for a soil element containing a number of particles. The volume increase per unit increase in shear strain is related to the dilatancy angle of the soil, as discussed in Chapter 4. The dilatancy angle is given by

$$\sin\psi = -\frac{d\varepsilon_v^p}{d\varepsilon_1^p - kd\varepsilon_3^p} = -\frac{d\varepsilon_1^p + kd\varepsilon_3^p}{d\varepsilon_1^p - kd\varepsilon_3^p} \tag{4.35}$$

where the superscript p indicates that the strain is a plastic strain, and k is given by

$$k = \begin{cases} 1 \text{ for plane-strain conditions} \\ 2 \text{ for triaxial conditions} \end{cases} \tag{4.36}$$

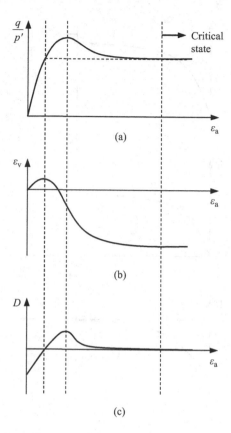

Figure 5.5 Triaxial test results for a dilative sand sample: (a) effective stress ratio q/p', (b) volumetric strain ε_v, and (c) dilatancy D versus axial strain ε_a.

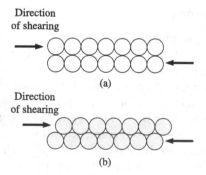

Figure 5.6 Shearing of (a) loose and (b) dense sands at the particle level.

Any increase in volume takes place with the opposition of an existing confining stress, implying that energy must be spent to do work against this confining stress. As the confining stress increases, so does the energy required for the dilation; thus, for very large confining stresses, it may be easier for particles to crush and shear through portions of other particles than to dilate. This alternative shearing mechanism, which involves more particle crushing and less dilation, will, for large confining stresses, require less energy than pure dilation.

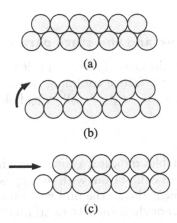

(a)

(b)

(c)

Figure 5.7 Particle climbing action necessary for the shearing of densely arranged particles in the absence of particle crushing: (a) ideally dense arrangement of equal-sized particles; (b) particle climbing action during shear; (c) end of climbing action, as shearing can now proceed without it.

The energy required to overcome interparticle friction and kinematic impediments related to particle morphology and to cause the soil to expand against the existing confining stress is supplied by the applied shear stress τ. This shear stress can be related to a friction angle ϕ. In the absence of volume change, this friction angle is known as the constant-volume or critical-state friction angle ϕ_c, which reflects primarily interparticle friction, with a small influence from the impediment to motion due to particle shape and angularity. No volume change at critical state implies zero dilatancy angle: $\psi=0$.

Referring to Figure 5.2, in which the principal effective stress ratio σ_1' / σ_3' is plotted versus axial strain for a drained triaxial compression test on a contractive sample of sand, we note that the principal effective stress ratio increases monotonically toward the critical-state shear strength. This means that the soil gets stronger during the test, a result of the fact that it becomes increasingly denser. Once the volume change stabilizes and the critical state is reached at very large axial strains, the mobilized friction angle[5] becomes equal to ϕ_c. Referring now to Figure 5.3, which shows the same type of test performed on an ultimately dilative sand sample, we see that there is initially contraction until point A is reached.[6] At point A, the sample is no longer contracting and expansion is about to start. This means that the sum of incremental elastic and plastic volumetric strains at that point is equal to zero. The plastic volumetric strain increment is not zero at point A, and thus the dilatancy angle, which is defined in terms of incremental plastic strains, is not zero either; however, it is possible, as a simple approximation, to assume it to be zero there, given that elastic strains are quite small.

Since ψ is approximately zero at A, conditions at A resemble the critical state, and the mobilized friction angle is approximately equal to the critical-state friction angle ϕ_c. Beyond point A, the soil starts dilating, ψ becomes positive, and ϕ rises above ϕ_c. This continues until the dilatancy angle peaks at point B, corresponding to conditions of maximum volume increase per unit increase in shear strain, at which point $\phi=\phi_p$, where ϕ_p is the peak friction angle. From that point on, expansion continues, but at a decreasing rate, and ϕ drops until it reaches ϕ_c at the critical state, point C, where the dilatancy angle becomes equal to zero.[7]

---

[5] Recall that ϕ can be calculated from σ_1' / σ_3' using Equation (5.1).

[6] The reason for soil that at the critical state will have a higher void ratio than it had initially to initially contract is related to the fabric of the soil and how it evolves during shearing. This goes beyond the scope of this text, but is an essential aspect of soil behavior that needs to be properly modeled in realistic sophisticated simulations.

[7] Note that what is plotted in the horizontal axis in the figure is the axial strain $\varepsilon_a = \varepsilon_1$. A more rigorous representation of shear strain in the sample would be to have the deviatoric strain ε_q in the horizontal axis.

5.2 CRITICAL STATE

5.2.1 The critical-state line and the state parameter

As seen in the previous section, the critical state is a *soil state* in which the soil is sheared at constant shear stress, constant effective confining stress, and constant volume (or void ratio). Mathematically, at the critical state, the following holds:

$$\dot{p}' = 0, \dot{q} = 0, \dot{e} = 0, \dot{\varepsilon}_q \neq 0 \qquad (5.2)$$

where the dot overlying the variables indicates a time rate. The critical state can be thought of as a state in which the soil is in equilibrium with the imposed stresses, no longer needing to either dilate or contract to shear (hence the null void ratio time rate). It is important to note that, given that stresses are not changing, elastic strain rates are also zero. Additionally, given that volume is not changing, not only the elastic, but also the plastic volumetric strain rate is zero.

The *critical-state line* (CSL) is the locus of points corresponding to critical states in void ratio–confining stress–shear stress space. Figure 5.8 shows a critical-state plot in p'–e space. This plot has a line in it separating states, as defined by the (p', e) pairs, in which the soil is dilative (located below the line) from those in which the soil is contractive (located above the line). For a dilative sample at a given initial void ratio e_0, the greater the confining stress p' at which it is tested, the lower the soil void ratio that is consistent with the stresses at critical state. So, for increasing p', fixing the void ratio, the sample becomes less dilative, will expand less during shearing, and will reach a lower critical-state void ratio. The vertical distance from the critical-state line can be seen to be an indicator of how much the soil will dilate or contract by the time it reaches critical state. This distance is represented by the state parameter ψ:

$$\psi = e - e_c \qquad (5.3)$$

where e_c is the void ratio on the critical line for the p' experienced by the soil when its void ratio is e.

Figure 5.9 illustrates the concept. The more positive the ψ is, the more ultimately contractive the soil is; conversely, the more negative it is, the more ultimately dilative the soil is. *Constitutive models* for soils, which are a set of relationships that allow prediction of soil

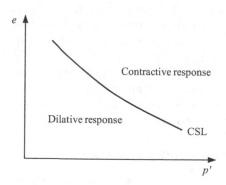

Figure 5.8 Critical-state plot: sand becomes less dilative or more contractive as either p' or e increases, all other things being equal.

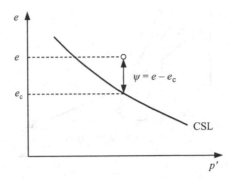

Figure 5.9 The state parameter is the difference between the soil's void ratio and the void ratio at the critical-state line for the same mean effective stress p'.

response under different loading conditions, are often developed using the state parameter as a state variable.

The critical-state line in p'–e space slopes downward, but is concave upward, as shown in Figure 5.8. The upward concavity reflects the fact that successive equal increments of confining stress p' make an increasingly smaller difference in terms of the critical-state void ratio that will ultimately be attained by the sand because the critical state corresponds to increasingly denser, stiffer particle arrangements. Dilatancy increases with increasing relative density; this can be inferred from the fact that, all other things being equal, the lower the void ratio, the more dilative the soil state is. If the mean effective stress p' is very large for the inherent sand particle strength, it inhibits dilation and keeps ϕ values at or near ϕ_c. This is so because, in the presence of a very large confining stress, particle crushing/breakage becomes a lower-energy path for the soil to follow than dilation. Regardless of particle crushing resistance, an increasing confining stress increasingly inhibits dilatancy, so that the contribution of dilatancy to the friction angle decreases with increasing confining stress.

Figure 5.15 shows the critical-state line in e–p'–q space. The CSL is a three-dimensional line that can be projected onto p'–e space (we have focused on this projection in our discussion of the CSL so far) and onto p'–q space (the slope of the CSL in p'–q space is M_c). The CSL encapsulates the effect of the key state variables (density and stress) on shear strength at large strains.

5.2.2 Shearing paths: all paths lead to the critical state*

Figure 5.10a shows the paths followed by initially loose (ultimately contractive) and initially dense (ultimately dilative) soil samples when loaded in triaxial compression under drained conditions, both starting from the same initial confining stress p_0'. Point A represents the initial state of the dilative sample. As the sample is sheared, note that its void ratio initially decreases slightly and then increases, as indicated by the path AC. The void ratio of the contractive sand sample decreases continuously from B to C, as shown in the figure. The paths for both samples converge on the same point C, which represents the critical state for both samples.

The initial decrease in volume of the dense sample is due to an initially weak fabric – that is, particle arrangement – in triaxial compression. Once particles reposition themselves and realign, the sample becomes stronger and dilation is then required for continuous shearing. The transition between contractive and dilative plastic response is known as *phase transformation*.

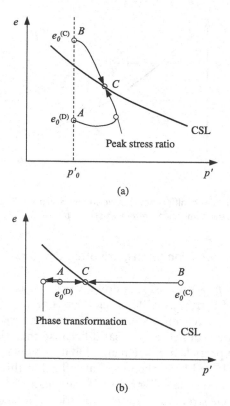

Figure 5.10 Paths followed in p'–e space by an ultimately contractive and an ultimately dilative sand sample in (a) drained and (b) undrained triaxial compression loadings.

The responses of two sand samples sheared under undrained conditions at the same initial void ratio, but different values of initial confining stress p' are shown in Figure 5.10b. The sand initially confined with a greater stress p' behaves as a contractive material: as the sand attempts to contract, which it cannot because drainage valves are closed and volume change is not possible, it squeezes the pore water, leading to an increase in pore pressure throughout the test. The other sample, the one subjected to a lower initial value of p', is also initially contractive, but, like the dense sand in Figure 5.3, soon undergoes phase transformation and turns dilative.

Undrained shearing is the best way to experimentally determine the phase transformation state or point. The phase transformation point is determined as the point at which the incremental change in p' (and thus the incremental change in pore pressure) is equal to zero. Since $dp'=0$, the elastic volumetric strain increment, which is dp' divided by the elastic bulk modulus of the soil, is equal to zero as well. Given that the total volumetric strain increment is zero, the incremental plastic volumetric strain must also be equal to zero, which implies that the dilatancy at phase transformation is zero. Both the dilative and contractive samples converge on the same critical state: p' increases sufficiently for the dilative sample and decreases sufficiently for the contractive sample for both to have, at critical state, the same value of p' and e.

Figure 5.11a and b shows the response of the contractive and dilative samples in Figure 5.10 in p'–q space. Both reach the critical-state line, whether under drained (Figure 5.11a) or undrained (Figure 5.11b) loading. The critical-state line is a straight line in this space because it represents purely frictional strength, as discussed earlier. The p' versus e plots in Figure 5.10 are shown together with the p'–q plots in Figure 5.11. We do this to show the correspondence between significant points during loading for both the drained and undrained loadings of

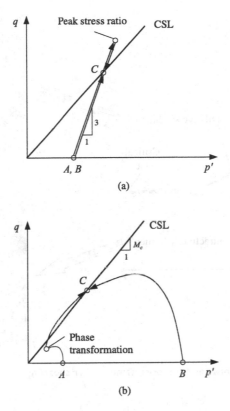

Figure 5.11 Paths followed in p′–q space by an ultimately contractive and an ultimately dilative sand sample in (a) drained and (b) undrained triaxial compression loadings.

the initially dense and loose samples. Phase transformation, peak stress ratio, and the critical state – all identified in the figure – are seen to occur for specific values of p' and so appear in the figure as aligned in the vertical direction.

In the case of the triaxial test, dense and loose samples look more or less as shown in Figure 5.12a and b, respectively, after shearing. In a dense sample, after shearing starts in a plane, this plane gets still weaker during shearing due to dilation, and shearing therefore continues along this same plane until the end of the test. This makes for a well-defined shear band or, in ideal terms, shear plane, as shown in Figure 5.12a. This process is called shear strain localization, and the shear "plane" was explored conceptually in Chapter 4 as a zero-extension surface. The loose sample, in contrast, bulges outward, and no distinct shear plane may be identified. Loose samples do not necessarily behave as the ideal contractive sample in Figure 5.2, which plots far above the CSL in p' versus e space and never experiences any dilation. They may contract during most of the test, but then start to dilate at larger axial strains, with the accumulated volumetric strain still positive, before they reach critical state. In samples that do not dilate or dilate to a limited extent only late in the shearing process, shearing ("failure") may start in one definite plane, but as soon as it does, that plane densifies (because the sample is contractive), the sand gets stronger there, and plastic failure then starts to develop in an adjacent plane. This process repeats itself until the sample is sheared throughout; as a result, instead of a well-defined shear plane, the whole sample can be said to have sheared, as shown in Figure 5.12b. Two facts follow: the dense sample expands locally, while the loose sample contracts almost as a whole. This has

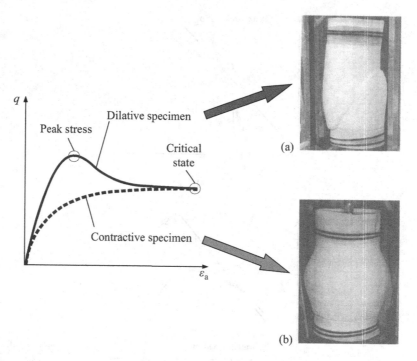

Figure 5.12 Appearance of (a) dense and (b) loose sample after shearing.

the following implication for the collapse of foundations and other geotechnical structures: "failure" or slip "surfaces" tend to be wider and more diffused in loose sandy soils and narrower and more localized in dense sandy soils.

Another consequence of shear localization is that we cannot find the critical-state void ratios by performing drained tests on dilative samples through measurement of the global void ratio for the sample because these samples shear along a well-defined shear band, and only the soil in the shear band expands to reach the critical-state void ratio e_c under the applied stress state. As a result, the overall void ratio of dilative samples is less than e_c. Contractive samples (loose samples under high confining stresses) can be used to obtain at least approximate values of e_c. Undrained tests on sand, which are discussed later in this chapter, can be more effectively used to obtain all the information needed to characterize critical-state behavior because sample volume is constant and only effective stresses change, and effective stresses tend to equalize throughout the sample.

The critical-state friction angle ϕ_c is the friction angle that a given soil has at critical state. The corresponding value of the stress ratio $\eta = M = q/p'$ is denoted by M_c. The difficulties in determining e_c do not apply to ϕ_c or M_c estimation; this is because the shear strength of the sample develops along the shear band (where critical state is reached) and does not depend, at least in any significant way, on the state of stress or the void ratio in the other parts of the sample. However, in triaxial tests, the large deformations required for critical state to be reached may lead to changes in sample geometry (bulging in a loose sample, for example) that may reduce accuracy in the estimation of stress at critical state. For our purposes in the context of this text, relatively simple ways of estimating ϕ_c when stabilization starts, at strains of the order of 20%, are satisfactory. Example 5.1 illustrates how to estimate ϕ_c from a triaxial test.

Example 5.1

Estimate the critical-state friction angle ϕ_c and the corresponding stress ratio M_c for the test sand used in the drained triaxial compression test in E-Figure 5.1. The effective confining stress is $\sigma_3' = 400$ kPa.

Solution:

From E-Figure 5.1, we see that the soil reaches its critical state for an axial strain between 25% and 30%. The deviatoric stress at that point is 720 kPa, that is:

$$\sigma_1' - \sigma_3' = 720 \text{ kPa}$$

Because $\sigma_3' = 400$ kPa at the critical state, the major principal stress will be:

$$\sigma_1' = 720 + 400 = 1120 \text{ kPa}$$

Using Equation (5.1) at critical state:

$$\sigma_1' = N_c \sigma_3' = \frac{1 + \sin\phi_c}{1 - \sin\phi_c}\sigma_3'$$

$$\frac{1 + \sin\phi_c}{1 - \sin\phi_c} = \frac{1120}{400} = 2.8$$

$$1 + \sin\phi_c = 2.8 - 2.8\sin\phi_c$$

$$\phi_c = 28.3° \quad \text{answer}$$

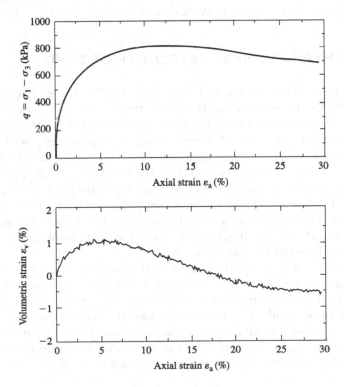

E-Figure 5.1 Results from a triaxial test on Ottawa Sand sample.

The stress ratio M_c at critical state can be calculated directly from the friction angle using Equation (4.49) as

$$M_c = \frac{6\sin\phi_c}{3 - \sin\phi_c} = \frac{6\sin 28.3°}{3 - \sin 28.3°} = 1.13 \text{ answer}$$

5.2.3 Critical-state friction angle

Critical-state friction angle values range from 28° to 36° for most clean silica sands. Poorly graded sands with more equidimensional, rounded particles tend to have ϕ_c values near the low end of that range, while well-graded sands with less equidimensional, more angular particles have ϕ_c values near the top of the range. Salgado et al. (2000) found that the addition of small amounts (5%–25%) of nonplastic silt to clean sand tends to increase ϕ_c by 1°–2°, but Carraro et al. (2003) found that the addition of small amounts (2%–10%) of low-plasticity clay to clean sand tends to reduce ϕ_c by 0.4°–1°.

Table 5.1 contains data on intrinsic variables, including morphology parameters, of 23 different sands. Based on these data, an equation may be proposed for the critical-state friction angle ϕ_c in triaxial compression in terms of particle size distribution and particle morphology parameters (Sakleshpur et al. 2021):

$$\phi_c(°) = 28.3\left(\frac{D_{50}}{D_{\text{ref}}}\right)^{\zeta}(C_U)^{2\zeta}(R)^{-3\zeta} \tag{5.4}$$

where D_{ref}=reference particle size = 1 mm, D_{50}=mean particle size, C_U=coefficient of uniformity, R=roundness, and ζ= 0.045.[8]

Figure 5.13 shows a plot of Equation (5.4) together with the data from Table 5.1. It is seen that the equation provides estimates of critical-state friction angle to within ±1°.

5.3 EVALUATION OF THE SHEAR STRENGTH OF SAND

The triaxial test (particularly the triaxial compression test) and the direct shear test are by far the most widely used tests to assess the shear strength of sandy soils in the laboratory. Triaxial tests for sand are usually of the *consolidated drained* (CD) type, although *consolidated undrained* (CU) tests are also used. Sandy soils drain fast, much faster than the loading rates of foundations, slopes, or retaining structures, so the use of CD tests is appropriate. The use of CD tests is also economical, as these tests are rapidly performed in sands.

Many strength tests performed in routine practice stop at the point corresponding to peak strength because they are performed in stress-controlled mode. To extend the test beyond peak strength, strain-controlled tests must be performed. Additional important information, notably an estimate of the critical-state friction angle, can be obtained only if a CD test is extended beyond the point of peak strength.

Laboratory testing can be used for estimating the peak friction angle ϕ_p of engineered fills, which are placed under specified conditions. Laboratory samples, in this case, should model anticipated field conditions. Compaction of the sample should simulate the compaction method used in the field. A common technique to prepare samples in those cases is the moist-tamping method, in which the sample is deposited in layers at a controlled water content and tamped by hand the number of times expected to produce the desired density. For evaluating the ϕ_p of natural sand deposits, laboratory tests are not a viable option

---

[8] These variables were defined and discussed in Section 3.3.

Table 5.1 Intrinsic parameters of 23 clean silica sands (after Sakleshpur et al. 2021)

Sand	Gradation		Morphology		Packing		Strength	
	D_{50} (mm)	C_U	R	S	e_{min}	e_{max}	ϕ_c (°)	Reference
FS Ohio 6–10	2.68	1.31	0.43	0.86	0.66	0.92	34.6	Han et al. (2018)
FS Ohio 10–16	1.59	1.30	0.44	0.83	0.65	0.92	33.7	Han et al. (2018)
FS Ohio 16–20	1.01	1.25	0.40	0.78	0.66	0.97	32.9	Han et al. (2018)
FS Ohio 20–40	0.63	1.42	0.39	0.82	0.62	0.91	31.8	Han et al. (2018)
FS Ohio 50–100	0.23	1.56	0.35	0.82	0.63	0.93	31.7	Han et al. (2018)
FS Ohio Coarse	1.50	2.00	—	—	0.45	0.72	33.6	Han et al. (2018)
FS Ohio Fine	0.35	2.00	—	—	0.48	0.72	33.4	Han et al. (2018)
FS Ohio SW	1.04	7.90	—	—	0.37	0.65	33.2[a]	Han et al. (2018)
Fontainebleau NE34	0.21	1.53	0.45	0.75[b]	0.51	0.90	30.0	Yang et al. (2010), Zheng and Hryciw (2016), Altuhafi et al. (2018)
Fraser River	0.30	2.40	0.43	0.83	0.68	1.00	33.0	Uthayakumar and Vaid (1998), Sukumaran and Ashmawy (2001), Gao et al. (2014)
Ham River	0.30	1.59	0.45	0.65[b]	0.59	0.92	32.0	Coop and Lee (1993), Jovicic and Coop (1997), Zheng and Hryciw (2016)
Lausitz	0.25	3.09	0.51	—	0.44	0.85	32.2	Herle and Gudehus (1999), Zheng and Hryciw (2016)
Leighton Buzzard	0.78	1.27	0.75	0.80[b]	0.51	0.80	30.0	Thurairajah (1962), Lings and Dietz (2004), Zheng and Hryciw (2016)
Longstone	0.15	1.43	0.30	0.65[b]	0.61	1.00	32.5	Tsomokos and Georgiannou (2010), Zheng and Hryciw (2016)
M31	0.28	1.54	0.62	0.70[b]	0.53	0.87	30.2	Tsomokos and Georgiannou (2010), Zheng and Hryciw (2016)
Monterey No. 0	0.38	1.58	—	0.89[c]	0.53	0.86	32.8	Riemer et al. (1990), Altuhafi et al. (2013)
Ohio Gold Frac	0.62	1.60	0.43	0.83	0.58	0.87	32.5	Han et al. (2018), Ganju et al. (2020)
Ottawa Graded	0.31	1.89	0.80	0.90	0.49	0.76	29.5	Carraro et al. (2009) and unpublished research
Ottawa 20–30	0.72	1.18	0.72	0.88	0.50	0.74	29.2	Han et al. (2018)
Q-Rok	0.63	1.50	0.40	0.73	0.70	1.03	33.0	Unpublished research
Sacramento River	0.30	1.80	—	0.88[c]	0.53	0.87	33.2	Riemer et al. (1990), Altuhafi et al. (2013)
Ticino	0.58	1.50	0.40	0.80[b]	0.57	0.93	33.0	Bellotti et al. (1996), Cho et al. (2006), Altuhafi et al. (2016)
Toyoura	0.17	1.70	0.35	0.65[b]	0.60	0.98	31.6	Verdugo and Ishihara (1996), Loukidis and Salgado (2009), Zheng and Hryciw (2016)

[a] Obtained from direct shear test results.
[b] Width-to-length ratio sphericity (Mitchell and Soga 2005; Zheng and Hryciw 2015).
[c] Perimeter sphericity (Altuhafi et al. 2013).
Note: D_{50}, mean particle size; C_U, coefficient of uniformity; e_{min}, minimum void ratio; e_{max}, maximum void ratio; R, roundness; S, diameter sphericity (unless otherwise indicated); and ϕ_c, critical-state friction angle of sand in triaxial compression (unless otherwise indicated).

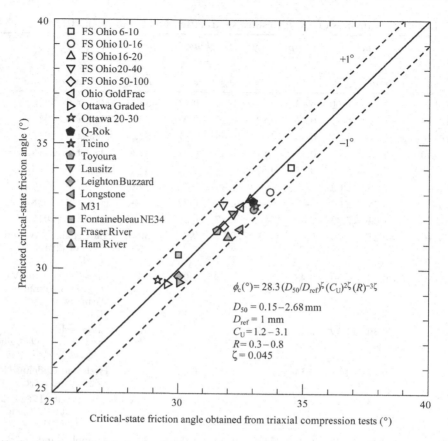

Figure 5.13 Comparison of critical-state friction angles obtained from Equation (5.4) and triaxial compression tests on poorly graded, clean silica sands (Sakleshpur et al. 2021).

except in projects with the budget to fund sophisticated methods of sample recovery from the project site. For the vast majority of projects, recovering undisturbed samples of sands is not possible: when samplers are introduced into the ground, loose sands contract, and dense sands dilate, so that the soil that enters the sampler is in a condition that differs from that in which it exists in the field. Any cementation the soil might have will also likely be completely or partially destroyed after sampling operations. It follows that, for natural sandy soils, the most practical approach is to estimate shear strength from the results of *in situ* tests, particularly the standard penetration test or the cone penetration test, both of which are discussed in Chapter 7.

5.4 SOURCES OF DRAINED SHEAR STRENGTH

5.4.1 Variables affecting the shear strength of sand

There are two types of variables on which the shear strength of soil depends: state variables and intrinsic variables. The soil state of a soil element is the physical state in which the soil exists. The variables that fully define the soil state for our purposes are the void ratio or relative density of the soil, the stress state, the soil fabric, and the interparticle cementation, if it exists. Intrinsic variables relate to soil composition, such as particle mineralogy, size, angularity, and particle surface roughness. We will discuss each of these in turn.

5.4.2 Soil State variables

5.4.2.1 Relative density or void ratio

Relative density, defined in Chapter 3 as

$$D_R = \frac{e_{max} - e}{e_{max} - e_{min}} \tag{3.6}$$

and expressed either as a number between 0 and 1, or in percentage terms between 0% and 100% is clearly the most important of the soil state variables. It is the best index available to characterize the influence of sand density on friction angle. Obviously, all other things being equal, ϕ_p increases with relative density. This increase is entirely due to increased dilatancy effects, as ϕ_c is independent of D_R.

The values of e_{max} and e_{min} are dependent on the method used to obtain them. Correlations based on relative density should be scrutinized for information on how e_{max} and e_{min} were determined since these numbers are critical to the evaluation of D_R. If either this information is not available or the methods used are not standard, the correlation should probably not be used. An alternative to relative density is the use of void ratio or the state parameter.

5.4.2.2 Effective confining stress

The peak friction angle decreases with increasing confining stress because dilatancy decreases with increasing confining stress. A larger amount of work needs to be done by the applied shear stress to expand the soil against a higher confining stress. This makes it gradually more attractive for particles to break and crush locally at particle contacts instead of climbing over each other for shearing to take place. So the contribution of dilatancy to the shear strength of a sand at a given void ratio decreases gradually with increasing confining stress until it completely disappears at a sufficiently high confining stress.

Figure 5.14 shows that, for a sand at a given value of relative density, the decreasing dilatancy with increasing confining stress leads to a curved peak strength envelope, corresponding to a decreasing peak friction angle. Recall that each point of the shear envelope represents the effective normal and shear stresses acting on the shear plane for that effective normal stress. The figure then shows how the peak friction angle ϕ_{p0} can be determined

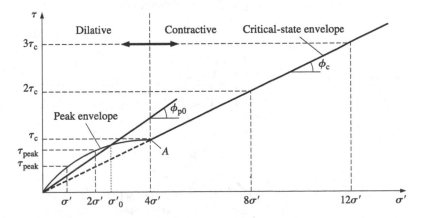

Figure 5.14 A shear envelope for a dilatant sand at a given value of relative density, in which the friction angle decreases with increasing confining stress until the critical-state line is reached.

for a given normal effective stress σ_0'. The shear envelope in Figure 5.14 contrasts with the straight Mohr–Coulomb envelope that a material with a constant ϕ would have (refer back to Figure 4.13). Note, however, that if the confining stress is increased to a sufficiently large value, the dilatancy is completely suppressed (point A in Figure 5.14), the dilatancy angle becomes zero from then on, and the failure envelope becomes a straight line with slope determined by the friction angle ϕ_c. This straight line is the representation of the CSL in $\sigma'-\tau$ space.

5.4.2.3 Soil fabric

The *soil fabric* is the way in which soil particles are arranged. The fabric of natural soils results from the way in which the soil has formed. River sand, for example, has a different fabric from wind-deposited sand. If fines are present in sufficient quantity to separate the sand particles, the soil behavior becomes very different, controlled by the fines. The fabric of such a soil can be referred to as a floating fabric (Salgado et al. 2000). Typically, fines contents greater than 15%–20% tend to float sand particles in a matrix of fine particles. Determination of the fines content of sandy soils is essential for determining whether the soil has a floating fabric, in which case great focus must be placed on characterizing the fines. This will be discussed further later in this section (Figure 5.15).

Soil fabric differs from soil structure. Soil structure is a term that encompasses both fabric and interparticle forces or interactions (Mitchell and Soga 2005). For example, two sands may have the same fabric, but one sand may be slightly cemented, so that the structure of these two soils is different. Obviously, the cemented sand tends to be stiffer and stronger. A feature of cementation is that it may break down during shearing, so any benefits to stiffness or strength that are available initially may be lost if the applied stresses (and the resulting strains) are sufficiently large.

Soil fabric evolves during loading of a soil element. Particles tend to reorient as the stress state in the soil element evolves. In modeling soil behavior, it is important to quantify fabric and to capture its evolution. Fabric can be described by a second-rank tensor: the *fabric*

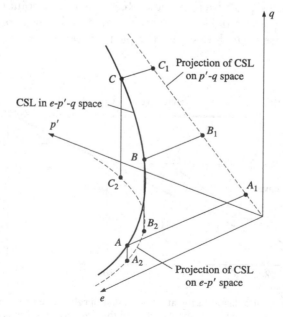

Figure 5.15 Critical-state line in e–p'–q space.

tensor. The fabric tensor for sand is usually defined in terms of the direction normal to the plane tangent to a contact between two particles (Oda 1982). If all the contact normals are known for all particle contacts within a soil element, then the fabric tensor can be computed (Loukidis and Salgado 2009; Dafalias et al. 2004). A soil that has a preponderance of vertical or subvertical contacts, for example, will have a different fabric tensor and will respond differently to loading from one in which such contacts are subhorizontal. For clays, the fabric tensor is usually expressed in terms of particle orientation, but it expresses the same general physical concept (Chakraborty et al. 2013; Loukidis and Salgado 2009; Woo and Salgado 2015).

5.4.2.4 Cementation

Cementation in sands is a result of precipitation of one or more of various cementing agents (including clays, oxides, and even amorphous silica) in the pores of the soil. Strongly cemented sand behaves more like rock, but lightly cemented sand is indistinguishable from a soil in that the same type of laboratory and *in situ* tests can be done on them as in uncemented sands and the same type of construction techniques can also be used. The most characteristic feature of cemented sand is cementation breakdown, which happens as a result of both shear and volumetric strains. So, while a cemented sand will have a true cohesive component of shear strength at the beginning of loading, this cohesive strength will degrade with increasing strain until, at large strains, the cementation is completely broken down and the sand behaves as if it were uncemented.

The breakdown of cementation in sand is not easy to model and therefore not easy to treat with routine approaches. We will not therefore discuss it further in any detail. We will just indicate that, at the same relative density, a cemented sand will have a much larger small strain stiffness and initial shear strength than an uncemented sand under the same initial confining stress. Both sands, however, will approach in essence the same critical state at large strains, when cementation is completely degraded. So the differences between cemented and uncemented sands may be large at small deformations, but are small to negligible at large deformations.

5.4.2.5 Aging

There is considerable evidence that freshly deposited soils have lower small-strain stiffness than older soils. This has been speculated to be the result of soil aging. While aging is not yet well understood, one possible aging mechanism would be local cementation at the contacts between soil particles due to the large stresses present there or very local, micro-crushing of contacts. The contact area between soil particles is very small; this leads to very large stresses even if the force between the particles is not large. A very large stress will cause plastic deformation and, as one hypothesis goes, some welding of one grain to the other, that is, cementation. Any such cementation happens faster initially and then decreases as the contact area increases. It imparts a cohesive strength component to the soil that is very brittle and usually very small.

5.4.3 Intrinsic factors: factors related to the nature and characteristics of the soil particles

5.4.3.1 Mineral composition

Mineral composition has an effect on ϕ_p both through dilatancy and through ϕ_c. Silica sands are sands composed mainly of particles made of silicate minerals. When silica sands are compared among themselves, mineral composition has a modest effect on shear strength.

For example, take feldspar and quartz. The interparticle friction angle ϕ_μ (as estimated from sliding one sheet of mineral on another sheet of the same mineral) for these minerals is (Mitchell and Soga 2005; Rowe 1962)

$$\phi_\mu = \begin{cases} 22° \text{ to } 24° \text{ for pure quartz} \\ 36° \text{ to } 38° \text{ for pure feldspar} \end{cases}$$

Although the difference in ϕ_μ for quartz and feldspar is 14°, the difference in ϕ_c between a pure quartz sand and a pure feldspar sand is only 2°–4°. This is so because particle morphology, the other major contributor to shear strength in the critical state, dominates over the pure, surface-to-surface friction. Additionally, the difference in rolling friction is not as much as in sheet-on-sheet friction. The difference of 2° to 4° is also observed in comparisons of peak friction angle ϕ_p for pure quartz and pure feldspar sands, suggesting that their dilatancy response is similar.

The presence of mica in sands can lower ϕ_p by as much as 4° compared with clean silica sand. Mica appears in soil as sand-sized platelets that facilitate particle sliding, thereby lowering ϕ_p. The presence of some mica particles is not uncommon in residual soils of granite and similar rocks.

Although we have seen that the differences are small when the sand is made of silicates, if we compare silica sands with sands with radically different mineral compositions, friction angles will likewise be quite different. For example, the nature of carbonate sand particles is such that interparticle friction, resistance to particle rearrangement, and crushability are all greater for carbonate sands than for silica sands. The larger interparticle friction and resistance to rearrangement imply a larger ϕ_c, which is indeed observed. While ϕ_c for silica sands is in the 28°–36° range, it is typically in the 37°–44° range for carbonate sands. The higher crushability of carbonate sands, in turn, implies a lower contribution of dilatancy to ϕ_p than is observed in silica sands. Despite the lower dilatancy, the peak friction angle of carbonate sands is still typically larger than that of a silica sand at the same density and confining stress because of the larger ϕ_c.

5.4.3.2 Particle morphology

As discussed in Chapter 3, particles of sandy soils exist in a variety of shapes. The main variables describing particle shape are sphericity and roundness. In general, the lesser the roundness and the lesser the sphericity, the greater the ϕ_c and the greater the ϕ_p. Angular particles resist rearrangement more than rounded particles. They also tend to interlock more, increasing the dilatancy component of ϕ_p for low to moderate confining stresses. At large confining stresses, however, there is more particle breakage in soils containing angular rather than rounded particles, reducing and even reversing the shear strength advantage of soils with angular particles.

Particle surface roughness is of limited importance in most cases. Soil particles are small and convex, the contact areas are small, and the opportunity for surface peaks and valleys of one particle to interlock with those of an adjoining particle is limited.

5.4.3.3 Particle size and soil gradation (grain size distribution)

Well-graded soils tend to be somewhat less compressible and more dilative than poorly graded soils, as small and large particles are more efficiently packed, with smaller particles occupying the voids and providing additional interlocking and support to larger particles.

Both ϕ_c and ϕ_p tend to be higher for well-graded soils. Salgado et al. (2000) showed that this holds true for silty sands with up to 10 to 15% silt content.[9] Addition of nonplastic silt to clean sand, leading to more well-graded soils, increases both ϕ_c and ϕ_p, as long as the silt particles do not float the sand particles. Plastic fines, however, may decrease both ϕ_p and ϕ_c by facilitating sliding.

The behavior of sandy soils containing fines is quite different if the fines content is sufficiently high to float the sand particles. The percentage of fines required for leading to such floating fabric depends on the relative density of the soil. Salgado et al. (2000) showed that a silty sand would have to be looser than $D_R = 3\%$ to have a floating fabric for a fines content of 5%, but that a $D_R < 37\%$ would produce a floating fabric for a fines content of 15%. The choice of values of shear strength to use in calculations for sands with more than 15% fines must be made with caution.

5.4.3.4 Presence of water

The presence of water does not affect the friction angle significantly in most cases. Exceptions, as noted in Lawton (2001), include soils in which the particles contain cracks or are brittle, in which case the presence of water can induce more crushing. It may have a very slight lubricant effect, but that is not detectable from comparisons of shear strength test data from saturated sand samples tested under CD conditions to data from tests on dry samples. Water does reduce the effective normal stress on the failure plane through pore pressures, which reduce shear strength; however, that is accounted for explicitly through the principle of effective stress, discussed in Chapter 3.

5.4.4 Loading path

The best way to visualize the importance of the loading path is to evaluate three shear strength tests that have been extensively discussed in the literature: the triaxial compression test, the triaxial extension test, and the plane-strain compression test. Figure 5.16 shows plan and side views of the triaxial and plane-strain test samples.

The confining stress, as expressed by the mean effective stress, is given by

$$p' = \sigma'_m = \frac{\sigma'_1 + \sigma'_2 + \sigma'_3}{3} \tag{5.5}$$

In the triaxial compression test, $\sigma'_1 > \sigma'_2 = \sigma'_3$, and

$$p' = \sigma'_m = \frac{\sigma'_1 + 2\sigma'_3}{3} \tag{5.6}$$

In the triaxial extension test, $\sigma'_1 = \sigma'_2 > \sigma'_3$ and

$$p' = \sigma'_m = \frac{2\sigma'_1 + \sigma'_3}{3} \tag{5.7}$$

In plane-strain loading, the soil element cannot deform in one of the principal directions, but can do so in the other two. In a plane-strain test, the stress in one of these two

---

[9] Some of the effect may be due also to increased angularity because the silt used in the study was more angular than the sand matrix.

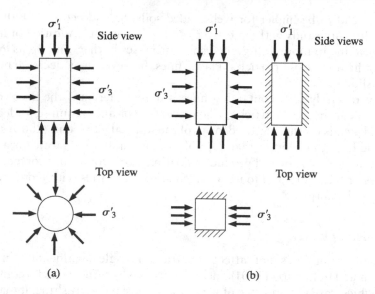

Figure 5.16 Plan view and cross section of (a) the triaxial test boundary conditions and (b) the plane-strain boundary conditions.

free directions (σ_3') remains the same throughout the test, whereas the axial stress (σ_1') is increased; as a result, the sample contracts vertically and tries to expand horizontally. It expands only in the direction of application of σ_3'. In the direction in which no normal deformation is allowed, the soil element tries to expand, but, since that is not possible, an increase in the normal stress results, so $\sigma_2' > \sigma_3'$.

Consider two samples of the exact same soil with exactly the same relative density, subjected to exactly the same initial stress σ_3' and to the same initial conditions. The peak friction angle ϕ_p is defined based on σ_1' and σ_3' at peak strength for both the triaxial and plane-strain compression tests using Equation (5.1). However, due to the larger σ_2' value for the plane-strain compression test, the mean effective stress σ_m' is larger under plain-strain conditions than under triaxial conditions for a given σ_3'. This means that the shear strength of the soil, the principal effective stress ratio σ_1' / σ_3', and consequently the value of ϕ_p calculated from σ_1' and σ_3' for the plane-strain test are higher than the corresponding values for the triaxial test. This difference increases with increasing dilatancy because the constraint in one of the principal directions in the plane-strain test makes it harder for the soil to expand, increasing the energy required for dilation.

In many soil mechanics problems, including the loading of strip footings, slopes, and retaining structures, boundary conditions can usually be best approximated as being plane-strain boundary conditions. This is because these structures are often long, and the normal strain perpendicular to any cross section located in central parts of these structures must therefore be equal to zero because of symmetry. When laboratory tests have been relied on for the estimation of friction angles, unconfined compression, direct shear, or triaxial compression tests have been used, ignoring the additional strength that is often available in these real-life problems that approximate plane-strain conditions. These tests are used simply as a consequence of the greater availability of equipment to perform them and the ease in performing them. Additionally, the resulting friction angles are smaller than plane-strain friction angles, so their use in problems that are in truth plane-strain problems is conservative.

5.5 REPRESENTATION OF DRAINED SHEAR STRENGTH OF SANDS

5.5.1 The Bolton correlation for the friction angle

We saw earlier that shear strength in sand has a component due to interparticle friction (the critical-state shear strength) and another due to dilatancy. The same, as a consequence, is true of the friction angle and of the flow number, calculated from the friction angle using Equation (4.45). De Josselin De Jong (1976) showed that this can be expressed mathematically for plane-strain conditions as

$$N = N_c N_d \tag{5.8}$$

where the dilatancy number N_d and the critical-state flow number N_c are given by

$$N_d = \frac{1+\sin\psi}{1-\sin\psi} \tag{5.9}$$

$$N_c = \frac{1+\sin\phi_c}{1-\sin\phi_c} \tag{5.10}$$

with the sine of the dilatancy angle given by Equation (4.35) with $k=1$, which becomes:

$$\sin\psi = -\frac{d\varepsilon_v}{d\varepsilon_1 - d\varepsilon_3}$$

Bolton (1986) examined a large number of triaxial compression and plane-strain compression tests and concluded that, for both types of loading, the following relationship held:

$$-\left(\frac{d\varepsilon_v}{d\varepsilon_1}\right)_p = 0.3I_R \tag{5.11}$$

where the subscript p indicates that the quantity is measured or calculated at the point at which peak strength is reached, and I_R is a variable called the relative dilatancy index, defined for peak strength as

$$I_R = I_D\left[Q - \ln\left(\frac{100\sigma'_{mp}}{p_A}\right)\right] - R_Q \tag{5.12}$$

where $I_D=D_R/100$=relative density, ranging from 0 to 1; Q, R_Q=fitting parameters that depend on the intrinsic characteristics of the sand; p_A=reference stress=100 kPa=0.1 MPa$\approx$1 tsf$\approx$2000 psf; and σ'_{mp}=mean effective stress at peak shear strength, given by

$$\sigma'_{mp} = \frac{1}{3}\left(\sigma'_{1p} + \sigma'_2 + \sigma'_3\right) \tag{5.13}$$

For a triaxial compression test, $\sigma'_2 = \sigma'_3 = \sigma'_c$ throughout the test, so

$$\sigma'_{mp} = \frac{1}{3}\left(\sigma'_{1p} + 2\sigma'_3\right) \tag{5.14}$$

For a plane-strain compression test, for which $\sigma_3' = \sigma_c'$ throughout the test but σ_2' varies, Davis (1968) proposed the following equation for σ_2':

$$\sigma_2' = \mu_{ps}\left(\sigma_{1p}' + \sigma_3'\right) \tag{5.15}$$

where:

$$\mu_{ps} = \frac{1}{2}\left(1 + \sin\phi_p \sin\psi_p\right) \tag{5.16}$$

Taking Equation (5.15) into Equation (5.13),

$$\sigma_{mp}' = \frac{1}{3}\left(1 + \mu_{ps}\right)\left(\sigma_{1p}' + \sigma_3'\right)$$

for plane-strain conditions. Note that μ_{ps} varies, in theory, from 0.5 to 1 (if both the sines in Equation (5.16) are equal to 1), but, for the maximum values of friction and dilatancy angles that may be found in realistic problems, it would not exceed 0.7 and would typically be close to 0.5. Some experimental evidence in fact exists that μ_{ps} may actually be slightly lower than 0.5.[10] So, in practice, if we take

$$\sigma_{mp}' = \frac{1}{2}\left(\sigma_{1p}' + \sigma_3'\right) \tag{5.17}$$

for calculating the mean effective stress at peak strength for plane-strain conditions, we will be close to the exact value of it.

Recalling from Chapter 4 that

$$k = \begin{cases} 1 \text{ for plane-strain conditions} \\ 2 \text{ for triaxial conditions} \end{cases} \tag{4.36}$$

we can consolidate the expressions for the mean effective stress into a single expression by using $k=1$ for plane-strain compression and $k=2$ for triaxial compression:

$$\sigma_{mp}' = \frac{1}{k+1}\left(\sigma_{1p}' + k\sigma_3'\right) \tag{5.18}$$

and we can also rewrite Equation (4.35) as

$$\sin\psi = -\frac{d\varepsilon_1 + k d\varepsilon_3}{d\varepsilon_1 - k d\varepsilon_3} = -\frac{d\varepsilon_v}{d\varepsilon_1 - k d\varepsilon_3} = -\frac{\dfrac{d\varepsilon_v}{d\varepsilon_1}}{2 - \dfrac{d\varepsilon_v}{d\varepsilon_1}} \tag{4.35}$$

The last equality in Equation (4.35) tells us that there is a unique relationship between ψ and $d\varepsilon_v/d\varepsilon_1$ regardless of whether we have plane-strain or triaxial compression. Taking Equation (5.11) into Equation (4.35) leads to:

$$\sin\psi_p = \frac{0.3 I_R}{2 + 0.3 I_R} = \frac{I_R}{6.7 + I_R} \tag{5.19}$$

---

[10] This is likely due to the restrictive assumption of perfect plasticity made by Davis (1968).

an expression also proposed by Schanz and Vermeer (1996) that shows that there is a one-to-one relationship between ψ_p and I_R regardless of the loading path (or, more specifically, regardless of whether we have triaxial compression or plane-strain compression conditions).

Bolton (1986) also codified the dependence of the peak friction angle ϕ_p on intrinsic soil variables and soil state variables for both triaxial and plane-strain compression conditions. He found that the following equation quantified the peak friction angle very well for both plane-strain and triaxial compression conditions:

$$\phi_p = \phi_c + A_\psi I_R = \phi_c + \left[5 - 2(k-1)\right]I_R \tag{5.20}$$

where, as before, $k=1$ for plane-strain and 2 for triaxial compression conditions, and so $A_\psi=3$ for triaxial conditions and $A_\psi=5$ for plane-strain conditions.[11]

Bolton (1986) also found that, for plane-strain conditions, he could use the following equation to calculate approximate values of ϕ_p:

$$\phi_p = \phi_c + 0.8\psi_p \tag{5.21}$$

What Equations (5.20) and (5.21), considered together, suggest is that, for plane-strain compression, $0.8\,\psi_p$ is equivalent to $\phi_p - \phi_c = A_\psi I_R$ with $A_\psi = 5$. If we now consider triaxial compression under conditions leading to the same value of I_R, knowing that $A_\psi = 3$, we should have the difference between ϕ_p and ϕ_c be 60% of the value of this difference for plane-strain compression, or $0.6 \times 0.8 = 0.48\psi_p$. So, for triaxial compression, instead of Equation (5.21), we use the following equation to approximate ϕ_p for a known peak dilatancy angle:

$$\phi_p = \phi_c + 0.48\psi_p \approx \phi_c + 0.5\psi_p \tag{5.22}$$

We can generalize:

$$\phi_p = \phi_c + B_\psi \psi_p = \phi_c + \left[0.8 - 0.3(k-1)\right]\psi_I \tag{5.23}$$

where $B_\psi = 0.5$ for triaxial conditions and 0.8 for plane-strain conditions. Between Equations (5.20) and (5.23), Equation (5.20) is preferable because it gives more accurate, reliable estimates of ϕ_p. If we need an estimate of ψ_p, then we should use Equation (5.19) directly.

Bolton (1986) found $Q=10$ and $R_Q=1$ produce peak friction angle values that match rather well the results of a large number of laboratory tests on many different clean silica sands. Figure 5.17 shows the scatter of experimental results around Equation (5.20) observed by Bolton (1986) for both triaxial and plane-strain compression using $Q=10$ and $R_Q=1$ for a variety of sands.

Chakraborty and Salgado (2010) extended the stress range of the Bolton peak friction angle correlation to very low stresses. If R_Q is set to 1, the values of Q tend to be lower at low confining stresses, meaning that the materials will still be more dilative than at higher confining stresses, but less so than would be predicted by the original Bolton (1986) correlation.

Salgado et al. (2000) and Carraro et al. (2003) found that the presence of a small amount of silt can increase dilatancy and peak friction angles and proposed the values of Q and R_Q presented in Table 5.2 for silty sands. Carraro et al. (2009) studied the drained response of clayey sand. Based on those studies, Q and R_Q are also presented in Table 5.2 for clayey sands. The table indicates that the fit using Equation (5.20) (or Equation (5.23)) is again very good (having high r^2 for all mixtures).

---

[11] Although there is experimental evidence that ϕ_c for plane-strain compression is greater than for triaxial compression (by as much as 5°), Bolton (1986) assumed the two values to be the same.

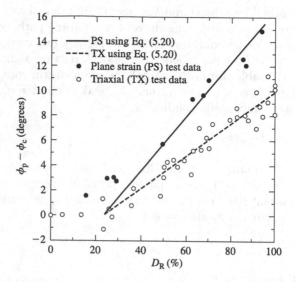

Figure 5.17 Illustration of fit of Equation (5.20) to triaxial and plane-strain compression test results. (Data from Bolton 1986.)

Table 5.2 Values of Q and R_Q for Ottawa Sand with various percentages by weight of nonplastic silt and kaolinite clay

Percentage of fines	Best fit		
	Q	R_Q	r^2
0%	9.9	0.86	0.95
5% Silt	9.1	−0.33	0.99
10% Silt	9.3	−0.30	0.98
2% Clay	12.1	2.78	0.96
5% Clay	11.7	3.17	0.95
10% Clay	10.9	3.43	0.80

r^2 = coefficient of determination.

Example 5.2

Estimate the peak friction angle of a clean sand with a relative density of 50% tested in triaxial compression under an effective consolidation stress of 100 kPa. The sand is known to have a critical-state friction angle of 30°, $Q=10$, and $R_Q=1$.

Solution:

We know $\phi_c = 30°$. Determining the peak friction angle is an iterative process until values converge. Given that the sand is medium dense and the confining stress is low, we expect a fair amount of dilatancy. Let's make an initial assumption that $\phi_p = 36°$. To use Bolton's equation, we must first know the stresses acting on the sample. We know $\sigma_3' = 100$ kPa, so we need to find σ_1' and σ_{mp}'.

Using Equation (4.45) for the peak friction angle:

$$N_p = \frac{1 + \sin\phi_p}{1 - \sin\phi_p} = \frac{1 + \sin 36°}{1 - \sin 36°} = 3.85$$

Now, using Equation (4.48),

$$\sigma'_{1p} = N_p \sigma'_3 = 3.85(100) = 385\,\text{kPa}$$

Equation (5.6) gives us σ'_{mp}:

$$\sigma'_{mp} = \frac{\sigma'_{1p} + 2\sigma'_3}{3} = \frac{385 + 2(100)}{3} = 195\,\text{kPa}$$

Substituting into Equation (5.12):

$$I_R = I_D\left[Q - \ln\left(\frac{100\sigma'_{mp}}{p_A}\right)\right] - R_Q$$

$$= 0.5\left[10 - \ln\left(\frac{100(195)}{100}\right)\right] - 1$$

$$= 1.36$$

Finally, we use Equation (5.20):

$$\phi_p = \phi_c + A_\psi I_R$$

$$= 30 + 3 \times 1.36$$

$$= 34.1°$$

We see that our result of 34.1° is not close to our initial assumption of $\phi_p = 36°$. We can use our result as our next assumption on the value of ϕ_p:

From Equation (4.45):

$$N_p = \frac{1 + \sin\phi_p}{1 - \sin\phi_p} = \frac{1 + \sin 34.1°}{1 - \sin 34.1°} = 3.55$$

From Equation (4.48):

$$\sigma'_{1p} = N_p \sigma'_3 = 3.55(100) = 355\,\text{kPa}$$

From Equation (5.6):

$$\sigma'_{mp} = \frac{\sigma'_{1p} + 2\sigma'_3}{2} = \frac{355 + 2(100)}{3} = 185\,\text{kPa}$$

From Equation (5.12):

$$I_R = I_D\left[Q - \ln\left(\frac{100\sigma'_{mp}}{p_A}\right)\right] - R_Q$$

$$= 0.5\left[10 - \ln\left(\frac{100(185)}{100}\right)\right] - 1$$

$$= 1.39$$

Using Equation (5.20):

$$\phi_p = \phi_c + A_\psi I_R$$

$$= 30 + 3 \times 1.39$$

$$= 34.2°$$

Our assumed and calculated values of ϕ_p are sufficiently close, so we can take our answer to be:

$$\phi_p = 34.2° \text{ }^{\text{answer}}$$

The calculation shown in Example 5.2 can be done for wide ranges of density and stress so that a chart can be prepared for ϕ_p as a function of D_R and consolidation stress σ'_3. Figure 5.18 shows charts that may be used to estimate ϕ_p given values of D_R and consolidation stress σ'_c. Figure 5.18a shows the values of ϕ_p in triaxial compression, and Figure 5.18b shows the values of ϕ_p in plane-strain compression. Note that ϕ_p may be less than ϕ_c in Figure 5.18. Negative values of $\phi_p - \phi_c$ mean the sand is very contractive, and a design value of ϕ would be taken as less than ϕ_c as an indirect way of preventing extremely large deformations (as suggested by Bolton 1986). However, it is also an option, and conceptually more correct, to make ϕ_c a floor for the value of ϕ_p, which means that, when the calculations or Figure 5.18 produces negative $\phi_p - \phi_c$, we would make $\phi_p = \phi_c$. In a design context, if one is checking for deformations using a separate analysis, then one could choose to not allow ϕ_p to fall below ϕ_c in stability calculations, as there is no need to indirectly limit deformations by using a lower ϕ_p value.

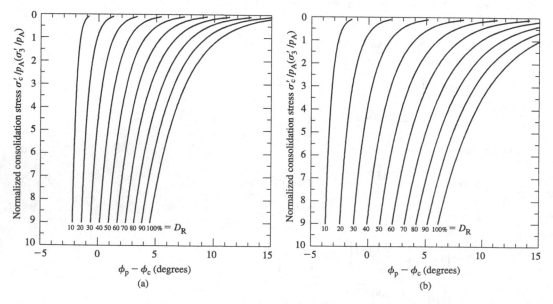

Figure 5.18 Peak friction angle in (a) triaxial compression and (b) plane-strain compression as a function of stress state, relative density, and critical-state friction angle using $Q=10$ and $R_Q=1$.

Example 5.3

Solve Example 5.2 using Figure 5.18.

Solution:

We know the confining stress (lateral effective stress) is 100 kPa. Given that $p_A = 100$ kPa, this corresponds to a normalized value, shown in the figure, of 1. The relative density is 50%. So we draw a horizontal line from a vertical axis value of 1 (corresponding to 100 kPa) until it intersects the $D_R = 50\%$ line in Figure 5.18 and project it down from there to the $\phi_p - \phi_c$ axis to read a value of approximately 4°.
So:

$$\phi_P - \phi_c = 4°$$

$$\phi_P = 30° + 4° = 34° \text{ answer}$$

which is consistent with the answer found analytically in Example 5.2.

Example 5.4

A level, clean sand deposit has an average unit weight of 19 kN/m³ in the upper 8 m. Estimate the peak friction angle at a depth of 8 m if the water table is located at a depth of 10 m. The relative density is estimated to be equal to 65% at 8 m depth, and the critical-state friction angle of the sand is 32°.

Solution:

The solution to this problem can be easily attained by using Figure 5.18.
The vertical effective stress is found to be:

$$8 \times 19 = 152 \text{ kPa}$$

Assuming K_0 to be 0.45, then the lateral effective stress follows directly:

$$\sigma'_h = 152 \times 0.45 = 68.4 \text{ kPa}$$

Using Figure 5.18, we determine where $\sigma'_h = 68$ kPa intersects the $D_R = 65\%$ line. Now we project that down to the $\phi_p - \phi_c$ axis and read off a value of approximately 6.8°.
So:

$$\phi_p - 32° = 6.8°$$

and

$$\phi_p = 38.8° \text{ answer}$$

A more interesting question is: if we could recover an "undisturbed" sample of sand and test it in the laboratory under triaxial compression, what stress would we reconsolidate it to? Would we reconsolidate it to $\sigma'_3 = \sigma'_c = \sigma'_h$ before we sheared it? Or to σ'_m or even σ'_v? We address this topic in detail when we discuss clays, where it is much more pertinent because of the possibility of recovering samples that are only moderately disturbed and testing them in the laboratory. Let us investigate the impact on the value of ϕ_p of these other two options.

Example 5.5

Redo Example 5.4 assuming $\sigma_3' = \sigma_v'$ and $\sigma_3' = \sigma_m'$.

Solution:

We carry forward, from Example 5.4, the following:

$$D_R = 65\%$$

$$\phi_c = 32°$$

$$\sigma_v' = 152\,kPa$$

$$\sigma_h' = 68.4\,kPa$$

Entering Figure 5.18 with $\sigma_3' = 152$ kPa: $\phi_p - \phi_c = 5.3°$, which leads to

$$\phi_p = 32° + 5.3° = 37.3° \approx 37°$$

Let us now repeat this calculation for σ_m':

$$\sigma_m' = \frac{\sigma_v' + 2\sigma_h'}{3} = \frac{152 + 2 \times 68.4}{3} = 96.3 \approx 96 \text{ kPa}$$

Entering Figure 5.18 with 96 kPa, we obtain $\phi_p - \phi_c = 6°$, from which:

$$\phi_p = 32° + 6° = 38° \text{ }^{\text{answer}}$$

From the example, we can see that the stress we use to calculate ϕ_p does make a difference, although not a large difference, in the result. We will show later for clays that we should consolidate triaxial test samples to the mean effective stress p' to approximate the friction angle the soil has *in situ*.[12]

5.5.2 Parameters c and ϕ from curve fitting

Some engineers request that a series of triaxial compression tests be performed for a mean effective stress range representative of the design problem at hand, plot the Mohr circles for each test, and draw a straight line approximately tangent to these circles as the strength envelope for the given stress range (Figure 5.19). From the strength envelope fitted to the circles, they will obtain a c and ϕ for the sand. If this is done for sand, it is possible that c takes a value greater than 0, but this does not mean that sand has cohesion (unless the sand is cemented); it means only that a graphical fit produced a positive number for a fitting parameter. These values of c and ϕ are then used in design, but they have no physical meaning. In contrast, ϕ_p as calculated using Equations (5.20) and (5.12) has a physical meaning. It is a secant friction angle, that is, the angle that a line extending from the origin of the $\sigma'-\tau$ space to the point of the peak strength envelope under consideration makes with the horizontal. Figure 5.20 shows that the friction angle a sand of a certain relative density has available at an effective normal stress on the shear plane equal to σ_A' is ϕ_{pA}; that drops to ϕ_{pB} at a greater normal stress $\sigma_B' > \sigma_A'$.

---

[12] We should also be aware that, if triaxial compression friction angles are used, even when doing plane strain analysis of long slopes, strip footings, and other structures for which plane-strain conditions are in effect, the results should be somewhat conservative.

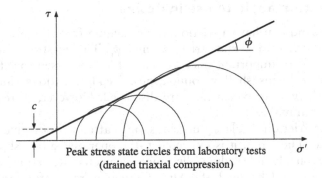

Figure 5.19 Curve-fitting approach to defining the peak-strength ("failure") envelope.

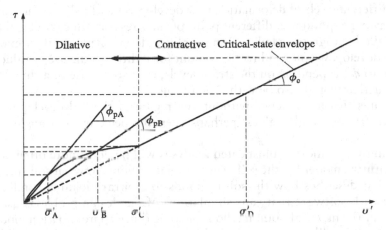

Figure 5.20 The peak friction angle is the secant angle to the stress level of interest. Note how the peak friction angle decreases with increasing confining stress until it eventually becomes equal to the critical-state friction angle.

A very important point regarding curve fitting is that not all geotechnical problems are well suited to it. A key point, as mentioned earlier, is that, in curve fitting, the c and ϕ must be representative of the stress range experienced by the soil. The stresses experienced by the soil near the base of a foundation loaded even to service load conditions can be very large and lie beyond the range of stresses that can be obtained in laboratory strength tests except in very few research laboratories; thus, foundation analysis and design should not, as a rule, be based on the values of c and ϕ curve-fitted to laboratory results. On the other hand, in slope or retaining structure stability limit states, stresses tend not to be large, being in a range easily reproducible in laboratory strength tests. Therefore, curve fitting may be acceptable, even if not preferable, as a way of determining input strength parameters for slope stability analysis or retaining structure analysis and design. The only care that needs to be taken is to assess whether the shear strength determined in this manner will be present for the useful life of the structure. As we will discuss in later chapters, seasonal changes in water content, cycles of freezing and thawing, and uneven mobilization of deformation along slip surfaces may lead to strength degradation; in that case, curve fits for peak-strength Mohr's circles may present risks.

5.5.3 Which friction angle to use in design?

We have seen that sand can have a friction angle ranging from the peak friction angle ϕ_p all the way down to the critical-state friction angle ϕ_c. The question of which ϕ to use in geotechnical design is an important one, which will be addressed specifically for each of the major classes of problems (shallow foundations, deep foundations, slopes, and retaining structures) in the chapters in which they are discussed. However, there are some general statements that we can make.

The critical-state friction angle ϕ_c develops only after considerable deformation has already taken place. Therefore, it is expected to be operative along slip surfaces (within shear bands) where considerable sliding (shearing) is expected. Referring back to the concept of limit states discussed in Chapter 2, if the limit state requires the formation of shear bands that are not constrained by relatively stable soil masses and along which, as a result, large shear strains develop, then the use of ϕ_c in design checks for that limit state is appropriate.

In cases in which a slip surface is extensive and is subjected to different conditions, it may happen that different levels of deformation may develop along the slip surface, in which case each point may correspond to a different point of the stress–strain curve for the sand. This condition is called "progressive failure." In simplified design methods, if we expect progressive failure to develop, we may indirectly account for it through the use of values within the range from ϕ_p to ϕ_c, depending on the strain levels, or even the use of a single "representative" value of ϕ that would also be within that range.

For limit states that are reached without any region of the soil developing large shear strains, higher friction angle values, perhaps even the peak friction angle ϕ_p, might be operative.

In the coming years, more sophisticated analyses will be done more often, and, in those cases, a constitutive model for the soil (a set of relationships between soil state and intrinsic variables that describes how the soil responds to arbitrary loading conditions) will be used together with software that runs simulations of the design problems using the finite element method, the material point method, or some other approximation model. Answers to questions like this will result automatically from the analyses; e.g., we will know where shear bands form (or not) and the operative friction angle there will be a result of the calculations.

5.6 UNDRAINED SHEAR STRENGTH*

Under certain conditions, particularly in the case of seismically loaded foundations or foundations of offshore structures (which can be subjected to loadings that are applied very fast), the foundations are loaded under undrained conditions. Slopes, retaining structures, and other geotechnical systems may also be loaded under undrained conditions. It is then necessary in design to consider the undrained stress–strain response of the sand and its undrained shear strength.

The undrained shearing of sand samples is typically studied in the laboratory using consolidated undrained triaxial (TXCU) tests. If we impose the same total stresses on two identical samples, allowing full drainage in one and no drainage in the other, the samples respond quite differently. When full drainage is allowed, water will either leave or enter the sample as the sample contracts or expands. In undrained loading, a sand tending to contract would experience an increase in pore pressure: the impossibility of a volume reduction loads the pore water in compression, increasing the pore pressure. In contrast, a sand tending to dilate would experience a pore pressure drop in an undrained test.

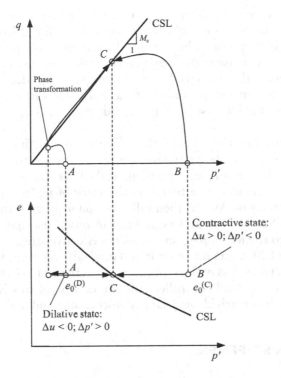

Figure 5.21 Stress path in p'–q space corresponding to the undrained shearing of fully contractive and fully dilative samples in TXCU tests.

Consider now two distinct samples (one fully contractive and one fully dilative) of the same sand, with the same void ratio but different consolidation stresses, sheared under undrained conditions. Figure 5.21 shows the effective stress paths (in p'–q space) of the two samples. Both paths end at the critical-state line, as we would expect. Moreover, because the two samples have the same void ratio, the mean effective stress p' for each sample changes (as a result of pore pressure changes) so that both samples achieve exactly the same critical state (same density and stress state) at the end of the tests. The way they get there, however, is far from the same. To understand the differences between the two effective stress paths, let us review first what makes a stress path move to the right or left. For that, recall the definitions of p, p', q, and q':

$$p = \sigma_m = \frac{1}{3}(\sigma_1 + 2\sigma_3) \tag{4.75}$$

$$p' = \sigma'_m = \frac{1}{3}(\sigma'_1 + 2\sigma'_3) \tag{4.76}$$

$$q = q' = \sigma_1 - \sigma_3 \tag{4.77}$$

The effective stress path moves to the left or right depending on the value of p', which in turn depends on the pore pressure generated by the applied axial stress and on the applied axial stress itself. The following simple rules apply to a triaxial compression test:

1. The effective stress path moves to the right if p' increases; p' increases if
   - the pore pressure decreases (which happens if the soil is dilative) and
   - the pore pressure increases but the total axial stress σ_1 increases more than sufficiently for the increase in σ_1' to exceed twice the decrease in σ_3'.
2. The effective stress path moves to the left if p' decreases; p' decreases if
   - the pore pressure increases, and the total axial stress σ_1 increases less than sufficiently to cause the increase in σ_1' to exceed twice the decrease in σ_3'.

Another factor determining the shape of the effective stress path is the evolution of $q'=q$ during the shearing process. As shown in Figure 5.21, q increases from the beginning to the end of the test on the dilative sample, exceeding, at the later stages of the test, the value of q the soil would have at critical state based on the value of p'. In contrast, for a continuously contractive sample, q peaks and then falls continuously until the stress path reaches the critical-state line (CSL). We can understand this by observing that the height of the CSL decreases as we move to the left in $p'-q$ space. As the effective stress path moves to the left, q rises initially, but the CSL bears down on it and gradually repels it, forcing q to peak and then drop until the effective stress path eventually ends at the CSL. Given that loading in real problems is virtually always load-controlled, the implication of this is that, once the peak undrained shear strength is reached, very large deformations and deflections will develop.

5.7 SMALL-STRAIN STIFFNESS

Soil deformations are recoverable and linear elastic so long as they are due to deformation of the particles themselves, resulting from the stresses experienced at particle contacts. Once the particles are crushed, broken, or dislodged from their original positions, the deformations that develop are irrecoverable or plastic.

Linear elastic soil behavior takes place only at very small deformation levels, corresponding to shear strains typically less than 10^{-5} ($10^{-3}\%$). The shear modulus and Poisson's ratio at these very low strain levels, under drained conditions, are denoted by G_0 and ν_0, respectively, and are given by

$$\nu_0 = 0.05 \text{ to } 0.2$$

and

$$\frac{G_0}{p_A} = C_g \frac{(e_g - e)^2}{1+e} \left(\frac{\sigma_m'}{p_A}\right)^{n_g} \tag{5.24}$$

where p_A is a reference stress ($= 100$ kPa$=0.1$ MPa≈ 1 tsf); $\sigma_m' = (\sigma_v' + 2\sigma_h')/3$ is the mean effective stress after consolidation but before any shearing has taken place; and C_g, e_g, and n_g are constants for a given sand. Equation (5.24) is in the form first proposed by Hardin and Black (1968). These constants can be taken as C_g=650, e_g=2.17, and n_g=0.45 for a typical clean silica sand. For sand containing nonplastic silt, G_0 drops by as much as 50% with respect to the value of G_0 for clean sand as the silt content is increased from 0% to 20% (Salgado et al. 2000). Determination of C_g, e_g, and n_g for a given sand can be done most economically by performing a series of bender element tests (Viggiani and Atkinson 1995a,b; Arulnathan et al. 1998; Salgado et al. 2000), but other methods include the use of strain gauges attached locally to triaxial soil samples or the use of less common devices, such as resonant columns.

Example 5.6

Determine the small-strain shear modulus and Young's modulus for a normally consolidated, typical silica sand at a depth of $10\,\mathrm{m}$ in a level soil deposit. Assume $\nu_0 = 0.15$. The water table is at a depth of $8\,\mathrm{m}$. The sand at that depth is estimated to have $D_R = 60\%$, $e_{max} = 0.8$, and $e_{min} = 0.5$.

Solution:

The void ratio at $10\,\mathrm{m}$ is calculated as

$$e = e_{max} - D_R\left(e_{max} - e_{min}\right) = 0.8 - 0.60(0.8 - 0.5) = 0.62$$

Assuming $K_0 = 0.4$ for normally consolidated conditions and a unit weight of $20\,\mathrm{kN/m^3}$:

$$\sigma'_v = (20)(10) - (2)(9.81) = 180\,\mathrm{kPa}$$

$$\sigma'_h = K_0\sigma'_v = (0.4)(180.4) = 72\,\mathrm{kPa}$$

$$\sigma'_m = 108\,\mathrm{kPa}$$

The small-strain shear modulus now follows directly from Equation (5.24):

$$\frac{G_0}{p_A} = C_g \frac{\left(e_g - e\right)^2}{1 + e}\left(\frac{\sigma'_m}{p_A}\right)^{n_g} = 650\frac{(2.17 - 0.62)^2}{1 + 0.62}\left(\frac{108}{100}\right)^{0.45} = 998$$

$$G_0 = 998 p_A = 998 \times 100 = 99800\,\mathrm{kPa} \approx 100\,\mathrm{MPa}\ ^{\text{answer}}$$

The small-strain Young's modulus can now be calculated as:

$$E_0 = 2\left(1 + \nu_0\right)G_0 = 2 \times 1.15 \times 100 = 230\,\mathrm{MPa}\ ^{\text{answer}}$$

5.8 CHAPTER SUMMARY

5.8.1 Main concepts

A shear stress will cause a sand to shear if it is sufficiently large to provide the energy required to overcome interparticle friction, rearrange the particles during shearing, and dilate against an existing confining stress. All other things being equal, the denser the soil is, the more closely packed the sand particles are, the greater the amount of dilation that must take place for particles to be able to climb over one another in the process of shearing is, and the greater the friction angle of the sand is.

As the **confining stress** increases, it becomes increasingly attractive for particles to crush and break at particle contacts, reducing the need for dilation. So the dilatancy-related component of shear strength increases with increasing relative density and decreases with increasing confining stress. If the confining stress increases far enough, it may completely suppress dilation, and a contractive sand results.

If triaxial tests are performed on two samples of sand at initially the same confining stress, one very dense and thus ultimately **dilative** and one very loose and thus ultimately **contractive**, by the end of the test the contractive sample will have contracted and the dilative sample will have dilated until both converge to the same void ratio at the **critical state**. On the way to the critical state, the dense sample may have initially contracted, becoming dilative only after going through the **phase transformation** point.

The critical state is an equilibrium state at which the sand shears at constant volume, constant shear stress, and constant confining stress. For each different initial confining stress, a different critical-state void ratio exists. These void ratios and their corresponding confining stresses define the **critical-state line** in void ratio versus confining stress space.

The stress–strain curve has notable points or ranges that are always useful in calculations. At very small strains (shear strains less than 10^{-5}), the soil is elastic and the elastic parameters G_0 (the **small-strain shear modulus**) and ν_0 (**small-strain Poisson's ratio**) can be defined. Referring to drained **triaxial compression** loading, as strains increase beyond the initial elastic range, the shear stress increases nonlinearly with shear strain. For relative densities usually observed in practice, the stress peaks at some value of strain and then drops until it reaches a plateau (at which point the sand is said to have reached the critical state). There are accordingly two important values of friction angle for sands: the peak and the critical-state friction angles. The **peak friction angle** is significantly affected by dilatancy, and thus by relative density and confining stress. The **critical-state friction angle** is an intrinsic property of the sand, being completely independent of density and stress.

Poisson's ratio in the linear elastic range is in the 0.05–0.2 range for silica sand, depending on the sand. G_0 can be estimated using the following equation:

$$\frac{G_0}{p_A} = C_g \frac{(e_g - e)^2}{1 + e} \left(\frac{\sigma'_m}{p_A} \right)^{n_g} \tag{5.24}$$

where $\sigma'_m = (\sigma'_v + 2\sigma'_h)/3$ and C_g, e_g, and n_g = constants for a given sand (values of 650, 2.17, and 0.45 can be used if specific information is not available).

The principal effective stress ratio can be expressed in terms of the friction and dilatancy angles as follows:

$$\frac{\sigma'_1}{\sigma'_3} = N = \frac{1 + \sin\phi}{1 - \sin\phi} = \tan^2 \left(45° + \frac{\phi}{2} \right) \tag{5.1}$$

The flow number N is given by

$$N = N_c N_d \tag{5.8}$$

with

$$N_d = \frac{1 + \sin\psi}{1 - \sin\psi} \tag{5.9}$$

The dilatancy angle given in terms of strain increments is:

$$\sin\psi = -\frac{d\varepsilon_v}{d\varepsilon_1 - kd\varepsilon_3} = -\frac{d\varepsilon_1 + kd\varepsilon_3}{d\varepsilon_1 - kd\varepsilon_3} \tag{4.35}$$

with

$$k = \begin{cases} 1 & \text{for plane-strain compression} \\ 2 & \text{for triaxial compression} \end{cases} \tag{4.36}$$

N_c is the value of N at critical state:

$$N_c = \frac{1 + \sin\phi_c}{1 - \sin\phi_c} \tag{5.10}$$

The critical-state friction angle ϕ_c of a sand is in the 28°–36° range. Sands with more rounded, smooth particles with a poorly graded particle size distribution have values near the low end of this range, while sands with angular, rough particles with a well-graded particle size distribution have values near the high end of the range. The peak friction angle ϕ_p may be estimated using:

$$\phi_p = \phi_c + A_\psi I_R = \phi_c + \left[5 - 2(k-1)\right]I_R \tag{5.20}$$

where:

$$I_R = I_D\left[Q - \ln\left(\frac{100\sigma'_{mp}}{p_A}\right)\right] - R_Q \tag{5.12}$$

$$\sigma'_{mp} = \sigma'_{1p} + \sigma'_{2p} + \sigma'_{3p}/3$$

or

$$\sigma'_{mp} = \frac{1}{k+1}\left(\sigma'_{1p} + k\sigma'_3\right)$$

for plane-strain ($k=1$) and triaxial compression ($k=2$), for which σ'_3 is the same throughout loading. For clean sand, Q can be taken as 10 and R_Q as 1.

5.8.2 Symbols and notations

Symbol	Quantity	US units	SI units
A_ψ, B_ψ	Parameters in Bolton's equation	Unitless	Unitless
e_c	Critical-state void ratio	Unitless	Unitless
N_d	Dilatancy number	Unitless	Unitless
N_c	Critical-state flow number	Unitless	Unitless
p'	Mean effective stress	psf	kPa
Q	Fitting parameter for Bolton's equation	Unitless	Unitless
R_Q	Fitting parameter for Bolton's equation	Unitless	Unitless
ε_a	Axial strain	Unitless	Unitless
σ'_c	Effective cell (confining) stress	psf	kPa
σ'_m	Mean effective stress	psf	kPa
τ_c	Critical-state shear strength	psf	kPa
ϕ_c	Critical-state friction angle	deg	deg
ϕ_p	Peak friction angle	deg	deg
ϕ_μ	Interparticle friction angle	deg	deg
ψ_p	Peak dilatancy angle	deg	deg

5.9 PROBLEMS

5.9.1 Conceptual problems

Problem 5.1 Define all the terms in bold contained in the chapter summary.

Problem 5.2* Discuss which of the following materials, when sheared in a triaxial test in exactly the same way, exhibit the greater amount of dilatancy and why. Consider shearing under both very small and very large confining pressures. The materials to consider are: (a) glass beads, (b) hollow glass beads, (c) sand particles, and (d) solid steel spheres.

Problem 5.3* Physically, in the absence of a peak in the stress–strain curve, local crushing and shearing of soil particles at contacts requires less energy than dilation. What does this imply about the physical meaning of the parameter Q in Equation (5.12)?

Problem 5.4 Discuss whether increases in the values of the following quantities have a positive, negative, or neutral impact on the peak friction angle of a sand:
  a. critical-state friction angle ϕ_c;
  b. relative density D_R;
  c. initial mean effective stress;
  d. median particle size D_{50}; and
  e. particle shape (rounded, subrounded, subangular, and angular).

Problem 5.5* Consider a soil with $D_R = 50\%$, $\phi_c = 30°$, $Q = 10$, and $R_Q = 1$. Using Equations (5.12) and (5.20), find the value of σ_1' in drained triaxial compression tests with $\sigma_3' = 100\,\text{kPa}$, $200\,\text{kPa}$, $500\,\text{kPa}$, $1\,\text{MPa}$, $2\,\text{MPa}$, and $3\,\text{MPa}$. Do not allow negative values of I_R. Using the stress analysis relationships of Chapter 4 or Mohr's circles, find the normal and shear stresses on the shear plane for each test. Then plot the shear envelope in $\sigma'-\tau$ space. Comment on the shape of the shear envelope.

Problem 5.6* In the beginning of an undrained triaxial compression test, the soil is predominantly elastic. In the near absence of plastic deformation, the pore pressure increases are due exclusively to increases in the total mean stress σ_m. Using this fact, show that, for an undrained triaxial compression test, the increase in pore pressure is one-third of the increase in σ_1.

Problem 5.7* We have seen that, in conditions found routinely in the practice of geotechnical engineering, the shear stress versus shear strain response of sand has a peak if strain is continuously increased (a process known as strain control). What would happen if stress, instead of strain, were increased all the way to the value corresponding to the peak observed for strain control? Would we still observe a peak? What would happen when we tried to increase the stress beyond the strain-controlled peak stress?

5.9.2 Quantitative problems

Problem 5.8 Two TXCD compression tests were performed on clean samples of Ottawa Sand. The samples were consolidated isotropically to $\sigma_c' = 100\,\text{kPa}$. Determine the peak friction angle and critical-state friction angle. Why does sample 1 have a larger shear strength than sample 2? The data from the tests are provided in P-Table 5.1 and P-Figure 5.1.

Problem 5.9 Three TXCD compression tests were performed on samples of Ottawa Sand with 5% silt by weight. The samples were consolidated isotropically to $\sigma_c' = 100\,\text{kPa}$, $200\,\text{kPa}$, and $400\,\text{kPa}$. (a) Estimate the critical-state friction angle for Ottawa Sand. (b)

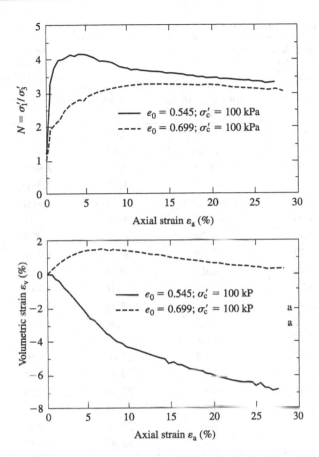

P-Figure 5.1 TXCD compression tests on samples 1 and 2 described in Problem 5.8.

Determine the peak friction angle for each sample. (c) Does the sample with the largest peak strength have the largest peak friction angle? Why or why not? The data from the tests are provided in P-Table 5.2 and P-Figure 5.2.

Problem 5.10 Considering that the minimum and maximum void ratios of the sand of Problem 5.8 are 0.48 and 0.78, use Equations (5.12) and (5.20) to calculate the difference between the peak and critical-state friction angles for the two tests. Do these estimates match well with the test results?

Problem 5.11 A deposit of clean sand has a unit weight equal to 22 kN/m³. The relative density increases approximately linearly from 60% at the surface of the deposit to 75% at a depth of 10 m. K_0 is 0.45 for this deposit. The sand can be assumed to have $\phi_c = 30°$, $Q = 10$, and $R_Q = 1$. Calculate and plot the values of ϕ_p under triaxial and plane-strain compression conditions between 0 and 10 m depth assuming $\sigma'_c = \sigma'_m$. Consider the water table to be very deep (deeper than 10 m).

Problem 5.12 For the deposit of Problem 5.11, calculate the values of the small-strain shear modulus G_0 of the sand and plot them versus depth for the 0–10 m depth range. Assume the sand to have $e_{max} = 0.8$; $e_{min} = 0.48$; and C_g, e_g, and n_g values equal to 650, 2.17, and 0.45, respectively.

Problem 5.13 Repeat Problem 5.11 for the case in which the water table is at the ground surface.

P-Table 5.1 TXCD compression tests on samples 1 and 2 described in Problem 5.8

Sample 1 $e_0=0.545$ $\sigma'_c = 100$ kPa			Sample 2 $e_0=0.699$ $\sigma'_c = 100$ kPa		
Axial strain (%)	Flow number N	Volumetric strain (%)	Axial strain (%)	Flow number N	Volumetric strain (%)
0.00	1.000	0.03	0.00	1.003	0.01
0.50	3.211	0.10	0.50	1.849	0.42
1.01	3.689	−0.20	0.99	2.048	0.61
1.50	3.898	−0.38	1.50	2.300	0.81
3.00	4.061	−1.32	3.00	2.600	1.16
4.50	4.093	−2.23	4.51	2.733	1.40
6.50	3.893	−3.20	6.51	2.976	1.54
8.50	3.772	−3.99	8.50	3.091	1.46
10.50	3.614	−4.43	10.50	3.158	1.34
12.50	3.571	−4.79	12.50	3.216	1.17
14.50	3.509	−5.25	14.50	3.197	1.03
16.50	3.473	−5.48	16.50	3.191	0.90
19.00	3.397	−5.89	19.00	3.165	0.74
22.01	3.332	−6.25	22.00	3.120	0.50
24.50	3.287	−6.43	24.50	3.074	0.46
27.01	3.234	−6.92	27.00	3.020	0.32
27.50	3.236	−6.84	27.51	3.032	0.26
—	—	—	28.00	3.020	0.31
—	—	—	28.47	2.987	0.29

Source: Carraro (2004).

Problem 5.14 Repeat Problem 5.12 for the case in which the water table is at the ground surface.

Problem 5.15 Consider the equation for the peak relative dilatancy index I_R, Equation (5.12). Let us imagine that triaxial compression tests on samples of a given sand with $D_R=40\%$ are performed at increasingly larger confining stresses and that, for $\sigma'_3=1.5$ MPa and $\sigma'_1=4.5$ MPa, the resulting stress–strain diagram does not have a peak (that is, the stress continues to increase until it reaches a plateau at large strains) and $I_R=0$. Considering that the sand has $R_Q=1$, what would be the value of Q implied by this result? What is the value of critical-state friction angle ϕ_c?

Problem 5.16 A drained triaxial compression test is performed on a sample of dry sand prepared in such way as to produce an extremely low relative density. Given the relative density of the sample and the confining stress applied (600 kPa) in the test, the sand is known to be in the contractive range. Knowing that the sample failed when the axial stress was equal to 2100 kPa, estimate the critical-state friction angle of the sand. What is the shear strength at failure?

Problem 5.17 A drained triaxial compression test is performed on a sample of dry sand prepared in such way as to produce a high relative density. Given the relative density of the sample and the confining stress applied (80 kPa) in the test, the sand is known to be in the dilative range. Knowing that the axial stress was equal to 300 kPa when the peak shear stress was observed, estimate the peak friction angle of the sand.

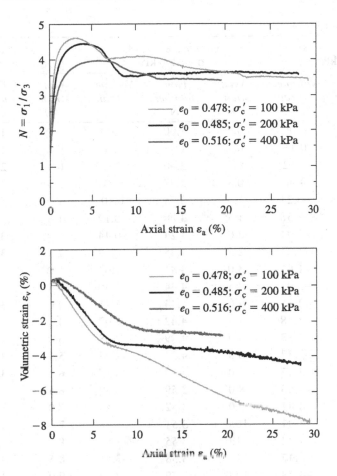

P-Figure 5.2 Flow number N and volumetric strain ε_v versus axial strain ε_a for the three samples of silty sand of Problem 5.9, which were consolidated to $\sigma_c' = 100$, 200, and 400 kPa.

Problem 5.18* A direct shear test was performed on a sample of clean sand at an initial relative density of $D_R = 80\%$. Conditions at peak strength are 100 kPa vertical stress and 90 kPa shear stress applied in the horizontal direction. The sample is dry.
a. What is the peak friction angle based on this test?
b. Determine the magnitudes and orientations of the principal stresses at failure.
c. Is the friction angle consistent with what the Bolton equation – Equation (5.20) – would predict for a ϕ_c of the order of 33°?

Problem 5.19 A fully drained triaxial test is performed on a dry sample of the same sand prepared to the same initial density ($D_R = 80\%$) as in Problem 5.18. At peak strength, the major principal (vertical) stress is 287 kPa and the minor principal (lateral) stress is 67 kPa.
a. What is the friction angle based on this test? Is it the same as in Problem 5.18? Does this appear reasonable?
b. Determine the orientation(s) of the shear planes, and the magnitudes of the normal stress and shear stress acting on the shear planes at peak shear strength.

P-Table 5.2 Data for Problem 5.9

Sample 1 $e_0=0.478$ $\sigma_c' = 100$ kPa			Sample 2 $e_0=0.485$ $\sigma_c' = 200$ kPa			Sample 3 $e_0=0.516$ $\sigma_c' = 400$ kPa		
Axial strain (%)	*Flow number N*	*Volumetric strain (%)*	*Axial strain (%)*	*Flow number N*	*Volumetric strain (%)*	*Axial strain (%)*	*Flow number N*	*Volumetric strain (%)*
0.0	1.00	0.00	0.0	1.00	0.00	0.0	1.00	0.00
0.2	2.25	0.16	0.2	1.86	0.16	0.2	1.68	0.18
0.3	3.03	0.12	0.3	2.40	0.27	0.3	2.00	0.18
0.5	3.57	0.06	0.5	2.97	0.31	0.5	2.48	0.28
1.0	4.19	−0.21	1.0	3.75	0.10	1.0	3.10	0.26
1.5	4.42	−0.58	1.5	4.08	−0.19	1.5	3.42	0.20
2.0	4.52	−0.91	2.0	4.24	−0.44	2.0	3.61	0.03
2.5	4.60	−1.31	2.5	4.35	−0.79	2.5	3.73	−0.13
3.0	4.60	−1.66	3.0	4.41	−1.04	3.0	3.81	−0.26
3.5	4.58	−2.03	3.5	4.44	−1.40	3.5	3.87	−0.49
4.0	4.53	−2.35	4.0	4.44	−1.75	4.0	3.90	−0.63
4.5	4.44	−2.68	4.5	4.42	−2.07	4.5	3.93	−0.88
5.0	4.37	−2.98	5.0	4.39	−2.38	5.0	3.94	−1.09
6.0	4.13	−3.32	6.0	4.24	−2.93	5.5	3.95	−1.28
7.0	4.03	−3.47	7.0	3.90	−3.27	6.0	3.95	−1.42
8.0	4.04	−3.63	8.0	3.59	−3.39	6.5	3.94	−1.66
9.0	4.07	−3.81	9.0	3.52	−3.45	7.0	3.92	−1.82
10.0	4.10	−4.01	10.0	3.51	−3.48	7.5	3.88	−1.99
11.0	4.08	−4.22	11.0	3.55	−3.50	8.0	3.80	−2.11
12.0	4.04	−4.47	12.0	3.58	−3.54	8.5	3.74	−2.22
13.0	3.98	−4.73	13.0	3.59	−3.62	9.0	3.67	−2.33
14.0	3.88	−5.02	14.0	3.59	−3.63	10.0	3.61	−2.56
15.5	3.79	−5.41	15.6	3.58	−3.71	11.0	3.56	−2.61
17.0	3.72	−5.79	17.0	3.60	−3.75	12.0	3.46	−2.72
18.5	3.68	−6.15	18.5	3.62	−3.84	13.0	3.42	−2.69
20.0	3.63	−6.51	20.0	3.62	−3.85	14.0	3.42	−2.80
22.0	3.51	−6.89	22.0	3.62	−4.11	15.0	3.42	−2.77
24.0	3.50	−7.18	24.0	3.62	−4.18	16.0	3.42	−2.82
26.0	3.49	−7.32	26.0	3.60	−4.43	17.0	3.42	−2.78
28.0	3.46	−7.68	28.0	3.56	−4.64	18.0	3.40	−2.88
29.0	3.38	−7.88	—	—	—	19.5	3.39	−2.96

Source: Carraro (2004).

5.9.3 Design problems

Problem 5.20 Consider a triaxial compression test on a sand sample with $D_R=60\%$ consolidated to σ_c':

a. What is the void ratio of the sand if $e_{min}=0.5$ and $e_{max}=0.9$?
b. What is the unit weight of the sand for $G_s=2.7$?

c. Considering that the sand is composed of well-rounded particles, what is an appropriate value to assume for the critical-state friction angle?

d. Using the Bolton equation – Equation (5.20) – what do you estimate the peak friction angle to be for $\sigma_c'=100$, 200, and 500 kPa?

e. What confining stress σ_c' would you need to apply to this sample to completely suppress dilatancy?

f. Assuming $K_0=0.5$, at what depth will dilatancy be suppressed?

g. Sketch the Mohr–Coulomb failure envelope using the calculated peak friction angles up to σ_c' values beyond the point of dilatancy suppression.

h. What is the initial, small-strain shear modulus for the sand before shearing?

Problem 5.21* A sample of sand with $D_R=50\%$ was tested in triaxial compression. The initial confining stress σ_3' was 200 kPa. For the sand of P-Figure 5.3:

a. Estimate the critical-state friction angle.

b. Plot the dilatancy angle versus the axial strain.

c. Discuss how well the principal effective stress ratio obtained from the dilatancy angle using Equation (5.8) compares with the observed values.

d. How well does Bolton's expression – Equation (5.20) – predict the peak friction angle for the dilatant sand?

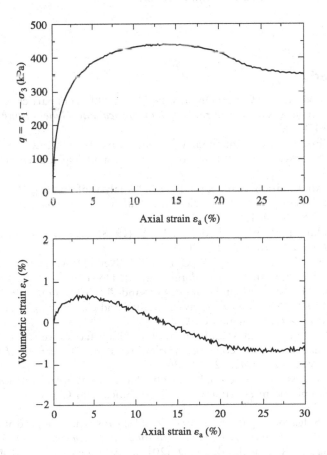

P-Figure 5.3 Deviatoric stress and volume change evolution with axial strain in a triaxial compression test.

P-Table 5.3 Sand properties for Problem 5.22

Maximum void ratio e_{max}	0.85
Minimum void ratio e_{min}	0.32
In situ void ratio e	0.43
Specific gravity G_s	2.65

Problem 5.22 A 2 m footing is placed in a sand deposit with critical-state friction angle $\phi_c = 30°$. The base of the footing is located at a depth of 1 m, and the water table is at a depth far below the footing. Estimate the peak friction angle ϕ_p at depths 0.5, 1, and 2 m below the base of the footing by assuming only an increase in the vertical effective stress at those depths (that is, consider the lateral stress to remain unchanged, as in a triaxial test, and the vertical stress to be increased up to a peak in the stress–strain relationship). Consider K_0 to be equal to 0.4. Use Equation (5.21) and the chart in Figure 5.18. Other required soil properties can be found in P-Table 5.3.

Problem 5.23* Discuss the reasonableness of your results for Problem 5.22. Are the ϕ_p estimates too high, too low, or just right? Does the lateral stress beneath a footing stay constant when the footing is loaded? In which zones beneath and around the footing would you expect it to increase and decrease?

REFERENCES

References cited

Altuhafi, F. N., Coop, M. R., and Georgiannou, V. N. (2016). "Effect of particle shape on the mechanical behavior of natural sands." *Journal of Geotechnical and Geoenvironmental Engineering*, 142(12), 04016071.

Altuhafi, F. N., Jardine, R. J., Georgiannou, V. N., and Moinet, W. W. (2018). "Effects of particle breakage and stress reversal on the behaviour of sand around displacement piles." *Géotechnique*, 68(6), 546–555.

Altuhafi, F., O'Sullivan, C., and Cavarretta, I. (2013). "Analysis of an image-based method to quantify the size and shape of sand particles." *Journal of Geotechnical and Geoenvironmental Engineering*, 139(8), 1290–1307.

Arulnathan, R., Boulanger, R. W., and Riemer, M. F. (1998). "Analysis of bender element tests." *Geotechnical Testing Journal*, 21(2), 120–131.

Bellotti, R., Jamiolkowski, M., Lo Presti, D. C. F., and O'Neill, D. A. (1996). "Anisotropy of small strain stiffness in Ticino sand." *Géotechnique*, 46(1), 115–131.

Bolton, M. D. (1986). "The strength and dilatancy of sands." *Géotechnique*, 36(1), 65–78.

Carraro, J. A. H. (2004). "Mechanical behavior of silty and clayey sands." Ph.D. Thesis, Purdue University, West Lafayette, Indiana, USA.

Carraro, J. A. H., Bandini, P., and Salgado, R. (2003). "Liquefaction resistance of clean and nonplastic silty sands based on cone penetration resistance." *Journal of Geotechnical and Geoenvironmental Engineering*, 129(11), 965–976.

Carraro, J. A. H., Prezzi, M., and Salgado, R. (2009). "Shear strength and stiffness of sands containing plastic or nonplastic fines." *Journal of Geotechnical and Geoenvironmental Engineering*, 135(9), 1167–1178.

Chakraborty, T., and Salgado, R. (2010). "Dilatancy and shear strength of sand at low confining pressures." *Journal of Geotechnical and Geoenvironmental Engineering*, 136(3), 527–532.

Chakraborty, T., Salgado, R., and Loukidis, D. (2013). "A two-surface plasticity model for clay." *Computers and Geotechnics*, 49, 170–190.

Cho, G.-C., Dodds, J., and Santamarina, J. C. (2006). "Particle shape effects on packing density, stiffness, and strength: natural and crushed sands." *Journal of Geotechnical and Geoenvironmental Engineering*, 132(5), 591–602.

Coop, M. R., and Lee, I. K. (1993). "The behaviour of granular soils at elevated stresses." *Predictive Soil Mechanics: Proceedings of Wroth Memorial Symposium*, Thomas Telford, London, UK, 186–198.

Dafalias, Y. F., Papadimitriou, A. G., and Li, X. S. (2004). "Sand plasticity model accounting for inherent fabric anisotropy." *Journal of Engineering Mechanics*, 130(11), 1319–1333.

Davis, E. H. (1968). "Theories of plasticity and failures of soil masses." In *Soil Mechanics, Selected Topics*, Lee, I. K. (ed.), Butterworth, London, 341–380.

Ganju, E., Han, F., Prezzi, M., Salgado, R., and Pereira, J. S. (2020). "Quantification of displacement and particle crushing around a penetrometer tip." *Geoscience Frontiers*, 11(2), 389–399.

Gao, Z., Zhao, J., Li, X.-S., and Dafalias, Y. F. (2014). "A critical state sand plasticity model accounting for fabric evolution." *International Journal for Numerical and Analytical Methods in Geomechanics*, 38(4), 370–390.

Han, F., Ganju, E., Salgado, R., and Prezzi, M. (2018). "Effects of interface roughness, particle geometry, and gradation on the sand-steel interface friction angle." *Journal of Geotechnical and Geoenvironmental Engineering*, 144(12), 04018096.

Hardin, B. O., and Black, W. L. (1968). "Vibration modulus of normally consolidated clay." *Journal of Soil Mechanics and Foundations Division*, 94(2), 353–369.

Herle, I., and Gudehus, G. (1999). "Determination of parameters of a hypoplastic constitutive model from properties of grain assemblies." *Mechanics of Cohesive-Frictional Materials*, 4(5), 461–486.

De Josselin De Jong, G. (1976). "Rowe's stress-dilatancy relation based on friction." *Géotechnique*, 26(3), 527–534.

Jovicic, V., and Coop, M. R. (1997). "Stiffness of coarse-grained soils at small strains." *Géotechnique*, 47(3), 545–561.

Lawton, E. C. (2001). "Soil improvement and stabilization- non-grouting techniques." In *Practical Foundation Engineering Handbook*, Brown, R. W. (ed.) Section 6A, 6.3–6.340, McGraw-Hill, New York.

Lings, M. L., and Dietz, M. S. (2004). "An improved direct shear apparatus for sand." *Géotechnique*, 54(4), 245–256.

Loukidis, D., and Salgado, R. (2009). "Modeling sand response using two-surface plasticity." *Computers and Geotechnics*, 36(1–2), 166–186.

Mitchell, J. K., and Soga, K. (2005). *Fundamentals of Soil Behavior*. Wiley, New York.

Oda, M. (1982). "Fabric tensor for discontinuous geological materials." *Soils and Foundations*, 22(4), 96–108.

Riemer, M. F., Seed, R. B., Nicholson, P. G., and Jong, H.-L. (1990). "Steady state testing of loose sands: limiting minimum density." *Journal of Geotechnical Engineering*, 116(2), 332–337.

Rowe, P. W. (1962). "The stress-dilatancy relation for static equilibrium of an assembly of particles in contact." *Proceedings of Royal Society of London. Series A. Mathematical and Physical Sciences*, 269(1339), 500–527.

Sakleshpur, V. A., Prezzi, M., Salgado, R., and Zaheer, M. (2021). *CPT-based geotechnical design manual–volume II: CPT-based design of foundations (methods)*. Joint Transportation Research Program, Purdue University, West Lafayette, IN, USA.

Salgado, R., Bandini, P., and Karim, A. (2000). "Shear strength and stiffness of silty sand." *Journal of Geotechnical and Geoenvironmental Engineering*, 126(5), 451–462.

Schanz, T., and Vermeer, P. A. (1996). "Angles of friction and dilatancy of sand." *Géotechnique*, 46(1), 145–151.

Sukumaran, B., and Ashmawy, A. K. (2001). "Quantitative characterisation of the geometry of discret particles." *Géotechnique*, 51(7), 619–627.

Thurairajah, A. (1962). "Some shear properties of kaolin and of sand." Ph.D. Thesis, University of Cambridge, Cambridge, UK.

Tsomokos, A., and Georgiannou, V. N. (2010). "Effect of grain shape and angularity on the undrained response of fine sands." *Canadian Geotechnical Journal*, 47(5), 539–551.

Uthayakumar, M., and Vaid, Y. P. (1998). "Static liquefaction of sands under multiaxial loading." *Canadian Geotechnical Journal*, 35(2), 273–283.

Verdugo, R., and Ishihara, K. (1996). "The steady state of sandy soils." *Soils and Foundations*, 36(2), 81–91.

Viggiani, G., and Atkinson, J. H. (1995a). "Interpretation of bender element tests." *Géotechnique*, 45(1), 149–154.

Viggiani, G., and Atkinson, J. H. (1995b). "Stiffness of fine-grained soil at very small strains." *Géotechnique*, 45(2), 249–265.

Woo, S. I., and Salgado, R. (2015). "Bounding surface modeling of sand with consideration of fabric and its evolution during monotonic shearing." *International Journal of Solids and Structures*, 63, 277–288.

Yang, Z. X., Jardine, R. J., Zhu, B. T., Foray, P., and Tsuha, C. H. C. (2010). "Sand grain crushing and interface shearing during displacement pile installation in sand." *Géotechnique*, 60(6), 469–482.

Zheng, J., and Hryciw, R. D. (2015). "Traditional soil particle sphericity, roundness and surface roughness by computational geometry." *Géotechnique*, 65(6), 494–506.

Zheng, J., and Hryciw, R. D. (2016). "Index void ratios of sands from their intrinsic properties." *Journal of Geotechnical and Geoenvironmental Engineering*, 142(12), 06016019.

Additional references

Alshibli, K. A., Batiste, S. N., and Sture, S. (2003). "Strain localization in sand: plane strain versus triaxial compression." *Journal of Geotechnical and Geoenvironmental Engineering*, 129(6), 483–494.

Alshibli, K., and Sture, S. (2000). "Shear band formation in plane strain experiments of sand." *Journal of Geotechnical and Geoenvironmental Engineering*, 126(6), 495–503.

Finno, R. J., and Rechenmacher, A. L. (2003). "Effects of consolidation history on critical state of sand." *Journal of Geotechnical and Geoenvironmental Engineering*, 129(4), 350–360.

Kulhawy, F. H., and Mayne, P. W. (1990). *Manual on Estimating Soil Properties for Foundation Design*. Electric Power Research Institute, Palo Alto, CA.

Lade, P. V., and Yamamuro, J. A. (1997). "Effects of nonplastic fines on static liquefaction of sands." *Canadian Geotechnical Journal*, 34(6), 918–928.

Mooney, M. A., Finno, R. J., and Viggiani, G. (1998). "A unique critical state for sand?" *Journal of Geotechnical and Geoenvironmental Engineering*, 124(11), 1100–1108.

Thevanayagam, S., Shenthan, T., Mohan, S., and Liang, J. (2002). "Undrained fragility of clean sands, silty sands, and sandy silts." *Journal of Geotechnical and Geoenvironmental Engineering*, 128(10), 849–859.

Relevant ASTM Standards

ASTM. (2015). Standard test method for modulus and damping of soils by fixed-base resonant column devices. ASTM D4015, American Society for Testing and Materials, West Conshohocken, Pennsylvania, USA.

ASTM. (2019). Standard test method for permeability of granular soils (constant head). ASTM D2434, American Society for Testing and Materials, West Conshohocken, Pennsylvania, USA.

Chapter 6

Consolidation, shear strength, and stiffness of clays

The advantages of a well-formulated theory are at least twofold. In the first place[,] it improves predictive power by extending an analytical treatment farther into the real world. The second advantage is the fact that often an advanced theory will show effects which are not present in the simplified approach.

Robert L. Schiffman

Clays differ from sands in one important way: clay particles are much, much smaller than sand particles. This simple fact has an important consequence: clays are much less pervious than sands. This means that saturated clays allow water to drain from its pores much more slowly than sands and consequently contract or expand also much more slowly than sands because volume change in a saturated soil requires water flow into or out of the clay's pores. Under the loading of a building, for example, clay deposits tend to compress, but will do so only slowly because the water in the clay pores must be expelled, and this requires time. This, in turn, has ramifications for the shear strength of the clay and thus for the bearing capacity of the foundations sustaining the building loads. Finally, another important way in which clay differs from sand is the shape of its particles, usually platy (but sometimes tubular or needle-like). A consequence of clay particles having such shapes is that prolonged shearing in a given direction will align the clay particles in that direction, leading to a lower shear strength than critical-state shear strength. In this chapter, we cover in detail these aspects of clay behavior and how to quantify them in a way that will be useful in design.

6.1 COMPRESSION AND CONSOLIDATION

6.1.1 Excess pore pressures

Let us consider a clay layer initially in equilibrium, as shown in Figure 6.1a. A free-draining sand layer overlays the clay layer. The conditions are hydrostatic; that is, the pore pressure increases linearly with depth from zero at the water table level. The hydraulic gradient is zero everywhere, as the elevation head decreases with depth at the same rate as the pressure head increases. Thus, there is no water flow.

Let us imagine now that we build a very extensive fill on top of the soil profile in Figure 6.1a and that we do so instantaneously. The surcharge due to this fill can be modeled as a uniform load p_v extending to infinity horizontally, as shown in Figure 6.1b. Because p_v is applied instantaneously and the clay cannot change volume immediately due to its low hydraulic conductivity, the soil skeleton carries no load and the applied surcharge is initially fully absorbed by the pore water. This creates, at time $t=0$, an excess pore pressure u in the pore water throughout the clay layer, as shown in the figure. No excess pore pressure is generated in the sand layer because of its large hydraulic conductivity, which means that all

DOI: 10.1201/b22079-6

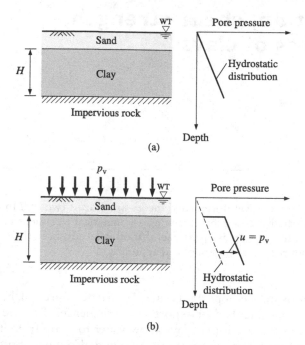

Figure 6.1 Clay layer: (a) under hydrostatic conditions; (b) with excess pore pressures.

of the surcharge is carried by the soil skeleton in the case of the sand layer. Now there is a nonzero hydraulic gradient exactly at the top of the clay layer. In fact, at the time the load is applied, the hydraulic gradient there is infinity (because there is a discontinuity in the pore pressure versus depth curve right at the drainage boundary). This causes an immediate flow of water from the clay to the sand at that depth. This means that the excess pore pressure dissipates instantaneously at the sand–clay interface, where drainage is immediate; however, at points that are distant from the draining layer, drainage and pore pressure dissipation will only occur with the passage of time. As the excess pore pressure dissipates toward zero and water is driven out of the clay pore space, the clay layer will compress until the soil skeleton develops enough stiffness to carry the entire applied load.

6.1.2 Soil compression

Compression is the reduction in volume experienced by soil upon an increase in stress. Consider, for example, what happens to clay deposited at the bottom of a lake, under very gentle, quiescent conditions. The soil at the top of the soil deposit (say at some point P, as shown in Figure 6.2) is under zero normal stress; that is, it is in effect in a slurry condition. However, over time, an increasing number of clay particles settle on top of that soil, so that after some time (as always when we deal with geology, even the geology of recent sediments, a considerable amount of time) it will be under the weight of a certain layer of soil. So our soil, which was under zero vertical effective stress initially (Figure 6.2a), now finds itself under some nonzero σ'_v (Figure 6.2b). Because it originally existed at a high void ratio (it was, after all, a slurry), it will have compressed considerably even for small increases in σ'_v. As the void ratio decreases with increasing accumulation of soil, the soil will be less compressible. This means that, for the same increase in stress, the void reduction will be less as σ'_v increases (this is shown in Figure 6.2c). If we plot void ratio versus the logarithm of σ'_v, we find that the relationship is approximately linear (Figure 6.2d).

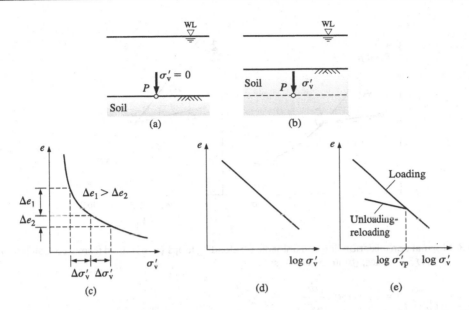

Figure 6.2 Compression of a soil deposit formed at the bottom of a lake: (a) recently deposited soil ($\sigma'_v = 0$ at point P); (b) some time later (with nonzero σ'_v at point P); (c) e versus σ'_v; (d) e versus $\log \sigma'_v$; (e) e versus $\log \sigma'_v$, including an unloading and reloading event.

If we removed some of the soil from above point P, in effect reducing σ'_v and unloading the soil, the void ratio would not go back to what it used to be for smaller σ'_v values. This is so because, as discussed in Chapter 3, the void ratio reduction when σ'_v increased came as a result of an inelastic, irrecoverable deformation associated with interparticle relative motion. This means that, on unloading, the void ratio e versus $\log \sigma'_v$ relationship is linear, but with a flatter slope, as shown in Figure 6.2e. Upon unloading, the vertical effective stress σ'_{vp} from which we started unloading becomes the preconsolidation pressure, discussed earlier in Chapter 4.

Soil compression and the time rate of compression (referred to as consolidation) are studied in the laboratory using consolidation tests, which were discussed in Chapter 4. In the consolidation test, a sample of clay with diameter usually equal to 70 mm (2.8 in.) and height equal to 25 mm (1 in.) is compressed one-dimensionally in the direction of its axis. The one-dimensional (1D) compression of a clay layer such as that in Figure 6.1b is thus simulated by increasing the total vertical stresses σ_v on the sample, allowing enough time for the excess pore pressures to dissipate fully, and, during this period, measuring the resulting compression at various times until the sample height stabilizes. Knowing the initial void ratio, it is possible to calculate the void ratio at the end of each loading step. The final void ratios can then be plotted versus the corresponding vertical effective stresses or their logarithms, as shown in Figure 6.3.

Assuming that we have determined the preconsolidation pressure σ'_{vp} of the soil, which is the maximum vertical effective stress ever experienced by the soil, by analysis of consolidation tests or other means (which include full knowledge of the geological history of the deposit), and that the vertical effective stress corresponding to point B in Figure 6.3 represents this preconsolidation pressure, line AB represents a recompression line. In $\log \sigma'_v$–e space,[1] the recompression line has a slope C_s, referred to as the swelling or recompression index, which is given by

---

[1] It is possible to use either decimal or natural logarithms of stress to define this space. What is essential is that one knows which one is used and that all quantities and parameters are consistently defined.

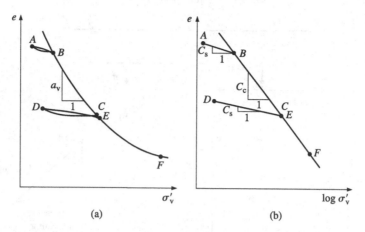

Figure 6.3 Normal consolidation line and unloading–reloading loops (as obtained from a consolidation test): (a) in σ'_v–e space; (b) in $\log \sigma'_v$–e space.

$$C_s = -\frac{de}{d \log \sigma'_v} = -2.303\sigma'_v \frac{de}{d\sigma'_v} \tag{6.1}$$

where σ'_v can be taken as the average of the initial and final vertical effective stresses for the vertical stress increment under consideration.[2]

During a consolidation test or even in the field, soil may be loaded and unloaded multiple times. For example, as consolidation of the soil continues beyond point *B* to point *C* in Figure 6.3, unloading the soil from point *C* to point *D* creates a new preconsolidation pressure σ'_{vp}, that corresponds to point *C*. Every time it is reloaded back to the previous largest stress, the path followed in the void ratio versus the log of stress space has a slope equal to C_s. And every time the consolidation progresses further and the stress is posteriorly reduced, a new state of overconsolidation, with a new preconsolidation pressure σ'_{vp}, is created.

The *BEF* line in Figure 6.3b is known as either the 1D *normal consolidation line* (NCL) or 1D virgin compression line (VCL). The *compression index* C_c is the slope of the NCL in the logarithm of stress versus void ratio space, as shown in Figure 6.3b:

$$C_c = -\frac{de}{d \log \sigma'_v} = -2.303\sigma'_v \frac{de}{d\sigma'_v} \tag{6.2}$$

Example 6.1

The data in E-Table 6.1 were obtained from a consolidation test on a clay, already corrected for sample disturbance. Determine the value of C_c.

E-Table 6.1 Corrected data from consolidation test

e	1.63	1.60	1.55	1.39	1.22	1.24	1.28	1.33	1.31	1.28	1.21	1.07
σ'_v (kPa)	50	100	200	400	800	400	200	100	200	400	800	1600

---

[2] Whether the initial, average, or final value of σ'_v is taken does not make any practical difference if the stress increment is very small.

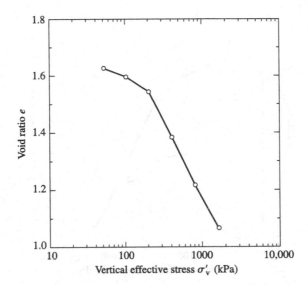

E-Figure 6.1 Compression of clay sample of Example 6.1.

Solution:

First, we plot the data in e–log σ'_v space (E-Figure 6.1). We note that there is a brief recompression segment, followed by normal consolidation. We can find the slope of the NCL. Analytically, using Equation (6.2):

$$C_c = -\frac{de}{d\log\sigma'_v}$$

Because the points are very well aligned, we can pick any pair of points to obtain a value for C_c:

$$C_c = \frac{-(1.07-1.22)}{\log 1600 - \log 800} = 0.50$$

A quick way of determining C_c is to simply subtract the void ratio at a vertical effective stress that lags another by one log cycle (say 100) from the void ratio at the stress one log cycle ahead (say 1000). We should only be careful that the plot is linear and representative of the sample behavior for the two points that we use in the calculation.

Example 6.2

Consider the unloading portion of the log σ'_v–e data for the soil of Example 6.1. Determine the value of C_s.

Solution:

We plot the data in E-Table 6.1 in log σ'_v versus e space, including the additional data for the unloading and reloading of the sample (E-Figure 6.2). We can find the slope of the swelling or recompression line using Equation (6.1):

$$C_s = \frac{-de}{d\log\sigma'_v}$$

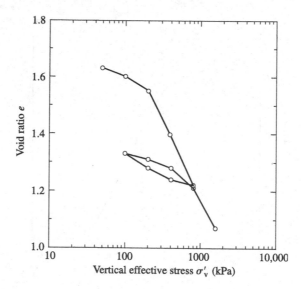

E-Figure 6.2 Compression, unloading, and recompression of clay sample of Example 6.1.

$$C_s = \frac{-(1.22 - 1.33)}{\log 800 - \log 100} = 0.12$$

The *coefficient of compressibility* a_v is defined in straight stress scale as the ratio of the change in void ratio per unit change in vertical effective stress:

$$a_v = -\frac{de}{d\sigma'_v} \tag{6.3}$$

The line *CDE* in Figure 6.3a and b represents unloading followed by reloading. This means that the vertical stress was reduced and then reapplied to the sample during the consolidation test. For practical purposes, the loading and unloading segments fall practically on top of each other and are represented in Figure 6.3b as a single line. Note that the slope of this line is the same as that of line *AB*. This is true of any unloading–reloading segment: the applicable *recompression index* is C_s, defined for the unloading–reloading loop by Equation (6.1).

Figure 6.4 shows the final, corrected compression plot for a clay sample extracted from some point within a soil profile. To obtain the idealized plot in Figure 6.4, we must first obtain the preconsolidation pressure σ'_{vp}, which may be done using the Casagrande (1936) graphical construction, and then correct the laboratory curve for disturbance, possibly using the Schmertmann (1953) graphical construction. Both of these graphical procedures are discussed in Appendix D. Once the final compression plot is ready, we can calculate the void ratio change associated with any vertical effective stress variation and, from the void ratio changes considered over the entire thickness of a given layer, the settlement of the layer.

The compression diagram in Figure 6.5 can be used to establish the link between void ratio reduction and 1D compression of a clay layer. Referring back to Chapter 4 for a definition of normal strain, the vertical normal strain at any depth in this layer can be calculated from the change in void ratio by considering the element of soil with initial void ratio e_0, unit cross-sectional area, and volume of solids equal to 1, as represented in Figure 6.5. As can be seen, the compression results from a decrease in the soil void ratio, so that the incremental vertical strain can be calculated as

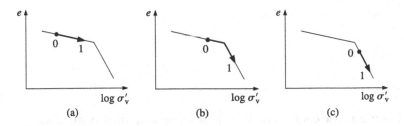

(a)   (b)   (c)

Figure 6.4 Calculation of change in void ratio as the vertical effective stress is increased (from point 0 to point 1 in the plots): (a) the stress increment is contained within the overconsolidated (OC) range; (b) the soil is initially overconsolidated, but becomes normally consolidated (NC) after the stress increment; and (c) the stress increment is contained within the NC range.

$$d\varepsilon_z = -\frac{1}{1+e_0} de \tag{6.4}$$

The change in void ratio (de) clearly depends on whether the initial vertical effective stress is equal to or greater than the preconsolidation stress σ'_{vp}. A soil for which the initial vertical effective stress $\sigma'_{v0} < \sigma'_{vp}$ is referred to as overconsolidated, while if $\sigma'_{v0} = \sigma'_{vp}$ the soil is said to be normally consolidated. For a vertical stress increment from an initial vertical effective stress σ'_{v0} to the current vertical effective stress σ'_v, the combination of Equation (6.4) with Equations (6.2) and (6.1) leads to the following equations for the compression $\Delta\varepsilon_z$ of normally consolidated (NC) and overconsolidated (OC) soils:

$$\Delta\varepsilon_z = \begin{cases} \dfrac{C_c}{1+e_0}\log\dfrac{\sigma'_v}{\sigma'_{v0}} & \text{if } \sigma'_{v0} = \sigma'_{vp} \text{ and } \sigma'_v \geq \sigma'_{vp} \text{ (NC)} \\[2ex] \dfrac{1}{1+e_0}\left[C_s\log\dfrac{\sigma'_{vp}}{\sigma'_{v0}} + C_c\log\dfrac{\sigma'_v}{\sigma'_{vp}}\right] & \text{if } \sigma'_{v0} < \sigma'_{vp} \leq \sigma'_v \text{ (OC then NC)} \\[2ex] \dfrac{C_s}{1+e_0}\log\dfrac{\sigma'_v}{\sigma'_{v0}} & \text{if } \sigma'_{v0} \leq \sigma'_{vp} \text{ and } \sigma'_v \leq \sigma'_{vp} \text{ (OC)} \end{cases} \tag{6.5}$$

Note that an increase in stress from σ'_{v0} to $\sigma'_v = \sigma'_{v0} + p_v$ leads to a positive (contractive) strain. The relationships expressed in Equation (6.5) are based on the void ratio versus vertical effective stress relationships illustrated in Figure 6.4.

Another expression for the vertical strain $d\varepsilon_z$ can be obtained in terms of a_v by combining Equations (6.3) and (6.4):

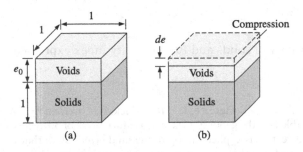

(a)   (b)

Figure 6.5 Phase diagram showing the relationship between the amount of compression and the void ratio change: (a) before compression and (b) after compression.

$$d\varepsilon_z = \frac{a_v}{1+e_0}d\sigma'_v \qquad (6.6)$$

which can be written as

$$d\varepsilon_z = m_v d\sigma'_v \qquad (6.7)$$

where m_v is known as the coefficient of volume compressibility, defined as

$$m_v = \frac{a_v}{1+e_0} \qquad (6.8)$$

Let us consider a layer of soil of total thickness H subdivided into n layers with thickness H_i, with $i = 1, ..., n$. The compression of clay layer i with an initial thickness H_i caused by an increment of strain can be calculated from:

$$dH_i = H_i d\varepsilon_{zi} \qquad (6.9)$$

For a larger strain increment $\Delta\varepsilon_{zi}$, we can write:

$$\Delta H_i = H_i \Delta\varepsilon_{zi} \qquad (6.10)$$

To obtain the total compression of the layer composed by the layers H_i, we simply integrate or sum up the individual compressions:

$$\Delta H = \sum_{i=1}^{n} H_i \Delta\varepsilon_{zi} \qquad (6.11)$$

with:

$$H = \sum_{i=1}^{n} H_i \qquad (6.12)$$

Ideally, the compression and recompression indexes would be determined from a suitable test. However, there are correlations that may be used to estimate both. One such correlation is that of Wroth and Wood (1978) for C_c; C_s is then 10%–20% of C_c. According to the Wroth and Wood (1978) correlation:

$$C_c = \frac{1}{200} G_s (PI)(\%) \qquad (6.13)$$

where G_s = specific gravity of solids and PI = plasticity index expressed as a percentage.

Example 6.3

An NC clay layer, shown in E-Figure 6.3, is 10 m thick. It is overlain by a 2-meter-thick sand layer. The water table is at the ground surface. A uniform load of 30 kPa extending to infinity in the horizontal direction is applied on top of the sand layer. Given that the plasticity index of the clay is 50% and that the water content at the center of the clay layer is equal to 70%, calculate the amount of surface settlement resulting from the application of the load. Take the unit weight of the sand to be 20 kN/m³ and the specific gravity of solids for the clay to be 2.67.

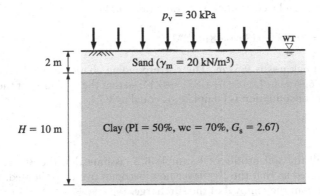

E-Figure 6.3 Soil profile and clay layer undergoing consolidation for Example 6.3.

Solution:

For now, we will use the center of the clay layer as a representative point. Using Equation (6.13), the compression index is calculated as

$$C_c = \frac{1}{200} G_s PI(\%) = \frac{1}{200} \times 2.67 \times 50 = 0.668$$

The void ratio and porosity follow from Equations (3.9) and (3.13):

$$e_0 = (wc)G_s = 0.7 \times 2.67 = 1.87$$

$$n_0 = \frac{e_0}{1+e_0} = \frac{1.87}{1+1.87} = 0.651$$

The unit weight of the clay, given that it is saturated, follows from Equation (3.17):

$$\gamma_m = \gamma_w \left[G_s (1 - n_0) + n_0 S \right]$$

$$\gamma_m = 9.81 \left[(1 - 0.651)2.67 + 0.651 \right] = 15.5 \text{ kN/m}^3$$

From the information on the soil profile and water table location, the initial vertical effective stress is equal to:

$$\sigma'_{v0} = 2 \times 20 + 5 \times 15.5 - 7 \times 9.81 = 48.9 \text{ kPa}$$

With the addition of the surcharge p_v, the current vertical effective stress, after complete pore pressure dissipation, becomes:

$$\sigma'_v = \sigma'_{v0} + p_v = 48.9 + 30 = 78.9 \text{ kPa}$$

The resulting strain of the clay layer can now be computed from the normally consolidated form of Equation (6.5):

$$d\varepsilon_z = \frac{C_c}{1+e_0} \log \frac{\sigma'_v}{\sigma'_{v0}} = \frac{0.668}{1+1.87} \log \frac{78.9}{48.9} = 0.048$$

The compression can be computed from Equation (6.9):

$$\Delta H = H\varepsilon_z = 10\,\text{m} \times 0.048 = 0.48\,\text{m} \quad \text{answer}$$

The new layer thickness is 9.52 m. The corresponding change in void ratio is obtained from Equation (6.4) as $(-0.048)(1+1.87)=-0.14$, so that the void ratio at the center of the clay layer, after consolidation is complete, is equal to 1.73.

Example 6.4

Considering still the soil profile of Example 6.3, assume that the 30 kPa surcharge is completely removed so that the clay layer now becomes overconsolidated. Calculate the heave resulting from the removal of the surcharge.

Solution:

Knowing that $C_c=0.668$ and that C_s is 10%–20% of C_c, we can take a conservative estimate of $C_s=0.2C_c=0.134$. The vertical strain can then be calculated using the form of Equation (6.5) suitable for a soil lying entirely in the overconsolidated range:

$$d\varepsilon_z = \frac{C_s}{1+e_0}\log\frac{\sigma'_v}{\sigma'_{v0}} = \frac{0.134}{1+1.73}\log\frac{48.9}{78.9} = -0.010$$

The heave is calculated as

$$\Delta H = -0.010 \times 9.52\,\text{m} = -0.095\,\text{m} \quad \text{answer}$$

Note that we used the new layer thickness (9.52 m) and new void ratio (1.73) in the calculation, but negligible error would have resulted from using 10 m. If we had assumed the original layer thickness and void ratio, we would have obtained –0.1 m, which is, given the level of precision we would work with in practice, the same result. Note that only 20% of the layer compression is recovered upon unloading; the other 80% is irrecoverable plastic deformation.

Example 6.5

Redo Example 6.3 by subdividing the clay layer into four sublayers, calculating the compression of each sublayer, and then adding the individual compressions to obtain the total clay layer compression. Discuss.

Solution:

When the clay layer is divided into four sublayers, the representative points for each sublayer are as indicated in E-Figure 6.4. The new initial void ratios reflect the fact that the midpoints of layers 1 and 2 have undergone less consolidation and the midpoints of layers 3 and 4 have undergone more consolidation than the midpoint of the entire layer. They can be calculated from the difference in effective stresses between the midpoint of the layer (let us call it point 0) and points A, B, C, and D. These stresses can be calculated by knowing the vertical distance between the representative points, each located at the center of the corresponding sublayer, and the unit weight of 15.5 kN/m³ obtained in the solution of Example 6.3, assumed to hold for the whole layer. Considering that all the layers are below the water table, we use the buoyant unit weight (= 15.5−9.81=5.7 kN/m³) to calculate the stresses as follows:

$$\sigma'_{vA} = \sigma'_{v0} + \gamma_b\Delta z_{0A} = 49 + 5.7 \times (-3.75) = 28\,\text{kPa}$$

$$\sigma'_{vB} = \sigma'_{v0} + \gamma_b\Delta z_{0B} = 49 + 5.7 \times (-1.25) = 42\,\text{kPa}$$

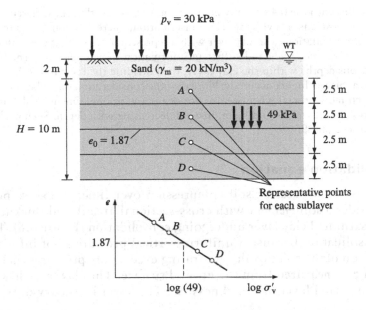

E-Figure 6.4 Soil profile, clay layer undergoing consolidation, and the four sublayers into which it is divided.

$$\sigma'_{vC} = \sigma'_{v0} + \gamma_b \Delta z_{0C} = 49 + 5.7 \times 1.25 = 56 \text{ kPa}$$

$$\sigma'_{vD} = \sigma'_{v0} + \gamma_b \Delta z_{0D} = 49 + 5.7 \times 3.75 = 70 \text{ kPa}$$

The initial void ratios at the center of each layer can now be calculated considering that the soil is NC and that, at 7 m depth, the void ratio is $e_0 = 1.87$ and the vertical effective stress $\sigma'_v = 49$ kPa:

$$\left(e_0 \right)_A = e_0 + C_c \log \frac{\sigma'_{v0}}{\sigma'_{vA}} = 1.87 + 0.668 \log \frac{49}{28} = 1.87 + 0.16 = 2.03$$

$$\left(e_0 \right)_B = e_0 + C_c \log \frac{\sigma'_{v0}}{\sigma'_{vB}} = 1.87 + 0.668 \log \frac{49}{42} = 1.87 + 0.04 = 1.91$$

$$\left(e_0 \right)_C = e_0 - C_c \log \frac{\sigma'_{vC}}{\sigma'_{v0}} = 1.87 - 0.668 \log \frac{56}{49} = 1.87 - 0.04 = 1.83$$

$$\left(e_0 \right)_D = e_0 - C_c \log \frac{\sigma'_{vD}}{\sigma'_{v0}} = 1.87 - 0.668 \log \frac{70}{49} = 1.87 - 0.10 = 1.77$$

The total compression of the clay layer is calculated using Equations (6.5) and (6.11):

$$\Delta H = 2.5 \times 0.668 \left[\begin{array}{l} \dfrac{1}{1+2.03} \log \dfrac{28+30}{28} + \dfrac{1}{1+1.91} \log \dfrac{42+30}{42} + \dfrac{1}{1+1.83} \log \dfrac{56+30}{56} \\[2mm] + \dfrac{1}{1+1.77} \log \dfrac{70+30}{70} \end{array} \right]$$

$$\Delta H = 2.5 \times 0.668 \left(0.104 + 0.0804 + 0.0658 + 0.0559 \right) = 0.51 \text{ m} \quad \text{answer}$$

Comparing this with the 0.48 m calculated previously, we see that the difference is small; therefore, in most cases involving 1D consolidation, there is enough accuracy if we assume in our calculations a single layer with a representative point at the center. If there are concerns about accuracy and sufficient information is available concerning the void ratio at various depths within the clay layer, then dividing the layer into sublayers is the best way to achieve a higher accuracy. Note that the consideration of sublayers in calculations is much more crucial when we are not dealing with one-dimensional compression and the stress increase is not uniform, instead decreasing with depth. Such problems are discussed in Chapter 9.

6.1.3 Consolidation equation

Consolidation is the process of soil compression over time, as excess pore pressures dissipate. Consider a soil element with cross-sectional area A and thickness dz located within a fully saturated clay layer undergoing consolidation (Figure 6.6). The clay layer undergoes consolidation because a uniform vertical pressure p_v of infinite extent was applied on the top of the soil profile, generating excess pore pressures that are then dissipated. Depth z is measured from the ground surface. The clay layer has thickness H, and the overlying sand layer has thickness H_0. The lower boundary of the clay layer is impervious.

The net volume of water dW expelled from the soil element in time dt is calculated as the difference between the volume of water leaving and the volume of water entering the element in time dt. This calculation can be done by considering the difference between the values of specific discharge v at z (the top of the element) and $v + (\partial v / \partial z)dz$ at $(z+dz)$ (the base of the element):

$$dW = \left(v + \frac{\partial v}{\partial z} dz \right) A dt - v A dt \tag{6.14}$$

Note that we draw the specific discharge in the positive z direction (downward) even though we can guess, because of our knowledge of the boundary conditions (that is, our knowledge that the top of the clay layer is a free-draining boundary and that the bottom of the clay layer is an impermeable boundary), that the flow will actually be directed upward. In formulating a solution to the problem, we do not allow our knowledge of the boundary conditions to interfere with the formulation of the physics of the process at the element level. Instead, we write equations for quantities with the correct positive signs or directions and allow mathematics to automatically take into account the boundary and initial conditions at the time of solution of a specific problem to produce positive or negative values of each quantity.

If the element has unit cross-sectional area ($A=1$):

$$dW = \frac{\partial v}{\partial z} dz\, dt \tag{6.15}$$

As the compression is one-dimensional and the element cross-sectional area can be assumed to be equal to one, the volume change dV of the soil element in time dt can be calculated from the incremental vertical strain in the soil element during time dt:

$$dV = -\frac{\partial \varepsilon_z}{\partial t} dt\, dz \tag{6.16}$$

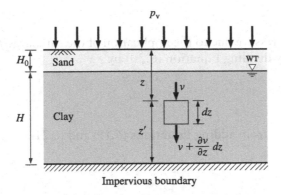

Figure 6.6 Depiction of an element of soil in a clay layer subjected to an increase in vertical stress p_v.

Now, we make three assumptions:

1. The soil is saturated.
2. Water is incompressible.
3. The soil particles are incompressible.

A direct consequence of these three assumptions is that the volume decrease in time dt is equal to the volume of water expelled in time dt, which is mathematically expressed as $-dV = dW$ or

$$dV + dW = 0$$

Using Equations (6.15) and (6.16), this equality results in:

$$\frac{\partial \varepsilon_z}{\partial t} = \frac{\partial v}{\partial z} \tag{6.17}$$

where both ε_z and v are functions of both time and depth: $\varepsilon_z = \varepsilon_z(t, z)$ and $v = v(t, z)$.

A fourth assumption, that Darcy's law holds, allows us to introduce the pore pressure u into Equation (6.17). According to Darcy's law, Equation (3.24):

$$v = -Ki = -K\frac{\partial h}{\partial z} \tag{6.18}$$

where $h = h(z, t)$ is the hydraulic head at depth z and time t.

If we take the datum to coincide with the lower boundary of the clay layer, the elevation head h_e can be calculated as

$$h_e = H + H_0 - z \tag{6.19}$$

The pressure head h_p is calculated from the total pore pressure. We can assume the water table to be anywhere above the top of the clay layer for our purpose, which is to analyze the volume change and pore pressure evolution within the saturated clay layer. Let us assume it coincides with the top boundary of the clay layer. For these conditions, the total pore pressure u_t at depth z is given by

$$u_t = (z - H_0)\gamma_w + u \tag{6.20}$$

where u=excess pore pressure due to the application of the surcharge p_v. The pressure head $h_p = u_t/\gamma_w$ is obtained by dividing Equation (6.20) by γ_w:

$$h_p = z - H_0 + \frac{u}{\gamma_w} \tag{6.21}$$

The total head results from adding Equations (6.19) and (6.21):

$$h = H + \frac{u}{\gamma_w} \tag{6.22}$$

Differentiating with respect to z and considering that H is a constant:

$$\frac{\partial h}{\partial z} = \frac{1}{\gamma_w} \frac{\partial u}{\partial z} \tag{6.23}$$

Taking Equation (6.23) into Equation (6.18):

$$v = -\frac{K}{\gamma_w} \frac{\partial u}{\partial z} \tag{6.24}$$

Equation (6.24) tells us that, for vertical flow, the specific discharge v depends only on the pore pressure gradient. This helps us understand the consolidation process. Initially, before the application of the uniform surcharge p_v, there was no flow because the pore pressure gradient was zero everywhere. After the application of p_v, flow takes place because of the difference in excess pore pressure created between the top and bottom of the clay layer. Because of the presence of the sand layer there, drainage at the top is immediate, with the excess pore pressure there going to zero immediately after the application of p_v, but it is delayed elsewhere, with the result that excess pore pressures are greater than zero. So water starts to flow from the bottom to the top, with higher flow rates[3] near the top, because hydraulic gradients there are higher.

Taking Equation (6.24) into Equation (6.17) gives:

$$\frac{\partial \varepsilon_z}{\partial t} = -\frac{\partial}{\partial z}\left(\frac{K}{\gamma_w} \frac{\partial u}{\partial z}\right) \tag{6.25}$$

in which we now have two variables: $\varepsilon_z = \varepsilon_z(z, t)$ and $u = u(z, t)$.

At this point, we make a fifth assumption: the additional total vertical stress $\Delta\sigma_v$ generated by the application of p_v is independent of z. For a surcharge p_v extending to infinity in all directions, this assumption holds exactly. This allows us to write that, for any depth z and time t:

$$\Delta\sigma_v = \Delta\sigma'_v + u = p_v = \text{constant} \tag{6.26}$$

Differentiating both sides of Equation (6.26), we get:

$$\frac{\partial \sigma'_v}{\partial z} + \frac{\partial u}{\partial z} = 0$$

---

[3] See Chapter 3 for a review of groundwater flow and definitions of terms such as specific discharge and flow rate.

which leads to

$$\frac{\partial \sigma_v'}{\partial z} = -\frac{\partial u}{\partial z} \tag{6.27}$$

Taking Equation (6.27) into Equation (6.25):

$$\frac{\partial \varepsilon_z}{\partial t} = \frac{\partial}{\partial z}\left(\frac{K}{\gamma_w} \frac{\partial \sigma_v'}{\partial z} \right) \tag{6.28}$$

The advantage of this last equation is that it allows us to relate σ_v' to ε_z, between which there is a known relationship, namely Equation (6.7). Taking Equation (6.7) into Equation (6.28):

$$\frac{\partial \varepsilon_z}{\partial t} = \frac{\partial}{\partial z}\left[c_v \frac{\partial \varepsilon_z}{\partial z} \right] \tag{6.29}$$

where c_v, called the *coefficient of consolidation*, is defined as

$$c_v = c_v(z,t) = \frac{K}{m_v \gamma_w} \tag{6.30}$$

As represented in Equation (6.30), the coefficient of consolidation c_v is not truly a constant. As time passes and consolidation progresses, the soil becomes denser, the soil particles are packed closer together, and the hydraulic conductivity K decreases. The soil also becomes stiffer, and the coefficient of volume compressibility m_v decreases as well. Since both numerator and denominator go down at presumably comparable rates, it is fair to state that c_v is "more constant" than either K or m_v. So we make the assumption, which is still acceptable for practical purposes in most cases, that c_v is constant, leading us to:

$$\frac{\partial \varepsilon_z}{\partial t} = c_v \frac{\partial^2 \varepsilon_z}{\partial z^2} \tag{6.31}$$

This equation is known in mathematics as the diffusion equation. It appears in many problems of interest, including heat flow and the pricing of financial derivatives. In soil mechanics, this equation was the root of a tragic controversy in the beginning of the 20th century, as detailed in this chapter's case history.

An alternative assumption would be to state that both K and m_v are constant, which is not as good as stating that c_v is constant, but is still reasonably realistic for most practical purposes. Based on this last assumption (that K and m_v are both constants), we take Equations (6.7) and (6.26) into Equation (6.28) to obtain:

$$\frac{\partial u}{\partial t} = c_v \frac{\partial^2 u}{\partial z^2} \tag{6.32}$$

which is known as the one-dimensional consolidation equation.

Summarizing our results, the consolidation process can be represented through the variation of either the vertical strain ε_z or the excess pore pressure u with depth and time, following either Equation (6.31) or (6.32). Equation (6.31) is more realistic than Equation (6.32), but Equation (6.32) is easier to use and is the one that is by far the most often used in practice.

The mathematical form of Equation (6.32) tells us some interesting facts. Recall from differential calculus that the second derivative of the excess pore pressure u with respect to z represents the curvature of the excess pore pressure distribution curve within the clay layer. Recall also that a positive second derivative at a point means that the slope of the function at that point is increasing. Using the consolidation equation, we can then conclude by inspection of the plot of u versus z whether u should increase, decrease, or remain constant with time t by observing whether its second derivative with respect to z is positive, equal to zero, or negative (see the conceptual problems at the end of this chapter).

6.1.4 Solution of the consolidation equation and the degree of consolidation

To solve Equation (6.32) to obtain $u=u(z, t)$ for a given problem, we need an initial condition and boundary conditions. The initial condition states that, at time $t=0$, the excess pore pressure is equal to p_v throughout the clay layer. The boundary conditions state that the drainage at the interface with the sand layer is immediate and that there is no flow across the impervious boundary. If we now assume z measured from the top of the clay layer, to simplify the notation, the 1D consolidation boundary-value problem can be expressed as follows:

Solve $\dfrac{\partial u}{\partial t} = c_v \dfrac{\partial^2 u}{\partial z^2}$ subject to:

1. $u=p_v$ for all values of z at $t=0$ (initial condition),
2. $u=0$ for all times at $z=0$ (immediate-drainage boundary condition), and
3. $\dfrac{\partial u}{\partial z} = 0$ at $z=H$ (no-flow boundary condition).

The solution of this boundary-value problem is:

$$\frac{u}{p_v} = \sum_{n=0}^{\infty} \frac{2}{N} \sin\left(\frac{Nz}{H}\right) \exp\left(-N^2 T_v\right)$$ (6.33)

where N contains the summation index n:

$$N = \frac{(2n+1)\pi}{2}$$

and

$$T_v = \frac{c_v t}{H^2}$$ (6.34)

is referred to as the time factor. Equation (6.33) can only be calculated numerically. A spreadsheet or software such as MATLAB® can be used for that purpose. We don't need many terms of the summation to obtain a very good approximation to u/p_v.

More often than calculating pore pressures, we are interested in assessing how much consolidation has taken place, of which the degree of consolidation U is an indicator. U represents the ratio of the current to the final total compression. It is defined as the ratio of the shaded area in Figure 6.7c, which represents an integral over depth of the excess pore pressure that has already dissipated, to the area representing the product of the initial excess pore pressure p_v by H; H is equal to the clay layer thickness in the case of single drainage (which we have been discussing) and half the clay layer thickness in the case of double drainage (the solution for which follows directly from the solution to the case of single drainage, as we will discuss later). According to this definition, $U=0$ implies that there has been no consolidation and $U=1$ implies that consolidation of the clay layer is complete. It is also common to express U in percentage units, in which case it varies between 0% and 100%. Mathematically, if U is expressed as a number in the [0, 1] range, it is given as

$$U = 1 - \frac{\int_0^H u \, dz}{H p_v} \tag{6.35}$$

where the integral corresponds to the nonshaded area below the curve $u=u(z)$ in Figure 6.7c.

Both the excess pore pressure u and the degree of consolidation U are determined numerically, but convenient chart solutions or approximate equations can be used in many practical problems. Following Terzaghi (1943), U can be calculated from approximate equations. For $0 \leq U \leq 0.526$ (or $0 \leq T_v \leq 0.217$):

$$U = \sqrt{\frac{4T_v}{\pi}} \tag{6.36}$$

or

$$T_v = \frac{\pi}{4} U^2 \tag{6.37}$$

For $0.526 < U \leq 1$ (or $0.217 < T_v \leq 1.781$):

$$U = 1 - 10^{-1.0718(T_v + 0.085)} \tag{6.38}$$

or

$$T_v = -0.933 \log(1 - U) - 0.085 \tag{6.39}$$

where the log is a logarithm of base 10.

Table 6.1 Values of the time factor T_v corresponding to various values of the degree of consolidation U expressed as a number

T_v	U	T_v	U
0.002	0.05	0.239	0.55
0.008	0.10	0.286	0.60
0.018	0.15	0.340	0.65
0.031	0.20	0.403	0.70
0.049	0.25	0.477	0.75
0.071	0.30	0.567	0.80
0.096	0.35	0.684	0.85
0.126	0.40	0.848	0.90
0.159	0.45	1.129	0.95
0.196	0.50	1.781	0.99

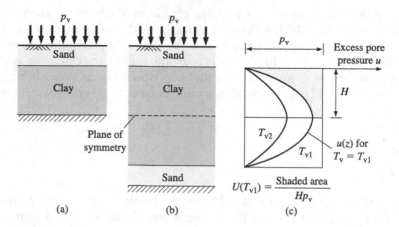

Figure 6.7 Drainage boundary conditions: (a) single drainage; (b) double drainage; and (c) definition of degree of consolidation.

The values of T_v are provided for simple reference in Table 6.1 for the entire range from $U=0$ to $U=1$.

Figure 6.7b shows a clay layer with thickness $2H$ and drainage at both the top and bottom of the layer. Excess pore pressures in the top half of the layer dissipate by upward water flow, and excess pore pressures in the lower half of the layer dissipate by downward water flow. Let us consider what happens to a drop of water located exactly at the center of the layer, thus located at a distance H from either boundary. Any water drop located there would be indifferent between moving to the top or bottom of the clay layer; thus, there is no flow across the midsection of the clay layer. So, in effect, the midsection of the clay layer functions much as the impervious boundary in Figure 6.7a. The problem of the consolidation of a clay layer with double drainage then reduces to the case of single drainage: what happens in each half of the double-drained clay layer is exactly what happens in the whole layer in the case of single drainage. In using the equations we derived earlier, H is half the thickness of the clay layer in the double-drainage case, whereas H is the full thickness of the clay layer in the single-drainage case. It is customary, therefore, to distinguish between the layer thickness and the "drainage path," which is the longest path a water drop would follow to a free-draining boundary during consolidation of a clay layer.

In the preceding discussion, we have tacitly assumed that the strains develop uniformly throughout the clay layer. In reality, for the single-drainage case, strains tend to be larger near the top and smaller near the bottom of the layer for a given soil type. This is so because void ratios will be higher, and thus the clay will be more compressible, near the top. Under these conditions, as discussed by Duncan (1993), consolidation happens faster than suggested by the relationships in Equations (6.36)–(6.39).

6.1.5 Estimation of the coefficient of consolidation

The coefficient of consolidation has the unit of L^2T^{-1}. It can be determined from the consolidation test by examining the evolution of void ratios with time for each load step. The two graphical methods commonly used for its determination are the square root of time method (Taylor 1942) and the logarithm of time method (Casagrande and Fadum 1940). The logarithm of time method is shown in Appendix D to illustrate the process. In practice, we often need to resort to correlations such as that in Figure 6.8 (NAVFAC 1982), which gives c_v values for overconsolidated, normally consolidated, and remolded (that is, extensively sheared

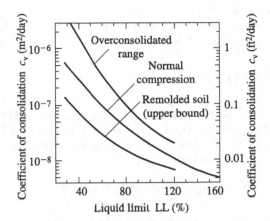

Figure 6.8 Coefficient of consolidation for normally consolidated, overconsolidated, and remolded clays as a function of the liquid limit. (After NAVFAC 1982.)

or disturbed) clays. Holtz and Kovacs (1981) compiled data from various clays and found values of c_v ranging approximately from 10^{-8} to $5 \times 10^{-7}\,m^2/s$.

Example 6.6

For the clay layer of Example 6.3, how long would it take for 50% consolidation to take place? What would your answer be if the clay layer were double drained? Take the LL as 80%.

Solution:

Entering Figure 6.8 with $LL=80\%$, $c_v=4.0\times10^{-8}\,m^2/s$ results.
$H=10\,m$ for the single-drainage case and $5\,m$ for the double-drainage case. The time factor corresponding to $U=50\%$ can be calculated from Equation (6.37):

$$T_v = \frac{\pi}{4}U^2 = \frac{\pi}{4}0.5^2 = 0.196$$

Now, the time t_{50} required for 50% consolidation can be calculated from Equation (6.34):

$$t_{50} = \frac{T_{v,50}H^2}{c_v} = \frac{0.196\times100}{4\times10^{-8}} = 4.91\times10^8 \text{ seconds} \approx 189 \text{ months} \quad \text{answer}$$

The time required for the double-drainage case is one-fourth the time required for the single-drainage case, or 47 months. answer

Example 6.7

For the clay layer of Example 6.3, how much consolidation takes place (1) after 1 year and (2) after 5 years? Assume the clay layer to be double-drained.

Solution:

First, we calculate the time factor values corresponding to 1 and 5 years using Equation (6.34):

$$T_v = \frac{c_v t}{H^2} = \frac{4\times10^{-8}\,\dfrac{\text{meter}^2}{\text{second}}\times1\text{ year}\times365\dfrac{\text{day}}{\text{year}}\times86,400\dfrac{\text{second}}{\text{day}}}{(5\text{ meters})^2} = 0.0505$$

for 1 year and five times that (0.252) for 5 years.

Now, we use Equation (6.36) for 1 year in the expectation that $U < 0.526$ for this time. If we are proven wrong, we should then use Equation (6.38) instead. We find:

$$U = \sqrt{\frac{4 \times 0.0505}{\pi}} = 0.25 \text{ answer}$$

which is less than 0.526, justifying the use of Equation (6.36). So 25% of the final consolidation of the clay layer is calculated to develop over the course of 1 year.

We use Equation (6.38) to calculate U for 5 years in the expectation that the resulting U value is larger than 0.526. We find:

$$U = 1 - 10^{-1.0718(T_v + 0.085)} = 1 - 10^{-1.0718(0.252 + 0.085)} = 0.56 \text{ answer}$$

justifying the initial assumption that $U > 0.526$, which means our use of Equation (6.38) is justified. So 56% of the final consolidation takes place after 5 years.

6.1.6 Secondary compression

It has been observed that compression may go on even after full excess pore pressure dissipation, and thus at constant effective stress. In mechanics, continuing deformation under constant stress (in this case, constant effective stress) is referred to as creep. *Secondary compression* is nothing more than creep, or the delayed deformation in response to a given increase in effective stress. It is believed that, at the particle level, secondary compression results from minor adjustments of the soil fabric. The change in void ratio associated with this process can be expressed as (Mesri and Castro 1987):

$$de = C_\alpha d(\log t) \tag{6.40}$$

where C_α is the *secondary compression index*.

Secondary compression is a continuation of the compression of the soil started by an increase in effective stress, which is calculated in terms of C_c. This initial compression, which occurs as the excess pore pressure dissipates, has been referred to as "primary" compression or consolidation, to contrast it with secondary compression or consolidation. The change in void ratio with time is shown in Figure 6.9 for a consolidation test. Also shown in the figure is a sketch of how C_α is determined from the test results.

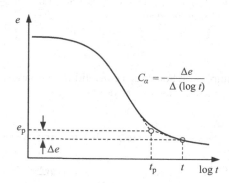

Figure 6.9 Evolution of void ratio with time during primary and secondary consolidation (t_p=time to completion of primary consolidation and e_p=void ratio at the end of primary consolidation).

It is reasonable to expect that a more compressible soil as expressed by its compression index will also be more compressible in secondary deformation. This suggests a relationship of proportionality between C_α and C_c. The C_α/C_c ratio tends to be higher for more plastic soils. It increases with organic content. The following values can be used as estimates of C_α/C_c values for various soils:

$$\frac{C_\alpha}{C_c} = \begin{cases} 0.03 \pm 0.01 & \text{for shale and mudstone} \\ 0.04 \pm 0.01 & \text{for inorganic clay and silt} \\ 0.05 \pm 0.01 & \text{for organic clay and silt} \\ 0.06 \pm 0.01 & \text{for peat} \end{cases}$$

6.1.7 Isotropic compression

Triaxial compression tests are often performed on isotropically consolidated samples. Isotropic compression follows the same pattern we have observed for one-dimensional compression: there are a normal consolidation line and much stiffer unloading–reloading loops. The isotropic NCL is roughly parallel to the 1D NCL, but it has a lower void ratio intercept (Figure 6.10). The lower intercept is due to the fact that the same value of consolidation (vertical effective) stress corresponds to a higher mean effective stress for isotropic consolidation (for which the ratio K of lateral to vertical effective stress is equal to 1) than for 1D conditions (for which $K = K_{0,NC}$). So, at any given value of stress, the void ratio of an isotropically consolidated NC sample is lower than that of a 1D consolidated NC sample.

Isotropic compression is physically more interesting because it is a pure expression of the soil's volume change response, with no shear stress or strain involved. It is customary, largely because of the influence of a school of soil mechanics known as critical-state soil mechanics (see Atkinson and Bransby 1978), to express the isotropic NCL in $\ln p'-e$ space, where $p' = \sigma'_m$ is the mean effective stress (see Chapter 4). The resulting equation is:

$$e = N - 1 - \lambda \ln p' \tag{6.41}$$

where $N - 1$ is the value of void ratio at $p' = 1\,\text{kPa}$ and the slope λ of the isotropic compression line, given that the isotropic and one-dimensional NCL can be taken as parallel, is related to C_c through:

$$C_c = 2.303\lambda \tag{6.42}$$

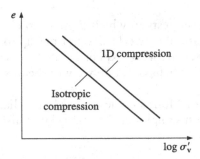

Figure 6.10 Isotropic and one-dimensional compression lines.

6.1.8 Large-strain consolidation analysis*

In the derivation of the consolidation equation, we used an infinitesimal strain formulation. If a 10 m layer of clay settles 1 m, that corresponds to an average strain of 10%, which exceeds the range in which the infinitesimal strain approximation would give sufficiently accurate results. Strains of this magnitude are not rare in the consolidation of soft clays,[4] so there has been a considerable amount of research focusing on obtaining a large-strain theory of consolidation (e.g., Gibson et al. 1967; Gibson et al. 1981; Lee and Sills 1981; Schiffman 1980; Fox and Berles 1997). In addition to quantifying strains in a more realistic manner, these theories also allow updating the hydraulic conductivity K and the coefficient of volume compressibility m_v as the soil compresses.

While the classical small-strain theory is acceptable in most routine cases in practice, in more sophisticated projects in which accuracy is important, the use of large-strain analysis (which requires a computer-based numerical analysis) is justified. As an alternative to computer analysis, Fox (1999) proposed charts that may be used for a preliminary assessment.

6.2 DRAINED SHEAR STRENGTH OF SATURATED CLAYS

The response of clays to drained loading is quite similar to that of sands. Figure 6.11 shows typical triaxial compression test[5] results for NC and OC clays. Note that the behavior of NC clays is qualitatively the same as that of contractive sands and that the behavior of OC clays is qualitatively the same as that of dilative sands.

The strength envelope of NC clays passes through the origin, just as the critical-state line of sands does. This is understood as follows: a saturated NC clay subjected to zero consolidation stress is nothing more than a clay slurry, with effectively zero shear strength. As the slurry consolidates, water is expelled from its pores, the void ratio of the slurry decreases, particles come into better contact and develop greater contact forces, and the shear strength increases. Accordingly, the higher the consolidation stress, the lower the void ratio, and the higher the shear strength. However, there are some quantitative differences:

1. NC clays have lower critical-state friction angles (typically in the 15°–30° range for clays, compared with 28°–36° for sands).
2. Clays are much less pervious; in most practical problems, clays are loaded much faster than the rate at which pore pressures can dissipate.
3. "Undisturbed" samples can be tested.
4. Upon large shear strains in a given direction, the friction angle of a clay will drop below the critical-state friction angle because of the alignment of clay particles in the direction of shearing; this friction angle ϕ is known as the residual friction angle ϕ_r.

In clays, drained triaxial tests (tests in which there is no change in pore pressure within the sample during load application) may take a long time (up to several days, or even weeks) to complete; consequently, they are very expensive and are seldom performed. However, they can help us understand clay behavior, and we will spend some time discussing them in this section.

Figure 6.12 shows a series of Mohr's circles in a Mohr's diagram, each one corresponding to a different triaxial test on samples of the same clay, consolidated to different effective

---

[4] A soft clay is a normally consolidated clay with s_u less than approximately 30 kPa.
[5] Refer to Chapter 4 for a discussion of triaxial tests.

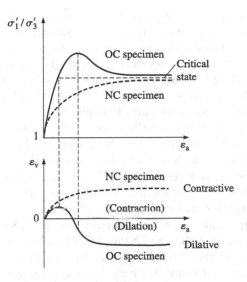

Figure 6.11 Principal effective stress ratio and volumetric strain versus axial strain for drained triaxial compression tests on two isotropically consolidated samples of the same clay. The samples are initially at the same initial effective consolidation stress $\sigma'_c = \sigma'_3$, but one sample is normally consolidated (ultimately contractive) and the other, overconsolidated (ultimately dilative).

consolidation stresses σ'_c. A strength envelope is drawn tangent to the circles. Observe that the envelope has a relatively sharp curvature change (a "bump"), with some Mohr's circles located to the left and some to the right of it. Those on the left correspond to OC samples, and those on the right, to NC samples. The question arises as to how to determine the preconsolidation pressure σ'_{vp} from such a plot. To find the preconsolidation pressure, we should consider circle 3, which is tangent to the envelope right at the "bump." The sample corresponding to circle 3 is borderline between NC and OC, and thus circle 3 can be considered to be associated uniquely with σ'_{vp}. The only question is whether σ'_{vp} is equal to σ'_3 or σ'_1. If it were equal to σ'_3, it would be possible for a slightly OC sample to have its strength controlled by the flatter strength envelope associated with overconsolidation even though the σ'_1 would be well into the NC range. Consequently, σ'_1 for this borderline sample is the preconsolidation pressure for it and for the samples corresponding to circles to the left of circle 3.

Let us now consider the results of a series of drained direct shear tests on eight different samples, each consolidated to a different normal effective stress acting on the horizontal slip plane through the sample, denoted by $\sigma' = \sigma'_v$. Four samples (Nos. 1–4) are

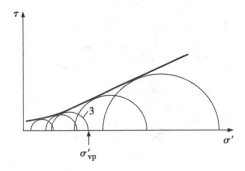

Figure 6.12 Mohr's circles corresponding to five drained triaxial tests in which samples were consolidated to different confining stresses.

normally consolidated (NC), and four (Nos. 5–8) are overconsolidated (OC) (Figure 6.13). It is observed that the shear strength of any of the OC samples is higher than that of an NC sample at the same normal stress. This means that stress history plays a role in the shear strength of OC clays. Referring to Figure 6.13, if a single sample started out with a void ratio and stress corresponding to point 1, were loaded to point 3, then unloaded to point 6, and finally reloaded all the way to point 4, its shear strength and overall mechanical behavior should conceptually be the same as that of sample 4. The effects of stress history cease to matter once the sample becomes normally consolidated again.

The eight direct shear tests show that clays have a larger shear strength at a given consolidation stress if previously subjected to a higher consolidation stress. This means that the shear strength of clay depends on stress history. We saw in Chapter 5 that the shear strength of sands depends on both effective stress and void ratio (or relative density). Naturally, we are inclined to ask the question as to whether the shear strength of clay can be explained likewise. To answer this question, we refer again to Figure 6.13. Since unloading takes place along a much flatter $\sigma_v' - e$ curve than the NCL, OC clays have lower void ratios than NC clays at the same $\sigma_v' = \sigma'$. So the "bump" on the strength envelope is due to the fact that an OC sample has a lower void ratio under the same confining stress than an NC sample and thus has a higher shear strength also. Moreover, the higher the overconsolidation ratio (OCR) is, the larger the difference between the void ratio of NC and OC samples at the same stress level is, and the greater the difference in shear strength is. The shear strength of clays is seen then as depending on both void ratio and effective consolidation stress, just as was true for sands. Differently from sands (at least for stress ranges within which particle

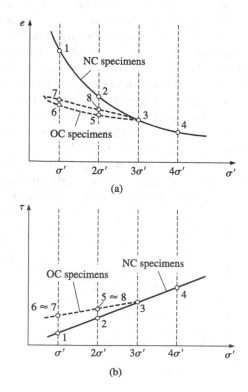

Figure 6.13 A series of eight direct shear tests showing the effects of previous loading to larger confining stresses on (a) void ratio and (b) shear strength of clay samples.

crushing is limited), clays can have different void ratios for the same confining stress (by having different preconsolidation pressures).

It is tempting to project the overconsolidated portion of the strength envelope in Figure 6.13b as a straight line to $\sigma' = 0$ so that it intercepts the shear stress axis at some value c', which would then be referred to as the "cohesion" of the clay. Schofield (1998) argues that this approach, although popular, is based on a view of the source of the excess shear strength exhibited by OC clays that is not correct. According to Schofield (1998), the excess shear strength of OC clays, as in sands, is due to dilatancy, and clays have truly no cohesion. We will discuss this in greater detail in a later section.

6.3 UNDRAINED SHEAR STRENGTH OF CLAYS

6.3.1 Consolidated undrained triaxial compression tests

The undrained shear strength of clay is typically obtained in the laboratory using the consolidated undrained triaxial compression test (TXCU). It helps us understand the results of such tests on clay samples if we examine the sequence of stress states experienced by a sample as it is extracted from the ground (using the techniques described in Chapter 7), taken to the laboratory, consolidated to an effective stress state in the triaxial test chamber, and then sheared under undrained conditions. This sequence is illustrated in Figure 6.14. The path followed by the sample in log σ'–e space as it is withdrawn from the ground and then reconsolidated in a triaxial chamber is shown in Figure 6.15. Note that the sample is subjected to an anisotropic effective stress state (point A in Figure 6.15), with $\sigma'_v \neq \sigma'_h$, and a corresponding total stress state initially in the ground. When recovered from the ground, the sample is initially contained within a sampling tube, but inevitably ends up subjected to zero total stresses. Yet, the sample does not crumble upon removal from the tube because the removal of the total stresses does not lead to an immediate change in the effective confining stress on account of its low permeability, leading instead to a negative pore pressure within the sample. Under conditions of minimum disturbance, the negative excess pore pressure that arises within the sample is approximately equal to the initial *in situ* mean effective stress σ'_{m0}:

$$u \approx -\sigma'_{m0} = -\frac{\sigma'_{v0} + 2\sigma'_{h0}}{3} \tag{6.43}$$

to be consistent with the effective stress principle, which in this case requires that

$$\sigma_{m0} = \sigma'_{m0} + u = 0$$

The effective stresses within the sample cannot immediately adjust to the new total stress state, but between sample recovery and testing there is usually time for the effective stress state within the sample to slowly change from $(\sigma'_{v0}, \sigma'_{h0})$ toward $(\sigma'_{m0}, \sigma'_{m0})$. This happens because the effective stress state seeks to converge toward the total stresses on the sample, which are isotropic. This process is one of the sources of sample disturbance. Due to disturbance related to the introduction of the sampling tube into the ground and further sample manipulation, the mean effective stress drops to a value $\sigma'_{mB} < \sigma'_{m0}$ (point B in Figure 6.15). The sample, protected by a latex membrane, is then placed within a triaxial test chamber. After that, the sample may be either anisotropically or isotropically consolidated. Because it is much more easily done, isotropic consolidation is the usual choice. Figure 6.15 shows the consolidation first to point C on the isotropic NCL, which happens along a recompression line, and then to point D further down the isotropic NCL. Samples recovered from the field and isotropically consolidated for triaxial testing are typically consolidated to the estimated mean effective stress they were subjected to originally in the field. We will discuss the reasons for this later in this chapter.

The *cell pressure* σ_c' is net of any back pressure. The *back pressure* is a pressure induced in the pores of the sample to dissolve any air bubbles that may still be left after the usual precautions to ensure sample saturation. The clay sample is allowed to consolidate under the action of the applied cell pressure, so that the effective consolidation stress itself will be equal to σ_c'. After the sample is consolidated to σ_c', the axial stress is increased by the application of a deviatoric stress $\sigma_d = \sigma_1 - \sigma_3$. This is done with the drainage valves closed, so that the shearing of the sample takes place under undrained conditions. As a result, pore pressures will develop within the sample. Skempton (1954) proposed the following equation to express the pore pressure change Δu induced within the sample by the application of principal stress increments $\Delta\sigma_1$ and $\Delta\sigma_3$:

$$\Delta u = B\left[\Delta\sigma_3 + A\left(\Delta\sigma_1 - \Delta\sigma_3\right)\right] \tag{6.44}$$

where A and B have become known as Skempton's pore pressure parameters (Figure 6.15).

It is customary to rewrite Equation (6.44) as

$$\Delta u = B\Delta\sigma_3 + \bar{A}\left(\Delta\sigma_1 - \Delta\sigma_3\right) \tag{6.45}$$

where the first term is the pore pressure increment due to the application of an isotropic stress state (equivalent to raising the cell pressure) and the second term is the pore pressure increment due to the deviatoric stress.

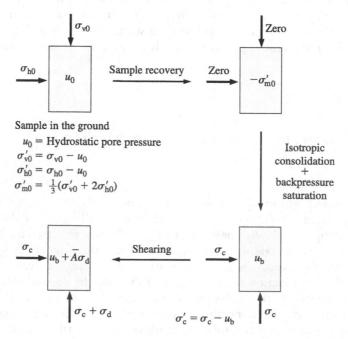

Figure 6.14 Sequence of stress states experienced by a sample as it is extracted from the ground, taken to the laboratory, and then sheared in a CU test (pore pressure values represented within samples).

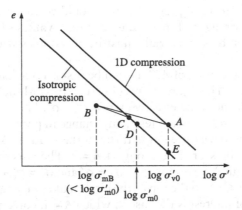

Figure 6.15 Path followed by clay sample from recovery from the ground to reconsolidation in the laboratory.

Let us use Equation (6.45) to arrive at what the value of B should be for different materials. Assuming full saturation and assuming that the stiffness of the soil particles is much greater than that of water, the sample volume V_{soil} will only contract by ΔV_{soil} if the pore water contracts by ΔV_{water}. If the isotropic compressibility of the soil skeleton and the water is b and b_w, respectively, this condition is expressed as

$$\Delta V_{soil} = \Delta V_{water}$$

$$V_{soil} b \Delta \sigma'_3 = n V_{soil} b_w \Delta u$$

where $n =$ porosity. In these equations, compressibility is used as the number that, multiplied by the stress increment, yields the volumetric strain. When multiplied by the volume, the volumetric strain yields the volume change. It follows that:

$$\Delta \sigma'_3 = n \frac{b_w}{b} \Delta u \tag{6.46}$$

Now, considering the principle of effective stress,

$$\Delta \sigma_3 = \Delta \sigma'_3 + \Delta u \tag{6.47}$$

and the definition of B in Equation (6.45) as $\Delta u / \Delta \sigma_3$, and taking Equation (6.46) into Equation (6.47), we obtain

$$B = \frac{\Delta u}{\Delta \sigma_3} = \frac{1}{1 + n \dfrac{b_w}{b}} \tag{6.48}$$

For soft to medium stiff clays (with s_u ranging roughly from 0 to 50 kPa), $b_w \ll b$, and B approaches 1 very closely. Sands and stiff clays are much less compressible than clays, but values of at least 0.98 would be required for full saturation. For very stiff soils, say sands at very large confining stresses, lower values (as low as 0.91) might be acceptable (Black and Lee 1973). The B value is an important check for engineers trying to assess the quality of any test that may have been solicited from a commercial laboratory. A low value of B in a laboratory test would lead to pore pressures that would be too low, and thus to unconservative estimates of shear strength. Table 6.2 gives target values of B for various materials. Quality laboratory tests on these materials must be on samples with B values at least equal to those values.

Now let us examine the meaning of $\bar{A}$. This can be best accomplished by referring to Figure 6.16 and to Equation (6.45). The figure shows effective stress Mohr's circles for triaxial compression test samples with the same effective stress $\sigma'_3 = \sigma'_c$. Since the total cell stress $\sigma_3 = \sigma_c$ remains constant in triaxial compression tests, any change in pore pressure must have resulted from the application of the deviatoric stress through the term $\bar{A}\,(\Delta\sigma_1 - \Delta\sigma_3)$. It follows that there is no change in pore pressure if and only if $\bar{A} = 0$. The effective stress Mohr's circle in this case (circle A in the figure) corresponds to $\sigma'_3 = \sigma'_c$ and $\sigma'_1 = \sigma'_c + (\Delta\sigma_1 - \Delta\sigma_3)$. Note how a line extending from point $(\sigma'_c, 0)$ to the top of the circle makes an angle of 45° (clockwise) with the σ' axis. Another case of interest is the case in which $\bar{A} = 1$, corresponding to circle B in the figure, for which the pore pressure increase Δu is equal to the deviatoric stress increase $\Delta\sigma_1 - \Delta\sigma_3$. In this case, $\sigma'_3 = \sigma'_c - (\Delta\sigma_1 - \Delta\sigma_3)$ and $\sigma'_1 = \sigma'_c$. Note how the line from point $(\sigma'_c, 0)$ to the top of the circle now makes an angle of 135° (or 45° measured from the other direction) with the σ' axis. Any other value of $\bar{A}$ will lead to a case between these two extremes. A circle corresponding to $\bar{A} = 0.5$ (which lies exactly halfway between $\bar{A} = 0$ and $\bar{A} = 1$) will have its center at $(\sigma'_c, 0)$. This means the line from point $(\sigma'_c, 0)$ to the top of circle C is exactly vertical, and the values of σ'_3 and σ'_1 are, respectively, $\sigma'_c - 0.5(\Delta\sigma_1 - \Delta\sigma_3)$ and $\sigma'_c + 0.5(\Delta\sigma_1 - \Delta\sigma_3)$.

Figure 6.17 shows plots of the results of two CU triaxial tests: one on an NC clay sample, and the other on an OC clay sample.[6] The plots show deviatoric stress $\sigma_d = \sigma_1 - \sigma_3$, generated pore pressure Δu, and Skempton's $\bar{A}$ parameter as a function of axial strain. Both samples

Table 6.2 Target B values for various materials

Material	B
Soft clay	1
Loose sand	>0.99
Dense sand	>0.95
Unsaturated soil	≪1
Concrete	≪1

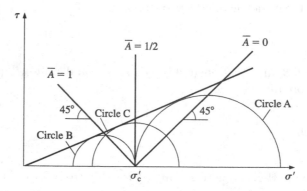

Figure 6.16 The meaning of $\bar{A}$: The three circles illustrate the cases of $\bar{A} = 0$, $\bar{A} = 1/2$, and $\bar{A} = 1$.

---

[6] The behaviors of the NC (soft) clay and of the OC (stiff) clay are qualitatively the same as those of the contractive and dilative sands loaded also under undrained conditions discussed in Chapter 5.

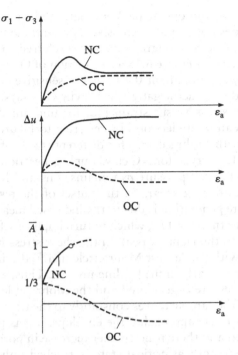

Figure 6.17 Results of CU triaxial tests on a normally consolidated clay sample and a heavily overconsolidated clay sample with the same void ratio.

have the same initial void ratio. The consolidation stress σ'_c for the NC sample is greater than that for the OC sample, as the OC sample is obtained by first consolidating it to a large effective stress and then allowing it to rebound to a void ratio matching that of the NC clay sample. As seen in Figure 6.11, NC clay contracts when sheared under drained conditions. In an undrained test, volume change is prevented and there is an increase in the pore pressure instead. The undrained behavior of the OC sample also mirrors its drained behavior. As no drainage is allowed during shearing, pore pressures start positive, then approach zero and, for samples with OCR[7] values greater than about 5, turn negative.

The value of $\bar{A}$ at zero axial strain is closely approximated by 1/3. The response of the soil sample at extremely small strains is linear elastic, meaning that the irrecoverable shear strains associated with large particle dislocations, which would lead to plastic volume change under drained conditions and to pore pressure changes under undrained conditions, do not take place at very small strains. It follows that pore pressure is generated only due to the increase in mean stress resulting from the application of an increment of axial stress σ_1. In essence, the increase in axial stress leads to an increase in total mean stress; however, the sample cannot change volume, so the effective mean stress must not change; in order for that to happen, the pore pressure needs to increase by the same amount as the total mean stress. We write this as

$$\Delta u = \Delta \sigma_m = \frac{1}{3}\left(\Delta \sigma_1 + \Delta \sigma_2 + \Delta \sigma_3\right) = \frac{1}{3}\Delta \sigma_1 = \frac{1}{3}\Delta \sigma_d$$

for $\Delta \sigma_2 = \Delta \sigma_3 = 0$ in the triaxial test. This result shows that the value of $\bar{A}$ at very small strains is indeed 1/3.

---

[7] Refer back to Chapter 4 for a definition and discussion of the overconsolidation ratio (OCR).

Of the other values of $\bar{A}$, the one at peak strength and the one at critical state are of interest. The literature does not contain extensive data on these values, but there are useful indications of the value at peak strength, sometimes referred to as the value "at failure," denoted by $\bar{A}_f$, which is given in Figure 6.18 as a function of OCR. For triaxial compression tests on NC clays, it ranges from about 0.7 to 1. For sensitive clays, it exceeds 1. It drops with increasing OCR and is in fact negative for heavily OC clays.

The peak in the shear stress versus strain plot for undrained shearing of an NC clay observed in Figure 6.17 can be understood by referring to Figure 6.19, in which the stress path is plotted together with Mohr's circles for different stages of the loading in $\sigma'-\tau$ space. In general, the values of $\bar{A}$ observed for NC clays imply generation of positive pore pressures throughout shearing. As the cell pressure σ_3 remains constant during the test, this implies σ'_3 continuously decreases during shear. At the outset of the test, the soil behaves elastically, and all pore pressure generation is due to the increased mean stress resulting from the increase in σ_1. This implies that $\bar{A} = 1/3$, which in turn implies, referring to Figure 6.19, that the stress path starts off to the right. A peak in the shear stress (the peak undrained shear strength s_{up}), associated with the largest Mohr circle (marked 2 in Figure 6.19) reached by the sample, follows relatively early in the loading process. Thus, as shearing proceeds, more and more positive pore pressure is generated and the Mohr circle representing the evolving state of effective stress in the sample moves constantly to the left, where it is bounded by the strength envelope. As the circle approaches the envelope, it gets progressively smaller until the critical state is reached and there is no further increase in pore pressure.

The undrained shear strength at critical state is reached only after large axial strains have developed. Given that loading in real problems is virtually always load-controlled, the implication of this is that, once the peak undrained shear strength is reached, very large deformations and deflections will develop in a soil mass that is part of a slope or supporting a structure.

An interesting question that arises for those planning a testing program on samples collected from the field is whether to consolidate the sample to $\sigma'_c = \sigma'_{v0}$ (initial *in situ* vertical effective stress), $\sigma'_c = \sigma'_{h0}$ (initial *in situ* horizontal effective stress), or $\sigma'_c = \sigma'_{m0}$ (initial *in situ*

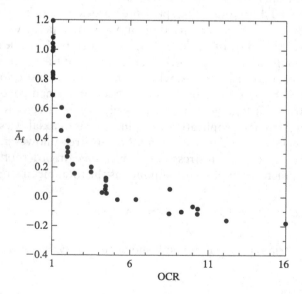

Figure 6.18 Values of $\bar{A}_f$ as a function of OCR for CU triaxial compression tests on isotropically consolidated clay. (Based on data from Mayne and Stewart (1988); with permission from ASCE.)

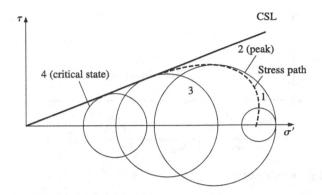

Figure 6.19 Undrained shearing of a normally consolidated clay. The evolution of the test can be understood by plotting Mohr's circles for key stages of the test. Note how, even before the first Mohr circle touches the critical-state line (Mohr's circle 3), the subsequent Mohr circles (such as 3 and 4) become progressively smaller.

mean effective stress). Because the isotropic NCL lies below the 1D NCL, it is clear that isotropic consolidation to $\sigma_c' = \sigma_{v0}'$ would lead to a void ratio (point E in Figure 6.15) that would be much lower than the *in situ* void ratio, leading to an unconservative estimate of shear strength. It turns out that isotropic consolidation to $\sigma_c' = \sigma_{m0}'$ leads to a combination of a more reasonable void ratio and a value of $\overline{A}_f$ that leads to Mohr's circles that are comparable in size to those the soil would have at its undisturbed *in situ* condition, and thus to comparable shear strength values. It is appropriate, therefore, unless a more refined (and complex) process is desired, to isotropically consolidate samples to $\sigma_c' = \sigma_{m0}'$. Naturally, to estimate σ_{m0}', we need to have an estimate of K_0. As shown in Figure 6.15, the isotropic reconsolidation path to σ_{m0}' may involve partial travel along the isotropic NCL.

6.3.2 Unconsolidated undrained tests

Imagine a direct shear test in which the confining force P and the shearing force S are applied so fast that: (1) the confining stress applied is carried entirely by the pore water pressure; (2) there is no change in effective stresses; and (3) there is no change in void ratio. In this type of test, no matter what consolidation or confining pressure is applied, there is no void ratio change and the shear strength will be the same at any load P. In other words, since the effective confining stress does not change, the shear strength is a function of the void ratio only and therefore does not change. Such tests are known as unconsolidated undrained (UU) tests.

If the results of UU tests are plotted in σ–τ (total stress) space, a horizontal strength envelope is obtained (Figure 6.20), indicating that, no matter what the total stress applied is, the shear strength remains the same. This envelope, obtained based on the total stresses applied to the sample, is referred to as the undrained failure (or strength) envelope. The figure shows three Mohr's circles that would have resulted from UU triaxial compression tests on samples consolidated to σ_3, $2\sigma_3$, and $3\sigma_3$.

The undrained strength envelope is used in total stress analyses, in which total stresses calculated for foundations, slopes, or retaining structures are compared with the undrained shear strength s_u given by

$$c = s_u = \frac{1}{2}(\sigma_1 - \sigma_3) \tag{6.49}$$

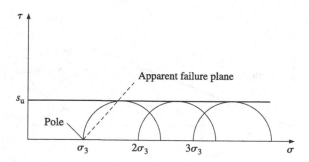

Figure 6.20 Undrained strength envelope.

with s_u being calculated usually either at peak or at the critical state. Some define a friction angle equal to zero to describe the fictitious, horizontal strength envelope in Figure 6.20. Both this friction angle and the failure envelope are artificial results of plotting total stress Mohr's circles for what is in effect the same effective stress state for all undrained tests. It is best therefore, not to generate unnecessary confusion, to avoid these terms.

A question that often arises is which of the two types of triaxial tests with undrained shearing – UU or CU – to use in practice. The knowledge of K_0 is required so that we can know which consolidation stress to impose on the sample in the case of a CU test. This is a disadvantage because K_0 can never be known with certainty. In a UU test, we count on the sample to retain most of the original effective confining stress by capillary action (which it will if handled with care) and we apply a total stress, not an effective stress. In a UU test, we don't need to know K_0 because the sample is not consolidated to different effective stresses from those it is able to retain through capillary action. One shortcoming of the UU test is precisely the lack of the consolidation stage, which is helpful in partially erasing disturbance effects.

6.3.3 Assessment of total stress analysis

Based on what we learned in Chapter 4, the total stress strength envelope in Figure 6.20 would suggest that the failure plane in a sample sheared in an undrained triaxial compression test would make an angle of 45° (because the total stress friction angle is zero) with the major principal plane π_1, which is horizontal. However, if we performed a CU or UU triaxial compression test on a clay sample, we would observe that the sample shears along a plane that makes an angle of roughly 60° with the major principal plane π_1. How can we explain this paradox?

The answer lies in the fact that the effective stresses, not the total stresses, are operational in determining the soil shear strength and stress response. The soil experiences only effective stresses (the effective stress circle at failure is shown in Figure 6.21), and thus the angle that the shear plane actually makes with π_1 is theoretically determined to be $45° + \phi/2$ or $45° + \psi/2$, depending on whether we use the stress- or strain-based way of estimating the direction of the shear plane (see Chapter 4 for details of the two approaches; the stress-based approach is illustrated in Figure 6.21). This illustrates yet again how the total stress envelope and the total stress ϕ should be avoided in our thinking and calculations because they do not relate to the reality of soil behavior. Figure 6.21 also shows that the actual shear strength τ_f along the real shear plane is lower than the undrained shear strength, raising the question of whether it would be unconservative to perform total stress analysis based on the undrained shear strength s_u to assess the stability of clay masses.

Despite the fact that total stresses do not represent the operative stress fields within a soil mass, total stress analysis offers an approximate, convenient way to analyze undrained

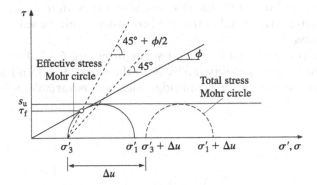

Figure 6.21 The soil experiences effective stress. The failure plane and shear strength are determined by the effective stress Mohr circle at failure, not the total stress ones.

stability problems. Unless one is performing an effective stress analysis using a constitutive model,[8] with the effective stresses following naturally from the constitutive model, total stress analysis offers the significant advantage of enabling easy calculations. The limitations of total stress analysis are alleviated by the following:

1. History and tradition (the use of s_u started before Terzaghi even proposed the concept of effective stress) and the resulting comfort engineers feel in using it.
2. Ease of use compared with effective stress analysis, for which a knowledge of the pore pressures within the soil mass (which requires a very good constitutive model and, for most problems, a numerical analysis framework, such as the finite element method) and better understanding of the mechanics of soil is required.
3. Effectiveness (it usually works well in most practical problems).
4. Laboratory compensation for unconservatism (sample disturbance reduces undrained shear strength by 10%–20%, which roughly compensates for the amount by which s_u exceeds τ_f).

6.4 CRITICAL STATE, RESIDUAL, AND DESIGN SHEAR STRENGTHS

6.4.1 Critical-state plots

To fully understand both the drained and undrained responses of clays, it is necessary to resort again to the critical-state framework. Recall that the critical state is a state in which the soil is consistent with the applied stress state and therefore does not need to change volume under drained conditions or adjust the effective stress state by changes in pore pressure under undrained conditions. When shearing of the soil starts, if it is too dense for the applied stress state, two outcomes are possible: (1) the soil dilates under drained conditions until the soil volume is consistent with the effective stress state or (2) the effective stresses increase (through drops in the pore pressure) under undrained conditions until the effective stress state is consistent with the soil volume. The process works similarly, but in reverse directions, for a soil that is initially too soft (with too high a void ratio) for the applied effective stress state. Such a soil would contract under drained conditions until reaching a void

---

[8] A constitutive model is a set of equations that describe the interrelationship of state variables of the soil; these equations allow us to apply loading on a soil element and calculate the corresponding strain, effective stress, and fabric tensors.

ratio that would be consistent with the effective stress state in it. Under undrained conditions, the mean effective stress would drop until consistent with the void ratio of the soil and the applied shear stress.

The critical-state line is shown in e–p'–q space in Figure 6.22. Figures 6.23 and 6.24 illustrate in p'–e and p'–q spaces separately, both the drained and undrained responses of clay in terms of the evolution of the Cambridge shear stress variable $q = \sigma_1 - \sigma_3 = \sigma_d$ with the

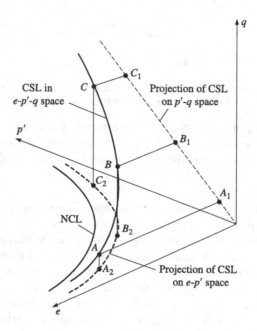

Figure 6.22 Critical-state line in e–p'–q space.

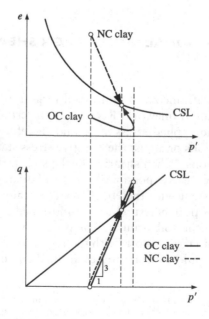

Figure 6.23 Critical-state plot for drained loading.

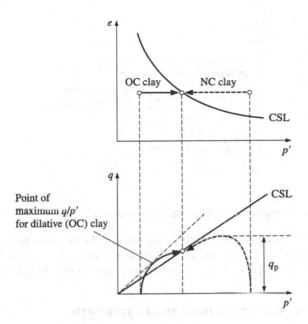

Figure 6.24 Critical-state plot for undrained loading.

mean effective stress $p' = \sigma'_m$ and the void ratio e of the soil. We can see in Figure 6.23 that the first time a sample of OC (dilative) clay passes through the value of p' it will ultimately have at critical state, it has not yet reached critical state. The clay is still too dense to be consistent with that value of p', needing to dilate further so the void ratio increases until it becomes consistent with p' further along the test (it is only at that point that the critical state is reached). During the dilation phase of the test, the drained shear strength of the OC clay develops a shear strength greater than the critical-state shear strength it will ultimately have because p' continues to increase (up to a point) before dropping back toward its critical-state value, while e remains lower than its critical-state value until critical state is reached. This response was also observed in Chapter 5 for sands.

6.4.2 Design shear strength

As shown in Figures 6.24 and 6.25, an NC clay sheared under undrained conditions first hits a peak deviatoric stress q_p and then reaches a maximum principal stress ratio when it reaches the critical-state line. This reverses for a heavily OC clay, which first hits a maximum principal stress ratio value and then a maximum deviatoric stress. The points corresponding to maximum principal stress ratio, maximum deviatoric stress, and critical-state strengths both for NC and OC clays are shown for the corresponding typical effective stress paths in Figure 6.25. A question that arises is: which one (maximum principal stress ratio or maximum deviatoric stress) should be used to define the design shear strength?

We can best answer this question by remembering that we design by avoidance of limit states. So, which shear strength to use is intimately connected with what the limit state is. If the limit state is one in which large shear strains are expected to develop in the soil, the appropriate shear strength to use is the critical-state shear strength. If the limit state generates only small shear strains in a given soil region, then the shear strength in that region could very well be the peak shear strength or values around it. We have to be very prudent if we use a shear strength greater than the critical-state shear strength though. All excess shear strength is due to interlocking, to a soil density that tends to degrade toward zero. This drop

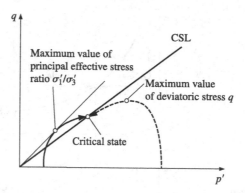

Figure 6.25 Design shear strengths for an NC clay and a heavily OC clay.

of the excess shear strength toward zero is seen in our laboratory tests as the shear strain increases, but it could also happen as a result of load cycles, freeze–thaw cycles, and other such loading patterns that the soil might experience during the design life of the structure.

6.4.3 Correlations for undrained shear strength

We sometimes rely on correlations of undrained shear strength with the vertical effective stress. In areas where extensive deposits of clays exist, these clays have typically been studied rather extensively, and there is good confidence on the likely value of s_u / σ_v'. Based on such estimates of s_u / σ_v', engineers can estimate design s_u values based on the knowledge of the soil profile.

For NC clays, Skempton (1957) proposed the following correlation for s_u / σ_v' in terms of plasticity index (PI):

$$\frac{s_u}{\sigma_v'} = 0.11 + 0.0037(\text{PI}) \tag{6.50}$$

Another correlation is available in terms of the critical-state friction angle ϕ_c of the clay (Wroth 1984):

$$\frac{s_u}{\sigma_v'} = \frac{\phi_c}{100} \tag{6.51}$$

As stated earlier, ϕ_c ranges from roughly $15°$ to as high as $30°$. This implies a 0.15–0.30 range for s_u/σ_v'. For OC clays, Ladd et al. (1977) proposed the following correlation:

$$\left(\frac{s_u}{\sigma_v'}\right)_{\text{OC}} = \left(\frac{s_u}{\sigma_v'}\right)_{\text{NC}} \text{OCR}^{0.8} \tag{6.52}$$

Having equations that tell us how much s_u varies with σ_v' and OCR allows us to plot profiles of s_u versus depth (as readers are asked to do in Problem 6.23). The OCR may be known from historic considerations (we may know, for example, that there were loadings due to soil that was removed or structures that were demolished at the site), or it may be deduced from geologic considerations or (and this is more difficult) from *in situ* test measurements. In connection with *in situ* tests, their measurements reflect the influence of the OCR, as do the s_u values calculated in Problem 6.23. Determining the OCR from the measurements is, however, not straightforward, given the dependence of the measurements on a myriad of other factors, not all of which are known well.

6.4.4 Residual shear strength

The critical-state friction angle ϕ_c in sands was a lower bound on the possible values of friction angle. In clays, a lower friction angle, called the residual friction angle ϕ_r, may develop. Recall that the critical state is a natural state of flow (shearing) resulting from the equilibrium between the applied confining stress, the shear stress, and the void ratio, all of which remain unchanged during shearing. This presupposes that the soil particles are in a somewhat random arrangement. This lack of order in the particle arrangement is natural in sands, which are made of bulky particles; however, in clays, if shearing proceeds in a given direction, the platy, tube-like, or needle-like clay particles will eventually align themselves with that direction, and a friction angle lower than ϕ_c – the *residual friction angle* – becomes possible.

The development of a residual friction angle requires very large shear strains (of 20% and even larger). The stress–strain response of a soil element sheared to such large strains would look like that shown in Figure 6.26. In the case depicted in the figure, an OC clay is sheared to very large strains. There is first a peak in shear stress due to soil dilatancy, followed by the critical-state plateau and finally by another plateau determined by the residual strength of the soil.

The relevance of residual friction angles is restricted to calculations involving preexisting slip surfaces or problems in which large shear strains are expected to develop. One case in practice in which it is relevant is in stability calculations for soil masses within which there has been sliding in the past; the shear strength along the former slip surface is calculated using the residual and not the critical-state shear strength. The installation of most piles, as discussed in greater detail in Chapter 13, also generates residual shear strength along the pile shaft. In the laboratory, the residual strength is typically studied using special laboratory devices such as the ring shear apparatus (e.g., Bishop et al. 1971; Ramsey et al. 1998; Stark and Eid 1994).

The residual shear strength is the product of the normal effective stress on the shearing plane by the tangent of the residual friction angle ϕ_r, which in turn decreases with increasing effective normal stress σ' on the plane of shearing. The drop in ϕ_r with increasing σ' occurs because a larger normal stress forces a greater alignment of particles as they are sheared. Figure 6.27 illustrates schematically the decrease in ϕ_r with increasing normal stress on the plane of shearing. At very large stresses, ϕ_r reaches an absolute minimum, denoted by $\phi_{r,min}$. For σ' on the shearing plane approaching zero, ϕ_r approaches the critical-state friction angle ϕ_c. The assumption that, at σ' equal to zero, $\phi_r = \phi_c$ is based on the expectation that there would be negligible reorientation of particles in the absence of a normal stress forcing this reorientation to happen. Following Maksimović (1989), ϕ_r can be expressed as

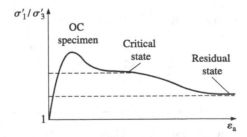

Figure 6.26 Drained stress–strain response of an overconsolidated clay: the peak is due to dilatancy; a first plateau is reached when the critical state is reached; and a second plateau forms when the clay particles have been aligned with the direction of shearing and only residual shear strength is then available.

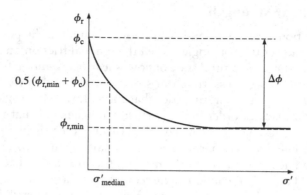

Figure 6.27 Variation of ϕ_r with normal stress perpendicular to the plane of shearing.

$$\phi_r = \phi_{r,min} + \frac{\phi_c - \phi_{r,min}}{1 + \dfrac{\sigma'}{\sigma'_{median}}} \tag{6.53}$$

where σ' is the effective normal stress on the plane of shearing and σ'_{median} is the value of σ' at which the friction angle is exactly equal to the average of the minimum residual friction angle $\phi_{r,min}$ and ϕ_c.

In addition to varying with pressure normal to the direction of shearing, ϕ_r also varies with sand content. Skempton (1985) showed that if the clay content exceeds 52%, ϕ_r is the same as for pure clay. For clay contents <25%, there is no residual friction angle that is lower

Table 6.3 Properties of Lower Cromer Till, Boston Blue Clay, San Francisco Bay Mud, London Clay, and Weald Clay

Soil	Mineralogy	CF (%)	PI (%)	A	ϕ_c (°)	$\phi_{r,min}$ (°)	References
Boston Blue Clay	Illite, quartz	35	13.1	0.37	32.4[a]	—	Ladd and Varallyay (1965)
London Clay	Kaolinite, illite, montmorillonite, quartz	53–62	42–45	0.73–0.79	21.3	9.4[b]	Bishop et al. (1971), Gasparre (2005), and Nishimura (2005)
Lower Cromer Till	Illite, calcite, quartz	14–20	10–12	0.60–0.71	30.0	—	Lupini et al. (1981), Gens (1982), and Dafalias et al. (2006)
San Francisco Bay Mud	Illite, montmorillonite	47	47	1.00	28.9[a]	16.2	Kirkgard and Lade (1991) and Meehan (2006)
Weald Clay	Illite, kaolinite, illite-montmorillonite, vermiculite	52	33	0.63	20.9	8.3	Parry (1960), Bishop et al. (1971), and Akinlotan (2017)

Source: Sakleshpur et al. (2021).

[a] Extrapolated value corresponding to 30% axial strain (Chakraborty 2009).

[b] Value corresponds to blue London Clay at Wraysbury (CF=57%, PI=43%, A=0.75). For brown London Clay at Walthamstow (CF=53%, PI=42%, A=0.79), $\phi_{r,min}$=7.5° (Bishop et al. 1971).

Note: CF, clay fraction; PI, plasticity index; A, activity; ϕ_c, critical-state friction angle (obtained from triaxial compression test results); and $\phi_{r,min}$, minimum residual friction angle (obtained from ring shear test results).

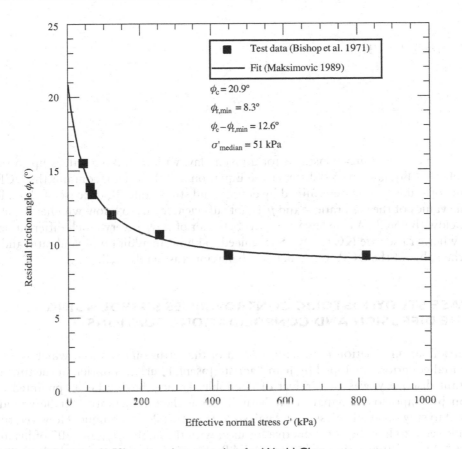

Figure 6.28 Fit of Equation (6.53) to ring shear test data for Weald Clay.

than ϕ_c. For clay content of 25%–52%, the lowest achievable ϕ transitions from ϕ_c to the ϕ_r of pure clay. Because silt particles are relatively small, they don't interfere significantly with the development of residual shear strength. In fact, many naturally occurring clays, such as lacustrine clays, have considerable silt content, and yet their friction angles exhibit significant drops toward residual upon normal effective stress increases in the direction normal to shearing.

Table 6.3 shows data on critical-state and residual shear strength for several clays. Figure 6.28 illustrates the fit of Equation (6.53) to data for a specific clay. The fit is done by first estimating the critical-state friction angle from the triaxial compression test data of Parry (1960) and then finding the values of σ'_{median} and $\phi_{r,min}$ that minimize the sum of least squares.

6.5 SMALL-STRAIN STIFFNESS

As for sands, the stress–strain response of clays is also strongly nonlinear from very small strains. However, elastic response does exist for shear strains smaller than ~10^{-5}. The small-strain shear modulus equation for clays takes the form (Hardin and Black 1968):

$$\frac{G_0}{p_A} = C_g F(e) \left(\frac{p'}{p_A} \right)^{n_g} OCR^{m_g} \tag{6.54}$$

where:

$$\sigma'_m = \left(\sigma'_v + 2\sigma'_h \right)/3;$$

$$F(e) = \frac{\left(e_g - e\right)^2}{1 + e} \qquad\qquad (6.55)$$

where C_g, e_g, n_g, and m_g=constants for a given clay, with m_g taking values up to roughly 0.5. Note that Equation (6.54) differs from Equation (5.24) in that it contains the OCR. The behavior of sands is fully determined by density and stress state. In the case of clays, having only the values of the void ratio e and p' is not sufficient for us to know whether the clay lies on or below the NCL. As we have seen, the behavior of clays is very much different depending on whether they are NC or OC. So we need to know, in addition to void ratio and stress state, the value of the OCR to predict clay behavior reasonably well.

6.6 CASE STUDY: HISTORIC CONTROVERSIES SURROUNDING THE DIFFUSION AND CONSOLIDATION EQUATIONS

The consolidation equation is actually a form of the diffusion equation, which is a partial differential equation developed by Jean Baptiste Joseph Fourier.[9] Fourier, at the time of this important discovery, was the Prefect of Grenoble, France, having been appointed to that position by Napoleon Bonaparte. He actually went there reluctantly, for he would have preferred to stay on as a Professor of Analysis at the Ecole Polytechnique. However, times in Grenoble seem to have been productive for him, with the publication in 1807 of his memoir entitled "On the Propagation of Heat in Solid Bodies," which was read to the Paris Institute on December 21, 1807.

A committee consisting of Lagrange, Laplace, Monge, and Lacroix, some of whom were Fourier's earlier Professors, was set up to analyze and report on this work. The committee and others raised a number of objections to Fourier's work, including his expansion of functions as trigonometric series (now known as Fourier series) and his derivation of the diffusion equation (this criticism was by Biot, who had published a paper on the subject, now known to be in error). Another committee, consisting of some of the same members, later evaluated the same work with a few additions entered by Fourier in a competition. Even though he won the competition, the final report of the committee was less than enthusiastic, containing the statement: "...the manner in which the author arrives at these equations is not exempt of difficulties, and his analysis to integrate them still leaves something to be desired on the score of generality and even rigor." Time would show the irony in this, for Fourier's contributions in this work are tremendous contributions that are routinely used today by mathematicians, scientists, and engineers and have served as the basis for much additional mathematical research.

---

[9] Largely based on: Herivel, J. (1975). *Joseph Fourier: The Man and the Physicist*. Clarendon Press, Oxford; Goodman, R. E. (1999). *Karl Terzaghi, The Engineer as Artist*, ASCE Press, Reston, VA; and de Boer, R., Schiffman, R. L., and Gibson, R. E. (1996). "The origins of the theory of consolidation: The Terzaghi-Fillunger dispute." *Geotechnique* 46(2), 175–186.

Unfortunately, critics insisted on questioning Fourier's theory and raising other obstacles to its publication and acceptance. For example, Biot claimed (unjustly) priority over Fourier, and Poisson attacked the theory and claimed to have an alternative, superior theory. About these claims, Fourier had the following to say:

Having contested the various results, Biot and Poisson now recognize that they are exact, but protest that they have invented another method of expounding them and that this method is excellent and the true one. If they had illuminated this branch of physics by important and general views and had greatly perfected the analysis of partial differential equations, if they had established a principal element of the theory of heat by fine experiments... they would have the right to judge my work and to correct it. I would submit with much pleasure... But one does not extend the bounds of science by presenting, in a form said to be different, results which one has not found oneself and, above all, by forestalling the true author in publication.

Controversies such as this one surrounding Fourier's brilliant work are not rare in academic and research circles, as was evidenced even more clearly by the injustice done to Wegener over a century later, who never saw his seminal contributions to what would later become tectonic plate theory become accepted and used in geology to explain so many important geologic phenomena, from earthquakes to the formation of igneous rocks. It is just natural then that there would be one more controversy involving the diffusion equation, now in its soil mechanics incarnation. This next controversy involved none other than the founder of soil mechanics, Karl Terzaghi.

In 1920, Terzaghi, then a Professor at Robert College in Turkey, was converging on what we now know as consolidation theory. Terzaghi had been deeply influenced by his mentor, Prof. P. Forchheimer, who was the first to recognize the analogy between electrical and hydraulic flows. Terzaghi had been able to use Forchheimer's findings to study groundwater flow through pervious soils and had afterward embarked on a quest to solve the problem of flow in comparatively impervious, compressible clays. In 1920, Terzaghi had been working with what is now known as the consolidometer to study clay compression. The mathematical formulation of the observations, however, eluded him until 1923, when he first formulated the principle of effective stress, according to which only the effective stresses are operative in bringing about compression or in determining mechanical behavior of soils. After that, by consulting books on heat conduction, he found that the diffusion equation proposed more than 100 years earlier by Fourier could be used for soil consolidation as well. In resorting to this analogy, Terzaghi followed Forchheimer, who had also used analogies between water flow and heat flow on earlier occasions. In 1925, Terzaghi published his findings in an *Engineering News Record* paper.

After a stint as an academic in the United States in the 1920s, Terzaghi went back to Austria, his home country, as a Professor. He continued to work on the development of soil mechanics. In 1936, he chaired the first International Conference on Soil Mechanics and Foundation Engineering (ICSMFE) at Harvard. Upon his return, he found that a colleague, Paul Fillunger, whose work concerning concrete gravity dams he had previously criticized, had started a massive campaign against soil mechanics and the theory of consolidation in particular. Fillunger had developed an alternative differential equation for the consolidation of clay based purely on mathematical analysis of the problem, an approach different from Terzaghi's, which was based on experimental observations, only later followed by the search for a suitable equation to explain the observations. The dispute between Terzaghi and Fillunger was ferocious, including the distribution of pamphlets by Fillunger and a letter-writing campaign by Terzaghi. A special committee was appointed to evaluate the merits of Terzaghi's research on consolidation and of Fillunger's criticism of it. The dispute had ups and downs, and Terzaghi believed at moments that he would not prevail. The committee at some point started leaning toward Terzaghi's side, after one of the committee members showed that Terzaghi's consolidation equation was in fact a special case of Fillunger's equation after "nonessential" factors were omitted. When it became clear to Fillunger that he was not going to come out the winner in the controversy, he and his wife committed suicide on February 14, 1937. Consolidation theory had a bloody start. Ironically, de Boer et al. (1996) showed that Fillunger's consolidation formulation was in fact rigorously correct, forming the basis of what we now call the mechanics of porous media.

6.7 CASE STUDY: THE LEANING TOWER OF PISA (PART II)

What lies behind the large settlements of the Tower of Pisa?[10] Naturally, a thick layer of clay. While we will examine the probable cause of the lean in Part III of the case history (at the end of Chapter 9), let us examine now the soil profile underneath the Tower and, in particular, some properties of the clay layer.

The soil profile is shown somewhat generally in Figure 6.29. It consist of Horizon A of mostly sand on the north side and sand mixed with clay and silt on the south side, Horizon B of clay (within which a relatively thin layer of sand exists), and Horizon C of dense silty sand. Geotechnical index properties are given in Table 6.4, and soil gradations, in Table 6.5. The water table is 1–2 m below the ground surface. In Problem 6.23, the reader is asked to use the index properties to arrive at an approximate shear strength profile for the clay. In Problem 6.24, the reader is asked to calculate the one-dimensional compression that would result from the application of a unit load equal to that on the base of the Tower's foundation on top of the soil profile at the Tower's site.

Table 6.4 Approximate soil profile at the Tower of Pisa site

Formation	Layer	Soil type	Sublayer	Elevation of top	Elevation of base	γ (kN/m³)	LL (%)	PI (%)	wc (%)	LI
A	A_1	Sandy–clayey silts	—	0.0	−5.4	18.9	35.0	12.9	30.9	0.68
	A_2	Silty sands	—	−5.4	−7.4	18.1	—	—	37.3	—
B	B_1	Upper clays	1	−7.4	−10.9	17.0	73.1	42.7	50.7	0.56
			2	−10.9	−12.9	17.5	59.2	32.7	46.5	0.73
			3	−12.9	−17.8	16.7	70.9	41.3	56.3	0.72
	B_2	Intermediate clays	4	−17.8	−19.0	19.5	53.2	33.3	27.7	0.27
			5	−19.0	−22.0	19.8	46.3	22.8	27.4	0.15
	B_3	Intermediate sands	6	−22.0	−24.4	19.1	33.0	8.6	29.4	—
	B_4	Lower clays	7	−24.4	−29.0	18.6	59.7	33.9	38.5	0.41
			8	−29.0	−30.4	18.4	48.9	22.6	37.1	0.57
			9	−30.4	−34.4	19.0	54.5	31.0	32.5	0.21
			10	−34.4	−37.0	19.4	51.3	29.0	31.2	0.34
C	—	Lower sands	—	−37.0	−68.0	20.5	—	—	20.6	—

Source: Data from Jamiolkowski (2006).

---

[10] All references for this case history are at the end of Part IV in Section 11.5.

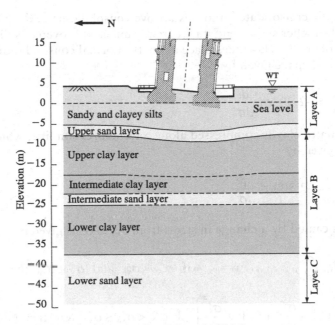

Figure 6.29 Soil profile at the Tower of Pisa site (Jamiolkowski 2006).

Table 6.5 Gradation of soil layers at the Tower of Pisa site

Formation	Layer	Soil type	Sublayer	Elevation of top	Elevation of base	Sand (%)	Silt (%)	Clay (%)
A	A_1	Sandy–clayey silts	—	0.0	−5.4	30.1	56.7	12.5
	A_2	Silty sands		−5.4	−7.4	56.4	30.3	13.2
B	B_1	Upper clays	1	−7.4	−10.9	14.2	32.2	53.0
			2	−10.9	−12.9	2.2	51.9	46.0
			3	−12.9	−17.8	2.0	37.7	60.0
	B_2	Intermediate clays	4	−17.8	−19.0	6.6	48.1	44.6
			5	−19.0	−22.0	16.3	51.4	33.0
	B_3	Intermediate sands	6	−22.0	−24.4	57.7	33.6	8.6
	B_4	Lower clays	7	−24.4	−29.0	2.8	50.0	47.3
			8	−29.0	−30.4	12.7	53.7	33.7
			9	−30.4	−34.4	6.4	62.7	31.5
			10	−34.4	−37.0	1.3	52.5	46.3
C		Lower sands	—	−37.0	−68.0	82.5	11.9	5.5

The case history continues at the end of Chapter 9, where we provide more information about the soil profile and explore the likely reason for the initiation of the tower lean.

6.8 CHAPTER SUMMARY

6.8.1 Main concepts and equations

A clay layer will undergo compression over time to dissipate positive **excess pore pressures**; it will expand to dissipate negative excess pore pressures. The excess pore pressures may appear as a result of the application or removal of loads on the surface of the soil deposit or as a result of changes in the groundwater regime.

A clay is **normally consolidated** and has an **overconsolidation ratio** (OCR)=1 if it has never experienced a higher stress than it is currently under. It is **overconsolidated** otherwise. The compression of an NC clay takes place along the **normal consolidation line**, which has a slope C_c in e–$\log \sigma_v'$ space given by

$$C_c = -\frac{de}{d\log\sigma_v'} = -2.303\sigma_v' \frac{de}{d\sigma_v'} \tag{6.2}$$

An **overconsolidated** clay is compressed along a **recompression line,** which has a slope C_s in e–$\log \sigma_v'$ space given by

$$C_s = -\frac{de}{d\log\sigma_v'} = -2.303\sigma_v' \frac{de}{d\sigma_v'} \tag{6.1}$$

The total strain caused by a change in stress from σ_{v0}' to σ_v' is given by

$$\Delta\varepsilon_z = \begin{cases} \dfrac{C_c}{1+e_0}\log\dfrac{\sigma_v'}{\sigma_{v0}'} & \text{if } \sigma_{v0}' = \sigma_{vp}' \text{ and } \sigma_v' \geq \sigma_{vp}' \text{ (NC)} \\[3mm] \dfrac{1}{1+e_0}\left[C_s \log\dfrac{\sigma_{vp}'}{\sigma_{v0}'} + C_c \log\dfrac{\sigma_v'}{\sigma_{vp}'} \right] & \text{if } \sigma_{v0}' < \sigma_{vp}' \leq \sigma_v' \text{ (OC then NC)} \\[3mm] \dfrac{C_s}{1+e_0}\log\dfrac{\sigma_v'}{\sigma_{v0}'} & \text{if } \sigma_{v0}' \leq \sigma_{vp}' \text{ and } \sigma_v' \leq \sigma_{vp}' \text{ (OC)} \end{cases} \tag{6.5}$$

The total compression ΔH of a soil layer with thickness H can be calculated by dividing the layer into thin sublayers, each with thickness H_i, calculating $\Delta\varepsilon_{zi}$ for each sublayer, and then using the following equation:

$$\Delta H = \sum_{i=1}^{n} H_i \Delta\varepsilon_{zi} \tag{6.11}$$

with

$$H = \sum_{i=1}^{n} H_i \tag{6.12}$$

The compression index C_c may be estimated using:

$$C_c = \frac{1}{200} G_s (\text{PI})(\%) \tag{6.13}$$

The recompression index varies from 10% to 20% of C_c.

Compression or swelling in clay takes time (sometimes, years) to happen because excess pore pressures require a long time to dissipate due to the very low hydraulic conductivity of the clay. The **consolidation equation** describes the evolution of the excess pore pressure in time and space:

$$\frac{\partial u}{\partial t} = c_v \frac{\partial^2 u}{\partial z^2} \tag{6.32}$$

where c_v is called the **coefficient of consolidation** and is defined as

$$c_v = c_v(z, t) = \frac{K}{m_v \gamma_w} \tag{6.30}$$

The most common initial and boundary conditions for the consolidation problem are:

$u = p_v$ for all values of z at $t = 0$ (initial condition),
$u = 0$ for all times at $z = 0$ (immediate-drainage boundary condition), and
$\partial u / \partial z = 0$ at $z = H$ (no-flow boundary condition).

The solution for the excess pore pressure in terms of position z and time t in that case is:

$$\frac{u}{p_v} = \sum_{n=0}^{\infty} \frac{2}{N} \sin\left(\frac{Nz}{H}\right) \exp\left(-N^2 T_v\right) \tag{6.33}$$

The **degree of consolidation** U, defined as the fraction of the total compression that has taken place, is usually of interest, instead of the values of pore pressure for every depth. It can be calculated at any given time as 1 minus the ratio of the integral of the excess pore pressures over depth to the integral of the initial excess pore pressures over depth. The following approximate equations are solutions to the boundary-value problem in terms of the degree of consolidation:

$$U = \sqrt{\frac{4 T_v}{\pi}} \tag{6.36}$$

or

$$T_v = \frac{\pi}{4} U^2 \tag{6.37}$$

for $0 \leq U \leq 0.526$, $0 \leq T_v \leq 0.217$, and

$$U = 1 - 10^{-1.0718(T_v + 0.085)} \tag{6.38}$$

or

$$T_v = -0.933 \log (1 - U) - 0.085 \tag{6.39}$$

for $0.526 < U \leq 1$ and $0.217 < T_v \leq 1.781$.

In these equations, T_v is the **time factor** for **one-dimensional compression**, which is given by

$$T_v = \frac{c_v t}{H^2} \tag{6.34}$$

where H is the layer thickness when a single-drainage boundary is present and half the layer thickness when double drainage is present.

Some compression may continue after consolidation is complete, that is, after $U = 1$. This is called **secondary compression** or **secondary consolidation**. It is an issue in practice mostly with very soft clays and organic soils.

Isotropic consolidation can be treated mathematically in the same way as one-dimensional consolidation. The isotropic NCL is parallel to the one-dimensional NCL, and the slope of recompression or swelling lines is also the same.

Clays take a long time to consolidate. Likewise, shear stresses would need to be applied very slowly for the drained shear strength of the clay to develop. The critical-state friction angle of clays is much lower than that of sand, falling in most cases in the 15°–30° range, versus 28°–36° for sand. Additionally, clays can develop a shear strength even lower than the critical-state shear strength due to the alignment of their platy, tube-like, or needle-like particles in the direction of shearing when shear strains are large, called the **residual shear strength**. This shear strength is that available against sliding in a preferential sliding direction, in contrast to the critical-state shear strength, which reflects a random particle arrangement during shearing. **Residual friction angles** can be as low as 5° or so for smectites.

In drained loading, stiff (highly overconsolidated) clay behaves very much like dense sands sheared under drained conditions. Likewise, soft clays (normally consolidated to lightly overconsolidated clays) behave much like loose (contractive) sands. This similarity between dense sands and stiff clays on the one hand and contractive sands and soft clays on the other holds also for undrained loading.

Most loads in engineering practice are applied much faster than there is time for excess pore pressure dissipation in clay. This means that in most cases we are interested in the **undrained shear strength** of clays. During undrained shearing of clays, pore pressures appear. These can be quantified using the following equation:

$$\Delta u = B\Delta\sigma_3 + \bar{A}(\Delta\sigma_1 - \Delta\sigma_3) \tag{6.45}$$

where $B \approx 1$ for fully saturated soils (with the approximation being excellent for soft clays and very good for loose sands, but less so for very stiff clays and dense sands). The values of $\bar{A}$ for an undrained triaxial test (CU or UU) are of the order of 0.3 at the start of loading and 0.5–1 at the peak shear stress for an NC clay. For an OC clay, $\bar{A}$ will decrease with increasing OCR and may reach 0 or even small negative values for a high OCR (typically exceeding 5).

The undrained shear strength of clays is denoted by s_u. It can be measured in the laboratory using **CU** or **UU** tests. If using CU tests, samples are consolidated to the estimated mean effective stress *in situ*. Correlations exist for the s_u of NC clays in terms of plasticity index and critical-state friction angle:

$$\frac{s_u}{\sigma'_v} = 0.11 + 0.0037(\text{PI}) \tag{6.50}$$

$$\frac{s_u}{\sigma'_v} = \frac{\phi_c}{100} \tag{6.51}$$

For OC soils, the following equation can be used:

$$\left(\frac{s_u}{\sigma'_v}\right)_{OC} = \left(\frac{s_u}{\sigma'_v}\right)_{NC} \text{OCR}^{0.8} \tag{6.52}$$

6.8.2 Notations and symbols

Symbol	Quantity	US units	SI units
A	Skempton's pore pressure parameter	Unitless	Unitless
$\bar{A}$	Skempton's pore pressure parameter $(= AB)$	Unitless	Unitless
$\bar{A}_f$	$\bar{A}$ at failure or at peak strength	Unitless	Unitless
a_v	Coefficient of compressibility	ft²/lb	m²/kN
B	Skempton's pore pressure parameter	Unitless	Unitless
B	Isotropic compressibility of soil skeleton	ft²/lb	m²/kN
b_w	Isotropic compressibility of water	ft²/lb	m²/kN
C_c	Compression index	Unitless	Unitless
C_s	Swelling or recompression index	Unitless	Unitless
c_v	Coefficient of consolidation	ft²/s	m²/s
C_α	Secondary compression index	Unitless	Unitless
e_0	Initial void ratio	Unitless	Unitless
H	Thickness of clay layer or maximum drainage path	ft	m
h	Hydraulic head	ft	m
h_e	Elevation head	ft	m
h_p	Pressure head	ft	m
i	Hydraulic gradient	Unitless	Unitless
K	Hydraulic conductivity	ft/s	m/s
k	Intrinsic permeability	in.²	mm²
m_v	Coefficient of volume compressibility	ft²/lb	m²/kN
OCR	Overconsolidation ratio	Unitless	Unitless
P	Confining force	lb	kN
p_v	Vertical pressure (surcharge)	psf	kPa
Q	Flow rate	ft³/s	m³/s
S	Shearing force	lb	kN
s_u	Undrained shear strength	psf	kPa
t	Time	s	s
T_v	Time factor	Unitless	Unitless
U	Degree of consolidation	%	%
V	Volume	ft³	m³
v	Specific discharge	ft/s	m/s
z	Depth	ft	m
ε_z	Vertical strain	in./in.	mm/mm
σ'_v	Vertical effective stress	psf	kPa
σ'_{v0}	Initial vertical effective stress	psf	kPa
σ'_{vp}	Preconsolidation pressure	psf	kPa
ϕ_r	Residual friction angle	deg	deg
$\phi_{r,min}$	Minimum residual friction angle	deg	deg

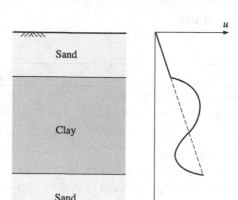

P-Figure 6.1 Pore pressure distribution inside a clay layer for Problem 6.2.

6.9 PROBLEMS

6.9.1 Conceptual problems

Problem 6.1 Define all the terms in bold contained in the chapter summary.

Problem 6.2* Simultaneous application of a surcharge at the top of a clay layer and a vacuum at its base produces, after some time, the pore pressure distribution shown in P-Figure 6.1. Sketch what the pore pressure distribution would look like (a) after a finite amount of time and (b) at the end of consolidation.

Problem 6.3 In a reclaimed site, samples of a soil taken to the laboratory and tested in the consolidometer give a preconsolidation pressure, using the Casagrande method, much less than the vertical effective stress one would expect based on calculations using the current soil profile and water table location. Discuss the reason for this.

6.9.2 Quantitative problems

Problem 6.4* Consider an ideal clay with activity $A=0.5$ and plasticity index $PI=50\%$. Its preconsolidation stress is equal to $200\,kPa$. Its recompression index is zero. Estimate ϕ_c for this clay, and plot the strength envelope in σ' versus τ space for this clay for values of $\sigma'=0\text{–}1000\,kPa$.

Problem 6.5 A consolidation test is performed on a sample of clay with a thickness of $25\,mm$. In one stage of the test, after the sample has been compressed by $2\,mm$, the stress was raised by $50\,kPa$. (a) Given that the final settlement under this stress increase was $1\,mm$, what is the value of m_v for this sample at this stage in the test? (b) Is m_v likely to be the same as the stress is further increased? (c) Given that c_v for this soil is known to be $2\,m^2/year$, how much time was required for the sample to settle by $0.25\,mm$? The sample is drained from both top and bottom. (d) How much settlement was observed after 3 minutes? (e) Comment on whether c_v may be taken as a constant throughout the consolidation test (and not for just this stage).

Problem 6.6 If the initial vertical effective stress for the stress increment analyzed in Problem 6.5 is $200\,kPa$ and the soil is NC with $e_0=1.3$, how much is the C_c of this soil?

Problem 6.7 The results of a consolidation test on a sample of clayey soil are shown in P-Table 6.1. For this test, (a) draw the e–$\log \sigma'_v$ curve, (b) compute the compression index, and (c) find the preconsolidation pressure using the Casagrande procedure.

P-*Table 6.1* Results of consolidation test pertaining
to Problem 6.7

σ'_v (kPa)	e
50	1.85
250	1.82
500	1.77
1,000	1.68
2,000	1.56
4,000	1.46
8,000	1.37
16,000	1.05
5,000	1.10
1,000	1.20
250	1.28
50	1.38

Problem 6.8 A sample of an NC clay was subjected to a stress-controlled UU tri-axial compression test. The total confining stress applied to the sample was equal to 60 kPa. This sample tested under a UU condition failed when the axial stress was equal to 135 kPa. A pore pressure of 30 kPa was measured at failure. (a) Determine the undrained shear strength of the clay. (b) Determine the effective mobilized friction angle. (c) Determine the effective shear strength at failure on the failure plane.

Problem 6.9 From the same soil of Problem 6.8, a sample was prepared and sheared under drained conditions. Discuss how you might use the results of Problem 6.8 to estimate the drained shear strength that would be observed in this drained test.

Problem 6.10* For a uniform clay layer, the groundwater table is known to have never been above a depth of 5 m. Presently, the water table is at a depth of 1 m. Calculate the OCR at different depths (1, 2, 3, 4, 5, 10, 20, and 30 m). The unit weight of saturated clay is known to be 17 kN/m³. Assume that water is capable of rising by capillary action more than 5 m so that the clay layer has always been saturated.

Problem 6.11 A soil profile is shown in P-Figure 6.2a. It is afterward excavated down to 2 m, with the resulting profile shown in P-Figure 6.2b. The clay layer was

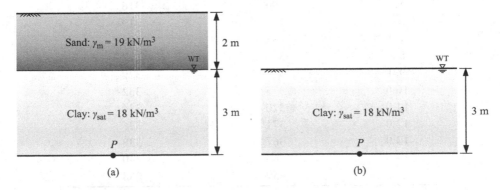

P-*Figure 6.2* Soil profile for Problem 6.11.

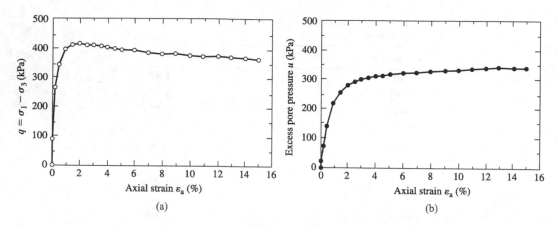

P-Figure 6.3 CU triaxial compression test on the sample of kaolinite–sand mixture of Problem 6.13: $q = \sigma_1 - \sigma_3$ and excess pore pressure u versus axial strain ε_a. (From Jafari and Shafiee (2004); reproduced with permission from the National Research Council Press 2006.)

P-Table 6.2 CU triaxial compression test described in Problem 6.13

Kaolinite 40%, sand 60%, $\sigma_c' = 500\,kPa$		
Axial strain ε_a (%)	Deviatoric stress q (kPa)	Excess pore pressure u (kPa)
0.0	0	0
0.1	87	24
0.2	266	71
0.5	341	140
1.0	395	216
1.5	410	254
2.0	414	278
2.5	410	290
3.0	410	299
3.5	406	304
4.0	402	309
4.5	398	310
5.0	394	316
6.0	391	319
7.0	383	322
8.0	378	327
9.0	379	329
10.0	375	332
11.0	370	334
12.0	371	336
13.0	367	339
14.0	363	340
15.0	359	338

Source: Data from Jafari and Shafiee (2004).

normally consolidated before the excavation. The critical-state friction angle for this clay is 25°. What is the undrained shear strength s_u of the clay at point P after the excavation?

Problem 6.12* A clay sample is isotropically consolidated to 300 kPa, at which point it is NC. Its critical-state friction angle is 18°. The overall total stress on the sample is suddenly reduced by 140 kPa; the sample is then sheared under undrained conditions, without any delay, in triaxial compression. Explain what happens to the sample, plotting total and effective stress Mohr's circles as needed.

Problem 6.13 A CU triaxial compression test was performed on an NC clay sample. The sample is a mixture of 40% kaolinite and 60% sand (the resulting soil behaves, for most purposes, like clay). The sample is isotropically consolidated under a pressure of 500 kPa before shearing. The results of the shearing stage of the test are shown in P-Table 6.2 and P-Figure 6.3. (a) Determine the undrained shear strength s_u both at the peak and at the end of the test. (b) Does the sample appear to be near or at the critical state at the end of the test? If so, what is your estimate of the critical-state friction angle for this soil? (c) Determine Skempton's $\bar{A}$ in the initial, peak, and final stages of the test. How does $\bar{A}_f$ compare with what you would expect for a pure clay?

Problem 6.14 A CU triaxial compression test was performed on an NC clay sample. The sample was isotropically consolidated to 1723 kPa before shearing. The results of the shearing stage of the test are shown in P-Figure 6.4 and P-Table 6.3. (a) Plot σ_1'/σ_3' versus axial strain for this sample. (b) Identify the points at which σ_1'/σ_3' and q peak. Based exclusively on this one test, what would be a reasonable value for the undrained shear strength s_u of this soil? (c) What is your estimate of critical-state friction angle for this clay? (d) What is the value of $\bar{A}_f$?

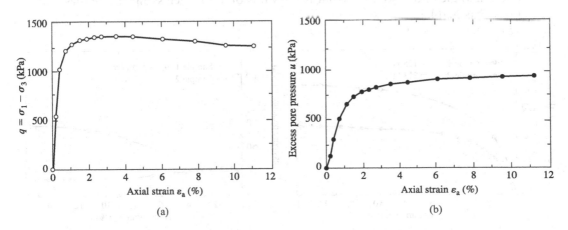

(a)   (b)

P-Figure 6.4 CU triaxial compression test described in Problem 6.14: $q = \sigma_1 - \sigma_3$ and excess pore pressure u versus axial strain ε_a. (Richardson and Whitman (1963); courtesy of the Institution of Civil Engineers and Thomas Telford Limited.)

P-Table 6.3 CU triaxial compression test described in Problem 6.14

Normally consolidated specimen, $\sigma'_c = 1723\,kPa$		
Axial strain ε_a (%)	Deviatoric stress q (kPa)	Excess pore pressure u (kPa)
0.0	0	0
0.2	552	122
0.4	1031	293
0.7	1220	502
1.1	1285	653
1.5	1325	735
1.9	1338	779
2.3	1357	805
2.7	1361	827
3.5	1364	858
4.4	1362	872
6.1	1334	901
7.8	1315	914
9.5	1268	926
11.1	1260	928

Source: Data from Richardson and Whitman (1963).

Problem 6.15 CU triaxial compression tests were performed on two samples of a clayey soil containing 60% kaolinite and 5% illite. The preconsolidation pressure for this soil was determined to be equal to 500 kPa. Additional information about the soil includes: $G_s = 2.63$; smectite content < 1%; nonclay material mostly composed of quartz and other silicates; and void ratio = 1.02–1.07. One sample was isotropically consolidated to 125 kPa; the other, to 313 kPa before shearing. The results of the shearing stage of the test are shown in P-Figure 6.5 and P-Table 6.4. (a) Determine the OCR for each sample. (b) Determine the undrained shear strength s_u. (c) Estimate the critical-state friction angle of this soil. (d) What is the value of $\bar{A}_f$ for each sample? Are these values of $\bar{A}_f$ reasonable?

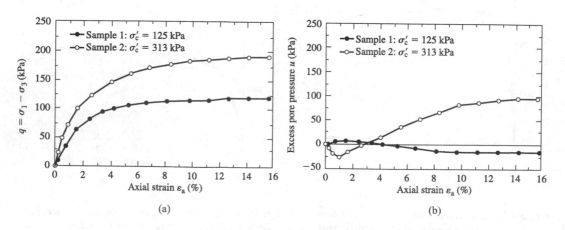

P-Figure 6.5 CU triaxial compression tests on the two samples described in Problem 6.15: (a) deviatoric stress change with respect to axial strain and (b) excess pore pressure change with respect to axial strain.

P-*Table 6.4* CU triaxial compression tests of samples 1 and 2 described in Problem 6.15

Sample 1			Sample 2		
Axial strain ε_a (%)	Deviatoric stress q (kPa)	Excess pore pressure u (kPa)	Axial strain ε_a (%)	Deviatoric stress q (kPa)	Excess pore pressure u (kPa)
0.0	0	0	0.0	0	0
0.2	11	2	0.3	24	−8
0.7	36	5	0.5	48	−19
1.5	63	7	0.9	71	−24
2.5	82	6	1.6	100	−17
3.4	95	3	2.6	124	−3
4.2	101	0	4.1	147	14
5.3	106	−3	5.5	161	36
6.6	110	−8	7.0	170	52
8.2	113	−13	8.4	178	67
9.7	114	−15	9.9	182	82
11.2	116	−15	11.3	185	87
12.8	118	−15	12.8	187	92
14.3	120	−15	14.2	190	95
15.8	119	−15	15.7	191	96

Source: Data from Li and Meissner (2002)

6.9.3 Design problems

Problem 6.16 Consider that the sample of Problems 6.5 and 6.6 was obtained from a representative point of a double-drained clay layer with a thickness of 3 m and that this clay layer will experience an increase in vertical stress from 200 to 250 kPa as a result of construction at the site. (a) Calculate the change in void ratio that this layer will experience. (b) Calculate the total settlement of the building due to the clay layer. (c) Calculate the times required for 30%, 60%, and 90% of the final consolidation settlement to take place. (d) Calculate the pore pressure in the middle of the layer for all three consolidation levels.

Problem 6.17 The total settlement of a building caused by compression of a double-drained, 5-meter-thick clay layer is estimated to be 100 mm. Assuming that the construction period is so short that the load application can be considered to have taken place instantaneously, calculate the times required for the building to settle 10, 50, and 80 mm given that the coefficient of consolidation is 25×10^{-8} m²/s.

Problem 6.18 A clay deposit with PI = 30% was at one time, long before the present time, subjected to a surcharge of 50 kPa. This surcharge no longer exists. The water table is at 2 m depth. The unit weight of the saturated clay is 16 kN/m³. The unit weight of the clay above the water table is about 15 kN/m³. Plot an estimated profile of undrained shear strength s_u with depth z down to 50 m. In particular, provide values for s_u at z = 0.5, 1, 2, 5, 10, 20, 30, 40, and 50 m.

Problem 6.19* A slip surface formed in the slope in P-Figure 6.6 in the past. There was considerable movement along this slip surface, which has continued slowly over the years, such that the residual strength is mobilized at each point of it. Considering that the silty clay soil above the slip surface has a unit weight of 20 kN/m³, that the clay in which the slip surface has formed has a critical-state friction angle of 15° and

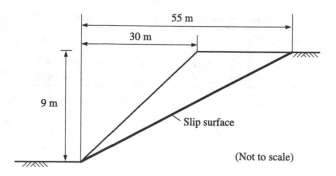

P-Figure 6.6 Slope in which a slip surface has formed in the past.

a minimum residual friction angle of 7°, and that the effective normal stress at which the residual friction angle is equal to 11° is 50 kPa, estimate the values of ϕ_r along the slip surface.

Problem 6.20 A 10-meter-thick layer of inorganic, overconsolidated clay (γ_{sat}=18.6 kN/m³) is overlain by 2 m of sand (γ_m=19.6 kN/m³) and underlain by impermeable, hard rock. The water table is located at the top of the clay layer, and a surcharge p_v of 50 kPa is applied on the sand surface. A 1D consolidation test was performed on a 2.5-centimeter-thick specimen extracted from the middle of the clay layer, and the following parameters were determined: e_0=2.0, σ'_{vp} = 100 kPa, C_c=1.0, and C_s=0.2. During the 1D consolidation test, when the stress conditions from the field (including the surcharge of 50 kPa) were applied to the specimen, it took 15 minutes for the specimen to reach 50% consolidation.

a. Determine the coefficient of consolidation c_v of the clay specimen.
b. Calculate the excess pore pressure u, the total pore pressure, and the vertical effective stress σ'_v at the middle of the clay layer for two cases: case 1 (before the surcharge is applied) and case 2 (immediately after the surcharge is applied).
c. Calculate the vertical effective stress σ'_v at the middle of the clay layer 4 years after the surcharge is applied.
d. Calculate the void ratio e_p at the end of primary consolidation.
e. Estimate the total primary consolidation settlement corresponding to 100% consolidation of the clay layer and the settlement corresponding to 80% consolidation.
f. Estimate the time required for the primary consolidation settlement to be 99% complete. Estimate the secondary consolidation settlement 20 years after the end of the primary consolidation. Assume that the time corresponding to U=99% ($\approx$ 100%) represents the end of primary consolidation.

Problem 6.21* A very wide fill is to be built on top of an NC clay layer underlain by a sand layer. The clay layer has a thickness of 5 m. From laboratory tests, we know that c_v=4×10⁻⁷ m²/s, C_c=0.32, and G_s=2.78. The water table is at the surface of the clay layer, and the saturated unit weight of the clay layer is 19 kN/m³. The initial void ratio at the middle of the clay layer is 0.85. After constructing the fill, which will have a height of 3 m, using a permeable fill material with a unit weight γ_{fill}=20 kN/m³, a building will be built on top of the fill. This building, whose area is very large, has an effect on the clay layer that may be approximated by a surcharge of 20 kPa. The building cannot tolerate total settlements greater than 10 cm. How long after the construction of the embankment can the building be constructed? Ignore the secondary compression of the clay layer.

Problem 6.22 Continue to study the viability of the foundation solution proposed for the building of Problem 4.28. The void ratios of the three samples collected from the clay layer shown in P-Figure 4.7 are 1.2, 1.15, and 1.12 at the depths of 5, 7.5, and 10 m. C_c for this clay is 0.5. The clay is normally consolidated. Estimate the amount of total compression of the clay layer using the following assumptions:

a. The uniform load of 100 kPa extends to infinity in the horizontal direction, and the clay layer is a single layer.

b. The uniform load of 100 kPa extends to infinity in the horizontal direction, and the clay layer is subdivided into 11 sublayers of 1 m.

c. The weighted average of the stresses is as computed in Problem 4.28 using 11 sublayers of 1 m (this will capture 3D effects that 1D compression cannot capture because of the uniformity of stress increase with depth). Discuss the results.

Problem 6.23 Plot an approximate profile of undrained shear strength with depth for the clay layers in the soil profile at the site of the Tower of Pisa. Either use the index properties in Table 6.4 and materials covered in this chapter to estimate C_c and C_s, or consult the references provided as part of the case history in this and other chapters to obtain more accurate values for these properties.

Problem 6.24 If the unit load obtained by dividing the weight of the Tower of Pisa by the plan area of the foundation were applied as an infinite load, we would have the case of one-dimensional compression. How much consolidation would take place? Either use the index properties in Table 6.4 and materials covered in this chapter to estimate C_c and C_s, or consult the references provided as part of the case history in this and other chapters to obtain more accurate values for these properties.

REFERENCES

References cited

Akinlotan, O. (2017). "Mineralogy and palaeoenvironments: The Weald Basin (Early Cretaceous), Southeast England." The Depositional Record, 3(2), 187–200.

Atkinson, J. H. and Bransby, P. L. (1978). The Mechanics of Soil: An Introduction to Critical State Soil Mechanics. McGraw-Hill, London, UK.

Bishop, A. W., Green, G. E., Garga, V. K., Andresen, A., and Brown, J. D. (1971). "A new ring shear apparatus and its application to the measurement of residual strength." Géotechnique, 21(4), 273–328.

Black, D. K. and Lee, K. L. (1973). "Saturating laboratory samples by back pressure." Journal of Soil Mechanics and Foundations Division, 99(SM1), 75–93.

Casagrande, A. (1936). "The determination of pre-consolidation load and it's practical significance." Proceedings of 1st International Conference on Soil Mechanics and Foundation Engineering, Harvard University, Cambridge, MA, 60–64.

Casagrande, A. and Fadum, R. E. (1940). Notes on soil testing for engineering purposes. Publication No. 8, Graduate School of Engineering, Harvard University, Cambridge, MA.

Chakraborty, T. (2009). "Development of a clay constitutive model and its application to pile boundary value problems." Ph.D. Thesis, Purdue University, West Lafayette, IN.

Dafalias, Y. F., Manzari, M. T., and Papadimitriou, A. G. (2006). "SANICLAY: Simple anisotropic clay plasticity model." International Journal for Numerical and Analytical Methods in Geomechanics, 30(12), 1231–1257.

de Boer, R., Schiffman, R. L., and Gibson, R. E. (1996). "The origins of the theory of consolidation: The Terzaghi–Fillunger dispute." Géotechnique, 46(2), 175–186.

Duncan, J. M. (1993). "Limitations of conventional analysis of consolidation settlement." Journal of Geotechnical Engineering, 119(9), 1333–1359.

Fox, P. J. (1999). "Solution charts for finite strain consolidation of normally consolidated clays." Journal of Geotechnical and Geoenvironmental Engineering, 125(10), 847–867.

Fox, P. J. and Berles, J. D. (1997). "CS2: A piecewise-linear model for large strain consolidation." *International Journal for Numerical and Analytical Methods in Geomechanics*, 21(7), 453–475.

Gasparre, A. (2005). "Advanced laboratory characterization of London Clay." Ph.D. Thesis, Imperial College London, London, UK.

Gens, A. (1982). "Stress–strain and strength of a low plasticity clay." Ph.D. Thesis, Imperial College London, London, UK.

Gibson, R. E., England, G. L., and Hussey, M. J. L. (1967). "The theory of one-dimensional consolidation of saturated clays. I. finite non-linear consolidation of thin homogeneous layers." *Géotechnique*, 17(3), 261–273.

Gibson, R. E., Schiffman, R. L., and Cargill, K. W. (1981). "The theory of one-dimensional consolidation of saturated clays. II. finite non-linear consolidation of thick homogeneous layers." *Canadian Geotechnical Journal*, 18(2), 280–293.

Hardin, B. O. and Black, W. L. (1968). "Vibration modulus of normally consolidated clay." *Journal of Soil Mechanics and Foundations Division*, 94(2), 353–369.

Holtz, R. D. and Kovacs, W. D. (1981). *An Introduction to Geotechnical Engineering*. Prentice-Hall, Upper Saddle River, NJ.

Jafari, M. K. and Shafiee, A. (2004). "Mechanical behavior of compacted composite clays." *Canadian Geotechnical Journal*, 41(6), 1152–1167.

Kirkgard, M. M. and Lade, P. V. (1991). "Anisotropy of normally consolidated San Francisco Bay Mud." *Geotechnical Testing Journal*, 14(3), 231–246.

Ladd, C. C. and Varallyay, J. (1965). "The influence of the stress system on the behavior of saturated clays during undrained shear." Research Report No. R65-11, Department of Civil Engineering, MIT, Cambridge, MA.

Ladd, C. C., Foott, R., Ishihara, K., Schlosser, F., and Poulos, H. G. (1977). "Stress-deformation and strength characteristics." *The 9th International Conference on Soil Mechanics and Foundation Engineering*, Tokyo-Japan, 2, 421–494.

Lee, K. and Sills, G. C. (1981). "The consolidation of a soil stratum, including self-weight effects and large strains." *International Journal for Numerical and Analytical Methods in Geomechanics*, 5(4), 405–428.

Li, T. and Meissner, H. (2002). "Two-surface plasticity model for cyclic undrained behavior of clays." *Journal of Geotechnical and Geoenvironmental Engineering*, 128(7), 613–626.

Lupini, J. F., Skinner, A. E., and Vaughan, P. R. (1981). "The drained residual strength of cohesive soils." *Géotechnique*, 31(2), 181–213.

Maksimović, M. (1989). "On the residual shearing strength of clays." *Géotechnique*, 39(2), 347–351.

Mayne, P. W. and Stewart, H. E. (1988). "Pore pressure behavior of K_0-consolidated clays." *Journal of Geotechnical Engineering*, 114(11), 1340–1346.

Meehan, C. L. (2006). "An experimental study of the dynamic behavior of slickensided surfaces." Ph.D. Thesis, Virginia Polytechnic Institute and State University, Blacksburg, VA.

Mesri, G. and Castro, A. (1987). "C_α/C_c concept and K_0 during secondary compression." *Journal of Geotechnical Engineering*, 113(3), 230–247.

NAVFAC. (1982). *Soil Mechanics Design Manual 7.1*. Naval Facility Engineering Command, Alexandria, VA.

Nishimura, S. (2005). "Laboratory study on anisotropy of natural London clay." Ph.D. Thesis, Imperial College London, London, UK.

Parry, R. H. G. (1960). "Triaxial compression and extension tests on remoulded saturated clay." *Géotechnique*, 10(4), 166–180.

Ramsey, N., Jardine, R., Lehane, B., and Ridley, A. (1998). "A review of soil-steel interface testing with the ring shear apparatus." *Proceedings of Conference on Offshore Site Investigation and Foundation Behaviour*, London, UK, 237–258.

Richardson, A. M. and Whitman, R. V. (1963). "Effect of strain-rate upon undrained shear resistance of a saturated remoulded fat clay." *Géotechnique*, 13(4), 310–324.

Sakleshpur, V. A., Prezzi, M., Salgado, R., and Zaheer, M. (2021). CPT-based geotechnical design manual–volume II: CPT-based design of foundations (methods). Joint Transportation Research Program, Purdue University, West Lafayette, IN.

Schiffman, R. L. (1980). "Finite and infinitesimal strain consolidation." *Journal of the Geotechnical Engineering Division*, 106(2), 203–207.

Schmertmann, J. H. (1953). "The undisturbed consolidation behavior of clay." *Transactions of the American Society of Civil Engineers*, 120, 1208–1216.

Schofield, A. N. (1998). "Mohr coulomb error correction." Ground Engineering, August, 30–32.

Skempton, A. W. (1954). "The pore-pressure coefficients A and B." *Géotechnique*, 4(4), 143–147.

Skempton, A. W. (1957). "The planning and design of the new Hong Kong airport." *Proceedings of the Institution of Civil Engineers*, London, UK, 305–307.

Skempton, A. W. (1985). "Residual strength of clays in landslides, folded strata and the laboratory." *Géotechnique*, 35(1), 3–18.

Stark, T. D. and Eid, H. T. (1994). "Drained residual strength of cohesive soils." *Journal of Geotechnical Engineering*, 120(5), 856–871.

Taylor, D. W. (1942). Research on consolidation of clays. Report No. 82, Department of Civil Engineering, Massachusetts Institute of Technology, Cambridge, MA.

Terzaghi, K. (1943). *Theoretical Soil Mechanics*. Wiley, New York.

Wroth, C. P. (1984). "The interpretation of in situ soil tests." *Géotechnique*, 34(4), 449–489.

Wroth, C. P. and Wood, D. M. (1978). "The correlation of index properties with some basic engineering properties of soils." *Canadian Geotechnical Journal*, 15(2), 137–145.

Additional references

Altschaeffl, A. G. (1965). "Interpretation of the consolidation test." Discussion on paper by C. B. Crawford, Journal of the Soil Mechanics and Foundations Division, 91(SM3), 146–147.

Hara, A., Ohta, T., Niwa, M., Tanaka, S., and Banno, T. (1974). "Shear modulus and shear strength of cohesive soil." *Soils and Foundations*, 14(3), 1–12.

Hvorslev, M. J. (1937). "Über die Festigheitseigen-Shaften Gestorter Bindinger Böden." (On the strength properties of remolded cohesive soils). Ingeniorvidenskabelige Skrifter, 45, Danmarks Naturvidenskabelige Samfund, Kovenhavn.

Kulhawy, F. H. and Mayne, P. W. (1990). *Manual on Estimating Soil Properties for Foundation Design*. Electric Power Research Institute, Palo Alto, CA.

Leonards, G. A. and Altschaeffl, A. G. (1964). "Compressibility of clay." *Journal of the Soil Mechanics and Foundations Division*, 90(SM5), 133–155.

Macari, E. J. and Arduino, P. (1995). "Overview of state-of-the practice modeling of overconsolidated soils." Transportation Research Record No. 1479. Engineering Properties and Practices in Overconsolidated Clays, 51–60.

Mesri, G. and Godlewski, P. M. (1977). "Time and stress-compressibility interrelationships." *Journal of the Geotechnical Engineering Division*, 103(5), 417–430.

Olson, R. E. and Ladd, C. C. (1979). "One-dimensional consolidation problems." *Journal of the Geotechnical Engineering Division*, 105(GT1), 55–60.

Schiffman, R. L., Chen, A. T., and Jordan, J. C. (1969). "An analysis of consolidation theories." *Journal of the Soil Mechanics Division*, 95(SM1), 285–312.

Schofield, A. N. and Wroth, C. P. (1968). *Critical State Soil Mechanics*. McGraw-Hill, New York.

Relevant ASTM standards

ASTM. (2011a). "Standard test method for one-dimensional consolidation properties of soils using incremental loading." ASTM D2435, American Society for Testing and Materials, West Conshohocken, PA.

ASTM. (2011b). "Standard test method for consolidated undrained triaxial compression test for cohesive soils." ASTM D4767, American Society for Testing and Materials, West Conshohocken, PA.

ASTM. (2015). "Standard test method for unconsolidated-undrained triaxial compression test on cohesive soils." ASTM D2850, American Society for Testing and Materials, West Conshohocken, PA.

ASTM. (2016a). "Standard test method for measurement of hydraulic conductivity of saturated porous materials using a flexible wall permeameter." ASTM D5084, American Society for Testing and Materials, West Conshohocken, PA.

ASTM. (2016b). "Standard test method for laboratory miniature vane shear test for saturated fine-grained clayey soil." ASTM D4648, American Society for Testing and Materials, West Conshohocken, PA.

ASTM. (2016c). "Standard test method for unconfined compressive strength of cohesive soil." ASTM D2166, American Society for Testing and Materials, West Conshohocken, PA.

ASTM. (2017). "Standard test method for consolidated undrained direct simple shear testing of fine grain soils." ASTM D6528, American Society for Testing and Materials, West Conshohocken, PA.

ASTM. (2020). "Standard test method for one-dimensional consolidation properties of saturated cohesive soils using controlled-strain loading". ASTM D4186, American Society for Testing and Materials, West Conshohocken, PA.

ASTM. (2021). "Standard test method for one-dimensional swell or collapse of soils." ASTM D4546, American Society for Testing and Materials, West Conshohocken, PA.

Chapter 7

Site exploration

The dispositions of a thoughtful commander ensue from correct decisions derived from correct judgments, which depend on a comprehensive and indispensable reconnaissance. The data gathered by observation and from reports are carefully appraised; the crude and false discarded; the refined and true retained.

Mao Tse-Tung

Invert, always invert.

Carl Jacobi

The solution to a foundation engineering problem requires the selection of a suitable foundation type (one satisfying constructability and performance criteria) and determination of the load capacities or resistances of the foundation elements (that is, the loads that they can take without reaching any limit state). Similarly, we are concerned with determining whether a slope or a retaining structure exists in a configuration that is stable. The data necessary for the engineer to make these determinations are obtained through the site investigation program. In this chapter, we discuss the planning of a site investigation program for a typical project. We discuss borings (and how they are drilled), sampling, and *in situ* testing. These form the core of a modern site investigation program, which is sometimes complemented by laboratory testing. By careful analysis of the data obtained during a site characterization program, the engineer can then discard the "crude and false" and retain the "refined and true."

7.1 GENERAL APPROACH TO SITE INVESTIGATION

The methods used for site investigation vary with the type of material present at the site and other site complexities, such as the nature of the soil (sandy versus clayey), its strength and stiffness (weak/soft/loose versus strong/stiff/dense), position of the water table (shallow versus deep), and the presence of rock. The choice of site investigation methods is also based on other factors, such as the type and method of foundation construction that will be used, and the type of analyses that will be used in design.

The ultimate aim of a site investigation program is always the same: to take a "peek" below the ground surface to be able to establish a model of the site profile to use in design. The site profile is described by the identification of the layering of soils or rock, the assignment of approximate values to the relevant soil properties (e.g., shear strength, stiffness, hydraulic conductivity, and coefficient of consolidation) for each layer, the assessment of the groundwater pattern at the site (location of water table and determination of whether hydrostatic or hydrodynamic conditions exist), and the presence of any defects in rock masses.

DOI: 10.1201/b22079-7

The most common way by which the soil profile is defined, at least in part, for a given site is through borings. Borings are small-diameter holes, usually vertical, drilled at the site observing appropriate spacings, from which disturbed soil samples or cuttings are obtained. Borings may be supplemented by the recovery of reasonably undisturbed samples for laboratory tests (from the borings themselves or from exposed soil cuts), by *in situ* testing, or by geophysical tests. The most common laboratory tests are the direct shear test, the triaxial compression test, and the consolidation test, all discussed in prior chapters. The most widely used *in situ* tests are the standard penetration test (SPT) and the cone penetration test (CPT). Detailed procedures for these tests are given in Sections 7.3 and 7.6. Most of our discussion will focus on the more common practices and procedures. We will cover the devices or practices that are infrequently used in a more summarized form.

The choice of methods and techniques used in a site investigation program and the extent of the program are largely a function of the general characteristics of the site and the type and importance of the structure to be supported. Even in the case of foundations for very small, rugged structures, at least a minimal number of borings are needed to provide the information that may allow the designer of the foundations to propose a more economical solution. There is, of course, an optimal level of site investigation for any given structure: do too little and you are left with far too many uncertainties that must be compensated for by a more conservative, expensive design; do too much and the reduction of uncertainties is not sufficient to lead to design economies that compensate for the additional site investigation costs! There is no formula for finding the right balance. The trick to deciding whether an additional boring or more testing is required is to assess whether the additional amount of work is likely to generate information that will lead to design economies that will more than pay for the additional cost.

A minimal site investigation in soil usually includes at least three borings, usually combined with SPTs. As an example of a commonsense approach to defining the number of borings, for a rectangular building, five borings would be enough to investigate the four corners and central area of the building. The number of borings would increase significantly with the plan area of the site, the size/weight of the structure, and the degree of complexity of the soil profile. Pits and trenches (shallow excavations) are also sometimes done to expose the subsoil at locations where this may be desired. When test pits are planned, the number of borings may be reduced. If it is known that the site is located in soils with erratic profiles, a larger number of borings will obviously be required. Because the number of borings depends on so many factors, it is preferable not to specify the spacing between borings or the number of borings per unit area.[1] It is better to start any site investigation planning with a minimum number of borings and increase that number as needed based on previous knowledge of the area, available geologic information, and considerations concerning the size of the structure and magnitude of the loads.

A trend over the last 30 years or so, as noted early by Mitchell (2000), has been the increasing reliance on *in situ* testing in place of laboratory testing. Mitchell (2000) attributes this to the expanded availability of more reliable *in situ* test interpretation methods today than in the past, making the less time-consuming, more economical *in situ* tests preferable to laboratory testing.

Site investigations vary significantly according to local preferences, and young engineers usually learn from more experienced colleagues about how things are done in their localities. Additionally, some local codes (New York City being a good example) are very specific about how site investigations should be conducted. Accordingly, we emphasize methods, concepts, and interpretation approaches that are universal in their applicability, leaving the more esoteric or specialized aspects of site investigation to be learned on the job, as needed.

---

[1] Note, however, that certain codes (an example being the New York City Code) do specify minimum numbers of borings for various situations.

Site investigations are extremely important for proper construction planning and geo-technical design. In fact, they are so important that these investigations are often at the core of one of the main objects of litigation: differing site conditions. Differing site conditions clauses are common in construction contracts. They protect contractors from bidding for work based on information (including site investigation reports) that differs in a significant way from the reality at the time of construction. In connection with site investigations, this would typically happen when the investigation misses a subsoil feature that ends up being an important or controlling feature in design or construction.

7.2 SOIL BORINGS

Soil borings are vertical holes whose purpose is (1) to establish the soil profile by observation of cuttings recovered during drilling or by obtaining disturbed soil samples, (2) to allow standard penetration testing at various depths during pauses in the drilling process, (3) to detect the presence of and estimate the depth of the water table, or (4) to install sensors that can provide useful data during and even after construction. There are different methods of soil drilling. The most common are hand augering, rotary drilling (with either drill bits or continuous flight augers), and wash boring. Rotary drilling with continuous flight augers is the most common method in the United States. Wash boring is the most common method in South America (notably Brazil) and parts of Asia and Africa. Hand augering is only suited to small projects, where exploration is required only to small depths.

Figure 7.1 shows a schematic representation of wash boring equipment. A short casing, with a length of 2–3 m, is first driven into the ground to prevent caving of surface soils. A chopping bit, connected to drill rods, is then lowered into the casing. It is raised, dropped, and rotated to weaken and loosen the soil. Water is pumped through the drill rods to the bit, where it aids in the drilling process by exiting through small holes at high velocity. The water then flows upward between the casing and the drill rods, carrying the soil cuttings with it. At the top, it flows into a container, where the cuttings settle and from which the water is pumped back into the drill rods; this way, water is continuously recycled. Examination of the cuttings brought to the surface by the water allows the operator to develop a log of the type of soil versus depth. This log is later confirmed or changed based on the soil samples obtained from the boring.

Rotary drilling combines downward pressure and rotation of a drill bit (Figure 7.2) to advance the boring in either soil or rock. In soils that are unable to stand when unsupported (and would cave into the borehole), bentonite slurry is circulated through the drill rods; it then exits the bit and flows upward between the rods and the borehole walls, carrying the soil or rock cuttings up to the surface and providing support for the borehole walls. Bentonite is a plastic clay that, when added to water, forms a slurry heavier than water. If the level of the slurry in the boring is above the water table, cave-ins are prevented. The slurry acts not only by hydrostatic pressure on the walls of the borehole, but also by forming a thin cohesive cake on the walls, which helps prevent even cohesionless soils from caving. We discuss bentonite slurry and its use in supporting the walls of excavations more extensively in Chapter 12, in which we cover the installation of drilled shafts (bored piles).

Rotary drilling with continuous flight augers (Figure 7.3) is the preferred method of advancing borings in North America. The augers come in 1 or 1.5 m (5 ft in the United States). Auger drilling works much the same way as a corkscrew: when rotated clockwise, the auger penetrates the ground. It is usual to use hollow-stem augers, as they allow the introduction of samplers into the stem either for undisturbed sampling or for performing SPT. The sequence would typically be as follows: the operator drills one segment into the ground, stops, inserts a sampler into the stem to either perform an SPT or collect a sample,

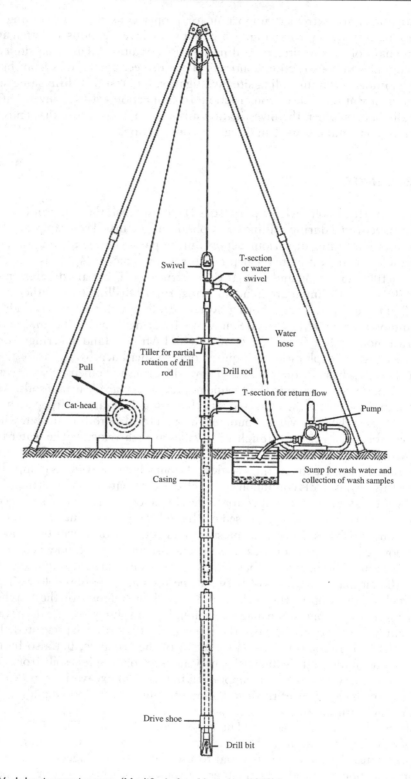

Figure 7.1 Wash boring equipment. (Modified after Hvorslev (1949).)

Figure 7.2 Drill bits for use with rotary drilling. (Courtesy of Ross Boulanger.)

(a)

(b)

Figure 7.3 Boring operations with continuous flight augers: (a) drilling rig and (b) solid-stem and hollow-stem augers. ((b) Courtesy of Ross Boulanger.)

then connects the next segment, and proceeds with the drilling. The tip of the auger may or may not be sealed by a suitable plug, depending on soil conditions. If a plug is used, it must be removed each time the SPT is performed or a sample is recovered. Upon removal of the plug, if the water table is located above the level of the auger tip, water will flow into the auger stem to reestablish a hydrostatic condition. This may be a problem in saturated sandy soils, when plug removal may induce quick groundwater flow into the auger stem, loosening the soil and invalidating any measurements made under such conditions.

Borings can be made in the middle of a lake, river, or ocean from a barge. It is important in those cases to carefully keep track of tidal fluctuations to obtain accurate depths for the loggings. The water level inside a boring can be measured by lowering a suitable indicator hanging from a cable into the borehole. When the indicator reaches the water, it will detect it and emit a signal. Although the water table can be estimated from such readings, this is not the most accurate way of assessing the water table. The main difficulty is the fact that equilibrium must be reestablished before the water elevation inside the boring will be reflective of the true water table elevation, which in soils of low permeability will require some time. When slurry is used in drilling, that further distorts the borehole water level. Waiting times might be several days for an accurate reading to be obtained. So an inspector should take several readings, minding each time to document the time of day and depth of any casing that may be in the hole. Borehole readings are best taken every morning prior to borehole continuation (if possible), after any interruptions to drilling (such as for equipment repair or even a lunch break), immediately after completion of drilling, and 24 hours after completion of drilling. If possible, readings should be continued for as long as practical.

A more accurate indication of water table elevation is obtained by using piezometers. A piezometer is a sensor that measures the hydraulic head at the point where it is placed. A simple piezometer is of the standpipe type, in which a filter tip, which is connected to a pipe that extends all the way to the ground surface, is placed at the location where the hydraulic head is desired. Above the tip, a bentonite seal is constructed; it is then topped by sand all the way to the surface. Groundwater will enter the piezometer through the filter tip and rise to a level h above the tip such that, following Equation (3.21):

$$h_{piezometer} = \frac{u}{\gamma_w} + \frac{v_t^2}{2g}$$

In a stationary groundwater condition, with $v_t = 0$ and no special geologic features, the water level inside the piezometer will directly yield the water table. There are other more sophisticated piezometers that provide a direct, electronic reading of the value of the hydraulic head or pore pressure at a given location.

Groundwater patterns that depart from the ordinary must be noted. Artesian conditions and perched water tables are good examples of this. Both are possible because of the existence of layers or masses of impervious material (clay and/or rock) and pervious material in the same profile. Artesian conditions exist when a relatively permeable soil mass confined by an impervious material is hydraulically connected to a zone with a higher piezometric surface than that occurring locally. This means that the water level inside a standpipe piezometer placed within the layer experiencing artesian conditions will rise higher than the one placed immediately above this layer.

A perched water table is one associated with a volume of water existing above a layer of impervious soil that may result from the accumulation of rainwater there. The water table elsewhere will be lower than the perched water table.

7.3 STANDARD PENETRATION TEST

7.3.1 Procedure

The SPT is still the most widely used *in situ* test in the world, although the use of CPT has been increasing steadily. The SPT is a relatively crude test, whose essence is to drive a standard sampler into the soil to a pre-defined distance by repeated blows of a standard hammer from a standard drop height. Naturally, the harder it is to advance the sampler, the stronger and stiffer the soil is supposed to be. The main criticism of the SPT is that, the name of the test notwithstanding, it is far from being standard.

The SPT has been codified in ASTM D1586 (ASTM 2018b). The test is performed at intervals varying typically from 2.5 to 5 ft, as shown in Figure 7.4. Drilling must be interrupted each time the test is performed. A sampler with standard geometry (Figure 7.5) is driven into the soil a total of 450 mm (18 in.) by blows of a hammer weighing 630 N (140 pounds) falling from a height of 760 mm (30 in.). The hammer falls on top of an anvil, which in turn transfers the energy down to the drill rods. Figure 7.6 shows an SPT in progress. The 450 mm (18 in.) penetration is divided into three separate advances of 150 mm (6 in.) each (Figure 7.7). The operator counts the number of blows of the hammer required to advance the sampler each of the 150 mm. The first count is usually discarded in data interpretation because it reflects the properties of soil at the bottom of the excavation, which is likely to be at least somewhat disturbed by the drilling process. The SPT blow count N_{SPT} is obtained by adding together the last two counts, representing the number of blows required for penetration of the sampler 300 mm (12 in.) into the soil.

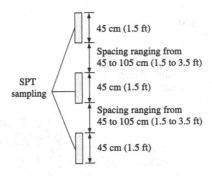

Figure 7.4 Frequency at which the SPT is performed.

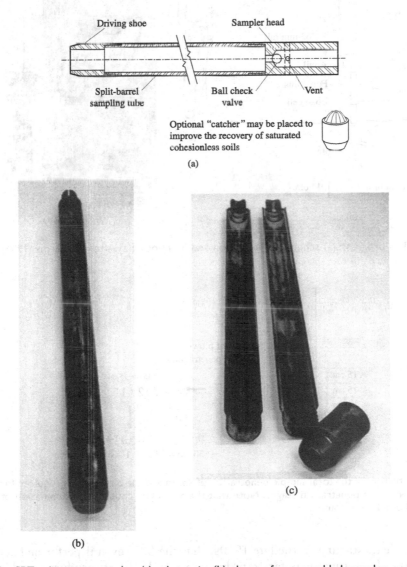

Figure 7.5 The SPT split spoon sampler: (a) schematic, (b) photo of an assembled sampler, and (c) photo of a sampler disassembled. ((a) After U.S. Army (1996).)

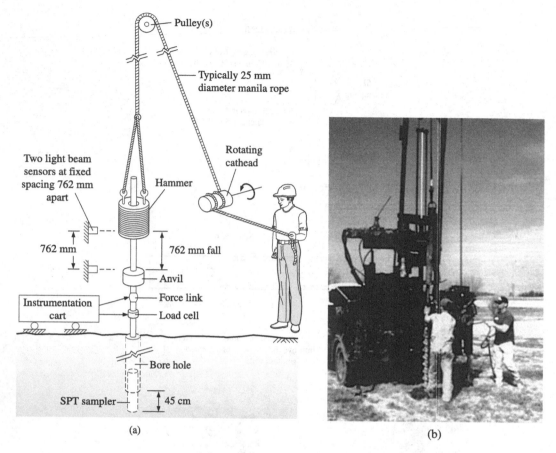

(a)   (b)

Figure 7.6 SPT in progress: (a) schematic illustration and (b) photo. ((a) After U.S. Army (1996).)

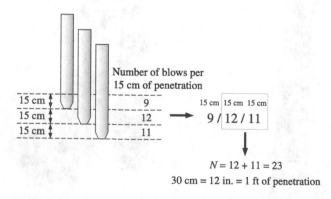

Figure 7.7 Illustration of the total sample penetration of 45 cm and the blow counts required for each of the three 15 cm penetration stages. Note that the SPT blow count is the summation of the second and third blow counts.

It became clear, starting in the late 1970s, that the SPT, even if performed according to the standards of the time, was far from being a standard test, producing wildly variable N_{SPT} values (Seed et al. 1985; Skempton 1986). The sources of uncertainties are many. For example, even though the weight of the hammer was standardized, there were (and still are)

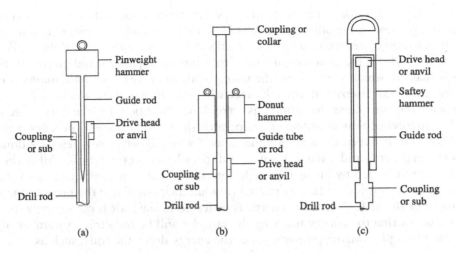

Figure 7.8 SPT hammers: (a) pin, (b) donut, and (c) safety hammers (U.S. Army 2001).

at least three different types of hammer in use (Figure 7.8). The safety and, more rarely, the donut hammers are used in North America, while the pin-weight hammer is most common in South America. The different geometries of these hammers result in different efficiencies in the impact of the hammer with the anvil, which means the amount of energy effectively transmitted to the rods varies with the hammer type.

The ratio of the energy effectively transferred to the rods to the maximum theoretical energy is simply referred to as the energy ratio ER (usually expressed as a percentage). The maximum theoretical energy ($m_{hammer}gh$) is the potential energy associated with the standard hammer mass m_{hammer} and standard fall height h. It is widely believed that the energy ratio is roughly 45% for the donut hammer and 60% for the safety hammer, but the only way to know ER with any certainty is to actually measure it.[2] There are indications that ER would be roughly 70% for the pin-weight hammer.

Additional factors influencing the energy ratio are the method of hammer release, quality of operation, and the condition of the hammer and anvil. There are two ways in which the method of hammer release may influence the ER. If the hammer is released via an automatic release mechanism, triggered whenever the hammer reaches the standard height, the actual initial energy is nearly the same for each blow, which is desirable. If the hammer release is manual, then the fall height is not always the same and could vary significantly from the standard fall height, depending on the operator. Another way in which the hammer release method can affect the ER is that it may initially slow down the hammer fall, effectively reducing the amount of input energy. This is the case with the rope and cathead method (Figure 7.6), in which the hammer is connected to one end of a rope that goes around a pulley located vertically above the hammer. The other end of the rope is then wrapped around a cathead and held by the operator. With a jerky motion, the operator quickly loosens the rope around the cathead, allowing the hammer to fall. There is, however, friction between the cathead and the rope, which reduces the fall velocity to below that of a true free fall. An attempt to standardize for this has been made by recommending that operators use only two turns of the rope around the cathead,[3] but some dependence on the operator clearly remains.

---

[2] SPT energy measurement is rarely done, although the techniques to do it have been available for many years. It could be argued that reliance on a truly standard test, such as the CPT, is superior to attempting to overcome the many limitations of the SPT with costly measures.

[3] There is a recent trend toward using three turns of the rope around the cathead to minimize the risk of shoulder injury.

A conscientious operator is important in the performance of the SPT. There are many ways in which an operator can produce erroneous N_{SPT} values. As an example, operators usually control the 450 mm of penetration by chalk marks they make on the rods. If the chalk marks are not made carefully, the penetration will differ from the required standard penetration. In an extreme case of which I am aware, the wrong scale to make these chalk marks was used (something that was discovered after the fact when the results were analyzed by someone familiar with the site where the soundings were done). Additionally, there have been cases of blunders involving blow counting. The operator should also strive to produce a vertical, centered collision between the anvil and hammer. Unfortunately, many tests are done with speed as the main goal and attention is not paid to such important details.[4] All of these can be avoided or minimized by hiring reputable companies with conscientious employees.

So far, we have considered factors related to what happens above the ground during the performance of the SPT. Even if the energy transferred to the rods is consistently the same, it does not mean that the energy reaching the sampler will be the same. A number of other factors come into play during propagation of the energy down the rods, such as

- the total string length;
- the tightness of the rod connections;
- the condition of the rods and the sampler;
- the sampler's ability to accommodate a liner; and
- the diameter of the borehole.

The length of the string of rods is a factor only if the SPT is done at a depth of 10 m or less. In this case, the wave generated by the hammer–anvil collision travels down the rods, reflects off the sampler, and returns to the anvil in time to interfere with the impact itself. For rod strings longer than 10 m, rod string length ceases to be a factor (although a long string would have a large weight, which could be a factor in soft clayey soils).

If the rods are not connected together tightly, there will be energy loss at the connections. Bent rods may touch the walls of the borehole, also affecting energy transfer. While this might at first glance suggest a large-diameter borehole would be desirable, too large a borehole actually creates conditions of low confinement at the base of the borehole, artificially reducing the N_{SPT} value (Figure 7.9). Yet another factor is the type of sampler and its condition. A worn-out sampler does not have a sharp cutting edge, thereby producing artificially high blow counts.

There are two types of samplers in use. One type, the ASTM sampler illustrated in Figure 7.10a, can accommodate a liner, whereas the so-called ISSMFE[5] sampler, illustrated in Figure 7.10b, cannot. If the ASTM sampler is used without the liner, lower blow counts result because of the larger internal diameter in the absence of the liner, which allows the soil sample to expand as it enters the sampler, reducing the internal friction between the soil and the sampler.

Figure 7.11 shows a sample SPT log, which contains the blow counts obtained for every depth at which the test was done. Ideally, these logs would also contain additional information that will allow corrections to be made to the blow count,[6] such as the type of hammer that was used and whether a sampler with or without a liner was used. Blow count corrections are discussed next.

---

[4] The basis on which contracts are awarded – price per foot of sounding, a lump sum, or day rates – also plays a role in the level of care with which soundings are done.

[5] International Society of Soil Mechanics and Foundation Engineering, known today as the International Society for Soil Mechanics and Geotechnical Engineering (ISSMGE).

[6] When not directly available, the information should be obtained from the firm in charge of performing the SPTs.

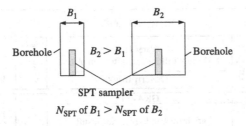

SPT sampler

N_{SPT} of B_1 > N_{SPT} of B_2

Figure 7.9 Illustration of borehole diameter effects: a larger borehole provides less confinement and thus leads to a lower blow count.

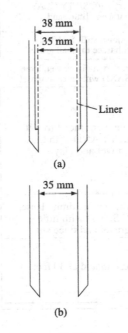

Figure 7.10 SPT split spoon samplers: (a) sampler (ASTM) that can accommodate the liner with the liner in place and (b) sampler (ISSMFE) that cannot accommodate a liner.

7.3.2 Blow count corrections

To be useful in design, the blow count must reflect closely the state of the soil. This means the effects of the operator, equipment, and test procedures on the SPT blow count N_{SPT} must be filtered out. Because the effect of some factors on N_{SPT} can be quantified, correction factors have been proposed to normalize the N_{SPT} with respect to these factors. The factors that cannot be quantified are dealt with by adhering to guidelines concerning the type and condition of the equipment and procedure used to perform the test. Table 7.1 summarizes the main recommendations of the ISSMGE concerning the SPT.

The factors that can be quantified approximately are the effects of the hammer type, rod length, sampler type, and borehole diameter. The corrected SPT blow count is the standard blow count, denoted as N_{60} and expressed as

$$N_{60} = C_h C_r C_s C_d N_{SPT} \qquad (7.1)$$

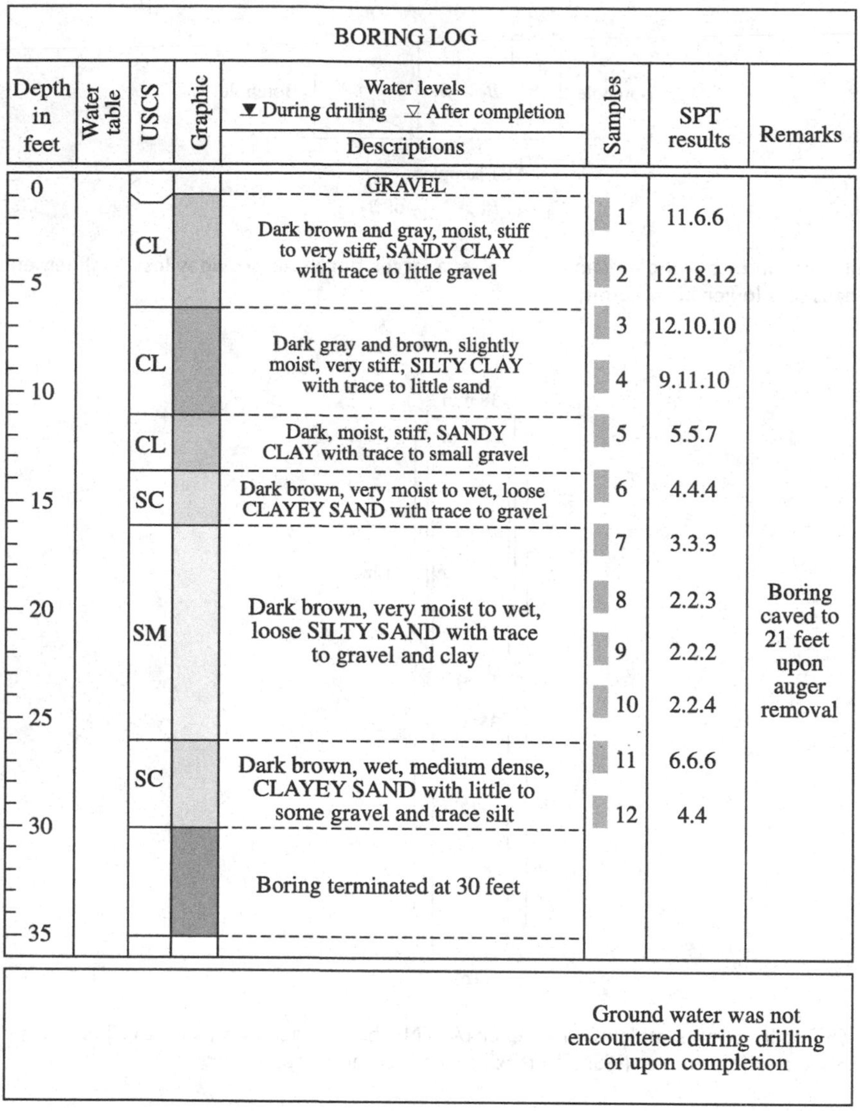

BORING LOG								
Depth in feet	Water table	USCS	Graphic	Water levels ▼ During drilling ▽ After completion		Samples	SPT results	Remarks
				Descriptions				
0				GRAVEL		1	11.6.6	
		CL		Dark brown and gray, moist, stiff to very stiff, SANDY CLAY with trace to little gravel		2	12.18.12	
5								
		CL		Dark gray and brown, slightly moist, very stiff, SILTY CLAY with trace to little sand		3	12.10.10	
10						4	9.11.10	
		CL		Dark, moist, stiff, SANDY CLAY with trace to small gravel		5	5.5.7	
15		SC		Dark brown, very moist to wet, loose CLAYEY SAND with trace to gravel		6	4.4.4	
						7	3.3.3	
20		SM		Dark brown, very moist to wet, loose SILTY SAND with trace to gravel and clay		8	2.2.3	Boring caved to 21 feet upon auger removal
						9	2.2.2	
25						10	2.2.4	
		SC		Dark brown, wet, medium dense, CLAYEY SAND with little to some gravel and trace silt		11	6.6.6	
30						12	4.4	
				Boring terminated at 30 feet				
35								

Ground water was not encountered during drilling or upon completion

Figure 7.11 Sample SPT log.

Table 7.1 Summary of ISSMGE guidelines for SPT testing

Borehole diameter	66–115 mm (2.5–4.5 in.)
Borehole support	Casing and/or drilling mud
Drilling	Wash boring with side discharge bit or rotary boring
Sampler	Outer diameter OD = 51 ± 1 mm (2 in.) Inner diameter ID = 35 ± 1 mm (1⅜ in.) Length = 457 mm (18 in.)
Blow count	Count the number of blows for each 150 mm (6 in.) of total 450 mm (18 in.) sample penetration into the soil; add the blow counts for the last two 150 mm of penetration to obtain the SPT blow count N_{SPT}

Table 7.2 Approximate values of energy ratio and correction factor C_h for common hammer types

Hammer	Approximate energy ratio (%)	C_h
Safety	60	1
Donut	45	0.75
Automatic trip	80	1.33
Pin-weight	72	1.2

where C_h = hammer correction, C_r = rod length correction, C_s = sampler correction, and C_d = borehole diameter correction.

The hammer correction factor C_h is given by

$$C_h = \frac{ER_{hammer}}{ER_{safety}} = \frac{ER_{hammer}}{60\%} \tag{7.2}$$

where ER_{hammer} = energy ratio for the hammer used in the test and $C_h = 1$ for the safety hammer, which is taken as the standard hammer. Approximate values of energy ratio and C_h are given in Table 7.2 for types of hammer that are most often used.

The sampler correction factor C_s refers to the occasional use of a liner sampler without the liner. The standard ISSMGE sampler, which does not accommodate a liner, has the same geometry as the ASTM sampler with the liner in place. The blow counts obtained when the standard sampler is used have been found to be 10%–30% higher than those obtained using the ASTM sampler without the liner. Accordingly, C_s is taken as 1 for the standard ISSMGE sampler or the liner sample with the liner in place, and $C_s = 1.2$ for the liner sampler without a liner.

The rod length[7] correction factor C_r is given by

$$C_r = \begin{cases} 0.75 & \text{if rod length} < 4\,\text{m} \\ 0.85 & \text{if } 4 \le \text{rod length} < 6\,\text{m} \\ 0.95 & \text{if } 6 \le \text{rod length} < 10\,\text{m} \\ 1 & \text{if rod length} \ge 10\,\text{m} \end{cases} \tag{7.3}$$

The borehole diameter correction factor C_d is equal to 1 (for the range specified in Table 7.1) or greater than 1 (for larger borehole diameters). It is taken as 1.05 and 1.15 for borehole diameters of 150 and 200 mm, respectively. Interpolation could be used for other values. C_d increases with borehole diameter because of the decreasing confinement that is associated with a larger diameter, leading to values of N_{SPT} that are too low.

Example 7.1

A soil boring was drilled in a sandy soil deposit shown in E-Figure 7.1. An SPT was also performed at intervals of one meter. A donut hammer and a liner sampler without the liner were used. Every care was taken to firmly connect the rod segments and to follow the standard procedure. If the SPT blow counts at 2 and 10 m were, respectively, 4 and 20, what would the corresponding blow counts be if a safety hammer had been used and if all recommended corrections were made?

---

[7] It is important to stress that this is the length of the rod string within the ground, that is, the depth at which the SPT sampler is when the blow counts are taken.

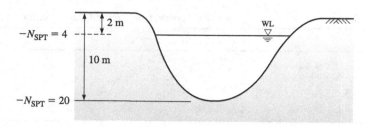

E-Figure 7.1 Soil deposit and SPT blow counts for Example 7.1.

Solution:

Using Equation (7.1), the blow count at 2 m becomes:

$$N_{60} = C_h C_d C_r C_s N_{SPT} = (45/60)(1)(0.75)(1.2)4 = 2.7$$

Note that $C_d = 1$ because standard procedures (including drilling with a certain diameter) were followed. The blow count might be rounded up to 3 or down to 2 on account of the impossibility of having fractional blow counts, but that is not strictly necessary, because the goal is to use the blow count in design calculations, and it matters little if a fractional number is used (so long as it is reasonable from a precision viewpoint).
For the blow count at 10 m:

$$N_{60} = C_h C_r C_s N_{SPT} = (45/60)(1)(1.2)20 = 18$$

Despite attempts at its standardization, the SPT cannot be relied on to give consistently the same blow counts and errors of ±50% are possible. This means a reported blow count of 50 might in truth be as low as 25 or as high as 75. Despite the considerable uncertainty regarding the SPT blow count value, the test continues to be used and is likely to remain in use. Engineers must therefore familiarize themselves with the test and its shortcomings to more effectively inspect SPT operations and interpret their results.

7.3.3 Interpretation of SPT results

There are two different approaches to *in situ* test interpretation for foundation design: direct and indirect interpretations. In direct interpretation, test results are directly related to foundation resistance or other variables of interest in design. In indirect interpretation, we first determine soil variables, such as relative density in sand or undrained shear strength in clay, and then evaluate foundation capacities from the estimated soil variables. We will emphasize the indirect interpretation of SPT results in this section. Direct interpretation is discussed in the chapters on shallow and deep foundation designs.

Assuming we are successful in normalizing the SPT blow count N_{SPT}, so that it reflects the state of the soil only and not the test procedure or equipment, then it is reasonable to assume that correlations can be established between N_{SPT} and various soil properties. We will present some of the more useful correlations for sands and clays.

7.3.3.1 Sand

For a given sand, it is intuitive that N_{SPT} increases with increasing relative density and increasing confining stress. Some correlations rely on the blow count N_1 normalized for the vertical effective stress. Normalization of blow counts with respect to stress requires that we know how N_{SPT} depends on stress. When we discuss the CPT, we will see that both theoretical and experimental results show that static penetration resistance varies approximately with the square root of the effective stress. Assuming the same holds true for dynamic penetration resistance, we can write

$$N_1 = N_{SPT}\sqrt{\frac{p_A}{\sigma'_v}} \tag{7.4}$$

where p_A = reference stress (= 100 kPa = 0.1 MPa). This relationship is due to Liao and Whitman (1986). Salgado et al. (1997a) showed that it is nearly exact for static penetration resistance in sand with $D_R = 50\%$; for other relative density values, it is a satisfactory approximation. It applies strictly to normally consolidated sands, for which K_0 is in the 0.38–0.50 range.

Figure 7.12 illustrates what we are attempting to do when we normalize blow count with respect to vertical effective stress. We take the number N_{SPT} measured at some depth for which the vertical effective stress is σ'_v, and we ask the question: can we obtain a value of blow count that reflects, for a given soil, the effect of soil density but not the stress? The way to do that is to normalize the blow count for the effect of stress. If we have all values of blow count normalized for the same stress, we can directly compare blow count values regardless of the depth at which they were obtained. A higher normalized blow count would always reflect a denser soil. The figure shows how this process can be visualized as taking the soil element for which a blow count N_{SPT} was obtained at a vertical effective stress σ'_v to the depth at which $\sigma'_v = 100$ kPa, where the blow count N_1 that would be obtained is different because of the different vertical effective stress.

Equation (7.4) does not normalize the N_{SPT} with respect to lateral stress effects if K_0 is not in the range for normally consolidated sand. Salgado et al. (1997a) proposed a general equation to normalize penetration resistance with respect to stress even when the sand is overconsolidated and thus has $K_0 > K_{0,NC}$:

$$N_1 = N_{SPT}\sqrt{\frac{p_A}{\sigma'_v}\frac{K_{0,NC}}{K_0}} \tag{7.5}$$

Equation (7.5) reduces to Equation (7.4) when the soil is normally consolidated and $K_0 = K_{0,NC}$.

It is now possible to speak of four different values of blow count for the same point in the soil profile: N_{SPT}, N_1, N_{60}, and $(N_1)_{60}$. N_{SPT} is the raw blow count as a function of depth as recorded during the test. N_1 is the blow count normalized with respect to vertical

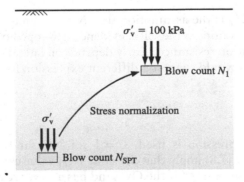

Figure 7.12 Illustration of stress normalization of blow counts.

effective stress. N_{60} is the blow count corrected for the type of hammer, sampler, rod length, and borehole diameter. $(N_1)_{60}$ is the blow count normalized with respect to vertical effective stress and corrected for the type of hammer, type of sampler, borehole diameter, and rod length. Which one should we use? It depends on the correlation or interpretation method. Certain correlations were proposed for specific types of hammers, without any normalization. If the same hammer is used in the investigation program for a given project, then the N_{SPT} values from the SPT report can be used directly in the correlations; otherwise, corrections to the N_{SPT} value are needed for use in the correlations. Other correlations were developed for the safety hammer without stress normalization. The appropriate blow count in this case would be N_{60}. Other correlations are in terms of N_1 or $(N_1)_{60}$. So we should normalize N_{SPT} according to the correlation or analysis we choose to use. Whenever we use a correlation with which we are not familiar, we should check whether the blow counts need to be corrected before using them in the correlation. If the basis for the correlation is unclear, we should not use that correlation.

For sands, the relative density D_R is a very useful concept. Strength and stiffness can be estimated based on D_R. Drawing from the work by Meyerhof (1957) and Skempton (1986), we can write the following general expression for relative density in terms of N_{60}:

$$\frac{D_R}{100\%} = \sqrt{\frac{N_{60}}{A + BC\dfrac{\sigma'_v}{p_A}}} \tag{7.6}$$

an expression applicable for $35\% \leq D_R \leq 85\%$ and $50\,\text{kPa} \leq \sigma'_v \leq 250\,\text{kPa}$, where A and B are correlation coefficients ($27 \leq A \leq 46$ and $B \approx 27$) and C is given by

$$C = \frac{1 + 2K_0}{1 + 2K_{0,NC}} \tag{7.7}$$

where K_0 = coefficient of lateral earth pressure at rest for the soil and $K_{0,NC}$ = value of this coefficient that the soil would have if it were normally consolidated. For an NC soil, $K_0 = K_{0,NC}$ and $C = 1$; for OC soils, $C > 1$. Refer to Chapters 4 and 5 for a detailed discussion of K_0.

Based on the limited data contained in Skempton (1986), the values of A range from 27 to 46 and B is ~27; however, these values should be used with caution, as Skempton observed great variability for the soils he studied. A common approach is to use the midpoint (36.5) of the 27–46 range for A. Once the relative density is estimated using Equation (7.6), it can be used in a correlation such as the Bolton (1986) correlation introduced in Chapter 5 to estimate the friction angle.

Implicit in Equation (7.7) is the assumption that N_{SPT} depends on mean effective stress. As we will discuss in the section on the CPT, evidence developed over the past 20–30 years demonstrates that penetration resistance clearly depends on lateral effective stress instead of mean effective stress. This would suggest a different expression for C:

$$C = \frac{K_0}{K_{0,NC}} \tag{7.8}$$

Regardless of which expression is used, $C = 1$ for NC sands, for which $K_0 = K_{0,NC}$. Equations (7.6) and (7.7) or (7.8) imply that the same N_{SPT} value can only be observed for an NC and an OC sand for the same σ'_v if the OC sand has a lower relative density to compensate for the larger lateral stress locked in within the overconsolidated soil.

There are variations of the Meyerhof–Skempton correlation available, as well as correlations of some other forms (Kulhawy and Mayne 1990). An interesting alternative to these correlations, though, is to convert the blow count to cone resistance and use CPT interpretation methods to estimate D_R and ϕ. This approach is discussed together with CPT interpretation in Section 7.6.

Example 7.2

Refer to Example 7.1 and E-Figure 7.1. The sandy deposit is located near a stream and is known to be normally consolidated. The average unit weight of the sand both above and below the water table is 20 kN/m^3. Obtain the stress-normalized blow count values at depths of 2 and 10 m.

Solution:

The energy-corrected blow counts obtained in Example 7.1 were 2.7 at 2 m and 18 at 10 m. In the absence of more specific information, we assume the groundwater table to be at the same level as the water level in the stream (2 m from the ground surface). The vertical effective stresses are as follows:
At 2 m:

$$\sigma'_v = 2 \times 20 = 40 \, \text{kPa}$$

And at 10 m:

$$\sigma'_v = 10 \times 20 - 8 \times 9.81 = 121.5 \, \text{kPa}$$

The stress-normalized blow counts are calculated using Equation (7.4):

$$(N_1)_{60} = 2.7 \sqrt{\frac{100}{40}} = 4.3 \approx 4$$

$$(N_1)_{60} = 18 \sqrt{\frac{100}{121.5}} = 16$$

7.3.3.2 Clay

Although there are SPT-based correlations for clays, these should be used with caution. There are many uncertainties in the SPT, which are aggravated by the dependence of the shear strength of clays – and thus of the blow counts – on the loading rate. Additionally, the blow count depends also on clay sensitivity, which expresses the drop in shear strength of the clay when disturbed or remolded.[8] Additionally, SPT blow counts in clays, except in very stiff clays, are low, potentially in the low single digits, which means the measurement lacks precision. In extremely soft soils, the SPT split spoon sampler may even go down under the weight of the rods, without any hammer blows. At great depths in soft clays, the weight of the rods may be such that artificially low blow counts result, which again raises the issue of lack of precision. The effects of rod length and weight on SPT blow count are discussed by Odebrecht et al. (2005).

---

[8] Formally, the sensitivity of a soil is its strength in an "undisturbed" state divided by its strength in a completely remolded state "at the same water content." For most soils, the sensitivity lies in the range from 1.5 to about 10.

While in sands it is common to relate N_{SPT} to relative density and stress state, in clays it is more common to directly relate it to undrained shear strength or compressibility. As an illustration of one such correlation, consider the correlations for undrained shear strength s_u and coefficient of compressibility m_v proposed by Stroud (1975):

$$\frac{s_u}{p_A} = f_1 N_{SPT} \tag{7.9}$$

and

$$m_v p_A = \frac{1}{f_2 N_{SPT}} \tag{7.10}$$

where p_A is the reference stress ($= 100$ kPa $= 0.1$ MPa ≈ 1 tsf); f_1 and f_2 are given in Figure 7.13.

Stroud (1975) based his correlations on relatively insensitive clays (with sensitivity <3). He also mentions that an automatic trip hammer was used. According to Clayton (1990), the average energy ratio for this type of hammer used in the United Kingdom is 73%. So, using N_{60}, the correlation Equations (7.9) and (7.10) would become:

$$\frac{s_u}{p_A} = 0.82 f_1 N_{60} \tag{7.11}$$

$$m_v p_A = 1.2 \frac{1}{f_2 N_{60}} \tag{7.12}$$

Example 7.3

For a clay at 10 m depth, an N_{60} count was determined to be equal to 30. This clay has $LL = 53$ and $PL = 20$. The unit weight of the clay is 16 kN/m³. Determine the undrained shear strength s_u at 10 m. What is the value of s_u/σ'_v at 10 m if the water table is at the surface?

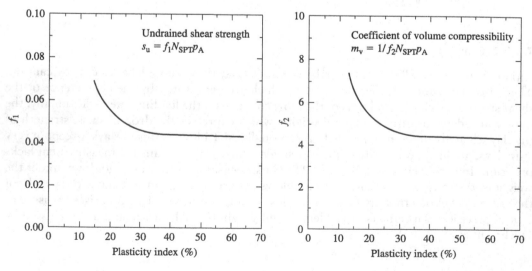

Figure 7.13 Values of f_1 and f_2 for use in Equations (7.9) and (7.10). (The plots are fits to the data by Stroud (1975).)

Solution:

Using Equation (7.11),

$$\frac{s_u}{p_A} = 0.82 f_1 N_{60}$$

The term f_1 is found from Figure 7.13. Our *PI* is 33, so our f_1 is found to be roughly 0.045. So:

$$s_u = (0.82)(0.045)(30)(100\,\text{kPa}) = 111\,\text{kPa} \quad \text{answer}$$

The vertical effective stress is:

$$\sigma'_v = (16 - 9.81) \times 10 = 62\,\text{kPa}$$

It follows that:

$$\frac{s_u}{\sigma'_v} = \frac{111}{62} = 1.8 \quad \text{answer}$$

Referring back to Chapter 6 and Equation (6.50), we see that:

$$\left(\frac{s_u}{\sigma'_v}\right)_{NC} = 0.11 + 0.0037 \times 33 = 0.23$$

The value of s_u / σ'_v we calculated is clearly higher than 0.23, suggesting the clay at 10 m seems to be overconsolidated. We can estimate the overconsolidation ratio using Equation (6.52):

$$\frac{\left(\dfrac{s_u}{\sigma'_v}\right)_{OC}}{\left(\dfrac{s_u}{\sigma'_v}\right)_{NC}} = \frac{1.8}{0.23} = 7.8 = (\text{OCR})^{0.8}$$

from which the OCR results equal to 13. Even though the number 13 is just a rough estimate for the OCR, it is nonetheless obvious that this clay is heavily overconsolidated. We should resist placing too much confidence on this number, given that, as stated earlier, SPT-based correlations for clay are somewhat crude. However, the result would give us a starting point for any calculations we might need to do. If necessary, we would still have the option of doing further testing using a method of higher quality than the SPT.

7.4 UNDISTURBED SOIL SAMPLING

An ideal undisturbed soil sample is one that preserves the soil state (density, stress state, and fabric) after sampling. Sampling procedures are naturally imperfect, and so the term "undisturbed" is better understood to mean high quality. The recovery of undisturbed clay samples is possible if proper care is taken in the sampling operations. However, upon any attempt to obtain samples of these soils using routine techniques, cohesionless soils will loosen considerably if initially dense and densify considerably if initially loose. It is also entirely possible that no recovery will be possible at all because the soil slides out of the sampler as the sampler is raised. There are some special procedures, expensive and rarely used, to obtain undisturbed samples of cohesionless soils, such as ground freezing. We will not discuss those, focusing instead on the undisturbed sampling of soils containing at least some clay so that it is possible to sample them with standard sampling tools.

The first care to take in sampling operations is the selection of a suitable tube sampler. The first feature such a sampler must have is relatively thin walls (with thickness of the order of 1.6–3.2 mm). Thick walls would cause considerable shearing in the soil, badly disturbing the sample. A low ratio of wall thickness to sample diameter is also desirable to minimize disturbance. The sampler must preferably be pushed into the soil, as opposed to driven. When the desired length of sample has been recovered, a slight rotation will shear the bottom of the sample from the underlying soil and the sampler may be brought up to the surface. The sample will be retained within the sampler by friction with the internal surface of the sampler. This common type of sampler is usually referred to as a Shelby tube sampler; it is shown in Figure 7.14.

As the sampler is lowered into the soil, it separates the soil flow into some soil that enters the sampler and becomes the sample, and some soil that flows around the sampler. This "split" in soil flow is partly responsible for the soil disturbance, as the sample is slightly compressed laterally to enter the sampler. In addition, the sample may be compressed vertically by the friction with the inner surface of the sampler as it flows into the sampler. Naturally, thin walls are helpful in reducing both of these effects. A significant amount of vertical compression of the sample is detected by a recovery ratio (length of the sample later extruded from the sampler to the sampler length) greater than one because the sample expands in the absence of the confinement provided by the sampler.

Special samplers have also been devised to minimize disturbance and facilitate total sample recovery. The piston sampler (Figure 7.15) is probably the most widely used of these special samplers. The piston sampler is introduced and advanced into the soil with the piston blocking the lower end of the sampler (see Figure 7.15). When the desired depth is reached, the thin-wall tube is pushed down into the soil while the piston remains stationary. At the end of sampling, the sampler and piston are raised together. The sample will be retained not only by friction, but also by suction that develops between the piston and sample.

An alternative to tube sampling is the collection of block samples from pits, trenches, and open excavations. There are procedures for removing such samples from the base and walls of excavations, by hand, such that disturbance is minimized. Theoretically, samples obtained in this way are those in which disturbance is the least. The application of this method, however, requires that excavations, pits, or trenches be made. These can be expensive to do at great depths specifically for the purpose of sampling, hence the primary reliance on tube sampling for "undisturbed" samples.

Figure 7.14 Shelby tube soil sampler.

Figure 7.15 Piston sampler. (Courtesy of Mueser Rutledge Consulting Engineers.)

7.5 ROCK SAMPLING*

7.5.1 Occurrence of rock

Rock may be present either at large depths, beneath the overlying soil, or in outcrops. When rock and soil coexist at a site, it often happens that boring is interrupted by the presence of rock at a shallower depth than that which the boring is required to reach. When this happens, it becomes important to ascertain whether an extensive rock mass has been reached. In some geologic settings, it is possible that either a large boulder in a soil matrix or a layer of rock underlain by soil is reached instead. Knowledge of the principles of geology and of the local geology is very helpful when facing situations of this type. If the rock is key to developing the necessary bearing capacity for the foundation elements, it may be necessary to core through it to sufficient depth. Coring at several locations is also helpful to assess the extent of a possible boulder.

7.5.2 Sampling operations

Rock may or may not be covered by soil layers. When it is, the soil is often derived from the rock itself and termed a residual soil. In the course of routine site investigations, bedrock is identified by difficulty in advancing borings using the normal procedures, by extremely high SPT blow counts (of the order of 100 per 150 mm penetration or greater), or even by bouncing of the SPT hammer on the anvil. When that happens, the question of whether more knowledge about the rock is required for the project at hand determines if borings are to be discontinued. Depending on the scale of the loads to be transferred to the ground, the rock mass may or may not be relevant for design of the foundations. If bedrock is located at relatively large depths, the stresses induced by shallow foundations will dissipate with depth and be minimal at the soil–rock interface. Likewise, pile foundations may be designed to bear on an intermediate soil layer with sufficient thickness and load-carrying capacity. In other cases, it will be important to learn more about the condition of the rock, its continuity, the presence of defects, and degree of weathering so that an intelligent estimate of its load-carrying capacity can be made. In these cases, it becomes necessary to extend the boring into the rock mass. The importance of knowledge of engineering geology and of the geology of the area cannot be overstated when rock behavior becomes important in a project.

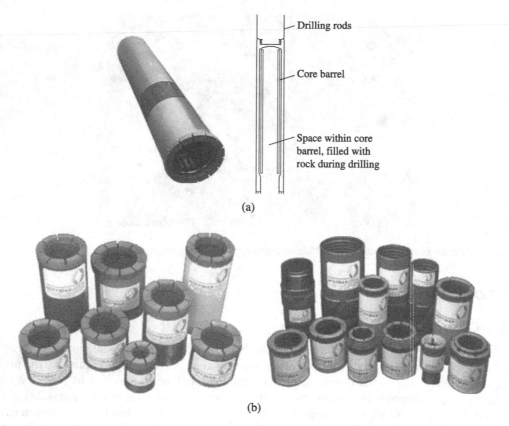

Figure 7.16 Rock coring tools: (a) core barrel and (b) diamond bits. (Courtesy of Hoffman Diamond Products.)

Drilling in rock requires different tools from those used to drill in soil. Figure 7.16 shows typical tools used to drill in rock and recover rock samples. The sampler is usually referred to as a core barrel, and the sample, a rock core. The coring bit is encrusted with diamonds for hardness. The core barrel, which is connected to a string of individual casing segments, is advanced by rotation and axial pressure, forming a groove around the core. The cylindrical sample of rock, the rock core, is the rock that has entered the core barrel during drilling.

7.5.3 Information from coring and rock testing

Two measures of rock quality and continuity provided by the coring are the recovery ratio and the rock quality designation (RQD). The recovery ratio is simply the ratio of the total length of rock recovered (whether continuous or not) to the total core length. The RQD is the ratio of the summation of the lengths of core pieces longer than 100 mm to the total core length (including pieces shorter than 100 mm). Clearly, poor recovery (recovery ratio of the order of 0.5) is an indication of very fractured or weak rock (or, alternatively, poor coring technique). A high RQD value is indicative of fairly continuous rock. The RQD is the basis for some failure criteria and bearing capacity assessments, as we will see when we discuss rock foundations in a later chapter. Site inspectors will also measure the rate of progress in coring as an additional, if imperfect (particularly because of rig and operator dependence), gauge of rock quality.

Rock cores are sometimes taken to the laboratory for additional testing, chiefly strength testing. Unconfined compression tests are fairly common, and rock shear strength is often expressed in terms of unconfined compressive strength. Unconfined compressive strength is obtained by compressing a cylinder of intact rock in the direction of its axis. Intact rock (that is, rock with no defects) can have extremely high shear strength. In general, igneous and metamorphic rocks tend to be stronger than sedimentary rocks, but some sedimentary rocks, such as quartzites (strongly cemented sandstones), are very strong as well (Goodman 1992).

Table 7.3 provides ranges of unconfined compressive strength, tensile strength, and shear strength for a number of rocks. The point load test (ISRM 1985)[9] may be used to provide an estimate of unconfined compressive strength. There are different ways to obtain tensile strength, including compressing the cylinder in the direction of one of its diameters, which generates tension along its vertical axis, and thus we provide a tensile strength reading while applying compression on the specimen. Shear strength can be obtained from the unconfined compression tests, direct shear tests, or triaxial compression tests. We can use Table 7.3 as a guide when selecting an appropriate value for rock strength. The ranges in Table 7.3 are rather wide, reflecting the variability of rock properties in nature as well as the effect that slight weathering, which was present in intact rock samples tested to develop the table, has on strength.

7.5.4 Rock mass strength

The term "rock mass" refers to the rock considered in its entirety, including the presence of defects of various kinds and their effect on load response. The rock mass is governed by discontinuities and defects to a greater extent than by its intact shear strength. Discontinuities are of different types and exist for a variety of reasons. The most important types of discontinuities are faults, shear zones, joints, schistosity planes, and bedding planes.

Table 7.3 Unconfined compressive strength q_u, tensile strength σ_t, and shear strength s of various types of intact rock

Rock	Stress-normalized unconfined compressive strength (q_u/p_A)	Stress-normalized tensile strength (σ_t/p_A)	Stress-normalized shear strength (s/p_A)
Basalt	780–4120	59–120	49–130
Diabase	1180–2450	59–127	59–98
Gabbro	1470–2940	49–80	39–83
Granite	980–2750	39–80	49–100
Dolomite	147–2450	25–60	25–69
Limestone	39–2450	10–70	15–70
Sandstone	95–1670	15–60	20–60
Shale	98–1000	20–98	29–110
Gneiss	780–2450	39–70	30–70
Marble	490–1960	49–80	35–80
Quartzite	850–3530	29–50	—
Slate	980–1960	69–196	—

Source: Based on data by Shroff and Shah (2003) and Széchy (1966).

$p_A = 0.1$ MPa ≈ 1 tsf.

---

[9] In the point load test, a sample of rock is positioned between two steel tips of approximately conical shape. The two tips are on the same vertical. One of the tips is pushed against the rock. The pressure required to fail or "break" the rock is converted back into an estimate of shear strength.

Discontinuities that are open, particularly if filled with plastic soil gouge, are most detrimental to shear strength and to stiffness. Faults, shear zones, and joints may fall in that category.

The characterization of discontinuities in rock should include observation and quantification of the following:

- average and minimum spacing between adjacent discontinuities;
- rugosity (roughness);
- width;
- degree of weathering;
- the presence of gouge material within the discontinuity (mainly with reference to faults); and
- three-dimensional pattern.

For routine foundation jobs, extensive mapping of discontinuities is typically not done or required. It suffices in these cases to make reasonable, conservative estimates of intact rock strength and use a simplified method for accounting for the effects of the discontinuities. A very simple method to account for the reduction in shear strength due to the presence of discontinuities is to apply a weakness coefficient c_w to rock intact strength so that the shear strength s_{rm} of the rock mass is given by

$$s_{rm} = c_w s_{ir} \tag{7.13}$$

where s_{ir} = shear strength of intact rock. The values of the weakness factor c_w are given in Table 7.4.

A more involved method for assessing the combined effects of discontinuities and intact rock strength on the overall behavior of a rock mass was developed by Bieniawski (1979, 1989); it is known as the rock mass rating (RMR) system. The RMR of a rock mass can be established based on Table 7.5.

Table 7.4 Coefficient of weakness c_w of rocks as a function of rock-jointing characteristics

	Coefficient of weakness c_w	
Joint characteristics in rock	Limiting values	Average value[a]
Dense network of fractures, in all directions, in layered rock or uncemented individual blocks	0.0–0.01	0.0005
Dense network of open fractures in all directions	0.001–0.02	0.005
Dense jointing	0.01–0.04	0.02
Above-average jointing	0.04–0.08	0.06
Average jointing (open/closed fractures every 20–30 cm)	0.08–0.12	0.1
Below-average jointing	0.12–0.9	0.2
Network of deep joints every 30–50 cm; insignificant number of open fractures	0.3–0.4	0.35
Little-jointed rocks; closed fractures	0.4–0.6	0.5
Microfractures almost absent	0.6–0.8	0.7
Monolithic rocks with no sign of jointing	0.8–1.0	0.9

Source: Modified after Komarnitskii (1968).

[a] Note that the average value is not the average of the two limiting values because distributions may be skewed.

Table 7.5 Rock mass rating for jointed rock

A. Classification parameters and their ratings

Parameter		Ranges of values					
1 Strength of intact rock material	Point load strength index (MPa)	>10	4–10	2–4	1–2	Point load test not indicated at this level	
	Uniaxial compressive strength (MPa)	>250	100–250	50–100	25–50	5–25 1–5	<1
Rating		15	12	7	4	2 1	0
2 Drill core quality RQD		90%–100%	75%–90%	50%–75%	25%–50%	<25%	
Rating		20	17	13	8	3	
3 Spacing of discontinuities (m)		>2	0.6–2	0.2–0.6	0.06–2	<0.06	
Rating		20	15	10	8	5	
4 Condition of discontinuities		• Very rough • Not continuous • Closed • Unweathered	• Slightly rough • Separation <1 mm • Slightly weathered walls	• Slightly rough • Separation <1 mm • Highly weathered walls	• Slickened surfaces or • Gouge <5 mm thick • Continuous separation of 1–5 mm	• Soft gouge >5 mm thick or • Continuous separation >5 mm	
Rating		30	25	20	10	0	
5 Ground water	General conditions	Completely dry	Damp	Wet	Dripping	Flowing	
Rating		15	10	7	4	0	

When calculating rock strength using Table 7.6, rating = 10; ground water pressures accounted for in stability analysis.

B. Rating adjustment for joint orientations

Strike and dip orientation of discontinuities	Very favorable	Favorable	Fair	Unfavorable	Very unfavorable
Adjustment for foundations	0	−2	−7	−15	−25
Adjustment for slopes	0	−5	−25	−50	−60

Source: After Bieniawski (1979, 1989).

When calculating rock strength using Table 7.6, adjustment = 0; joint orientation accounted for in stability analysis.

Example 7.4

An open cut in a site where basalt outcrops shows the basalt to have 5 to 10 fissures per meter with a horizontal direction (zero dip). Although borings for this specific location are not available, typical RQD values for this rock are in the 25%–50% range. The basalt as observed in this cut is slightly weathered. Its unconfined compressive strength is 80 MPa. Find the RMR and the Hoek–Brown parameters for this rock mass. The effect of water is negligible, and the joint orientation (horizontal) is neutral.

Solution:

According to Table 7.5, the RMR is obtained as follows:

RMR = 7 for rock with $50 \le q_u < 100$ MPa

+8 for RQD in the 25% – 50% range

+8 for discontinuity spacing in the 0.06 – 0.2 range

+25 (joints in good condition)

+10 (water to be accounted for in analysis)

+0 (joint orientation to be accounted for in analysis)

= 58

Consulting Table 7.6 with an RMR = 58 and the rock description we have, the best match would be a fine-grained igneous rock with fair quality rock mass, leading to

$m = 1.395$

and

$s = 0.00293$

We could also interpolate between these values and those immediately in Table 7.6 ($m = 0.311$ and $s = 0.00009$), but given the subjectivity of this approach and the fact that we are close to the 65 RMR corresponding to the values of s and m we selected, it is not strictly necessary to do that. Note also that the descriptions of rock mass quality are given in the first column of Table 7.6 to provide an idea of what a rock with a particular RMR would resemble, but the computed RMR is what we would go by to determine the values of m and s.

In Chapter 10, we will continue this example by calculating the bearing capacity of a footing in this rock using the estimated rock strength.

7.6 CONE PENETRATION TEST: CONE PENETROMETER, TYPES OF RIG, AND QUANTITIES MEASURED

7.6.1 Cone penetrometer and CPT rigs

The CPT was developed much more recently than the SPT, but it is now widely used for geotechnical site characterization and *in situ* determination of soil properties. Among its advantages are simplicity, speed, and quasi-continuous profiling. The test is performed by pushing a penetrometer with a conical tip and standard geometry (Figure 7.17) vertically into the ground at a standard rate of 20 mm/s. Penetration at this rate is fully drained for sands with nonplastic silt contents up to 10% (Carraro et al. 2003) and fully undrained for most clays, but it is partially drained for soils intermediate between these two extremes.

Originally, the cone penetrometer (often referred to as just the "cone") was used to measure only the tip or cone resistance q_c, defined as the vertical force acting on the tip of the penetrometer divided by the base area of the tip (1000 mm² for the standard penetrometer, with a diameter of 35.7 mm). Over the years, sensors have been incorporated into the cone to measure the friction along a lateral sleeve (also of standard dimensions), the arrival of a seismic shear wave (which allows determining shear wave velocities), pore pressure, and other quantities (Mitchell 1988). This versatility has further enhanced the use of the CPT. Additionally, an important advantage of the CPT is that the penetration process is amenable to theoretical modeling.

Table 7.6 Approximate relationship between rock mass quality and the Hoek–Brown failure criterion parameters *m* and *s*

Rock description and rock mass rating	Yield criterion parameters	Carbonate rocks with well-developed crystal cleavage – dolomite, limestone, and marble	Lithified argillaceous rocks – mudstone, siltstone, shale, and slate (normal to cleavage)	Arenaceous rocks with strong crystals and poorly developed crystal cleavage – sandstone and quartzite	Fine-grained polymineralic igneous crystalline rocks – andesite, dolerite, diabase, and rhyolite	Coarse-grained polymineralic igneous and metamorphic crystalline rocks – amphibolite, gabbro, gneiss, granite, norite, and quartz-diorite
Intact rock samples Laboratory size specimens free from discontinuities RMR = 100	*m*	7.00	10.00	15.00	17.00	25.00
	s	1.00	1.00	1.00	1.00	1.00
Very good quality rock mass Tightly interlocking undisturbed rock with unweathered joint at 1–3 m RMR = 85	*m*	2.40	3.43	5.14	5.82	8.56
	s	0.082	0.082	0.082	0.082	0.082
Good quality rock mass Fresh to slightly weathered rock, slightly disturbed with joints at 1–3 m RMR = 65	*m*	0.575	0.821	1.231	1.395	2.052
	s	0.00293	0.00293	0.00293	0.00293	0.00293
Fair quality rock mass Several sets of moderately weathered joints spaced at 0.3–1 m RMR = 44	*m*	0.128	0.183	0.275	0.311	0.458
	s	0.00009	0.00009	0.00009	0.00009	0.00009
Poor quality rock mass Numerous weathered joints at 30–500 mm, some gouge. Clean compacted waste rock RMR = 23	*m*	0.029	0.041	0.061	0.069	0.102
	s	0.000003	0.000003	0.000003	0.000003	0.000003
Very poor quality rock mass Numerous heavily weathered joints spaced <50 mm with gouge. Waste rock with fines. RMR = 3	*m*	0.007	0.010	0.015	0.017	0.025
	s	0.0000001	0.0000001	0.0000001	0.0000001	0.0000001

Figure 7.17 Cone penetrometer.

CPT rigs for onshore testing are of three types: truck-mounted, crawler-mounted, and trailer-mounted (Figure 7.18). Each is suited to a different terrain. The trailer-mounted rig is quite light, relying mostly on anchors to develop the reaction required for pushing the cone. The cone is located right at the center of gravity of truck- and crawler-mounted rigs, so the

Figure 7.18 CPT rigs: (a) truck-mounted, (b) crawler-mounted, and (c) trailer-mounted. (Courtesy of ISMES.)

weight of the rig can be relied on to provide most of the reaction required for testing, although anchoring remains a source of additional reaction. Total reaction capacity ranges from as little as 5 tons for trailer-mounted systems to as much as 40 tons for large truck-mounted systems.

7.6.2 Measurements made during a CPT

Figure 7.19 shows the cross sections of the electrical cone penetrometer. The figure shows a possible configuration of load cells within the cone to measure tip resistance and sleeve friction. As stated earlier, cone resistance is the ratio of the force opposing penetration acting only on the cone tip divided by the projected area of the tip (1000 mm² in the standard cone). Another measurement that is always made is the sleeve friction f_s, defined as the ratio of the shear force acting along the cylindrical friction sleeve located above the cone tip to the area of the sleeve (15,000 mm² in the standard cone). Sleeve friction was originally envisioned as useful for the estimation of pile shaft resistance (an application discussed in

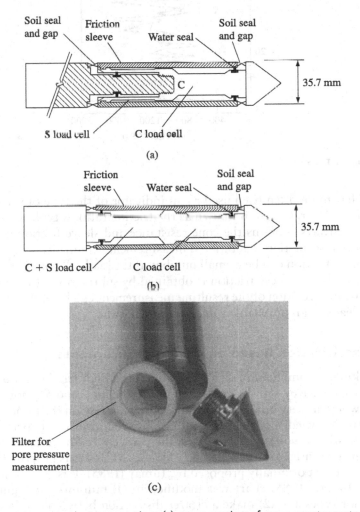

Figure 7.19 Cone penetrometer instrumentation: (a) cross section of cone penetrometer with a load cell scheme in which cone resistance and sleeve friction are measured separately; (b) cross section of cone penetrometer with a load cell scheme in which cone resistance plus sleeve resistance are measured by one load cell and cone resistance is measured by another load cell; and (c) disassembled cone penetrometer. ((a) and (b) Courtesy of ASTM D5778 (ASTM 2020a).)

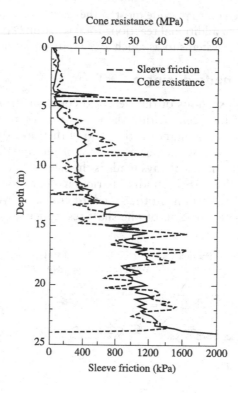

Figure 7.20 Sample CPT log.

Chapter 13); it has more often been used as an indicator of the type of soil through which the cone is pushed. The ratio f_s/q_c is known as the friction ratio. A typical CPT log is shown in Figure 7.20. It always contains the cone resistance and sleeve friction versus depth and may contain more information if additional measurements are made.

Given that sleeve friction can be a small number, if the load cell configuration in the cone penetrometer is such that sleeve friction is obtained by subtraction of the values measured by two load cells, the precision of the resulting measurement can be of the order of the sleeve friction, which leads to an unreliable value of sleeve friction.

7.6.3 Soil classification based on CPT measurements

Generally speaking, a combination of low q_c values and high friction ratio (f_s/q_c) suggests the soil is a clay or a clayey soil. In sands, q_c tends to be high and f_s/q_c low. All other soils would be somewhere in between. It is more appropriate to rely on the information from soil borings or on the knowledge of the site stratigraphy from previous experience in the area of the project to estimate soil types. However, there are charts that allow estimation of soil type based on the measured values of f_s and q_c. One such chart was proposed by Ganju et al. (2017) based on a chart originally proposed by Tumay (1985). It is shown in Figure 7.21.

The original Tumay (1985) chart was modified to (1) minimize ambiguities associated with soil behavior types and (2) make a clearer distinction between soil intrinsic variables (related closely to soil composition) and soil state variables, such as relative density, stress state, and fabric. In general, a combination of low q_c/p_A (<10) and high f_s/q_c values (>4%) suggests a clayey soil, whereas a combination of high q_c/p_A (>50) and low f_s/q_c values (<2%) suggests a sandy soil, where p_A = reference stress (= 100 kPa or 14.5 psi).

Table 7.7 Soil behavior types associated with the modified Tumay (1985) chart

Soil type	Soil description
1	Sensitive clay
2	Very soft clay
3	Soft clay
4	Medium stiff clay
5	Stiff clay
6	Very stiff clay
7	Sandy clay or silty clay
8	Clayey silty sand
9	Clayey sand or silt
10	Clayey silt
11	Very dense sand or silty sand
12	Dense sand or silty sand
13	Medium-dense sand or silty sand
14	Loose sand or silty sand
15	Very loose sand or silty sand

Table 7.7 summarizes the soil behavior types associated with the modified Tumay (1985) chart. Each soil behavior type that appears in Figure 7.21 is assigned an index. For instance, indices 1 to 7 correspond to clays of different stiffnesses, 8 corresponds to sands containing fines,

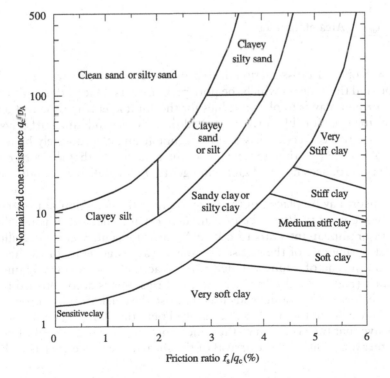

Figure 7.21 Soil type from cone resistance, pore pressure, and sleeve friction. (After Ganju et al. (2017), with permission from Elsevier Science and Technology Journals.)

and 9 and 10 correspond to clayey sand or silt and clayey silt, respectively. Ganju et al. (2017) classified soils falling in the "clean sand or silty sand" region of the chart using five indices (zones 11–15 in Table 7.7) based on the relative density (see Table 3.10), which can be estimated from CPT data using Equation (7.20) proposed by Salgado and Prezzi (2007).

Mitchell and Brandon (1998) suggested that there are limitations associated with the methods that have been proposed to estimate soil type based on measured q_c and f_s, so we should use Figure 7.21 with caution.

7.6.4 Measurement of pore pressures and shear wave velocity

The cone can also be instrumented to measure pore pressures during penetration and shear wave velocity. It is sometimes called a piezocone when it is enabled to measure pore pressures and a seismic cone when it can detect the arrival of shear waves. The measured pore pressure depends on the location of the pore pressure sensor. The three possible locations are (1) behind the friction sleeve, (2) in front of the friction sleeve and just behind the cone face, and (3) on the cone face. Of the three, the best location is just behind the cone face. The main reason for this is the existence of a small area behind the cone face on which the pore pressure pushes down on the cone, aiding penetration. The implication of this is that we must subtract the downward force resulting from the pore pressure acting on this small area behind the cone tip from the upward force caused by pore pressure acting on the projected area of the cone. If the pore pressure behind the cone face is measured, a corrected, total cone resistance q_t can be calculated by taking into account the unbalanced pore pressure force:

$$q_t = q_c + (1-a)u_{\mathrm{meas}} \tag{7.14}$$

where

$$a = \frac{A_c - A_{\mathrm{diff}}}{A_c} \approx \frac{\text{Area of load cell}}{A_c} \tag{7.15}$$

in which A_c = projected cross-sectional area of the cone ($10\,\mathrm{cm^2}$ in standard cones) and A_{diff} = area behind the cone on which the pore pressure acts. One of the goals of cone design and manufacturing today is to obtain values for the parameter a as close to 1 as possible, so as to minimize the need for this correction. Additionally, in sands and stiff clays, in which q_c takes high values, this correction is small to negligible and can routinely be ignored. In the use we make of the CPT in this text, we make the assumption that the correction has been made whenever it produces nonnegligible changes to q_c. We will accordingly not distinguish q_c from q_t.

A seismic sensor may be placed in the cone to allow the detection of the arrival of shear waves. Shear wave velocity V_s can be measured as often as needed. It is usual to stop penetration at $1\,\mathrm{m}$ or similar intervals to take a V_s measurement. Figure 7.22 illustrates how V_s is measured during one of the stops. The shear wave is generated by a horizontal blow of a hammer on the end of a long wooden plank placed on and pressed against the ground for good contact (Figure 7.22b). In truck-mounted rigs, this is accomplished by placing the wooden plank below the hydraulic pads used to raise the truck from the ground and level it. A portion of the weight of the truck is thus applied onto the plank. The hammer blow generates a shear wave that travels down to the cone approximately in a straight line.[10] The blow triggers time measurement, and the arrival of the shear wave at the cone is picked up by the

---

[10] This is the model we make for the wave propagation; in fact, small differences in soil stiffness along the way may cause some bending of the travel path. This can typically be neglected without significant loss of accuracy.

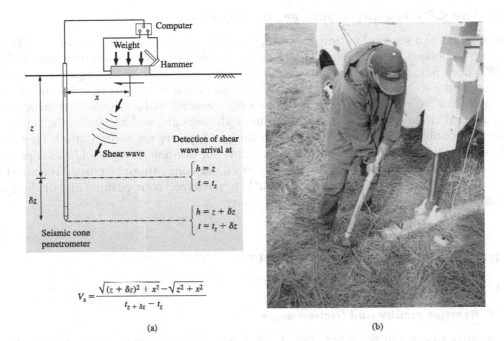

Figure 7.22 Performance of seismic CPT and measurement of shear wave velocity using a truck-mounted rig: (a) schematic and (b) photo.

seismic sensor. Because both the distance from the wooden plank to the cone and the arrival time are known, the average shear wave velocity along the travel path can be calculated. However, it is of much greater interest to estimate V_s approximately at the depth at which the cone is located. Even though V_s cannot be determined continuously, seismic measurements can be taken at a spacing δz within a range of 0.5–1 m or so without making the test excessively expensive.

To illustrate how V_s is calculated from each pair of consecutive measurements, let us consider that measurements are taken at depths z and $z+\delta z$, the corresponding arrival times are t_z and $t_{z+\delta z}$, and the distance from the axis of cone penetration to the wooden plank is x (Figure 7.22a). The shear wave velocity within the δz interval is given by the ratio of the difference in travel distances to the difference in travel times:

$$V_s = \frac{\sqrt{\left(z+\delta z\right)^2 + x^2} - \sqrt{z^2 + x^2}}{t_{z+\delta z} - t_z} \tag{7.16}$$

From the knowledge of V_s at a given depth, we can immediately obtain the small-strain shear modulus there from the following equation from elasticity theory:

$$G_0 = \rho_m V_s^2 = \frac{\gamma_m}{g} V_s^2 \tag{7.17}$$

where ρ_m is the material (moist) mass density of the soil at depth z and γ_m is the material unit weight there.

In interpreting the results of seismic measurements, it is customary to look for layers of similar soil with either relatively constant or slightly increasing V_s values. By combining this information with the cone resistance versus depth log, it is possible to obtain a reasonable model for the soil profile at any given location.

7.6.5 The CPT in a site investigation program

In the context of a site exploration program, it may be advantageous to substitute CPTs for some of the borings because the CPT is a more reliable test and provides information that the SPT does not. Another factor in deciding the extent of the site investigation program is how well the soils in the area of the project are known. In urban areas, it is common for an engineer to have been involved in the design of foundations for neighboring structures. If the general characteristics of the soil profile in the area are well known to the engineer, it may be possible to rely entirely on CPT testing and specify no borings at all. However, CPTs may be limited by features of the site profile. For example, if a gravel layer is present at a site, and information below this layer is desired, the cone cannot be pushed through the gravel. In cases such as these, borings with SPTs or combined cone pushing and drilling are typically specified.

7.7 INTERPRETATION OF CPT RESULTS

7.7.1 Sands

7.7.1.1 Relative density and friction angle

It is intuitive that it will be increasingly hard to push a cone penetrometer into sand as its relative density increases. The dependence of penetration resistance on stress is best understood by viewing the penetrometer as creating and expanding a cylindrical cavity in the soil as it is pushed through it. Accordingly, there is a relationship between cone resistance and the pressure required to expand a cylindrical cavity in the soil from zero initial radius. The cylindrical cavity expansion pressure is a function not only of the relative density of the soil, but also of the initial lateral effective stress $\sigma'_h = K_0\sigma'_v$ in the soil. The larger this stress is, the harder it is for the cone to push aside the soil in its path, and the higher the q_c value will be. Sands with a given vertical effective stress can have different lateral effective stresses, and, therefore, different q_c values, if the overconsolidation ratios (and therefore K_0) are different. The relationship between cone resistance and the soil state variables (relative density and stress state) can be expressed as

$$q_c = q_c\left(D_R, \sigma'_h\right) \tag{7.18}$$

indicating that q_c is a function of D_R and σ'_h. Intrinsic soil variables, related to particle size distribution and particle morphology, are implicit in the function of Equation (7.18); the most important of those is the critical-state friction angle ϕ_c of the soil.

Salgado and Prezzi (2007) proposed a single equation to estimate q_c as a function of D_R and σ'_h, obtained by doing a regression on the results of a rigorous cavity expansion analysis. The equation is:

$$\frac{q_c}{p_A} = 1.64\exp\left[0.1041\phi_c + \left(0.0264 - 0.0002\phi_c\right)D_R\right]\left(\frac{\sigma'_h}{p_A}\right)^{0.841-0.0047D_R} \tag{7.19}$$

where p_A = reference stress (= 100 kPa ≈ 1 tsf). Salgado and Prezzi (2007) also provided charts for $\phi_c = 29°$–$36°$. Four of these are presented in Figure 7.23; another four, in Figure 7.24.

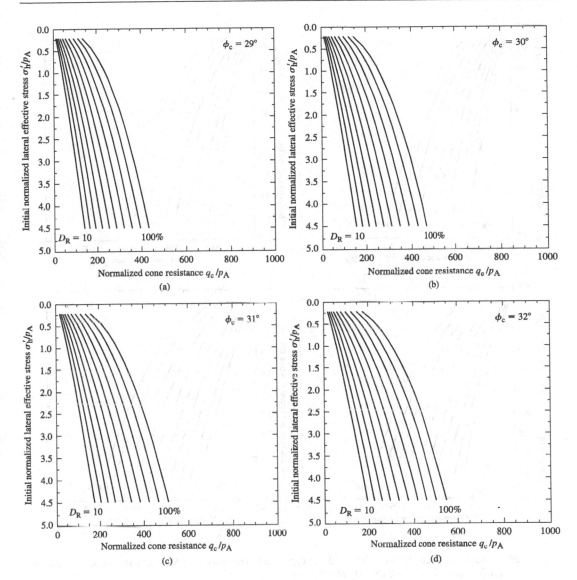

Figure 7.23 Cone resistance charts, calculated for typical intrinsic variables for silica sand and ϕ_c values of (a) 29°, (b) 30°, (c) 31°, and (d) 32°.

In interpreting the CPT results in sand, we are concerned with estimating relative density. It is convenient therefore to write D_R in terms of q_c:

$$D_R = \frac{\ln\left(\dfrac{q_c}{p_A}\right) - 0.4947 - 0.1041\phi_c - 0.841\ln\left(\dfrac{\sigma'_h}{p_A}\right)}{0.0264 - 0.0002\phi_c - 0.0047\ln\left(\dfrac{\sigma'_h}{p_A}\right)} \tag{7.20}$$

with $0\% \leq D_R \leq 100\%$.

Salgado et al. (1997a,b) analyzed the variability in cone resistance measurements and cone resistance calculations using the program CONPOINT. Figure 7.25 illustrates the scatter in terms of normalized lateral effective stress for four relative density ranges. They concluded

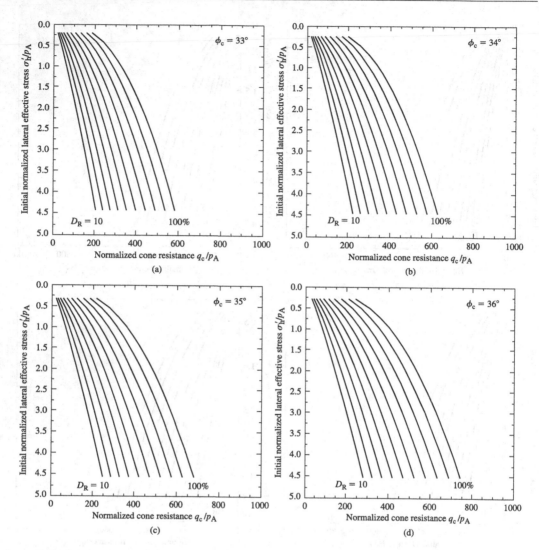

Figure 7.24 Cone resistance charts, calculated for typical intrinsic variables for silica sand and ϕ_c values of (a) 33°, (b) 34°, (c) 35°, and (d) 36°.

that cone resistance could be computed using a relationship such as (7.19) to within ±30%. Given the exponential dependence of q_c on D_R, Equation (7.20) allows a tighter estimation of D_R from q_c, with D_R determinable to within 10% (as units of relative density; that is, if we compute $D_R = 80\%$, it would with good reliability be within the 70%–90% range if the input parameters we use are realistic). Calculations leading to very large values of D_R, approaching 100%, suggest that either the sand is extremely dense sand or the value of ϕ_c of the sand used in the calculations is too low and not realistic.

In some correlations, a stress-normalized q_{c1} may be used. This is common, for example, in correlations for liquefaction resistance assessment. Stress normalization for q_c is the same as for the N_{SPT}. Mathematically:

$$q_{c1} = q_c \sqrt{\frac{p_A}{\sigma_v'} \frac{(K_0)_{NC}}{K_0}} \qquad (7.21)$$

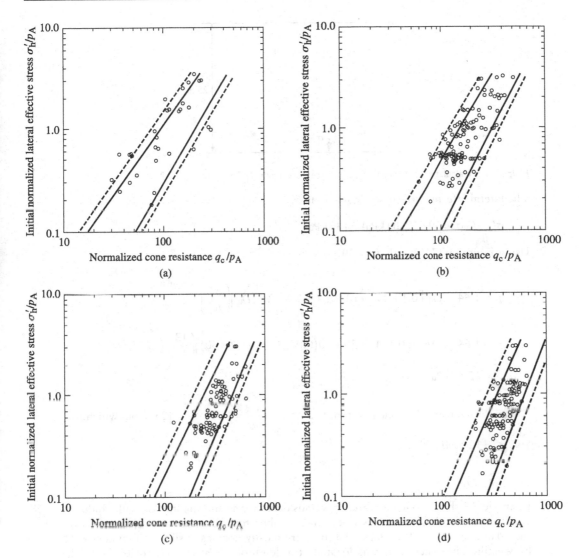

Figure 7.25 Cone resistance versus relative density from CPTs performed in calibration chamber samples of sands with various properties for (a) D_R 20%–40%, (b) D_R 40%–60%, (c) D_R 60%–80%, and (d) D_R 80%–100%. The solid lines represent a range of calculated q_c values for the test sands, and the dashed lines, 80% and 120% of these q_c values. (Based on data from Salgado et al. (1997b).)

Example 7.5

The cone resistance is to be estimated at a depth of 25 m within a clean sand deposit with water table at a depth of 5 m. The sand is assumed to be normally consolidated with $K_0 = 0.45$. The relative density is 50%. The critical-state friction angle estimated from triaxial compression tests is 30°. The average unit weight of the sand over the 25 m is 20 kN/m³.

Solution:

Looking at E-Figure 7.2, the vertical effective stress is calculated as

$$\sigma'_v = \gamma_m z - \gamma_w (z - z_w) = (25\,\text{m})(20\,\text{kN/m}^3) - (20\,\text{m})(9.81\,\text{kN/m}^3) = 304\,\text{kPa}$$

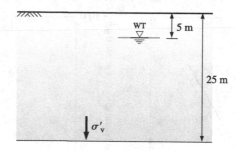

E-Figure 7.2 Calculation of vertical effective stress for Example 7.5.

The lateral effective stress is calculated as

$$\sigma_h' = K_0\sigma_v' = (0.45)(304\,kPa) = 137\,kPa$$

Using Equation (7.19) with $\phi_c = 30°$:

$$q_c = 1.64 p_A \exp\left[0.1041\,\phi_c + (0.0264 - 0.0002\phi_c)D_R\right]\left(\frac{\sigma_h'}{p_A}\right)^{0.841 - 0.0047 D_R}$$

$$= 1.64 \times 0.1 \exp\left[0.1041 \times 30 + 50(0.0264 - 0.0002 \times 30)\right]\left(\frac{137}{100}\right)^{0.841 - 0.0047 \times 50}$$

$$= 12.5\,MPa$$

Alternatively, using the cone resistance charts, for $\dfrac{\sigma_h'}{p_A} = 1.37$, $\dfrac{q_c}{p_A} = 125$, from which:

$$q_c = 12.5\,MPa \quad ^{answer}$$

Example 7.6

Example 7.3 shows cone resistance q_c versus depth for a sand site in Evansville, Indiana. Soil from a depth of 14 m was tested in the laboratory. The average unit weight over the 14 m is equal to 19 kN/m³. The fines are mostly nonplastic silt, with content <7% by weight. The water table was located at a depth of 9 m at the time of the test. The critical-state friction angle is ~33°. From the knowledge of geologic conditions of the area, the deposit is a young normally consolidated sand with low fines content throughout. Estimate the relative density and the peak friction angles at 14 m.

Solution:

Using 19 kN/m³ for the soil unit weight, the vertical effective stress can be estimated as

$$\sigma_v' = 19 \times 14 - 9.81 \times 5 = 217\,kPa$$

A K_0 of 0.4, consistent with a young NC deposit, gives

$$\sigma_h' = K_0 \times \sigma_v' = 0.4 \times 217 = 87\,kPa$$

E-Figure 7.3 gives $q_c = 16\,MPa$ at 14 m. Entering Figure 7.24a with $q_c = 16\,MPa$ and $\sigma_h' = 87\,kPa$, we get $D_R = 60\%$. Alternatively, using Equation (7.20), we obtain $D_R = 62\%$. The peak friction angles calculated using (5.12) and (5.20) are $\phi_p = 37.9°$ for $D_R = 60\%$ and $\phi_p = 38°$ for $D_R = 62\%$.

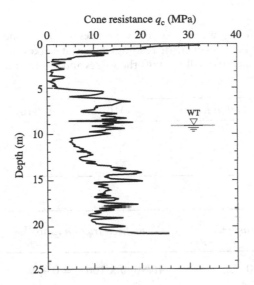

E-Figure 7.3 CPT log from a site in Evansville, Indiana.

Example 7.7

Given the CPT log in E-Figure 7.4, determine the cone resistance and friction ratio at depths of 3, 5, and 7 m. What type of material is likely to be present at these depths? What is the relative density of the soil at these depths? Assume unit weight $\gamma - 18 \text{ kN/m}^3$, critical-state friction angle $\phi_c = 33°$, and coefficient of lateral earth pressure $K_0 = 0.45$.

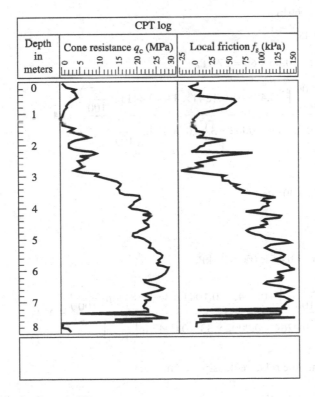

E-Figure 7.4 CPT log for Example 7.7.

Solution:

By examining the CPT log, we can see that the values for both cone resistance and local friction will need to be smoothened; readings taken at spikes are not representative. For the sake of simplicity, we will read off the values and put them in a table:

Depth (m)	q_c (MPa)	f_s (kPa)
3	9	40
5	22	105
7	22	140

Depth (m)	Normalized cone resistance $\dfrac{q_c}{p_A}$	Normalized friction ratio $\dfrac{f_s}{q_c}$	Type of material
3	$\dfrac{9000}{100} = 90$	$40/9000 \times 100\% = 0.44\%$	Clean sand or silty sand
5	$\dfrac{22{,}000}{100} = 220$	$105/22{,}000 \times 100\% = 0.48\%$	Clean sand or silty sand
7	$\dfrac{22{,}000}{100} = 220$	$140/22{,}000 \times 100\% = 0.64\%$	Clean sand or silty sand

Relative density at 3 m:

$$\sigma'_v = 20 \times 3 = 60\,\text{kPa}$$

Using a K_0 of 0.45:

$$\sigma'_h = K_0 \times \sigma'_v = 0.45 \times 60 = 27\ \text{kPa}$$

$$D_R = \frac{\ln\left(\dfrac{9000}{100}\right) - 0.4947 - 0.1041 \times 33 - 0.841 \ln\left(\dfrac{27}{100}\right)}{0.0264 - 0.0002 \times 33 - 0.0047 \ln\left(\dfrac{27}{100}\right)} = 64\%$$

Relative density at 5 m:

$$\sigma'_v = 20 \times 5 = 100\,\text{kPa}$$

Using a K_0 of 0.45:

$$\sigma'_h = K_0 \times \sigma'_v = 0.45 \times 100 = 45\,\text{kPa}$$

$$D_R = \frac{\ln\left(\dfrac{22{,}000}{100}\right) - 0.4947 - 0.1041 \times 33 - 0.841 \ln\left(\dfrac{45}{100}\right)}{0.0264 - 0.0002 \times 33 - 0.0047 \ln\left(\dfrac{45}{100}\right)} = 91\%$$

Now we calculate the relative density at 7 m:

$$\sigma'_v = 20 \times 7 = 140\ \text{kPa}$$

Using a K_0 of 0.45:

$$\sigma'_h = K_0 \times \sigma'_v = 0.45 \times 140 = 63 \text{ kPa}$$

$$D_R = \frac{\ln\left(\dfrac{22{,}000}{100}\right) - 0.4947 - 0.1041 \times 33 - 0.841 \ln\left(\dfrac{63}{100}\right)}{0.0264 - 0.0002 \times 33 - 0.0047 \ln\left(\dfrac{63}{100}\right)} = 84\%$$

From the calculated relative densities D_R, the degree of compactness of material from Table 3.10 is medium dense at 3 m, very dense at 5 m, and dense at 7 m.

Unit weights assumed in the calculations can be refined after the relative densities are calculated, but this usually does not change the results by much.

Example 7.8

The cone resistance for a clean sand at 9 m has been measured at 14.2 MPa. Estimate the relative density of the sand. The water table is 3 m below the surface.

Solution:

Referring to Equation (7.20), we see that we need the values of ϕ_c and σ'_h. For a clean sand, we can estimate ϕ_c to be 33°. To find σ'_h, we must first find σ'_v, so we estimate γ to be 20 kN/m³. This gives us:

$$\sigma'_v = \gamma h - \gamma_w h_w = 20 \times 9 - 9.81 \times 6 = 121 \text{ kPa}$$

We will assume the sand is normally consolidated, so we use $K_0 = 0.45$ to estimate σ'_h:

$$\sigma'_h = K_0 \sigma'_v = 0.45(121) = 54.5 \text{ kPa}$$

Using Equation (7.20), we obtain $D_R = 68\%$. **answer**

Figure 7.23 may also be used to estimate D_R by finding the intersection of a horizontal line at $\sigma'_h = 54.5$ kPa with a vertical line at $q_c = 14.2$ MPa. The point falls between two lines; interpolating between the two curves, we find D_R to be equal to 66%.

7.7.1.2 Shear modulus

The small-strain shear modulus G_0 of sand can be calculated directly from the shear wave velocity V_s determined with the seismic cone. As seen earlier, the following equation from elasticity theory can be used:

$$G_0 = \rho_m V_s^2 = \frac{\gamma_m}{g} V_s^2 \tag{7.17}$$

There are correlations between soil stiffness and cone resistance. The use of such correlations is not indicated, however, because different soil states (that is, different combinations of density, stress, and soil structure) can produce the same cone resistance, but different values of shear modulus. When soil stiffness is an important determination, one way to determine it is to measure V_s and compute G_0 using (7.17). Another is to use correlations for G_0 directly (Equations (5.24) and (6.54)). The stiffness at higher strain levels, usually of interest in routine design, can then be estimated from nonlinear stress–strain relationships.

7.7.2 Clays

7.7.1.3 Undrained shear strength

The undrained shear strength of clay can be estimated from cone resistance q_c through an equation of the form:

$$q_c = N_k s_u + \sigma_v \tag{7.22}$$

This equation is in essence the bearing capacity equation (discussed in Chapter 10) with $q_0 = \sigma_v$, as shown in Figure 7.26. N_k is called the *cone factor*. There is not a single value of cone factor because there is not a single value of undrained shear strength one might be interested in. The three loading paths of potential interest are triaxial compression, triaxial extension, and direct simple shear (refer to Section 4.5 for a discussion of this). When calculating the unit shaft resistance of pile in clay, we might be interested in the s_u value corresponding to direct simple shear, but when calculating the unit base resistance, we are likely to be more interested in the s_u for triaxial compression.

Cavity expansion analyses (Yu and Mitchell 1998) as well as large-strain finite element analyses (Yu et al. 2000) suggest that N_k should be of the order of 10. More recent work using the material point method (e.g., Bisht et al. 2021) found N_k to be equal to 12.2 for triaxial compression, 17.6 for triaxial extension, and 13.2 for simple shear; 14.3 is the average of these.

The literature contains values determined based on experimental data as high as 24.5, as shown in Table 7.8. The apparently wide range of N_k values has been a source of confusion to engineers, who often face the question of which value of N_k to use for a specific project. If we examine Table 7.8 carefully, it is apparent that the high N_k values are associated with soils containing low clay content and high sand content, for which penetration at the standard rate of 20 mm/s is not fully undrained or with rather unusual situations (such as extremely high OCR values). When partially drained conditions are in force, q_c is determined by soil shear strength that is higher than s_u; this, in turn, means that q_c values are higher than the values that would result from undrained penetration. So when q_c obtained from partially drained penetration is divided by a value of s_u determined under undrained conditions, a larger N_k value results. This N_k value is artificial, relating q_c measured under partially drained conditions to shear strength s_u measured under fully undrained conditions, and therefore must not be used. An approach that is correct but simple enough for practical purposes is to perform the cone tests at a rate of penetration that is sufficiently high to ensure undrained penetration, and to use N_k values as discussed earlier.

A higher-quality, more recent compilation was done by Low et al. (2010). These authors suggest that N_k values are in the range of 10–14 for s_u from triaxial compression tests, with an average of 12, and 11.5–15.5 for s_u from all types of tests, including triaxial compression and simple shear tests, with an overall average of 13.5. These values are largely in line with the values obtained theoretically by Bisht et al. (2021).

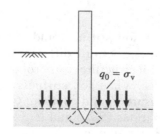

Figure 7.26 Cone penetrometer and vertical stress acting at the level of the cone tip.

Table 7.8 N_k *values from experimental research reported in the literature*

References	Soil description	Method of s_u determination	OCR	Clay (%)	LL (%)	PI (%)	N_k
Rad and Lunne (1988)	Clayey silt	TXCAU	1.45	23	37	12.5	8.5
	Silty clay	TXCAU	1.85	29	33	18	8.7
	Lean clay	TXCAU	1.35	48	40	26	9.5
	Silty quick clay	TXCAU	1.93	—	28.3	6.5	11.3
	Plastic clay	TXCAU	1.45	57.3	69.5	42.5	12.5
	Plastic OC clay	TXCAU	4.5	54.3	48.3	21.3	12.6
	Plastic clay	TXCAU	1.6	31.8	42.4	23.4	13.0
	Unweathered stony clay	TXCIU	3.15	—	34.5	18.8	15.7
	Silty sandy clay with gravel	TXCAU	6.1	—	33	18	15.7
	Clay	TXCIU	35.3	58.7	28.3	46.7	24.5
La Rochelle et al. (1988)	Silty clay with pine seams of silt, sensitive	Field vane	1.7–2.6		66	40	11
	Gray silty clay	Field vane	1.1–1.3		37–50	16–26	12.5
		Field vane	25–50		31	8	16
Stark and Juhrend (1989)	Soft to medium silty clay with low plasticity	TXUU	1–2		40	20	11
		Field vane	1–2		40	20	13
Lunne et al. (1986)	Silty clay	TXCAU	5	15–40	41–50	22–28	15
	Stiff clay	TXCAU	1–2	23–35	34–46	18–25	12
	Onshore quick clay	TXCAU					10.5
Denver (1988)	Glacial meltwater clay						9.5
	Lake glacial clay				25–35	10–15	7.3
Luke (1995)		Triaxial test	8–11	62	65	39	9.9
			3–4	15	22	7	12.2 10.5
			2–3	83	174	137	9.9
			1–2	21	37	15	10.6
			10–11	32	23	10	8.5 10.6
			1–2	35	97	53	8.4 9.5
Anagnostopoulos et al. (2003)		TXUU					18.9
		TXUU					17.2
Jamiolkowski et al. (1982)	Very young silty clay of medium plasticity	Field vane					11 ± 3
	Cohesive deposit of hard marine clay	TXCK$_0$U	2.5–4				9 ± 1
	Very stiff nonfissured clay		15–25				16 ± 2
Carpentier (1982)	Boom clay	Field vane					9 ± 1.5
Tani and Craig (1995)	Remolded glacial clay till	TXUU and field vane	1			23	12.5
Almeida et al. (1996)	Brown London Clay	TXCU	17–25				19–21
	Stiff clay	TXUU	4–7.5		40	18	20
	Stiff clay	TXUU	2–3		35	17	13.2
Van Impe (2004)	Boom clay	TXUU	14–30		65–71	40–50	13–24

TX= triaxial test; CU = consolidated undrained; CIU = isotropically consolidated undrained; CAU = anisotropically consolidated undrained; CK$_0$U = K$_0$-consolidated undrained; UU = unconsolidated undrained.

Figure 7.27 shows how q_c varies with normalized penetration rate (the penetration rate v multiplied by the cone diameter d_c divided by the coefficient of consolidation c_v of the soil). For small normalized penetration rate, the value of q_c is large, resulting from penetration under drained conditions. For the standard cone pushed at the standard penetration rate of 2 cm/s, penetration is drained for soils with large c_v values (clean sands as well as sands with relatively small contents of silt and very small contents of clay). As normalized penetration rate vd_c/c_v increases, q_c drops until a new plateau is reached at a sufficiently large value of vd_c/c_v, from which point penetration is undrained. Research suggests that penetration is fully drained at penetration rates less than about $0.1c_v/d_c$ to c_v/d_c and fully undrained at penetration rates of at least $10c_v/d_c$ to $20c_v/d_c$ (Kim et al. 2008, 2010; Salgado et al. 2013; Salgado and Prezzi 2014).

Another source of inaccuracy in the values in Table 7.8 is the method used to determine s_u. In some cases, the vane shear test (discussed later in this chapter) was used. In others, s_u was obtained in the laboratory.

7.7.1.4 Compressibility and rate of consolidation*

The compressibility of clay cannot be accurately determined from the cone resistance only, although many correlations have been proposed for this purpose. As was true of sands, the measurement of shear wave velocity, from which the small-strain shear modulus can be calculated, is helpful in calculating the initial elasticity modulus and in estimating the initial void ratio. We should not rely on q_c to estimate the compression index.

The coefficient of consolidation of clay can be estimated from dissipation tests. In these tests, penetration is halted at the depth where the value of the coefficient of consolidation is desired, and dissipation of the pore pressure generated during penetration is observed. Dissipation takes place predominantly in the lateral direction, which justifies the use of cylindrical cavity expansion analysis to develop the theoretical basis for the interpretation of dissipation tests. Carter et al. (1979) and Randolph and Wroth (1979) developed analyses of this type. Figure 7.28 shows the charts obtained with this analysis.

The normalized pore pressure, which is useful in interpreting dissipation tests, is defined as

$$u_n = \frac{u - u_0}{u_i - u_0} \tag{7.23}$$

where u = pore pressure at time t, u_0 = pore pressure due to hydrostatic conditions before cone penetration, and u_i = measured pore pressure at the beginning of the dissipation test.

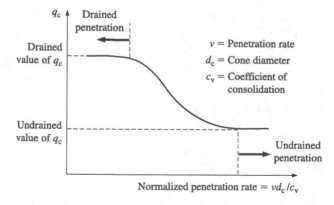

Figure 7.27 Cone resistance versus normalized rate of penetration.

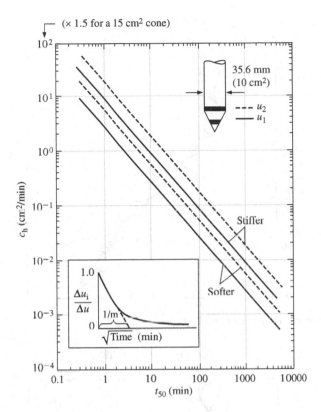

Figure 7.28 Chart for finding the horizontal coefficient of consolidation c_h in terms of the time t_{50} for 50% consolidation. (Lunne et al. (1997); Courtesy of Thomson Publishing Services, on behalf of Taylor & Francis Books.)

A dissipation test is done by allowing the value of u_n to drop from its initial value of 1. The value of the horizontal coefficient of consolidation c_h is determined using Figure 7.28 from the time t_{50} required for u_n to drop to 0.5. In this figure, relationships are provided for piezocone configurations with the pore pressure sensor located both on the face and behind the tip. Soils with a high stiffness-to-strength ratio are represented by the upper lines; those with low stiffness-to-strength ratio, by the lower lines.

Example 7.9

The data in E-Figure 7.5 are for a clay site. (1) Estimate the profile of undrained shear strength with depth. Consider the water table to be at 2 m depth, but the clay to be fully saturated due to capillary rise. (2) What is the profile of implied OCR with depth? The unit weight of the clay is 18 kN/m³, the plasticity index is 30%, and the cone factor is estimated to be around 12.

Solution:

First, we smoothen the CPT log using an appropriate technique (such as a rolling average). Both the smoothed and original logs are superimposed in E-Figure 7.6.

By rearranging terms in Equation (7.22), we obtain the undrained shear strength as a function of the cone resistance and the total vertical stress:

$$q_c = N_k s_u + \sigma_v \Rightarrow s_u = \frac{q_c - \sigma_v}{N_k} \Rightarrow s_u = \frac{q_c - 18z}{12} \text{ (in kPa)}$$

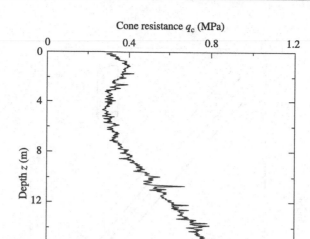

E-Figure 7.5 CPT log at clay site for Example 7.9.

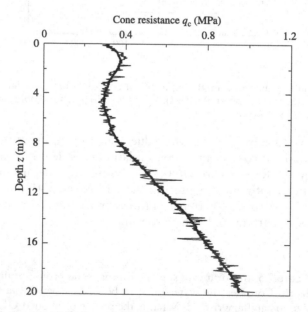

E-Figure 7.6 CPT log at clay site for Example 7.9 including smoothened plot.

Given that the plasticity index is 30, the undrained shear strength of the clay in its normally consolidated state can be estimated using the Skempton (1957) correlation, Equation (6.50):

$$\left(\frac{s_u}{\sigma_v'}\right)_{NC} = 0.11 + 0.0037PI = 0.11 + 0.0037 \times 30 \Rightarrow s_{u,NC} = 0.221\sigma_v'$$

$$\sigma_v' = 18z \quad \text{for} \quad z \le 2\,\text{m}$$

and

$$\sigma_v' = 36 + (18 - 9.81)(z - 2) \quad \text{for} \quad z > 2\,\text{m}$$

E-Figure 7.7 shows both the undrained shear strength calculated using Equation (7.22) and the undrained shear strength the soil would have if it were normally consolidated, calculated using Equation (6.50) and the above equations of vertical effective stress versus depth.

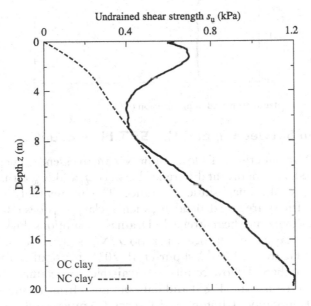

E-Figure 7.7 Undrained shear strength of the clay in its *in situ*, OC state (full line) and undrained strength the clay would have if it were NC according to Equation (6.50).

Finally, the overconsolidation ratio can be estimated using Equation (6.52):

$$\left(\frac{s_u}{\sigma_v'}\right) = \left(\frac{s_u}{\sigma_v'}\right)_{NC} \text{OCR}^{0.8}$$

from which:

$$\text{OCR} = \left(\frac{s_u}{s_{u,NC}}\right)^{1.25}$$

The resulting OCR versus depth plot is shown in E-Figure 7.8.

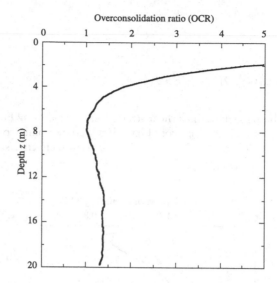

E-Figure 7.8 OCR versus depth estimated using Equation (6.52).

7.7.3 Correlation between q_c and the SPT blow count

Because both the CPT q_c and the SPT blow count N_{60} are in essence penetration resistances, they are closely related. An important difference between q_c and N_{60} is that N_{60} is associated with a dynamic instead of a quasi-static resistance. This means that, for materials whose shear strength and stiffness are rate dependent, such as clay, N_{60} does not reflect static shear strength, but a higher dynamic shear strength. Drainage conditions during penetration also differ for the two tests. As a consequence, the ratio q_c/N_{60} is higher for sands than for clays.

The correlation in Figure 7.29 (Sakleshpur et al. 2021) is useful in case we wish to use a CPT method when only SPT blow counts are available, for example. The chart includes data reported by Robertson et al. (1983) and data obtained from 15 sites in Indiana (2 sites each in Hamilton, Tippecanoe, Clinton, and Greene Counties, and 1 site each in Jasper, Lake, Newton, Knox, Starke, Dubois, and Carroll Counties). Starke, Newton, Jasper, and Lake Counties are located in northern Indiana; Hamilton, Tippecanoe, Carroll, and Clinton Counties are in central Indiana; and Greene, Knox, and Dubois Counties are in southern Indiana. However, as with any correlation involving the SPT blow count, it should be used with caution. There is an additional error introduced by the transformation from N_{SPT} to q_c.

The q_c versus N_{60} relationship, as represented by the trend defined by the 98 data points plotted in Figure 7.29, is well captured by this equation (Sakleshpur et al. 2021):

$$\frac{q_c}{p_A N_{60}} = 6.95 \left(\frac{D_{50}}{D_{ref}} \right)^{0.25} - 0.18 \quad \text{for} \quad 0.001 \leq \frac{D_{50}}{D_{ref}} \leq 10 \tag{7.24}$$

where p_A = reference stress (= 100 kPa or 14.5 psi), D_{50} = mean particle size, and D_{ref} = reference particle size (= 1 mm or 0.0394 in.). The coefficient of determination r^2 and the standard error (SE) of the regression are 0.886 and 0.774, respectively.

7.7.4 Cemented sands

In some cases in which sand is lightly cemented, a cone will not be stopped by the cementation, but the cone resistance will reflect the greater shear strength of the soil. Puppala et al.

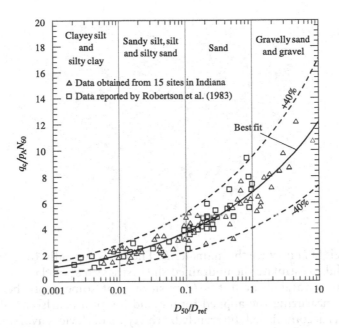

Figure 7.29 Correlation between CPT cone resistance and SPT blow count. (After Sakleshpur et al. (2021).)

(1995) showed that the cone resistance measured in calibration chambers increased by a factor of as much as 2 if 1% Portland cement was added to a medium-dense sand to create an initial cohesive intercept c of 10 kPa and by as much as 4 if 2% Portland cement was added to create an initial cohesive intercept c in a dense sand. These factors were observed at an overburden (vertical effective) stress of 100 kPa. At greater confining stresses, the increase was found not to be as drastic, presumably because of a higher rate of cementation breakdown, but it was still present.

If we perform a CPT in a lightly cemented sand, unless the geology of the site is known or samples are collected, we may not know that we are pushing the cone through a cemented soil. Identification of the material using the combined values of friction ratio and cone resistance, which provides at least clues as to what types of soils a cone may be traversing in the case of uncemented soils, is of no help in the case of cemented soils (Schnaid and Consoli 1996; Puppala et al. 1996). We may then misinterpret the higher cone resistance values as indicative of high relative densities. In some cases, this faulty interpretation may lead to a defective design. The reason is the very different behavior of cemented and uncemented soils. As discussed in Chapter 5, cementation breaks down during shearing or compression, and so the strength derived from cementation will be available for small strains, but not necessarily for moderate to large strains. Additionally, while dense soil will dilate, a cemented soil misidentified as dense may even be contractive. Even though this is rarely a serious issue in routine projects, the distinction is important in some cases in which performance is critical and analysis must be precise.

7.8 OTHER *IN SITU* TESTS

7.8.1 Vane shear test

The *shear vane* is a device composed of four mutually perpendicular cutting blades connected to a cylindrical rod. The blades may be rectangular or tapered (Figure 7.30). The blades are usually 1.95 mm thick, the diameter B of the circumscribed circle ranges from roughly 40 to

Figure 7.30 Field vane.

90 mm, and the height H is twice the diameter B. The rod is typically 16–20 mm in diameter. The device is useful to estimate the undrained shear strength s_u of clay.

The objective of a vane shear test is to position it at some depth below the soil and to rotate it while measuring the applied torque and rotation (Carlson 1948), as shown in Figure 7.31. This is accomplished differently by the types of devices available in the market. In one common approach, the device is lowered into the soil with the torsion rod within an external casing. At a depth where a test is desired, the torsion rod is pushed down 0.5 m and the vane is rotated. In this case, the measured torque is only due to resistances to the vane rotation. In another approach, the vane is either pushed or driven through the ground with the rod without any encasement. In this case, the torques needed to rotate both the vane

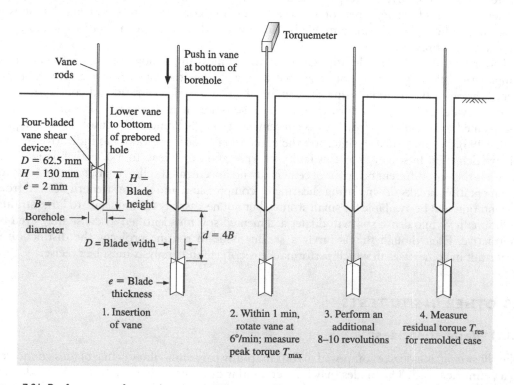

Figure 7.31 Performance of vane shear test. (Courtesy of Paul Mayne.)

and rod and the rod alone are measured. The torque required for vane rotation can then be calculated by subtracting one from the other.

For the rectangular vane, the resistance to vane rotation is provided by the cylindrical surface with diameter B and height $H = 2B$ along which the edges of the four blades will move. Additional resistance comes from shearing also on the top and bottom of this cylinder, if the vane is fully inserted in the soil where s_u is to be estimated, or from the bottom only, otherwise. Assuming the shear strength to be the same at the top, bottom, and lateral surfaces of the shear cylinder, the torque T is given by

$$T = 2\pi B^2 s_u \frac{B}{2} + n \int_0^{\frac{B}{2}} (s_u)(2\pi r) r\, dr$$

where $n = 1$ for the case where the vane is partially inserted and 2 when it is fully inserted into the soil for which s_u is desired. After performing the integration and grouping terms:

$$T = \frac{1}{12} \pi B^3 s_u (12 + n)$$

Considering that vanes are manufactured with $H = 2B$, the undrained shear strength $(s_u)_{FV}$ estimated from the field vane shear test can finally be written as

$$(s_u)_{FV} = \frac{12T}{(12 + n)\pi B^3} \tag{7.25}$$

The undrained shear strength measured by the vane is, of course, affected by the soil disturbance caused by the insertion of the vane. A waiting period will allow the material to heal to some degree, increasing $(s_u)_{FV}$. Rate effects[11] appear if the vane is rotated at different speeds, with $(s_u)_{FV}$ increasing with increasing angular velocity [see ASTM D2573 (ASTM 2018a) for recommendations on which angular velocity to test at]. There are also three-dimensional end effects that have not been accounted for in the derivation of Equation (7.25). Bjerrum (1972, 1973) suggested that the design shear strength s_u be obtained by multiplying $(s_u)_{FV}$ by a correction factor λ approximately given as

$$\lambda = 1.18 - 0.0107(PI) + 0.0000513(PI)^2 \le 1 \tag{7.26}$$

so that

$$s_u = \lambda (s_u)_{FV} \tag{7.27}$$

Similar equations can be derived for the tapered vane by considering the additional triangular-shaped areas of the blades, which generate a truncated cone where the vane joins the rod and approximately a cone at the bottom. Bowles (1996) provides the equation for the undrained shear strength in terms of the torque T as

$$s_u = \frac{0.3183T}{1.354B^3 + 0.354\left(B_1 B^2 - BB_1^2\right) + 0.2707B_1^3} \tag{7.28}$$

where B_1 = vane shaft diameter (usually in the 12–22 mm range).

---

[11] Under undrained conditions, the shear strength of clay tends to increase with loading rate (that is, clay exhibits some degree of "viscosity").

7.8.2 Pressuremeter test

The *pressuremeter* was developed by Menard (1956). That model of the pressuremeter is known today as the Menard pressuremeter. It consists of three separate cylindrical rubber membranes, one on top of each other, as shown in Figure 7.32. The upper and lower membranes are referred to as the top and bottom guard cells; the middle one, the measuring cell. The pressuremeter is lowered into a preexisting borehole (making it difficult to use in sands and nonplastic silts). The measuring cell contains water connected to a pressure source, so that the pressure in the cell can be increased hydraulically. The cell expansion is obtained by measuring the volume of water going into the cell. By increasing the pressure in the guard cells (inflated by gas, typically CO_2) as well, it is possible for the measuring cell to expand essentially in the radial direction. This is important, meaning that cylindrical cavity expansion analysis applies directly to the Menard pressuremeter with very good approximation.

The pressuremeter expansion curve (Figure 7.33) is usually plotted as pressure in the vertical axis versus volumetric or radial strain in the horizontal axis. Its initial portion is not representative of the soil properties because of the poor initial fit between the membrane and the borehole as well as because of soil disturbance in the vicinity of the borehole wall (a problem that is most serious in soft clays, sands, and soils containing gravel). This part of the curve is clearly identifiable, as the slope of the curve is quite small and usually increases with strain until the curve starts better reflecting the soil properties, when the slope then starts decreasing with increasing strain. Some pressuremeter test (PMT) interpretation methods are based on calculating the Young modulus so that it is consistent with the pressure versus strain relationship at some value of strain and relating that to a design modulus to use in various soil analyses (such as the calculation of shallow foundation settlements or the calculation of the deflections of laterally loaded piles). At large-strain values, the pressure approaches a limit value, the limit pressure p_L. This pressure, if well defined (which usually requires some extrapolation), can be taken as a cylindrical cavity limit pressure. Cylindrical cavity expansion analysis can then be used to estimate the shear strength of the soil from p_L. Alternatively, the limit pressure can be defined, conventionally, as the pressure corresponding to the point in the test where the measuring cell volume has doubled.

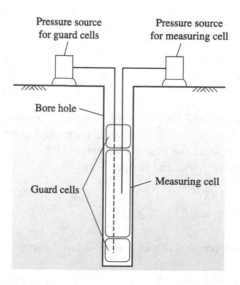

Figure 7.32 The Menard pressuremeter.

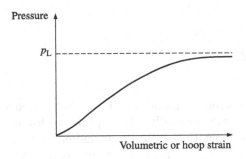

Figure 7.33 Pressuremeter expansion curve.

Another type of pressuremeter, the self-boring pressuremeter (SBP), has a cutting tool at the end that is used to drill the hole as the pressuremeter is advanced in the ground (Denby and Clough 1980). It has some advantages over the Menard pressuremeter, such as the lower degree of soil disturbance because of the way it is installed and the possibility of performing the test effectively in caving soils (sands, silts, and soft clays under the water table). One of its disadvantages is the much greater degree of difficulty in performing the test, which results in the unavailability of qualified personnel to perform the test on a commercial basis in most places. Additionally, in contrast to the Menard pressuremeter, the SBP is composed of a single cell, which curves when expanded, deviating from a cylindrical cavity expansion mode and thus being less suited to rigorous interpretation.

Interpretation of the pressuremeter test is beyond the scope of this text. Much experience with the pressuremeter has been accumulated in France, and the French Standard (1991) contains useful information on test execution and interpretation. The interested reader is also referred to Briaud (1992) and Yu et al. (2000) for more information on the pressuremeter test and its interpretation.

7.9 GEOPHYSICAL EXPLORATION

Geophysics developed in the first half of the 20th century and came into full use in the second half of that century in the exploration for oil. The techniques were gradually transferred to civil engineering and mining applications. These techniques are mostly based on the principles of wave propagation through media and their reflection or refraction (meaning crossing of) boundaries between media of different elastic properties. In the last few years, there has been significant progress in the interpretation of the results of geophysical tests. Treatment of this topic in any detail is beyond the scope of the present text. References under Additional references at the end of this chapter can be consulted.

7.10 SUBSURFACE EXPLORATION REPORT AND GEOTECHNICAL REPORT

The subsurface exploration report is usually brief, but can be lengthy if conditions at the site prove to be complex or unanticipated, requiring expanded discussion. It is often done after the conclusion of the site investigation program and any laboratory testing that may have been requested. The scope of the report is dependent on the extent of the investigation, the types of tests done, and the type and size of the project. In routine jobs, the report may be quite simple, containing the results of the site investigation and nothing more.

The geotechnical report is broader in scope. An example of how a report might be organized would be as follows:

1. Letter of transmittal.
2. Cover page.
3. Table of contents.
4. Introduction. It frames the work done and how it helps model the subsurface for design purposes. It includes any potentially serious problems found in the course of the work.
5. Description of site investigation program, including:
   - the date(s) during which the procedures were performed;
   - the name of the engineering inspector and drilling company;
   - the plan of borings;
   - a description of how the borings and SPT testing were performed, if done;
   - the elevation of the ground surface at each boring, with respect to a permanent benchmark;
   - the elevation of the water table;
   - the soil profile at each location according to the inspector's soil identification and spoon samples collected;
   - SPT blow counts for each boring;
   - similar details on *in situ* tests done instead of or in addition to the SPT;
   - similar details on any laboratory testing done; and
   - any other significant facts.
6. Additional information. A more complete geotechnical report would typically contain the interpretation of the results of the exploratory program, the proposed values for the stiffness and shear strength of the soil, the bearing capacity of the soil, and the allowable bearing pressures to prevent excessive settlement. It may contain additional items if deep foundations and other geotechnical work are recommended. To complete this part of the report, the engineer needs to perform calculations that are discussed in subsequent chapters.
7. Summary.

7.11 CHAPTER SUMMARY

7.11.1 Main concepts

To construct the most economically efficient foundations and other geotechnical structures, the knowledge of the subsurface is required. This knowledge cannot be perfect, but a combination of the knowledge of the local geology with a suitable site investigation program usually produces a model of the subsurface that is adequate for design. Routine site investigation programs include a number of **soil borings** across the site. The **standard penetration test** (SPT) is usually performed together with the borings. The **cone penetration test** (CPT) is often performed as well. Other *in situ* tests are not as frequently used. Usually, there are not hard and fast rules about the number of soil borings or CPTs as a function of the size of the site and the magnitude of the loads to be supported by the foundations, but clearly there is an optimum number. Too few borings will not allow a sufficiently accurate estimation of design parameters across the site. Too many borings will cost too much relative to the information they provide.

The SPT is performed by driving a **sampler** with standard geometry (Figure 7.5) into the soil a total of 450 mm (18 in.) by blows of a hammer weighing 630 N (140 pounds) falling from a height of 760 mm (30 in.). The hammer falls on top of an anvil, which in turn transfers the energy down to the drill rods. The 450 mm (18 in.) penetration is divided into three separate penetration intervals of 150 mm (6 in.) each. The operator counts the number of blows of the hammer required to advance the boring each of the 150 mm. The SPT **blow count** N_{SPT} is the sum of the numbers of blows for each of the last two 150 mm penetrations.

The SPT blow count is normalized for the type of hammer, rod string length, type of sampler, and borehole diameter as follows:

$$N_{60} = C_h C_r C_s C_d N_{SPT} \tag{7.1}$$

where C_h = hammer correction, C_r = rod length correction, C_s = sampler correction, and C_d = borehole diameter correction.

The hammer correction is given by

$$C_h = \frac{ER_{hammer}}{ER_{safety}} = \frac{ER_{hammer}}{60\%} \tag{7.2}$$

where ER_{hammer} = energy ratio for the hammer used in the test, given in Table 7.2.

The sampler correction C_s is equal to 1 for the standard ISSMGE sampler or the liner sampler with the liner in place, and $C_s = 1.2$ for the liner sampler when used without the liner.

The rod length correction factor C_r is given by

$$C_r = \begin{cases} 0.75 & \text{if rod length} < 4\,\text{m} \\ 0.85 & \text{if } 4 \le \text{rod length} < 6\,\text{m} \\ 0.95 & \text{if } 6 \le \text{rod length} < 10\,\text{m} \\ 1 & \text{if rod length} \ge 10\,\text{m} \end{cases} \tag{7.3}$$

The borehole diameter correction factor C_d is equal to 1.05 and 1.15 for borehole diameters equal to 150 and 200 mm, respectively.

The SPT blow count is often used to estimate the relative density of sand. The following equation can be used for that purpose:

$$\frac{D_R}{100\%} = \sqrt{\frac{N_{60}}{A + BC\dfrac{\sigma'_v}{p_A}}} \tag{7.6}$$

This expression is applicable for $35\% \le D_R \le 85\%$ and $50\,\text{kPa} \le \sigma'_v \le 250\,\text{kPa}$, where A and B are correlation coefficients ($27 \le A \le 46$ and $B \approx 27$), and C is given by

$$C = \frac{K_0}{K_{0,NC}} \tag{7.8}$$

where K_0 = coefficient of lateral earth pressure at rest for the soil as it is, and $K_{0,NC}$ = value of this coefficient that the soil would have if it were normally consolidated.

SPT correlations for clay are not very reliable and should be used only when something better (such as laboratory tests or CPTs) is not available.

The CPT is performed by pushing a **penetrometer** with a conical tip and standard geometry (Figure 7.17) vertically into the ground at a standard rate of 20 mm/s. Penetration at this rate is fully drained for sands with nonplastic silt contents up to 10% and fully undrained for most clays, but partially drained for soils intermediate between these two extremes.

The most important quantity measured in a CPT is the **cone resistance** q_c, which is the ratio of the vertical force resisting penetration on the tip of the cone to the projected area of the cone. It is usually provided as a plot of q_c versus depth. This plot may have relatively sharp peaks and valleys, particularly in sandy and gravelly soils. It is important to read off q_c values that are representative of the range of depths under consideration, as opposed to reading a single value from an unrealistic peak or valley.

Figures 7.23 and 7.24 are very useful in interpreting the results of CPTs in sand. It provides cone resistance q_c as a function of relative density D_R and lateral effective stress σ'_h. Once an estimate of K_0 and thus of σ'_h is made, the chart can be entered with q_c for an estimate of D_R. D_R may also be estimated from:

$$D_R = \frac{\ln\left(\dfrac{q_c}{p_A}\right) - 0.4947 - 0.1041\,\phi_c - 0.841\ln\left(\dfrac{\sigma'_h}{p_A}\right)}{0.0264 - 0.0002\,\phi_c - 0.0047\ln\left(\dfrac{\sigma'_h}{p_A}\right)} \leq 100\% \tag{7.20}$$

In contrast to the SPT, the CPT can be used very effectively to estimate undrained shear strength in clay. The equation for that is:

$$q_c = N_k s_u + \sigma_v \tag{7.22}$$

where N_k is the **cone factor**, which is no less than 10 for soils with a large clay content, in which penetration at 20 mm/s is a fully undrained process.

7.11.2 Notations and symbols

Symbol	Quantity represented	US units	SI units
C_d	Borehole diameter correction factor for SPT	Unitless	Unitless
C_h	Hammer correction factor for SPT	Unitless	Unitless
C_r	Rod length correction factor for SPT	Unitless	Unitless
C_s	Sampler correction factor for SPT	Unitless	Unitless
d_c	Cone diameter (CPT)	in.	mm
ER	Energy ratio for SPT hammers	%	%
G_0	Small-strain shear modulus	psf	MPa
$(N_1)_{60}$	Corrected blow count normalized by the vertical effective stress	Unitless	Unitless
N_1	Blow count normalized with respect to vertical effective stress	Unitless	Unitless
N_{60}	Corrected blow count	Unitless	Unitless
N_k	Cone factor	Unitless	Unitless
N_{SPT}	SPT blow count	Unitless	Unitless
q_c	Cone penetration resistance	psf	MPa
q_t	Cone resistance corrected for pore pressure area effects	psf	MPa
u_n	Normalized pore pressure	Unitless	Unitless
u_i	Initial pore pressure	psf	kPa
u_0	Pore pressure due to hydrostatic conditions	psf	kPa
V_s	Shear wave velocity	ft/s	m/s

7.12 PROBLEMS

7.12.1 Conceptual problems

Problem 7.1 Define all the terms in bold contained in the chapter summary.

Problem 7.2 A 55-story residential building will be built at a 60 m × 40 m beachfront property. The subsurface is typical of barrier island geology in this area: a very loose sand layer on top of sandstone. Foundations usually consist of piles or piled rafts with the piles bearing on the sandstone. Develop a site investigation plan for this site.

Problem 7.3 A 30-story office building will be built at a 30 m × 40 m property. The geology of the area is residual soil of gneiss extending to depths ranging from 10 to 20 m (this depth can vary significantly across short distances because of the nature of the banding in gneiss). Sound rock (gneiss) is located at that depth. There are occasionally large boulders found at shallow depths. For large buildings, piles to rock are usually used. Develop a site investigation plan for this site.

7.12.2 Quantitative problems

Problem 7.4 A vane shear test was performed at a point within a clay layer. The maximum moment required to rotate the vane, which had a diameter of 60 mm and a height of 120 mm, was measured as 70 N m. The vane was fully inserted in the soil. What is the undrained shear strength of the clay?

Problem 7.5 If a cone penetration test were to be performed next to the vane shear test of Problem 7.4, which was performed at a depth of 5 m, what value of cone resistance would you expect? The water table is at the surface, and the clay has a unit weight of 17 kN/m^3. Use $N_k = 12$.

Problem 7.6 A CPT was performed in a deposit of soft clay with the water table at a depth of 1 m. The cone resistance at a depth of 10 m was equal to 0.6 MPa. What are the minimum and maximum values of s_u of the clay at that depth that you would expect based on the range of values possible for N_k? The unit weight of this clay is 17 kN/m^3.

Problem 7.7 The results of SPTs performed with an automatic trip hammer using the standard ASTM split spoon sampler with a liner are shown in the table below. The borehole diameter was within the recommended range. The soil profile consists of a normally consolidated sand with a unit weight equal to 20 kN/m^3. The water table is located at 4.5 m below the ground surface. Use Equation (7.6) with Equation (7.8) to estimate the relative density D_R at depths where SPT measurements are available. You may use the following table to guide your calculations.

Depth (m)	N_{SPT}	C_b	C_r	N_{60}	σ'_v (kPa)	D_R (%)
4	25					
5	30					
6	35					

Problem 7.8 P-Table 7.1 has SPT blow counts obtained at intervals of 1 m at a sandy site. A donut hammer and a liner sampler without the liner were used. Every care was taken to connect the rod segments firmly and to follow standard procedure. The water table is at a depth of 3 m, and the site is lightly overconsolidated because ~2 m of soil of unit weight ≅17 kN/m^3 was removed before the SPT was performed. The K_0 of this soil

in a normally consolidated state would be 0.48. Calculate the corresponding stress-normalized, energy-corrected blow counts $(N_1)_{60}$.

P-Table 7.1 N_{SPT} for Problem 7.8

Depth (m)	SPT blow count
1	15
2	18
3	22
4	23
5	25
6	28

Problem 7.9 The cone resistance for a clean sand at 6 m has been measured as 11 MPa. The average total unit weight of the soil column above 6 m is 21 kN/m³. The water table is 3 m below the surface. The soil is normally consolidated, with $K_0 = 0.45$. The soil has $\phi_c = 30°$. Estimate the relative density of the sand at 6 m.

Problem 7.10 For the sand deposit and conditions of Problems 5.11 and 5.12, estimate and plot the cone resistance q_c as a function of depth for the 0–10 m depth range. Plot also the ratio of the small-strain shear modulus G_0 to q_c.

Problem 7.11 For the sand deposit and conditions of Problems 5.13 and 5.14, estimate and plot the cone resistance q_c as a function of depth for the 0–10 m depth range. Plot also the ratio of the small-strain shear modulus G_0 to q_c.

7.12.3 Design problems

Problem 7.12 An SPT log is given in P-Figure 7.1. The SPT was performed with a safety hammer using the standard ASTM split spoon sampler with a liner. The borehole diameter is within the recommended range. The sand is normally consolidated (with a K_0 of ~0.45), and the water table is very deep. The critical-state friction angle of this sand is ~32°. The unit weights of the sandy clay, silty clay, and sand are equal to 17, 15, and 20 kN/m³, respectively. For the sand layer extending from 8.5 to 21 ft, estimate the relative density D_R and the peak friction angle ϕ_p that would be obtained in triaxial compression tests performed on ideal, "undisturbed" samples recovered from the following depths where SPT measurements are available: 9.3, 11.5, 14, 16.5, and 19 ft. Use Equation (7.6) to estimate D_R and Equations (5.12) and (5.20) to estimate ϕ_p. Assume that you are estimating the ϕ_p that would be obtained in the laboratory under triaxial compression of a sample consolidated isotropically to the *in situ* mean effective stress.

Boring log for Problem 7-12							
Depth (ft)	Water table	USCS	Graphic	Water levels ▼ During drilling ▽ After completion / Descriptions	Samples	SPT results	Remarks
0				GRAVEL	1	3,4,4	
		CL		Dark brown, moist, medium stiff, SANDY CLAY	2	3,3,3	
5							
		CL		Brown, very moist, soft, SILTY CLAY with sand seams	3	2,1,3	
					4	7,6,6	
10							
				Brown, slightly moist to moist, medium dense, fine to medium grained, SAND with trace silt and gravel	5	10,12,14	
					6	7,7,10	
15		SP			7	7,8,12	
					8	6,6,7	Boring caved to 18 ft upon auger removal
20					9	7,15,15	
		SP		Brown, moist, medium dense, fine to medium grained, SAND with little gravel	10	9,15,13	
25							
		SM		Brown, slightly moist, medium dense, SILTY SAND	11	9,10,11	
30					12	5,7,8	
				Boring terminated after 30 ft			
35							
Ground water was not encountered during drilling or upon completion							

P-Figure 7.1 SPT log for Problem 7.12.

Problem 7.13 Redo Problem 7.12, part (a), assuming consolidation of the sample to the *in situ* vertical effective stress.

Problem 7.14 Redo Problem 7.12, part (a), assuming consolidation of the sample to the *in situ* horizontal effective stress.

Problem 7.15 An SPT log is given in P-Figure 7.2. The SPT was performed with a safety hammer using the standard ASTM split spoon sampler with a liner. The borehole diameter is within the recommended range. The sand is normally consolidated, and the water table is very deep. The unit weights of the sandy clay, silty clay, and sand are equal to 17, 15, and 20 kN/m³, respectively. The sand has $\phi_c = 30°$. For the sand layer extending from 11 to 21 ft, estimate the relative density D_R and the peak friction angle ϕ_p that would be obtained in triaxial compression tests performed on ideal, undisturbed samples recovered from the following depths where SPT measurements are available: 11.5, 14, 16.5, and 19 ft. Use Equation (7.6) to estimate D_R and Equations (5.12) and (5.20) to estimate ϕ_p. Assume that you are estimating the ϕ_p that would be obtained in the laboratory under triaxial compression of a sample consolidated isotropically to the *in situ* mean effective stress.

Boring log for Problem 7-15							
Depth (ft)	Water table	USCS	Graphic	Water levels ▼ During drilling ▽ After completion Descriptions	Samples	SPT results	Remarks
0				GRAVEL			
		CL		Brown, moist, medium stiff, SANDY CLAY with trace gravel (Possible fill)	1	2,3,4	
5					2	3,3,2	
		SP		Brown, moist, very loose coarse grained SAND with trace clay (Possible fill)	3	8,3,6	
10		CL		Brown, moist, medium stiff, SILTY CLAY with SAND and gravel seams	4	8,3,19	
					5	20,14,16	
15				Brown, slightly moist, dense to very dense, fine to medium grained, SAND with trace to little gravel	6	8,17,22	
		SP			7	14,21,29	
20					8	24,25,27	
					9	13,16,14	Boring caved to 22.5 ft upon auger removal
25		SP		Brown, slightly moist, medium dense to dense, fine grained, SAND with trace silt	10	8,12,11	
					11	10,14,15	
30					12	12,16,23	
		SP		Brown, slightly moist, dense to very dense, fine to medium grained, SAND with little to some gravel	13	11,12,19	
35					14	50	
				Auger refusal at 35 ft			
40							

Ground water was not encountered during drilling or upon completion

P-Figure 7.2 SPT log for Problem 7.15.

Problem 7.16* Two CPTs and one SPT were performed in close proximity. Results are in P-Tables 7.2 and 7.3.

a. For the data given, prepare plots of q_c, f_s, and f_s/q_c versus depth.

b. Estimate the relative density D_R using CPT-based methods. The coefficient of lateral earth pressure is equal to 0.4, and the critical-state friction angle is equal to 36°. You may use the charts in Figure 7.24. The sand is normally consolidated, and the water table is very deep. Estimate also the peak friction angle ϕ_p that would be obtained in triaxial compression tests performed on ideal, undisturbed samples recovered from depths equal to 6.1, 7.6, and 9.10 m and reconsolidated to the respective mean effective stresses p' at those depths.

c. Estimate the relative density D_R and the peak friction angle ϕ_p of the sand at depths equal to 6.1, 7.6, and 9.10 m using an SPT-based method. Use Equation (7.6) to estimate D_R and Equations (5.20) and (5.12) to estimate ϕ_p. The SPT was performed with a safety hammer using the standard ASTM split spoon sampler with a liner. The borehole diameter is within the recommended range. The peak friction angle ϕ_p estimate should be for the friction angle that would be obtained in triaxial compression tests performed on ideal, undisturbed samples recovered from depths equal to 6.1, 7.6, and 9.10 m and reconsolidated to the respective mean effective stresses p' at those depths.

d. Compare the results obtained in parts (b) and (c).

P-Table 7.2 SPT data for Problem 7.16

Depth (m)	Soil type (from borings)	Unit weight (kN/m³)	Depth (m)	N_{SPT}
0–1.5	Clayey silt	14	6.1	27
1.5–4.3	Sand	15	7.6	25
4.3–5.2	Silty clay	14.5	9.1	40
5.2–14.3	Sand	19		
14.3–16.8	Clayey silt	15.5		

P-Table 7.3 CPT data for Problem 7.16

Depth (m)	q_c (MPa)	f_s (kPa)	f_s/q_c (%)	q_c (MPa)	f_s (kPa)	f_s/q_c (%)
0.05	1.29	55.87	4.3	1.5	38.39	2.6
0.1	1.66	60.83	3.7	1.3	51.98	4.0
0.15	1.37	59.92	4.4	1.3	46.35	3.7
0.2	1.37	52.73	3.9	1.3	45.94	3.7
0.25	1.61	53.39	3.3	1.3	46.45	3.5
0.3	1.62	57.92	3.6	1.4	53.51	3.8
0.35	1.86	64.49	3.5	1.4	60.4	4.2
0.4	1.86	70.35	3.8	1.5	64.53	4.3
0.45	1.88	72.81	3.9	1.5	68.28	4.6
0.5	2.14	79.71	3.7	1.6	72.51	4.5
0.55	2.43	87.22	3.6	1.9	95.97	5.1
0.6	2.73	95.76	3.5	2.3	85.23	3.7
0.65	2.94	105.75	3.6	2.1	81.87	3.9
0.7	2.88	113.71	3.9	2.3	83.62	3.7
0.75	2.82	120.79	4.3	2.1	89.88	4.3
0.8	2.67	114.1	4.3	2.0	89.42	4.4
0.85	2.57	108.38	4.2	2.0	89.8	4.5
0.9	2.50	109.09	4.4	2.1	92.35	4.5
0.95	2.57	115.72	4.5	1.9	93.2	4.9
1	2.51	116.5	4.6	1.9	90.43	4.7
1.05	2.48	114.24	4.6	2.0	96.19	4.9
1.1	2.37	117.15	4.9	2.0	103.78	5.2
1.15	2.40	123.25	5.1	2.0	108.85	5.4
1.2	2.36	137.13	5.8	2.0	114.32	5.8
1.25	2.32	148.17	6.4	2.0	123.76	6.2
1.3	2.37	160.71	6.8	2.2	143.25	6.6
1.35	2.37	165.26	7.0	2.3	155.05	6.8
1.4	2.31	152.95	6.6	2.3	160.34	7.1
1.45	2.32	147.83	6.4	2.4	167.56	7.0

(Continued)

P-Table 7.3 (Continued) CPT data for Problem 7.16

Depth (m)	q_c (MPa)	f_s (kPa)	f_s/q_c (%)	q_c (MPa)	f_s (kPa)	f_s/q_c (%)
1.5	2.32	145.35	6.3	2.5	162.33	6.5
1.55	2.46	141.5	5.8	2.8	149.58	5.4
1.6	2.83	129.19	4.6	3.3	132.36	4.0
1.65	3.28	87.32	2.7	3.8	107.26	2.8
1.7	3.76	56.72	1.5	3.7	67.42	1.8
1.75	3.51	39.06	1.1	3.5	39.27	1.1
1.8	3.25	41.36	1.3	3.3	24.8	0.7
1.85	3.35	47.63	1.4	3.2	23.44	0.7
1.9	3.33	40.95	1.2	3.5	38.68	1.1
1.95	3.38	43.6	1.3	4.1	34.12	0.8
2	4.49	33.41	0.7	3.8	29.54	0.8
2.05	4.60	19.17	0.4	3.4	45.43	1.3
2.1	4.55	48.52	1.1	3.3	45.9	1.4
2.15	3.62	51.84	1.4	2.9	58.8	2.1
2.2	1.33	51.23	3.8	1.2	50.82	4.3
2.25	0.89	25.43	2.9	0.9	41.48	4.7
2.3	1.79	18.96	1.1	1.4	40.1	2.9
2.35	1.95	17.92	0.9	1.8	35.54	2.0
2.4	1.93	31.86	1.7	1.8	30.25	1.7
2.45	1.82	25.59	1.4	1.9	43.09	2.2
2.5	1.64	25.76	1.6	1.7	39.29	2.3
2.55	1.37	18.05	1.3	1.6	22.73	1.4
2.6	1.39	12.53	0.9	1.7	25.09	1.5
2.65	1.43	8.02	0.6	2.0	56.54	2.9
2.7	1.85	1.04	0.1	2.6	7.39	0.3
2.75	2.08	22.01	1.1	2.8	65.69	2.4
2.8	2.91	22.89	0.8	2.0	41.71	2.1
2.85	2.25	27.81	1.2	2.8	32.53	1.2
2.9	2.31	26.98	1.2	3.0	6.23	0.2
2.95	3.16	33.59	1.1	3.4	25.43	0.8
3	4.06	35.36	0.9	4.2	29.46	0.7
3.05	4.22	30.76	0.7	4.2	36.64	0.9
3.1	4.13	21.95	0.5	4.8	32.69	0.7
3.15	3.83	23.21	0.6	4.9	37.05	0.8
3.2	4.15	23.95	0.6	5.0	39.63	0.8
3.25	4.75	27.59	0.6	5.4	37.33	0.7
3.3	5.67	31.23	0.6	6.0	39.37	0.7
3.35	5.70	34.04	0.6	6.4	44.96	0.7
3.4	4.99	28.83	0.6	6.3	50.39	0.8
3.45	4.13	20.18	0.5	5.5	30.07	0.5
3.5	3.45	17.88	0.5	4.2	28.97	0.7
3.55	2.52	19.08	0.8	3.0	18.86	0.6
3.6	1.88	11.21	0.6	2.3	17.44	0.8
3.65	1.43	8.22	0.6	1.9	16.42	0.9
3.7	1.37	7.79	0.6	2.1	20.2	1.0

(Continued)

P-Table 7.3 (Continued) CPT data for Problem 7.16

Depth (m)	q_c (MPa)	f_s (kPa)	f_s/q_c (%)	q_c (MPa)	f_s (kPa)	f_s/q_c (%)
3.75	1.55	7.95	0.5	3.1	27.22	0.9
3.8	1.97	9.87	0.5	4.5	34.22	0.8
3.85	2.65	12.8	0.5	5.0	32.74	0.7
3.9	3.48	10.62	0.3	4.8	29.58	0.6
3.95	3.51	13.63	0.4	4.3	23.58	0.5
4	3.02	10.74	0.4	3.4	24.11	0.7
4.05	2.61	12.11	0.5	2.9	21.48	0.7
4.1	2.31	24.01	1.0	2.6	33.37	1.3
4.15	2.38	26.57	1.1	3.7	37.37	1.0
4.2	2.70	25.7	1.0	5.7	38.61	0.7
4.25	5.25	37.92	0.7	6.1	85.16	1.4
4.3	4.06	52.78	1.3	3.2	68.95	2.2
4.35	2.12	45.37	2.1	1.9	44.41	2.3
4.4	1.76	49.74	2.8	1.7	37.11	2.1
4.45	3.07	52.31	1.7	2.4	72	3.0
4.5	2.27	62.89	2.8	2.1	76.5	3.7
4.55	2.19	49.26	2.2	2.4	62.01	2.5
4.6	1.93	50.62	2.6	2.0	51.57	2.6
4.65	2.80	76.29	2.7	5.8	56.99	1.0
4.7	3.96	59.79	1.5	3.8	56.7	1.5
4.75	2.49	55.36	2.2	2.3	45.21	2.0
4.8	1.87	37.82	2.0	2.0	32.53	1.7
4.85	1.85	36.54	2.0	1.9	31.27	1.6
4.9	2.01	35.32	1.8	2.1	36.48	1.7
4.95	2.22	51.11	2.3	2.5	69.11	2.7
5	3.37	61.63	1.8	3.0	74.36	2.5
5.05	3.50	92.06	2.6	3.6	72.49	2.0
5.1	5.28	74.73	1.4	16.5	78.55	0.5
5.15	14.03	77.09	0.5	21.6	98.08	0.5
5.2	17.80	83.25	0.5	24.6	121.15	0.5
5.25	22.74	139.02	0.6	27.6	157.1	0.6
5.3	24.36	141.93	0.6	29.3	157.23	0.5
5.35	23.90	130.53	0.5	28.7	143.57	0.5
5.4	22.81	111.92	0.5	28.9	150.45	0.5
5.45	22.65	76.4	0.3	27.9	111.37	0.4
5.5	19.51	92.57	0.5	26.8	94.36	0.4
5.55	22.27	79	0.4	22.6	106.73	0.5
5.6	21.65	72.79	0.3	26.1	138.04	0.5
5.65	16.93	102.34	0.6	26.0	168.5	0.6
5.7	17.07	115.93	0.7	23.7	138.04	0.6
5.75	17.80	109.62	0.6	23.1	105.55	0.5
5.8	19.35	95.36	0.5	23.5	94.06	0.4
5.85	18.30	93.97	0.5	24.0	94.97	0.4
5.9	17.72	68.66	0.4	22.5	106.38	0.5
5.95	17.31	64.68	0.4	20.4	48.48	0.2

(*Continued*)

P-Table 7.3 (Continued) CPT data for Problem 7.16

Depth (m)	q_c (MPa)	f_s (kPa)	f_s/q_c (%)	q_c (MPa)	f_s (kPa)	f_s/q_c (%)
6	17.44	70.66	0.4	19.9	65.65	0.3
6.05	17.84	76.01	0.4	19.9	73.28	0.4
6.1	18.68	91.9	0.5	22.2	95.48	0.4
6.15	21.38	113.2	0.5	23.3	112.89	0.5
6.2	21.37	118.71	0.6	23.9	125.65	0.5
6.25	19.14	132.24	0.7	22.7	138.18	0.6
6.3	18.35	112.98	0.6	21.8	146.2	0.7
6.35	17.55	111.63	0.6	21.0	150.07	0.7
6.4	16.97	110.94	0.7	20.6	154.4	0.8
6.45	17.02	110.6	0.6	20.4	−4.29	0.0
6.5	16.92	115.62	0.7	20.3	115.56	0.6
6.55	17.29	97.41	0.6	6.9	139.63	2.0
6.6	17.93	114.6	0.6	19.7	151.12	0.8
6.65	17.65	126.44	0.7	21.3	162.76	0.8
6.7	17.54	123.35	0.7	21.8	167.72	0.8
6.75	16.73	120.34	0.7	21.2	163.98	0.8
6.8	16.06	114.03	0.7	20.6	164.23	0.8
6.85	15.40	106.59	0.7	19.1	148.72	0.8
6.9	14.03	98.82	0.7	16.5	132.57	0.8
6.95	13.00	88.79	0.7	15.9	96.37	0.6
7	13.64	93.93	0.7	15.9	125.85	0.8
7.05	12.81	158.43	1.2	12.4	228.13	1.8
7.1	7.13	180.16	2.5	7.2	212.36	3.0
7.15	3.87	125.51	3.2	3.6	113.34	3.2
7.2	3.12	59.61	1.9	8.3	102.46	1.2
7.25	9.75	48.26	0.5	15.3	105.39	0.7
7.3	13.79	62.58	0.5	17.4	97.15	0.6
7.35	15.19	65.67	0.4	17.6	107.91	0.6
7.4	15.60	83.15	0.5	20.8	119.47	0.6
7.45	17.91	78.71	0.4	22.3	105.65	0.5
7.5	17.03	82.09	0.5	23.0	135.78	0.6
7.55	16.38	78.23	0.5	25.8	151.27	0.6
7.6	17.67	74.14	0.4	25.2	148.78	0.6
7.65	19.09	98.12	0.5	22.6	193.03	0.9
7.7	15.89	105.08	0.7	19.5	190.37	1.0
7.75	13.73	98.14	0.7	17.2	173.36	1.0
7.8	14.03	93.2	0.7	15.9	182.84	1.2
7.85	12.85	157.33	1.2	16.3	189.98	1.2
7.9	5.97	146.85	2.5	14.2	272.85	1.9
7.95	3.74	97.6	2.6	6.1	247.86	4.1
8	3.38	58.84	1.7	3.4	164.55	4.9
8.05	3.88	80.91	2.1	3.4	75.85	2.2
8.1	12.28	89.76	0.7	5.0	98.57	2.0
8.15	15.27	83.11	0.5	17.4	106.55	0.6
8.2	15.61	84.84	0.5	17.1	89.09	0.5
8.25	14.78	84.59	0.6	16.4	91.47	0.6

(Continued)

P-Table 7.3 (Continued) CPT data for Problem 7.16

Depth (m)	q_c (MPa)	f_s (kPa)	f_s/q_c (%)	q_c (MPa)	f_s (kPa)	f_s/q_c (%)
8.3	16.35	91.04	0.6	15.0	88.01	0.6
8.35	17.78	103.6	0.6	14.1	85.21	0.6
8.4	18.35	111.25	0.6	14.0	88.68	0.6
8.45	16.68	120.83	0.7	15.8	62.09	0.4
8.5	16.19	108.05	0.7	17.5	73.95	0.4
8.55	14.91	101.44	0.7	16.0	97.88	0.6
8.6	14.79	136.64	0.9	17.2	115.82	0.7
8.65	13.59	101.58	0.7	17.3	107.89	0.6
8.7	15.29	91.33	0.6	17.7	86.59	0.5
8.75	19.52	105.59	0.5	19.0	92.12	0.5
8.8	22.41	115.62	0.5	21.0	68.01	0.3
8.85	25.06	113.12	0.5	24.6	−4.39	0.0
9	31.94	206.81	0.6			
9.05	34.32	187.46	0.5			
9.1	34.64	211.51	0.6			
9.15	37.37	315.14	0.8			
9.2	42.11	344.06	0.8			
9.25	38.23	341.31	0.9			
9.3	33.55	256.45	0.8			
9.35	36.35	195.58	0.5			
9.4	34.96	120.52	0.3			
9.45	35.90	126.46	0.4			
9.5	36.37	124	0.3			
9.55	35.85	168.5	0.5			
9.6	34.06	293.97	0.9			
9.65	34.10	253.82	0.7			
9.7	29.22	207.44	0.7			
9.75	29.14	145.16	0.5			
9.8	32.31	181.05	0.6			
9.85	30.29	199.02	0.7			

Problem 7.17 To estimate the undrained shear strength of a normally consolidated soft clay deposit, vane shear tests were performed at four different depths. In these tests, a rectangular vane with 60 mm diameter and 120 mm height was used. Both ends of the vane were inserted in the soil. The plasticity index of the clay is equal to 65%. The unit weight of the clay is 16 kN/m³, and the water table is at the ground surface. The results are shown in P-Table 7.4.

P-Table 7.4 Vane shear test results for Problem 7.17

Depth (m)	Maximum torque (N.m)
3	7
5	11.1
8	18.5
10	22.4

a. Estimate the undrained shear strength for each depth and develop a plot of the design undrained shear strength versus depth (depth on the vertical axis and undrained shear strength on the horizontal axis).

b. Estimate the *in situ* undrained shear strength using the correlation of Equation (6.50). Compare the result with the result from (a) and discuss.

REFERENCES

References cited

Almeida, M. S., Danziger, F. A., and Lunne, T. (1996). "Use of the piezocone test to predict the axial capacity of driven and jacked piles in clay." *Canadian Geotechnical Journal*, 33(1), 23–41.

Anagnostopoulos, A., Koukis, G., Sabatakakis, N., and Tsiambaos, G. (2003). "Empirical correlations of soil parameters based on cone penetration tests (CPT) for Greek soils." *Geotechnical and Geological Engineering*, 21(4), 377–387.

Bieniawski, Z. T. (1979). "The geomechanics classification in rock engineering applications." *Proceedings of 4th International Congress on Rock Mechanics*, Montreux, Vol. 2, 41–48.

Bieniawski, Z. T. (1989). *Engineering Rock Mass Classifications : A Complete Manual for Engineers and Geologists in Mining, Civil, and Petroleum Engineering*. Wiley, New York.

Bisht, V., Salgado, R., and Prezzi, M. (2021). "Analysis of cone penetration using the material point method." *Proceedings of 16th International Conference of the International Association for Computer Methods and Advances in Geomechanics*, Turin, Italy, 765–771.

Bjerrum, L. (1972). "Embankments on soft ground." *Proceedings of ASCE Conference on Performance of Earth and Earth-Supported Structures*, Purdue University, Vol. 2, 1–54.

Bjerrum, L. (1973). "Problems of soil mechanics and construction on soft clays." *Proceedings of 8th International Conference on Soil Mechanics and Foundation Engineering*, Moscow, Vol. 3, 111–159.

Bolton, M. D. (1986). "The strength and dilatancy of sands." *Géotechnique*, 36(1), 65–78.

Bowles, J. E. (1996). *Foundation Analysis and Design*. 5th Edition, McGraw-Hill, New York.

Briaud, J. L. (1992). *The Pressuremeter*. 1st Edition, Taylor & Francis, Oxford, UK.

Carlson, L. (1948). "Determination in situ of the shear strength of undisturbed clay by means of a rotating auger." *Proceedings of 2nd International Conference on Soil Mechanics and Foundation Engineering*, Rotterdam, 1, 265–270.

Carpentier, R. (1982). "Relationship between the cone resistance and the undrained shear strength of stiff fissured clays." *Proceedings of 2nd European Symposium on Penetration Testing*, Amsterdam, 519–528.

Carraro, J. A. H., Bandini, P., and Salgado, R. (2003). "Liquefaction resistance of clean and non-plastic silty sands based on cone penetration resistance." *Journal of Geotechnical and Geoenvironmental Engineering*, 129(12), 965–976.

Carter, J. P., Randolph, M. F., and Wroth, C. P. (1979). "Stress and pore pressure changes in clay during and after the expansion of a cylindrical cavity." *International Journal for Numerical and Analytical Methods in Geomechanics*, 3(4), 305–322.

Clayton, C. R. I. (1990). "SPT energy transmission: Theory, measurement and significance." *Ground Engineering*, 23(10), 35–43.

Denby, G. M. and Clough, G. W. (1980). "Self-boring pressuremeter tests in clay." *Journal of Geotechnical and Geoenvironmental Engineering*, 106(12), 1369–1387.

Denver, H. (1988). "CPT and shear strength of clay." *Proceedings of 1st International Symposium on Penetration Testing*, Orlando, FL, 2, 723–727.

French Standard. (1991). NF P94-110-1: Essi Pressiometrique Menard, AFNOR.

Ganju, E., Prezzi, M., and Salgado, R. (2017). "Algorithm for generation of stratigraphic profiles using cone penetration test data." *Computers and Geotechnics*, 90, 73–84.

Goodman. (1992). *Engineering Geology*. Wiley, New York.

Hoek, E. and Brown, E. T. (1980). "Empirical strength criterion for rock masses." *Journal of the Geotechnical Engineering Division*, 106(9), 1013–1035.

Hoek, E. and Brown, E. T. (1988). "The Hoek-Brown failure criterion–a 1988 update." *Proceedings of 15th Canadian Rock Mechanics Symposium*, Toronto, Canada, 31–38.

Hvorslev, M. J. (1949). Subsurface exploration and sampling of soils for civil engineering purposes. Report on a Research Project of the Committee on Sampling and Testing, Soil Mechanics and Foundations Division, American Society of Civil Engineers.

ISRM. (1985). "Suggested method for determining point load strength." *International Journal of Rock Mechanics*, 22(2), 53–60.

ISSMGE (International Society of Soil Mechanics and Geotechnical Engineering). "Standard penetration test (SPT): International Reference Test Procedure." International Society of Soil Mechanics and Foundation Engineering Technical Committee on Penetration Testing, International Symposium on Penetration Testing (ISOPT-1), Orlando, FL, Vol. 1, 3–26.

Jamiolkowski, M., Lancellotta, R., Tordella, L., and Battaglio, M. (1982). "Undrained strength from CPT." *Proceedings of 2nd European Symposium on Penetration Testing*, Amsterdam, 2, 599–606.

Kim, K., Prezzi, M., Salgado, R., and Lee, W. (2008). "Effect of penetration rate on cone penetration resistance in saturated clayey soils." *Journal of Geotechnical and Geoenvironmental Engineering*, 134(8), 1142–1153.

Kim, K., Prezzi, M., Salgado, R., and Lee, W. (2010). "Penetration rate effects on cone resistance measured in a calibration chamber." *Proceedings of 2nd International Symposium on Cone Penetration Testing*, Huntington Beach, CA.

Komarnitskii, N. I. (1968). *Zones and Planes of Weakness in Rocks and Slope stability*. Consultants Bureau, New York.

Kulhawy, F. H. and Mayne, P. W. (1990). *Manual on Estimating Soil Properties for Foundation Design*. Report No. EPRI-EL-6800, Electric Power Research Institute, Palo Alto, CA.

La Rochelle, P., Zebdi, M., Leroueil, S., Tavenas, F., and Virely, D. (1988). "Piezocone tests in sensitive clays of eastern Canada." *Proceedings of 1st International Symposium on Penetration Testing*, Orlando, FL, 831–841.

Liao, S. C. and Whitman, R. V. (1986). "Overburden correction factors for SPT in sand." *Journal of Geotechnical Engineering*, 112(3), 373–377.

Low, H. E., Lunne, T., Andersen, K. H., Sjursen, M. A., Li, X. and Randolph, M. F. (2010). "Estimation of intact and remoulded undrained shear strengths from penetration tests in soft clays." *Geotechnique*, 60(11), 843–859.

Luke, K. (1995). "The use of Cu from Danish triaxial tests to calculate the cone factor." *Proceedings of International Symposium on Cone Penetration Testing: CPT'95*, Linkoping, Sweden, 209–214.

Lunne, T., Eidsmoen, T., Gillespie, D., and Howland, J. D. (1986). "Laboratory and field evaluation of cone penetrometers." *Proceedings of Use of in situ Tests in Geotechnical Engineering: In situ'86*, Blacksburg, VA, 714–729.

Lunne, T., Powell, J. J. M., and Robertson, P. K. (1997). *Cone Penetration Testing in Geotechnical Practice*. Blackie Academic and Professional, New York.

Menard, L. (1956). "An apparatus for measuring the strength of soils in place." Ph.D. Thesis, University of Illinois, Illinois.

Meyerhof, G. G. (1957). "Discussion on soil properties and their measurement–session I." *Proceedings of 4th International Conference of Soil Mechanics and Foundation Engineering*, London, 3, 10.

Mitchell, J. K. (1988). "New developments in penetration tests and equipment." *Proceedings of International Symposium on Penetration Testing*, Rotterdam, 1, 245–261.

Mitchell, J. K. (2000). "Education in geotechnical engineering: Its evaluation, current status and challenges for the 21st century." *Proceedings of 11th Panamerican Conference on Soil Mechanics and Geotechnical Engineering*, Foz do Iguaçu, Brazil, 4, 167–174.

Mitchell, J. K. and Brandon, T. L. (1998). "Analysis and use of CPT in earthquake and environmental engineering." *Proceedings of 1st International Conference on Site Characterization*, Rotterdam, 69–114.

Odebrecht, E., Schnaid, F., Rocha, M. M., and de Paula Bernardes, G. (2005). "Energy efficiency for standard penetration tests." *Journal of Geotechnical and Geoenvironmental Engineering*, 131(10), 1252–1263.

Puppala, A. J., Acar, Y. B., and Tumay, M. T. (1995). "Cone penetration in very weakly cemented sand." *Journal of Geotechnical Engineering*, 121(8), 589–600.

Puppala, A. J., Acar, Y. B., and Tumay, M. T. (1996). "Cone penetration in very weakly cemented sand: Closure." *Journal of Geotechnical Engineering*, 122(11), 948–949.

Rad, N. S. and Lunne, T. (1988). "Direct correlations between piezocone test results and undrained shear strength of clay." *Proceedings of 1st International Symposium on Penetration Testing*, Orlando, FL, 2, 911–917.

Randolph, M. F. and Wroth, C. P. (1979). "An analytical solution for the consolidation around a driven pile." *International Journal for Numerical and Analytical Methods in Geomechanics*, 3, 217–229.

Robertson, P. K., Campanella, R. G., and Wightman, A. (1983). "SPT-CPT correlations." *Journal of Geotechnical Engineering*, 109(11), 1449–1459.

Sakleshpur, V. A., Prezzi, M., Salgado, R., and Zaheer, M. (2021). "CPT-based geotechnical design manual–volume II : CPT-based design of foundations (methods)." Joint Transportation Research Program, Purdue University, West Lafayette, IN.

Salgado, R. and Prezzi, M. (2007). "Computation of cavity expansion pressure and penetration resistance in sands." *International Journal of Geomechanics*, 7(4), 251–265.

Salgado, R. and Prezzi, M. (2014). "Penetration rate effects on cone resistance: Insights from calibration chamber and field testing." *Soils and Rocks*, 37(3), 233–242.

Salgado, R., Boulanger, R. W., and Mitchell, J. K. (1997a). "Lateral stress effects on CPT liquefaction resistance correlations." *Journal of Geotechnical and Geoenvironmental Engineering*, 123(8), 726–735.

Salgado, R., Mitchell, J. K., and Jamiolkowski, M. (1997b). "Cavity expansion and penetration resistance in sand." *Journal of Geotechnical and Geoenvironmental Engineering*, 123(4), 344–354.

Salgado, R., Prezzi, M., Kim, K., and Lee, W. (2013). "Penetration rate effects on cone resistance measured in a calibration chamber." *Proceedings of 4th International Conference on Site Characterization*, Pernambuco, Brazil, 1, 1025–1030.

Schnaid, F. and Consoli, N. C. (1996). "Discussion of 'cone penetration in very weakly cemented sand.'" *Journal of Geotechnical Engineering*, 122(11), 948–948.

Seed, H. B., Tokimatsu, K., Harder, L. F., and Chung, R. M. (1985). "Influence of SPT procedures in soil liquefaction resistance evaluations." *Journal of Geotechnical Engineering*, 111(12), 1425–1445.

Shroff, A. V. and Shah, D. L. (2003). *Soil Mechanics and Geotechnical Engineering*. Taylor & Francis, Oxford, UK.

Skempton, A. W. (1957). "The planning and design of the new Hong Kong airport." Institution of Civil Engineers, London, 305–307.

Skempton, A. W. (1986). "Standard penetration test procedures and the effects in sands of overburden pressure, relative density, particle size, ageing and overconsolidation." *Géotechnique*, 36(3), 425–447.

Stark, T. D. and Juhrend, J. E. (1989). "Undrained shear strength from cone penetration tests." *Proceedings of 12th International Conference on Soil Mechanics and Foundation Engineering*, Rio de Janeiro, Brazil, 327–330.

Stroud, M. A. (1975). "The standard penetration test in insensitive clays and soft rocks." *Proceedings of 1st European Symposium on Penetration Testing*, Stockholm, Sweden, 2, 367–375.

Széchy, K. (1966). The art of tunnelling. Translated from the Hungarian by Denis Széchy and others, Akadémiai Kiadó, Budapest.

Tani, K. and Craig, W. H. (1995). "Bearing capacity of circular foundations on soft clay of strength increasing with depth." *Soils and Foundations*, 35(4), 21–35.

Tumay, M. T. (1985). Field calibration of electric cone penetrometers in soft soils: Executive summary. Report No. FHWA/LA/LSU-GE-85/02, Washington, DC.

U.S. Army. (1996). Engineering manual: Engineering and design–soil sampling. Publication No. EM 1110-1-1906.

U.S. Army. (2001). Engineering and design–geotechnical investigations. Publication No. EM 1110-1-1804.

Van Impe, W. F. (2004). Two decades of full scale research on screw piles. An overview. Published by The Laboratory of Soil Mechanics, Ghent University, Belgium.

Yu, H. S. and Mitchell, J. K. (1998). "Analysis of cone resistance: Review of methods." *Journal of Geotechnical and Geoenvironmental Engineering*, 124(2), 140–149.

Yu, H. S., Herrmann, L. R., and Boulanger, R. W. (2000). "Analysis of steady cone penetration in clay." *Journal of Geotechnical and Geoenvironmental Engineering*, 126(7), 594–605.

Additional references

Baldi, G. (1985). "Laboratory validation of *in situ* tests." *Proceedings of 11th International Conference on Soil Mechanics and Foundation Engineering*, San Francisco, CA.

Baligh, M. M. (1985). "Strain path method." *Journal of Geotechnical Engineering*, 111(9), 1108–1136.

Baligh, M. M., Azzouz, A. S., and Chin, C. T. (1987). "Disturbances due to 'ideal' tube sampling." *Journal of Geotechnical Engineering*, 113(7), 739–757.

Bieniawski, Z. T. (1976). "Rock mass classification in rock engineering applications." *Proceedings of Symposium on Exploration for Rock Engineering*, Johannesburg, South Africa, 97–106.

Brown, E. T. (1970). "Strength of models of rock with intermittent joints." *Journal of Soil Mechanics and Foundations Division*, 96(SM6), 1935–1949.

Burns, S. E. and Mayne, P. W. (2002). "Analytical cavity expansion-critical state model for piezocone dissipation in fine-grained soils." *Soils and Foundations*, 42(2), 131–137.

Durgunoglo, H. T. and Mitchell, J. K. (1975). "Static penetration resistance of soils I-analysis." *Proceedings of Specialty Conference in in-situ Measurements of Soil Properties*, Raleigh, NC, 151–171.

Durgunoglu, H. T. and Mitchell, J. K. (1975). "Static penetration resistance of soils II: Evaluation of theory and implications for practice." *Proceedings of Specialty Conference in in-situ Measurements of Soil Properties*, Raleigh, NC, 172–189.

Puppala, A. J., Acar, Y. B., and Tumay, M. T. (1992). "Miniature CPT tests in dense monterey No. 0/30 sand in a flexible double-walled calibration chamber." *Proceedings of 1st International Symposium on Calibration Chamber Testing*, Potsdam, NY, 339–350.

Salgado, R., Mitchell, J. K., and Jamiolkowski, M. (1998). "Calibration chamber size effects on penetration resistance in sand." *Journal of Geotechnical and Geoenvironmental Engineering*, 124(9), 878–888.

Teh, C. I. and Houlsby, G. T. (1991). "An analytical study of the cone penetration test in clay." *Géotechnique*, 41(1), 17–34.

Relevant ASTM standards

ASTM. (2015). "Standard practice for thin-walled tube sampling of fine-grained soils for geotechnical purposes." ASTM D1587, American Society for Testing and Materials, West Conshohocken, PA.

ASTM. (2016). "Standard test method for mechanical cone penetration testing of soils." ASTM D3441, American Society for Testing and Materials, West Conshohocken, PA.

ASTM. (2017). "Standard practice for classification of soils for engineering purposes (unified soil classification system)." ASTM D2487, American Society for Testing and Materials, West Conshohocken, PA.

ASTM. (2018a). "Standard test method for field vane shear test in saturated fine-grained soils." ASTM D2573, American Society for Testing and Materials, West Conshohocken, PA.

ASTM. (2018b). "Standard test method for penetration test and split-barrel sampling of soils." ASTM D1586, American Society for Testing and Materials, West Conshohocken, PA.

ASTM. (2020a). "Standard test method for electronic friction cone and piezocone penetration testing of soils." ASTM D5778, American Society for Testing and Materials, West Conshohocken, PA.

ASTM. (2020b). "Standard test methods for prebored pressuremeter testing in soils." ASTM D4719, American Society for Testing and Materials, West Conshohocken, PA.

Chapter 8

Shallow foundations in soils: types of shallow foundations and construction techniques

> Every one therefore that hears these words of mine, and does them, will be like a wise man, who built his house upon the rock: and the rain descended, and the floods came, and the winds blew, and beat upon that house; and it did not fall: for it was founded upon the rock. And every one that hears these words of mine, and does not do them, will be like a foolish man, who built his house upon the sand: and the rain descended, and the floods came, and the winds blew and struck that house; and it fell: and great was the fall thereof.
>
> KJV (Matthew 7:24–7:27)

Shallow foundations transfer structural loads to soils at relatively small depths in the ground. They range from isolated foundations, each carrying its own column load, to elements carrying several columns, walls, or even all of the loads for a given structure or building. Shallow foundations are easy to build, requiring little to no specialized equipment. In this chapter, we discuss the various types of shallow foundations and how they are constructed.

8.1 TYPES OF SHALLOW FOUNDATIONS AND THEIR APPLICABILITY

8.1.1 Applicability of shallow foundations

As discussed in Chapter 2, one of the main strategies of load transfer to the ground is the use of *shallow foundations*, in which the loads are spread horizontally at the contact between the foundation element and the ground. In this chapter, we discuss shallow foundations bearing on soil (foundations in or on rock are discussed in later chapters). For shallow foundations in soil, the load transfer from the foundation element to the soil takes place predominantly through the base of the element, with only a small fraction of the load transferred through the sides of the element (which is accordingly disregarded in load capacity computations). The Sermon on the Mount – epigraph to this chapter – notwithstanding, it is often possible and economical to use shallow foundations in sand, and even in clay. We examine the main types of shallow foundations, their construction, and their construction inspection.

The most important advantages of shallow foundations over deep foundations are that they don't require the use of special equipment for their construction, they tend to be more economical if competent soil is available near the ground surface, and they allow inspection of the base soil before concrete placement. This last advantage means that any design assumptions made at the design stage about the base soil can be checked at least in some simple way. Shallow foundations also have some disadvantages. They require stable excavations in which to form the foundation element. This sometimes requires the use of struts

DOI: 10.1201/b22079-8

and other ground-supporting elements (discussed in Chapter 16) and may require the use of a dewatering system if the water table is too high. Such measures increase costs and lead to construction difficulties that may make deep foundations a preferable alternative.

Shallow foundations are not well suited for very compressible soils, such as normally consolidated clay and peat, nor for expansive soils. Problems with excessive total settlements would develop in the case of very compressible soils, and with differential movements in the case of expansive soils. Uncontrolled fills should also be avoided as a support material for shallow foundations because large differences in soil density that may exist from location to location will inevitably lead to large differential settlements. Footings near the property line are also not advised if future excavations or significant construction activities are foreseen in the adjacent lot. Even though it would be the responsibility of those involved in future foundation construction not to damage existing foundations, it may be possible to prevent future disputes and the associated litigation cost by selecting deep foundations in such cases, even at a higher foundation cost.

8.1.2 Types of shallow foundations

We will now discuss the main types of shallow foundations, which are summarized in Figure 8.1. For very lightly loaded foundations, unreinforced concrete or masonry footings are possible; however, shallow foundations are most of the time built of reinforced concrete. When a foundation element supports a single column load, it is referred to as an *isolated footing*. When it supports two columns, it is called a combined footing. When supporting a line of columns or a load-bearing wall, it is called a *strip footing*. If it supports a large number of columns that are not in one line, it is referred to as a *mat* or *raft*. When two isolated footings are connected by a beam, so that they work as a unit, the term *strap footing* is used to refer to them.

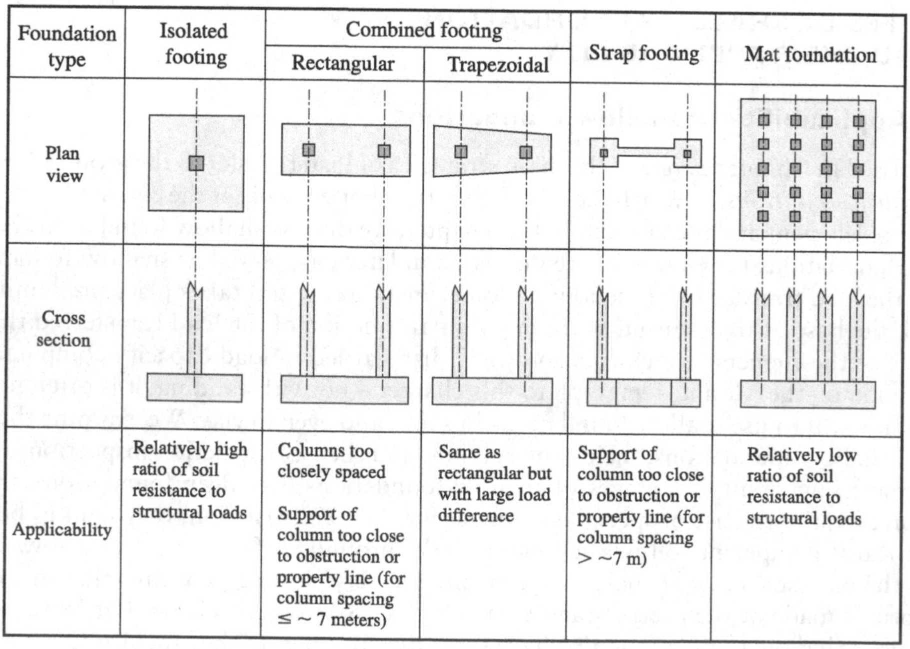

| Foundation type | Isolated footing | Combined footing | | Strap footing | Mat foundation |
		Rectangular	Trapezoidal		
Plan view					
Cross section					
Applicability	Relatively high ratio of soil resistance to structural loads. Support of column too close to obstruction or property line (for column spacing ≤ ~ 7 meters)	Columns too closely spaced. Support of column too close to obstruction or property line (for column spacing ≤ ~7 m)	Same as rectangular but with large load difference	Support of column too close to obstruction or property line (for column spacing > ~7 m)	Relatively low ratio of soil resistance to structural loads

Figure 8.1 Summary of types of shallow foundations.

If the load from the superstructure is small (typically less than about 100 kN, or roughly 10 tons), vertical, and centered, an unreinforced concrete footing is a possible foundation solution (Figure 8.2). This is because the shear forces and bending moments within the footing are moderate and the stresses are mostly compressive, eliminating the need for reinforcement. If the loads are large, inclined, or off-center, isolated footings are reinforced (Figure 8.3). Circular footings are used for special applications, such as machine and tank foundations. For buildings and infrastructure systems, footings usually have plan shapes that are square, rectangular, or trapezoidal, which are easier to form. Square footings are the natural choice for the cases where a single column load is to be supported; however, it is not always possible to use them. There may be obstructions or design-imposed boundaries that constrain the engineer from extending the footing beyond these boundaries. For example, a footing cannot extend beyond the property line, nor can it extend into the area where an elevator shaft has been planned. For this or other similar reasons, a rectangular footing may be used instead of a square footing.

Footings must sometimes absorb moment loadings. These are often due to wind, but some structures (such as cranes and sign poles) do impose permanent moments on foundations. One way to handle permanent moments is to extend the footing in the direction of the moment (Figure 8.4), in effect displacing the centroid of the footing so that the soil resistance at the base of the footing counterbalances the moment from the superstructure.

In some cases, when two columns are near one another and isolated footings for each column would overlap or be excessively close, a combined footing (Figure 8.5) may be used.

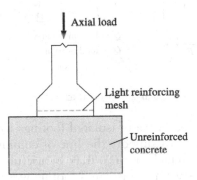

Figure 8.2 Unreinforced concrete footing.

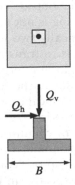

Figure 8.3 Isolated reinforced concrete footing (reinforcement not shown in the figure).

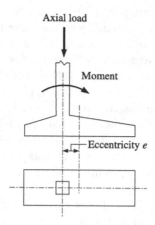

Figure 8.4 Footing with eccentric column to counterbalance the applied moment.

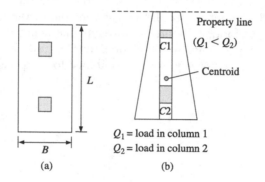

Figure 8.5 Combined footings: (a) rectangular and (b) trapezoidal.

A rule of thumb is that if the total area of isolated footings would exceed 50% of the area that a strip footing would have to support the line of columns, a strip footing is preferable. Another case where combined footings may be useful is when a column is located very close to an obstruction; whereas an isolated footing would have to extend beyond the obstruction, it may be possible to design a combined footing supporting this and another column located next to it that will occupy only the space available. This is possible because the support of the soil located between the two columns reduces the need to extend the plan area of the footing much beyond the outside column. When a column is immediately next to an obstruction, a combined footing is not appropriate and strap footings may be a superior solution. Strap footings are basically two footings connected by means of a beam (Figure 8.6). The function of the beam is to transfer loads between the two footings (the external footing loaded eccentrically and the internal one loaded centrically) so as to create a pair that will balance the moment due to the eccentricity of the outside column. Strap footings are also preferable to (more economical than) combined footings when the distance between the two columns is large (typically in excess of 7 m).

When the load/spacing ratio for a line of columns is high, isolated footings would be too close to one another, making it more economical to build a single footing to support all columns. Such a footing is known as a strip or "continuous" footing (shown in plan view and cross section in Figure 8.7 and in perspective in Figure 8.8). Strip footings are also used to support load-bearing walls. Strip footings are long (that is, have large values of L/B,

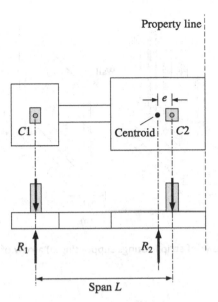

Figure 8.6 Strap footing.

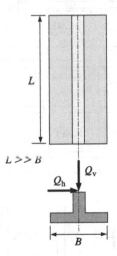

Figure 8.7 Strip footing: plan view and cross section.

exceeding, say, 7, where B is the width and L is the length of the foundation) and may overlie soils of variable stiffness. In order for a strip footing to be able to bridge over weaker soils, it is essential that the footing be reinforced.

A similar situation to those requiring the use of strip footings exists when isolated, combined, and strip footings would occupy a large percentage (more than 50%) of the plan area of a building or structure. In this case, it is usually more economical (because of savings in such items as excavation time and formwork) to build a single foundation covering the whole area of the foundation. This foundation is known as a mat or raft foundation. Figure 8.9 shows three different types of mats. The first type (Figure 8.9a) is a reinforced

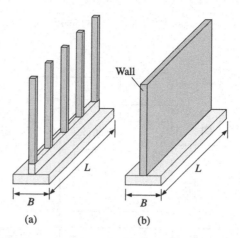

Figure 8.8 Three-dimensional views of strip footings supporting (a) a row of columns and (b) a load-bearing wall.

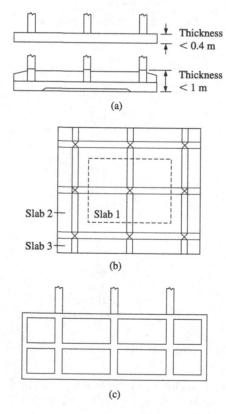

Figure 8.9 Mats: (a) reinforced concrete mat with uniform thickness across the site; (b) reinforced concrete mat with stiffening beams along the lines of columns; and (c) cellular mat.

concrete slab of uniform thickness (with thickness typically in the 0.75–2 m range), appropriate for column loads that don't vary excessively. The slab is reinforced at both the top and bottom with negative and positive steels. The second type (Figure 8.9b) is a more structured mat where beams traverse the slab along the lines of loads, providing extra reinforcement at those locations. The third type, shown in (Figure 8.9c), is a cellular mat, used when it would be very expensive to use a reinforced concrete slab, even one with beams running across. These mats are usually used where the volume of excavation is large, which has the added benefit of reducing the net load imposed on the foundation soil.

8.2 CONSTRUCTION OF SHALLOW FOUNDATIONS

8.2.1 Basic construction methods

The first step in the construction of shallow foundations is soil excavation. This is usually accomplished by the use of a backhoe or similar equipment; manual excavation is sometimes also possible. Excavation is easy in clayey soils, but becomes quite difficult in sandy/gravelly soil, particularly below the water table. Under right conditions, dewatering (lowering of the water table) can be done to allow soil excavation; it is discussed later in this chapter.

After excavation is complete, the soil exposed at the base of the excavation must be inspected for an assessment to be made of whether it has the properties required in the construction specifications and assumed in design. The base of the footing should be cleaned of any loose material that may have been left on it after excavation. The base soil, if sandy, may benefit from a moderate amount of compaction (Figure 8.10). A thin (50–75 mm) coat of lean concrete (concrete with a higher water–cement ratio than usual), often called a blinding layer, is sometimes applied to the base soil to protect the soil from degradation that may result from stress relief and exposure to rain and other weather actions. This layer also offers a better working platform than simply compacted soils. If the base soil is found to be satisfactory, construction can proceed. In some cases, when the soil is good and excavations in it will remain stable, no formwork is necessary and concrete is placed directly into the excavation. This is not common in foundations for reinforced concrete buildings, for which pedestals rising from the foundation elements are used to connect the foundations to the supported columns, as discussed later. The construction of these pedestals requires formwork.

Figure 8.10 Compaction of footing base.

The reinforcing mesh (Figure 8.11) and the formwork, if used, are prepared and positioned inside the excavation at that point (Figure 8.12). Figure 8.12 also shows the four dowels that will later be used to connect the steel columns to the footing. Concrete placement is the next step (Figure 8.13). The required cover ranges from ~40 mm (1½ in.) when a lean concrete coat is present at the base of the footing to ~75 mm (3 in.) in the absence of any coat. If the soil is not aggressive (corrosive), a typical concrete mix would have a compressive strength of the order of 25–30 MPa (3500–4000 psi). Concrete placement is followed, after curing, by backfilling above the footing up to the level of the ground surface. This backfill is usually compacted.

For frame buildings, attachment of the column to the footing may precede backfilling. Although there is little hesitation as to where the columns should be positioned on square or other symmetric cross sections, for L-shaped columns (which are found in reinforced concrete structures), there might be some doubt. The rule is that the centroid of an isolated footing must coincide with the centroid of the column it supports.

Figure 8.12a shows that for a reinforced concrete structure, the formwork for the footing already has an extension (a pedestal) from which the reinforced concrete column will later continue. In this case, backfilling precedes the attachment of the column to the footing. A steel column may or may not be attached to a pedestal. If not, then the column is first connected to the footing and then backfilling is done.

There is a difference in the means by which the column is connected to the footing depending on whether the structure is a steel or reinforced concrete frame structure. For steel columns, a number of anchor bolts (typically four) are positioned partway into the concrete at the time of concrete placement (Figures 8.12b and 8.14a). These are later used to fix a steel base plate to the footing, to which the column is then bolted. Alternatively, if a base plate is already welded to the column, the base of the column is bolted to the footing. If a reinforced concrete pedestal stems from the footing, the bolting of the plate and column will be done at the top of the pedestal. The base plate (or column base) has a number of holes corresponding to the number of anchor bolts used. These holes are slightly larger than the bolts to accommodate any misalignment. The column is lowered onto the footing using a crane (Figure 8.14b), and the column base is bolted to the footing (Figure 8.14c). Even though every attempt is made to level the base plate (or column base) on top of the footing, it is

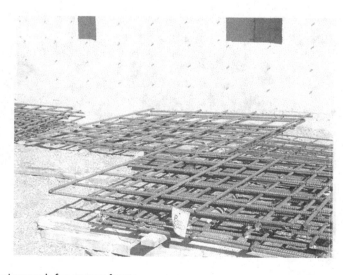

Figure 8.11 Reinforcing mesh for square footing.

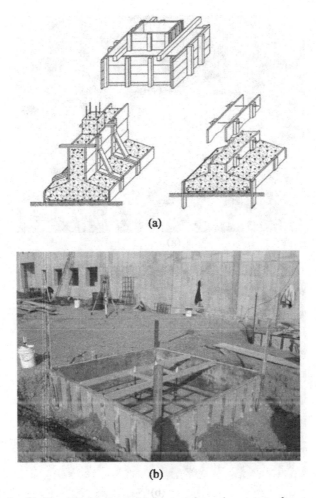

(a)

(b)

Figure 8.12 Formwork for shallow foundations: (a) for reinforced concrete frame structures and (b) for steel frame structures.

Figure 8.13 Concrete placement.

(a)

(b)

(c)

Figure 8.14 Attachment of steel columns to reinforced concrete footings by the use of bolts: (a) finished footing with four bolts sticking out, (b) column being lowered on the footing, and (c) tightening of bolts for secure connection of column to footing.

almost always necessary to grout between the base plate and the footing or pedestal surface to fill the remaining voids. Several design variations are possible and may be required in the case of foundations for cellular phone or energy transmission towers.

8.2.2 Basic construction specifications and items for inspection

There are three construction requirements that must be addressed in the construction specifications for footings in soil: (1) minimum depth to base of footing, (2) certain minimum distances with respect to buried utilities, cavities, or other foundation elements, and (3) sequence of construction of individual footings.

Specification of a minimum depth prevents potential problems related to the weakening of the support for the footing that could result from any type of excavation, erosion, or ground loss in the immediate neighborhood of the footing. It is usually sufficient to require 0.75 m as a minimum. An exception occurs near the property line, where it is prudent to adopt a minimum depth of 1.50 m. A minimum depth is also essential in areas where the ground freezes during the winter and thaws in the summer. It is important to place the footing on soil that will not be subjected to freeze and thaw. This depth varies from area to area, but, in the United States, nowhere is it greater than about 1.5 m. In Chicago, for example, the local code requires a minimum depth of 3.5 ft (roughly 1 m).

In the presence of a buried cavity or utility line, a line extending from the edge of the footing, tangent to the contour of the cavity or utility line, should make an angle of at least 45° with the vertical (Figure 8.15). Underlying this rule is the understanding that load is transferred primarily within this region delimited by 45° orientation. A larger angle may be warranted in extremely weak soils. The same rule can be used when the bases of the footings are placed at different levels within the ground, as would be common in sloping ground. In this case, no footing should be placed in the zone defined by a 45° line extending down from the edge of any footing (Figure 8.16).

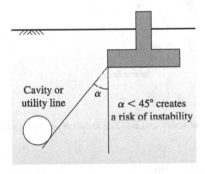

Figure 8.15 Footing placement guidelines in the presence of buried utility lines or cavities.

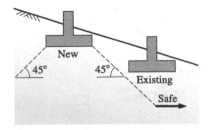

Figure 8.16 Footing placement guidelines for sloping ground conditions.

Example 8.1

You are designing a square footing near the property line of an adjacent structure (E-Figure 8.1). The local municipality owns an easement along the property line for a sewer pipe with a diameter equal to 0.5 m and an invert depth of 4 m below grade. At what distance from the property line and at what depth would you consider placing your footing?

Solution:

Let's look at the pipe first. Referring to Figure 8.15, we know we need the pipe to fall outside a 45° line extending from the base of the footing. Using the invert as the "tangent," we know that our base must be on the outside of the 45° line.

Now we can look at depth. Since we are near the property line, we must consider using a depth of 1.5 m. However, because the sewer restricts our proximity to the property line, we can go back to the minimum depth of 0.75 m. E-Figure 8.2 shows the 45° line and depth of 0.75 m. Using geometry, we see that the distance from the property line is:

3.25m at 0.75m deep [answer]

The sequence of execution is also quite important. Obviously, project management requirements play a major role, but there is at least one obvious constraint imposed by mechanics. If footings are to be built at different depths (the case of sloping ground, such as in Figure 8.16), deeper footings should be built first. This assures that there will be no

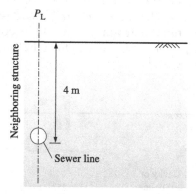

E-Figure 8.1 Schematic diagram of sewer and building location pertinent to Example 8.1.

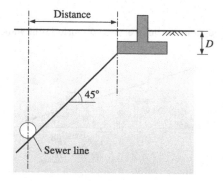

E-Figure 8.2 Location of footing so as not to stress a sewer line excessively.

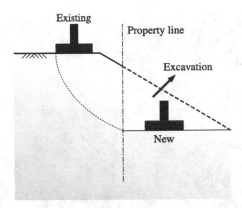

Figure 8.17 Removal of support from an existing footing by the construction of a new footing at a lower elevation.

removal of support from the footings located further above. When new foundations are constructed next to existing structures, it is essential not to remove support from the foundations of the neighboring structures (this is illustrated in Figure 8.17).

Lastly, the time during which the excavation remains open and unsupported must be minimized, as soil degradation may result, particularly if the soil is exposed to rain during this period. Some engineers favor using a lean layer of concrete or even laying down a concrete slab on the base of the footing if it will stay open overnight to prevent soil degradation.

8.2.3 Construction inspection

There are seven checks that an inspector should consider, although not all of them are required in every project. These are (1) base elevation; (2) nature and type of soil at the base of the excavation; (3) base cleaning before concrete placement; (4) dimensions and cross sections of the footings, grade beams, and other foundation elements; (5) concrete placement; (6) time for concrete placement; and (7) footing integrity after concrete placement.

The first check is an obvious one: the footing is to be placed at the design depth. The depth of the excavation should be measured to guarantee that it is as per design. There are also design assumptions made about the soil at the base of the footing. The soil actually uncovered by the excavation needs to match the design expectations; if it does not, calculated bearing capacities and settlements will be in error. Additionally, any loose soil or other debris needs to be removed from the base before construction can proceed. If specified, compaction of the base soil must also be done in an effective way. The dimensions of the excavation and the formwork need to be consistent with the specified dimensions for the foundation elements and the structural elements connecting with them. With respect to cross sections, the rebars must be of the right diameter and need to be positioned correctly, with the correct cover.

The concrete must be of the right quality and strength (some concrete cylinders must be tested, something usually done by the supplier of the concrete if ready-mix concrete is used) and must be properly compacted. Simple checks, such as checking the delivery documentation and doing slump tests, allow verification that the concrete delivered meets specifications. The time during which the excavation remains open and unsupported should be controlled, unless measures have been taken to prevent degradation of the soil at the footing base.

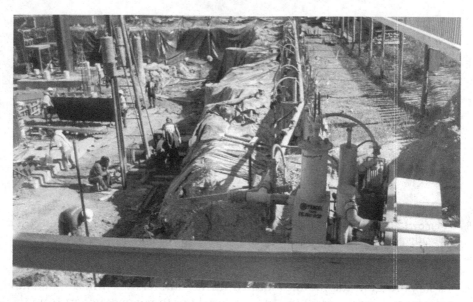

Figure 8.18 Dewatering by the wellpoint method.

8.2.4 Dewatering

Dewatering for foundation construction is mostly accomplished by one of two methods: sump pumping or wellpoints. Sump pumping can, under certain conditions, be effective. In this method, a pump is placed inside a sump, usually partly backfilled with gravel to prevent the development of a quick condition. Groundwater accumulates within the sump as it surfaces, from which it is pumped out. The sump can be located in the plan area of the footing or immediately next to it. Sometimes, a ditch may be excavated around several or many footings, to lower the water table for the construction of all these footings, and sumps are excavated within the ditch. The draining with the sump pumping method is relatively slow, the effective lowering of the water table is only of the order of 1–2 m, and the sump may be in the way of the contractor's operations.

A more effective method, the use of wellpoints, is illustrated in Figure 8.18. Wellpoints are small-diameter (5–8 cm) wells usually installed in series along the borders of the zone to be dewatered. This method is effective for relatively clean sand (containing <15% fines by weight). A higher fines content may lead to wellpoint clogging. Poorly conceived wellpoint systems may give a false sense of confidence.

A detailed treatment of the design of dewatering systems is found in Powers (1992).

8.3 CHAPTER SUMMARY

Shallow foundations may be **isolated** (when one foundation element supports one column), **combined** (when one foundation element supports more than one column), **strap** (when two foundation elements are linked by a beam to absorb a large load eccentricity if present in one of the foundation elements), **strip** (when one foundation element supports a wall or line of columns), or **mat foundations** (when a single, usually slab-like element supports the whole building or structure). Shallow foundations are typically used when the soil near the surface is reasonably competent. Mat foundations may be used when the shallow soils are too compressible or weak for the imposed loads.

The typical construction sequence of a shallow foundation is to first excavate (usually using a backhoe), then to compact the foundation soil (if the soil is sandy or gravelly), lay a lean concrete base (when necessary), position any reinforcement and formwork that may be required, place the concrete, and finally connect the footing with the column. Attention to the effects of foundation construction on neighboring facilities is required. If the groundwater table is high, **dewatering** is necessary for excavation to be done. Items for inspection are related to ascertaining that the quality of the bearing soil is as assumed in design, that the location and geometry of the foundation and its components are correct, and that the materials used are of sufficient quality.

8.4 PROBLEMS

8.4.1 Conceptual problems

Problem 8.1 Define all the terms in bold contained in the chapter summary.

Problem 8.2 What items would you check as a foundations inspector controlling the construction of shallow foundations?

Problem 8.3 When a column exists near an obstruction (such as an elevator shaft or the property line), what alternatives are there to support this column load? Upon which factors do you base your choice of solution?

8.4.2 Design problems

Problem 8.4 You are designing the foundations for a residential building next to an adjacent structure built a long time ago, with foundations flush with the property line (P-Figure 8.1). You are considering using square footings. At what distance from the property line and at what depth would you consider placing your peripheral footings?

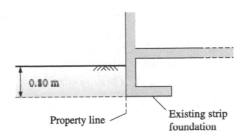

P-Figure 8.1 Existing foundations next to site under development for which you are asked to design foundations in Problem 8.4.

Problem 8.5* You are designing the foundations for a building that will be constructed on sloping natural ground with footings spaced, on average, 5 m from each other. The natural ground slope varies between 20° and 35°. Consider an embedment of 0.5 m and footing widths of 1.5–2 m. Will you be able to maintain every footing outside the zone of load transfer from the footings above it? Comment on how you could adapt your design if this were not possible.

REFERENCE

Powers, J. P. (1992). *Construction Dewatering: New Methods and Applications*. Wiley, New York.

Chapter 9

Shallow foundation settlement

[A]ll subsidence estimates, including [my] own[,] are inevitably based on simplifying assumptions, and[,] as additional facts become available[,] the estimates may require revision.

Karl Terzaghi

As discussed extensively in Chapter 2, excessive foundation settlements can impair the serviceability and even the safety of a structure. Chapter 2 focused on establishing criteria for tolerable settlements. This chapter concentrates on the analyses and methods used to estimate the settlement of shallow foundations. These settlements are due to either volume change or distortion of the soil and can take place immediately or over a time that could be measured in years. Immediate settlements are estimated using equations from elasticity theory, while long-term settlements are calculated using the concepts of consolidation and secondary compression in clays explored in Chapter 6. An essential design step consists of the comparison of these estimates with the tolerable settlements: a well-designed foundation will settle less than what is deemed tolerable.

Our ability to perform settlement computations has improved considerably since Terzaghi cautioned us against over-reliance on settlement predictions and about the need for updating them as needed. However, to different degrees, it is always true that a theory is based on simplifying assumptions. Unless the assumptions cross some boundary that separates what is realistic from what is not, that is not fatal to a given theory. The quote to some extent reflects the reality of the time: limited modeling capability, a material that is more variable than practically any other a civil engineer must contend with, and still incipient site investigation methods. It does, however, remind us to keep track of how realistic our calculations are and where uncertainties are concentrated.

9.1 TYPES OF SETTLEMENT

Consider the case of one-dimensional consolidation discussed in Chapter 6 for clays, where a distributed load p_v extending to infinity horizontally is applied on top of a soil mass (Figure 6.1b). If the soil is saturated, vertical settlements due to the load p_v will develop only if the soil contracts and water is expelled from the soil pores. This type of settlement is referred to as *consolidation settlement*.

A strip load of finite width, such as that shown in Figure 9.1, does generate an immediate settlement even in the absence of soil compression because the soil distorts under the applied load. This type of settlement is accordingly referred to as *distortion settlement*. Distortion settlement is not synonymous with *immediate settlement*. Sands (or any free-draining soils) experience settlement caused by both distortion and volume change (compression or

DOI: 10.1201/b22079-9

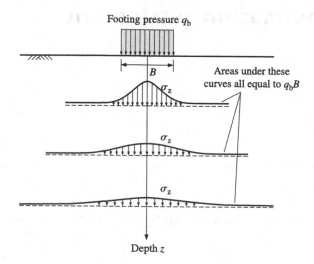

Figure 9.1 Loading q_b, with finite width, acting on top of soil mass and the stress distributions at various depths due to q_b.

consolidation) immediately after load application. In addition to consolidation and distortion settlements, some soils, particularly those with significant organic content, experience nonnegligible *secondary compression settlement*. This can be summarized through a simple equation:

$$w = w_d + w_c + w_s \tag{9.1}$$

where w=total settlement, w_d=distortion settlement, w_c=consolidation settlement, and w_s=secondary compression settlement.[1]

Another way to decompose the settlement experienced by a foundation is to write it in terms of the immediate (or short-term) and delayed (or long-term) settlements w_{st} and w_{lt}:

$$w = w_{st} + w_{lt} \tag{9.2}$$

In saturated clays, the immediate settlement is due purely to distortion, while the component of settlement that takes place over time is due to consolidation and secondary compression. Consolidation settlement is calculated as discussed in Chapter 6 for one-dimensional consolidation, but a correction is necessary for most foundations because conditions deviate significantly from one-dimensional conditions. In sands, in contrast, immediate settlement is due to both distortion and consolidation[2], while delayed settlement is caused by secondary compression. Delayed settlement in sand is often, but not always, neglected in settlement estimations. Table 9.1 summarizes the terms we have discussed in this section. Figure 9.2 illustrates the different stages of settlement of a footing in clay over time.

---

[1] A word on notation: we will use w without a subscript when we focus on the settlement (vertical displacement) of a foundation or surface of a soil mass. For the more general concept of displacement, we will also use w, but typically with a subscript to indicate the direction of the displacement. For example, in cylindrical coordinates, the displacement would have components w_r, w_z, and w_θ.

[2] Some authors associate the term "consolidation" with time-dependent volume change in clay. Physically, consolidation is simply change in volume. The fact that it takes place faster for some soils than for others (and practically immediately for clean sands) does not change the nature of the process.

Table 9.1 Sources of settlement of foundations

	Type of soil	
Settlement	Clay	Sand
Short-term or immediate	Distortion	Distortion Consolidation
Long-term or delayed	Consolidation Secondary compression	Secondary compression (small to negligible)

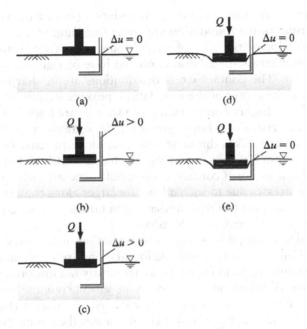

Figure 9.2 Settlement of a footing over time: (a) footing before loading; (b) immediate settlement and generation of excess pore pressures upon loading; (c) consolidation settlement in progress; (d) end of consolidation settlement (note that the excess pore pressures have completely dissipated); and (e) secondary compression.

9.2 INFLUENCE OF FOUNDATION STIFFNESS

Foundations are neither perfectly rigid, nor ideally flexible. Most isolated reinforced concrete foundations are designed in such a way that they are stiff, closer to rigid than flexible. Foundations supporting more than one column, on the other hand, particularly mats, are more flexible than isolated foundations. However, it helps us understand the problem if we consider the two extremes: perfectly rigid and perfectly flexible foundations.

Rigid foundations settle uniformly; that is, every point of the base of the foundation settles the same. Flexible foundations settle differently. Consider two identical flexible foundations bearing on sand in one case and clay in another. The load applied on them, which they transfer to the ground, is in both cases a uniformly distributed load. The settlement patterns for perfectly flexible foundations resting on saturated clay and sand are illustrated in Figure 9.3. Settlement in sand tends to be larger near the edges of the footing, as the confining stress and thus the stiffness there are less than those near the center. In clays, the stiffness at the moment of load application is the same across the entire plan of the foundation.

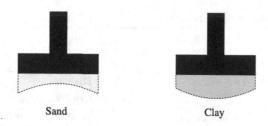

Sand   Clay

Figure 9.3 Settlement patterns of flexible footings on sand and clay.

Settlements are larger toward the center of the foundation because the stresses that develop at depth below the center of the foundation are larger (see Figure 9.1).

The stresses that develop at the base of a rigid foundation are not uniform. Figure 9.4a and b show the contact stress distributions on the base of rigid foundations resting on sand and saturated clay. The contact stress distributions are in sharp contrast. In sands, stresses are higher near the center of the foundation, peaking exactly at the center, because sand is stiffer there due to higher confinement. The stresses are higher at the center because the larger stiffness there requires a larger stress for the settlement to be the same as near the edges.[3] In clays, the stiffness is the same anywhere along the base of the foundation. In fact, the soil stiffness is roughly the same at any given depth, although it does increase with depth. This means that a uniform contact stress would generate larger displacements near the center because the stresses due to such a load are larger along the axis of the foundation (Figure 9.1). It follows that, for uniform settlement to take place, the contact stresses must not be uniform and must be larger near the edges.

In some types of calculations, it is necessary to know the contact stress distribution at the base of foundations. Considering that realistic foundations are somewhat rigid and that the contact stress distributions for rigid foundations are clearly not uniform, the question arises as to what form of stress distribution to assume in settlement computations. It is appropriate to consider a uniform distribution for a centrally loaded foundation (Figure 9.4c) because the fact that the foundation is not perfectly rigid attenuates the nonuniformity of the distribution. The nonuniformity is also reduced when soils are intermediate between the extreme cases of clean sand and pure clay, which is often the case. In any case, the error made in making the assumption of uniform distribution is rather small. This assumption is made throughout the text.

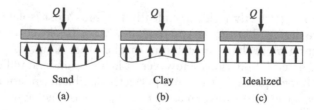

Sand   Clay   Idealized
(a)   (b)   (c)

Figure 9.4 Contact stress distribution: (a) beneath rigid foundation on sand, (b) beneath rigid foundation on clay, and (c) uniform, which is the idealized pressure distribution assumed in most settlement calculations.

---

[3] It should be noted that the sand yields (enters the plastic range) at the edges of a footing on the surface of a soil deposit due to the low strength associated with the low to zero confinement there.

9.3 APPROACHES TO SETTLEMENT COMPUTATION

Most settlement analyses are based on equations from elasticity theory. These equations are developed for finding both the stresses and displacements at every point of an elastic half-space caused by distributed loads of finite size that are applied on the boundary of the elastic half-space. An elastic half-space is a semi-infinite medium with elastic properties E and v. It represents the soil mass, with its free boundary representing the ground surface.

Some methods of displacement calculation are based on equations for the displacements at the base of the applied loads as a function of the magnitude of the applied load and its extent (that is, whether it is a point load or a distributed load, and over what area it is distributed). In such cases, we calculate the displacements of foundations directly. In the vast majority of cases, we are interested in the vertical displacement of the foundations, which we refer to as settlements. It is necessary in this approach to have representative values of the elastic constants E and v to calculate accurate settlements. Much of the engineering challenge in settlement computations resides in selecting values for E and v.

In other methods, settlements are obtained by first computing the stresses at various depths within the ground by using elastic solutions for the stresses as a function of the magnitude and size of the applied loads. From these stresses, the strains at the corresponding depths are calculated; this requires that we establish the values of the elastic constants at those depths. Again, selecting appropriate values for the elastic modulus and Poisson's ratio is the most challenging task of the whole procedure. We will discuss how to find these values when we examine specific settlement computation methods. The displacements are finally obtained by integration of the strains across the depth of influence of the load. This depth of influence is the depth within which most deformation takes place. We neglect the contribution of deformation at greater depths to foundation settlement because these are very small.

9.4 SETTLEMENT EQUATIONS FROM ELASTICITY THEORY

9.4.1 General form of the equations

The following general expression for the displacement w at the base of a distributed load q_b with a characteristic dimension B is obtained in the solution to elastic boundary-value problems:

$$w = I_1 I_2 \frac{q_b B}{E} \tag{9.3}$$

where I_1 is an influence factor dependent on the geometry of the load and the point where the displacement w is desired, I_2 is a function of the Poisson's ratio v of the soil, and E is the elastic modulus of the soil. The size B of the load may be the diameter of a circle, the width of a strip load, or the width of a rectangle with length L. We will always use the term settlement to mean the vertical displacement of a loaded area, of the base of a foundation, or of the ground surface.

9.5 SETTLEMENT OF FLEXIBLE FOUNDATIONS

9.5.1 Point load

Using Saint-Venant's principle, discussed in Chapter 4, it is sometimes possible to approximate a column load applied on a footing, for example, by a point load. The three components of the displacement at a point represented by its radial r, vertical z, and hoop θ coordinates due to a point load applied on the surface of a semi-infinite soil mass (see Figure 4.28) were found by Boussinesq to be given by

$$w_r = \frac{Q}{4\pi G\sqrt{r^2+z^2}}\left[\frac{rz}{r^2+z^2} - \frac{(1-2v)r}{\sqrt{r^2+z^2}+z}\right] \tag{9.4}$$

$$w_z = \frac{Q}{4\pi G\sqrt{r^2+z^2}}\left[2(1-v) + \frac{z^2}{r^2+z^2}\right] \tag{9.5}$$

$$w_\theta = 0 \tag{9.6}$$

where

$$R^2 = r^2 + z^2 \tag{9.7}$$

The settlement w of any point on the surface of the half-space is obtained by making $z=0$ in Equation (9.5), leading to:

$$w = \frac{Q(1-v^2)}{\pi ER} = \frac{Q(1-v^2)}{\pi Er} \tag{9.8}$$

The only point where we cannot use Equation (9.8) is the point of application of the load, where the equation gives settlement equal to infinity, a result that is not useful in a practical world where ideal point loads do not really exist.

Example 9.1

A 500 kN load is applied at a point on the surface of a homogeneous soil deposit. Find the vertical displacement (a) at a depth of 3 m and a horizontal distance of 4 m from the load and (b) at the surface 4 m away from the load. This soil has elastic constants $v=0.15$ and $E=10\,\text{MPa}$.

Solution:

a. The distance R from the point where the vertical displacement is to be calculated to the point of application of the load is:

$$R = \sqrt{r^2+z^2} = \sqrt{4^2+3^2} = 5$$

Using Equation (9.5) with $G = E/[2(1+v)]$:

$$w = \frac{Q(1+v)}{2\pi ER}\left[2(1-v) + \frac{z^2}{R^2}\right]$$

$$= \frac{500(1+0.15)}{2\pi(10000)5}\left[2(1-0.15) + \frac{3^2}{5^2}\right] = 0.0038\,\text{m} = 3.8\,\text{mm} \quad \text{answer}$$

b. In this case, because the point of interest is on the surface, we can use Equation (9.8):

$$w = \frac{Q(1-v^2)}{\pi ER} = \frac{500(1-0.15^2)}{\pi(10000)4} = 3.9 \times 10^{-3} \text{ m} = 3.9 \text{ mm} \quad \text{answer}$$

9.5.2 Uniform circular load

The circular load may be viewed as a collection of point loads, so integration of the Boussinesq problem, as discussed already in Chapter 4 (refer back to Figure 4.30), leads to the solution to the problem of the circular loads as well. Because of symmetry, the radial displacement w_r at any point beneath the center of the loaded area is equal to zero. The vertical displacement beneath the center of the loaded area can be expressed as a function of z as

$$w_z\big|_{r=0} = \frac{q_b B}{4G}\left[\frac{\sqrt{B^2+4z^2}}{B} - \frac{2z}{B}\right]\left[2(1-v) + \frac{2z}{\sqrt{B^2+4z^2}}\right] \tag{9.9}$$

Equation (9.9) may be rewritten in terms of nondimensional depth ζ as:

$$w = \frac{q_b B(1-v^2)}{E}\left(\sqrt{1+\zeta^2} - \zeta\right)\left(1 + \frac{\zeta}{2(1-v)\sqrt{1+\zeta^2}}\right) \tag{9.10}$$

where

$$\zeta = \frac{2z}{B} \tag{9.11}$$

At the base of the loaded area, that is, at $z=0$, we get the following for the displacement at the center of the loaded area:

$$w = w_z\big|_{r=0;z=0} = \frac{q_b B(1-v)}{2G} = \frac{q_b B(1-v^2)}{E} \tag{9.12}$$

The vertical displacement at the edge of the loaded area is:

$$w_z\big|_{r=\frac{B}{2};z=0} = \frac{2q_b B(1-v^2)}{\pi E} \tag{9.13}$$

The average vertical displacement underneath the loaded area is:

$$w_z\big|_{\text{avg},z=0} = \frac{0.85 q_b B(1-v^2)}{E} \tag{9.14}$$

Recalling that the strain is given by differentiation of Equation (9.9) according to:

$$\varepsilon_z = -\frac{dw_z}{dz}$$

and that G and E are related through

$$G = \frac{E}{2(1+v)}$$

we obtain the following equation for the vertical strain along the axis of the loaded area:

$$\left.\varepsilon_z\right|_{r=0} = \frac{q_b}{E}\left\{1-v(1+2v)+\frac{4v(1+v)\dfrac{z}{B}}{\sqrt{1+4\left(\dfrac{z}{B}\right)^2}}-\frac{8(1+v)\left(\dfrac{z}{B}\right)^3}{\left[1+4\left(\dfrac{z}{B}\right)^2\right]^{3/2}}\right\} \tag{9.15}$$

which is often expressed in terms of a strain influence factor I_z as

$$\left.\varepsilon_z\right|_{r=0} = \frac{q_b}{E}I_z \tag{9.16}$$

with

$$I_z = 1-v(1+2v)+\frac{4v(1+v)\dfrac{z}{B}}{\sqrt{1+4\left(\dfrac{z}{B}\right)^2}}-\frac{8(1+v)\left(\dfrac{z}{B}\right)^3}{\left[1+4\left(\dfrac{z}{B}\right)^2\right]^{3/2}} \tag{9.17}$$

The strain influence factor I_z is plotted in Figure 9.5 as a function of normalized depth for various values of Poisson's ratio. Note how the plots come together at around $z/B=1$. They differ in the value of I_z at $z/B=0$, which increases with decreasing Poisson's ratio from

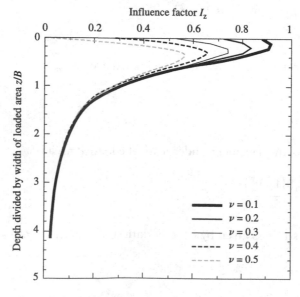

Figure 9.5 Strain influence factor I_z for a circular loaded area.

a value of 0 for $v=0.5$ to almost 1 for $v=0.1$, and in the peak value of I_z, which develops at $z/B=0.2$–0.4 approximately. These plots provide the conceptual basis for a much used method of foundation settlement calculation in sand that we will discuss later in this chapter.

Example 9.2

Determine the settlement of a flexible circular foundation of diameter 2 m and the vertical displacement of a point located below the center of the foundation at a depth of 5 m. The soil has properties $v=0.45$ and $E=10.5$ MPa. The axial load is 1200 kN.

Solution:

Because we have a circular foundation, we must use Equation (9.10) to calculate the vertical displacement of a point vertically below the center of the foundation:

$$w = \frac{q_b B\left(1-v^2\right)}{E}\left(\sqrt{1+\zeta^2}-\zeta\right)\left(1+\frac{\zeta}{2(1-v)\sqrt{1+\zeta^2}}\right)$$

$B = 2\,\mathrm{m}$

$$q_b = \frac{Q}{\pi\left(\dfrac{B}{2}\right)^2} = \frac{1200}{\pi\left(1\right)^2} = 382 \text{ kPa}$$

From Equation (9.11), for $z=5$ m:

$$\zeta = \frac{2z}{B} = \frac{2(5)}{2} = 5$$

Substituting ζ back into the equation for w:

$$w_{5\,\mathrm{m}} = \frac{382(2)\left(1-0.45^2\right)}{10500}\left(\sqrt{1+5^2}-5\right)\left(1+\frac{5}{2(1-0.45)\sqrt{1+5^2}}\right)$$

$$= (0.058)(0.099)(1.89)$$

$$= 10.9\times10^{-3}\,\mathrm{m} = 10.9 \text{ mm } ^{\text{answer}}$$

The same calculation with $z=0$ gives us the settlement of the foundation itself:

$$w = \frac{q_b B\left(1-v^2\right)}{E} = \frac{382\times2\times\left(1-0.45^2\right)}{10500} = 0.058 \text{ m} \approx 58 \text{ mm } ^{\text{answer}}$$

9.5.3 Rectangular load

Rectangular loads can often be used to approximate not only the loads on foundation elements, but also the loads of entire buildings. The settlement under the corner of a flexible rectangular load is given by:

$$w = I\frac{q_b B\left(1-v^2\right)}{E} \tag{9.18}$$

where I =influence factor given by:

$$I = \frac{1}{2\pi}\left\{m\left[\ln\left(\frac{\sqrt{1+m^2}+1}{\sqrt{1+m^2}-1}\right)\right]+\ln\left(\frac{\sqrt{1+m^2}+m}{\sqrt{1+m^2}-m}\right)\right\} \tag{9.19}$$

where $m=L/B$, in which B, L =dimensions of the rectangular load.

Example 9.3

For a flexible foundation with sides of 2 m and 3 m and an axial load of 750 kN, determine the settlement under one of the corners. The soil has properties $v=0.37$ and $E=9.8\,$MPa.

Solution:

To find the settlement under the corner of a rectangular foundation, we can use Equations (9.18) and (9.19). The settlement is given by:

$$w = I\frac{q_b B\left(1-v^2\right)}{E}$$

where the influence factor I is calculated as a function of m:

$$I = \frac{1}{2\pi}\left\{m\left[\ln\left(\frac{\sqrt{1+m^2}+1}{\sqrt{1+m^2}-1}\right)\right]+\ln\left(\frac{\sqrt{1+m^2}+m}{\sqrt{1+m^2}-m}\right)\right\}$$

$$m = \frac{L}{B} = \frac{3}{2} = 1.5$$

$$I = \frac{1}{2\pi}\left(1.875+2.390\right) = 0.679$$

The unit load q_b is given as:

$$q_b = \frac{Q}{A} = \frac{750}{2\times 3} = 125 \text{ kPa}$$

So

$$w = \frac{0.679\times 125\times 2\times\left(1-0.37^2\right)}{9800} = 14.9\times 10^{-3} \text{ m} \approx 15\text{mm} \quad \text{answer}$$

The settlement at the center of a flexible rectangular load can be calculated by superposing the settlements at the corners of four rectangles with total area equal to the area of the rectangle for which the settlement is desired. Equation (9.12) may also be used to estimate the settlement at the center of both square and rectangular loaded areas. To do that, we need to use an equivalent circle with the same area as the square or rectangle.

The average vertical displacement at the base of a rectangular loaded area with plan dimensions B and L is:

$$w_{\text{avg}} = C\frac{Q\left(1-v^2\right)}{E\sqrt{BL}}$$

Table 9.2 Coefficient C for the calculation of average settlement of a rectangular loaded area

Circle	L/B =						
	1	1.5	2	3	5	10	100
0.96	0.95	0.94	0.92	0.88	0.82	0.71	0.37

Modified after Saada (2009).

where C is a coefficient whose values are given in Table 9.2.

9.5.4 Settlement of rigid foundations

The settlement of a rigid cylinder under a load Q is:

$$w = w_z|_{z=0} = \frac{Q(1-v^2)}{BE} \tag{9.20}$$

Equation (9.20) may also be written in terms of the average distributed load q_b:

$$w = \frac{\pi}{4} \frac{q_b B}{E} (1 - v^2) \tag{9.21}$$

The settlement of other rigid areas can be estimated from the settlements calculated at the center and corner or edge of equivalent flexible areas as

$$w_{rigid} = \frac{1}{2} \left(w_{center} + w_{edge} \right)_{flexible} \tag{9.22}$$

for circular or strip foundations, and

$$w_{rigid} = \frac{1}{3} \left(2w_{center} + w_{corner} \right)_{flexible} \tag{9.23}$$

for square foundations.

9.6 SETTLEMENT OF SHALLOW FOUNDATIONS ON SAND

9.6.1 SPT-based methods

There are several methods available for the calculation of footing settlements using SPT results. Most of these methods use equations that are based on elasticity theory.

9.6.1.1 Meyerhof's method

Meyerhof (1965) suggested a relationship for settlements of spread footings on sand that is presented next in terms of energy-corrected SPT blow counts and normalized quantities:

$$\frac{w}{L_R} = \frac{0.152}{\min\left(1 + \dfrac{D}{3B}, 1.33\right) N_{60}} \left(\frac{q_b - \sigma'_{vp}|_{z_f=0}}{p_A} \right) \tag{9.24}$$

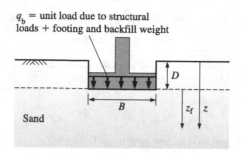

q_b = unit load due to structural
loads + footing and backfill weight

Figure 9.6 Depth measured from footing base level.

for $B \le 1.2L_R$, and

$$\frac{w}{L_R} = \frac{0.229}{\min\left(1+\dfrac{D}{3B},\ 1.33\right)N_{60}}\left(\frac{q_b - \sigma'_{vp}\big|_{z_f=0}}{p_A}\right)\left(\frac{B}{B+0.305L_R}\right)^2 \tag{9.25}$$

for $B > 1.2L_R$, where w=footing settlement, z_f=depth measured from the level of the base of the footing (see Figure 9.6), q_b=gross unit load at the base of the footing (including both structural loads and the weight of the backfill and foundation element), $\sigma'_{vp}\big|_{z_f=0}$=maximum previous vertical effective stress experienced by the soil at the footing base level, N_{60}=average SPT blow count at 60% energy ratio over a depth of $1B$ below the footing base for square footings and $2B$ below the footing base for strip footings, B=footing width, L_R=reference length=1 m=3.281 ft=39.37 in., and p_A=reference stress=100 kPa $\approx$ 1 tsf. In Equations (9.24) and (9.25), the SPT N values are not corrected for the water table or overburden pressure and the $\min[1+D/(3B), 1.33]$ term is a depth factor that attempts to account for reduced settlement when the footing is embedded in the soil a depth equal to D, all else being the same. There is a small discontinuity at $B=1.2L_R$. Additionally, Equations (9.24) and (9.25) were intended for the calculation of settlements under working loads, so they should not be expected to be accurate for loads that are either much smaller than the limit bearing resistance (the load at which the footing will plunge into the ground, which we will discuss in Chapter 10), or close to it. The equations therefore are supposed to be used in design, when we are looking for the size of the footing that will give us a settlement that is close to the tolerable values we discussed in Chapter 2.

Many settlement estimation methods subtract from the gross (total) unit load q_b on the base of the footing the maximum stress ever experienced by the soil at that depth or some number related to that. The purpose is to account for the fact that the footing will settle as a result of stress that exceeds a stress the soil was already subjected to in the past (much like the concept of compressing a soil element along a reloading curve onto the normal consolidation line, as discussed in Chapter 6).

9.6.1.2 Peck and Bazaraa's method

Peck and Bazaraa (1969) proposed the following relationship, a modification of the Meyerhof method, for estimating the settlements of footings in sand:

$$w = C_{gw}\left(\frac{0.229L_R}{N_B}\right)\left(\frac{q_b - \sigma'_{vp}\big|_{z_f=0}}{p_A}\right)\left(\frac{B}{B+0.305L_R}\right)^2 \tag{9.26}$$

where q_b=gross unit load at the base of the footing (including both structural loads and the weight of the backfill and foundation element), $\sigma'_{vp}\big|_{z_f=0}$=maximum previous vertical effective stress at footing base level, B=footing width, L_R=reference length=1 m=3.281 ft=39.37 in., and p_A=reference stress=100 kPa $\approx$ 1 tsf.

The stress-normalized SPT N value N_B is given by

$$N_B = \left(\frac{3N_{60}}{1+4\dfrac{\sigma'_v}{p_A}} \right) \frac{\sigma'_v}{p_A} \tag{9.27}$$

for $\sigma'_v \leq 0.75p_A$, and

$$N_B = \left(\frac{3N_{60}}{3.25+\dfrac{\sigma'_v}{p_A}} \right) \frac{\sigma'_v}{p_A} \tag{9.28}$$

for $\sigma'_v > 0.75p_A$, where N_{60}=representative SPT N value at 60% energy ratio. One way of obtaining N_{60} is by taking an average of the energy-corrected blow counts over a representative depth below the footing base, as we did for the Meyerhof (1965) method. Equation (9.26) was based on the assumption that settlements predicted by the Terzaghi and Peck (1967) correlation would produce excessively conservative results (that is, excessively large settlements).

The groundwater correction factor C_{gw} is simply the ratio of the vertical effective stress at a depth of $B/2$ below the base of the footing when the water table is deep to that when it is within the zone of influence for the footing:

$$C_{gw} = \frac{\left(\sigma'_v\big|_{z_f=\frac{B}{2}} \right)_{\text{without WT}}}{\left(\sigma'_v\big|_{z_f=\frac{B}{2}} \right)_{\text{with WT}}} \tag{9.29}$$

The Peck and Bazaraa (1969) method should be corrected for embedment, although exactly what this correction should be was not elaborated on by the authors of the equations. Presumably, the same correction suggested by Meyerhof (1965) would be appropriate.

9.6.1.3 Burland and Burbidge's method

Another slightly more modern method for estimating settlements of footings in sand or gravel was proposed by Burland and Burbidge (1985):

$$\frac{w}{L_R} = 0.10 f_s f_L f_t I_c \frac{q_b - \frac{2}{3}\sigma'_{vp}\big|_{z_f=0}}{p_A} \left(\frac{B}{L_R} \right)^{0.7} \tag{9.30}$$

where f_s=shape factor, f_L=layer thickness factor, f_t=time factor, q_b=gross unit load at the footing base (including both structural loads and the weight of the backfill and foundation element), $\sigma'_{vp}\big|_{z_f=0}$ =maximum previous vertical effective stress at footing base level, z_f=depth measured from foundation base level, B=footing width, I_c=compressibility index, L_R=reference length=1 m=3.281 ft=39.37 in., and p_A=reference stress=100 kPa ≈ 1 tsf.

The use of $\sigma'_{vp}\big|_{z_f=0}$ in Equations (9.24)–(9.26) and (9.30) allows for the fact that any measurable settlement will only develop after the vertical effective stress exceeds the maximum vertical effective stress ever experienced by the soil below the foundation base.

Burland and Burbidge (1985) proposed the values of the compressibility index I_c as a function of SPT blow count:

$$I_c = \frac{1.71}{\overline{N}^{1.4}} \tag{9.31}$$

where $\overline{N}$=average blow count over the depth of influence z_{f0} below the footing base. The authors are not explicit as to whether to normalize blow counts, but we may assume $\overline{N}$ to be an energy-corrected, standardized blow count N_{60}. The depth of influence can be calculated from:

$$\frac{z_{f0}}{L_R} = \left(\frac{B}{L_R}\right)^{0.79} \tag{9.32}$$

The blow counts are corrected as follows:

1. For gravel or sandy gravel, increase the measured blow counts by 25%.
2. For sands below the water table, any excess of the measured blow count over 15 blows should be halved (for example, 21 would really be 15+6/2=15+3=18).

The shape factor is given by:

$$f_s = \left(\frac{1.25\dfrac{L}{B}}{\dfrac{L}{B}+0.25}\right)^2 \tag{9.33}$$

The sand layer thickness factor f_L is used to account for cases in which the depth of influence z_{f0} is larger than the actual thickness H of the sand layer. This means that, if a very hard layer (typically bedrock) exists at a depth smaller than the depth of influence, the settlement should be less; this result is obtained in the method by using f_L. This factor is given as

$$f_L = \begin{cases} \dfrac{H}{z_{f0}}\left(2-\dfrac{H}{z_{f0}}\right) & \text{if } H \leq z_{f0} \\ 1 & \text{if } H > z_{f0} \end{cases} \tag{9.34}$$

The time factor is based on the observations of increasing settlement with time by Burland and Burbidge (1985). Some like to refer to this as a creep factor, but this factor is also intended to capture the effect, particularly on tall structures (such as tall chimneys, silos, and high-rise buildings) of wind loads, which is not creep in the strict definition of the term. The repeated application of wind loads can always generate some additional settlement.

Over time, these can add up to a nonnegligible amount. Burland and Burbidge (1985) proposed the following expression for f_t:

$$f_t = \left(1 + R_3 + R_t \log\frac{t}{3}\right) \tag{9.35}$$

where t=time in years, R_3=ratio of settlement developing over a period of 3 years to the immediate settlement, and R_t=ratio of settlement developing over a log cycle of time to the immediate settlement. It is not necessarily easy to estimate the values of R_3 and R_t because precise measurements of immediate versus long-term settlement are not available. A design based on f_t=1 is typically acceptable, although Burland and Burbidge give values of R_t=0.2 and R_3=0.3.

Example 9.4

A 6 ft square footing (see E-Figure 9.1), which is backfilled, is embedded 2.5 ft into a sand. SPT N_{60} values for the sand are given in E-Table 9.1. Estimate the settlement of this footing caused by a column load of 182 tons. The sand is approximately 40 ft thick and has an average unit weight of 125 pcf. It is underlain by a shale bedrock. Consider that the worst-case scenario is for the water table to be located at the level of the base of the footing, as shown in the figure. Example 10.15 will examine the limit bearing capacity of this same footing.

Solution:

We will use the Burland and Burbidge method, for which we use Equation (9.30):

$$\frac{w}{L_R} = 0.10 f_s f_L f_t I_c \frac{q_b - \frac{2}{3}\sigma'_{vp}\big|_{z_f=0}}{p_A}\left(\frac{B}{L_R}\right)^{0.7}$$

E-Table 9.1 Corrected SPT blow counts, normalized for 60% energy ratio for Example 9.4

Depth (ft)	N_{60}
2.5	3
5	5
7.5	12
10	14
12.5	15
15	14
17.5	12
20	16
22.5	16
25	18
27.5	13
30	15
32.5	17
35	19
37.5	20

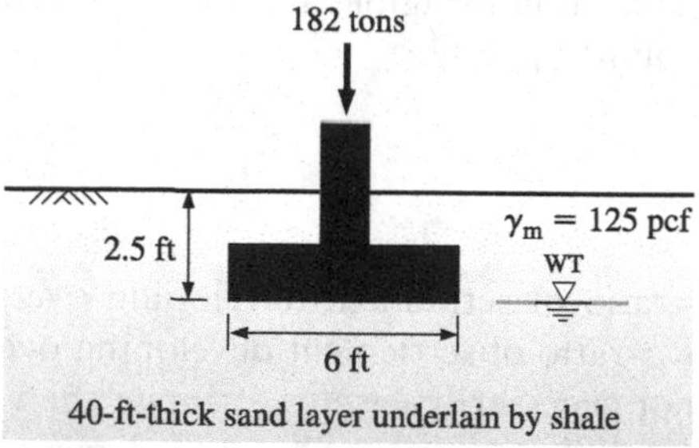

E-Figure 9.1 Footing for Examples 9.4 and 10.15.

The correction factors are all equal to 1: the shape factor is 1 because the footing is square, as seen from Equation (9.33); the layer thickness factor is 1 because the sand layer is sufficiently deep, as seen from Equation (9.34); and the time factor is 1 because we are ignoring settlement increases over time.

The depth of influence is calculated using Equation (9.32):

$$\frac{z_{f0}}{L_R} = \frac{z_{f0}}{3.281} = \left(\frac{B}{L_R}\right)^{0.79} = \left(\frac{6}{3.281}\right)^{0.79} = 1.61$$

from which

$$z_{f0} = 1.61 \times L_R = 1.61 \times 3.281 = 5.3 \text{ ft}$$

The blow counts must thus be averaged between a depth of 2.5 and 2.5 + 5.3 ft or 2.5 and approximately 8 ft. This gives:

$$\bar{N} = \frac{1}{3}(3+5+12) = 6.7$$

The factor I_c is calculated using Equation (9.31) as

$$I_c = \frac{1.71}{\bar{N}^{1.4}} = \frac{1.71}{6.7^{1.4}} = 0.12$$

The maximum previous vertical effective stress at the footing base level is:

$$\sigma'_{vp}\big|_{z_f=0} = 2.5 \times \left(125 \text{ pcf} \times \frac{1 \text{ ton}}{2000 \text{ lb}}\right) = 0.156 \text{ tsf}$$

and the gross unit load at the base of the footing is:

$$q_b = \frac{Q}{BL} + \sigma'_{vp}\big|_{z_f=0} = \frac{182 \text{ tons}}{36 \text{ ft}^2} + 0.156 = 5.2 \text{ tsf}$$

So we are finally able to calculate the footing settlement:

$$\frac{w}{3.281 \text{ ft}} = 0.10 \times 1 \times 1 \times 1 \times 0.12 \times \frac{5.2 - \frac{2}{3}(0.156)}{1} \times \left(\frac{6}{3.281}\right)^{0.7} = 0.093$$

This gives:

$$w = 0.093 \times 3.281 = 0.305 \text{ ft} = 3.7 \text{ in.}^{\text{answer}}$$

This value of settlement is considered excessive in most applications. The implication of this result is that 182 tons is clearly impractical for this footing and this soil profile. We will revisit this example as two end-of-chapter problems and in Chapters 10 (as an additional example) and 11 (in the form of two problems).

Example 9.5

A square footing carrying a load of 600 kN is to be built on sand and embedded 1 m below the ground surface. To ensure a limit state is not reached, the settlement is limited to 25 mm. An SPT was conducted with a safety hammer with the following results:

Depth (m)	N_{60}
1	14
2	18
3	17
4	19
5	22
6	24
7	26

Determine an appropriate footing width.

Solution:

We will use the Meyerhof (1965) method. We must start with an estimated size. Let us choose $B = 2$ m and proceed from there. Because the depth of influence for this method is B below the embedded depth, we need to calculate the average blow count values from depths 1–3 m:

$$N_{60} = \frac{14 + 18 + 17}{3} = 16.3 \approx 16$$

Using our estimated area, the added stress at the base of the foundation is:

$$q_b - \sigma'_{vp}\big|_{z_f=0} = \frac{Q}{A} = \frac{600}{2 \times 2} = 150 \text{ kPa}$$

Since our $B > 1.2 L_R$, we use Equation (9.25):

$$\frac{w}{L_R} = \frac{0.229}{\min\left(1 + \dfrac{D}{3B}, \, 1.33\right) N_{60}} \left(\frac{q_b - \sigma'_{vp}\big|_{z_f=0}}{p_A}\right)\left(\frac{B}{B + 0.305 L_R}\right)^2$$

$$= \frac{0.229}{\min\left(1 + \dfrac{1}{3 \times 2}, \, 1.33\right) \times 16} \times \left(\frac{150}{100}\right) \times \left(\frac{2}{2 + 0.305 \times 1}\right)^2$$

$$= \frac{0.229}{16 \times 1.167} \times \left(\frac{150}{100}\right) \times \left(\frac{2}{2 + 0.305}\right)^2$$

$$= 14 \times 10^{-3}$$

This corresponds to a settlement of 14 mm. Therefore, our footing is overdesigned at 2 m. Let us try a smaller dimension, say, $B = 1$ m. Now $q_b = 600$ kN/1 m² = 600 kPa. Again, because we are looking at B, or 1 m below the embedded depth, we take the corresponding average blow count value:

$$N_{60} = \frac{14 + 18}{2} = 16$$

This time, since $B < 1.2 L_R$, we use Equation (9.24):

$$\frac{w}{L_R} = \frac{0.152}{\min\left(1 + \dfrac{D}{3B}, 1.33\right) N_{60}} \left(\frac{q_b - \sigma'_{vp}\big|_{z_f = 0}}{p_A}\right)$$

$$= \frac{0.152}{\min\left(1 + \dfrac{1}{3 \times 1}, 1.33\right) \times 16} \times \left(\frac{600}{100}\right)$$

$$= \frac{0.152}{16 \times 1.33} \times 6$$

$$= 43 \times 10^{-3}$$

Continuing with the iterations, we would find that $B = 1.35$ m for the 25 mm maximum settlement.

9.6.2 CPT-based methods

9.6.2.1 Schmertmann's method

The method of choice for the calculation of the settlement of shallow foundations on sand is that of Schmertmann (1970) and Schmertmann et al. (1978). The fundamental basis for the method is the fact that strains start at some value at the base of the loaded area, peak at some depth below the loaded area, and then approach zero, as we observed earlier in this chapter for a circular load on the surface of a soil mass (Figure 9.5).

The strain influence diagrams for strip, square, and circular footings have a shape that resembles that of Figure 9.5, but it is useful in design to obtain simplified, generalized influence factor diagrams. These may be obtained from two fundamental postulates regarding the soil deformations beneath the footing. The first postulate is that the settlement of footings is mostly due to deformations within an influence depth z_{f0}, measured from the base of the footing. This depth z_{f0} is taken as $2B$ for square or circular footings and $4B$ for footings with L equal to $10B$ or greater (strip footings), where B is the width or diameter of the footing and L is the length of a rectangular footing. The depth z_{f0} is interpolated for rectangular footings with $1 \le L/B \le 10$ using the following equation:

$$\frac{z_{f0}}{B} = 2 + 0.222\left(\frac{L}{B} - 1\right) \le 4 \tag{9.36}$$

The second postulate is that the deformations start small at the footing base, peak at a depth equal to one-fourth the depth of influence z_{f0}, and then decrease gradually to zero at z_{f0}. An influence diagram, developed based on this observation, is shown in Figure 9.7. The influence diagram provides values of the influence factor I_z as a function of depth. Note how the influence factor starts from I_{z0} at the footing base, increases linearly to its peak value I_{zp},

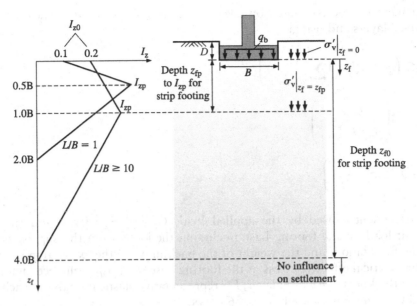

Figure 9.7 Influence diagram for settlement computations.

and then drops to zero at depth z_{f0}. The values of the influence factor I_{z0} at the base of the foundation, of the depth z_{fp} below the foundation at which the influence factor peaks, and of the peak influence factor I_{zp} are given by

$$I_{z0} = 0.1 + 0.0111\left(\frac{L}{B} - 1\right) \le 0.2 \tag{9.37}$$

$$\frac{z_{fp}}{B} = 0.5 + 0.0555\left(\frac{L}{B} - 1\right) \le 1 \tag{9.38}$$

$$I_{zp} = 0.5 + 0.1\sqrt{\frac{q_b - \sigma'_v|_{z_f=0}}{\sigma'_v|_{z_f=z_{fp}}}} \tag{9.39}$$

where the initial vertical effective stress σ'_v (before footing construction) is considered at the level of the foundation base $\sigma'_v|_{z_f=0}$ and at the depth at which the influence factor peaks $\sigma'_v|_{z_f=z_{fp}}$.

The influence factor I_z at any depth can now be calculated as

$$I_z = I_{z0} + \frac{z_f}{z_{fp}}\left(I_{zp} - I_{z0}\right) \tag{9.40}$$

for depths z_f below the foundation base such that $z_f < z_{fp}$, and

$$I_z = \frac{z_{f0} - z_f}{z_{f0} - z_{fp}} I_{zp} \tag{9.41}$$

for $z_{fp} \le z_f \le z_{f0}$.

The settlement can be computed by discretizing the soil profile underneath the footing into several sublayers and using:

$$w = C_1 C_2 \left(q_b - \sigma'_v \big|_{z_f=0} \right) \sum \left(\frac{I_{zi} \Delta z_i}{E_i} \right) \tag{9.42}$$

$$C_1 = 1 - 0.5 \left(\frac{\sigma'_v \big|_{z_f=0}}{q_b - \sigma'_v \big|_{z_f=0}} \right) \tag{9.43}$$

$$C_2 = 1 + 0.2 \log \left(\frac{t}{0.1 t_R} \right) \tag{9.44}$$

where w=settlement caused by the applied load, C_1 and C_2=depth and time factors, q_b=gross unit load on the footing base (including the loads from the superstructure, the weight of the foundation, and the weight of the backfill when the excavation is backfilled), $\sigma'_v \big|_{z_f=0}$=initial vertical effective stress at the footing base level, I_{zi}=influence factor for each sublayer, Δz_i=thickness of each sublayer, E_i=representative elastic modulus of each sublayer, t=time, and t_R=reference time=1 year=365 days.

The depth factor accounts for the effects of embedment on footing settlement, compared with what it would be if the footing were located on the ground surface. Naturally, the settlement decreases as the embedment depth and thus $\sigma'_v \big|_{z_f=0}$ increase. The time factor attempts to account for observations that the settlement would tend to increase with time, even in sands. Short-term settlement, in this context, would be the settlement observed about a month (0.1 year) after the application of the load to the footing. Note that there is no need for a shape factor to account for L/B ratios between 1 (circular or square footings) and 10 (strip footings). The interpolated values of z_{fp}, z_{f0}, and I_{zp} account for that.

In the model load tests discussed by Schmertmann et al. (1978), the settlement of strip foundations is greater than that of circular or square foundations under the same unit load q_b. This results from the larger influence depth for strip foundations, which outweighs the slightly higher moduli of the supporting soil in the case of strip foundations resulting from the larger stresses generated by a strip load for the same value of q_b.

The soil modulus for each sublayer is determined based on the cone resistance q_c. The original correlations between the elastic modulus E_i and the cone resistance q_{ci} for any sublayer i were taken as (Schmertmann et al. 1978; Robertson and Campanella 1989)[4]:

$$E_i = \begin{cases} 2.5 q_{ci} & \text{for young normally consolidated silica sand} \\ 3.5 q_{ci} & \text{for aged normally consolidated silica sand} \\ 6.0 q_{ci} & \text{for overconsolidated silica sand} \end{cases} \tag{9.45}$$

The values of E_i proposed in Equation (9.45) are not a function of the footing settlement. In reality, the values of E_i drop as settlement increases, due to the nonlinearity of soil stress–strain relationships (refer to Chapter 5). Because it is based on linear elasticity, the load–settlement response obtained by Schmertmann's method with E_i given by Equation (9.45) leads to unconservative results if the load or settlement increases beyond the range to which the assumed elasticity modulus applies.

---

[4] It is acceptable to use other correlations that the engineer feels comfortable with; there has been progress in the estimation of soil stiffness from *in situ* tests since these original correlations were proposed.

9.6.2.2 Lee et al.'s method

A more general expression for E_i in terms of cone resistance is given by Lee and Salgado (2002):

$$E_i = \varphi_i q_{ci} \tag{9.46}$$

where φ_i = a parameter for sublayer i depending on the footing settlement, the footing geometry, and the relative density; and q_{ci} = representative cone resistance of sublayer i.

As discussed in Chapter 2, a limiting settlement for footings in sand of 25 mm ($\approx$1 in.) has traditionally been used in foundation design practice. The parameter φ_i allows flexibility in setting tolerable settlements. We can now stipulate a tolerable settlement less than or greater than 25 mm and calculate either the load on a given footing or the area of the footing for a given load corresponding to that settlement.

Lee et al. (2008) perfected the Lee and Salgado (2002) method with the aid of nonlinear finite element analyses. The analyses revealed that the influence diagrams from Schmertmann et al. (1978) must be revised. The revised equations for the depth to the peak influence factor and to the bottom of the influence zone are (Sakleshpur et al. 2021):

$$\frac{z_{f0}}{B} = 2 + 0.4\left[\min\left(\frac{L}{B}, 6\right) - 1\right] \tag{9.47}$$

$$\frac{z_{fp}}{B} = 0.5 + 0.1\left[\min\left(\frac{L}{B}, 6\right) - 1\right] \tag{9.48}$$

The parameter φ_i for sublayer i can be determined using the following equation, which was proposed by Sakleshpur et al. (2021) based on the work by Lee and Salgado (2002):

$$\varphi_i = \lambda \left(\frac{w}{L_R}\right)^{-0.285} \left(\frac{B}{L_R}\right)^{0.4} \left(\frac{D_R}{100}\right)^{-0.65} \tag{9.49}$$

where

$$\lambda = \begin{cases} 0.38 & \text{for young normally consolidated silica sand} \\ 0.53 & \text{for aged normally consolidated silica sand} \\ 0.91 & \text{for overconsolidated silica sand} \end{cases} \tag{9.50}$$

where w = footing settlement, B = width or diameter of the footing, D_R = relative density (expressed as a percentage), and L_R = reference length = 1 m = 3.281 ft = 39.37 in.

The value of φ_i varies according to the footing settlement level, footing size, and relative density. The larger the footing size is, the larger the value of φ_i will be due to the higher confinement imposed by a more extensive foundation. As the relative density increases, the rate of modulus degradation intensifies and φ_i drops accordingly. It is interesting to note that Lee and Salgado (2002) found that, for the assumed strain influence diagrams proposed by Schmertmann et al. (1978), the values of φ_i for strip footings are almost exactly the same as for isolated footings. This means the elasticity modulus E_i for each sublayer is the same whether the footing has $L/B=1$ or a very large L/B.[5]

The appropriate value of φ_i for a given settlement can be determined for both strip and square footings using Equations (9.49) and (9.50). It can then be used for footing settlement checks in sand following these steps:

1. Define the tolerable settlement.
2. Calculate the depth of influence using Equation (9.47).
3. Divide the influence zone in sublayers with similar q_c values.
4. From the CPT cone resistance profile, estimate the representative value of cone resistance q_{ci} for each sublayer.
5. Calculate I_{z0} using Equation (9.37), z_{fp} using Equation (9.48), I_{zp} using Equation (9.39), and I_z using either Equation (9.40) or Equation (9.41).
6. Estimate a representative value of relative density for each sublayer (refer to Chapter 7 for how to estimate D_R from q_c).
7. For each sublayer, determine φ_i for the given D_R, B, and tolerable settlement using Equations (9.49) and (9.50).
8. Determine the elastic modulus E_i (from q_{ci}) for each sublayer using Equation (9.46).
9. Calculate the settlement corresponding to the load acting on the footing using Equation (9.42). Compare with the tolerable settlement established in step (1).

Example 9.6

A square footing is embedded 3 m into a young, normally consolidated silica sand with CPT values given in E-Figure 9.2. The unit weight of the sand is 21 kN/m³, and the critical-state friction angle is 33°. Assume that the coefficient of earth pressure at rest is 0.4 and that the worst-case scenario is for the water table to be located at the level of the base of the footing. Considering a column load of 1200 kN, find the area of the footing so that the settlement will not be more than 38 mm.

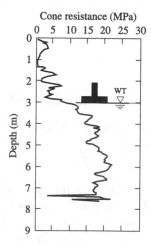

E-Figure 9.2 Cone resistance versus depth for Example 9.6.

---

[5] Note that these values of modulus are simply stiffness indices to be used in connection with the assumed strain influence diagrams; the soil, in reality, should be stiffer under plane-strain conditions (that is, for strip footings).

Solution:

We will follow the procedure described previously in this section. Note that the ultimate goal in such problems is to find the summation:

$$\sum \left(\frac{I_{zi} \Delta z_i}{E_i} \right)$$

to use it in Equation (9.42):

$$w = C_1 C_2 \left(q_b - \sigma_v' \big|_{zf=0} \right) \sum \left(\frac{I_{zi} \Delta z_i}{E_i} \right)$$

Therefore, it is useful to create a table that will include all the necessary elements to end up with the preceding summation. Such a table would look like the following:

Sublayer	z_{top}	z_{bottom}	z_{middle}	Δz_i	q_{ci}	σ_v'	σ_h'	D_{Ri}	φ_i	E_i	I_{zi}	$\dfrac{I_{zi} \Delta z_i}{E_i}$
1	...	...	...	...	...	...	...	...	...	...	...	...
...	...	...	...	...	...	...	...	...	...	...	...	...
i	...	...	...	...	...	...	...	...	...	...	...	...
...	...	...	...	...	...	...	...	...	...	...	...	...
...	...	...	...	...	...	...	...	...	...	...		...

$$\sum \left(\frac{I_{zi} \Delta z_i}{E_i} \right) = \dots$$

The first step in finding the solution to such problems is to determine the tolerable settlement. In our case, it is already given:

$$w_{max} = 38 \, \text{mm}$$

The next step is to divide the influence zone into sublayers with similar q_c values. To do this, we need to make an assumption regarding the width of the footing. We assume that it is:

$$B = L = 2 \, \text{m}$$

Note that we assume that a footing with a width equal to this value of B is acceptable with respect to bearing capacity. This would of course have to be checked separately. Using Equation (9.47) with $B=L$, the depth of influence is:

$$z_{f0} = B \left\{ 2 + 0.4 \left[\min \left(\frac{L}{B}, 6 \right) - 1 \right] \right\} = 2 \times \left\{ 2 + 0.4 \left[\min \left(\frac{2}{2}, 6 \right) - 1 \right] \right\} = 4 \, \text{m}$$

We are therefore interested in the region between 3 m (that is, the base of the footing) and 7 m (that is, 3 m plus the influence depth of 4 m starting from the base of the footing). Considering E-Figure 9.2, we divide the influence zone into five sublayers, each one with its own approximate CPT value. Note that we divide the influence depth so that the CPT values within each layer are either approximately constant or approximately linear with depth so that a representative value may be easily calculated. We start filling the table as follows:

Sublayer i	z_{top} (m)	z_{bottom} (m)	Δz_i (m)	q_{ci} (MPa)
1	3.0	3.5	0.5	12
2	3.5	4.0	0.5	15
3	4.0	5.0	1.0	17
4	5.0	6.0	1.0	20
5	6.0	7.0	1.0	18

Now we need to estimate a representative value of relative density for each sublayer. We can do this by using either Figure 7.24 or Equation (7.20), as discussed in Chapter 7. To use the figure, we need to calculate the lateral effective stresses. We do so at the midpoint of each sublayer, because there we have the representative value of relative density for each sublayer. For the first sublayer, we have:

$$\sigma'_{v(1)} = 21 \times 3 + (21 - 9.81) \times 0.25 = 66 \text{ kPa}$$

So

$$\sigma'_{h(1)} = K_0 \sigma'_{v(1)} = 0.4 \times 66 \text{ kPa} = 26 \text{ kPa}$$

Following the same process for all sublayers, we find that:

$$\sigma'_{h(2)} = 29 \text{ kPa}$$

$$\sigma'_{h(3)} = 32 \text{ kPa}$$

$$\sigma'_{h(4)} = 36 \text{ kPa}$$

$$\sigma'_{h(5)} = 41 \text{ kPa}$$

We are now ready to estimate the relative density in the same way we did in Chapter 7. For each sublayer, we find the following representative relative density:

$$D_{R(1)} = 76\%$$

$$D_{R(2)} = 83\%$$

$$D_{R(3)} = 86\%$$

$$D_{R(4)} = 90\%$$

$$D_{R(5)} = 84\%$$

The next step is to determine φ_i for each of the five sublayers, for the given tolerable settlement, using Equations (9.49) and (9.50). For a 2 m footing in a young, normally consolidated silica sand and a tolerable settlement of 38 mm, we have:

$$\varphi_{(1)} = \lambda \left(\frac{w}{L_R} \right)^{-0.285} \left(\frac{B}{L_R} \right)^{0.4} \left(\frac{D_{R(1)}}{100} \right)^{-0.65} = 0.38 \left(\frac{38}{1000} \right)^{-0.285} \left(\frac{2}{1} \right)^{0.4} \left(\frac{76}{100} \right)^{-0.65} = 1.52$$

$$\varphi_{(2)} = 1.44$$

$$\varphi_{(3)} = 1.40$$

$$\varphi_{(4)} = 1.36$$

$$\varphi_{(5)} = 1.43$$

Using Equation (9.46), we can now determine the elastic modulus E_i for each sublayer:

$$E_{(1)} = \varphi_{(1)} q_{c(1)} = 1.52 \times 12 = 18.2 \text{ MPa}$$

$$E_{(2)} = 1.44 \times 15 = 21.6 \text{ MPa}$$

$$E_{(3)} = 1.40 \times 17 = 23.8 \text{ MPa}$$

$$E_{(4)} = 1.36 \times 20 = 27.2 \text{ MPa}$$

$$E_{(5)} = 1.43 \times 18 = 25.7 \text{ MPa}$$

The last thing that remains to be calculated is the influence factor I_{zi}. We do so using Equation (9.40) or Equation (9.41), depending on depth; that is:

$$I_z = I_{z0} + \frac{z_f}{z_{fp}}\left(I_{zp} - I_{z0}\right) \text{ for } z_f \leq z_{fp}$$

or

$$I_z = \frac{z_{f0} - z_f}{z_{f0} - z_{fp}} I_{zp} \text{ for } z_{fp} \leq z_f \leq z_{f0}$$

Using Equation (9.37), we get:

$$I_{z0} = 0.1 + 0.0111\left(\frac{L}{B} - 1\right) = 0.1 + 0.0111\left(\frac{2}{2} - 1\right) = 0.1$$

Using Equation (9.48), we get:

$$z_{fp} = B\left\{0.5 + 0.1\left[\min\left(\frac{L}{B}, 6\right) - 1\right]\right\} = 2 \times \left\{0.5 + 0.1\left[\min\left(\frac{2}{2}, 6\right) - 1\right]\right\} = 1 \text{ m}$$

Using Equation (9.39), we get:

$$I_{zp} = 0.5 + 0.1\sqrt{\frac{q_b - \sigma_v'|_{z_f=0}}{\sigma_v'|_{z_f=z_{fp}}}} = 0.5 + 0.1\sqrt{\frac{363 - 63}{74}} = 0.701$$

because

$$q_b = \frac{1200}{4} + 3 \times 21 = 363 \text{ kPa}$$

$$\sigma_v'\big|_{z_f=0} = 3 \times 21 = 63 \text{ kPa}$$

$$\sigma_v'\big|_{z_f=z_{fp}} = 3 \times 21 + 1 \times (21 - 9.81) = 74 \text{ kPa}$$

So for $z_f < z_{fp}$:

$$z_f = 0.25 \text{ m}: \quad I_{z(1)} = 0.1 + \frac{0.25}{1}(0.701 - 0.1) = 0.250$$

$$z_f = 0.75 \text{ m}: \quad I_{z(2)} = 0.1 + \frac{0.75}{1}(0.701 - 0.1) = 0.551$$

For $z_{fp} \leq z_f \leq z_{f0}$:

$$z_f = 1.5 \text{ m}: \quad I_{z(3)} = \frac{4 - 1.5}{4 - 1} \times 0.701 = 0.584$$

$$z_f = 2.5 \text{ m}: \quad I_{z(4)} = \frac{4 - 2.5}{4 - 1} \times 0.701 = 0.351$$

$$z_f = 3.5 \text{ m}: \quad I_{z(5)} = \frac{4 - 3.5}{4 - 1} \times 0.701 = 0.117$$

So, finally:

$$\frac{I_{z(1)} \Delta z_{(1)}}{E_{(1)}} = \frac{0.250 \times 0.5}{18.2} = 0.0069 \text{ mm/kPa}$$

$$\frac{I_{z(2)} \Delta z_{(2)}}{E_{(2)}} = \frac{0.551 \times 0.5}{21.6} = 0.0128 \text{ mm/kPa}$$

$$\frac{I_{z(3)} \Delta z_{(3)}}{E_{(3)}} = \frac{0.584 \times 1}{23.8} = 0.0245 \text{ mm/kPa}$$

$$\frac{I_{z(4)} \Delta z_{(4)}}{E_{(4)}} = \frac{0.350 \times 1}{27.2} = 0.0129 \text{ mm/kPa}$$

$$\frac{I_{z(5)} \Delta z_{(5)}}{E_{(5)}} = \frac{0.117 \times 1}{25.7} = 0.0046 \text{ mm/kPa}$$

The summation of these five numbers is 0.0617 mm/kPa. Now our table will look like this:

Sublayer i	z_{top} (m)	z_{bottom} (m)	Δz_i (m)	q_{ci} (MPa)	σ'_v (kPa)	σ'_h (kPa)	D_{Ri} (%)	φ_i	E_i (MPa)	I_{zi}	$\dfrac{I_{zi}\Delta z_i}{E_i}$ (mm/kPa)
1	3.0	3.5	0.5	12	66	26	76	1.52	18.2	0.250	0.0069
2	3.5	4.0	0.5	15	71	29	83	1.44	21.6	0.551	0.0128
3	4.0	5.0	1.0	17	80	32	86	1.40	23.8	0.584	0.0245
4	5.0	6.0	1.0	20	91	36	90	1.36	27.2	0.351	0.0129
5	6.0	7.0	1.0	18	102	41	84	1.43	25.7	0.117	0.0046

$$\sum \frac{I_{zi}\Delta z_i}{E_i} = 0.0617$$

Using Equation (9.42), we can calculate the total immediate settlement:

$$w = C_1 C_2 \left(q_b - \sigma'_v \big|_{z_f=0} \right) \sum \left(\frac{I_{zi}\Delta z_i}{F_i} \right)$$

The depth factor is calculated using Equation (9.43):

$$C_1 = 1 - 0.5 \left(\frac{\sigma'_v \big|_{z_f=0}}{q_b - \sigma'_v \big|_{z_f=0}} \right) = 1 - 0.5 \left(\frac{63}{363 - 63} \right) = 0.895$$

We will neglect the small creep effect that may be present to some extent in sand and take the time factor as equal to 1. This leaves us with:

$$w = 0.895 \times 1 \times (363 - 63) \times 0.0617 = 16.6 \text{ mm} \approx 1.7 \text{ cm}$$

This means that the assumption of $B=2\,m$ was conservative. We would now need to reduce B to obtain a settlement that is closer to the tolerable settlement. This is an exercise proposed in Problem 9.18.

9.7 SETTLEMENT OF SHALLOW FOUNDATIONS ON CLAY

9.7.1 Immediate settlement

9.7.1.1 Christian and Carrier's method

Janbu et al. (1956) proposed a form of Equation (9.3) for the average settlement under a flexible foundation embedded D into a half-space with elastic constants E and v as

$$w = I_0 I_1 \frac{q_b B}{E} \left(1 - v^2 \right) \tag{9.51}$$

where I_0 and I_1 are the influence factors related to the depth D from the ground surface to the base of the foundation, and the depth H from the foundation base to a very stiff ("rigid") layer, respectively. It is practical to rewrite Equation (9.51) in this form:

$$w_i = I_0 I_1 \frac{q_b^{net} B}{E_u} \tag{9.52}$$

where q_b^{net} is the net unit load on the footing base (that is, the excess of vertical normal stress at the footing base over the initial *in situ* vertical effective stress), I_0 and I_1 are given in Figure 9.8, and E_u is a representative elasticity modulus.

Most elasticity-based equations, such as Equation (9.52), are developed based on the assumption of a uniform elastic half-space. In reality, each point of the soil mass beneath the footing has a different initial modulus and will be subjected to different strains for any given applied load (with the consequent different degrees of modulus degradation). Additionally, the degradation of the modulus at any given point increases (and the modulus decreases) as the load on the footing increases. The greatest challenge in the use of Equation (9.52) therefore is in the selection of a representative modulus to use in Equation (9.52) that will produce the correct result.

Researchers and engineers have approached this problem in two different ways: by back-calculating modulus values from both full-scale and model load tests on footings and by performing realistic numerical analyses and calculating the value of E_u that would produce results consistent with the numerical results for a given foundation load. The modulus E_u can be expressed as a multiple of the undrained shear strength s_u:

$$E_u = K s_u \tag{9.53}$$

The correlation in Equation (9.53) is empirical. The values of K range from 100 for very soft soils to as high as 1500 for very stiff clays. Most design charts available in the literature make no reference to the settlement level. The values proposed in Figure 9.9 may be used for K provided that we recognize that these values are subject to significant uncertainties.

9.7.1.2 Foye et al.'s method

An improved procedure would take account of the level of settlement and the strains imposed on the soil. Such a procedure would recognize that the higher the strain level, the

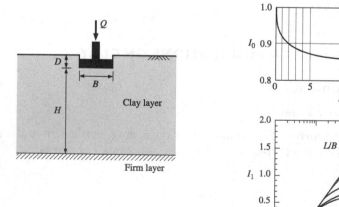

Figure 9.8 Values of I_0 and I_1 in Equation (9.52). (From Christian and Carrier 1978; reproduced with permission from the National Research Council Press 2006.)

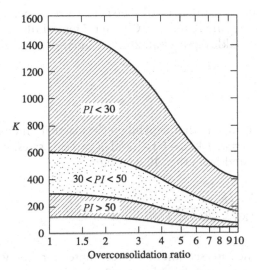

Figure 9.9 Value of $K=E_u/s_u$ in terms of *PI* and *OCR*. (Duncan and Buchignani 1987; Courtesy of J.M. Duncan.)

higher the degree of modulus degradation, and the lower the representative modulus to use in Equation (9.52).

One such method was proposed by Foye et al. (2008). The method relies on the calculation of settlement using an equation that is essentially the same as Equation (9.52), but written in terms of the small-strain Young's modulus:

$$w_i = I_q \frac{q_b^{net} B}{\bar{E}_0} \tag{9.54}$$

The representative small-strain Young's modulus $\bar{E}_0$ of clay below the footing base is given by

$$\bar{E}_0 = 2(1+v)\bar{G}_0 \tag{9.55}$$

where v=Poisson's ratio=0.5 for undrained conditions.

The representative small-strain shear modulus $\bar{G}_0$ is calculated as the weighted average of the values of G_0 within an influence depth $z_{\bar{G}_0}$ measured from the base of the footing (Sakleshpur et al. 2021):

$$\bar{G}_0 = \frac{\displaystyle\sum_{i=1}^{n} G_{0,i}^{avg} H_i}{\displaystyle\sum_{i=1}^{n} H_i} \tag{9.56}$$

where $G_{0,i}^{avg}$=average small-strain shear modulus of layer i, H_i=thickness of layer i, and n=number of clay layers within the influence depth $z_{\bar{G}_0}$ below the footing base, which is given by

$$\frac{z_{\bar{G}_0}}{B} = 1+0.111\left(\frac{L}{B}-1\right) \leq 2 \tag{9.57}$$

In the absence of direct measurement of the shear wave velocity, from which the small-strain shear modulus can be calculated (see Chapter 7 for details), it may be estimated using the following correlation (Sakleshpur et al. 2021; Viggiani and Atkinson 1995):

$$\frac{G_0}{p_A} = C_g \left(\frac{100\sigma'_m}{p_A} \right)^{n_g} R_0^{m_g} \tag{9.58}$$

where C_g, n_g, and m_g = parameters that depend on the plasticity index PI; σ'_m = mean effective stress; p_A = reference stress = 100 kPa ≈ 1 tsf; and R_0 = mean stress-based overconsolidation ratio, which can be estimated from the OCR using:

$$R_0 = \frac{p'_p}{p'} = OCR \left(\frac{1 + 2K_{0,NC}}{1 + 2K_{0,NC}\sqrt{OCR}} \right) \tag{9.59}$$

where p'_p = value of p' at the intersection of the recompression line with the normal consolidation line in v–$\ln p'$ space, v = specific volume (= $1+e$), and $K_{0,NC}$ = coefficient of lateral earth pressure at rest for normally consolidated soils (≈ 0.5–0.75 for NC clays).

The parameters C_g, n_g, and m_g can be calculated using (Foye et al. 2008; Viggiani and Atkinson 1995):

$$C_g = 37.9\exp(-0.045PI) \quad \text{for } PI > 5\% \tag{9.60}$$

$$n_g = 0.109\ln(PI) + 0.4374 \quad \text{for } PI > 5\% \tag{9.61}$$

$$m_g = 0.0015PI + 0.1863 \quad \text{for } PI > 5\% \tag{9.62}$$

The net unit load q_b^{net} on the footing base is given by:

$$q_b^{net} = q_b - \gamma_m D \tag{9.63}$$

where q_b = gross unit load on the footing base (including the loads from the superstructure, the weight of the foundation, and the weight of the backfill when the excavation is backfilled) and $\gamma_m D$ = total overburden stress at the footing base level.

Lastly, the influence factor I_q is given in Figure 9.10, where H = thickness of the clay layer below the footing base, B = footing width, and $\bar{s}_u$ = representative undrained shear strength, which is calculated as the average of the values of s_u over depth B below the footing base. For circular footings, an equivalent footing width may be obtained by equating the cross-sectional area of the footing with that of an equivalent square.

Example 9.7

A 1.5 m square footing is founded on an NC clay with LL=70%, PL=40%, and unit weight equal to 18 kN/m³. The base of the footing is placed at a depth of 1 m. The groundwater table may rise as high as 1 m below the ground surface. The clay layer is 7 m thick. It is underlain by a very stiff layer. Calculate the immediate settlement of this footing under a 60 kN column load using the method by Christian and Carrier (1978). The excavation is backfilled.

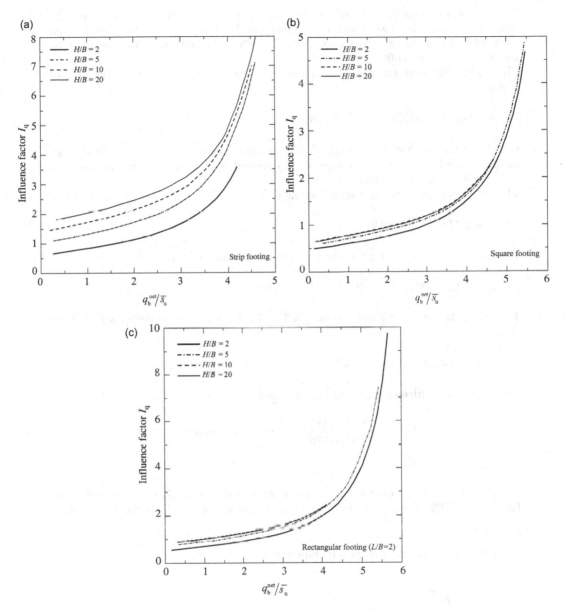

Figure 9.10 Influence factor I_q as a function of $q_b^{net}/\bar{s}_u$ and H/B for (a) strip footings, (b) square footings, and (c) rectangular ($L/B=2$) footings. (Foye et al. 2008; with permission from ASCE.)

Solution:

The immediate settlement is given by Equation (9.52):

$$w_i = I_0 I_1 \frac{q_b^{net} B}{E_u}$$

For this footing, the distance H from the base of the footing to the top of the stiff layer is 6 m, so that $H/B = 6/1.5 = 4$. The relative embedment is $D/B = 1/1.5 = 0.67$. Entering the charts in Figure 9.8 with these values, we find the influence factors I_0 and I_1:

$$I_0 = 0.92 \quad \text{and} \quad I_1 = 0.62$$

The modulus is determined from Equation (9.53). To use this equation, we require the values of K and s_u. Both are determined in terms of the plasticity index PI, which is given as $70-40=30\%$. The value of K corresponding to $PI=30\%$ and OCR=1 is read from the chart in Figure 9.9 as 600.

The undrained shear strength s_u is calculated using Equation (6.50) as a function of depth:

$$s_u = \left[0.11+0.0037(30)\right]\sigma'_v = 0.22\sigma'_v$$

Given that the increase in s_u with depth is linear, the representative s_u to use in these calculations can be taken as the average of s_u calculated at the base of the footing and that at depth B below the footing base.

The undrained shear strength at the footing base is:

$$s_u = 0.22 \times 18(1) = 4.0 \text{ kPa}$$

The undrained shear strength at depth B below the footing base is:

$$s_u = 0.22 \times \left[18(2.5) - 9.81(2.5-1)\right] = 6.7 \text{ kPa}$$

Therefore, the representative value of s_u is 5.3 kPa. The value of the modulus E_u follows directly:

$$E_u = Ks_u = 600 \times 5.3 = 3180 \text{ kPa}$$

The immediate settlement can now be calculated:

$$w_i = 0.92 \times 0.62 \times \frac{60/(1.5 \times 1.5) \text{ kPa} \times 1.5 \text{ m}}{3180 \text{ kPa}} = 0.0072 \text{ m} \approx 7 \text{ mm} \quad \text{answer}$$

Example 9.8

Calculate the immediate settlement of the footing in Example 9.7 using the method by Foye et al. (2008). The coefficient of lateral earth pressure at rest K_0 of the clay is 0.6.

Solution:

The immediate settlement is given by Equation (9.54):

$$w_i = I_q \frac{q_b^{net}B}{\bar{E}_0}$$

The influence depth below the base of the footing for the calculation of small-strain shear modulus is:

$$z_{\bar{G}_0} = B\left[1 + 0.111\left(\frac{L}{B}-1\right)\right] = 1.5 \times \left[1 + 0.111\left(\frac{1.5}{1.5}-1\right)\right] = 1.5 \text{ m}$$

The parameters C_g, n_g, and m_g are calculated from the plasticity index PI using Equations (9.60), (9.61), and (9.62), respectively:

$$C_g = 37.9\exp(-0.045PI) = 37.9\exp(-0.045 \times 30) = 9.83$$

$$n_g = 0.109\ln(PI) + 0.4374 = 0.109\ln(30) + 0.4374 = 0.81$$

$$m_g = 0.0015PI + 0.1863 = 0.0015(30) + 0.1863 = 0.23$$

Because the plasticity index of the clay layer is uniformly equal to 30%, the mean effective stress σ'_m is calculated at the midpoint of the depth of influence, that is, at a depth of $1 + (1.5/2) = 1.75$ m from the ground surface:

$$\sigma'_v = 18(1.75) - 9.81(1.75 - 1) = 24.1 \text{ kPa}$$

$$\sigma'_h = K_0\sigma'_v = 0.6 \times 24.1 = 14.5 \text{ kPa}$$

$$\sigma'_m = \frac{1}{3}(\sigma'_v + 2\sigma'_h) = \frac{1}{3}[24.1 + (2 \times 14.5)] = 17.7 \text{ kPa}$$

In the absence of shear wave velocity data, the small-strain shear modulus is calculated using Equation (9.58):

$$G_0 = C_g\left(\frac{100\sigma'_m}{p_A}\right)^{n_g} R_0^{m_g} p_A = 9.83 \times \left(\frac{100 \times 17.7}{100}\right)^{0.81} \times (1)^{0.23} \times 100 = 10,079 \text{ kPa}$$

The representative small-strain shear modulus is calculated using Equation (9.56):

$$\bar{G}_0 = \frac{\sum_{i=1}^{n} G_{0,i}^{avg} H_i}{\sum_{i=1}^{n} H_i} = \frac{10,079 \times 1.5}{1.5} = 10,079 \text{ kPa}$$

which gives the representative small-strain Young's modulus through Equation (9.55) as

$$\bar{E}_0 = 2(1 + v)\bar{G}_0 = 2 \times (1 + 0.5) \times 10079 = 30237 \text{ kPa}$$

From Example 9.7, we know that the representative undrained shear strength $\bar{s}_u$ is 5.3 kPa. The ratio $q_b^{net}/\bar{s}_u$ is calculated as

$$\frac{q_b^{net}}{\bar{s}_u} = \frac{60/(1.5 \times 1.5)}{5.3} = 5$$

For this footing, the distance H from the base of the footing to the top of the stiff layer is 6 m, so that $H/B = 6/1.5 = 4$ and $q_b^{net}/\bar{s}_u = 5$. Entering the chart in Figure 9.10b for square footings with these values, we find the influence factor I_q to be 3.0. The immediate settlement can now be calculated:

$$w_i = I_q\frac{q_b^{net}B}{\bar{E}_0} = 3 \times \frac{[60/(1.5 \times 1.5)] \times 1.5}{30,237} = 0.00397 \text{ m} \approx 4 \text{ mm} \quad \text{answer}$$

9.7.2 Consolidation settlement

We already know how to calculate one-dimensional consolidation settlement from Chapter 6. In calculating the settlement of a footing of finite size (circular, square, or strip) acted upon by a column or wall load, however, there are three-dimensional effects that must be accounted for. The differences with respect to 1D consolidation settlement are that (1) the initial excess pore pressure generated at any given depth is not equal to the initial vertical stress there and (2) the initial vertical stress increment at any given depth is not equal to the applied stress at the top of the layer.

Skempton and Bjerrum (1957) proposed an approximate solution to the problem of the consolidation settlement of a footing (or any loaded area of finite size) based on Skempton's pore pressure parameters A and B (Skempton 1954) and on the elastic solutions to the values of the principal stress increments within the clay caused by the application of the loadings. The analysis is performed for points along the vertical axis of the foundation element passing through its center; it is implicitly assumed to hold for other points of the soil mass. The pore pressure increment due to increases in the major and minor principal stresses along the axis of the foundation can be assumed to be given by Equation (6.44). After dissipation of this excess pore pressure, the final consolidation settlement will have fully developed. The final consolidation settlement will be the integral of Equation (6.7) over the thickness H of the clay layer, with the final vertical effective stress increments equal to the initial pore pressure increments:

$$w_c = \int_0^H d\varepsilon_z = \int_0^H m_v \Delta\sigma_1 \left[A + \frac{\Delta\sigma_3}{\Delta\sigma_1}(1-A) \right] dz \tag{9.64}$$

If we completely ignored the effects of the differences between the major and minor principal stress increments, we would calculate a value of settlement just as if we had 1D consolidation. Let us call that settlement w_{c1D} and write it as

$$w_{c1D} = \int_0^H m_v \Delta\sigma_1 \, dz \tag{9.65}$$

It is now possible to write w_c in terms of w_{c1D} as

$$w_c = \mu w_{c1D} \tag{9.66}$$

where

$$\mu = A + \alpha(1-A) \tag{9.67}$$

and

$$\alpha = \frac{\int_0^H \Delta\sigma_3 \, dz}{\int_0^H \Delta\sigma_1 \, dz} \tag{9.68}$$

Table 9.3 Values of α for the computation of consolidation
settlement (Skempton and Bjerrum 1957)

H/B	Circular footing	Strip footing
0	1.00	1.00
0.25	0.67	0.74
0.5	0.50	0.53
1	0.38	0.37
2	0.30	0.26
4	0.28	0.20
10	0.26	0.14
∞	0.25	0

The values of α for circular and strip footings as a function of the ratio of the clay layer thickness H to the footing width B are given in Table 9.3. For square footings, the values for circular footings with the same area as the square footing can be used. For rectangular footings, interpolation is required.

Ideally, this procedure would take into account the strain level for the determination of the appropriate value of the pore pressure parameter A and of the coefficient of volume compressibility m_v. Because the stress path of the soil during consolidation settlement does not approach failure conditions or critical state, the value of A for normally consolidated clay is less than 1.[6] However, there is enough plastic shearing during consolidation settlement for the value to be greater than the 0.33 associated with elastic deformation. The value of A is most likely in the 0.50–0.75 range for normally consolidated clays. There is no detailed information in the literature as to a way to refine the estimate of A to use in this type of calculation. Chapter 6 discusses the values of A for triaxial tests and also has information on the values of m_v.

Example 9.9

A square footing embedded 2 m into an NC clay is to support a load of 200 kN. The water table is at a depth of 2 m. The clay layer thickness is 10 m. It overlies a very firm sand layer. The structure to be supported can tolerate as much as 80 mm of total settlement. What is the required width of the footing that can tolerate that settlement? The excavation is backfilled. The following data are available for the clay layer: $\gamma = 18$ kN/m³, $C_c = 0.2$, $LL = 70\%$, $PL = 20\%$, and $s_u/\sigma'_v = 0.3$. At $z = 6$ m, $e_0 = 0.8$. Assume that, for the loads and spans involved, any interaction with the other footings that are part of the same structure can be ignored.

Solution:

The maximum tolerable total settlement is:

$$w = 80 \text{ mm}$$

The total settlement is resolved into an immediate settlement and a consolidation settlement. So we must have:

$$w = w_i + w_c \leq 80 \text{ mm}$$

---

[6] To follow this argument, you must be acquainted with the material presented in Chapter 6.

We assume that the width of the footing is:

$$B = 4 \text{ m}$$

We will calculate first the immediate settlement. This is given by Equation (9.52):

$$w_i = I_0 I_1 \frac{q_b^{net} B}{E_u}$$

The distance H from the base of the footing to the bottom of the clay layer is 8 m, so $H/B = 8/4 = 2$. The relative embedment is $D/B = 2/4 = 0.5$. Entering the charts in Figure 9.8 with these values, we find the influence factors I_0 and I_1:

$$I_0 = 0.93 \quad \text{and} \quad I_1 = 0.52$$

The modulus E_u is determined from Equation (9.53). To use this equation, we need the values of K and s_u. The value of the plasticity index PI is $LL - PL = 70\% - 20\% = 50\%$. The value of K corresponding to $PI = 50\%$ and $OCR = 1$ (because the clay is an NC clay) is read from the chart in Figure 9.9 as 300.
Given that the increase in s_u with depth is linear ($s_u/\sigma'_v = 0.3$), the representative s_u to use in these calculations can be taken as the average of s_u calculated at the base of the footing and that at depth B below the footing base.
The undrained shear strength at the footing base is:

$$s_u = 0.3 \times 18(2) = 10.8 \text{ kPa}$$

The undrained shear strength at depth B below the footing base is:

$$s_u = 0.3 \times \left[18(6) - 9.81(6-2) \right] = 20.6 \text{ kPa}$$

Therefore, the representative value of s_u is 15.7 kPa. The value of the modulus E_u follows directly:

$$E_u = K s_u = 300 \times 15.7 = 4710 \text{ kPa}$$

The immediate settlement can now be calculated:

$$w_i = 0.93 \times 0.52 \times \frac{\left(\dfrac{200}{4 \times 4} \right) \times 4}{4710} = 0.0051 \text{ m} \approx 5 \text{ mm}$$

Now we must calculate the consolidation settlement. This is given by Equation (9.66):

$$w_c = \mu w_{c1D}$$

where, according to Equation (9.67):

$$\mu = A + \alpha(1 - A)$$

For a square footing, we can use the value of α for circular footings from Table 9.3 with $H/B = 2$:

$$\alpha = 0.30$$

Recall from Chapter 6 that the value of A at peak stress for NC clays is 1. For initial loading, it is 1/3. Since, for settlement of the order of the tolerable values, footings are rather far from failure conditions, it is reasonable to assume that a representative value for A is 2/3. So

$$\mu = \frac{2}{3} + 0.30\left(1 - \frac{2}{3}\right) \approx 0.77$$

The 1D consolidation settlement is calculated using Equations (6.5), (6.10), and (6.11) with the stress increase estimated using the 2-to-1 stress distribution rule, expressed using Equation (4.107). Because the footing lies on 8 m of clay, what we wish is to compute the compression of these 8 m. If we divide them into four sublayers of 2 m, we can calculate the increase in vertical stress at the center of each sublayer as

$$\Delta\sigma_v = \frac{Q}{(B+z)(L+z)} = \frac{200}{(4+z)^2} = \begin{cases} \dfrac{200}{(4+1)^2} = 8 \text{ kPa} \\[2mm] \dfrac{200}{(4+3)^2} = 4.1 \text{ kPa} \\[2mm] \dfrac{200}{(4+5)^2} = 2.5 \text{ kPa} \\[2mm] \dfrac{200}{(4+7)^2} = 1.7 \text{ kPa} \end{cases}$$

The vertical effective stress at the center of the clay layer is:

$$\sigma_v'(6 \text{ m}) = 6 \times 18 - 4 \times 9.81 = 69 \text{ kPa}$$

Given that the clay is normally consolidated, the initial void ratio at the center of each sublayer can be computed from the known value at z=6 m using the equation for the NC line:

$$e_0(z) = e_0(6 \text{ m}) + C_c \log\left(\frac{\sigma_v'(6 \text{ m})}{\sigma_v'(6 \text{ m}) - \gamma_b\Delta z}\right)$$

$$= 0.8 + 0.2\log\left(\frac{69}{69 - 8\Delta z}\right) = \begin{cases} 0.8 + 0.2\log\left(\dfrac{69}{69 - 8 \times 3}\right) = 0.84 \text{ for } z = 3 \text{ m} \\[2mm] 0.8 + 0.2\log\left(\dfrac{69}{69 - 8 \times 1}\right) = 0.81 \text{ for } z = 5 \text{ m} \\[2mm] 0.8 + 0.2\log\left(\dfrac{69}{69 + 8 \times 1}\right) = 0.79 \text{ for } z = 7 \text{ m} \\[2mm] 0.8 + 0.2\log\left(\dfrac{69}{69 + 8 \times 3}\right) = 0.77 \text{ for } z = 9 \text{ m} \end{cases}$$

We can now calculate the total one-dimensional compression by adding the compressions of the sublayers. For that, we need to calculate the initial vertical effective stresses at the centers of the sublayers, and we also need the previously calculated stress increases at the same points. The one-dimensional compression is:

$$
w_{c1D} = \sum \frac{C_c}{1+e_0} \log\left(\frac{\sigma'_{v0} + \Delta\sigma_v}{\sigma'_{v0}}\right) \times \Delta H_0
$$

$$
= C_c \Delta H_0 \sum \frac{1}{1+e_0} \log\left(\frac{\sigma'_{v0} + \Delta\sigma_v}{\sigma'_{v0}}\right)
$$

$$
= 0.2 \times 2\ \text{m} \times \left(\begin{array}{c} \dfrac{1}{1+0.84}\log\left(\dfrac{45+8}{45}\right) + \dfrac{1}{1+0.81}\log\left(\dfrac{61+4.1}{61}\right) \\[3mm] + \dfrac{1}{1+0.79}\log\left(\dfrac{77+2.5}{77}\right) + \dfrac{1}{1+0.77}\log\left(\dfrac{93+1.7}{93}\right) \end{array} \right)
$$

$$
= 0.027\ \text{m}
$$

So

$$
w_c = 0.77 \times 0.027 = 0.021\ \text{m}
$$

Finally,

$$
w_i + w_c = 0.005 + 0.021 = 0.026\ \text{m} = 26\ \text{mm} < 80\ \text{mm}
$$

So a width of 4 m is sufficiently large for the footing to settle no more than 26 mm. Because there is a tolerance of 80 mm, there is room for reducing the size of this foundation, subject to the bearing capacity check (see Chapter 10). Note how the immediate settlement is a small fraction (in this case, of the order of 25%) of the consolidation settlement.

9.8 CASE STUDY: THE LEANING TOWER OF PISA (PART III) AND THE LEANING BUILDINGS OF SANTOS

Anyone familiar with the material in Chapter 6 might naturally suspect an uneven clay layer thickness as the cause of the initial lean of the Tower of Pisa or other structures in the same situation.[7] However, as shown by Mitchell et al. (1977), the clay layer underneath the Tower is actually fairly uniform. The same cannot be stated regarding the sand with silt and clay of Horizon A[8]. It is very likely that the lean to the south of the Tower of Pisa is in fact due to the differences in the soil in Horizon A on the north and south sides. On the north side, near the ground surface, the soil is mostly sand, whereas on the south side, the soil ranges from yellow silty sand to yellow clayey silt. From the material covered in Chapters 5 and 6, we would expect the soil to be weaker and more compressible if the fines content were high. This reflects on the values of cone resistance measured on the south and north sides of the Tower, shown in Figure 9.11. Note how the values of q_c are lower on the south side. Jamiolkowski (2006) showed more recent data (Figure 9.12) that demonstrate this point more clearly. The CPTs on the south side of the Tower (DH4 and DH5) actually suggest the

---

[7] All references for this case history are at the end of Part IV in Section 11.5.
[8] For a complete description of the soil profile underneath the tower, please refer to Section 6.7.

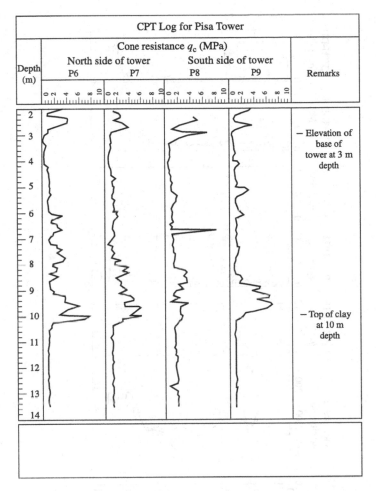

Figure 9.11 CPTs on the north and south sides of the Tower of Pisa. (Data from Mitchell et al. 1977.)

complete disappearance of the sand layer on the south side, while it is clearly visible in the other CPTs. This observation would be consistent with an initial lean to the south, which would then tend to intensify for reasons we will discuss in a continuation of this case history in Chapter 11. While this hypothesis cannot be definitively proven to be the reason for the initial lean of the Tower, it is the hypothesis that best fits all that has been learned about the Tower. For completeness, Figure 9.13 provides the OCR values for the entire soil profile and SPT blow counts for the lower sand layer in addition to a representative CPT log.

As seen in Part I of the case history, in Chapter 2, the foundation of the Tower of Pisa is a spread foundation in the form of a hollow cylinder with an outer diameter equal to 19.58 m and an inner diameter equal to 4.47 m (Mitchell et al. 1977). The hollow space appears to have been filled with rubble and mortar at the time of construction (Burland et al. 2003). The depth of embedment was 3 m. In Problem 9.9, the reader is asked to estimate the contribution of the sand layer to the tower settlement using the data in Figures 9.11 and 9.12.

It is interesting that nine centuries later, a fairly similar situation developed not once, but many times in the coastal city of Santos, Brazil. Until 1968, the local building code had no restrictions whatsoever on the type of foundation that could be used for multi-story buildings. Many buildings were constructed on shallow foundations; the result can be seen in

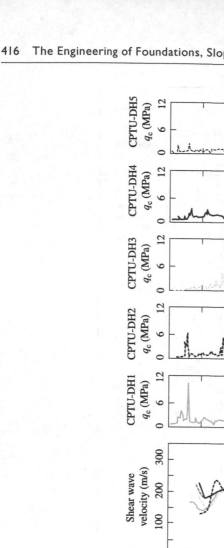

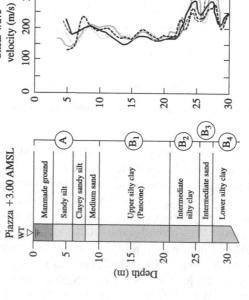

Figure 9.12 Recent CPTs at the tower site (Jamiolkowski 2006).

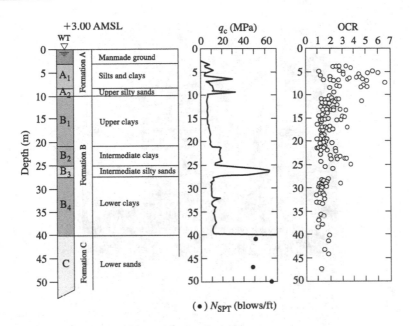

Figure 9.13 Results from CPT and SPT as well as OCR values at the site of the tower (Jamiolkowski 2006).

Figures 9.14–9.16. After the leanings in the first buildings became visible, there was a realization that the practice of placing tall buildings on shallow footings could not continue, and a requirement was added to the Santos building code to use deep foundations for buildings 11 stories tall or taller. This limit has recently been lowered to 5 stories.

Rock or strong residual soils at the sites of these buildings, mostly clustered together in a particular oceanfront neighborhood, are at depths of 40–60 m. Above rock, there is a layer of marine clay extending 8–12 m to rock. The similarity with the profile on which the Tower of Pisa rests is also observed with respect to the presence of an 8- to 12-meter-thick sand layer at the top and even an intermediate sand layer mixed in the marine clay layer. Local variations in the properties of the upper sand layer could have played a role in the initial lean of the buildings, just as in the case of the Tower of Pisa, although there has been limited scientific work published on the subject.

An aggravating factor in the case of the buildings of Santos is that many are located close to one another, so the settlement caused by buildings constructed next to existing buildings most likely did contribute in some cases to the lean. The proximity of the buildings led the late Prof. Milton Vargas to state, somewhat controversially, that the buildings near the beautiful beach area in Santos are like dominoes, and that if one building collapsed, all the others would also collapse. Although people continue to live in the leaning buildings and the danger may have been somewhat overstated, attention is being paid to the potential structural problems. In 2001, there was an intervention in one of the buildings, the 55-meter-high Nuncio Malzoni building, which was leaning 2.2° in 1998. This lean corresponded to a lateral overhang of the building of the order of 2 m. The maximum settlement of the building was almost 1 m.

The intervention in the Nuncio Malzoni was conducted by a team of professors from the University of São Paulo (Carlos Eduardo Maffei, Heloísa Helena Gonçalves, and Paulo Pimenta) over the course of 2 years and ending in 2001. Sixteen 57-meter-long drilled shaft foundations with diameters ranging from 1 to 1.4 m were installed all the way to rock.

Figure 9.14 Leaning building in Santos, Brazil. (Courtesy of Fernando Schnaid.)

Figure 9.15 Leaning building in Santos, Brazil. (Courtesy of Fernando Schnaid.)

Figure 9.16 Leaning building in Santos, Brazil. (Courtesy of Fernando Schnaid.)

Fourteen hydraulic jacks were used to restore the building to plumb by pushing up on reinforced concrete beams designed to support all the affected columns and have the building move as much as a rigid body as possible. The hydraulic jacks were supported by caps resting on the drilled shafts. The total cost of the intervention was R$1.5 million, which at the time was roughly equivalent to US$1.5 million. It is reported that this is the first successful intervention of this type anywhere.

Residents of the approximately 100 buildings in Santos reported to have some degree of lean have learned to cope, but there has been economic loss associated with the leaning buildings of Santos. Not surprisingly, the prices of condo units in these buildings plummeted after the leans became visible to the naked eye many years ago. Much engineering work has gone into understanding the situation, assessing danger, and, in at least one known case, correcting a potentially dangerous condition. But people have become used to the situation and live normally in these buildings. This allows us to comment also on the light side of the story. The buildings, like the Tower of Pisa, are somewhat of a tourist attraction. News media have covered the peculiarities of life in the buildings, such as the facts that balls released on the floor on the "high" end of a room will roll across the room and that one can never fill a cup of coffee all the way to the top. One of the buildings (officially, the Excelsior) received the nickname Torto (which means crooked in English). A cafe opened in the Excelsior, called the Bar do Torto, where conversations about the leaning buildings of Santos are, naturally, unavoidable.

End of Part III. The Tower of Pisa's case history continues at the end of Chapter 11, where we discuss how the tower came close to collapse in the early 1990s.

9.9 CHAPTER SUMMARY

9.9.1 Main concepts and equations

It is often necessary, mostly due to the need to calculate **foundation settlement**, to calculate the stresses induced at depth in a soil mass by surface loads of various geometries and magnitudes. Equations derived from elasticity theory are often used for this purpose. These were discussed in Chapter 4. In this chapter, we examined equations that allow direct calculation of displacements and settlements of foundation elements from applied boundary loads.

Settlements are sometimes associated with the mechanism by which they develop. **Distortion settlement** is settlement due purely to shear strains in the soil, with no volume change taking place. **Consolidation settlement** is due to consolidation of the soil, and **secondary compression** settlement is due to the enduring distortions of the soil fabric after primary consolidation is over, which lead to additional volumetric change.

Another way to classify settlement is to refer to settlement that happens shortly after load application as **immediate settlement**, and the remainder as **long-term settlement**. In sands, this makes more sense, because both consolidation and distortion settlements happen immediately or in the very short term, and it is impossible to separate them in any practical way.

Let us now summarize the equations for the calculation of immediate settlement in sands and clays and consolidation settlement in clays.

9.9.2 Equations for the calculation of settlement of shallow foundations in sand using the SPT

The settlement w according to Meyerhof (1965) is given by

$$\frac{w}{L_R} = \frac{0.152}{\min\left(1 + \dfrac{D}{3B}, 1.33\right) N_{60}} \left(\frac{q_b - \sigma'_{vp}\big|_{z_f=0}}{p_A}\right) \tag{9.24}$$

for $B \le 1.2L_R$, and

$$\frac{w}{L_R} = \frac{0.229}{\min\left(1 + \dfrac{D}{3B}, 1.33\right) N_{60}} \left(\frac{q_b - \sigma'_{vp}\big|_{z_f=0}}{p_A}\right)\left(\frac{B}{B + 0.305L_R}\right)^2 \tag{9.25}$$

for $B > 1.2L_R$. In these equations, N_{60} is the energy-corrected SPT blow count averaged over a depth from the foundation base to $1B$ and $2B$ below the base for square and strip footings, respectively.

The settlement w according to Peck and Bazaraa (1969) is given by

$$w = C_{gw}\left(\frac{0.229L_R}{N_B}\right)\left(\frac{q_b - \sigma'_{vp}\big|_{z_f=0}}{p_A}\right)\left(\frac{B}{B + 0.305L_R}\right)^2 \tag{9.26}$$

where

$$N_B = \left(\frac{3N_{60}}{1 + 4\dfrac{\sigma'_v}{p_A}}\right)\frac{\sigma'_v}{p_A} \tag{9.27}$$

for $\sigma'_v \leq 0.75 p_A$, and

$$N_B = \left(\frac{3 N_{60}}{3.25 + \dfrac{\sigma'_v}{p_A}} \right) \frac{\sigma'_v}{p_A} \tag{9.28}$$

for $\sigma'_v > 0.75 p_A$.

The groundwater correction factor is:

$$C_{gw} = \frac{\left(\sigma'_{vp} \big|_{z_f = \frac{B}{2}} \right)_{\text{without WT}}}{\left(\sigma'_{vp} \big|_{z_f = \frac{B}{2}} \right)_{\text{with WT}}} \tag{9.29}$$

The settlement w according to Burland and Burbidge (1985) is given by

$$\frac{w}{L_R} = 0.10 f_s f_L f_t I_c \frac{q_b - \dfrac{2}{3} \sigma'_{vp} \big|_{z_f = 0}}{p_A} \left(\frac{B}{L_R} \right)^{0.7} \tag{9.30}$$

where

$$I_c = \frac{1.71}{\overline{N}^{1.4}} \tag{9.31}$$

$$\frac{z_{f0}}{L_R} = \left(\frac{B}{L_R} \right)^{0.79} \tag{9.32}$$

$$f_s = \left(\frac{1.25 \dfrac{L}{B}}{\dfrac{L}{B} + 0.25} \right)^2 \tag{9.33}$$

$$f_L = \begin{cases} \dfrac{H}{z_{f0}} \left(2 - \dfrac{H}{z_{f0}} \right) & \text{if } H \leq z_{f0} \\ 1 & \text{if } H > z_{f0} \end{cases} \tag{9.34}$$

$$f_t = \left(1 + R_3 + R_t \log \frac{t}{3} \right) \tag{9.35}$$

9.9.3 Equations for the calculation of settlement of shallow foundations in sand using the CPT

The depth of influence z_{f0} – measured from the foundation base – is given by

$$\frac{z_{f0}}{B} = 2 + 0.4 \left[\min\left(\frac{L}{B}, 6\right) - 1 \right] \tag{9.47}$$

The influence factor I_{z0} at the base of the foundation is given by

$$I_{z0} = 0.1 + 0.0111\left(\frac{L}{B} - 1\right) \le 0.2 \tag{9.37}$$

The depth z_{fp} – measured from the foundation base – at which the influence factor peaks is given by

$$\frac{z_{fp}}{B} = 0.5 + 0.1 \left[\min\left(\frac{L}{B}, 6\right) - 1 \right] \tag{9.48}$$

The peak influence factor I_{zp} is given by

$$I_{zp} = 0.5 + 0.1 \sqrt{\frac{q_b - \sigma_v'\big|_{z_f=0}}{\sigma_v'\big|_{z_f=z_{fp}}}} \tag{9.39}$$

The influence factor I_z at depth z_f measured from the base of the foundation is given by

$$I_z = I_{z0} + \frac{z_f}{z_{fp}}\left(I_{zp} - I_{z0}\right) \tag{9.40}$$

for $z_f < z_{fp}$, and

$$I_z = \frac{z_{f0} - z_f}{z_{f0} - z_{fp}} I_{zp} \tag{9.41}$$

for $z_{fp} \le z_f \le z_{f0}$.

The elastic modulus of sublayer i is a multiple φ of the cone resistance:

$$E_i = \varphi_i q_{ci} \tag{9.46}$$

where

$$\varphi_i = \lambda \left(\frac{w}{L_R}\right)^{-0.285} \left(\frac{B}{L_R}\right)^{0.4} \left(\frac{D_R}{100}\right)^{-0.65} \tag{9.49}$$

Now the settlement w can be calculated from:

$$w = C_1 C_2 \left(q_b - \sigma_v'\big|_{z_f=0}\right) \sum \left(\frac{I_{zi} \Delta z_i}{E_i}\right) \tag{9.42}$$

where

$$C_1 = 1 - 0.5 \left(\frac{\sigma_v'\big|_{z_f=0}}{q_b - \sigma_v'\big|_{z_f=0}}\right) \tag{9.43}$$

$$C_2 = 1 + 0.2 \log\left(\frac{t}{0.1 t_R}\right) \tag{9.44}$$

9.9.4 Equations for the calculation of immediate settlement of shallow foundations in clay

The immediate settlement w_i according to Christian and Carrier (1978) is given by

$$w_i = I_0 I_1 \frac{q_b^{net} B}{E_u} \tag{9.52}$$

where I_0 and I_1 are given in Figure 9.8 and E_u is a representative small-strain Young's modulus. The Young's modulus can be expressed as a multiple of the undrained shear strength s_u:

$$E_u = K s_u . \tag{9.53}$$

where K is given in Figure 9.9.

The immediate settlement w_i according to Foye et al. (2008) is given by

$$w_i = I_q \frac{q_b^{net} B}{E_0} \tag{9.54}$$

In this equation, I_q is given by Figure 9.10 and $\bar{E}_0$ is a representative small-strain Young's modulus:

$$\bar{E}_0 = 2(1+v)\bar{G}_0 \tag{9.55}$$

where v is the Poisson's ratio and $\bar{G}_0$ is a representative small-strain shear modulus. This modulus – $\bar{G}_0$ – is calculated as

$$\bar{G}_0 = \frac{\sum\limits_{i=1}^{n} G_{0,i}^{avg} H_i}{\sum\limits_{i=1}^{n} H_i} \tag{9.56}$$

where $G_{0,i}^{avg}$ is the average small-strain shear modulus of layer i, H_i is the thickness of layer i, and n is the number of clay layers within the influence depth $z_{\bar{G}_0}$ below the footing base. This influence depth is calculated using:

$$\frac{z_{\bar{G}_0}}{B} = 1 + 0.111\left(\frac{L}{B} - 1\right) \le 2 \tag{9.57}$$

9.9.5 Equations for the calculation of consolidation settlement of shallow foundations in clay

The consolidation settlement w_c is written in terms of w_{c1D} as

$$w_c = \mu w_{c1D} \tag{9.66}$$

where

$$w_{c1D} = \int_0^H m_v \Delta\sigma_1 \, dz \tag{9.65}$$

$$\mu = A + \alpha(1 - A) \tag{9.67}$$

and

$$\alpha = \frac{\displaystyle\int_0^H \Delta\sigma_3 \, dz}{\displaystyle\int_0^H \Delta\sigma_1 \, dz} \tag{9.68}$$

The values of α for circular and strip footings as a function of the ratio of the clay layer thickness H to the footing width B are given in Table 9.3.

9.9.6 Symbols and notations

Symbol	Quantity represented	US units	SI units
A, B	Skempton's pore pressure parameters	Unitless	Unitless
C_1	Depth factor	Unitless	Unitless
C_2	Time factor	Unitless	Unitless
C_{gw}	Groundwater correction factor	Unitless	Unitless
D	Depth of embedment of footing	ft	m
E	Young's modulus	psf	kPa
f_L	Layer thickness factor	Unitless	Unitless
f_s	Shape factor	Unitless	Unitless
f_t	Time factor	Unitless	Unitless
G	Shear modulus	psf	kPa
H	Clay layer thickness	ft	m
I	Influence factor	Unitless	Unitless
I_0	Influence factor related to the depth from ground surface to the base of foundation	Unitless	Unitless
I_1	Influence factor related to the depth from foundation base to a very stiff layer	Unitless	Unitless
I_2	Influence factor related to Poisson's ratio of soil	Unitless	Unitless
I_c	Compressibility index	Unitless	Unitless
I_q	Influence factor accounting for footing shape, clay layer thickness, and settlement level	Unitless	Unitless
I_z	Strain influence factor	Unitless	Unitless
I_{z0}	Initial influence factor	Unitless	Unitless
I_{zp}	Peak influence factor	Unitless	Unitless
L_R	Reference length (=1 m = 3.281 ft)	ft	m
N_{60}	SPT blow count corresponding to 60% energy ratio	Unitless	Unitless
p_A	Reference stress (=100 kPa ≈ 1 tsf = 2000 psf)	psf	kPa
q_b	Unit load at the base of footing	psf	kPa

(*Continued*)

Symbol	Quantity represented	US units	SI units
q_b^{net}	Net unit load on the footing base	psf	kPa
R_0	Mean stress-based overconsolidation ratio	Unitless	Unitless
R_3	Ratio of settlement developing over a period of 3 years to the immediate settlement	Unitless	Unitless
R_t	Ratio of settlement developing over a log cycle of time to the immediate settlement	Unitless	Unitless
t	Time	year	year
t_R	Reference time (=1 year)	year	year
w	Total settlement	in.	mm
w_c	Consolidation settlement	in.	mm
w_{c1D}	1D consolidation settlement	in.	mm
w_d	Distortion settlement	in.	mm
$w_{flexible}$	Settlement of flexible foundation	in.	mm
$w_i = w_{st}$	Immediate (short-term) settlement	in.	mm
w_{lt}	Delayed (long-term) settlement	in.	mm
w_r	Radial displacement	in.	mm
w_{rigid}	Settlement of rigid foundation	in.	mm
w_s	Secondary compression settlement	in.	mm
w_z	Vertical displacement	in.	mm
z_f	Depth measured from the base of the footing	ft	m
z_{f0}	Depth of influence	ft	m
z_{fp}	Depth to the peak influence factor	ft	m
Δz_i	Thickness of each sublayer	ft	m
v	Poisson's ratio	Unitless	Unitless
$\sigma_v'\big|_{z_f=0}$	Initial effective overburden pressure at foundation base level	psf	kPa
$\sigma_{vp}'\big|_{z_f=0}$	Maximum previous vertical effective stress at foundation base level	psf	kPa
φ	Ratio of elastic modulus to cone resistance; it is a function of settlement, footing geometry, relative density, and geologic history	Unitless	Unitless
ζ	Nondimensional depth	Unitless	Unitless

9.10 PROBLEMS

9.10.1 Conceptual problems

Problem 9.1 Define all the terms in bold contained in the chapter summary.

Problem 9.2* Discuss possible applications of Saint-Venant's principle in the estimation of displacements caused by foundation loads.

Problem 9.3 Discuss all the factors involved in settlement computations. Discuss which pertain to the soil and which pertain to the foundation element and the superstructure.

Problem 9.4* When calculating the area that a shallow foundation must have in order for it to sustain a given load with a settlement not exceeding a limiting value, we increase the plan area of the footing if calculations at first produce a settlement that is too large. For this to work, the factors that would lead to a decrease in the settlement with increasing plan area of the footing dominate over those that would lead to an increase in the settlement with increasing footing area. Discuss which factors belong to each of the two groups and the magnitude by which they change when the plan area of the footing changes.

9.10.2 Quantitative problems

Problem 9.5 A 700 kN load is applied at a point on the surface of a homogeneous soil deposit. Find the vertical displacement (a) at a depth of 1 m and a horizontal distance of 1 m from the load and (b) at the surface 3 m away from the load. This soil has elastic constants $v=0.2$ and $E=12$ MPa.

Problem 9.6 For a foundation with sides that measure 2 m and 3 m and an axial load of 1100 kN, determine the settlement under one of the corners. The soil has properties $v=0.2$ and $E=14$ MPa.

Problem 9.7 A rectangular footing 2 m wide and 4 m long is placed in clay with an average undrained shear strength $s_u=20$ kPa (see P-Figure 9.1). The embedment depth is 1 m (which is where the water level is as well). As shown in the figure, the excavation is not backfilled. The clay layer extends 7 m below the footing base and is slightly overconsolidated, with an average OCR=2.5 and an average $K_0=0.8$. The unit weight is 16 kN/m³. Atterberg limits were found to be $PL=30\%$ and $LL=70\%$. Use the method by Christian and Carrier (1978) in conjunction with Figures 9.8 and 9.9 and the method of Foye et al. (2008) to estimate the immediate settlement given that the footing will carry a column load of 320 kN. Which method do you expect to be more accurate? Why?

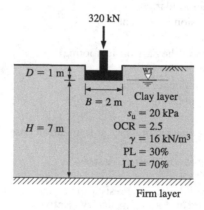

P-Figure 9.1   Footing on clay for which settlement is to be calculated in Problem 9.7.

Problem 9.8 The 3 m square footing in P-Figure 9.2 bears on a clayey sand underlain by a clean sand layer. What is the settlement of this footing when acted upon by a 2100 kN load? Take the unit weights of the sand and clayey sand to be 20 and 18 kN/m³, respectively.

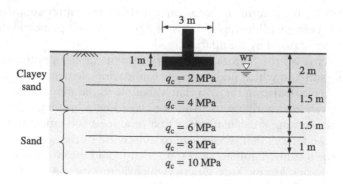

P-Figure 9.2   Footing of Problem 9.8.

Problem 9.9 Using the CPTs in Figure 9.11: (a) calculate the settlement of the Tower of Pisa due to the sand layer if the soil conditions over the entire plan area of the foundation were represented by the CPT logs obtained on the south side of the Tower; (b) repeat your calculations using the CPT logs from the north side of the Tower; (c) based on these calculations, provide an estimate of the initial differential settlement due to the difference in sand properties between the north and south sides. Treat the soil as if it were a pure sand. Use both the original Schmertmann's method and Lee et al.'s method, accounting for the size of the foundation using adjustments as needed. Discuss the potential sources of error in your estimate.

9.10.3 Design problems

Problem 9.10* What are the immediate and consolidation settlements of a 9 ft square footing placed at a depth of 3 ft in the soil profile in P-Figure 9.3? The load on the footing is 200 kN. The footing will be backfilled after construction. The soil has a unit weight of 118 pcf (in Problem 10.21, you will continue to work with this footing and soil profile). Assume a depth of influence based on your expectation of the depth at which the deformations become negligible and adjust if necessary.

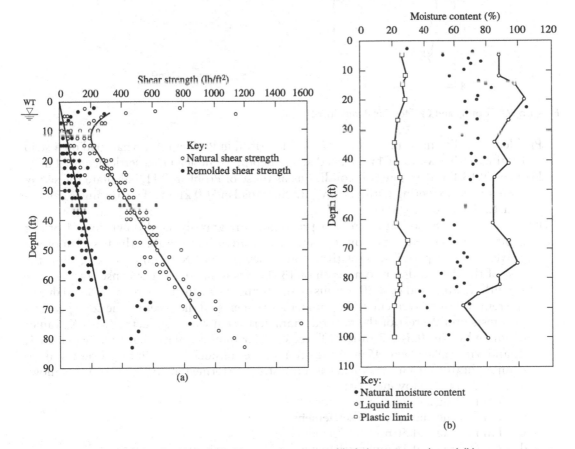

P-Figure 9.3 Soil profile for Problem 9.10: (a) natural and remolded shear strength, and (b) water content, liquid limit, and plastic limit versus depth. (Golder (1957); courtesy of the Institution of Civil Engineers and Thomas Telford Limited.)

Problem 9.11 A 2-meter-wide square footing is placed in sand (see P-Figure 9.4). The footing, which is going to be backfilled, is embedded 1.35 m into the sand layer. The soil unit weight is 19 kN/m³ above the water table (which is at 1.35 m depth) and 20 kN/m³ below the water table. Calculate the footing settlement for a column load of 1500 kN using both Schmertmann's method and Lee et al.'s method. Would this footing pass the serviceability limit state check? The sand deposit is young and normally consolidated ($K_0 = 0.45$ and $\phi_c = 32°$). Neglect any time effects on the settlement.

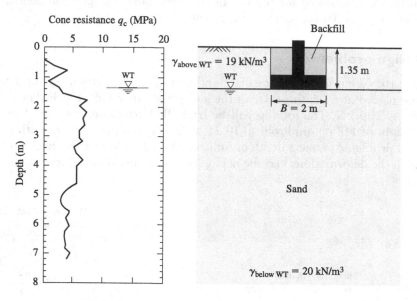

P-Figure 9.4 Footing and CPT log for Problem 9.11.

Problem 9.12 For the same soil conditions described in Problem 9.7, what load on a strip footing with a width of 1 m, placed at 1 m depth, would lead to a settlement of 1 in.?

Problem 9.13 For the same soil conditions as those of Problem 9.11, calculate the settlement of a strip footing with 1.5 m width for a load of 800 kN/m. The footing is embedded 1.35 m into the sand.

Problem 9.14* The soil at a site is a predominantly gravelly sand down to a depth of around 13–14 m. Results from the site investigation indicate that the first 3 m of the gravelly sand deposit is in a relatively loose state ($\gamma \approx 19$ kN/m³ and $\phi_c \approx 30°$), while the rest of the deposit down to a depth of 13–14 m is in dense to very dense states ($\gamma \approx 21$ kN/m³, $K_0 = 0.5$, and $\phi_c \approx 30°$). Soils near the surface are likely overconsolidated due to the removal of a pavement and fill materials that existed there before the cone penetration tests and the rest of the ground characterization were conducted (with K_0 values estimated in the 0.5–0.7 range). Three circular footings, with diameters equal to 1, 2, and 3 m, embedded 0.45 m in the ground, are considered. P-Figure 9.5 shows the q_c profile obtained from cone penetration testing performed at the site. Find the allowable load in the following cases:

a. 1 m footing and 25 mm settlement
b. 1 m footing and 37.5 mm settlement
c. 1 m footing and 50 mm settlement
d. 2 m footing and 25 mm settlement

e. 2 m footing and 37.5 mm settlement
f. 2 m footing and 50 mm settlement
g. 3 m footing and 25 mm settlement
h. 3 m footing and 37.5 mm settlement
i. 3 m footing and 50 mm settlement.

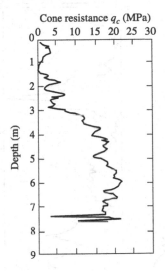

P-Figure 9.5 CPT log for Problem 9.14.

Problem 9.15 Compute the immediate settlement of a footing placed in the soil profile with the CPT log of P-Figure 9.6. The footing is a 1.5 m square footing placed at a depth of 1.5 m. The sand is reasonably well graded, with subangular particles, gravelly in some places and clayey in others. The unit weight of the sand for use in your calculations is 20 kN/m³. The sand is normally consolidated, with $K_0=0.4$. The critical-state friction angle is 30°. A clay layer starts at 3.5 m and extends to roughly 6.5 m. The water table is at a depth of 3 m. The load on the footing is 2000 kN. Is this settlement likely to be tolerable? In Problem 10–25, you will investigate the stability of this footing.

Problem 9.16 Recalculate the settlement of the footing of Example 9.4 using (a) Peck and Bazaraa's method and (b) Meyerhof's method.

Problem 9.17* Convert the SPT blow counts to cone penetration resistance q_c and recalculate the settlement of the footing of Example 9.4 using Lee et al.'s method. Assume the sand to be normally consolidated with $D_{50}=0.23$ mm and $\phi_c=30°$.

Problem 9.18* Complete the design of Example 9.6 by recalculating the footing width B until the settlement is just under the tolerable settlement of 38 mm. The footing will be backfilled after construction.

Problem 9.19* In Problem 6.24, you calculated the one-dimensional compression caused by a unit load numerically equal to that acting on the base of the Tower of Pisa. Now, estimate the total consolidation settlement taking full account of the three-dimensional (axisymmetric) nature of the problem. Use only the information on the index properties of the clay given in Table 6.4 and your knowledge of the material covered in Chapters 6 and 9.

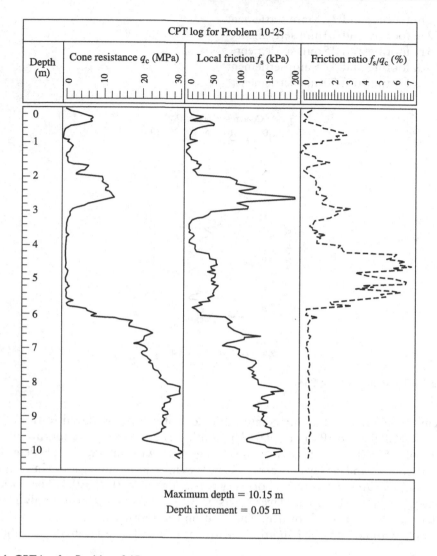

CPT log for Problem 10-25			
Depth (m)	Cone resistance q_c (MPa)	Local friction f_s (kPa)	Friction ratio f_s/q_c (%)

Maximum depth = 10.15 m

Depth increment = 0.05 m

P-Figure 9.6 CPT log for Problem 9.15.

Problem 9.20 The soil profile at a site consists of 5–7 m of a poorly graded, medium-dense, quartz dune sand ($K_0=0.43$, $\gamma_m=16.4$ kN/m³, and $\phi_c=32°$) overlying a weakly cemented limestone. The sand layer is of Holocene age and was formed from the chemical weathering (dissolution) of limestone with subsequent erosion, transportation, and re-deposition by wind. The water table is typically located at a depth of about 5.5 m, just above the limestone layer. P-Figure 9.7 shows the q_c profile obtained from cone penetration testing performed at the site. A 1.5-meter-wide and 1.0-meter-thick square footing is embedded 1.0 m into the sand layer. The footing is not backfilled. Neglect the effect of creep and take the time factor as equal to 1. Using both Schmertmann's method and Lee et al.'s method[9]:

a.   Plot the load–settlement curve of the footing up to a settlement of 100 mm.

---

[9] This problem is in part based on real data obtained by Lehane et al. (2008).

b. Compute the net unit load at the base of the footing for settlements of 25 and 50 mm.

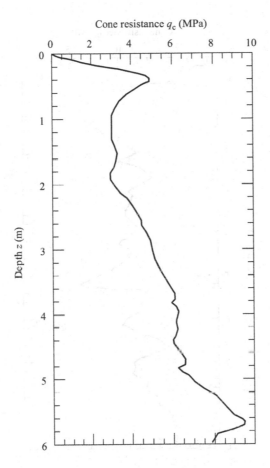

P-Figure 9.7 CPT log for Problem 9.20. (Data from Schneider 2007.)

Problem 9.21 The soil profile at a site consists predominantly of a medium-dense, silty, fine silica sand ($K_{0,NC}=0.45$ and $\phi_c=34.2°$) of Pleistocene age up to a depth of 11 m. The fines content at the site generally varies from about 8% near the ground surface to 35% at a depth of 9 m; the fines are nonplastic. The sand layer is overconsolidated due to the desiccation of the fines and the removal of about 1 m of overburden prior to the construction of the footing. Below the sand layer, there is a very stiff, marine clay deposit of Eocene age extending down to a depth of about 33 m. The liquid limit LL and plasticity index PI of the clay layer are 40% and 21%, respectively. The unit weight of the sand layer is 15.5 kN/m³ above the water table (which is at 4.9 m depth) and 20.5 kN/m³ below the water table. P-Figure 9.8 shows the q_c profile obtained from cone penetration testing performed at the site. A 3.0-meter-wide and 1.2-meter-thick square footing is embedded 0.75 m into the sand layer. The footing is not back-filled. Neglect the effect of creep and take the time factor as equal to 1. Using both Schmertmann's method and Lee et al.'s method[10]:

---

[10]This problem is in part based on real data obtained by Briaud and Gibbens (1997).

a. Plot the load–settlement curve of the footing up to a settlement of 100 mm.
b. Compute the net unit load at the base of the footing for settlements of 25 and 50 mm.

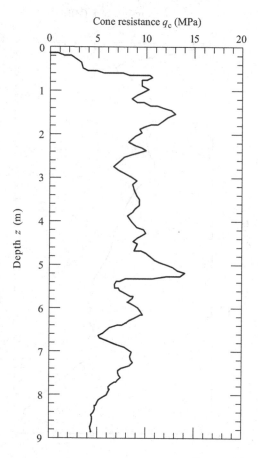

P-Figure 9.8 CPT log for Problem 9.21. (Modified from Briaud and Gibbens 1997.)

Problem 9.22 The soil profile at a site consists of a 1.5-meter-thick stiff crust of clay followed by soft, normally consolidated, estuarine clay ($G_s = 2.70$ and $K_0 = 0.63$) down to a depth of 10–15 m. Dense gravel lies below the clay. P-Figure 9.9 shows the depth profiles of net cone resistance $q_t - \sigma_v$, plastic limit PL, water content wc, liquid limit LL, and undrained shear strength s_u. A 10-meter-long, 5-meter-wide, and 175-millimeter-thick rectangular footing is embedded 0.175 m into the clay crust. The load on the footing is 1100 kN, and the footing is not backfilled. Assume the water table to be at the ground surface, and take $\gamma_{sat} = 16.5$ kN/m³ for both the clay crust and clay layers.[11]

a. Calculate the immediate settlement w_i of the footing using the methods by Christian and Carrier (1978) and Foye et al. (2008).
b. Calculate the consolidation settlement w_c of the footing.
c. Calculate the total settlement w of the footing.
d. Would this footing pass the serviceability limit state check?

---

[11] This problem is in part based on real data obtained by Schnaid et al. (1993).

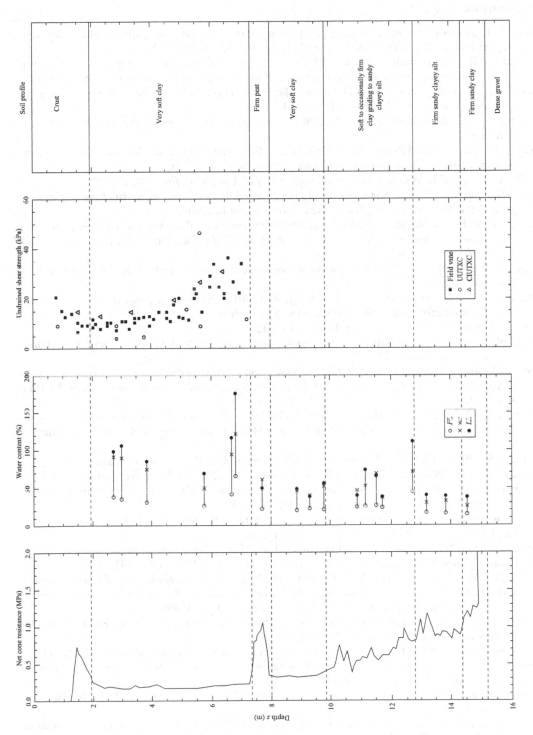

P-Figure 9.9 CPT log and soil profile for Problem 9.22. (Modified from Schnaid et al. 1993.)

REFERENCES

References cited

Briaud, J.-L. and Gibbens, R. (1997). *Large-scale load tests and data base of spread footings on sand.* Report No. FHWA-RD-97-068, Federal Highway Administration, Virginia, USA.

Burland, J. B. and Burbidge, M. C. (1985). "Settlement of foundations on sand and gravel." *Proceedings of Institution of Civil Engineers. Part 1: Design and Construction*, 78, 1325–1381.

Christian, J. T. and Carrier III, W. D. (1978). "Janbu, Bjerrum and Kjaernsli's chart reinterpreted." *Canadian Geotechnical Journal*, 15(1), 123–28.

Duncan, J. M. and Buchignani, A. L. (1987). *An Engineering Manual for Settlement Studies*, Blacksburg, VI.

Foye, K. C., Basu, P., and Prezzi, M. (2008). "Immediate settlement of shallow foundations bearing on clay." *International Journal of Geomechanics*, 8(5), 300–310.

Golder, H. Q. (1957). "A note on piles in sensitive clays." *Géotechnique*, 7(4), 192–195.

Janbu, N., Bjerrum, L., and Kjaernsli, B. (1956). "Veiledning ved losning av fundamenteringsoppgaver." *Publication No. 16, Norwegian Geotechnical Institute*.

Lee, J., Eun, J., Prezzi, M., and Salgado, R. (2008). "Strain influence diagrams for settlement estimation of both isolated and multiple footings in sand." *Journal of Geotechnical and Geoenvironmental Engineering*, 134(4), 417–427.

Lee, J. and Salgado, R. (2002). "Estimation of footing settlement in sand." *International Journal of Geomechanics*, 2(1), 1–28.

Lehane, B. M., Doherty, J. P., and Schneider, J. A. (2008). "Settlement prediction for footings on sand." *Proceedings of 4th International Symposium on Deformation Characteristics of Geomaterials*, Atlanta, Georgia, USA, 1, 133–150.

Meyerhof, G. G. (1965). "Shallow foundations." *Journal of the Soil Mechanics and Foundations Division*, 91(SM2), 21–31.

Peck, R. B. and Bazaraa, A. R. (1969). "Discussion on settlements of spread footings on sand by D'Appolonia and Brisette." *Journal of the Soil Mechanics and Foundations Division*, 95(SM3), 905–909.

Robertson, P. K. and Campanella, R. G. (1989). *Guidelines for Geotechnical Design Using the Cone Penetrometer Test and CPT with Pore Pressure Measurement.* Hogentogler and Co., Columbia, MD.

Saada, A. S. (2009). *Elasticity Theory and Applications.* J. Ross Publishing, Plantation, FL.

Sakleshpur, V. A., Prezzi, M., Salgado, R., and Zaheer, M. (2021). *CPT-based geotechnical design manual–volume II: CPT-based design of foundations (methods).* Joint Transportation Research Program, Purdue University, West Lafayette, IN, USA.

Schmertmann, J. H. (1970). "Static cone to compute static settlement over sand." *Journal of the Soil Mechanics and Foundations Division*, 96(SM3), 1011–1042.

Schmertmann, J. H., Brown, P. R., and Hartman, J. P. (1978). "Improved strain influence factor diagrams." *Journal of the Geotechnical Engineering Division*, 104(GT8), 1131–1135.

Schnaid, F., Wood, W. R., Smith, A. K. C., and Jubb, P. (1993). "An investigation of bearing capacity and settlements of soft clay deposits at Shellhaven." *Predictive Soil Mechanics: Proceedings of Wroth Memorial Symposium*, St. Catherine's College, Oxford, 609–627.

Schneider, J. A. (2007). "Analysis of piezocone data for displacement pile design." Ph.D. Thesis, The University of Western Australia, Perth, Australia.

Skempton, A. W. (1954). "The pore-pressure coefficients A and B." *Géotechnique*, 4(4), 143–147.

Skempton, A. W. and Bjerrum, L. (1957). "A contribution to the settlement analysis of foundations on clay." *Géotechnique*, 7(4), 168–178.

Terzaghi, K. and Peck, R. B. (1967). *Soil Mechanics in Engineering Practice.* 2nd Edition, Wiley, New York.

Viggiani, G. and Atkinson, J. H. (1995). "Stiffness of fine-grained soil at very small strains." *Géotechnique*, 45(2), 249–265.

Additional references

Taylor, B. B. and Matyas, E. L. (1983). "Influence factors for settlement estimates of footings on finite layers." *Canadian Geotechnical Journal*, 20(4), 832–835.

Timoshenko, S. P. and Goodier, J. N. (1970). *Theory of Elasticity.* 3rd Edition, McGraw-Hill, New York.

Chapter 10

Shallow foundations: limit bearing capacity

Theory is the language by means of which lessons of experience can be clearly expressed. When there is no theory, there is no collected wisdom, merely incomprehensible fragments.

Karl Terzaghi

The action was brought to recover damages for a failure of defendants to erect and complete a building on a lot of plaintiffs, on Minnesota street, between Third and Fourth streets, in the city of St. Paul, which ... the defendants had agreed to build, erect, and complete, according to plans and specifications annexed to and made part of the agreement. The defendants commenced the construction of the building, and had carried it to the height of three stories, when it fell to the ground. The next year, 1869, they began again and carried it to the same height as before, when it again fell to the ground, whereupon defendants refused to perform the contract.

Stees v. Leonard, 20 Minn. 494, 449 (1874).

As the load on a foundation element increases from zero initial value, the element first experiences some settlement, which can be estimated using analyses discussed in Chapter 9. After the load exceeds a threshold value, some regions of the soil enter the plastic range. Further increases in load enlarge these plastic zones within the soil mass to the extent that they may coalesce and reach a free boundary (typically, the ground surface). Extremely large settlements are then possible with small to no further increases in load. A foundation element is said to have failed in bearing capacity or to have reached the bearing capacity limit state when it undergoes such very large settlements.

While not recognized as such at the time, it is reasonable to conclude that the contractor in *Stees v. Leonard* (the epigraph contains an extract from the corresponding court opinion) twice brought the building foundations to their limit bearing capacity. The magnitude of the load causing a foundation element to fail in this manner (referred to as the limit bearing capacity, limit resistance, or limit load capacity of the footing) depends on a variety of factors, including the plan dimensions of the element, the depth at which it is placed in the ground, and the shear strength of the soil. Bearing capacity checks are an essential part of foundation design: a foundation element must not give under the application of the structural loads. While some questions remain unsolved, there is now a solid theoretical basis for the calculation of shallow foundation limit bearing capacity. In this chapter, we will examine this theory: the mechanics of bearing capacity failures and the analyses available for calculating the limit bearing capacity of a shallow foundation.

DOI: 10.1201/b22079-10

10.1 THE BEARING CAPACITY EQUATION FOR STRIP FOOTINGS

10.1.1 Bearing capacity failure and the bearing capacity equation

Figure 10.1a shows the force equilibrium for a vertically loaded shallow foundation element. The water table is assumed to be deep. Any side resistance is neglected, and the weight W_{ftg} of the footing, the weight W_{fill} of the backfill, and the applied load Q are carried only by the mobilized *base resistance* Q_b. In Figure 10.1b, the water table is located above the footing base, and the pore pressure at the footing base also helps support the applied load and weight of footing and backfill. As Q increases, the mobilized gross base resistance Q_b also increases. The increase in Q_b will be capped at the limit load Q_{bL}, the value of the *gross resistance* at which the footing will plunge into the ground. This is referred to as the limit bearing capacity failure. The difference between Q_{bL} and the pore water force at the footing base is the effective limit base resistance or effective limit bearing capacity of the footing.

The limit bearing capacity is usually calculated from the geometry of the footing and the strength profile of the ground in which it rests. Figure 10.2 shows a cross section of a strip footing with width B and base placed at depth D within a soil mass. In the most general set of conditions and most basic analysis of this problem, the soil is assumed to have unit weight γ, to be isotropic in terms of its shear strength response, and to be linear elastic, perfectly plastic. This means that the soil response is elastic before the yield criterion is satisfied, and then deformation takes place without a change in shear strength.

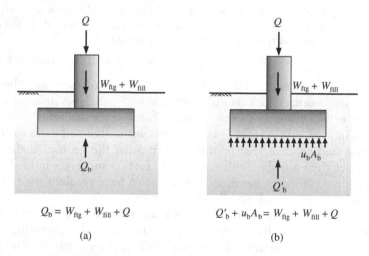

$$Q_b = W_{ftg} + W_{fill} + Q$$

$$Q'_b + u_b A_b = W_{ftg} + W_{fill} + Q$$

(a)   (b)

Figure 10.1 Vertical force equilibrium for a shallow foundation element: (a) without pore pressure at base; (b) with pore pressure at base.

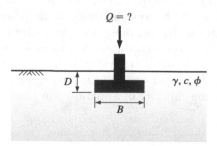

Figure 10.2 The bearing capacity problem.

Traditional bearing capacity analysis has been performed using the Mohr–Coulomb yield criterion. The shear strength parameters are then the cohesive intercept c and friction angle ϕ. In practice, uncemented soil (whether sandy or clayey) does not have a nonzero cohesive intercept. The bearing capacity of a footing in sand is calculated using drained analysis based on a single shear strength parameter: the friction angle ϕ. However, the bearing capacity of saturated soils loaded under undrained conditions can be calculated using total stresses, a nonzero c value, and ϕ = zero.[1]

For now, let us ignore that there may be a water table present. If the vertical (axial) load on this footing is continually increased, the footing will first settle by increasing amounts. For values of load below some limit, plasticity does not develop to any significant extent within the soil (Figure 10.3a) and foundation settlements can be estimated using elasticity theory (see Chapter 9). There will be one value of load for which, at some point or points in the soil, the soil first enters the plastic range (that is, the stresses at those points satisfy the operative yield criterion; see Footnote 1). Figure 10.3b illustrates the typical case in which the soil at the edges of the footing enters the plastic range first. Upon further increases in load, other points, first adjacent to the initial points and then further away, will also enter the plastic range (Figure 10.3c). The plastic zone thus formed increases in size with increasing load (Figure 10.3d), but continues at all times to be confined, surrounded by soil in the elastic range. For soils that are not excessively contractive, at some value of load,

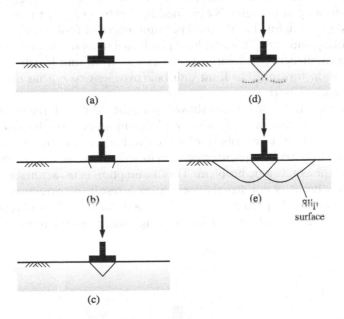

Figure 10.3 Effects of increasing the load on a shallow foundation element: (a) elastic settlement; (b) footing experiences further settlement as the soil enters the plastic range at the footing edges; (c) settlement continues as the plastic zone increases; (d) settlement may become large as the plastic zone expands, but flow continues to be constrained by the surrounding elastic zone; and (e) unrestrained flow occurs when the plastic zone extends to the surface of the soil mass (this is referred to as the classical or limit bearing capacity failure). Evolution of settlement is not shown.

---

[1] As discussed in Chapter 5, sands have been modeled using the Mohr–Coulomb criterion with $c = 0$ and $\phi > 0$. In Chapter 6, we saw that, in total stress analysis of clays, they can be modeled as having $c = s_u$ and $\phi = 0$. Nonzero c and ϕ values are specified for an uncemented soil typically only when both are curve-fitting parameters, but in Chapter 5, we saw that curve fitting is not the best approach for finding shear strength parameters for foundation analysis. A soil with interparticle cementation will have nonzero values for both c and ϕ.

which we will call the limit load, the plastic zone extends to the soil mass boundary (the ground surface), at which point there is no material still in the elastic range constraining it (Figure 10.3e). Very large displacements and foundation settlement can then occur because the plastic zone expansion is no longer constrained. We can state that, at that point, a limit bearing capacity failure has occurred.

The limit bearing capacity problem can be concisely stated as follows: determine the limit load Q_L that, when applied to the footing, causes it to plunge into the ground, that is, causes it to settle by very large amounts. There is a difference between the load applied to the footing by the superstructure and the load transferred to the soil across the base of the footing. This difference is equal to the weight of the footing and any backfill soil on top of the footing.[2]

It is useful at this point to define two terms often used in the context of bearing capacity calculations: unit gross resistance and unit net resistance. Unit gross resistance is the stress that must be applied at the soil–foundation interface to reach the critical ultimate limit state for the foundation. Some of this stress will be caused by the application of a superstructure load to the foundation element, but some will be due simply to the weight of the foundation. The unit net resistance is the stress that must be applied in excess of any stress due to foundation or backfill loads. It is the net resistance that is available to sustain superstructure loads, but it is often gross resistance that is calculated using design equations.

At the plunging point, the load transferred to the soil across the footing base is denoted by Q_{bL}; it may be referred to as the limit bearing capacity of the footing. The unit load q_{bL} at the base of the footing at plunging is obtained by dividing Q_{bL} by the area of the footing. For strip footings, Q_{bL} is defined as the load per unit length of footing, so q_{bL} is equal to Q_{bL} divided by the footing width B. This unit load is referred to as the limit unit gross resistance (sometimes, for convenience, just limit unit resistance), limit unit bearing capacity or limit bearing pressure of the footing. The limit unit bearing capacity q_{bL} has units of stress: load per unit length per unit length.

Figure 10.4 portrays the same footing shown in Figure 10.2 with the overburden soil (the soil above the level of the base of the footing) replaced by an equivalent overburden unit load $q_0 = \gamma D$. If we assume that the contribution of the overburden soil (that is, the soil above the base of the footing) to bearing capacity stems exclusively from its weight, then Figures 10.2 and 10.4 represent the exact same problem. This assumption is in fact made to develop bearing capacity theory, but it is not strictly correct; as a result, a correction factor has to be introduced into the bearing capacity equation to account for this. We will return to this point later. We will now focus on finding the limit bearing capacity of the footing in Figure 10.4.

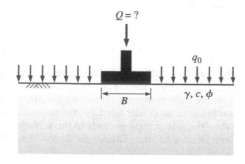

Figure 10.4 The bearing capacity problem with surcharge substituted for the overburden soil.

---

[2] This difference is in most cases small, but in other cases can be important. We will discuss this in greater detail later.

Example 10.1

What is the equivalent soil surcharge q_0 for the situation shown in E-Figure 10.1?

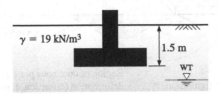

$\gamma = 19 \text{ kN/m}^3$

1.5 m

WT

E-Figure 10.1 Footing for Example 10.1.

Solution:

The surcharge q_0 is calculated by calculating the weight of a unit slice of soil above the base of the footing. So,

$$q_0 = \gamma D = 19 \times 1.5 = 28.5 \text{ kPa} \approx 28 \text{ kPa}$$

To identify the sources of bearing capacity for this footing, we need to understand the process by which it would plunge into the soil. For the footing to move down substantially, a "mechanism" must form. This mechanism becomes possible as slip surfaces (shear surfaces or, in more realistic terms, shear bands) develop in the soil, so that blocks of relatively rigid soil can move with respect to one another, allowing the footing to move down. An example of a possible mechanism is shown in Figure 10.5a. Observe how the downward movement of the footing implies a downward movement of block A, a lateral, rotational movement of block T, and an upward movement of block B, as seen in Figure 10.5b. The elements that resist the downward movement of the footing are the same that resist movement of the whole mechanism, starting with the surcharge q_0, which counters the upward movement of block B. The shear strength along the slip surfaces, which is a function of the soil unit weight γ

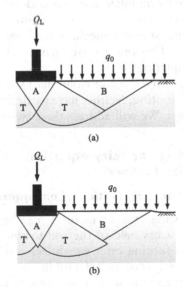

Q_L

q_0

A   B

T   T

(a)

Q_L

q_0

A   B

T   T

(b)

Figure 10.5 Simplified bearing capacity mechanism: (a) before and (b) after sliding.

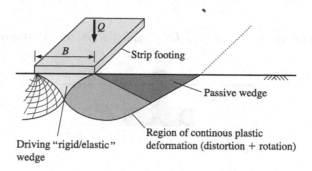

Figure 10.6 Slip mechanism for a strip footing.

and the Mohr-Coulomb parameters ϕ and c, also opposes the movement. Figure 10.6 shows a three-dimensional perspective of the slip mechanism, with shapes that are less schematic and more consistent with more realistic representations of bearing capacity failure.

Summarizing the discussion so far, the resistance of the footing to plunging (that is, its bearing capacity) increases with increasing q_0, c, and γ for a given friction angle ϕ. The bearing capacity also increases with increasing ϕ. An equation for bearing capacity can be proposed, based on these observations, in terms of the three sources of bearing capacity we have identified for the footing: the surcharge q_0, any cohesive component c of shear strength, and a representative soil unit weight γ below the base of the foundation. This can be expressed through an equation called the *bearing capacity equation* (Brinch Hansen 1970; Meyerhof 1951, 1963; Terzaghi 1943):

$$q_{bL} = cN_c + q_0 N_q + \frac{1}{2}\gamma B N_\gamma \tag{10.1}$$

where N_c, N_q, and N_γ = bearing capacity factors multiplying terms, respectively, containing (i) the cohesive intercept c, (ii) the surcharge q_0 acting at the level of the base of the footing, and (iii) a unit weight γ representative of the soil below the base of the footing, respectively.

The presence of the foundation width B in the third term of Equation (10.1) deserves further discussion. Wider footings mobilize larger and deeper shear mechanisms. Deeper mechanisms imply greater effective stresses. Greater effective stresses imply, for soils with nonzero ϕ and γ, greater frictional shear strength. Accordingly, wider footings have greater limit unit bearing capacities q_{bL}. This result is mathematically captured through the presence of B in the third term of Equation (10.1).

Recall that, so far, we have ignored the contribution of shear strength of the soil located above the level of the base of the footing. Neither have we derived an equation valid for footings with shapes other than strips. We will address these two issues later.

10.1.2 Derivation of bearing capacity equation and bearing capacity factors*

10.1.2.1 Frictional, weightless soil: derivation of an equation for N_q

Following Bolton (1979), the bearing capacity factor N_q can be determined using a relatively simple analysis that relates the stress field in the soil to the boundary loads that it must support. Consider the case of a footing on a weightless soil ($\gamma = 0$) with $c = 0$, shown in Figure 10.7. The major principal stress σ_1^A is constant in magnitude (equal to q_{bL}) and direction (vertical) within zone A. In zone B, q_0 is a principal stress because it acts on a plane

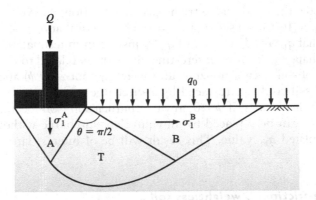

Figure 10.7 Stress rotation beneath shallow foundation.

where the shear stress is equal to zero. We also know that q_0 is the minor principal stress, for zone B is being compressed laterally. The major principal stress in zone B is horizontal and related to q_0 through Equation (4.48):

$$\sigma_1^B = Nq_0 \tag{10.2}$$

where the flow number N is given by Equation (4.45), repeated here:

$$N = \frac{1 + \sin\phi}{1 - \sin\phi} \tag{10.3}$$

Zones A and B are separated by a radial transition zone T. Considering that $\sigma_1^A = q_{bL}$ is vertical and σ_1^B, given by Equation (10.2), is horizontal, there is a stress rotation of 90° degrees as the major principal stress goes from vertical below the footing in zone A to horizontal in zone B. This means the angle θ between the boundaries of zones A and B must equal $\pi/2$ (as shown in the figure) in order for the stress field to be continuous. It is possible to show (see Appendix E) that the relationship between the major principal stress σ_1^A in zone A and the major principal stress σ_1^A in zone B is given by

$$\sigma_1^A = \sigma_1^B e^{2\theta\tan\phi} \tag{10.4}$$

Making the appropriate substitutions in Equation (10.4):

$$q_{bL} = Nq_0 e^{\pi\tan\phi} \tag{10.5}$$

which is the bearing capacity equation for a purely frictional, weightless soil. It follows that N_q is given by:

$$N_q = \frac{1 + \sin\phi}{1 - \sin\phi} e^{\pi\tan\phi} \tag{10.6}$$

So now:

$$q_{bL} = q_0 N_q \tag{10.7}$$

Let us now consider the ideal and extreme case of a frictionless, weightless soil.[3] By making $\phi = 0$ in Equation (10.6), we see that $N_q = 1$. This mean that $q_{bL} = q_0$ for this hypothetical soil, meaning that q_{bL} needs to exceed q_0 by just an infinitesimal amount for a bearing capacity failure to happen. Another interesting observation related to this case is that, if we had a thickness D of soil with a nonzero unit weight γ (but zero ϕ) above the level where q_{bL} is applied, that soil could be replaced by a surcharge $q_0 = \gamma D$, and we would still have the same result. In other words, if $\phi = 0$, any overburden soil (soil located above the level at which q_{bL} is applied) can be replaced by an equivalent surcharge without introducing any error into the calculated q_{bL} value. This result will be of interest later when we examine depth factors.

10.1.2.2 Cohesive-frictional, weightless soil

The bearing capacity factor N_c can be easily obtained from N_q by making use of Caquot's stress similarity principle. According to this principle, which was introduced in Chapter 4, every equation involving normal stresses for a purely frictional soil (with $c = 0$ and nonzero ϕ) is also valid for a c–ϕ soil so long as the term $c.\cot\phi$ is added to every normal stress in the equation. The principle is easily proven, for adding $c.\cot\phi$ to normal stresses is equivalent to shifting the vertical axis of a Mohr's diagram (Figure 4.15) to the left by an amount equal to $c.\cot\phi$, so that the intercept of the Mohr–Coulomb envelope drops from a nonzero c to $c = 0$. Applying this principle to an ideal, weightless c–ϕ soil requires that we add $c.\cot\phi$ to both q_{bL} and q_0 in Equation (10.5), yielding:

$$q_{bL} + c.\cot\phi = N\left(q_0 + c.\cot\phi\right)e^{\pi\tan\phi} \tag{10.8}$$

which can be rewritten to become:

$$q_{bL} = Nq_0 e^{\pi\tan\phi} + \left(Ne^{\pi\tan\phi} - 1\right)c.\cot\phi \tag{10.9}$$

from which, by comparison with Equation (10.1), it is apparent that N_c is given by:

$$N_c = \left(\frac{1+\sin\phi}{1-\sin\phi}e^{\pi\tan\phi} - 1\right)\cot\phi \tag{10.10}$$

or, simply,

$$N_c = \left(N_q - 1\right)\cot\phi \tag{10.11}$$

10.1.2.3 Soil with self-weight: expressions for N_γ for associative materials

Equations (10.6) and (10.11), obtained earlier for N_q and N_c, are mathematically rigorous for a material that follows an associated flow rule. As discussed in Chapter 4, this means that the dilatancy and friction angles are identical. They were first obtained a century ago by Prandtl (1920) and Reissner (1924) using a more advanced analysis known as the method of characteristics or slipline method (Sokolovski 1965). N_γ has been a more elusive quantity, but exact solutions are now available for N_γ (Martin 2005; Lyamin et al. 2007), also for an associated flow rule. According to these solutions, based on limit analysis and the method of characteristics,

---

[3] A frictionless, weightless soil is obviously an ideal material that does not exist, but one that can teach us something about the workings of the equations we are deriving.

the following expression for N_γ (due to Brinch Hansen 1970) approximates rigorous results for purely frictional soils (soils with nonzero ϕ and $c = 0$) fairly well up to ϕ values not exceeding 40°:

$$N_\gamma = 1.5\left(N_q - 1\right)\tan\phi \tag{10.12}$$

Lyamin et al. (2007) and Martin (2005) arrived independently at exact solutions for N_γ. The following equation fits almost perfectly the exact values of N_γ, even for very low values of friction angle (Loukidis and Salgado 2011):

$$N_\gamma = \left(N_q - 0.6\right)\tan\left(1.33\phi\right) \tag{10.13}$$

N_γ calculated using this expression is plotted together with N_c and N_q in Figure 10.8.

Example 10.2

Calculate the following: (a) N_q and N_γ for a friction angle of 30°; (b) N_c for total stress analysis of clay.

Solution:

a. We know from Equation (10.6) that:

$$N_q = \frac{1 + \sin\phi}{1 - \sin\phi}e^{\pi\tan\phi} = \frac{1 + \sin 30°}{1 - \sin 30°}e^{\pi\tan 30°} = 18.4$$

Using Equations (10.12) and (10.13) in sequence:

$$N_\gamma = 1.5\left(N_q - 1\right)\tan\phi = 1.5\left(18.4 - 1\right)\tan 30° = 15.1$$

$$N_\gamma = \left(N_q - 0.6\right)\tan\left(1.33\phi\right) = \left(18.4 - 0.6\right)\tan\left(1.33 \times 30°\right) = 14.9$$

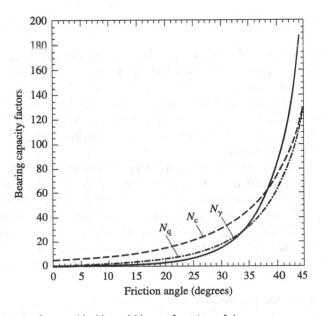

Figure 10.8 Bearing capacity factors N_c, N_q, and N_γ as a function of ϕ.

b. For total stress analysis in clay, $\phi = 0°$. Plugging this value of ϕ into Equation (10.11) leads to an undefined number. So we need to take the limit of N_c as ϕ approaches zero. The application of L'Hospital's rule produces the exact limit: $2 + \pi$. Although not very elegant from a mathematical standpoint, we can also make ϕ very small and calculate N_c. When we do that, we find that N_c is approximately 5.14. Let us illustrate this by making ϕ equal to 0.1°:

$$N_q = \frac{1+\sin\phi}{1-\sin\phi}e^{\pi\tan\phi} = \frac{1+\sin 0.1°}{1-\sin 0.1°}e^{\pi\tan 0.1°} = 1.003497 \times 1.005498 = 1.009014$$

$$N_c = (N_q - 1)\cot\phi = (1.009014 - 1) \times 572.96 = 5.165$$

Repetition of these calculations for $\phi = 0.01°$ would give 5.144. In fact, there is an exact value for N_c: $2 + \pi$.

Although it is not theoretically correct to combine the effects of c, q_0, and γ as is done in Equation (10.1), this equation has been used in practice for decades. It is possible, however, to develop the bearing capacity equation in such a way that there is no error due to this combination. This was done by Lyamin et al. (2007) for sands by developing expressions for the shape and depth factors that multiply a single term (not the separate q_0 and γ terms of the traditional bearing capacity equation).

10.1.3 The bearing capacity equation for materials following a nonassociated flow rule*

The derivation of the bearing capacity equation done earlier presupposes that the soil follows an associated flow rule. The flow rule is a rule from plasticity theory that links the principal directions of plastic strain increments to those of the principal stresses. When the material follows an associated flow rule, the plastic strain rates follow directly from the yield function of the material by differentiation:

$$\dot{\varepsilon}_{ij}^p = \dot{\lambda}\frac{\partial F}{\partial \sigma_{ij}} \tag{10.14}$$

where, as discussed in Chapter 4, i and j are indices taking the values 1, 2, or 3.

For a Mohr–Coulomb material, the use of an associated flow rule implies that the dilatancy angle is equal to the friction angle of the material: $\psi = \phi$. As seen in Chapter 4, this is not verified experimentally for a real sand. If the sand is modeled with a set value of ϕ and a set value of ψ, then realistic modeling requires $\phi > \psi$. This means that the calculation of bearing capacity using Equation (10.13) or any other equation derived based on associated flow would lead to incorrect results.

Salgado (2020) shows how the use of an associated flow rule leads to the calculation of bearing capacity factors that are up to 30% greater than they in reality would be. Figure 10.9 shows the comparison for realistic pairings of $\phi > \psi$. Loukidis and Salgado (2009) performed finite element (FE) simulations of strip and circular footings on sand using an elastic–perfectly plastic constitutive model following the Mohr–Coulomb yield criterion considering nonassociated flow, and obtained equations for the bearing capacity factors in terms of both ϕ and ψ. Equation (10.13) can again be used for N_γ, but N_q is now calculated using:

$$N_q = \frac{1+\sin\phi}{1-\sin\phi}\exp\left[F(\phi,\psi)\pi\tan\phi\right] \tag{10.15}$$

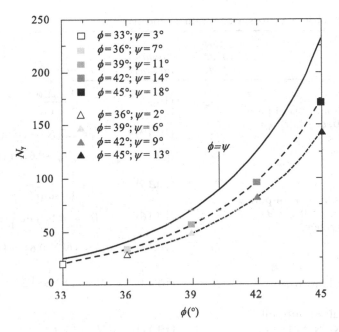

Figure 10.9 Comparison of values of bearing capacity factor N_γ calculated based on the assumption of associated flow ($\psi = \phi$) with values calculated based on nonassociated flow ($\psi < \phi$).

where

$$F(\phi,\psi) = 1 - \tan\phi \left[\tan\left(0.8(\phi - \psi)\right) \right]^{2.5} \qquad (10.16)$$

The difficulty in using the bearing capacity equation for materials modeled using non-zero ϕ and ψ is that there is not a single friction or dilatancy angle that represents the soil around the footing. Every point in the soil below and around the footings will be subjected to a different stress state, will have deformed to different degrees, and accordingly will have different values of mobilized friction and dilatancy angles. Regardless of which equations are used for the calculation of the bearing capacity factors, the question remains of which value of friction angle and, in the case of Equations (10.15) and (10.16), dilatancy angle to use to produce the correct value of bearing capacity. This is an important question in the calculation of the bearing capacity of shallow foundations in sands, so we return to it when we focus our discussion on sand.

10.1.4 Using the bearing capacity equation

We have seen in the preceding subsections that the limit unit bearing capacity of a shallow foundation element can be calculated using Equation (10.1). The bearing capacity factors can be determined from equations repeated in Table 10.1 or from Figure 10.8. Bearing capacity factors derived using different techniques by different authors based on different assumptions must not be used together. We can identify three groups of factors in the table:

1. Factors derived based on an assumption that a Mohr–Coulomb strength envelope applies with $\psi = \phi$ (that is, Mohr–Coulomb materials with an associated flow rule);

Table 10.1 Bearing capacity factors used for sands and clays

Soil/analysis	Bearing capacity factor	Equation number	Equation
Sand: effective stress analysis of purely frictional soil ($c = 0$, $\phi > 0$, $\psi = \phi$) Associated flow rule	N_q	(10.6)	$N_q = \dfrac{1 + \sin\phi}{1 - \sin\phi} e^{\pi\tan\phi}$
	N_γ	(10.12)	$N_\gamma = 1.5(N_q - 1)\tan\phi$
	N_γ	(10.13)	$N_\gamma = (N_q - 0.6)\tan(1.33\phi)$
Sand: effective stress analysis of purely frictional soil ($c = 0$, $\phi > 0$, $\psi < \phi$) Nonassociated flow rule	N_q	(10.15)	$N_q = \dfrac{1 + \sin\phi}{1 - \sin\phi} \exp\left[F(\phi,\psi)\pi\tan\phi \right]$
		(10.16)	$F(\phi,\psi) = 1 - \tan\phi\left[\tan\left(0.8(\phi - \psi)\right) \right]^{2.5}$
	N_γ	(10.13)	$N_\gamma = (N_q - 0.6)\tan(1.33\phi)$
Sand: N_γ based on relative density	N_γ	(10.43)	$N_\gamma = 2.82\exp\left(3.64\dfrac{D_R}{100\%} \right)\left(\dfrac{\gamma B}{p_A} \right)^{-0.4}$
Clay: total stress analysis with soil modeled as purely cohesive ($c = s_u$ and $\phi = 0$)	N_c	(10.11)	$N_c = 2 + \pi$

2. Factors derived based on an assumption that a Mohr–Coulomb strength envelope applies with $\psi < \phi$ (that is, Mohr–Coulomb materials with a nonassociated flow rule);
3. Factors derived based on numerical analyses using a refined constitutive model for the soil. These factors can be expressed directly in terms of state variables such as initial relative density. Equation (10.43), appearing in the table, will be discussed more completely in a later section.

As a rule, factors from one group must not be used with factors from another group.

Example 10.3

For the strip footing with $B = 2\,\mathrm{m}$, shown in E-Figure 10.2, determine the unit limit load and the load on the footing right before plunging. The soil is a clay with design parameters $\gamma = 17\,\mathrm{kN/m^3}$ and $s_u = 26\,\mathrm{kPa}$. Ignore for now the effects of the shear strength along the slip surfaces extending above the footing base level.

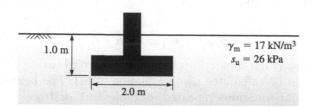

1.0 m

$\gamma_m = 17\ \mathrm{kN/m^3}$
$s_u = 26\ \mathrm{kPa}$

2.0 m

E-Figure 10.2 Strip footing for Example 10.3.

Solution:

According to Equation (10.1), the bearing capacity q_{bL} of the footing is:

$$q_{bL} = cN_c + q_0 N_q + \frac{1}{2}\gamma B N_\gamma$$

For a clay, given that $\phi = 0°$, $N_c \approx 5.14$, $N_q = 1$, and $N_\gamma = 0$. The surcharge in clay is a total stress, so it is calculated using $q_0 = \gamma_{m0} D$, where γ_{m0} is the total unit weight of the soil above the base of the foundation:

$$q_0 = 17 \times 1 = 17 \text{ kPa}$$

Equation (10.1) now becomes:

$$q_{bL} = s_u N_c + q_0 = 26 \times 5.14 + 17 = 151 \text{ kPa} \text{ }^{\text{answer}}$$

The base limit load Q_{bL} is given as

$$Q_{bL} = B q_{bL} = 2 \times 151 = 302 \text{ kN/m}$$

Not all of that load needs to be applied to the footing to get it to plunge. If we subtract the weight of the footing and backfill soil from the calculated Q_{bL}, we obtain an estimate of the structural load that would cause plunge, the limit load Q_L. If we assume a footing thickness of 0.4 m, we have 0.4 m of concrete and 0.6 m of soil. So we can write:

$$Q_L = Q_{bL} - A\left[D_{footing}\gamma_{concrete} + \left(D - D_{footing}\right)\gamma_{m0} \right]$$

$$= 302 - 2\left[0.4 \times 24 + 0.6 \times 17\right]$$

$$= 262 \text{ kN/m} \text{ }^{\text{answer}}$$

Example 10.4

For the same strip footing of Example 10.3, shown in E-Figure 10.2, determine the unit limit load right before failure if the soil is now a sand with design parameters $\gamma = 20$ kN/m³ and $\phi = 35°$. Ignore for now the effects of the shear strength along the slip surfaces extending above the footing base level. Consider the water table to be deep.

Solution:

For a sand, $c = 0$. The surcharge q_0 in sand is an effective stress. Because the water table is deep, $q_0 = \gamma_{m0} D$:

$$q_0 = 20 \times 1 = 20 \text{ kPa}$$

Using Equation (10.6), we find N_q:

$$N_q = \frac{1 + \sin\phi}{1 - \sin\phi} e^{\pi\tan\phi} = \frac{1 + \sin 35°}{1 - \sin 35°} e^{\pi\tan 35°} = 33.3$$

Using Equations (10.12) and (10.13), we find the following two values of N_γ:

$$N_\gamma = 1.5\left(N_q - 1\right)\tan\phi = 1.5\left(33.3 - 1\right)\tan 35° = 33.9$$

$$N_\gamma = \left(N_q - 0.6\right)\tan\left(1.33\phi\right) = \left(33.3 - 0.6\right)\tan\left(1.33 \times 35°\right) = 34.5$$

So, for $\phi = 35°$, the difference between the two values of N_γ is small. We will use the value from Eq. (10.13).

Finally, substituting N_γ into Equation (10.1):

$$q_{bL} = cN_c + q_0 N_q + \frac{1}{2}\gamma B N_\gamma$$

$$= q_0 N_q + \frac{1}{2}\gamma B N_\gamma = 20 \times 33.3 + \frac{1}{2} \times 20 \times 2 \times 34.5$$

So:

$$q_{bL} = 1356 \text{ kPa} \quad \text{answer}$$

If we use the expressions resulting from the assumption of a nonassociated flow rule (that is, $\psi < \phi$), then the calculations proceed as follows.

The surcharge q_0 in sand is an effective stress. As the water table is deep, $q_0 = \gamma_{m0} D$:

$$q_0 = 20 \times 1 = 20 \text{ kPa}$$

Using Equation (10.16), we find the value of the function F by assuming a value of $\psi = 5°$ for a 35° friction angle. The Bolton (1986) relationship for the friction angle, discussed in Chapter 5, is a practical tool to make this determination. The dilatancy angle ψ must be reasonable given the peak and critical-state friction angles. Thus, the value of F is:

$$F(\phi, \psi) = 1 - \tan\phi \left[\tan\left(0.8(\phi - \psi)\right) \right]^{2.5} = 1 - \tan 35° \left[\tan\left(0.8(35° - 5°)\right) \right]^{2.5}$$

$$F(\phi, \psi) = 0.9074$$

Using Equations (10.15) and (10.13), we find the following values of N_γ and N_q:

$$N_q = \frac{1 + \sin\phi}{1 - \sin\phi} \exp\left[F(\phi, \psi) \pi \tan\phi \right] = \frac{1 + \sin 35°}{1 - \sin 35°} e^{0.9074 \times \pi \tan 35°} = 27.2$$

$$N_\gamma = \left(N_q - 0.6\right) \tan(1.33\phi) = (27.2 - 0.6)\tan(1.33 \times 35°) = 28.1$$

Finally, substituting N_γ into Equation (10.1):

$$q_{bL} = cN_c + q_0 N_q + \frac{1}{2}\gamma B N_\gamma$$

$$= q_0 N_q + \frac{1}{2}\gamma B N_\gamma = 20 \times 27.2 + \frac{1}{2} \times 20 \times 2 \times 28.1$$

So:

$$q_{bL} = 1106 \text{ kPa} \quad \text{answer}$$

Up to this point, we have not considered the shear strength of the soil mobilized along the part of the slip surface extending above the level of the base of the footing (Figure 10.10). We will do this in the next sections. Aside from this, Equation (10.1) applies to strip footings, which are idealized as extending to infinity in the direction perpendicular to the plane of the

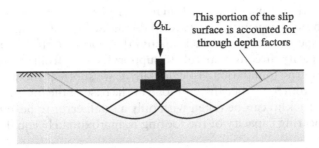

Figure 10.10 Contribution to the footing bearing capacity of the shear strength along the portion of the slip surface located above the base level of the footing.

footing cross section. This assumption is quite satisfactory in practice, and obviously very realistic near the center of the strip footing. This type of condition, in which the element we analyze extends to infinity in one direction and is loaded in directions that are normal to this first direction, is called a plane-strain condition, since the strains normal to the footing cross section are zero. This assumption is good for strip footings, but it does not apply to footings of finite size, such as square or rectangular footings. Equation (10.1) is still used in practice to estimate the bearing capacity of these footings, but it requires the use of correction factors referred to as shape factors. We will revisit this point when we discuss square, circular, and rectangular footings in both clay and sand later in this chapter.

10.2 THE BEARING CAPACITY OF SATURATED CLAYS

10.2.1 The bearing capacity equation for clays

The rate of construction of a building, bridge, and most structures is much higher than the rate of consolidation of clays, so loading conditions for clays are much closer to undrained conditions than they are to drained conditions. It is usually realistic, therefore, and in any case not unconservative,[4] to assume the loading of foundations in clay to take place under undrained conditions. Accordingly, the soil is modeled as a material with $c = s_u$ and $\phi = 0$, where s_u = the undrained shear strength of the clay, assumed for now to be constant with depth. It follows that $N_q = 1$, $N_\gamma = 0$, and Equation (10.1) reduces to

$$q_{bL} = s_u N_c + q_0 \tag{10.17}$$

As shown in Example 10.2, if we take the limit of Equation (10.10) as ϕ approaches zero, we find that $N_c = 2 + \pi \approx 5.14$. Advanced methods of analysis, such as limit analysis or the slipline method,[5] show that this is indeed the exact value of N_c. So Equation (10.17) can be taken as

$$q_{bL} = 5.14 s_u + q_0 \tag{10.18}$$

---

[4] An exception to this would be the rare situation when negative pore pressures would develop in extremely dilative clays, in which case the bearing capacity in the short term could be higher than in the long run. This is a rare situation because foundation loads impose a considerable amount of compressive stresses on the soil, as opposed to shear stresses predominantly, and compressive stresses generate positive pore pressures even in dilative soils. If a long-term check is required, the equations for drained loading can be used.

[5] The exact value for N_c in purely cohesive materials $(2 + \pi)$ was found by Prandtl (1920, 1921) using the slipline method.

The limit unit bearing capacity q_{bL} of Equation (10.18) is a gross unit bearing capacity, meaning that it is the unit bearing capacity available at the base of the footing for supporting the load from the superstructure, the backfill, and the footing itself. The net bearing capacity is the bearing capacity directly available to support the load from the superstructure; it is the useful capacity of the foundation. To calculate it, we subtract the weight of the footing and the soil backfill from q_{bL}. Since the unit load at the base of the footing due to the footing self-weight plus the backfill can be taken with only a small error to be approximately equal to q_0, the net unit bearing capacity of the footing is approximately equal to:

$$q_{bL}^{net} = 5.14 s_u \tag{10.19}$$

Example 10.5

A saturated clay with $\gamma = 19 \text{ kN/m}^3$ and $s_u = 25 \text{ kPa}$ is the soil into which a foundation is placed 1.5 m below the surface. Determine the gross limit unit bearing capacity for this soil.

Solution:

We must first determine the surcharge q_0:

$$q_0 = \gamma_{m0} D = 19 \times 1.5 = 28.5 \text{ kPa}$$

Using Equation (10.18), we can find q_{bL}:

$$q_{bL} = 5.14 s_u + q_0 = 5.14 \times 25 + 28.5 = 157 \text{ kPa}$$

For sensitive clays, which have an undrained shear stress–strain relationship with a sharp peak, the shear strength drops sharply after it is first mobilized. In these cases, it is sometimes convenient to take the load at which the peak strength first mobilizes anywhere in the soil as the bearing capacity of the footing. For these conditions, it can be shown that N_c can be taken as π. This is an approximate result that follows from using elasticity equations for a strip load on the soil surface (see Section 4.6) to calculate the maximum shear stress in the soil mass for a given footing load. When this shear stress is made equal to s_u, that point of the soil yields. It is shown that this happens for $q_b = \pi s_u$. The reader is asked to show this in Problem 10.9.

10.2.2 Shape, depth, and load inclination factors for footings in clay

Equation (10.18) is used for strip footings, for which plane-strain conditions apply. We can use Equation (10.18) to compute the bearing capacity of square, circular, or rectangular footings if we introduce into the equation a correction factor known as a *shape factor* s_{su}:

$$q_{bL} = 5.14 s_{su} s_u + q_0 \tag{10.20}$$

For clay deposits with uniform shear strength throughout, shape factor values are always greater than 1. This is easy to understand. Because plane-strain conditions no longer apply, we can expect that slip mechanisms will develop in all directions around the footing when it is loaded to failure. So, as opposed to having only two sets of potential slip surfaces per unit length of footing into the plane of representation, we now have four: one on each of the four sides of the footing. The larger area of slip surface per unit length of foundation leads to a larger bearing capacity than that of strip footings. It follows that the shape factor is obviously a function of the aspect ratio B/L of a rectangular footing, with $B/L \leq 1$. The lower the value of B/L, the closer we are to plane-strain conditions, and the closer s_{su} is to 1.

In addition to shape factors, we use depth factors to account for the portion of the slip mechanism extending above the level of the base of the footing, as discussed briefly earlier (refer to Figure 10.10). This is needed because we initially made the assumption that the soil located above the base of the footing could be replaced by a surcharge q_0, and this assumption ignores the contribution of soil shear strength along that portion of the slip surface. The bearing capacity equation for clay now becomes:

$$q_{bL} = 5.14 s_{su} d_{su} s_u + q_0 \tag{10.21}$$

where d_{su} is the *depth factor*. Depth factors are obviously greater than 1.

A question that is natural at this point is why we have included the shape and depth factors in the first, but not in the second (the surcharge) term of (10.21). The best way to understand this is to consider, as we did already once before, the case where the base of the footing rests on a weightless, frictionless, cohesionless soil, but the soil above the level of the base of the footing, although also having $c = 0$ and $\phi = 0°$, does have a nonzero unit weight. The soil above the footing base level, having zero friction angle, can be replaced by an equivalent surcharge q_0 without introducing any error into the calculation. This means that there is no need for depth or shape factors multiplying the surcharge term of the bearing capacity equation. This, if we consider it carefully, is an intuitive result: we would expect that, whatever the shape of the footing, we would be able to get the footing to move down into the soil deposit by simply exceeding slightly the value of q_0 if the soil has zero c and ϕ. This shows that the q_0 term does not require a shape factor.

In the case of clays and the applicable form of the bearing capacity equation, Equation (10.18), the soil does have a nonzero c (equal to s_u), but continues to have zero ϕ, so the same reasoning applies to the surcharge term of Equation (10.18). Because the soil above the level of the footing does have a nonzero c, though, it is not perfectly equivalent to a surcharge applied at the footing base level. That must be reflected somewhere in the equation, and the natural place for that is the first term of the equation, that containing c.

The shape and depth factors historically used in practice are mostly based on experimental data due to Meyerhof (1951, 1963) and Skempton (1951). Meyerhof suggested:

$$s_{su} = 1 + 0.2 \frac{B}{L} \tag{10.22}$$

$$d_{su} = 1 + 0.2 \frac{D}{B} \tag{10.23}$$

for $D/B < 2.5$.

Brinch Hansen (1970) recommended a different set of expressions for the depth factor:

$$d_{su} = \begin{cases} 1 + 0.4 \dfrac{D}{B} & \text{for } \dfrac{D}{B} \leq 1 \\[3mm] 1 + 0.4 \tan^{-1} \dfrac{D}{B} & \text{for } \dfrac{D}{B} > 1 \end{cases} \tag{10.24}$$

Today, the effects of shape and depth factor are better quantified. Salgado et al. (2004), based on the results of rigorous limit analysis, proposed the following expressions for the shape and depth factors:

$$s_{su} = 1 + C_1\frac{B}{L} + C_2\sqrt{\frac{D}{B}} \qquad (10.25)$$

$$d_{su} = 1 + 0.27\sqrt{\frac{D}{B}} \qquad (10.26)$$

where C_1 and C_2 are functions of B/L and are given in Table 10.2.

The fact that sound analysis of the bearing capacity problem shows the shape factor s_{su} to depend on D/B highlights the deficiency of the traditionally made assumption that shape and depth factors are independent. Assuming independence of the shape and depth factors produces conservative results in practical problems. Figure 10.11 shows that the shape factor of Equation (10.22) leads to an overprediction of the bearing capacity of square or circular footings on the surface of the soil deposit. The same equation, however, leads to underprediction of the bearing capacity as soon as D/B is allowed to take even small positive values; this underprediction increases with increasing D/B and becomes substantial for values of D/B that are common in practice.

Table 10.2 Regression constants in Equation (10.25) for the shape factor

B/L	C_1	C_2
1 (circle)	0.163	0.210
1 (square)	0.125	0.219
0.50	0.156	0.173
0.33	0.159	0.137
0.25	0.172	0.110
0.20	0.190	0.090

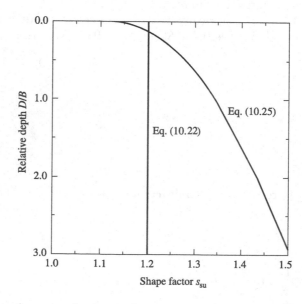

Figure 10.11 Shape factor for square footings in clays.

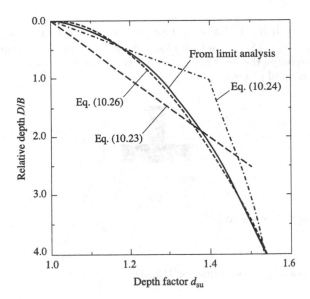

Figure 10.12 Depth factor relationships for footings in clays.

Figure 10.12 shows the depth factors given by Equations (10.23), (10.24), and (10.26) plotted together. It is clear that, of the traditional expressions for the depth factor, Brinch Hansen's expression would be more appropriate for low values of D/B, and Meyerhof's, for higher values of D/B. It can be seen that Meyerhof's depth factors are generally conservative within the range of depths ($D/B = 0$–2.5) for which they were defined, underpredicting the exact depth factors by as much as 9% for $0 < D/B < 2$ and overpredicting them slightly for $2 < D/B < 2.5$. Brinch Hansen tends to be unconservative for $D/B > 0.5$.

An additional correction factor is necessary if the load acting on the footing is inclined, in which case its horizontal component is denoted by Q_h and its vertical component is denoted by Q_v. Two separate checks are required: one with respect to bearing capacity, in which the vertical component Q_v of the load is compared with a properly reduced vertical load capacity, and one in which the horizontal load is compared with the sliding resistance at the base of the footing. The presence of the horizontal load has a negative impact on the vertical bearing capacity of the footing because it induces additional shearing. The greater the Q_h/Q_v ratio is, the less the bearing capacity of the footing will be. This effect can be captured by inserting one additional factor, a load inclination factor i_{su}, into Equation (10.21):

$$q_{bL} = 5.14 s_{su} d_{su} i_{su} s_u + q_0 \tag{10.27}$$

Based on the Meyerhof (1953) experimental results, the load inclination factor i_{su} can be approximated by

$$i_{su} = 1 - 1.3 \frac{Q_h}{Q_v} \tag{10.28}$$

for $Q_h/Q_v \le 0.4$, where Q_h = horizontal load and Q_v = vertical load acting on the footing. Shallow foundations in clay are not practical for cases where $Q_h/Q_v > 0.4$.

Example 10.6

A square footing with $B = 1.75\,\text{m}$ is designed as shown in E-Figure 10.3. The soil is a clay with $\phi = 0$, $c = s_u = 25\,\text{kPa}$, and $\gamma = 17\,\text{kN/m}^3$. The footing is to be embedded $1.25\,\text{m}$ below grade. Determine the limit bearing capacity for this location considering that the horizontal load is equal to zero.

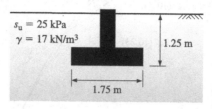

E-Figure 10.3 Footing for Example 10.6.

Solution:

The bearing capacity equation, Equation (10.21), is to be used:

$$q_{bL} = 5.14 s_{su} d_{su} s_u + q_0$$

The surcharge is

$$q_0 = \gamma_{m0} D = 17 \times 1.25 = 21.2 \approx 21\,\text{kPa}$$

We note there are many equations to determine the shape and depth factors. We will use the "classic" Meyerhof approach first, and then we will show the solution using a more modern approach.

Using Equation (10.22), $s_{su} = 1 + 0.2\dfrac{B}{L} = 1 + 0.2\left(\dfrac{1.75}{1.75}\right) = 1.20$

Using Equation (10.23), $d_{su} = 1 + 0.2\dfrac{D}{B} = 1 + 0.2\left(\dfrac{1.25}{1.75}\right) = 1.14$

$$q_{bL} = 5.14 s_{su} d_{su} s_u + q_0 = 5.14 \times 1.2 \times 1.14 \times 25 + 21 = 197\,\text{kPa}$$

Now, using the Salgado et al. (2004) equations, Equation (10.26) gives

$$d_{su} = 1 + 0.27\sqrt{\dfrac{D}{B}} = 1 + 0.27\sqrt{\dfrac{1.25}{1.75}} = 1.23$$

Referring to Table 10.2, considering that we have a square footing with $B/L = 1$:

$$C_1 = 0.125 \quad \text{and} \quad C_2 = 0.219.$$

So, using Equation (10.25):

$$s_{su} = 1 + 0.125\dfrac{B}{L} + 0.219\sqrt{\dfrac{D}{B}}$$

$$= 1 + 0.125\dfrac{1.75}{1.75} + 0.219\sqrt{\dfrac{1.25}{1.75}} = 1.31$$

Now we can calculate the limit unit base resistance:

$$q_{bL} = 5.14 s_{su} d_{su} s_u + q_0 = 5.14 \times 1.31 \times 1.23 \times 25 + 21.2 = 228 \text{ kPa}$$

The limit base resistance Q_{bL} is:

$$Q_{bL} = q_{bL} A = 228 \times 1.75^2 = 698 \text{ kN} \quad \text{answer}$$

10.2.3 Bearing capacity of footings in clay with strength increasing with depth

10.2.3.1 Surface, strip footings

In Chapter 6, we saw that the undrained shear strength increases with depth in normally consolidated soil deposits. Davis and Booker (1973) showed, by using the method of characteristics, that the bearing capacity of footings in such deposits can be significantly different from what we would expect based on an idealization of the deposit as having uniform shear strength with depth. Figure 10.13a–c shows idealized strength profiles below the level of the foundation base in which the strength increases with depth. They are given in terms of s_u (plotted in the horizontal axis) and the depth z_f measured from the foundation base (plotted in the vertical axis).

Figure 10.13a applies to the idealized case (and not very realistic case for onshore foundations in typical applications) of a footing placed directly on a recently deposited clay, which has not undergone any consolidation at the surface (hence the zero shear strength at $z_f = 0$). Figure 10.13b can be used to model the case of a footing embedded sufficiently deep in a clay deposit that the soil located below the base of the footing is normally consolidated and the shear strength profile beneath the footing is linear. Clay deposits frequently have a crust of hardened clay due to overconsolidation, the action of plant roots, desiccation, etc. Figure 10.13c is an idealized strength profile that may be used to model a footing bearing on the overconsolidated crust of such a deposit. In Figure 10.13c, the strength remains equal to s_{u0} down to the depth

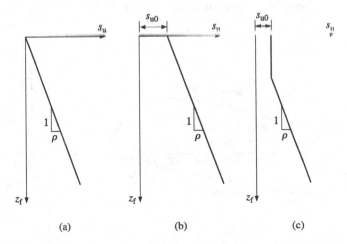

Figure 10.13 Idealized shear strength versus depth profiles (in terms of s_u versus depth z_f measured from the footing base) for saturated clay deposits: (a) normally consolidated with $s_u = 0$ at footing base, (b) normally consolidated with nonzero s_u at the footing base, and (c) normally consolidated below a certain depth, with the footing base resting on an overconsolidated crust for which the strength is constant with depth.

required for it to meet with a line starting from $s_u = 0$ at the level of the base of the foundation. Note that s_{u0} in the case of both Figure 10.13b and c is the shear strength at the level of the base of the foundation.

Using the notation in Figure 10.13, Davis and Booker (1973) expressed the limit unit bearing capacity of a strip footing on the surface of a deposit with increasing strength with depth as

$$q_{bL} = F\left[s_{u0}N_c + \frac{1}{4}\rho B\right] \tag{10.29}$$

where F is a correction factor that is given in Figure 10.14, s_{u0} is the undrained shear strength of the clay at depth $z_f = 0$ below the base of the foundation, and $\rho = ds_u/dz$ = rate of increase in the undrained shear strength s_u with depth. The correction factor is given for footings with both smooth and rough bases. In foundation engineering practice, concrete footings can be assumed to have rough bases. Even steel foundations will be sufficiently corroded that their interface response will be that of a rough footing for clay and even silts and fine sand.

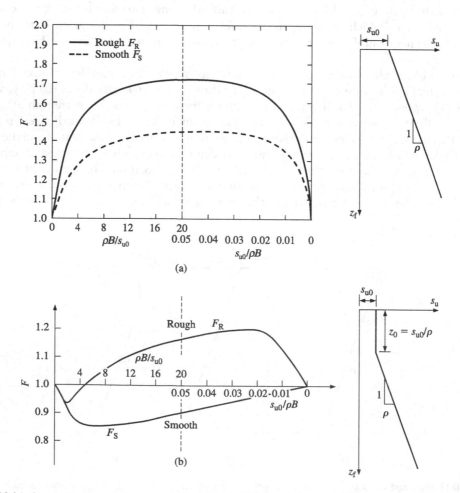

Figure 10.14 Correction factor F for use in Equation (10.29) in two cases: (a) the case of Figure 10.13b and (b) the case of Figure 10.13c. F_R is the correction factor F for a rough footing base, and F_S is the correction factor F for a smooth footing base. (Davis and Booker 1973; courtesy of the Institution of Civil Engineers and Thomas Telford Limited.)

Using the ABC program (Martin 2004) or FE limit analysis (Sloan 2013), we can update the values originally provided by Davis and Booker (1973) for the correction factor to use in Equation (10.29) for the case shown in Figure 10.13b, of which the case shown in Figure 10.13a is a special case. The updated values are given in Figure 10.14a. Figure 10.14b gives the correction factor F obtained by Davis and Booker (1973) to use in Equation (10.29) when the undrained shear strength below the footing base stays constant with depth up to a certain point and then increases at a rate ρ with depth (the case shown in Figure 10.13c).

If we rewrite Equation (10.29) as

$$q_{bL} = F\left[1 + \frac{\rho B}{4 s_{u0} N_c}\right] s_{u0} N_c \tag{10.30}$$

it may be compared with Equation (10.17) with $q_0 = 0$. This comparison between Equations (10.30) and (10.17) implies that the use of Equation (10.17) for a deposit with increasing shear strength would be correct only if we used an equivalent $s_{u,eq}$ given by

$$s_{u,eq} = F\left[1 + \frac{\rho B}{4 s_{u0} N_c}\right] s_{u0} = F\left[s_{u0} + \frac{1}{4 N_c}\rho B\right] \tag{10.31}$$

When attempting to determine a value of s_u to use in a bearing capacity calculation, an engineer may be tempted to average s_u values obtained from *in situ* or laboratory tests over the depth of the slip mechanism beneath the footing. The depth of the mechanism might, in error, be assumed to be about B below the base of the footing. For the case of strength increasing linearly with depth, this would imply that the representative value of s_u would be the value at a depth of $B/2$ below the base of the footing. In reality, a shallower mechanism develops in the case of strength increasing with depth because the strong soil at greater depths repels the slip surfaces, diverting them back up. The following example illustrates this point.

Example 10.7

Let us examine the case of a strip footing on top of a "slurry," a normally consolidated soil deposit with $s_{u0} = 0$. This is something one might face in trying to walk on a muddy deposit, where one's foot would be the "footing," with the difference that feet are not "infinitely long." Some offshore foundations (spudcans) are installed by sinking what amounts to a steel base into ocean floor soil. The force required to sink into soil could be estimated using this approach (but, again, a shape factor would be required, because such foundations tend to be circular). Considering that $F = 1$ and that N_c is approximately 5.14, Equation (10.31) becomes:

$$s_{u,eq} = \frac{\rho B}{20.6}$$

The depth at which $s_{u,eq}$ must be calculated can now be determined from the knowledge that the strength increases linearly at a rate ρ from 0 at the surface:

$$z_{eq} = \frac{s_{u,eq}}{\rho} = \frac{B}{20.6}$$

This shows that taking a reading at $B/2$, as one might be tempted to do if we assumed the depth of influence to be equal to B, would be very much in error (besides being clearly unconservative).

We can generalize the result of Example 10.7 for the case shown in Figure 10.13b. The depth at which $s_{u,eq}$ would need to be calculated for the traditional bearing capacity equation, which assumes constant strength with depth, to generate the same bearing capacity as that calculated taking the increase in s_u with depth into account can be determined from Equation (10.31) as

$$z_{f,eq} = \frac{s_{u,eq} - s_{u0}}{\rho} = \frac{(F-1)s_{u0}}{\rho} + \frac{BF}{4N_c} \tag{10.32}$$

For the case shown in Figure 10.13c, we would first need to make a determination of the depth to which s_u remains constant and equal to s_{u0}.

10.2.3.2 Footings with finite dimensions embedded in soil with increasing strength with depth

Although Davis and Booker (1973) restricted their analysis to strip footings on the surface of a soil deposit, if it is to be useful in practice, we must be able to introduce shape, depth, and inclination factors into Equation (10.30). We must also be able to superpose q_0 to it. This leaves us with the equation[6]:

$$q_{bL} = Fs_{su}d_{su}i_{su}\left[1 + \frac{\rho B}{4s_{u0}N_c}\right]s_{u0}N_c + q_0 \tag{10.33}$$

Comprehensive, rigorous solutions to the problem of embedded footings in soil with increasing strength with depth are still not available, so the shape, depth, and inclination factors appearing in Equation (10.33) are not yet available in definitive form.

Salençon and Matar (1982) and Houlsby and Wroth (1983) observed that the shape factor of a footing in a soil with increasing strength with depth decreases with increasing strength gradient ρ. Randolph et al. (2004), using the results from Davis and Booker (1973) and Martin (2001), arrived at values of shape factors for circular footings that are reproduced in Table 10.3. Table 10.4 shows calculations made with the ABC program (Martin 2004). A least-squares fit to the values of shape factor s_{su}, given in the last column, leads to the following equation:

$$s_{su} = 1 + 0.176\frac{B}{L}\left(\frac{2.3}{\exp\left[0.353\left(\dfrac{\rho B}{s_{u0}}\right)^{0.509}\right]} - 1.3\right)$$

where $B/L = 1$ for circular footings. Note that the $s_{su} - 1$ is equal to 0.176 when $\rho = 0$, a value that is slightly greater than the C_1 values in Table 10.2. The difference is not relevant from a practical point of view, and we could easily replace 0.176 by C_1 in all that follows.

Table 10.3 Shape factor for surface footings as a function of $\rho B/s_{u0}$

$\rho B/s_{u0}$	0	1	2	3	6	10
s_{su}	1.18	1.05	1.00	0.98	0.93	0.90

---

[6] The use of s_{su}, d_{su}, and i_{su} in Equation (10.33) was not proposed by Davis and Booker (1973) and has not been validated in research, so it should be used with caution.

Table 10.4 Shape factor for circular footings in clays with increasing strength with depth

$\rho B/s_{u0}$	$q_{bL,strip}$	F	$q_{bL,circle}$	$s_{su}F$	s_{su}
0.0	5.14	1.00	6.05	1.18	1.18
0.1	5.33	1.03	6.15	1.19	1.15
0.2	5.51	1.06	6.26	1.21	1.14
0.5	5.97	1.13	6.54	1.24	1.09
1.0	6.61	1.23	6.95	1.29	1.05
2.0	7.60	1.35	7.63	1.35	1.00
4.0	9.13	1.49	8.74	1.42	0.96
6.0	10.42	1.57	9.69	1.46	0.93
8.0	11.58	1.62	10.56	1.48	0.91
10.0	12.66	1.66	11.37	1.49	0.90
12.0	13.69	1.68	12.13	1.49	0.89
15.0	15.14	1.70	13.21	1.49	0.87
20.0	17.40	1.72	14.89	1.47	0.86
22.5	18.48	1.72	15.69	1.46	0.85
25.0	19.54	1.72	16.47	1.45	0.84
30.0	21.58	1.71	17.98	1.42	0.83
35.0	23.54	1.70	19.43	1.40	0.83
50.0	29.16	1.65	23.54	1.34	0.81
60.0	32.72	1.63	26.14	1.30	0.80
100.0	46.17	1.53	35.88	1.19	0.78
150.0	61.97	1.45	47.17	1.11	0.76

This equation so far applies only to surface footings. For values of D/B common in foundation engineering practice (say $D/B < 0.5$) and values of $\rho B/s_{u0}$ for typical soils and foundations (that is, values less than about 2), we can retain the same dependence of the shape factor on depth as we had earlier without incurring significant error.[7] This leads us to the following expression for the shape factor:

$$s_{su} = 1 + C_1 \frac{B}{L}\left(\frac{2.3}{\exp\left[0.353\left(\frac{\rho B}{s_{u0}}\right)^{0.509}\right]} - 1.3\right) + C_2\sqrt{\frac{D}{B}} \tag{10.34}$$

where, for C_1, we may use the value 0.176 from the ABC program or the values from Table 10.2 and C_2 is the same as presented in Table 10.2. These values all apply to footings with a rough base, a realistic assumption for concrete footings, as discussed earlier in this chapter.

Example 10.8

For a typical value of $ds_u/d\sigma'_v = 0.22$, a unit weight of 17 kN/m³, and water table at the soil surface, under what conditions can we expect the bearing capacity of a circular footing bearing on the surface of a clay with undrained shear strength at the surface equal to s_{u0} to be roughly the same as that of a strip footing bearing on the same soil?

---

[7] There are not sufficient results yet in the literature that allow us to test this assumption, but it is not expected that the difference in how the shape factor varies with D/B for different values of $\rho B/s_{u0}$ will generate large errors.

Solution:

According to Table 10.4, the value of $\rho B/s_{u0}$ for which the circular and strip footings have the same bearing capacity (and thus a shape factor of 1) is 2. For the water table flush with the soil surface, the gradient ρ can be computed as

$$\rho = \frac{ds_u}{dz} = \gamma_b \frac{ds_u}{d(\gamma_b z)} = \gamma_b \frac{ds_u}{d\sigma_v'} = 0.22(17 - 9.81) = 1.6 \text{ kPa/m}$$

Based on Table 10.4, this gives us the following relationship between B and s_{u0}:

$$B = 1.25 s_{u0}$$

Considering a wide range of reasonable s_{u0} values, the values of B shown in the following table can then be obtained:

s_{u0} (kPa)	5	10	20	50
B (m)	6.25	12.5	25	62.5

These limiting values of B are quite large, meaning that, for typical design situations, the shape factor will certainly be greater than 1.

With the shape factor determined, we turn our attention to the depth factor. An analysis of the results of Martin (2001) for circular footings embedded to various degrees in soil with strength increasing with depth shows that for values of D/B common in foundation engineering practice (say $D/B < 0.5$) and values of $\rho B/s_{u0}$ for typical soils and foundations (that is, values less than about 2), the depth factor is not significantly different from that given by Equation (10.26). For cases with particularly high D/B or $\rho B/s_{u0}$ values, Equation (10.26) may overpredict by as much as 10% the value of the depth factor and, thus, of the bearing capacity of the footing.

Example 10.9

Consider a soil deposit with $ds_u/d\sigma_v' = 0.3$, $\gamma_b = 10$ kN/m³, and water table at the surface. Consider two situations: a 3 m square footing is located at a depth of 1 m, from which depth the shear strength increases linearly with depth, and a 3 m strip footing under the same conditions. The undrained shear strength s_{u0} at a depth of 1 m is equal to 20 kPa. What is the limit bearing capacity of the strip and square footings?

Solution:

Let us start by calculating the values of q_0 and ρ. Irrespective of the location of the water table, we use the total unit weight in the calculation of the surcharge because we have a clay layer, and capillary rise assures full saturation of the clay above the water table. The values of q_0 and ρ are:

$$q_0 = 19.8 \text{ kN/m}^3 \times 1 \text{ m} = 19.8 \text{ kPa}$$

$$\rho = \frac{ds_u}{dz} = \gamma_b \frac{ds_u}{d(\gamma_b z)} = \gamma_b \frac{ds_u}{d\sigma_v'} = 0.3\gamma_b = 0.3 \times 10 = 3 \text{ kPa/m}$$

This leads us to:

$$\frac{s_{u0}}{\rho B} = \frac{20}{3 \times 3} = 2.2 \quad \text{or} \quad \frac{\rho B}{s_{u0}} = \frac{9}{20} = 0.45$$

The value of F for a rough strip footing is read from Figure 10.14 as 1.12. The depth factor is calculated from Equation (10.26):

$$d_{su} = 1 + 0.27\sqrt{\frac{D}{B}} = 1 + 0.27\sqrt{\frac{1}{3}} = 1.16$$

The bearing capacity of the strip footing is now calculated as

$$q_{bL} = Fd_{su}i_{su}\left[1 + \frac{\rho B}{4s_{u0}N_c}\right]s_{u0}N_c + q_0$$

$$= 1.12 \times 1.16 \times 1 \times \left[1 + \frac{0.45}{4 \times 5.14}\right] \times 20 \times 5.14 + 19.8$$

$$= 136.5 + 19.8$$

$$= 136 + 20 = 156 \text{ kPa}$$

So, the limit bearing capacity of the strip footing is

$$Q_{bL} = q_{bL}A = 156 \times 3 = 468 \text{ kN/m} \quad \text{answer}$$

The limit bearing capacity of the square footing is obtained by multiplying the 136.5 kPa by the shape factor given by Equation (10.34):

$$s_{su} = 1 + C_1\frac{B}{L}\left(\frac{2.3}{\exp\left(0.353\left(\frac{\rho B}{s_{u0}}\right)^{0.509}\right)} - 1.3\right) + C_2\sqrt{\frac{D}{B}}$$

$$= 1 + 0.125 \times \frac{3}{3}\left(\frac{2.3}{\exp\left(0.353(0.45)^{0.509}\right)} - 1.3\right) + 0.219 \times \sqrt{\frac{1}{3}}$$

$$= 1.19$$

So the limit bearing capacity of the square footing is:

$$Q_{bL} = Aq_{bL} = 3 \times 3 \times (136.5 \times 1.19 + 19.8) = 1640 \text{ kN} \quad \text{answer}$$

Example 10.10

Consider a soil deposit with $ds_u/d\sigma_v' = 0.3$, $\gamma_b = 10$ kN/m³, and the water table located permanently 1.5 m below the footing base. A 3 m square footing is located at a depth of 1 m. What is the limit bearing capacity of this footing?

Solution:

This example illustrates how a nonzero s_{u0} may develop because of capillary stresses. Let us start by calculating the values of q_0 and ρ:

$$q_0 = 19.8 \text{ kN/m}^3 \times 1 \text{ m} = 19.8 \text{ kPa}$$

$$\rho = \frac{ds_u}{dz} = \gamma_b \frac{ds_u}{d(\gamma_b z)} = \gamma_b \frac{ds_u}{d\sigma'_v} = 0.3\gamma_b = 0.3 \times 19.8 = 3 \text{ kPa/m}$$

To calculate s_{u0}, we need to calculate the effective stresses at the base of the footing:

$$\sigma'_v = 19.8 \times 1 - (-9.81 \times 1.5) = 34.5 \text{ kPa}$$

$$s_{u0} = 0.3 \times 34.5 = 10.35 \text{ kPa} \approx 10 \text{ kPa}$$

This leads us to:

$$\frac{s_{u0}}{\rho B} = \frac{10}{3 \times 3} = 1.1 \quad \text{or} \quad \frac{\rho B}{s_{u0}} = \frac{3 \times 3}{10} = 0.9$$

The value of F for a rough footing is read from Figure 10.14 as 1.21. The depth factor is calculated from Equation (10.26):

$$d_{su} = 1 + 0.27\sqrt{\frac{D}{B}} = 1 + 0.27\sqrt{\frac{1}{3}} = 1.16$$

The shape factor is calculated from Equation (10.34):

$$s_{su} = 1 + C_1 \frac{B}{L} \left(\frac{2.3}{\exp\left[0.353\left(\frac{\rho B}{s_{u0}}\right)^{0.509}\right]} - 1.3 \right) + C_2\sqrt{\frac{D}{B}}$$

$$= 1 + 0.125 \frac{3}{3} \left(\frac{2.3}{\exp\left(0.353(0.9)^{0.509}\right)} - 1.3 \right) + 0.219 \times \sqrt{\frac{1}{3}}$$

$$= 1.17$$

where C_1 and C_2 come from Table 10.2.
The limit unit bearing capacity is now calculated as

$$q_{bL} = Fs_{su}d_{su}s_{u0}\left[1 + \frac{\rho B}{4s_{u0}N_c}\right]N_c + q_0$$

$$= 1.21 \times 1.17 \times 1.16 \times 10\left[1 + \frac{0.9}{4 \times 5.14}\right]5.14 + 19.8$$

$$= 88 + 19.8$$

$$\approx 108 \text{ kPa}$$

The limit bearing capacity of the footing can now be calculated from:

$$Q_{bL} = A \times q_{bL} = 3 \times 3 \times 108 = 972 \text{ kN} \approx 970 \text{ kN} \text{ answer}$$

Example 10.11

Consider again the footing of Example 9.7. The 1.5 m square footing is founded on a clay with $LL = 70\%$, $PL = 40\%$, and unit weight equal to 18 kN/m³. The base of the footing is placed at a depth of 1 m. The groundwater table may rise as high as 1 m below the ground surface. The clay layer is 7 m thick. It is underlain by a very stiff layer. Calculate the limit bearing capacity of this footing.

Solution:

We wish to use Equation (10.33):

$$q_{\text{bL}} = F s_{su} d_{su} i_{su} \left[1 + \frac{\rho B}{4 s_{u0} N_c} \right] s_{u0} N_c + q_0$$

Considering the water table at 1 m below the ground surface as a worst-case scenario and considering that $\gamma = 18$ kN/m³, the surcharge is:

$$q_0 = \gamma_{m0} D = 18 \times 1 = 18 \text{ kPa}$$

As discussed earlier, irrespective of the location of the water table, we use the total unit weight in the calculation of the surcharge because we have a clay layer, and capillary rise assures full saturation of the clay above the water table.
Using Equation (10.26):

$$d_{su} = 1 + 0.27 \sqrt{\frac{D}{B}} = 1 + 0.27 \sqrt{\frac{1}{1.5}} = 1.22$$

The rate of increase in the undrained shear strength s_u with respect to the vertical effective stress increase is estimated from Equation (6.50):

$$ds_u = \left[0.11 + 0.0037(PI) \right] d\sigma_v'$$

The plasticity index is calculated as $PI = LL - PL = 70 - 40 = 30\%$. We must now calculate s_{u0} and ρ. Let's first calculate σ_v' at the level of the footing base (at $z_f = 0$):

$$\sigma_v' = 18 \times 1 = 18 \text{ kPa}$$

So,

$$s_{u0} = \left[0.11 + 0.0037(30) \right] 18 = 4 \text{ kPa}$$

The strength increase rate ρ is given as

$$\rho = \frac{ds_u}{dz_f} = \gamma_b \frac{ds_u}{d\sigma_v'} = \gamma_b \left[0.11 + 0.0037(PI) \right] = 8.2 \left[0.11 + 0.0037(30) \right] = 1.8 \text{ kPa/m}$$

This leads to:

$$\frac{s_{u0}}{\rho B} = \frac{4}{1.8 \times 1.5} = 1.5 \quad \text{or} \quad \frac{\rho B}{s_{u0}} = \frac{1.8 \times 1.5}{4} = 0.675$$

and to an F value of 1.18 from Figure 10.14.

From Table 10.2, we see that $C_1 = 0.125$ and $C_2 = 0.219$ because we have a square footing. Using Equation (10.34):

$$s_{su} = 1 + C_1 \frac{B}{L} \left(\frac{2.3}{\exp\left(0.353 \left(\frac{\rho B}{s_{u0}} \right)^{0.509} \right)} - 1.3 \right) + C_2 \sqrt{\frac{D}{B}}$$

$$= 1 + 0.125 \frac{1.5}{1.5} \left(\frac{2.3}{\exp\left(0.353 (0.675)^{0.509} \right)} - 1.3 \right) + 0.219 \times \sqrt{\frac{1}{1.5}}$$

$$= 1.23$$

We can now calculate q_{bL} as

$$q_{bL} = F s_{su} d_{su} i_{su} \left[1 + \frac{\rho B}{4 s_{u0} N_c} \right] s_{u0} N_c + q_0$$

$$= 1.18 \times 1.23 \times 1.22 \times 1 \times \left[1 + \frac{0.675}{4 \times 5.14} \right] \times 4 \times 5.14 + 18$$

$$= 37.6 + 18 \approx 56 \text{ kPa}$$

It is interesting to compare this with the limit unit bearing capacity that would be calculated if an average s_u of $5.3\,\text{kPa}$ (the value of s_u at $B/2$ below the base of the footing) were used in the bearing capacity equation:

$$q_{bL} = 5.14(1.30 \times 1.22)5.3 + 18 = 61 \text{ kPa}$$

The estimated q_{bL} would have been about 10% too high. Note the larger shape factor (1.30), which is in part responsible for the difference in load capacities, that results when $\rho = 0$.

The limit bearing capacity of the footing is:

$$Q_{bL} = 1.5 \times 1.5 \times 56 = 126\,\text{kN} \quad \text{answer}$$

In Chapter 11, which deals with shallow foundation design, we discuss in detail how to combine our knowledge of shear strength of clay (Chapter 6), site exploration (Chapter 7), and limit bearing capacity (this chapter) to determine the values of limit bearing capacity of footings in clay for use in design.

10.3 BEARING CAPACITY OF FOOTINGS IN SAND

10.3.1 The bearing capacity equation for sands

One of the main distinctions between sands and clays lies in their very different drainage rate: clays drain slowly; sands drain very rapidly. Because the construction rate of most civil engineering projects is quite fast when compared to the slow drainage/consolidation rate of clays, we assumed that the loading of foundations in clays takes place under undrained conditions, an assumption that is quite satisfactory for most clays. Any deviations from this assumption cause calculation results to be slightly conservative, as partial drainage would

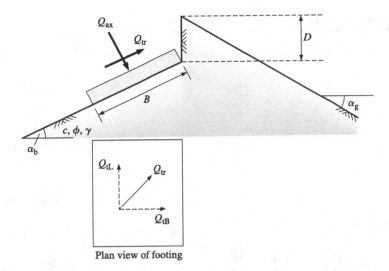

Figure 10.15 A foundation of arbitrary shape with inclined base under the most general conditions, embedded D into the soil, subjected to inclined loading, and supported by sloping ground.

increase shear strength and thus bearing capacity. A rare exception might be found in highly overconsolidated clays. In contrast, for foundations in sand, we assume perfectly drained conditions because structural loads are applied sufficiently slowly for full drainage to be possible in sands. We saw in Chapter 5 that the shear strength of uncemented sands is best modeled, particularly for foundation analysis, with $c = 0$ and nonzero ϕ.

Figure 10.15 shows the general case of a foundation with base with dimensions B and L inclined at α_b with respect to the horizontal near a slope with ground inclination α_g (the following constraint applies to these angles: $\alpha_b + \alpha_g \leq 90°$). The foundation is subjected to an inclined load with axial and transverse components Q_{ax} and Q_{tr}. The foundation depth D is measured from the side of the foundation that is at the highest elevation. Note from the figure that the transverse component of the load must point upward, because the only reason to incline a footing is to make the footing absorb the applied load in a direction that is closer to perpendicular to the base of the footing; note also that it would not make sense to incline the footing past the point at which the resultant load is perpendicular to the base of the footing.

Equation (10.1) can be generalized for the case shown in Figure 10.15 by multiplying each term by suitable shape, depth, load inclination, base inclination, and ground inclination factors (denoted, respectively, by the letters s, d, i, b, and g with subscripts indicating whether they apply to the c, q_0, or γ term of the bearing capacity equation):

$$q_{bL} = \left(s_c d_c i_c b_c g_c \right) c N_c + \left(s_q d_q i_q b_q g_q \right) q_0 N_q + \frac{1}{2}\left(s_\gamma d_\gamma i_\gamma b_\gamma g_\gamma \right)\gamma B N_\gamma \qquad (10.35)$$

For uncemented sands, $c = 0$, and the equation reduces to:

$$q_{bL} = \left(s_q d_q i_q b_q g_q \right) q_0 N_q + \frac{1}{2}\left(s_\gamma d_\gamma i_\gamma b_\gamma g_\gamma \right)\gamma B N_\gamma \qquad (10.36)$$

Table 10.5 contains expressions commonly used for the bearing capacity correction factor for footing shape: shape factor, for short. They are due to Meyerhof (1963), Brinch

Table 10.5 Commonly used expressions for shape factors[a]

q_0 term	γ term
Meyerhof (1963)[b]	
$s_q = 1 + 0.1N\dfrac{B}{L}$	$s_\gamma = 1 + 0.1N\dfrac{B}{L}$
Brinch Hansen (1970)[c]	
$s_q = 1 + \dfrac{B}{L}\sin\phi$	$s_\gamma = 1 - 0.4\dfrac{B}{L} \geq 0.6$
Vesic (1973)	
$s_q = 1 + \dfrac{B}{L}\tan\phi$	$s_\gamma = 1 - 0.4\dfrac{B}{L} \geq 0.6$
Lyamin et al. (2007)[d]	
$s_q = 1 + \left(0.098\phi - 1.64\right)\left(\dfrac{D}{B}\right)^{0.7-0.01\phi}\left(\dfrac{B}{L}\right)^{1-0.16\left(\frac{D}{B}\right)}$	$s_\gamma = 1 + \left(0.0336\phi - 1\right)\dfrac{B}{L}$
Loukidis and Salgado (2009)	
$s_q^{circ} = 1 + 2.9\tan^2\phi$	$s_\gamma^{circ} = 1 + \left(0.26\dfrac{1+\sin\phi}{1-\sin\phi} - 0.73\right)$

[a] Use nominal values of B and L for the Vesic factors and effective values of B and L for the Brinch Hansen and Lyamin et al. factors.
[b] N is the flow number, given by Equation (4.45).
[c] Brinch Hansen (1970) s_c derived from s_q through Equation (10.66).
[d] ϕ measured in degrees. For circular footings, the s_q and s_γ equations should be multiplied by an additional term equal to $1 + 0.0025\phi$ and $1 + 0.002\phi$, respectively.

Hansen (1970), and Vesic (1973). The table also contains the expressions of Lyamin et al. (2007) for the shape and depth factors, calculated based on limit analysis, and the factors derived by Loukidis and Salgado (2009) for a material following a nonassociated flow rule. The Loukidis and Salgado (2009) factors apply to a circular footing in sand with $\psi < \phi$ with N_γ calculated using Equation (10.13). The factors are:

$$s_q^{circ} = 1 + 2.9\tan^2\phi \tag{10.37}$$

$$s_\gamma^{circ} = 1 + \left(0.26\frac{1+\sin\phi}{1-\sin\phi} - 0.73\right) \tag{10.38}$$

No shape factor equations are available in Loukidis and Salgado (2009) for rectangular footings, but the shape factors for circular footings can be used to approximate those of square footings.

Brinch Hansen (1970) and Vesic (1973) have the same expressions for the shape and depth factors, with the difference that the values used for B and L should be the nominal footing dimensions in the Vesic (1973) equations, but should be the effective dimensions B_{eff} and L_{eff} in the Brinch Hansen (1970) equations. The effective dimensions were defined by Meyerhof (1953) for a footing loaded eccentrically as the dimensions of a rectangular portion of the foundation with center coinciding with the point of application of the load. It follows that the effective dimensions are equal to $B_{eff} = B - 2e_B$ and $L_{eff} = L - 2e_L$ for a footing with load eccentricities e_B and e_L in the B and L directions, respectively, where B and L are the nominal dimensions. Load eccentricity is discussed in greater detail in Section 10.6.

10.3.2 Estimation of ϕ value to use in bearing capacity equation

Chapter 5 examines in detail how to determine the peak friction angle ϕ_p of sand sheared in triaxial compression or plane strain. In the bearing capacity problem, however, we need to obtain a friction angle value to plug into the bearing capacity equation that is not the value at a point, but rather a value representative of the entire slip mechanism that forms beneath the footing. Figure 10.16 illustrates this point by showing that the slip surface is made up of an infinite number of elements or points subjected to different loading paths. To provide a sense of this loading direction, soil elements are positioned in such way that the slip surface makes an angle of $45° + \phi/2$ with the major principal plane for each element in the figure. Each of these elements would have their own friction angle on the verge of bearing capacity failure because of the different loading paths that they experience.[8] The friction angle would tend to be lower near the footing (near element E_1), where the mean effective stress is larger, and higher away from the footing, say near element E_2, where the mean effective stress is smaller.

If there is a friction angle for each point of the slip surface, but a single friction angle is to be used in the bearing capacity equation, then this representative friction angle should be one that, when plugged into the bearing capacity equation, produces the correct or true value of the bearing capacity. It must therefore account for the contribution from each and every element along the slip surface to the overall bearing capacity of the footing. We will estimate the value of ϕ to use in the bearing capacity equation using the Bolton (1986) correlation, which consists of Equations (5.20) and (5.12), with a representative value of the mean effective stress σ'_{mp} to use in Equation (5.12). Meyerhof (1950) suggested that a representative value of σ'_{mp} to use in bearing capacity calculations would be of the order of 10% of q_{bL}. Three different equations are given in Table 10.6 for calculating σ'_{mp}. Of the three, Equations (10.39) and (10.41) are the most current. Perkins and Madson (2000) proposed an additional correction factor to use with Equation (10.41) to account for progressive failure[9] and other effects (mostly the fact that real sands have $\phi > \psi$, where ψ is the dilatancy angle). Those effects are already accounted for when Equation (10.39) is used with the Lyamin et al. (2007) shape and depth factors.

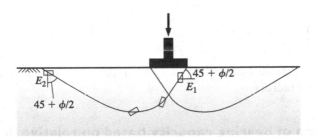

Figure 10.16 Soil elements along a slip surface positioned in such a way that the slip surface makes an angle of $45° + \phi/2$ with the major principal plane. Note how the major principal plane is horizontal near E_1, rotates along the slip mechanism, and becomes vertical near E_2.

---

[8] Extracting a number of samples from the ground with the orientations shown in Figure 10.16, even if we knew a priori the location of the slip surface, is clearly not a practical possibility. As a mental exercise, however, we can imagine that if we indeed extracted them and tested them under plane-strain conditions (for strip footings) or in triaxial compression (for square or circular footings), following the same stress path experienced by them in the ground during loading of the footing, we would have a response in the laboratory that would closely reproduce what happens in the field, at least in those cases where shearing is narrowly located along clearly identifiable slip surfaces. At the end of each of our imaginary tests, a slip plane would form through the sample with its usual orientation with respect to the principal planes, representing the slip surface that forms underneath the footing.

[9] "Progressive failure" is the drop of the friction angle below the peak friction angle in portions of the soil or slip surfaces because of uneven shear strains within the soil and thus uneven mobilization of shear strength.

Table 10.6 Equations for the representative mean effective stress for use in Equation (5.12) for the estimation of ϕ_p in bearing capacity calculations

Equation	Equation number	Reference
$\sigma'_{mp} = 20 p_A \left(\dfrac{\gamma B}{p_A} \right)^{0.7} \left(1 - 0.32 \dfrac{B}{L} \right)$	(10.39)	
$\sigma'_{mp} = \dfrac{1}{\left(1 + \tan^2 \phi_p \right)\left(1 + \sin \phi_p \right)} \dfrac{q_{bL} + 3 q_0}{4}$	(10.40)	De Beer (1970)
$\sigma'_{mp} = \dfrac{1}{6} \left(0.52 - 0.04 \dfrac{L}{B} \right) q_{bL} \geq \dfrac{q_{bL}}{25}$	(10.41)	Perkins and Madson (2000)

Note: L and B = length and width of footing, respectively.

Equation (10.39), used together with the Lyamin et al. (2007) shape and depth factors, produces the value of mean effective stress that, once used to calculate the corresponding friction angle, produces the correct value of q_{bL} for sand. The equation is based on both FE analysis using a realistic soil model (Loukidis and Salgado 2011) and experimental results (Ueno et al. 1998; Tatsuoka et al. 1991). It accounts for soil anisotropy, for the difference between friction and dilatancy angles, and for the evolution of soil properties (and thus also for "progressive failure") during loading toward limit bearing capacity failure.

Let us now examine in more detail the procedure that must be followed to use Equation (10.39) to calculate footing limit bearing capacity:

1. Calculate σ'_{mp} using Equation (10.39).
2. Calculate ϕ_p using Equations (5.20) and (5.12). If the footing is square or circular, the value of $A_\psi = 3$ for triaxial conditions is used. For a strip footing (one with $L/B \geq 7$), the value of $A_\psi = 5$ for plane-strain conditions is used. For $1 \leq L/B < 7$, A_ψ is interpolated between 3 and 5:

$$A_\psi = \frac{1}{3} \left(\frac{L}{B} + 8 \right) \tag{10.42}$$

3. Calculate q_{bL} using the bearing capacity formulation using the Lyamin et al. (2007) bearing capacity factors.

10.3.3 Estimation of bearing capacity based on relative density

A possible way to circumvent the difficulty in selecting a value of ϕ to use in bearing capacity calculations is to perform advanced FE simulations with realistic constitutive models to obtain expressions for N_γ directly in terms of initial relative density. Such models capture key features of the soil mechanical response – phase transformation, peak stress ratio, critical state, and nonassociativity – that the use of merely a Mohr–Coulomb shear strength envelope cannot capture. This is what Loukidis and Salgado (2011) did, obtaining the following equation for N_γ based on the analysis of the bearing capacity of a footing on the surface of the soil deposit:

$$N_\gamma = 2.82 \exp \left(3.64 \frac{D_R}{100\%} \right) \left(\frac{\gamma B}{p_A} \right)^{-0.4} \tag{10.43}$$

Loukidis and Salgado (2011) also obtained a shape factor for circular footings, which can also be used for square footings:

$$s_\gamma^{circ} = 1 - 0.23 \frac{D_R}{100\%} \tag{10.44}$$

10.3.4 Some perspective on the depth factor

The depth factor is a factor that corrects for the fact that the soil located above the level of the base of the footing also contributes to its capacity. The material above the footing base is not in a state of compression like the material below the base of the footing, so whether a material follows an associated or nonassociated flow rule becomes of modest influence at best. This means that depth factors such as those of Lyamin et al. (2007) in Table 10.7 could be used even with bearing capacity factors derived based on a nonassociated flow rule.

10.3.5 Some perspective on the shape factor

Notably in Table 10.5, the shape factor s_γ of both Meyerhof (1963) and Lyamin et al. (2007) is greater than 1, but those of Brinch Hansen (1970) and Vesic (1973) are less than 1. How can we reconcile these conflicting results? Surely, if the shear strength has a greater slip surface area to operate on, as it does for a square footing compared with a strip footing, the bearing capacity must be greater for square footings than for strip footings and the shape factor must be accordingly greater than 1 (see both the discussion earlier in this chapter and Figure 10.17). But this only holds if the friction angle used in the calculation of the bearing capacity of both the strip and the square footings is the same. In other words, it holds if we are comparing the bearing capacity of a square footing calculated with a friction angle appropriate for the square footing with the bearing capacity of a strip footing calculated with the same friction angle, not with the value of friction angle appropriate for a strip footing.

For the same initial relative density and lateral effective stress, the friction angle to use in the calculation of the bearing capacity of a strip footing (a plane-strain problem) must be greater than that used for a square footing (for which the friction angle from a triaxial test would be appropriate).[10] The use of the shape factors proposed by Brinch Hansen (1970) and Vesic (1973) can, in this context, be seen as an empirical way of accounting for the absence of plane-strain conditions, and thus the existence of lower ϕ values, in footings of finite size. Otherwise, if a triaxial ϕ is used for square or circular footings, then the Meyerhof (1963)

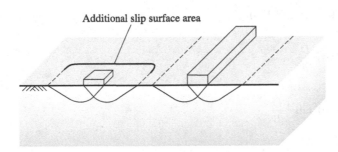

Additional slip surface area

Figure 10.17 Illustration of the greater slip surface area for a square footing compared with a strip footing.

---

[10]The reader should refer to the discussion in Chapter 5 on triaxial versus plane-strain friction angles. In that chapter, the reason for the higher plane-strain ϕ values and the use of Equations (5.20) and (5.12) to calculate both triaxial and plane-strain ϕ values are discussed.

or the Lyamin et al. (2007) shape factor, in particular, is more appropriate. Interpolation is needed for rectangular footings.

10.3.6 Load, base, and ground inclinations

Table 10.7 provides equations used for the depth factor, and Table 10.8 provides equations for the load, ground, and footing base inclination factors. If we consider the possibility that the transverse load Q_{tr} may have components Q_{tB} in the B direction and Q_{tL} in the L direction, the m values appearing in the Vesic factors given in Table 10.8 are calculated as follows:

$$m = \sqrt{m_B^2 + m_L^2} \tag{10.45}$$

where:

$$m_B = \begin{cases} \dfrac{2+B/L}{1+B/L} & \text{if } Q_{tB} > 0 \\ 0 & \text{otherwise} \end{cases} \tag{10.46}$$

$$m_L = \begin{cases} \dfrac{2+L/B}{1+L/B} & \text{if } Q_{tL} > 0 \\ 0 & \text{otherwise} \end{cases} \tag{10.47}$$

Note that B and L in these equations are nominal plan dimensions of the footing.

Except for the Lyamin et al. (2007) equations, the equations in Tables 10.5, 10.7, and 10.8 are semi-empirical because the theoretical basis for calculating collapse loads for footings under conditions other than plane-strain and with nonzero embedment was not available at

Table 10.7 Commonly used expressions for depth factors[a]

q_0 term	γ term
Meyerhof (1963)[b]	
$d_q = 1 + 0.1\sqrt{N}\dfrac{D}{B}$	$d_\gamma = 1 + 0.1\sqrt{N}\dfrac{D}{B}$
Brinch Hansen (1970)[c], Vesic (1973)	
$D/B \leq 1$	
$d_q = 1 + 2\tan\phi(1-\sin\phi)^2\dfrac{D}{B}$	$d_\gamma = 1$
$D/B > 1$	
$d_q = 1 + 2\tan\phi(1-\sin\phi)^2\tan^{-1}\dfrac{D}{B}$	$d_\gamma = 1$
Lyamin et al. (2007)	
$d_q = 1 + (0.0036\phi + 0.393)\left(\dfrac{D}{B}\right)^{-0.27}$	$d_\gamma = 1$

[a] Use nominal values of B and L for the Vesic factors and effective values of B and L for the Brinch Hansen and Lyamin et al. factors.
[b] N is the flow number, given by Equation (4.45).
[c] Brinch Hansen (1970) d_c derived from d_q through Equation (10.67).

Table 10.8 Commonly used load, base, and ground inclination factors[a]

Factors	q_0 term	γ term
Load inclination[b]	Meyerhof (1963)	
	$$i_q = \left[1 - \frac{\arctan\left(\dfrac{Q_{tr}}{Q_{ax}}\right)}{90°} \right]^2$$	$$i_\gamma = \left[1 - \frac{\arctan\left(\dfrac{Q_{tr}}{Q_{ax}}\right)}{\phi} \right]^2$$
	Brinch Hansen (1970)	
	$$i_q = \max\left[\left(1 - \frac{0.5 Q_{tr}}{Q_{ax} + Ac\cot\phi} \right)^5, 0 \right]$$	$$i_\gamma = \max\left[\left(1 - \frac{0.7 Q_{tr}}{Q_{ax} + Ac\cot\phi} \right)^5, 0 \right]$$
	Vesic (1973)	
	$$i_q = \max\left[\left(1 - \frac{Q_{tr}}{Q_{ax} + Ac\cot\phi} \right)^m, 0 \right]$$	$$i_\gamma = \max\left[\left(1 - \frac{Q_{tr}}{Q_{ax} + Ac\cot\phi} \right)^{1+m}, 0 \right]$$
	Loukidis et al. (2008)	
	—	$$i_\gamma = \left(1 - 0.94 \frac{\tan\alpha}{\tan\phi} \right)^{(1.5\tan\phi + 0.4)^2}$$
Base inclination	Brinch Hansen (1970)	
	$$b_q = \exp(-0.035\alpha_b \tan\phi)$$	$$b_\gamma = \exp(-0.047\alpha_b \tan\phi)$$
	Vesic (1973)[c]	
	$$b_q = \left(1 - \frac{\alpha_b \tan\phi}{57°} \right)^2$$	$$b_\gamma = \left(1 - \frac{\alpha_b \tan\phi}{57°} \right)^2$$
Ground inclination	Brinch Hansen (1970)	
	$$g_q = (1 - 0.5\tan\alpha_g)^5$$	$$g_\gamma = (1 - 0.5\tan\alpha_g)^5$$
	Vesic (1973)	
	$$g_q = (1 - \tan\alpha_g)^2$$	$$g_\gamma = (1 - \tan\alpha_g)^2$$

[a] m values are calculated from Equations (10.45)–(10.47) for the Vesic factors; these values may also be used for the Brinch Hansen factors if they fall in the $2 < m < 5$ range.
[b] A = area of footing base.
[c] α_b and α_g are measured in degrees.

the time they were proposed. Without 3D analyses, it could not be known by which factors to multiply the terms in the original bearing capacity equation (applicable to strip footings at the surface) to obtain the bearing capacity of embedded circular, square, or rectangular footings with inclined load, base, or ground surface. Without 3D analyses, it was also not possible to investigate the interdependence of the different terms of the bearing capacity equation. The experimental data on which these equations are based are mostly due to Meyerhof (1951, 1953, 1963), who tested both prototype and model foundations.

The choice of the value for the angle of inclination of the ground surface to use in the ground inclination factor calculation is made based on the direction the load (that is, the resultant of all the loadings on the footing) points to with respect to the ground. The ground inclination angle used in the equations should be the ground inclination on the side of the footing where most of the slip mechanism develops (that is, the side of the footing toward which the load on the footing is directed). In the case of a centered, vertical load applied on a level footing, the load clearly points down the slope, and that is the direction in which the slip mechanism will primarily develop. Additionally, base inclination must also be considered. Figure 10.15 represents the geometry of a very general foundation/load configuration.

When the base of the foundation is inclined, base inclination factors are used. The values of limit bearing capacity calculated in this manner are axial (normal to the foundation base), not vertical capacity.

Example 10.12

A square footing with side $B = 2\,m$ is to be constructed well above the water table in a clean sand with $\phi = 40°$. The ground is level, and the load is to be applied axially. The base of the footing will be placed at a depth of $0.75\,m$. The sand has $\gamma = 20\,kN/m^3$. Determine the limit bearing capacity for this situation.

Solution:

We will use the traditional methods of Meyerhof and Brinch Hansen. We start by calculating the values of the flow number N using Equation (4.45), N_q using Equation (10.6), and N_γ using Equation (10.12):

$$N = \frac{1+\sin\phi}{1-\sin\phi} = \frac{1+\sin 40°}{1-\sin 40°} = 4.6$$

$$N_q = \frac{1+\sin\phi}{1-\sin\phi}e^{\pi\tan\phi} = \frac{1+\sin 40°}{1-\sin 40°}e^{\pi\tan 40°} = 64.2$$

$$N_\gamma = 1.5(N_q - 1)\tan\phi = 1.5(64.2 - 1)\tan 40° = 79.5$$

The surcharge is:

$$q_0 = \gamma D = 20 \times 0.75 = 15 \text{ kPa}$$

We proceed to calculate the Meyerhof shape and depth factors:

$$s_q = 1 + 0.1N\frac{B}{L} = 1 + 0.1 \times 4.6 \times \frac{2}{2} = 1.46$$

$$s_\gamma = 1 + 0.1N\frac{B}{L} = 1 + 0.1 \times 4.6 \times \frac{2}{2} = 1.46$$

$$d_q = 1 + 0.1\sqrt{N}\frac{D}{B} = 1 + 0.1 \times \sqrt{4.6} \times \frac{0.75}{2} = 1.08$$

$$d_\gamma = 1 + 0.1\sqrt{N}\frac{D}{B} = 1 + 0.1 \times \sqrt{4.6} \times \frac{0.75}{2} = 1.08$$

Taking these into Equation (10.36):

$$q_{bL} = \left(s_q d_q\right) q_0 N_q + \frac{1}{2}\left(s_\gamma d_\gamma\right)\gamma B N_\gamma$$

$$= (1.46 \times 1.08) \times 15 \times 64.2 + \frac{1}{2}(1.46 \times 1.08) \times 20 \times 2 \times 79.5$$

$$q_{bL} = 4025 \text{ kPa}$$

The limit resistance of this footing is then calculated as

$$Q_{bL} = B^2 q_{bL} = 2^2 \times 4025 = 16,100 \text{ kN} \text{ }^{\text{answer}}$$

Let us now repeat the calculations using Brinch Hansen's shape and depth factors:

$$s_q - 1 + \frac{B}{L}\sin\phi - 1 + \frac{2}{2}\sin 40° = 1.64$$

$$s_\gamma - 1 - 0.4\frac{B}{L} - 1 - 0.4 \times \frac{2}{2} = 0.6$$

$$d_q = 1 + 2\tan\phi(1 - \sin\phi)^2 \frac{D}{B} = 1 + 2\tan 40°(1 - \sin 40°)^2 \frac{0.75}{2} = 1.08$$

$$d_\gamma = 1$$

Taking these into Equation (10.36):

$$q_{bL} = \left(s_q d_q\right) q_0 N_q + \frac{1}{2}\left(s_\gamma d_\gamma\right)\gamma B N_\gamma$$

$$= (1.64 \times 1.08) \times 15 \times 64.2 + \frac{1}{2}(0.6 \times 1) \times 20 \times 2 \times 79.5$$

$$q_{bL} = 2660 \text{ kPa}$$

The limit resistance of this footing is then calculated as

$$Q_{bL} = B^2 q_{bL} = 2^2 \times 2660 = 10,640 \text{ kN} \text{ }^{\text{answer}}$$

The fairly large difference between the two results, particularly in the N_γ term, illustrates the fact that we still find methods in the bearing capacity literature and in practice that can produce very different results. Most of the difference can be attributed to a difference in the N_γ term appearing due to the difference in s_γ values, which is striking in that $s_\gamma > 1$ according to the Meyerhof equation, but $s_\gamma < 1$ according to the Brinch Hansen equation.

Example 10.13

A rectangular footing with sides $B = 2$ m and $L = 3$ m is to be constructed well above the water table in a clean sand with $D_R = 70\%$. The ground is level, and the load is to be applied axially. The base of the footing will be placed at a depth of 0.75 m. The sand has $\gamma = 20$ kN/m³. Samples of the well-graded sand show its particles to be between subangular and subrounded. Determine the limit bearing capacity for this situation.

Solution:

In the absence of shear strength tests, we must estimate the critical-state friction angle ϕ_c. Because we have a clean sand that is not particularly angular, we estimate ϕ_c as being 32°. We must start by calculating the mean effective stress using Equation (10.39):

$$\sigma'_{mp} = 20p_A \left(\frac{\gamma B}{p_A}\right)^{0.7} \left(1 - 0.32\frac{B}{L}\right)$$

$$= 20 \times 100 \left(\frac{20 \times 2}{100}\right)^{0.7} \left(1 - 0.32\frac{2}{3}\right) = 828 \text{ kPa}$$

The value of A_ψ is obtained from Equation (10.42):

$$A_\psi = \frac{1}{3}\left(\frac{L}{B} + 8\right) = \frac{1}{3}\left(\frac{3}{2} + 8\right) = 3.17$$

The friction angle follows:

$$\phi_p = \phi_c + A_\psi \left\{ I_D \left[Q - \ln\left(\frac{100\sigma'_{mp}}{p_A}\right) \right] - R_Q \right\}$$

$$= 32° + 3.17 \left\{ 0.7 \left[10 - \ln\left(\frac{100 \times 828}{100}\right) \right] - 1 \right\}$$

$$= 32° + 4.1°$$

$$= 36.1°$$

The bearing capacity factors N_q and N_γ are calculated using Equations (10.6) and (10.13):

$$N_q = \frac{1 + \sin\phi_p}{1 - \sin\phi_p} e^{\pi\tan\phi_p} = \frac{1 + \sin 36.1°}{1 - \sin 36.1°} e^{\pi\tan 36.1°} = 3.87 \times 9.9 = 38.3$$

$$N_\gamma = (N_q - 0.6)\tan(1.33\phi_p) = (38.3 - 0.6)\tan(1.33 \times 36.1°) = 41.9$$

With the water table not being a factor, the surcharge q_0 is also easily calculated:

$$q_0 = \gamma_{m0}D = 20 \times 0.75 = 15 \text{ kPa}$$

Now we must determine the depth and shape factors. Since we are taking due consideration of confinement effects (that is, we are accounting for how close the footing is to plane-strain versus triaxial conditions) by using Equation (10.40), we use the Lyamin et al. (2007) shape factors. From Table 10.5,

$$s_q = 1 + (0.098\phi_p - 1.64)\left(\frac{D}{B}\right)^{0.7 - 0.01\phi_p} \left(\frac{B}{L}\right)^{1 - 0.16\left(\frac{D}{B}\right)}$$

$$= 1 + (0.098 \times 36.1 - 1.64)\left(\frac{0.75}{2}\right)^{0.7 - 0.01 \times 36.1} \left(\frac{2}{3}\right)^{1 - 0.16\left(\frac{0.75}{2}\right)}$$

$$= 1.93$$

$$s_\gamma = 1 + (0.0336\phi_p - 1)\frac{B}{L}$$

$$= 1 + (0.0336 \times 36.1 - 1)\frac{2}{3} = 1.14$$

The depth factors are:

$$d_\gamma = 1$$

$$d_q = 1 + (0.0036\phi_p + 0.393)\left(\frac{D}{B}\right)^{-0.27}$$

$$= 1 + (0.0036 \times 36.1 + 0.393)\left(\frac{0.75}{2}\right)^{-0.27} = 1.68$$

We can now calculate the limit bearing capacity:

$$q_{bL} = (s_q d_q)q_0 N_q + \frac{1}{2}(s_\gamma d_\gamma)\gamma B N_\gamma$$

$$= 1.93 \times 1.68 \times 15 \times 38.3 + \frac{1}{2} \times 1.14 \times 1 \times 20 \times 2 \times 41.9$$

$$q_{bL} = 1863 + 955 \approx 2818 \text{ kPa}$$

$$Q_{bL} = q_{bL}A = q_{bL}BL = 2818 \times 2 \times 3 \approx 16,908 \text{ kN}^{\text{answer}}$$

This bearing capacity is obviously very large. This load would cause collapse (extremely large settlements) of the footing. We would not be able to rely on it to support actual superstructure loads. In Chapter 9, we saw how to calculate shallow foundation settlement, which controls design in most cases. In Chapter 11, we combine what we have learned in both Chapters 9 and 10 to design footings that will be serviceable and safe during the design life of the superstructure for the given super-structure loads.

10.4 GENERAL SHEAR, LOCAL SHEAR, AND PUNCHING BEARING CAPACITY FAILURE MODES

According to a traditional view of the bearing capacity failure of footings, three modes of failure are possible: general shear, local shear, and punching shear (Figure 10.18). General shear has been our focus so far. In this mode of failure, slip surfaces extend all the way to the soil surface. Accordingly, most of the movement of the footing as it enters a bearing capacity failure mode is due to slippage between adjacent masses of soil upon development of the slip surfaces. In local shear, slip surfaces develop in the neighborhood of the base of the footing, but don't reach the surface. In this case, the footing settlement is due partly to slipping along slip surfaces and partly to contraction of the soil around the slip zone. In "punching shear," slip surfaces don't even develop, and settlement is due almost exclusively to volumetric contraction and widespread and diffuse (as opposed to localized) distortion of the soil, except very near the footing.

The concepts of punching and local shears were proposed originally to avoid excessive settlements. The procedure, which recommended a reduction in the shear strength parameters of the soil used in bearing capacity calculations, was rationalized on the basis that loose sands would not develop slip surfaces or would develop them only partially. Although still taught and covered in textbooks, these concepts have outlived their usefulness. We know today that even relatively loose sands at the confining stresses operative in foundation engineering problems are somewhat dilative, and, consequently, slip surfaces (or, more

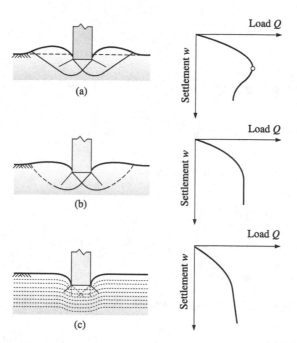

Figure 10.18 Modes of bearing capacity failure: (a) general, (b) local, and (c) punching shear modes. (Modified after Vesic (1973); with permission from ASCE.)

precisely, bands of intense shearing) do tend to form in them in most cases. The slip surfaces develop more completely and shearing concentrates more intensely along these shear bands as L/B increases (Loukidis 2006). The modern equations that we saw earlier, with an equivalent friction angle that reflects the relative density and representative mean effective stress beneath the footing, already reflect the mobilization of shear bands, full or partial, that occurs in the soil. In any case, the settlement check using the concepts of Chapters 2 and 9 will guarantee that we keep the allowable load at a limit that will not lead to excessive settlements.

Figure 10.19, proposed by Vesic (1973), provides some indication of which failure mode would, according to the traditional view, develop, depending on relative density and depth to the footing base. Naturally, high relative density and low confining stress (conditions that cause a sandy soil to be dilative) would lead to general shear modes. Since $D/B < 1$ for most shallow foundations, the figure suggests that, for $D_R > 67\%$, the foundation would fail in general shear. For $33 < D_R < 67\%$, the foundation would fail in local shear, and, for $D_R < 33\%$, it would fail in punching shear. The original use of Figure 10.19 was to ascertain for which conditions (those leading to punching or local shear) we would correct the shear strength of the soil by multiplying it by a factor of 2/3 before calculating bearing capacity. This led to an awkward discontinuity in bearing capacity, so that a shallow footing in sand with $D_R = 66\%$ would have a radically lower bearing capacity than the same footing in sand with $D_R = 68\%$. We will not be using reduced shear strength parameters to limit settlements indirectly, but we can still use Figure 10.19 for rough guidance on when settlement is likely to control design – when the chart suggests that local and punching shear would develop – instead of classical bearing capacity failure (which would be observed in the general shear area of that chart).

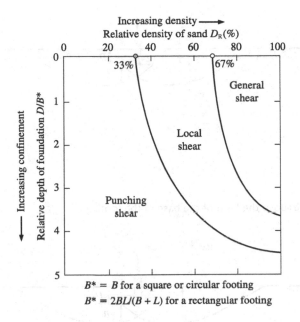

Figure 10.19 Bearing capacity failure modes versus relative density and depth to footing base for foundations in sands. (Modified after Vesic (1973); with permission from ASCE.)

10.5 FOOTINGS IN SAND: EFFECTS OF GROUNDWATER TABLE ELEVATION

Consider the water table to be at a depth z_w. If the water table is located at an elevation such that part of the slip mechanism for the footing would be below the water table (Figure 10.20), higher pore pressures will exist in that zone, implying lower shear strength and consequently lower bearing capacity. The effect of the water table is best accounted for by suitably choosing the value of the unit weight for use in the bearing capacity equation as the buoyant unit weight γ_b, the wet (or material) unit weight γ_m, or some number in between.

The buoyant unit weight γ_b must be used if the groundwater table is at or above the base of the footing (that is, $z_w \leq D$, as shown in Figure 10.21), for in this case the whole of the potential slip mechanism is under water. Since the depth of the slip mechanism is of the order of B below the base of the foundation (that is, a total depth of $D + B$), if the water table

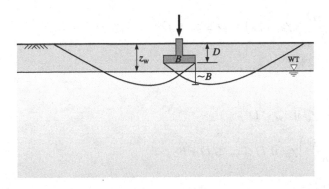

Figure 10.20 Footing bearing on soil with groundwater table within its influence zone.

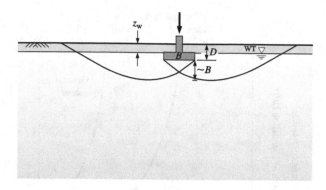

Figure 10.21 Water table above the level of the base of the footing.

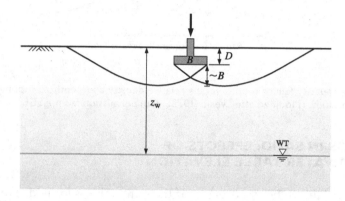

Figure 10.22 Water table below the lowest level to which the slip mechanism extends.

is below this depth (that is, $z_w \geq D + B$, as shown in Figure 10.22), the material moist unit weight γ_m must be used. A simple interpolation can be used for water table depths between D and $D + B$ (for simplicity, γ_m may also be taken as equal to γ_{sat} above the water table if interpolation is required). Taking z_w to be the depth to the water table, the value of the unit weight γ to use in the bearing capacity equation can be written as

$$\gamma = \begin{cases} \gamma_b = \gamma_{sat} - \gamma_w \text{ if } z_w < D \\ \gamma_b + \left(\dfrac{z_w - D}{B}\right)(\gamma_m - \gamma_b) \text{ if } D \leq z_w \leq D + B \\ \gamma_m \text{ if } z_w > D + B \end{cases} \tag{10.48}$$

which reduces to

$$\gamma = \begin{cases} \gamma_b = \gamma_{sat} - \gamma_w \text{ if } z_w < D \\ \gamma_b + \left(\dfrac{z_w - D}{B}\right)\gamma_w \text{ if } D \leq z_w \leq D + B \\ \gamma_m \text{ if } z_w > D + B \end{cases} \tag{10.49}$$

when γ_m is assumed to be the same as γ_{sat} (because of saturation above the water table due to capillary rise).

The surcharge q_0 also depends on the position of the water table. It is simply the effective stress at the level of the footing base. So, for $z_w < D$, it can be written as

$$q_0 = \gamma_m z_w + (D - z_w)\gamma_b \qquad (10.50)$$

Example 10.14

Redo Example 10.13 considering the water table to be located at a depth of 2 m. Consider the saturated unit weight of the soil to be equal to 22 kN/m³.

Solution:

The depth to water table (2 m) is such that the water table is located below the level of the base of the footing (0.75 m), but above the approximate depth of the bottom of the slip mechanism at $0.75 + 2 = 2.75$ m. So q_0 does not change from the value we calculated in Example 10.13, but the value of γ to use in the γ term of the bearing capacity equation changes. Equation (10.48) for $D \le z_w \le D + B$ gives:

$$\gamma = \gamma_b + \left(\frac{z_w - D}{B}\right)(\gamma_m - \gamma_b)$$

$$= (22 - 9.81) + \left(\frac{2 - 0.75}{2}\right)(20 - 22 + 9.81)$$

$$= 17.1 \text{ kN/m}^3$$

The calculation sequence is the same as in Example 10.13. We simply calculate the same quantities as before in succession. We start calculating the mean effective stress:

$$\sigma'_{mp} = 20 p_A \left(\frac{\gamma B}{p_A}\right)^{0.7}\left(1 - 0.32\frac{B}{L}\right)$$

$$= 20 \times 100\left(\frac{17.1 \times 2}{100}\right)^{0.7}\left(1 - 0.32\frac{2}{3}\right) = 742 \text{ kPa}$$

The friction angle follows:

$$\phi_p = \phi_c + A_\psi \left\{ I_D\left[Q - \ln\left(\frac{100\sigma'_{mp}}{p_A}\right)\right] - R_Q \right\}$$

$$= 32° + 3.17\left\{0.7\left[10 - \ln\left(\frac{100 \times 742}{100}\right)\right] - 1\right\}$$

$$= 32° + 4.4°$$

$$= 36.4°$$

The bearing capacity factors N_q and N_γ are calculated using Equations (10.6) and (10.13):

$$N_q = \frac{1 + \sin\phi_p}{1 - \sin\phi_p}e^{\pi\tan\phi_p} = \frac{1 + \sin 36.4°}{1 - \sin 36.4°}e^{\pi\tan 36.4°} = 3.92 \times 10.1 = 39.6$$

$$N_\gamma = (N_q - 0.6)\tan(1.33\phi_p) = (39.6 - 0.6)\tan(1.33 \times 36.4°) = 43.9$$

$$q_0 = \gamma_{m0}D = 20 \times 0.75 = 15 \text{ kPa}$$

We use the Lyamin et al. (2007) shape factors. From Table 10.5,

$$s_q = 1 + (0.098\phi_p - 1.64)\left(\frac{D}{B}\right)^{0.7-0.01\phi_p}\left(\frac{B}{L}\right)^{1-0.16\left(\frac{D}{B}\right)}$$

$$= 1 + (0.098 \times 36.4 - 1.64)\left(\frac{0.75}{2}\right)^{0.7-0.01\times36.4}\left(\frac{2}{3}\right)^{1-0.16\left(\frac{0.75}{2}\right)}$$

$$= 1.94$$

$$s_\gamma = 1 + (0.0336\phi_p - 1)\frac{B}{L}$$

$$= 1 + (0.0336 \times 36.4 - 1)\frac{2}{3} = 1.14$$

The depth factors are:

$$d_\gamma = 1$$

$$d_q = 1 + (0.0036\phi_p + 0.393)\left(\frac{D}{B}\right)^{-0.27}$$

$$= 1 + (0.0036 \times 36.4 + 0.393)\left(\frac{0.75}{2}\right)^{-0.27} = 1.68$$

We are now ready to compute the limit bearing capacity:

$$q_{bL} = (s_q d_q)q_0 N_q + \frac{1}{2}(s_\gamma d_\gamma)\gamma B N_\gamma$$

$$= 1.94 \times 1.68 \times 15 \times 39.6 + \frac{1}{2} \times 1.14 \times 1 \times 17.1 \times 2 \times 43.9$$

$$q_{bL} = 1936 + 856 \approx 2792 \text{ kPa}$$

$$Q_{bL} = q_{bL}A = q_{bL}BL = 2792 \times 2 \times 3 \approx 16,752 \text{ kN} \text{ answer}$$

Note that this bearing capacity is approximately 0.9% lower than the capacity available in the case of Example 10.13, for which the water table is deep. The reduction in the unit weight below the foundation base reduced the N_γ term, but also increased slightly the effective friction angle – two offsetting effects.

Example 10.15

Calculate the limit bearing capacity of the 6 ft square footing (see Example 9.4) shown in E-Figure 10.4. The footing is embedded 2.5 ft into a sand with SPT N_{60} given in E-Table 10.1. The sand is approximately 40 ft thick and has an average unit weight of 125 pcf. It is underlain by a shale bedrock. Consider that the worst-case scenario is for the water table to be located at the level of the base of the footing.

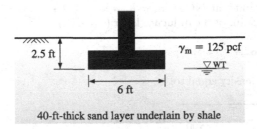

E-Figure 10.4 Footing for Examples 10.15 and 9.4.

E-Table 10.1 Corrected SPT blow counts, normalized for 60% energy ratio, for Example 10.15

Depth (ft)	N_{60}
2.5	3
5	5
7.5	12
10	14
12.5	15
15	14
17.5	12
20	16
22.5	16
25	18
27.5	13
30	15
32.5	17
35	19
37.5	20

Solution:

The depth of influence for this footing is equal to B below the base of the footing, that is, $2.5 + 6 = 8.5$ ft. From E-Table 10.1 we see that the average SPT blow counts normalized for 60% energy ratio down to this depth is:

$$\bar{N}_{60} = \frac{3+5+12}{3} = 7$$

Using Equation (7.6):

$$\frac{D_R}{100\%} = \sqrt{\frac{\bar{N}_{60}}{A + BC\dfrac{\sigma'_v}{p_A}}}$$

where the coefficient A varies between 27 and 46 (we take an average value equal to 36.5), $B = 27$, and C is given by Equation (7.8):

$$C = \frac{K_0}{K_{0,NC}}$$

We assume that the sand is an NC sand, so $K_0 = K_{0,NC} = 0.4$ and $C = 1$. The vertical effective stress at the midpoint of the influence layer ($z = 5.5$ ft) is:

$$\sigma'_v = 125 \times 5.5 - 62.4 \times 3 = 500 \text{ psf}$$

This gives a relative density equal to:

$$\frac{D_R}{100\%} = \sqrt{\frac{7}{36.5 + 27 \times 1 \times \dfrac{500}{2000}}} = 0.402 \approx 0.40$$

or

$$D_R = 40\%$$

Given that there is no information provided regarding the value of ϕ_c, we will, conservatively, assume $\phi_c = 30°$. We then follow the now familiar sequence of calculations (already used in Examples 10.13 and 10.14):

$$\gamma = \gamma_b + \left(\frac{z_w - D}{B}\right)(\gamma_m - \gamma_b)$$

$$= (125 - 62.4) + \left(\frac{2.5 - 2.5}{2}\right)(125 - 125 + 62.4)$$

$$= 62.6 \text{ pcf}$$

$$\sigma'_{mp} = 20p_A\left(\frac{\gamma B}{p_A}\right)^{0.7}\left(1 - 0.32\frac{B}{L}\right)$$

$$= 20 \times 2000\left(\frac{62.6 \times 6}{2000}\right)^{0.7}\left(1 - 0.32\frac{6}{6}\right) = 8440 \text{ psf}$$

The friction angle follows:

$$\phi_p = \phi_c + A_\psi\left\{I_D\left[Q - \ln\left(\frac{100\sigma'_{mp}}{p_A}\right)\right] - R_Q\right\}$$

$$= 30° + 3\left\{0.4\left[10 - \ln\left(\frac{100 \times 8440}{2000}\right)\right] - 1\right\}$$

$$= 30° + 1.7°$$

$$= 31.7°$$

The bearing capacity factors N_q and N_γ are calculated using Equations (10.6) and (10.13):

$$N_q = \frac{1 + \sin\phi_p}{1 - \sin\phi_p}e^{\pi\tan\phi_p} = \frac{1 + \sin 31.7°}{1 - \sin 31.7°}e^{\pi\tan 31.7°} = 3.21 \times 6.96 = 22.3$$

$$N_\gamma = (N_q - 0.6)\tan(1.33\phi_p) = (22.3 - 0.6)\tan(1.33 \times 31.7°) = 19.6$$

$$q_0 = \gamma_{m0}D = 125 \times 2.5 \approx 312 \text{ psf}$$

We use the Lyamin et al. (2007) shape factors from Table 10.5:

$$s_q = 1 + \left(0.098\phi_p - 1.64\right)\left(\frac{D}{B}\right)^{0.7-0.01\phi_p}\left(\frac{B}{L}\right)^{1-0.16\left(\frac{D}{B}\right)}$$

$$= 1 + \left(0.098 \times 31.7 - 1.64\right)\left(\frac{2.5}{6}\right)^{0.7-0.01\times31.7}\left(\frac{6}{6}\right)^{1-0.16\left(\frac{2.5}{6}\right)}$$

$$= 2.04$$

$$s_\gamma = 1 + \left(0.0336\phi_p - 1\right)\frac{B}{L}$$

$$= 1 + \left(0.0336 \times 31.7 - 1\right)\frac{6}{6} = 1.06$$

The depth factors are:

$$d_\gamma = 1$$

$$d_q = 1 + \left(0.0036\phi_p + 0.393\right)\left(\frac{D}{B}\right)^{-0.27}$$

$$= 1 + \left(0.0036 \times 31.7 + 0.393\right)\left(\frac{2.5}{6}\right)^{-0.27} = 1.64$$

We are now ready to compute the limit bearing capacity:

$$q_{bL} = \left(s_q d_q\right)q_0 N_q + \frac{1}{2}\left(s_\gamma d_\gamma\right)\gamma B N_\gamma$$

$$= 2.04 \times 1.64 \times 312 \times 22.3 + \frac{1}{2} \times 1.06 \times 1 \times 62.6 \times 6 \times 19.6$$

$$q_{bL} = 23,277 + 3902 \approx 27,179 \text{ psf}$$

$$Q_{bL} = q_{bL} A = q_{bL} BL = 27,179 \times 6 \times 6 \approx 9.78 \times 10^5 \text{ lbs} \approx 489 \text{ tons}^{\text{answer}}$$

Footings on loose sand, such as this one, will likely undergo very large settlements by the time they approach the limit load divided by the factor of safety. So we will, in these cases, have a settlement-controlled design. In Example 9.4, we saw that 182 tons, which corresponds to a ratio of 489/182 = 2.7, produces 3.7 in. of settlement. Obviously, settlement will control design in this case, as 3.7 in. is too large a settlement. In problems at the end of Chapter 11, we will explore what the settlement would be for a factor of safety of 5 (Problem 11.17) and then do a design of a suitable footing for these soil conditions (Problem 11.18).

10.6 FOUNDATIONS SUBJECTED TO LOAD ECCENTRICITY

10.6.1 Idealized distributions of pressure at foundation base

Sometimes, because of space constraints, a column may not be located at the center of the footing. This gives rise to a geometric load eccentricity, shown in Figure 10.23a. On other occasions, the column is centered on the footing, but a moment (associated with a lateral

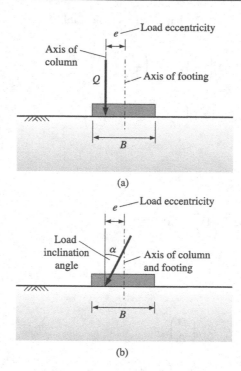

Figure 10.23 Foundation with eccentric load: (a) geometric eccentricity and (b) mechanical eccentricity.

load) is transferred from the column to the footing, giving rise to a mechanical load eccentricity defined mathematically as the ratio of the applied moment to the applied vertical load. This second situation is shown in Figure 10.23b, in which the load is shown inclined because it has both a vertical and a lateral component. Both types of eccentricities are treated in the same way in calculations of bearing capacity. The eccentricity in the B direction is denoted as e_B, and that in the L direction, as e_L.

The analytical treatment of a rectangular foundation subjected to an eccentric load is almost the same as that of the rectangular cross section of a beam subjected to both a normal load and a bending moment. Figure 10.24 shows that the eccentric load may be converted to a statically equivalent system consisting of a moment and a centered load. The only difference with respect to what we normally do for a beam cross section is that, typically, no tensile stresses are allowed to develop at the soil–foundation interface. To quantify the compressive stresses at the base of the footing, we establish a system of reference defined by the two axes passing through the center of the footing base: x_B parallel to the B direction and x_L parallel to the L direction. We set the positive directions of x_B and x_L so that we have a right-hand reference system assuming z to be positive down into the soil (see Figure 10.25a). If Q is applied in the quadrant where both x_B and x_L are positive (the first quadrant), its moments with respect to the x_B and x_L axes will be positive as shown in Figure 10.25a. These positive moments generate positive (compressive) stress in the first quadrant.

So the contact stress q_b normal to the foundation base at a point at a distance x_L from the x_B axis and at a distance x_B from the x_L axis is given by

$$q_b\left(x_B, x_L\right) = \frac{Q}{A} + \frac{M_L x_B}{I_L} + \frac{M_B x_L}{I_B} \tag{10.51}$$

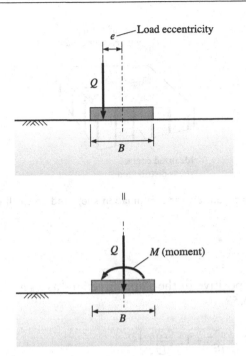

Figure 10.24 Equivalence between an eccentric load and a centered load combined with a moment.

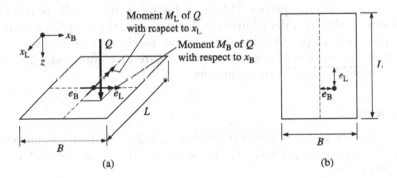

Figure 10.25 Eccentric force Q acting on a footing: (a) 3D view and reference system and (b) plan view of load and eccentricities with respect to axes x_B and x_L.

$$q_b\left(x_B, x_L\right) = \frac{Q}{BL} + 12\frac{M_L x_B}{LB^3} + 12\frac{M_B x_L}{BL^3} \tag{10.52}$$

where Q = axial load acting on the foundation; I_B, I_L = moments of inertia of the plan area of the footing with respect to the x_L (L) and x_B (B) axes, respectively; and M_L, M_B = moments in the L and B directions (that is, with respect to the x_L and x_B axes, respectively) of the footing. Using the right-hand rule, it is seen that M_L creates a stress gradient in the B direction, and M_B, in the L direction.

Let us now write the moments in terms of the load and the load eccentricities in the B and L directions:

$$M_B = Qe_L \tag{10.53}$$

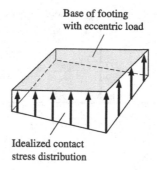

Base of footing
with eccentric load

Idealized contact
stress distribution

Figure 10.26 Contact stress distribution for a foundation subjected to small eccentricities in the *B* and *L* directions.

$$M_L = Q\, e_B \tag{10.54}$$

where both e_B and e_L are positive in the same directions as x_B and x_L, respectively. We can now rewrite Equation (10.52) as

$$q_b\left(x_B, x_L\right) = \frac{Q}{BL}\left(1 + 12\frac{e_B x_B}{B^2} + 12\frac{e_L x_L}{L^2}\right) \tag{10.55}$$

The contact stress distribution for an eccentrically loaded footing with eccentricities in two directions is seen from Equation (10.55) to be linear in both the *B* and *L* directions. Figure 10.26 shows what this distribution looks like when the eccentricities are small and the contact stresses are all positive. In reality, soil is not elastic and, as a result, the contact stress distributions are not linear as represented in Figure 10.26 for working loads and much less so for loads near the limit bearing capacity.

10.6.2 The kern

The maximum contact stress develops at point ($B/2$, $L/2$), and the minimum contact stress is found at ($-B/2$, $-L/2$) for positive eccentricities. Taking these values of x_B and x_L into Equation (10.55) gives:

$$q_{b,\text{max}/\text{min}} = \frac{Q}{BL}\left(1 \pm 6\frac{e_B}{B} \pm 6\frac{e_L}{L}\right) \tag{10.56}$$

If the eccentricities take certain limiting values, the minimum contact stress becomes zero at one of the corners of the foundation. These limiting values e_{B0} and e_{L0} of the eccentricities can be calculated by making $q_{b,\text{min}} = 0$ in Equation (10.56), which leads to:

$$\pm\frac{e_{B0}}{B} \pm \frac{e_{L0}}{L} = \frac{1}{6} \tag{10.57}$$

Equation (10.57) describes four straight lines that meet at four points: ($\pm 1/6\,B$, $\pm 1/6\,L$). These four lines form a rhombus that bounds what is usually referred to as the core or kern of the foundation (Figure 10.27). If the load is applied at any point contained within the kern, the contact stresses are compressive everywhere. Obviously, when the load is located outside of the kern, tensile stresses (negative contact stresses) would tend to develop. Soil cannot take tensile stresses, and a redistribution of stresses occurs instead; as a result, the

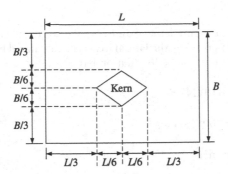

Figure 10.27 The four straight lines define the locus of points of application of the axial load that lead to positive contact stress between the footing and the soil at all points except at one of the footing corners or one of its edges, where the contact stress is equal to zero. The enclosed area is the kern of the foundation; loads applied inside the kern produce only positive contact stresses.

stresses in the tensile zone become zero and the compressive stresses all increase to help balance the moment, compensating for the absence of tensile stresses.

We would not normally engineer a foundation in such a way that a portion of its base is not in contact with the soil because this would clearly be a waste of material. Instead, we do everything we can to keep the eccentricities small or, when that is not possible, use alternative schemes, such as combined or strap footings.

Example 10.16

A bridge pier is supported on a shallow foundation as shown in E-Figure 10.5. The total vertical force, including the self-weight of the pier and footing, is 2000 kN. The lateral forces are 85 kN and 50 kN. Determine the distribution of contact stresses at the base of the footing, which has plan dimensions of 2.3 m in the direction of the smaller lateral load and 3.5 m in the direction of the larger lateral load.

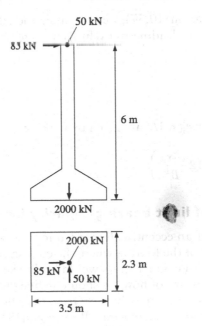

E-Figure 10.5 Bridge foundation supporting vertical and lateral loads.

Solution:

As always, we take B as the smaller side and L as the larger side of the plan area of the footing. The moments imposed by the lateral loads are calculated by multiplying them by the leverage arm with respect to the foundation base:

$$M_B = 85\,\text{kN} \times 6\,\text{m} = 510\,\text{kN} \cdot \text{m}$$

$$M_L = 50\,\text{kN} \times 6\,\text{m} = 300\,\text{kN} \cdot \text{m}$$

The eccentricities are given as

$$e_L = 510/2000 = 0.26\,\text{m}$$

$$e_B = 300/2000 = 0.15\,\text{m}$$

Using Equation (10.57), we can show that the resultant is in the core (kern) of the foundation:

$$\frac{0.26\,\text{m}}{3.5\,\text{m}} + \frac{0.15\,\text{m}}{2.3\,\text{m}} = 0.14 < \frac{1}{6}$$

So we can now calculate the contact stresses at the corners of the footing using Equation (10.56):

$$q_{b,\max/\min} = \frac{2000}{3.5 \times 2.3}\left(1 \pm 6\frac{0.26}{3.5} \pm 6\frac{0.15}{2.3}\right)$$

which gives 456, 262, 235, and 40 kPa. If we plot in three dimensions these stresses perpendicularly to the footing base and join the four resulting points, we obtain a plane that represents the stress distribution for the footing.

10.6.3 Eccentricity in one direction

If we make $e_{L0} = 0$ in Equation (10.57), we see that, in the case of eccentricity in one direction only, the largest eccentricity leading to exclusively compressive contact stresses at the base of the footing is given by

$$\frac{e_{B0}}{B} = \pm\frac{1}{6} \tag{10.58}$$

The contact stresses for $e_B < 1/6$ are given by making $e_L = 0$ in Equation (10.55):

$$q_b\left(x_B, x_L\right) = \frac{Q}{BL}\left(1 + 12\frac{e_B x_B}{B^2}\right) \tag{10.59}$$

10.6.4 Calculation of limit bearing capacity for eccentric loads

It stands to reason that, if an eccentrically loaded footing is to fail in bearing capacity, this should start at the corner of the footing where the contact stress is maximum. The question that follows is what value of contact stress would lead to bearing capacity failure. The question may be rephrased in terms of how we should define the unit limit bearing capacity of a footing for which the unit load varies linearly from one edge to the other(s).

An answer to this question is sketched by Peck et al. (1953) for footings with eccentricity in one direction. They point out that the right comparison would be between the maximum stress at the toe of the footing and the limit unit bearing capacity for a much smaller effective width of the footing (an "effective" area near the toe). This is conceptually attractive because it recognizes that any bearing capacity failure will indeed start at or near the toe of the footing. However, the unit bearing capacity near the toe is not the same as that calculated for the whole footing in the case of a centrally loaded footing, for which the whole width B is operative. We cannot possibly estimate the value of effective width to use to calculate this hypothetical bearing capacity.

Another possible answer lies is the Meyerhof (1953) proposal to define the effective area of the footing as the area with respect to which the load is centered. This means that we neglect a portion of the footing in the calculation of bearing capacity. The effective dimensions of the footing then become (see Figure 10.28 for one-way eccentricity and Figure 10.29 for two-way eccentricity):

$$B_{eff} = B - 2e_B \tag{10.60}$$

$$L_{eff} = L - 2e_L \tag{10.61}$$

Once the effective dimensions B_{eff} and L_{eff} are determined, the bearing capacity of the footing is fully calculated with reference to B_{eff} and L_{eff}. The various correction factors are calculated as prescribed in Tables 10.5, 10.7, and 10.8. The effective width B_{eff} is used in place of B in the self-weight (γ) term of the bearing capacity equation. The limit load is then calculated as

$$Q_{bL} = B_{eff} L_{eff} q_{bL} \tag{10.62}$$

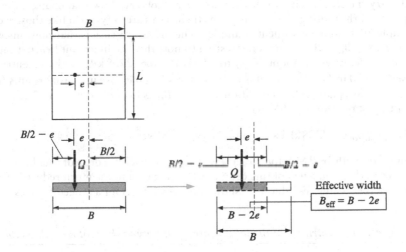

Figure 10.28 Effective width for one-way eccentricity.

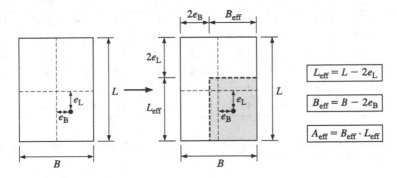

Figure 10.29 Effective dimensions for two-way eccentricity.

Based on the results obtained from FE analyses of strip footings subjected to combined eccentric and inclined loading resting on a purely frictional (sandy) soil following the Mohr–Coulomb failure criterion, Loukidis et al. (2008) proposed a more accurate expression for the effective width B_{eff} of strip footings as a function of load eccentricity e:

$$B_{eff} = B\left(1 - 2.273\frac{e}{B}\right)^{0.8} \tag{10.63}$$

This equation is always used with the following load inclination factor:

$$i_\gamma = \left(1 - 0.94\frac{\tan\alpha}{\tan\phi}\right)^{(1.5\tan\phi + 0.4)^2} \tag{10.64}$$

where α = load inclination with respect to the vertical direction.

Example 10.17

Study the effect of one-way load eccentricity in the direction of the smaller footing dimension on the bearing capacity of the footing of Example 10.13. Consider the eccentricity here to be physical (not due to a moment created by a lateral force)[11].

Solution:

We will vary the eccentricity all the way to $B/6$ and observe how that changes the bearing capacity of the footing. The calculations themselves are very much like those we used in Example 10.13 and subsequent examples. The only difference is that now, instead of using B, we use $B_{eff} = B - 2e_B$. It is interesting to note that the limit unit bearing capacity falls only moderately (by about 10%) to 2503 kPa from 2788 kPa as the eccentricity is increased from 0 to $B/6$. In part, this is due to the slight increase in ϕ_p that results from a transition toward plane-strain conditions, with L/B_{eff} being greater than L/B. The limit resistance for $e_B = B/6$ is calculated as

$$Q_{bL} = q_{bL}B_{eff}L = 2538\,\text{kPa} \times 1.33\,\text{m} \times 3\,\text{m} = 10.1\,\text{MN}$$

This compares with 16.9 MN for the case with no eccentricity, a drop of roughly 40%. As we have seen, three-quarters of this drop can be attributed to the much reduced effective plan area of the footing. E-Table 10.2 shows the calculations for a range of values of e_B/B.

E-Table 10.2 *Effect of eccentricity on the limit bearing capacity of the footing of Example 10.17*

e_B/B	σ'_{mp} (kPa)	A_ψ	ϕ_p (°)	N_{peak}	N_q	N_γ	s_q	s_γ	d_q	q term (kPa)	γ term (kPa)	q_{bL} (kPa)
0	828	3.17	36.1	3.87	38.3	41.9	1.93	1.14	1.68	1862	956	2818
1/24	797	3.21	36.3	3.89	39.0	42.9	1.89	1.13	1.67	1844	892	2736
1/18	786	3.23	36.3	3.90	39.2	43.4	1.88	1.13	1.66	1839	871	2710
1/12	762	3.27	36.4	3.92	39.9	44.3	1.86	1.12	1.65	1831	830	2660
1/9	737	3.31	36.6	3.95	40.6	45.4	1.83	1.12	1.64	1825	789	2615
1/6	680	3.42	36.9	4.01	42.4	48.2	1.78	1.11	1.61	1827	711	2538

---

[11] In Problem 10.23, the reader is asked to consider the case when the eccentricities are caused by increasing values of a lateral load applied at a set distance from the footing base.

10.7 CALCULATION OF BEARING CAPACITY USING CURVE-FIT c AND ϕ PARAMETERS

Although fundamentally incorrect and difficult to use because of uncertainty as to how to estimate the shear strength parameters, some engineers still use curve-fit c and ϕ parameters to calculate bearing capacity. For completeness, we will provide here the bearing capacity correction factors necessary for these calculations.

We saw earlier that N_c and N_q are related through Equation (10.11). The five correction factors (for shape, depth, load inclination, base inclination, and ground inclination, given in Table 10.9) for the c and q terms are related as well. To see why, consider the case of the shape factor. For a weightless, cohesionless soil, the limit bearing capacity is:

$$q_{bL} = s_q q_0 N_q \tag{10.65}$$

We consider now a weightless soil with (instead of without) cohesion. We can introduce the cohesion c by using Caquot's principle:

$$q_{bL} + c \cot\phi = s_q N_q \left(q_0 + c \cot\phi \right)$$

Simple reordering gives:

$$q_{bL} = s_q N_q q_0 + c \cot\phi \left(s_q N_q - 1 \right)$$

which can finally be rewritten as

$$q_{bL} = s_q N_q q_0 + c s_c N_c$$

where

$$s_c = \frac{s_q N_q - 1}{N_q - 1} \tag{10.66}$$

Table 10.9 Bearing capacity correction factors for the c term of the equation

Factor	Meyerhof (1963)	Brinch Hansen (1970)	Vesic (1973)
Shape	$s_c = 1 + 0.2N\dfrac{B}{L}$	$s_c = 1 + \cos\phi\,\dfrac{N_q}{N_c}\dfrac{B}{L}$	$s_c = 1 + \dfrac{N_q}{N_c}\dfrac{B}{L}$
Depth	$d_c = 1 + 0.2\sqrt{N}\dfrac{D}{B}$	$d_c = \begin{cases} 1 + 2(1-\sin\phi)^2\dfrac{N_q}{N_c}\dfrac{D}{B} & \text{for } \dfrac{D}{B} \le 1 \\[2mm] 1 + 2(1-\sin\phi)^2\dfrac{N_q}{N_c}\tan^{-1}\dfrac{D}{B} & \text{for } \dfrac{D}{B} > 1 \end{cases}$	$d_c = \begin{cases} 1 + 0.4\dfrac{D}{B} & \text{for } \dfrac{D}{B} \le 1 \\[2mm] 1 + 0.4\tan^{-1}\dfrac{D}{B} & \text{for } \dfrac{D}{B} > 1 \end{cases}$
Load inclination	$i_c = \left[1 - \dfrac{\arctan\left(\dfrac{Q_t}{Q_a}\right)}{90°}\right]^2$	$i_c = i_q - \dfrac{1-i_q}{N_q -}$	$i_c = i_q - \dfrac{1-i_q}{N_q - 1}$
Base inclination	NA	$b_c = b_q - \dfrac{1-b_q}{N_q - 1}$	$b_c = 1 - \dfrac{2\alpha_b}{(2+\pi)\tan\phi}$
Ground inclination	NA	$g_c = g_q - \dfrac{1-g_q}{N_q - 1}$	$g_c = i_q - \dfrac{1-i_q}{(2+\pi)\tan\phi}$

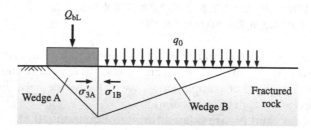

Figure 10.30 Limit bearing capacity mechanism for a footing on fractured rock.

Likewise:

$$d_c = \frac{d_q N_q - 1}{N_q - 1}$$

(10.67)

10.8 LIMIT BEARING CAPACITY OF SHALLOW FOUNDATIONS IN ROCKS

The limit bearing capacity of intact rocks is usually quite large: in hard rocks, the structural capacity of the footing is likely to control. If the rock contains structural defects, however, the bearing capacity of the rock may control. An intensely fractured rock, for example, behaves somewhat like a dilative soil. Following Wyllie (1992), a simple mechanism may be used to estimate the bearing capacity of a strip footing in a fractured rock. This mechanism assumes that an active wedge forms immediately below the footing; this active wedge pushes laterally on a passive wedge extending to the side (Figure 10.30).

Using the Hoek–Brown criterion, the limit unit bearing capacity q_{bL} is given as

$$q_{bL} = \sigma'_{3A} + q_u \sqrt{m \frac{\sigma'_{3A}}{q_u} + s}$$

where σ'_{3A} is the minor principal stress in the active wedge A, which in limit equilibrium matches the major principal stress σ'_{1B} in the passive wedge B. This, in turn, is given by

$$\sigma'_{1B} = \sigma'_{3B} + q_u \sqrt{m \frac{\sigma'_{3B}}{q_u} + s}$$

where σ'_{3B} is obviously equal to the surcharge q_0 acting on the passive wedge. So we can write:

$$q_{bL} = q_0 + q_u \left\{ \sqrt{m \frac{q_0}{q_u} + s} + \sqrt{m \frac{q_0 + q_u \sqrt{m \frac{q_0}{q_u} + s}}{q_u} + s} \right\}$$

(10.68)

where:
  m, s = Hoek–Brown strength parameters (refer to Chapters 4 and 7);
  q_u = unconfined compressive strength of the rock; and
  q_0 = surcharge at the level of base of foundation.

Table 10.10 Shape factors for footings in rocks

Foundation shape	Shape factor C_s
Strip ($L/B > 6$)	1
Rectangular with $L/B = 5$	1.05
Rectangular with $L/B = 2$	1.12
Square	1.25
Circular	1.20

If the surcharge q_0 is due to embedment of the footing in the rock (the usual scenario), then a depth factor is required. Likewise, if we have a square or rectangular footing instead of a strip footing, a shape factor is also required. Unfortunately, these factors have not been studied as well in rocks as in soils. For simplicity, we can take the depth factor as equal to 1 and use the values in Table 10.10, proposed by Sowers and Sowers (1970), for the shape factor.

Example 10.18

Find the limit unit load for a 1.5 m square footing bearing directly on a basalt, with little embedment, that will support a centered, vertical column load. An open cut near the location of the footing shows the basalt to have 5–10 fissures per meter with a horizontal direction (zero dip). Although borings for this specific location are not available, typical RQD values for this rock are in the 25%–50% range. The basalt, as observed in this cut, is slightly weathered. Its compressive strength is 80 MPa.

Solution:

If we are to use Equation (10.68) to compute q_{bL} for this footing, we need to find the values of the parameters m and s. We already did this back in Chapter 7 when we solved Example 7.4, but we will repeat our steps here for clarity. We start by obtaining the rock mass rating (RMR) from Table 7.5 and then m and s from Table 7.6. According to Table 7.5, the RMR is obtained as follows:

RMR = 7 for rock with $50 < q_u < 100$ MPa

$\qquad$ + 8 for RQD in the 25% – 50% range

$\qquad$ + 8 for discontinuity spacing in the 0.1 – 0.2 range

$\qquad$ + 25 (for slightly weathered walls)

$\qquad$ + 10 (water to be accounted for in analysis)

$\qquad$ + 0 (joint orientation to be accounted for in analysis)

$\qquad$ = 58

We can assume the effect of water to be negligible and the joint orientation (horizontal) to be neutral. Entering Table 7.6 with an RMR = 58 and the rock description we have, the best match would be a fine-grained igneous rock with fair quality rock mass, leading to

$\qquad m = 1.395$

and

$\qquad s = 0.00293$

In the absence of any surcharge, $q_0 = 0$, Equation (10.68) reduces to:

$$q_{bL} = q_u \sqrt{s} \left\{ 1 + \sqrt{\frac{m}{\sqrt{s}} + 1} \right\}$$

Plugging in the values of q_u, m, and s, we find:

$$q_{bL} = 80\sqrt{0.00293} \left\{ 1 + \sqrt{\frac{1.395}{\sqrt{0.00293}} + 1} \right\} = 26.7 \text{ MPa}^{\text{answer}}$$

10.9 CHAPTER SUMMARY

10.9.1 Main concepts and equations

10.9.1.1 Bearing capacity equation

The **bearing capacity equation** gives the value of the **limit unit bearing capacity** q_{bL} of the footing, which, multiplied by the plan area of the footing, gives the **limit bearing capacity** or **limit load capacity** or **limit resistance** Q_{bL} of the footing load. Q_{bL} is the load that will cause the foundation to plunge into the ground. This means that when the load on a foundation (including its own weight and the weight of any backfill) is increased gradually up to the limit bearing capacity, the foundation first undergoes moderate settlement and then settles by large amounts for small additional load increments as the limit bearing capacity is approached. For this to happen, the foundation element must push aside a mass of soil by forming a mechanism that may or may not reach the ground surface. When it does reach the ground surface, it is referred to as **general shear**; otherwise, if the mechanism is not fully developed, it has traditionally been known as **local shear** or **punching shear**, depending on the extent of the slip mechanism.

The sources of resistance to the formation of the slip mechanism are the soil self-weight, the weight of the soil above the base of the footing, and any cohesive shear strength that may be present. The three terms of the bearing capacity equation correspond to the three sources of bearing capacity. The general bearing capacity equation is written as

$$q_{bL} = \left(s_c d_c i_c b_c g_c \right) c N_c + \left(s_q d_q i_q b_q g_q \right) q_0 N_q + 0.5 \left(s_\gamma d_\gamma i_\gamma b_\gamma g_\gamma \right) \gamma B N_\gamma \qquad (10.36)$$

The unit bearing capacity q_{bL} is the **gross unit bearing capacity**. When multiplied by the footing plan area, it yields the load capacity available at the base of the foundation; it must be sufficient to carry the load on the foundation in addition to the foundation self-weight and the weight of any backfill on top of the foundation. When these are subtracted from $Q_{bL} = A \, q_{bL}$, and the result is again divided by the plan area of the footing, we obtain the **net unit bearing capacity** $q_{bL}{}^{\text{net}}$, which is the capacity net of the requirements to support the foundation and backfill and is thus the capacity directly available to support loads from the superstructure.

10.9.1.2 Calculation of bearing capacity in clays

Bearing capacity in clays is calculated assuming that loads are applied rapidly compared with the drainage rate of clays and that the short term is the critical loading condition. Loading therefore takes place under undrained conditions, and the footing fails in the general shear mode.

The bearing capacity equation for a clay deposit with strength s_u increasing linearly with depth z at a rate $\rho = ds_u/dz$ is:

$$q_{bL} = Fs_{su}d_{su}i_{su}\left[s_{u0} + \frac{\rho B}{4N_c}\right]N_c + q_0 \tag{10.33}$$

where F (which is equal to 1 for $\rho = 0$ throughout the depth of influence for the foundation) is a correction factor that is given in Figure 10.14, and s_{u0} is the undrained shear strength of the clay at depth $z_f = 0$ below the base of the foundation.

Making $\rho = 0$ in Equation (10.33) leads to the traditional bearing capacity equation for a soil with uniform strength $s_u = s_{u0}$ throughout:

$$q_{bL} = 5.14s_{su}d_{su}i_{su}s_u + q_0 \tag{10.27}$$

The bearing capacity factor of 5.14 in Equation (10.27) results from taking the limit of

$$N_c = (N_q - 1)\cot\phi \tag{10.11}$$

as ϕ approaches zero.

The shape, depth, and load inclination factors are:

$$s_{su} = 1 + C_1\frac{B}{L}\left(\frac{2.3}{\exp\left[0.353\left(\frac{\mu B}{s_{u0}}\right)^{0.509}\right]} - 1.3\right) + C_2\sqrt{\frac{D}{B}} \tag{10.34}$$

which reduces to Equation (10.25) when $\rho = 0$; C_1 and C_2 are given in Table 10.2.

$$i_{su} = 1 - 1.3\frac{Q_h}{Q_v} \tag{10.28}$$

$$d_{su} = 1 + 0.27\sqrt{\frac{D}{B}} \tag{10.26}$$

10.9.1.3 Calculation of bearing capacity in sands

Equation (10.36), with $c = 0$, is directly used to calculate bearing capacity in sands. Bearing capacity in sands is calculated assuming drained conditions. The terms in the bearing capacity equation for sands are given by

$$N_q = \frac{1 + \sin\phi}{1 - \sin\phi}e^{\pi\tan\phi} \tag{10.6}$$

$$N_\gamma = (N_q - 0.6)\tan(1.33\phi) \tag{10.13}$$

Shape factors are given in Table 10.5. Depth factors are given in Table 10.7. Load inclination, base inclination, and ground inclination factors are given in Table 10.8.

Because, in realistic foundation design problems, the friction angle for use in the bearing capacity equation is not given, a procedure is needed to determine it. Equation (10.39) can be used for this purpose. It gives the value of the mean effective stress that is representative of the bearing capacity problem; this stress can be used in Equations (5.21) and (5.12) for calculating the friction angle.

10.9.1.4 Load eccentricity

A load applied off-center generates a (theoretical) contact stress distribution on the footing base that is planar but not uniform. If the eccentricity is large, this plane cuts across the plane of the footing base, meaning that a portion of the footing will not be in contact with the soil. This will not happen, and the entire base will be subjected to compressive stresses only, if the load lies within a central area of the footing base known as the **kern**. For two-way eccentricity, the kern is a rhombus defined by joining with straight lines the four points defined by ($\pm B/6, \pm L/6$) with respect to the center of the footing. For one-way eccentricity, if we slice the footing in three equal rectangles in the direction normal to the eccentricity, the kern is the middle rectangle, the so-called middle third of the footing.

We account for load eccentricity by using a fictitious effective plan area for the footing, such that the load is centered with respect to this area. This means the effective dimensions B_{eff} and L_{eff} are used in equations instead of the true footing dimensions B and L. These effective widths are defined as

$$B_{eff} = B - 2e_B \tag{10.60}$$

$$L_{eff} = L - 2e_L \tag{10.61}$$

10.9.2 Notations and Symbols

Symbol	Quantity	US units	SI units
B	Foundation width	ft	m
b_c	Base inclination factor for the c term of bearing capacity equation	Unitless	Unitless
B_{eff}	Effective footing dimension	ft	m
B_n	Nominal footing dimension	ft	m
b_q	Base inclination factor for the q term of bearing capacity equation	Unitless	Unitless
b_γ	Base inclination factor for the γ term of bearing capacity equation	Unitless	Unitless
d_c	Depth factor for the c term of bearing capacity equation	Unitless	Unitless
d_q	Depth factor for the q term of bearing capacity equation	Unitless	Unitless
d_{su}	Depth factor for saturated clay	Unitless	Unitless
d_γ	Depth factor for the γ term of bearing capacity equation	Unitless	Unitless
e	Eccentricity	ft	m
e_B	Eccentricity in the B direction	ft	m

(*Continued*)

Symbol	Quantity	US units	SI units
e_L	Eccentricity in the L direction	ft	m
F	Correction factor for bearing capacity in soil with increasing strength with depth	Unitless	Unitless
g_c	Ground inclination factor for the c term of bearing capacity equation	Unitless	Unitless
g_q	Ground inclination factor for the q term of bearing capacity equation	Unitless	Unitless
g_γ	Ground inclination factor for the γ term of bearing capacity equation	Unitless	Unitless
i_c	Load inclination factor for the c term of bearing capacity equation	Unitless	Unitless
i_q	Load inclination factor for the q term of bearing capacity equation	Unitless	Unitless
i_{su}	Load inclination factor for saturated clay	Unitless	Unitless
i_γ	Load inclination factor for the γ term of bearing capacity equation	Unitless	Unitless
L_{eff}	Effective footing dimension	ft	m
L_n	Nominal footing dimension	ft	m
m	Exponent appearing in equations for load inclination factors	Unitless	Unitless
M_B	Moment in the B direction	ft·lb	N·m
M_l	Moment in the L direction	ft·lb	N·m
N_c	Bearing capacity factor for the c or s_u term	Unitless	Unitless
N_q	Bearing capacity factor for soil surcharge	Unitless	Unitless
N_γ	Bearing capacity factor for unit weight term	Unitless	Unitless
q_0	Soil surcharge	psf	kPa
Q_a	Axial component of load Q	lb	kN
q_{bL}	Limit unit load	psf	kPa
Q_L	Limit load	lb	kN
Q_{tB}	Transverse load in the B direction	lb	kN
Q_{tL}	Transverse load in the L direction	lb	kN
Q_{tr}	Transverse component of load Q	lb	kN
s_c	Shape factor for the c term of bearing capacity equation	Unitless	Unitless
s_q	Shape factor for the q term of bearing capacity equation	Unitless	Unitless
s_{su}	Shape factor for saturated clay	Unitless	Unitless
s_u	Undrained shear strength of clay	psf	kPa
s_{u0}	Undrained shear strength of clay at the level of foundation base	psf	kPa
s_γ	Shape factor for the γ term of bearing capacity equation	Unitless	Unitless
α_b	Inclination angle of foundation base	deg	deg
α_g	Inclination angle of ground	deg	deg
γ_m	Soil unit weight	pcf	kN/m^3
ρ	Strength gradient with depth for soil with strength increasing linearly with depth	psf/ft	kPa/m
σ_1^A	Major principal stress in zone A	psf	kPa
σ_1^B	Major principal stress in zone B	psf	kPa

10.10 PROBLEMS

10.10.1 Conceptual problems

Problem 10.1 Define all the terms in bold contained in the chapter summary.

Problem 10.2 The surcharge q_0 for the calculation of bearing capacity of a footing in clay is the total stress at the level of the base of the footing, while it is the effective stress there for a footing in sand. Explain why.

Problem 10.3 In this chapter, we developed an equation for the limit bearing capacity of a footing in sand (P-Figure 10.1a) and one for a footing in clay (P-Figure 10.1b). Knowing that there is no specific equation for either the case shown in P-Figure 10.1c or that shown in P-Figure 10.1d, how would you calculate bearing capacity for these two cases in a way that would be conservative?

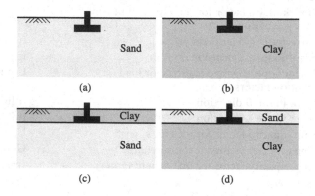

P-Figure 10.1   The bearing capacity equation has been developed for (a) footings in sands and (b) footings in clays, but not for mixed-soil conditions (c) and (d).

Problem 10.4 Consider a footing in a deep clay deposit. Assuming drainage only at the surface and excess pore pressures to develop mostly within the top $D + 2B$ of the clay layer, where B is the width of the footing, discuss how long it would take for a load applied instantaneously at a given time to dissipate. For typical clay properties and typical construction times (considering also that loads are applied over the construction time instead of all at the same instant of time), what range of values would you expect for the degree of consolidation at the end of construction? What is the degree of conservatism in calculating bearing capacities using total stress analysis and an undrained shear strength?

Problem 10.5 Does the limit unit bearing capacity q_{bL} of a surface, strip footing increase with B? If so, is this increase linear? How does the limit load Q_{bL} vary with B for the same footing? Provide reasons for all your answers.

Problem 10.6 Does the limit unit bearing capacity q_{bL} of a square footing increase with B? If so, is this increase linear? How does the limit load Q_{bL} vary with B for the same footing? Provide reasons for all your answers.

Problem 10.7 Plot the shape factor for footings in clays and sands versus B/L using the various equations discussed in this chapter. Explain why the shape factor varies in this manner with B/L.

Problem 10.8 With reference to Example 10.9, at what depth would we need to calculate an equivalent s_u to obtain the correct result with Equation (10.21)? What result would we have obtained had we used the traditional bearing capacity equation with an average s_u taken over a depth B below the footing?

Problem 10.9 Show that the first-yield bearing capacity factor N_c for a sensitive clay is equal to π by imposing that the maximum shear stress caused by the unit load q_b be equal to s_u. Refer to Problem 4.29 for the expression of the maximum shear stress beneath a strip footing in clay.

Problem 10.10 Discuss the effect on q_{bL} of both small and large oscillations in the water table for (a) footings in sands and (b) footings in saturated clays.

10.10.2 Quantitative problems

Problem 10.11 What is the limit bearing capacity of a strip footing with width $B = 1$ m on the surface of a sand deposit with $D_R = 70\%$? The sand has a unit weight equal to 19 kN/m³ and a critical-state friction angle ϕ_c of 31°. Assume the water table to be deep.

Problem 10.12 Consider that the footing of Problem 10.11 is embedded 0.6 m in the ground, that is, $D = 0.6$ m. What is the limit resistance of the footing now?

Problem 10.13 Consider a 1 m square instead of a strip footing on the surface of the soil deposit of Problem 10.11. What is the limit resistance of the footing?

Problem 10.14 Consider that the footing of Problem 10–13 is embedded 0.6 m in the ground, that is, $D = 0.6$ m. What is the limit resistance of the footing now?

Problem 10.15 Study the bearing capacity of a square footing with width $B = 2$ m placed at a depth of 0.75 m in sand with $\gamma = 20$ kN/m³ and ϕ ranging from 35° to 45°. Consider the water table to range from the ground surface to deep enough not to affect the results of the calculations.

Problem 10.16 Redo Example 10.13 considering the water table to be located at a depth of 0.3 m. Consider the saturated unit weight of the soil to be equal to 22 kN/m³.

Problem 10.17 Consider a soil deposit with $ds_u/d\sigma'_v = 0.3$, $\gamma_b = 10$ kN/m³, and water table at the surface. A 2 m square footing is located at a depth of 0.75 m, as shown in P-Figure 10.2. The undrained shear strength s_{u0} is equal to 30 kPa down to a depth of 3 m. It increases linearly with depth from 3 m down. What is the limit bearing capacity of this footing? Use Equations (10.25) and (10.26) for the shape and depth factors.

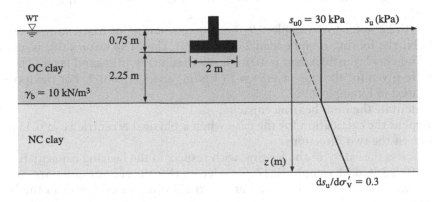

P-Figure 10.2 Footing and soil profile for Problem 10.17.

Problem 10.18 Consider the square footing with cross section shown in P-Figure 10.3. Calculate the limit resistance (the limit load capacity) for this footing in the absence of any lateral load. The sand has D_R = 73%, with a unit weight equal to 19 kN/m³ and a critical-state friction angle ϕ_c of 30°. The footing dimensions are B = 3 m and D = 1.25 m. The water table is very deep and has no bearing on the calculations.

Problem 10.19 Consider now, for the footing of Problem 10.18, that the load has a lateral component of 20 kN (in addition to the 300 kN vertical load) in a direction parallel to one of the footing sides.

a. Compute the maximum and minimum soil normal pressures at the base of the footing. Neglect the footing weight.
b. Using the Vesic expression for the load inclination factor, calculate the limit load for vertical bearing.
c. What is the sliding resistance? Neglect passive resistance.

Problem 10.20 Repeat Problem 10.19 for the case in which the load has 20 kN components in both directions parallel to the footing sides.

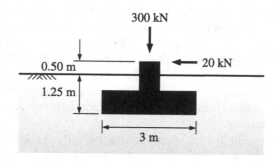

P-Figure 10.3 Footing for Problems 10.18, 10.19, and 10.20.

Problem 10.21 What is the limit bearing capacity of a 9 ft square footing placed at a depth of 3 ft in the soil profile shown in P-Figure 10.4? The footing will be backfilled after construction. The soil has a unit weight of 118 pcf. Use Equations (10.26) and (10.34).

Problem 10.22 Refer to Example 9.9. The square footing of that example will support 200 kN; the footing is embedded 2 m in an NC clay. The water table is at a depth of 2 m. The clay layer thickness is 10 m. It overlies a very firm sand layer. The following data are given for the clay layer: γ = 18 kN/m³ and s_u/σ'_v = 0.3. For the footing width calculated in Example 9.9:

a. Calculate the limit bearing capacity.
b. Repeat the calculation for the case when a physical eccentricity of 0.15 m exists in one of the two directions.
c. Discuss the safety of the footing with respect to the bearing capacity limit state.

Problem 10.23 Redo Example 10.17 considering that the eccentricities are caused by an increasing horizontal load applied at a vertical distance of 1 m from the base of the footing.

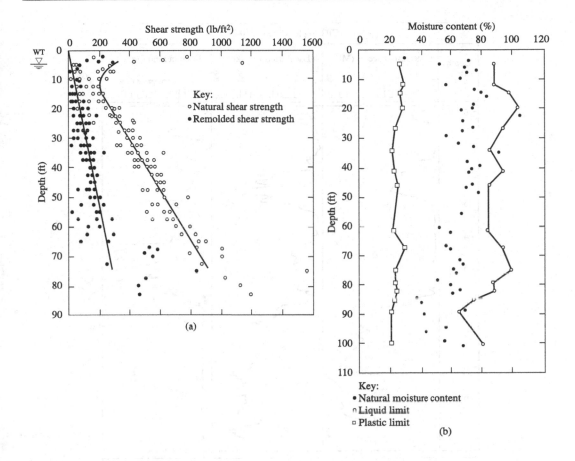

P-Figure 10.4 Soil profile for Problem 10.21: (a) natural and remolded shear strength and (b) water content, liquid limit, and plastic limit versus depth. (Golder 1957; courtesy of the Institution of Civil Engineers and Thomas Telford Limited.)

Problem 10.24 Study the changes in the limit bearing capacity of the footing of Example 10.12 if the following changes are made in sequence:
a. a horizontal load of 1500 kN is applied at a vertical distance of 1 m from the base of the footing (0.25 m above ground level);
b. the ground surface is inclined at an angle of 10°; and
c. the footing is placed at an angle of 10°.

10.10.3 Design problems

Problem 10.25 For the CPT log provided in P-Figure 10.5, determine the limit bearing capacity of a 1.5 m square footing placed at a depth of 1.5 m. The sand is reasonably well graded, with subangular particles, gravelly in some places and clayey in others. The unit weight of the sand for use in your calculations is 20 kN/m³. The sand is normally consolidated, with $K_0 = 0.4$. The critical-state friction angle is 30°. A clay layer starts at 3.5 m, extending to roughly 6.5 m. The water table is at a depth of 3 m.

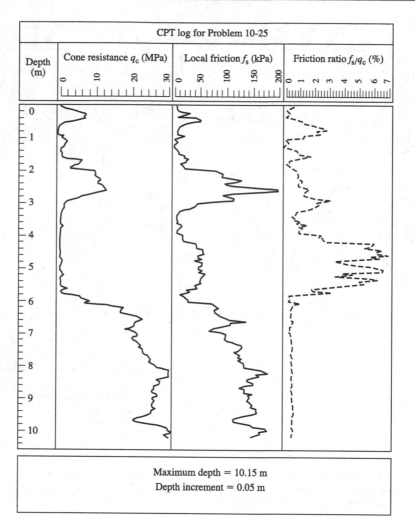

CPT log for Problem 10-25		
Depth (m)	Cone resistance q_c (MPa)	Local friction f_s (kPa)	Friction ratio f_s/q_c (%)

Maximum depth = 10.15 m
Depth increment = 0.05 m

P-Figure 10.5 CPT log for Problem 10.25.

Problem 10.26 Referring to the parameters defined in Figure 10.15, a square footing with $B = 1.5$ m rests on soil with a footing base slope $\alpha_b = 20°$. The natural ground slope is $\alpha_g = 10°$ in the opposite direction. The footing is embedded $D = 0.75$ m in the ground. The soil is a sand with small amounts of clay that has a unit weight of 21 kN/m³, $\phi_c = 30°$, and $D_R = 75\%$. Find the limit load normal to the base of the footing for the following cases of load inclination: (a) resultant load normal to the footing, (b) resultant load at 5° with normal to the footing, (c) resultant load at 10° with normal to the footing, and (d) resultant load at 30° with normal to the footing.

Problem 10.27 A footing 1.5 m wide and 2.0 m long is placed in sand. The relative density in the depth range of interest is estimated (from SPT data) to be $D_R = 75\%$. Laboratory tests show that the critical-state friction angle ϕ_c is 33°. The depth of embedment is 0.6 m, and the unit weight of the sand is 20 kN/m³. Assume that the water table is at the level of the base of the footing. Find the bearing capacity of the footing.

Problem 10.28 A square footing ($B = 2.0$ m) is placed in sand. The relative density within the depth of interest is estimated (from SPT data) to be $D_R = 70\%$. Laboratory tests show that the critical-state friction angle ϕ_c is 32°. The depth of embedment is 1 m, and the unit weight of the sand is 20 kN/m³. Assume that the water table is at 0.5 m below the base of the footing. Find the limit bearing capacity of the footing.

Problem 10.29 Calculate the limit bearing capacity for the footing of Problem 9.20.

Problem 10.30 Calculate the limit bearing capacity for the footing of Problem 9.21.

Problem 10.31 Calculate the limit bearing capacity for the footing of Problem 9.22.

REFERENCES

References cited

De Beer, E. E. (1970). "Experimental determination of the shape factors and the bearing capacity factors of sand." *Géotechnique*, 20(4), 387–411.

Bolton, M. D. (1979). *A Guide to Soil Mechanics*. 1st Edition, MacMillan Education, London, UK.

Bolton, M. D. (1986). "The strength and dilatancy of sands." *Géotechnique*, 36(1), 65–78.

Brinch Hansen, J. (1970). "A revised and extended formula for bearing capacity." *Bulletin of the Danish Geotechnical Institute*, 28, 5–11.

Davis, E. H. and Booker, J. R. (1973). "The effect of increasing strength with depth on the bearing capacity of clays." *Géotechnique*, 23(4), 551–563.

Golder, H. Q. (1957). "A note on piles in sensitive clays." *Géotechnique*, 7(4), 192–195.

Houlsby, G. T. and Wroth, C. P. (1983). "Calculation of stresses on shallow penetrometers and footings." *Proceedings of International Union of Theoretical and Applied Mechanics (IUTAM)/ International Union of Geodesy and Geophysics (IUGG) Symposium on Seabed Mechanics*, Newcastle, UK, 107–112.

Loukidis, D. (2006). "Advanced constitutive modeling of sands and applications to foundation engineering." PhD Thesis, Purdue University, West Lafayette, Indiana, USA.

Loukidis, D., Chakraborty, T., and Salgado, R. (2008). "Bearing capacity of strip footings on purely frictional soil under eccentric and inclined loads." *Canadian Geotechnical Journal*, 45, 768–787.

Loukidis, D. and Salgado, R. (2009). "Bearing capacity of strip and circular footings in sand using finite elements." *Computers and Geotechnics*, 36(6), 871–879.

Loukidis, D. and Salgado, R. (2011). "Effect of relative density and stress level on the bearing capacity of footings on sand." *Géotechnique*, 61(2), 107–119.

Lyamin, A. V, Salgado, R., Loan, S. W. S., and Prezzi, M. (2007). "Two-and three-dimensional bearing capacity of footings in sand." *Géotechnique*, 57(8), 647–662.

Martin, C. M. (2001). "Vertical bearing capacity of skirted circular foundations on Tresca soil." *Proceedings of 15th International Conference on Soil Mechanics and Geotechnical Engineering*, Istanbul, 743–746.

Martin, C. M. (2004). *User guide for ABC: analysis of bearing capacity*. Department of Engineering Science, University of Oxford, Oxford, UK.

Martin, C. M. (2005). "Exact bearing capacity calculations using the method of characteristics." *Proceedings of 11th International Conference on Computer Methods and Advances in Geomechanics*, Turin, Italy, 4, 441–450.

Meyerhof, G. G. (1950). "A general theory of bearing capacity." *Building Research Station Note No. C143*, Watford, England.

Meyerhof, G. G. (1951). "The ultimate bearing capacity of foudations." *Géotechnique*, 2(4), 301–332.

Meyerhof, G. G. (1953). "The bearing capacity of foundations under eccentric and inclined loads." *Proceedings of 3rd International Conference on Soil Mechanics and Foundation Engineering*, Switzerland, 440–445.

Meyerhof, G. G. (1963). "Some recent research on the bearing capacity of foundations." *Canadian Geotechnical Journal*, 1(1), 16–26.

Peck, R. B., Hanson, W., and Thornburn, T. H. (1953). *Foundation Engineering*. Wiley, New York.

Perkins, S. W. and Madson, C. R. (2000). "Bearing capacity of shallow foundations on sand: a relative density approach." *Journal of Geotechnical and Geoenvironmental Engineering*, 126(6), 521–530.

Prandtl, L. (1920). "Über die Härte plastischer Körper." *Nachrichten von der Gesellschaft der Wissenschaften zu Göttingen, Mathematisch-Physikalische Klasse,* 12, 74–85.

Prandtl, L. (1921). "Uber die eindringungsfestigkeit (Harte) plastischer baustoffe und die festigkeit von schneiden." *Zeitschrift für Angewandte Mathematik und Mechanik,* 1(1), 15–20.

Randolph, M. F., Jamiolkowski, M. B., and Zdravković, L. (2004). "Load carrying capacity of foundation." *Proceedings of the Skempton Conference: In Advances in Geotechnical Engineering,* Royal Geographical Society, London, UK, 1, 207–240.

Reissner, H. (1924). "Zum erddruckproblem." *Proceedings of the 1st International Congress for Applied Mechanics,* Delft, Netherlands, 295–311.

Salençon, J. and Matar, M. (1982). "Capacité portante des fondations superficielles circulaires." *Journal de Mécanique Théorique et Appliquée,* 1(2), 237–267.

Salgado, R. (2020). "Forks in the road: rethinking modeling decisions that defined the teaching and practice of geotechnical engineering." *Keynote Geotechnical Lecture in International Conference on Geotechnical Engineering Education,* Athens, Greece.

Salgado, R., Lyamin, A. V., Sloan, S. W., and Yu, H. S. (2004). "Two- and three-dimensional bearing capacity of foundations in clay." *Géotechnique,* 54(5), 297–306.

Skempton, A. W. (1951). "The bearing capacity of clays." *Proceeding of Building Research Congress,* London, UK, 180–189.

Sloan, S. W. (2013). "Geotechnical stability analysis." *Géotechnique,* 63(7), 531–572.

Sokolovski, V. V. (1965). *Statics of Granular Media.* 1st Edition, Pergamon Press, London, UK.

Sowers, G. B. and Sowers, G. F. (1970). *Introductory Soil Mechanics and Foundations.* 3rd Edition, McMillan, New York.

Tatsuoka, F., Okahara, M., Tanaka, T., Tani, K., Morimoto, T., and Siddiquee, M. S. (1991). "Progressive failure and particle size effect in bearing capacity of footing on sand." *Proceedings of Geotechnical Engineering Congress,* Geotechnical Special Publication No. 27, 788–802.

Terzaghi, K. (1943). *Theoretical Soil Mechanics.* Wiley, New York.

Ueno, K., Miura, K., and Maeda, Y. (1998). "Prediction of ultimate bearing capacity of surface footings with regard to size effects." *Soils and Foundations,* 38(3), 165–178.

Vesic, A. S. (1973). "Analysis of ultimate loads of shallow foundations." *Journal of the Soil Mechanics and Foundations Division,* 99(SM1), 45–73.

Wyllie, D. C. (1992). *Foundations on Rock.* E. and F.N. Spon, London, UK.

Additional references

Asaoka, A. and Ohtsuka, S. (1987). "Bearing capacity analysis of a normally consolidated clay foundation." *Soils and Foundations,* 27(3), 58–70.

Balla, A. (1962). "Bearing capacity of foundations." *Journal of the Soil Mechanics and Foundations Division,* 88(5), 13–34.

Baracos, A. (1957). "The foundation failure of the Transcona grain elevator." *The Engineering Journal,* 40(7), 973–990.

Booker, J. R. (1969). "Applications of theories of plasticity to cohesive frictional soils." Ph.D. Thesis, Sydney University, Australia.

Bozozuk, M. (1972). "Foundation failure of the Vanleek hill tower silo." *Proceedings of Specialty Conference on Performance of Earth- Supported Structures,* Lafayette, Indiana, 2(2), 885–902.

Bozozuk, M. (1977). "Evaluating strength tests from foundation failures." *Proceedings of 9th International Conference on Soil Mechanics and Foundation Engineering,* Tokyo, Japan, 1, 55–59.

Brown, J. D. and Meyerhof, G. G. (1969). "Experimental study of bearing capacity in layered clays." *Proceedings of 7th International Conference on Soil Mechanics and Foundation Engineering,* Mexico City, Mexico, 2, 45–51.

Brown, J. D. and Paterson, W. G. (1964). "Failure of an oil storage tank founded on a sensitive marine clay." *Canadian Geotechnical Journal,* 1(4), 205–214.

Button, S. J. (1953). "The bearing capacity of footing on a two-layer cohesive subsoil." *Proceedings of 3rd International Conference on Soil Mechanics and Foundation Engineering,* Switzerland, 1, 332–335.

Caquot, A. and Kerisel, J. (1953). "Sur le terme de surface dans le calcul des fondations en milieu pulverent." *Proceedings of 3rd International Conference on Soil Mechanics and Foundation Engineering*, Switzerland, 1, 336–337.

Chen, W. F. and Davidson, H. L. (1973). "Bearing capacity determination by limit analysis." *Journal of the Geotechnical Engineering Division*, 99(SM6), 433–449.

DeBeer, E. E. (1965). "Bearing capacity and settlement of shallow foundation on sand." *Proceedings of Symposium on Bearing Capacity and Settlement of Foundations*, Duke University, Durham, North Carolina, USA, 15–34.

DeBeer, E. E. (1965). "The scale effect on the phenomenon of progressive rupture in cohesionless soils." *Proceedings of 6th International Conference on Soil Mechanics and Foundation Engineering*, Montreal, Canada, 2, 13–17.

Drucker, D. C., Greenberg, H. J., and Prager, W. (1951). "The safety factor of an elastic-plastic body in plane strain." *Journal of Applied Mechanics – Transactions of the ASME*, 18(4), 371–378.

Eden, W. J. and Bozozuk, M. (1962). "Foundation failure of a silo on varved clay." *The Engineering Journal*, 45(9), 54–57.

Golder, H. Q., Fellenius, W., Kogler, F., Meischeider, H., Krey, H., and Prandtl, L. (1941). "The ultimate bearing pressure of rectangular footings." *Journal of the Institution of Civil Engineers*, 17(2), 161–174.

Griffiths, D. V. (1982). "Computation of bearing capacity factors using finite elements." *Géotechnique*, 32(3), 195–202.

Brinch Hansen, J. (1961). "A general formula for bearing capacity." *Danish Geotechnical Institute Bulletin No. 11*, 38–46.

Brinch Hansen, J. (1970). "A revised and extended formula for bearing capacity." *Danish Geotechnical Institute Bulletin No. 28*, 5–11.

Hu, G. C. Y. (1964). "Variable-factors theory of bearing capacity." *Journal of the Soil Mechanics and Foundations Division*, 90(SM4), 85–95.

Karafiath, L. L. (1970). "Shape factors in bearing capacity equation." *Journal of the Soil Mechanics and Foundations Division*, 9(SM4), 1493–1497.

Ko, H. Y. and Davidson, L. W. (1973). "Bearing capacity of footings in plane strain." *Journal of the Soil Mechanics and Foundation Division*, 99(SM1), 1–23.

Ko, H. Y. and Scott, R. F. (1973). "Bearing capacities by plasticity theory." *Journal of the Soil Mechanics and Foundations Division*, 99(SM1), 25–43.

Lundgren, H. and Mortensen, K. (1953). "Determination by the theory of plasticity of the bearing capacity of continuous footings on sand." *Proceedings of 3rd International Conference on Soil Mechanics and Foundation Engineering*, Switzerland, 1, 409–412.

Meyerhof, G. G. (1965). "Shallow foundations." *Journal of the Soil Mechanics and Foundations Division*, 91(SM2), 21–31.

Meyerhof, G. G. (1957). "The ultimate bearing capacity of foundations on slopes." *Proceedings of 4th International Conference on Soil Mechanics and Foundation Engineering*, London, UK, 1, 384–386.

Meyerhof, G. G. (1974). "Ultimate bearing capacity of footings on sand layer overlying clay." *Canadian Geotechnical Journal*, 11, 223–229.

Michalowski, R. L. and Shi, L. (1995). "Bearing capacity of footings over two-layer foundation soils." *Journal of Geotechnical Engineering*, 121(5), 421–428.

Milovic, D. M. (1965). "Comparison between the calculated and experimental values of the ultimate bearing capacity." *Proceedings of the 6th International Conference on Soil Mechanics and Foundation Engineering*, Montreal, Canada, 2, 142–144.

Narita, K. and Yamaguchi, H. (1992). "Three-dimensional bearing capacity analysis of foundations by use of a method of slices." *Soils and Foundations*, 32(4), 143–155.

Nordlund, R. L. and Deer, D. U. (1970). "Collapse of the Fargo grain elevator." *Journal of the Soil Mechanics and Foundations Division*, 96(SM12), 585–607.

Peck, R. B. and Bryant, F. G. (1953). "The bearing capacity failure of the Transcona elevator." *Géotechnique*, 3, 201–208.

Prakash, S. and Saran, S. (1971). "Bearing capacity of eccentrically loaded footings." *Journal of the Soil Mechanics and Foundations Division*, 97(SM1), 95–117.

Skempton, A. W. (1942). "An investigation of the bearing capacity of a soft clay soil." *Journal of the Institution of Civil Engineering*, 18, 307–321.

Stuart, J. G. (1962). "Interference between foundations with special reference to surface footings in sand." *Géotechnique*, 12(1), 15–22.

Tanaka, A., Bauer, G., and Queiroz De Carvalho, J. B. (1989). "Failure and analysis of a concrete silo." *Proceedings of 12th International Conference on Soil Mechanics and Foundation Engineering*, Rio De Janeiro, Brazil, 1, 345–348.

White, L.S. (1953). "Transcona elevator failure: eye-witness account." *Géotechnique*, 3(5), 209–214.

Wilson, G. (1941). "The calculation of the bearing capacity of a soft clay soil." *Journal of the Institution of Civil Engineers*, 17, 87–96.

Chapter 11

Shallow foundation design

> The secret of getting ahead is getting started. The secret of getting started is breaking your complex[,] overwhelming tasks into small manageable tasks, and then starting on the first one.
>
> Mark Twain

Foundation design requires the identification of possible limit states, followed by a series of checks to guarantee that no limit state will be reached by the foundations. We covered the analyses required to check the main limit states in Chapters 9 and 10. Shallow foundation design consists in using these analyses in a proper sequence to determine the dimensions and location of the foundation element and any reinforcement needed to prevent the limit states. In this chapter, we discuss in greater detail the shallow foundation design process using both the working stress design (WSD) and the load and resistance factor design (LRFD) approaches.

11.1 THE SHALLOW FOUNDATION DESIGN PROCESS

11.1.1 The design problem

The geotechnical design problem for a shallow foundation can be concisely stated as follows: given one or more loadings (which structural engineers calculate from consideration of the superstructure as prescribed by codes) and information about the subsoil, define the location within the ground where the foundation should be placed and the most economical dimensions of the foundation such that the foundation will support the loadings safely and without excessive deflections for the lifetime of the structure. The quantities to define are B, L, and D, as shown in Figure 11.1.

The foundation loads are determined as discussed in Chapter 2. It is usually desirable for structural engineers to provide the foundation engineers with the values of dead, live, wind, seismic, snow, earth pressure, and any other loads that the structural engineers may have calculated. This becomes essential for LRFD of the foundations, as will be clear later in this chapter.

The subsoil information is collected as discussed in Chapter 7. Given that collapse mechanisms may extend to depths in excess of the width of the foundation below the foundation base (Chapter 10) and that the calculation of settlements (Chapter 9) considers the deformation of soil extending down to twice the width of the foundation (for square or circular footings) and as much as four times the width of the foundation (for strip footings), it is necessary to characterize the subsoil at least down to a depth that is consistent with these numbers. If soft clay layers are present, characterization of the clay along the entire layer

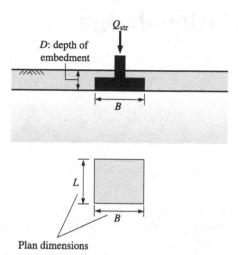

Figure 11.1 Part of foundation design is to determine the location of and proportion individual foundation elements. For a footing, that calls for determining B, L, and D, given the loading.

thickness is needed, except if the layer is so deep that induced vertical stresses in it can be neglected. The quantities that are needed to characterize the soil depend on whether the soil is sand or clay and on which analyses are performed. Sands and clays were extensively discussed in Chapters 5 and 6, and the settlement and bearing capacity analyses used in shallow foundation design were discussed in Chapters 9 and 10.

Once the loads and subsoil conditions are known, the shallow foundation is positioned at a certain depth (which must not be less than the minimum depth set forth in Chapter 8) and proportioned to ensure safety and adequate performance at the lowest possible cost. The requirements of safety and performance are better expressed through the concept of limit states, as discussed in Chapter 2.

11.1.2 Limit states design of shallow foundations

It is helpful to design foundations referring to the limit states the design must prevent. A limit state, as discussed extensively in Chapter 2, is a set of conditions or outcomes that must not be reached. Limit states in which the safety of the structure is jeopardized are known as ultimate limit states. Limit states in which the structure fails to perform its intended function or to be of service are known as serviceability limit states. In this chapter, we are only interested in limit states achieved due to inadequate foundation design. The limit states for shallow foundations are those in Table 11.1.

Three of the limit states listed in Table 11.1 result from excessive settlement. In Chapter 2, we discussed in detail what constitutes excessive settlement. The ultimate limit states are identified by the numerals I and III. Limit state II is a serviceability limit state caused by excessive settlement. In most cases, we design foundations so that their total settlement is not excessive, indirectly limiting differential settlement. This means that preventing limit state II is in most cases sufficient to prevent limit state IB, an ultimate limit state caused by excessive differential settlement. Limit state III is a special case. This limit state is reached if a global instability problem develops. The most obvious example of this is the case of a building that rests on top of a slope (Figure 11.2). If the building foundations are properly designed for all limit states, but the slope is unsafe, the overall geotechnical design is inadequate. A second example is a building that stands next to a slope and may potentially be

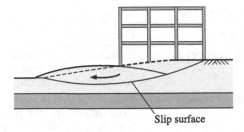

Figure 11.2 Overall stability failure caused by loss of stability of a slope on which the building stands.

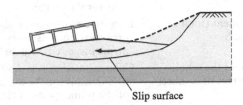

Figure 11.3 Overall stability failure caused by loss of stability of a slope next to which the building stands.

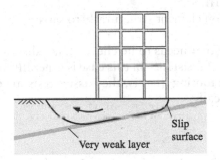

Figure 11.4 Overall stability failure due to the presence of a very weak layer in the soil profile.

sitting on the landslide, if it develops (Figure 11.3). An example of this happened in Kuala Lumpur, Malaysia, as discussed in the case history at the end of Chapter 17. Figure 11.4 illustrates another situation in which general instability is possible. In this case, a layer of weak soil with the "right" inclination below the foundations may facilitate collapse of all or part of the building foundations.

A logical sequence for the design of shallow foundations is as follows:

1. Calculate a first approximation for the plan area of each foundation element by requiring that the foundation element passes a check for limit state IA-1.
2. Check for limit states II and IB; this usually consists of sizing the foundation element such that a total tolerable settlement is not exceeded, but it may also involve the calculation of the settlements of adjacent elements and the requirement that the angular distortion be less than a limiting value. If excessively large plan areas for the foundation elements are required to keep the settlement at acceptable levels, shallow foundations on the natural soil are not an acceptable solution. Pile foundations are usually indicated. Other alternatives, involving ground modification before placement of shallow foundations, may also be possible.

Table 11.1 Limit states for shallow foundation design

Limit state	Nature of limit state	Consequences
IA-1	Classical bearing capacity failure (ULS)	Excessive movement/collapse of foundation causes serious damage (partial collapse or complete collapse of structure).
IA-2	Structural failure of foundation element (ULS)	Column or wall is inadequately supported by foundation element, punching through it; this causes serious damage (partial or complete collapse of superstructure).
IB	Excessive differential foundation settlement (ULS)	Excessive differential settlement creates excessive additional loads in the structure, leading to structural damage.
II	Excessive settlement (total or differential) (SLS)	Excessive settlement leads to serviceability problems, such as access difficulties, drainage problems, or damage to architectural finishing.
III	Stability failure of the whole foundation system or a subset thereof	Collapse mechanism develops that encompasses the foundations for the building or structure or a part of the foundations (a classical example is the stability failure of a slope on top of which the building is founded).

3. Check for limit state III.
4. Design each foundation element structurally to satisfy limit state IA-2.

An additional limit state that needs to be checked, not discussed previously, is resistance against sliding. The source of resistance in this case is generally assumed to be only the shear stress along the base of the footing (that is, the passive resistance due to embedment is usually neglected). Mathematically,

$$\mu Q_v = (FS)(Q_h) \qquad (11.1)$$

where the factor of safety (FS) is taken to be at least 1.5 and μ is a coefficient of friction given by

$$\mu = \tan \delta \qquad (11.2)$$

where δ is equal to ϕ_c for concrete or masonry footings (and as low as 0.8 ϕ_c for steel footings). For clays, excavation of the footing may have generated a layer of clay at the base of the footing whose shear strength is the residual shear strength. So it is safer to use δ equal to ϕ_r for clays. Residual friction angles in these cases may be quite low (as is the case for smectites), so μ values will also be low for clays.

11.2 LIMIT STATE IA-1 CHECK

11.2.1 Working stress design of shallow foundations

The gross bearing capacity of the footing needs to be sufficient to support the column (or wall) load Q_{str}, the weight of the footing itself, and the weight of the backfill. The relationship between load and resistance at limit bearing capacity is:

$$A q_{bL} = Q_L + W_{ftg} + W_{fill} \qquad (11.3)$$

In working stress design (WSD), a limit bearing capacity failure is prevented by using a safety factor. Equation (11.3), therefore, corresponds to a footing on the verge of collapse ($FS=1$). In that case, the gross unit resistance q_{bL} will relate to the applied load and the weight of the footing and backfill through:

$$q_{bL} = \frac{Q_{str} + W_{ftg} + W_{fill}}{A} \tag{11.4}$$

Let us now consider clays and sands separately. For clays,[1] Equation (11.4) becomes:

$$q_{bL} = s_u N_c s_{su} d_{su} i_{su} + q_0 = \frac{Q_{str} + W_{ftg} + W_{fill}}{A} \approx \frac{Q_{str}}{A} + q_0 \tag{11.5}$$

The approximation made in Equation (11.5) is that the weight of the footing and backfill is considered to be the same as the weight of the soil. Now the surcharge q_0 appears on both sides of the equation, and subtraction from both sides of Equation (11.5) makes it go away. It is appropriate that we do this before we introduce the safety factor because the value of q_0 is not subject to the same uncertainty as the other terms in the bearing capacity equation. Now, with the use of a safety factor FS, Equation (11.5) can be rewritten in terms of the limit unit net resistance:

$$q_{bL}^{net} = s_u N_c s_{su} d_{su} i_{su} = FS \frac{Q_{str}}{A} \tag{11.6}$$

For sands, we can rewrite Equation (11.4) in much the same way as we did for clays:

$$q_{bL} = s_q d_q i_q b_q g_q q_0 N_q + \frac{1}{2} \gamma B N_\gamma s_\gamma d_\gamma i_\gamma b_\gamma g_\gamma \approx \frac{Q_{str}}{A} + q_0 \tag{11.7}$$

As before, accepting the approximation as an equality, subtracting q_0 from both sides of the equation, and considering a safety factor FS:

$$q_{bL}^{net} = q_0 \left(s_q d_q i_q b_q g_q N_q - 1 \right) + \frac{1}{2} \gamma B N_\gamma s_\gamma d_\gamma i_\gamma b_\gamma g_\gamma = FS \frac{Q_{str}}{A} \tag{11.8}$$

The area of the footing can be expressed in terms of the allowable net unit bearing capacity and the applied load as follows:

$$A = \frac{Q_{str}}{q_{b,all}^{net}} \tag{11.9}$$

where the allowable net unit bearing capacity is given by

$$q_{b,all}^{net} = \frac{q_{bL}^{net}}{FS} \tag{11.10}$$

Appropriate values of the safety factor are given in Table 2.2. Typically, for footings, safety factors will be in the 2–4 range.

Because the limit bearing capacity of footings is affected by their geometry, footing design is an iterative process. Plan dimensions must first be assumed so that the net limit bearing

---

[1] Here, we will use the form of the bearing capacity equation for strength not varying with depth, but the same argument could be made using the equation for strength increasing linearly with depth.

capacity can be computed. Then the area A can be calculated by using Equations (11.9) and (11.10). This area and the corresponding plan dimensions are then fed back into the calculation of the net limit bearing capacity. This process is repeated until reasonable convergence is obtained.

In the case of sands, it makes very little difference whether one uses net or gross bearing capacity in Equation (11.9) because the gross bearing capacity is calculated using $s_q d_q i_q b_q g_q N_q$ and the net bearing capacity is calculated using $s_q d_q i_q b_q g_q N_q - 1$. Realistically, $s_q d_q i_q b_q g_q N_q \approx s_q d_q i_q b_q g_q N_q - 1$. In contrast, for clays, the difference between using net and gross bearing capacity can be substantial. The use of the safety factor on the gross bearing capacity and not on the net bearing capacity would subject the q_0 that appears on the left-hand side of Equation (11.5) to a safety factor, but it would not affect the one on the right-hand side. Although this approach is recommended by many textbooks, it is conceptually incorrect, because it implies that the same quantity is subjected to different levels of uncertainty depending on which side of the equation it is in.

If the footing is not backfilled, $W_{\text{fill}} = 0$. Again, in sands, this makes no difference for practical purposes, but for clays, we must rewrite Equation (11.5) as

$$q_{\text{bL}}^{\text{gross}} = s_u N_c s_{su} d_{su} i_{su} + q_0 = \frac{Q_{\text{str}} + W_{\text{ftg}}}{A} \tag{11.11}$$

There are two possibilities here: either we divide the gross bearing capacity by the factor of safety FS (which is the same as multiplying $Q_{\text{str}} + W_{\text{ftg}}$ by FS), or we multiply only Q_{str} by FS on the basis that W_{ftg} has very little uncertainty compared with Q_{str}. We favor this second approach, which leads to:

$$A = \frac{FS Q_{\text{str}} + W_{\text{ftg}}}{q_{\text{bL}}^{\text{gross}}} \tag{11.12}$$

Example 11.1

Does the footing of Examples 9.7 and 10.11 pass the limit states design checks?

Solution:

In Example 10.11, we calculated the limit unit base resistance as

$$q_{\text{bL}} = 56\,\text{kPa}$$

The surcharge q_0 was calculated as being $18\,\text{kPa}$. Therefore, the factor of safety FS on bearing capacity, considering that the applied load is $60\,\text{kN}$ (see Example 9.7), is:

$$FS = \frac{B \times L(q_{\text{bL}} - q_0)}{Q} = \frac{1.5 \times 1.5(56 - 18)}{60} = 1.43$$

which means that the footing does not pass the bearing capacity check.

In Example 9.7, we found that the estimated immediate settlement of the same footing was

$$w = 7\,\text{mm}$$

In the absence of consolidation settlement, this would be acceptable because it is much lower than the 25 mm usually taken as the serviceability limit. In this case, bearing capacity would control the design of the footing, and a new design with a larger plan area would be required. However, large consolidation settlements (several times larger than the immediate settlement) are likely to develop in a soft clay, as the one in this example is. So settlement may still control design. As an exercise, the reader may wish to explore this further by proportioning the foundation such that FS be equal to 2.5 and then checking the resulting settlement.

Example 11.2

Verify that the footing of Example 9.9 passes the bearing capacity ultimate limit state check.

Solution:

In summary, the data for the footing of Example 9.9 are:

$Q=200$ kN
NC clay
$D=2$ m
$z_w=2$ m
$H=10$ m
$\gamma=18$ kN/m^3
$s_u/\sigma'_v=0.3$

The equivalent surcharge q_0 and the shear strength s_{u0} at the footing base level are:

$$q_0 = \gamma D = 18 \times 2 = 36\,\text{kPa}$$

$$s_{u0} = 0.3 \times 36 = 11\,\text{kPa}$$

Now, to determine whether the footing passes the limit bearing capacity check, we will use Equation (10.33):

$$q_{bL} = F s_{su} d_{su} i_{su} \left[s_{u0} + \frac{\rho B}{4 N_c} \right] N_c + q_0$$

To determine F and s_{su}, we need to first determine the strength gradient ρ:

$$\rho = \frac{ds_u}{dz} = \gamma_b \frac{ds_u}{d(\gamma_b z)} = \gamma_b \frac{ds_u}{d\sigma'_v} = 0.3 \times (18 - 9.81) = 2.5\,\text{kPa/m}$$

This leads us to:

$$\frac{s_{u0}}{\rho B} = \frac{11}{2.5 \times 4} = 1.1 \quad \text{and} \quad \frac{\rho B}{s_{u0}} = \frac{2.5 \times 4}{11} = 0.91$$

The value of F for a rough footing is read from Figure 10.14 as 1.22. Recalling that a width of 4 m was tried in Example 9.9, we now use Equation (10.26), Table 10.2, and Equation (10.34):

$$d_{su} = 1 + 0.27\sqrt{\frac{D}{B}} = 1 + 0.27\sqrt{\frac{2}{4}} = 1.19$$

$$s_{su} = 1 + C_1 \frac{B}{L} \left\{ \frac{2.3}{\exp\left[0.353\left(\frac{\rho B}{s_{u0}}\right)^{0.509}\right]} - 1.3 \right\} + C_2 \sqrt{\frac{D}{B}}$$

where for C_1 we may use 0.125 and $C_2 = 0.219$ (as presented in Table 10.2). So

$$s_{su} = 1 + 0.125\frac{4}{4} \left\{ \frac{2.3}{\exp\left[0.353(0.91)^{0.509}\right]} - 1.3 \right\} + 0.219\sqrt{\frac{2}{4}}$$

$$= 1 + 0.043 + 0.155$$

$$= 1.2$$

Now we can calculate q_{bL} as

$$q_{bL} = F s_{su} d_{su} i_{su} \left[s_{u0} + \frac{\rho B}{4N_c} \right] N_c + q_0$$

$$= 1.22 \times 1.2 \times 1.19 \times 1 \times \left(11 + \frac{2.5 \times 4}{4 \times 5.14} \right) \times 5.14 + 36$$

$$= 139\,\text{kPa}$$

Assuming $FS = 3$:

$$Q_{all} = \frac{(q_{bL} - q_0)A}{FS} = \frac{(139 - 36) \times 4^2}{3} = 549\,\text{kN} > 200\,\text{kN}$$

So the footing passes the limit bearing capacity check because $FS > 3$. As indicated in Example 9.9, there is room to reduce the size B of the footing if a more refined design is desired. This is left as an exercise in Problem 11.12.

Example 11.3

Consider the two footings in E-Figure 11.1, each supporting a column, with loads as given in the figure. A surcharge of 0.75 m of sand will be placed around each footing. The unit weight of this fill may be taken as 20 kN/m³. A 300 kN horizontal load is applied 1 m above the base of the footing A. The soil at the site is a gravelly sand deposit extending to a depth of 15 m, at which bedrock is found. The critical-state friction angle for this sand can be taken as 33°. Four SPT logs are available, which apply to the footings in question. The N_{60} values are given in E-Table 11.1. The unit weight of the sand can be taken to be 20 kN/m³. Consider the water table to be flush with the level of the bases of the footings. Calculate the plan areas of the footings leading to an FS equal to 3. Assume no load transfer between the two footings.

Solution:

The size B of each footing is not yet known. Starting with footing A and assuming $B = 3$ m, the slip mechanism will extend from 3 m to approximately 6 m. E-Table 11.2 shows the lowest SPT values for depths in the range of 3–6 m, the vertical effective stresses, and the relative densities.

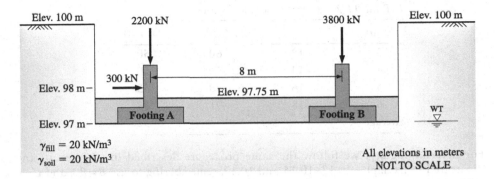

E-Figure 11.1 Footings of Example 11.3.

E-Table 11.1 SPT N_{60} values from four borings pertaining to Example 11.3

Depth (m)	SPT 01	SPT 02	SPT 03	SPT 04
1	2	3	4	6
2	5	5	6	7
3	18	13	13	23
4	23	15	14	25
5	35	16	19	31
6	29	18	20	32
7	19	13	18	25
8	23	17	14	26
9	23	17	15	24
10	27	18	19	31
11	29	13	18	22
12	28	15	15	19
13	31	18	19	25
14	32	19	18	29
15	29	20	20	26
16	34	22	21	33

Following Equation (7.6) proposed by Skempton (1986):

$$\frac{D_R}{100\%} = \sqrt{\frac{N_{60}}{A + BC\dfrac{\sigma_v'}{p_A}}}$$

with $A=36$ (median value of the 27–46 range recommended by Skempton (1986)), $B=27$, and $C=1$.

Footing A is subject to an eccentricity calculated as

$$e_B = \frac{M}{Q_v} = \frac{hQ_h}{Q_v} = \frac{1\ m \times 300\ kN}{2200\ kN} = 0.136\ m$$

This eccentricity is not excessive, so the Meyerhof effective dimension approach can be used to calculate the effective value of B:

$$B_{eff} = B - 2e_B = 3 - 2 \times 0.136 = 2.73\ m$$

E-Table 11.2 Summary of calculations

z (m)	N_{60}	σ_v' (kPa)	D_R (%)
3	13	60	50
4	14	70	50
5	16	80	53
6	18	91	55
Average =			52

For determining q_{bL}, we follow the same procedure described in Chapter 10 and illustrated in Examples 10.13–10.15 and 10.17, remembering to use $B = B_{eff}$ and $L = B$ in the calculations. We must start by calculating the mean effective stress using Equation (10.40):

$$\sigma_{mp}' = 20 p_A \left(\frac{\gamma B}{p_A} \right)^{0.7} \left(1 - 0.32 \frac{B}{L} \right)$$

$$= 20 \times 100 \left(\frac{10.19 \times 2.73}{100} \right)^{0.7} \left(1 - 0.32 \frac{2.73}{3} \right) = 579 \text{ kPa}$$

The value of A_ψ is obtained from Equation (10.42):

$$A_\psi = \frac{1}{3} \left(\frac{L}{B} + 8 \right) = \frac{1}{3} \left(\frac{3}{2.73} + 8 \right) = 3.03$$

The friction angle follows:

$$\phi_p = \phi_c + A_\psi \left\{ I_D \left[Q - \ln \left(\frac{100 \sigma_{mp}'}{p_A} \right) \right] - R_Q \right\}$$

$$= 33° + 3.03 \left\{ 0.52 \left[10 - \ln \left(\frac{100 \times 579}{100} \right) \right] - 1 \right\}$$

$$= 33° + 2.7°$$

$$= 35.7°$$

We also must use load inclination factors on both the N_q and N_γ terms, using the Meyerhof (1963) equations in Table 10.8:

$$i_q = \left[1 - \frac{\arctan\left(\frac{Q_{tr}}{Q_{ax}} \right)}{90°} \right]^2$$

and

$$i_\gamma = \left[1 - \frac{\arctan\left(\frac{Q_{tr}}{Q_{ax}} \right)}{\phi} \right]^2$$

In these equations, we use $Q_{tr}=300$ kN and $Q_{ax}=2200$ kN. Finally, we obtain $q_{bL}=2048$ kPa.

For an assumed footing height of 0.7 m and concrete unit weight $\gamma_{concrete}=25$ kN/m^3, the safety factor is:

$$FS = \frac{Aq_{bL}^{gross} - W_{ftg}}{Q_{str}} = \frac{8.2 \times (2048 - 25 \times 0.7)}{2200} = 7.6$$

For the next iteration, as we reduce the size of the footing, the value of gross unit limit bearing capacity would tend to decrease because of the reduction in B, but increase because the effective L/B increases. Assuming no change in q_{bL} due to these offsetting effects, we would have:

$$A_{eff} = \frac{(FS)Q_{str} + W_{ftg}}{q_{bL}^{gross}} \approx \frac{(FS)Q_{str}}{q_{bL}^{gross}}$$

which leads to $A \approx 3.2$ m^2. This leads to B given by

$$(B - 2e_B)B = (B - 2 \times 0.136)B = 3.2$$

a second-degree algebraic equation that leads to $B=1.93$ m. As q_{bL} decreases with the reduction in B, our next iteration should be slightly greater than 1.93 m, say 2 m; consequently, $B_{eff}=1.73$ m and $L=2$ m. Calculations with these new effective dimensions for the footing give, in sequence:

$q_{bL} = 2085$ kPa

$FS = 3.3$

$A = 3.5$ m^2

$B = 2$ m

Footing B is not subjected to a lateral load and supports a vertical load of 3800 kN. To arrive at our first guess of what the area of the footing should be, we can divide 3800 kN by 695 kPa (approximately 2085 kPa for q_{bL} divided by an $FS=3$), obtaining 5.5 m^2. Given that this footing is loaded without any eccentricity, the corresponding width of the footing is approximately 2.3 m. We can now go through the bearing capacity calculations using this initial estimate of footing size. We obtain $q_{bL}=2775$ kPa and $B=2.1$ m corresponding to $FS=3.2$. The reader is asked in Problem 11.13 to reproduce these results.

A check on the settlements of these footings under the imposed loads is now in order. If these settlements exceed tolerable limits, resizing of the footings is required. Design of both the footings by considering also whether the settlements they will undergo under the imposed loads are tolerable is started in Example 11.6. Completion of the design is left as an exercise (see Problem 11.15).

11.2.2 Load and resistance factor design of shallow foundations

11.2.2.1 The fundamental design inequality

In load and resistance factor design (LRFD), a failure in limit bearing capacity is prevented by requiring that the final foundation design resistance R_d be at least equal to the corresponding design load Q_d. Mathematically:

$$R_d \geq Q_d \tag{11.13}$$

or, alternatively:

$$(RF)\,R_n \geq \sum (LF)_i\, Q_{ni} \tag{11.14}$$

where RF and $(LF)_i$ = resistance and load factors, respectively; R_n and Q_{ni} = nominal resistance and loads (including the footing weight), respectively. The resistance and loads refer to a specific loading mode, such as vertical or lateral. Nominal loads (also known as characteristic loads) are defined in very specific ways by the codes, as discussed in Chapter 2. Likewise, nominal (or characteristic) resistances must be defined in a standard way for the successful use of LRFD. We discuss this in the next subsection.

It is interesting to contrast Equation (11.14) with the WSD inequality:

$$R_n \geq (FS)\sum Q_{ni} \tag{11.15}$$

In WSD, dead and live loads (and other types of loads, if present) are combined into one total applied nominal load, which is multiplied by a safety factor and then compared with the nominal resistance.

Load factors are given in Table 2.3. Typically, we would use load factors stipulated by the code or set of recommendations governing a specific type of geotechnical design. For example, AASHTO (2020) would be used for bridge foundations, but ACI (2002) would be used for residential or commercial building foundations with reinforced concrete structures. This ensures consistency between structural and foundation designs.

11.2.2.2 Nominal resistances and resistance factors

Nominal resistances (or characteristic resistances) are the load capacities of the foundation calculated for either the vertical or horizontal load directions using an appropriate model or analysis, reduced in some specified manner to account for uncertainties in the ways in which the shear strength is measured or estimated. For example, the bearing capacity equation is used to produce a vertical resistance, given the values of the shear strength parameters (either ϕ for sands or s_u for clays). Usually, reduced values of the shear strength parameters are used in the calculations, to produce a nominal bearing capacity q_{bLn} and nominal resistance $R_n = q_{bLn}A$.

Foye et al. (2006a,b) proposed a design sequence for LRFD of shallow foundations. With modifications, it consists of the following steps:

1. Select the worst applicable boring (or CPT/SPT log)[2]; combine two or more borings/loggings if deemed appropriate.
2. Establish the dependence (if it exists) of the measured variable (for example s_u, N_{SPT}, and q_c) on depth within the depth of influence for limit bearing capacity calculations; separate in one or more layers if needed.
3. Either graphically or mathematically, find the relationship of the quantity in question with depth that is exceeded by 80% of the measurements.[3]

---

[2] This is the simplest and most practical way of accounting for horizontal site variability in most projects.
[3] It was found that the 90 or 95% exceedance used with materials such as concrete or steel is too strict for soils, which have large variations in shear strength. The 80% value captures best the variability in soil properties.

4. Either find the nominal value of s_u or ϕ to use in the bearing capacity equation from the 80% exceedance profile established in Step 3, or in the case of clays with increasing strength with depth, find a depth gradient ρ of strength that guarantees that strength is exceeded by 80% of the measurements.

5. Substitute the nominal value of s_u, ρ, or ϕ in the bearing capacity equation to determine q_{bLn}.

An assessment of the 80% exceedance values can be made using statistical methods. For a normal distribution, the nominal value x_n of the variable x under consideration corresponding to the value exceeded by 80% of the values is given by

$$x_n = E(x) - 0.84\sigma_x \tag{11.16}$$

where $E(x)$ is the expected value[4] of x; σ_x is the standard deviation of x; and x represents cone resistance q_c, SPT blow count N, undrained shear strength s_u, or any other measurement used in the calculation of bearing capacity and nominal resistance. The standard deviation of x can be estimated based on the limited number of measurements (the statistical sample) of x using the process proposed by Tippett (1925). According to Tippett (1925), the range of values contained in the sample, that is, the difference between the maximum and minimum values of x, corresponds to a certain number of standard deviations of x that depends on the number of measurements of x available (that is, the size of the sample). The ratio of the range of x to the standard deviation of x is given in Table 11.2 as a function of sample size. It is clear that there is a benefit to be gained from having a larger sample (more data) from the site exploration (or the reverse. a penalty to be paid for having too few data). The larger the number n of data points, the larger the ratio of the range of the data to the standard deviation, the lower the standard deviation for a given range, and thus the larger the nominal value of x_n.[5] A larger value of x_n (the nominal value of s_u, N_{SPT}, q_c, or any input variable into the calculation of resistance) will obviously reflect in a larger nominal resistance, calculated from Equation (11.16).

Table 11.2 Ratio of the range of the statistical sample to the number of standard deviations of the population

n	$\dfrac{(x_{max} - x_{min})_{sample}}{\sigma_x}$	n	$\dfrac{(x_{max} - x_{min})_{sample}}{\sigma_x}$	n	$\dfrac{(x_{max} - x_{min})_{sample}}{\sigma_x}$
2	1.128379	12	3.258457	100	5.0152
3	1.692569	13	3.335982	200	5.492108
4	2.058751	14	3.406765	300	5.755566
5	2.325929	15	3.471828	400	5.936396
6	2.534413	16	3.531984	500	6.073445
7	2.704357	17	3.587886	600	6.183457
8	2.847201	18	3.640066	700	6.275154
9	2.970027	19	3.688965	800	6.353645
10	3.077506	20	3.734952	900	6.422179
11	3.172874	50	4.498153	1000	6.482942

---

[4] The expected (or mean) value of x is typically obtained from the trend line through the x (q_c, N_{SPT}, or s_u) values with depth.

[5] Note how, for a large sample size, this ratio is approximately 6 times the standard deviation.

Table 11.3 Resistance factors for footings in sands (for $D/B \leq 1$)

Design case	ASCE-7 LFs	AASHTO (1998) LFs
Strip footings using CPT	0.23	0.26
Strip footings using SPT	0.17	0.20
Rectangular footings using CPT	0.33	0.36
Rectangular footings using SPT	0.20	0.23

Table 11.4 Resistance factors for footings in clays

Design case	ASCE-7 LFs	AASHTO (1998) LFs
Strip footings using Equation (10.26) to calculate the depth factor	0.70	0.73
Strip footings using Equation (10.24) to calculate the depth factor	0.60	0.63
Rectangular footings using Equations (10.26) and (10.34) to calculate the depth and shape factors	0.75	0.78
Rectangular footings using Equations (10.22) and (10.23) to calculate the depth and shape factors	0.70	0.73

Foye et al. (2006a,b) also performed reliability analyses to determine resistance factors consistent with the nominal resistance determined based on the 80% exceedance criterion. These values are given in Tables 11.3 and 11.4. The tables refer to the values of load factors put forth by ASCE, which were adopted in the ACI 318 reinforced concrete design code and AASHTO (1998). Both of these sets of load factors were discussed in Chapter 2. Table 11.4 also refers to the shape and depth factors of Meyerhof (1963) and Salgado et al. (2004), given, respectively, by Equations (10.22), (10.23), (10.26), and (10.34).

11.2.3 Relationship between resistance factors, load factors, and the factor of safety

As the profession transitions to LRFD, the question of what equivalent safety factor is produced by the design will appear frequently. The relationship between the safety factor, the load and resistance factors, and the live-to-dead load ratio can be obtained by referring to Equations (11.14) and (11.15). Let us assume that our factored resistance is exactly equal to our factored load in the case of LRFD and that the mean resistance is exactly equal to the mean load times the safety factor in the case of WSD. The nominal resistance in LRFD is a conservatively assessed mean, which is less than the resistance used in WSD, which is the mean resistance. A bias factor b_R, the ratio between mean and nominal resistances, is defined as

$$b_R = \frac{R}{R_n}$$

Let us assume further that we have only dead and live loads. It follows that:

$$(RF) R_n = \sum (LF)_i Q_{ni} = (LF)_{DL} (DL) + (LF)_{LL} (LL)$$

or

$$R_n = \frac{(LF)_{DL} (DL) + (LF)_{LL} (LL)}{RF}$$

for LRFD, and

$$b_R R_n = (FS)(DL + LL)$$

for WSD. If we now combine the two expressions, we get the following equation for the factor of safety FS:

$$FS = b_R \frac{(LF)_{DL} + (LF)_{LL}\left(\dfrac{LL}{DL}\right)}{\left(\dfrac{LL}{DL} + 1\right)(RF)} \qquad (11.17)$$

The resistance bias factor b_R is typically not the same for every design situation. So we could, to obtain a quick estimate of the equivalent FS, assume it to be 1 in Equation (11.17). We will discuss the possible error resulting from this in the context of Example 11.5.

It is important to understand that LRFD is a more evolved design method than WSD. If we trust the way in which resistance factors have been developed by authors or code developers and then follow the design procedure closely, there will be no need to calculate safety factors (which would be, in a way, "second-guessing" the design produced by LRFD). Additionally, no unique FS will result from Equation (11.17). A low RF value, arrived at by code developers because of greater uncertainties in material properties, testing methods, or analyses, would naturally lead to higher factors of safety. The live-to-dead load ratio would also have an impact on the value of FS corresponding to a set of RF and LF values.

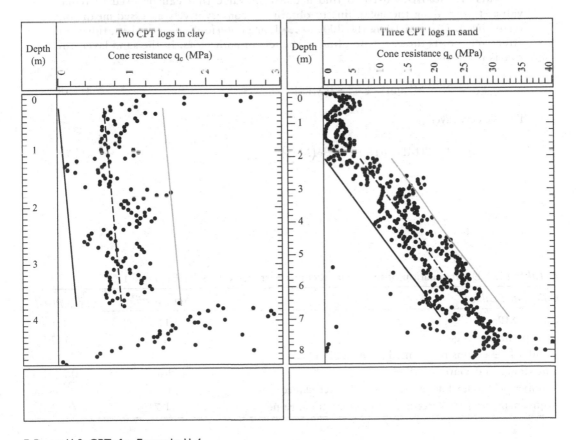

E-Figure 11.2 CPTs for Example 11.4.

Example 11.4

Two buildings with the exact same design are to be founded at two different sites – one with a primarily sand subsurface profile and the other with a clay profile. E-Figure 11.2 shows the cone resistance profiles obtained from a number of CPT soundings performed at each of the sites. The loads for one of the columns, to be supported by a square footing, are 440 kN live load and 600 kN dead load. The basement is to extend to a depth of 1 m. The water table is very deep. Based on the available logs, a reasonable foundation should be possible at a depth of 2 m (1 m below basement elevation). Find the plan area of the footing in both cases that will pass a check of limit state IA-1 following the LRFD approach.

Solution:

Determination of factored load

Using live load and dead load factors of 1.6 and 1.2 (ASCE-7), respectively, the factored load is 1424 kN.

Determination of characteristic cone resistances

For simplicity, cone resistance q_c is assumed to vary linearly with depth for both the clay and sand sites. E-Figure 11.2 shows linear regression lines for both the clay and sand sites. Lines can also be drawn bounding the q_c data points, representing the entire range of q_c data for those depths. These range lines can be used to determine the 80% exceedance criterion values of q_c (denoted by $q_{c,CAM}$, where CAM stands for conservatively assessed mean) for design using Table 11.2. E-Table 11.3 presents the statistics used to find a constant value that can be used to reduce values found along the mean line to obtain the conservatively assessed mean cone resistance $q_{c,CAM}$ satisfying the 80% exceedance criterion. Once applied, this value effectively shifts the regression line to the left in both plots. In the sand layer, this line is given by

$$q_{c,CAM} = 2.75(\text{MPa/m}) \times z - 1.62(\text{MPa}) \tag{11.18}$$

For the clay layer:

$$q_{c,CAM} = 0.049(\text{MPa/m}) \times z + 0.236(\text{MPa}) \tag{11.19}$$

E-Table 11.3 Determination of the nominal cone resistance values for Example 11.4

Calculated values	Sand profile	Clay profile
Range (MPa)	11.7	1.2
Data points in range	294	142
Standard deviations represented (from Table 11.2)	5.743	5.261
One standard deviation (MPa)	2.037	0.2281
Number of standard deviations for 80% exceedance	0.84	0.84
Adjustment for 80% exceedance (MPa): nominal value	1.71	0.192

E-Table 11.4 Summary of equations used to calculate the nominal unit bearing capacity

Factor	Equation	Equation or table number
N_q	$N_q = \dfrac{1+\sin\phi}{1-\sin\phi}e^{\pi\tan\phi}$	Equation (10.6)
N_γ	$N_\gamma = \left(N_q - 0.6\right)\tan\left(1.33\phi\right)$	Equation (10.13)
Shape	$s_q = 1 + \left(0.098\phi - 1.64\right)\left(\dfrac{D}{B}\right)^{0.7-0.01\phi}\left(\dfrac{B}{L}\right)^{1-0.16\left(\frac{D}{B}\right)}$	Table 10.5
Shape	$s_\gamma = 1 + \left(0.0336\phi - 1\right)\dfrac{B}{L}$	Table 10.5
Depth	$d_q = 1 + \left(0.0036\phi + 0.393\right)\left(\dfrac{D}{B}\right)^{-0.27}$	Table 10.7
Depth	$d_\gamma = 1$	Table 10.7

Design in sand

Considering the footing in sand, with the base at a depth of 2 m (and $D=1$ m, given the 1 m basement) and a trial footing width of 1.5 m, we can, for simplicity, determine the representative relative density as that at a depth of 2.75 m ($0.5B$ below the footing base). To use the charts in Figure 7.23 to interpret q_c, a value for the horizontal effective stress σ'_h must be found. Assuming a unit weight and lateral earth pressure coefficient at rest K_0 of 20 kN/m^3 and 0.45, respectively, a depth of 2.75 m gives $\sigma'_h = 25$ kPa. Using Equation (11.18), the CAM of q_c at 2.75 m is 5.9 MPa. Assuming a critical-state friction angle of 33°, Equation (7.19) yields a relative density of 50%.

The procedure outlined in Chapter 10 and illustrated in Examples 10.13–10.15 will be used to calculate q_{bL} using Equation (10.36). The values of N_q, N_γ, s_q, d_q, and s_γ for use in Equation (10.36) are computed using Equations (10.6) and (10.13) and the equations from Tables 10.5 and 10.7, repeated here in E-Table 11.4 for convenience. Working with a unit weight of 20 kN/m^3, the resulting value of q_{bL} is 3551 kPa.

Using the trial footing area, 2.25 m^2, the design resistance R_n for this trial design is $3551 \times 2.25 = 7990$ kN. The appropriate resistance factor RF from Table 11.3 is 0.33. Thus, the factored resistance for this trial design $(RF)R_n$ is approximately 2637 kN, a value much greater than the factored load of 1424 kN. Further iterations lead us to $B = 1.1$ m and $q_{bL} = 3679$ kPa, giving us a value of $(RF)R_n$ of 1469 kN, just above the required value.

Design in clay

We will use Equation (10.33). A foundation design will be attempted with a base depth of 2 m and an initial trial width of 1.5 m. The depth at which s_u will be determined is found using the method after Davis and Booker (1973) discussed in Chapter 10 (Equation 10.32). From Equations (11.19) and (7.22) with $N_k = 10$ and a soil unit weight of 16 kN/m^3:

$$q_{c,\text{CAM}} = 0.049\left(\text{MPa/m}\right) \times z + 0.236\left(\text{MPa}\right)$$

$$s_u = \frac{q_c - \sigma_v}{N_k} = \frac{49\left(\text{kPa/m}\right) \times z + 236\left(\text{kPa}\right) - 16\left(\text{kPa/m}\right) \times z}{10} = \left(3.3z + 24\right)\text{kPa}$$

The strength gradient with depth and the undrained shear strength at the footing base level are:

$$\rho = 3.3\,\text{kPa/m} \quad \text{and} \quad s_{u0} = 31\,\text{kPa}$$

Thus,

$$\frac{\rho B}{s_{u0}} = 0.16$$

yielding a value of $F = 1.08$ from Figure 10.14a.

Using a value of 5.14 for N_c and Equations (10.26) and (10.34) for the depth and shape factors, we can now calculate q_{bL}^{net} using Equation (10.33) without including q_0:

$$q_{bL}^{net} = F s_{su} d_{su} i_{su} \left[s_{u0} + \frac{\rho B}{4 N_c} \right] N_c$$

$$= 1.08 \times 1.27 \times 1.22 \times 1 \times \left[31 + \frac{3.3 \times 1.5}{4 \times 5.14} \right] \times 5.14$$

$$= 269\,\text{kPa}$$

Applying the trial footing area, $2.25\,\text{m}^2$, the design resistance R_n is 605 kN. The appropriate resistance factor from Table 11.4 is 0.75. Thus, the factored resistance $(RF)R_n$ is 454 kN, well below the required 1424 kN. Further iterations lead us to the final answer: $B = 2.8\,\text{m}$, giving us a value of $(RF)R_n$ of 1470 kN, just above the required value.

Example 11.5

Calculate the factors of safety corresponding to the designs in sand and clay carried out in Example 11.4.

Solution:

Factor of safety in sand

From Equation (11.17) with $b_R = 1$:

$$FS = \frac{(LF)_{DL} + (LF)_{LL}\left(\dfrac{LL}{DL}\right)}{\left(\dfrac{LL}{DL} + 1\right)(RF)} = \frac{1.2 + 1.6\left(\dfrac{440}{600}\right)}{\left(\dfrac{440}{600} + 1\right)(0.33)} = 4.1$$

Factor of safety in clay

$$FS = \frac{1.2 + 1.6\left(\dfrac{440}{600}\right)}{\left(\dfrac{440}{600} + 1\right)(0.75)} = 1.8$$

Now let us examine the true safety factor, given that we can redo our calculations not with the CAM of q_c, but with the mean q_c, from which the mean D_R will result. At 2.75 m, referring to E-Figure 11.2, $q_c = 9\,\text{MPa}$. Taking that into Equation (7.20), we obtain

$D_R = 66\%$. For this relative density and $B = 1.1\,\text{m}$ from the design obtained in Example 11.4, we get $q_{bL} = 5040\,\text{kPa}$, $b_R = 1.37$, and $FS = 5.6$, which is greater than the value of 4.1 we estimated by a factor of 1.37 (which, of note, is equal to the resistance bias factor).

11.3 SETTLEMENT CHECK

There are a variety of ways in which engineers try to ensure that settlements will not be excessive. The most common ones are:

1. Total settlements are kept smaller than some value, indirectly limiting also the differential settlements to a tolerable value.
2. Differential settlements and angular distortions are calculated for each pair of footings (considering the possible variability of soil resistances beneath each footing) and compared with the Skempton and MacDonald's (1956) 1/500 limit on angular distortion discussed in Chapter 2 or other such values determined by the structural engineer.
3. Differential settlements are calculated (considering the possible variability of soil resistances beneath each footing) and provided to the structural engineer, who then calculates the overloading on structural members; if this overloading is excessive, footing sizes are increased or other measures (such as the use of grade beams or even a change to deep foundations) are taken.

Spatial variability of soil density (and properties, such as shear strength, that depend on density) can only be estimated approximately based on the relatively few borings that are economical to sink for any given project. With approach 1, it is typical to use the worst applicable sounding to design each footing (that is, we take, of the available SPT or CPT near the footing, the one with the lowest values). With approaches 2 and 3, we are trying to capture the possibility that one footing rests on a relatively weak part of the soil profile, while the other rests on much better soil, as illustrated in Figure 11.5, which shows the footing bearing on soil with lower q_c values settling more than the one bearing on firmer soil. These approaches capture in a practical, but approximate way the spatial heterogeneity of the soil.

Example 11.6

With reference to the two footings of Example 11.3, check if the angular distortion is less than 1/250, the tolerable value.

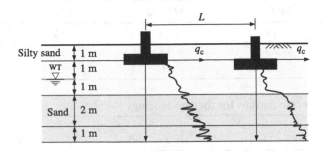

Figure 11.5 Illustration of differential settlement caused by spatial variability of soil properties at a site (WT = groundwater level).

Solution:

There are different ways in which this problem could be solved. In Chapter 2, we saw a variety of correlations between angular distortion or differential settlement and total settlement with an implied variability for the soil. In this case, we have four SPT records that are applicable to the part of the site where these footings are located. We then assume the worst sounding to be under one of the two footings and the best sounding to be under the other (a pessimistic assumption), neglect the effects of the lateral load applied on one of the footings, and calculate the settlement of each footing. Next, we divide the differential settlement by the distance between the two footings to get the angular distortion. We will use the Burland and Burbidge method to estimate the settlement of each footing. Tables are used to facilitate the presentation of the results. The reader should refer to Equations (9.30)–(9.35) to follow the solution. Given the effective footing sizes (1.73 m and 2.1 m), the depth range of interest is determined from:

$$\log \frac{z_{f0}}{L_R} = 0.79 \log \frac{B}{L_R}$$

$$= 0.79 \log \frac{1.73}{1} = 0.19 \text{ for } B_{eff} = 1.73 \text{ m}$$

$$= 0.79 \log \frac{2.1}{1} = 0.25 \text{ for } B = 2.1 \text{ m}$$

giving

$$1.54 \text{ m} < z_{f0} < 1.80 \text{ m}$$

which corresponds to a depth of influence from approximately 3 to 5 m. For this range, the worst sounding is SPT-02 and the best is SPT-04. The worst result is obtained when the worst sounding is placed at the location of the more heavily loaded footing.

E-Table 11.5 shows the SPT blow counts used to obtain the average blow count within the zone of influence of each footing, and E-Table 11.6 shows the complete results of the calculations.

Given the span of 8 m between the two footings, the angular distortion follows directly from the numbers in the last column of E-Table 11.6:

$$\alpha = \frac{56 \text{ mm} - 17 \text{ mm}}{8000 \text{ mm}} = 0.005 = \frac{1}{200}$$

which is greater than the tolerable value. So, in this case, settlement controls.

E-Table 11.5 SPT blow counts for use in the calculation of settlement of the two footings of Example 11.6

Depth (m)	SPT 01	SPT 02	SPT 03	SPT 04
3	18	13	13	23
4	23	15	14	25
5	35	16	19	31

E-Table 11.6 Results of calculations for the two footings

B_{eff} (m)	L (m)	z_{f0} (m)	$\bar{N}$	I_c	f_s	f_L	f_t	q_b (kPa)	σ'_{vp} (kPa)	w (mm)
1.73	2	1.54	26.3	0.0176	1.06	1	1	656	60	17
2.1	2.1	1.80	14.7	0.0397	1	1	1	879	60	56

11.4 STRUCTURAL CONSIDERATIONS

11.4.1 Interaction with the structural engineer

It is customary practice for the geotechnical/foundation engineer (GE) and the structural engineer (SE) to interact in the solution to foundation problems. The SE calculates the loads that must be sustained by the foundations. What happens afterward varies greatly depending on the engineers involved and indeed on the country and locality where they practice. In one model, the SE looks for the GE to provide the SE with a single allowable unit bearing resistance. The SE then uses this value of unit bearing resistance to size the footings. This is a flawed approach because the bearing resistance of a footing depends on the size and shape of this footing, among other things. An improved approach calls for the GE to assign the footings to different groups based on size ranges; the GE would then provide the SE with one value of bearing resistance for each of these groups.

Another approach would call for the SE to provide the GE with the loads. If LRFD is used, this information must include type and duration of loads, in addition to their magnitude. The GE must decide how much of the live load and of other temporary loads should be considered for settlement computations. Usually, maximum wind loads, for example, would be appropriate for the calculation of settlement in sands, but not in clays.

The SE must also provide the GE with a foundation plan of the building or structure, where the loads are indicated. This allows the GE to account for space limitations so that combined and strap footings may be considered where isolated footings are not possible. Based on such considerations, the GE and the SE arrive, working both independently and together, at suitable values of design bearing resistance for final footing sizing. The end product of the GE's work may still be values of design bearing resistance, but the greater the level of interaction between the GE and the SE, the better targeted these values can be to various conditions, particularly footing size.

11.4.2 Location, configuration, and flexibility of the structure

In foundation engineering, resources can be saved by having a broad view of the engineering issues relevant to each project. If foundation engineers take a narrow view, they may end up designing expensive foundations when an alternative solution involving some change to the structure or its location may be possible. These actions can produce significant savings, but they can only be pursued in cooperation with the architect or structural engineer, who can make certain decisions that foundation engineers cannot make.

Take the case of a site containing an area with weak, soft soil. Instead of designing and building expensive foundations to support the portion of the structure located over the soft soil, it may be possible to move, reorient, or reconfigure the structure so that no foundation element will have to be supported by the soil of poor quality. Reconfiguration of the structure, in case it cannot be moved, may call for the elimination of columns and redistribution of loads so that larger spans may result. The question then becomes whether the savings in the foundations justify such changes. This can be decided only on a case-by-case basis.

The rigidity/flexibility of the structure can also be adjusted to fit soil conditions. A rigid structure is one in which the structure moves as a rigid body, with no possibility of relative displacements between points internal to the structure (that is, there is no possibility of structural deformation). A high degree of rigidity may be convenient in highly heterogeneous, locally weak soils, for the structure can then bridge over weak zones. Conversely, a more flexible structure may be more economical for sites with good, strong soils. Well-located construction joints can be useful in accommodating excessive deformation due to temperature changes and shrinkage effects, as well as foundation movements. In soft soils, which will consolidate by large amounts during the useful life of the structure, jacks have occasionally been used between foundation elements and the supported structure to allow periodic readjustment of the elevation of columns or walls.

11.4.3 Sizing of rectangular and trapezoidal combined footings

Consider two neighboring columns C_1 and C_2 with loads Q_1 and Q_2. The center-to-center distance between the two columns is L'. If the loads Q_1 and Q_2 are such that isolated footings designed to support these loads would either overlap or be very close to each other, then a combined footing is often used. The idea is to size the footing so that the pressure at the footing base is as uniform as possible.

A relatively uniform bearing pressure is accomplished for rectangular footings by requiring that the centroid of the area of the footing coincide with the location of the resultant of the loads. When $Q_2 \gg Q_1$, this may not be possible, for it would then be necessary to extend the rectangular footing much beyond the location of column C_2. This is not only inefficient, but often impossible to do because of space limitations. A trapezoidal footing could be used in such instances, as discussed in Chapter 8. The footing would have a larger width near C_2 and a smaller one near C_1 so that its centroid would coincide with the location of the resultant of the loads.

A trapezoidal footing is shown in Figure 11.6. Let a be the small side and b the large side of the trapezoidal footing. Let $L=L'+d_1+d_2$ be its length, where d_1 and d_2 are the distances from the center of each column to the closest edge of the footing. The footing is sometimes flush with the column, in which case d_1 is the half-width of column C_1. The plan area A of the footing and the distance x_c from the large side of the footing to the centroid of the footing are given by

$$A = \frac{1}{2}(a+b)L \tag{11.20}$$

$$x_c = \frac{1}{3}L\frac{2a+b}{a+b} \tag{11.21}$$

If the allowable bearing pressure, after consideration of all the pertinent limit states (using the analyses covered in Chapters 9 and 10, and this chapter), is $q_{b,all}$, the area A is calculated as in Equation (11.9) or (11.12), depending on whether the excavation for the footing is backfilled, using the following for the structural load Q_{str}:

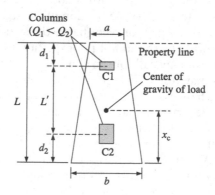

Figure 11.6 Analysis of trapezoidal footing.

$$Q_{str} = Q_1 + Q_2 \tag{11.22}$$

The centroidal distance x_c must coincide with the distance of the resultant load $Q_1 + Q_2$ from the large side of the trapezoid, which leads us to the following expression for x_c:

$$x_c = \frac{d_2 Q_2 + (d_2 + L') Q_1}{Q_1 + Q_2} \tag{11.23}$$

Taking Equations (11.22) and (11.23) into Equations (11.20) and (11.21), we can solve for a and b:

$$a = \frac{2A}{L}\left(\frac{3x_c}{L} - 1\right) \tag{11.24}$$

$$b = \frac{2A}{L} - a \tag{11.25}$$

With the geometry of the trapezoidal footing fully known, it is possible to obtain shear and bending moment diagrams and to carry out the structural design.

The procedure we have just outlined for the proportioning of trapezoidal footings aims at obtaining a relatively uniform contact pressure at the base of the footing. Consider now what happens at the time of structural design of the footing if the ratio of dead-to-live load in the two columns is different. Since LRFD is used for the structural design, a nonuniform contact stress is more likely to result for the dimensions determined using WSD. However, if LRFD had been used in the geotechnical design as well, this would likely not occur. This illustrates one of the advantages of having geotechnical and structural engineers use the same design framework.

11.4.4 Sizing of strap footings

As seen in Chapter 8, a strap footing is a combination of two isolated footings, each supporting a column, connected by a beam (the strap). Strap footings are used when one of the columns is near an obstruction or near the property line and combined footings are not economical because the distance between the two columns is large (typically, larger than 7 m or so). The outer column (let us call it C_1; see Figure 11.7) clearly is not centered over its supporting footing, but the moment associated with this eccentricity can be efficiently absorbed by the two footings working together. The function of the strap becomes to transfer this moment from the outer to the internal footing.

Figure 11.7 shows that the analysis of strap footings is relatively simple. The problem is modeled as a simply supported beam where one of the column loads (Q_1) is cantilevered with respect to the outer support. Each support corresponds to the center of the corresponding foundation. If $L' + e$ is the distance between the columns and e is the load eccentricity for the outer footing, equilibrium yields the following loads for each footing:

$$R_1 = Q_1\left(1 + \frac{e}{L'}\right) \tag{11.26}$$

and

$$R_2 = Q_2\left(1 - \frac{Q_1}{Q_2}\frac{e}{L'}\right) \tag{11.27}$$

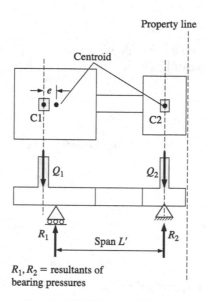

Figure 11.7 Analysis of strap footing.

We can now size the footings for loads R_1 and R_2 so that no serviceability or ultimate limit states are attained. The structural design involves design not only of the footings, but of the beam as well. This foundation arrangement is quite dependent on the structural design, which must be done with utmost care.

11.4.5 Analysis and structural design of mat foundations

Structural design of mats is commonly done using the finite element method. The mat is divided into finite elements typically smaller than 1 m in size. Each element is supported at its corners by springs. Each spring is shared by all the elements having a node at the spring location. The spring is supposed to carry loads applied on the fractions of the areas of the elements it helps support. The contributing area for the spring is the summation of these fractional areas. For square finite elements, the contributing areas would be the areas of four squares, each with one-fourth the area of each element.

Each spring stiffness[6] is the product of the contributing area for that spring by the unit resistance the soil is capable of offering at some level of deflection divided by that value of deflection. The unit resistance offered by the soil per unit of deflection is known as the modulus of subgrade reaction k of the soil.[7] The modulus of subgrade reaction k is also known as Winkler's spring constant in honor of the originator of the concept Winkler (1867). When loads are applied on the mat, they transfer down to the soil by means of contact stresses between the mat and the soil. The stiffer the soil is, the greater the value of k, the smaller the deflections experienced by the mat, and the larger the spring constants must be to accurately represent the problem.

In routine mat design, the geotechnical work consists mostly of arriving at reasonable values of the modulus of subgrade reaction. Structural engineers use these values to multiply contributing areas to obtain spring constants. They then select appropriate stiffness

---

[6] A spring stiffness is a ratio of force to displacement (the force required to either compress or extend a spring by a unit length).

[7] We will run into the modulus of subgrade reaction concept again when we study axially and laterally loaded piles.

values for the mat elements and run finite element analyses, from which the stresses, strains, bending moments, shear forces, and displacements in the mat are obtained. If these are all acceptable, the design is complete; otherwise, an adjustment is made to either the thickness or reinforcement of the mat foundation, and a new set of calculations are performed. The process is repeated until the results are satisfactory. Best designs are achieved by adjusting the value of k for each spring as a function of settlement at that location, given that soil has a nonlinear load–deflection response.

11.5 CASE STUDY: THE LEANING TOWER OF PISA (PART IV)

On Friday, April 5, 1991, a headline on the *Corriere della Sera*, the Milan newspaper, read: "*Pisa chiede una riunione per l'emergenza Torre.*" The article reports the alarm that had taken over the population of Pisa as word of the latest measurements of the lean of the tower started spreading. The tower had been closed since January 6, 1990, and the population was caught off guard by the perceived difficulty of stopping the ongoing rotation of the tower toward the south and the acceleration of the lean detected by instrumentation. While in 1991 it was still believed that it would take tens of years for the tower to reach a state of collapse, we now know that the geotechnical experts on the commission in charge of stabilizing the tower came to believe in subsequent years that there were moments of grave danger of collapse. In fact, according to Potts (2003), the tower was likely dangerously close to collapse in the 1990–1991 period, during which the acceleration in the rotation of the tower was detected. Given that Mitchell et al. (1977) showed that the safety factor against bearing capacity failure appeared to be sufficient at all times of the Tower's existence (even if it is not settled by how much the safety factor exceeded 1 at certain times of the Tower's life), what was the nature of this collapse that the population of Pisa feared and the experts now confirm was indeed a possibility?

The beginnings of an answer lie in the observation by Mitchell et al. (1977) that the maximum shear stress at points or within zones in the clay layer matched or exceeded the soil shear strength, even if the safety factor against bearing capacity failure was greater than 1. This indicated that a plastic zone had started to form below the south end of the tower. Potts (2003) provided a simple example that illustrates the type of collapse that the Tower of Pisa could have experienced. This type of collapse, known as leaning instability, is closely associated not only with the shear strength of the soil, which we use in our bearing capacity calculations, but also with the soil stiffness, as represented by its shear modulus. Figure 11.8 shows a tower (with geometry quite similar to that of the Tower of Pisa) on top of a clay with $s_u = 80\,\text{kPa}$. The tower is built with an initial inclination of $0.5°$. A finite element analysis then simulates the soil–tower response as the weight of the tower is gradually increased for three different soil stiffnesses: $G = 10s_u$, $G = 100s_u$, and $G = 1000s_u$. Note that, since the value of s_u is the same in all three cases, the weight at which collapse would occur would be the same if bearing capacity were the mechanism of collapse. Instead, as shown in Figure 11.9, collapse takes place for a much lower weight in the case of low stiffness ($G/s_u = 10$). It also happens quite suddenly. Why the difference?

Figure 11.10 shows a plot of incremental displacements at the last loading increment (just before collapse) for the case of low stiffness. It also shows that the shear stress in the soil becomes equal to the shear strength (forming a plastic zone, represented as a shaded zone) in only a portion of the soil deposit, and a plastic mechanism does not form. So how does collapse occur? It is, in essence, an overturning failure. The decreasing stiffness of the soil as it enters the plastic range does not provide enough support beneath the right edge of the tower after a plastic zone forms there to balance the moment of the weight of the tower with

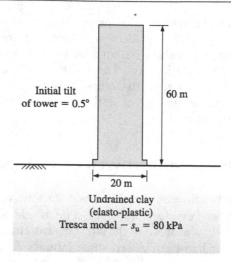

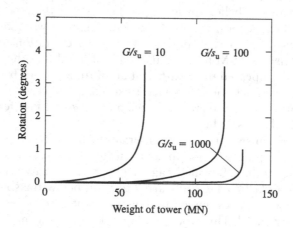

Figure 11.8 Tower built with an initial inclination of 0.5° on top of a clay with s_u=80 kPa; the shear modulus G of this clay is not fixed, and cases with both low and high G/s_u ratios are analyzed. (After Potts 2003; courtesy of the Institution of Civil Engineers and Thomas Telford Limited.)

Figure 11.9 Plot of tower weight versus angle of rotation for G/s_u= 10, 100, and 1000. (After Potts 2003; courtesy of the Institution of Civil Engineers and Thomas Telford Limited.)

respect to the center of the foundation. Leaning instability is a second-order phenomenon: as the rotation increases because of insufficient soil stiffness, the driving moment increases, which in turn leads to further rotation, and so on. Contrast that with the full plastic mechanism that forms below the tower in the high-stiffness-ratio case, shown in Figure 11.11. Here, clearly, we have a bearing capacity failure, which is a very different mechanism from leaning instability, hence the difference in tower weights for which these two failures are observed.

Understanding the potential mechanism of instability for the Tower of Pisa was not merely an academic exercise. For example, using lead ingots on the north side of the tower to stabilize it, as described in Part I of the case history in Chapter 2, would only work, as argued by Potts (2003), if the prevailing mechanism was leaning instability, in which case the lean would reduce upon placement of the ingots. If a bearing capacity failure had been in place (which, however, previous studies suggested was not the case), the ingots would actually precipitate failure.

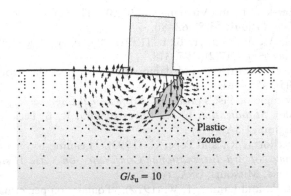

Figure 11.10 Incremental displacement field underneath the tower just before collapse for low (10) G/s_u ratio; the shaded zone is the zone where the shear stresses are equal to the shear strength of the soil s_u. (After Potts 2003; courtesy of the Institution of Civil Engineers and Thomas Telford Limited.)

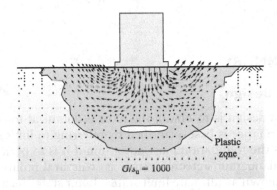

Figure 11.11 Incremental displacement field underneath the tower just before collapse for high (100) G/s_u ratio; the shaded zone is the zone where the shear stresses are equal to the shear strength of the soil s_u. (After Potts 2003; courtesy of the Institution of Civil Engineers and Thomas Telford Limited.)

Now that we have covered the main facts about the Tower of Pisa, readers are asked in Problem 11.20 to design the foundations for the Tower with current design and foundation technology (they may also wait until they have studied pile foundations in later chapters to investigate solutions using piles, which is what is asked in Problem 15.16). We are optimistic that, after nine centuries of progress, we will be at some advantage to our predecessors in Pisa, who, dealing with processes they could not at the time suspect, set off a chain of events that produced a fascinating engineering failure, an ingenious engineering intervention, and an incredible success for the tourism industry in Italy.

11.5.1 References for "Case History: The Leaning Tower of Pisa" – Parts I–IV

Burland, J. B. (2000). "Ground-structure interaction: does the answer lie in the soil?" *Structural Engineer*, 78(23/24), 42–49.

Burland, J. B. (2002). "The stabilization of the leaning tower of Pisa." *Journal of Architectural Conservation*, 8(3), 7–23.

Burland, J. B., Jamiolkowski, M., and Viggiani, C. (2003). "The stabilization of the leaning tower of Pisa." *Soils and Foundations*, 43(5), 63–80.

Cheney, J. A., Abghari, A., and Kutter, B. L. (1991). "Stability of leaning towers." *Journal of Geotechnical Engineering*, 117(2), 297–318.

Foti, S. (2003). "Small-strain stiffness and damping ratio of Pisa clay from surface wave tests." *Géotechnique*, 53(5), 455–461.

Greek, D. (1998). "New angle on Pisa's problem." *Professional Engineering*, 11(3), 14–15.

Jamiolkowski, M. (2006). *Stabilization of the tower of Pisa.* R. B. Peck Lecture, Geo-Institute, ASCE, Reston, Virginia, USA.

Jamiolkowski, M., Lancellotta, R., and Pepe, C. (1993). "Leaning tower of Pisa—updated information." *Proceedings of 3rd International Conference on Case Histories in Geotechnical Engineering*, St. Louis, Missouri, USA, 1–6.

Mitchell, J. K., Vivatrat, V., and Lambe, T. W. (1977). "Foundation performance of tower of Pisa." *Journal of the Geotechnical Engineering Division*, 103(GT3), 227–249.

Morley, J. (1996). "Acts of God: the symbolic and technical significance of foundation failures." *Journal of Performance of Constructed Facilities*, 10(1), 23–31.

Potts, D. M. (2003). "Numerical analysis: a virtual dream or practical reality?" *Géotechnique*, 53(6), 535–573.

Ranzini, S. M. T. (2001). "Stabilisation of leaning structures: the tower of Pisa case." *Géotechnique*, 51(7), 647–648.

11.6 CHAPTER SUMMARY

11.6.1 Foundation design

The goal of foundation design is to obtain the most economical foundation that will perform adequately. Performance is best quantified by referring to limit states. A limit state is a set of conditions that the foundation must not attain if it is to perform adequately. Limit states associated with dangerous outcomes, such as structural failures and loss of stability in various forms, are known as ultimate limit states. Limit states related to functionality of the structure are called serviceability limit states.

Foundation design consists mostly of limit state checks. These checks must reflect the fact that we do not know all the variables involved with certainty. For ultimate limit states, either an FS or a combination of load and resistance factors is used. **Working stress design** requires that the resistance divided by the FS be greater than or equal to the applied load. **Load and resistance factor design** requires that the product of the **resistance factor** by a **nominal (characteristic) resistance** be equal to or greater than the summation of the **nominal (characteristic) loads** (dead load, live load, and loads due to earthquakes, earth pressure, snow, and wind) multiplied by suitable load factors.

11.6.2 Bearing capacity check using working stress design

The following equations can be used to determine the net unit bearing capacity of a footing that has been backfilled:

$$q_{bL}^{net} = s_u N_c s_{su} d_{su} i_{su} = FS \frac{Q_{str}}{A} \tag{11.6}$$

for clays with uniform strength with depth,

$$q_{bL}^{net} = Fs_{su}d_{su}i_{su}\left[s_{u0} + \frac{\rho B}{4N_c}\right]N_c = FS\frac{Q_{str}}{A} \tag{10.33}$$

for clays with increasing strength with depth, and

$$q_{bL}^{net} = q_0\left(s_q d_q i_q b_q g_q N_q - 1\right) + \frac{1}{2}\gamma B N_\gamma s_\gamma d_\gamma i_\gamma b_\gamma g_\gamma = FS\frac{Q_{str}}{A} \tag{11.8}$$

for sands.

The area of the footing can be expressed in terms of the allowable net unit bearing capacity and the applied load as follows:

$$A = \frac{Q_{str}}{q_{b,all}^{net}} \tag{11.9}$$

where the allowable net unit bearing capacity is given by

$$q_{b,all}^{net} = \frac{q_{bL}^{net}}{FS} \tag{11.10}$$

The following equation can be used to find the plan area of a footing that is not backfilled:

$$A = \frac{FSQ_{str} + W_{ftg}}{q_{bL}^{gross}} \tag{11.12}$$

11.6.3 Bearing capacity check using LRFD

In the LRFD approach, the design inequality is:

$$(RF)R_n \geq \sum (LF)_i Q_{ni} \tag{11.14}$$

The resistance factors RF are given for vertically loaded footings in Tables 11.3 and 11.4. The load factors are the same as those used in structural design of the footings. The characteristic load is provided by the structural engineer, just as in the case of WSD.

The characteristic resistance (bearing capacity) is calculated from the nominal value x_n of cone resistance q_c, SPT blow count N, undrained shear strength s_u, or any other measurement of strength used in the calculation of bearing capacity. This nominal value x_n is defined as the value exceeded by 80% of the values of x and is given by

$$x_n = E(x) - 0.84\sigma_x \tag{11.16}$$

where $E(x)$ is the expected value of x, and σ_x is the standard deviation of x. The standard deviation of x can be estimated from the range of values contained in the sample, that is, from the difference between the maximum and minimum values of x. This difference corresponds to a certain number of standard deviations of x that depends on the number of measurements of x available (that is, the size of the sample). The ratio of the range of x to the standard deviation of x is given in Table 11.2 as a function of sample size.

The design procedure consists then in determining the most economical plan area of the footing that will satisfy the inequality in Equation (11.14).

11.6.4 Settlement check

For settlement checks, uncertainties have traditionally been handled by imposing limits on settlement that are stricter than those known to lead to attainment of limit states. For example, the value of angular distortion used to derive common values of tolerable settlements is 1/500, much smaller than the 1/300 known to cause architectural damage and the 1/170 known to lead to damage of structural frames. Settlements are calculated from the values of soil modulus and compared directly with tolerable settlements. Alternative approaches are based on calculating the differential settlement between adjacent footings and requiring that these differential settlements either are below a tolerable value or do not overload structural elements.

11.6.5 Symbols and notations

Symbol	Quantity represented	US units	SI units
A	Area of the footing	ft^2	m^2
B	Width of the footing	ft	m
b_q, b_γ	Base inclination factors	Unitless	Unitless
b_R	Resistance bias factor	Unitless	Unitless
D	Embedment depth of the footing	ft	m
d_q, d_γ, d_{su}	Depth factors	Unitless	Unitless
D_R	Relative density	%	%
g_q, g_γ	Ground inclination factors	Unitless	Unitless
i_q, i_γ, i_{su}	Load inclination factors	Unitless	Unitless
K_0	Coefficient of lateral earth pressure at rest	Unitless	Unitless
L_R	Reference length (= 1 m = 3.281 ft)	ft	m
N_c, N_q, N_γ	Bearing capacity factors	Unitless	Unitless
N_{SPT}	SPT blow count	Unitless	Unitless
p_A	Reference stress (= 100 kPa)	psf	kPa
q_0	Surcharge unit load	psf	kPa
q_{bL}^{gross}	Gross bearing capacity of the footing	psf	kPa
q_c	Cone resistance	psf	kPa
$q_{c,CAM}$	Conservatively assessed mean of cone resistance	psf	kPa
Q_d	Design load	ton or lb	kN
Q_h	Horizontal load	ton or lb	kN
$q_{b,all}^{net}$	Allowable net unit bearing capacity	psf	kPa
Q_{ni}	Nominal load	ton or lb	kN
Q_{str}	Column (wall) load	ton or lb	kN
Q_v	Vertical load	ton or lb	kN
R_d	Design resistance	lb	kN
R_n	Nominal resistance	lb	kN
s_q, s_γ, s_{su}	Shape factors	Unitless	Unitless
s_u	Undrained shear strength	psf	kPa
w	Settlement	ft	m
W_{fill}	Weight of backfill	ton or lb	kN

(Continued)

Symbol	Quantity represented	US units	SI units
W_{ftg}	Weight of footing	ton or lb	kN
x_c	Centroidal distance	ft	m
z_{f0}	Depth from foundation base level	ft	m
α	Angular distortion	deg	deg
γ	Unit weight of soil	pcf	kN/m^3
μ	Coefficient of friction	Unitless	Unitless
ϕ_c	Critical-state friction angle	deg	deg
ϕ_p	Peak friction angle	deg	deg
ϕ_r	Residual friction angle	deg	deg
ρ	Strength gradient with depth	psf/ft	kPa/m
σ'_{mp}	Peak mean effective stress	psf	kPa

11.7 PROBLEMS

11.7.1 Conceptual problems

Problem 11.1 Define all the terms in bold contained in the chapter summary.

Problem 11.2 Discuss the use of net versus gross unit bearing capacity to size shallow foundations.

Problem 11.3 A single value of unit bearing capacity is often used to size all the footings for a given building. Discuss whether this is conceptually correct. Discuss also under what conditions (such as magnitude of the project, magnitude of the loads, or type of soil) the practice might lead to excessively conservative designs and thus costly solutions.

Problem 11.4 Consider a centrally, vertically loaded square footing on sand supporting live and dead loads that, added together, are equal to 1000 kN. Considering the load and resistance factors given in Tables 2.3 and 11.3 as well as the definition of characteristic resistance, study the dependence of the implied safety factor on the dead-to-live load ratio. Work with ASCE (ACI) load factors.

11.7.2 Quantitative problems

Problem 11.5 If the allowable unit load for the soil is 0.25 MPa, determine the dimensions into the plane of the paper of the strap footings shown in P-Figure 11.1. The vertical loads on the two footings are 450 and 900 kN, respectively.

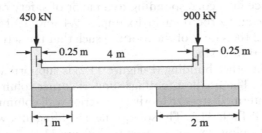

P-Figure 11.1 Footings for Problem 11.5.

Problem 11.6 For a strip footing with a width of 1 m embedded 0.75 m in clay with a representative undrained shear strength equal to 40 kPa, plot the maximum live load for various values of dead load that would be allowable for this footing. Use Equation (10.26) to calculate the depth factor.

Problem 11.7 For a strip footing with a width of 1 m embedded 0.75 m in clay with a representative undrained shear strength equal to 40 kPa, plot the maximum live load for various values of dead load that would be allowable for this footing. Use Equation (10.22) to calculate the depth factor.

Problem 11.8 For a square footing with a width of 2 m embedded 0.75 m in clay with a representative undrained shear strength equal to 40 kPa, plot the maximum live load for various values of dead load that would be allowable for this footing. Use Equations (10.26) and (10.34) to calculate the depth and shape factors.

Problem 11.9 For a square footing with a width of 2 m embedded 0.75 m in clay with a representative undrained shear strength equal to 40 kPa, plot the maximum live load for various values of dead load that would be allowable for this footing. Use Equations (10.22) and (10.23) to calculate the depth and shape factors.

11.7.3 Design problems

Problem 11.10 A reinforced concrete column with square cross section and side equal to 300 mm is located such that one of its faces is aligned with the property line. The load on this column is 500 kN. An interior, square column supporting a load of 1000 kN (with sides equal to 400 mm) exists at a center-to-center distance of 5 m from the exterior column. Study the use of combined footings (both rectangular and trapezoidal) and strap footings in this case, given that the allowable soil unit load is in the 0.1–0.2 MPa range.

Problem 11.11 Complete the design of Example 11.1. Assume that large total settlements can be tolerated. Work with a factor of safety of 3.

Problem 11.12 Complete the design of Example 11.2.

Problem 11.13 Complete Example 11.3 by finding the size of the second footing such that the factor of safety is 3. The solution has been sketched as part of the example.

Problem 11.14 Check if the footings of Example 11.3 designed with $FS = 3$ pass the ultimate limit state check using LRFD.

Problem 11.15 Consider the two footings of Example 11.3. What would the new plan areas of the two footings be if calculations were made for each footing based on a tolerable settlement of 25 mm (1 in.)? For 38 mm (1.5 in.)?

Problem 11.16 Increase the areas of footings A and B of Example 11.6 so that (a) both have the same factor of safety and (b) the angular distortion between them becomes less than the tolerable value.

Problem 11.17 For the footing of Examples 9.4 and 10.15, calculate the settlement associated with the service load corresponding to a factor of safety of 5.

Problem 11.18 Referring to the data in Examples 9.4 and 10.15, proportion a square footing (that is, find the side B of the footing) such that the settlement of the footing is approximately equal to 1.5 in.

Problem 11.19* A light office building (P-Figure 11.2) is supported on footings with three different plan areas. The smallest footings support corner columns; the largest, interior columns; and the intermediate-size footings, exterior wall columns. The loads to be supported are given in P-Table 11.1. The soil profile, including unit weight, index properties for the clay, and SPT blow counts, is given in P-Figure 11.2. For the clay layer, $C_c = 0.28$ and OCR =1. The initial void ratio at the middle of the clay layer is equal to 0.9.

a. Considering both immediate and consolidation settlements, proportion the footings such that the angular distortion between every pair of footing is no greater than 1/300. Assume that the total settlement will be accommodated by architectural solutions so long as it does not exceed 150 mm.

b. Estimate the evolution of the settlements of the three classes of footings with time.

c. Check the three types of footings for limit bearing capacity failure using both the WSD and the LRFD approaches.

Problem 11.20 We have studied the case history of the Leaning Tower of Pisa in Parts I (Chapter 2), II (Chapter 6), III (Chapter 9), and IV (Chapter 11). Provide a design for the foundations of the Tower of Pisa using shallow foundations. Make whatever assumptions may be necessary, or consult the papers we have referenced in the case history.

Problem 11.21 Check if the footing of Problem 9.20 passes the ultimate limit state check using both WSD and LRFD. The nominal dead and live loads on the footing are 500 kN each.

Problem 11.22 Check if the footing of Problem 9.21 passes the ultimate limit state check using both WSD and LRFD. The nominal dead and live loads on the footing are 3000 kN each.

P-Table 11.1 Loads for Problem 11.19

Column	Dead load (kN)	Live load (kN)
Corner columns	450	530
Exterior wall columns	550	620
Interior columns	650	730

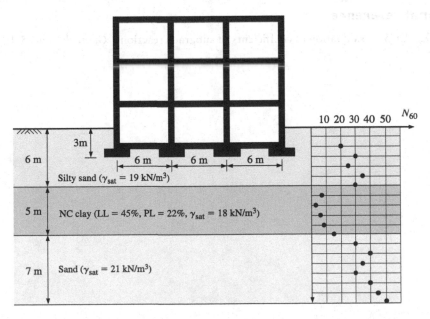

P-Figure 11.2 Cross-sectional view of building supported on three types of footings, to be designed as part of Problem 11.19.

REFERENCES

References cited

AASHTO. (1998). AASHTO LRFD bridge design specifications. 2nd Edition, American Association of State Highway and Transportation Officials, Washington, D.C., USA.

AASHTO. (2020). AASHTO LRFD bridge design specifications. 9th Edition, American Association of State Highway and Transportation Officials, Washington, D.C., USA.

ACI. (2002). Building code requirements for structural concrete (318-02) and commentary (318R-02). American Concrete Institute, Detroit, Michigan, USA.

Davis, E. H. and Booker, J. R. (1973). "The effect of increasing strength with depth on the bearing capacity of clays." *Géotechnique*, 23(4), 551–563.

Foye, K. C., Salgado, R., and Scott, B. (2006a). "Assessment of variable uncertainties for reliability-based design of foundations." *Journal of Geotechnical and Geoenvironmental Engineering*, 132(9), 1197–1207.

Foye, K. C., Salgado, R., and Scott, B. (2006b). "Resistance factors for use in shallow foundation LRFD." *Journal of Geotechnical and Geoenvironmental Engineering*, 132(9), 1208–1219.

Meyerhof, G. G. (1963). "Some recent research on the bearing capacity of foundations." *Canadian Geotechnical Journal*, 1(1), 16–26.

Salgado, R., Lyamin, A. V., Sloan, S. W., and Yu, H. S. (2004). "Two- and three-dimensional bearing capacity of foundations in clay." *Géotechnique*, 54(5), 297–306.

Skempton, A. W. (1986). "Standard penetration test procedures and the effects in sands of overburden pressure, relative density, particle size, ageing and overconsolidation." *Géotechnique*, 36(3), 425–447.

Skempton, A. W. and MacDonald, D. H. (1956). "The allowable settlements of buildings." *Proceedings of Institution of Civil Engineers, Part III*, 727–768.

Tippett, L. H. C. (1925). "On the extreme individuals and the range of samples taken from a normal population." *Biometrika*, 17(3/4), 364–387.

Winkler, E. (1867). *Die lehre von elastizitat und festigkeit (on elasticity and fixity)*. H. Dominicus, Prague.

Additional reference

Terzaghi, K. (1955). "Evaluation of coefficients of subgrade reaction." *Géotechnique*, 5(4), 297–326.

Chapter 12

Types of piles and their installation

Do not design on paper what you have to wish into the ground.

Karl Terzaghi

Piles are long, slender elements made of concrete, steel, timber, or polymers used to support structural loads. Historically, piles have been used to transfer structural loads to deeper rock or firm soil layers at sites where soft clays or loose sands exist at shallow depths. In recent decades, their uses have expanded to absorbing tensile and lateral loads, to supporting loads by shaft resistance, and to reducing the settlement of mat foundations. In this chapter, we discuss the main pile types, their applications, and their installation.

12.1 PILE FOUNDATIONS: WHAT ARE THEY AND WHEN ARE THEY REQUIRED?

Piles are deep foundation elements made of concrete, steel, timber, polymers, or a combination of two or more of these materials. Their obvious use is when surface soils are too loose or soft to support shallow foundations safely and economically. However, as Figure 12.1 illustrates, there are many other uses for deep foundations. The choice of pile foundations is largely a design decision, as discussed earlier in Chapter 2.

While the construction of shallow foundations is relatively simple, the installation of piles is not. It requires piling rigs that are sometimes very sophisticated and expensive (costing in excess of a million US dollars) and involves techniques that depend on soil conditions and other factors. Additionally, there are a large number of pile types to choose from. The performance of pile foundations is very much dependent on the type of pile and the method of pile installation. Thus, the choice of pile type and decisions regarding installation techniques are an integral part of the design process, and engineers must understand them to design deep foundations effectively.

12.2 CLASSIFICATIONS OF PILE FOUNDATIONS

12.2.1 Classification based on the method of fabrication and installation process

Pile response to loading is very much dependent on the method of pile installation. We can identify two types of piles that are on the opposite ends of the spectrum of the changes caused in the soil by pile installation: *nondisplacement piles* and *displacement piles*. Nondisplacement piles are piles that are cast *in situ* in the space left after a volume of soil is removed from the ground.

DOI: 10.1201/b22079-12

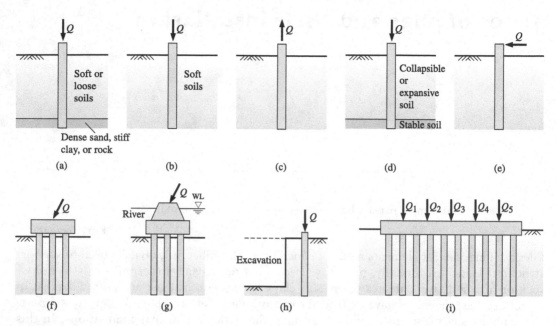

Figure 12.1 Several uses of pile foundations: (a) end-bearing pile used to cross a soft or loose soil deposit; (b) floating pile in soft soils; (c) pile used to develop sufficient tensile load capacity; (d) pile foundation used to extend beyond collapsible or swelling soil; (e) pile used to develop sufficient lateral load capacity; (f) group of piles used to support a large inclined load; (g) piles used to extend foundations below maximum probable depth of scour; (h) pile foundation used when it is known that a deep excavation will be done next to the foundation; and (i) piled mat (piled raft).

Soil elements around the pile are not pushed away or displaced from its original positions with this method of pile installation. Displacement piles, on the other hand, are piles that are inserted into the soil by either driving (most commonly) or jacking (more infrequently) without any soil removal before the piles are inserted. In the installation of displacement piles, soil elements originally located where the pile will be installed or near it undergo large displacements. Between the two extremes, there is a growing number of pile types that are neither full-displacement piles nor nondisplacement piles. They are most commonly referred to as partial-displacement piles; most are installed using some type of auger. The terminology for these piles is still evolving and differs substantially across the Atlantic. We will adopt those terms that we find most descriptive, regardless of whether they are used most often in Europe or the United States. Where appropriate, we will point to alternative terminology.

Nondisplacement piles are cast in place. The various types of nondisplacement piles differ based on the equipment and method by which the soil is excavated and the concrete or other materials placed into the excavation. Nondisplacement piles can be classified as

- Percussion bored (Strauss) piles
- Rotary bored
  - In soil
  - In rock
- Bored piles (drilled shafts)[1]

---

[1] It is important to emphasize that drilled shafts, also referred to in earlier times as drilled piers or drilled caissons, are piles. Traditionally, the term pile was used for smaller-diameter, typically driven piles. A notion developed that drilled shafts needed to be treated in a different way, and this notion still survives to some extent. We will see in this chapter and the next that no such distinction is needed to understand how they respond to loading, to analyze this response, or to design them.

- Straight shaft
- Underreamed (base-enlarged)
  Partial-displacement piles include the following:
- Small-displacement piles
  - H-piles[2]
  - Open-ended pipe piles without soil plugging
- Continuous flight auger (CFA) piles
  - Made with grout
  - Made with concrete
- Drilled displacement piles
  - Atlas
  - Auger pressure grouted displacement (APGD)
  - Fundex
  - Omega

Full-displacement piles can be either pre-fabricated and later transported to the site for installation, or cast in place. Cast-in-place piles are usually made of concrete, grout, or cement paste. Displacement piles can be classified as

- Cast-in-place
  - Closed-ended concrete/steel pipe driven and filled with concrete
  - Raymond piles
  - Franki piles
- Pre-fabricated
  - Concrete
    - precast
    - full-length reinforced
    - prestressed
    - reinforced with connections ready for extension
    - tubular section
  - Steel
    - H-piles (when soil adheres to pile between flanges)
    - pipe piles (closed-ended or small-diameter piles, for which significant soil plugging occurs during installation)
    - other cross sections

This classification of piles as nondisplacement, partial-displacement, or full-displacement is the most important consideration from a pile design perspective, because these classes of piles behave radically differently when loaded, particularly under axial load. This difference in load response is mostly due to the state of the soil around the pile after installation. Both the density and stress state of the soil around a displacement pile and beneath its base change significantly (Figure 12.2). Displacement piles may be seen as having preloaded the soil during the installation process. The installation of nondisplacement piles, on the other hand, preserves the soil density and state to a significant degree. When we discuss the design of axially loaded piles in the next chapter, we will see in detail the influence that the installation process has on pile load response.

---

[2] The load response of these piles depends on the extent of soil plug formation during installation. We discuss soil plug formation and the conditions under which they occur later in Chapter 13.

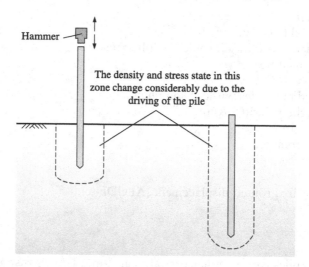

Figure 12.2 Change in soil state around a displacement pile.

12.2.2 Classification based on pile material

Pre-fabricated piles have traditionally been made using one of the three materials: timber, concrete, or steel. Composite piles, made of different materials, have occasionally been used. More recently, polymer piles are being produced. These are usually tubular, filled with concrete. They are still rarely used. Accordingly, we will focus our discussion on timber, concrete, and steel piles.

12.2.2.1 Timber piles

Timber piles have been used since antiquity. The natural cross section of these piles is approximately circular, with diameter varying slightly from one end to the other of the pile. Timber piles have varying diameter because tree diameters tend to be larger near the roots and smaller near the branches. Pile diameter and length range from 150 to 400 mm (6–16 in.) and from 6 to 20 m (20–66 ft), respectively. Less common than circular piles, square piles are also available. These are cut out of the natural circular tree trunks and are less effective and durable than circular piles because the outer, natural protective layers of the pile are removed.

There are certain requirements for timber piles to be acceptable. The diameter of timber piles should vary gradually and uniformly from the top to the bottom of the pile. Additionally, the pile should be reasonably straight. The pile should have no defects (such as decayed wood and discontinuities of various types) anywhere.

When timber piles are exposed to water, they may be subject to decay by the action of fungi. This is possible only if both water and a supply of oxygen are available, so damage tends to happen at the water table level for an onshore pile and around the water level in piles used in docks, bridges, or other waterfront structures. Figure 12.3 shows the seriousness of this type of decay for a timber pile in the docking area of the Vizcaya Museum and Gardens (formerly the winter residence of 1920s tycoon and International Harvester vice president, James Deering, in Miami, Florida). Note the much reduced cross section of the pile. Other organisms can also attack wooden piles, including insects and marine borers. Although piles can be treated with creosote, oils, and other products, such treatment is 100% effective only above the permanent water table. All of these limitations should be taken into account when planning the use of timber piles.

Figure 12.3 Seriously decayed timber piles used in the docking area of what today is the Vizcaya Museum and Gardens in Miami, Florida.

12.2.2.2 Steel piles

The most common types of steel piles are pipe piles and H-piles, although other types of cross sections are occasionally used. Advantages of steel piles include their high resistance to driving and handling, as well as their large lateral stiffness. Pipe piles, in particular, offer an excellent combination of lateral stiffness and ideal shape to resist current and wave loadings, as well as ship collisions, in both waterfront and offshore applications. An additional advantage is the ease with which these piles are cut off or extended. Extension is usually done by welding. There are some patented splicing methods that may be indicated under some circumstances. Figure 12.4 illustrates the general approach to splicing.

The obvious shortcoming of steel piles is their susceptibility to corrosion in marine environments. Corrosion of steel piles on shore is rarely a problem (Manning and Morley 1981; Romanoff 1962, 1969; Wong and Law 1999). The reason appears to be the absence in soils of a sufficient supply of oxygen for corrosion to develop to a meaningful extent, even in fairly aggressive soils.

Pipe piles are commercially available usually in the cross sections shown in Table 12.1, and H-piles, in those shown in Table 12.2. Figure 12.5 illustrates the meaning of the cross-sectional dimensions given in these tables. The range of pipe pile diameters is 50–4000 mm

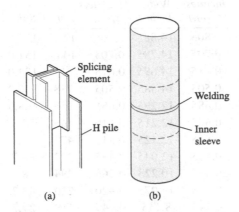

Figure 12.4 Illustration of splicing of (a) H-piles and (b) pipe piles.

Table 12.1 Typical pipe pile cross sections

Outer diameter (in.)	Wall thickness (in.)	Weight per foot (lb)	Area of steel (in.²)	Internal area (in.²)
10.75	0.219	24.63	7.242	83.5
10.75	0.250	28.04	8.243	82.5
10.75	0.312	34.78	10.226	80.5
10.75	0.365	40.48	11.902	78.8
12.00	0.219	27.55	8.101	104.9
12.00	0.250	31.37	9.224	103.8
12.00	0.312	38.95	11.450	101.6
12.00	0.375	46.56	13.688	99.4
12.75	0.219	29.31	8.617	119.0
12.75	0.250	33.38	9.813	117.8
12.75	0.312	41.45	12.185	115.4
12.75	0.375	49.56	14.572	113.0
14.00	0.250	36.71	10.794	143.1
14.00	0.312	45.61	13.410	140.5
14.00	0.375	54.57	16.043	137.8
14.00	0.500	72.09	21.195	132.7
16.00	0.250	42.05	12.364	188.6
16.00	0.312	52.27	15.369	185.6
16.00	0.375	62.58	18.389	182.6
16.00	0.500	82.77	24.335	176.6
18.00	0.375	70.59	18.408	182.7
18.00	0.500	93.45	24.347	176.7
20.00	0.375	78.60	23.120	291.0
20.00	0.500	104.13	30.631	283.5
24.00	0.375	94.62	27.832	424.6
24.00	0.500	125.49	36.913	415.5
24.00	0.625	156.03	45.897	406.5

Table 12.2 Typical H-pile cross sections and properties

Designation (in. × lb/ft)	Area A (in.²)	Depth d (in.)	Web thickness t_w (in.)	Flange Width b_f (in.)	Flange Thickness t_f (in.)	Cross-sectional properties X–X I (in.⁴)	S (in.³)	r (in.)	Y–Y I (in.⁴)	S (in.³)	r (in.)
HP 14 × 117	34.60	14.21	0.805	14.885	0.805	1230	173.0	5.96	443.0	59.50	3.58
×102	30.20	14.01	0.705	14.785	0.705	1060	151.0	5.92	380.0	51.40	3.55
×89	26.30	13.83	0.615	14.695	0.615	913	132.0	5.89	326.0	44.30	3.52
×73	21.60	13.61	0.505	14.585	0.505	738	108.0	5.84	261.0	35.80	3.48
HP 12 × 84	24.60	12.28	0.685	12.295	0.685	650	106.0	5.14	213.0	34.50	2.94
×74	21.80	12.13	0.605	12.215	0.610	569	93.8	5.11	186.0	30.30	2.92
×63	18.40	11.94	0.515	12.125	0.515	472	79.1	5.07	153.0	25.20	2.89
×53	15.50	11.78	0.435	12.045	0.435	393	66.7	5.04	127.0	21.00	2.86
HP 10 × 57	16.70	9.99	0.565	10.225	0.565	294	58.7	4.19	101.0	19.70	2.46
×42	12.30	9.70	0.415	10.075	0.420	210	43.3	4.14	71.7	14.20	2.41
HP 8 × 36	10.50	8.02	0.445	8.155	0.445	119	29.7	3.37	40.3	9.88	1.96

I, moment of inertia; *S*, section modulus; *r*, radius of gyration.

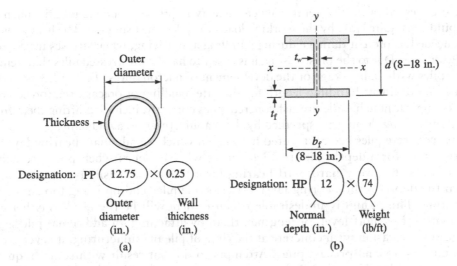

Figure 12.5 Cross sections of (a) pipe and (b) steel H-piles.

(the large diameters being used in offshore piling, and the smaller ones as mini-piles) and as large as 10 m for monopiles, which are commonly used to support wind turbines in offshore environments. Wall thickness ranges from 4 to 150 mm. Length is not a constraint for steel piles because pile sections can easily be welded or cut to obtain the desired length. Piles can be as short as 3 m onshore and as long as 500 m offshore. Pipe pile steel can be of grades 1–3, with corresponding yield strengths of 205, 230, and 310 MPa [ASTM A252 (ASTM 2019b)].

12.2.2.3 Concrete piles

12.2.2.3.1 Concrete properties

Concrete is used in both pre-fabricated and cast-in-place piles. Pre-fabricated piles are manufactured in plants and transported to the job site. In very large projects, temporary plants may be set up at the site to manufacture the piles. Cast-in-place piles are installed by first creating a void in the soil and then filling it with concrete. In the case of micropiles, cement paste or high-strength grout may be used in place of concrete.

The compressive strength of concrete ranges from 25 to 40 MPa (3600–5800 psi) for precast concrete piles and from 11 to 25 MPa (1600–3600 psi) for cast-in-place piles. Concrete strength is the primary consideration for both precast and cast-in-place piles; workability can also be very important for certain types of cast-in-place piles, as we will see later. For nondisplacement piles (such as drilled shafts) and augered partial-displacement piles (such as CFA piles), workability is just as important as strength, for reasons that will be discussed in more detail later.

12.2.2.3.2 Precast, reinforced concrete piles

Precast concrete piles can be made with any cross section. Square cross sections are most common commercially, particularly in short lengths. Piles with other cross sections, typically hexagonal, octagonal, or circular, are also available from some manufacturers in some countries. Concrete pipe piles also exist, prestressed versions of which are sometimes used in the offshore industry.

Precast concrete piles can be of two types: reinforced or prestressed concrete piles. The reinforcement elements in precast piles are regular rebars, while tendons or cables are normally used in prestressed piles. The reinforcement density in precast concrete piles is much higher than would be required just by the working loads the piles must support. The heavy reinforcement is needed to prevent damage during handling and driving, when stresses usually exceed working stresses. Due to the prestressing, it is easier to handle prestressed piles than reinforced concrete piles without breakage or the development of microfissures. Table 12.3 gives the values of the approximate handling length for the safe handling of precast, reinforced concrete piles. It is important to handle precast concrete piles correctly, including lifting them from the proper points along the pile, as specified by the manufacturer, to avoid damage.

Precast concrete piles are best suited for sites in which an adequate bearing layer exists at relatively uniform depth. Figure 12.6 shows what happens when piles are driven to refusal (that is, to a very hard, stiff bearing layer that is not easily penetrated) at a site in which the depth to the bearing layer varies: some piles end up being too short; others are too long. This is highly undesirable because there will be the need to either cut the piles down to the cutoff level, or lengthen them by attaching an additional pile segment. Unless joints are cast in with concrete at the time of pile manufacturing, it is very difficult to both cut and extend precast piles. Attempts to do that result with some frequency in damage to the piles. Figure 12.7 shows the over-cutting of precast piles in a pile group; labor-intensive work is required to fix such problems.

Figure 12.8 shows a few types of joints used to extend precast concrete piles. When using precast concrete piles in a soil profile where the depth to the bearing layer varies, the use of

Table 12.3 Approximate maximum handling length for safe handling of square precast reinforced concrete piles

Pile cross section (mm)	Maximum length for safe handling (m)
250	12
300	15
350	18
400	21
450	25

Figure 12.6 Precast concrete piles driven to refusal at site where bearing layer is located at uneven depths.

Figure 12.7 Damage to concrete piles caused during or after installation.

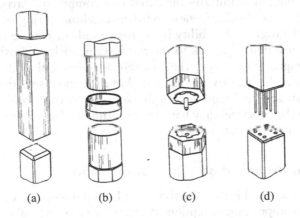

(a)   (b)   (c)   (d)

Figure 12.8 Types of splices used for prestressed concrete piles: (a) sleeve; (b) connector ring; (c) mechanical; and (d) dowel (after PCI 1993; with permission from PCI).

jointed piles (piles with joints built in at the time of pile casting) greatly simplifies piling. Pile sections ranging from 2.5 to 12.5 m are typically available commercially in most locations. The economics of the operations require that sections be supplied to the site in the longest practical length. When driving a pile that is already a composite of two segments, the joints must fit well to avoid energy loss and the creation of tensile stresses during driving. They must also be stronger than the pile and must be properly aligned, so that the pile remains straight and on plumb.

When jointless piles are used, then both cutting and splicing may be necessary. A simple way to cut the excess pile above the ground level is to first expose the rebars by breaking the corners of the pile at the desired cutoff level and then cut the rebars. Pulling on the pile head (or pile top), given the absence of reinforcement and the reduced concrete cross section at the cutoff level, breaks the pile at that location.

Pile splicing is accomplished by breaking away the concrete to expose a length of approximately 40 rebar diameters, and then lapping and welding the rebars along the full 40 diameters. Formwork is then positioned, and the concrete for the pile extension placed. Connection of the pile with the pile cap also requires exposure of the reinforcement. The pile cap is a block of reinforced concrete used to make a pile group work as a whole.

12.2.2.4 Precast, prestressed concrete piles

There are two methods for prestressing the piles: pre-tension (the most common) and post-tension (used in piles for special applications, particularly in the offshore piling industry). In pre-tension, wires are placed in forms and stressed by jacking and the concrete is then placed into the forms and vibrated. After the concrete hardens, the wire ends are cut and their load is transferred to the concrete. As a result, the wires are in tension and the concrete is in compression. Piles manufactured in this manner are available in solid square cross sections with side up to 400 mm.

Prestressed piles manufactured using the post-tension method are occasionally employed in offshore piling projects (Fleming et al. 1992a). They are made from large-diameter cylindrical pipe pile segments. These segments have several holes that extend throughout their length. The segments are juxtaposed, the holes of each segment are carefully aligned, and wires (or cables) are thread through each series of holes. These wires are then tensioned by jacking, and grout is injected into the holes under pressure and allowed to harden. The tension in the wires is then released, at which point the load in each wire is transferred to the pile.

The fact that the concrete is initially subjected to a compressive stress means prestressed piles are less subject to the development of hairline cracking during transporting and handling. This ensures the highest durability in corrosive soils or marine environments, where any crack is a path for potential attack to the reinforcements by corrosive agents. The presence of an initial compressive stress in the concrete also means the pile is not as easily subjected to tensile stresses during driving, but it makes them more sensitive to the compressive waves caused by obstructions found during driving. As another consequence of the initial compressive stress in the concrete, they have lower ultimate compressive load capacities than corresponding reinforced concrete piles.

12.2.3 Classification based on pile loading mode

Piles may be subjected to axial loads, transverse loads, and moments. Axially loaded piles can be subjected to either compression or tension. Piles can be used very effectively to resist vertical and lateral loads, as well as moments. In docking applications, bridges, and offshore platforms, where lateral loads are large, battered (inclined) piles have traditionally been used. Even though it may still make sense to use battered piles in these applications, vertical piles can resist large lateral loads quite effectively, depending on the soil profile, and this lateral resistance can be estimated with sufficient accuracy with current methods. In applications where vertical loads are large and sustained, and lateral loads are not sustained loads, it is not advantageous to use battered piles because they will be permanently subjected to bending moments and shear, while vertical piles would be subjected most of the time to axial loads only.

Moments are usually absorbed by piles groups, although some large-diameter piles do have considerable moment-resisting capacity. The piles in a pile group are unified by means of a pile cap (Figure 12.9), which is a reinforced concrete block that envelops the pile heads, absorbing the moments and transferring almost exclusively transverse or axial loads to the piles.

Figure 12.9 Pile cap formwork in preparation.

If the wind-related moment on a pile group is large and the vertical compressive forces on the piles are small, some piles will be subjected to tensile forces when gusts of wind blow on the superstructure. This would happen, for example, in energy transmission towers or more notably in cellular phone transmission towers, which are light structures that may be acted upon by strong winds. There are other situations where piles may be permanently subjected to tension, such as when supporting a cantilevered structure with a large permanent moment.

12.3 NONDISPLACEMENT PILES

12.3.1 Drilled shafts (bored piles)

12.3.1.1 Basic idea

Drilled shafts are installed by first removing a volume of soil from the ground by drilling and then filling the resulting cylindrical void left in the ground with concrete. Drilled shafts are known in the British Commonwealth as bored piles. In the United States, some refer to them as drilled piers or drilled caissons, but these terms are not as descriptive and may in fact convey an idea that these are large-diameter piles, when the diameters can in fact be as small as 300 mm (1 ft) or so.

12.3.1.2 Equipment

Drilling rigs may be crane-, truck-, or crawler-mounted (Figure 12.10). The type of mounting depends on the site conditions. Truck-mounted rigs are adequate for good, dry terrain, but cannot be used in wet, soft, or loose terrain. Crane-mounted rigs can be used in such conditions. If the terrain is very rugged and cannot be easily leveled, then crawler-mounted rigs, which have a lower center of gravity and much more mobility, are indicated.

Drilling is usually done with augers (Figure 12.11), but buckets (Figure 12.12) are occasionally used, particularly for sands under the water table, when soil would not adhere to the augers. The drilling principle is the same as that used in corkscrews: rotate clockwise and push down to insert the auger into the soil, pull it up, move the auger away from the hole, and then rotate counterclockwise to expel the soil from the auger by centrifugal action (Figure 12.13).

Figure 12.10 Different drilling rig mountings: (a) crane-mounted rig; (b) crawler-mounted rig; (c) truck-mounted rig.

Figure 12.11 Augers used in soil drilling.

Figure 12.12 Soil buckets used in drilling in sands under the water table.

The capacity of the drilling rig is a function of the crowd (the vertical, downward force that can be applied to the auger or bucket) and the torque. Drilling rigs can be mechanical or hydraulic. The distinction is in how the crowd and the torque are applied. Hydraulic rigs are superior in every respect, from capacity to durability.

The steel column to which the auger or bucket is connected is called the kelly. The kelly can be a single piece, or it may be telescoped, with up to three pieces, each fitting inside the next. Telescoped kellies allow depths of up to 45 m to be reached.

Steel casings are used to support caving ground. A pipe called a tremie tube or tremie pipe is used for placing concrete when drilling mud is used. The way in which these are used is discussed later.

12.3.1.3 Procedures

The method of drilled shaft installation depends on the nature of the ground. The simplest approach, known as the dry method, applies to self-supporting soil, such as stiff clays, residual soils of most types, and very lightly cemented soils. For caving soils, either a casing or

Figure 12.13 Removing soil from an auger by centrifugal action.

drilling mud must be used to support the walls of the excavation. The procedures associated with each approach are sometimes referred to as the casing and the wet (or slurry) methods. Different contractors will use different variations of these procedures, so the descriptions of the procedures we provide next are general and not necessarily strictly followed.

12.3.1.3.1 Dry method

In the dry method, the rig is positioned near the location where the drilled shaft is to be installed and the auger is positioned as exactly as possible above the location where the shaft is to be installed. Drilling is started and continued in increments until a cylindrical hole has been created with the desired depth. It is necessary to introduce the auger into the hole and remove it with soil cuttings many times until drilling is finished. As to auger positioning, construction tolerances are typically of the order of 1% of the shaft diameter, and verticality must also be insured by the use of bubble levels strategically placed on the equipment.

Because the soil is self-supporting, there is no need to resort to excavation support techniques, except in cases in which a very thin layer of surface soil requires the use of a short casing (of the order of 1 m long) to avoid cave-ins. After the excavation is complete, concrete is poured into the hole. It is important that the concrete falls without hitting the sides of the excavation, as this might lead to concrete segregation, which would in turn lead to decreased durability and strength. This is easily accomplished by using a centering funnel through which the concrete is poured, as shown in Figure 12.14.

The concrete must be quite workable, with a slump of the order of 200–220 mm. The main reason for this requirement is that concrete with this much fluidity will fully occupy the space available and indeed push against the walls of the excavation, which helps to preserve the lateral effective stress in the soil around the shaft, thus improving shaft resistance. Additionally, there is some slight penetration of cement paste into the soil, which also enhances shaft resistance. Concrete compressive strengths as low as 11 MPa (1600 psi) have been used for drilled shafts, but they are more commonly in the 15–25 MPa range (2200–3600 psi). Larger strengths are required for shafts that are more heavily loaded.

Drilled shafts, when subjected to compressive loads and small lateral loads, are unreinforced, except for a short (about 1–2 m in length) reinforcement cage placed at the head of the shaft. This light reinforcement at the top aids the shaft in resisting small loadings associated with eccentricities that fall within the construction tolerance limits, with construction traffic before the concrete has hardened, and with other minor sources of moments or lateral loads.

Figure 12.14 Funnel for concrete placement in the dry method. (Courtesy of ADSC.)

Drilled shafts can be installed adjacent or secant to each other to form a drilled shaft wall, which serves as an earth-retaining structure (see Chapter 16). In this function, they require either partial-length or full-length reinforcement. Another application in which a drilled shaft requires reinforcement is when the shaft is subject to a large lateral load (as those caused by strong wind or earthquakes). The reinforcement cage is typically made with circular ties (or spiral transverse reinforcement for higher confinement in seismic areas), as shown in Figure 12.15. The reinforcement is placed into the excavated hole using a suitable winch and cable system. It is important to be careful in inserting the reinforcing cage into the hole without hitting against the walls, which would cause soil to cave in. For this reason, the reinforcing cage should not be excessively flexible. To center and properly align the reinforcing cage within the hole, as well as to guarantee minimum cover both at the base and sides, spacers such as those shown in Figure 12.15 are recommended.

12.3.1.3.2 Casing method of drilling

The casing method is used, ideally, when a layer of caving soil (typically, a sandy or gravelly soil under the water table) is underlain by a clayey, relatively impervious soil. In such cases, it is possible to drill at the same time as the casing is advanced into the excavation.

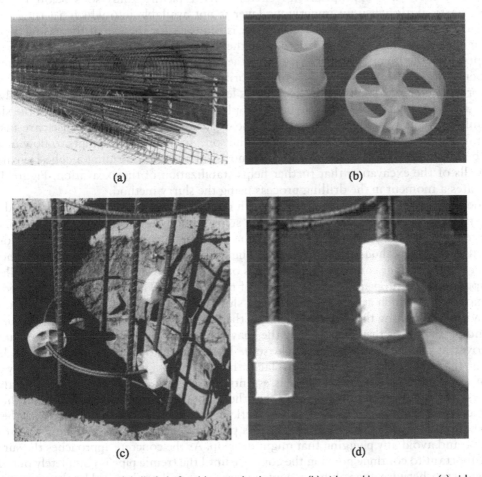

Figure 12.15 Reinforcement of drilled shafts: (a) typical rebar cage; (b) side and base spacers; (c) side spacer in use; (d) base spacer in use. [Parts (c) and (d) courtesy of Foundation Technologies, Inc.]

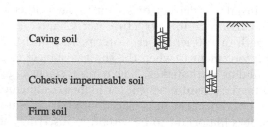

Figure 12.16 The casing method.

The casing can be either pushed, driven, or vibrated into the ground. Slurry may be required during this initial phase of drilling (see the discussion of the slurry method next). Once the casing reaches the impervious layer, it is effectively sealed into it. If slurry was used, then it is at this point pumped or bucketed out of the excavation. From this point on, the drilling can proceed as in the dry method (see Figure 12.16).

12.3.1.3.3 Wet or slurry method

The slurry method is appropriate for caving soils (typically sandy soils below the water table), particularly when an impervious layer is not available into which a casing can be introduced to provide the water sealing that is required for the use of the casing method. In the slurry method, drilling proceeds at the same rate as drilling slurry is pumped or dumped into the hole. The slurry is, in most cases, a mixture of water and approximately 5% bentonite, which is a very plastic montmorillonite clay. The slurry is used for excavation support, which is accomplished by keeping its level in the excavation above the groundwater table. This differential of hydraulic head, amplified by the larger unit weight of the slurry, not only prevents water from flowing into the excavation, which would cause cave-ins, but also pushes the slurry a short distance into the soil. As the slurry attempts to flow into the soil, bentonite particles adhere to the outermost soil particles, forming a cohesive cake on the walls of the excavation that further helps stabilization of the excavation. Figure 12.17 illustrates a moment in the drilling process using the slurry method.

When the final depth is reached, full-length reinforcement is introduced, if required. The next step is the placement of concrete. These steps are shown in Figure 12.18. The concrete is placed with a long pipe called a tremie tube or tremie pipe. This pipe is sealed by one of several available methods at the lower end. The simplest of such methods is using a spherical rubber plug. Another common method relies on a hinged cover which can be controlled by the operator or can open under the weight of the concrete. The tremie pipe is lowered into the excavation, which at this point is full of slurry. Once it reaches the bottom of the hole (but without touching the soil at the base of the hole), it is filled with concrete. The plug at the end of the pipe is then displaced by the heavy concrete column above it (or, alternatively, the cover at the end of the pipe is released). The concrete then exits the pipe and, being heavier than the slurry, completely displaces the slurry from the hole. To prevent any mixing of the concrete with the slurry, it is essential to keep the lower end of the pipe always immersed in the concrete. This requires that the tremie pipe be raised no faster than the rate at which concrete ascends within the excavation. The word "tremie" follows from the fact that the tremie pipe is vibrated as concrete is poured to keep the concrete flowing through the pipe and avoid any plugging that might develop. As the concrete approaches the surface, it is important to continue pouring the concrete until the tremie pipe is completely out of the excavation; otherwise, the volume occupied by the pipe will not be filled by the concrete and the pile head may result lower than the appropriate cutoff level.

Figure 12.17 Drilling using the slurry method.

12.3.1.3.4 Base enlargement

When using the dry or casing methods, it is possible and occasionally desirable to enlarge the base of the drilled shafts. This is most often the case when the base capacity is a significant contributor to the total capacity. The procedure involves lowering a base-enlarging tool, called underreamer, to the base of the shaft when the final depth has been reached. This tool goes into the hole closed (Figure 12.19a); when it reaches the base of the hole, the operator starts rotating it and opening its cutting arms at the same time (Figure 12.19b), thereby forming a conical base. The final shape of the shaft will be as in Figure 12.19c.

12.3.2 Barrette piles

Barrette piles are installed using a clam shell (Figure 12.20), the same tool used for building slurry walls (see Chapter 16). They are essentially bored piles with a rectangular cross section. Their load response is comparable to that of drilled shafts. As with drilled shafts, they may be installed using the dry method if the soil is self-supporting or slurry is used to support the excavation if needed.

12.3.3 Strauss piles

Strauss piles are small-diameter bored piles installed to relatively short depths (less than 10 m or so) using relatively simple equipment, typically a tripod similar to that used for soil boring and standard penetration testing in many countries (Figure 12.21). Diameters start at around 300 mm. A casing is typically advanced with the excavation to support the soil (this is particularly needed in sandy soils).

Figure 12.18 Concrete placement in the slurry method. (a) A rebar cage, if required, is lowered into the excavation, which is filled with drilling mud. (b) The tremie pipe is lowered into the excavation. (c) Concrete is placed through the tremie pipe into the excavation. (d) The pipe is raised at a rate no faster than the rate at which concrete rises inside the excavation to avoid contamination of the concrete with slurry. (e) Enough concrete must be placed to avoid having a deficit of concrete associated with the volume occupied by the tremie pile while it was still inside the excavation.

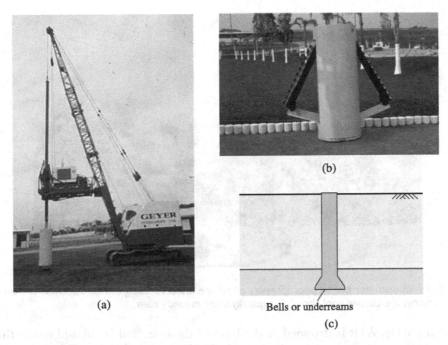

(a)

(b)

Bells or underreams

(c)

Figure 12.19 Underreamer (a) goes into the hole in closed configuration and (b) is opened at the base of the hole to enlarge the base of the shaft, producing, at the end, a drilled shaft with the geometry shown in (c).

In clays, either a cylinder with a cutting edge or a cutting tool with a cross section resembling a cross is typically used to advance the excavation. The soil is retained by adhesion with the surface of the tool. Addition of water is usually necessary to soften the material. In sands, a special shell, shown in Figure 12.21, is used. This shell has a cutting edge and

Figure 12.20 Clam shell.

Figure 12.21 Small-diameter bored piles (Strauss piles) are installed using a tripod such as the one shown here; the cutting tool shown is typically used in sandy soils.

a valve at the base. As it is dropped at the base of the hole, soil is cut and enters the shell. When the shell is raised, the valve closes back, holding the soil inside. To overcome harder soils or obstructions, a heavy I-section chisel is used.

When the desired depth is reached, concrete is placed. It is also possible to reinforce these piles, in which case the reinforcement cage is placed in the hole before concrete placement.

12.4 AUGER PILES

12.4.1 Types of auger piles

The piling industry has made an effort to develop auger piles installed by displacing instead of removing soil. The motivation has been more often than not the avoidance of the need to remove soil cuttings from the site, the disposal of which can be very expensive in countries or localities with stringent regulations. A beneficial side effect is the greater pile capacities that are obtained. We will start with the discussion of auger piles with lowest capacities (CFA piles) and then move on to piles with greater capacities (such as the Atlas and Omega piles).

12.4.2 Continuous flight auger piles (augercast piles)

12.4.2.1 Equipment

Continuous flight auger (CFA) piles are also known as augercast piles. These piles are installed in a way that reminds us of drilled shafts, but there are important differences between the two types of piles. CFA piles are installed using rigs that are usually crawler-mounted for stability. The rigs are usually quite tall so that deep piles can be installed. The maximum length of CFA piles that can be installed with commercially available rigs is of the order of 30 m, so rigs are slightly taller than that (see, for example, Figure 12.22).

A concrete pump is required for the installation of CFA piles. Hollow-stem augers are used; the concrete for the pile is pumped through the hollow space. Modern rigs are usually equipped with sensors that allow monitoring of concrete placement.

Figure 12.22 CFA pile rig.

12.4.2.2 Procedures

CFA piles are installed by first positioning the auger above the desired pile location, establishing auger verticality, and then drilling to the final depth. During drilling, the soil is transported up to the surface along the auger flights. Drilling should proceed steadily and, for best pile capacity, at a high rate. Studies suggest that the bearing capacity of CFA piles installed faster than a threshold rate approaches that of displacement piles, while if they are installed too slowly, their bearing capacity will be more like that of nondisplacement piles.

After the desired depth is attained, concrete is pumped through the hollow stem of the auger; the seal at the lower end of the auger, whose function is to prevent soil from entering the hollow stem, is released by the action of the concrete. As concrete is pumped, the auger is pulled up without rotation or, in the case of sandy soils, with very slow clockwise rotation (rotation in the direction used for drilling). The rate of ascension of the auger must be compatible with the pumping rate of the concrete to ensure continuity of the pile. In modern rigs, the operator has access to a computer that compares the volume of concrete pumped into the excavation with the theoretical volume of the excavation, aiding the operator with setting the appropriate rate of auger ascension. The volume of concrete pumped per unit of time can be estimated by counting pump strokes (a method that is sometimes unreliable because pump strokes do not necessarily imply that a corresponding volume of concrete has been pumped) or by directly measuring concrete flow rate using appropriate sensors placed just before the concrete enters the auger.

Reinforcement, if necessary, is introduced after the concrete has been placed. This is done by either pushing it down into the concrete manually or, if that proves too difficult, by vibrating or driving it down.

12.4.3 Prepakt piles

These piles, more commonly used in the past, are similar in principle to CFA piles, but they are of smaller diameter and length. Drilling is done with a common soil auger, with grout pumped through a pipe going through the hollow stem of the auger at a rate compatible with the rate of ascension of the auger.

12.4.4 Drilled displacement piles

12.4.4.1 Common features of drilled displacement piles

Drilled displacement pile is the term currently favored in North America for piles that are installed with augers specifically designed to displace the soil laterally during pile installation (several types are shown in Figure 12.23). One of the results of installing piles in this manner is that their resistance will be greater than that of both nondisplacement piles and CFA piles, in some cases approaching the resistance of full-displacement piles.

Prezzi and Basu (2005) reviewed the installation of drilled displacement piles in detail. Typically, the drilling tool contains some or all of the following components: (1) a soil displacement body, (2) a helical, partial-flight auger segment, and (3) a specially designed sacrificial tip attached to the bottom of the drilling tool. The displacement body is cylindrical and may contain single or multiple helices. The drilling tool is connected to a casing (or drill stem) of diameter smaller than or equal to the diameter of the pile. At the end of drilling, the sacrificial tip (if used) is released from the displacement body. Concrete or grout is then placed through the casing as the drilling tool and casing are withdrawn from the ground. Reinforcement is inserted either before or after concrete placement. A nearly smooth pile shaft is obtained if the drilling tool is rotated clockwise as it is withdrawn from the ground (for example, in the cases

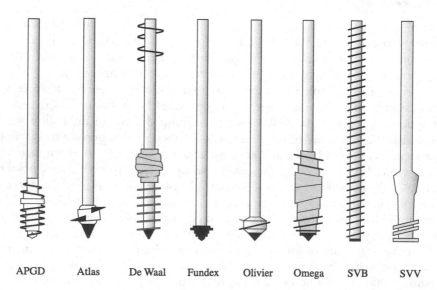

APGD   Atlas   De Waal   Fundex   Olivier   Omega   SVB   SVV

Figure 12.23 Drilling tools for the installation of drilled displacement piles. (https://engineering.purdue. edu/~mprezzi/DrilledDisplacementPiles.html; Courtesy of Monica Prezzi.)

of the APGD and Omega piles). If the displacement body is rotated counterclockwise (in the case of the Atlas pile) during withdrawal, then a screw-shaped shaft is obtained.

We will next discuss briefly the main features and installation of the drilled displacement piles that are most often used.

12.4.4.2 Omega pile

Figure 12.24 shows the steps in the installation of an Omega pile. A displacement auger, which is closed at the bottom and has a sacrificial tip, is used. The displacement auger is connected to a casing. Concrete is injected under pressure into the casing as the desired depth is approached. After reaching the required depth, the sacrificial tip is released and the auger is slowly rotated clockwise and pulled up. The withdrawal of the auger with a clockwise rotation produces a nearly smooth shaft. The reinforcement cage is vibrated into the fresh concrete.

12.4.4.3 Atlas pile

The Atlas pile is a drilled, cast-in-place concrete pile that displaces soil laterally during both drilling and extraction of the auger. The drilling rig has two hydraulic rams that can work independently (one taking over from the other after its full stroke is achieved) to allow a continuous drilling operation. In the case of hard soils, the two hydraulic rams can work simultaneously.

The Atlas pile is installed using a sacrificial tip connected to a displacement body, which is in turn attached to a steel casing (Figure 12.25). The combined action of the torque and crowd forces the casing down into the ground with a continuous, clockwise movement.

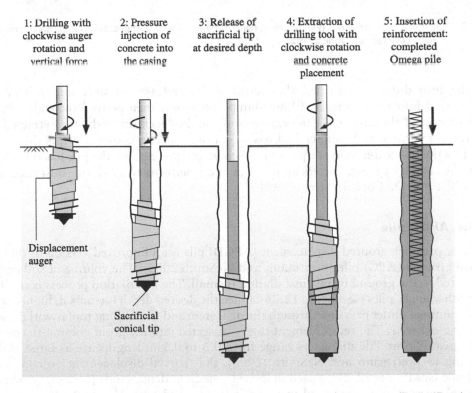

Figure 12.24 Installation of an Omega pile. (https://engineering.purdue.edu/~mprezzi/DrilledDisplacement Piles.html; Courtesy of Monica Prezzi.)

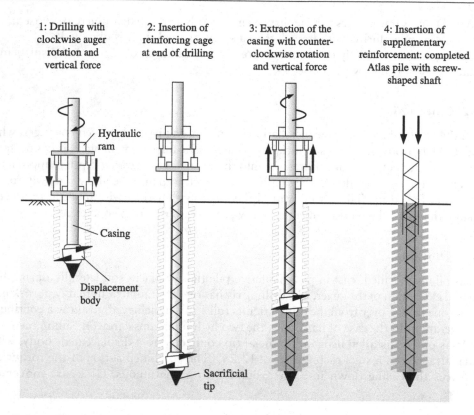

Figure 12.25 Installation of an Atlas pile. (https://engineering.purdue.edu/~mprezzi/DrilledDisplacementPiles. html; courtesy of Monica Prezzi.)

After the desired depth is reached, the sacrificial tip is released. A steel reinforcing cage is then inserted into the casing, and high-slump concrete is then poured through a hopper placed on top of the casing. As the casing and the displacement body are extracted by a vertical pulling force and counterclockwise rotation, concrete completely fills the opening formed by the upward-moving displacement screw. For this reason, the pile has the shape of a screw. Atlas piles have diameters up to 0.8 m and lengths up to 22–25 m (Bustamante and Gianeselli 1998; De Cock and Imbo 1994).

12.4.4.4 APGD pile

The auger pressure grouted displacement (APGD) pile is an improved version of the auger pressure grouted (APG) pile (Brettmann and NeSmith 2005). The volume of soil cuttings transported to the ground by pile installation is small. The installation process is similar to that of other auger piles (see Figure 12.26). Once the desired depth is reached, high-strength grout is pumped under pressure through the drill stem and the drilling tool is withdrawn as it rotates clockwise. The reinforcement cage is inserted into the grout column to complete the pile installation. Pile diameters range from 0.3 to 0.5 m; lengths are as large as 24 m. According to Brettmann and NeSmith (2005), the "partial-displacement" version of the APGD pile (see Figure 12.27) is used in loose to medium-dense sands ($N_{SPT} < 25$), while the "full-displacement" piles are used in loose to dense sand with $N_{SPT} < 50$ (Brettmann and NeSmith 2005).

1: Drilling with clockwise auger rotation and vertical force

2: Pressure injection of grout after reaching the desired depth

3: Extraction of drilling tool with clockwise rotation

4: Insertion of reinforcement: completed APGD pile

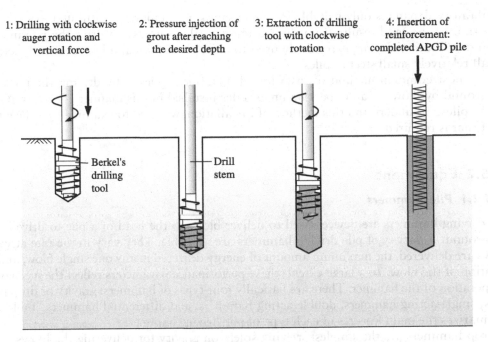

Berkel's drilling tool

Drill stem

Figure 12.26 Installation of an APGD pile. (https://engineering.purdue.edu/~mprezzi/DrilledDisplacement Piles.html; courtesy of Monica Prezzi.)

Figure 12.27 Installation of a partial-displacement APGD pile. (Courtesy of Tracy Brettmann.)

12.5 DISPLACEMENT PILES

12.5.1 Installation methods

There are three methods of installing displacement piles: jacking, vibratory driving, and driving. The first two, jacking and vibratory driving, are comparatively rare. The reaction needed to push a pile into the ground is equal to the limit pile capacity (a term we will define precisely in Chapter 13), which can be a very large load. Until recently, this made jacking suitable only for small piles; large, heavy rigs are now available that can jack normal-sized piles for onshore applications (this is relatively common in Hong Kong, where H-piles are often jacked instead of being driven into the ground).

Vibratory driving is only suitable for loose sands, particularly if saturated, because liquefaction of the sand results from the vibration, making it easy to drive the pile into the ground. Vibratory driving is routinely used to drive sheet piles and less frequently used to install relatively small steel H-piles.

The most common method of installing displacement piles is by driving the piles into the ground by blows of an impact hammer. Piles installed in this manner are referred to as driven piles. To understand this method of installation, we need to examine first the equipment that is required.

12.5.2 Equipment

12.5.2.1 Pile hammers

Pile driving hammers are devices used to deliver blows to the head of a pile to drive it into the ground. A variety of pile driving hammers are available. They vary in the rate at which blows are delivered, the maximum amount of energy delivered in any one single blow, and the duration of the blow. To a large extent, these performance parameters reflect the mechanism of operation of the hammer. There are basically four types of hammers: gravity or drop hammers, single-acting hammers, double-acting hammers, and differential hammers. Table 12.4 summarizes the main features of each type of pile driving hammer.

Drop hammers are the simplest, relying solely on gravity for delivering the blows to the pile head. A drop hammer is usually made of a single block or a system of steel blocks, which may be removed or added as needed. Drop hammer weights are typically in the 10–50 kN range (1–5 tons). Since the weight of the hammer is usually fixed during the driving of any given pile, the only variable available to the operator for adjusting the energy delivered by hammer blows is the drop height. There is an implied danger when driving concrete piles through hard, strong soils (under the so-called hard driving conditions): the operator may drop the hammer from too great a height, generate damaging stresses in the pile, and damage it. We will examine this problem in greater detail in Chapter 14.

In single-acting hammers, the ram is connected to a piston located within a cylinder (Figure 12.28). The piston is lifted by either steam or compressed air (which is called the upstroke) and then allowed to fall by the action of gravity (the downstroke). When it does, the ram impacts the head of the pile, driving it some distance into the ground. The weight of

Table 12.4 General characteristics of different types of pile hammers (Fleming et al. 1992; U.S. Army 1991)

Hammer	Ram weight (kN)	Stroke (m)	Maximum strike rate (bpm)	Conditions under which use indicated	Caution
Drop	10–50 kN (½ to 2 times pile weight)	Wide range	5–10	Noise restrictions	Possible damage during hard driving of concrete piles
Single-acting	20–150 kN	<1.5	40–60		
Double-acting/ differential	0.5–180 kN		90–300	Underwater operations Sheet pile driving	
Diesel	10–150 kN		40–100	All types of piles (with diameter up to 2.2 m) in most soil conditions	Soft clays (where combustion may not occur)

bpm, blows per minute.

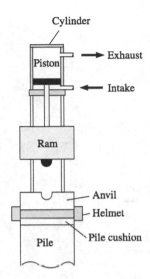

Figure 12.28 Single-acting hammer.

a single-acting hammer (in the 20–150 kN range) is much larger than that of a drop hammer, but the fall height (stroke) is much smaller (up to 1.5 m, typically).

In a double-acting hammer (Figure 12.29), either steam or compressed air is used to increase the pressure on the piston both on its way up and on its way down, so that the ram impacts the pile with a greater force and higher velocity and does so more times per minute. This is possible because these hammers are closed at the top. The stroke is typically less than that for the single-acting hammer, resulting in a higher production rate. The stroke cannot be controlled visually because the hammer is closed. The differential hammer is much like the double-acting hammer, but it relies on the different areas of the upper and lower parts of the piston to generate the repeating up and down strokes.

Hydraulic hammers are moved by oil pressure and can be of the single- or double-acting varieties. Their principle of operation is essentially the same as that of other single- and double-acting hammers. Diesel hammers, such as the one shown in Figure 12.30, also come in both the single- and double-acting varieties, but they differ from other hammers in one important aspect. In single-acting hammers, extra "zip" is added to the blow by combustion

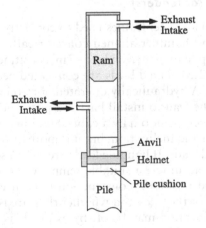

Figure 12.29 Double-acting hammer.

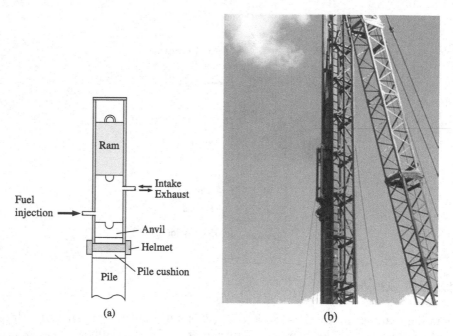

Figure 12.30 Diesel hammer: (a) sketch and (b) photo.

of fuel injected before the downstroke is completed. In a double-acting hammer, a bounce chamber is present in the upper part of the hammer, providing quicker and stronger rebound from the upstroke. These hammers tend to be smaller and lighter than other hammers, as the extra energy and blow duration obtained from the fuel combustion make them very efficient.

The amount of fuel injected into the chamber of diesel hammers can be controlled, allowing adjustment for lighter or harder driving conditions. However, in soils alternating loose/soft layers with extremely hard layers, the bounce of the ram will vary from low to high, which may be damaging to concrete piles. Diesel hammers may be attached directly to the pile head, not strictly requiring the use of leads for their operation.

12.5.2.2 Pile driving leads (or leaders)

Pile driving leads (or leaders) are steel frames used to correctly position the pile for driving and to keep the pile head and hammer aligned concentrically during driving. Leads, with length exceeding that of the pile to be driven by 5–7 m, are attached to a crane in one of the two ways shown in Figure 12.31. Fixed leads are connected near the top with a horizontal hinge at the tip of the boom. A hydraulically operated horizontal brace allows the operator to adjust the inclination of the lead to install battered piles and adjust verticality. Hanging leads are suspended from the crane boom by a cable. Stabbing points at the base of the lead allow adjustments in position and inclination, but it is more difficult to position the pile with hanging leads than with fixed leads. If hanging leads are to be used to drive piles that require a high degree of positioning accuracy, a suitable template should be provided to maintain the leads in a steady or fixed position. Construction tolerances on positioning depend to some extent on the diameter of the piles and whether they are isolated piles or are part of a group. For group piles, pile location may be off by as much as 75 mm, and deviation from vertical as large as in 1 in 25 may be acceptable.

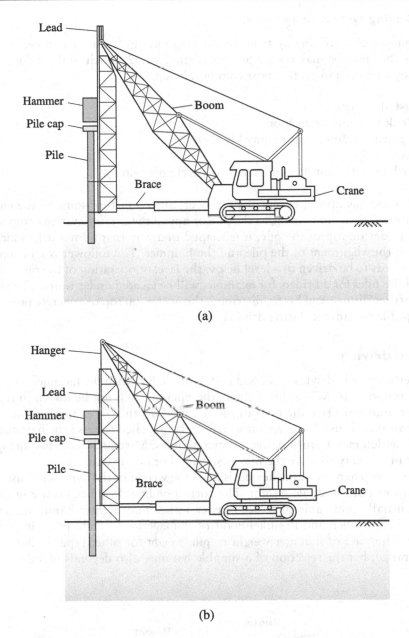

Figure 12.31 Crane-mounted leads: (a) fixed and (b) hanging leads.

Leads that are not properly restrained may cause pile damage, particularly to concrete piles. When driving long, slender piles, the use of intermediate pile supports in the leads may be necessary to prevent pile damage that may be caused by long unbraced pile lengths. Leads are not absolutely necessary for every pile driving operation, but they are normally used to maintain concentric alignment of the pile and hammer, and to obtain the required accuracy of pile position and alignment while driving the pile, especially for battered piles. Even if leads are not used, it is highly advisable to use a template to maintain the pile at the right location throughout driving.

12.5.2.3 Driving system components

The components of a driving system are the hammer itself and a number of additional components that may or may not be present (Figure 12.32). Each of these components are referred to by various names, the most common being:

- anvil (striker plate);
- cap block or hammer cushion;
- driving head (helmet, cap, or anvil block);
- follower; and
- pile cushion (used for driving precast concrete piles only).

All of the elements above, except the follower, aim to diffuse some of the energy from the hammer blow to avoid damage to the pile or any of the driving system components. The driving head goes on top of the pile; it is shaped in a way that allows it to slide along the leads, forcing the alignment of the pile and the hammer. The follower is an extension used when a pile needs to be driven to a level below the level of operation of the rig, such as when the heads of the piles for a bridge, for example, will be located under water. The pile cushion is used to further diffuse and better distribute the energy on top of concrete piles, which are more susceptible to damage during driving.

12.5.3 Pile driving

The key to efficient pile driving is a good match of the pile with the hammer and other driving system components. When this is done, the operator will not be forced to try anything out of the ordinary to drive the piles to the required depth. For example, if an excessively light drop hammer[3] is used, the operator may feel compelled to raise the hammer to excessive heights, which may in turn damage concrete piles. Mismatches of this sort quite often result either in inability to drive the pile as specified or in pile damage.

When using drop hammers, the key decision is the weight of the hammer. It usually ranges from one half to twice the pile weight. The corresponding drop heights are in the 0.2–2 m range. It is usually preferable to select heavier rather than lighter hammers, as the drop heights are then smaller and the likelihood of damage to concrete piles, in particular, is much lower. The ratio of hammer weight to pile weight for other types of hammers lies in the 0.25–1 range, but the selection of a suitable hammer also depends on other factors and

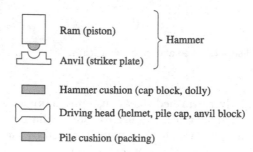

Ram (piston)
Anvil (striker plate) } Hammer

Hammer cushion (cap block, dolly)

Driving head (helmet, pile cap, anvil block)

Pile cushion (packing)

Figure 12.32 Driving system components.

---

[3] As seen earlier, a drop hammer is typically a cylindrical weight that is raised to a certain height using a winch and dropped on top of the pile.

is best done with the aid of computer-based drivability analysis (discussed in Chapter 14). For example, the energy delivered by diesel hammers to piles increases with the driving resistance; in fact, if the driving is too easy, as in the first few meters in soft clays, there may be no ignition at all in the hammer, which would make it very inefficient.

The driving of precast concrete piles is probably the most challenging. Concrete piles may be damaged when driven through soft or loose soils, something which is not possible for either timber or steel piles. This is so because tensile stresses may develop in the pile under these conditions,[4] and concrete is weak in tension. In general, in going through soft/loose soil layers, the operator should use light hammer blows to avoid this. As a general rule, light blows are always used when driving resistance is small.

Immediately after driving, the pile resistance may be either higher or lower than the resistance it will ultimately have. The process by which pile resistance increases with time after driving is referred to as setup or freeze. When pile resistance decreases with time after driving, the process is referred to as relaxation. At least approximate estimation of the rate at which these processes take place is important to plan continuing pile driving around previously driven piles, to plan and perform load tests, and to take account in design of the real, long-term resistance of the pile.

12.5.4 Franki piles (Pressure-injected footings)

Franki piles, as they are more commonly known,[5] are cast-in-place concrete piles that may or may not be reinforced. They are installed with the aid of a steel casing. The casing is positioned vertically, a plug of dry concrete is introduced near the lower end of the casing, and a drop hammer is used to pound on the concrete. Because the concrete is quite dry, its friction with the steel casing is sufficient for the blows to drag the casing down into the ground. This continues until the desired depth is reached. If necessary, more dry concrete is placed along the way, so that there is sufficient concrete within the casing at all times. When the desired depth is reached, concrete with a much wetter mix is introduced and the casing is pulled up as the concrete is driven out of the casing by the hammer blows. As the casing is pulled up, the level of concrete within the casing must be kept above the lower end of the casing at all times to guarantee a continuous pile. The process is illustrated in Figure 12.33.

12.5.5 Raymond piles

Raymond piles are cast-in-place concrete piles installed by the driving of a thin, corrugated, usually tapered shell into the ground with a closed end, followed by concrete placement. These piles were common in the past, but are less used today, given the many new types of piles that have been developed. The shells are available in lengths up to 12 m or so.

12.6 PILING IN ROCK*

12.6.1 Rock sockets

When loads are large, embedding the pile into rock may be better than simply resting its base on top of the rock. The length of pile embedded into the rock is called a rock socket. The length of rock sockets does not exceed roughly two pile diameters. The same modern, powerful drilling rigs used to install drilled shafts in soil can frequently be used to do the drilling in rock.

---

[4] For why this happens, see Chapter 14.
[5] The term originally proposed by the Franki Company, pressure-injected footings, is not often used anymore; it is also not descriptive of the method of installing these piles.

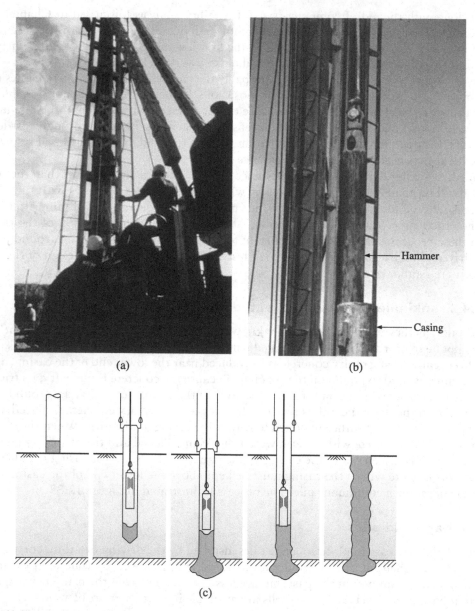

(a)

(b)

Hammer

Casing

(c)

Figure 12.33 Installation of Franki piles: (a) Franki pile rig; (b) hammer pounding on the concrete inside the casing; (c) installation sequence.

12.6.2 Micropiles

Micropiles have become common in the past 15 years or so. These are piles with diameters ranging from 100 mm (4 in.) to 300 mm (12 in.) that are installed in rock by drilling followed by the placement of cement paste, grout, or high-strength concrete in the cylindrical hole created by the drilling.

The drilling is done using a rig such as that shown in Figure 12.34 and a downhole hammer that is attached to the lowest of a string of drilling rods. This downhole hammer (Figure 12.35) has an eccentric, swing-out moving part on which a string of casings rests.

Figure 12.34 Micropile drilling rig.

Figure 12.35 Downhole hammer with eccentric bit moved out of alignment.

As the hole is formed, the casing moves down with the hammer, supporting the excavation. The drilling takes place by repeated, quick blows of the downhole hammer – which has a bit made of tungsten, diamond, or any other very hard material – to the rock, essentially grinding it (the sizes of the resulting particles range from silt to particles as large as a small gravel). The excavation is cleaned by blowing compressed air through the hammer, which forces the flow of rock fragments, particles, and dust up to the surface through the space between the drilling rods and hammer.

Upon completion of drilling, the movable part of the hammer is retracted so that the whole hammer fits within the casing, and the hammer is withdrawn from the excavation. The reinforcing element is now introduced into the hole. This can be a rebar cage, a pipe, a bundle of rebars, and even an I-beam or H-pile. The rebar cage has the distinct advantage that it makes for a more effective composite material with the concrete, grout, or cement paste that is placed into the hole next. This placement is normally done under pressures of up to four atmospheres (roughly 400 kPa) to ensure good bonding with the rock. Figure 12.36 shows completed micropiles.

Micropiles are a convenient way of underpinning or reinforcing existing foundations. Figure 12.37 shows drilling through an existing footing. At the end of installation of the micropiles, a cap is built incorporating the existing footing and the heads of all the micropiles.

Figure 12.36 Completed micropiles.

Figure 12.37 Drilling through an existing footing.

12.7 CHAPTER SUMMARY

Piles are foundation elements that can be used to transfer superstructure loads to depth in a soil profile when shallow foundations are not feasible or economical. Piles can be made of a variety of materials, most commonly of concrete, steel, and timber. If piles are installed by **driving** or **jacking,** they are referred to as **displacement** piles. These piles are installed by pushing the soil mainly radially away from the pile. If piles are installed by pre-removal of soil from the ground, followed by placement of concrete or grout in the resulting excavation, they are known as **nondisplacement** piles.

Displacement piles typically have a higher load capacity than nondisplacement piles, all other things being equal. The main types of displacement piles are **precast and prestressed concrete piles, closed-ended steel pipe piles, H-piles, timber piles,** and **Franki piles.**

Partial-displacement piles are installed by partly replacing soil by the pile and partly pushing the soil laterally away from the pile. They include **open-ended pipe piles, CFA piles,** and **drilled displacement piles.** The main types of drilled displacement piles are **Atlas, Omega,** and **APGD piles.** Some drilled displacement piles are considered to come close to being full-displacement piles.

The main types of nondisplacement piles are **drilled shafts (bored piles).** A nondisplacement pile generates the most spoils (soil cuttings). They also impose the least change to the state of the soil around the pile, which typically means that their unit load capacity tends to be smaller than that of partial- or full-displacement piles.

12.8 WEBSITES OF INTEREST

www.dfi.org
   Deep Foundations Institute
http://www.adsc-iafd.com/
   The International Association of Foundation Drilling

12.9 PROBLEMS

12.9.1 Conceptual problems

Problem 12.1 Define all the terms in bold contained in the chapter summary.

Problem 12.2 Explain in detail and very clearly the installation process of the following piles: precast concrete piles, Franki pile, CFA pile, and drilled shaft (both dry and slurry methods). Indicate whether each would be appropriate for the following sites: sand with high water table, stiff clay, deep soft clay over a very stiff formation, and residual soil of granite.

Problem 12.3 In contaminated sites, it is desirable to minimize the amount of spoils resulting from pile installation, for it is expensive to dispose of these spoils in a lawful way. Comment on which piles would be best and which worst when the aim is to minimize spoils.

Problem 12.4 In certain urban areas, there are strict limits on noise that may rule out the use of certain piles. Under such conditions, comment on which piles would likely be affected by such regulations.

Problem 12.5 Describe in detail the equipment that is needed to drive piles, including the components of the driving system.

Problem 12.6 What is the difference between the driving system for a steel pile and a concrete pile?

REFERENCES

References cited

Brettmann, T. and NeSmith, W. (2005). "Advances in auger pressure grouted piles: design, construction and testing." *Advances in Designing and Testing Deep Foundations, Geotechnical Special Publication No.129*, American Society of Civil Engineers, Reston, Virginia, USA, 1–13.

Bustamante, M. and Gianeselli, L. (1998). "Installation parameters and capacity of screwed piles." *Proceedings of 3rd International Geotechnical Seminar on Deep Foundations on Bored and Auger Piles–BAPIII*, Ghent, Belgium, 95–108.

De Cock, F. and Imbo, R. (1994). "Atlas screw pile: a vibration-free, full displacement, cast-in-place pile." *Transportation Research Record*, 1447, 49–62.

Fleming, W. G. K., Weltman, A. J., Randolph, M. F. and Elson, W. K. (1992). *Piling Engineering*. 2nd Edition, Wiley, New York.

Manning, J. T. and Morley, J. (1981). "Corrosion of steel piles." *Piles and Foundations*, The Institute of Civil Engineers, London, UK, 223–229.

PCI. (1993). "Recommended practice for design, manufacture, and installation of prestressed concrete piling." *PCI Journal*, 38(2), 14–41.

Prezzi, M. and Basu, P. (2005). "Overview of construction and design of auger cast-in-place and drilled displacement piles." *Proceedings of 30th Annual Conference on Deep Foundations*, Chicago, Illinois, USA, 497–512.

Romanoff, M. (1962). "Corrosion of steel pilings in soils." *National Bureau of Standards Monograph 58, U.S. Department of Congress*, Gaithersburg, Maryland, USA.

Romanoff, M. (1969). "Performance of steel pilings in soil." *Proceeding of 25th Conference of National Association of Corrosion Engineers*, Houston, Texas, USA, 14.

U.S. Army. (1991). *Engineering and design-design of pile foundations*. Engineering Manual No. EM 1110-2-2906, U.S Army Corps of Engineers, Washington D.C., USA.

Wong, I. H. and Law, K. H. (1999). "Corrosion of steel H piles in decomposed granite." *Journal of Geotechincal and Geoenvironment Engineering*, 125(6), 529–532.

Additional references

ADSC/DFI. (1989). *Drilled shaft inspector's manual*. The International Association of Foundation Drilling and DFI, Dallas, Texas, USA.

Prakash, S., and Sharma, H. D. (1990). *Pile Foundations in Engineering Practice*. 1st Edition, Wiley, New York.

Turner, J. P. (1992). "Constructability for drilled shafts." *Journal of Construction Engineering and Management*, 118(1), 77–93.

Relevant ASTM standards

ASTM. (2019a). *Standard specification for carbon structural steel*. ASTM A36, American Society for Testing and Materials, West Conshohocken, Pennsylvania, USA.

ASTM. (2019b). *Standard specification for welded and seamless steel pipe piles*. ASTM A252, American Society for Testing and Materials, West Conshohocken, Pennsylvania, USA.

ASTM. (2020). *Standard test method for determining density of construction slurries*. ASTM D4380, American Society for Testing and Materials, West Conshohocken, Pennsylvania, USA.

Chapter 13

Analysis and design of single piles

When utilizing past experience in the design of a new structure, we proceed by analogy, and no conclusion by analogy can be considered valid unless all the vital factors involved in the cases subject to comparison are practically identical. Experience does not tell us anything about the nature of these factors, and many engineers who are proud of their experience do not even suspect the conditions required for the validity of their mental operations. Hence our practical experience can be very misleading, unless it combines with it a fairly accurate conception of the mechanics of the phenomena under consideration.

Karl Terzaghi

Pile foundation design has been an area of geotechnical design in which "a fairly accurate conception of the mechanics of the phenomena under consideration," to use Terzaghi's words, has become clear only recently. Thus, it has been an area in which empiricism has controlled. Piles must be able to sustain axial (typically, vertical) and lateral (typically, horizontal) loads without suffering structural damage, without failing in bearing capacity, and without undergoing excessive settlements or lateral deflections. In this chapter, we examine both the mechanics of pile loading and the analyses that are available for us to check whether piles respond satisfactorily with respect to these three requirements and the specific ways in which these analyses are used in the design of single piles.

13.1 RESPONSE OF SINGLE PILES TO AXIAL LOAD

A pile derives its load-carrying capacity from friction along the pile shaft with the surrounding soil and/or from compressive resistance at the contact of the pile base with the underlying soil. The load Q_{str} at the head of a single pile, which is the load coming from the superstructure, and the weight W_{pile} of the pile are balanced by the mobilized base resistance Q_b and the mobilized shaft resistance Q_s (see Figure 13.1):

$$Q_s + Q_b = W_{pile} + Q_{str} \tag{13.1}$$

In Figure 13.1a, the mobilized base resistance Q_b can be calculated as

$$Q_b = A_b q_b \tag{13.2}$$

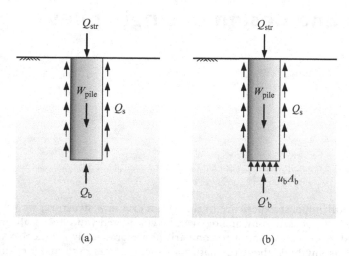

Figure 13.1 Vertical equilibrium of a pile loaded in compression: (a) formulation in terms of total stresses and (b) formulation in terms of effective stresses.

where A_b is the area of the pile base and q_b is the mobilized unit base load, a total stress. In Figure 13.1b, Q_b is broken down into an effective stress times the pile base area plus the pore pressure at the base times the pile base area.

The mobilized shaft resistance is given by

$$Q_s = \sum_{i=1}^{n} q_{si} A_{si} \tag{13.3}$$

where the summation is over all n soil layers crossed by the pile; q_{si} = unit shaft resistance within a soil layer i; A_{si} = pile shaft area interfacing with layer i; and $i = 1, 2, ..., n$ represents any one of these layers. Here there is no difference between total and effective shaft resistances because q_s is a shear stress.

Continuous penetration of a pile into a homogeneous soil mass without the need for any increase in load at the pile head would characterize the classical concept of bearing capacity failure, corresponding to the plunging of the pile. Plunging would of course be a ULS.[1] We will refer to the plunging load as the *limit load* Q_L of the pile. The unit loads at the base and shaft of the pile in this case are referred to as the limit *unit base resistance* q_{bL} and limit *unit shaft resistance* q_{sL}. The classical bearing capacity failure of a pile may be observed in a load test on a pile in a soft clay or loose sand, but is otherwise usually not attainable in routine load tests because of equipment limitations. In most conventional load tests, determination of the limit load would require a reaction capacity of the reaction system that would be cost-prohibitive.

Figure 13.2 shows a conceptual representation of how a pile absorbs an increasing vertical load applied at its head until it no longer can take up any additional load and fails in the plunging mode. In this figure, the load at the pile head (or pile top) is gradually increased from a very small value all the way to the limit load Q_L required for continuing pile penetration into the ground. In Figure 13.2a, axial load is plotted versus depth z. This type of plot shows the load carried by the pile cross section at each depth. The small load Q_1 applied at the head of the pile (shown in Figure 13.2a) is fully absorbed by friction along an upper

---

[1] Limit states are discussed in Chapters 2 and 11.

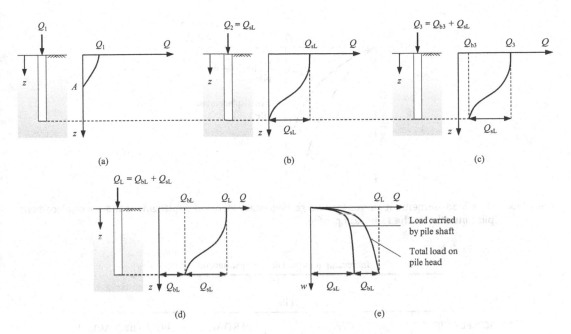

Figure 13.2 Pile axial load transfer: (a) a small axial load is fully taken by shaft resistance; (b) the shaft resistance is fully mobilized at small pile settlement levels; (c) any additional load is taken by base resistance; (d) the base resistance is fully mobilized at large pile settlements; (e) the load response of the pile is expressed through the shaft and total load–settlement curves.

section of the shaft. Point A indicates the depth or pile cross section below which no load is transferred from the pile shaft to the soil. As the load increases, point A moves down until shaft resistance is fully mobilized (Figure 13.2b) and some load starts being transferred to the base of the pile (Figure 13.2c). Shaft resistance is fully mobilized for small pile displacements: locally, at a given cross section, the displacement required to fully mobilize q_{sL} can be as low as a few millimeters. This would correspond to relative displacements at the pile head as low as 0.25% of the pile diameter, depending on pile length and diameter. We will discuss the reason for this later. After full mobilization of shaft resistance, any load incre ment applied at the pile head is taken up by the soil at the pile base. At the limit load Q_L, the pile plunges (Figure 13.2d). The shape of the load transfer curve in Figure 13.2 shows a changing slope near the pile base that reflects the possibility of the complex stress and strain field around the pile base locally reducing q_{sL}. Figure 13.2e shows a typical pile load–settlement (Q–w) curve, for which Q_L would be reached only at very large pile displacements. It also shows a schematic load–settlement curve for the pile shaft only, which reaches its plateau at a small pile settlement. The difference between the curves is equal to the base resistance Q_b.

While shaft capacity is mobilized for very small pile settlements, mobilization of pile base resistance may require very large pile settlements. The rate of mobilization (not necessarily the magnitude) of base resistance per unit pile settlement tends to be higher for clays than for sands, for softer clays than for stiffer clays, for looser sands than for denser sands, and for displacement piles than for nondisplacement piles. The physical processes at play are complex; the extent of strain localization below the pile base is a key factor. The greater the volume of soil over which strain exceeds some threshold value, the more displacement is needed to mobilize the base load.

Figure 13.3 shows the base load–settlement curve for a nondisplacement pile and a displacement pile with the same diameter installed in the same sand to the same depth.

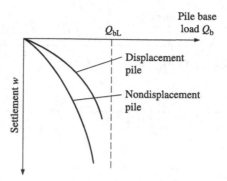

Figure 13.3 Base load–settlement curves for two geometrically identical displacement and nondisplacement piles installed in the same soil profile.

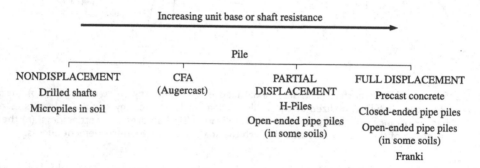

Figure 13.4 Spectrum of soil displacement caused by pile installation and its relationship to bearing capacity.

Generally speaking, both would eventually approach a similar limit base load Q_{bL}, but the rate of progress toward that load is much higher for the displacement pile. The best way to understand this is to realize that the installation of a displacement pile preloads the soil. The installation of a displacement pile, typically by driving and occasionally by jacking, is itself a form of loading that, in nondisplacement piles, will have to be accomplished by static loading of the pile, hence the much greater pile settlements required for mobilization of a given load. Partial-displacement piles (CFA piles, drilled displacement piles, and, in some cases, open-ended pipe piles) have load–settlement curves between the two extremes. Figure 13.4 shows the range from nondisplacement to full-displacement piles and the relationship of a pile type's position in the displacement spectrum to the pile bearing capacity.

13.2 DESIGN OF SINGLE, AXIALLY LOADED PILES

13.2.1 Design process

The process of deep foundation design for axial loads consists of a succession of intermediate goals, typically the following:

1. Selection of piles over other types of foundations.
2. Selection of pile type considering local practice (which types of piles are used locally), local economics (even if not usual, some pile types may still be economical and thus

selected for use), constructability (a pile type that is impossible or difficult to install, given the site conditions, must not be selected), and range of column loads to be supported by the piles (as much as possible, we should select a pile type with structural load capacity consistent with the column loads to minimize the use of pile groups and the additional pile cap costs).

3. Decision on the pile length based on the soil profile. Usually, pile foundations are best designed by first finding a suitable bearing layer for end-bearing piles or by deciding on a practical length for floating piles,[2] and then by determining the cross-sectional dimensions required to develop the necessary load capacity. In the case of floating piles, fixing the cross section and varying pile length is also a possibility. In establishing pile length, we should keep in mind that, according to Randolph (1983), for axial loading, long, slender floating piles are more efficient than short, stubby floating piles on two accounts: load capacity per unit volume installed (and thus per unit cost) and axial stiffness.

4. Determination of the cross section of each pile based on static analysis of pile capacity. We can use dynamic (or wave equation) analysis to verify whether the pile can be driven into a given soil profile, without damaging the pile, to the load capacity predicted by the static analysis. If a driving system can be found that can drive the pile into the ground successfully, then a correlation between the driving resistance and the static resistance (or, alternatively, between the set, which is the pile penetration resulting from one blow, and the static resistance) can also be determined using wave equation analysis. This is discussed in detail in Chapter 14.

5. Selection of a driving system if the piles are driven piles. It is best to use wave equation analysis (discussed in Chapter 14) to make an informed decision about this.

6. Specification of a minimum pile length. We require a minimum length because calculations assume certain embedment of the pile into the ground. A short pile (that is, a pile that cannot be installed all the way down to the design depth), even if embedded into the intended strong layer (which would in this case be at a higher elevation), may not develop the expected shaft capacity, leading to an excessively low safety factor. The pile may also be short because it did not reach the bearing layer, but rather an obstruction or intermediate layer that cannot be relied on to provide the required end-bearing resistance.

7. For driven piles, setting of a minimum driving resistance (in blows per unit penetration depth) below which a pile will not be acceptable. This is typically done based on the results of wave equation analysis (discussed in Chapter 14). For nondisplacement piles, selection of the installation method based on the knowledge of the soil profile and groundwater pattern at the site.

13.2.2 Limit states

The limit states of Chapter 11 can be adapted for piles as follows:

(IA): bearing capacity failure of a single pile, which may correspond to either
    (IA-1): classical bearing capacity failure (plunging) of the pile, or
    (IA-2): crushing or yield of the pile cross section upon loading (more of a possibility for cast-*in situ* concrete piles installed in soil with large shear strength and less so for precast concrete or steel piles);
(IB): collapse or severe damage to the superstructure due to foundation movement;
(II): loss of functionality or serviceability of the superstructure due to foundation movement;

---

[2] An ideal floating pile is a pile deriving all of its resistance from shaft resistance, but a pile with small nonzero base resistance is also referred to as a floating pile.

(III): overall stability failure, consisting of the development of a failure mechanism enveloping the pile foundations or a part thereof (as illustrated in Figures 11.2–11.4).

In foundation design problems, limit state (LS) IB or II controls design in most cases. This is so because, by the time a pile has actually failed as in LS IA-1, either the serviceability or stability of the supported structure in all likelihood will have already been jeopardized (see Figure 13.5). It would require a comparatively small limit (plunging) load to have plunging occur before serviceability or other ultimate limit states. This is possible, as an example, in soft clays. As to LS IA-2, it may be the controlling limit state for small-diameter (20–50 cm), relatively long bored piles (drilled shafts), made of concrete with characteristic strengths of the order of 10–15 MPa, installed in strong, stiff soil. Otherwise, failure of well-constructed piles following LS IA-2 is rare. It is of course still possible in poorly constructed piles.

LS II can be achieved not only as a result of differential settlements, but also because of total settlements, when these cause problems at the interface of the building with its surroundings (such as impairment of access, drainage problems, or shearing of utility lines). There are cases, however, of facilities that undergo very large total settlements and remain serviceable; Mexico City and some European cities provide good examples of this. If the costs and inconveniences associated with repair and maintenance are seen as preferable to more expensive foundations, total settlements without a strong distortion component may be considered not to represent loss of serviceability. There are modern facilities, such as the Kansai International Airport in Japan, built on reclaimed land founded on very compressible soils, where jacks between the structure and the foundations are periodically adjusted to account for ongoing settlements. So, clearly, there is some flexibility both in defining serviceability limit states (SLSs) and in devising strategies to deal with them.

13.2.3 Design ultimate limit state

Figure 13.6a shows the balance of forces in terms of total stresses, and Figure 13.6b breaks down the base resistance into an effective base resistance $q'_{b,ult} A_b$ and a water force $u_b A_b$ at the base. The equilibrium between the applied load, the pile weight, and the base and shaft resistances is of greatest interest at the design ULS. If we rewrite Equation (13.1) for the ULS for a fully or partially submerged pile and subtract the $u_b A_b$ term from both sides of the equation for sand only, we obtain the following expressions for the ultimate load Q_{ult}:

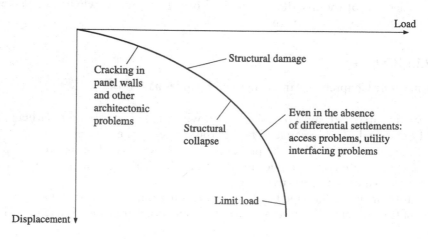

Figure 13.5 Load–settlement curve showing that SLSs very often precede ULSs, including plunging of the pile (which corresponds to the classical ULS of bearing capacity failure).

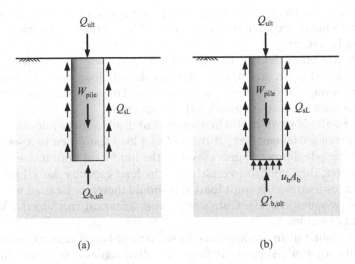

Figure 13.6 Vertical equilibrium of a pile loaded in compression at the ULS: (a) formulation in terms of total stresses and (b) formulation in terms of effective stresses.

$$Q_{ult} = \begin{cases} Q_{sL} + Q'_{b,ult} - W'_{pile} \text{ for sand} \\ Q_{sL} + Q_{b,ult} - W_{pile} \text{ for clay} \end{cases} \tag{13.4}$$

where Q_{sL} is the limit shaft resistance, $Q_{b,ult}$ and $Q'_{b,ult}$ are the total and effective ultimate base resistances, W_{pile} is the weight of the pile, and $W'_{pile} = W_{pile} - u_b A_b$ is the buoyant weight of the pile. W'_{pile}, as defined, correctly accounts for full, as well as partial submergence of the pile. For piles in clay, we work directly in terms of total stresses.

The design ULS is the one chosen or determined by the designer as most critical. For example, piles in soft clay tend to plunge at relatively low pile settlements, and plunging is often the ULS underlying design methods. In other cases, the critical ULS will be associated with a settlement of the pile required to cause structural distress. We expand this discussion next.

13.3 ULTIMATE LOAD

13.3.1 Ultimate load: What is it?

As discussed earlier, the limit load or limit resistance is associated with the plunging of the pile through the soil. It is often unattainable, particularly for piles bearing in dense sand. Load test reaction systems will often not have the capacity to take a pile all the way to plunging, so a limit load often cannot be determined in the field. A limit load is clearly an ultimate load in that it will lead to a ULS, for any portion of the structure whose stability depends on the bearing capacity of the pile will be compromised. Ultimate load and ultimate resistance are often used as synonyms.

Given the difficulty in determining the limit load and given also the fact that excessive settlements of the pile before plunging is reached can also lead to ULSs, the concept of an ultimate load or ultimate resistance is more general than that of the limit load. An ultimate load is associated with a ULS. Plunging of the pile (in which case the ultimate load is equal to the limit load) is not the only possible ULS. Superstructure damage due to excessive differential settlement is frequently the basis for ultimate load criteria. However, there are also many ultimate load criteria that have been established rather arbitrarily in the geotechnical engineering literature.

Ultimate loads are often defined based on load–settlement curves. Many ultimate load criteria were developed in an attempt to interpret the pile load capacity from pile load test results. Some are extrapolation criteria, some are based on a set value of relative or absolute settlement, and some are based on graphical constructions.

Extrapolation methods work best when applied to displacement piles because the load on these piles for a given settlement (say, at the end of a load test taken to a settlement equal to some fraction of the pile diameter) gets closer to the limit load than it does for nondisplacement piles. However, they tend to overestimate the load capacity for all pile types because they are aiming at estimating the limit load and should therefore be used with some caution. We will examine two such criteria: Chin's hyperbolic criterion and Van der Veen's exponential extrapolation criterion.

It is rational to define ultimate loads using settlement-based criteria because, in the vast majority of situations, a pile-supported structure often achieves serviceability and ultimate limit states long before limit loads are reached on the piles. The most widely used settlement-based criterion defines the ultimate load based on a settlement equal to 10% of the pile diameter. Davisson's criterion – still commonly used in US practice although sometimes overly conservative – is also a settlement-based criterion.

The criteria based on graphical constructions explore changes in the curvature of the load–settlement curve plotted in a variety of ways. They are generally arbitrary and bear little relationship to the pile-loading process and its effect on the superstructure. We will discuss, as an example of graphical criteria, the criterion of De Beer (1968).

Except for the extrapolation criteria, which should in theory lead to the limit load (which conceptually should be about the same for geometrically identical nondisplacement and displacement piles installed in the same soil profile), all other ultimate load criteria will give an ultimate load that is greater for displacement piles than for nondisplacement piles. This is due to the stiffer response of displacement piles due to the significant amount of soil preloading during their installation. Design methods should reflect this fact.

13.3.2 Ultimate load criteria

13.3.2.1 Chin's criterion

The method of Chin (1970) is based on the assumption that the pile load–settlement (Q–w) curve is hyperbolic:

$$Q = \frac{w}{C_1 w + C_2} \tag{13.5}$$

where C_1 and C_2 are constants with very specific physical meanings, as we will see.

If we rewrite Equation (13.5) as

$$Q = \frac{w}{C_1 w + C_2} = \frac{1}{C_1 + \dfrac{C_2}{w}}$$

we see that, as the settlement w approaches infinity, Q approaches $1/C_1$. But the load at infinite settlement is, by definition, the limit load Q_L, so it follows that:

$$Q_L = \frac{1}{C_1}$$ (13.6)

If we rewrite Equation (13.5) yet again as

$$K_t = \frac{Q}{w} = \frac{1}{C_1 w + C_2}$$ (13.7)

we obtain an equation for the pile head stiffness K_t. Making $w = 0$, we see that C_2 is the inverse of the initial pile head stiffness:

$$K_t\big|_{w=0} = \frac{1}{C_2}$$ (13.8)

Usually, we are interested in determining Q_L; occasionally, we are also interested in determining the initial pile head stiffness. In either case, it is useful to use Equations (13.6) and (13.8) to rewrite Equation (13.5) as

$$\frac{w}{Q} = \frac{1}{Q_L} w + \frac{1}{K_t\big|_{w=0}}$$ (13.9)

Equation (13.9) has the implication that the limit load Q_L is the inverse of the slope of the straight line fit through load test results in w/Q–w space (see Figure 13.7). The intercept of this line for $w = 0$ is the initial pile head stiffness.

13.3.2.2 Van der Veen's criterion

Van der Veen's criterion (Van der Veen 1953) is based on the assumption that the load approaches asymptotically the limit load, following:

$$Q = Q_L \left(1 - e^{-\frac{w}{\beta}} \right)$$ (13.10)

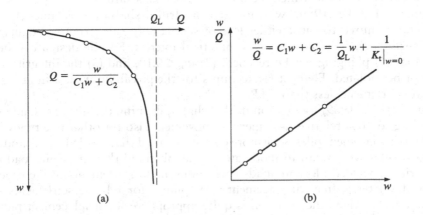

Figure 13.7 Chin's hyperbolic criterion: (a) load–settlement curve and (b) replotted results for the determination of C_1 and Q_L.

where β is a positively valued fitting parameter.

Although it is now used only sparingly, the Van der Veen criterion was used to define the ultimate load in several legacy pile design methods developed from direct comparison of pile load test results with *in situ* test results.

13.3.2.3 Ultimate load based on 10% relative settlement

One of the best measures of structural distortion at the foundation level is the angular distortion, as discussed in Chapter 2. Maximum tolerable values for the angular distortion can be arrived at for LS IB and LS II. In connection with piles, there is a fundamental reason for linking an ultimate load to relative settlement that can be best understood by considering the case of frame structures with columns supported on single drilled shafts. Larger-diameter piles are used to carry heavier columns loads. Such columns have larger cross-sectional areas and are often associated with larger spans. It can therefore be assumed that the larger the pile diameter is, the larger the distance between adjacent piles is and, for a fixed limiting angular distortion, the larger the tolerable differential settlements will be. Because excessive differential settlements are proportional to total settlements and are thus usually prevented through containment of total settlements, it follows that the total allowable settlement of a pile increases with pile diameter. A practical way to prevent attainment of limit states associated with excessive settlements is then to establish a limiting value of the relative settlement w_R that is not to be exceeded.

The relative settlement w_R is defined as the ratio of the pile head settlement w to the pile diameter B:

$$w_R = \frac{w}{B} \tag{13.11}$$

The value of w_R for LS IA can be very large. In most cases, the structure either will have collapsed or will no longer be of service once a certain level of settlement is achieved, with loss of serviceability typically preceding structural collapse. The proportionality between tolerable settlement and pile diameter forms the basis for a relative settlement-based criterion. A practical way to define a certain limit state for piles is then to establish the value for the relative settlement w_R at which the limit state will be attained.

According to Franke (1989), who focused on drilled shafts (bored piles),[3] the relative displacements required to cause either loss of serviceability or structural damage will be greater than $w_R = 0.10$ in most cases of practical interest. So pile design can be done by requiring that (1) plunging not be reached, (2) $w_R \leq 0.10$, and (3) the integrity of the pile material not be violated. Element (1) accounts for the possibility that a limit load may be achieved for settlements less than $0.1B$.

Although Franke (1989) focused on drilled shafts, defining the ultimate load as the load corresponding to 10% relative settlement is convenient also for other pile types. Up to the 1970s, when displacement piles were more commonly used than drilled shafts and the diameters of most piles were small to moderate, it was observed that most pile load test interpretation criteria yielded ultimate loads corresponding to settlements of the order of 10% of the pile diameter. Often, displacement piles plunge for relative settlements not much greater than 10%. Thus, the criterion is quite appropriate for displacement piles as well.

---

[3] The different types of piles are described in detail in Chapter 12.

This criterion was adopted in the British Code and is today the standard way of defining ultimate load internationally.

As pointed out by Salgado et al. (2011a, 2013) and Basu and Salgado (2014), the choice of 10% relative settlement as the ultimate pile capacity in routine design is logically correct and widely accepted (BSI 1986; Fleming et al. 2009; Franke 1989, 1993; Ghionna et al. 1994; Mayne and Harris 1993; Briaud et al. 2000; Jardine et al. 2005; Kim et al. 2009; Lee and Salgado 1999; McCabe and Lehane 2006; Paik and Salgado 2003; Paik et al. 2003; Randolph 2003; de Sanctis and Mandolini 2006; Seo et al. 2009; Skempton 1959; Xu et al. 2008; Tomlinson and Woodward 2008). There are certain cases in which the 10% settlement criterion may not control design. Examples include drilled shafts in strong soil or rock and extremely large-diameter piles. Another notable example is piles in soft clays, which respond as "floating piles" and may plunge for relative settlements substantially less than 10%. In these cases, plunging may control design.

13.3.2.4 Davisson's criterion

The Davisson ultimate load criterion (Davisson 1972, 1975) for the ultimate load of a pile may be expressed in the following form:

$$w_{ult} = 0.004 L_R + \frac{B}{120} + \frac{Q_{ult} L}{A_p E_p} \tag{13.12}$$

where:

Q_{ult} = ultimate load;
L_R = reference length = 1 m = 3.281 ft;
B = pile diameter (or width) in the same unit as L_R;
w_{ult} = settlement (in the same unit as L_R) observed for the pile when $Q = Q_{ult}$;
A_p = cross-sectional area of the pile (in unit of L_R^2);
E_p = pile Young's modulus (in units consistent with those of load and length); and
L = pile length in the same unit as L_R.

The Davisson ultimate load may be visualized graphically in Figure 13.8. First, we plot a straight line parallel to the elastic compression line (the dashed line in the figure) for a free-standing column with the same geometry as the pile. Note that the dashed line in

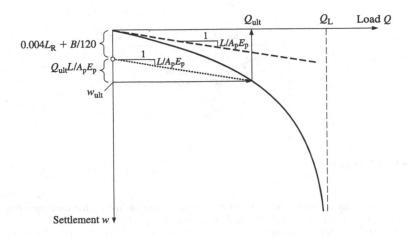

Figure 13.8 Graphical representation of Davisson's criterion.

Figure 13.8 does not provide the elastic compression of the pile under a given load, but that of a hypothetical free-standing column geometrically identical to the pile subjected to the same load as the pile. The dashed line obviously passes through the origin because the column's elastic compression is zero at $Q = 0$. The dotted line plotted in the graph is parallel to the dashed line, with an intercept equal to $0.004L_R + B/120$. The ultimate load corresponds to the point at which this line intersects the load–settlement curve.[4]

Davisson's criterion was proposed for driven piles with diameters of the order of 1 ft (12 in. or ~300 mm) (Davisson 1975). In fact, originally, the second term in Equation (13.12) was simply made equal to 0.1 in., the value that was often assumed in the pile foundations literature of the time to approximate the settlement required to mobilize most of the pile base resistance, a settlement known as the "quake."[5] Note that 0.1 in. is equal to 120th of a foot, which is what we obtain from the second term of Equation (13.12) if we make $B = 12$ in.

The Davisson criterion suffers from a number of shortcomings. It is often too conservative and is based on empirical considerations and the very approximate notion of a perfectly plastic soil response and the related soil "quake." It was proposed for very specific conditions (driven piles with relatively small diameters). Despite these limitations, the criterion is still used in the United States. Recognizing the conservatism of the criterion, some engineers have engaged in the practice of correcting for this conservatism by increasing the first term of Equation (13.12). Instead of continuing to modify a criterion that was proposed for specific conditions and is approximate in nature, it may be more appropriate to use the 10% relative settlement criterion instead, unless one of the situations discussed earlier applies. In those cases, the Davisson method would not be applicable either.

13.3.2.5 De Beer's criterion

The criterion proposed by De Beer (1968) requires that pile load test data be plotted on log Q–log w space. According to De Beer, a point is identified where the curvature is maximum (Figure 13.9). The ultimate load is the load corresponding to this point.

13.3.2.6 Which criterion to use?

The explicit use of an ultimate load criterion is only necessary when interpreting pile load tests, something that is discussed in a later chapter. The other way in which an ultimate load criterion "appears" in the design process is in the choice of a method of pile resistance

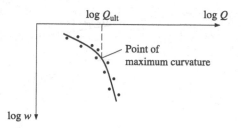

Figure 13.9 De Beer's criterion.

---

[4] Readers may recognize this approach based on an offset from an elastic line as that used sometimes in strength of materials to define the yield stress of steel based on tension tests.

[5] For more on the "quake," refer to Chapter 14.

estimation. These methods are developed or should be developed based on a definition of ultimate load or ultimate resistance, so they have or should have a definition of ultimate resistance embedded in them. In this case, the developer of the method will have used a definition of ultimate resistance in the development of the method's equations. Thus, when a user calculates an ultimate resistance using the method, there is an implicit reliance on this definition of ultimate load, but that is all in the background. When using a given pile design method, it is necessary to know on which ultimate load criterion it is based. A pile design method should not be used if there is no documentation on how ultimate loads were defined by the proponents of the method.

The most appropriate ultimate resistance criterion is the one based on pile head settlement equal to 10% of the pile diameter. The criterion applies equally to displacement, partial-displacement, and nondisplacement piles and is firmly associated with the concepts of serviceability and ultimate limit states. Although its adoption leads to acceptable designs in most cases, it is possible that a stricter limit on settlement will be needed for certain projects with special requirements. If this is found to be the case, more detailed, specific settlement analyses are required. We discuss these analyses toward the end of the chapter. The criterion will also not apply for floating piles, which tend to plunge for settlements less than $0.1B$, or for very large-diameter piles, when it could lead to excessively large absolute settlements.

As to other possible criteria, the Davisson criterion was developed for very specific conditions and its application to piles in general produces ultimate resistances that are lower than those deriving from the 10% relative settlement criterion. As far as extrapolation criteria are concerned, it can be useful to have an estimate of the limit load from an extrapolation criterion, but that requires sufficient development of the load–settlement curve for the extrapolation to be accurate. The availability of extrapolation criteria notwithstanding, the use of the limit load as an ultimate load is fraught with risks, as illustrated in Figure 13.5. All the limit states we wish to prevent through sound design of our foundations usually happen for settlements smaller, and in many cases much smaller, than the settlement required to reach the limit load. An exception is piles in soft clay, for which the limit load is often the ultimate load because it is reached for relatively small settlements.

Example 13.1

Determine the ultimate load of a closed-ended pipe pile using the static load test data in E-Table 13.1. Use the Chin, Davisson, and 10% relative settlement criteria. The outside diameter of the pile is 355 mm (14 in.), and the total length is 8.24 m (27 ft). The steel

E-Table 13.1 Static load test data for Example 13.1

Axial load (kN)	Settlement (cm)
0	0
268	0.091
401	0.161
535	0.264
669	0.372
803	0.494
936	0.651
1070	0.891
1204	1.405
1338	2.800
1472	6.735

cross-sectional area is $136.8\,\text{cm}^2$ ($21.2\,\text{in}^2$). Use $200\,\text{GPa}$ ($29,000$ ksi) for the Young's modulus of the steel pipe pile.

Solution:

Chin's method makes use of the assumption that a pile settlement curve is hyperbolic and can be represented, in w/Q versus w space, by a straight line given by Equation (13.9):

$$\frac{w}{Q} = \frac{1}{Q_\text{L}}w + \frac{1}{K_\text{t}\big|_{w=0}}$$

E-Figure 13.1a shows a least-squares regression on the data in E-Table 13.1 that gives Equation (13.9) as

$$\frac{w}{Q} = 6.357 \times 10^{-4}\frac{1}{\text{kN}} \times w + 2.958 \times 10^{-4}\frac{\text{cm}}{\text{kN}}$$

from which the limit load is found to be:

$$Q_\text{L} = \frac{1}{C_1} = \frac{1}{6.357 \times 10^{-4}}\,\text{kN} = 1573 \approx 1570\,\text{kN}$$

The Davisson criterion for ultimate load is represented in the form of Equation (13.12):

$$w_\text{ult} = 0.004 L_\text{R} + \frac{B}{120} + \frac{Q_\text{ult}L}{A_\text{p}E_\text{p}}$$

The ultimate load is that corresponding to the settlement w_ult calculated using this equation; that is, one reads from the load–settlement curve obtained from the load test the load corresponding to w_ult. E-Figure 13.1b shows the plots of the load versus settlement calculated using Equation (13.12) and load versus observed settlement. The ultimate load is found by selecting the intersection of the two lines and is observed to be 1100 kN for this pile.

The 10% relative settlement criterion states that the ultimate load is reached when the pile head settlement reaches 10% of the outside diameter of the pile. For this pile, the outside diameter is 355 mm and 10% of that is 35.5 mm, or about 3.5 cm. E-Figure 13.1c shows the load–settlement curve for this pile with the 3.5 cm limit plotted. From this figure, the ultimate load based on the 10% relative settlement criterion is found to be 1360 kN.

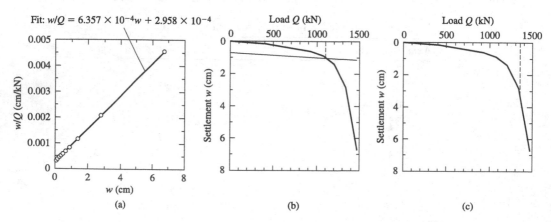

E-Figure 13.1 Interpretation of the load–settlement response from a pile load test: (a) using Chin's method; (b) using Davisson's criterion; and (c) using a relative settlement equal to 10% as a criterion.

So our estimated ultimate loads are:

1100, 1360, and 1570 kN ^{answers}

for the Davisson, 10% relative settlement, and Chin methods, respectively.

13.4 CALCULATION OF PILE RESISTANCE

13.4.1 General framework

The ultimate load Q_{ult} of a single pile is the load, net of the pile weight, that the pile is able to sustain at the ULS. It may be expressed in terms of the sum of the ultimate base resistance $Q_{b,ult}$ and the limit shaft resistance Q_{sL} by referring back to Equation (13.4) – and Equations (13.2) and (13.3) – and written for the ULS:

$$Q_{ult} + W_{pile} = Q_{sL} + Q_{b,ult} = q_{b,ult} A_b + \sum_{i=1}^{n} q_{sLi} A_{si} \qquad (13.13)$$

where Q_{ult} = ultimate resistance of the pile net of the pile weight; $q_{b,ult}$ = ultimate unit base resistance; q_{sLi} = limit unit shaft resistance along the interface of the pile with soil layer i; A_b = area of pile base; A_{si} = pile shaft area interfacing with layer i; and n = number of soil layers crossed by the pile. In most pile design scenarios, the pile weight and the pore water force at the pile base become negligible next to the pile resistances and applied load. Thus, generally, $q_{b,ult}$ is approximately equal to $q'_{b,ult}$, and we will use $q_{b,ult}$ and $Q_{b,ult}$ to refer to effective resistances hereafter, and will neglect the pile weight when designing piles in sand. However, for long, larger-diameter piles, particularly if the piles are in soft clay, explicit consideration of the pile weight in calculations is necessary.

The key assumption behind Equation (13.13) is that the limit shaft resistance has already been fully mobilized by the time the ultimate base load starts mobilizing. This is consistent with our previous discussion of the very small settlements required for shaft resistance mobilization. The reason small settlements are sufficient for full mobilization of shaft resistance is the development of critical state or, for clay, a residual strength state next to the pile shaft. As discussed extensively in earlier chapters, shear strains in most soils tend to localize in narrow bands called shear bands. In the case of piles, a shear band of the order of five to ten times D_{50} for sands and a few millimeters for clays forms next to the pile shaft upon loading. This means that a settlement equal to the shear band thickness – that is, a settlement as small as a few millimeters – is sufficient to cause shear strains of the order of 100% in the soil next to the pile shaft. Consequently, the critical-state shear strength in sand or residual shear strength in clay is mobilized next to the shaft, and the unit shaft resistance of the pile stays at these values with any further pile settlement.

The ultimate unit base resistance $q_{b,ult}$ is determined according to the criterion used as a basis for developing a specific design method. For example, for the 10% relative settlement criterion, the ultimate unit base load corresponds to that for which the pile settlement is 10% of the pile diameter. When we calculate an ultimate resistance using a given design method, we calculate a resistance corresponding to the ultimate resistance criterion based on which the method was developed.

There are two general types of methods to design piles for axial loads: those based on soil state variables and those based on *in situ* tests. There are also some mixed methods, with both soil state variables and *in situ* test measurements appearing in their relationships. Soil property-based methods arrive at estimates of $q_{b,ult}$ and q_{sL} based on the values of fundamental soil state variables for the soil layers crossed by the pile. For example, these are usually the relative density and initial stress state in the case of sands and the undrained shear strength and the plasticity index in the case of clays. The *in situ* test-based methods correlate $q_{b,ult}$ and q_{sL} directly with either CPT cone resistance q_c or SPT blow counts N_{SPT}. The correlation of $q_{b,ult}$ with q_c is particularly natural for a pile with a solid cross section, as q_{bL} for such a pile (the unit base resistance when the pile is plunging through the soil) is approximately equal to the cone resistance q_c under the same soil conditions.

13.4.2 Factor of safety and allowable load

The design of axially loaded piles requires first the calculation of an ultimate resistance, defined as in Equation (13.13). For that, we calculate the ultimate base resistance and the limit shaft resistance separately using a suitable design method. In the working stress design (WSD) framework, we add them together and divide them by a factor of safety *FS*. The resulting reduced resistance must exceed the load coming from the superstructure added to the pile weight. In the load and resistance factor design (LRFD) framework, we consider the ultimate base resistance separately from the individual shaft resistance contributions from layers constituting the soil profile crossed by the pile, each multiplied by suitable resistance factors. The resulting reduced resistance must exceed a magnified applied load. The pile weight may, in such calculations, be magnified together with the superstructure load as a dead load.

The allowable load of a pile is the load that can be transmitted to the pile from the superstructure without leading to either serviceability or ultimate limit states. This load must be arrived at by taking the uncertainties in measurements and calculations fully into account. In working stress design, the allowable load is simply equal to the ultimate load divided by a suitable factor of safety. In the majority of cases, we can safely ignore the pile weight.

In WSD, so long as the structural load applied on the pile is less than or at most equal to the allowable load, the design of the pile is a safe design. Factors of safety for pile design tend to be slightly lower than for shallow foundation design, being typically in the 2–3 range.

The LRFD of piles relies on the design inequality

$$(RF)_s Q_{sL} + (RF)_b Q_{b,ult} \geq \sum_i (LF)_i Q_i \tag{13.14}$$

where Q_{sL} and $Q_{b,ult}$ are the limit shaft and ultimate base resistances; $(RF)_s$ and $(RF)_b$ are the shaft and base resistance factors; $(LF)_i$ is the load factor for load Q_i; and Q_i is any of the dead or live loads that must be transferred to the ground, including the pile weight. The resistance factors depend on the load factors that are used, on the type of pile, and on the design method. The literature on this subject is still evolving, and at least several years will be required until the profession settles on suitable values of resistance factors for piles.

13.4.3 "Floating piles" and "end-bearing piles"

Some engineers design piles by fixing the pile cross section and calculating the length required to reach the target capacity, but that is usually not the best approach to design piles. It is usually more economical to embed a pile into a strong bearing layer if one is available at a practical depth. Doing so also practically eliminates any risk that future construction on adjacent property might in some way undermine the foundations. Additionally, placing the tips of the piles in the same layer reduces the possibility of development of excessive differential settlements.

There are many examples, the most recent being the Millennium Tower in San Francisco, CA, that illustrate how not following the sound principle of placing piles on an intermediate layer proven by sound analysis to be a capable bearing layer for the structure's foundations can lead to significant problems. The Millennium Tower was founded on a piled mat, with piles bearing on an intermediate sand layer (the "Colma Sand"). The Colma sand, which is underlain by an overconsolidated clay layer (the "Old Bay Clay") and then bedrock, has often been used to support buildings in San Francisco in the past. The Millennium Tower, however, is taller and heavier than the typical San Francisco building. Stresses induced by its foundations on Old Bay Clay pushed the clay into its normal consolidation range. The Tower settled approximately 0.5 m, with a significant lean (Engineering News Record 2016). As is expected in these situations, a dispute developed about causation, with the suggestion made that construction activities, including dewatering, in its vicinity may have aggravated the situation. Local practice after this design failure appears to have shifted away from placing piles on intermediate layers and taking them to bedrock. This may not be necessary for lighter buildings, but the lesson offered by the Millennium Tower case is the same as that offered by many other cases before it: intermediate layers must not be used for supporting pile foundations unless very careful analysis has demonstrated that settlement – total or differential – will not be excessive. This case also illustrates the risk inherent to operating outside the range of recorded experience.

When designing a pile, if we wish to derive the full benefit of a bearing layer for the base resistance of the pile, it is not sufficient to place the pile base at the top of this layer. It is good practice to have it embedded in the bearing layer to ensure that the benefits of that layer's large resistance are achieved (Figure 13.10 illustrates this point). Depending on how strong the bearing layer is, an embedment of the order of a diameter would be adequate and likely achievable with routine drilling or driving equipment. Embedment in rock is only achievable if rock drilling rigs are used and is only necessary if the rock is the primary source of pile capacity.

Piles that have their tips bearing in a strong layer are sometimes called "end-bearing piles." This is a misnomer because these piles also derive their capacity from shaft resistance. Indeed, when in service, it is possible, depending on the soil profile, that a pile will derive most of its mobilized capacity from shaft resistance. Another misnomer is that of "floating piles" or, equivalently, "friction piles." These two terms are used for piles whose tips are not placed in a strong bearing layer. Every pile derives its capacity from shaft resistance ("friction") to some extent, so the term does not effectively distinguish piles in which shaft resistance is a significant fraction of the total resistance from other piles. Additionally, the piles are not floating, having some, even if small, base capacity. These piles are used when a strong bearing layer is not within practical reach.

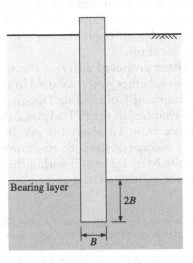

Bearing layer

2B

B

Figure 13.10 Need for embedment of end-bearing piles into the bearing layer.

Whether we refer to a pile as either end-bearing or floating, in practice, the ratio of ultimate unit base resistance to limit unit shaft resistance in clay is in the 10–20 range, which is much less than the 50–200 range for sand. This means shaft resistance is comparatively more important in clay than in sand. In soft clay, the base resistance may be as low as one-tenth of the shaft resistance.

13.4.4 Calculation of pile resistance from CPT or SPT results

As discussed in Chapter 7, the CPT was, in its origins, viewed as a "scaled-down pile load test." While our understanding of the CPT and our capabilities to interpret CPT results are today much broader than this rather narrow view of the test, the reality is that cone penetration is a process that is very similar to the plunging of a pile, and so the CPT is ideal for soil characterization for pile design.

The cone resistance q_c is a reasonably good approximation to the limit unit pile base resistance q_{bL}: CPT results yield, with little processing, the pile limit unit base resistance directly. The only caution necessary is to recognize that the cone is much smaller in diameter than virtually every pile. This means that the "sensing" depth of the cone is much smaller than that of the pile. The sensing depth is controlled by the depth of the slip mechanism or plastic zone forming below the level of the cone or pile base as it moves through the soil. This depth is of the order of two to five diameters for limit bearing capacity calculations, depending on whether density increases or decreases with depth below the base (Tehrani et al. 2018). Thus, the mechanical properties of soil located outside this range have little to no impact on the value of q_c. Design methods also consider the spatial variability of CPT results across the site: the pile will not be installed at the exact same location as a given CPT log. So a representative cone resistance q_{cb} at the pile base level must also reflect the horizontal variability of the site.

When designing a pile, if we wish to derive the full benefit of a bearing layer for the base resistance of the pile, it is not sufficient to place the pile base at the top of this layer. It is required to have it penetrate the layer some distance if the benefits of that layer's large resistance are to be fully developed. Accordingly, the estimation of q_{bL} is that it would be equal to an "average"[6] or representative value of q_c over a distance of one to two diameters from the pile base. This can be done in a simple way as follows: first, we use always the worst applicable CPT log to calculate pile capacity (this is conservative with respect to horizontal variability of the site if a sufficient number of borings exist); second, in obtaining q_{cb}, we may also consider values not only below, but also above the pile base, as shown in Figure 13.11. The worst applicable log is simply the one with lower q_c values of all the logs in the neighborhood of the pile location.

Different method developers have proposed different averaging schemes for the cone resistance at the base. We discuss two schemes next: one used in the Purdue Pile Design Method (PPDM) and the other in the Imperial College Pile Design Method (ICPDM), both to be discussed in detail later in this chapter. In the PPDM, we calculate the average of the cone resistance over a vertical distance from $1B$ above the pile base to $2B$ below the pile base. In the ICPDM, we calculate the average of the cone resistance over a vertical distance from $1.5B$ above to $1.5B$ below the pile base. If the soil within the averaging zone is clay, we use the corrected total cone resistance q_t, instead of q_c.

---

[6] Several averaging schemes have been proposed in the literature (e.g., Nordlund 1963; Bustamante and Gianeselli 1982) for determining the representative q_c. So long as the averaging does not take place over a too wide range of depths or overemphasizes the soil above at the expense of the soil below the pile base level, these averaging schemes are all acceptable. For belled drilled shafts, the diameter B_b of the bell should be used in these schemes.

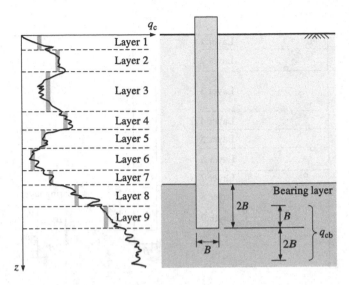

Figure 13.11 Pile design using the CPT: discretization of the soil profile for shaft resistance calculation and averaging the cone resistance for base resistance calculation.

The rationale for averaging q_c over an interval that includes points located both below and above the pile base is to account for the possibility that a lower q_c value appearing above the pile base at one location might have appeared below it if the log had been done at a location shifted horizontally by a few meters. These are simple, but effective ways to reduce the risk associated with site variability.

While the SPT bears no resemblance to pile loading, the SPT blow count is affected by the same factors as q_c; thus, the representative SPT blow count N_b at the pile base must, as the CPT cone resistance q_{cb}, be representative of the zone around the pile base. The only difference is that while q_c values are available almost continuously, SPT values are obtained at discrete depths spaced no closer than 2.5 ft. This means that, for small-diameter piles, only two or, at most, three N_{SPT} values will be available for averaging in the calculation of N_b.

When using CPT or SPT based methods for calculating shaft resistance, it is practical to divide the soil profile into layers, as illustrated in Figures 13.11 and 13.12. These layers should be constituted of the same soil type (that is, a layer will be a clay layer or a sand layer, for example, but not a layer containing both clay and sand). Moreover, if thick soil layers exist in the profile, they can be broken down into sublayers; these sublayers should preferably have similar SPT/CPT measurements. The values of the limit unit shaft resistance q_{sLi} for each soil layer i and of the ultimate unit base resistance $q_{b,ult}$ are determined based on available CPT or SPT results.

The general forms of the CPT-based equations for estimating $q_{b,ult}$ and q_{sLi} are:

$$q_{b,ult} = c_b q_{cb} \tag{13.15}$$

$$q_{sLi} = c_{si} q_{ci} \tag{13.16}$$

where c_b and c_{si} = constants or functions of soil type, soil state, and pile type, q_{cb} = representative cone resistance at pile base level; and q_{ci} = representative cone resistance for layer i. We will discuss specific methods and the values of functional forms they propose for c_b and c_{si} later.

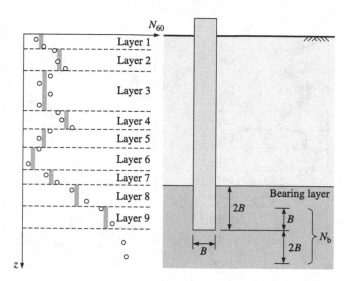

Figure 13.12 Pile design using the SPT: discretization of the soil profile for shaft resistance calculation and averaging the blow counts for base resistance calculation.

The general forms of the SPT-based prediction methods for estimating $q_{b,ult}$ and q_{sLi} are:

$$\frac{q_{b,ult}}{p_A} = n_b \left(N_b \right)_{60} \tag{13.17}$$

and

$$\frac{q_{sLi}}{p_A} = n_{si} \left(N_{si} \right)_{60} \tag{13.18}$$

where n_{si} and n_b are constants that depend on soil type, pile type, and the method used to calculate the unit resistances, and N_b and N_{si} are representative blow counts at the pile base and for layer i, respectively. To facilitate calculations, in this chapter we have attempted to present the methods in terms of N_{60} (the subscript 60 indicates that the blow counts are corrected for 60% energy ratio). In the literature, there typically is not enough information as to whether other corrections discussed in Chapter 7 (type of sampler, borehole diameter, and string length) were made in the development of the methods. This introduces additional uncertainty into these empirical design methods.

Researchers in this area have tended to consolidate their analytical, computational, and experimental work into methods of design that reflect what they have learned in research. Because there has been considerable research on CPT-based methods since the 1980s, more recently developed methods are considerably superior to more traditional methods, particularly if they were developed for the SPT. We will discuss several methods, including traditional, empirical CPT- and SPT-based methods, in this chapter, with focus on three methods: the PPDM, the ICPDM, and the University of Western Australia Pile Design Method (UWAPDM). The coverage of these methods is not identical, with some methods covering pile types and design scenarios that other methods do not cover.

With regard to pile design based on the SPT, a possible path is to convert SPT blow counts to cone resistance, as discussed in Chapter 7, and use the resulting estimated cone resistance values in a CPT-based method. We do this by referring to Figure 7.29 or using the equation fit to the data plotted in the figure. This approach takes advantage of the progress that has taken place in the development of CPT-based methods, but it introduces into the results an error associated with the conversion from SPT to CPT.

13.4.5 The sources of ultimate shaft and base resistance in piles

13.4.5.1 Shaft resistance

When a pile is loaded, shearing develops between the pile shaft and the soil. It is easier to understand the nature of the processes that develop along the shaft if we focus on sand. The sand in the immediate neighborhood of the pile, if dilative, will attempt to expand when the pile is pushed down with respect to it, but is constrained from doing so on one side by the presence of the pile shaft and on the other side by the presence of more soil. As the soil is constrained radially by the presence of the pile, the normal stress between the pile and sand increases when shearing is underway. The resulting shaft resistance may be seen as the product of this effective normal stress by the tangent of the friction angle between the pile and the soil. A similar process would be observed for clay under drained conditions. Under undrained conditions, the thinking shifts, with negative or positive excess pore pressures developing upon shearing, depending on the overconsolidation ratio of the clay.

The shear strain that develops in the soil along the pile shaft at a ULS is quite large, exceeding 100%. This means that sand reaches a critical state and clay reaches a residual state along most, if not all, of the length of the pile at the ultimate limit states we consider in design. We must distinguish conditions at a ULS from those in service. The unit shaft resistance in end-bearing piles may still be fully mobilized along most of the pile in service conditions. It is the base resistance that is typically far from fully mobilized to guarantee a sufficient factor of safety. But, in floating piles in service conditions, the shear stress along the pile shaft is less than the limit unit shaft resistance for a sufficiently long segment of the pile to ensure the required safety factor.

13.4.5.2 Base resistance

It might be tempting to think of the base resistance of piles in terms of the bearing capacity equation seen in Chapter 10 for shallow foundations, but that equation does not apply to the calculation of the base resistance of piles. The key difference between the base resistance of piles and footings is this: in footings, there is some soil that is below and to the side of the footings that is pushed laterally and then up; in piles, the soil below and to the side of the pile is pushed away laterally. This fundamental difference in mechanism means that pile base resistance has to be calculated using a different approach.

The base resistance of piles develops gradually with increasing base settlement. This is particularly true of nondisplacement piles. This means that large settlements are needed for the base resistance to achieve its limit value. The limit value is, accordingly, rarely of direct interest in design, for limit states will typically occur before the pile plunges. Instead, design is usually based on a conventional ultimate load corresponding to a settlement equal to 10% of the pile diameter. This covers a wide range of design situations, but, in some situations, the critical ULS may be defined by a different value of settlement (Basu and Salgado 2014). In others, such as floating piles in clay, the limit load is directly of interest because plunging would occur after relatively small settlements.

13.4.6 Treatment of sands, silts, and clays

When a pile crossing or bearing on a sand layer is loaded, a fraction of the load is transferred to that sand layer. That transfer can be assumed to be immediate, as sands consolidate extremely fast. The sand layer will carry the load to the extent of its capacity to do so.

In clays, consolidation is a factor. Although construction of many types of structures is rather slow, consolidation is even slower. Thus, it is very likely that a clay will need to sustain a substantial fraction of the applied loads in undrained or substantially undrained loading mode. It is also possible to calculate a long-term pile unit resistance contributed by a clay layer, but that is rarely critical in design.

With regard to silts, they are treated as sands if nonplastic and as clays if plastic.

13.4.7 Design methods

With the growing understanding of the pile resistance mobilization mechanisms and the accumulation of high-quality, full-scale pile load test case histories, researchers have proposed and improved pile design methods to estimate the ultimate resistances (those corresponding to a pile head settlement of $0.1B$) of closed-ended and open-ended driven pipe piles.

We will focus on four methods for displacement piles: the PPDM, the ICPDM, the UWAPDM, and the Norwegian Geotechnical Institute Pile Design Method (NGIPDM). The PPDM has developed as a result of research involving both computational simulations (Loukidis and Salgado 2008; Han et al. 2017a; Chakraborty et al. 2013; Basu et al. 2014) and extensive small-scale model pile and full-scale, field pile load testing (Bica et al. 2014; Ganju et al. 2020; Han et al. 2017b, 2018b, 2019b, 2020; Kim et al. 2009; Lee et al. 2003; Paik et al. 2003; Paik and Salgado 2003). The ICPDM, UWAPDM, and NGIPDM have a similar foundation, but without a comparable emphasis on numerical analyses.

Of the modern design methods that we will cover, only the PPDM has design equations for nondisplacement piles. All four methods have design equations for other pile types. We will also see some equations from empirical methods.

The design methods do not usually specify whether their equations for unit shaft and base resistances were developed accounting for the contact stresses between the pile and soil after pile installation. In nondisplacement piles, such stresses result almost exclusively from the balancing of the weight of the pile by the mobilized resistances. But in driven piles, these stresses result from the locking in of stresses generated during driving. The usual distribution of stresses is such that stresses at the base are compressive, with the shaft being partly in tension and partly in compression. These stresses have been referred to as "residual stresses," and their measurement before pile load tests is essential for correct interpretation of pile load tests and informed development or validation of design methods using pile load tests (Fellenius 2002a,b; Fellenius et al. 2004; Ganju et al. 2020; Han et al. 2017b, 2019b, 2020; Kim et al. 2009; Paik et al. 2003; Seo et al. 2009). In the absence of better information, they can be assumed based on one's best understanding of the driving process.

13.4.8 Special considerations for drilled shafts

It has been suggested that the ultimate unit base resistance $q_{b,ult}$ of drilled shafts be capped. A cap of 5 MPa has been proposed. This limit on unit base resistance provides insurance against a number of things that could go wrong in the process of installation (such as the accumulation of loose cuttings on the base of the shaft). It also accounts for some limitations of design methods. For example, most CPT- or SPT-based methods use a constant base-to-cone resistance ratio $q_{b,ult}/q_c$ for each soil type. In sands, this means that $q_{b,ult}/q_c$ used in design is always the same, regardless of the relative density of the sand. Lee and Salgado (1999) showed that $q_{b,ult}/q_c$ actually decreases with increasing relative density. For design methods that do not take into account the effect of relative density on $q_{b,ult}/q_c$, the limit on $q_{b,ult}$ indirectly accounts for the excessively high values of $q_{b,ult}$ that might result when relative density, and thus q_c or N_{SPT}, is very large.

13.4.9 Special considerations for belled drilled shafts

Although not as common as in the past,[7] belled drilled shafts are still occasionally used. The belled shaft is first sized as a structural element. This is done by calculating the minimum shaft diameter B_s for which LS IA-2 (concrete crushing) is not achieved. The design compressive strength f_{cd} of the concrete is obtained from the characteristic compressive strength f_c' through:

$$f_{cd} = \frac{f_c'}{(FS)_{IA-2}} \tag{13.19}$$

where the safety factor $(FS)_{IA-2}$ is in the 2.5–4 range. Alternatively, we may follow the LRFD-based ACI-318 reinforced concrete design procedure to achieve the same goal.

Given the column load Q_{col}, the shaft diameter is obtained from:

$$B_s = \sqrt{\frac{4Q_{col}}{\pi f_{cd}}} \tag{13.20}$$

Once B_s is determined, the limit shaft capacity Q_{sL} can be determined using Equation (13.3), where q_{sL} is estimated using the methods applicable to drilled shafts. No load transfer is assumed along the shaft one shaft diameter above the top of the truncated cone that forms the base of the belled shaft because instrumented load tests and numerical analyses have indicated that minimal load transfer is expected in that zone.

The base must support the excess of the column load (magnified by a suitable global safety factor FS) over the available shaft capacity Q_{sL}. The ultimate unit base resistance $q_{b,ult}$ is calculated using the methods for drilled shafts, described later in this chapter. The base diameter B_b is then determined from:

$$B_b = \sqrt{\frac{4\left[(FS)Q_{col} - Q_{sL}\right]}{\pi q_{b,ult}}} \tag{13.21}$$

The angle of the lateral surface of the bell depends on the characteristics of the under-reamer (the tool used to enlarge the base). An angle of 30° with the vertical is convenient because it dispenses with the need for elaborate calculations to check for structural failure, as there is virtually no possibility of punching of the shaft through the base for an angle of 30° if the concrete is of good quality.

---

[7] The infrequent use of belled shafts is for good reason, given the construction difficulties, the difficulty with inspection, and the possible danger if workers are sent in for either construction or inspection of the base of the shaft.

13.4.10 Special considerations for steel pipe piles

In certain areas of the United States (such as the Midwest) and throughout the world, pipe piles are often used in piling practice. Pipe piles are often closed at the base by welding of a thick steel plate or a conical tip. Closed-ended pipe piles are large displacement piles and should be designed as such. Open-ended piles cause less change in the soil state around the pile when driven than closed-ended piles with the same diameter, but more change than nondisplacement piles with the same diameter.

A major difference between closed- and open-ended piles is the possible formation of a "soil plug" inside the open-ended pile during driving, as shown in Figure 13.13. If no soil entered the pile during installation, open-ended piles would behave exactly as closed-ended piles. But, in reality, as the pile is driven into the ground, soil does enter the pile. As the soil column (or plug) forms inside the pile, frictional resistance is gradually mobilized between the soil and the inner surface of the pile. The soil plug is compressed as it is pushed from below by the soil entering the pile and from above by the frictional resistance between the upper part of the soil plug and the inner pile surface. Two conditions are necessary for soil to stop entering the pile: (1) sufficient friction needs to develop between the soil plug and the inner pile surface, and (2) the plug needs to become sufficiently stiff to no longer undergo significant compression.

The response of open-ended piles is generally intermediate between that of nondisplacement and displacement piles. The length of the soil column inside the pile may be equal to or less than the pile driving length. If it is the same as the pile driving length, the pile has been driven in a fully coring or unplugged mode, which is illustrated in Figure 13.14a, throughout driving. If driving takes place in a fully plugged mode from the beginning (clearly not a realistic possibility), the pile looks like the pile shown in Figure 13.14b, into which no soil has entered during driving. If driving takes place in a partially plugged mode or a fully plugged mode during part of the driving, the length of the soil plug within the pile will be greater than zero, but less than the length of pile penetration into the soil. It may be possible to observe all three driving modes (fully coring, partially plugged, or fully plugged) during the driving of a single pile (Paikowsky et al. 1989).

Although the installation of an open-ended pile imparts less change to the surrounding soil than that of a closed-ended or full-displacement pile, the soil conditions are certainly different from those before installation (Randolph et al. 1979; Nauroy and Le Tirant 1983). The unplugged or fully coring mode is commonly observed during the initial stages of pile

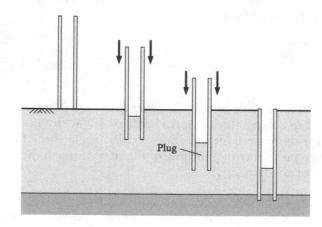

Figure 13.13 Formation of soil plug during driving of an open-ended pipe pile.

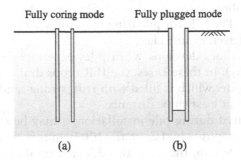

Figure 13.14 Two extreme, ideal conditions for an open-ended pipe pile: (a) unplugged and (b) perfectly plugged.

driving. As penetration and formation of the soil plug continue, internal frictional resistance mobilizes between the inner pile surface and the soil plug, densifying the lower part of the soil plug. However, some soil continues to enter the pile, characterizing partially plugged driving. Finally, with further driving, soil intrusion is prevented by the now sufficiently large interface shear strength between the soil plug and the inner pile surface and by the large soil plug stiffness. Driving is said to take place under fully plugged conditions while that remains true. The base response of the open-ended pile installed to this stage approaches that of a closed-ended pile.

In order for a soil plug to form during driving, the shear strength between the plug and the inner area of the pile must be sufficient to overcome the dynamic base resistance of the pile base as well as the inertia of the plug itself.[8] As the diameter of the pile increases, inertia and bearing capacity tend to exceed the internal shaft resistance, and driving takes place in the fully coring mode.

The incremental filling ratio (IFR), defined as

$$\text{IFR} = \frac{dL_p}{dL} \tag{13.22}$$

where L_p = plug length and L = pile penetration length, can be used to quantify the plugging response of open-ended pipe piles. In the fully coring mode, soil enters the pile at the same rate as the pile moves down, with an implied incremental filling ratio (IFR) of 1 (or 100%). A pile installed under fully coring conditions has a load response that is only slightly stiffer than that of a nondisplacement pile. If plugging develops during driving, the installation increasingly resembles the installation of a full-displacement pile (for which IFR = 0), and the pile capacity will reflect that.

In practice, a value of zero for the IFR is not commonly achieved by pipe piles. In sands, as discussed by Lee et al. (2003), the IFR decreases with increasing relative density, increasing lateral effective stress (which generally increases with depth), and decreasing pile diameter. Increasing density and stress lead to greater soil plug stiffness, and a decreasing diameter

---

[8] Inertia is the tendency of a body to retain its state of motion. In this case, the soil "would prefer" to stay where it is instead of being dragged down by the pile. One way to express this physically is to state that, if an acceleration is imposed on the soil, then an inertial force appears which is the product of the soil mass – itself the product of volume by the mass density of the soil – by the acceleration. If the soil volume is large, therefore, it will be difficult for the pile to drag the plug down with it, the soil will likely stay put, and the driving will take place in a fully coring mode.

increases the ratio of the shear force that develops between the plug and the inner pile surface to the inertial forces. The IFR for pile driving in soft clay is typically close to 1, dropping somewhat for piles driven in stiff clay.

Another situation sometimes develops when piles penetrate through strong soil (dense sand) into weaker soil (clay). In these cases, the IFR drops dramatically as the weaker clay soil cannot push into the pile, which is filled with stiff, strong sand. The pile then advances in a fully plugged state for at least some distance.

If the IFR is not measured during pile installation, it may be estimated for sands using the chart in Figure 13.15 proposed by Lee et al. (2003). In this chart, which is applicable to piles of short to moderate length, the normalized incremental filling ratio (NIFR) and the normalized pile driving depth D_n are defined as

$$\text{NIFR} = 100\frac{\text{IFR}}{D_n} \tag{13.23}$$

$$D_n = \frac{\text{Driving depth}}{\text{Inner pile diameter}} \tag{13.24}$$

where IFR is given as a number between 0 and 1.[9]

Equations (13.23) and (13.24) and the chart in Figure 13.15 may be used for a sand layer overlain by clay so long as there is enough pile penetration into the sand layer. It has been observed through instrumentation of open-ended piles that the internal transfer of stresses between the sand column within the pipe and the internal area of the pipe takes place along a length measured from the base of the pile of approximately six diameters according to Murff et al. (1990), Randolph et al. (1991), and O'Neill and Raines (1991) and as much as 10–15 diameters according to Han et al. (2020). The active length of the soil plug may depend on certain factors, including the density and mean effective stress within the plug.

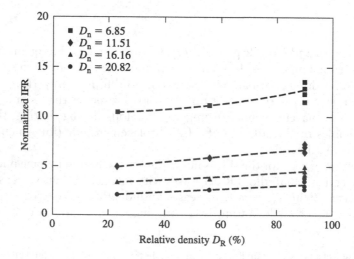

Figure 13.15 Chart for estimating the incremental filling ratio. (Modified after Lee et al. (2003); with permission from ASCE.)

---

[9] Whenever the IFR appears in equations, it will always be a number between 0 and 1.

It does not matter if there is soil within the pipe above the active length of the plug or if any such soil is sand or clay.

When the pile is loaded statically, after installation is completed, its capacity will depend on the response of the soil plug, in addition to the resistances mobilized at the pile annulus and along the pile shaft. Thus, for open-ended pipe piles, Equation (13.1) applies, with the base capacity defined as

$$Q_b = Q_{plug} + Q_{ann} \tag{13.25}$$

where Q_b = base capacity, Q_{plug} = soil plug capacity, and Q_{ann} = annulus capacity.

The resultant unit base resistance of an open-ended pile with outside diameter B and annulus thickness t is a weighted average of the unit plug resistance q_{plug} acting on the plug area (the internal cross-sectional area of the pile) and the unit annulus resistance q_{ann} acting on the annulus. This can be expressed mathematically as the weighted average:

$$q_b = \frac{q_{plug}B_i^2 + 4q_{ann}Bt}{B^2} \tag{13.26}$$

where B_i = inner pile diameter.

Roughly speaking, at plunging or large relative settlements, q_{ann} approaches the cone resistance value q_c obtained from a CPT performed at the same location. This is true for sand or clay and possibly even for fine gravel, for the dimensions of the cone tip are of the order of those of the pile wall. The challenge lies in determining q_{plug}, which is quite dependent on the degree of plug formation during driving and of the resulting soil plug stiffness (it is closely related to the value of the IFR at the end of driving). The driving of large-diameter pipe piles (say, with diameters of 500 mm or larger) into sands tends to take place under fully coring (unplugged) conditions. While essentially inactive during driving, the soil plug does contribute to base capacity during static loading. Analyzing a large dataset of the pile capacity of open-ended piles in sand, Lehane and Randolph (2002) recommend that the ultimate unit base capacity $q_{b,ult}$ of these piles be estimated, conservatively, using the values of Lee and Salgado (1999) for nondisplacement piles.[10] They note that, while $q_{plug,ult}$ would tend to be lower than the $q_{b,ult}$ of nondisplacement piles due to plug compression during loading, the annulus resistance fully compensates for plug compressibility, with $q_{b,ult}$ of pipe piles resulting at least as large as and probably slightly greater than that of nondisplacement piles.

For smaller-diameter piles, a plug tends to form during driving at larger depths. However, because the IFR rarely reaches zero, penetration does not typically take place under fully plugged conditions. Thus, base capacities tend to be somewhere between the base capacities of nondisplacement and displacement piles. Methods of design of open-ended pipe piles must reflect this.

13.4.11 Special considerations for steel tapered piles

These piles are rarely used nowadays. Their shaft resistance tends to be larger than straight-shaft piles. Following Nordlund (1963), q_{sL} for these piles can be taken as (1 + *taper*) times the value of q_{sL} for a straight-shaft pile with equivalent geometry, where *taper* is the taper (slope) of the lateral surface of the pile.

---

[10]These values can be calculated using Equation (13.31) provided later in this chapter.

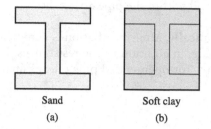

Sand
(a)

Soft clay
(b)

Figure 13.16 Areas for the calculation of base and shaft resistances for an H-pile in (a) sand and (b) clay. Solid lines indicate the area for the calculation of shaft resistance; shaded area is the area for the calculation of base resistance.

13.4.12 Special considerations for steel H-section piles

A plug does not form when an H-pile in sand is loaded statically. Thus, the shaft capacity of an H-pile in sand should be calculated by considering the full steel–soil interface area, which includes not only the outside and inside of the flanges and their tips, but the web as well, as shown in Figure 13.16a. The base capacity is calculated with basis on the actual H-pile cross-sectional area.

When an H-pile is installed in soft clay, it is possible that the "plug" occupying the space between the flanges of the pile will be activated when the pile is loaded statically. It may be appropriate in this case to assume the outside (circumscribed) perimeter of the pile in the calculation of shaft capacity, as shown in Figure 13.16b. Likewise, the gross cross-sectional area should be used for base capacity calculations. Although some authors have questioned whether this applies also to stiff clays on the basis of any separation between pile and soil that might develop, such separation at depth under pressure is not likely.

According to Tomlinson (1987), the soil plug along a considerable length in the upper part of the pile may crack and detach from the pile in stiff clays. This would imply that a relatively loose volume of soil would develop within the pile flanges that would not be contributing to shaft resistance. Although this may be somewhat speculative, if we wish to account for it, we would not consider the outside perimeter of the cross section in the calculation of shaft capacity. Rather, only the external surface area associated with the two flanges would be considered as contributing to shaft capacity. The gross area should be used in calculating the base capacity of these piles in stiff clay.

> Example 13.2
>
> For an HP 14×117 H-pile, find the appropriate values of the lateral surface area per unit pile length and of the base area in sand and clay for use in the calculation of pile axial load capacity.
>
> *Solution:*
>
> The typical H-pile cross sections and properties are given in Table 12.2. For an HP 14×117 H-pile, we have the dimensions shown in E-Figure 13.2.
>
> Sand
>
> There is no formation of plug. Referring to Figure 13.16a, the lateral surface area per unit pile length is given by

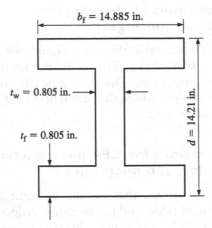

E-Figure 13.2 Dimensions of the H-pile of Example 13.2.

$$A_s = \left(2 \times 14.885 + 2 \times (14.885 - 0.805) + 2 \times (14.21)\right) \text{in}^2/\text{in}$$

$$A_s = 86.35 \text{ in}^2/\text{in} = 2.19 \text{ m}^2/\text{m}$$

The base area is the actual pile cross-sectional area:

$$A_b = \left(2 \times (14.885 \times 0.805) + (14.21 - 2 \times 0.805) \times 0.805\right) \text{in}^2$$

$$A_b = 34.1 \text{ in}^2 = 0.022 \text{ m}^2$$

Clay

There is plug formation and therefore the lateral surface area per unit pile length is given by assuming the outside perimeter of the pile in the calculation, as shown in Figure 13.16b. It is calculated as

$$A_s = \left(2 \times 14.885 + 2 \times 14.21\right) \text{in}^2/\text{in} = 58.2 \text{ in}^2/\text{in} = 1.48 \text{ m}^2/\text{m}$$

The base area is the gross cross-sectional area:

$$A_b = 14.885 \times 14.21 \text{ in}^2 = 211.5 \text{ in}^2 = 0.136 \text{ m}^2$$

As summarized in E-Table 13.2, the areas used to calculate base and shaft resistance for an H-pile depend on the soil type.

E-Table 13.2 Areas for the calculation of base and shaft resistances for the H-pile of Example 13.2

Soil profile	A_s (m²/m)	A_b (m²)
Sand	2.19	0.022
Clay	1.48	0.136

13.4.13 Special considerations for Franki piles ("pressure-injected footings")

As discussed in Chapter 12, the lateral surface of a Franki pile is quite rough, and its base has an expanded size with a bulb shape. That notwithstanding, the dimensions used in design are the nominal pile dimensions. This is marginally conservative, except in some CPT- or SPT-based design methods, where the multipliers of q_c or N_{SPT} account indirectly for the enlarged dimensions of the pile.

13.4.14 Special considerations for CFA piles, partial-displacement piles, and micropiles

There has been some question as to whether continuous flight auger piles respond more as displacement or nondisplacement piles, and the literature reflects two different viewpoints. Research by Viggiani (1989) in Italy, as discussed by Mandolini et al. (2002), suggests that, if CFA piles are installed quickly, they do not allow soil unloading to any significant extent and the resulting piles are closer in mechanical behavior to displacement piles. In contrast, if installation is slow, the piles behave more like nondisplacement piles. The rate (velocity) of installation below which the resulting pile behaves as a nondisplacement pile is given as

$$ v_{crit} = \omega\lambda\left(1 - \frac{B}{B_h}\right) \tag{13.27} $$

where ω is the rate of revolution of the auger, λ is the pitch of the auger, and B and B_h are the diameters of the auger and the hollow stem, respectively. Given the uncertainties concerning CFA resistance mobilization, it may be advisable not to use unit shaft resistance values substantially greater than those used for nondisplacement piles.

Micropiles in soil can be designed using the same numbers as CFA piles, as they are usually installed in ways that affect similarly the surrounding soil. Prezzi and Basu (2005) cover in detail the construction and design of partial-displacement piles that have been developed in recent years (notably "drilled displacement" piles, such as the Omega pile, the Atlas pile, and the APGD pile).

13.5 CALCULATION OF THE ULTIMATE RESISTANCE OF NONDISPLACEMENT PILES

13.5.1 The relationship of pile installation to pile load response

As discussed in detail in Chapter 12, a nondisplacement pile is installed by the removal of soil from the ground without allowing the soil to cave in and then by filling the void created by the material that will constitute the pile, which is usually concrete or grout. This means that the soil state – density and effective stress – is largely preserved both around the pile and below the pile base. Any limited unloading that may take place due to excavation is corrected by the placement of the concrete, which has enough fluidity before setting to apply a normal stress on the soil sufficient to reestablish the original lateral stress (in fact, Fleming et al. (1992) argued that the placement of concrete with high fluidity would even lead to an initial K slightly greater than K_0). Concrete placed using a tremie will also remove, as it flows upward, any cake that may have formed on the walls of the excavation and any superficial layer of soil that may have been disturbed by auger action. The shaft resistance of piles installed using the slurry method and polymer slurry has been found to be no less (Lam et al. 2014) or even modestly greater than (Mobley et al. 2018) that of piles installed using bentonite slurry.

Loading of a nondisplacement pile can then be considered to take place under "fresh" conditions. This contrasts with displacement piles, which are forced into the ground, significantly shearing the soil along the pile shaft and compressing and shearing the soil below the base during installation. This produces significant changes in the soil state next to the pile even before it is loaded. This means that there is significant difference in response between a nondisplacement and a displacement pile.

13.5.2 Nondisplacement piles in sandy soil

13.5.2.1 Shaft resistance

Research performed at Purdue University since 1993 has led to a relatively comprehensive pile design method referred to, in some papers, as the PPDM (Han et al. 2017a, 2018c; Loukidis and Salgado 2008; Salgado et al. 2017; Salgado 2006a,b). In this method, the limit unit shaft resistance q_{sL} is calculated as

$$q_{sL} = K\sigma'_v \tan\delta_c \qquad (13.28)$$

$$K = \frac{K_0}{e^{0.3\sqrt{K_0 - 0.4}}} 0.67 \exp\left\{\frac{D_R}{100}\left[1.5 - 0.35 \ln\left(\frac{\sigma'_v}{p_A}\right)\right]\right\} \qquad (13.29)$$

where δ_c is the pile–soil interface friction angle, K_0 is the initial coefficient of lateral earth pressure, and D_R is the relative density.

The relative density can be calculated from the cone resistance q_c using Equation (7.20), proposed by Salgado and Prezzi (2007), repeated here:

$$D_R(\%) = \frac{\ln\left(\frac{q_c}{p_A}\right) - 0.4947 - 0.1041\phi_c - 0.841 \ln\left(\frac{\sigma'_h}{p_A}\right)}{0.0264 - 0.0002\phi_c - 0.0047 \ln\left(\frac{\sigma'_h}{p_A}\right)} \leq 100\%$$

Equation (13.28) expresses a simple relationship: the shear strength along the pile – that is, the limit unit shaft resistance q_{sL} – is equal to the normal effective stress acting on the pile shaft times the tangent of the pile–soil interface friction angle. The challenge in using this equation has been to answer two questions:

1. What value of friction angle should we use?
2. What is the operative effective stress and thus K?

Research has shown that soil particle sliding along the pile–soil interface is minimal because the outer surface of nondisplacement piles is sufficiently rough and thus shearing mainly occurs within the soil immediately next to the pile shaft. It follows that the interface friction angle is in essence equal to the friction angle of the soil. Additionally, the shear strain that develops along the pile shaft has been shown to be quite large. This is so because the shearing localizes in a shear band (refer to Chapters 5 and 6) with thickness of the order of a few millimeters. So even a small relative motion of the pile shaft with respect to the outer boundary of the shear band results in shear strains exceeding 100%. The consequence of this is that the sand immediately next to the pile is in the critical state and δ_c takes the value of the internal critical-state friction angle ϕ_c. That answers question (1): the appropriate friction angle to use is the critical-state friction angle.

Question (2) cannot be as simply answered. When the pile is loaded, shearing develops between the pile and the sand. The sand, if dilative, will attempt to expand, but is constrained from doing so on one side by the presence of the pile shaft and on the other side by the presence of more soil. As the sand is prevented from expanding by the presence of the pile, the normal stress between the pile and sand increases. The result is that, for contractive or only slightly dilative sands, K is close in value to K_0, but, for moderately to strongly dilative (dense to very dense) sands, K may be significantly greater than K_0. This has been shown by finite element analyses using a realistic constitutive model. Salgado (2006a,b) first proposed an equation for the calculation of K, the updated version of which is Equation (13.29). Equation (13.29) is based on the results of finite element analyses (Han et al. 2017a; Loukidis and Salgado 2008; Salgado 2006a,b). Figure 13.17 shows the values of K produced by the equation as a function of relative density D_R and vertical effective stress σ'_v.

13.5.2.2 Base resistance

Using the PPDM (Han et al. 2017a), we can compute the ultimate unit base resistance $q_{b,ult}$ as

$$\frac{q_{b,ult}}{p_A} = 62 \left(\frac{D_R}{100} \right)^{1.83} \left(\frac{\sigma'_h}{p_A} \right)^{0.4} \tag{13.30}$$

This equation is applicable for $L/B < 50$. The same method allows the calculation of $q_{b,ult}$ from CPT cone resistance using the following equation (Salgado et al. 2011b):

$$q_{b,ult} = 0.23 e^{-0.66(D_R/100)} q_{cb} \tag{13.31}$$

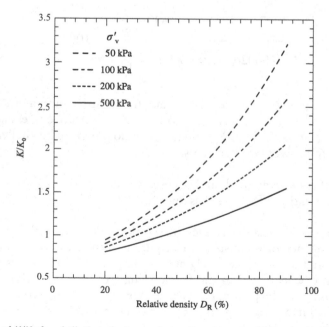

Figure 13.17 Values of K/K_0 for drilled shafts in sand as a function of relative density and vertical effective stress.

where q_{cb} is the cone resistance averaged within a distance from $1B$ above the pile base to $2B$ below the pile base.

The choice of which of these two equations to use hinges on the availability of CPT results. If calculations are based on CPT results, the use of Equation (13.31) is more direct. Both equations are based on sophisticated finite element analyses and have been validated using pile load test data.

13.5.3 Nondisplacement piles in clayey soil

13.5.3.1 Shaft resistance

In drilled shafts, the most ubiquitous nondisplacement piles, augering causes clay particles along the walls of the excavation and future shaft of the pile to realign, leading to a drop in friction angle to the residual friction angle value.[11]

Additionally, augering generates excess pore pressures at small distances into the excavation walls. After that, concrete is placed, reloading the soil in the horizontal direction. After hardening of the concrete, the pile is loaded as the structure is erected on top of it. During the period between excavation and loading of the pile, the excess pore pressures generated by installation dissipate. Axial loading of the pile then generates some additional pore pressure, leading to a lateral effective stress σ'_{hds} in the disturbed soil at a given depth down the pile that is of course dependent on the initial lateral effective stress σ'_h there and the value of pore pressure generated by the installation and loading of the pile. The limit unit shaft resistance q_{sL} of the pile can be calculated from the knowledge of this effective stress on the pile shaft through

$$q_{sL} = s_{ur} = \sigma'_{hds} \tan\phi_r = K\sigma'_v \tan\phi_r \tag{13.32}$$

where s_{ur} is the residual shear strength of the soil, ϕ_r is the residual friction angle of the soil, σ'_v is the initial vertical effective stress, and K is the ratio of the normal lateral effective stress on the pile shaft at the ULS to σ'_v, all at the depth where q_{sL} is to be calculated.

Instead of proposing equations for K for use in the effective stress-based Equation (13.32) to calculate the limit unit shaft resistance q_{sL} in clay, Salgado (2006a) proposed the values of α to use in the total stress-based equation:

$$q_{sL} = \alpha s_u \tag{13.33}$$

The values of α have since been updated and incorporated into the PPDM. The equation for α in the PPDM is (Chakraborty et al. 2013):

$$\alpha = \left(\frac{s_u}{\sigma'_v}\right)^{-0.05} \left\{ A_1 + (1-A_1)\exp\left[-\left(\frac{\sigma'_v}{p_A}\right)(\phi_c - \phi_{r,min})^{A_2} \right] \right\} \tag{13.34}$$

where $A_1 = 0.75$ for $\phi_c - \phi_{r,min} \leq 5°$, 0.4 for $\phi_c - \phi_{r,min} \geq 12°$, and a linearly interpolated value for $5° \leq \phi_c - \phi_{r,min} \leq 12°$; and

$$A_2 = 0.4 + 0.3\ln\left(\frac{s_u}{\sigma'_v}\right)$$

These equations resulted from finite element analyses using a realistic constitutive model for clay. The simulations were done for various clays and validated against the results of pile load tests. In these simulations, a key insight is that the shear strength that develops along the pile shaft at the ULS is the residual shear strength (see Salgado 2006a). In sands, the friction angle

---

[11] Burland and Twine (1988) and Skempton (1959) appear to have been the first to propose that a residual friction angle is appropriate for the calculation of shaft resistance of drilled shafts in clay.

along the pile shaft is the critical-state friction angle. In clays, in contrast, the particles tend to align with the direction of shearing and a residual state is reached. Accordingly, in Equation (13.33), $\phi_{r,min}$ is the minimum residual-state friction angle. This does not mean that the minimum residual friction angle is reached everywhere on the shaft of the pile, but simply that the equation was developed with $\phi_{r,min}$ as a parameter. For example, at shallow depths, realignment will be limited and the residual friction angle ϕ_r will be higher than $\phi_{r,min}$ and may even be equal to the critical-state friction angle because the normal effective stress perpendicular to the direction of shear is small. The analyses do capture and reflect these responses.

The value of α also depends on the presence of any sand. Salgado (2006a) provides a plot of α versus undrained shear strength for various values of the clay fraction – that is, the percentage of clay – in mixtures of sand and clay (see Figure 13.18).

13.5.3.2 Base resistance

We use the following equation to calculate the ultimate unit base resistance $q_{b,ult}$ using the PPDM (Salgado 2006a):

$$q_{b,ult} = 9.6 s_u$$

where s_u is the undrained shear strength at the pile base.

13.5.4 Piles in tension

The shaft resistance of nondisplacement piles loaded in tension is the same as that of piles loaded in compression. This has been shown through calibration chamber testing of nondisplacement piles (Galvis-Castro et al. 2019a). In these tests, model piles pre-installed in sand samples prepared in a large chamber were loaded in tension and compression, with no difference in measured shaft resistance.

The reason for this is that, for a given fabric of the prepared sand sample, it does not make any difference whether the pile is moved up or down with respect to the soil. This is true for first loading, but, if there are load reversals, then there will be a difference in the shaft

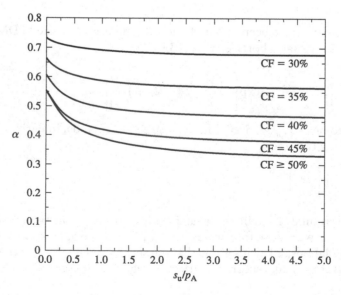

Figure 13.18 Values of α for clay with OCR in the 3–5 range as a function of clay fraction CF.

resistance. This occurs because, if the pile has already been moved in one direction, it has realigned the fabric of the soil next to the shaft in a way consistent with that load direction. On load reversal, the fabric will be mechanically weaker to resist the movement of the pile and thus the shaft resistance will be lower.

13.5.5 Examples of calculations of the ultimate resistance of nondisplacement piles

Example 13.3

A 15-meter-long drilled shaft is installed through a sand layer with $D_R = 50\%$, $\gamma_{sat} = 22$ kN/m³, and $\phi_c = 30°$ extending down to a depth of 10 m. The water table is at the surface of the deposit. The deposit is normally consolidated, having a K_0 of 0.4. Calculate and plot the values of K and q_{sL} versus depth for this drilled shaft.

Solution:

The solution is straightforward. The pile here is a nondisplacement pile, a drilled shaft, crossing a 10-meter-thick sand layer. We simply calculate K using Equation (13.29) for various depths and plot it versus depth (see E-Figure 13.3). For example, for $z = 5$ m, we write

$$K = \frac{K_0}{e^{0.3\sqrt{K_0-0.4}}} 0.67 \exp\left\{\frac{D_R}{100}\left[1.5-0.35\ln\left(\frac{\sigma'_v}{p_A}\right)\right]\right\}$$

$$= \frac{0.4}{e^{0.3\sqrt{0.4-0.4}}} 0.67 \exp\left\{\frac{50}{100}\left[1.5-0.35\ln\left(\frac{(22\times 5 - 9.81\times 5)}{100}\right)\right]\right\}$$

$$= 0.62$$

For $\delta_c = \phi_u$, the limit unit shaft resistance can be computed as follows:

$$q_{sL} = K\sigma'_v \tan\delta_c$$

$$= 0.62\times\left[(22-9.81)\times 5\right]\times\tan 30°$$

$$= 21.8\text{ kPa}$$

Example 13.4

A soil profile consists of an 8-meter-thick layer of a normally consolidated clay overlying a stiffer clay. The s_u/σ'_v of the NC clay is 0.3, while that of the stiff clay is 1. The unit weights are 16 and 18 kN/m³, respectively. The clay has a critical-state friction angle $\phi_c = 25°$ and a minimum residual-state friction angle $\phi_{r,min} = 13°$. Estimate the ultimate resistance of a drilled shaft with 300 mm diameter and 10 m length, bearing on the stiff clay. Consider the water table to be at the ground surface.

Solution:

The undrained shear strength in the NC clay layer is calculated using

$$s_u = 0.3\sigma'_v$$

and in the stiff OC clay layer using

$$s_u = \sigma'_v$$

If we divide the 8-meter-thick NC clay layer into eight layers of 1 m thickness, we can consider, as an illustration, the midpoint of the fifth layer, with depth equal to 4.5 m:

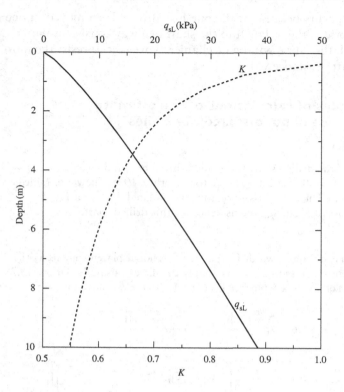

E-Figure 13.3 K and q_{sL} versus depth for the drilled shaft of Example 13.3.

$$\sigma'_v = (16 - 9.81) \times 4.5 = 27.9\,\text{kPa}$$

$$s_u = 0.3\sigma'_v = 0.3 \times 27.9 = 8.4\,\text{kPa}$$

The value of α is calculated as

$$\alpha = \left(\frac{s_u}{\sigma'_v}\right)^{-0.05}\left[A_1 + (1 - A_1)e^{-\left(\frac{\sigma'_v}{p_A}\right)(\phi_c - \phi_{r,\min})^{A_2}}\right]$$

Because $\phi_c - \phi_{r,\min} = 25° - 13° = 12°$, the value of A_1 is equal to 0.4. The value of A_2 is calculated using

$$A_2 = 0.4 + 0.3\ln\left(\frac{s_u}{\sigma'_v}\right) = 0.4 + 0.3\ln(0.3) = 0.04$$

So

$$\alpha = (0.3)^{-0.05}\left[0.4 + (1 - 0.4)e^{-\left(\frac{27.9}{100}\right)(12)^{0.04}}\right] = 0.89$$

The limit unit shaft resistance follows directly:

$$q_{sL5} = 0.89 \times 8.4 = 7.5\ \text{kPa}$$

This can now be multiplied by the shaft area of the pile segment in contact with this layer to calculate the contribution of the layer to the total shaft resistance:

$$Q_{sL5} = q_{sL5}A_{s5} = 7.5(\pi \times 0.3 \times 1) = 7.1 \text{ kN}$$

This calculation now needs to be repeated for all eight layers of the NC layer and the two additional layers into which the 2 m OC layer is divided. The results are summarized in E-Table 13.3, showing a total limit shaft resistance Q_{sL} of 106 kN.
The pile base capacity is mobilized in the stiff layer. The ultimate unit base resistance of the pile is calculated as

$$q_{b,ult} = 9.6s_u = 9.6\sigma'_v$$

$$\sigma'_v = [(16 - 9.81) \times 8] + [(18 - 9.81) \times 2] = 65.9 \text{ kPa}$$

Therefore,

$$q_{b,ult} = 9.6 \times 65.9 = 633 \text{ kPa}$$

The pile base area is given by

$$A_b = \pi\left(\frac{B}{2}\right)^2 = 3.14 \times 0.15^2 = 0.071 \text{ m}^2$$

The ultimate base resistance of the pile is obtained by multiplying the ultimate unit base resistance with the pile base area:

$$Q_{b,ult} = q_{b,ult}A_b = 633 \times 0.071 = 45 \text{ kN}$$

The ultimate resistance of the pile is then obtained by adding together the limit shaft resistance and the ultimate base resistance:

$$Q_{ult} = Q_{sL} + Q_{b,ult} = 106 + 45 = 151 \text{ kN} \quad \text{answer}$$

E-Table 13.3 Calculations of shaft resistance of the drilled shaft of Example 13.4

Depth (m)	s_u (kPa)	α	q_{sLi} (kPa)	$Q_{sLi} = q_{sLi}A_{si}$ (kN)
0.5	0.9	1.04	1.0	0.9
1.5	2.8	1.00	2.8	2.6
2.5	4.6	0.96	4.5	4.2
3.5	6.5	0.93	6.0	5.7
4.5	8.4	0.89	7.5	7.0
5.5	10.2	0.86	8.8	8.3
6.5	12.1	0.83	10.1	9.5
7.5	13.9	0.81	11.2	10.6
8.5	53.6	0.54	29.0	27.3
9.5	61.8	0.51	31.7	29.9
				$\Sigma = 106$ kN

Example 13.5

Calculate the ultimate resistance and the allowable load for a drilled shaft with diameter $B = 0.4$ m and length $L = 12$ m, with the base embedded 3 m into a layer of sand with critical-state friction angle equal to 30°, relative density equal to 70%, and unit weight equal to 19 kN/m³. A 9-meter-thick normally consolidated clay layer with unit weight equal to 17 kN/m³ and $s_u/\sigma_v' = 0.25$ is located above the sand layer. The clay has a critical-state friction angle $\phi_c = 25°$ and a minimum residual-state friction angle $\phi_{r,min} = 13°$. The water table is located 2 m below the ground surface.

Solution:

The ultimate unit base resistance of the drilled shaft is calculated using Equation (13.30):

$$\frac{q_{b,ult}}{p_A} = 62\left(\frac{D_R}{100}\right)^{1.83}\left(\frac{\sigma_h'}{p_A}\right)^{0.4}$$

Assuming that the sand is normally consolidated, we can take K_0 as 0.5. The vertical and horizontal effective stresses at a depth of 12 m are:

$$\sigma_v' = (17 \times 9) + (19 \times 3) - (9.81 \times 10) = 111.9 \text{ kPa}$$

$$\sigma_h' = 0.5 \times 111.9 = 56 \text{ kPa}$$

It follows that

$$q_{b,ult} = 62\left(\frac{D_R}{100}\right)^{1.83}\left(\frac{\sigma_h'}{p_A}\right)^{0.4} p_A = 62 \times \left(\frac{70}{100}\right)^{1.83} \times \left(\frac{56}{100}\right)^{0.4} \times 100 = 2560 \text{ kPa}$$

The pile base area is given by

$$A_b = \pi\left(\frac{B}{2}\right)^2 = \pi\left(\frac{0.4}{2}\right)^2 = 0.126 \text{ m}^2$$

The ultimate base resistance of the pile is then calculated as

$$Q_{b,ult} = q_{b,ult} A_b = 2560 \times 0.126 = 323 \text{ kN}$$

The limit unit shaft resistance is calculated in three different layers: in the clay above the water table (calculations at $z = 1$ m), in the clay below the water table (calculations at $z = 5.5$ m), and in the sand layer (calculations at $z = 10.5$ m).

For the first layer ($z = 0$–2 m):

The vertical effective stress at the center of the layer is

$$\sigma_v' = 17 \times 1 = 17 \text{ kPa}$$

Because $\phi_c - \phi_{r,min} = 25° - 13° = 12°$, the value of A_1 is equal to 0.4. The value of A_2 is calculated using

$$A_2 = 0.4 + 0.3 \ln\left(\frac{s_u}{\sigma_v'}\right) = 0.4 + 0.3 \ln(0.25) = -0.016$$

The value of α is calculated using

$$\alpha = \left(\frac{s_u}{\sigma'_v}\right)^{-0.05}\left[A_1 + (1-A_1)e^{-\left(\frac{\sigma'_v}{p_A}\right)(\phi_c-\phi_{r,min})^{A2}}\right]$$

$$= (0.25)^{-0.05}\left[0.4 + (1-0.4)e^{-\left(\frac{17}{100}\right)(12)^{-0.016}}\right]$$

$$= 0.97$$

The undrained shear strength at the center of the layer is given by

$$s_u - 0.25\sigma'_v - 0.25 \times 17 - 4.25 \text{ kPa}$$

The limit unit shaft resistance of the pile segment in contact with the first layer is calculated using Equation (13.33):

$$q_{sL1} = 0.97 \times 4.25 = 4.1 \text{ kPa}$$

For the second layer ($z = 2$–9 m):
The vertical effective stress at the center of the layer is:

$$\sigma'_v - (17 \times 5.5) \ (9.81 \times 3.5) - 59.2 \text{ kPa}$$

The values of A_1 and A_2 are unchanged because the critical-state friction angle, the minimum residual-state friction angle, and the undrained strength ratio have not changed, so the value of α is

$$\alpha = \left(\frac{s_u}{\sigma'_v}\right)^{-0.05}\left[A_1 + (1-A_1)e^{-\left(\frac{\sigma'_v}{p_A}\right)(\phi_c-\phi_{r,min})^{A2}}\right]$$

$$= (0.25)^{-0.05}\left[0.4 + (1-0.4)e^{-\left(\frac{59.2}{100}\right)(12)^{-0.016}}\right]$$

$$= 0.79$$

The undrained shear strength at the center of the layer is given by

$$s_u = 0.25\sigma'_v = 0.25 \times 59.2 = 14.8 \text{ kPa}$$

It follows that

$$q_{sL2} = 0.79 \times 14.8 = 11.7 \text{ kPa}$$

For the third layer ($z = 9$–12 m):
We will use Equations (13.28) and (13.29) to estimate the limit unit shaft resistance of the pile segment in contact with the sand layer:

$$q_{sL} = K\sigma'_v \tan\delta_c$$

$$K = \frac{K_0}{e^{0.3\sqrt{K_0 - 0.4}}} 0.67 \exp\left\{\frac{D_R}{100}\left[1.5 - 0.35 \ln\left(\frac{\sigma_v'}{p_A}\right)\right]\right\}$$

The vertical and horizontal effective stresses at the center of the layer are:

$$\sigma_v' = (17 \times 9) + (19 \times 1.5) - (9.81 \times 8.5) = 98.1 \text{ kPa}$$

$$\sigma_h' = 0.5 \times 98.1 = 49.1 \text{ kPa}$$

Therefore:

$$K = \frac{0.5}{e^{0.3\sqrt{0.5 - 0.4}}} 0.67 \exp\left\{\frac{70}{100}\left[1.5 - 0.35 \ln\left(\frac{98.1}{100}\right)\right]\right\} = 0.875$$

and

$$q_{sL3} = K\sigma_v' \tan\delta_c = 0.875 \times 98.1 \times \tan 30° = 49.6 \text{ kPa}$$

The limit shaft resistance of the pile is then calculated as

$$Q_{sL} = \sum\left(q_{sLi}A_{si}\right)$$

$$= q_{sL1}A_{s1} + q_{sL2}A_{s2} + q_{sL3}A_{s3}$$

$$= 4.1\left(\pi \times 0.4 \times 2\right) + 11.7\left(\pi \times 0.4 \times 7\right) + 49.6\left(\pi \times 0.4 \times 3\right)$$

$$= 300 \text{ kN}$$

The ultimate resistance of the pile is obtained by adding together the ultimate base resistance and the limit shaft resistance:

$$Q_{ult} = Q_{b,ult} + Q_{sL} = 323 + 300 = 623 \text{ kN}$$

Assuming that the factor of safety FS is equal to 3, the allowable load for the pile is:

$$Q_{all} = \frac{Q_{ult}}{FS} = \frac{623}{3} = 208 \text{ kN} \quad \text{answer}$$

13.6 CALCULATION OF THE ULTIMATE RESISTANCE OF DISPLACEMENT PILES

13.6.1 The relationship of pile installation to pile load response

In simple terms, the driving process imposes the following three types of motion on the soil around the pile: (1) relatively large-magnitude vertical shearing along the pile shaft, (2) vibration of the soil due to the hammer blows, and (3) radial compression of the soil around the pile. Vibration is associated with load reversals, which have an effect on soil contractiveness. It is easier to understand the effects that these processes have on q_{sL} if we focus our discussion on sand, but similar processes, though under undrained conditions, are at play in clays.

The value of K – the ratio of the lateral effective stress operative at the ULS to the initial vertical effective stress – depends on the tendency of the sand to contract or expand during driving and upon loading. For sand, this depends on the relative density of the sand. Aside from vibration, which tends to cause some soil densification, even in dense sands, the other two processes tend to produce dilation in dilative sands and densification in contractive sands. In an ideally contractive soil, we would not expect large increases of K over K_0 because the contractive soil would not "push back" against the pile as much due to its contractiveness upon shearing. Indeed, for loose sands, K values can be materially lower than K_0 at shallow depths (due to shaft resistance degradation, discussed in more detail next) and not materially greater than it at greater depths. On the other hand, when piles are driven into relatively dense sands, whose natural tendency is to dilate, dilatancy generates large effective normal stresses against the pile shaft, even in the presence of vibration, because volume change is somewhat constrained. Additionally, after installation, on loading, shearing develops between the pile shaft and the soil. A dilative sand will still be dilative even after driving; given that it is constrained from dilating, it will generate additional normal compressive stresses against the pile shaft. As a result, the values of K can be significantly greater than K_0 (and in excess of 1 even at relatively large depths) for very dense sands. Example 13.6 illustrates this using one of the methods of analysis for piles in sand. Unfortunately, rigorous analyses of partial- and full-displacement piles are not yet possible because modeling the effects of driving, jacking, or drilling a pile into place on the surrounding soil is not yet possible from both a fundamental and computational perspective. So, for these piles, we still rely on relationships resulting mostly from experimental research coupled with experience.

13.6.2 Unit shaft resistance degradation

Randolph (2003) discusses in detail how the unit shaft resistance at a fixed depth degrades as a pile is driven deeper into the ground. This would imply that the longer the pile is, the less the local unit shaft resistance available at a fixed depth is. The reason for this degradation in unit shaft resistance seems to be the progressive contraction of the soil due to exposure to a larger number of vibration cycles as the pile is driven deeper and deeper. As a result of the soil's contraction, it retracts and "moves away" from the pile, leading to a decrease in the value of K. Randolph (2003) provides an equation for K as a function of depth z:

$$K = K_{\min} + \left(K_{\max} - K_{\min}\right)\exp\left(-\alpha\frac{h}{B}\right) \qquad (13.35)$$

where $h = L - z$; B, L = pile diameter and length; α = coefficient of degradation rate = 0.05; $0.2 \leq K_{\min} \leq 0.4$; and $K_{\max}$ can be taken as 1%–2% of q_c/σ_v'. The variable h is a measure of how many loading cycles (how many hammer blows) the pile has experienced.

The assumption behind this mathematical form for K is that vibrations caused by the hammer blows degrade the unit shaft resistance due to the soil's contractive response when subjected to vibration. At the pile base, $h = 0$, and the soil will have experienced only one blow, and there will have been no degradation. The greater the value of h, the greater the degradation and the smaller the value of q_{sL} at a given depth.

The shaft resistance calculations in the modern pile design methods that we consider in this chapter all incorporate pile shaft resistance degradation in some form. It is possible that the normalization of h by B is not justified. Degradation would be a function of the number of loading cycles and would not be impacted by the pile diameter. The PPDM accordingly has a different normalization for it.

13.6.3 Variation of driven pile resistance with time

Seed and Reese (1955) observed a sixfold increase in the capacity of a pile driven in clay. This has been observed many times since then, and not exclusively in clays. The term used for it is "pile setup" or "ground freeze," indicating an increase in bearing capacity over time. A decrease in bearing capacity over time is referred to as relaxation.

The phenomenon is reasonably well understood, at least in qualitative terms. It is mostly due to the dissipation of the excess pore pressures generated during driving and to the healing of remolded soil near the pile over time. Randolph and Wroth (1979) proposed a procedure, based on theoretical considerations, to estimate the amount of setup. In design, it is desirable to account for setup; otherwise, we could underestimate significantly the pile capacities in soils where setup is considerable. However, confidence in estimating setup is not yet such that it can be routinely considered in calculations. There is considerable doubt, for example, as to whether dynamic load tests, discussed in the next chapter, can be used to accurately account for pile capacity gains with time. It is particularly uncertain whether capacity gains in sand can be as large as some data show. The nature of the sand (e.g., carbonate versus silica sand), for example, could be a factor. Additionally, if setup in clay will take a long time, that capacity gain would not be helpful if the pile will be loaded earlier than the gain in capacity can be achieved.

Design methods will ideally provide some indication of how long after installation the calculated pile resistances apply. The methods emphasized in this chapter have been validated by comparison to the results of static pile load tests performed from several days to a few weeks after installation.

13.6.4 Displacement piles in sandy soil

Tables 13.1–13.4 show the equations for the PPDM, ICPDM, UWAPDM, and NGIPDM for calculating both limit unit shaft resistance and ultimate unit base resistance for displacement piles in sand. The authors of the ICPDM, UWAPDM, and NGIPDM joined efforts to propose a "unified" method for driven piles in sand, which we refer to as the UPDM, given in Table 13.5.

The five methods account for shaft resistance degradation due to driving. As discussed earlier, shaft resistance decreases, all other things being equal, with increasing distance from the pile base. Shaft resistance degradation is calculated by having the limit unit shaft resistance q_{sL} be a function of the distance h from the pile base to the cross section at which it is calculated. Differently from other methods, normalization of h is done in the PPDM with respect to $L_R = 1$ m or the equivalent in other units. This is done because degradation is related to the number of blows, and thus approximately to the distance from the cross section to the pile base, but that does not depend on pile diameter.

The PPDM relies on the fundamental shaft resistance equation, already discussed in connection with nondisplacement piles. This amounts to setting q_{sL} equal to the lateral effective stress, which is normal to the pile shaft, multiplied by the tangent of the interface friction angle. The ICPDM and UWAPDM rely on equations that directly relate q_{sL} to q_c, but the equation in the ICPDM and one of the UWA unit limit shaft resistance equations are in truth effective stress-based equations, with the effective stress related to q_c and that multiplied by the tangent of the critical-state friction angle.

The critical-state interface friction angle δ_c can be determined for use in the PPDM, and possibly in the other methods as well, as follows:

1. For precast concrete piles, $\delta_c/\phi_c = 0.95$.
2. For cast-in-place concrete piles, $\delta_c/\phi_c = 1.00$.
3. For steel piles, we can obtain the value of δ_c/ϕ_c from Figure 13.19. If the D_{50} and C_U values of the sand are unknown, then we can conservatively assume $\delta_c/\phi_c = 0.80-0.85$.

A significant difference between the methods is in the way they handle open-ended pipe piles. With respect to this, the NGIPDM's equal treatment of open-ended and closed-ended pipe piles in terms of their shaft resistance estimation is a shortcoming of the method.

For base resistance calculations, all five methods average cone resistance around the pile base according to some formula, and all relate the ultimate unit base resistance to cone resistance, which serves as a proxy for the limit unit base resistance of the pile.

Table 13.1 PPDM equations for the unit base and shaft resistances for displacement piles driven in sand (Han et al. 2019a)

Pile type	Limit unit shaft resistance q_{sL}	Ultimate unit base resistance $q_{b,ult}$
Closed-ended pipe pile	$q_{sL} = K\sigma_v' \tan\delta_c$	$q_{b,ult} = (1 - 0.0058 D_R)q_{cb}$
	$K = K_{min} + (K_{max} - K_{min})\exp\left(\dfrac{-\alpha h}{L_R}\right)$	
	$K_{min} = 0.2; \; K_{max} = 0.01(q_c/p_A)/\sqrt{\sigma_h'/p_A}$	
	$\alpha = 0.14$	
Open-ended pipe pile	$q_{sL} = K(1 - 0.66\text{PLR})\sigma_v' \tan\delta_c$	$q_{b,ult} = 0.21(\text{IFR})^{-1.2} q_{cb} \leq 0.6 q_{cb}$
	$K = K_{min} + (K_{max} - K_{min})\exp\left(\dfrac{-\alpha h}{L_R}\right)$	$\text{IFR} = \min\left[1;\left(\dfrac{B_i}{1.5 L_R}\right)^{0.2}\right]$
	$K_{min} = 0.2; \; K_{max} = 0.01(q_c/p_A)/\sqrt{\sigma_h'/p_A}$	
	$\alpha = 0.14$	

Notes:

1. The exponential term in the equation for K accounts for shaft resistance degradation due to driving.
2. For open-ended pipe piles, the plug length ratio (PLR) used in the equation for q_{sL} is that measured at the specific depth where q_{sL} is calculated. If the PLR is not measured, it can be approximated using the same equation provided for the IFR.
3. IFR is the incremental filling ratio averaged over the last $3B$ of pile driving; if not measured, it can be estimated using the equation provided, which was proposed by Lehane et al. (2005).
4. Notation: σ_v' = initial vertical effective stress at the depth being considered; δ_c = critical-state interface friction angle (determined from Figure 13.19); B_i = inner diameter of open-ended pipe pile; L_R = reference length (= 1 m); h = distance from the pile base to the depth being considered; q_c = cone resistance; and q_{cb} = value of q_c for pile base resistance calculation, estimated as the average of q_c from $1B$ above to $2B$ below the pile base.

Table 13.2 ICPDM equations for the unit base and shaft resistances for displacement piles driven in sand (Jardine et al. 2005; Yang et al. 2015)

Pile type	Limit unit shaft resistance q_{sL}	Ultimate unit base resistance $q_{b,ult}$
Closed-ended pipe pile	$q_{sL} = \begin{cases} (\sigma_{rc}' + \Delta\sigma_{rd}')\tan\delta_c & \text{for compression} \\ (0.8\sigma_{rc}' + \Delta\sigma_{rd}')\tan\delta_c & \text{for tension} \end{cases}$	$q_{b,ult} = \max\left[0.3, 1 - 0.5\log_{10}(B/d_c)\right]q_{cb}$
	$\sigma_{rc}' = 0.029 q_c \left(\dfrac{\sigma_v'}{p_A}\right)^{0.13}\left(\max\left[\dfrac{h}{R},8\right]\right)^{-0.38}$	
	$\Delta\sigma_{rd}' = 2G\Delta r/R$	
	$G = q_c\left[0.0203 + 0.00125\eta - 1.216 \times 10^{-6}\eta^2\right]^{-1}$	
	$\eta = q_c(p_A\sigma_v')^{-0.5}$	
	$\Delta r = 0.02\,\text{mm}$ for lightly rusted steel piles	

(Continued)

Table 13.2 (Continued) ICPDM equations for the unit base and shaft resistances for displacement piles driven in sand (Jardine et al. 2005; Yang et al. 2015)

Pile type	Limit unit shaft resistance q_{sL}	Ultimate unit base resistance $q_{b,ult}$
Open-ended pipe pile	Use the same equations as for closed-ended pipe pile, but with an equivalent pile radius R given by $$R = \sqrt{R_o^2 - R_i^2}$$ For piles in tension, the value of q_{sL} is decreased further by 10%	Pile response during static loading as an unplugged pile if: $B_i/L_R \geq 0.02(D_R - 30)$ or $$\frac{B_i}{d_c} \geq 0.083 \frac{q_{cb}}{p_A}$$ $$q_{b,ult} = \max\left[0.15, 0.5 - 0.25\log_{10}(B/d_c),\right.$$ $$\left.\left(1 - \frac{R_i^2}{R_o^2}\right)\right] q_{cb}$$ for plugged piles $$q_{b,ult} = q_{cb}\left(1 - \frac{R_i^2}{R_o^2}\right)$$ for unplugged piles
H-pile	Use the same equations as for closed-ended pipe pile, but with an equivalent pile radius R given by $$R = \sqrt{\frac{A_b}{\pi}}$$ $A_b = 2b_f t_f + (2X_p + t_w)(d - 2t_f)$ $X_p = b_f/8$ if $b_f/2 < (d - 2t_f) < b_f$ $X_p = b_f^2/[16(d - 2t_f)]$ if $(d - 2t_f) \geq b_f$	$q_{b,ult} = q_{cb}$
Square or rectangular pile	Use the same equations as for closed-ended pipe pile, but with an equivalent pile radius R given by $$R = \sqrt{\frac{A_b}{\pi}}$$ where $A_b = B_w B_l$, and B_w and B_l = width and length of the pile cross section (in plan)	$q_{b,ult} = 0.7 q_{cb}$

Notes:

1. The method is intended to predict the pile capacity 10 days after driving for "virgin" piles (that is, piles that have not been load-tested).
2. The pile radius R for H-piles and open-ended pipe piles is an equivalent radius.
3. The representative cone resistance q_{cb} for base resistance calculations is q_c averaged from $1.5B$ above to $1.5B$ below the pile base.
4. Notation: σ'_{rc} = local radial effective stress acting on the pile segment after installation; $\Delta\sigma'_{rd}$ = increase in local radial effective stress associated with constrained dilation during pile loading; σ'_v = initial vertical effective stress at the depth being considered; δ_c = critical-state interface friction angle; B = pile diameter; B_i = inner diameter of open-ended pipe pile; d_c = cone diameter; R = pile radius; h = distance from the pile base to the depth being considered; q_c = cone resistance; R_o = outer radius of open-ended pipe pile; R_i = inner radius of open-ended pipe pile; A_b = area of pile base; G = shear modulus; Δr = radial displacement of soil during loading; b_f = width of flange; d = depth of H-section; t_f = thickness of flange; and t_w = thickness of web.

Table 13.3 UWAPDM equations for the unit base and shaft resistances for displacement piles driven in sand (Lehane et al. 2005)

Pile type	Limit unit shaft resistance q_{sL}	Ultimate unit base resistance $q_{b,ult}$
Closed-ended pipe pile	$$q_{sL} = \frac{f}{f_c}(\sigma'_{rc} + \Delta\sigma'_{rd})\tan\delta_c$$ $$\sigma'_{rc} = 0.03q_c\left[\max\left(\frac{h}{B},2\right)\right]^{-0.5}$$ $$\Delta\sigma'_{rd} = \frac{4G\Delta r}{B}$$ $$\frac{G}{q_c} = 185\left[\frac{q_c/p_A}{(\sigma'_v/p_A)^{0.5}}\right]^{-0.75}$$ $\Delta r = 0.02\,\text{mm}$; $f/f_c = 1$ for compression and 0.75 for tension	$q_{b,ult} = 0.6q_{cb}$
Open-ended pipe pile	$$q_{sL} = \frac{f}{f_c}(\sigma'_{rc} + \Delta\sigma'_{rd})\tan\delta_c$$ $$\sigma'_{rc} = 0.03q_c(A_{rs})^{0.3}\left[\max\left(\frac{h}{B},2\right)\right]^{-0.5}$$ $$\Delta\sigma'_{rd} = \frac{4G\Delta r}{B}; \quad A_{rs} = 1 - \text{IFR}\left(\frac{B_i}{B}\right)^2$$ $$\frac{G}{q_c} = 185\left[\frac{q_c/p_A}{(\sigma'_v/p_A)^{0.5}}\right]^{-0.75}$$ $\Delta r = 0.02\,\text{mm}$; $f/f_c = 1$ for compression and 0.75 for tension. IFR is the average incremental filling ratio measured over the final $20B$ of pile driving; when plug length measurement is not available, it can be estimated using: $$\text{IFR} = \min\left[1; \left(\frac{B_i}{1.5L_R}\right)^{0.2}\right]$$	$q_{b,ult} = (0.15 + 0.45A_{rb})q_{cb}$ $$A_{rb} = 1 - \text{FFR}\left(\frac{B_i}{B}\right)^2$$ FFR is the average final filling ratio measured over the final $3B$ of pile driving; if not measured, it can be roughly approximated by using the same equation for IFR

Notes:

1. The method is intended to predict the pile capacity measured 10–20 days after driving.
2. The representative cone resistance q_{cb} for base resistance calculation is q_c averaged using the Dutch technique: $q_{cb} = 0.5(q_{c1} + q_{c2})$, with $q_{c1} = 0.5(q_{c1a} + q_{c1b})$, q_{c1a} = average of the q_c values over a vertical distance of λB below the pile base, q_{c1b} = average of the q_c values over a vertical distance of λB below the pile base following a minimum path rule, and q_{c2} = average of the q_c values over a vertical distance of $8B$ above the pile base following a minimum path rule. The value of q_{c1} is calculated for different λ values ranging from 0.7 to 4.0, and the minimum value of q_{c1} obtained is used in the calculation of q_{cb}. Additional information about the computation of q_{c1} and q_{c2} can be found in Schmertmann (1978). For open-ended pipe piles, B is replaced by B_{eff} [$= B(A_{rb})^{0.5}$] in the calculation of q_{cb}.
3. Notation: σ'_{rc} = local radial effective stress acting on the pile segment after installation; $\Delta\sigma'_{rd}$ = increase in local radial effective stress associated with constrained dilation during pile loading; σ'_v = initial vertical effective stress at the depth being considered; δ_c = critical-state interface friction angle; A_{rs} = effective shaft area ratio; A_{rb} = effective base area ratio; B = pile diameter; B_i = inner diameter of open-ended pipe pile; L_R = reference length (= 1 m); h = distance from the pile base to the depth being considered; q_c = cone resistance; FFR = final filling ratio; IFR = incremental filling ratio; G = shear modulus; and Δr = radial displacement of soil during loading.

Table 13.4 NGIPDM equations for the unit base and shaft resistances for displacement piles driven in sand (Clausen et al. 2005)

Pile type	Limit unit shaft resistance q_{sL}	Ultimate unit base resistance $q_{b,ult}$
Closed-ended pipe pile	$q_{sL} = \max\left[p_A \left(\dfrac{z}{L}\right)\left(\dfrac{\sigma'_v}{p_A}\right)^{0.25} F_{DR}F_{tip}F_{load}F_{mat}, \; 0.1\sigma'_v \right]$ $F_{DR} = 2.1\left[D_R^* - 0.1\right]^{1.7}$ $D_R^* = 0.4 \ln\left[\dfrac{q_c}{22\left(\sigma'_v p_A\right)^{0.5}}\right]$ $F_{tip} = 1.6; F_{load} = 1.3$ for compression and 1.0 for tension; and $F_{mat} = 1.0$ for steel and 1.2 for concrete pile	$q_{b,ult} = \dfrac{0.8 q_{cb}}{1 + \left\{0.4\ln\left[\dfrac{q_{cb}}{22\left(\sigma'_{vb}p_A\right)^{0.5}}\right]\right\}^2}$
Open-ended pipe pile	Use the same equations as for closed-ended pipe pile, but with $F_{tip} = 1.0$	$q_{b,ult} = \dfrac{0.7 q_{cb}}{1 + 3\left\{0.4\ln\left[\dfrac{q_{cb}}{22\left(\sigma'_{vb}p_A\right)^{0.5}}\right]\right\}^2}$

Notes:

1. The $q_{b,ult}$ equation for an open-ended pipe pile describes the response of a plugged pile under static loading.

2. Notation: z = depth below the ground surface; L = embedded length of the pile; D_R^* = nominal relative density, which may be greater than 100%; p_A = reference stress (= 100 kPa); σ'_v = initial vertical effective stress at the depth being considered; q_c = cone resistance; q_{cb} = representative cone resistance at the pile base level (the method does not specify how to calculate q_{cb}); and σ'_{vb} = initial vertical effective stress at the pile base level.

Example 13.6

For a 15-meter-long pile, use the PPDM to calculate the profile of lateral effective stress ratio K on the shaft of the pile at full unit shaft resistance mobilization.

Solution:

The equation for the calculation of K can be obtained from Table 13.1:

$$K = K_{min} + \left(K_{max} - K_{min}\right)\exp\left(\frac{-\alpha h}{L_R}\right)$$

$$K_{max} = 0.01\left(q_c/p_A\right)/\sqrt{\sigma'_h/p_A}$$

$K_{min} = 0.2$ and $\alpha = 0.14$.

We can calculate q_c as a function of depth using Equation (7.19):

$$\frac{q_c}{p_A} = 1.64\exp\left[0.1041\phi_c + \left(0.0264 - 0.0002\phi_c\right)D_R\right]\left(\frac{\sigma'_h}{p_A}\right)^{0.841 - 0.0047 D_R}$$

Successive application of these equations to increasing depths produces the chart in E-Figure 13.4. Note how there are two processes at play: the increase in K due to constrained dilatancy and its drop due to degradation caused by vibrations in the soil due to hammer blows. These are competing processes that generate the pattern we see in the figure.

Table 13.5 UPDM equations for the unit base and shaft resistances for displacement piles driven in sand (Lehane et al. 2020)

Pile type	Limit unit shaft resistance q_{sL}	Ultimate unit base resistance $q_{b,ult}$
Closed-ended pipe pile	$q_{sL} = \dfrac{f_t}{f_c}(\sigma'_{rc} + \Delta\sigma'_{rd})\tan\delta_c$ $\sigma'_{rc} = \dfrac{q_c}{44}\left[\max\left(\dfrac{h}{B},1\right)\right]^{-0.4}$ $\Delta\sigma'_{rd} = 0.1q_c\left(\dfrac{q_c}{\sigma'_v}\right)^{-0.33}\left(\dfrac{d_c}{B}\right)$ $f_t/f_c = 1$ for compression and 0.75 for tension	$q_{b,ult} = 0.5q_{cb}$
Open-ended pipe pile	$q_{sL} = \dfrac{f_t}{f_c}(\sigma'_{rc} + \Delta\sigma'_{rd})\tan\delta_c$ $\sigma'_{rc} = \dfrac{q_c}{44}(A_{rs})^{0.3}\left[\max\left(\dfrac{h}{B},1\right)\right]^{-0.4}$; $A_{rs} = 1 - PLR\left(\dfrac{B_i}{B}\right)^2$ $\Delta\sigma'_{rd} = 0.1q_c\left(\dfrac{q_c}{\sigma'_v}\right)^{-0.33}\left(\dfrac{d_c}{B}\right)$ $f_t/f_c = 1$ for compression and 0.75 for tension PLR is the plug length ratio; when plug length measurements are not available, it can be estimated using: $PLR \approx \tanh\left[0.3\left(\dfrac{B_i}{d_c}\right)^{0.5}\right]$	$q_{b,ult} = (0.12 + 0.38A_{rb})q_{cb}$ $A_{rb} = 1 - FFR\left(\dfrac{B_i}{B}\right)^2$ FFR is the final filling ratio, which is defined as the average incremental filling ratio measured over the final $3B$ of pile driving; if not measured, it can be roughly approximated by using the same equation for the PLR

Notes:

1. The method predicts the ultimate load capacity Q_{ult} of the pile corresponding to a pile base settlement equal to 10% of the pile diameter. In addition, the method is intended to predict the pile capacity measured 14 days after driving.
2. For piles installed in relatively homogeneous sands, the representative cone resistance q_{cb} for base resistance calculation is q_c averaged from $1.5B$ above to $1.5B$ below the pile base. For piles installed in highly variable soil profiles (that is, when q_c varies significantly in the vicinity of the pile base), q_{cb} can be either taken as $1.2q_{c,Dutch}$, or estimated using the procedure developed by Boulanger and DeJong (2018); $q_{c,Dutch} = q_c$ averaged using the Dutch technique (Schmertmann 1978). For open-ended pipe piles, B is replaced by B_{eff} [$= B(A_{rb})^{0.5}$] in the calculation of q_{cb}.
3. Notation: UPDM = Unified Pile Design Method, σ'_{rc} = local radial effective stress acting on the pile segment after installation, $\Delta\sigma'_{rd}$ = increase in local radial effective stress associated with constrained dilation during pile loading, σ'_v = *in situ* vertical effective stress at the depth being considered, δ_c = critical-state interface friction angle (= 29° in the absence of laboratory interface shear test results), A_{rs} = effective shaft area ratio, A_{rb} = effective base area ratio, B_i = inner diameter of open-ended pipe pile, B_{eff} = effective pile diameter, d_c = cone diameter, h = vertical distance from the pile base to the depth being considered, and q_c = cone resistance.

13.6.5 Displacement piles in clayey soil

The unit base resistance of full-section displacement and nondisplacement piles in clays is almost as large as the cone resistance, except in stiff clays, when $q_{b,ult}/q_c$ may be somewhat less than 1. Since cone resistance is low in soft clays, base resistance is often not a major component of the ultimate resistance of piles in these soils. In soft clay, piles tend to be floating piles.

Tables 13.6–13.9 have the equations for the calculation of both shaft and base unit resistance for the PPDM, ICPDM, UWAPDM, and NGIPDM. The equations are of the same general form as those for nondisplacement piles in clay, with the same key factors appearing. For example, in the PPDM, the undrained shear strength, the critical-state friction angle, and the minimum residual friction angle are the key soil parameters. Other quantities appearing in the equation are either soil variables or equation parameters.

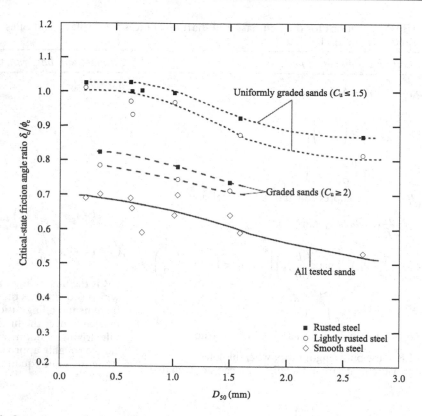

Figure 13.19 Critical-state friction angle ratio δ_c/ϕ_c versus mean particle size D_{50} for silica sands tested against smooth, lightly rusted, and rusted steel surfaces (Han et al. 2018a, 2019a). Interpolation can be used for $1.5 < C_U < 2$.

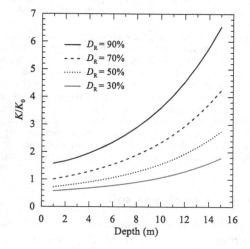

E-Figure 13.4 Coefficient K of lateral earth pressure used to calculate the limit unit shaft resistance q_{sL} of displacement of piles in sand using the PPDM.

The ICPDM relies on some key quantities: the residual interface friction angle δ_r, the sensitivity S_t of the clay, and the OCR of the clay. For piles in clay, δ_r can be taken as approximately equal to the residual friction angle ϕ_r of the clay. Recall from Chapter 6 that ϕ_r for clays can be as low as single digits and that it depends on the normal effective stress on the pile shaft, and thus on depth.

Table 13.6 PPDM equations for the unit base and shaft resistances for displacement piles driven in clay (Basu et al. 2009)

Time after installation	Limit unit shaft resistance q_{sL}	Ultimate unit base resistance $q_{b,ult}$
Short term (several days)	$q_{sL} = \alpha s_u$ $\alpha = A_1 + (1 - A_1)\exp\left[-\left(\dfrac{\sigma_v'}{p_A}\right)(\phi_c - \phi_{r,min})^{A_2}\right]$ $A_1 = 0.75$ for $\phi_c - \phi_{r,min} \leq 5°$, 0.43 for $\phi_c - \phi_{r,min} \geq 12°$, and a linearly interpolated value for $5° < \phi_c - \phi_{r,min} < 12°$ $A_2 = 0.55 + 0.43\ln\left(\dfrac{s_u}{\sigma_v'}\right)$	$q_{b,ult} = 10 s_u$
Long term (several months)	$q_{sL} = \alpha s_u$ $\alpha = 1.28\left(\dfrac{s_u}{\sigma_v'}\right)^{-0.05}\left\{A_1 + (1 - A_1)\exp\left[-\left(\dfrac{\sigma_v'}{p_A}\right)(\phi_c - \phi_{r,min})^{A_2}\right]\right\}$ $A_1 = 0.75$ for $\phi_c - \phi_{r,min} \leq 5°$, 0.43 for $\phi_c - \phi_{r,min} > 12°$, and a linearly interpolated value for $5° \leq \phi_c - \phi_{r,min} \leq 12°$ $A_2 = 0.64 + 0.4\ln\left(\dfrac{s_u}{\sigma_v'}\right)$	$q_{b,ult} = 12 s_u$

Notes:

1. The method is intended to estimate the shaft resistance after dissipation of the excess pore pressure generated during pile installation.
2. Notation: ϕ_c = critical-state friction angle; $\phi_{r,min}$ = minimum residual-state friction angle; s_u = undrained shear strength; σ_v' = initial vertical effective stress at the depth being considered, and p_A = reference stress ($\approx 100\,\text{kPa}$).

Table 13.7 ICPDM equations for the unit base and shaft resistances for displacement piles driven in clay (Jardine et al. 2005)

Pile type	Limit unit shaft resistance q_{sL}	Ultimate unit base resistance $q_{b,ult}$
Closed-ended pipe pile	$q_{sL} = 0.8K\sigma_v'\tan\delta_r$ $K = \left[2.2 + 0.016\,\text{OCR} - 0.87\Delta I_{vy}\right]\text{OCR}^{0.42}\left(\max\left[\dfrac{h}{R}, 8\right]\right)^{-0.20}$ $\Delta I_{vy} = \log_{10} S_t$ $S_t = \dfrac{s_u}{s_{ur}}$	$q_{b,ult} = 0.8q_{cb}$ for undrained loading $q_{b,ult} = 1.3q_{cb}$ for drained loading
Open-ended pipe pile	Use the same equations as for closed-ended pipe pile, but with an equivalent pile radius R given by $R = \sqrt{R_o^2 - R_i^2}$	The pile responds as a plugged pile during static loading if: $\dfrac{B_i}{d_c} + 0.45\dfrac{q_{cb}}{p_A} < 36$ *Response as a plugged pile during static loading:*

(Continued)

Table 13.7 (Continued) ICPDM equations for the unit base and shaft resistances for displacement piles driven in clay (Jardine et al. 2005)

Pile type	Limit unit shaft resistance q_{sL}	Ultimate unit base resistance $q_{b,ult}$
		$q_{b,ult} = 0.4q_{cb}$ for undrained loading $q_{b,ult} = 0.65q_{cb}$ for drained loading *Response as an unplugged pile during static loading:* Calculate base resistance using the annulus area multiplied by $q_{ann,ult}$ given by $q_{ann,ult} = q_{cb}$ for undrained loading $q_{ann,ult} = 1.6q_{cb}$ for drained loading
H-pile	Use the same equations as for closed-ended pipe pile, but with an equivalent pile radius R given by $$R = \sqrt{\frac{A_b}{\pi}}$$ $A_b = 2b_f t_f + (2X_p + t_w)(d - 2t_f)$ $X_p = b_f/8$ if $b_f/2 < (d - 2t_f) < b_f$ $X_p = b_f^2/[16(d - 2t_f)]$ if $(d - 2t_f) \geq b_f$	$q_{b,ult} = q_{cb}$
Square or rectangular pile	Use the same equations as for closed-ended pipe pile, but with an equivalent pile radius R given by $$R = \sqrt{\frac{A_b}{\pi}}$$ where $A_b = B_w B_l$, and B_w and B_l = width and length of the pile cross section (in plan)	$q_{b,ult} = 0.7q_{cb}$

Notes:

1. The method is intended to estimate the shaft resistance after dissipation of the excess pore pressure generated during pile installation.
2. The pile radius R for H-piles and open-ended pipe piles is an equivalent radius.
3. The representative cone resistance q_{cb} for base resistance calculations is q_c averaged from $1.5B$ above to $1.5B$ below the pile base.
4. The method does not account for the dependence of the residual interface friction angle δ_r on the effective normal stress acting on the pile–soil interface during shearing.
5. Notation: σ'_v = initial vertical effective stress at the depth being considered; δ_r = residual interface friction angle (determined from interface ring shear test results); A_b = area of pile base; d_c = cone diameter; R = pile radius; h = distance from the pile base to the depth being considered; OCR = overconsolidation ratio; s_u = undrained shear strength; ΔI_{vy} = relative void index at yield in e–$\log\sigma'_v$ space; S_t = sensitivity; s_{ur} = remolded undrained shear strength = $0.017[10^{2(1-LI)}]p_A$ (Randolph and Wroth 1979); p_A = reference stress (= 100 kPa); LI = liquidity index; q_{cb} = representative cone resistance at pile base level; R_o = outer radius of open-ended pipe pile; R_i = inner radius of open-ended pipe pile; b_f = width of flange; d = depth of H-section; t_f = thickness of flange; and t_w = thickness of web.

Table 13.8 UWAPDM equations for the unit base and shaft resistances for displacement piles driven in clay (Lehane 2019; Lehane et al. 2013)

Pile type	Limit unit shaft resistance q_{sL}	Ultimate unit base resistance $q_{b,ult}$
Closed-ended pipe pile	$$q_{sL} = \dfrac{0.23q_t\left[\max\left(\dfrac{h}{R},1\right)\right]^{-0.2}}{\left(q_t/\sigma_v'\right)^{0.15}}\tan\delta_r$$ or $$q_{sL} = 0.055q_t\left[\max\left(\dfrac{h}{R},1\right)\right]^{-0.2}$$	$q_{b,ult} = 0.5q_{cb}$ for undrained loading
Open-ended pipe pile	Use the same equations as for closed-ended pipe pile, but with an equivalent pile radius R given by $$R = \sqrt{R_o^2 - R_i^2}$$	*Response as a plugged pile during static loading:* $q_{b,ult} = 0.5q_{cb}$ for undrained loading

Notes:

1. The method is intended to estimate the shaft resistance after dissipation of the excess pore water pressure generated during pile installation (Lehane 2019; Lehane et al. 2017).
2. Two equations were proposed for the limit unit shaft resistance q_{sL}, and the second one was reported by Lehane et al. (2013) to be slightly more reliable than the first. The second equation for q_{sL} can be used in the absence of measurement or other estimations of δ_r.
3. The method does not account for the dependence of the residual interface friction angle δ_r on the effective normal stress acting on the pile–soil interface during shearing.
4. Notation: δ_r = residual interface friction angle; R = pile radius; h = distance from the pile base to the depth being considered; $q_t = q_c + (1-a)u_2$, where q_c = cone resistance, a = cone area ratio, and u_2 = pore pressure measured at the shoulder position behind the cone face; σ_v' = initial vertical effective stress at the depth being considered; R_o = outer radius of open-ended pipe pile; and R_i = inner radius of open-ended pipe pile.

Table 13.9 NGIPDM equations for the unit base and shaft resistances for displacement piles driven in clay (Karlsrud et al. 2005)

Pile type	Limit unit shaft resistance q_{sL}	Ultimate unit base resistance $q_{b,ult}$
Closed-ended pipe pile	For NC clays with $s_u/\sigma_v' < 0.25$: $q_{sL} = \alpha s_u$ $\alpha = 0.32(PI - 10)^{0.3}$ for $0.20 \leq \alpha \leq 1.0$ For OC clays with $s_u/\sigma_v' > 1.0$: $q_{sL} = \alpha s_u F_{tip}$ $\alpha = 0.5\left(\dfrac{s_u}{\sigma_v'}\right)^{-0.3}$ $F_{tip} = 0.8 + 0.2\left(\dfrac{s_u}{\sigma_v'}\right)^{0.5}$ for $1.0 \leq F_{tip} \leq 1.25$ For clays with $0.25 < s_u/\sigma_v' < 1.0$: α is determined by interpolation between the above two cases	$q_{b,ult} = 9s_u$
Open-ended pipe pile	Use the same equations as for closed-ended pipe pile, but with $F_{tip} = 1.0$	*Response as a plugged pile:* $9s_u$ *Response as an unplugged pile:* NA

Notes:

1. The method is intended to predict the pile capacity measured 100 days after driving.
2. The undrained shear strength s_u is recommended to be determined from unconsolidated undrained (UU) triaxial compression test results.
3. Notation: σ_v' = initial vertical effective stress at the depth being considered, and PI = plasticity index.

Like all methods that incorporate shaft friction degradation, the degradation term is expressed as a function of the distance h of the pile cross section where q_{sL} is to be calculated to the pile base (often normalized by pile diameter B). As indicated earlier, it is likely though that the degradation depends only on h, not on h normalized with respect to radius or diameter of the pile.

Both the UWAPDM and the ICPDM have equations for q_{sL} relying on the use of a residual interface friction angle δ_r for the clay. According to the ICPDM, we should determine the residual interface friction angle δ_r of the sublayer from the results of ring shear interface tests performed for the applicable value of normal effective stress (Ramsey et al. 1998). In the absence of that, it is possible to estimate δ_r by recognizing that it varies with normal effective stress acting on the pile shaft. Recognizing that production piles are typically rough, so that the interface friction angle is approximately equal to the soil's residual friction angle, we can use Equation (6.53), reproduced below, to calculate δ_r along the pile length:

$$\delta_r \approx \phi_r = \phi_{r,\min} + \frac{\phi_c - \phi_{r,\min}}{1 + \dfrac{\sigma'}{\sigma'_{\text{median}}}}$$

In this equation, in the context of pile shaft resistance calculation, $\sigma' = \sigma'_{hds}$, which is the effective lateral stress on the pile operative at the time of shearing and can be obtained from the method equations themselves. From the equation for q_{sL} given in Table 13.7, we can infer

$$\sigma'_{hds} = 0.8 K \sigma'_v$$

where

$$K = \left[2.2 + 0.016\,\text{OCR} - 0.87\Delta I_{vy}\right]\text{OCR}^{0.42}\left(\max\left[\frac{h}{R}, 8\right]\right)^{-0.20}$$

for the ICPDM and, from Table 13.8,

$$\sigma'_{hds} = \frac{0.23 q_t \left[\max\left(\dfrac{h}{R}, 1\right)\right]^{-0.2}}{\left(q_t/\sigma'_v\right)^{0.15}} \tag{13.36}$$

for the UWAPDM.

The remaining variables were introduced earlier in Section 6.4, but are repeated here for convenience: $\phi_{r,\min}$ is the absolute minimum residual friction angle, achieved only for sufficiently large effective normal stress on the plane of shearing (here, the pile shaft–soil interface), and σ'_{median} is the effective normal stress at which the residual friction angle ϕ_r is equal to the average of the critical-state friction angle ϕ_c and $\phi_{r,\min}$.

13.6.6 Piles in tension

The shaft resistance of a displacement pile installed in sand loaded in tension is 40 to 50% lower than that of the same displacement pile loaded in compression. This was shown through the calibration chamber testing performed by Galvis-Castro et al. (2019b). In these tests, model piles were first installed by multi-stroke jacking in sand samples prepared in a large chamber and then loaded in tension and compression. Shaft resistance was consistently lower in tension.

The reason for this is that the soil fabric next to the pile shaft aligns with the direction of installation, which is down. So, the fabric is mechanically strong for piles loaded in compression after installation, but weaker when loaded in tension.

The same is observed to a lesser extent in clay, but the ratio of q_{sL} in tension to q_{sL} in compression has not been sufficiently studied. Some methods have a factor to account for that based on a limited number of load tests performed in the field.

13.6.7 Examples of calculations of the ultimate resistance of displacement piles

Example 13.7

A 15-meter-long precast concrete pile having a diameter $B = 0.3$ m is driven through a sand layer ($\gamma_{sat} = 22$ kN/m^3 and $\phi_c = 30°$) with uniform relative density extending down to a depth of 10 m. The q_c profile for the sand layer is given in E-Figure 13.5. The water table is at the surface of the deposit. Calculate and plot the values of K and q_{sL} versus depth for this pile using the PPDM. The deposit is normally consolidated, having a K_0 of 0.4.

Solution:

We will illustrate the calculation of K and q_{sL} by performing these calculations for a depth of 5 m. The value of q_c at a depth of 5 m, as read from E-Figure 13.5, is 4392 kPa. The coefficient of lateral earth pressure is calculated using

$$K = K_{min} + \left(K_{max} - K_{min}\right)\exp\left(\frac{-\alpha h}{L_R}\right)$$

We now set about calculating the quantities that appear in this equation:

$$K_{max} = 0.01\left(q_c/p_A\right)/\sqrt{\sigma'_h/p_A} = 0.01\left(4392/100\right)/\sqrt{0.4 \times (22 - 9.81) \times 5/100} = 0.89$$

The default values of K_{min} and α are 0.2 and 0.14, respectively (Table 13.1). Now we can calculate K:

$$K = K_{min} + \left(K_{max} - K_{min}\right)\exp\left(\frac{-\alpha h}{L_R}\right)$$

$$= 0.2 + \left(0.89 - 0.2\right)\exp\left[\frac{-0.14 \times (15 - 5)}{1}\right]$$

$$= 0.37$$

Finally, the limit unit shaft resistance is calculated as follows:

$$q_{sL} = K\sigma'_v \tan\delta_c = 0.37 \times \left[(22 - 9.81) \times 5\right] \times \tan(0.95 \times 30°) = 12.2 \text{ kPa}$$

E-Figure 13.6 shows the calculation results for the length of pile crossing the sand layer.

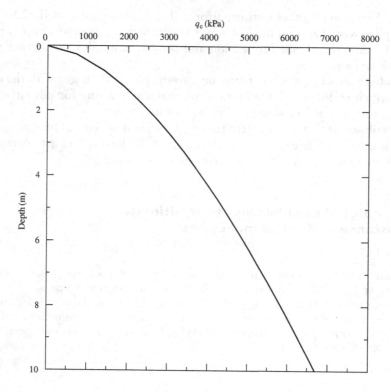

E-Figure 13.5 q_c versus depth profile for Example 13.7.

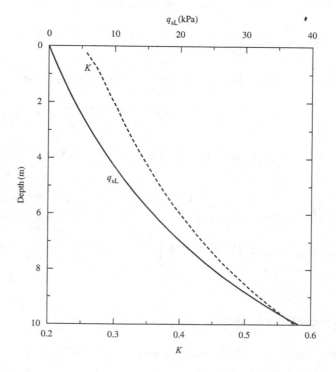

E-Figure 13.6 K and q_{sL} versus depth for the precast concrete pile of Example 13.7.

Example 13.8

Estimate the ultimate resistance of a closed-ended steel pipe pile with 300 mm diameter and 10 m length driven through the same soil profile of Example 13.4.

Solution:

We will use the PPDM (Table 13.6) to estimate the ultimate resistance of the pile. The undrained shear strength in the NC and OC clay layers is calculated using

$$s_u = 0.3\sigma'_v \quad \text{for the NC clay}$$

$$s_u = \sigma'_v \quad \text{for the OC clay}$$

If we divide the 8-meter-thick NC clay layer into eight layers of 1 m thickness, we can consider, as an illustration, the midpoint of the fifth layer, with depth equal to 4.5 m:

$$\sigma'_v = (16 - 9.81) \times 4.5 - 27.9 \text{ kPa}$$

$$s_u = 0.3\sigma'_v = 0.3 \times 27.9 = 8.4 \text{ kPa}$$

The value of α is calculated as

$$\alpha = 1.28 \left(\frac{s_u}{\sigma'_v} \right)^{-0.05} \left[A_1 + (1 - A_1)e^{\left(\frac{\sigma'_v}{p_A} \right)(\phi_c - \phi_{r,min})^{A_2}} \right]$$

Because $\phi_c - \phi_{r,min} = 25° - 13° = 12°$, the value of A_1 is equal to 0.43. The value of A_2 is calculated using

$$A_2 = 0.64 + 0.4 \ln \left(\frac{s_u}{\sigma'_v} \right) = 0.64 + 0.4 \ln(0.3) = 0.16$$

Thus:

$$\alpha = 1.28(0.3)^{-0.05} \left[0.43 + (1 - 0.43)e^{-\left(\frac{27.9}{100} \right)(12)^{0.16}} \right] = 1.10$$

The limit unit shaft resistance follows directly:

$$q_{sL5} = 1.1 \times 8.4 = 9.2 \text{ kPa}$$

This can now be multiplied by the shaft area of the pile segment in contact with this layer to calculate the contribution of the layer to the total shaft resistance:

$$Q_{sL5} = q_{sL5}A_{s5} = 9.2(\pi \times 0.3 \times 1) = 8.6 \text{ kN}$$

This calculation now needs to be repeated for all eight layers of the NC layer and the two additional layers into which the 2 m OC layer is divided. The results are summarized in E-Table 13.4, showing a total limit shaft resistance Q_{sL} of 124 kN.

E-Table 13.4 Calculations of shaft resistance of the closed-ended pipe pile of Example 13.8

Depth (m)	s_u (kPa)	α	q_{sLi} (kPa)	$Q_{sLi} = q_{sLi}A_{si}$ (kN)
0.5	0.9	1.32	1.2	1.2
1.5	2.8	1.26	3.5	3.3
2.5	4.6	1.20	5.6	5.3
3.5	6.5	1.15	7.5	7.0
4.5	8.4	1.10	9.2	8.6
5.5	10.2	1.05	10.7	10.1
6.5	12.1	1.01	12.2	11.5
7.5	13.9	0.97	13.6	12.8
8.5	53.6	0.60	32.3	30.5
9.5	61.8	0.59	36.2	34.1
				$\Sigma = 124$ kN

The pile base capacity is mobilized in the stiff layer. The ultimate unit base resistance of the pile is calculated as

$$q_{b,ult} = 10s_u = 10\sigma'_v$$

$$\sigma'_v = [(16 - 9.81) \times 8] + [(18 - 9.81) \times 2] = 65.9 \text{ kPa}$$

Therefore,

$$q_{b,ult} = 10 \times 65.9 = 659 \text{ kPa}$$

The pile base area is given by

$$A_b = \pi \left(\frac{B}{2}\right)^2 = 3.14 \times 0.15^2 = 0.071 \text{ m}^2$$

The ultimate base resistance of the pile is obtained by multiplying the ultimate unit base resistance with the pile base area:

$$Q_{b,ult} = q_{b,ult}A_b = 659 \times 0.071 = 47 \text{ kN}$$

The ultimate resistance of the pile is then obtained by adding together the limit shaft resistance and the ultimate base resistance:

$$Q_{ult} = Q_{sL} + Q_{b,ult} = 124 + 47 = 171 \text{ kN} \text{ }^{\text{answer}}$$

Example 13.9

Calculate the ultimate resistance and the allowable load for a prestressed square concrete pile with $B = 0.3$ m and length $L = 12$ m, with the base embedded 3 m into a layer of sand with critical-state friction angle equal to $30°$, relative density equal to 70%, and unit weight equal to 19 kN/m³. The q_c profile for the sand layer is given in E-Figure 13.7. A 9-meter-thick normally consolidated clay layer with unit weight equal to 17 kN/m³ and $s_u/\sigma'_v = 0.25$ is located above the sand layer. The clay has a critical-state friction angle $\phi_c = 25°$ and a minimum residual-state friction angle $\phi_{r,min} = 13°$. The water table is located 2 m below the ground surface.

Solution:

The PPDM is used to calculate the ultimate unit base resistance of the pile in the sand layer (Table 13.1):

$$q_{b,ult} = (1 - 0.0058D_R)q_{cb}$$

where q_{cb} is q_c averaged from $1B$ above to $2B$ below the pile base.

Averaging the cone resistance values between $1B$ (0.3 m) above and $2B$ (0.6 m) below the pile base, we obtain $q_{cb} = 11.6\,\text{MPa}$.

The ultimate unit base resistance of the pile will therefore be:

$$q_{b,ult} = [1 - 0.0058(70)] \times 11.6 = 6.9 \text{ MPa}$$

The pile base area is given by

$$A_b = B^2 = 0.3^2 = 0.09 \text{ m}^2$$

The ultimate base resistance of the pile will be:

$$Q_{b,ult} = q_{b,ult}A_b = 6.9 \times 0.09 = 0.621 \text{ MN} = 621 \text{ kN}$$

The limit shaft resistance is calculated in three different layers: in the clay above the water table (calculations at $z = 1\,\text{m}$), in the clay below the water table (calculations at $z = 5.5\,\text{m}$), and in the sand layer (calculations at $z = 10.5\,\text{m}$). The PPDM is used to calculate the limit unit shaft resistance in the clay layers using the equations given in Table 13.6.

<u>For the first layer ($z = 0$–$2\,\text{m}$):</u>

The vertical effective stress at the center of the layer is

$$\sigma'_v = 17 \times 1 = 17 \text{ kPa}$$

Because $\phi_c - \phi_{r,min} = 25° - 13° = 12°$, the value of A_1 is equal to 0.43. The value of A_2 is calculated using

$$A_2 = 0.64 + 0.4 \ln\left(\frac{s_u}{\sigma'_v}\right) = 0.64 + 0.4 \ln(0.25) = 0.085$$

The value of α is calculated using

$$\alpha = 1.28\left(\frac{s_u}{\sigma'_v}\right)^{-0.05}\left[A_1 + (1 - A_1)e^{-\left(\frac{\sigma'_v}{P_A}\right)(\phi_c - \phi_{r,min})^{A_2}}\right]$$

$$= 1.28(0.25)^{-0.05}\left[0.43 + (1 - 0.43)e^{-\left(\frac{17}{100}\right)(12)^{0.085}}\right]$$

$$= 1.22$$

The undrained shear strength at the center of the layer is given by

$$s_u = 0.25\sigma'_v = 0.25 \times 17 = 4.25 \text{ kPa}$$

The limit unit shaft resistance of the pile segment in contact with the first layer is

$$q_{sL1} = 1.22 \times 4.25 = 5.2 \text{ kPa}$$

<u>For the second layer ($z = 2$–$9\,\text{m}$):</u>

The vertical effective stress at the center of the layer is

$$\sigma'_v = (17 \times 5.5) - (9.81 \times 3.5) = 59.2 \text{ kPa}$$

The values of A_1 and A_2 are unchanged because the critical-state friction angle, the minimum residual friction angle, and the undrained strength ratio have not changed, so the value of α is

$$\alpha = 1.28 \left(\frac{s_u}{\sigma'_v}\right)^{-0.05} \left[A_1 + (1 - A_1)e^{-\left(\frac{\sigma'_v}{p_A}\right)\left(\phi_c - \phi_{r,min}\right)^{A_2}} \right]$$

$$= 1.28 (0.25)^{-0.05} \left[0.43 + (1 - 0.43)e^{-\left(\frac{59.2}{100}\right)(12)^{0.085}} \right]$$

$$= 0.97$$

The undrained shear strength at the center of the layer is given by

$$s_u = 0.25\sigma'_v = 0.25 \times 59.2 = 14.8 \text{ kPa}$$

It follows that

$$q_{sL2} = 0.97 \times 14.8 = 14.3 \text{ kPa}$$

For the third layer ($z = 9–12$ m):

The PPDM equation given in Table 13.1 is used to calculate the limit unit shaft resistance of the pile segment in contact with the sand layer:

$$q_{sL} = K\sigma'_v \tan \delta_c$$

with

$$K = K_{min} + (K_{max} - K_{min})\exp\left(\frac{-\alpha h}{L_R}\right)$$

where K_{min} and α are equal to 0.2 and 0.14, respectively, and

$$K_{max} = 0.01(q_c/p_A)/\sqrt{\sigma'_h/p_A}$$

The vertical and horizontal effective stresses at the center of the layer are

$$\sigma'_v = (17 \times 9) + (19 \times 1.5) - (9.81 \times 8.5) = 98.1 \text{ kPa}$$

$$\sigma'_h = 0.5 \times 98.1 = 49.1 \text{ kPa}$$

The representative cone resistance of the layer is read as 10.8 MPa from E-Figure 13.7. It follows that

$$K_{max} = 0.01(q_c/p_A)/\sqrt{\sigma'_h/p_A} = 0.01(10.8 \times 1000/100)/\sqrt{49.1/100} = 1.54$$

$$K = K_{min} + (K_{max} - K_{min})\exp\left(\frac{-\alpha h}{L_R}\right)$$

$$= 0.2 + (1.54 - 0.2)\exp\left[\frac{-0.14 \times (12 - 10.5)}{1}\right]$$

$$= 1.29$$

and

$$q_{sL3} = K\sigma'_v \tan\delta_c = 1.29 \times 98.1 \times \tan(0.95 \times 30°) = 68.7 \text{ kPa}$$

We are now ready to use Equation (13.3) to obtain the limit shaft resistance of the pile:

$$\begin{aligned}
Q_{sL} &= \sum(q_{sLi}A_{si}) \\
&= q_{sL1}A_{s1} + q_{sL2}A_{s2} + q_{sL3}A_{s3} \\
&= 5.2(4 \times 0.3 \times 2) + 14.3(4 \times 0.3 \times 7) + 68.7(4 \times 0.3 \times 3) \\
&= 380 \text{ kN}
\end{aligned}$$

The ultimate resistance of the pile is then obtained by adding together the ultimate base resistance and the limit shaft resistance:

$$Q_{ult} = Q_{b,ult} + Q_{sL} = 621 + 380 = 1001 \text{ kN}$$

Assuming that the factor of safety *FS* is equal to 3, the allowable load for the pile is

$$Q_{all} = \frac{Q_{ult}}{FS} = \frac{1001}{3} = 334 \text{ kN} \text{ answer}$$

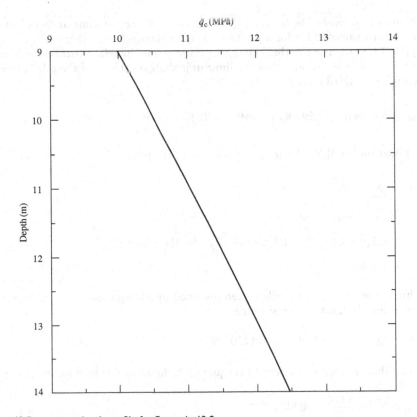

E-Figure 13.7 q_c versus depth profile for Example 13.9.

Example 13.10

For the soil profile given in Example 13.9, calculate the ultimate resistance and the allowable load for a closed-ended steel pipe pile with diameter $B = 0.4\,$m and length $L = 12\,$m.

Solution:

We will use the PPDM (Table 13.1) to compute the unit shaft and base resistances of the pile. The pile in this example has the same length as that in Example 13.9, but a different B, so the value of q_{cb} would not be strictly the same as in that example. Given, however, the smooth, quasi-linear nature of the q_c profile, the value results the same: $q_{cb} = 11.6\,$MPa.

The ultimate unit base resistance of the pile will be

$$q_{b,ult} = (1 - 0.0058 D_R)q_{cb} = \left[1 - 0.0058(70)\right] \times 11.6 = 6.9\ \text{MPa}$$

with the pile base area now given by

$$A_b = \pi r^2 = \pi \left(\frac{0.4}{2}\right)^2 = 0.126\ \text{m}^2$$

The ultimate base resistance of the pile is then obtained by multiplying the ultimate unit base resistance with the pile base area:

$$Q_{b,ult} = q_{b,ult} A_b = 6.9 \times 0.126 = 0.869\ \text{MN} = 869\ \text{kN}$$

For the first two layers, the limit unit shaft resistances are the same as those obtained previously in Example 13.9 for a prestressed square concrete pile, that is, $q_{sL1} = 5.2\,$kPa and $q_{sL2} = 14.3\,$kPa. For the third layer, the critical-state interface friction angle is calculated using $\delta_c/\phi_c = 0.8$, and thus the limit unit shaft resistance of the pile segment in contact with the third layer is

$$q_{sL3} = K\sigma_v' \tan\delta_c = 1.29 \times 98.1 \times \tan(0.8 \times 30°) = 56.3\ \text{kPa}$$

Using Equation (13.3), the limit shaft resistance of the pile is

$$\begin{aligned}
Q_{sL} &= \sum (q_{sLi} A_{si}) \\
&= q_{sL1} A_{s1} + q_{sL2} A_{s2} + q_{sL3} A_{s3} \\
&= 5.2(\pi \times 0.4 \times 2) + 14.3(\pi \times 0.4 \times 7) + 56.3(\pi \times 0.4 \times 3) \\
&= 351\ \text{kN}
\end{aligned}$$

The ultimate resistance of the pile is then obtained by adding together the ultimate base resistance and the limit shaft resistance:

$$Q_{ult} = Q_{b,ult} + Q_{sL} = 869 + 351 = 1220\ \text{kN}$$

Assuming that the factor of safety *FS* is equal to 3, the allowable load for the pile is

$$Q_{all} = \frac{Q_{ult}}{FS} = \frac{1220}{3} = 407\ \text{kN}\ \text{answer}$$

Example 13.11

E-Figure 13.8 shows the CPT log and soil profile at a site. Estimate the ultimate resistance of a 356-millimeter-diameter closed-ended steel pipe pile driven to a depth of 15.4 m. The water table is located at a depth of 4.3 m from the ground surface. The silty clay layer has the following properties: $\phi_c = 24°$ and $\phi_{r,min} = 12°$. The critical-state friction angle of the layers containing sand and sand–silt mixtures is 33° (the silt is nonplastic).[12]

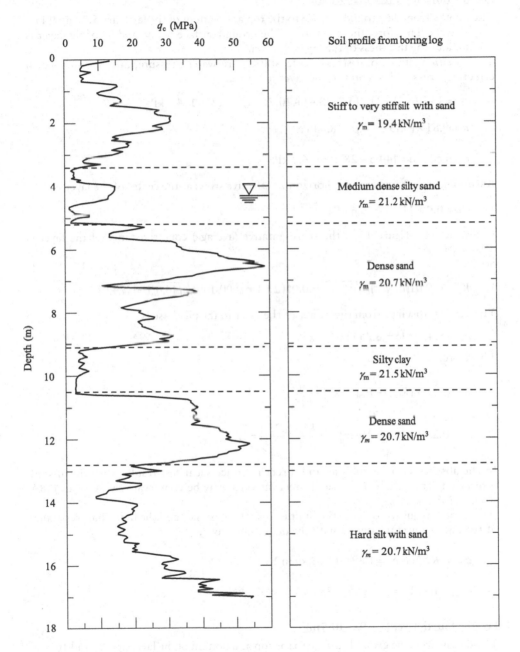

E-Figure 13.8 CPT log and soil profile for Example 13.11.

---

[12]This problem is in part based on real data obtained by Han et al. (2017b).

Solution:

We will use the PPDM (Tables 13.1 and 13.6) to solve the problem. E-Table 13.5 summarizes the shaft resistance calculations for the six layers that are in contact with the pile shaft. Example calculations for the third (dense sand) and fourth (silty clay) layers are presented below.

For the third layer ($z = 5.2$–9.1 m):

The depths from the ground surface to the top and bottom of the layer are 5.2 and 9.1 m, respectively. Therefore, the depth from the ground surface to the middle of the layer is 7.15 m, and the thickness of the layer is 3.9 m.

The *in situ* vertical total stress, hydrostatic pore water pressure, and *in situ* vertical effective stress at the center of the layer are given by

$$\sigma_v = 19.4(3.4) + 21.2(5.2 - 3.4) + 20.7(7.15 - 5.2) = 144.5 \text{ kPa}$$

$$u = 9.81 \times (7.15 - 4.3) = 28.0 \text{ kPa}$$

$$\sigma'_v = \sigma_v - u = 144.5 - 28.0 = 116.5 \text{ kPa}$$

Assuming $K_0 = 0.4$, the *in situ* horizontal effective stress at the center of the layer is

$$\sigma'_h = 0.4 \times 116.5 = 46.6 \text{ kPa}$$

Referring to E-Figure 13.8, the representative (average) cone resistance of the layer is 29.4 MPa.

So

$$K_{max} = 0.01(q_c/p_A)/\sqrt{\sigma'_h/p_A} = 0.01(29.4 \times 1000/100)/\sqrt{46.6/100} = 4.31$$

The vertical distance from the center of the layer to the pile base is

$$h = 15.4 - 7.15 = 8.25 \text{ m}$$

Therefore,

$$K = K_{min} + (K_{max} - K_{min})\exp\left(\frac{-\alpha h}{L_R}\right)$$

$$= 0.2 + (4.31 - 0.2)\exp\left(\frac{-0.14 \times 8.25}{1}\right) = 1.49$$

In the absence of sieve analysis and direct interface shear test results, the critical-state interface friction angle δ_c for sand–steel interface may be conservatively taken as $0.8\phi_c$ ($= 0.8 \times 33° = 26.4°$).

We are now ready to compute the limit unit shaft resistance and the limit shaft resistance of the pile segment in contact with the dense sand layer:

$$q_{sL3} = K\sigma'_v \tan\delta_c = 1.49 \times 116.5 \times \tan 26.4° = 86.2 \text{ kPa}$$

$$Q_{sL3} = q_{sL3}A_{s3} = 86.2(\pi \times 0.356 \times 3.9) = 376 \text{ kN}$$

For the fourth layer ($z = 9.1$–10.5 m):

The depths from the ground surface to the top and bottom of the layer are 9.1 and 10.5 m, respectively. Therefore, the depth from the ground surface to the middle of the layer is 9.8 m, and the thickness of the layer is 1.4 m.

The *in situ* vertical total stress, hydrostatic pore water pressure, and *in situ* vertical effective stress at the center of the layer are given by

$$\sigma_v = 19.4(3.4) + 21.2(5.2 - 3.4) + 20.7(9.1 - 5.2) + 21.5(9.8 - 9.1) = 200 \text{ kPa}$$

$$u = 9.81 \times (9.8 - 4.3) = 54 \text{ kPa}$$

$$\sigma'_v = \sigma_v - u = 200 - 54 = 146 \text{ kPa}$$

Referring to E-Figure 13.8, the representative cone resistance of the layer is about 2.6 MPa. Assuming a cone factor N_k of 12, the undrained shear strength at the center of the layer is calculated using Equation (7.22):

$$s_u = \frac{q_c - \sigma_v}{N_k} = \frac{2600 - 200}{12} = 200 \text{ kPa}$$

Because $\phi_c - \phi_{r,min} = 24° - 12° = 12°$, the value of A_1 is equal to 0.43. The value of A_2 is calculated using

$$A_2 = 0.64 + 0.4\ln\left(\frac{s_u}{\sigma'_v}\right) = 0.64 + 0.4\ln\left(\frac{200}{146}\right) = 0.77$$

So

$$\alpha = 1.28\left(\frac{s_u}{\sigma'_v}\right)^{-0.05}\left[A_1 + (1 - A_1)e^{-\left(\frac{\sigma'_v}{p_A}\right)(\phi_c - \phi_{r,min})^{A_2}}\right]$$

$$= 1.28\left(\frac{200}{146}\right)^{-0.05}\left[0.43 + (1 - 0.43)e^{-\left(\frac{146}{100}\right)\times 12^{0.77}}\right]$$

$$= 0.542$$

We are now ready to compute the limit unit shaft resistance and the limit shaft resistance of the pile segment in contact with the silty clay layer:

$$q_{sL4} = \alpha s_u = 0.542 \times 200 = 108.4 \text{ kPa}$$

$$Q_{sL4} = q_{sL4}A_{s4} = 108.4(\pi \times 0.356 \times 1.4) = 170 \text{ kN}$$

E-Table 13.5 summarizes the shaft resistance calculations for the six layers that are in contact with the pile shaft. The coefficient of lateral earth pressure at rest K_0 is taken as 0.45 for layer 2 and 0.40 for layers 1, 3, 5, and 6. Note that the layers consisting of sand–silt mixtures are treated as "sand" for the purpose of pile shaft capacity analysis because the silt is nonplastic. The limit shaft resistance of the pile is obtained by summing the last column of E-Table 13.5:

$$Q_{sL} = \sum_{i=1}^{6} Q_{sLi} = 49 + 33 + 376 + 170 + 626 + 464 = 1718 \text{ kN}$$

E-*Table 13.5* Calculation of limit shaft resistance of the closed-ended pipe pile of Example 13.11

Layer	Soil type	z_{top} (m)	z_{bottom} (m)	Δz (m)	z_{middle} (m)	q_c (MPa)	σ'_v (kPa)	σ'_h (kPa)	h (m)	K	s_u (kPa)	α	q_{sLi} (kPa)	Q_{sLi} (kN)
1	Sand	0.0	3.4	3.4	1.7	15.4	33.0	13.2	13.7	0.79	—	—	13.0	49
2	Sand	3.4	5.2	1.8	4.3	6.7	85.0	38.3	11.1	0.39	—	—	16.4	33
3	Sand	5.2	9.1	3.9	7.2	29.4	116.5	46.6	8.3	1.49	—	—	86.2	376
4	Clay	9.1	10.5	1.4	9.8	2.6	145.9	—	—	—	200	0.54	108.4	170
5	Sand	10.5	12.8	2.3	11.7	39.5	166.7	66.7	3.8	2.94	—	—	243.4	626
6	Sand	12.8	15.4	2.6	14.1	17.2	193.3	77.3	1.3	1.66	—	—	159.6	464

The representative cone resistance q_{cb} for use in pile base capacity calculation is obtained by averaging the q_c values between $1B$ $(0.356\,\text{m})$ above and $2B$ $(0.712\,\text{m})$ below the pile base, corresponding to a 15.0–$16.1\,\text{m}$ depth range. This yields $q_{cb} = 23.7\,\text{MPa}$.

The bearing layer for the pile base, which consists of hard silt with sand, is treated as "sand" for the purpose of pile base capacity analysis because the silt is nonplastic. The *in situ* vertical total stress, hydrostatic pore water pressure, and *in situ* vertical and horizontal effective stresses at depth $B/2$ $(0.178\,\text{m})$ below the pile base are given by

$$\sigma_v = 19.4(3.4) + 21.2(5.2 - 3.4) + 20.7(9.1 - 5.2) + 21.5(10.5 - 9.1) + 20.7(15.6 - 10.5)$$

$$= 320.5\,\text{kPa}$$

$$u = 9.81 \times (15.6 - 4.3) = 110.8\,\text{kPa}$$

$$\sigma_v' = \sigma_v - u = 320.5 - 110.8 = 209.7\,\text{kPa}$$

$$\sigma_h' = 0.4 \times 209.7 = 83.9\,\text{kPa}$$

The relative density of the bearing layer is calculated using Equation (7.20):

$$D_R = \frac{\ln\left(\dfrac{q_{cb}}{p_A}\right) - 0.4947 - 0.1041\phi_c - 0.841\ln\left(\dfrac{\sigma_h'}{p_A}\right)}{0.0264 - 0.0002\phi_c - 0.0047\ln\left(\dfrac{\sigma_h'}{p_A}\right)}$$

$$= \frac{\ln\left(\dfrac{23.7 \times 1000}{100}\right) - 0.4947 - 0.1041(33) - 0.841\ln\left(\dfrac{83.9}{100}\right)}{0.0264 - 0.0002(33) - 0.0047\ln\left(\dfrac{83.9}{100}\right)}$$

$$= 81.7\%$$

We are now ready to compute the ultimate unit base resistance and the ultimate base resistance of the pile:

$$q_{b,ult} = (1 - 0.0058 D_R)q_{cb}$$

$$= [1 - 0.0058(81.7)] \times 23.7 = 12.5\,\text{MPa}$$

$$Q_{b,ult} = q_{b,ult}A_b = 12.5 \times \left(\frac{\pi \times 0.356^2}{4}\right) = 1.244\,\text{MN} = 1244\,\text{kN}$$

The ultimate resistance of the pile is the summation of the limit shaft resistance and the ultimate base resistance:

$$Q_{ult} = Q_{sL} + Q_{b,ult} = 1718 + 1244 = 2962\,\text{kN} \quad \text{answer}$$

This example is based on real data (Han et al. 2017b), which included not only the CPT soundings, but also a static pile load test. The calculated value of Q_{ult} is conservative compared to that obtained from the static load test (3275 kN), underestimating it by about 10%. Note that the ultimate load capacity of the pile calculated using the PPDM corresponds to a pile head settlement of 35.6 mm, which is equal to 10% of the pile diameter. In this example, we divided the soil profile into six layers for the estimation of the limit shaft resistance of the pile. However, a more refined shaft resistance analysis is possible by further subdividing the soil profile into additional sublayers. This is an exercise proposed in Problem 13.31. Problem 13.32 asks the reader to recalculate the ultimate resistance of the pile using both the ICPDM and the UWAPDM.

13.6.8 Examples of calculations of the ultimate resistance of open-ended pipe piles and H-piles

Example 13.12

Estimate the ultimate resistance and the allowable load for the following piles: (a) open-ended steel pipe pile with $B = 0.406\,\text{m}$ and $0.635\,\text{cm}$ wall thickness, and (b) HP 14×117 H-pile. Consider the piles to have the same length $L = 12\,\text{m}$ and installed in the same soil profile as given in Example 13.9. The sensitivity S_t of the clay layer is 4.

Solution:

(a) Open-ended pipe pile

We will use the ICPDM (Tables 13.2 and 13.7) to estimate the pile shaft and base resistances. The outer diameter and wall thickness of the open-ended pipe pile are $0.406\,\text{m}$ and $0.635\,\text{cm}$, respectively. Therefore, the inner diameter of the pile is

$$B_i = 0.406 - (2 \times 0.00635) = 0.393\,\text{m}$$

The equivalent pile radius is given by

$$R = \sqrt{R_o^2 - R_i^2} = \sqrt{\left(\frac{0.406}{2}\right)^2 - \left(\frac{0.393}{2}\right)^2} = 0.05\,\text{m}$$

The limit unit shaft resistance of the pile is calculated in three different layers: in the clay above the water table (calculations at $z = 1\,\text{m}$), in the clay below the water table (calculations at $z = 5.5\,\text{m}$), and in the sand layer (calculations at $z = 10.5\,\text{m}$).

<u>For the first layer ($z = 0$–$2\,\text{m}$):</u>

The values of σ_v', ΔI_{vy}, and K are calculated using

$$\sigma_v' = 17 \times 1 = 17\,\text{kPa}$$

$$\Delta I_{vy} = \log S_t = \log(4) = 0.60$$

$$K = \left[2.2 + 0.016\,\text{OCR} - 0.87\Delta I_{vy}\right]\text{OCR}^{0.42}\left(\max\left[\frac{h}{R}, 8\right]\right)^{-0.20}$$

$$= \left[2.2 + 0.016(1) - 0.87(0.60)\right] \times \left(\max\left[\frac{11}{0.05}, 8\right]\right)^{-0.20}$$

$$= 1.694 \times (220)^{-0.20}$$

$$= 0.58$$

The effective lateral stress $\sigma'\,(= \sigma_h')$ on the pile operative at the time of shearing is given by

$$\sigma_h' = 0.8 K \sigma_v' = 0.8 \times 0.58 \times 17 = 7.9\,\text{kPa}$$

For $\sigma_{median}' = 50\,\text{kPa}$, the residual interface friction angle δ_r is calculated using

$$\delta_r \approx \phi_r = \phi_{r,min} + \frac{\phi_c - \phi_{r,min}}{1 + \dfrac{\sigma'}{\sigma_{median}'}} = 13° + \frac{25° - 13°}{1 + \dfrac{7.9}{50}} = 23.4°$$

The limit unit shaft resistance of the pile segment in contact with the first layer is

$$q_{sL1} = 0.8 K \sigma_v' \tan \delta_r = 0.8 \times 0.58 \times 17 \times \tan 23.4° = 3.4 \text{ kPa}$$

For the second layer ($z = 2$–9 m):

The value of ΔI_{vy} is the same as that calculated for the first layer. The values of σ_v', K, σ' ($= \sigma_h'$), and δ_r are given by

$$\sigma_v' = (17 \times 5.5) - (9.81 \times 3.5) = 59.2 \text{ kPa}$$

$$K = \left[2.2 + 0.016 \text{OCR} - 0.87 \Delta I_{vy} \right] \text{OCR}^{0.42} \left(\max \left[\frac{h}{R}, 8 \right] \right)^{-0.20}$$

$$= \left[2.2 + 0.016(1) - 0.87(0.60) \right] \times \left(\max \left[\frac{6.5}{0.05}, 8 \right] \right)^{-0.20}$$

$$= 1.694 \times (130)^{-0.20}$$

$$= 0.64$$

$$\sigma' = \sigma_h' = 0.8 K \sigma_v' = 0.8 \times 0.64 \times 59.2 = 30.3 \text{ kPa}$$

$$\delta_r \approx \phi_r = \phi_{r,min} + \frac{\phi_c - \phi_{r,min}}{1 + \dfrac{\sigma'}{\sigma'_{median}}} = 13° + \frac{25° - 13°}{1 + \dfrac{30.3}{50}} = 20.5°$$

It follows that

$$q_{sL2} = 0.8 K \sigma_v' \tan \delta_r = 0.8 \times 0.64 \times 59.2 \times \tan 20.5° = 11.3 \text{ kPa}$$

For the third layer ($z = 9$–12 m):

The representative cone resistance of the layer is read as 10.8 MPa from E-Figure 13.7. Taking that into the appropriate ICPDM equations:

$$\sigma_{rc}' = 0.029 q_c \left(\frac{\sigma_v'}{p_A} \right)^{0.13} \left(\frac{h}{R} \right)^{-0.38}$$

$$= 0.029 \times 10.8 \times 1000 \times \left(\frac{98.1}{100} \right)^{0.13} \left(\frac{1.5}{0.05} \right)^{-0.38}$$

$$= 85.8 \text{ kPa}$$

$$\eta = q_c \left(p_A \sigma_v' \right)^{-0.5} = 10.8 \times 1000 \times (100 \times 98.1)^{-0.5} = 109$$

$$G = q_c \left[0.0203 + 0.00125 \eta - 1.216 \times 10^{-6} \eta^2 \right]^{-1}$$

$$= 10.8 \times 1000 \left[0.0203 + (0.00125 \times 109) - (1.216 \times 10^{-6} \times 109^2) \right]^{-1}$$

$$= 76,001 \text{ kPa}$$

$$\Delta \sigma_{rd}' = 2 G \Delta r / R = (2 \times 76,001 \text{ kPa} \times 0.02 \text{ mm}) / (0.05 \text{ m} \times 1000) = 60.8 \text{ kPa}$$

and for $\delta_c/\phi_c = 0.8$:

$$q_{sL3} = (\sigma'_{rc} + \Delta\sigma'_{rd})\tan\delta_c = (85.8 + 60.8)\tan(0.8 \times 30°) = 65.3 \text{ kPa}$$

We are now ready to use Equation (13.3) to obtain the limit shaft resistance of the pile:

$$Q_{sL} = \sum(q_{sLi}A_{si})$$

$$= q_{sL1}A_{s1} + q_{sL2}A_{s2} + q_{sL3}A_{s3}$$

$$= 3.4(\pi \times 0.406 \times 2) + 11.3(\pi \times 0.406 \times 7) + 65.3(\pi \times 0.406 \times 3)$$

$$= 359 \text{ kN}$$

To calculate the ultimate unit base resistance of the pile, we must first check if the pile is unplugged using the relation $B_i/L_R \geq 0.02(D_R - 30)$ from Table 13.2. As $0.02(70 - 30) = 0.8$ and $B_i/L_R = 0.393/1 = 0.393 < 0.8$, the pile is plugged, and we will accordingly use the following equation to calculate $q_{b,ult}$:

$$q_{b,ult} = \max\left[0.15, 0.5 - 0.25\log(B/d_c), \left(1 - \frac{R_i^2}{R_0^2}\right)\right]q_{cb}$$

Averaging the cone resistance values between $1.5B$ (0.61 m) above and $1.5B$ (0.61 m) below the pile base, we obtain $q_{cb} \approx 11.5 \text{ MPa}$.

$$\left[0.5 - 0.25\log(B/d_c)\right] = \left[0.5 - 0.25\log(0.406/0.036)\right] = 0.24$$

$$1 - \frac{R_i^2}{R_0^2} = 1 - \frac{0.393^2}{0.406^2} = 0.06$$

Therefore,

$$q_{b,ult} = \max[0.15, 0.24, 0.06]q_{cb} = 0.24q_{cb} = 0.24 \times 11,500 = 2760 \text{ kPa}$$

The ultimate base resistance of the pile is calculated using

$$Q_{b,ult} = q_{b,ult}A_b = 2760 \times \frac{\pi \times 0.406^2}{4} = 357 \text{ kN}$$

The ultimate resistance of the pile is then obtained by adding together the limit shaft resistance and the ultimate base resistance:

$$Q_{ult} = Q_{sL} + Q_{b,ult} = 359 + 357 = 716 \text{ kN}$$

Assuming that the factor of safety FS is equal to 3, the allowable load for the pile is

$$Q_{all} = \frac{Q_{ult}}{FS} = \frac{716}{3} = 239 \text{ kN} \text{ answer}$$

(b) HP 14 × 117 H-Pile

We will use the ICPDM (Tables 13.2 and 13.7) to estimate the pile shaft and base resistances. The limit shaft resistance of the H-pile is estimated following the same procedure as done previously for the open-ended pipe pile, except for the equivalent pile radius R, which is now calculated using

$$R = \sqrt{\frac{A_b}{\pi}}$$

According to the ICPDM, the pile base area A_b for an H-pile is given by

$$A_b = 2b_f t_f + (2X_p + t_w)(d - 2t_f)$$

and

$$X_p = b_f/8 \text{ if } b_f/2 < (d - 2t_f) < b_f$$

$$X_p = b_f^2 / [16(d - 2t_f)] \text{ if } (d - 2t_f) \geq b_f$$

From Table 12.2, for HP 14×117 H-pile, we obtain:

$$d = 0.36 \text{ m}, t_w = 0.02 \text{ m}, b_f = 0.378 \text{ m, and } t_f = 0.02 \text{ m}$$

Since $b_f/2 < (d - 2t_f) < b_f = 0.189 < 0.32 < 0.378$, the value of X_p is

$$X_p = b_f / 8 = 0.378 / 8 = 0.047 \text{ m}$$

By substituting these values, the pile base area is obtained as

$$A_b = 2 \times 0.378 \times 0.02 + (2 \times 0.047 + 0.02)(0.36 - 2 \times 0.02) = 0.052 \text{ m}^2$$

and

$$R = \sqrt{\frac{0.052}{\pi}} = 0.13 \text{ m}$$

For the first layer ($z = 0–2$ m):

The values of σ'_v and ΔI_{vy} are the same as those calculated previously for the open-ended pipe pile. The values of K and δ_r are given by

$$K = \left[2.2 + 0.016 OCR - 0.87 \Delta I_{vy}\right] OCR^{0.42} \left(\max\left[\frac{h}{R}, 8\right]\right)^{-0.20}$$

$$= \left[2.2 + 0.016(1) - 0.87(0.60)\right] \times \left(\max\left[\frac{11}{0.13}, 8\right]\right)^{-0.20}$$

$$= 1.694 \times (84.6)^{-0.20}$$

$$= 0.70$$

$$\sigma' = \sigma'_h = 0.8 K \sigma'_v = 0.8 \times 0.70 \times 17 = 9.5 \text{ kPa}$$

$$\delta_r \approx \phi_r = \phi_{r,min} + \frac{\phi_c - \phi_{r,min}}{1 + \dfrac{\sigma'}{\sigma'_{median}}} = 13° + \frac{25° - 13°}{1 + \dfrac{9.5}{50}} = 23.1°$$

The limit unit shaft resistance of the pile segment in contact with the first layer is

$$q_{sL1} = 0.8 K \sigma'_v \tan \delta_r = 0.8 \times 0.70 \times 17 \times \tan 23.1° = 4.1 \text{ kPa}$$

and the corresponding pile shaft area is

$$A_{s1} = 2 \times (0.378 + 0.36) \times 2 = 2.95 \text{ m}^2$$

For the second layer ($z = 2$–9 m):
The values of σ'_v and ΔI_{vy} are the same as those calculated previously for the open-ended pipe pile. The values of K and δ_r are given by

$$K = \left[2.2 + 0.016 \text{OCR} - 0.87 \Delta I_{vy} \right] \text{OCR}^{0.42} \left(\max \left[\frac{h}{R}, 8 \right] \right)^{-0.20}$$

$$= \left[2.2 + 0.016(1) - 0.87(0.60) \right] \times \left(\max \left[\frac{6.5}{0.13}, 8 \right] \right)^{-0.20}$$

$$= 1.694 \times (50)^{-0.20}$$

$$= 0.77$$

$$\sigma' = \sigma'_h = 0.8 K \sigma'_v = 0.8 \times 0.77 \times 59.2 = 36.5 \text{ kPa}$$

$$\delta_r \approx \phi_r = \phi_{r,min} + \frac{\phi_c - \phi_{r,min}}{1 + \dfrac{\sigma'}{\sigma'_{median}}} = 13° + \frac{25° - 13°}{1 + \dfrac{36.5}{50}} = 19.9°$$

It follows that

$$q_{sL2} = 0.8 K \sigma'_v \tan \delta_r = 0.8 \times 0.77 \times 59.2 \times \tan 19.9° = 13.2 \text{ kPa}$$

and

$$A_{s2} = 2 \times (0.378 + 0.36) \times 7 = 10.33 \text{ m}^2$$

For the third layer ($z = 9$–12 m):
The values of q_c, η, G, and δ_c are the same as those calculated previously for the open-ended pipe pile. Taking them into the appropriate ICPDM equations:

$$\sigma'_{rc} = 0.029 q_c \left(\frac{\sigma'_v}{p_A} \right)^{0.13} \left(\frac{h}{R} \right)^{-0.38}$$

$$= 0.029 \times 10.8 \times 1000 \times \left(\frac{98.1}{100} \right)^{0.13} \left(\frac{1.5}{0.13} \right)^{-0.38}$$

$$= 123.3 \text{ kPa}$$

$$\Delta \sigma'_{rd} = 2 G \Delta r / R = (2 \times 76{,}001 \text{ kPa} \times 0.02 \text{ mm})/(0.13 \text{ m} \times 1000) = 23.4 \text{ kPa}$$

It follows that

$$q_{sL3} = (\sigma'_{rc} + \Delta\sigma'_{rd})\tan\delta_c = (123.3 + 23.4)\tan(0.8\times 30°) = 65.3 \text{ kPa}$$

and

$$A_{s3} = 2\times(0.378 + 0.36)\times 3 = 4.43 \text{ m}^2$$

Using Equation (13.3), the limit shaft resistance of the pile is

$$\begin{aligned}
Q_{sL} &= \sum(q_{sLi}A_{si}) \\
&= q_{sL1}A_{s1} + q_{sL2}A_{s2} + q_{sL3}A_{s3} \\
&= 4.1(2.95) + 13.2(10.33) + 65.3(4.43) \\
&= 438 \text{ kN}
\end{aligned}$$

The ultimate unit base resistance $q_{b,ult}$ of the H-pile is equal to q_{cb}. For the purpose of calculating q_{cb}, the equivalent diameter B of the H-pile is obtained by equating the gross cross-sectional area $(b_f\times d)$ of the H-pile to an equivalent circular area. Averaging the cone resistance values over a distance of $1.5B$ (0.62 m) above and $1.5B$ (0.62 m) below the pile base, we obtain $q_{cb}\approx 11.5$ MPa.

$$q_{b,ult} = q_{cb} = 11.5 \text{ MPa}$$

The ultimate base resistance of the pile is calculated using

$$Q_{b,ult} = q_{b,ult}A_b = 11,500\times 0.052 = 598 \text{ kN}$$

The ultimate resistance of the pile is then obtained by adding together the limit shaft resistance and the ultimate base resistance

$$Q_{ult} = Q_{sL} + Q_{b,ult} = 438 + 598 = 1036 \text{ kN}$$

Assuming that the factor of safety FS is equal to 3, the allowable load for the pile is

$$Q_{all} = \frac{Q_{ult}}{FS} = \frac{1036}{3} = 345 \text{ kN} \quad \text{answer}$$

Example 13.13

An open-ended pipe pile with an outer diameter of 36 in. and a wall thickness of 0.75 in. is driven to a depth of 100 ft from the ground surface. The soil profile consists of normally consolidated, sensitive clay with the following properties: $\gamma_{sat} = 18$ kN/m³, $s_u/\sigma'_v = 0.22$, $S_t = 3$, $\phi_c = 22°$, $\phi_{r,min} = 12°$, and $\sigma'_{median} = 50$ kPa. The water table is at the ground surface. E-Figure 13.9 shows the cone resistance profile. Do the following:

    a. Plot the values of q_{sL} versus depth for the pile using the ICPDM and the UWAPDM.
    b. Calculate the limit shaft capacity Q_{sL}, the ultimate base capacity $Q_{b,ult}$, and the ultimate load capacity Q_{ult} of the pile using these methods.

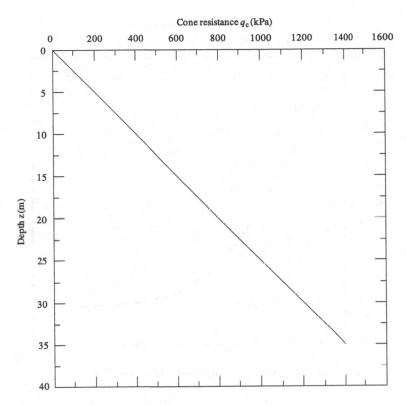

E-Figure 13.9 CPT log for Example 13.13.

Solution:

(a) **Depth profiles of limit unit shaft resistance**

The pile properties are the following:
Outer diameter B of the pile = 36 in. = 0.9144 m.
Wall thickness of the pile = 0.75 in. = 19.05 mm.
Inner diameter B_i of the pile = 0.9144 – 2(0.01905) = 0.8763 m.
Embedded length L of the pile = 100 ft = 30.48 m.
E-Figure 13.10 shows the q_{sL} profiles obtained using the ICPDM and the UWAPDM. Calculations of the limit unit shaft resistance are illustrated below for a depth z of 50 ft (= 15.24 m):

Imperial College Pile Design Method (ICPDM)

The vertical effective stress σ'_v at a depth of 15.24 m is equal to $(18 - 9.81) \times 15.24 = 124.8$ kPa.
The vertical distance from the pile base to the depth of 15.24 m is:

$$h = L - z = 30.48 - 15.24 = 15.24 \text{ m}$$

The equivalent pile radius is:

$$R = \sqrt{R_o^2 - R_i^2} = \sqrt{\left(\frac{0.9144}{2}\right)^2 - \left(\frac{0.8763}{2}\right)^2} = 0.1306 \text{ m}$$

$$\Delta I_{vy} = \log_{10} S_t = \log_{10}(3) = 0.48$$

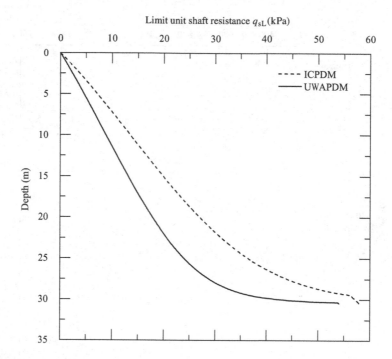

E-Figure 13.10 q_{sL} versus depth for the open-ended pipe pile of Example 13.13.

$$K = \left[2.2 + 0.016\mathrm{OCR} - 0.87\Delta I_{vy}\right]\mathrm{OCR}^{0.42}\left(\max\left[\frac{h}{R}, 8\right]\right)^{-0.20}$$

$$= \left[2.2 + 0.016(1) - 0.87(0.48)\right] \times 1^{0.42} \times \left(\max\left[\frac{15.24}{0.1306}, 8\right]\right)^{-0.20}$$

$$= 0.69$$

The effective lateral stress σ' ($= \sigma'_h$) on the pile operative at the time of shearing is given by

$$\sigma'_h = 0.8K\sigma'_v = 0.8 \times 0.69 \times 124.8 = 68.9 \text{ kPa}$$

The residual interface friction angle δ_r is calculated using

$$\delta_r \approx \phi_r = \phi_{r,\min} + \frac{\phi_c - \phi_{r,\min}}{1 + \dfrac{\sigma'}{\sigma'_{\mathrm{median}}}} = 12° + \frac{22° - 12°}{1 + \dfrac{68.9}{50}} = 16.2°$$

The limit unit shaft resistance is given by

$$q_{sL} = 0.8K\sigma'_v \tan\delta_r = 0.8 \times 0.69 \times 124.8 \times \tan 16.2° = 20 \text{ kPa}$$

University of Western Australia Pile Design Method (UWAPDM)
The value of q_t at a depth of 15.24 m is read from E-Figure 13.9 as 604 kPa.

The effective lateral stress σ' $(=\sigma'_h)$ on the pile operative at the time of shearing is given by

$$\sigma'_h = \frac{0.23 q_t \left[\max\left(\frac{h}{R}, 1\right)\right]^{-0.2}}{(q_t/\sigma'_v)^{0.15}}$$

$$= \frac{0.23 \times 604 \times \left[\max\left(\frac{15.24}{0.1306}, 1\right)\right]^{-0.2}}{(604/124.8)^{0.15}} = 42.3 \text{ kPa}$$

The residual interface friction angle δ_r is calculated using

$$\delta_r \approx \phi_r = \phi_{r,\min} + \frac{\phi_c - \phi_{r,\min}}{1 + \dfrac{\sigma'}{\sigma'_{median}}} = 12° + \frac{22° - 12°}{1 + \dfrac{42.3}{50}} = 17.4°$$

The limit unit shaft resistance is given by

$$q_{sL} = \frac{0.23 q_t \left[\max\left(\frac{h}{R}, 1\right)\right]^{-0.2}}{(q_t/\sigma'_v)^{0.15}} \tan \delta_r$$

$$= \frac{0.23 \times 604 \times \left[\max\left(\frac{15.24}{0.1306}, 1\right)\right]^{-0.2}}{(604/124.8)^{0.15}} \tan 17.4° = 13.3 \text{ kPa}$$

(b) Calculation of pile shaft and base capacities

E-Table 13.6 summarizes the limit shaft capacity Q_{sL}, the ultimate base capacity $Q_{b,ult}$, and the ultimate load capacity Q_{ult} of the pile using the ICPDM and the UWAPDM. The limit shaft capacity of the pile was calculated in the usual manner by multiplying the value of q_{sL} computed at each depth by the corresponding pile perimeter and integrating that over the embedded length of the pile. Calculation of the ultimate base capacity of the pile using the ICPDM is shown below.

Imperial College Pile Design Method (ICPDM)

The representative cone resistance q_{cb} near the pile base is obtained by taking the average of the q_c values over a vertical distance of $1.5B$ ($= 1.37$ m) above and below the pile base. Referring to E-Figure 13.9, the value of q_{cb} is 1208 kPa.

Now we check whether the pile behaves as a fully plugged pile during static loading. We do that by calculating the plugging index for the pile:

$$\frac{B_i}{d_c} + 0.45 \frac{q_{cb}}{p_A} = \frac{0.8763}{0.036} + 0.45\left(\frac{1208}{100}\right) = 29.8 < 36$$

E-Table 13.6 Values of Q_{sL}, $Q_{b,ult}$, and Q_{ult} for the open-ended pipe pile of Example 13.13

Method	Limit shaft capacity Q_{sL} (kN)	Ultimate base capacity $Q_{b,ult}$ (kN)	Ultimate load capacity Q_{ult} (kN)
ICPDM	1957	317	2274
UWAPDM	1300	not calculated	not calculated

Since the plugging index is less than 36, the pile behaves as a fully plugged pile during static loading. For undrained loading, the ultimate unit base resistance of a plugged pile is given by

$$q_{b,ult} = 0.4q_{cb} = 0.4 \times 1208 = 483 \text{ kPa}$$

The ultimate base capacity of the pile is calculated using:

$$Q_{b,ult} = q_{b,ult}A_b = q_{b,ult}\frac{\pi B^2}{4} = 483 \times \frac{\pi \times 0.9144^2}{4} = 317 \text{ kN}$$

Note that the base capacity of the pile is substantially less than the shaft capacity, which means that, for practical purposes, this is a floating pile. The ultimate load accounting for the pile weight, which is 127 kN, would be 2147 kN.

The UWAPDM for open-ended pipe piles in clay does not provide a criterion to evaluate whether the pile behaves as a fully plugged pile during static loading. However, based on the preceding calculations for the ICPDM, we may assume that the pile in this example behaves as a fully plugged pile during static loading for the estimation of pile base resistance using the UWAPDM. This calculation is left as an exercise for the reader.

13.7 OTHER SPT AND CPT DESIGN CORRELATIONS

13.7.1 Form of the correlations

In this section, we provide values for the coefficients in Equations (13.15)–(13.18) for the calculation of pile unit resistances from CPT cone resistance or SPT blow count. These values are mostly empirical. One of the shortcomings of these methods for driven piles is the lack of a shaft resistance degradation factor. This may lead to unconservative results, particularly for long piles.

The SPT has many shortcomings, as discussed in Chapter 7. The SPT, as a rule, should not be used for the design of piles in clay. One of the reasons for that is the low blow count number, particularly in soft clay, when variations from 1 to 2 or 3 blow counts are significant percentage variations, and yet the soil state may be approximately the same. In other words, the SPT lacks resolution in clay. Nonetheless, some legacy correlations used for this purpose are provided in this section for completeness.

13.7.2 Sands

13.7.2.1 Base resistance

Table 13.10 has the values of the coefficients c_b and n_b proposed by various authors for displacement piles. The methods are mostly empirical.

Table 13.11 contains the values of c_b and n_b for use for nondisplacement piles. Most methods have been developed for drilled shafts, but can be used for Strauss piles, micropiles in soil, and other nondisplacement piles installed in a way that produces similar effects on the soil as drilled shaft installation. The methods are mostly empirical.

13.7.2.2 Shaft resistance

The values of c_s and n_s found in the literature for driven piles in sand are provided in Tables 13.12 and 13.13, respectively. The values are based on pile load tests performed either in the

Table 13.10 Design values of $c_b = q_{b,ult}/q_{cb}$ and $n_b = q_{b,ult}/[p_A(N_b)_{60}]$ for displacement (driven) piles in sand

Source	Equation or value	Basis and applicability[a]
Chow (1997)	$c_b = 0.35–0.50$	Database of high-quality pile load tests
Randolph (2003)	$c_b = 0.40$	Reinterpretation of the Chow (1997) database
Meyerhof (1983)	$n_b = 4$	Experience; not clear how ultimate load was defined
Aoki and Velloso (1975)	$c_b = \begin{cases} 0.4 & \text{for Franki piles} \\ 0.57 & \text{for precast concrete piles} \end{cases}$	Database of noninstrumented pile load tests. Ultimate load defined as the limit load obtained by extrapolation using Van der Veen's criterion. The independence of c_b from soil type means some caution is needed in using these numbers
Aoki and Velloso (1975)	$n_b = \begin{cases} 4.8 & \text{for clean sand} \\ 3.8 & \text{for silty sand} \\ 3.3 & \text{for silty sand with clay} \\ 2.4 & \text{for clayey sand with silt} \\ 2.9 & \text{for clayey sand} \end{cases}$ Franki piles: multiply the number above by 0.7	Database of noninstrumented pile load tests. Ultimate load defined as the limit load obtained by extrapolation using Van der Veen's criterion
Lee et al. (2003)	$c_b = 0.52 - 0.4(\text{IFR})$ IFR = incremental filling ratio	Database of calibration chamber and field load tests. Applicable to open-ended steel pipe piles only

[a] Applicable to displacement piles and ultimate load corresponding to 10% relative settlement, unless otherwise indicated.

Table 13.11 Design values of $c_b = q_{b,ult}/q_{cb}$ and $n_b = q_{b,ult}/[p_A(N_b)_{60}]$ for nondisplacement piles (drilled shafts) and CFA piles in sand

Source	Equation or value	Basis and applicability[a]
Franke (1989)	$c_b = 0.2$	Experience with load tests on drilled shafts
Ghionna et al. (1994)	$c_b = 0.13 \pm 0.02$	Database of calibration chamber load tests
Salgado (2006a) based on Lee and Salgado (1999)	$c_b = 0.23\exp(-0.0066 D_R)$	Finite element analysis verified using calibration chamber tests
Lopes and Laprovitera (1988)	$n_b = \begin{cases} 0.82 & \text{for clean sand} \\ 0.72 & \text{for sand with silt or clay} \end{cases}$	Database of noninstrumented pile load tests. Ultimate load = the limit load estimated by extrapolation using the Van der Veen criterion
Reese and O'Neill (1989)	$n_b = 0.6$ $\dfrac{q_{b,ult}}{p_A} \leq 45$	Experience and load tests on drilled shafts[b]. Ultimate resistance mobilized at $w/B = 0.05$
Neely (1991), after Roscoe (1983) and Meyerhof (1976)	$n_b = \begin{cases} 1.9 & \text{for CFA piles} \\ 1.2 & \text{for drilled shafts} \end{cases}$	CFA piles and drilled shafts. Load tests and experience. Ultimate load criterion not clearly defined

[a] Applicable to the calculation of the ultimate load corresponding to 10% relative settlement for a drilled shaft or similarly constructed nondisplacement pile, unless otherwise indicated.
[b] The authors indicated that a linear reduction in n_b with the pile diameter was necessary for diameters greater than 1.2 m (4 ft), but, in light of the current knowledge of pile response, this reduction is not necessary.

Table 13.12 Design values of $c_s = q_{sL}/q_c$ for piles in sand

Source	Equation or value	Basis and applicability[a]
Schmertmann (1978)	$c_s = \begin{cases} 0.008 & \text{for open-end steel pipe piles} \\ 0.012 & \text{for precast concrete and} \\ & \text{closed-ended steel pipe piles} \\ 0.018 & \text{for Franki and timber piles} \end{cases}$	Applicable to pipe piles and precast concrete piles
Lee et al. (2003)	$c_s = \begin{cases} 0.004-0.006 & \text{for } D_R \leq 50\% \\ 0.004-0.007 & \text{for } 50\% < D_R \leq 70\% \\ 0.004-0.009 & \text{for } 50\% < D_R \leq 90\% \end{cases}$	Database of calibration chamber pile load tests Applicable to closed-ended pipe piles
Lee et al. (2003)	$c_s = \begin{cases} 0.0015-0.003 & \text{for IFR} \leq 0.60 \\ 0.0015-0.004 & \text{for } 0.60 < \text{IFR} \leq 1 \end{cases}$ IFR = incremental filling ratio	Database of calibration chamber pile load tests Applicable to open-ended pipe piles
Aoki and Velloso (1975) and Aoki et al. (1978)	$c_s = \begin{cases} 0.0040 & \text{for clean sand} \\ 0.0057 & \text{for silty sand} \\ 0.0069 & \text{for silty sand with clay} \\ 0.0080 & \text{for clayey sand with silt} \\ 0.0086 & \text{for clayey sand} \end{cases}$ Franki piles: multiply the number above by 0.7 Drilled shafts: multiply the number above by 0.5	Database of noninstrumented pile load tests Applicable to driven piles; for Franki piles and drilled shafts, multiply by the coefficient provided
Lopes and Laprovitera (1988)	$c_s = \begin{cases} 0.0027 & \text{for clean sand} \\ 0.0037 & \text{for silty sand} \\ 0.0046 & \text{for silty sand with clay} \\ 0.0054 & \text{for clayey sand with silt} \\ 0.0058 & \text{for clayey sand} \end{cases}$	Database of noninstrumented pile load tests Applicable to nondisplacement piles
Eslami and Fellenius (1997)[b]	$c_s = \dfrac{q_{sL}}{q_c - u} = 0.0034 - 0.006$ u = pore pressure at depth corresponding to q_c	Pile load test database Steel and concrete piles with full cross section

[a] Applicable to displacement piles, unless otherwise indicated.
[b] This method uses a corrected q_c. The pore pressure at the level of the cone is subtracted from the cone resistance to obtain a new, corrected value of q_c. This is in addition to the correction discussed in Chapter 7 to account for the difference in area behind and in front of the cone acted on by the pore pressure.

field or in calibration chambers. Considering the value of c_s for precast concrete and closed-ended steel pipe piles, it can be seen that the values of Schmertmann seem to be too high. The values of Aoki and Velloso (1975) appear to be high for sand. It does seem that a lower bound for c_s is of the order of 0.4% of q_c for these piles.

Table 13.13 Design values of $n_s = q_{sL}/[p_A(N_s)_{60}]$ for piles in sand

Source	Equation or value	Basis and applicability[a]
Thorburn and MacVicar (1971)	$n_s = 0.02$	
Meyerhof (1976, 1983)	$n_s = \begin{cases} 0.02 & \text{for full-displacement piles} \\ 0.01 & \text{for H-piles} \end{cases}$	Experience
Aoki and Velloso (1975) and Aoki et al. (1978)	$n_s = \begin{cases} 0.033 & \text{for sand} \\ 0.038 & \text{for silty sand} \\ 0.040 & \text{for silty sand with clay} \\ 0.033 & \text{for clayey sand with silt} \\ 0.043 & \text{for clayey sand} \end{cases}$ Franki piles: multiply the number above by 0.7 Drilled shafts: multiply the number above by 0.5	Database of noninstrumented pile load tests Applicable to driven piles; for Franki piles and drilled shafts, multiply by the coefficient provided
Lopes and Laprovitera (1988)	$n_s = \begin{cases} 0.014 & \text{for sand} \\ 0.016 & \text{for silty sand} \\ 0.020 & \text{for silty sand with clay} \\ 0.024 & \text{for clayey sand with silt} \\ 0.026 & \text{for clayey sand} \end{cases}$	Database of noninstrumented pile load tests Applicable to nondisplacement piles

[a] Applicable to displacement piles, unless otherwise indicated.

The values of c_s for open-ended pipe piles are lower than the values for closed-ended pipe piles, precast piles, and other full displacement piles. Again, the values of Schmertmann (1978) appear high. A reasonable lower bound on c_s would be of the order of 0.15% of q_c, with values as high as 0.4% of q_c for piles with low IFR.

There is a scarcity of direct CPT or SPT correlations for the shaft resistance of drilled shafts in the literature. Table 13.12 contains the numbers put forth by Aoki et al. (1978) for these piles. It is interesting to note that the value of c_s for clean sand is practically the same as the lower-end value for open-ended pipe piles of Lee et al. (2003). This suggests that, perhaps, as remarked by Lehane and Randolph (2002) for base resistance, the unit resistance of drilled shafts may serve as a reasonable approximation of the unit shaft resistance of open-ended pipe piles in sand. It would likely not be a lower bound on shaft resistance because of shaft degradation.

13.7.3 Clays

13.7.3.1 Base resistance

The unit base resistance of full-section displacement and nondisplacement piles in clays is almost as large as the cone resistance, except in stiff clays, when $q_{b,ult}/q_c$ may be somewhat less than 1 (see Table 13.14). Unfortunately, this is an under-researched area. Nonetheless,

Table 13.14 Design values of $c_b = q_{b,ult}/q_{cb}$ and $n_b = q_{b,ult}/[p_A(N_b)_{60}]$ for piles in clay

Source	Equation or value		Basis and applicability[a]
State of the art	$c_b = 0.9{-}1$		Soft to lightly overconsolidated clays
Price and Wardle (1982)	$c_b = \begin{cases} 0.35 \\ 0.30 \end{cases}$	for driven piles for jacked piles	Pile load tests and CPTs in London Clay Stiff clays
Aoki and Velloso (1975) and Aoki et al. (1978)	$n_b = \begin{cases} 0.95 \\ 1.05 \\ 1.57 \\ 1.43 \\ 1.67 \end{cases}$	for pure clay for silty clay for silty clay with sand for sandy clay with silt for sandy clay	Database of noninstrumented pile load tests Applicable to driven piles; for Franki piles and drilled shafts, multiply by the coefficient provided Ultimate load = the limit load estimated by extrapolation using the Van der Veen criterion
	Franki piles: multiply the number above by 0.7 Drilled shafts: multiply the number above by 0.5		
Lopes and Laprovitera (1988)	$n_b = \begin{cases} 0.34 \\ 0.41 \\ \\ 0.66 \end{cases}$	for pure clay and silty clay for silty clay with sand and sandy clay with silt for sandy clay	Database of noninstrumented pile load tests Applicable to nondisplacement piles Ultimate load = the limit load estimated by extrapolation using the Van der Veen criterion

[a] Applicable to displacement piles and ultimate load corresponding to 10% relative settlement, unless otherwise indicated.

Table 13.14 provides some guidance on base resistance calculation in clay. SPT-based correlations in clay are not very reliable, but, for completeness, Table 13.14 contains some guidance on how to obtain estimates of base resistance directly from the representative SPT blow count at the pile base level.

13.7.3.2 Shaft resistance

Table 13.15 provides some values of c_s and n_s that may be used in design. At first glance, the Aoki and Velloso (1975) c_s values appear to be significantly lower than those of Eslami and Fellenius (1997). However, an important difference is that the Eslami and Fellenius (1997) values are for a corrected cone resistance, from which the pore pressure during penetration at the level of the cone is subtracted, which would explain the higher values.

13.7.4 Silts

The literature contains even less in terms of data and correlations for silts than for clays and sands. It does not even clearly distinguish between plastic silts (which behave more like clays) and nonplastic silts (which behave more like sands). Nonetheless, some numbers for n_b, c_s, and n_s are given in Tables 13.16 and 13.17.

Table 13.15 Design values of $c_s = q_{sL}/q_c$ and $n_s = q_{sL}/[p_A(N_s)_{60}]$ for piles in clay

Source	Equation or value	Basis and applicability
Eslami and Fellenius (1997)[a]	$c_s = \begin{cases} 0.074 - 0.086 & \text{for sensitive clay} \\ 0.046 - 0.056 & \text{for soft clay} \\ 0.021 - 0.028 & \text{for silty clay or stiff clay} \end{cases}$	Pile load test database Driven piles Use of modified cone resistance
Thorburn and MacVicar (1971)	$c_s = 0.025$	Displacement piles
Aoki and Velloso (1975) and Aoki et al. (1978)	$c_s = \begin{cases} 0.017 & \text{for pure clay} \\ 0.011 & \text{for silty clay} \\ 0.0086 & \text{for silty clay with sand} \\ 0.0080 & \text{for sandy clay with silt} \\ 0.0069 & \text{for sandy clay} \end{cases}$ Franki piles: multiply the number above by 0.7 Drilled shafts: multiply the number above by 0.5	Database of noninstrumented pile load tests Applicable to driven piles; for Franki piles and drilled shafts, multiply by the coefficient provided
Lopes and Laprovitera (1988)	$c_s = \begin{cases} 0.012 & \text{for pure clay} \\ 0.011 & \text{for silty clay} \\ 0.010 & \text{for silty clay with sand} \\ 0.0087 & \text{for sandy clay with silt} \\ 0.0077 & \text{for sandy clay} \end{cases}$	Database of noninstrumented pile load tests Applicable to nondisplacement piles
Aoki and Velloso (1975) and Aoki et al. (1978)	$n_s = \begin{cases} 0.029 & \text{for clay} \\ 0.021 & \text{for silty clay} \\ 0.024 & \text{for silty clay with sand} \\ 0.020 & \text{for sandy clay with silt} \\ & \text{and sandy clay} \end{cases}$ Franki piles: multiply the number above by 0.7 Drilled shafts: multiply the number above by 0.5	Database of noninstrumented pile load tests Applicable to driven piles; for Franki piles and drilled shafts, multiply by the coefficient provided
Lopes and Laprovitera (1988)	$n_s = \begin{cases} 0.024 & \text{for clay} \\ 0.022 & \text{for silty clay} \\ 0.024 & \text{for silty clay with sand} \\ 0.022 & \text{for sandy clay with silt} \\ 0.031 & \text{for sandy clay} \end{cases}$	Database of noninstrumented pile load tests Applicable to nondisplacement piles

[a] This method uses a corrected q_c. The pore pressure at the level of the cone is subtracted from the cone resistance to obtain a new, corrected value of q_c. This is addition to the correction discussed in Chapter 7 to account for the difference in area behind and in front of the cone acted on by the pore pressure.

Table 13.16 Design values of $c_b = q_{b,ult}/q_{cb}$ and $n_b = q_{b,ult}/[p_A(N_b)_{60}]$ for piles in silt

Source	Equation or value	Basis and applicability
Meyerhof (1976, 1983)	$n_b = 2.7$	Experience Ultimate load criterion not explicitly discussed
Aoki and Velloso (1975) and Aoki et al. (1978)	$n_b = \begin{cases} 1.9 & \text{for silt} \\ 2.6 & \text{for sandy silt} \\ 2.1 & \text{for sandy silt with clay} \\ 1.2 & \text{for clayey silt with sand} \\ 1.1 & \text{for clayey silt} \end{cases}$ Franki piles: multiply the number above by 0.7 Drilled shafts: multiply the number above by 0.5	Database of noninstrumented pile load tests Applicable to driven piles; for Franki piles and drilled shafts, multiply by the coefficient provided Ultimate load = the limit load estimated by extrapolation using the Van der Veen criterion
Lopes and Laprovitera (1988)	$n_b = \begin{cases} 0.66 & \text{for silt and sandy silt} \\ 0.52 & \text{for sandy silt with clay} \\ & \text{and clayey silt with sand} \\ 0.41 & \text{for clayey silt} \end{cases}$	Database of noninstrumented pile load tests Applicable to nondisplacement piles Ultimate load = the limit load estimated by extrapolation using the Van der Veen criterion

Table 13.17 Design values of $c_s = q_{sL}/q_c$ and $n_s = q_{sL}/[p_A(N_s)_{60}]$ for piles in silt

Source	Equation or value	Basis and applicability
Eslami and Fellenius (1997)[a]	$c_s = \begin{cases} 0.087 - 0.0134 & \text{for sand/silt mixtures} \\ 0.0206 - 0.0280 & \text{for clay/silt mixtures} \end{cases}$	Pile load test database Driven piles Use of modified cone resistance
Aoki and Velloso (1975) and Aoki et al. (1978)	$c_s = \begin{cases} 0.0086 & \text{for silt} \\ 0.0063 & \text{for sandy silt} \\ 0.0080 & \text{for sandy silt with clay} \\ 0.0086 & \text{for clayey silt with sand} \\ 0.0097 & \text{for clayey silt} \end{cases}$ Franki piles: multiply the number above by 0.7 Drilled shafts: multiply the number above by 0.5	Database of noninstrumented pile load tests Applicable to driven piles; for Franki piles and drilled shafts, use the coefficient provided
Lopes and Laprovitera (1988)	$c_s = \begin{cases} 0.0058 & \text{for silt, sandy silt, sandy silt with clay} \\ & \text{and clayey silt with sand} \\ 0.0034 & \text{for clayey silt} \end{cases}$	Database of noninstrumented pile load tests Applicable to nondisplacement piles
Aoki and Velloso (1975) and Aoki et al. (1978)	$n_s = \begin{cases} 0.029 & \text{for silt and sandy silt} \\ 0.030 & \text{for sandy silt with clay} \\ 0.018 & \text{for clayey silt with sand} \\ & \text{and clayey silt} \end{cases}$ Franki piles: multiply the number above by 0.7 Drilled shafts: multiply the number above by 0.5	Database of noninstrumented pile load tests Applicable to driven piles; for Franki piles and drilled shafts, use the coefficient provided

(Continued)

Table 13.17 (Continued) Design values of $c_s = q_{sL}/q_c$ and $n_s = q_{sL}/[p_A(N_s)_{60}]$ for piles in silt

Source	Equation or value		Basis and applicability
Lopes and Laprovitera (1988)	$n_s = $	0.023 for silt and sandy silt	Database of noninstrumented pile load tests
		0.018 for sandy silt with clay	
		and clayey silt with sand	Applicable to nondisplacement piles
		0.016 for clayey silt	

[a] This method uses a corrected q_c. The pore pressure at the level of the cone is subtracted from the cone resistance to obtain a new, corrected value of q_c. This is addition to the correction discussed in Chapter 7 to account for the difference in area behind and in front of the cone acted on by the pore pressure.

13.8 LOAD AND RESISTANCE FACTOR DESIGN PROCEDURE FOR SINGLE PILES

Due to the complexity of LRFD factor development for piles, only limited high-quality research has been done in that area. Thus, factors are available only for some methods and some pile and soil types. For the PPDM, the resistance factors, RF_s and RF_b, for the pile shaft and base resistances, respectively, can be obtained from Table 13.18 based on the selected pile type and the predominant soil type at the site. Resistance factors RF_s and RF_b for axially loaded, driven H-piles in sand for the ICP method can be obtained from Table 13.19.

Having the resistance factors, we verify that the following LRFD inequality is satisfied (Basu and Salgado 2012; Foye et al. 2009):

$$RF_s Q_{sL}^n + RF_b Q_{b,ult}^n \geq LF_{DL} DL^n + LF_{LL} LL^n \tag{13.37}$$

where DL^n and LL^n = nominal dead and live loads, respectively; LF_{DL} and LF_{LL} = load factors for dead load and live load, respectively; Q_{sL}^n and $Q_{b,ult}^n$ = nominal limit shaft and

Table 13.18 Resistance factors for axially loaded drilled shafts and closed-ended pipe piles in sand and clay based on the PPDM

Pile type	Predominant soil type at the site	$p_{f,T} = 10^{-1}$	
		RF_b	RF_s
Drilled shaft	Sand	0.70	0.65
Drilled shaft	Clay	0.65	0.70
Closed-ended pipe pile	Sand	0.30	0.60
Closed-ended pipe pile	Clay	0.65	0.65

Source: Modified from Han et al. (2015).

Note: $p_{f,T}$, target probability of failure. For layered clay deposits (that is, soft over stiff layers), the values of RF_b and RF_s should be decreased by 25% and 20%, respectively.

Table 13.19 Resistance factors for axially loaded, driven H-piles in sand based on the ICPDM

Pile type	Predominant soil type at the site	$p_{f,T} = 2 \times 10^{-4}$ (or $\beta_T = 3.5$)	
		RF_b	RF_s
H-pile	Sand	0.56	0.45

Source: Modified from Kim et al. (2011).

Note: $p_{f,T}$, target probability of failure and β_T, target reliability index. The resistance factors listed in Table 13.19 are the lowest values reported by Kim et al. (2011) for different combinations of DL^n/LL^n and $Q_{b,ult}^n/Q_{sL}^n$.

ultimate pile base resistances, respectively; and RF_s and RF_b = resistance factors for pile shaft and base resistances, respectively.

If the pile design fails the inequality test, then a redesign is needed.

If we wish to obtain the factor of safety FS resulting from the LRFD design, we can use the following equation (Han et al. 2015):

$$FS = \frac{C^n}{D^n} = \frac{Q_{sL}^n + Q_{b,ult}^n}{DL^n + LL^n}$$

(13.38)

where C^n = nominal capacity and D^n = nominal demand.

13.9 CALCULATION OF SETTLEMENT OF PILES SUBJECTED TO AXIAL LOADINGS*

13.9.1 Nature and applicability of the analysis

When the scale and nature of the project justify it, a settlement analysis for each pile used as a foundation element may allow more economical designs. The settlement w_t at the pile head is of course a function of the compressibility of the pile itself and of how the axial load is distributed along the shaft and between the shaft and base of the pile. As the pile is pushed down by the load applied at its head, it in turn transfers this load to the surrounding soil, which settles along with the pile at a rate that decreases as the distance from the pile increases (see Figure 13.20). The stiffer the soil, the more it will resist deforming and, as a consequence, the less the pile itself will settle. To develop an accurate analysis to relate pile load to pile settlement, it is therefore essential to model pile and soil stiffness and their interplay as realistically as possible.

We follow here the approaches of Randolph and Wroth (1978) and Mylonakis and Gazetas (1998), which are based on linear elasticity and require some judgment in the selection of values for the elastic constants so that the results of the calculations are realistic. The limitation of these methods is that they do not model soil response in a realistic manner and do not capture the shear strain localization that takes place along the piles. They are valuable to set up the conceptual basis for settlement analyses, and we examine them here with that purpose.

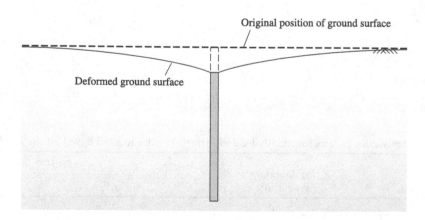

Figure 13.20 Settlement pattern around an axially loaded pile.

13.9.2 Basic differential equation of pile compression

The change dQ in axial load Q along a small distance dz along a circular pile is given by the mobilized unit shaft resistance (shear stress) q_s along the pile shaft multiplied by the lateral surface of the pile for the segment dz:

$$dQ = -\pi B q_s dz \qquad (13.39)$$

where B = pile diameter. Depending on the distribution of q_s along the pile, different load transfer curves would result (Figure 13.21). Load transfer curves show the load experienced by every pile cross section. Except in the cases where a mass of soil undergoing consolidation exists around the pile, the load carried by the cross section of the pile always decreases with depth (or, at least, stays the same, in the absence of any shaft resistance along some portion of the pile). The shape of load transfer curves can be more complex than those shown in Figure 13.21 if the pile goes through many soil layers with very different properties.

The axial strain experienced by the small pile element dz can be obtained from the same relationship applied to rods subjected to axial loads (a topic studied in basic strength of materials):

$$\frac{dw(z)}{dz} = -\varepsilon_z = -\frac{Q(z)}{(EA)_p} \qquad (13.40)$$

where $(EA)_p$ is the axial stiffness of the pile. We use here, as in previous chapters, the geomechanics definition of strain (positive in contraction) and also explicitly indicate that both w and Q are functions of z (we will do this for emphasis where desirable).

Combination of Equations (13.39) and (13.40) leads to:

$$\frac{d^2w}{dz^2} = \frac{\pi B}{(EA)_p} q_s \qquad (13.41)$$

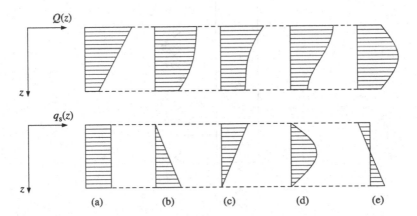

Figure 13.21 Load transfer curves corresponding to different shaft unit load distributions: (a) uniform; (b) increasing linear; (c) decreasing linear; (d) "parabolic"; and (e) distribution in the presence of negative side resistance (negative skin friction). (From Vesic (1977), Figure 28, p. 27. Copyright, National Academy of Sciences. Reproduced with permission from the Transportation Research Board).

This differential equation – Equation (13.41) – can be rewritten in terms of a Winkler constant $k = k(z)$ varying with depth z:

$$(EA)_\text{p} \frac{d^2w(z)}{dz^2} - k(z)w(z) = 0 \tag{13.42}$$

where:

$$k(z) = \frac{\pi B q_\text{s}}{w(z)} \tag{13.43}$$

is the Winkler constant for depth z, defined as the ratio of the load per unit pile length at depth z to the displacement there.[13]

The concept of the Winkler constant (also called a modulus of subgrade reaction) can better be understood by referring to Figure 13.22. In this figure, the soil is modeled as a series of springs positioned along the pile shaft. This may be visualized best by thinking of each

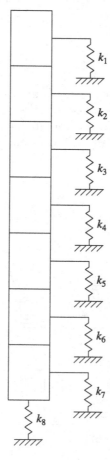

Figure 13.22 Model for axial soil–pile interaction by discretization of the soil by Winkler springs positioned along the pile shaft.

---

[13]Note that k is not a spring constant, which has the unit of force per length.

spring as connected to a separate pile element. If the pile is discretized using n elements, there would be n springs. In a limiting case, for an element of length dz, the force transferred to the soil by the pile element is given by Equation (13.39). Since k is defined as

$$k(z) = -\frac{dQ / dz}{w(z)} \qquad (13.44)$$

it may be seen that Equation (13.43) follows directly from Equation (13.39). Note that the negative sign reflects the fact that the load Q in the pile decreases with increasing depth due to the existence of a shaft resistance dQ_s along dz, while k is a positive number because a positive load must be applied to the soil to get it to settle by an amount w. Note that if the spring spacing is made equal to 1, then the modulus of subgrade reaction is numerically (but not dimensionally) equal to the spring constant.

The differential Equation (13.42) ignores the contribution of soil compression around the pile shaft in the vertical direction. The correct differential equation is

$$-\left[(EA)_p + 2t\right]\frac{d^2w}{dz^2} + kw = 0$$

where the coefficient t represents the soil compressive resistance in the vertical direction. This compressive resistance is mobilized because the load transfer to the soil does not occur in the idealized way we assumed thus far, relying solely on vertical shear stress mobilized on cylindrical surfaces concentric with the pile. In reality, some load is transferred by compressive stresses "emanating" from the shaft of the pile. This could be loosely referred to as an "arching" process. For simplicity, we will continue to assume $t = 0$, which takes us back to Equation (13.42).

The solution to the differential Equation (13.42) gives the pile vertical displacement as a function of depth, which by differentiation leads to the pile load at each cross section. Numerical solution to Equation (13.41) or (13.42) is required if the soil profile is complex (with k varying with depth in some way) or consideration of nonlinearity in the stress–strain response is required; however, for the simple case of a pile installed in a homogeneous, elastic soil (with elastic properties as shown in Figure 13.23), analytical solutions are possible.

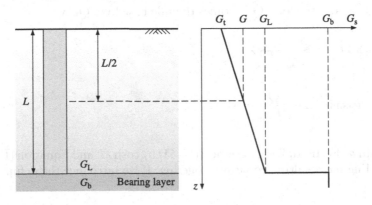

Figure 13.23 Pile in an elastic soil.

13.9.3 Pile compression in homogeneous, elastic soil

For a homogeneous, elastic soil, the modulus of subgrade reaction k is the same everywhere along the pile. This means Equation (13.42) is a second-order, homogeneous, linear differential equation with general solution:

$$w(z) = C_1 \cosh \lambda z + C_2 \sinh \lambda z \tag{13.45}$$

where C_1 and C_2 are arbitrary constants and λ is given by

$$\lambda = \sqrt{\frac{k}{E_p A_p}} \tag{13.46}$$

Differentiation of Equation (13.45) leads to

$$\frac{dw(z)}{dz} = -\varepsilon_z(z) = -\frac{Q(z)}{E_p A_p} = \lambda \left(C_1 \sinh \lambda z + C_2 \cosh \lambda z \right) \tag{13.47}$$

where $Q(z)$ is the load carried by the pile and $\varepsilon_z(z)$ is the axial normal strain in the pile at depth z.

We note from making $z = 0$ in Equation (13.45) that C_1 is the pile settlement, that is, the pile top (head) displacement w_t. Similarly, making $z = 0$ in Equation (13.47) gives us that C_2 is directly related to the load Q_t at the top of the pile through:

$$\frac{Q_t}{\lambda E_p A_p} = -C_2 \tag{13.48}$$

We can now rewrite Equations (13.45) and (13.47) as follows:

$$w(z) = w_t \cosh \lambda z - \frac{Q_t}{\lambda E_p A_p} \sinh \lambda z \tag{13.49}$$

$$\frac{Q(z)}{E_p A_p} = -\lambda \left(w_t \sinh \lambda z - \frac{Q_t}{\lambda E_p A_p} \cosh \lambda z \right) \tag{13.50}$$

Making $z = L$ in Equations (13.49) and (13.50) so that the settlement w becomes the pile base settlement w_b and the load Q becomes the pile base load Q_b, we get:

$$w_b = w_t \cosh \lambda L - \frac{Q_t}{\lambda E_p A_p} \sinh \lambda L \tag{13.51}$$

$$-\frac{Q_b}{\lambda E_p A_p} = w_t \sinh \lambda L - \frac{Q_t}{\lambda E_p A_p} \cosh \lambda L \tag{13.52}$$

We can obtain w_t by multiplying Equation (13.51) by $\cosh \lambda L$ and Equation (13.52) by $-\sinh \lambda L$, and by adding the resulting equations together. If we introduce the definitions

$$\Omega = \frac{K_b}{\lambda E_p A_p} \tag{13.53}$$

with Ω being the dimensionless pile base stiffness, and

$$K_b = \frac{Q_b}{w_b} \tag{13.54}$$

with K_b being the pile base stiffness, the equation for the pile head settlement can be written as

$$w_t = C_1 = w_b \cosh \lambda L (1 + \Omega \tanh \lambda L) \tag{13.55}$$

Equation (9.20) can be used to calculate the settlement of a rigid circular foundation bearing on soil with modulus G_b and Poisson's ratio ν. Recalling Equation (9.20):

$$w = \frac{Q(1 - \nu^2)}{BE}$$

Considering now that

$$E = 2G(1 + \nu)$$

and equating w and Q to the settlement w_b and load Q_b at the base of the pile, we get:

$$w_b = \frac{Q_b}{G_b r_s} \frac{1 - \nu}{4} \tag{13.56}$$

where $r_s = B/2$ is the radius of the pile.

Taking Equation (13.56) into Equation (13.54), we obtain an expression for the pile base stiffness K_b:

$$K_b = \frac{4 G_b r_s}{1 - \nu} = \frac{2 G_b B}{1 - \nu} \tag{13.57}$$

We can obtain Q_t by multiplying Equation (13.51) by $\sinh \lambda L$ and Equation (13.52) by $-\cosh \lambda L$, and by adding the resulting equations together:

$$\frac{Q_t}{\lambda E_p A_p} = w_b \sinh \lambda L + \frac{Q_b}{\lambda E_p A_p} \cosh \lambda L \tag{13.58}$$

The substitution of Equations (13.53) and (13.54) into Equation (13.58), after some rearrangement, gives:

$$\frac{Q_t}{\lambda E_p A_p} = w_b \cosh \lambda L (\tanh \lambda L + \Omega) \tag{13.59}$$

The pile head stiffness K_t can now be determined from Equations (13.55) and (13.59) as

$$K_t = \frac{Q_t}{w_t} = \lambda E_p A_p \frac{\Omega + \tanh \lambda L}{1 + \Omega \tanh \lambda L} \tag{13.60}$$

We can obtain an expression for Ω that is sometimes more practical to use from the combination of Equations (13.53) and (13.57):

$$\Omega = \frac{4G_b r_s}{1-v}\frac{1}{\lambda E_p A_p} \tag{13.61}$$

13.9.4 Limiting cases: Ideal floating, infinitely long, and end-bearing piles

Let us examine what Equation (13.60) teaches us about the special cases of ideal floating piles (piles with $\Omega = 0$), infinitely long piles (piles with $\lambda L \to \infty$), and perfect end-bearing piles (piles with $\Omega \to \infty$).

For $\Omega = 0$, Equation (13.60) reduces to

$$K_t = \lambda E_p A_p \tanh \lambda L \tag{13.62}$$

Figure 13.24 shows pile head stiffness K_t versus normalized length λL plots for an ideal floating pile ($\Omega = 0$) as well as piles with small Ω values. Although the figure was prepared for a specific set of soil properties and pile properties (both material properties and cross-sectional parameters), conclusions about the shape of the plots are general. The examination of the portion of the curve for small deflections, shown in Figure 13.24a, shows that the pile head stiffness K_t varies practically linearly with pile length L for all values of Ω; that holds if λL is less than approximately 0.50 for $\Omega = 0$ and 0.30 for $\Omega = 0.6$. This interesting result can be obtained mathematically for an ideal floating pile by recalling that $\tanh \lambda L$ is defined in terms of the exponentials $e^{\lambda L}$ and $e^{-\lambda L}$, which can be expanded using the Maclaurin series. For small values of λL, the higher-order terms of these expansions can be dropped, and $\tanh \lambda L$ results equal λL. This means that Equation (13.62) reduces to:

$$K_t\big|_{\lambda L \to 0} = \lambda^2 E_p A_p L \tag{13.63}$$

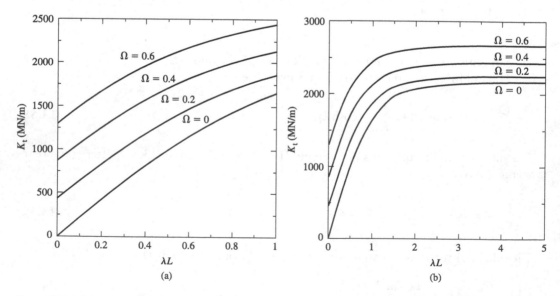

Figure 13.24 Pile head stiffness versus normalized length λL for floating piles (piles with $\Omega = 0$–0.6): (a) $0 \le \lambda L \le 1$ and (b) $0 \le \lambda L \le 5$.

Equation (13.63) clearly establishes that the pile head stiffness K_t varies linearly with pile length L for a relatively short floating pile. To understand this physically, let us assume that we will make Q_t always equal to the total pile resistance available, which is entirely due to shaft resistance because $\Omega = 0$ for an ideal floating pile. The shaft resistance is directly proportional to the lateral surface and thus to the length of the pile. This means Q_t increases linearly with pile length L. So, as the pile length increases, the pile top settlement w_t, equal to the ratio of Q_t to K_t (both of which increase in linear proportion to L), remains unchanged. Another way to obtain the same result is to substitute Equation (13.64) for λ in Equation (13.65), which leads to the pile head stiffness being simply kL. This means that the pile head stiffness has ceased to depend on $E_p A_p$, being a function only of the modulus of subgrade reaction k and the pile length L. Furthermore, if Q_t is increased proportionally with L, the settlement does not change, being a function only of k. This type of response is what we would expect from a free-standing rigid column, for which there would be no additional displacement at the top regardless of the pile length.

Let us contrast this response of a floating pile with what would happen if we had a free-standing, elastic column with the same geometry and material properties as the pile. In that case, the stiffness of the column, given by $E_p A_p/L$, would decrease with pile length, so the same load increment that produces no settlement for the floating pile (so long as we increase its length, as discussed previously) would lead to some compression and thus some settlement at the top of the free-standing column.

For an infinitely long pile, $\lambda L \to \infty$ and $\tanh \lambda L$ approaches 1. For an infinitely long pile, the pile base stiffness is irrelevant. Taking the limit of Equation (13.66) as $\lambda L \to \infty$, we obtain the following for the pile head stiffness:

$$K_t\big|_{\lambda L \to \infty} = \lambda E_p A_p \qquad (13.67)$$

This means the pile head stiffness is constant for an infinitely long pile: if the load is increased by always the same amount, that leads always to the same increase in deflection.

K_t of a pile with finite length approaches the value of K_t for an infinitely long pile with the same λ, E_p, and A_p asymptotically and monotonically, as shown in Figure 13.24b. An interesting question, which can be answered by the observation of Figure 13.24b, is: when does it cease to be beneficial to increase the pile length of a floating pile? Obviously, if pile head stiffness stops increasing once some value of λL is reached, then nothing is gained from further increases in pile length. Figure 13.24b shows that the curve for an ideal floating pile ($\Omega = 0$) becomes horizontal at a value of λL of ~2. This value is slightly smaller for small positive values of Ω. A more effective way to answer this question is to set a practical criterion for when the behavior of the floating pile is essentially the same as that of an infinitely long pile. We can do that by stating that when the pile head stiffness reaches 90% of the stiffness of an equivalent infinitely long pile, then there is little to gain from making it longer. This critical length can then be calculated by setting:

$$\frac{\lambda E_p A_p \tanh \lambda L}{\lambda E_p A_p} \geq 0.90$$

which leads to a requirement that $\lambda L > 1.50$ for a floating pile (a pile with $\Omega = 0$) to behave as an infinitely long pile.

A final case to consider is that of an ideal end-bearing pile (one for which $w_b = 0$ and $\Omega \to \infty$). By considering the limit of Equation (13.60) as $\Omega \to \infty$, we obtain:

$$K_t\big|_{\Omega \to \infty} = \frac{\lambda E_p A_p}{\tanh \lambda L} \qquad (13.68)$$

By comparing Equation (13.68) with Equations (13.62) and (13.67), we can see that the stiffness of an infinitely long pile, given by Equation (13.67), is the geometric mean of the stiffness of the ideal floating pile, given by Equation (13.62), and that of the ideal end-bearing pile, given by Equation (13.68).

In real problems, G, G_L, G_b, k, λ, and all derivative variables change when the pile length L changes. As a consequence, the K_t versus λL curves will not asymptotically approach a limit. This point will be illustrated in Example 13.14.

13.9.5 Application to real problems

The largest difficulty in applying pile settlement analysis of the type we described to real problems lies in choosing appropriate values of soil moduli to use for the shaft and for the base of the pile and then to determine how these translate into a suitable, representative value for the modulus of subgrade reaction k. Elasticity-based analyses always present this problem: how do you reduce an inelastic, nonlinear problem to an elastic representation of it? It is important to remember that the modulus values, particularly the value for the pile base, must reflect the level of settlement expected.

Figure 13.23 shows a depth profile of soil shear modulus G_s that would be typically assumed in this type of analysis. Three values of soil shear modulus are of interest: the average or representative shear modulus G along the shaft (this would be the value used in the linear elastic, homogeneous approximation we have been working with), the value of G_s at the level of the base of the pile (denoted by G_L), and a value of G_s representative of the pile base resistance (that is, a shear modulus G_b that would be representative of the soil below the pile base). G_L and G_b would typically be the same in a floating pile, that is, a pile such that the soil above and below its base is essentially the same. If the pile is embedded in a bearing layer, then G_b could be substantially greater than G_L. With respect to G, its lowest value will typically be half of G_L (for the case of G increasing linearly with depth from zero at the ground surface); it is commonly greater than that.

Now let us tackle the issue of how to arrive at a representative value of k. We will present two different ways: one appropriate for floating piles, and another for end-bearing piles.

13.9.5.1 Floating piles or piles with limited base resistance

Randolph and Wroth (1978) considered concentric cylindrical surfaces through the soil deposit, all with axes coincident with the pile axis. For vertical equilibrium to hold, the shear stress τ_v acting on the lateral area of these cylindrical surfaces must decrease with the distance from the pile axis according to:

$$\tau_v = q_s \frac{r_s}{r} \tag{13.69}$$

where r_s = radius of pile shaft (equivalent radius in the case of noncircular cross section); r = radius of cylindrical surface; and q_s = mobilized unit load (shear stress) on the pile shaft. Note that q_s is not necessarily the limit unit shaft resistance because, at working loads, q_{sL} may not be completely mobilized along the entire pile length.[14]

---

[14] The factor of safety of 2 to 3 we apply to the ultimate load calculated for 10% relative settlement guarantees the working load on the pile will generate a settlement that is less than $0.1B$. How much less depends on how accurate our estimate of Q_{ult} is. If we apply a safety factor of 3 on a conservative estimate of Q_{ult}, the actual settlement on the pile could be very small.

The distortion around the pile is mostly vertical. Because of this, we can disregard the component of shear strain associated with the change in radial displacement over increments of depth z, and we can write the following equation for the engineering shear strain at radius r:

$$\gamma = -\frac{dw}{dr} \tag{13.70}$$

where we recognize that w decreases with increasing r, hence the minus sign. The corresponding shear stress is then:

$$\tau = G\gamma = -G\frac{dw}{dr} = q_s\frac{r_s}{r} \tag{13.71}$$

Rewriting Equation (13.71) in terms of dw gives:

$$dw = -\frac{q_s r_s}{G}\frac{dr}{r} \tag{13.72}$$

and integrating with the boundary condition that $w = 0$ at a "magical" radius[15] r_m gives:

$$w = \frac{q_s r_s}{G}\ln\frac{r_m}{r} \tag{13.73}$$

The magical radius r_m at which the displacement w goes to zero can be estimated using the following equation (Fleming et al. 1992), which was based on extensive numerical simulations of the settlement in the soil surrounding an axially loaded pile:

$$r_m = \left\{0.25 + \left[2.5\left(1-v\right)\frac{G}{G_L} - 0.25\right]\frac{G_L}{G_b}\right\}L \tag{13.74}$$

The displacement at the pile shaft (at $r = r_s$) will now be given as

$$w = \frac{q_s r_s}{G}\ln\frac{r_m}{r_s} \tag{13.75}$$

This allows us to write the modulus of subgrade reaction k as

$$k = -\frac{dQ/dz}{w} = \frac{2\pi G}{\ln\dfrac{r_m}{r_s}} \tag{13.76}$$

Equation (13.76) can be used for floating piles or, in more practical terms, for piles in which the base resistance is not much larger than the shaft resistance.

13.9.5.2 End-bearing piles

Mylonakis (2001) found a very good, while still approximate, solution to Equation (13.42) with k variable with depth for the case of $w_b = 0$ (ideal end-bearing pile). Using this solution, he then determined equivalent values of k, assumed constant with depth, that would

---

[15] The term "magical" was used by the proponents of the analysis.

produce the same pile head stiffness as that obtained for the case with k varying with depth. The following approximate equation captures the effects of the key parameters:

$$\frac{k}{G} \approx 1.3\left(\frac{E_p}{E}\right)^{-\frac{1}{40}}\left[1+7\left(\frac{L}{B}\right)^{-0.6}\right]$$ (13.77)

where $E = 2(1+\nu)G$. So, for end-bearing piles, that is, piles with significant end-bearing resistance, Equation (13.77) may be used.

13.9.5.3 Piles with noncircular cross sections

Finally, when applying this analysis to both pipe and H-piles, an equivalent pile with a solid circular cross section is used. The equivalency is on the basis of having the same axial stiffness A_pE_p as the original pile. This is usually done by finding an equivalent pile radius r_s such that:

$$r_s = \sqrt{\frac{A_p}{\pi}}$$ (13.78)

Example 13.14

Precast concrete piles with a 0.4 m square cross section are being considered to support a structure as floating piles in a soft clay deposit. Consider the modulus of the concrete to be 25 GPa. The top of the soil deposit is slightly overconsolidated, and the soil shear modulus increases approximately as $G/p_A = 150 + 15(z/L_R)$, with p_A and L_R being the reference pressure (100 kPa = 0.1 MPa = 1 tsf) and reference length (1 m = 3.281 ft), respectively. What is the optimum length of these piles for efficient control of settlement?

Solution:

We need to obtain a figure similar to Figure 13.24, but, in this case, k and λ will vary with pile length. We will demonstrate the settlement calculations for a length of 10 m and then show a table and figure with results for a wide range of pile lengths. We need to start by finding an equivalent radius using Equation (13.78):

$$r_s = \sqrt{\frac{A_p}{\pi}} = \sqrt{\frac{0.40^2}{\pi}} = 0.225 \text{ m}$$

We now calculate the soil modulus values at the elevation of the top and base of the pile, and then go on to calculate the average modulus G:

$$G\big|_{z=0} = 0.1\left(150 + 15 \times \frac{0}{1}\right) = 15 \text{ MPa}$$

$$G\big|_{z=10\,m} = 0.1\left(150 + 15 \times \frac{10}{1}\right) = 30 \text{ MPa}$$

This gives $G = 22.5$ MPa and $G/G_L = 22.5/30 = 0.75$. Note that, for floating piles, $G_b = G_L$.

We now have what we need to calculate the "magical" radius using Equation (13.74) with $v = 0.5$ (because we have a pile in clay and, as usual, we are interested in the undrained response):

$$r_m = \left\{ 0.25 + \left[2.5(1-v)\frac{G}{G_L} - 0.25 \right]\frac{G_L}{G_b} \right\} L$$

$$= \left\{ 0.25 + \left[2.5(1-0.5)0.75 - 0.25 \right]1 \right\}10 = 9.4 \text{ m}$$

For floating piles, we use Equation (13.76) to calculate k, so:

$$k = -\frac{dQ/dz}{w} = \frac{2\pi G}{\ln\frac{r_m}{r_s}} = \frac{2\pi \times 22.5}{\ln\frac{9.4}{0.225}} = 37.9 \frac{\text{MN/m}}{\text{m}}$$

We can now get λ from Equation (13.46):

$$\lambda = \sqrt{\frac{k}{E_p A_p}} = \sqrt{\frac{37.9}{25,000 \times 0.16}} = 0.097 \text{ m}^{-1}$$

As discussed previously, the optimal length should correspond to a normalized length λL no greater than roughly 1.5. This suggests a critical length of no more than 16 m. This calculation assumes that the k we calculated for 10 m would hold for $L = 16$ m. Let us proceed further by taking due account of the effect of L on G, G_L, k, and, through k, on λ and then K_t.

We continue by calculating K_t for $L = 10$ m. We must first obtain K_b and then Ω from K_b. We use Equations (13.57) and (13.53):

$$K_b = \frac{4G_b r_s}{1-v} = \frac{4 \times 30 \times 0.225}{1-0.5} = 54 \text{ MN/m}$$

$$\Omega = \frac{K_b}{\lambda E_p A_p} = \frac{54}{0.097 \times 25,000 \times 0.16} = 0.139$$

Now the pile head stiffness K_t can be calculated using Equation (13.60):

$$K_t = \lambda E_p A_p \frac{\Omega + \tanh \lambda L}{1 + \Omega \tanh \lambda L}$$

$$= 0.097 \times 25,000 \times 0.16 \times \frac{0.139 + \tanh(0.097 \times 10)}{1 + 0.139 \times \tanh(0.097 \times 10)} = 312 \text{ MN/m}$$

Repeating the calculations for $L = 1$–100 m, we obtain the pile stiffness values shown in E-Figure 13.11. Note how, at a pile length around 10 m ($\lambda L \approx 1$), the rate of increase in pile head stiffness starts dropping at an increasing rate and how, at around 20 m ($\lambda L \approx 2$), the slope of the curve reaches a more stable lower value. The optimum pile length, without considering economic aspects, would be ~20 m. Note that this is not necessarily the most economical length. That would require minimizing costs, which is a more comprehensive criterion. But it appears, based on these calculations, that it would be uneconomical to extend the piles beyond 20 m.

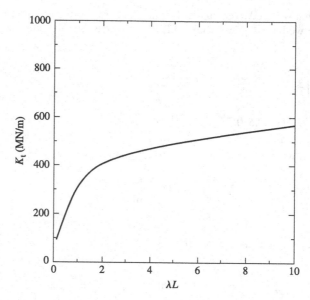

E-Figure 13.11 Pile head stiffness K_t versus λL for Example 13.14.

13.9.6 Negative skin friction

While the usual settlement analysis problem involves moving the pile down with respect to the soil (with the soil, therefore, opposing the pile movement), the opposite occasionally happens in soft clays, a phenomenon known as negative skin friction or downdrag. The classical case where this happens is the case when a fill is placed on top of soft clay. The piles are then installed through the fill and the soft clay before the soft clay has finished consolidating. As a result, the soft clay will be moving down with respect to the pile, applying a downward shear stress on it. Any time a pile is placed in a consolidating soil, downdrag will be observed. The consolidation may even be induced by pile driving itself, when it generates large pore pressures around the pile, which then dissipate as the soil consolidates.

Downdrag and the force it applies to the pile shaft (referred to as dragload) can be very significant, potentially leading to serious problems (see, for example, Chellis 1961; Brand and Luangdilok 1974; Acar et al. 1994). There are three approaches to dealing with downdrag: (1) design the pile for the added load imposed by negative skin friction; (2) design the pile for the structural load only, allowing the piles to settle more uniformly with the soil surrounding the building or supported structure; and (3) install the piles with a coat of bitumen or other materials that will demobilize the shear stress between the pile and the consolidating soil.

The first approach requires first that the motion of the consolidating soil in the absence of piles (its free-field motion) be understood and quantified. Typically, free-field soil settlement will decrease with depth because of either increasing soil stiffness with depth or the presence of a bearing layer (in which the piles are embedded), which is a nearly rigid boundary for the consolidation process. Then the motion of the pile needs to be quantified. It is, except in the case of rigid piles, greater near the pile head, from which it will decrease toward the base. This decrease in pile movement will depend on many factors, as we have already discussed in detail in this section. The next step is the identification of the depth at which the relative motion between the pile and the free-field soil reverses (let us call that the reversal depth), if it exists. It may be that the pile is an end-bearing pile with the entire soil mass around it settling, with therefore no reversal until the very tip of the pile. The pile tip will then support the entire load applied on the pile plus any loads due to downdrag. All of the shear stress developed along the segment of the shaft of the pile through soil with free-field motion greater than the pile is then considered an added load to be supported by the working part of the pile, located below the reversal depth.[16] Along the portion of the pile subject to downdrag, it is then necessary to locate the point above which there is slip between pile and soil. At that point, the difference between the soil free-field settlement and the local pile settlement (the settlement of the pile cross section) reaches the value required for full mobilization of shaft resistance (which, as discussed earlier in this chapter, is of the order of 0.25%–1% of the pile diameter). Above that point, the downdrag stress can be calculated, for example, using the soil properties approach:

$$q_{s,downdrag} = K_0 \sigma'_v \tan \delta \tag{13.79}$$

where K_0 is used for K because the soil is undergoing consolidation and δ is slightly less than or equal to ϕ_c, as discussed earlier.

Below the point demarcating the limit at which slip is observed, the consolidating soil still applies a downward shear stress on the pile, but the relative motion between the pile and the free-field soil is not sufficiently large for there to be slip and the behavior can then be modeled as elastic. A simple approach to quantifying this shear stress is to estimate the difference between the soil free-field settlement and the local pile settlement and linearly interpolate with respect to the value calculated using Equation (13.79) with the assumption that the value calculated using Equation (13.79) is reached at a value of the order of 0.25%–1% of the pile diameter.

Other analyses, with various degrees of sophistication, have been proposed. Some are elastic and cannot account for pile–soil slip (e.g., Poulos and Davis 1980; Chow et al. 1990; Lee 1993; Teh and Wong 1995); others are elastic–plastic and are able to account for pile–soil slip (e.g., Lee and Ng 2004).

The second approach to dealing with downdrag is not to proportion the piles for it at all. This approach is discussed by Fleming et al. (1992) as being applicable to many situations where downdrag is present. The rationale for this approach is that serviceability problems (discussed in Chapter 2) will develop at the interface of the building with the surrounding soil if the piles are designed to carry the downdrag loads and, as a result, settle considerably less than the surrounding ground and any adjacent structure not supported on piles. It is best, then, to design the piles for the superstructure loads only, obviously making sure that a sufficient factor of safety against classical bearing capacity failure exists and that angular distortions in the superstructure will be well within the safe range. This will result in settlements of the supported structure that will transition smoothly to the surrounding ground.

The last approach, using bitumen to reduce downdrag, has been in use for some time. One of the first times it appeared in the technical literature is in Bjerrum et al. (1969). The successful use of bitumen requires that steps be taken to ensure that the bitumen adheres to the pile and is not removed during driving. Removal is likely if a layer of sandy or gravelly soil exists near the ground surface (Fellenius 1999). Discussion on the aspects of selection of bitumen and its effective use to reduce downdrag can be found in Briaud (1997) and Fellenius (1999). The environmental impact of bitumen is discussed by Yeung et al. (1996).

---

[16] The term neutral plane is used to refer to the horizontal plane at the depth of reversal in soil–pile relative motion.

13.10 PILING IN ROCK*

13.10.1 Rock sockets and micropiles in rock

When a rock mass is relatively close to the ground surface, it becomes economical, for large loads, to socket piles into rock. When only the bottom one or two diameters are embedded into the rock, we call that a rock socket. Rock sockets can be installed using modern drilling rigs that have the power to drill into rock. In design, we are interested in determining both the shaft and the base capacity of the rock socket if the rock socket is short (typically less than twice its diameter). For longer rock sockets, the entire load will be carried by shaft resistance, and determination of the base capacity becomes less important. Whatever base capacity may be present would provide in this case an extra margin of safety.

Equipment exists also to install full-length piles into rock. The piles installed in rock are typically micropiles, with diameters in the range of 100–300 mm (4–12 in.). Micropiles develop virtually all of their load capacity from shaft resistance on account of their small diameter and considerable length.

13.10.2 Estimation of base resistance

The same somewhat simplistic analysis done for footings on rock in Chapter 10 can be used for finding the limit unit base capacity q_{bL} of rock sockets. Figure 13.25 shows the rock socket, the mechanism (consisting of failure wedges A and B), and the relevant quantities. We repeat Equation (10.68) here for convenience:

$$q_{bL} = q_0 + q_u \left\{ \sqrt{m \frac{q_0}{q_u} + s} + \sqrt{m \frac{q_0 + q_u \sqrt{m \frac{q_0}{q_u} + s}}{q_u} + s} \right\} \tag{13.80}$$

where:

m, s = Hoek–Brown strength parameters (refer to Chapter 4);
q_u = unconfined compressive strength of the intact rock; and
q_0 = surcharge at the level of the base of the rock socket.

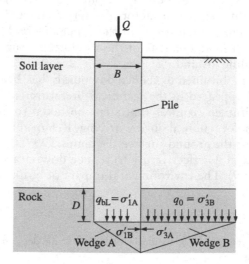

Figure 13.25 Bearing capacity analysis of rock socket: slip mechanism assumed to consist of blocks A and B.

AASHTO (1989) has an empirical procedure for the determination of an ultimate unit bearing capacity $q_{b,ult}$ that takes into account both the unconfined compressive strength of the rock and its rock mass quality:

$$q_{b,ult} = N_{ms}q_u \tag{13.81}$$

where N_{ms} is an empirical factor that depends on rock type, jointing and weathering.

The Canadian Foundation Engineering Manual (CGS 1985) has another empirical procedure for estimating $q_{b,ult}$:

$$q_{b,ult} = 3K_{sp}q_uD \tag{13.82}$$

where:

$$K_{sp} = \frac{3 + \dfrac{s}{B}}{10\sqrt{1 + 300\dfrac{g}{s}}} \tag{13.83}$$

$$D = 1 + 0.4\frac{L_{RS}}{B} \le 3.4 \tag{13.84}$$

B = pile diameter;
L_{RS} = rock socket length;
s = discontinuity spacing; and
g = discontinuity aperture.

The limits of applicability of this procedure are:

$$0.05 \le \frac{s}{B} \le 2$$

and

$$0 \le \frac{g}{s} \le 0.02$$

Yet another method for calculating $q_{b,ult}$ was proposed by Zhang and Einstein (1998), which we write below in nondimensional form:

$$\frac{q_{b,ult}}{p_A} = C_B\sqrt{\frac{q_u}{p_A}} \tag{13.85}$$

where:

$$C_B = \begin{cases} 9.5 & \text{for RMR} \approx 40 \\ 15 & \text{for RMR} \approx 65 \\ 21 & \text{for RMR} \approx 100 \end{cases} \tag{13.86}$$

and RMR is the rock mass rating, discussed in Chapter 7.

13.10.3 Estimation of shaft resistance

There are two methods to calculate the limit unit shaft resistance q_{sL} of piles in rock. The first method is based on the concept that q_{sL} is some fraction of the unconfined compressive strength q_u that the rock would have if it were intact (that is, had no discontinuities). The better the rock quality and the greater the q_u of the intact rock are, the greater is the q_{sL}. However, no matter how good the rock is, the shaft resistance cannot develop if the quality of the pile material is not at least as good. No interface can have more strength than the weakest of the two materials it separates. Thus, the calculated value of q_{sL} can be no higher than the shear strength of the concrete, grout, or cement paste used to construct the rock socket or micropile. A value of 5% of the characteristic (compressive) strength of the pile material has been found to be a conservative estimate of the upper limit on the interface shear strength between the pile material and the rock.

This is expressed mathematically as

$$q_{sL} = \min\left(\frac{1}{2}c_w q_u, 0.05 f_c'\right) \tag{13.87}$$

where c_w is the coefficient of weakness given in Table 7.4 as a function of discontinuity spacing, q_u is the unconfined compressive strength of the intact rock, and f_c' is the characteristic (nominal) strength of the concrete, cement paste, or grout used to construct the micropile or rock socket. The coefficient of weakness does not account for degree of weathering, so Equation (13.87) applies strictly only to unweathered or slightly weathered rock. Smaller values of q_{sL} must be assumed in case of a more severe weathering condition.

A second method assumes a nonlinear relationship between q_{sL} and the unconfined compressive strength q_u (Zhang and Einstein 1998):

$$\frac{q_{sL}}{p_A} = \min\left(C_S\sqrt{\frac{q_u}{p_A}}, 0.05\frac{f_c'}{p_A}\right) \tag{13.88}$$

where $C_S = 1.25$ (for a "smooth" interface) to 2.5 (for a rough interface); p_A = reference stress = 100 kPa = 0.1 MPa; and we have added the limit of 5% of the characteristic strength of the pile material to the original equation.

13.10.4 Estimation of structural capacity

Another particularity of piles in rock is the fact that the cross-sectional strength of the pile itself may control the design, a condition that is infrequent in soils. Either the pile can be designed as a short column following ACI 318-19, or for simplicity, a factor of safety of 2 can be applied to the cross-sectional strength calculated from the characteristic compressive strengths of the concrete (or grout or cement paste) and reinforcing steel.

The axial load applied on the pile must not exceed the structural capacity Q_0 of the pile. According to reinforced concrete design guidelines by the American Concrete Institute (ACI), the maximum axial load the pile can take (limiting concrete compressive strength to 85% of the unconfined compressive strength) is given by

$$Q_0 = 0.85 f_c'\left(A_g - A_{st}\right) + f_y\left(A_{st}\right)$$

where A_g is the gross cross-sectional area of the pile, A_{st} is the cross-sectional area of the steel reinforcement, and f_y is the yield stress of the steel.

Example 13.15

Find the contribution of a rock socket to the total capacity of a drilled shaft with 500 mm diameter installed through 8 m of a clayey soil. The rock in which the pile is socketed, described in Examples 7.4 and 10.18, is a slightly weathered basalt with five to ten horizontal (zero-dip) fissures per meter with typical RQD values in the 25%–50% range and a compressive strength (for the intact rock) of 80 MPa. Concrete with f'_c = 20 MPa is used. Consider two lengths for the rock socket: (a) 500 mm and (b) 1000 mm.

Solution:

In Example 10.18, we determined the Hoek–Brown strength criteria parameters m and s as 1.395 and 0.00293, respectively, for this rock.

We can use Equation (13.80) to estimate the base capacity of the rock socket. To determine the surcharge q_0 at the base of the rock socket, which appears in Equation (13.80), we will assume a conservative unit weight of 16 kN/m³ for the overlying clay soil and 28 kN/m³ for the basalt. Our surcharge is then:

$$q_0 = 8\ \text{m} \times 16\ \frac{\text{kN}}{\text{m}^3} + 0.5\ \text{m} \times 28\ \frac{\text{kN}}{\text{m}^3} = 142\ \text{kPa}$$

for the 500-millimeter-long socket and

$$q_0 = 8\ \text{m} \times 16\ \frac{\text{kN}}{\text{m}^3} + 1\ \text{m} \times 28\ \frac{\text{kN}}{\text{m}^3} = 156\ \text{kPa}$$

for the 1000-millimeter-long socket.

We may now calculate the unit base resistance of the two rock sockets:

$$q_{bL} = q_0 + q_u \left[\sqrt{m\frac{q_0}{q_u} + s} + \sqrt{m\frac{q_0 + q_u\sqrt{m\dfrac{q_0}{q_u} + s}}{q_u} + s} \right]$$

In this equation, the following term appears twice:

$$\sqrt{m\frac{q_0}{q_u} + s} = \sqrt{1.395\frac{142}{80,000} + 0.00293} = 0.074$$

for the 500-millimeter-long socket and

$$\sqrt{m\frac{q_0}{q_u} + s} = \sqrt{1.395\frac{156}{80,000} + 0.00293} = 0.075$$

for the 1000-millimeter-long socket.
We can now write:

$$q_{bL} = 142 + 80,000 \left[0.074 + \sqrt{1.395\frac{142 + 80,000 \times 0.074}{80,000} + 0.00293} \right]$$

$$q_{bL} = 32,430\ \text{kPa}$$

for the 500-millimeter-long socket and

$$q_{bL} = 156 + 80,000 \left[0.075 + \sqrt{1.395 \frac{156 + 80,000 \times 0.075}{80,000} + 0.00293} \right]$$

$$= 32,722 \text{ kPa}$$

for the 1000-milllimeter-long socket.

The shaft resistance may be determined by using Equation (13.87). Focusing on the rock first:

$$q_{sL} = \frac{1}{2} c_w q_u$$

The coefficient of weakness c_w may be found in Table 7.4. For our case, with five to ten joints per meter (every 10–20 cm) we find the coefficient of weakness to be:

$$c_w = 0.06$$

The shaft resistance of our rock socket would be:

$$q_{sL} = \frac{1}{2} 0.06 \times 80 \text{ MPa} = 2400 \text{ kPa}$$

However, q_{sL} cannot exceed 5% of the compressive strength of the cement paste, $0.05 f_c'$. This will ensure the transfer of load from the drilled shaft to the surrounding rock. In this case, the pile material strength controls, so:

$$q_{sL} = 0.05 f_c' = 0.05 \times 20 \text{ MPa} = 1000 \text{ kPa}$$

For the 500-millimeter rock socket:

$$Q_{bL} = \frac{\pi}{4}(0.5)^2 \times 32,430 = 6368 \text{ kN}$$

$$Q_{sL} = \pi(0.500 \text{ m}) \times 0.500 \text{ m} \times 1000 \text{ kPa} = 785 \text{ kN}$$

$$Q_L = 6368 + 785 = 7153 \text{ kN} \text{ answer}$$

For the 1000-millimeter rock socket:

$$Q_{bL} = \frac{\pi}{4}(0.5)^2 \times 32,722 = 6425 \text{ kN}$$

$$Q_{sL} = \pi(0.500 \text{ m}) \times 1.000 \text{ m} \times 1000 \text{ kPa} = 1571 \text{ kN}$$

$$Q_L = 6425 + 1571 = 7996 \text{ kN} \text{ answer}$$

Example 13.16

Assuming the same rock as that of Examples 7.4 and 10.18, find the allowable capacity of a micropile with diameter equal to 200 mm (8 in.). Cement paste with $f_c' = 27$ MPa is used. Consider the limitations of the cross section of the pile to calculate the allowable load; then determine the pile length required to carry the allowable load.

Solution:

For micropiles, we consider only shaft resistance when estimating design capacity. We do this because the base capacity of the micropile is typically small, given its small diameter and the much smaller settlement required for shaft resistance mobilization compared with base resistance. We estimate shaft resistance using Equation (13.87):

$$q_{sL} = \frac{1}{2}c_w q_u$$

where the coefficient of weakness c_w may be found from Table 7.4. For our case, with five to ten joints per m (every 10–20 cm), we find the coefficient of weakness to be:

$$c_w = 0.06$$

The shaft resistance of our micropile would be equal to:

$$q_{sL} = \frac{1}{2}0.06 \times 80 \text{ MPa} = 2400 \text{ kPa}$$

However, this value exceeds the upper limit of 5% of the compressive strength of the cement paste. The upper limit on q_{sL} is:

$$q_{sL} = 0.05f_c' = 0.05 \times 27 \text{ MPa} = 1350 \text{ kPa}$$

The shaft capacity per unit length of our micropile is then:

$$Q_{sL} = 1350 \text{ kPa} \times (0.200\pi) = 848 \frac{\text{kN}}{\text{m}}$$

We must also consider the structural capacity of the micropile cross section. For this, we will consider the cement paste strength in combination with reinforcement consisting of four steel bars with 20 mm diameter. From ACI reinforcing steel specifications, we know that:

$$f_y = 420 \text{ MPa}$$

According to ACI reinforced concrete design guidelines, the maximum axial load (limiting concrete compressive strength to 85% of the unconfined compressive strength) is given by

$$Q_0 = 0.85f_c'\left(A_g - A_{st}\right) + f_y\left(A_{st}\right)$$

The gross area A_g of our micropile cross section is:

$$A_g = \frac{\pi}{4}(0.200 \text{ m})^2 = 0.0314 \text{ m}^2$$

For reinforcement, we will use four 20 mm rebars, with steel area equal to:

$$A_{st} = 4 \times \frac{\pi}{4}(0.020 \text{ m})^2 = 0.001257 \text{ m}^2$$

The cross-sectional load capacity can now be calculated from the compressive strength of the cement paste of 27 MPa and the yield strength of the steel of ~420 MPa. This value is:

$$Q_0 = 0.85 \times 27 \text{ MPa}\left(0.0314 \text{ m}^2 - 0.001257 \text{ m}^2\right) + 420 \text{ MPa}\left(0.001257 \text{ m}^2\right) = 1220 \text{ kN}$$

Using a factor of safety of 2, this sets the maximum allowable axial load at:

$$Q_{allow} = \frac{1220 \text{ kN}}{2} = 610 \text{ kN}$$

To minimize the cost of our micropiles, we would like to select a micropile length that develops as much capacity as the structural capacity of the micropile allows. We will use a factor of safety of 3. The required structural resistance, with a factor of safety of 3, of the micropile is:

$$Q_{req} = 3 \times 610 \text{ kN} = 1830 \text{ kN}$$

and the required minimum length of our micropile is:

$$L = \frac{1830 \text{ kN}}{848 \dfrac{\text{kN}}{\text{m}}} = 2.2 \text{ m}$$

It is desirable, for safety reasons, to use minimum lengths for the micropiles. A safe minimum length would be 3 m, so we would in this case simply specify a 3-meter-long micropile.

13.11 LATERALLY LOADED PILES*

13.11.1 The design problem

The problem of a laterally loaded pile can be represented as shown in Figure 13.26a for the case of the pile base embedded in a strong layer and Figure 13.26b for a pile installed in soil without a clear firm layer within reach. At its head, the pile is subjected to loading that includes a lateral load and may also include an axial load and an applied moment. The applied moment usually results from the load being applied some distance above the pile head; this distance is often referred to as the load eccentricity. For small deflections, the effects of the two loads and moments can be considered to be independent, but for larger deflections (often of interest in practical problems), they truly aren't. While some methods of analysis and software will account for structural dependence (mainly the dependence of lateral deflection on the presence of an axial load because of the moment the axial load starts having with respect to a given pile cross section as it displaces laterally), the dependence of the compressive resistance of soil on any vertical shearing caused by the vertical load and, in

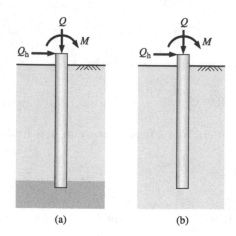

(a)   (b)

Figure 13.26 Vertical pile subjected to axial (vertical), transverse (horizontal), and moment loadings: (a) "fixed" base and (b) free base.

Table 13.20 Serviceability limit state (SLS) and ultimate limit state (ULS) criteria for laterally
            loaded monopiles

References	SLS	ULS
Luo et al. (2018)		$\theta = 2°$
Klinkvort and Hededal (2013)		$\theta = 4°$
Ahmed and Hawlader (2016)		$\theta = 5°$
Zdravkovic et al. (2015)		$y_{ml} = 10\% B$
Doherty and Gavin (2012)	$\theta = 0.5°$	
Arany et al. (2017)	$\theta = 0.5°$ or $y_{ml} = 0.2\,m$	

Note: θ, pile rotation at the mudline; y_{ml}, lateral pile deflection at the mudline.

turn, the dependence of the shearing resistance of the soil on the compressive action caused
by the lateral load are not well studied.

The design goal, as always, is to prevent achievement of limit states. ULSs may be of two
types: large deflections that lead either to structural damage of the supported structure or
slippage of the pile with respect to the soil (slippage may be of the upper part of the pile with
respect to the soil in case of pile breakage). An SLS for a laterally loaded pile corresponds
to a lateral deflection or rotation of the pile head that is sufficiently large to cause nonstruc-
tural damage to the superstructure or impair the functionality of the superstructure.

One of the main tasks of the design engineer is to establish the value of the tolerable
deflection to use in design. There are not as much data on tolerable lateral displacements as
there are for vertical displacements. One of the exceptions is the work of Bozozuk (1978) for
bridges, in which significant structural damage to bridges and "rough ride" conditions were
matched to the foundation movements and plotted as shown in Figure 2.18. For bridges,
Figure 2.18 can be used as a reference. For wind turbines, engineers will often look to the
manufacturer for guidance. These turbines today are most often supported by monopiles
(large-diameter pipe piles). Table 13.20 lists the values found in the literature for the angle of
rotation or displacement at the mudline for monopiles. For buildings and other structures,
it is more difficult to make this assessment, and the structural engineer should be engaged
in this decision as much as possible. The tolerable lateral deflection engineers work with is
typically in the 5–25 mm range.

13.11.2 Pile lateral load response

As discussed earlier, piles are better suited than shallow foundations to support large lateral
loads imposed by, for example, traffic, collisions, wind, earthquakes, and earth pressures.
This superior lateral capacity results from their embedment into the ground and from the
soil resistances that can thus be mobilized. Soil resistance p is mobilized in the direction
opposite to that of the lateral deflection y.[17] These resistances are transferred to the pile
through two mechanisms: (1) side resistance due to friction and adhesion between the soil
and the pile and (2) normal stresses between the soil and the pile.

---

[17] This notation for the lateral deflection is unfortunate, particularly because of the possible confusion with a
Cartesian coordinate y, but also because y is not used to denote displacement anywhere in the mechanics literature. We have decided to retain it because of its common use in practice. The notation for the soil resistance is
equally undesirable because lowercase letters such as the letter p are usually associated with pressure and thus
would typically have the unit of stress. Additionally, p is used in critical-state soil mechanics to note the total
mean stress, as seen earlier in this text (in Chapters 4–6). It seemed, however, that using a different notation
in this text would have been more confusing than helpful, as the p–y method, to be discussed later in greater
detail, is firmly established in practice with precisely this notation.

In the so-called *p–y method*, discussed in greater detail later, it is customary to combine the side and passive resistances into a single unit resistance p in units of load per unit length of pile. The best way to understand p is to think of it as a normal stress equivalent to the side and passive resistance effects multiplied by the pile diameter B. To obtain a force acting on a segment of the pile, we then multiply p by the length of the pile segment.

As illustrated in Figure 13.27, the side resistance is the predominant component of the unit lateral resistance p at small lateral deflections, whereas the passive resistance is more significant at large lateral deflections. This is due to the fact that, as seen earlier for axially loaded piles, the rate of mobilization with respect to deflection of frictional resistance far exceeds that of compressive, passive resistance.

As in the case of axially loaded piles, load is transferred down the pile (and, naturally, from the pile to the soil) as the load at the pile head (and thus the deflection) is increased. Considering a given pile cross section, the soil next to it yields after some level of deflection is reached and can take no additional load (this is shown in Figure 13.27d as the attainment of the limit resistance p_L); this means that additional load must be transferred down the pile, where the resistance of soil can be mobilized. If the load is increased up to a sufficiently large value (which we will call the ultimate lateral load), a ULS associated with pile instability will eventually be reached. The pile response to the applied lateral load depends on the following factors: (1) pile length; (2) pile bending stiffness $E_p I_p$; (3) soil stiffness; and (4) degree of fixity of the head and base of the pile. The pile bending stiffness is the product of the pile's Young's modulus E_p and the moment of inertia I_p.

A long pile is a pile that is sufficiently embedded in the soil for its base to be essentially motionless and experience no moment, shear force, rotation, or deflection. How long the pile has to be depends on soil and pile properties. For example, if the soil is very stiff, the pile embedment does not need to be very large for the pile to be "long." The embedment of a short pile, in contrast, is insufficient to prevent pile base motion. As a result, pile

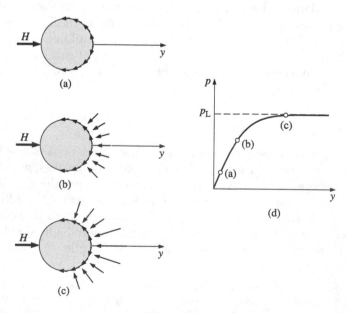

Figure 13.27 Representation of the shear force H and lateral resistance p corresponding to a given pile cross section: (a) soil–pile interface side resistance (friction + adhesion), (b) soil–pile interface compressive (normal stress) resistance, (c) unit lateral pile resistance p reflecting both the compressive and shear resistances of the soil, and (d) p as a function of lateral displacement y.

deformation is negligible compared with the movement of the whole pile as a rigid body. Accordingly, a short pile fails as a rigid body in either rotation (Figure 13.28a) or translation (Figure 13.28b), depending on the boundary conditions. Single piles are loaded under free-head conditions and will rotate freely, while piles connected to a pile cap or restrained in some manner are subjected to conditions that can typically be approximated as fixed-head conditions, so the pile head will not rotate with respect to the pile cap, moving instead with it (and predominantly in horizontal translation) under the action of a lateral load.

A long, flexible pile will not move as a rigid body when subjected to an increasing lateral load, but rather deform until one or more plastic hinges finally form at one or two cross sections for which the bending moments exceed the cross-sectional plastic moment. Free-headed long piles will form one plastic hinge (Figure 13.29a), while fixed-headed piles will typically form a hinge first at the connection with the pile cap and then possibly another at the depth corresponding to the maximum bending moment on the pile after the formation of the first hinge (Figure 13.29b). The load corresponding to either outcome is an ultimate load. The ultimate load in this case is associated with the structural failure of the pile and/or cap due to excessive movement of the pile, which is then followed by very large deflections. Furthermore, before this ultimate load is reached, it is very possible that pile movement will cause damage to the superstructure, which would be associated with its own ULS and hence its own ultimate load.

For applied loads less than the ultimate load, the pile head will deflect as a result of either pile deformation (in the case of long piles) or rigid body motion (in the case of short piles). It is usually more important for us to be able to estimate lateral displacements than to calculate ultimate loads because SLSs are more commonly critical. This can be done in one of two ways: (1) by treating the soil mass around the pile as a continuum (Figure 13.30a) or (2) by

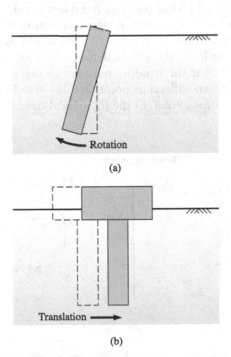

(a)

(b)

Figure 13.28 Failure modes for short piles: (a) rotation for free-head boundary conditions and (b) translation for fixed-head boundary conditions.

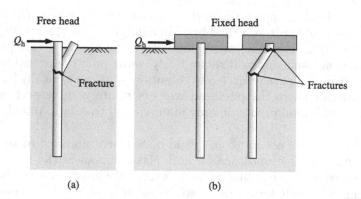

Figure 13.29 Failure modes for long piles: (a) formation of one plastic hinge for the free-head boundary conditions and (b) formation of two plastic hinges for the fixed-head boundary conditions.

substituting springs located at discrete points along the pile for the soil (Figure 13.30b). This second approach, when nonlinear springs are used, is generally known as the p–y method, because each spring is represented by a specific p–y curve, as shown in Figure 13.30c. The p–y method is common in current practice, despite shortcomings that we discuss later. We will accordingly focus on it in this chapter. But before we introduce the methods of analysis used for predicting the deflections of laterally loaded piles, it is worthwhile exploring what the deflected shape of the pile and the shapes of the shear force and bending moment diagrams along the pile axis are like for various conditions.

Figure 13.31 shows, qualitatively, the deflected shape, bending moments, and shear forces for moderately long piles with the following combinations of boundary conditions: (a) fixed head and base; (b) free head and fixed base; (c) fixed head and free base; and (d) free head and base. The figure shows also a consistent sign convention for the pile deflection y, the bending moment M and shear force H. What is of note here is that, for all boundary conditions and a sufficiently large load, the pile is deformed along practically its entire length. Another important point is that the bending moment changes sign (which happens when the distorted pile shape has an inflection point) for the fixed head–fixed base boundary conditions. It also does, to some extent, in the fixed head–free base case; the reason for the

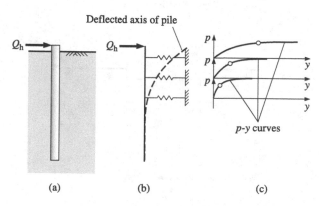

Figure 13.30 Modeling the soil–pile system for load–deflection analysis of laterally loaded piles: (a) pile in a soil continuum; (b) model using Winkler springs; (c) deflected pile shape when nonlinear springs described by p–y curves are used to model the soil.

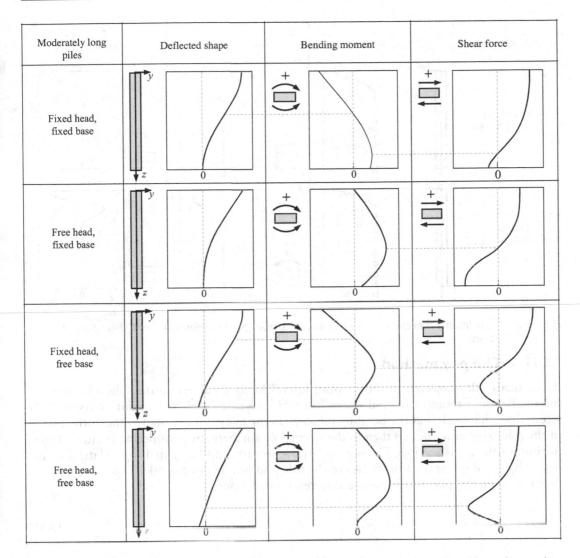

Moderately long piles	Deflected shape	Bending moment	Shear force
Fixed head, fixed base			
Free head, fixed base			
Fixed head, free base			
Free head, free base			

Figure 13.31 Qualitative representation of deflected shape, bending moments, and shear force for moderately long piles.

presence of an inflection point in this case (near the pile base) is that the pile is, by definition, a long pile, so it does develop fixity. The fact that the inflection point is near the pile base is related to the fact that the pile is moderately long (that is, long but not sufficiently long). All the curves shown do exhibit some pile deformation all the way down to the pile base or to depths near the pile base. We can contrast that with what we observe in Figure 13.32 for long piles. In that case, the pile's deflected shape and the bending moment and shear force along its axis are insensitive to the pile base boundary condition. This is shown in the figure for both fixed-head and free-head conditions by using symbols for fixed base and a line for free base. We can see that the lines track the symbols perfectly. We can also see that through the fact that a relatively extended length of the pile above the pile base is undeformed in all cases. In the figure, horizontal lines suggest the relationship that exists for key features in the plots between the three quantities. The quantities are obtained by differentiation or integration from each other, and that is the source of the relationship.

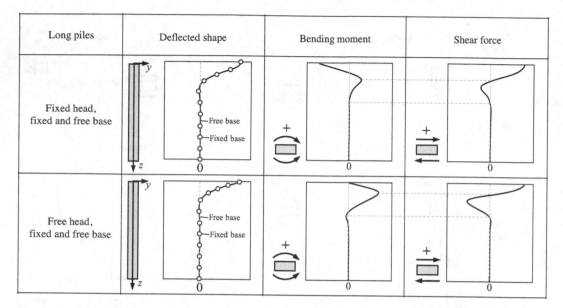

Long piles	Deflected shape	Bending moment	Shear force
Fixed head, fixed and free base			
Free head, fixed and free base			

Figure 13.32 Qualitative representation of deflected shape, bending moments, and shear force for long piles.

13.11.3 The p–y method

A (vertical) pile subjected to (horizontal) lateral loads is subject to the following internal forces: bending moment M and shear force H. We may recall from your coursework in strength of materials that the bending moment in the pile can be related to the normal stress in the pile cross section, and then to the normal strain there, by considering rotational equilibrium of the cross section. The normal strain, in turn, can be related to the lateral deflection y by considering the kinematics of the pile and bending of the pile. This gives us, for a depth z down the pile, the following differential equation:

$$M = E_p I_p \frac{d^2 y}{dz^2}$$

(13.89)

where E_p is the Young's modulus of the pile, I_p is the moment of inertia of the pile cross section, and $E_p I_p$ is the bending stiffness of the pile. This differential equation is not yet useful because the bending moment profile is not known. What is required is a relationship between pile deflection and soil stiffness, so we need to modify Equation (13.89) with that goal in mind.

The change in shear force across the element dz (that is, between depths z and $z + dz$) is due to the unit lateral soil resistance p. The soil resistance p is a continuous, distributed force (per unit length, so that p has the unit of force per length[18]) acting along the pile shaft in the negative direction (that is, opposing the lateral deflection $y(z)$). There are two sources of soil resistance: (1) compressive resistance and (2) shear resistance (see Figure 13.33). If we picture the soil mass as composed of an infinite number of horizontal layers (crossed by the pile) of infinitesimal thickness dz, compressive resistance follows from the reaction of these

[18] This is not a good notation because lowercase letters such as the letter p are usually associated with pressure and thus would typically have units of stress. Additionally, p is used in critical state soil mechanics to note the total mean stress, as seen earlier in this text (in Chapters 4–6). It seemed, however, that using different notation here would have been more confusing than helpful, as the p-y method is firmly established in practice with precisely this notation.

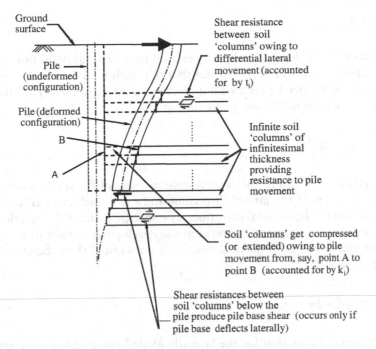

Ground surface

Pile (undeformed configuration)

Pile (deformed configuration)

B

A

Shear resistance between soil 'columns' owing to differential lateral movement (accounted for by t_i)

Infinite soil 'columns' of infinitesimal thickness providing resistance to pile movement

Soil 'columns' get compressed (or extended) owing to pile movement from, say, point A to point B (accounted for by k_i)

Shear resistances between soil 'columns' below the pile produce pile base shear (occurs only if pile base deflects laterally)

Figure 13.33 The lateral resistance p to lateral movement of a pile results from compressive resistance and shear resistance within the soil. (Basu et al. (2009); courtesy of the Institution of Civil Engineers and Thomas Telford Limited.)

horizontal layers of soil to being compressed as the pile pushes against them. Shear resistance follows from the fact that the lateral deflection of the pile varies with depth, imposing shearing between the infinitesimally thin soil layers (that is, causing a relative horizontal motion between two adjacent horizontal layers), which is also resisted by the soil. The total soil resistance p can then be seen as having contributions from both the soil compressive resistance p_c (because any horizontal layer of soil around the pile is compressed as the pile presses against it) and the soil shear resistance p_s (because the varying lateral deflection with depth causes a measure of horizontal shearing – that is, shearing between adjacent horizontal layers – in the soil). Thus, at a depth z, we get:

$$p = p_c + p_s \tag{13.90}$$

The total soil resistance p balances the change in pile shear force dH over the infinitesimal length dz and keeps the pile element in equilibrium. Considering the equilibrium of the pile element in contact with the soil layer with thickness dz at depth z, we obtain

$$dH = -pdz \tag{13.91}$$

The relationship between the change in the bending moment M across a differential depth dz and the shear force H at depth z is obtained from the consideration of moment equilibrium:

$$dM = Hdz \tag{13.92}$$

Using Equations (13.91) and (13.92), Equation (13.89) can be rewritten as

$$E_p I_p \frac{d^4 y}{dz^4} = -p \tag{13.93}$$

The soil resistance p at depth z is directly related to the lateral deflection y, as discussed earlier. The greater the value of y, the greater that of p, although the increase is not linear in realistic problems. So we need a relationship between p and y to obtain a differential equation in terms of the lateral deflection y. We now write

$$p = p_c + p_s = ky - 2t \frac{d^2 y}{dz^2} \tag{13.94}$$

where k is a coefficient related to the compressive stiffness of the soil (known as the modulus of subgrade reaction in the lateral direction) and t is a coefficient related to the shear stiffness of the soil in the horizontal direction. The specific form of this relationship follows from more involved considerations concerning the development of shearing along horizontal planes. Note that k has the unit of stress, but t does not. Taking Equation (13.94) into Equation (13.93), we get

$$E_p I_p \frac{d^4 y}{dz^4} - 2t \frac{d^2 y}{dz^2} + ky = 0 \tag{13.95}$$

which is the differential equation for the laterally loaded pile problem in terms of the coefficients k and t.

In the p–y method of analysis, the pile is assumed to be pushing against springs positioned at unit spacing with respect to one another, as represented in Figure 13.30b. No shear interaction between the springs is considered, so the relationship between the unit soil resistance p and the lateral deflection y follows from Equation (13.94) by making $t = 0$:

$$p = ky \tag{13.96}$$

where the modulus of subgrade reaction k is assumed to vary with the deflection y to account for the modulus degradation experienced by the soil with increasing strain. The differential equation follows from Equation (13.95) also by making $t = 0$:

$$E_p I_p \frac{d^4 y}{dz^4} + ky = 0 \tag{13.97}$$

Because the differential Equation (13.97) is typically solved numerically, it is possible and common for curves of p versus y, known as p–y curves, to be used instead of Equation (13.96) to represent p.

13.11.4 Limit unit lateral resistance p_L

An analysis of the limit[19] unit lateral resistance p_L of piles must account for the proximity to the ground surface of the cross section where p_L is desired to the ground surface. Where shallow conditions exist, a slip mechanism develops more easily and the limit pressure is

---

[19] The term "ultimate" instead of limit is widely used in publications and software. We prefer "limit" because it is not ambiguous, indicating clearly that p_L is associated with very large lateral sliding of the pile with respect to the soil.

accordingly lower. According to API (1993), the limit unit resistance p_L for piles in sand can be calculated from:

$$p_L = (C_1 z + C_2 B)\sigma'_v \tag{13.98}$$

for shallow conditions, and

$$p_L = C_3 B \sigma'_v \tag{13.99}$$

for deep conditions, where z = depth, B = pile diameter, and C_1, C_2, and C_3 are dimensionless functions of the relative density or peak friction angle. Stewart (2000) provides the following approximation for these dimensionless functions:

$$C_1 = 0.115 \times 10^{0.0405\phi_p} \tag{13.100}$$

$$C_2 = 0.571 \times 10^{0.022\phi_p} \tag{13.101}$$

$$C_3 = 0.646 \times 10^{0.0555\phi_p} \tag{13.102}$$

In calculating p_L in sand, we use both Equations (13.98) and (13.99) and take the lower of the two values.

According to API (1993), p_L/B for piles in clay transitions from $3s_u$ at the ground surface to $9s_u$ at a depth z_R[20]; for greater depths, it stays constant. This can be written mathematically as follows:

$$p_L = B(3s_u + \sigma'_v) + Js_u z \tag{13.103}$$

for $z \leq z_R$, where J is a dimensionless empirical parameter typically ranging from 0.25 for medium-stiff clays to 0.5 for soft clays, and

$$p_L = 9Bs_u \tag{13.104}$$

for $z > z_R$.

The depth z_R needed for full lateral resistance mobilization can be obtained from Equation (13.103) by requiring that p_L calculated from Equation (13.103) be equal to $9Bs_u$. If we can express the vertical effective stress as $\sigma'_v = \gamma_b z$, the resulting equation for z_R is:

$$z_R = \frac{6B}{B\dfrac{\gamma_b}{s_u} + J} \geq 2.5B \tag{13.105}$$

13.11.5 Long piles

As discussed previously, for long piles, in contrast to short piles, there is no load that, applied to the head, would cause the pile to move through the soil as a whole. This load, if it existed, would correspond to the concept of a limit load. There is instead a structural ultimate load, which results from full plastic yield of one (for the free-head condition) or

---

[20] The similarity with the ratio of q_{bL} for a pile to s_u (see discussions of axially loaded piles earlier in this chapter) is not accidental. Remember that a ratio of approximately nine has traditionally been used for axial loading as well.

two (for the fixed-head condition) cross sections. The formation of a plastic hinge at a pile cross section happens when the moment acting at the cross section equals the plastic hinge moment (or simply plastic moment) M_L of that cross section. The plastic moment is the moment for which the entire cross section of the pile becomes plastic, so that it will flow (rotate) unless the moment is reduced.

At this point, it may be helpful to review cross-sectional properties in bending. If the material of which the pile is made can be modeled as elastic–plastic, we can define the elastic section modulus S as the ratio of the bending moment at the cross section to the normal stress σ_{max} on the edge of the cross section (that is, the maximum normal stress for the cross section). When an element on the periphery of the cross section starts to yield (that is, when $\sigma_{max} = f_y$, where σ_{max} is obviously on the edge of the cross section and f_y is the yield stress or strength of the material), we say the bending moment at that point is the yield moment M_y of the cross section. Since f_y is still on the limit of elastic behavior, we can write:

$$M_y = S f_y$$

For reference, for rectangular cross sections, the elastic section modulus is $bh^2/6$ and the plastic modulus is $1.5M_y$. However, this is not usually very helpful because piles with rectangular cross sections are usually reinforced concrete piles, and reinforced concrete is not a homogeneous material; thus, M_y is calculated differently.

The plastic section modulus Z is defined as

$$Z = \frac{M_L}{f_y}$$

If Z is provided for a cross section, as is often done for rolled-steel cross sections, calculating M_L is very simple. The ratio of Z to S is 1.5 for rectangular cross sections, but is much less for, say, H-piles (for which we can take 1.1 as an approximate value).

For concrete piles, there is no symmetry between rupture in the compressive and tensile ranges, so the calculation changes somewhat. For reinforced concrete, the contribution of concrete on the side of the cross section in tension is disregarded. For unreinforced concrete, it is taken as between 0.05 and $0.1f_c'$. ACI (2019) offers more guidance in this connection. The plastic moment should be an input to lateral load versus lateral deflection analysis because we must not extend the analysis beyond the ultimate load for the pile (when hinges form and the pile ceases to work as a whole), which, for long piles, obviously depends on the plastic moment. The analysis for long piles using the p–y method is described later in this chapter.

13.11.6 Limit resistance of short piles

For short piles, which move as rigid bodies, the limit lateral load applied at the pile head can be calculated relatively simply. Broms (1964a,b) defined short piles as those satisfying the following criterion:

$$2 \geq \begin{cases} \dfrac{L}{T} & \text{for sandy soils} \\[2ex] \dfrac{L}{R} & \text{for clayey soils} \end{cases} \tag{13.106}$$

where L is the pile length;

$$T = \left[\frac{E_p I_p}{k_g} \right]^{\frac{1}{5}}$$
(13.107)

for sands, and

$$R = \left[\frac{E_p I_p}{k_0} \right]^{\frac{1}{4}}$$
(13.108)

for clays, where $k_g = dk_0/dz$ is the depth gradient of the initial subgrade reaction modulus k_0[21] in sands (given in Figure 13.34 as a function of the *in situ* initial relative density D_R of the soil and the position of the water table; note that variations of the initial modulus of subgrade reaction of sands with stress are much more pronounced for higher than for lower relative densities); k_0 in clays is assumed to be uniform with depth. In both equations, $E_p I_p$ is the pile bending stiffness.

In practice, k_0 of clay may vary slightly with depth, so we may use an average value of k_0 over the length of the pile as the value of k_0 in Equation (13.108). An estimate of k_0 can be

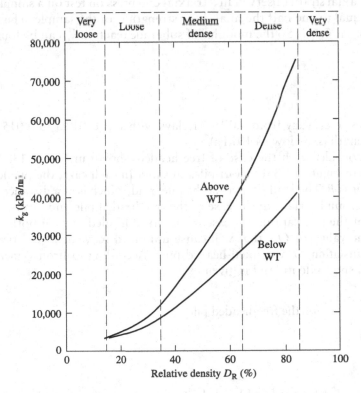

Figure 13.34 Gradient k_g of the initial modulus of subgrade reaction as a function of initial relative density according to API (1993). (Courtesy of API, RP 2A, Figure 31, 20th edition.)

---

[21] Remember that k should decrease with increasing deflection y (we will see how to quantify that decrease in the next section, when we discuss p–y curves), hence the term initial modulus.

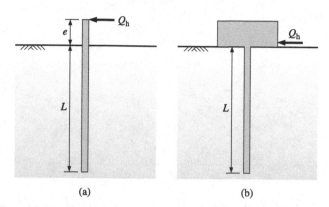

Figure 13.35 Laterally loaded short pile: (a) free-headed, with e = leverage arm or eccentricity, and (b) fixed-headed.

made by relying on the information contained in API (1993), which suggests we may assume the lateral deflection y_{50} corresponding to a soil resistance equal to 50% of the limit resistance to be roughly equal to

$$y_{50} = 2.5\varepsilon_{50}B$$

where ε_{50} is the axial strain observed in a triaxial compression test on a sample of the clay at a shear stress equal to one half the final shear strength s_u of the sample. The value of p corresponding to y_{50} is $0.5p_L$. So the modulus of subgrade reaction k_0 can be found by dividing $0.5p_L$ by y_{50}:

$$k_0 = \frac{p_L}{5\varepsilon_{50}B} \tag{13.109}$$

where the values of ε_{50} vary from 0.005 for clays with $s_u \geq 100\,\text{kPa}$ to 0.015 for clays with small shear strength (s_u as low as 10 kPa).

Let us now consider both the case of free-headed (shown in Figure 13.35a) and fixed-headed (shown in Figure 13.35b) short piles in sand. In each case, the pile length is L and the pile diameter is B. The load Q_h acts at the pile head, which is at a distance e (the leverage arm) from the ground surface in the case of the free-headed pile and flush with the ground (and the base of the pile cap) in the case of the fixed-headed pile. In both cases, we wish to determine the values of Q_{hL} that will cause unlimited movement (rotation for the free-headed pile; translation for the fixed-headed pile). According to Broms' method, the limit lateral load of a short pile in sand is given by

$$Q_{hL} = \begin{cases} \dfrac{1}{2}\dfrac{\gamma NL^3}{\dfrac{e}{B}+\dfrac{L}{B}} & \text{for the free-headed pile} \\[2em] \dfrac{3}{2}\gamma NBL^2 & \text{for the fixed-headed pile} \end{cases} \tag{13.110}$$

where γ is the material (wet) unit weight above the water table and the buoyant unit weight below the water table, and N is the flow number, given by Equation (4.45):

$$N = \frac{1+\sin\phi}{1-\sin\phi}$$

As always, a decision needs to be made as to what value of N should be used. Given that this limit state involves significant pile movement with respect to the soil, a conservative calculation would assume $\phi = \phi_c$.

Example 13.17

A drilled shaft with 1 m of diameter is to be installed in sand with $D_R = 70\%$, unit weight equal to 22 kN/m^3, and $\phi = 38°$. The drilled shaft has a length of 3 m. The water table is deeper than 3 m. The concrete modulus of elasticity is 23 GPa. Is this a short shaft from the point of view of lateral response? The leverage arm for the lateral load is 0.4 m. If so, calculate the limit lateral resistance of this shaft.

Solution:

Let us start by calculating the moment of inertia of the pile cross section:

$$I_p = \frac{1}{4}\pi r^4 = \frac{1}{64}\pi B^4 = \frac{1}{64}\pi (1)^4 = 0.04909 \text{ m}^4$$

The subgrade reaction modulus gradient is 50,000 kPa/m (from Figure 13.34). Now we can check whether the pile is short using Equations (13.106) and (13.107):

$$T = \left[\frac{E_p I_p}{k_g}\right]^{\frac{1}{5}} = \left[\frac{23 \times 10^6 \times 0.04909}{50,000}\right]^{\frac{1}{5}} = 1.9 \text{ m}$$

$$\frac{L}{T} = \frac{3}{1.9} < 2$$

so the pile is indeed short. The flow number N for 38° is 4.2. Now we can calculate the limit lateral resistance:

$$Q_{hL} = \frac{1}{2}\frac{\gamma N L^3}{\dfrac{e}{B}+\dfrac{L}{B}} = \frac{1}{2}\frac{22 \times 4.2 \times 3^3}{\dfrac{0.4}{1}+\dfrac{3}{1}} = 367 \text{ kN} \quad \text{answer}$$

We can find an expression for Q_{hL} for a short pile in clay by considering the limit equilibrium of the pile under the forces shown in Figure 13.36. The limit pressure is assumed on both sides of the pile as shown. Resistance at the top αB of the pile is considered not mobilized (Broms, for example, suggested that, because of pile–soil detachment, the top $1.5B$ should not be considered to contribute to lateral resistance). Taking moments around point O and requiring that they add up to zero and requiring also that there is lateral equilibrium (or, equivalently, that Q_{hL} be minimum), we arrive at the following expression for Q_{hL}:

$$Q_{hL} = L p_L \left[-\left(\alpha\frac{B}{L} + 2\frac{e}{L} + 1\right) + \sqrt{2\alpha^2\left(\frac{B}{L}\right)^2 + 4\left(\frac{e}{L}\right)^2 + 4\alpha\left(\frac{e}{L}\right)\left(\frac{B}{L}\right) + 4\left(\frac{e}{L}\right) + 2} \right] \quad (13.111)$$

Proceeding similarly for a fixed-headed pile in clay, we obtain:

$$Q_{hL} = p_L (L - \alpha B) \quad (13.112)$$

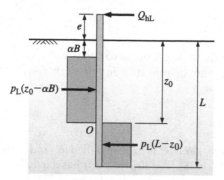

Figure 13.36 Limit resistance Q_{hL} of a short pile in clay.

which gives the same solution as that proposed by Broms if we take $\alpha = 1.5B$ and $p_L = 9s_u B$. However, we now know (see previous discussion on variation of p_L with depth according to the American Petroleum Institute) that Broms' original assumption of $p_L = 9Bs_u$ was not appropriate. A conservative solution would be based on $p_L = 3Bs_u$.

Example 13.18

The same drilled shaft as in Example 13.17 is to be installed in a clay with $s_u = 50\,kPa$. The geometry of the problem remains the same. Determine whether it is still considered a short pile, and, if so, determine its limit lateral resistance.

Solution:

To verify that the drilled shaft is a short pile, we need to first estimate the value of k_0 for this clay using Equation (13.109). We will be using $p_L = 3Bs_u$ in our solution because it will produce a conservative estimate of Q_{hL}. So we will use it also in calculating k_0. Given the moderately low shear strength of the clay of 50 kPa, we can take $\varepsilon_{50} = 0.01$ as a good approximation. This gives us:

$$k_0 = \frac{p_L}{5\varepsilon_{50}B} = \frac{3Bs_u}{5\varepsilon_{50}B} = \frac{3 \times 50}{5 \times 0.01} = 3000 \text{ kPa}$$

$$R = \left[\frac{E_p I_p}{k_0}\right]^{\frac{1}{4}} = \left[\frac{23 \times 10^6 \times 0.04909}{3000}\right]^{\frac{1}{4}} = 4.4 \text{ m}$$

So L/R (= 3/4.4) is indeed much smaller than 2, and the pile may be treated as a short pile. We can now calculate the limit lateral resistance of the free-headed pile as

$$Q_{hL} = Lp_L\left[-\left(\alpha\frac{B}{L}+2\frac{e}{L}+1\right)+\sqrt{\left\{2\alpha^2\left(\frac{B}{L}\right)^2+4\left(\frac{e}{L}\right)^2+4\alpha\left(\frac{e}{L}\right)\left(\frac{B}{L}\right)+4\left(\frac{e}{L}\right)+2\right\}}\right]$$

$$= 3 \times 3 \times 1 \times 50\left[-\left(1.5 \times \frac{1}{3}+2\frac{0.4}{3}+1\right)+\sqrt{2 \times 1.5^2\left(\frac{1}{3}\right)^2+4\left(\frac{0.4}{3}\right)^2+4 \times 1.5\left(\frac{0.4}{3}\right)\left(\frac{1}{3}\right)+4\left(\frac{0.4}{3}\right)+2}\right]$$

$$= 450 \times (-1.77+1.84)$$

$$= 32 \text{ kN}$$

$Q_{hL} = 32 \text{ kN}$ [answer]

Not surprisingly, the limit lateral resistance is much smaller than that available for the same drilled shaft in sand.

13.11.7 p–y Curves

A variety of approaches have been proposed to develop p–y curves based on load tests, *in situ* tests, analyses of different types, and combinations of these. However, examination of the literature shows frequent suggestions that site-specific validation of any correlation or method is advisable, underscoring the concern that any specific p–y correlation may not be generally applicable. The p–y curve depends not only on the soil and depth, but also on the pile type (because of both pile material and effects of pile installation on the surrounding soil) and several other factors.

To illustrate the use of the p–y method, we will introduce only the industry-standard API (1993) method for driven piles in sand and soft clay and another method for stiff clay. These API methods were developed for driven steel pipe piles with diameters no greater than 0.6 m. Whereas the API methods are generally found to be acceptable for driven piles, there has been some question regarding their applicability to drilled shafts. In general, the body of literature on lateral loading of drilled shafts provides some support for the use of API curves for drilled shafts. However, the amount and scope of research on this topic are very limited, and it is reasonable to expect that in reality there are differences between the p–y curves for displacement and nondisplacement piles. The body of research on the lateral load response of piles other than driven piles is certainly not sufficient to establish a satisfactory framework for the estimation of p–y curves for these piles. Whatever differences there may be in the details of the p–y curves become more or less relevant depending on what level of lateral deflection the pile is loaded to and on factors specific to the case being analyzed. More research is needed on this topic.

The API (1993) models for sand and clay (Matlock 1970; Reese et al. 1974) attempt to capture the most important factors affecting the p–y relationship. For sands, the following p–y relationship is used:

$$p = Cp_L \tanh\left(\frac{k_g z y}{Cp_L}\right) \tag{13.113}$$

where:
  p_L = limit unit resistance at depth z in the unit of force per length;
  $k_g = dk_0/dz$ = gradient of initial modulus of subgrade reaction k_0 with depth; and

$$C = \begin{cases} 3 - 0.8\dfrac{z}{B} \geq 0.9 & \text{for static loading} \\[2mm] 0.9 & \text{for cyclic loading} \end{cases} \tag{13.114}$$

For soft clays, the p–y curves are given in table form in Tables 13.21 and 13.22 and in graph form in Figures 13.37 and 13.38. In these tables, the quantities z_R and y_{50} appear normalizing other quantities. The quantity y_{50} was discussed earlier; it is defined as follows:

$$y_{50} = 2.5\varepsilon_{50}B \tag{13.115}$$

Table 13.21 API (1993) soft clay p–y relationships for static loading

$\dfrac{p}{p_L}$	$\dfrac{y}{y_{50}}$
0	0
0.5	1
0.72	3
1	8
1	∞

Table 13.22 API (1993) soft clay p–y relationships for cyclic loading

$z \leq z_R$		$z \geq z_R$	
$\dfrac{p}{p_L}$	$\dfrac{y}{y_{50}}$	$\dfrac{p}{p_L}$	$\dfrac{y}{y_{50}}$
0	0	0	0
0.5	1	0.5	1
0.72	3	0.72	3
$0.72\dfrac{z}{z_R}$	15	0.72	∞
$0.72\dfrac{z}{z_R}$	∞		

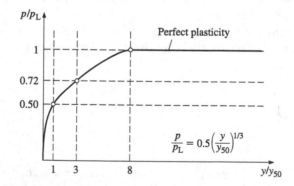

Figure 13.37 Normalized API p–y for a pile in clay under static loading.

where ε_{50} = axial strain in an undrained triaxial compression test corresponding to a shear stress equal to 0.5 times the maximum undrained shear strength. The values of ε_{50} vary from 0.005 for clays with $s_u \geq 100\,\text{kPa}$ to 0.015 for clays with a small shear strength (s_u as low as 10 kPa). The quantity z_R was discussed previously; it is the limiting depth defining the zone within which the limit unit resistance of laterally loaded piles is reduced due to the proximity of the free surface. Note that, in this model, perfect plasticity (unchanged value of p as y is increased further) is established once a given value of y/y_{50} (for example, 8 for static loading) is reached.

The API criteria for clays strictly apply to clay below the water table, but they may be used, with caution, for clay above the water table so long as it is saturated. Their use would

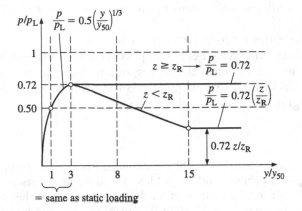

Figure 13.38 Normalized API p–y for a pile in clay under cyclic loading.

tend to be conservative in conditions of partial saturation. The literature contains a number of p–y curves that target this and other specific conditions.

For stiff clays below the water table, the model of Dunnavant and O'Neill (1986) may be used. According to this model, p for static loading is given as

$$p = \begin{cases} 1.02p_L \tanh\left[0.537\left(\dfrac{y}{y_{50}}\right)^{0.7}\right] & \text{for } y \le 8y_{50} \\[2ex] p_L & \text{for } y > 8y_{50} \end{cases} \tag{13.116}$$

where:

$$p_L = \left(2 + \frac{\sigma_v'}{s_{u,avg}} + 0.4\frac{z}{B}\right)s_u B \le 9Bs_u \tag{13.117}$$

$$y_{50} = 0.0063\varepsilon_{50}B\left(K_R\right)^{-0.875} \tag{13.118}$$

$$K_R = \frac{E_p I_p}{E_s L^4} \le 3B\left(\frac{E_p I_p}{E_s B^4}\right)^{0.286} \tag{13.119}$$

$s_{u,avg}$ = average s_u from the surface to the depth z;
L = pile length; and
E_s = Young's modulus of the soil at depth z.

13.11.8 Use of computer programs

It is not practical to do realistic calculations for laterally loaded piles by hand. Computer programs that analyze the lateral load response of piles typically use the p–y method. The main steps in the analysis of a laterally loaded pile problem using a computer program (which may differ in detail and ease, but not in substance) are the following:

1. Define the pile: (a) define the cross section of the pile; (b) define the bending stiffness $E_p I_p$ of the pile; (c) define the plastic moment of the cross section; and (d) define the pile length.

2. Define the soil profile: (a) define how many layers there are and (b) define the thickness of each layer.
3. Define layer properties (or p–y curves) for each layer.
4. Define the loading: (a) choose between applying a displacement or a load and a rotation or a moment and (b) define the corresponding increments.
5. Specify the results desired (for example, load–deflection curve, bending moment and shear force diagrams, and deflected pile shape).

Example 13.19

A 20-meter-long drilled shaft with 500 mm diameter (shown in E-Figure 13.12) is to be installed to support a column load of 500 kN. Extensive *in situ* testing has been done that provides us with some information on the soil properties at the site. The first 4 m of soil is a single clay layer ($\gamma_m = 16$ kN/m³). Vane shear tests conducted at the top and bottom of this first layer yield undrained strengths of 10 and 30 kPa, respectively. CPT results from the site show four distinct sand layers below the top (clay) layer: from 4 to 10 m, from 10 to 15 m, from 15 to 18 m, and from 18 to 40 m ($\gamma_m = 18$ kN/m³ and $\phi_c = 33°$ for all layers). The CPT data indicate average q_c values of 9 MPa for the first sand layer, 20 MPa for the second, and 30 MPa for the third. The cone was unable to penetrate the fourth sand layer. The water table is at 18 m. Use a suitable computer program for laterally loaded pile load–deflection analysis to estimate the horizontal deflection for a lateral load of 50 kN. If this load induces a deflection of more than 10 mm, find the load that results in a deflection that is smaller than this limit. Use the values of 80,000 kN m² for E_pI_p of the pile and 200 kN m for the plastic moment of its cross section. In solving this example problem, assume that the API (1993) p–y curves have been shown to apply.

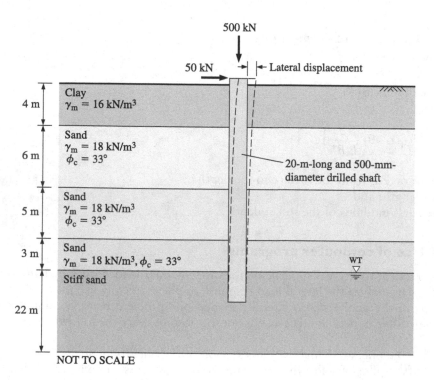

E-Figure 13.12 Drilled shaft of Example 13.19.

Solution:

The solution uses the program PYGMY. Readers will benefit the most if they follow the solution to this example at the same time as they attempt to reproduce the results using a program for the analysis of laterally loaded piles they may have access to. Our first task is to define the soil layers to use in the analysis. We know of five distinct layers within the soil profile located beneath our site. We may define the soil layers as follows: first layer (a clay layer) from 0 to 4 m, second layer (a sand layer) from 4 to 10 m, third layer (a sand layer) from 10 to 15 m, fourth layer (a sand layer) from 15 to 18 m, and fifth layer (a sand layer) from 18 to 40 m. There are no specific guidelines for selecting the depth to which consideration must be given, but twice the pile length should suffice.

Our first task is to define the soil layers to use in the analysis. We know of five distinct layers within the soil profile located beneath our site. We may define the soil layers as follows: first layer (a clay layer) from 0 to 4 m, second layer (a sand layer) from 4 to 10 m, third layer (a sand layer) from 10 to 15 m, fourth layer (a sand layer) from 15 to 18 m, and fifth layer (a sand layer) from 18 to 40 m. There are no specific guidelines for selecting the depth to which consideration must be given, but twice the pile length should suffice. We start by entering in the program all of the required inputs related to the pile:

$B = 0.4$ m (pile diameter)
$L = 20$ m (pile length)
$E_p I_p = 80,000$ kN m^2 (bending stiffness)
$M_L = 200$ kN m (plastic hinge moment)

We then proceed to define individual soil layer properties. We will use the API (1993) soil models (API soft clay for the clay layers and API sand for the sand layers) discussed earlier. These should be built in any software for the analysis of laterally loaded piles (or, alternatively, the program must have the facility of accepting manual input of p–y curves). For the clay soil, we need the undrained shear strength at the top of the layer. We know this to be 10 kPa from the vane shear test results. The undrained strength gradient is the rate of increase in undrained strength with depth. Since we know the undrained strength at the bottom of the first layer, we may calculate this gradient as

$$\frac{s_{u,bottom} - s_{u,top}}{\Delta z} = \frac{30 - 10 \text{ kPa}}{4 \text{ m}} = 5 \frac{\text{kPa}}{\text{m}}$$

We will use 16 kN/m^3 for the clay unit weight. The section on lateral load analysis in this chapter gives us some information regarding the parameters J and ε_{50}. We use an average value of 0.375 for J (which typically ranges from 0.25 to 0.5) and 0.015 for ε_{50} (typical for clays of small undrained shear strength). The program may require that we enter a p multiplier f for the pile. For our situation, with a single pile, f is 1 by definition.[22]

Slightly more work is required for obtaining the input parameters for the sand layers. Two of the important modeling parameters for the sand layers are the peak friction angle ϕ_p and the initial stiffness gradient k_g. Both of these parameters can be related to relative density, and we can estimate that from our CPT results. D_R can be expressed according to Equation (7.20):

$$D_R = \frac{\ln\left(\dfrac{q_c}{p_A}\right) - 0.4947 - 0.1041\phi_c - 0.841\ln\left(\dfrac{\sigma_h'}{p_A}\right)}{0.0264 - 0.0002\phi_c - 0.0047\ln\left(\dfrac{\sigma_h'}{p_A}\right)}$$

---

[22] In Chapter 15, we will discuss p multipliers in detail, as they are needed to calculate the lateral resistance of pile groups.

First, we calculate the lateral effective stresses at the centers of the top three layers:

$$\sigma'_{h1} = K_0 \sigma'_v = 0.45(4 \times 16 + 3 \times 18) = 0.45 \times 118 = 53 \text{ kPa}$$

$$\sigma'_{h2} = K_0 \sigma'_v = 0.45(4 \times 16 + 6 \times 18 + 2.5 \times 18) = 0.45 \times 217 = 98 \text{ kPa}$$

$$\sigma'_{h3} = K_0 \sigma'_v = 0.45(4 \times 16 + 6 \times 18 + 5 \times 18 + 1.5 \times 18) = 0.45 \times 289 = 130 \text{ kPa}$$

We obtain the following for the top three sand layers:

$$D_{R,1} = \frac{\ln\left(\dfrac{9}{0.1}\right) - 0.4947 - 0.1041 \times 33 - 0.841\ln\left(\dfrac{53}{100}\right)}{0.0264 - 0.0002 \times 33 - 0.0047\ln\left(\dfrac{53}{100}\right)} = 48\%$$

$$D_{R,2} = \frac{\ln\left(\dfrac{20}{0.1}\right) - 0.4947 - 0.1041 \times 33 - 0.841\ln\left(\dfrac{98}{100}\right)}{0.0264 - 0.0002 \times 33 - 0.0047\ln\left(\dfrac{98}{100}\right)} = 70\%$$

$$D_{R,3} = \frac{\ln\left(\dfrac{30}{0.1}\right) - 0.4947 - 0.1041 \times 33 - 0.841\ln\left(\dfrac{130}{100}\right)}{0.0264 - 0.0002 \times 33 - 0.0047\ln\left(\dfrac{130}{100}\right)} = 84\%$$

For the bottom sand layer, we do not have cone resistance readings. However, we may safely assume that the sand has at least $D_R = 90\%$. We may now use Figure 5.18 to find ϕ_p given the relative density and lateral effective stress of the four layers: 38.2°, 39.9°, 40.8°, and 41.5°.

Finally, we may use Figure 13.34 to predict the initial stiffness gradients k_g for each of the sand layers, based on the relative density of the layer and the relative location of the water table. The initial stiffness gradients are 22,500, 49,000, 74,000, and 45,000 kPa/m for the first, second, third, and fourth sand layers, respectively (the drop for the fourth layer occurs because it lies below the water table).

Lastly, we must enter some of the information about the loading conditions of the pile. Initially, we may compute the horizontal deflections resulting from a lateral load of 50 kN. The results of the analysis may be viewed in E-Figures 13.13–13.16, which show the profile of lateral deflection, rotation, bending moment, and shear force versus depth for the drilled shaft. From this analysis, we find that a lateral load of 50 kN yields a horizontal deflection of 17.8 mm at the top of the pile, more than our 10 mm limit.

To find the load required to induce a horizontal deflection of 10 mm, we specify displacement increments instead of load. Again, we perform an analysis, finding that a lateral load of 28 kN produces a horizontal deflection of 10 mm. To prevent a horizontal deflection exceeding 10 mm, we should ensure that the lateral load does not exceed 28 kN, clearly a small load, indicating the 10 mm requirement to be rather stringent for the soil conditions at this site. The deflected shape of the pile is shown in E-Figure 13.17. Note how the clay layer dominates the pile response.

The reader may easily modify this example for further practice by doing any of the following: replacing the drilled shafts by driven piles, assuming the water table to be at the ground surface, or assuming a soil profile made entirely of the sand.

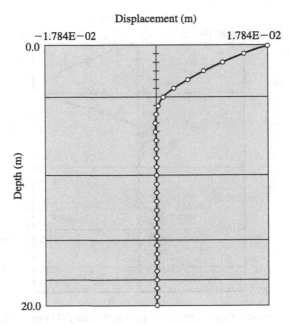

E-Figure 13.13 Lateral displacements along laterally loaded drilled shaft of Example 13.19.

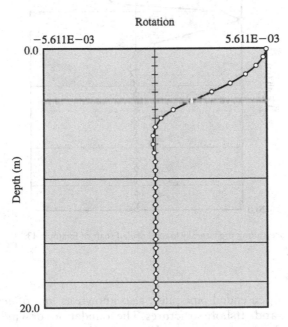

E-Figure 13.14 Rotation along the laterally loaded drilled shaft of Example 13.19.

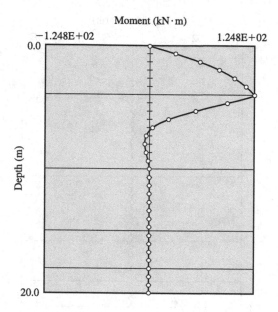

E-Figure 13.15 Bending moment along the laterally loaded drilled shaft of Example 13.19.

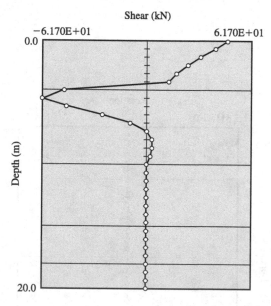

E-Figure 13.16 Shear force along the laterally loaded drilled shaft of Example 13.19.

13.11.9 Monopiles

Single, large-diameter, open-ended pipe piles, also known as *monopiles*, are widely used as foundations for onshore and offshore structures. The foundations of wind turbines in most new offshore windfarm developments, for example, are largely monopiles. The state of practice in the design of these piles is to use the *p–y* method, but there has been recent progress in this area that may enable engineers to perform more accurate analyses and better designs. Based on the results of realistic FE analyses, Hu et al. (2021a,b) proposed design equations – summarized in

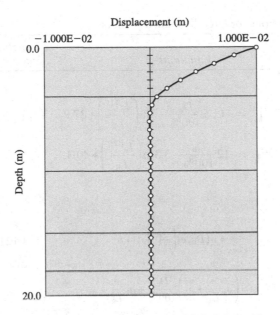

E-Figure 13.17 Lateral displacements along the laterally loaded drilled shaft of Example 13.19 for a deflection of 10 mm at the top.

Table 13.23 – that can be used to obtain the lateral capacity and lateral load–pile head rotation relationship for monopiles in uniform and layered sand profiles.

The lateral capacity Q_h of a monopile, for a fixed diameter, increases with increasing pile length L until L reaches a limiting value: the critical pile length L_{crit}. At $L = L_{crit}$, Q_h reaches it maximum value: $Q_{h,crit}$. The critical slenderness ratio L_{crit}/B decreases with increasing pile diameter B and also with increasing sand relative density D_R. The critical slenderness ratio L_{crit}/B decreases from about 11 to 6 when the pile diameter increases from 1 to 10 m. The lateral capacities $Q_h|_{\theta=0.5°}$ and $Q_h|_{\theta=1°}$ of a monopile in uniform sand – defined for pile cross-sectional rotation $\theta = 0.5°$ and $\theta = 1°$ at the mudline – can be calculated using Equations (13.120)–(13.123) from Table 13.23.

In a two-layer (sand A over sand B) sand profile, the lateral capacity $Q_{h,A-B}$ of a monopile transitions from the lateral capacity $Q_{h,A}$ for uniform sand A to the capacity $Q_{h,B}$ for uniform sand B as the thickness d_1 of the top layer (sand A) increases. Equation (13.124) captures this transition from $Q_{h,A}$ to $Q_{h,B}$. The top $5B$ of the soil profile has the greatest impact on the lateral load response of the monopile.

Considering three-layer profiles, the presence of a thin, loose middle sand layer – one with a thickness equal to B – in a dense sand profile has a small impact on the lateral capacity of a monopile. The lateral capacity $Q_{h,A-B-C}$ in a three-layered sand (A–B–C) profile can be estimated as a linear combination of the lateral pile capacities of the same pile in one uniform and two two-layered sand profiles using Equation (13.125). Using Equation (13.126), we can obtain the load–rotation curve for a monopile for a rotation level ranging from 0° to 2°.

The equations in Table 13.23 are applicable to a range of pile diameters B from 1 to 10 m, slenderness ratios L/B from 3 to 20, wall thickness-to-diameter ratios t_w/B from 1:100 to 1:50, two types of sand (Toyoura Sand and Ottawa Sand) with different particle morphologies, relative densities from 40% to 90%, and load eccentricities from 15 to 30 m. To use the table, we need to find which of the two sands for which the equations were developed – Toyoura Sand or Ottawa Sand – has properties that best approximate those of the sand of interest (Table 13.24).

Table 13.23 Design equations for laterally loaded monopiles

Calculation step	Equations	Equation number	Notes
Uniform sand			
Step 1: Calculate the critical pile length L_{crit} for $\theta = 0.5°$ and $1°$	$L_{crit}/I_{\mathrm{p}}^{0.25}\big|_{0.5°} = -12.5\dfrac{D_{\mathrm{R}}}{100\%} - 3\ln\left(\dfrac{I_{\mathrm{p}}^{0.25}}{L_{\mathrm{R}}}\right) + 37.9$ $L_{crit}/I_{\mathrm{p}}^{0.25}\big|_{1°} = -12\dfrac{D_{\mathrm{R}}}{100\%} - 3.3\ln\left(\dfrac{I_{\mathrm{p}}^{0.25}}{L_{\mathrm{R}}}\right) + 40.3$	(13.120)	• L_{crit} is the critical pile length beyond which the pile lateral capacity no longer increases • θ is the pile rotation at the mudline • $I_{\mathrm{p}} = \pi(B^4 - B_{\mathrm{i}}^4)/64$
Step 2: Calculate the critical pile lateral capacity $Q_{h,crit}$ for $\theta = 0.5°$ and $1°$	$\dfrac{Q_{h,crit}}{p_{\mathrm{A}}L_{\mathrm{R}}^2} = a(D_{\mathrm{R}},e)\left(\dfrac{I_{\mathrm{p}}}{L_{\mathrm{R}}^4}\right)^{0.72} + b$ $a(D_{\mathrm{R}},e) = \left(a_1\dfrac{e}{L_{\mathrm{R}}} + a_2\right)\dfrac{D_{\mathrm{R}}}{100\%} + a_3\dfrac{e}{L_{\mathrm{R}}} + a_4$	(13.121)	• e is the load eccentricity: the distance from the point of lateral load application to the mudline • The values of the coefficients a_1, a_2, a_3, a_4, and b are provided in Table 13.24
Step 3: Calculate the lateral capacity ratio $Q_h/Q_{h,crit}$ for $\theta = 0.5°$ and $1°$	$Q_h/Q_{h,crit} = \begin{cases} \left[\dfrac{1}{2}\sin\left(\pi\dfrac{L}{L_{crit}} - \dfrac{\pi}{2}\right) + 0.5\right]^{\beta} & \text{for } L < L_{crit} \\ 1 & \text{for } L \geq L_{crit} \end{cases}$ $\beta(D_{\mathrm{R}},I_{\mathrm{p}})\big|_{0.5°} = 0.084\left(L_{crit}/I_{\mathrm{p}}^{0.25}\right) - 1.2$ $\beta(D_{\mathrm{R}},I_{\mathrm{p}})\big|_{1°} = 0.08\left(L_{crit}/I_{\mathrm{p}}^{0.25}\right) - 1.25$	(13.122)	
Step 4: Calculate the lateral capacity Q_h in a uniform sand for $\theta = 0.5°$ and $1°$	$Q_h\big|_{0.5°} = \dfrac{Q_h}{Q_{h,crit}}\bigg|_{0.5°} Q_{h,crit}\big|_{0.5°}$ $Q_h\big|_{1°} = \dfrac{Q_h}{Q_{h,crit}}\bigg|_{1°} Q_{h,crit}\big|_{1°}$	(13.123)	• The critical capacity $Q_{h,crit}$ is obtained from Equation (13.121) • The lateral capacity ratio $Q_h/Q_{h,crit}$ is obtained from Equation (13.122)
Two-layer sand profile			
Calculate the pile lateral capacity $Q_{h,A\text{-}B}$ for a two-layer (A over B) sand			• d_1 is the thickness of the top layer • $Q_{h,A}$ and $Q_{h,B}$ are the lateral capacities of the pile in uniform sand A and B, respectively (obtained from Steps 1–4)

(Continued)

Table 13.23 (Continued) Design equations for laterally loaded monopiles

Calculation step	Equations	Equation number	Notes
	$$\frac{Q_{h,A\text{-}B}}{Q_{h,B}} = \frac{Q_{h,A}}{Q_{h,B}} + \left(1 - \frac{Q_{h,A}}{Q_{h,B}}\right)\exp\left[\alpha\left(d_1/B\right)^{1.5}\right]$$	(13.124)	
	$$\alpha = \begin{cases} -0.23 & \text{for } L/B \geq 5 \\ 1 \big/ \left(-1.07\dfrac{L}{B}+1\right) & \text{for } L/B < 5 \end{cases}$$		

Three-layer sand

| Calculate the pile lateral capacity $Q_{h,A\text{-}B\text{-}C}$ for a three-layer (A–B–C) sand | $$Q_{h,A\text{-}B\text{-}C} = Q_{h,A\text{-}A\text{-}A} - Q_{h,B\text{-}A\text{-}A} + Q_{h,B\text{-}B\text{-}C}$$ 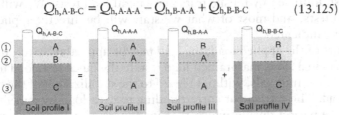 | (13.125) | • $Q_{h,A\text{-}A\text{-}A}$ is the pile lateral capacity for uniform sand A
• $Q_{h,B\text{-}A\text{-}A}$ and $Q_{h,B\text{-}B\text{-}C}$ are the pile lateral capacities for two-layer sand profile (obtained from Equation (13.124)) |

Load–rotation relationship for profiles with one, two, or three sand layers

| Calculate the relationship between the lateral load capacity Q_h versus pile rotation θ at the mudline | $$Q_h = \frac{\theta}{k + \eta\theta}$$ $$\eta = 2/Q_h|_{1^\circ} - 1/Q_h|_{0.5^\circ}$$ $$k = 1/Q_h|_{0.5^\circ} - 1/Q_h|_{1^\circ}$$ | (13.126) | • $Q_h|_{0.5^\circ}$ and $Q_h|_{1^\circ}$ are the pile lateral capacities obtained for $\theta = 0.5^\circ$ and 1°
• Pile lateral capacity Q_h for any rotation level θ ($<2^\circ$) can be estimated from Equation (13.126) |

Table 13.24 Values of the coefficients in Equation (13.121) to calculate the critical lateral capacity $Q_{h,crit}$ for Ottawa Sand and Toyoura Sand for pile rotation $\theta = 0.5^\circ$ and $\theta = 1^\circ$ at the mudline

Case	a_1	a_2	a_3	a_4	b
Ottawa Sand for $\theta = 0.5^\circ$	−0.95	49	−0.34	25.8	−0.01
Ottawa Sand for $\theta = 1^\circ$	−2.05	107	−0.35	33.5	0.07
Toyoura Sand for $\theta = 0.5^\circ$	−0.85	32	−0.35	33.5	−0.15
Toyoura Sand for $\theta = 1^\circ$	−1.30	65	−0.42	43.0	−0.02

13.12 STATIC LOAD TESTS

13.12.1 Definition and classification

The engineering of pile foundations involves more than predicting pile response and performance. We would also like to be as sure as possible that the piles will indeed perform adequately after installation. Static load tests on piles are often performed as a means of verification that the piles will support the load that they are designed to support. They are sometimes also done before the start of piling operations to obtain additional information for use in design.

Conceptually, a *static load test* is simply the successive application of load increments to the pile and the measurement of the resulting settlement (for an axial load test) or lateral deflection (for a lateral load test) for each load increment. Static load tests may be classified according to (1) the nature of the loading, (2) the rate of application of the load, (3) the method of load application, and (4) the presence or absence of pile instrumentation.

13.12.2 Type of loading and rate of load application

With respect to the nature of the load, the most common type of pile load test is the axial compressive load test, followed by tensile tests and lateral load tests. We will focus mostly on compressive load tests, and most of what we state will be directly applicable to these tests, unless otherwise indicated.

Whatever the nature of the applied load, the load may be applied relatively quickly, as in a quick test, or slowly, as in a slow, maintained-load test. In the quick test, load increments are applied at prescribed time intervals, without regard to the stabilization of the pile deflection under the applied load. The settlement corresponding to each load increment is the settlement at the end of the period during which that load increment is applied and sustained. The load is then increased, and the new settlement will be the settlement at the end of this subsequent load increment. ASTM D1143 (ASTM 2020) prescribes waiting periods of not less than 4 minutes and no more than 15 minutes for each applied load increment.

A slow, maintained-load test is a test in which the load increment is applied and enough time is allowed for most of the settlement the pile will eventually experience to develop (see an illustration of plots of load and settlement versus time for this type of loading in Figure 13.39). To standardize this wait period, stabilization criteria are set forth in the applicable standards, such as ASTM D1143 (ASTM 2020). Usually, such criteria are based on the time rate of settlement falling to some level equated with stabilization. After this criterion is satisfied, the next load increment is applied and again time is allowed for settlement stabilization. Slow, maintained-load tests are much superior to quick tests because they minimize rate effects, which

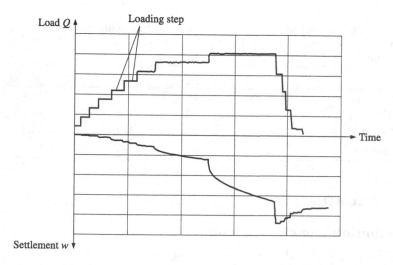

Figure 13.39 Load and settlement versus time for a slow, maintained-load test.

might distort the true bearing capacity of the pile. In soils with a substantial clay fraction, it is necessary to accept that long-term settlements will by necessity exceed any settlement measured over a period of only many hours or even days. Other test protocols exist (such as the constant rate of penetration or CRP test), but they are not as commonly used.

In the planning and performance of a pile load test, the magnitude of the load increments and the maximum load to which the tests are taken must also be decided beforehand. These are also set forth in applicable standards. For example, ASTM D1143 (ASTM 2020) prescribes load increments of 5%–10% of the design load. Tests used primarily for the verification of piling jobs are carried out to twice the design load in most cases, ensuring a safety factor of 2. However, whenever possible, it is desirable to take the pile load test all the way to the limit load or, at least, to a conventional or nominal ultimate load. This is particularly important if the test is done before the production piling to provide information that will be used in design.

13.12.3 Source of reaction

In a load test, loads are applied on the pile using a hydraulic jack, which pushes against something that can provide a suitable reaction.[23] There are two basic types of reaction systems: dead-weight and beam–pile or frame–pile reaction systems. Figure 13.40 shows the first type of system, in which the hydraulic jack placed on top of the pile pushes against a platform on top of which a large mass constituted of blocks of concrete, a water tank, a sand box, or whatever else may be available to provide a stable, large weight. Figure 13.41 shows the second type of system in which a beam (or block) (in Figure 13.41a) or frame (in Figure 13.41b) is connected to two or more reaction piles that are loaded in tension when the hydraulic jack pushes against the beam (frame) and the pile head.

In beam–pile or frame–pile reaction systems, it is important that the reaction piles be located at sufficiently large distances from the test pile. According to analysis done by Poulos and Davis (1990), for the reaction piles not to interfere with the load transfer in the test pile, the distance between each reaction pile and the test pile must be at least equal to about 5 pile

Figure 13.40 Dead-weight reaction system.

---

[23]Technically, there is a load cell between the hydraulic jack and the reaction system to measure the applied load.

(a)   (b)

Figure 13.41 Pile-beam reaction system: (a) beam or block reaction system; (b) frame reaction system.

diameters for a test pile with a significant fraction of its capacity coming from end bearing to as much as 15 pile diameters for floating piles. In ASTM D1143 (ASTM 2020), this minimum distance is prescribed to be the maximum of 5 pile diameters and 2.5 m. An unpublished research by the author and his research group at Purdue University suggests that, for piles in ordinary circumstances, 6 to 7 diameters should be sufficiently large to minimize or even eliminate test and reaction pile interaction.

13.12.4 Measurements and instrumentation

In a routine axial load test, only the load at the pile head and the deflection there are measured. In instrumented load tests, the strains, stresses, and forces along the pile are also determined. Figure 13.42 shows a schematic representation of all the elements involved in a pile load test. Pile head deflections are measured on both sides of the test pile (to account for any possible pile head rotation) using deflectometers placed on reference beams. These reference beams must be placed on supports that are sufficiently removed from the test pile not to settle themselves and thus influence the readings. Theoretical results suggest this distance to be at least 0.3–0.5 pile lengths away from the test pile for end-bearing piles and at least 1 pile length away for a floating pile (Poulos and Davis 1990). Distances of the order of one pile length may not be practical and we may need to accept some effect of the pile settlement on the supports for the reference beam. As long as we understand the magnitude of this effect, and can show it to be small, the load test will produce results that are adequate.

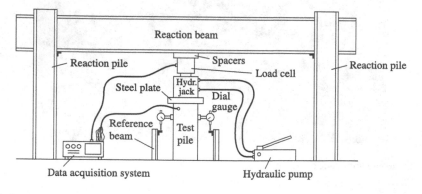

Figure 13.42 Schematic of a pile load test.

The load is measured using a load cell, which is placed between the hydraulic jack and the reaction system. Two undesirable, but not uncommon, outcomes of load tests are the failure of the reaction system (either bending of steel beams, or cracking of a reinforced concrete reaction block) or the inability of the jack to take the load test to the desired load on the pile. Both of these outcomes can be easily prevented by adequate planning of the load test and careful design of all components.

The combination of the load measured with the load cell and the corresponding deflection, calculated as the average of the deflections measured with the deflectometers, gives us a load versus settlement curve at the pile head. While this may be sufficient for the verification of pile performance, information about how the load is distributed along the pile is necessary if the goal is to refine design or do research. An instrumented load test allows us to obtain such information, making it by far the most valuable form of load test.

In an instrumented load test, strain gauges are placed along the pile. These strain gauges – which may be either vibrating-wire or electrical-resistance strain gauges – allow measurement of strain, which can be used to calculate stress using the modulus of elasticity of the pile. The stress, multiplied by the pile cross-sectional area, yields the force at the cross section. This gives us a load transfer curve along the pile (such as those in Figure 13.2). Instrumented tests permit separation of base from shaft resistance, as well as determination of the values of unit shaft resistance at various depths.

13.12.5 Interpretation of pile load tests

The most basic question that we wish to answer with a load test is: how much load can the pile take without impairing the performance or safety of the structure it will help to support? This takes us back to the discussion of Section 13.3. Since not every pile is taken to a clear limit (plunging) load, an ultimate load criterion is usually necessary. The ultimate load criterion that is currently the most accepted internationally is that which states the following: the ultimate load is the load corresponding to a settlement equal to 10% of the pile diameter. If a limit load is desired, but has not developed during the load test, the Chin (1970) criterion can be used to estimate it by extrapolation. Both of these criteria have been discussed in this chapter. In the problems at the end of the chapter, we will have a chance to practice the estimation of ultimate loads using these criteria on real load test data.

In connection with driven piles, the driving process often causes severe changes in the soil surrounding the pile. This is particularly true in clays, where the soil is severely remolded and large pore pressures are generated. This means that, if a load test is performed soon after pile driving, the measured load capacity will likely be a low estimate of that which will be available when the pile is put into service. As the pore pressures dissipate and the soil heals, the pile load capacity will increase to its final value, in a process that is called soil setup. The reverse process, by which pile load capacity decreases with time after pile driving, is known as soil relaxation. Although there have been efforts to develop analyses to approximately quantify soil setup and relaxation effects, it is often better in practice to perform load tests several days (10 days being a recommended minimum) after pile driving to obtain an estimate of pile capacity that is closer to the final capacity.

13.13 CHAPTER SUMMARY

Piles derive their axial load capacity from a combination of **shaft resistance** and **base resistance**. Shaft resistance mobilizes completely at small settlements (on the order of several millimeters, corresponding to 0.25%–1% of the pile diameter). Substantial mobilization of base resistance requires much larger settlements; if it mobilizes fully, the pile plunges. The **plunging load** is also referred to as the **limit load**. This load is usually very large and thus unreachable by pile load tests except for relatively small-diameter piles in weak soils. So pile load capacity is more usually defined in terms of a conventional **ultimate load**. The most common and most practical definition of ultimate load is that of the load corresponding to a settlement equal to 10% of the pile diameter.

Ultimate vertical (axial) resistances can be calculated from soil properties or directly from *in situ* test measurements (mainly the CPT and SPT penetration resistances). The unifying equation for all of these methods is:

$$Q_{ult} + W_{pile} = A_b q_{b,ult} + \sum_{i=1}^{n} A_{si} q_{sLi}$$

where n is the number of soil layers, q_{sLi} is the **limit unit shaft resistance** for soil layer i,[24] A_{si} is the lateral area of the pile in contact with soil layer i, $q_{b,ult}$ is a conventional **ultimate unit base resistance** defined according to a specific criterion, A_b is the base area of the pile, and W_{pile} is the weight of the pile.

There are several methods for the calculation of $q_{b,ult}$ and q_{sL} from soil properties or *in situ* test results. The most current methods are those known as the **Purdue Pile Design Method (PPDM)**, the **Imperial College Pile Design Method (ICPDM)**, the **University of Western Australia Pile Design Method (UWAPDM)**, the **Unified Pile Design Method (UPDM)**, and the **Norwegian Geotechnical Institute Pile Design Method (NGIPDM)**. These methods are discussed in Sections 13.5 and 13.6, and the corresponding equations are given in Tables 13.1–13.9.

Piles are also used in rock. One way in which this is done is to embed the pile into rock. The portion of the pile embedded in the rock is called a **rock socket**. The rock socket, if shorter than 1 diameter, derives its capacity from both base and side resistances. For longer rock sockets, side resistance predominates. **Micropiles** in rock are also used often. Their resistance comes exclusively from the shaft.

The limit unit shaft resistance q_{sL} of piles in rock can be calculated using one of two methods. In the first:

$$q_{sL} = \min\left(\frac{1}{2} c_w q_u, 0.05 f_c' \right) \tag{13.87}$$

where c_w is the coefficient of weakness given in Table 7.4 as a function of discontinuity spacing, q_u is the unconfined compressive strength of the intact rock, and f_c' is the characteristic (nominal) strength of the concrete, cement paste, or grout used to construct the micropile. The coefficient of weakness does not account for degree of weathering, so Equation (13.87) applies strictly only to unweathered or slightly weathered rock. Smaller values of q_{sL} must be assumed in case of a more severe weathering condition.

A second method assumes a nonlinear relationship between q_{sL} and the unconfined compressive strength (Zhang and Einstein 1998):

$$\frac{q_{sL}}{p_A} = \min\left(C_S \sqrt{\frac{q_u}{p_A}}, 0.05 \frac{f_c'}{p_A} \right) \tag{13.88}$$

---

[24]The shaft resistance may be assumed in most cases to reach a limit value in ultimate limit states and sometimes even under working load conditions if the base resistance is sufficiently large to cover the margin of safety considered in the design.

where $C_S = 1.25$ (for a "smooth" interface) to 2.5 (for a rough interface) and $p_A =$ reference stress = 100 kPa = 0.1 MPa.

Piles derive resistance to lateral loads from the soil compressive and shear stiffnesses. The design of laterally loaded piles is carried out in most cases using the *p–y* method. The practical use of the method is only possible using a computer. The main idea of the method is that the pile is divided into elements and each element is connected to a **nonlinear spring** representing the **unit lateral soil resistance** *p*. The unit lateral soil resistance *p* is related to the lateral displacement *y* through a known nonlinear relationship, the *p–y* **curve**. This relationship is a function of the soil state, that is, the soil density and depth of the pile element. The end result of the analysis is a lateral load versus lateral displacement relationship at the pile head as well as information along the pile, such as horizontal load (shear force), bending moment, and slope of the pile axis.

Static **load tests** are used to verify the vertical or lateral load capacity of piles. In these tests, loads are applied according to some protocol. Different protocols can be found in standards from ASTM and other organizations. Preferably, loads should be applied slowly to better reflect the load capacity that will ultimately be available to support the structure, and tests should be extended until the ultimate load has been reached. Instrumented load tests provide details of the load transfer curve along the pile, which is very helpful to determine the profile of unit shaft resistance versus depth and the unit base resistance acting on the pile.

13.13.1 Symbols and notations

Symbol	Quantity represented	US units	SI units
A_b	Area of pile base	ft^2	m^2
A_p	Cross-sectional area of pile	ft^2	m^2
A_s	Pile shaft area	ft^2	m^2
B	Diameter of pile	ft	m
B_b	Base diameter of belled drilled shaft	ft	m
B_h	Diameter of hollow stem of auger	ft	m
B_i	Inner pile diameter	ft	m
B_s	Belled drilled shaft diameter	ft	m
C_1, C_2	Constants appearing in Chin's load–settlement criterion	Unitless	Unitless
D_n	Normalized pile driving depth	Unitless	Unitless
E_p	Young's modulus of the pile	psf	kN/m^2
f'_c	Characteristic compressive strength	psf	kN/m^2
f_{cd}	Design compressive strength	psf	kN/m^2
FS	Factor of safety	Unitless	Unitless
H	Shear force in laterally loaded pile	ton	kN
IFR	Incremental filling ratio	Unitless	Unitless
k	Modulus of subgrade reaction	tsf	kPa
K_t	Pile head (top) stiffness	lb/ft	kN/m
L	Pile length	ft	m
L_p	Plug length	ft	m
L_R	Reference length (= 1 m = 3.281 ft)	ft	m
NIFR	Normalized incremental filling ratio	Unitless	Unitless

(*Continued*)

Symbol	Quantity represented	US units	SI units
p	Unit lateral resistance	ton/ft	kN/m
p_L	Limit unit lateral resistance	ton/ft	kN/m
Q	Vertical load on pile	lb	kN
Q_{ann}	Annulus capacity	lb	kN
Q_b	Load carried by base	lb	kN
q_b	Unit base resistance	psf	kN/m²
q_{bL}	Limit unit base resistance	psf	kN/m²
Q_{col}	Column load	lb	kN
Q_h	Lateral load on pile	lb	kN
Q_L	Limit load	lb	kN
Q_{plug}	Soil plug capacity	lb	kN
Q_s	Load carried by shaft	lb	kN
q_s	Unit shaft resistance	psf	kN/m²
q_{sL}	Limit unit shaft resistance	psf	kN/m²
s_q, d_q, N_q	Shape, depth, and bearing capacity factors	Unitless	Unitless
t	Parameter for laterally loaded piles representing additional stiffness due to vertical compressive resistance in vertically loaded piles and horizontal shearing resistance in laterally loaded piles	ton	kN
v_{crit}	Velocity of pile installation below which the pile behaves as a nondisplacement pile	ft/s	m/s
w	Settlement	ft	m
w_R	Relative settlement	Unitless	Unitless
w_{ult}	Settlement observed for pile when $Q = Q_{ult}$	ft	m
y	Lateral deflection	in.	mm
y_{50}	Lateral deflection corresponding to $p = 0.5p_L$ for analysis of laterally loaded piles in clay	in.	mm
β	Constant appearing in Van der Veen's load–settlement extrapolation method	ft	m
ε_{50}	Axial strain corresponding to a shear stress equal to 50% of the peak shear strength in an undrained triaxial compression test on a clay sample	Unitless	Unitless
λ	Pitch of auger	ft	m
ω	Rate of revolution of auger	rad/s	rad/s

13.14 PROBLEMS

13.14.1 Conceptual problems

Problem 13.1 Define all the terms in bold contained in the chapter summary.

Problem 13.2 What is the main difference between a pile and a footing?

Problem 13.3 Discuss the concepts of limit load and ultimate load in connection with the design of axially loaded piles.

Problem 13.4* Discuss the physical mechanism responsible for the mobilization of shaft resistance at small pile settlements. Contrast that with the physical mechanisms involved in base resistance mobilization, which requires much larger settlements.

Problem 13.5 Obtain the ratio q_{sL}/q_c for drilled shafts and full-displacement piles in a sand with $\phi_c = 30°$ using Figure 7.23b. Use the Purdue Pile Design Method (PPDM) with $K_0 = 0.45$ for $D_R = 70\%$ and $\sigma'_v = 200$ kPa. For the full-displacement pile, assume the length and diameter of the pile to be 15 and 0.5 m, respectively, and the unit weight of the sand to be 20 kN/m³.

Problem 13.6 Plot the ratio q_{sL}/q_c for drilled shafts and full-displacement piles in sand versus relative density for σ'_v ranging from 50 to 200 kPa. The sand has $\phi_c = 30°$ and $K_0 = 0.45$. Use Figure 7.23b and the Purdue Pile Design Method (PPDM). For the full-displacement pile, assume the length and diameter of the pile to be 15 and 0.5 m, respectively, and the unit weight of sand to be 20 kN/m³.

Problem 13.7 Using the Purdue Pile Design Method (PPDM), plot the ratio q_{sL}/q_c for drilled shafts and full-displacement piles in clay versus depth. The clay is overconsolidated due to the removal of 2 m of overburden prior to the installation of the piles. Assume that $s_u/\sigma'_v = 0.3$ for NC clay, $\phi_c = 20°$, $\phi_{r,min} = 8°$, $\gamma_{sat} = 16$ kN/m³, and $N_k = 10$.

Problem 13.8 What is the difference between a "short" and a "long" pile from the point of view of axial loading? How do they respond differently?

Problem 13.9 What is the difference between a "short" and a "long" pile from the point of view of lateral loading? How do they respond differently?

13.14.2 Quantitative problems

Problem 13.10 Consider the soil profile in P-Figure 13.1 consisting of intermixed clay and sand layers. A 3-meter-thick sand layer extends from 9.5 to 12.5 m. Results from SPTs and CPTs at the site suggest a relative density of ~70% for this layer. The sand is subrounded with ϕ_c of the order of 31°. The average unit weight of the soil above the layer is 19 kN/m³, the unit weight of the sand itself is 22 kN/m³, and the water table is flush with the ground surface. What is the value of shaft load ΔQ_{sL} carried by this sand layer for (a) a drilled shaft with diameter $B = 0.4$ m and (b) a 17-meter-long precast concrete pile with square cross section and side $B = 0.4$ m? Assume $K_0 = 0.4$ and use the Purdue Pile Design Method (PPDM). Divide the sand layer into three sublayers of equal thickness to facilitate your calculations.

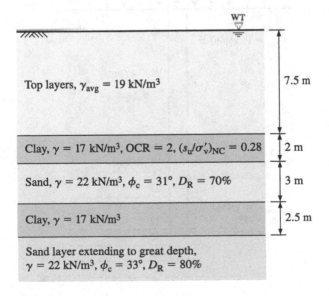

P-Figure 13.1 Soil profile for Problem 13.10.

Problem 13.11 One of the clay layers in the soil profile of Problem 13.10 is 2 m thick and has an average OCR of 2 and s_u/σ'_v of 0.28. The depth to the top of the clay layer is 7.5 m. What is the value of q_{sL} for (a) a drilled shaft and (b) a precast concrete pile through this clay layer? Assume a critical-state friction angle of 18°, a minimum residual-state friction angle of 9°, and unit weight of the saturated clay of 17 kN/m³.

Problem 13.12 A typical bearing layer in the soil profile of Problem 13.10 is an 80% relative density sand layer with $\phi_c = 33°$. The depth to the top of this layer is 15 m. What are the values of $q_{b,ult}$ for (a) a drilled shaft and (b) a precast concrete pile bearing in this sand layer located at a depth of 17 m?

Problem 13.13 Both closed-ended steel pipe piles and drilled shafts are under consideration in a project. As shown in P-Figure 13.2, the piles will be installed through a 10-meter-thick natural clay layer with $(s_u/\sigma'_v)_{NC} = 0.25$ into a sand layer with energy-corrected, normalized blow counts N_{60} of 25, 29, 32, 40, 42, and 45 for the first 6 m into the layer. The sand has unit weight equal to 20 kN/m³, and the clay has unit weight equal to 17 kN/m³. The critical-state friction angle of the sand is 32° and $K_0 = 0.45$. The water table is at the surface. In the past, the soil profile had been subjected to uniform surcharge of 50 kPa applied on the surface of the soil deposit, which was later removed, causing both the sand and clay layers in their current states to be overconsolidated. Calculate, for a 15-meter-long, 500-millimeter drilled shaft and a geometrically identical closed-ended steel pipe pile, the following: (a) the shaft capacity due to the clay layer (divide the clay layer into ten sublayers of equal thickness in your calculations); (b) the shaft capacity due to the sand layer; (c) the total shaft capacity; (d) the ultimate base capacity; (e) the ultimate load capacity of the pile; (f) the allowable load based on a suitable factor of safety (without consideration of the strength of the pile cross section); and (g) the allowable load if the compressive strength of the concrete is 15 MPa.

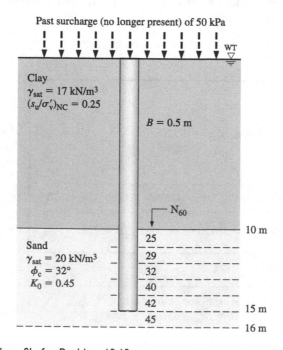

P-Figure 13.2 Pile and soil profile for Problem 13.13.

Problem 13.14 A precast concrete pile with square cross section (0.3 m×0.3 m) is going to be installed in the profile shown in P-Figure 13.3. The cone penetration resistance values are given in the figure. The water table is at 3 m depth below the surface. Determine the allowable vertical load capacity of the pile using a safety factor equal to 2.5. Use the Purdue Pile Design Method (PPDM) to obtain unit base and unit shaft resistance values. Assume $N_k = 10$ for the clay at this site. The shear strength gradient of this clay in a normally consolidated state is $(s_u/\sigma'_v)_{NC} = 0.3$. For the sand layers, a K_0 of 0.45 may be assumed. For the clay layers, assume a critical-state friction angle of 18° and a minimum residual-state friction angle of 9°.

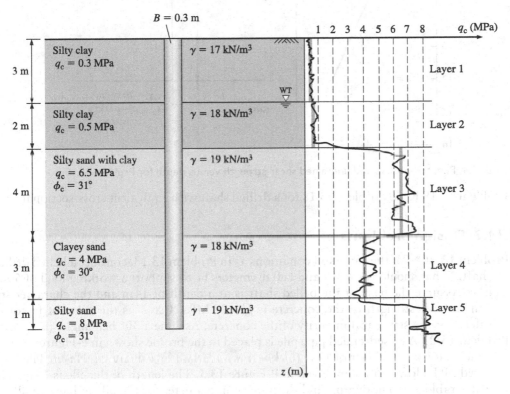

P-Figure 13.3 Pile and soil profile for Problem 13.14.

Problem 13.15 Redo Problem 13.14 using the Imperial College Pile Design Method (ICPDM) and the University of Western Australia Pile Design Method (UWAPDM) to obtain unit base and unit shaft resistance values. Assume clay sensitivity to be equal to 1. The interface friction angle is 0.95 times the critical-state friction angle.

Problem 13.16 For the same profile as in Problem 13.14, calculate the ultimate load capacity of a drilled shaft with a diameter of 0.3 m and the same length as the precast concrete pile.

Problem 13.17 Calculate the limit unit shaft resistance of the micropile in Example 13.16 using Equation (13.88).

Problem 13.18 Estimate the ultimate load capacity of a square prestressed concrete pile placed in the profile shown in P-Figure 13.4. The pile is 13.5 m long and has 0.4 m×0.4 m square cross section. The water table is at 2 m depth. Assume that the ratio $(s_u/\sigma'_v)_{NC} = 0.25$ for all clay layers. The profile of the undrained shear strength is given in P-Figure 13.4.

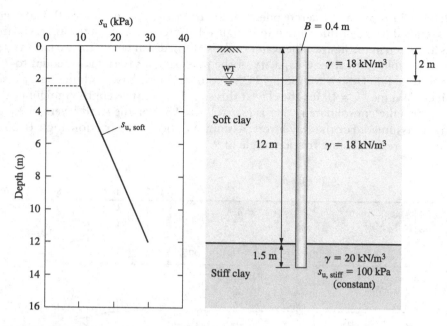

P-Figure 13.4 Pile, soil profile, and undrained shear strength versus depth for Problem 13.18.

Problem 13.19 Redo Problem 13.18 for a drilled shaft with equivalent cross-sectional area.

13.14.3 Design problems

Problem 13.20* For the same soil conditions as in Problem 13.13, consider a belled drilled shaft. What should the shaft and bell diameters be to support a working load of 5000 kN? Assume the base of the belled shaft is at a depth of 14 m and the characteristic compressive strength of the concrete is $f'_c = 20$ MPa. Use 3 as the value of factor of safety for the structural integrity of the concrete. Assume a 30° bell angle.

Problem 13.21 A closed-ended pipe pile is placed in the profile shown in P-Figure 13.5, in which a loose sand layer of 12 m thickness is underlain by a dense sand layer. The measured SPT blow counts are given in P-Figure 13.5. The length of the pile is 14 m. The water table is at 1 m depth. Find the minimum diameter for the pile to have an allowable capacity of at least 500 kN using a safety factor equal to 3. Assume $D_{50} = 0.4$ mm, $\phi_c = 30°$, and $K_0 = 0.45$ for both sand layers. Obtain the relative density of the sands using Skempton's (1986) equation. Use a soil property-based method. This problem is best solved using a spreadsheet (P-Table 13.1).

P-Table 13.1 SPT blow counts for Problem 13.21

Depth (m)	Blow count (blows per foot)
0.5	2
1.5	2
2.5	3
3.5	1
4.5	4

(Continued)

P-Table 13.1 (Continued) SPT blow counts for Problem 13.21

Depth (m)	Blow count (blows per foot)
5.5	5
6.5	7
7.5	8
8.5	9
9.5	12
10.5	13
11.5	15
12.5	48
13.5	50
14.5	50

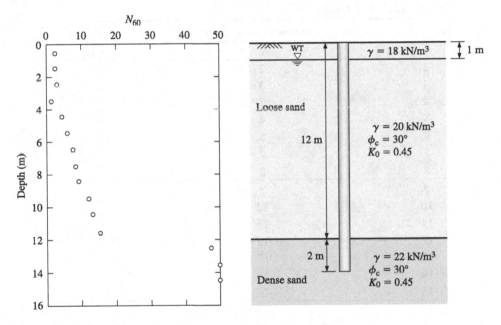

P-Figure 13.5 Pile, soil profile, and SPT blow counts for Problem 13.21.

Problem 13.22 Redo Problem 13.21 for the case of a drilled shaft.

Problem 13.23* Drilled shaft foundations will be installed at a site to support a building with column loads ranging from 250 to 6000 kN. The diameters available for the drilled shafts range from 0.30 to 0.70 m every 0.05 m and from 0.8 to 1.6 m every 0.10 m. Representative energy-corrected, normalized SPT blow counts are given in P-Table 13.2, together with soil type. The water table is at around 8 m depth.

a. At what depth would you place the base of the drilled shafts? Would this depth be the same for all shafts? Discuss how your decision might affect the total settlement of the supported building, which considers the combined effects of all the drilled shafts that will be used.

b. Calculate the drilled shaft axial resistance for each drilled shaft diameter. Consider all pertinent limit states, including structural failure of the individual shafts. Use a safety factor of 2.5 on the ultimate load.

c. Select drilled shaft diameters for column loads of 3750 and 5980 kN.

d. For a structure completely separate from the building, 300- and 400-millimeter-diameter drilled shafts with the bases located at a depth of 15 m will be used to carry the loads. (i) What are the ultimate and the allowable axial resistances of the 300 and 400 mm drilled shafts? (ii) How much of the load is due to shaft and how much to base capacity in each case? (iii) What are the compressive stresses at the base and the head of the drilled shaft in each case? Are they different? Why or why not?

P-Table 13.2 SPT results

Depth (m)	Blow count (blows per foot)	Soil type
1	2	Clay
2	2	
3	3	
4	1	
5	4	Sand
6	5	
7	7	
8	8	
9	9	
10	12	
11	13	
12	15	Silty, clayey sand
13	17	
14	18	
15	22	
16	23	
17	25	
18	20	
19	19	
20	25	Sand
21	33	
22	35	
23	50	
24	60	Sand and gravel
25	Refusal	

Problem 13.24* The soil profile at a site consists of a 1-meter-thick uncontrolled fill underlain by a stiff clay. The uncontrolled fill offers no shear resistance. A simple structure is to be supported by drilled shafts at this site, with a maximum column load of 900 kN. The drilled shaft has a length of 15 m, a diameter of 360 mm, and a Young's modulus of 25,000 MPa. If the undrained shear strength at the depth of 1 m is 50 kPa and at the depth of 15 m is 200 kPa, calculate the settlement that the pile will undergo. Make the necessary assumptions wherever it is needed.

Problem 13.25* A micropile has the following characteristics:

Diameter = 200 mm (8 in.)
Total length = 12 m (39.3 ft)
Length in residual soil = 7 m
Length in rock = 5 m
Grout injected by gravity only
Compressive strength f_c' of grout = 30 MPa (4170 psi)

The micropile is reinforced using a cage of 3 longitudinal rebars with diameter of 25 mm (1 in.) with 6.3 mm (¼ in.) ties each 125 mm (4.9 in.). The steel has a yield strength of 500 MPa. The casing used to install the pile is not left in the ground.

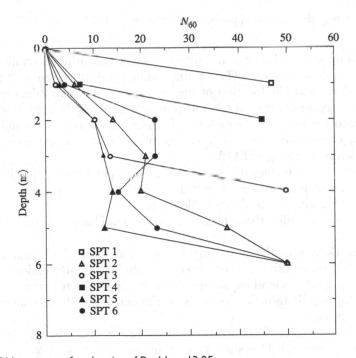

P-Figure 13.6 SPT blow counts for the site of Problem 13.25.

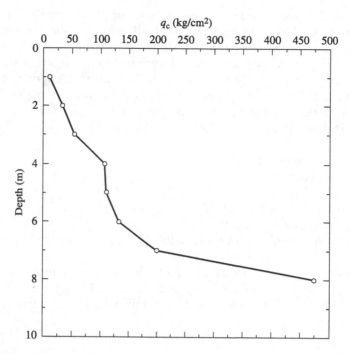

P-Figure 13.7 CPT q_c near the location of the micropile of Problem 13.25.

The micropile is installed at a site where a slightly weathered granite is overlain by residual soil of granite. SPTs for the residual soil are given in P-Figure 13.6. One CPT done near the location of the micropile is given in P-Figure 13.7. Rock is found at varying depths. At the location of the micropile, rock is at a depth of 7 m. The recovery ratio for the rock was found to be between 25% and 100%. The unconfined compressive strength was determined from one cylinder without any apparent defects as being 131 MPa.

a. determine the resistance of the micropile developed within the soil layer;
b. find the resistance of the micropile developed within the rock;
c. calculate the ultimate and allowable loads;
d. discuss the possibility that a buckling failure could happen before the ultimate load is reached.

Problem 13.26 Determine the lateral capacities of the 300 and 400 mm drilled shafts of Problem 13.23 using a suitable computer program for laterally loaded pile load–deflection analysis. Find the load capacities for 5, 10, and 25 mm lateral deflection. (Hint: break the soil profile into five layers, with the bottom of the last layer 40 m beneath the ground surface).

Problem 13.27 Two drilled shafts were installed in a site where the soil is a (mostly sandy) residual soil of gneiss.[25] One shaft was drilled to a depth of 16.9 m, and the other all the way to rock (at a depth of 21.4 m). The groundwater level is located between the depths of 16 and 19 m. For these two shafts (P-Figures 13.8 and 13.9):

a. calculate the unit shaft resistance along the shaft;

---

[25]The two drilled shafts were installed at a site on the Georgia Tech campus and were load-tested axially. Details can be found in FHWA Technical Report No. 41-30-2175, prepared by Georgia Tech Research Corporation, Paul W. Mayne (Research Director); Dean E. Harris (Research Engineer).

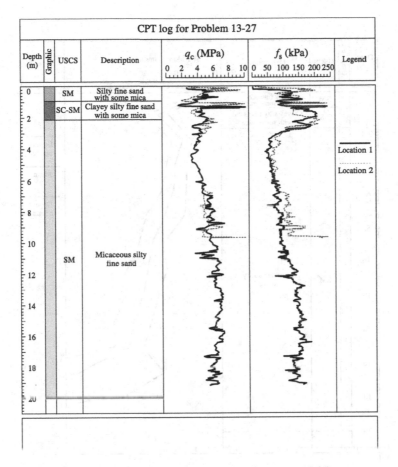

P-Figure 13.8 CPTs near the location of the two drilled shafts of Problem 13.27.

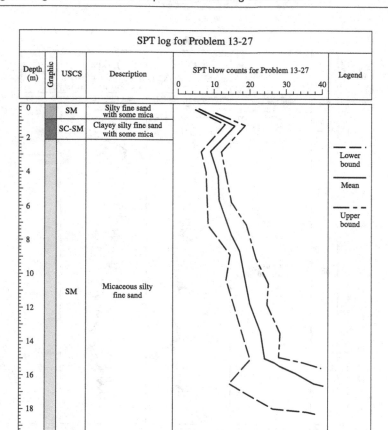

P-Figure 13.9 Range of SPT blow count values near the two drilled shafts of Problem 13.27.

    b. calculate the unit base resistance;
    c. calculate the total axial resistance;
    d. calculate the lateral resistance for a 30 mm pile head deflection (you will need a specific program for this).

Problem 13.28* Design a foundation solution using closed-ended, driven steel pipe piles bearing in the lower sand layer for supporting the building of Problem 11.19. Compare your design with the design using footings, done as part of Problem 11.19, for prevailing prices in your area of practice.

Problem 13.29* Design a foundation solution using CFA piles bearing in the lower sand layer for supporting the building of Problem 11.19. Compare your design with the design of Problem 13.28 using pipe piles.

Problem 13.30 For the soil profile shown in P-Figure 13.10 – consisting of normally consolidated silty sand, sand, and gravelly sand – and a pipe pile with outer diameter

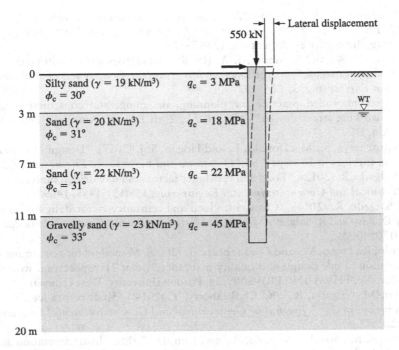

P-Figure 13.10 Soil profile and pipe pile for Problem 13.30.

equal to 20 in. (508 mm), wall thickness equal to 0.5 in. (12.7 mm), and length equal to 39.4 ft (12 m), determine the lateral load required to cause a pile head lateral deflection of 0.25, 0.5, and 1 in. (6.35, 12.7, and 25.4 mm). The axial load on the pile is equal to 62 tons (550 kN). Ignore the effects of any moment at the pile head. The plastic moment for this pile, which is filled with concrete after driving, is 950 kN m, and its bending stiffness $E_p I_p$ is 170,000 kN m².

Problem 13.31 Recalculate the ultimate load capacity of the pile of Example 13.11 by dividing the soil profile into additional sublayers.

Problem 13.32 Redo Problem 13.31 using the ICPDM and the UWAPDM. The values of OCR, S_t, and σ'_{median} for the silty clay layer are 5.3, 2.6, and 50 kPa, respectively.

REFERENCES

References cited

AASHTO. (1989). *Standard Specifications for Highway Bridges.* 14th Edition, American Association of State Highway and Transportation Officials, Washington, DC.

Acar, Y. B., Avent, R. R., and Taha, M. R. (1994). "Downdrag on friction piles: A case history." *Proceedings of Settlement 94: Vertical and Horizontal Deformations of Foundations and Embankment,* College Station, TX, 986–999.

ACI. (2019). "Building code requirements for structural concrete." ACI 318-19. American Concrete Institute, Farmington Hills, MI.

Ahmed, S. S. and Hawlader, B. (2016). "Numerical analysis of large-diameter monopiles in dense sand supporting offshore wind turbines." *International Journal of Geomechanics,* 16(5), 04016018.

Aoki, N. and Velloso, D. D. A. (1975). "An approximate method to estimate the bearing capacity of piles." *Proceedings of 5th Pan-American Conference of Soil Mechanics and Foundation Engineering*, Buenos Aires, Argentina, 1, 367–376.

Aoki, N., Velloso, D. A., and Salamoni, J. A. (1978). "Fundações para o silo vertical de 100000t no porto de paranaguá." *Proceedings of 6th Brazilian Conference of Soil Mechanics and Foundation Engineering*, 3, 125–132.

API. (1993). "Recommended practice for planning, designing, and constructing fixed offshore platforms-working stress design." API RP-2A, 20th Edition, American Petroleum Institute, Washington, DC.

Arany, L., Bhattacharya, S., MacDonald, J., and Hogan, S. J. (2017). "Design of monopiles for offshore wind turbines in 10 steps." *Soil Dynamics and Earthquake Engineering*, 92, 126–152.

Basu, D. and Salgado, R. (2012). "Load and resistance factor design of drilled shafts in sand." *Journal of Geotechnical and Geoenvironmental Engineering*, 138(12), 1455–1469.

Basu, D. and Salgado, R. (2014). "Closure to 'load and resistance factor design of drilled shafts in sand' by D. Basu and R. Salgado." *Journal of Geotechnical and Geoenvironmental Engineering*, 140(3), 07014002.

Basu, P., Salgado, R., Prezzi, M., and Chakraborty, T. (2009). "A method for accounting for pile setup and relaxation in pile design and quality assurance." Joint Transportation Research Program Publication No. FHWA/IN/JTRP-2009/24, Purdue University, West Lafayette, IN.

Basu, P., Prezzi, M., Salgado, R., and Chakraborty, T. (2014). "Shaft resistance and setup factors for piles jacked in clay." *Journal of Geotechnical and Geoenvironmental Engineering*, 140(3), 04013026.

Bica, A. V., Prezzi, M., Seo, H., Salgado, R., and Kim, D. (2014). "Instrumentation and axial load testing of displacement piles." *Proceedings of Institution of Civil Engineers-Geotechnical Engineering*, 167(3), 238–252.

Bjerrum, L., Johannessen, I. J., and Eide, O. (1969). "Reduction of negative skin friction on steel piles to rock." *Proceedings of 7th International Conference on Soil Mechanics and Foundation Engineering*, Mexico City, Mexico, 27–34.

Boulanger, R. W. and DeJong, J. T. (2018). "Inverse filtering procedure to correct cone penetration data for thin-layer and transition effects." *Proceedings of 4th International Symposium on Cone Penetration Testing*, CPT 2018, Delft, Netherlands, 25–44.

Bozozuk, M. (1978). "Bridge foundations move." Transportation Research Record No. 678, Transportation Research Board, Washington, DC, 17–21.

Brand, E. W. and Luangdilok, N. (1974). "A long term foundation failure caused by dragdown on piles." *Proceedings of 4th Southeast Asian Conference on Soil Engineering*, Malaysia, Kuala Lumpur, 4, 15–24.

Briaud, J.-L. (1997). "Bitumen selection for reduction of downdrag on piles." *Journal of Geotechnical and Geoenvironmental Engineering*, 123(12), 1127–1134.

Briaud, J.-L., Ballouz, M., and Nasr, G. (2000). "Static capacity prediction by dynamic methods for three bored piles." *Journal of Geotechnical and Geoenvironmental Engineering*, 126(7), 640–649.

Broms, B. B. (1964a). "Lateral resistance of piles in cohesive soils." *Journal of the Soil Mechanics and Foundations Division*, 90(SM2), 27–64.

Broms, B. B. (1964b). "Lateral resistance of piles in cohesionless soils." *Journal of the Soil Mechanics and Foundations Division*, 90(SM3), 123–158.

BSI. (1986). *Code of Practice for Foundations: CP 8004*. British Standards Institution, London, UK.

Burland, J. B. and Twine, D. (1988). "The shaft friction of bored piles in terms of effective strength." *Proceedings of 1st International Geotechnical Seminar: Deep Foundations on Bored and Auger Piles*, Rotterdam, Netherland, 411–420.

Bustamante, M. and Gianeselli, L. (1982). "Pile bearing capacity prediction by means of static penetrometer CPT." *Proceedings of 2nd European Symposium on Penetration Testing*, Amsterdam, Netherland, 493–500.

CGS. (1985). *Canadian Foundation Engineering Manual*. 2nd Edition. Canadian Geotechnical Society, Toronto, Canada.

Chakraborty, T., Salgado, R., Basu, P., and Prezzi, M. (2013). "Shaft resistance of drilled shafts in clay." *Journal of Geotechnical and Geoenvironmental Engineering*, 139(4), 548–563.

Chellis, R. D. (1961). *Pile Foundations*. 2nd Edition, McGraw-Hill, New York.

Chin, F. K. (1970). "Estimation of the ultimate load of piles from tests not carried to failure." *Proceedings of 2nd Southeast Asian Conference on Soil Engineering*, Singapore, 81–90.

Chow, F. C. (1997). "Investigations into the behaviour of displacement pile for offshore foundations." Ph.D. Thesis, Imperial College London, UK.

Chow, Y. K., Chin, J. T., and Lee, S. L. (1990). "Negative skin friction on pile groups." *International Journal for Numerical and Analytical Methods in Geomechanics*, 14(2), 75–91.

Clausen, C. J. F., Aas, P. M., and Karlsrud, K. (2005). "Bearing capacity of driven piles in sand; the NGI approach." *Proceedings of 1st International Symposium on Frontiers in Offshore Geotechnics: ISFOG*, University of Western Australia, Perth, Australia, 677–681.

Davisson, M. T. (1972). "High capacity piles." *Proceedings of Lecture Series on Innovations in Foundation Construction*, Chicago, IL, 81–112.

Davisson, M. T. (1975). "Pile load capacity." *Proceedings of Conference on Design Construction and Performance of Deep Foundations*, Berkeley, CA.

De Beer, E. E. (1968). "Proefondervindelijke bijdrage tot de studie van het grensdraagvermogen van zand onder funderingen op staal." Tijgshift der Openbar Verken van Belgie (No. 6, 1967 and 4, 5, 6, 1968).

de Sanctis, L. and Mandolini, A. (2006). "Bearing capacity of piled rafts on soft clay soils." *Journal of Geotechnical and Geoenvironmental Engineering*, 132(12), 1600–1610.

Doherty, P. and Gavin, K. (2012). "Laterally loaded monopile design for offshore wind farms." *Proceedings of Institution of Civil Engineers-Energy*, 165(1), 7–17.

Dunnavant, T. W. and O'Neill, M. W. (1986). "Evaluation of design-oriented methods for analysis of vertical pile groups subjected to lateral loads." Numerical Methods in Offshore Piling, Institute Francais du Petrole, Laboratoire Central de Ponts et Chausses, 303–316.

Engineering News Record (2016). "Signs of Millennium tower's settling, tilting are subtle but visible." https://www.enr.com/articles/40364-signs-of-millenium-towers-settling-tilting-are-subtle-but-visible. Date published: September 28, 2016. Last accessed: January 3, 2022.

Eslami, A. and Fellenius, B. H. (1997). "Pile capacity by direct CPT and CPTu methods applied to 102 case histories." *Canadian Geotechnical Journal*, 34(6), 886–904.

Fellenius, B. H. (1999). "Discussion on bitumen selection for reduction of downdrag on piles." *Journal of Geotechnical Engineering*, 125(4), 341–344.

Fellenius, B. H. (2002a). "Determining the resistance distribution in piles. part 1: Notes on shift of no-load reading and residual load." *Geotechnical News Magazine*, 20(2), 35–38.

Fellenius, B. H. (2002b). "Determining the resistance distribution in piles. part 2: Method for determining the residual load." *Geotechnical News Magazine*, 20(3), 25–29.

Fellenius, B. H., Harris, D. E., and Anderson, D. G. (2004). "Static loading test on a 45 m long pipe pile in Sandpoint, Idaho." *Canadian Geotechnical Journal*, 41(4), 613–628.

Fleming, W. G. K., Weltman, A. J., Randolph, M. F., and Elson, W. K. (1992). *Piling Engineering*. Surrey University Press, Surrey, UK.

Fleming, M., Weltman, A., Randolph, M. F., and Elson, K. (2009). *Piling Engineering*. 3rd Edition, Taylor & Francis, London, UK.

Foye, K. C., Abou-Jaoude, G., Prezzi, M., and Salgado, R. (2009). "Resistance factors for use in load and resistance factor design of driven pipe piles in sands." *Journal of Geotechnical and Geoenvironmental Engineering*, 135(1), 1–13.

Franke, E. (1989). "Co-report to discussion, session 13: Large diameter piles." *Proceedings of 12th International Conference on Soil Mechanics and Foundation Engineering*, Rio de Janeiro, Brazil.

Franke, E. (1993). "Design of bored piles, including negative skin friction and horizontal loading." *Proceedings of 2nd International Geotechnical Seminar: Deep Foundations on Bored and Auger Piles*, Ghent, Belgium, 43–57.

Galvis-Castro, A. C., Tovar-Valencia, R. D., Salgado, R., and Prezzi, M. (2019a). "Compressive and tensile shaft resistance of nondisplacement piles in sand." *Journal of Geotechnical and Geoenvironmental Engineering*, 145(9), 04019041.

Galvis-Castro, A. C., Tovar-Valencia, R. D., Salgado, R., and Prezzi, M. (2019b). "Effect of loading direction on the shaft resistance of jacked piles in dense sand." *Géotechnique*, 69(1), 16–28.

Ganju, E., Han, F., Prezzi, M., and Salgado, R. (2020). "Static capacity of closed-ended pipe pile driven in gravelly sand." *Journal of Geotechnical and Geoenvironmental Engineering*, 146(4), 04020008.

Ghionna, V. N., Jamiolkowski, M., Pedroni, S., and Salgado, R. (1994). "The tip displacement of drilled shafts in sands." *Proceedings of Settlement 94: Vertical and Horizontal Deformations of Foundations and Embankments*, 2, 1039–1057.

Han, F., Lim, J., Salgado, R., Prezzi, M., and Zaheer, M. (2015). Load and resistance factor design of bridge foundations accounting for pile group–soil interaction. Joint Transportation Research Program Report No. FHWA/IN/JTRP-2015/24, Purdue University, West Lafayette, IN.

Han, F., Salgado, R., Prezzi, M., and Lim, J. (2017a). "Shaft and base resistance of non-displacement piles in sand." *Computers and Geotechnics*, 83, 184–197.

Han, F., Prezzi, M., Salgado, R., and Zaheer, M. (2017b). "Axial resistance of closed-ended steel-pipe piles driven in multilayered soil." *Journal of Geotechnical and Geoenvironmental Engineering*, 143(3), 04016102.

Han, F., Ganju, E., Salgado, R., and Prezzi, M. (2018a). "Effects of interface roughness, particle geometry, and gradation on the sand-steel interface friction angle." *Journal of Geotechnical and Geoenvironmental Engineering*, 144(12), 04018096.

Han, F., Prezzi, M., and Salgado, R. (2018b). "Static and dynamic pile load tests on closed-ended driven pipe pile." *Proceedings of International Foundation Congress and Equipment Expo*, Orlando, FL, 496–506.

Han, F., Salgado, R., and Prezzi, M. (2018c). "Numerical and experimental study of axially loaded non-displacement piles in sand." *Proceedings of International Conference on Deep Foundations and Ground Improvement*, Rome, Italy, 221–229.

Han, F., Ganju, E., Prezzi, M., and Salgado, R. (2019a). "Closure to 'effects of interface roughness, particle geometry, and gradation on the sand–steel interface friction angle' by Fei Han, Eshan Ganju, Rodrigo Salgado, and Monica Prezzi." *Journal of Geotechnical and Geoenvironmental Engineering*, 145(11), 07019017.

Han, F., Ganju, E., Salgado, R., and Prezzi, M. (2019b). "Comparison of the load response of closed-ended and open-ended pipe piles driven in gravelly sand." *Acta Geotechnica*, 14(6), 1785–1803.

Han, F., Ganju, E., Prezzi, M., Salgado, R., and Zaheer, M. (2020). "Axial resistance of open-ended pipe pile driven in gravelly sand." *Géotechnique*, 70(2), 138–152.

Hu, Q., Han, F., Prezzi, M., and Salgado, R. (2021a). "Finite-element analysis of the lateral load response of monopiles in layered sand." *Journal of Geotechnical and Geoenvironmental Engineering* (in review).

Hu, Q., Han, F., Salgado, R., and Prezzi, M. (2021b). "Lateral load response of large-diameter mono-piles in sand." *Géotechnique*, 1–16. DOI:10.1680/jgeot.20.00002.

Jardine, R., Chow, F., Overy, R., and Standing, J. (2005). *ICP Design Methods for Driven Piles in Sands and Clays*. 2nd Edition, Thomas Telford, London, UK.

Karlsrud, K., Clausen, C. J. F., and Aas, P. M. (2005). "Bearing capacity of driven piles in clay, the NGI approach." *Proceedings of 1st International Symposium on Frontiers in Offshore Geotechnics: ISFOG*, University of Western Australia, Perth, Australia, 775–782.

Kim, D., Bica, A. V. D., Salgado, R., Prezzi, M., and Lee, W. (2009). "Load testing of a closed-ended pipe pile driven in multilayered soil." *Journal of Geotechnical and Geoenvironmental Engineering*, 135(4), 463–473.

Kim, D., Chung, M., and Kwak, K. (2011). "Resistance factor calculations for LRFD of axially loaded driven piles in sands." *KSCE Journal of Civil Engineering*, 15(7), 1185–1196.

Klinkvort, R. T., and Hededal, O. (2013). "Lateral response of monopile supporting an offshore wind turbine." *Proceedings of Institution of Civil Engineers-Geotechnical Engineering*, 166(2), 147–158.

Lam, C., Jefferis, S. A., and Martin, C. M. (2014). "Effects of polymer and bentonite support fluids on concrete–sand interface shear strength." *Geotechnique*, 64(1), 28–39.

Lee, C. Y. (1993). "Pile groups under negative skin friction." *Journal of Geotechnical Engineering,* 119(10), 1587–1600.

Lee, C. J. and Ng, C. W. W. (2004). "Development of downdrag on piles and pile groups in consolidating soil." *Journal of Geotechnical and Geoenvironmental Engineering,* 130(9), 905–914.

Lee, J. H. and Salgado, R. (1999). "Determination of pile base resistance in sands." *Journal of Geotechnical and Geoenvironmental Engineering,* 125(8), 673–683.

Lee, J., Salgado, R., and Paik, K. (2003). "Estimation of load capacity of pipe piles in sand based on cone penetration test results." *Journal of Geotechnical and Geoenvironmental Engineering,* 129(5), 391–403.

Lehane, B. M. (2019). "CPT-based design of foundations." *Australian Geomechanics Journal,* 54(4), 23–48.

Lehane, B. M. and Randolph, M. F. (2002). "Evaluation of a minimum base resistance for driven pipe piles in siliceous sand." *Journal of Geotechnical and Geoenvironmental Engineering,* 128(3), 198–205.

Lehane, B. M., Li, Y., and Williams, R. (2013). "Shaft capacity of displacement piles in clay using the cone penetration test." *Journal of Geotechnical and Geoenvironmental Engineering,* 139(2), 253–266.

Lehane, B. M., Lim, J. K., Carotenuto, P., Nadim, F., Lacasse, S., Jardine, R. J. and van Dijk, B. F. J. (2017). "Characteristics of unified databases for driven piles." Proceedings of 8th International Conference on Offshore Investigation and Geotechnics: Smarter Solutions for Offshore Developments, Society for Underwater Technology (SUT), London, UK, Vol. 1, 162–194.

Lehane, B. M., Schneider, J. A., and Xu, X. (2005). "The UWA-05 method for prediction of axial capacity of driven piles in sand." *Proceedings of International Symposium on Frontiers in Offshore Geotechnics,* Perth, Western Australia, 683–689.

Lehane, B., Liu, Z., Bittar, E., Nadim, F., Lacasse, S., Jardine, R., Carotenuto, P., Rattley, M., Jeanjean, P., Gavin, K., Gilbert, R., Bergan-Haavik, J., and Morgan, N. (2020). "A new 'unified' CPT-based axial pile capacity design method for driven piles in sand." *Proceedings of 4th International Symposium on Frontiers in Offshore Geotechnics (ISFOG 2020),* Austin, TX, 462–477.

Lopes, F. R. and Laprovitera, H. (1988). "On the prediction of the bearing capacity of bored piles from dynamic penetration tests." *Proceedings of 1st International Geotechnical Seminar: Deep Foundations on Bored and Auger Piles,* Ghent, Belgium, 537–540.

Loukidis, D. and Salgado, R. (2008). "Analysis of the shaft resistance of non-displacement piles in sand." *Géotechnique,* 58(4), 283–296.

Luo, R., Yang, M., and Li, W. (2018). "Numerical study of diameter effect on accumulated deformation of laterally loaded monopiles in sand." *European Journal of Environmental and Civil Engineering,* 24(14), 2440–2452.

Mandolini, A., Ramondini, M., Russo, G., and Viggiani, C. (2002). "Full scale loading tests on instrumented CFA piles." *Proceedings of Deep Foundations 2002: An International Perspective on Theory, Design, Construction, and Performance,* Orlando, FL, 1088–1097.

Matlock, H. (1970). "Correlations for design of laterally loaded piles in soft clay." *Proceedings of 2nd Annual Offshore Technology Conference,* Houston, TX, 577–594.

Mayne, P. W. and Harris, D. E. (1993). Axial load-displacement behavior of drilled shaft foundations in piedmont residuum. FHWA Reference No. 41-30-2175, Report to ADSC/ASCE by Georgia Tech, School of Civil & Environmental Engineering, Atlanta, GA, 172 p.

McCabe, B. A. and Lehane, B. M. (2006). "Behavior of axially loaded pile groups driven in clayey silt." *Journal of Geotechnical and Geoenvironmental Engineering,* 132(3), 401–410.

Meyerhof, G. G. (1976). "Bearing capacity and settlement of pile foundations." *Journal of the Geotechnical Engineering Division,* 102(3), 197–228.

Meyerhof, G. G. (1983). "Scale effects of ultimate pile capacity." *Journal of Geotechnical Engineering,* 109(6), 797–806.

Mobley, S., Costello, K., and Mullins, G. (2018). "The effect of slurry type on drilled shaft cover quality." *DFI Journal – The Journal of the Deep Foundations Institute,* 11(2–3), 91–100.

Murff, J. D., Raines, R. D., and Randolph, M. F. (1990). "Soil plug behavior of piles in sand." *Proceedings of 22nd Offshore Technology Conference*, Houston, TX, 25–32.

Mylonakis, G. (2001). "Winkler modulus for axially loaded piles." *Géotechnique*, 51(5), 455–461.

Mylonakis, G. and Gazetas, G. (1998). "Settlement and additional internal forces of grouped piles in layered soil." *Géotechnique*, 48(1), 55–72.

Nauroy, J. F. and Le Tirant, P. (1983). "Model tests of piles in calcarecus sands." *Proceedings of Conference on Geotechnical Practice in Offshore Engineering*, Austin, TX, 356–369.

Neely, W. J. (1991). "Bearing capacity of auger-cast piles in sand." *Journal of Geotechnical Engineering*, 117(2), 331–345.

Nordlund, R. L. (1963). "Bearing capacity of piles in cohesionless soils." *Journal of the Soil Mechanics and Foundations Division*, 89(SM3), 1–36.

O'Neill, M. W. and Raines, R. D. (1991). "Load transfer for pipe piles in highly pressured dense sand." *Journal of Geotechnical Engineering*, 117(8), 1208–1226.

Paik, K. and Salgado, R. (2003). "Determination of bearing capacity of open-ended piles in sand." *Journal of Geotechnical and Geoenvironmental Engineering*, 129(1), 46–57.

Paik, K., Salgado, R., Lee, J., and Kim, B. (2003). "Behavior of open-and closed-ended piles driven into sands." *Journal of Geotechnical and Geoenvironmental Engineering*, 129(4), 296–306.

Paikowsky, S. G., Whitman, R. V., and Baligh, M. M. (1989). "A new look at the phenomenon of offshore pile plugging." *Marine Georesources and Geotechnology*, 8(3), 213–230.

Poulos, H. G. and Davis, E. H. (1980). *Pile Foundation Analysis and Design*. Wiley, New York.

Poulos, H. G. and Davis, E. H. (1990). *Pile Foundation Analysis and Design*. Robert E. Krieger Publishing Company, Malabar, FL.

Prezzi, M. and Basu, P. (2005). "Overview of construction and design of auger cast-in-place and drilled displacement piles." *Proceedings of 30th Annual Conference on Deep Foundations*, Chicago, IL, 497–512.

Price, G. and Wardle, I. F. (1982). "A comparison between cone penetration test results and the performance of small diameter instrumented pile in stiff clay." *Proceedings of 2nd European Symposium on Penetration Testing, ESOPT II*, Balkema, Amsterdam, 775–780.

Ramsey, N., Jardine, R., Lehane, B., and Ridley, A. (1998). "A review of soil-steel interface testing with the ring shear apparatus." *Proceedings of Conference on Offshore Site Investigation and Foundation Behaviour*, London, UK, 237–258.

Randolph, M. F. (1983). "Discussion: The influence of shaft length on pile load capacity in clays, by H. G. Poulos." *Géotechnique*, 33(1), 75–76.

Randolph, M. F. (2003). "Science and empiricism in pile foundation design." *Géotechnique*, 53(10), 847–875.

Randolph, M. F. and Wroth, C. P. (1978). "Analysis of deformation of vertically loaded piles." *Journal of the Geotechnical Engineering Division*, 104(12), 1465–1488.

Randolph, M. F. and Wroth, C. P. (1979). "An analytical solution for the consolidation around a driven pile." *International Journal for Numerical and Analytical Methods in Geomechanics*, 3(3), 217–229.

Randolph, M. F., Steenfeld, J. S., and Wroth, P. (1979). "The effect of pile type on design parameters for driven piles." *Proceedings of 7th European Conference on Soil Mechanics and Foundation Engineering*, Brighton, UK, 2, 107–114.

Randolph, M. F., Leong, E. C., and Houlsby, G. T. (1991). "One-dimensional analysis of soil plugs in pipe piles." *Géotechnique*, 41(4), 587–598.

Reese, L. C. and O'Neill, M. W. (1989). "New design method for drilled shafts from common soil and rock tests." *Proceedings of Foundation Engineering: Current Principles and Practices*, Evanston, IL, 2, 1026–1039.

Reese, L. C., Cox, W. R., and Koop, F. D. (1974). "Analysis of laterally loaded piles in sand." *Proceedings of 6th Annual Offshore Technology Conference*, Houston, TX, 473–483.

Roscoe, G. H. (1983). "The behavior of flight auger bored piles in sand." *International Conference on Advances in Piling and Ground Treatment for Foundations*, London, UK, 283.

Salgado, R. (2006a). "The role of analysis in non-displacement pile design." *Springer Proceedings in Physics 106: Modern Trends in Geomechanics*, Berlin, Heidelberg, 521–540.

Salgado, R. (2006b). "Analysis of the axial response of non-displacement piles in sand." *Proceedings of 2nd Japan–U.S. Workshop on Testing, Modeling and Simulation, American Society of Civil Engineers*, Reston, VA, 427–439.

Salgado, R. and Prezzi, M. (2007). "Computation of cavity expansion pressure and penetration resistance in sands." *International Journal of Geomechanics*, 7(4), 251–265.

Salgado, R., Kim, D., Prezzi, M., Bica, A. V., and Lee, W. (2011a). "Closure to 'Load testing of a closed-ended pipe pile driven in multilayered soil' by Daehyeon Kim, Adriano Virgilio Bica, Rodrigo Salgado, Monica Prezzi, and Wonje Lee." *Journal of Geotechnical and Geoenvironmental Engineering*, 137(10), 986–988.

Salgado, R., Woo, S. I., and Kim, D. (2011b). "Development of load and resistance factor design for ultimate and serviceability limit states of transportation structure foundations." Joint Transportation Research Program Publication No. FHWA/IN/JTRP-2011/03, Purdue University, West Lafayette, IN.

Salgado, R., Zhang, Y., Dai, G., and Gong, W. (2013). "Reply to the discussion by Fellenius on 'Load tests on full-scale bored pile groups.'" *Canadian Geotechnical Journal*, 50(4), 454–455.

Salgado, R., Han, F., and Prezzi, M. (2017). "Axial resistance of non-displacement piles and pile groups in sand." *Rivista Italiana di Geotecnica*, 51(4), 35–46.

Schmertmann, J. H. (1978). "Guidelines for cone penetration test : Performance and design." Report No. FHWA-TS-78-209, U.S. Department of Transportation, Federal Highway Administration, Washington, DC.

Seed, H. B. and Reese, L. C. (1955). "The action of soft clay along friction piles." *Transactions of the American Society of Civil Engineers*, 122(1), 731–754.

Seo, H., Yildirim, I. Z., and Prezzi, M. (2009). "Assessment of the axial load response of an H pile driven in multilayered soil." *Journal of Geotechnical and Geoenvironmental Engineering*, 135(12), 1789–1804.

Skempton, A. W. (1959). "Cast *in situ* bored piles in London clay." *Géotechnique*, 9(4), 153–173.

Skempton, A. W. (1986). "Standard penetration test procedures and the effects in sands of overburden pressure, relative density, particle size, ageing and overconsolidation." *Géotechnique*, 36(3), 425–447.

Stewart, D. P. (2000). "User manual for program PYGMY." The University of Western Australia.

Teh, C. I. and Wong, K. S. (1995). "Analysis of downdrag on pile groups." *Géotechnique*, 45(2), 191–207.

Tehrani, F. S., Arshad, M. I., Prezzi, M., and Salgado, R. (2018). "Physical modeling of cone penetration in layered sand." *Journal of Geotechnical and Geoenvironmental Engineering*, 144(1), 04017101.

Thorburn, S. and MacVicar, R. S. L. (1971). "Pile load tests to failure in the Clyde alluvium." *Proceedings of Conference on Behavior of Piles*, The Institution of Civil Engineers, London, UK, 1–7.

Tomlinson, M. J. (1987). *Pile Design and Construction Practice*. 3rd Edition, Viewpoint Publications, London, UK.

Tomlinson, M. and Woodward, J. (2008). *Pile Design and Construction Practice*. 5th Edition, Taylor & Francis, London, UK.

Van der Veen, C. (1953). "The bearing capacity of a pile." *Proceedings of 3rd International Conference on Soil Mechanics and Foundation Engineering*, Switzerland, 2, 84–90.

Vesic, A. S. (1977). Design of pile foundations. NCHRP Synthesis of Highway Practice No.42. Transportation Research Board, Washington, DC.

Viggiani, C. (1989). "Influenza dei fattori tecnologici sul comportamento dei pali." *Proceedings of 17th Convegno di Geotecnica*, Taormina, 2, 83–91.

Xu, X., Schneider, J. A., and Lehane, B. M. (2008). "Cone penetration test (CPT) methods for end-bearing assessment of open-and closed-ended driven piles in siliceous sand." *Canadian Geotechnical Journal*, 45(8), 1130–1141.

Yang, Z. X., Guo, W. B., Zha, F. S., Jardine, R. J., Xu, C. J., and Cai, Y. Q. (2015). "Field behavior of driven prestressed high-strength concrete piles in sandy soils." *Journal of Geotechnical and Geoenvironmental Engineering*, 141(6), 04015020.

Yeung, A. T., Viswanathan, R., and Briaud, J.-L. (1996). "Field investigation of potential contamination by bitumen-coated piles." *Journal of Geotechnical Engineering*, 122(9), 736–744.

Zdravkovic, L., Taborda, D. M. G., Potts, D. M., Jardine, R. J., Sideri, M., Schroeder, F. C., and Skov Gretlund, J. (2015). "Numerical modelling of large diameter piles under lateral loading for offshore wind applications." *Proceedings of 3rd International Symposium on Frontiers in Offshore Geotechnics*, 1, 759–764.

Zhang, L. and Einstein, H. H. (1998). "End bearing capacity of drilled shafts in rock." *Journal of Geotechnical and Geoenvironmental Engineering*, 124(7), 574–584.

Additional references

Bazaraa, A. R. and Kurkur, M. M. (1986). "N-values used to predict settlements of piles in Egypt." *Proceedings of Specialty Conference, In Situ '86: In Use of In Situ Tests in Geotechnical Engineering*, Blacksburg, VA, 462–474.

Brand, E. W. and Luangdilok, N. (1975). "A long term foundation failure caused by dragdown on piles." *Proceedings of 4th Southeast Asian Conference on Soil Engineering*, Kuala Lumpur, 4.15–4.24.

Briaud, J. L. and Tucker, L. (1984). "Piles in sand: A method including residual stresses." *Journal of Geotechnical Engineering*, 110(11), 1666–1680.

Broms, B. (1965). "Design of laterally loaded piles." *Journal of Soil Mechanics and Foundation Division*, 91(3), 79–99.

Burns, S. E. and Mayne, P. W. (1999). "Pore pressure dissipation behavior surrounding driven piles and cone penetrometers." Transportation Research Record No.1675, Transportation Research Board, Washington, DC, 17–23.

De Kuiter, J. and Beringen, F. L. (1979). "Pile foundations for large north sea structures." *Marine Georesources and Geotechnology*, 3(3), 267–314.

Decourt, L. (1982). "Prediction of the bearing capacity of piles based exclusively on N values of the SPT." *Proceedings of 2nd European Symposium on Penetration Testing*, Amsterdam, Netherland, 29–34.

Evans, L. T. and Duncan, J. M. (1982). "Simplified analysis of laterally-loaded piles." Report No. UCB/GT/82-04, Department of Civil Engineering, U.C. Berkeley, CA.

Fragaszy, R. J., Argo, D., and Higgins, J. D. (1989). "Comparison of formula predictions with pile load tests." Transportation Research Record No.1219, Transportation Research Board, Washington, DC, 1–12.

Franke, E. (1993). "Design of bored piles, including negative skin friction and horizontal loading." *Proceedings of 2nd International Geotechnical Seminar: Deep Foundations on Bored and Auger Piles*, Ghent, Belgium, 43–57.

Frost, J. D. and DeJong, J. T. (2005). "*In situ* assessment of role of surface roughness on interface response." *Journal of Geotechnical and Geoenvironmental Engineering*, 131(4), 498–511.

Ghionna, V. N., Jamiolkowski, M., Lancellotta, R., and Pedroni, S. (1993). "Base capacity of bored piles in sands from *in situ* tests." *Proceedings of 2nd International Geotechnical Seminar: Deep Foundations on Bored and Auger Piles*, Ghent, Belgium, 67–74.

Gwizdala, K. and Tejchman, A. (1995). "Pile settlement analysis using CPT and load-transfer functions t-z and q-z." *Proceedings of International Symposium on Cone Penetration Testing*, Linkoping, Sweden, 2, 4–5.

Hunt, C. E., Pestana, J. M., Bray, J. D., and Riemer, M. (2002). "Effect of pile driving on static and dynamic properties of soft clay." *Journal of Geotechnical and Geoenvironmental Engineering*, 128(1), 13–24.

Jamiolkowski, M. and Lancellotta, R. (1988). "Relevance of *in situ* test results for evaluation of allowable base resistance of bored piles in sands." *Proceedings of 1st International Geotechnical Seminar: Deep Foundations on Bored and Auger Piles*, Ghent, Belgium, 107–120.

Klotz, E. U. and Coop, M. R. (2001). "An investigation of the effect of soil state on the capacity of driven piles in sands." *Géotechnique*, 51(9), 733–751.

Lawton, E. C., Fragaszy, R. J., Higgins, J. D., Kilian, A. P., and Peters, A. J. (1986). "Review of methods for estimating pile capacity." Transportation Research Record No.1105, Transportation Research Board, Washington, DC, 32–40.

Lehane, B. M., Jardine, R. J., Bond, A. J., and Frank, R. (1993). "Mechanisms of shaft friction in sand from instrumented pile tests." Journal of Geotechnical Engineering, 119(1), 19–35.

Matlock, H. and Reese, L. C. (1960). "Generalized solutions for laterally loaded piles." Journal of the Soil Mechanics and Foundations Division, 86(5), 63–92.

McVay, M. C., Townsend, F. C. and Williams, R. C. (1992). "Design of socketed drilled shafts in limestone." Journal of Geotechnical Engineering, 118(10), 1626–1637.

Meyerhof, G. G. (1956). "Penetration tests and bearing capacity of cohesionless soils." Journal of the Soil Mechanics and Foundations Division, 82(1), 1–19.

Nogami, T., Otani, J., Konagai, K., and Chen, H. L. (1992). "Nonlinear soil-pile interaction model for dynamic lateral motion." Journal of Geotechnical Engineering, 118(1), 89–106.

Ochiai, H., Otani, J., and Matsui, K. (1994). "Performance factor for bearing resistance of bored friction piles." International Journal of Structural Safety, 14, 103–130.

Paik, K. H. and Lee, S. R. (1993). "Behavior of soil plugs in open-ended model piles driven into sands." Marine Georesources and Geotechnology, 11(4), 353–373.

Paikowsky, S. G. and Whitman, R. V. (1990). "The effects of plugging on pile performance and design." Canadian Geotechnical Journal, 27(4), 429–440.

Pestana, J. M., Hunt, C. E., and Bray, J. D. (2002). "Soil deformation and excess pore pressure field around a closed-ended pile." Journal of Geotechnical and Geoenvironmental Engineering, 128(1), 1–12.

Philipponnat, G. (1980). "Méthode pratique de calcul d'un pieu isolé, à l'aide du pénétromètre statique." Revue Francaise de Geotechnique, 10, 55–64.

Poulos, H. G. (1971). "Behavior of laterally loaded piles: I-single piles." Journal of the Soil Mechanics and Foundations Division, 97(5), 711–731.

Poulos, H. G. (1972). "Behavior of laterally loaded piles: III-socketed piles." Journal of the Soil Mechanics and Foundations Division, 98(4), 341–360.

Prakash, S. and Sharma, H. D. (1990). Pile Foundations in Engineering Practice. Wiley, New York.

Puppala, A. J. and Moalim, D. (2002). "Evaluation of driven pile load capacity using CPT based LCPC and European interpretation methods." In Deep Foundations 2002: An International Perspective on Theory, Design, Construction, and Performance, Orlando, FL, 931–943.

Rollins, K. M., Peterson, K. T., and Weaver, T. J. (1998). "Lateral load behavior of full-scale pile group in clay." Journal of Geotechnical and Geoenvironmental Engineering, 124(6), 468–478.

Rollins, K. M., Clayton, R. J., Mikesell, R. C., and Blaise, B. C. (2005) "Drilled shaft side friction in gravelly soils." Journal of Geotechnical and Geoenvironmental Engineering, 131(8), 987–1003.

Salgado, R., Mitchell, J. K., and Jamiolkowski, M. (1997). "Cavity expansion and penetration resistance in sand." Journal of Geotechnical and Geoenvironmental Engineering, 123(4), 344–354.

Shioi, Y. and J. Fukui (1982). "Application of N-value to design of foundations in Japan." Proceedings of 2nd European Symposium on Penetration Testing, Amsterdam, Netherland, 1, 159–164.

Tejchman, A. and Gwizdala, K. (1988). "Comparative analysis of bearing capacity of large diameter bored pile." Proceedings of 1st International Geotechnical Seminar: Deep Foundations on Bored and Auger Piles, Ghent, Belgium, 553–558.

Wesselink, B. D., Murff, J. D., Randolph, M. F., Nunez, I. L., and Hyden, A. M. (1988). "Analysis of centrifuge model test data from laterally loaded piles in calcareous sand." Proceedings of International Conference on Calcareous Sediments, Perth, Western Australia, 2, 261–270.

Winkler, E. (1867). "Die Lehre von der Elasticitaet und Festigkeit: mit besonderer Rücksicht auf ihre Anwendung in der Technik, für polytechnische Schulen." Bauakademien, Ingenieure, Maschinenbauer, Architecten, Dominicus, Prague, Czech Republic.

Yang, Z. and Jeremić, B. (2003). "Numerical study of group effects for pile groups in sands." International Journal for Numerical and Analytical Methods in Geomechanics, 27(15), 1255–1276.

Yang, Z. and Jeremić, B. (2005). "Study of soil layering effects on lateral loading behavior of piles." Journal of Geotechnical and Geoenvironmental Engineering, 131(6), 762–770.

Relevant ASTM standards

ASTM. (2007a). "Standard test methods for deep foundations under static axial tensile load." ASTM D3689, American Society for Testing and Materials, West Conshohocken, PA.

ASTM. (2007b). "Standard test methods for deep foundations under lateral load." ASTM D3966, American Society for Testing and Materials, West Conshohocken, PA.

ASTM. (2018). "Standard test methods for individual piles in permafrost under static axial compressive load." ASTM D5780, American Society for Testing and Materials, West Conshohocken, PA.

ASTM. (2020). "Standard test methods for deep foundation elements under static axial compressive load." ASTM D1143, American Society for Testing and Materials, West Conshohocken, PA.

Chapter 14

Pile driving analysis and quality control of piling operations

> Reality isn't the way you wish things to be, nor the way they appear to be, but the way they actually are.
>
> Robert J. Ringer

In previous chapters, we examined the static methods of analysis and design of axially loaded piles. In this chapter, we study the dynamics of pile driving and of the response of piles to dynamic loading. Several analyses and techniques exist based on pile dynamics that allow us to (1) predict how hard it is to drive a pile to some given static resistance and what stresses develop in the pile during driving, (2) detect pile anomalies due to pile installation based on measurements of velocity near the pile head, and (3) estimate static bearing capacity based on measurements of force and velocity near the pile head during driving. We will cover these three pile dynamics applications in this chapter. We will also discuss a new generation of pile driving formulas, based on pile driving analysis, that can be helpful in the inspection of the driving of piles.

14.1 APPLICATIONS OF PILE DYNAMICS

We can use the principles of dynamics to analyze the response of the soil–pile system to dynamic loads. This is usually accomplished by taking advantage of the fact that the pile is a slender body surrounded by material with much lower stiffness. Under these conditions, we assume that any dynamic load applied at the pile head will move down the pile as a wave or pulse following the one-dimensional wave equation.

The design of piles for axial loads, discussed in Chapter 13, is based on estimating ultimate axial resistance using static analysis and obtaining an allowable load by suitable reduction in the ultimate resistance through division by a factor of safety. In choosing the type of pile used in a given project, consideration must be given to constructability, which was discussed in detail for many types of piles in Chapter 12. One component of constructability that can be better assessed using pile dynamic analyses is drivability, which we consider in this chapter. Dynamic analysis also allows us to establish a relationship between driving resistance and static resistance for a given site and a given pile and driving system; knowing this relationship, it becomes much easier to inspect piling operations.

Additionally, even if we are careful with pile type selection and design, without techniques to tell us that the piles will perform the way we designed them to perform, excessive uncertainty remains. Static load tests can be used to reduce this uncertainty, but they are typically expensive to perform. Dynamic load tests, done during pile driving, are more economical and can be used to estimate static pile capacity, although with less accuracy than provided by a static load test. These tests are based on measuring the force

DOI: 10.1201/b22079-14

and velocity near the pile head after each hammer blow and then matching these records by assigning values to the pile resistance both along the shaft and at base. When we obtain a match, we can claim that the assigned resistances are good estimates of the real resistances. This analysis is referred to as a *signal matching analysis*. The so-called *Case method* – a simpler form of signal matching in which total pile resistance is assumed to be lumped at the pile base, as opposed to estimating the shaft resistance along the pile and the base resistance separately – has become a common quality-assurance tool in connection with pile driving. It provides a simple means to estimate static pile capacity during pile driving. It also gives indications as to any pile damage that may have developed during driving.

Finally, pile installation can sometimes lead to defects, as discussed in Chapter 12. A third pile dynamics application that has become somewhat popular is *pile integrity testing*, in which a small hammer is used to impact the head of a previously installed pile. An interpretation of the measured velocity signal near the pile head is then made to assess whether there is the likelihood of defects along the pile shaft.

In the remainder of this chapter, we will present the theory of pile dynamics (largely based on one-dimensional wave mechanics) and its applications to signal matching analysis for dynamic test interpretation, to integrity testing, and to wave equation analysis of pile driving.

14.2 WAVE MECHANICS*

14.2.1 Wave equation

Because a pile with uniform cross section is a slender element surrounded by material with much lower stiffness, any mechanical impact on the pile head will travel primarily down the pile as a pulse or wave. Considering the distance z measured from the pile head down, we consider an infinitesimal element of the pile with length dz bounded by sections S_1 and S_2 (Figure 14.1). Consider what happens as a compressive pulse traveling in the positive (downward) z direction arrives at the element S_1–S_2. A compressive (thus, positive) force Q acts at z (section S_1). Let us say that a force Q $(z+dz)$ (infinitesimally more compressive than Q by dQ) acts at $z+dz$ (section S_2). We can also assume that the element S_1–S_2 is subject to an external shear stress q_s applied along its shaft pointing down (that is, assumed positive when pointing down, that is, when pointing in the positive z direction). This would be the unit shaft resistance at depth z. It is important to stress that we are simply setting up the differential equation for this problem at this stage. We are not asserting that any of the forces will necessarily be positive in their assumed positive directions or that compressive forces with increase with depth. Noting that, the axial displacement w being positive downward, the resultant force is also assumed positive downward, the application of Newton's second law to this element yields:

$$Q(z) - Q(z+dz) + q_s a_s dz = \rho A_p dz \frac{\partial^2 w}{\partial t^2} \tag{14.1}$$

where ρ is the mass density of the pile, A_p is its cross-sectional area, a_s is its lateral surface area per unit length (or its perimeter), and w is the axial displacement of the cross section at z – all functions of z in the most general scenario.

Equation (14.1), when properly rewritten and with $dQ = \left(\dfrac{\partial Q}{\partial z} \right) dz$, produces the equation of motion for our problem:

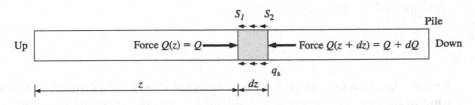

Figure 14.1 Pile element S_1–S_2 subject to a compressive pulse traveling in positive z direction.

$$-dQ + q_s a_s dz = \rho A_p dz \frac{\partial^2 w}{\partial t^2}$$

$$-\frac{\partial Q}{\partial z} dz + q_s a_s dz = \rho A_p dz \frac{\partial^2 w}{\partial t^2}$$

$$-\frac{\partial Q}{\partial z} + q_s a_s = \rho A_p \frac{\partial^2 w}{\partial t^2} \tag{14.2}$$

Using the elastic relationship for stresses and strains ($\sigma = E_p \varepsilon$, where E_p is the elastic modulus of the pile), the soil mechanics definition of axial strain ($\varepsilon = -\partial w/\partial z$) and the relationship between force and strain in the cross-section of the pile ($Q = \sigma A_p$), we can rewrite Equation (14.2) as

$$E_p A_p \frac{\partial^2 w}{\partial z^2} + q_s a_s = \rho A_p \frac{\partial^2 w}{\partial t^2} \tag{14.3}$$

Assuming that there is no external force acting on the pile element (thus, assuming $q_s = 0$), Equation (14.3) can be rewritten as

$$\frac{\partial^2 w}{\partial t^2} - c^2 \frac{\partial^2 w}{\partial z^2} = 0 \tag{14.4}$$

where

$$c = \sqrt{\frac{E_p}{\rho}} \tag{14.5}$$

Equation (14.4) is the one-dimensional *wave equation*, which is the governing differential equation for a wave of any shape traveling in the z direction. If we make the following change in variables:

$$\varphi = z - ct \tag{14.6}$$

and

$$\eta = z + ct \tag{14.7}$$

the wave equation reduces to:

$$\frac{\partial^2 w}{\partial \varphi \, \partial \eta} = 0 \tag{14.8}$$

where w is now a function of φ and η.[1] We now search for the form of $w(\varphi,\eta)$ such that the wave equation in the form Equation (14.8) is satisfied. The wave equation in this form states that the function $w=w(\varphi,\eta)$ has mixed second derivative equal to zero. This means that w must be of the form:

$$w(\varphi, \eta) = w_1(\varphi) + w_2(\eta) \tag{14.9}$$

Equation (14.9) must be the form of the solution w to Equation (14.8) because any other form would have a mixed second derivative equal at least to a nonzero constant, and not to zero, as required. Reversing back to the original variables z and t, we have then shown that the solution w to the wave equation is of the form

$$w(z,t) = w_1(z - ct) + w_2(z + ct) \tag{14.10}$$

The first term in Equation (14.10) represents a wave traveling in the positive z direction, that is, down the pile, with speed c, while the second term represents a pulse or wave traveling up the pile with speed c.[2] We will use this conclusion, that the displacement at any cross section is always the sum of a component associated with an upward-traveling wave and another associated with a downward-traveling wave, later in this chapter in the interpretation of dynamic pile tests.

14.2.2 Relationship between force and particle velocity

The velocity v imposed on the particles of a given cross section upon passage of the wave or pulse described by Equation (14.9) or Equation (14.10) is obtained by differentiating the displacement w with respect to time t. Operating on Equation (14.9):

$$v = \frac{\partial w}{\partial t} = \frac{\partial w_1}{\partial \varphi}\frac{\partial \varphi}{\partial t} + \frac{\partial w_2}{\partial \eta}\frac{\partial \eta}{\partial t}$$

$$= \frac{\partial w_1}{\partial \varphi}\frac{\partial (z - ct)}{\partial t} + \frac{\partial w_2}{\partial \eta}\frac{\partial (z + ct)}{\partial t}$$

which gives us the expression:

$$v = c\left(-\frac{\partial w_1}{\partial \varphi} + \frac{\partial w_2}{\partial \eta}\right) \tag{14.11}$$

Equation (14.11) can be interpreted to mean that the downward-propagating compressive wave and the upward-propagating compressive wave impose velocities of different signs on

---

[1] In Problem 14.2, the reader is asked to make the change of variables and show that Equation (14.8) does indeed follow from Equation (14.4).

[2] In Problem 14.3, the reader is asked to show that this important conclusion is indeed correct.

a cross section as they cross it. Equation (14.11) can be rewritten more compactly and more generally as

$$v(z,t) = v_1(z - ct) + v_2(z + ct) \tag{14.12}$$

where:

$$v_1 = -c \frac{\partial w_1}{\partial \varphi} \tag{14.13}$$

and

$$v_2 = c \frac{\partial w_2}{\partial \eta} \tag{14.14}$$

The strain is obtained by differentiation of w with respect to z:

$$\varepsilon = -\frac{\partial w}{\partial z} = -\left(\frac{\partial w_1}{\partial \varphi} + \frac{\partial w_2}{\partial \eta} \right) \tag{14.15}$$

so that the force F is given by

$$F = A_p E_p \varepsilon = A_p E_p \left(-\frac{\partial w_1}{\partial \varphi} - \frac{\partial w_2}{\partial \eta} \right) \tag{14.16}$$

Using Equations (14.13) and (14.14), we can rewrite Equation (14.16) as

$$F = \frac{A_p E_p}{c} \left(v_1 - v_2 \right) \tag{14.17}$$

When a single pulse is traveling down the pile (in the positive z direction), $v_2 = 0$. Substituting that into Equation (14.17) leads to the conclusion that force and velocity are directly proportional and have the same sign. On the other hand, if a single pulse travels up the pile (in the negative z direction), $v_1 = 0$, which, in light of Equation (14.17), means that force and velocity are directly proportional, but have opposite signs.

The quantity $\frac{E_p A_p}{c}$ represents the force carried by the pile for a unit particle velocity and is thus known as the pile impedance Z. The pile impedance represents the total axial force experienced by a pile cross section when it is subjected to a unit velocity $v = 1$. Given Equation (14.5), Z is also given by

$$Z = \frac{E_p A_p}{c} = \rho A_p c \tag{14.18}$$

Another way to look at impedance is the following: in general terms, for a given fall height, a hammer blow imposes a certain initial velocity at the pile head. This initial velocity tends to depend mostly on the hammer, driving system, and pile properties. The height from which the ram falls, how much energy is lost during the fall, and how much energy is lost upon energy transfer to the pile are the main factors determining the initial velocity imposed on the pile. For the same velocity imposed on a pile, the force that is sustained by the pile is directly proportional to the pile impedance. This means that a pile with larger impedance (because it has either a larger cross section or a larger product of mass density by wave speed c) is able to take a larger force and can be driven harder by the same hammer and driving system. This means it can be driven through harder soil layers and thus develop a larger static resistance.

Repeating Equations (14.12) and (14.17) in terms of the newly defined impedance Z:

$$F = Zv_1 - Zv_2 = F_d + F_u \tag{14.19}$$

$$Zv = Zv_1 + Zv_2 = Zv_d + Zv_u \tag{14.20}$$

Solving the resulting system of equations for the upward-traveling force $F_u = -Zv_u$ and the downward-traveling force $F_d = Zv_d$, in turn, we get:

$$F_u = -Zv_2 = \frac{1}{2}(F - Zv) \tag{14.21}$$

$$F_d = Zv_1 = \frac{1}{2}(F + Zv) \tag{14.22}$$

Equations (14.21) and (14.22) will be useful later when we discuss the interpretation of the Case method and of the use of signal matching to estimate pile resistance.

14.2.3 Boundary conditions

14.2.3.1 Types of boundary conditions

There are four boundary conditions that are of interest in pile dynamics:

1. wave approaching a free end;
2. wave approaching a fixed end;
3. prescribed force at a given cross section along the pile; and
4. prescribed displacement at the pile end.

It is useful to examine each boundary condition from the view point of what happens to the force and velocity when a pulse reaches it.

14.2.3.2 Wave approaching free end

Consider a pile with length L without any external force acting on it anywhere. Consider a velocity pulse $v_d = v_0(z - ct)$ approaching the free end of the pile (an end where the stress is zero). A free boundary can be thought of as a sudden discontinuity in inertia. When the pulse reaches the free end, there is no longer any inertia to overcome and the energy cannot continue to travel in the positive z direction; however, the energy cannot disappear either, so it reflects back in the negative z direction. At the exact moment it reflects back, the velocity at the free end doubles and the force there goes to zero. Let us examine in greater detail how this happens.

On reaching the end, the pulse must result in zero stress (or zero force, which is the same). Since there is a force associated with the pulse $v_d = v_0(z - ct)$, equilibrium requires that an

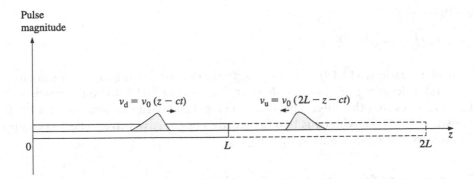

Figure 14.2 Reflection off a free end viewed as the crossing of two mirror pulses at the fixed end at the same time.

opposite force be generated at the free end exactly at the moment $v_0(z-ct)$ reaches it. A compressive pulse, on arrival at the free end, generates a tensile pulse that will now travel up the pile.[3] The pulse with particle velocity v_u that is thus generated now travels in the direction opposite that of $v_d=v_0(z-ct)$, so we say there is a reflection. The reflected pulse has a velocity of the same magnitude and sign as the incident pulse to have the cancellation of forces at the free end.[4] The best way to handle this reflection mathematically is to consider an imaginary mirror image of the pile extending beyond the free end. Consider now an imaginary pulse $v_u=v_0(2L-z-ct)$ with the same magnitude as v_d approaching the free end of the pile from the imaginary mirror image of the pile (Figure 14.2). At time $t=0$, $z=0$ for the downward pulse traveling in the real pile and $z=2L$ for the upward pulse traveling in the imaginary pile, so $v_u=v_0(2L-z-ct)$ and $v_d=v_0(z-ct)$ are exactly symmetric with respect to the pile end. Given the arguments $(2L-z-ct)$ of v_u and $(z-ct)$ of v_d, the two pulses arrive at the pile end at exactly the same time $(t=L/c)$, but from opposite directions.

After they pass each other at $z=L$, the pulse $v_d=v_0(z-ct)$ will continue to travel beyond the end of the pile into the imaginary pile extension and the pulse $v_u=v_0(2L-z-ct)$ travels up the real pile. From Equation (14.17), we know that the force associated with the pulse $v_d=v_0(z-ct)$, traveling in the positive z direction, has the same sign as the velocity and the force associated with $v_u=v_0(2L-z-ct)$, traveling in the negative z direction, has the opposite sign. For our purposes, this can be viewed as a reflection of the incident pulse v_d off the free end as a pulse with the same magnitude, but opposite sign for the force and with equal magnitude and sign for the velocity. This implies, as we have already seen, that, at the exact moment the downward pulse arrives at the pile base, the force there becomes zero and the velocity doubles.[5]

14.2.3.3 Wave approaching fixed end

The reasoning for a fixed end is similar to the one we went through for a free end, except that now the velocity (or displacement), not the force, must be zero at the fixed end. Considering Equation (14.12), we must now have the following relationship between the pulses approaching the fixed end from the real and from the imaginary side of the fixed end:

---

[3] Conversely, a tensile pulse, on arriving at a free end, generates a compressive pulse in the opposite direction.

[4] Note that the force associated with v_u is opposite in sign to that of v_d because force is proportional to the negative of velocity when the pulse travels in the negative z direction, as we saw earlier.

[5] This also has the important practical implication that when the pile base is in very soft soil (with conditions approaching those for a free end), a compressive wave reflects off the base as a tensile wave, which can be very damaging for a concrete pile.

$$v_d = v_0 (z - ct) \tag{14.23}$$

$$v_u = -v_0 (2L - z - ct) \tag{14.24}$$

In light of Equations (14.19) and (14.20), Equation (14.24) implies zero velocity at the fixed end and a doubling of the force there at the moment $t = L/c$ the pulse reaches the pile end. It also implies that the pulse reflects off the fixed end with the same sign for the force F and opposite sign for the velocity v. In other words, compressive waves reflect as compressive waves, and tensile waves, as tensile waves.

14.2.3.4 Prescribed force at a point along the pile

Consider now a force $R(z^*)$ applied at a point $z = z^*$ along the pile (Figure 14.3). Continuity requires that the velocity (or displacement) on the left of z^* (let us call that z_L^*) be the same as on the right of z^* (which we will call z_R^*). This can be written as

$$\left. v \right|_{z_L^*} = \left. v \right|_{z_R^*} \tag{14.25}$$

If the orientation of force $R(z^*)$ is as in Figure 14.3 (in the negative z direction), equilibrium requires that the force be balanced by the sum of a stress increase (that is, an increase in compression, and thus positive) above the point of application of the force and a stress decrease (an increase in tension, and thus negative) below the point of application of the force. It can be shown mathematically that these two forces are of equal magnitude.[6] The particle velocities of these two pulses generated by $R(z^*)$ point in the same direction as $R(z^*)$ and thus have the same sign, with magnitude $\dfrac{\frac{1}{2} c R(z^*)}{E_p A_p}$.

Let us examine what happens as a downward-traveling compressive pulse, as shown in Figure 14.4, arrives at a point of the pile where a soil resistance R_i can be developed. So long as the full resistance R_i is mobilized, the incident pulse generates an upward-traveling compressive wave with magnitude $R_i/2$ and a downward-traveling tensile wave also of magnitude $R_i/2$ upon reaching the force R_i. Note that the incident pulse needs to be strong enough to mobilize R_i fully.

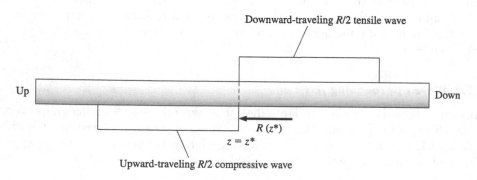

Figure 14.3 Compression and tension waves generated by sudden application of a force R at a point along the pile shaft.

---

[6] Indeed, there is no physical reason why the magnitudes of the two forces that are generated to balance the applied force $R(z*)$ at $z*$ would be different.

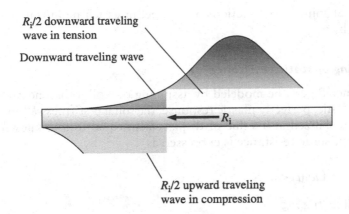

Figure 14.4 Interaction between wave and a force R_i applied at a given pile cross section.

When the force R_b is applied at the base of the pile, according to the preceding discussion, both an upward-traveling pulse and a downward-traveling pulse are generated, but, in this case, the downward-traveling pulse is immediately reflected up. The resulting velocity and force are given by

$$v(z,t) - \frac{1}{Z} R_b \tag{14.26}$$

$$F(z,t) = R_b \tag{14.27}$$

14.2.3.5 Prescribed velocity at pile top

Imposing a velocity v_t at the pile top generates a single pulse that will move down the pile. Until there is a reflection, this pulse is the single wave in the pile, and force and velocity are therefore proportional. So we can simply multiply the velocity at the pile top by the pile impedance to obtain the force there, and this boundary condition reduces to the case of the previous section, but with the force now applied not on the base but on the top of the pile. We have:

$$R_t = \frac{E_p A_p}{c} v_t \tag{14.28}$$

14.2.4 Modeling of soil resistances

14.2.4.1 Decomposition in static and dynamic components

In the analysis of wave propagation in piles, soil resistances R_i are sometimes assumed to act at a finite number of equally spaced points along the pile. A resistance $R_i(t)$ acting at z_i may be assumed to consist of a static component $R_{si}(t)$ and a dynamic component $R_{di}(t)$:

$$R_i(t) = R_{si}(t) + R_{di}(t) \tag{14.29}$$

The resistances at point z_i are functions of time because their mobilization depends on the passage of waves by z_i.

14.2.4.2 Modeling of static resistance

The static resistance R_{si} can be modeled by assuming the soil to be linear elastic, perfectly plastic. In linear elastic, perfectly plastic response, the soil has stiffness K_{si} until a limit static resistance R_{sLi} is mobilized at a value of displacement equal to w_{qi}, known as the *quake*. Mathematically, the static resistance is expressed as

$$R_{si}(t) = \begin{cases} K_{si}w_i(t) \text{ if } w_i(t) < w_{qi} \\ R_{sLi} \text{ if } w_i(t) \geq w_{qi} \end{cases} \tag{14.30}$$

or, alternatively,

$$R_{si}(t) = \min\left(K_{si}w_i(t), R_{sLi}\right) \tag{14.31}$$

in loading, and

$$R_{si}(t) = R_{sLi} - K_{si}\left(w_{i,\max} - w_i\right) \tag{14.32}$$

in unloading, where $w_{i,\max}$ is simply the maximum displacement reached by pile element i during loading and

$$K_{si} = \frac{R_{sLi}}{w_{qi}} \tag{14.33}$$

This model is represented schematically by a spring associated with a stick-slip support, which gives way when the force acting on the spring reaches R_{sLi} (Figure 14.5).

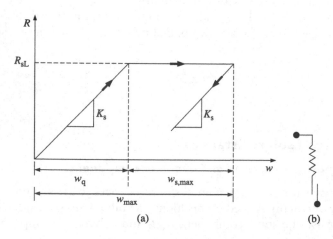

(a)   (b)

Figure 14.5 Model for static resistance: (a) linear elastic, perfectly plastic force–displacement relationship in both loading and unloading and (b) spring and stick-slip representation of this relationship.

A simpler model, useful in the more basic methods of analyzing dynamic tests, is based on rigid-plastic response, expressed mathematically as

$$R_{si}(t) = \begin{cases} 0 \text{ if } w_i(t) = 0 \\ R_{sLi} \text{ if } w_i(t) > 0 \end{cases} \tag{14.34}$$

14.2.4.3 Modeling of dynamic resistance

It is known that soils offer larger resistance to loads applied quickly than to loads applied slowly. Dynamic resistance forces in soil are usually linked to velocity by use of a suitable damping factor or coefficient. In the usual formulations of soil dynamics, the dynamic resistance R_{di} is written as

$$R_{di}(t) = d_i v_i(t) \tag{14.35}$$

where d_i=damping factor. Equation (14.35) can be represented by a dashpot model as shown in Figure 14.6.

There are two other ways of expressing soil damping: one based on the Case damping factor j_c (after Case Western Reserve University, where the Case method, discussed subsequently, originated) and the other based on the Smith damping factor J_s. The dynamic resistance force is expressed as follows, for each case, respectively:

$$R_{di}(t) = j_c Z v_i(t) \tag{14.36}$$

where the Case damping factor j_c is clearly a dimensionless number, and

$$R_{di}(t) = J_s R_{sLi} v_i(t) \tag{14.37}$$

where J_s is called the Smith damping factor.

The implication of Equations (14.35), (14.36), and (14.37) is that, if the velocity is known at a cross section, the dynamic force is known as well. Assume, for example, that the velocity at $z=z_i$, where a resistance $R_i=R_{ai}+R_{di}$ exists, is $v_{ia}(t)$ if the resistance is not mobilized (that is, $R_i=0$). The mobilization of R_i generates two pulses, one up and one down the pile, each with velocity opposite in sign to that of the incident wave. The actual velocity at the point under consideration will thus be:

$$v_i(t) = v_{ia}(t) - \frac{1}{2Z} R_i(t) \tag{14.38}$$

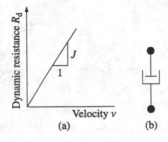

(a)   (b)

Figure 14.6 Model for dynamic soil resistance: (a) force versus velocity relationship and (b) dashpot representation of such relationship.

where R_i is a function of v_i through its dynamic component R_{di}. It is possible to solve Equation (14.38) for v_i and then to calculate R_{di}.

Let's now consider the pile base. The base resistance R_b, once mobilized, creates two pulses with force magnitude equal to $\frac{1}{2}R_b$ and velocity equal to $-\frac{1}{2}R_b/Z$. However, the down pulse is reflected immediately upward. So, for the pile base, Equation (14.38) becomes

$$v_b(t) = v_{ba}(t) - \frac{1}{Z} R_b(t) \tag{14.39}$$

14.3 ANALYSIS OF DYNAMIC PILE LOAD TESTS

14.3.1 The Case method

The Case method requires measurement of force and velocity near the pile head as a hammer strikes the pile. Accelerations are measured using an accelerometer; they are integrated to give velocities. Strain gauges are used to measure strains, from which we can obtain the corresponding force in the pile. Figure 14.7 shows the two accelerometers and one strain gauge attached to a pipe pile at the usual distance of about two diameters from the pile head. Figure 14.8 shows a typical record obtained from such sensors. The simplest interpretation of such records leads to what is known as the Case formula.

The Case formula assumes that:

1. the static component of the resistance is rigid-plastic, as in Equation (14.34),
2. the dynamic resistance follows Equation (14.36), and
3. during the time required for the wave generated by the hammer blow to travel down the pile and back, the pile is still moving down as a result of the hammer blow.

As a consequence of the last assumption, the soil resistances are still mobilized in the same direction: pointing up. Recall that an upward resistance along the shaft generates an upward

Figure 14.7 Accelerometers and strain gauge placed at a distance equal to roughly two pile diameters from the pile head.

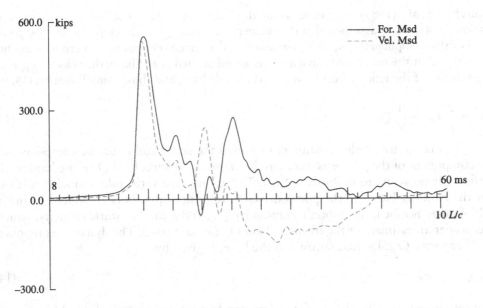

Figure 14.8 Force and velocity record from a dynamic test on a pile.

compressive wave with force equal to one half the resistance force and another tensile wave moving down with equal magnitude.

Now let $Q_{s,total}$ be the total mobilized shaft resistance (including static and dynamic components) and $Q_{b,total}$ be the total mobilized base resistance of the pile (including static and dynamic components) under a given hammer blow that imposed a downward force $F_d(t_0)$ at the pile head at time t_0. Consider now a time t_0+2L/c, where L is the pile length. The applied force F_d reflects at the pile base as $-F_d$. One half the resistance forces along the shaft are mobilized as upward-moving compressive waves, being responsible for the arrival of a compressive force equal to $Q_{s,total}/2$ at the pile top. The downward tensile waves generated by the shaft resistances reflect off the pile base as compressive waves, being responsible for the arrival at the top of a compressive force also equal to $Q_{s,total}/2$. Finally, the base capacity $Q_{b,total}$ also arrives at the top as a compressive wave. So the up-wave F_u at time $t=t_0+2L/c$ adds up to:

$$F_u\left(t_0 + \frac{2L}{c}\right) = Q_{b,total} + \frac{1}{2}Q_{s,total} + \frac{1}{2}Q_{s,total} - F_d(t_0)$$

or, simply,

$$F_u\left(t_0 + \frac{2L}{c}\right) = Q_{total} - F_d(t_0) \tag{14.40}$$

In this scenario, when the up-force arrives at the top, it contains not only the reflected wave at the base, but also the summation Q_{total} of all the resistance forces along the shaft and base. Thus, we can write the total pile resistance as a function of the down-force and up-force, measured at times t_0 and t_0+2L/c, respectively, as

$$Q_{total} = F_u\big|_{(t_0+2L/c)} + F_d\big|_{t_0} \tag{14.41}$$

which is the Case formula.

Rausche et al. (1985) proposed a method for separating the static capacity out of Equation (14.41). They assumed that the entire pile capacity Q_{total} developed at the pile base and that all of the damping effects, as a consequence, developed there as well (a crude assumption). Considering that the entire pile capacity is assumed located at the base, the velocity v_{ba} is equal to the velocity of the reflected down wave off the pile base, and thus, using Equation (14.39):

$$v_b = \frac{2}{Z} F_d \bigg|_{t_0} - \frac{1}{Z} Q_{total} \qquad (14.42)$$

where Q_{total} is the total pile resistance, including the static and dynamic components. The static component of the pile resistance can be taken as the limit load Q_L if the hammer blow is sufficiently powerful to mobilize it fully. This is only true if the pile is moving sufficiently under the hammer blow. As the pile is driven to refusal, displacements induced by the hammer blow may not be large enough to mobilize Q_L, and a smaller static capacity, which we will consider an estimate of the ultimate load Q_{ult}, is mobilized. The dynamic component of the pile capacity Q_{db}, located entirely at the base, is given by

$$Q_{db} = j_c Z v_b \qquad (14.43)$$

where the suggested values of the Case damping factor j_c are given in Table 14.1.

Substituting Equation (14.42) into Equation (14.43):

$$Q_{db} = j_c \left(2F_d \big|_{t_0} - Q_{total} \right) \qquad (14.44)$$

The static pile resistance Q_{ult} is obtained by subtracting Q_{db} from Q_{total}; in light of Equation (14.44), Q_{ult} is given by

$$Q_{ult} = Q_{total} - j_c \left(2F_d \big|_{t_0} - Q_{total} \right) \qquad (14.45)$$

We now substitute Equation (14.41) into Equation (14.45):

$$Q_{ult} = F_u \big|_{(t_0 + 2L/c)} + F_d \big|_{t_0} - j_c \left(F_d \big|_{t_0} - F_u \big|_{(t_0 + 2L/c)} \right)$$

which is easily rewritten as

$$Q_{ult} = \left(1 + j_c \right) F_u \big|_{(t_0 + 2L/c)} + \left(1 - j_c \right) F_d \big|_{t_0} \qquad (14.46)$$

While F_d is simply equal to whatever force is measured at time t_0 (because in the beginning of the record there is only a downward-propagating wave), F_u is not readily available from the record. So it is not practical to work with Equation (14.46), as it is expressed as

Table 14.1 Suggested values of the Case damping factor j_c

Soil type in bearing layer	Range of j_c values
Sand	0.05–0.20
Silty sand/sandy silt	0.15–0.30
Silt	0.20–0.45
Silty clay/clayey silt	0.40–0.70
Clay	0.60–1.10

Based on Pile Dynamics, Inc (2002).

a function of F_u. Equations (14.21) and (14.22) can be used to rewrite Equation (14.46) in terms of F and v directly, which leads us to the final form of the Case formula:

$$Q_{ult} = \frac{1}{2}\left[(F - Zv)(1 + j_c)\Big|_{(t_0 + 2L/c)} + (F + Zv)(1 - j_c)\Big|_{t_0} \right] \tag{14.47}$$

One of the limitations of the Case method has to do with the way in which damping is defined. As pointed out by Chow et al. (1988), j_c is not a soil property. It is, in a way, a soil–structure interaction parameter, with different values of it required to model different piles driven in the same soil. However, the Case method has the important advantage that it is quite simple and can thus be used to provide an estimate of pile capacity at the time of pile driving. This estimate, obtained using Equation (14.47), is somewhat crude, given all the simplifying assumptions, but is still useful to have it at the time of pile driving as a quality-assurance tool. It can be used to focus attention on piles that are showing estimates of Q_{ult} that differ significantly from the values obtained at the design stage.

Example 14.1

Using the strike and restrike force and velocity records (shown in E-Figure 14.1) obtained from dynamic tests on a closed-ended steel pipe pile, calculate the ultimate pile resistance with and without the damping effects at the base. The pile was driven in a gravelly sand.

Solution:

Having the force and velocity records, we notice that the velocity is represented in units of force. This means that the velocity values shown on the record are in fact velocity multiplied by the impedance. These records are typically available electronically and numbers given with a precision that goes beyond 1 kip. To be consistent with the calculations that are done automatically using these numbers, we will work with a precision of 0.1 kip and only round the final numbers to a 1-kip precision.

Let us first work with the strike record. We start by finding the values of F and Zv at time t_0 and $t_0 + \frac{2L}{c}$. Notice that the x axis of the record is in terms of $\frac{L}{c}$. The time t_0 is taken as the time when the force is maximum. The corresponding values of F and v are:

$$F_{t_0} = 500 \text{ kips}$$

$$Zv_{t_0} = 457.1 \text{ kips}$$

At $t_0 + \frac{2L}{c}$:

$$F_{t_0 + \frac{2L}{c}} = -35 \text{ kips}$$

$$Zv_{t_0 + \frac{2L}{c}} = 335.7 \text{ kips}$$

The total pile resistance Q_{total} is calculated using Equation (14.41):

$$Q_{total} = \frac{1}{2}\left[(F - Zv)_{(t_0 + 2L/c)} + (F + Zv)_{t_0} \right]$$

$$Q_{total} = \frac{1}{2}\left[(-35 - 335.7) + (500 + 457.1) \right]$$

$$Q_{total} = 293.2 \approx 293 \text{ kips}$$

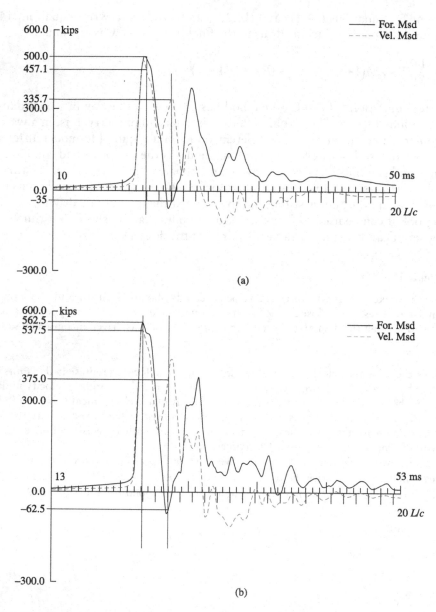

E-Figure 14.1 Dynamic tests on steel pipe pile: (a) driving records and (b) restrike records.

From Table 14.1, we see that the Case damping factor j_c for sand ranges from 0.05 to 0.2. Taking an average value $j_c = 0.125$, we can now find the ultimate static pile resistance Q_{ult} using Equation (14.46):

$$Q_{ult} = \frac{1}{2}\left[(F - Zv)(1 + j_c)\Big|_{(t_0 + 2L/c)} + (F + Zv)(1 - j_c)\Big|_{t_0}\right]$$

$$Q_{ult} = \frac{1}{2}\left[(-35 - 335.7)(1 + 0.125) + (500 + 457.1)(1 - 0.125)\right]$$

$$Q_{ult} = 210.2 \text{ kips} \approx 210 \text{ kips}^{\text{answer}}$$

We will now work with the restrike record. As before, we start by finding the values of F and Zv at time t_0 and $t_0 + \frac{2L}{c}$. At time t_0 (taken to be the time when the force is maximum):

$$F_{t_0} = 562.5 \text{ kips}$$

$$Zv_{t_0} = 537.5 \text{ kips}$$

At $t_0 + \frac{2L}{c}$:

$$F_{t_0+\frac{2L}{c}} = -62.5 \text{ kips}$$

$$Zv_{t_0+\frac{2L}{c}} = 375 \text{ kips}$$

Using Equation (14.41) to calculate the total pile resistance Q_{total}:

$$Q_{\text{total}} = \frac{1}{2}\left[(F - Zv)_{(t_0+\frac{2L}{c})} + (F + Zv)_{t_0}\right]$$

$$Q_{\text{total}} = \frac{1}{2}\left[(-62.5 - 375) + (562.5 + 537.5)\right]$$

$$Q_{\text{total}} = 331.3 \approx 331 \text{ kips}$$

From Table 14.1, we see that the Case damping factor j_c for sand ranges from 0.05 to 0.2. Taking an average value $j_c - 0.125$, we can now find the ultimate static pile resistance using Equation (14.47):

$$Q_{\text{ult}} = \frac{1}{2}\left[(F-Zv)(1+j_c)\big|_{(t_0+\frac{2L}{c})} + (F+Zv)(1-j_c)\big|_{t_0}\right]$$

$$Q_{\text{ult}} = \frac{1}{2}\left[(-62.5 - 375)(1+0.125) + (562.5 + 537.5)(1-0.125)\right]$$

$$Q_{\text{ult}} = 235.2 \text{ kips} \approx 235 \text{ kips}^{\text{ answer}}$$

These results show that the pile–soil system tested during initial driving and restrike is not the same. Some setup – the gain of bearing capacity with the passage of time, a phenomenon discussed in Chapter 13 – may have occurred. It is also possible that the pile advanced into firmer ground.

Example 14.1 illustrates an important point about dynamic tests. As with static load tests, it is important to obtain estimates of load capacity some time after pile driving, so that relaxation or setup (see Chapter 13) has had time to take effect. So measurements are often made again days after pile installation.[7] These measurements are commonly referred to as measurements taken on restrike or redriving.

14.3.2 Signal matching

An improved method for estimating static pile capacity from measurements of force and velocity during driving would no longer make the assumption of lumping all the pile resistance at the base of the pile, as done in the Case method. It is better to estimate shaft and base capacities using a more sophisticated analysis that attempts to closely match calculated

---

[7] In some cases, much longer waiting times are required for setup or relaxation effects to develop fully.

force and velocity at the top of the pile to the recorded force and velocity signals. This type of analysis is known as *signal matching*.

In signal matching analysis, the shaft resistance distribution along the pile and the pile base resistance are estimated by considering soil resistances along the shaft at some spacing and calculating their effects on the pile head by using superposition of effects (this was explored in depth by Rausche 1970). Instead of using rigid-plastic behavior, linear elastic, perfectly plastic models with consideration of unloading or even more sophisticated models can be used. The aim is of course to obtain the same velocity and force values measured over the time $2L/c$ after the hammer blow. This requires that the values of the resistances be adjusted until the match between calculated and measured values is satisfactory.

The matching is done by a suitable program. The iterative matching sequence is as follows:

1. select a record with appropriate energy and data quality;
2. build the pile model;
3. create or update the soil resistance distribution along the pile;
4. apply hammer blow;
5. check match quality: if not satisfactory, return to Step 3;
6. end.

Naturally, the pile resistance estimate using signal matching tends to be superior to that obtained from the Case method, as the assumptions in the Case method are rather limiting.

14.3.3 Pile integrity testing

When we discussed design in earlier chapters, we saw that design should not attempt to account for unquantifiable uncertainties, including defective installation. It is important to have quality-control/quality-assurance processes in place that will prevent installation of defective piles or, if they fail to do that, at least identify them when they exist. Figure 14.9 illustrates two of the most common defects in steel piles. Figure 14.9a shows a pile that was overdriven, compressing the tip of the pile as opposed to increasing pile penetration. Figure 14.9b shows a bent pipe pile. This was one of several with the same problem. The piles were installed in the early 1960s for a Boston Post Office parking garage. A 75- to 100-foot-long portion of the garage had to be demolished for the construction of one of the I-90 to I-93 tunnel ramps (part of the "Big Dig" project). The area had to be treated with deep soil mixing before a 50- to 60-foot-deep excavation for tunnel construction could be done, so the piles had to be pulled out. The piles were all 16-inch steel pipes, filled with concrete, originally driven 110–120 ft to glacial till (J. Lambrechts, personal communication). There was no suspicion that these piles had bent in this manner at the time of installation or, for that matter, afterward, indicating that they still performed well enough even if their configuration was not exactly what the designer had in mind. This suggests that the factor of safety used more than covered for the detrimental effects of these defects.

Pile integrity is a concern not only with pre-fabricated driven piles, which may bend, compress, or crack as a result of excessive compressive or tensile stresses during driving, but also with cast-*in situ* piles (drilled shafts, CFA piles, drilled displacement piles, and Franki piles) which may present discontinuities, soil inclusions, necking, and other anomalies that may reduce the load capacity of the pile. In piles cast in *in situ*, defects are most often associated with not placing enough concrete to fill the void left by excavated soil. This often results from pulling a tremie pipe, hollow-stem auger, or Franki-pile casing too fast. Figure 14.10 shows a case of necking, which left rebars completely exposed, observed in a CFA pile. Defects of this type, and also those in driven steel and concrete piles, are often not detected at the time of installation. They develop underground and the operator cannot see what is happening below the surface, so experience and a certain feel for "feedback" from the drilling equipment during installation are desirable on the part of the operator. New rigs also have sensors of various types that may help prevent installation defects.

If defects do make it past quality-assurance checks (that is, checks that a contractor and inspector are supposed to do to have confidence that the correct pile installation processes are in place), we need quality-control checks on the final product itself. Several techniques exist for checking pile integrity. We will focus on a technique that is simple to perform, based on the use of a light (0.5–5 kg) hammer (Figure 14.11). This hammer is used to strike the pile head (Figure 14.11a), which generates a pulse that travels down the pile and reflects back off any defect or anomaly it finds. The reflected signal is detected by an accelerometer on its return, allowing us to obtain velocity by integration, which is shown on a screen in real time (Figure 14.11b) and stored for analysis in the office.

In the interpretation of pile integrity test results, we use the fact that the wave reflected off a free end has force with the opposite sign as the force of the incident wave, but velocity that has the same sign as the velocity of the incident wave, and vice versa for a fixed end. In the absence of shaft resistance and energy loss along the shaft, we would obtain signals such as those shown in Figure 14.12. Note first that the velocity would have the

(a)   (b)

Figure 14.9 Defects in steel piles: (a) pile with compressed tip and (b) bent pipe pile. ((a) Courtesy of GRL Engineers, Inc.; (b) courtesy of James Lambrechts.)

Figure 14.10 Defect in a CFA pile detected by pile integrity testing. (Courtesy of GRL Engineers, Inc.)

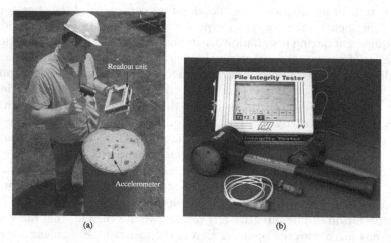

Figure 14.11 Pile integrity testing: (a) light hammer and read-out unit during a test and (b) details of the equipment with read-out unit showing a curve obtained from a test. (Courtesy of GRL Engineers, Inc.)

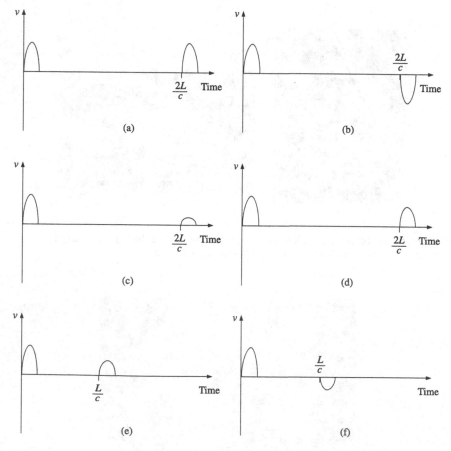

Figure 14.12 Schematic velocity signals for various pile conditions: (a) free-end condition at pile base, (b) fixed-end condition at pile base, (c) primarily end-bearing pile, (d) pile with very small base resistance, (e) pile with severe necking or discontinuity at $L/2$ (midway between head and base), and (f) diameter enlargement at $L/2$.

same signal and the same magnitude as the signal generated at the pile head in the case of a free-end condition at the pile base (Figure 14.12a). In the case of a fixed-end condition at the pile base, the velocity associated with the reflected wave would have the opposite sign and the same magnitude as the velocity generated at the pile head (Figure 14.12b). Even in end-bearing piles, upon arrival of the wave at the pile base, there is significant loss of impedance, and velocity signals would have small reflections with the same sign as the generated signal (Figure 14.12c). When pile base resistance is extremely small (such as a drilled shaft in which soil caving took place and soft or loose material accumulated at the base), the conditions at the base of the pile resemble those at a free end, hence the form of the signal in Figure 14.12d. In Figure 14.12e, we see that a reduction in diameter at the pile midpoint generates a velocity change of the same sign as the input signal because it represents a reduction in impedance. The opposite is observed in Figure 14.12f in the case of diameter enlargement, which represents an increase in impedance. Shaft resistance would appear in the same way as a diameter enlargement would: as an increase in impedance and thus a reduction in velocity.

Plots such as that in Figure 14.12 are often presented with the time axis transformed into a space axis. This can be done because there is a one-to-one correspondence between the depth z of any section of the pile and the time it takes for the wave to travel from the pile head to that section and return to the top ($2z/c$). So the results can be plotted as measured particle velocity versus z with scale from 0 to L (pile length). Another way in which results can be presented is by plotting estimated pile diameter versus depth along the pile. Where a reduction in impedance is detected, a reduction in diameter (which in truth can be a crack or gap in the pile) is plotted.

In reality, there is energy loss in a pile integrity test. As a result, the signals must be amplified and processed before interpretation can be made. This is potentially a problem, because, if the processing of the measured signals is not done correctly, interpretation becomes more difficult and mistakes can be made. There are other potential limitations of pile integrity testing. For example, Chow et al. (2003) show that, although one-dimensional wave propagation is assumed, three-dimensional effects may produce signals that could be (mistakenly) interpreted as anomalies depending on the diameter of the pile and where the hammer hits the pile head. Other factors also exist that could lead to false positives (that is, results that would suggest a defect that does not exist). Thus, much caution is needed before piles are condemned based exclusively on pile integrity testing.

So long as the possibility of false positives is not dismissed, the pile integrity test can be very useful. Figure 14.13 illustrates that fact for a case where a five-story expansion of an existing building was planned, requiring an assessment of the conditions and load capacities of the existing piles. Pile integrity testing showed a sudden drop of stiffness at 1.5 m, which, as we have seen, could be associated with necking or a discontinuity in the pile. Excavation around the pile confirmed the existence of the defect.

14.4 WAVE EQUATION ANALYSIS

14.4.1 Wave equation analysis and its applications

As we saw in connection with the Case method and the signal matching technique, it is possible to estimate the bearing capacity of a pile from how hard it is to drive it, which in turn reflects on the values of force and velocity measured near the pile head. The signal matching analysis is done by modeling the pile either as a continuous elastic rod or as pile segments connected by springs (and possibly dashpots) and modeling the surrounding soil as springs

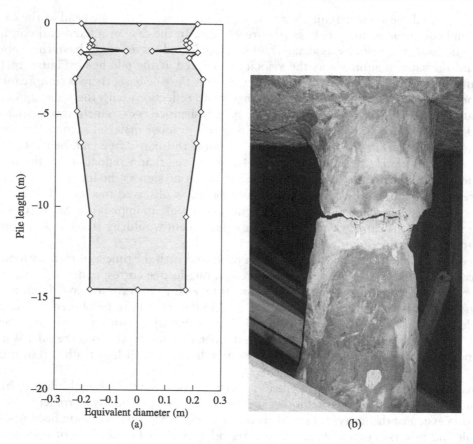

Figure 14.13 Defect in existing pile: (a) test result plotted as theoretical pile diameter versus depth suggesting a sharp drop in pile stiffness at a depth of approximately 1.5 m and (b) a view of the pile after excavation for confirmation of the presence of the defect. (Courtesy of John S. Higgs, Integrity Testing Pty Ltd., Victoria, Australia.)

and dashpots. If, instead of measuring the force and velocity generated by a hammer blow, as is done in the Case or signal matching methods, we model the entire system – pile, soil, and driving system (hammer, hammer cushion, and pile cushion) – we perform an analysis that is customarily referred to as *wave equation analysis*.

The solution of the wave equation for a pile subjected to a pulse traveling along the axis of the pile is the axial displacement of each pile cross section as a function of time. With knowledge of the displacement everywhere in the pile at a given time, we can compute the strain everywhere in the pile. Knowing the strain, we can calculate stress and thus the force experienced by every pile cross section. Additionally, by considering the accumulation of displacements at the pile base during one given hammer blow, we can calculate the pile penetration (the term used for that in piling jargon is "set") in the soil for that hammer blow. This allows us to establish the relationship between static pile axial resistance and set. An alternative is to plot static pile axial resistance versus driving resistance in blows per unit length of penetration. Piles deriving most of their capacity from side resistance reach their static capacity at lower dynamic driving resistances (less than approximately 4 blows/cm or 10 blows/in.) than end-bearing piles (between 4 and 10 blows/cm or 25 blows/in.). Knowledge of the relationship between the driving resistance (expressed in blows per unit length of penetration) and the pile capacity is helpful to piling inspectors, who can use it as a means to verify that piles are indeed driven to a depth at which the required capacity is available.

The other useful result produced by wave equation analysis is the maximum compressive and tensile stresses experienced at any pile cross section for the hammer blow considered. Knowledge of the relationship between pile axial resistance and set and knowledge of the maximum compressive and tensile stresses experienced by the pile enable engineers to select driving system components so that the piles can be driven to the desired capacity in an economical way without damage to the pile.

In summary, some of the useful determinations that can be made using wave equation analysis are the following:

1. the best hammer for the pile;
2. the thickness of the pile cushion for a given concrete pile;
3. the possibility of driving the pile through an intermediate layer of stronger soil without damaging the pile; and
4. installation time (as expressed by the number of hammer blows required to drive the pile to the required capacity).

14.4.2 Pile and soil model

14.4.2.1 Overview

In "wave equation analysis" of pile driving, the wave equation is not directly solved. Rather, the pile is discretized as shown in Figure 14.14. Pile inertia is represented through mass elements, and its stiffness by springs between mass elements. The shaft resistance applied to

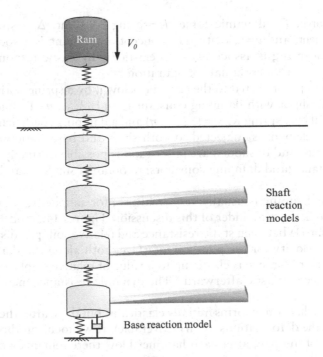

Figure 14.14 Pile discretization for wave equation analysis. (After Salgado et al. (2017); with permission from Elsevier Science and Technology Journals.)

each element and the base resistance applied to the bottom element of the pile can be modeled in different ways, with different degrees of complexity.

The hammer impact is represented by the instantaneous application of an initial velocity to the topmost mass (the ram mass). This initial velocity will generate compression of the spring between the ram and the first element of the pile and a force there. This force in turn will cause the acceleration of the first element of the pile, which will get it to move down. Any movement of pile elements will mobilize soil resistance. As time progresses, these effects will propagate down the pile.

We will first introduce the pile resistance model proposed by Smith (1960), which is the simplest that could be used in a wave equation analysis. We will then discuss a more advanced, more realistic model.

14.4.2.2 Smith model

As seen in Figure 14.15, the pile is discretized as n elements with weight W_i ($i=1, 2,..., n$). The imposed velocity at the top of the pile is propagated down the pile numerically, as proposed by Smith (1960). Each element of the pile and the soil around the pile can transmit force either statically by deforming some amount (as represented by a stiffness constant), or dynamically by deforming at some rate (as represented by a suitable damping coefficient). The equations for static and dynamic force transmissions are:

$$F_s = K\Delta w \tag{14.48}$$

and

$$F_d = JF_s v \tag{14.49}$$

where F_s=static force, F_d=dynamic force, K=spring constant, Δw=spring compression, J=damping coefficient, and v=velocity. The concept of damping is associated with energy dissipation, while a spring is associated with elasticity. In elastic response, force removal always returns a spring to its original configuration.

Each pile element i is connected to the element below it by a spring with stiffness K_i ($i=1, 2,..., n-1$) and a dashpot with damping constant J_i ($i=1, 2,..., n-1$). Each element is also connected to the soil by a spring K_{si} ($i=1, 2,..., n$) and a damping coefficient J_{si} ($i=1, 2,..., n$). The pile base (last) element is subjected to both shaft and base static and dynamic resistances. As the springs (and dashpots) are in series, it is possible to simply add the shaft and base stiffness constants (and damping constants) to obtain a single K_s and a single J_s for the last pile element.

The internal pile damping is usually very small, so, for simplicity, we will assume the J_i values equal to zero in the remainder of this discussion. Figures 14.5 and 14.6 show, graphically, the relationship (1) between static resistance and relative soil–pile displacement and (2) between force and velocity for the pile–soil interface, both along the shaft and at the base of the pile. The static resistance is elastic up to a value of relative displacement equal to w_q, the quake, and perfectly plastic afterward.[8] The dynamic resistance increases linearly with velocity.

Figure 14.5 shows that soil deforms initially elastically, but then, after the quake is reached, plastically (that is, the deformations are irrecoverable). If we focus on the last pile element, located at the base of the pile, after each hammer blow there remains a certain amount of

---

[8] Perfectly plastic response is that leading to increasing displacement in the face of a constant force.

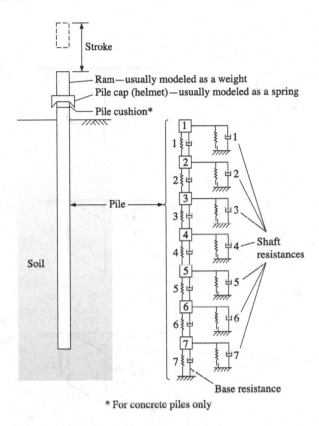

Figure 14.15 Modeling of pile and soil for wave equation analysis.

relative displacement there after unloading. There will also be a permanent displacement at the pile head, called the *pile set*, after unloading. The set is the pile penetration per hammer blow. When the pile meets refusal, the set will approach zero.

Figure 14.16 shows what happens as a static load is applied to a pile element initially at rest (Figure 14.16a) somewhere along the pile shaft (that is, the element is not the lowest pile element and is therefore subject only to a shaft resistance). Initially, there is no slip and the soil undergoes only elastic deformation (Figure 14.16b and c). After the quake w_q is reached, the pile slips with respect to the soil (Figure 14.16d); this continues until unloading starts (Figure 14.16e), at which point the displacement of the pile with respect to the soil (denoted by w_s) will have reached a maximum value $w_{s,max}$. Figure 14.17 shows what happens to the pile element and the soil during the unloading process, where some of the relative slip between pile and soil developed during loading may be reduced. Note that, in the dynamic loading of a pile segment, both the static response depicted in these two figures and the dynamic response discussed earlier develop at the same time. This leads to nonlinear load–deflection curves, as will be illustrated in an example later.

The pile base element responds somewhat differently from other pile elements. The main difference is that, on unloading, suction between the pile and the soil would develop (up to a limit). So the recovery of relative slip in this case would require that the suction be overcome, leading to a gap between the pile base and the soil, something that is not observed under realistic loading conditions. So, an indication of refusal occurs when the pile base element ceases to go down or shows evidence of a possible rebound (that is, recovery of pile-to-soil slip, represented by reduction in w_s from $w_{s,max}$); it can, at this point, be used as a criterion for stopping computations.

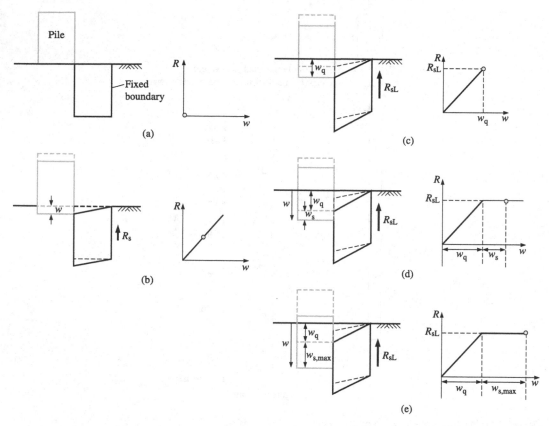

Figure 14.16 Relative motion of the pile with respect to the soil during the loading stage: (a) initial configuration, (b) elastic deformation of the soil, (c) maximum elastic soil deflection at the moment the quake is reached, (d) slip (irrecoverable, plastic) of pile with respect to the soil, and (e) maximum displacement of the pile with respect to the soil reached for the applied load increment (this happens typically as unloading starts).

Let us now examine how an engineer selects the appropriate values of the pile and soil parameters for use in analysis, starting with the pile spring stiffness. From basic strength of materials, we know that the force F required to compress a pile element with length L_e and constant cross section by an amount ΔL_e is given by

$$F = \frac{\left(E_p A_p\right) \Delta L_e}{L_e} \tag{14.50}$$

where $E_p A_p$=pile axial stiffness. Comparing Equations (14.50) with (14.48) and recognizing that $\Delta L_e = \Delta w$ leads to the following expression for the spring stiffness K:

$$K = \frac{E_p A_p}{L_e} \tag{14.51}$$

The critical step of wave equation analysis is the selection of the values of the soil resistance parameters. Usually, the limit soil resistance is established beforehand for the entire pile based on static analysis (such as the analyses we have seen in Chapter 13) or other considerations, so that the value of the limit static resistance would be known for each pile element, including the base element. The values of quake w_q and damping J_s are then selected

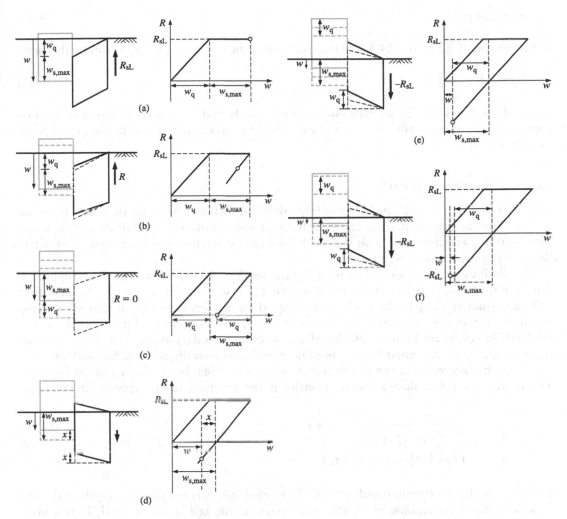

Figure 14.17 Relative motion of the pile with respect to the soil during the unloading stage: (a) initial configuration is the state of maximum relative motion between the pile and the soil reached during loading, (b) elastic unloading of the soil reduces the distortion in the soil, but leads to no slip between the pile and the soil, (c) complete unloading takes the soil element to its unstrained configuration, (d) load reversal starts distorting the soil in the other direction, (e) the quake is reached in the upward direction, and (f) slip of the pile with respect to the soil in the upward direction (thereby reducing some of the slip that developed during loading) takes place.

for each pile element. The values of these parameters are selected in today's practice empirically at best and frequently based on site-dependent correlations. The soil stiffness constant follows from the value of R_{sL} and w_q:

$$K_s = \frac{R_{sL}}{w_q} \tag{14.52}$$

The dynamic resistance can be expressed either in terms of the current, mobilized static resistance or in terms of the limit static resistance. Referring to Equation (14.49), the approach originally proposed by Smith (1960) expressed dynamic resistance in terms of the current static resistance, which would be written as

$$R_d = J_s R_s v \tag{14.53}$$

but a version of Equation (14.53), based on limit static resistance, is more commonly used:

$$R_d = J_s R_{sL} v = d_s v \tag{14.54}$$

where $d_s = J_s R_{sL}$ is the viscous damping constant of the soil. This second approach is more convenient mathematically because it allows for a constant damping coefficient during loading of the soil.

14.4.2.3 Advanced model*

Salgado et al. (2015) developed a set of improved soil shaft and base reaction models for use in one-dimensional pile driving analysis that take soil nonlinearity and hysteresis explicitly into account, are consistent with the mechanics of pile driving, and have input parameters that are physically meaningful.

The shaft reaction model, shown in Figure 14.18, is based on a continuum approach (Holeyman 1985) and has three components: (1) a soil disk representing the near-field soil surrounding the pile shaft; (2) a rheological model representing the thin shear band forming at the soil–pile interface located at the inner boundary of the soil disk; and (3) far-field-consistent boundaries placed at the outer boundary of the soil disk. The continuum approach accommodates a nonlinear soil stress–strain relationship and provides solutions that are valid for the highly transient motion caused by the impact of the hammer. The shear stress versus shear strain relationship in the near-field disk is expressed in rate form:

$$\dot{\tau} = \frac{G_0}{\left(1 + b_f \dfrac{|\tau - L_i \cdot \tau_{rev}|}{(L_i + 1) \times |\mathrm{sgn}(\dot{\gamma}) \cdot \tau_f - \tau|}\right)^2} \dot{\gamma} \tag{14.55}$$

where G_0 is the maximum (small-strain) shear modulus, τ_f is the shear strength of the soil in simple shear conditions, τ_{rev} is the shear stress at the last stress reversal, L_i is a loading index parameter that takes the values 0 for virgin loading (loading starting from zero shear stress) and 1 for subsequent unloading and reloading, and the variable $sgn(\dot{\gamma})$ is the

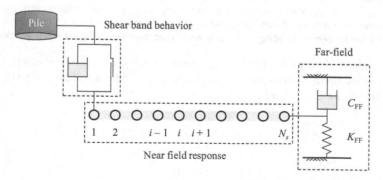

Figure 14.18 Shaft resistance model used in the Salgado et al. (2015) analysis and its three components: a rheological model for the shear band, a continuous near-field model beyond the shear band, and a far-field-consistent spring and radiation dashpot. (After Salgado et al. (2017); with permission from Elsevier Science and Technology Journals.)

sign of the shear strain rate $\dot{\gamma}$. The loading index L_i follows from the second Masing rule (Randolph and Simons 1986); it helps achieve closed and symmetrical stress–strain loops during cycling. The parameter b_f controls the rate of degradation of the shear modulus.

The rheological model was first proposed by Randolph and Simons (1986). It consists of a plastic slider and a viscous dashpot connected in parallel. The strength of the plastic slider is set equal to the static unit limit shaft resistance q_{sL}. If the stress τ_s is less than the unit limit shaft resistance q_{sL}, no sliding occurs, and τ_s is transferred to the pile. Sliding starts once $\tau_s = q_{sL}$. The viscous dashpot is then activated. The reaction of the viscous dashpot is described by a power function of the relative velocity across the shear band. The total (static+viscous) unit resistance of the rheological model at depth z along the pile is expressed as

$$\tau_{sf}(z) = q_{sL}\left(1 + m_s\left(\frac{\dot{w}(z) - \dot{w}_1(z)}{v_{ref}}\right)^{n_s}\right) \tag{14.56}$$

where $\dot{w}(z)$ is the velocity of the pile at depth z, $\dot{w}_1(z)$ is the velocity of the soil at the first node of the disk model used to model pile–soil interaction, m_s and n_s are parameters, and v_{ref} is a reference relative velocity (1 m/s in SI units).

As in Berghe and Holeyman (2002) and Charue (2004), a radiation dashpot with constant damping coefficient c_{FF} – following Novak et al. (1978) – at the outer boundary of the near-field models the loss of energy through wave propagation away from the pile. The damping coefficient c_{FF} is given by

$$c_{FF} = \rho V_s \tag{14.57}$$

where ρ is the soil mass density and V_s is the soil shear wave velocity. The absorbing boundary formulations commonly employed in finite difference or finite element analysis of seismic problems consist only of radiation dashpots. In the pile driving problem, however, we should recover the static solution after all vibrations have stopped. There will be a permanent soil displacement with magnitude decreasing with increasing radial distance from the pile axis that will be consistent with the presence of residual shear stresses on the pile shaft. The static solution is recovered by the addition of a linear spring with stiffness k_{FF} at the outer boundary of the soil disk. The static solution derived by Randolph and Wroth (1978) leads to a spring constant k_{FF} given by

$$k_{FF} = \frac{G_0}{r_f}\frac{1}{\ln(r_m/r_f)} \tag{14.58}$$

where r_f is the radius of the outer boundary of the near-field soil disk. The parameter r_m is the radius of pile influence (the so-called magical radius), seen earlier in Chapter 13. It represents the radial distance from the pile axis at which the displacement is zero (or negligible, for practical purposes) and can be estimated by the following equation (Randolph and Wroth 1978):

$$r_m = 2.5L(1-v) \tag{14.59}$$

where L is the length of the pile and v is the soil Poisson's ratio.

The base reaction model, shown in Figure 14.19, takes into account the nonlinear soil response under the pile base and the loading rate effect on base resistance; it also distinguishes between the different types of damping. The total base reaction is the sum of the spring reaction $R_b^{(S)}$ and the radiation dashpot reaction $R_b^{(D)}$:

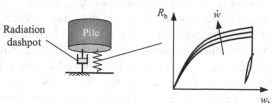

Figure 14.19 Base resistance model: nonlinear spring and dashpot. (After Salgado et al. (2017); with permission from Elsevier Science and Technology Journals.)

$$R_b = R_b^{(S)} + R_b^{(D)}$$

(14.60)

The nonlinear spring incremental load–settlement relationship is described by

$$\dot{R}_b^{(S)} = \frac{K_{b,max}}{\left(1 + b_{fb} \dfrac{\left| R_b - L_{ib} \cdot R_{b,rev} \right|}{(L_{ib}+1) \times \left| sgn(\dot{w}_b) \cdot R_{bL}^* - R_b \right|}\right)^2} \dot{w}_b$$

(14.61)

where $K_{b,max}$ is the maximum (small-strain) base spring stiffness, R_{bL}^* is the limit base capacity including any dynamic resistance, $R_{b,rev}$ is the spring reaction $R_b^{(S)}$ at the last displacement reversal, and L_{ib} is a loading index parameter that takes the values 0 for virgin loading and 1 for subsequent unloading and reloading. The variable $sgn(\dot{w}_b)$ is the sign of the base velocity $\dot{w}_b$, and b_{fb} is a model parameter that controls the rate of degradation of the base spring stiffness. Energy loss is captured through the variables L_{ib}, $sgn(\dot{w}_b)$, and $R_{b,rev}$. $R_b^{(S)}$ does not take negative values (no tension is allowed between the pile base and soil).

The rate dependence of soil strength is included in the model through R_{bL}^*, which is set to be a function of the base velocity:

$$R_{bL}^* = Q_{bL} \left(1 + m_b \left(\frac{\dot{w}_b}{v_{ref}} \right)^{n_b} \right)$$

(14.62)

where m_b and n_b are input parameters. For zero base velocity, R_{bL} is equal to Q_{bL}, which is the limit base capacity under quasi-static loading.

The radiation dashpot reaction $R_b^{(D)}$ in Equation (14.60) is given by the following equation:

$$R_b^{(D)} = \left(c_{Lysm} c_{emb} c_{hys} \right) \dot{w}_{b,el}$$

(14.63)

where c_{Lysm}, c_{emb}, and c_{hys} are the Lysmer dashpot coefficient (Lysmer and Richart 1966), depth factor for the radiation damping, and the hysteretic damping in the far-field soil, respectively, and

$$\dot{w}_{b,el} = \frac{1}{\left(1 + b_{fb} \dfrac{\left| R_b - L_{ib} \cdot R_{b,rev} \right|}{(L_{ib}+1) \times \left| sgn(\dot{w}_b) \cdot R_{bL} - R_b \right|}\right)^2} \dot{w}_b$$

(14.64)

is the elastic component of the pile base velocity, which can be calculated with reference to Equation (14.61). The damping coefficients in Equation (14.63) can be calculated using:

$$c_{\text{Lysm}} = \frac{3.4R^2}{1-v}\sqrt{\rho G_0} = \frac{3.4R^2}{1-v}\rho V_{s,\text{max}} \tag{14.65}$$

$$c_{\text{emb}} = 1.15 + \frac{2\left(\dfrac{D}{B}\right) - 0.15}{\exp\left(\dfrac{D}{B}\right)} \tag{14.66}$$

$$c_{\text{hys}} = \left[1 + \frac{2K_{b,\text{max}} / \left(c_{\text{Lysm}} c_{\text{emb}}\right)}{\omega}\xi\right] \tag{14.67}$$

where $R = B/2$ is the radius of the pile, D is the pile embedment depth, B is the pile diameter, $K_{b,\text{max}}$ is the maximum (small-strain) spring stiffness at the pile base, ω is the fundamental period of the pile–soil system, and ξ is the hysteretic damping ratio.

14.4.3 Modeling of driving system

As seen in Figure 12.32, the driving system components are the hammer itself (composed of the ram and the anvil or striker plate), the hammer cushion (cap block), the helmet (driving head, anvil block, or cap), and, for concrete piles, the pile cushion. Each of these is modeled using springs and masses as shown in Figure 14.15 and Table 14.2. For impact hammers, the driving of the pile through the soil starts with an impact velocity of the ram on the anvil, which is transmitted through the remainder of the driving system components and then to the pile.

The hammer blow for an impact hammer can be modeled by inputting a stroke (fall height) for the hammer and an efficiency. The efficiency is less than one due to internal frictional losses, external frictional losses, and the effects of hammer and driving system geometries on the quality of the impact. According to Pile Dynamics, Inc. (2002), the value of efficiency for single-acting hammers operated by air or steam can be taken as 0.67. The value for hydraulic hammers is 0.80. Diesel hammers are more complicated to model, since pile driving is accomplished not only by ram impact, but also by the pressure generated by the ignition that develops before impact.

Table 14.2 Modeling of driving system components

Component	Model
Ram (piston)	Mass
Anvil (striker plate)	Nonlinear spring
Hammer cushion (cap block)	Nonlinear spring
Driving head (cap, helmet)	Mass
Pile cushion	Spring

14.4.4 Analysis

To understand how a wave equation analysis operates, it is useful to track calculations for three elements during two time increments for the Smith model, which is the simplest. Consider three elements $i-1$, i, and $i+1$ of the pile and times t_{j-1}, t_j, and t_{j+1} such that:

$$\Delta t = t_{j+1} - t_j = t_j - t_{j-1} \tag{14.68}$$

If the stress pulse from the hammer blow is already traveling through the pile, element i will have been displaced already by some value $w_i(t_{j-1})$ and will be moving at a velocity $v_i(t_{j-1})$ at the end of time t_{j-1}. The new displacement at the end of time t_j will be:

$$w_i(t_j) = w_i(t_{j-1}) + v_i(t_{j-1})\Delta t \tag{14.69}$$

Similarly, for elements $i+1$ and $i-1$:

$$w_{i+1}(t_j) = w_{i+1}(t_{j-1}) + v_{i+1}(t_{j-1})\Delta t \tag{14.70}$$

$$w_{i-1}(t_j) = w_{i-1}(t_{j-1}) + v_{i-1}(t_{j-1})\Delta t \tag{14.71}$$

The pile spring i, which expresses the pile stiffness between pile elements i and $i+1$, is compressed by

$$\Delta w_i = w_i(t_j) - w_{i+1}(t_j) \tag{14.72}$$

at time t_j, generating a force

$$F_i = K_i \Delta w_i \tag{14.73}$$

in the pile spring i. Similarly, for spring $(i-1)$ connecting pile element i to pile element $i-1$, the compression can be written as

$$\Delta w_{i-1} = w_{i-1}(t_j) - w_i(t_j) \tag{14.74}$$

at time t_j, generating a force

$$F_{i-1} = K_{i-1}\Delta w_{i-1} \tag{14.75}$$

The acceleration $a_i(t_j)$ acting on element i at the end of time t_j can now be computed using Newton's second law from the resultant force acting on the element i:

$$\frac{a_i}{g} = \frac{F_{i-1} - F_i - R_i}{W_i} \tag{14.76}$$

where g=acceleration of gravity and R_i=total (static plus dynamic) resistance acting on element i.

The acceleration a_i can now be used to calculate the new velocity v_i of element i:

$$v_i(t_j) = v_i(t_{j-1}) + a_i(t_{j-1})\Delta t \tag{14.77}$$

and this new velocity, in turn, can be used to calculate the new displacement $w_i(t_{j+1})$ at time t_{j+1} using Equation (14.69) rewritten for the new time interval:

$$w_i(t_{j+1}) = w_i(t_j) + v_i(t_j)\Delta t \tag{14.78}$$

and the calculation proceeds as before. To calculate the progression of displacement, velocity, and acceleration of a given pile element, we need information about the element and its two neighboring elements for the current time increment and the one just before it.

The total soil resistance that appears in Equation (14.76) can be written as

$$R_i = K_{si}(w_i - w_{si}) + J_{si}R_{sLi}v_i \tag{14.79}$$

and the dynamic resistance is written as in (14.54). Consider the case in which there is no dynamic resistance. Equation (14.79) then reduces to:

$$R_i = K_{si}(w_i - w_{si}) \tag{14.80}$$

Consider that any displacement of the pile with respect to the surrounding soil in excess of the quake becomes an irrecoverable, plastic displacement of the pile with respect to the soil. Consider the case of loading from zero deflection between the pile and the surrounding soil. According to Equation (14.80), R_i is simply the elastic resistance $K_{si}w_{si}$ until w becomes equal to w_q. From that point on, the displacement w_{si} increases at the same rate as w_i, so relative pile–soil displacement occurs, but there is no further change in R_i, which becomes constant and equal to $R_{sLi} = K_{si}w_q$.

More generally, the value of w_s at the outset of the calculations is simply the accumulated value of the irrecoverable displacement of the pile element up to that point. This can best be visualized by referring to Figure 14.5, which shows that, when the load is completely removed, the component of the displacement that is in excess of the quake is not recovered. Mathematically, this imposes the following constraint on w_s:

$$w_s \geq w - w_q \tag{14.81}$$

The inequality in Equation (14.81) states that, on first loading, any relative displacements between pile and soil develop only after the quake is surpassed, and, at that point, w_s is equal to the difference between w and w_q. If there is then unloading, the relative displacement will not change until the quake is mobilized in the opposite direction, which means that w will now decrease, so $w - w_q$ will decrease, but w_s will remain unchanged. Thus, w_s is now greater than $w - w_q$.

Upon complete removal of the load, the settlement w will be equal to w_s. If a load in the other direction (up the pile) is applied, the relative settlement w_s will remain the same only up the point when there has been a displacement w_q in the other direction. At that point, $w = w_s - w_q$ (or $w_s = w + w_q$). If the element moves beyond that point, w_s will decrease; that is, it will become less than $w + w_q$. Based on these considerations, for unloading, the following condition exists on w_s:

$$w_s \leq w + w_q \tag{14.82}$$

For the pile base, where no rebound takes place, only condition Equation (14.81) needs to be enforced. For shaft elements, the conditions expressed in inequalities (14.81) and (14.82) are checked for every time increment and w_s is updated (increased on loading after the quake is reached in the loading direction or decreased on unloading after the quake is reached in the unloading direction). For the base element, w_s is increased on loading after the quake is reached in the loading direction. If the base segment of the pile shows signs of rebound, then the calculation is terminated, because we do not wish to allow the pile to "remove itself" from the ground, something that clearly is not realistic.

In simple terms, as we can see, wave equation analysis is the repetition of relatively straightforward calculations for every pile element. Calculations start at the top of the pile and progress downward. It will take time for the effects of the hammer blow to reach the lower elements of the piles, so calculations initially involve only the top pile elements, and then more elements until all elements have experienced the effects of the hammer blow. Dissipation of energy through soil damping should then lead to reduction in the effects for all elements. Calculations terminate when there is no more penetration of the pile into the ground.

A word about the time interval Δt is in order. The pulse created by the hammer blow travels down the pile with speed c, as seen earlier. The time it takes for it to travel through a pile element is $\Delta t_{crit} = L_e / c$, where L_e is the pile element length. Obviously, Δt must be selected to be no greater than Δt_{crit} or else the analysis will miss completely the effects of the pulse on any individual element.

14.4.5 Results of the analysis

For each hammer blow, the energy transferred from the hammer to the pile via an imposed velocity travels down through the pile, generating displacements, velocities, accelerations, and forces. By keeping track of the maximum and minimum forces experienced by every pile element for each hammer blow, it is possible to obtain an estimate of the maximum compressive stresses developed in the pile during driving, as well as the appearance of any tensile stress. This is useful because the development of tensile stresses or excessively large compressive stresses may lead to pile damage and thus indicate that the driving system is not adequate for driving the selected pile through the soil profile at hand.

For each hammer blow, a final set is calculated. The inverse of the set gives the number of blows required to drive the pile a unit length through the soil (say, 1 in. or 1 cm). A useful analysis involves considering the pile to have penetrated the soil down to some depth. For the pile at this depth, a range of credible values of static resistances on the shaft and base are considered, for each of which a dynamic analysis is performed. From the dynamic analysis, we have either a dynamic resistance in blows per unit length of penetration, or a set (penetration per blow). It is then possible to obtain, for that depth, a plot of static capacity versus dynamic resistance, which is known as a bearing graph. This is useful to an inspector attempting to determine whether a pile has been driven to the required capacity for a given depth based on the observation of its response during driving.

A drivability analysis consists in setting unit shaft and base resistances for the various soil layers the pile will be driven through, and then considering several different depths for the pile on the way to its final penetration depth. At each of these depths, a bearing graph is obtained.

Example 14.2

Consider a closed-ended pipe pile that is driven in soil using an ICE 42S single-acting diesel hammer. The ram weight is 18.2 kN. The maximum hammer stroke is 3.13 m, and the maximum driving energy is 56.8 kN·m. The pile has an outer diameter of 0.356 m, thickness of 0.0127 m, and length of 8.24 m. It is driven into the ground to a depth of 6.87 m.

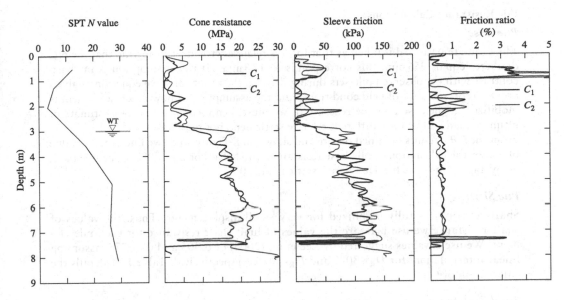

E-Figure 14.2 Site investigation results. (Modified after Kim et al. 2002.)

The efficiency of the hammer is 67%. The soil at the site is predominantly gravelly sand. The first 3 m of the deposit is in a loose state (D_R=30%), and the rest of the deposit is dense to very dense (D_R=80%), as illustrated by the results of CPT and SPT performed before driving, shown in E-Figure 14.2. Analyze the response of the pile to a single blow of the hammer.

As discussed in Chapter 13, $c_b=q_b/q_c$. For D_R=80% at the pile base, an average value of c_b is 0.35 for a relative settlement w_R=5% (Lee and Salgado 1999). According to Lee et al. (2003), $c_s=q_{sL}/q_c$=0.005 and 0.00775 for D_R=30% and D_R=80%, respectively.

Hammer properties for the ICE 42S hammer, according to manufacturer specifications, are as follows:

Efficiency=67%
Ram weight=18.2 kN
Fall height=3.13 m (maximum)
Hammer width=510 mm
Hammer depth=740 mm
Ram area=0.3774 m²
Ram length=4.9 m

Solution:

Static resistance calculations

Pile Dimensions

$$A_p = \frac{\pi \times 0.356^2}{4} - \frac{\pi(0.356 - 0.0127 \times 2)^2}{4} = 0.0137\,\text{m}^2$$

L=8.24 m

Pile Resistance Calculations

Pile Base

The ultimate base load is taken as that for 5% relative settlement. Knowing that the pile diameter equals 356 mm, this corresponds to 17.8 mm.[9] If this assumption is in sharp contrast with the observed pile sets during driving, the results of wave equation analysis are not representative of field conditions, and the assumption of how much settlement is mobilized at the base must be reassessed; we must then recognize that our estimate of ultimate load will be for a different relative settlement, not for 5%.

From the CPT results obtained on site and shown in E-Figure 14.2, we find that the value of q_{cb} for the calculation of pile base resistance is 15 MPa. For $D_R = 80\%$ at the pile base, an average value of c_b for 5% relative settlement is 0.35.

Pile Shaft

Shaft resistance is fully mobilized for very small displacements. Thus, the values of shaft resistance we use here are the values of limit shaft resistance for each pile element. We use c_s values suggested by Lee et al. (2003) of 0.005 and 0.00775 (assuming linear interpolation) for $D_R = 30\%$ and $D_R = 80\%$, respectively. E-Table 14.1 details the calculation of Q_{sL}.

Total Resistance

The resistances are:

$$Q_{b,ult(w_R=5\%)} = c_b q_{cb} A_b$$

$$Q_{sL} = \sum_{i=1}^{n} q_{sLi} A_{s_i}$$

with:

$$q_{b,ult(w_R=5\%)} = c_b q_{cb} = 0.35 \times 15 = 5.25 \text{ MPa}$$

E-Table 14.1 Details of calculation of Q_{sL}

z (m)	q_{ci} (MPa)	q_{sLi} (MPa)	Q_{sL} (MN)
0–1	2.5	0.0125	0.014
1–2	2.5	0.0125	0.014
2–3	4.5	0.0225	0.025
3–4	12	0.093	0.104
4–5	16	0.124	0.139
5–6	20	0.155	0.174
6–7	22	0.171	0.192
		Total Q_{sL}	0.662

[9] It is important to understand how we do this. We in essence make an assumption regarding the maximum base displacement that the hammer blow will be able to generate. If, in doing the analysis, we find that the assumption is not acceptable, then we recalculate the pile base resistance. This example is a real case (see Paik et al. (2003) for details), and the permanent displacement observed per blow of the pile varied between 9 and 15 mm. This information, of course, would not be available before pile driving.

$$A_{si} = 2\pi r h_i = 2\pi \times \frac{0.356}{2} \times 1 = 1.12 \text{ m}^2$$

$$A_b = \frac{\pi \times 0.356^2}{4} = 0.0995 \text{ m}^2$$

This gives us the following:

$$Q_{b,\text{ult}(w_R=5\%)} = 5.25 \times 0.0995 = 0.522 \text{ MN}$$

$$Q_{sL} = 0.662 \text{ MN}$$

$$Q_{\text{ult}} = 0.522 + 0.662 = 1.184 \text{ MN}$$

Pile and driving system model

We now divide the pile into N segments and consider separately the section that has no soil resistance (above ground level). If we take $N=8$ segments, we obtain:

$$dL_1 = 1.37 \text{ m}$$

$$dL_2 = dL_3 = dL_4 = dL_5 = dL_6 = dL_7 = 1 \text{ m}$$

$$dL_8 - 0.87 \text{ m}$$

Steel modulus: 210 GPa.
Steel mass density: 7900 kg/m³.
Mass of ram m_r and of pile segments m_i:

$$m_r = \frac{W}{g} = \frac{18.2 \times 10^3}{9.81} = 1856 \text{ kg}$$

$$m_1 - A dL_1 \rho = 0.0137 \times 1.37 \times 7900 = 148 \text{ kg}$$

$$m_2 = \ldots = m_7 = 0.0137 \times 1 \times 7900 = 108 \text{ kg}$$

$$m_8 = 0.0137 \times 0.87 \times 7900 = 94.2 \text{ kg}$$

Spring stiffness of segments:

$$K_1 = \frac{E_p A_p}{dL_1} = \frac{210 \times 10^9 \times 0.0137}{1.37} = 2.1 \times 10^9 \text{ N/m} = 2100 \text{ kN/mm}$$

$$K_2 = \ldots = K_7 = \frac{E_p A_p}{dL_{2\ldots7}} = \frac{210 \times 10^9 \times 0.0137}{1} = 2.877 \times 10^9 \text{ N/m} = 2877 \text{ kN/mm}$$

$$K_8 = \frac{E_p A_p}{dL_8} = \frac{210 \times 10^9 \times 0.0137}{0.87} = 3.307 \times 10^9 \text{ N/m} = 3307 \text{ kN/mm}$$

In the modeling of the pile and driving system, we consider a single spring to model the combined stiffness of the hammer and first pile segment. From the specifications for the ICE 42S single-acting diesel hammer, we know that the width of the hammer is 510 mm and its depth is 740 mm. Thus, the area of the ram is equal to $0.3774\,m^2$. The length of the ram is also given; it is equal to 4.9 m. Therefore, the stiffness of the ram is

$$K_{ram} = \frac{E_{ram} A_{ram}}{L_{ram}} = \frac{210 \times 10^9 \times 0.3774}{4.9} = 1.617 \times 10^{10}\ N/m\ =\ 16{,}170\ kN/mm$$

The ram and pile helmet are located on top of the pile at the time of collision. We will represent the stiffness of the ram and pile helmet together as a "ram spring." According to the spring–dashpot model shown in Figure 14.11, we see that the springs of the ram and the first pile segment are in series. We need to find the equivalent spring constant for these two springs using Kirchhoff's law:

$$\frac{1}{K_{eq}} = \frac{1}{K_{ram}} + \frac{1}{K_1} = 5.38 \times 10^{-4}\ mm/kN$$

$$K_{eq} = 1859\ kN/mm$$

The wave speed in steel is:

$$c = \sqrt{\frac{E_p}{\rho}} = \sqrt{\frac{210 \times 10^9}{7900}} = 5155\ m/s$$

The critical time is found using the shortest pile segment:

$$\frac{dL}{c} = \frac{0.87}{5155} = 0.000168\ s = 0.168\ ms$$

We make the time increment equal to one half the critical time:

$$dt = \frac{1}{2}\frac{dL}{c} = 0.0844\ ms$$

We now choose the quake values for shaft and base resistances, which should be different. For shaft resistance, we take the quake $w_q = 2.5$ mm, and for base resistance, we take $w_q = 3.0$ mm. These numbers are empirical; experience with wave equation analysis suggests these numbers typically produce acceptable results. We take the damping constant equal to $J_s = 0.2$ s/m for shaft resistance damping and $J_b = 0.5$ s/m for base resistance damping. The resistances for each segment are:

Pile Segment 1: no soil resistance $\Rightarrow R_{s1} = 0$.

Pile Segment 2 to 7: resistances are equal to the values for each layer as calculated above.

Pile Segment 8: neglect shaft resistance; assume it subject to only base resistance

$\Rightarrow R_{s8} = Q_{b,ult(s/B=5\%)} = 522\ kN$

Computations for first time increment: $\Delta t = 0.0844$ ms

Ram Impact velocity

$$v_{ram}(t_0) = \sqrt{2gh \times e_h} = \sqrt{2 \times 9.81 \times 3.13 \times 0.67} = 6.414\ m/s$$

Predicted Displacement

The ram has an initial velocity because of the conversion of part of the potential energy from the fall height to kinetic energy. At that velocity, in the time increment $\Delta t = 0.084$ ms, it undergoes the following displacement:

$$w_{ram}(t_1) = v_{ram}(t_0)\Delta t = 6.414 \times 0.0000844 = 5.413 \times 10^{-4} \text{ m}$$

The pile segments have not yet been affected by the hammer blow, so their displacements are all zero: $w_1(t_1) = w_2(t_1) = \ldots = w_8(t_1) = 0$.

Spring Compression and Forces

The compression of a pile segment is better visualized with respect to Figure 14.15 as compression of the spring between the segment and the preceding segment. Recall that the stiffness of Pile Segment 1 is combined with that of the ram, so the compression for Pile Segment 1 is the difference between the displacement of the ram and the displacement of Segment 1:

$$\Delta w_1 = w_{ram}(t_1) - w_1(t_1) = 5.413 \times 10^{-4} \text{ m}$$

Hence, the force on Segment 1 is:

$$F_1 = K_{eq} \times \Delta w_1 = 1859 \times 5.413 \times 10^{-1} = 1006 \text{ kN}$$

For Pile Segments 2–8:

$$\Delta w_2 = \Delta w_3 = \Delta w_4 = \Delta w_5 = \Delta w_6 = \Delta w_7 = \Delta w_8 = 0$$
$$F_2 = F_3 = F_4 = F_5 = F_6 = F_7 = F_8 = 0$$

Relative Displacement of the Pile with Respect to the Soil

Plastic deflections occur when the displacement of the pile segments with respect to the soil is greater than the corresponding quake. The basic condition is that the displacement at any time t for the pile segment should be greater than the quake in order for plastic deflections to occur.

According to Smith (1960), the relative deflection of any pile segment with respect to the soil remains constant (starting at zero) unless one of the conditions in Equations (14.81) and (14.82) kicks in. These can be restated as follows:

1. On loading, w_{si} cannot be less than $w_i - w_q$.
2. On unloading, w_{si} cannot be less than $w_i + w_q$.

As for the base, the permanent set also remains constant (starting at zero) unless changed by Equation (14.81); that is, w_{sb} cannot be less than $w_b - w_q$. The analysis should make these comparisons for each time interval and adjust the values of the plastic deflection of each segment accordingly. For this time step, all plastic deflections are still zero.

Shaft and Base Resistances

$$R_i = K_{si}(w_i - w_{si}) + J_{si}R_{sLi}v_i$$

where:

$$K_{si} = \frac{R_{sLi}}{w_q}$$

No resistance has developed for any pile segment at this time since the displacements of all segments are still zero.

Newton's Second Law
Calculation of force:
We will take

$$F_{Ri} = \frac{W_i a_i}{g} = F_{i-1} - F_i - R_i$$

Note that

$$F_{i-1} = F_i^{top} \quad and \quad F_i = F_i^{bottom}$$

$$F_{Ri} = F_i^{top} + W_i - F_i^{bottom} - R_i$$

$$F_{R,ram} = F_{ram}^{top} + W_{ram} - F_{ram}^{bottom} = 0 + 18.2 - 1006 \approx -988 \text{ kN}$$

$$F_{R1} = F_1^{top} + W_1 - F_1^{bottom} - R_1 = 1006 + 1.455 - 0 - 0 \approx 1007 \text{ kN}$$

$$F_{R2} = F_2^{top} + W_2 - F_2^{bottom} - R_2$$

$$= 0 + 1.0617 - 0 - 0 = 1.0617 \text{ kN} \approx 1.1 \text{ kN} = F_{R3} = ... = F_{R7}$$

$$F_{R8} = F_8^{top} + W_8 - R_8 = 0.924 \text{ kN} \approx 1 \text{ kN}$$

Calculation of velocities:

$$v_i(t_j) = v_i(t_{j-1}) + a_i(t_{j-1})\Delta t = v_i(t_{j-1}) + F_{Ri}\frac{g}{W_i}\Delta t$$

$$v_{ram}(t_1) = v_{ram}(t_0) + F_{R,ram}\frac{g}{W_{ram}}\Delta t$$

$$= 6.414 + (-988.08) \times \frac{9.806}{18.2} \times 0.0000844 = 6.369 \text{ m/s}$$

$$v_1(t_1) = v_1(t_0) + F_{R1}\frac{g}{W_1}\Delta t = 0 + 1007.74 \times \frac{9.806}{1.455} \times 0.0000844 = 0.573 \text{ m/s}$$

$$v_2(t_1) = v_2(t_0) + F_{R2}\frac{g}{W_2}\Delta t$$

$$= 0 + 1.0617 \times \frac{9.806}{1.0617} \times 0.0000844 = 8.28 \times 10^{-4} \text{ m/s} = ... v_8(t_1)$$

A comment is in order about the inclusion of the pile weight in the calculations. Some soil resistance (typically just a small fraction of total resistance) is required to support the pile, and we wish our calculations to account for that, even if in problems of practical interest the inclusion of the pile weight makes only a small difference.

<u>Calculations for second time increment:</u> $\Delta t = 0.0844$ ms

Starting with the calculated velocities at time t_1, we follow the same procedure as above to calculate the new displacements and forces in the segments as the stroke propagates downward. Those velocities are:

$$v_{ram}(t_1) = 6.369 \text{ m/s}$$

$$v_1(t_1) = 0.573 \text{ m/s}$$

$$v_2(t_1) = 8.28 \times 10^{-4} \text{ m/s} = ... = v_8(t_1)$$

Predicted Displacements

Ram

$$w_{ram}(t_2) = w_{ram}(t_1) + v_{ram}(t_1)\Delta t = 5.413 \times 10^{-4} + 6.369 \times 0.0000844 = 1.079 \times 10^{-3} \text{ m}$$

Pile Segments

$$w_1(t_2) = w_1(t_1) + v_1(t_1)\Delta t = 0 + 0.573 \times 0.0000844 = 4.836 \times 10^{-5} \text{ m}$$

$$w_2(t_2) = w_2(t_1) + v_2(t_1)\Delta t = 0 + 8.28 \times 10^{-4} \times 0.0000844 = 6.988 \times 10^{-8} \text{ m} = ... = w_8(t_2)$$

Spring Compression and Forces

Compression of Pile Segments

$$\Delta w_1 = w_{ram}(t_2) - w_1(t_2) = 1.079 \times 10^{-3} - 4.836 \times 10^{-5} = 1.031 \times 10^{-3} \text{ m}$$

$$\Delta w_2 = w_1(t_2) - w_2(t_2) = 4.836 \times 10^{-5} - 6.988 \times 10^{-8} = 4.829 \times 10^{-5} \text{ m}$$

There is no compression for deeper pile segments.

Forces in Segments

$$F_1 = K_{eq} \times \Delta w_1 = 1859 \times 1.031 \times 10^{-3} = 1917 \text{ kN}$$

$$F_2 = K_2 \times \Delta w_2 = 2877 \times 4.829 \times 10^{-5} = 139 \text{ kN}$$

$$F_3 = K_3 \times \Delta w_3 = 0 = ... = F_8$$

Relative Pile–Soil Deflections

The displacements of all pile segments are still less than the quake; therefore, all plastic deflections are still zero.

Shaft and Base Resistances

$$R_i = K_{si}(w_i - w_{si}) + J_{si}R_{sLi}v_i$$

where:

$$K_{si} = \frac{R_{sLi}}{w_q}$$

$$R_2 = K_{s2}(w_2 - w_{s2}) + J_{s2}R_{sL2}v_2$$

$$\Rightarrow R_2 = \frac{14}{0.00254} \times \left(6.988 \times 10^{-8} - 0\right) + 0.2 \times 14 \times 8.28 \times 10^{-4}$$

$$\Rightarrow R_2 = 2.704 \times 10^{-3} \text{ kN} \approx 0 \text{ kN}$$

Similar calculations for the shaft resistances of other pile segments are done using the limit resistance values R_{sL} calculated in E-Table 14.1. The results are:

$$R_3 = 2.704 \times 10^{-3} \text{ kN}$$

$$R_4 = 4.828 \times 10^{-3} \text{ kN}$$

$$R_5 = 2.008 \times 10^{-2} \text{ kN}$$

$$R_6 = 2.684 \times 10^{-2} \text{ kN}$$

$$R_7 = 3.708 \times 10^{-2} \text{ kN}$$

$$R_8 = 2.283 \times 10^{-1} \text{ kN}$$

which are all, for practical purposes, equal to zero.

Newton's Second Law

Calculation of force:

$$F_{R,ram} = F_{ram}^{top} + W_{ram} - F_{ram}^{bottom} = 0 + 18.2 - 1916.63 \approx -1899 \text{ kN}$$

$$F_{R1} = F_1^{top} + W_1 - F_1^{bottom} - R_1 = 1916.63 + 1.455 - 138.93 - 0 \approx 1779 \text{ kN}$$

$$F_{R2} = F_2^{top} + W_2 - F_2^{bottom} - R_2 = 138.93 + 1.0617 - 0 - 2.704 \times 10^{-3} \approx 140 \text{ kN}$$

Similarly,

$$F_{R3} = 1.059 \text{ kN}$$

$$F_{R4} = 1.057 \text{ kN}$$

$$F_{R5} = 1.042 \text{ kN}$$

$$F_{R6} = 1.035 \text{ kN}$$

$$F_{R7} = 1.025 \text{ kN}$$

$$F_{R8} = 0.696 \text{ kN}$$

which are all approximately 1 kN (corresponding approximately to the pile weight and essentially zero for practical purposes).

Calculation of velocities:

$$v_{ram}(t_2) = v_{ram}(t_1) + F_{R,ram}\frac{g}{W_{ram}}\Delta t = 6.283\,\text{m/s}$$

$$v_1(t_2) = 1.585\,\text{m/s}$$

$$v_2(t_2) = 0.110\,\text{m/s}$$

$$v_3(t_2) = 1.654 \times 10^{-3}\,\text{m/s}$$

$$v_4(t_2) = 1.652 \times 10^{-3}\,\text{m/s}$$

$$v_5(t_2) = 1.640 \times 10^{-3}\,\text{m/s}$$

$$v_6(t_2) = 1.635 \times 10^{-3}\,\text{m/s}$$

$$v_7(t_2) = 1.627 \times 10^{-3}\,\text{m/s}$$

$$v_8(t_2) = 1.452 \times 10^{-3}\,\text{m/s}$$

Calculations for third time increment: $\Delta t = 0.0844$ ms

Starting with the calculated velocities at time t_2, we follow the same procedure as before to calculate the new displacements and forces in the segments as the stroke propagates downward. We do this for subsequent time increments as well. The typical resistance versus displacement response is shown for the base segment in E-Figure 14.3 and for a shaft segment in E-Figure 14.4.

The maximum set for this analysis is 13.3 mm for a stroke height of 3.13 m. This falls within the range of permanent set observed during driving. The maximum base displacement is 16.3 mm, just shy of the 17.8 mm that would correspond to 5% relative settlement, so the value of c_b used in calculations is satisfactory.

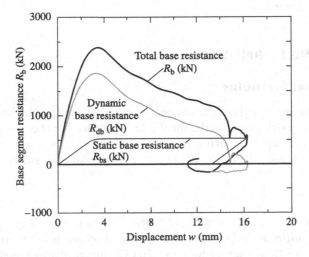

E-Figure 14.3 Base resistance versus displacement.

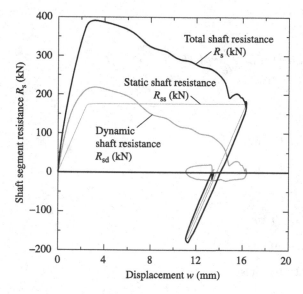

E-Figure 14.4 Shaft resistance versus displacement for Segment 7.

14.4.6 Importance of choice of soil resistance models

As discussed extensively in Chapter 13, pile resistance is determined most of the time based on pile settlement. If pile resistance is to be inferred from a wave equation analysis, then the static components of the pile resistance models for shaft and base need to be closely related to what these resistances truly will be. In other words, if we performed a static load test on the pile, we would obtain a load–resistance curve that resembles the one we would get using the assumed soil resistance models.

By this criterion, the Smith model falls short. The linear elastic, perfectly plastic responses assumed for the shaft and base resistances are not realistic. The values of quakes assumed are usually small and bear no relation to the values of settlement usually required to achieve ultimate limit states, particularly for large-diameter piles, which require considerable displacement to fully mobilize the plunging resistance, a requirement for the pile to move down as a result of the hammer blow.

14.5 PILE DRIVING FORMULAS

14.5.1 Traditional formulas

Pile driving formulas are based on the concept that the energy of the hammer (ram) before impact is equal to the work done by the total pile resistance for the observed pile set after a blow plus the energy dissipated during the blow and the energy lost at impact between the hammer and the pile head. The following equation expresses this mathematically:

$$e_h W_H H = Q_{ult} \left(w_s + w_{sc} \right) \tag{14.83}$$

where W_H is the hammer (ram) weight, H is the hammer drop height, e_h is the hammer efficiency, Q_{ult} is the ultimate pile capacity, w_s is the observed pile set, and w_{sc} is an empirical constant (called the "lost" set (Allin et al. 2015)) that captures the effects of the energy losses discussed earlier and the energy stored temporarily inside the pile due to elastic compression during the blow. We solve the equation for Q_{ult} for a given pile set w_s.

In the absence of better information, the following values can be assumed for the hammer efficiency e_h:

- 0.38 and 0.25 for diesel hammers acting on steel and concrete piles, respectively (Rausche 2000);
- 0.54 and 0.40 for single-acting air/steam hammers on steel and concrete piles, respectively (Rausche 2000); and
- 0.55 for drop hammers acting on either steel or concrete piles (Lam 2007; Lim and Broms 1990; Mostafa 2011; Allen 2005).

Table 14.3 Traditional pile driving formulas

Formula	Equations[a]	Notes
Gates formula (Gates 1957)	$Q_u = a\sqrt{e_h E_h}\left(b - \log(w_s)\right)$	w_s in mm, and E_h in kJ $a = 104.5$ $b = 2.4$
Modified ENR (ENR 1965)	$Q_u = \left(\dfrac{1.25 e_h F_h}{w_s + C}\right)\left(\dfrac{W_H + n^2 W_P}{W_H + W_P}\right)$	$C = 0.0025\,\mathrm{m}$ $n = 0.5$ for steel-on-steel anvil on steel or concrete piles
Danish formula (Olson and Flaate 1967)	$Q_u = \dfrac{e_h E_h}{w_s + C_1}$ $C_1 = \sqrt{\dfrac{e_h E_h L}{2 A_p E_p}}$	
Pacific Coast Uniform Building Code (PCUBC) formula (Bowles 1996)[b]	$Q_u = \dfrac{e_h E_h C_1}{w_s + C_2}$ $C_1 = \dfrac{W_H + k W_P}{W_H + W_P}$ $C_2 = \dfrac{Q_u L}{A_p E_p}$	$k = 0.25$ for steel piles and 0.1 for all other piles
Janbu (Bowles 1996)	$Q_u = \dfrac{e_h E_h}{K_u w_s}$ $C_d = 0.75 + 0.15\dfrac{W_P}{W_H}$ $K_u = C_d\left(1 + \sqrt{1 + \dfrac{\lambda}{C_d}}\right)$ $\lambda = \dfrac{e_h E_h L}{A_p E_p w_s^2}$	

[a] Q_u, the predicted pile capacity; e_h, the hammer efficiency; E_h, the maximum driving energy of the hammer; w_s, the observed pile set; W_H, the weight of the ram; W_P, the weight of the pile; L, the length of the pile; A_p, the cross-sectional area of the pile; E_p, the Young's modulus of the pile material.

[b] Calculations using the PCUBC formula are iterative.

The simplicity of these formulas and limited quality-assurance and quality-control budgets has led to their significant use in practice. Pile driving formulas include the Gates formula (Gates 1957), the modified ENR formula (ENR 1965), the Danish formula (Olson and Flaate 1967), the Janbu formula (Olson and Flaate 1967), the Pacific Coast Uniform Building Code (PCUBC) formula (Bowles 1996), and the Canadian National Building Code formula (Bowles 1996).

Table 14.3 lists five formulas that Salgado et al. (2017) have evaluated, together with a more recent set of formulas discussed next.

14.5.2 Modern formulas

The static capacity equations of the Purdue Pile Design Method (Chapter 13) and the one-dimensional dynamic pile driving analysis of Salgado et al. (2015) discussed in Section 14.4 can be used to produce pairs of values of total static pile capacity Q_{ult} and final pile set s for a wide range of values for each variable involved in the pile driving problem. A formula can then be fit to these pairs. This is in essence what Salgado et al. (2017) did for two generic soil profiles involving sand and clay. These profiles and the pile positioned in them are shown in Figure 14.20, and the values considered for the soils in the profiles are provided in Table 14.4.

The pile driving formulas are expressed as products of power functions of dimensionless variables and parameters c_1–c_5. For each of the five cases, they have the following forms:

$$\frac{Q_{10\%}}{p_A L_R^2} = c_1 \left(\frac{e_h E_h}{W_R L_R} \right)^{c_2} \exp\left(c_3 \frac{D_R}{100\%} \right) \left(\frac{w_s}{L_R} \right)^{c_4} \left(\frac{W_P}{W_R} \right)^{c_5} \tag{14.84}$$

for piles in a uniform sand layer, piles crossing a normally consolidated clay layer and resting on a dense sand layer, and end-bearing piles in sand; and

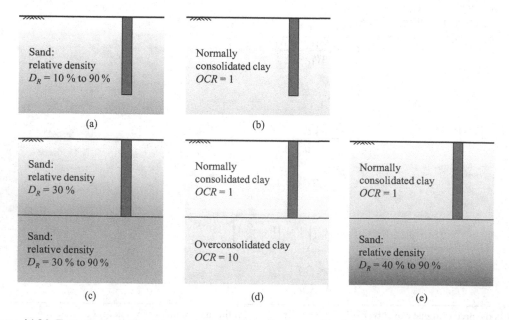

Figure 14.20 Typical pile–soil profile systems found in pile design: (a) piles in sand of uniform density, (b) a floating pile in clay, (c) an end-bearing pile in sand, (d) an end-bearing pile in clay, and (e) a pile crossing clay resting on sand.

Table 14.4 Soil parameters for typical soil profiles

Case	Shaft parameter	Values of shaft parameter	Base parameter	Values of base parameter
Piles in sand of uniform density	Relative density D_R (%)	10, 20, 30, 40, 50, 60, 70, 80 and 90	Relative density D_R (%)	Same as for shaft
End-bearing in sand	Relative density D_R (%)	30	Relative density D_R (%)	40, 50, 60, 70, 80, and 90
Floating piles in clay	Overconsolidation ratio OCR	1	Overconsolidation ratio OCR	1
	s_u/σ_v'	0.2, 0.23, 0.25, 0.28, and 0.3	s_u/σ_v'	Same as for shaft
End-bearing piles in clay	Overconsolidation ratio OCR	1	Overconsolidation ratio OCR	10
	s_u/σ_v'	0.2, 0.23, 0.25, 0.28, and 0.3	$(s_u/\sigma_v')_{NC}$	Same as for shaft
Piles crossing clay resting on sand	Overconsolidation ratio OCR	1	Relative density D_R (%)	40, 50, 60, 70, 80, and 90
	s_u/σ_v'	0.25		

$$\frac{Q_L}{p_A L_R^2} = c_1 \left(\frac{e_h E_h}{W_R L_R} \right)^{c_2} \exp\left(c_3 \frac{s_u}{\sigma_v'} \right) \left(\frac{w_s}{L_R} \right)^{c_4} \left(\frac{W_P}{W_R} \right)^{c_5} \qquad (14.85)$$

for floating piles in clay and end-bearing piles in clay. In all these equations, $Q_{10\%}$ is the ultimate load defined as the static load causing a pile to settle by 10% of the pile diameter, and Q_L is the limit load.

The proposed formulas can be rewritten in the general form of the pile driving formula proposed by Janbu (listed in Table 14.3):

$$Q_{10\%} = \frac{e_h L_h}{\left[\dfrac{W_R L_R}{p_A L_R^3} \dfrac{1}{c_1} \left(\dfrac{e_h E_h}{W_R L_R} \right)^{1-c_2} \exp\left(-c_3 \dfrac{D_R}{100} \right) \left(\dfrac{w_s}{L_R} \right)^{-c_4-1} \left(\dfrac{W_P}{W_R} \right)^{-c_5} \right] \times w_s} = \frac{e_h L_h}{A \times w_s} \qquad (14.86)$$

and

$$Q_L = \frac{e_h E_h}{\left[\dfrac{W_R L_R}{p_A L_R^3} \dfrac{1}{c_1} \left(\dfrac{e_h E_h}{W_R L_R} \right)^{1-c_2} \exp\left(-c_3 \dfrac{s_u}{\sigma_v'} \right) \left(\dfrac{w_s}{L_R} \right)^{-c_4-1} \left(\dfrac{W_P}{W_R} \right)^{-c_5} \right] \times w_s} = \frac{e_h E_h}{B \times w_s} \qquad (14.87)$$

where A and B are compact notations for the term multiplying the set w_s in the denominator of these expressions.

These formulas contain, as variables, soil properties: D_R for sands and s_u for clays. The values for these variables are arrived at by considering the results of field investigations, which normally precede pile installation. The values of these coefficients for each typical soil profile are listed in Table 14.5 for closed-ended steel pipe piles and Table 14.6 for concrete piles.

Table 14.5 Values of parameters for closed-ended steel pipe piles

Soil profile	c_1	c_2	c_3	c_4	c_5	R^2
Piles in sand of uniform density	14.97	0.33	1.04	−0.41	0.89	0.91
End-bearing piles in sand	8.11	0.41	0.74	−0.53	0.69	0.87
Floating piles in clay	0.72	0.71	1.30	−0.76	0.30	0.97
End-bearing piles in clay	1.57	0.59	1.51	−0.67	0.46	0.96
Piles crossing clay and resting on sand	5.77	0.36	0.12	−0.52	0.71	0.95

After Salgado et al. (2017).

Table 14.6 Values of parameters for precast concrete piles

Soil profile	c_1	c_2	c_3	c_4	c_5	R^2
Piles in sand of uniform density	19.97	0.07	1.73	−0.07	0.79	0.99
End-bearing piles in sand	22.05	0.08	1.13	−0.08	0.76	0.99
Floating piles in clay	0.96	0.69	1.82	−0.56	0.28	0.91
End-bearing piles in clay	3.01	0.24	3.10	−0.19	0.75	0.96
Piles crossing clay and resting on sand	9.50	0.08	1.56	−0.09	0.77	0.99

After Salgado et al. (2017).

Example 14.3

A soil profile consists of loose gravelly sand with $D_R=30\%$ down to 3 m, followed by dense gravelly sand with $D_R=80\%$. The groundwater table is 3 m below the ground surface. A closed-ended steel pipe pile, 8.249 m long, with an outer diameter equal to 356 mm and a wall thickness of 12.7 mm is driven by an ICE 42S single-acting diesel hammer down to 6.87 m depth. The ram weight is 18.2 kN, and the maximum hammer stroke during driving was 3.12 m. The rated maximum driving energy E_h was 56.8 kN m. According to the driving log, the observed final pile set was 10 mm. The pile base was located in the dense sand layer. Estimate the ultimate capacity $Q_{10\%}$ of the pile using a pile driving formula.

Solution:

Given that the pile base is embedded in the dense sand layer, the pile driving formula for end-bearing piles in sand (Equation (14.84)) is used:

$$\frac{Q_{10\%}}{p_A L_R^2} = c_1 \left(\frac{e_h E_h}{W_R L_R} \right)^{c_2} \exp\left(c_3 \frac{D_R}{100\%} \right) \left(\frac{w_s}{L_R} \right)^{c_4} \left(\frac{W_P}{W_R} \right)^{c_5}$$

The values of the c_1–c_5 parameters in the formula come from Table 14.5:

$c_1 = 8.11, c_2 = 0.41, c_3 = 0.74, c_4 = -0.53$ and $c_5 = 0.69$

The values of the reference stress, force, and length are:

$p_A = 100\,\text{kPa}$

$W_R = 100\,\text{kN}$

$L_R = 1\,\text{m}$

For diesel hammers acting on steel pipe piles:

$$e_h = 0.38$$

The weight of the pile is calculated from its volume (considering the annulus steel area to calculate the volume) as follows:

$$W_P = 78.5\ \frac{kN}{m^3} \times 8.24\ m \times \frac{\pi}{4}\left(0.356^2 - 0.3306^2\right) m^2$$

$$= 8.86\ kN$$

Substituting these values and the pile set of 10 mm into the pile driving formula yields:

$$\frac{Q_{10\%}}{100\ kPa \times 1^2\ m^2} = 8.11\left(\frac{0.38 \times 56.8\ kNm}{100\ kN \times 1\ m}\right)^{0.41} \exp\left(0.74 \times \frac{80}{100}\right)\left(\frac{0.01\ m}{1\ m}\right)^{-0.53}\left(\frac{8.86\ kN}{100\ kN}\right)^{0.69}$$

$$Q_{10\%} = 1686\ kN \quad \text{answer}$$

The ultimate capacity $Q_{10\%}$ of the pile using the pile driving formula is thus 1686 kN. This example is a real case (see Paik et al. (2003) for details) where a load test was performed on the pile at a site with this exact description, and $Q_{10\%}$ was measured as 1499 kN.

14.6 CHAPTER SUMMARY

Pile dynamics is the analysis of piles subjected to dynamic loads. These dynamic loads are applied either theoretically, as in the wave equation analysis, by actual blows to the pile head applied by a hammer during pile driving or by a handheld hammer to a pile in the ground as part of a quality-assurance test. The three main varieties of pile dynamics analysis are useful in (1) establishing the relationship between **driving resistance** and static pile resistance, (2) confirming the availability of pile capacity based on how hard it is to drive the pile, (3) assisting in the choice of pile driving system components, and (4) ensuring **pile integrity** after installation.

Underlying all types of pile dynamics analysis is the equation:

$$E_p A_p \frac{\partial^2 w}{\partial z^2} + q_s = \rho A_p \frac{\partial^2 w}{\partial t^2} \tag{14.3}$$

where E_p=pile material modulus of elasticity, A_p=pile cross-sectional area, ρ=pile material mass density, q_s=external unit force (shear stress) along pile shaft, w=pile cross section displacement, and t=time. Assuming that there is no q_s acting on the pile, Equation (14.3) can be rewritten as the wave equation:

$$\frac{\partial^2 w}{\partial t^2} = c^2 \frac{\partial^2 w}{\partial z^2} \tag{14.4}$$

where

$$c = \sqrt{\frac{E_p}{\rho}} \tag{14.5}$$

is the **velocity of wave propagation** in the pile.

The solution to the wave equation, any solution, can be expressed as the sum of an upward-traveling and a downward-traveling wave:

$$w(z,t) = w_1(z - ct) + w_2(z + ct) \tag{14.10}$$

The force on any cross section can be expressed in terms of the velocities associated with the upward-traveling and downward-traveling pulses:

$$F = Z(v_1 - v_2) \tag{14.17}$$

where Z=pile impedance, given by

$$Z = \frac{E_p A_p}{c} = \rho A_p c \tag{14.18}$$

For a single traveling pulse, that is, when either $v_1 = 0$ or $v_2 = 0$, Equation (14.17) implies that force and velocity are directly proportional, having the same sign if the pulse travels in the positive (down the pile) z direction and opposite signs if it travels in the negative (up the pile) z direction. It follows that, when a hammer hits a pile, generating a wave, and there has been no time for the wave to reflect back to the pile head, force and velocity at a cross section near the pile head are directly proportional. After the reflected waves arrive back at the top, the following equations are used to calculate the forces associated with the upward-traveling and downward-traveling waves, respectively:

$$F_u = -Zv_2 = \frac{1}{2}(F - Zv) \tag{14.21}$$

$$F_d = Zv_1 = \frac{1}{2}(F + Zv) \tag{14.22}$$

Equations (14.21) and (14.22) are the basis for the **Case method**. The Case formula for the total resistance (including dynamic and static components) is:

$$Q_{total} = F_u\big|_{(t_0 + 2L/c)} + F_d\big|_{t_0} \tag{14.41}$$

where t_0=time immediately after hammer blow and L=pile length (or, more precisely, the distance from the cross section where the force and velocity are measured to the pile base).

The derivation of Equation (14.41) is based on the assumption that, when the up-force arrives at the top, it contains not only the reflected wave at the base, but also the summation Q_{total} of all the resistance forces along the shaft and base.

The dynamic version of the formula is:

$$Q_{ult} = \frac{1}{2}\left[(F - Zv)(1 + j_c)\big|_{(t_0 + 2L/c)} + (F + Zv)(1 - j_c)\big|_{t_0}\right] \tag{14.47}$$

where j_c=Case damping factor, values for which are given in Table 14.1. In the derivation of this equation, the damping (energy loss) is assumed to take place all at the base of the pile.

Wave equation analysis is basically the solution of Equation (14.3) using the finite difference method given initial conditions imposed by a hammer blow. An alternative way to visualize the analysis is to discretize the pile in many elements, each of which interfaces with the soil through a spring and a dashpot or another similar model that can capture both static and dynamic soil resistances. A hammer blow is imposed on the top element, and the propagation of the pulse down the pile is analyzed.

Wave equation analysis cannot be done by hand. A computer program is required. Programs allow the evaluation of proposed driving systems. For example, the program can keep track of the forces generated during the propagation of a hammer blow down the pile and compare maximum tensile stresses and maximum compressive stresses with the compressive and tensile strengths of the pile material. A driving system that imposes excessive stresses on the pile to drive it to the required resistance is obviously undesirable. Wave equation programs also allow that, for the pile installed to a certain depth, a relationship between pile static resistance and pile driving resistance (in blows per unit penetration) be established. This allows an informed decision as to when pile driving can be stopped because the pile has reached a depth at which the necessary static load capacity will be available.

A **pile driving formula** provides an estimate of pile static capacity based on input data that include the driving resistance measured in blows per unit length of penetration of the pile into the ground. Current formulas allow input of general data about the soil profile – which can be obtained from soundings – into the formula to obtain more accurate estimates.

14.6.1 Symbols and notations

Symbol	Quantity represented	US units	SI units
a_i	Acceleration of element i	ft/s^2	m/s^2
c	Velocity of wave propagation	ft/s	m/s
d_i	Damping factor at pile element i	lb/ft/s	kN/m/s
d_s	Viscous damping constant of the soil	lb·s/ft	kN·s/m
E_p	Young's modulus of the pile	psf	kPa
F	Force	lb or ton	kN
F_d	Force due to downward-traveling wave	lb or ton	kN
F_d	Dynamic force	lb or ton	kN
F_s	Static force	lb or ton	kN
F_u	Force due to upward-traveling wave	lb or ton	kN
g	Acceleration due to gravity	ft/s^2	m/s^2
j_c	Case damping factor	Unitless	Unitless
J_s	Smith damping factor	s/ft	s/m
K	Spring constant	lb/in.	kN/mm
K_{si}	Soil stiffness	lb/in.	kN/mm
L	Pile length	ft	m
L_i	Loading index	Unitless	Unitless
L_e	Pile element length	ft	m
p_A	Reference stress	psf	kPa
Q	Compressive force	lb or ton	kN
$Q_{b,total}$	Total mobilized base capacity	lb or ton	kN
Q_{db}	Dynamic component of the pile capacity	lb or ton	kN
Q_L	Limit load	lb or ton	kN
$Q_{s,total}$	Total mobilized shaft capacity	lb or ton	kN
Q_{total}	Total pile resistance	lb or ton	kN
Q_{ult}	Ultimate load	lb or ton	kN
q_s	Unit shaft resistance	psf	kPa

(Continued)

Symbol	Quantity represented	US units	SI units
R_b	Force at the base of pile	lb or ton	kN
R_{di}	Dynamic soil resistances at pile element i	lb or ton	kN
R_i	Soil resistances at pile element i	lb or ton	kN
R_{si}	Static soil resistances at pile element i	lb or ton	kN
R_{sL}	Limit soil static resistance	lb	kN
R_{sLi}	Limit static soil resistance at pile element i	lb or ton	kN
R_t	Force at the pile top	lb or ton	kN
$R(z*)$	Force at a point $z*$	lb or ton	kN
v	Particle velocity	ft/s	m/s
$v\|_{z*L}$	Velocity just to the left of $z*$	ft/s	m/s
$v\|_{z*R}$	Velocity just to the right of $z*$	ft/s	m/s
v_1, v_d	Downward-traveling wave velocity	ft/s	m/s
v_2, v_u	Upward-traveling wave velocity	ft/s	m/s
v_b	Velocity at the base	ft/s	m/s
v_t	Velocity at the pile top	ft/s	m/s
w	Axial displacement at any cross section	in.	mm
w_{qi}	Quake at pile element i	in.	mm
w_s	Pile set	in.	mm
Z	Pile impedance	lb/ft/s	kN/m/s
Δt_{crit}	Critical time interval	s	s
Δw_s	Relative displacement	in.	mm
ε	Axial strain	in./in.	mm/mm
ρ	Density	pcf	kg/m³
σ	Axial stress	psf	kPa

14.7 PROBLEMS

14.7.1 Conceptual problems

Problem 14.1 Define all the terms in bold contained in the chapter summary.

Problem 14.2 Show that the change in variables expressed by Equations (14.6) and (14.7) does indeed lead from the form Equation (14.4) to the form Equation (14.8) of the one-dimensional wave equation.

Problem 14.3 Show that the term $w_1(x - ct)$ of Equation (14.10) does indeed represent a wave moving with speed c in the positive z direction. Hint: evaluate the value of w_1 for positions x_1 and x_2 for times t_1 and t_2, respectively, and impose that the value of w_1 be the same for both cases (indicating that the wave traveled from x_1 to x_2 during time t_2-t_1).

Problem 14.4 During most of 20th century practice, engineers would use pile driving formulas to estimate pile static resistance from driving resistance. In these formulas, the pile is considered a rigid body. The kinetic energy input into the pile by the hammer blow (a hammer with mass m_h falling from a height h_f and efficiency e_h) is converted into work, which is used to overcome the resistance R offered by the soil, allowing the pile to penetrate an amount w_s into the soil. Develop a simple pile driving formula by expressing R in terms of m_h, h_f, e_h, and w_s.

Problem 14.5 In what fundamental ways do the pile driving formulas of Problem 14.4 differ from the Case method? What errors present in the driving formulas are avoided with the Case method?

14.7.2 Quantitative problems

Problem 14.6 The results of a pile integrity test are given in P-Figure 14.1 in terms of velocity v versus time t for a CFA pile with diameter equal to 600 mm and recorded length equal to 23 m. The measured wave velocity was 3900 m/s. Does the plot in P-Figure 14.1 suggest the existence of any anomaly? If so, where is it located and what is its nature?

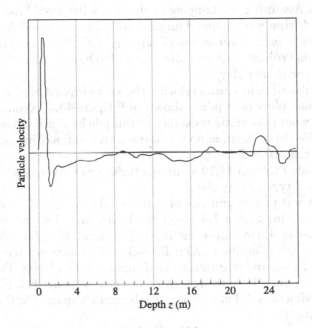

P-Figure 14.1 Results from pile integrity test for Problem 14.6.

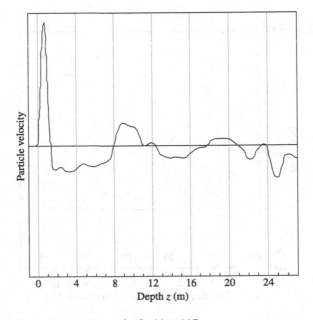

P-Figure 14.2 Results from pile integrity test for Problem 14.7.

Problem 14.7 The results of a pile integrity test are given in P-Figure 14.2 in terms of velocity v versus time t for design diameter equal to 600 mm and recorded length equal to 23 m. The measured wave velocity was 3900 m/s. Does the plot in P-Figure 14.2 suggest the existence of any anomaly? If so, where is it located and what is its nature?

Problem 14.8 A record of force and velocity (shown multiplied by impedance Z) obtained during a hammer blow on a pile that was being driven into silty clay is shown in P-Figure 14.3. Assuming no damping in the soil, what would the static resistance of this pile be if driving were stopped immediately after the blow? Assume no pile setup or relaxation (that is, no gain or loss of capacity with time after driving).

Problem 14.9 Redo Problem 14.8 assuming the soil to have a value of damping consistent with the soil type (a silty clay).

Problem 14.10 A record of force and velocity (shown multiplied by impedance Z) obtained during a hammer blow on a pile is shown in P-Figure 14.4. Assuming no damping in the soil, what would the static resistance of this pile be if driving were stopped immediately after the blow? Assume no pile setup or relaxation (that is, no gain or loss of capacity with time after driving).

Problem 14.11 Redo Problem 14.10 assuming the soil to have a value of damping consistent with the soil type (a silty clay).

Problem 14.12 A soil profile consists of dense sand intermixed with silt to a depth of approximately 25 m, with a 1.4-meter-thick silty layer located at a depth of 9.1 m. The water table is 4.3 m below the ground surface. The test pile was driven by a single-acting impact hammer (APE D30–32). The hammer has a weight of 29.4 kN, stroke of 3.2 m, and maximum rated energy of 94.1 kNm. The pile embedment depth is 15.4 m, and the total length of the pile is 16 m. The observed pile set at the end of driving was 6.4 mm. Estimate the static capacity of this pile using a pile driving formula.

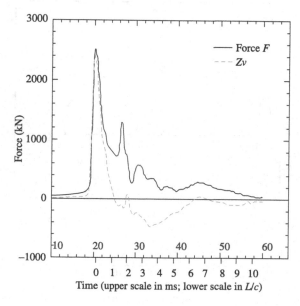

P-Figure 14.3 Force and velocity record for Problem 14.8.

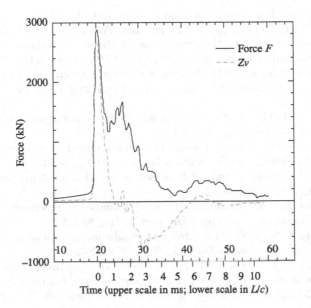

P-*Figure 14.4* Force and velocity record for Problem 14.10.

REFERENCES

References cited

Allen, T. M. (2005). *Development of the WSDOT pile driving formula and its calibration for load and resistance factor design.* Report No. WA-RD 610, Washington State Department of Transportation, Olympia, Washington, USA.

Allin, R., Likins, G., and Honeycutt, J. (2015). "Pile driving formulas revisited." *Proceedings of International Foundations Congress and Equipment Expo*, San Antonio, Texas, 1052–1063.

Berghe, J. V. and Holeyman, A. E. (2002). "Application of a hypoplastic constitutive law into a vibratory pile driving model." *Proceedings of International Conference on Vibratory Pile Driving and Deep Soil Compaction*, Louvain la Neuve, Belgium, 61–68.

Bowles, L. E. (1996). *Foundation Analysis and Design.* McGraw-Hill, New York.

Charue, N. (2004). "Loading rate effects on pile load-displacement behaviour derived from back-analysis of two load testing procedures." Ph.D. Thesis, Université Catholique de Louvain, Belgium.

Chow, Y. K., Phoon, K. K., Chow, W. F., and Wong, K. Y. (2003). "Low strain integrity testing of piles: three-dimensional effects." *Journal of Geotechnical and Geoenvironmental Engineering*, 129(11), 1057–1062.

Chow, Y. K., Wong, K. Y., Karunaratne, G. P., and Lee, S. L. (1988). "Wave equation analysis of piles a rational theoretical approach." *Proceedings of 3rd International Conference on Applications of Stress–Wave Theory to Piles*, Canada, 208–218.

ENR. (1965). "Michigan pile test program test results are released." *Engineering News Record*, May 20, 33–34.

Gates, M. (1957). "Empirical formula for predicting pile bearing capacity." *Civil Engineering*, 27(3), 65–66.

Holeyman, A. E. (1985). "Dynamic non-linear skin friction of piles." *Proceedings of International Symposium on Penetrability and Driveability of Piles*, San Francisco, California, 173–176.

Kim, K., Salgado, R., Lee, J., and Paik, K. (2002). *Load tests on pipe piles for development of CPT-based design method.* Joint Transportation Research Program Publication No. FHWA/IN/JTRP-2002/4, Purdue University, West Lafayette, Indiana, USA.

Lam, J. (2007). "Termination criteria for high-capacity jacked and driven steel H-piles in Hong Kong." Ph.D. Thesis, The University of Hong Kong, Pokfulam, Hong Kong.

Lee, J. H. and Salgado, R. (1999). "Determination of pile base resistance in sands." *Journal of Geotechnical and Geoenvironmental Engineering*, 125(8), 673–683.

Lee, J., Salgado, R., and Paik, K. (2003). "Estimation of load capacity of pipe piles in sand based on cone penetration test results." *Journal of Geotechnical and Geoenvironmental Engineering*, 129(5), 391–403.

Lim, P. C. and Broms, B. B. (1990). "Influence of pile driving hammer performance on driving criteria." *Geotechnical Engineering*, 21(1), 63–69.

Lysmer, J. F. E. R. and Richart, F. E. (1966). "Dynamic response of footings to vertical loading." *Journal of Soil Mechanics and Foundation Division*, 92(1), 65–91.

Mostafa, Y. E. (2011). "Onshore and offshore pile installation in dense soils." *The Journal of American Science*, 7(7), 549–563.

Novak, M., Aboul-Ella, F., and Nogami, T. (1978). "Dynamic soil reactions for plane strain case." *Journal of the Engineering Mechanics Division*, 104(4), 953–959.

Olson, R. E. and Flaate, K. S. (1967). "Pile–driving formulas for friction piles in sand." *Journal of the Soil Mechanics and Foundations Division*, 93(6), 279–296.

Paik, K., Salgado, R., Lee, J., and Kim, B. (2003). "Behavior of open-and closed-ended piles driven into sands." *Journal of Geotechnical and Geoenvironmental Engineering*, 129(4), 296–306.

Pile Dynamics, Inc. (2002). *Wave equation analysis of pile driving*. GRLWEAP Manual. Cleveland, Ohio. USA.

Randolph, M. F. and Simons, H. A. (1986). "An improved soil models for one-dimensional pile driving analysis." *Proceedings of 3rd International Conference of Numerical Methods in Offshore Piling*, Cambridge, UK, 3–17.

Randolph, M. F. and Wroth, C. P. (1978). "Analysis of deformation of vertically loaded piles." *Journal of the Geotechnical Engineering Division*, 104(12), 1465–1488.

Rausche, F. (1970). "Soil response from dynamic analysis and measurements on piles." Ph.D. Thesis, Case Western Reserve University, Cleveland, Ohio, USA.

Rausche, F. (2000). "Pile driving equipment: capabilities and properties." *Proceedings of 6th International Conference on the Application of Stress–Wave Theory to Piles*, Sao Paulo, Brazil, 11–13.

Rausche, F., Goble, G. G., and Likins, G. E. (1985). "Dynamic determination of pile capacity." *Journal of Geotechnical Engineering*, 111(3), 367–383.

Salgado, R., Loukidis, D., Abou-Jaoude, G., and Zhang, Y. (2015). "The role of soil stiffness non-linearity in 1D pile driving simulations." *Géotechnique*, 65(3), 169–187.

Salgado, R., Zhang, Y., Abou-Jaoude, G., Loukidis, D., and Bisht, V. (2017). "Pile driving formulas based on pile wave equation analyses." *Computers and Geotechnics*, 81, 307–321.

Smith, E. A. L. (1960). "Pile driving analysis by the wave equation." *Journal of the Soil Mechanics and Foundations Division*, 86(4), 35–64.

Additional references

Coyle, H. M. and Gibson, G. C. (1970). "Empirical damping constants for sands and clays." *Journal of the Soil Mechanics and Foundations Division*, 96(3), 949–965.

Deeks, A. J. and Randolph, M. F. (1995). "A simple model for inelastic footing response to transient loading." *International Journal for Numerical and Analytical Methods in Geomechanics*, 19(5), 307–329.

Gassman, S. L. and Finno, R. J. (2000). "Anomaly detection in drilled shafts." *Geotechnical Special Publication*, (93), 221–234.

Goble, G. G. and Rausche, F. (1981). "Wave equation analysis of pile driving: WEAP program." Report FHWA-IP-76-14, Federal Highway Administration, Washington, D.C., USA.

Issacs, D. V. (1931). "Reinforced concrete pile formulas." *Transactions of the Institute of Engineers*, 12, 305–323.

Lawton, E. C., Fragaszy, R. J., Higgins, J. D., Kilian, A. P., and Peters, A. J. (1986). "Review of methods for estimating pile capacity." *Transportation Research Record*, 1105, 32–40.

Lee, S. L., Chow, Y. K., Karunaratne, G. P., and Wong, K. Y. (1988). "Rational wave equation model for pile-driving analysis." *Journal of Geotechnical Engineering*, 114(3), 306–325.

Novak, M. (1977). "Vertical vibration of floating piles." *Journal of the Engineering Mechanics Division*, 103(1), 153–168.

Randolph, M. F. (2003). "Science and empiricism in pile foundation design." *Géotechnique*, 53(10), 847–875.

Randolph, M. F. (1992). "Dynamic and static soil models for axial pile response." *Proceedings of 4th International Conference on Application of Stress-Wave Theory to Piles*, The Hague, Netherlands, 3–14.

Simons, H. A. and Randolph, M. F. (1985). "A new approach to one dimensional pile driving analysis." *Proceedings of 5th International Conference on Numerical Methods in Geomechanics*, Nagoya, Japan, 3, 1457–1464.

Svinkin, M. R. and Abe, S. (1992). "Relationship between case and hysteric damping." *Proceedings of 4th International Conference on Application of Stress-Wave Theory to Piles*, The Hague, Netherlands, 175–181.

Relevant ASTM standards

ASTM. (2016a). Standard test method for low strain impact integrity testing of deep foundations. ASTM D5882, American Society for Testing and Materials, West Conshohocken, Pennsylvania, USA.

ASTM. (2016b). Standard test method for integrity testing of concrete deep foundations by ultrasonic crosshole testing. ASTM D6760, American Society for Testing and Materials, West Conshohocken, Pennsylvania, USA.

ASTM. (2017). Standard test method for high-strain dynamic testing of deep foundations. ASTM D4945, American Society for Testing and Materials, West Conshohocken, Pennsylvania, USA.

Chapter 15

Pile groups and piled rafts

Despite adverse comments by some of the pioneers of soil mechanics, there is a significant role for scientific methods in pile design. If we are to continue to attract the best of each new generation of engineers, then we must incorporate such science in our teaching and our practice, using empirical approaches to validate and calibrate, but not replace scientific theory.

Mark F. Randolph

In many cases, structural loads from one or more columns are much larger than the resistance of one pile, requiring multiple piles to support them. These piles are connected at the top by means of either a pile cap (when a single column is supported) or a piled mat or raft (when more than one column and usually all of the structural loads are supported). An additional distinction between what we usually refer to as a pile cap and what we call a piled raft is the reliance (in the case of piled rafts) on the soil-bearing resistance at the base of the raft, while, in pile caps, the total resistance has traditionally been assumed to be due exclusively to the resistances of the individual piles. In this chapter, we cover the calculation of the load capacity and design of both pile groups with a pile cap and piled mats.

15.1 USE OF PILE GROUPS, PILE CAPS, AND PILED RAFTS

When loads are large, it is often necessary to use more than one pile to support them. The piles acting together to support one or more columns are known as a *pile group*. The pile group is typically capped by a reinforced concrete block known as the *pile cap*, which is responsible for making the piles work as a unit. Figure 15.1 shows a schematic diagram of a pile group and the overlying pile cap. Note that the center-to-center pile spacing s_p may be different in the x and y directions. Figures 15.2 and 15.3 show piles in what will be two large groups; in both cases, the concrete for the pile cap has not yet been placed. The minimum pile embedment into the pile cap for effective load transmission (particularly in the case of lateral loads) is of the order of 150–300 mm according to AASHTO (2020); it is 150 mm for steel piles and 100 mm for concrete piles according to CSRI (1992). The pile cap can be assumed to be rigid so long as the distance from the face of peripheral piles to the edge of the cap does not exceed 2.2 times the pile cap thickness (Duan and McBride 1995). Assuming the pile group is adequately designed structurally, our design problem reduces to checking geotechnical serviceability and ultimate limit states.

Pile spacing within the group is one of the critical design decisions. Excessive spacing makes the pile caps too costly. When piles are too closely spaced, they transfer an increasing portion of their loads not to the foundation soil, but to neighboring piles, reducing the net load that each pile can carry and thus *pile group efficiency*. The optimal spacing depends on various

DOI: 10.1201/b22079-15

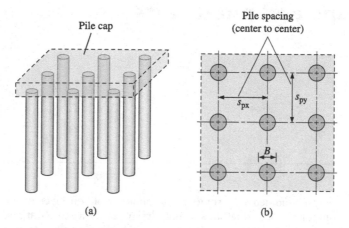

Figure 15.1 A group of piles supports a pile cap, which in turn supports the superstructure loads: (a) three-dimensional view of piles and pile cap and (b) plan view of pile arrangement and spacings.

Figure 15.2 Pile group that is part of the foundation of a high-rise building.

factors, being typically in the range from 2 to 5 pile diameters, center-to-center. Greater spacing increases efficiency by minimizing pile interaction, but increases the pile cap costs. Another factor is the overall size of the group. A large number of piles tend to respond as if they were a block foundation, so individual pile response within this block becomes different, perhaps very different, from that of an isolated pile. So the type of structure, magnitude of the loads, and number of piles in the group all influence what the optimum spacing should be.

Traditionally, design has been based on the assumption of no contact between the pile cap and the soil, which is a conservative assumption, given that many caps do bear on the soil and thus develop some bearing capacity in addition to the pile load capacities. A recent trend, starting in the late eighties, is to take full consideration of this additional bearing capacity using extensive pile groups, often encompassing the whole building. These new foundations are known as *piled rafts* or *piled mats* and are discussed in Section 15.3.

Figure 15.3 A group of APGD piles. (Courtesy of Tracy Brettmann.)

15.2 VERTICALLY LOADED PILE GROUPS

15.2.1 Definition

A vertically loaded pile group is a collection of piles connected by a pile cap (a reinforced concrete block) subjected to a vertical load or loads and perhaps moments, but with a negligible lateral load. The nominal load to be carried by each pile in the absence of moments is given as the total load to be supported by the pile group divided by the number of piles; this is of course based on the assumption that the pile cap distributes the loads uniformly to each pile.

For pile groups in which there is a moment applied, due either to eccentricity of the vertical load or to relatively small lateral loads that produce comparatively large moments, the pile loads can be obtained from consideration of the moment of inertia of the pile group, in a way similar to what was done for shallow foundations subjected to moments in Chapter 10. Consider the case shown in Figure 15.4 of a pile group with n_p piles subjected to a vertical load Q and a moment that we decompose in the x and y directions (so we have moments M_x with respect to the x axis and M_y with respect to the y axis with the directions shown in Figure 15.4). For a rigid cap, the load Q_i acting on the ith pile of the group is given by

$$Q_i = \frac{Q}{n_p} + \frac{M_y x_i}{\sum\limits_{j=1}^{n_p} x_j^2} + \frac{M_x y_i}{\sum\limits_{j=1}^{n_p} y_j^2} \tag{15.1}$$

where x_i, y_i = x and y coordinates of the center of pile i with respect to the centroid of the pile group (for the pile group centroid, $x_i = y_i = 0$).

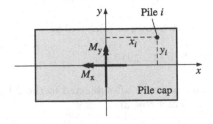

Figure 15.4 Pile group subjected to vertical load Q and moments M_x and M_y with respect to the x and y axes.

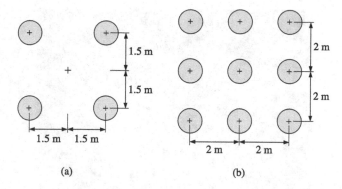

E-Figure 15.1 Two alternative pile distributions to absorb the loading of Example 15.1: (a) four-pile group and (b) nine-pile group.

Example 15.1

Study pile group arrangements to absorb a vertical load of 21,000 kN with a moment of 7000 kN·m due to wind action acting in only one of the directions at any given time. Assume that the lateral loads are easily absorbed by the piles and can thus be neglected in calculations. The piles are drilled shafts with 1 m diameter.

Solution:

Consider the use of four piles spaced at 3 diameters (E-Figure 15.1). Since the moment acts in only one direction at any given time, but can act in any direction, we will focus on the x direction only and consider either the maximum compression load or, if the moment is large enough to overcome the applied vertical load, tension load for each pile. We have

$$\sum x_i^2 = 4 \times 1.5^2 = 9 \, \text{m}^2$$

No pile in this arrangement is subjected to a tensile load. The largest compressive load given by Equation (15.1) is:

$$Q_i = 21,000/4 + (7000)(1.5)/9 = 6417 \, \text{kN}$$

Let's assume now that the soil profile is not sufficiently strong for each drilled shaft to support a 6417 kN compressive load (which may be a problem structurally as well). Then we could try a group with nine piles spaced at 2 diameters, which will keep the pile cap size to a still manageable size, although with some loss of efficiency due to the close pile proximity. In this second alternative, assuming once more drilled shafts with 1 m diameter, we have:

$$\sum x_i^2 = 6 \times 2^2 = 24 \, \text{m}^2$$

The load per pile for the outer two rows of piles will be:

$$Q_{t,\text{outer}} = 21,000/9 + (7000)(2)/24 = 2917 \, \text{kN}$$

Note that the outer piles will never be all subjected to the 2917 kN at the same time. The loads for the piles in the center row are:

$$Q_{t,\text{central}} = 21,000/9 = 2333 \, \text{kN}$$

The loads we calculated for the nine-pile arrangement are now considerably smaller, and if the soil profile is such that the drilled shafts can absorb them, the problem is solved. In Problem 15.6, the reader is asked to solve this problem with loadings that will lead to tension in the outer piles.

The key assumptions made in connection with Equation (15.1) are that there is no interaction between the piles and that the centroid of the pile group is at $x = y = 0$. State-of-the-art practice requires consideration of pile interaction. Pile interaction implies that piles connected via a stiff cap will be subjected to different loads and will thus work with different factors of safety; on the other hand, if the pile cap is flexible, piles will settle differentially. This has the following implication for design: should piles be designed (1) so that an average factor of safety is achieved for the entire group or (2) so that each has approximately the same factor of safety? The analyses necessary to deal with this question are dealt with in subsequent sections. A more important question is: which limit state tends to control pile group design? We will see as the chapter develops that the overall ultimate capacity of the pile group is very large, so settlement-based limit states, most often serviceability limit states, tend to control.

15.2.2 Ultimate bearing capacity

The first limit state that we need to check for a vertically loaded pile group is the ultimate bearing capacity failure of the whole pile group (Figure 15.5a) or a subset thereof (usually a row of piles). In this limit state, all of the piles, or some number of them, fail in bearing capacity as a block. A quick rule is that block failure is possible only for very close pile spacing (less than 2–3 pile diameters). The likelihood of a block ultimate limit state failure increases for: (1) long, slender piles at a given spacing, as opposed to short, stubby piles; (2) piles in clay, as opposed to piles in sand; and (3) closer spacings for a given diameter. Note that the common thread in the three cases is that the lower the contribution of the base of the pile group toward total bearing capacity, the higher the likelihood of block failure. Note also that if there are different spacings in the two directions within a pile group, only failure of a row of piles in the direction with closer spacing is possible.

In sands, block failure is practically impossible because the unit base resistance of the block is much larger than the individual pile unit base resistances (due to the influence of the larger B term of the bearing capacity equation on the base capacity of the pile group), and so individual piles would fail first (see Figure 15.5b). For clays, block failure is possible, and we must check the bearing capacity of (1) individual piles, (2) the whole group, and (3) pile rows in the direction of the smaller pile spacing, if applicable. To check the bearing capacity of a

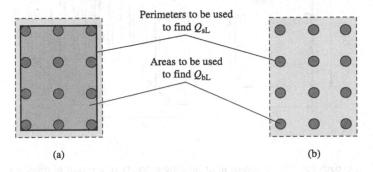

Figure 15.5 Failure modes: (a) block failure and (b) failure of individual piles within the group.

pile group or pile row, assume it to work as a single foundation element, take the unit shaft resistance as $q_{sL} = s_u$, and calculate the unit base capacity in a suitable manner. This procedure was proposed by Skempton (1951), according to whom q_{bL} should be computed using

$$q_{bL} = 5s_u\left(1 + 0.2\frac{B_g}{L_g}\right)\left(1 + \frac{L}{12B_g}\right) \tag{15.2}$$

where B_g, L_g = plan dimensions of the group (circumscribed area of all the piles in the group), L = embedded pile length, and s_u is a representative undrained shear strength (see Figure 15.6).

The side resistance q_{sL} is calculated as for piles. The total limit capacity of the pile group will be the summation of the base with the side resistance:

$$Q_L = B_g L_g q_{bL} + 2\left(B_g + L_g\right)L q_{sL} \tag{15.3}$$

Example 15.2

Calculate the total limit bearing capacity of a group of nine piles disposed symmetrically in the two directions for two cases: center-to-center pile spacing of (a) 2 diameters and (b) 3 diameters. The piles have a diameter of 500 mm and length equal to 10 m. For simplicity, assume the pile head elevation is flush with the ground surface. The piles are floating piles embedded in a deposit of normally consolidated clay with the water table at the surface and $s_u/\sigma'_v = 0.3$. The unit weight of the clay is 16 kN/m³.

Solution:

The pile group has plan dimensions given by

$$B_g = L_g = 2s_p + B = \begin{cases} 5B = 2.5\text{m} & \text{for} \quad s_p = 2B \\ 7B = 3.5\text{m} & \text{for} \quad s_p = 3B \end{cases}$$

where s_p is the pile spacing.
The average unit shaft resistance is equal to the average undrained shear strength for the 10-meter-long piles:

$$q_{sL} = \bar{s}_u = 0.3 \times 5 \times (16 - 9.81) = 9.3 \text{ kPa}$$

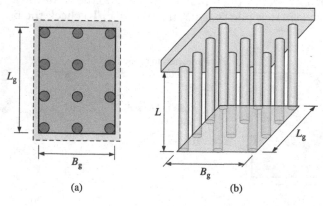

(a)   (b)

Figure 15.6 Block approach for calculation of limit bearing capacity of a group of piles in clay: (a) plan view; (b) 3D view.

The representative s_u for base resistance calculation may be obtained as for a footing with shear strength increasing with depth. The shear strength s_u at one-third of B_g below the base of the piles serves as a reasonable approximation[1]:

$$s_u = 0.3\left(10 + \frac{1}{3}B_g\right)(16 - 9.81) = \begin{cases} 20.1\,\text{kPa for } s_p = 2B \\ 20.7\,\text{kPa for } s_p = 3B \end{cases}$$

We can use, conservatively, $s_u = 20\,\text{kPa}$ for both cases. The unit base resistance follows from Equation (15.2):

$$q_{bL} = \begin{cases} 5s_u(1+0.2)\left(1 + \dfrac{10}{12 \times 2.5}\right) = 8s_u = 160\,\text{kPa} \quad \text{for} \quad s_p = 2B \\[2mm] 5s_u(1+0.2)\left(1 + \dfrac{10}{12 \times 3.5}\right) = 7.4s_u = 148\,\text{kPa} \quad \text{for} \quad s_p = 3B \end{cases}$$

The total limit resistances of the pile group for 2- and 3-diameter spacings are, respectively:

$$Q_L = 160 \times 2.5^2 + 9.3 \times 4 \times 2.5 \times 10 = 1930\,\text{kN} \quad \text{answer}$$

$$Q_L = 148 \times 3.5^2 + 9.3 \times 4 \times 3.5 \times 10 = 3115\,\text{kN} \quad \text{answer}$$

15.2.3 Pile group settlement

Settlement-based limit states are usually critical in connection with pile groups. If we assume the soil to have an elastic response, we can calculate group settlement by considering superposition of effects. The settlement of each pile is the summation of settlements that the pile would undergo if each pile of the group were loaded one at a time. Consider two piles i and j. If pile j is subjected to some load, pile i settles even if the load on it is zero. In this case, the settlement of pile i results from the effect pile j has on the surrounding soil, which moves down as a result of dragging by pile j. When the soil moves down, it takes pile i with it. This can be visualized in Figure 15.7 for an idealized pile with stiffness identical to the stiffness of the surrounding soil.

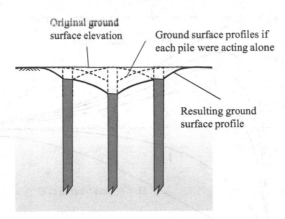

Figure 15.7 Pile group settlement: the settlement of a given pile is due not only to the load it is itself carrying, but also to the effects the other piles in the group have on it.

---

[1] Chapter 10 discusses in detail the bearing capacity of a footing in soil with strength increasing with depth. For linear variation of shear strength s_u with depth, the value of s_u at a representative depth below the footing base in the $0.25B$ to $0.33B$ range can be taken as representative for bearing capacity calculations.

In reality, piles have a stiffness that is much greater than that of the surrounding soil. This means that the piles will not move down with the soil without offering any resistance. This reinforcing effect was studied by Mylonakis and Gazetas (1998). They proposed that the head displacement of a pile i is a composition of the effects of the loads on adjacent piles with due consideration of the reinforcing effect:

$$w_i = \sum_{j=1}^{n_p} \alpha_{ij} \frac{Q_j}{K_{pj}} \tag{15.4}$$

where Q_j = load acting on pile j, K_{pj} = stiffness of pile j (in the sense of how much load is required to have unit pile head settlement), α_{ij} = influence factor between piles i and j (which depends on the proximity of the two piles), and w_i = settlement of pile i. Note that $\alpha_{ij} = 1$ when $i = j$, and the pile settles under its own load.

The interaction factors are given by

$$\alpha_{ij} = \Lambda(\lambda L, \Omega) \frac{\ln(r_m / s_{pij})}{\ln(2r_m / B)} \geq 0 \tag{15.5}$$

where:

r_m = "magical" radius of Randolph and Wroth (1978), at which the settlement w of the ground surface becomes zero;

s_{pij} = spacing between the pair of piles considered (pile i and pile j);

B = pile diameter; and

$\Lambda = \Lambda(\lambda L, \Omega)$ = factor containing the pile-reinforcing effect, given in Figure 15.8.

Note that s_{pij} is not the spacing s_p between adjacent piles in the group, but the center-to-center spacing or distance between the two interacting piles.

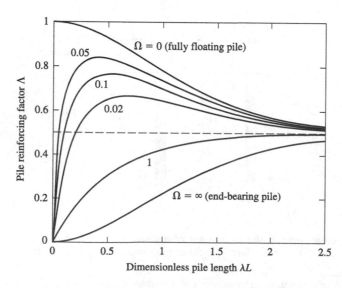

Figure 15.8 Pile-reinforcing factor for the calculation of influence factor for group pile settlement calculation. (Randolph 2003a; Courtesy of the Institution of Civil Engineers and Thomas Telford Limited.)

In these equations, λ is the Winkler factor defined in Equation (13.46) and Ω is the dimensionless base stiffness defined as in Equation (13.53) or (13.61); both λ and Ω are repeated here for convenience:

$$\lambda = \sqrt{\frac{k}{E_p A_p}} \tag{15.6}$$

$$\Omega = \frac{Q_b}{w_b E_p A_p \lambda} = \frac{4 G_b r_s}{1-v} \frac{1}{\lambda E_p A_p} \tag{15.7}$$

If, as part of a group, each pile settles more than if it were alone, then the pile group as a whole settles more for the same load per pile than an individual pile. This means that the stiffness of the group is less than that of the individual pile. Traditionally, this result has been expressed for identical piles through the concept of the group efficiency η_g. The overall pile group stiffness K_g can then be written as

$$K_g = \eta_g n_p K_t \tag{15.8}$$

where η_g = group efficiency (a number less than one), n_p = number of piles in the group, and K_t = individual pile stiffness (the subscript t, as seen in Chapter 13, represents the stiffness at the top or head of the pile). The group efficiency depends on the pile spacing, the number of piles, and the pile and soil elastic properties. The way to calculate K_g or η_g for a group of identical piles is to first calculate K_t using one of the analyses discussed in Section 13.9 and then to use Equation (15.4) with Equations (15.5)–(15.7) to calculate the settlement that each pile would have as part of the pile group if the foundation were flexible. We then take the average w_{avg} of those settlements to calculate K_g:

$$K_g = \frac{Q_{total}}{w_{avg}} \tag{15.9}$$

where Q_{total} is the total load applied on the pile group. The ratio of K_g to the stiffness of all piles in the absence of group effects yields η_g.

Randolph (2003a) prepared a chart of $K_g/B_g G_L$, where B_g is the width of the pile group and G_L is the shear modulus at the level of the pile base,[2] for square arrays of piles ranging from 2×2 up to 30×30 with spacing-to-diameter ratios s_p/B ranging from 2 to 10 (see Figure 15.9). The calculations were done for piles with length equal to 25 diameters, the ratio of pile modulus E_p to the soil shear modulus at the level of the pile base G_L equal to 1000, and the ratio of average shear modulus G (over the pile length) to the shear modulus at the level of the pile base G_L equal to 0.75. It is noteworthy how the normalized group stiffness decreases with increases in either pile spacing s_p or group width B_g, all else being the same. Note also that, if B_g/L is large (say, greater than 1), the group stiffness becomes of the same order as the stiffness of an unpiled mat (raft). Randolph (2003a) also makes the observation that the stiffness of the pile group could be approximated conservatively using 80% of the stiffness of an equivalent cylindrical pier with area equivalent to the area of the group (see Figure 15.10).

---

[2] G_L appears in the settlement analysis discussed in Chapter 13.

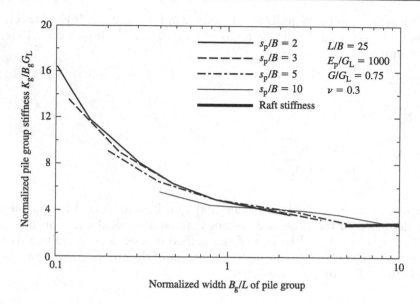

Figure 15.9 Normalized pile group stiffness as a function of pile spacing and normalized pile group width. (Randolph 2003a; Courtesy of the Institution of Civil Engineers and Thomas Telford Limited.)

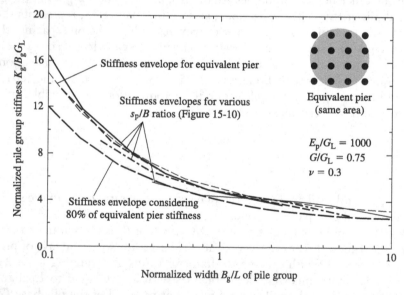

Figure 15.10 Normalized pile group stiffness estimated using the equivalent pier approach. (Randolph 2003a; Courtesy of the Institution of Civil Engineers and Thomas Telford Limited.)

Example 15.3

Calculate the group efficiency for a 2×2 pile group arrangement with drilled shafts with 600 mm diameter spaced at 3 pile diameters center-to-center installed in a 15 m clay layer bearing in a strong sandstone. The piles are 15 m long. The elastic modulus E_p of these piles is 25 GPa. Assume undrained conditions. Assume also that the soil shear modulus within the clay layer is defined by the equation $G = 12 + 1.25z$, where z is in meters and G is in MPa. The sandstone shear modulus G_b at the elevation of the pile base is 310 MPa.

Solution:

Our first tasks are to determine the average shear modulus over the pile length, calculate the base shear modulus, and find the ratio of the two. The shear modulus at the top of the piles is:

$$G_t = 12\,\text{MPa}$$

That at the bottom is:

$$G_L = 12 + 1.25 \times 15 = 30.75\,\text{MPa}$$

Since the shear modulus increases linearly with depth, the average shear modulus over the length of the piles is:

$$G = \frac{1}{2}(12 + 30.75) = 21.38\,\text{MPa}$$

from which we can compute the ratio

$$\rho = \frac{G}{G_L} = \frac{21.38\,\text{MPa}}{30.75\,\text{MPa}} = 0.695$$

Now we must find the magical radius r_m, as defined by Equation (13.74):

$$r_m = \left\{ 0.25 + \left[2.5(1-v)\rho - 0.25 \right] \frac{G_L}{G_b} \right\} L$$

The Poisson's ratio is equal to 0.5 because we are considering undrained conditions. The ratio of G_L to G_b is the ratio of shear modulus at the level of the base of the pile to that we would use to calculate base resistance. Because the piles bear in sandstone with $G_b = 310\,\text{MPa}$ and G_L is almost $31\,\text{MPa}$, this ratio is very nearly 0.1 in our case. It follows that:

$$r_m = \left\{ 0.25 + 0.1 \left[2.5(1-0.5)0.695 - 0.25 \right] \right\} 15\,\text{m} = 0.31 \times 15\,\text{m} = 4.7\,\text{m}$$

Given that the piles are clearly end-bearing piles, we may use Equation (13.77) to find k. First, however, we must determine the Young's modulus of the soil. For that, we use the relation between elastic and shear modulus:

$$E_s = 2G(1+v) = 2 \times 21.38 \times (1+0.5) = 64.14\,\text{MPa}$$

Then, using Equation (13.77), we find:

$$k = 1.3 \left(\frac{E_s}{E_p} \right)^{\frac{1}{40}} \left[1 + 7\left(\frac{L}{B} \right)^{-0.6} \right] G$$

$$= 1.3 \left(\frac{64.14}{25 \times 10^3} \right)^{\frac{1}{40}} \left[1 + 7\left(\frac{15}{0.60} \right)^{-0.6} \right] \times 21.38 = 48.24\,\text{MPa}$$

We must now determine the parameters λ and Ω given by Equations (15.6) and (15.7):

$$\lambda = \sqrt{\frac{k}{E_p A_p}} = \sqrt{\frac{48.24}{25 \times 10^3 \times \frac{\pi}{4}(0.60)^2}} = 0.083\,\text{m}^{-1}$$

$$\Omega = \frac{Q_b}{w_b \lambda E_p A_p} = \frac{K_b}{\lambda E_p A_p} = \frac{4 G_b r_s}{1-v} \frac{1}{\lambda E_p A_p}$$

$$\Omega = \frac{4 \times 310 \times 0.30}{1-0.5} \frac{1}{0.083 \times 25 \times 10^3 \times \frac{\pi}{4}(0.60)^2} = 1.27 \approx 1.3$$

The final step is to calculate the individual interaction factors, α_{ij}, using Equation (15.5):

$$\alpha_{ij} = \Lambda(\lambda L, \Omega) \frac{\ln(r_m / s_{pij})}{\ln(2 r_m / B)}$$

The term Λ is a function of the dimensionless pile length λL and the unitless term Ω and can be found from Figure 15.8. For λL equal to $(0.083\,\mathrm{m^{-1}})(15\,\mathrm{m}) = 1.25$ and Ω equal to 1.3, Λ is equal to 0.43.

Each pile of our group of four identical piles has four interaction factors: 1 (with itself), $\alpha_{12} = \alpha_{13}$ with respect to the piles next to it (which are the same because of symmetry), and α_{14} with respect to the pile diagonally opposite to it. The settlement w_g that any of the piles (and thus that the group) would have for a given load Q per pile is given by Equation (15.4):

$$w_g = \sum_{j=1}^{n} \alpha_{ij} \frac{Q_j}{K_{pj}} = \left(1 + 2\alpha_{12} + \alpha_{14}\right) \frac{Q}{K_t}$$

where K_t is the individual pile stiffness and Q/K_t would be the settlement of each pile in isolation.

The spacing between piles 1 and 2 is $3B$ or $3 \times 0.60\,\mathrm{m} = 1.80\,\mathrm{m}$. The spacing between piles 1 and 4 is $3\sqrt{2}B$ or $3\sqrt{2} \times 0.60\,\mathrm{m} = 2.55\,\mathrm{m}$.

We may now proceed to calculate the individual interaction factors:

$$\alpha_{12} = 0.43 \times \frac{\ln(4.7 / 1.80)}{\ln(2 \times 4.7 / 0.60)} = 0.15$$

$$\alpha_{14} = 0.43 \frac{\ln(4.7 / 2.55)}{\ln(2 \times 4.7 / 0.60)} = 0.1$$

The settlement of the pile group is:

$$w_g = \left(1 + 2\alpha_{12} + \alpha_{14}\right) \frac{Q}{K_t} = \left(1 + 2 \times 0.15 + 0.1\right) \frac{Q}{K_t} = 1.4 \frac{Q}{K_t}$$

The pile group stiffness is given by

$$K_g = \frac{4Q}{w_g} = \frac{4Q}{1.4 \frac{Q}{K_t}} = 2.86 K_t$$

The efficiency of the pile group can now be calculated by entering the number of piles $n_p = 4$ and the above value of K_g into Equation (15.8):

$$\eta_g = \frac{K_g}{n_p K_t} = \frac{2.86}{4} = 0.72 \quad \text{answer}$$

15.2.4 Impact of soil constitutive model used in the analyses*

The assumption that the soil is a linear elastic material is a strong assumption. As discussed extensively in previous chapters, shear strain tends to localize in soils. This means that deformation in vertically loaded piles will tend to localize along their shaft, reducing or even completely eliminating the interaction between piles that linear elasticity would suggest there is. This means that the analyses we have just seen tend to overestimate group pile interaction and underestimate pile group load capacity for a set value of settlement, at least for pile groups that are of small to moderate size (Han et al. 2019; Salgado 2020).

To provide an illustration of this, Figure 15.11 shows the significant difference in pile interaction and pile response within a group resulting from finite element analyses assuming a linear elastic soil, an elasto-plastic soil with a Mohr–Coulomb yield criterion, and a realistic sand model with an appropriately fine finite element mesh. These results show clearly that shear strain localization cannot be ignored if we desire accurate, realistic solutions to geotechnical boundary-value problems. In coming years, improved quantification of pile interaction within a group and group effects is likely to become available.

15.3 PILED MATS

15.3.1 The concept of piled mat foundations

Piled mats or piled rafts, as the name suggests, are hybrid foundations consisting of a raft and piles working together. First, the piles are installed; then a mat is built enveloping the pile heads (see Figure 15.12). Piled rafts were proposed as early as four or five decades ago, but they have been used more commonly only in the last two decades or so. Today, several of the high-rise buildings built in the United States and abroad are built on this type of foundation. The concept of a piled raft seems to have originated from the desire to use rafts (mats) to support tall and heavy structures in less than ideal soil conditions. The settlement that would result from unpiled mats would be excessive, and so the piles were thought of as settlement-reducing elements.

One important difference between the raft in this case and a pile cap is that the contact between the raft and the soil is taken into consideration, while pile caps have not traditionally been assumed to be in contact with the soil and to develop their own bearing resistance. Another important distinction is that pile caps are assumed and designed to be "rigid" so that differential settlement between piles in the same group is negligible. The piles in this case are used to guarantee safety against an ultimate bearing capacity limit state as well as to limit deflections. In a piled raft, the raft is large enough that classical bearing capacity failure is not possible even without the piles, which are then used for the purpose of either reducing differential settlements or bending moments in the raft.

15.3.2 Design

Design methods for piled rafts have not been developed to the point that charts are available for every possible soil and loading condition. We will only summarize the main goals of the design of these foundations and indicate general strategies. The key design questions in connection with the performance of a piled raft are the following:

1. What proportion of the total structural load is carried by the piles?
2. How do the piles reduce total and differential settlements?
3. How much are the bending moments in the raft reduced by the presence of the piles?

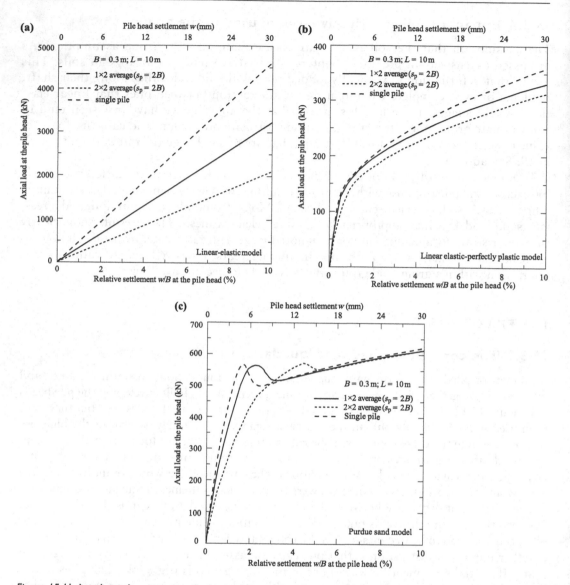

Figure 15.11 Load–settlement curves obtained for a specific combination pile length and diameter and pile spacing from analyses using: (a) a linear elastic model; (b) a linearly elastic–perfectly plastic model with the Mohr–Coulomb yield criterion; and (c) the Purdue sand model (Loukidis and Salgado 2009) – a bounding surface constitutive model that produces realistic simulations of soil mechanical response – for relative density = 80%. (Modified after Han et al. 2019; with permission from ASCE.)

When we design a piled raft, we specify the following:

1. the raft (mat) thickness and reinforcement;
2. the number of piles to be used;
3. the location of each pile; and
4. the pile diameter and length.

For any given structure, we will be interested in varying all of the variables listed above (raft thickness, the number and configuration of piles, and pile diameter and length) to obtain an optimal, most economical foundation system.

Figure 15.12 Piles for piled mat foundation.

There is evidence (Randolph 2003a; Reul and Randolph 2004) that the best pile configuration is not necessarily equally spaced piles throughout the raft. It appears, for example, that piles located in an area around the center of the raft lead to the most reduction in differential settlement with respect to those experienced by an unpiled raft, if such a configuration is possible. In a very stiff raft, which would not have significant differential settlements even in the absence of the piles, the bending moments are reduced most effectively by piles located in a central area of the raft. This conclusion appears to hold even in buildings where much of the load is applied near the edges of the raft, where sagging moments drop with the side effect that hogging moments increase. The overall result is still beneficial.

The piled raft performance may be quantified by using a coefficient α_{pr} (Reul and Randolph 2004) to express how much the piles are contributing to the total load-carrying capacity of the piled raft:

$$\alpha_{pr} = \frac{\sum Q_i}{Q} \tag{15.10}$$

where Q_i are the loads on the individual piles and Q is the total load supported by the piled raft. This number is naturally equal to 1 in the case of a traditional pile group (or, more correctly, that is the assumption we make when we design pile groups), and it is 0 in the case of an unpiled raft. It is somewhere in between for a piled raft.

No simplified procedures are available that can be used for the design of piled rafts. The analysis of piled rafts requires a large number of calculations and can be done effectively only by computer, typically using a finite element approach. Even then, the topic is in continuous evolution because much research is underway on how best to design these foundations. While we use finite elements to model mats and rely on the Winkler spring foundation approach to model the soil resistance, in the case of piled rafts, it is advantageous to use finite elements to model the soil as well, making design of these foundations relatively advanced compared with most designs done in geotechnical engineering.

15.4 LATERALLY LOADED PILE GROUPS

15.4.1 Design approaches

Pile groups subjected to lateral loads will not reach a block-failure ultimate limit state caused by the lateral loads, except in very unusual circumstances. However, serviceability limit states and ultimate limit states associated with superstructure failure are possible; these limit states require estimation of deflections under working loads. Considering the complexities of the interactions of the piles, particularly in the presence of vertical loads, these analyses must be done using computer programs specifically written for pile group analysis, such as PIGLET (Randolph 2003b), if reasonably accurate results are desired. In such computer programs, the interactions between the piles and the pile cap are automatically accounted for, dispensing with the need to make certain assumptions about, for example, pile fixity that are required to use simpler methods.

15.4.2 Simplified design approach

15.4.2.1 Pile head fixity

The two classical boundary conditions for the pile head are free and fixed. Assuming free-head conditions for piles joined by a pile cap is far too conservative because the predicted lateral deflections will be far in excess of the actual deflections. Fixed-head conditions work better because the true conditions are closer to a fixed-head condition than to a free-head condition, as shown by Mokwa and Duncan (2003). The slight deviation from fixity is mostly due to the fact that the pile cap itself rotates as a lateral load is applied to the cap. This happens as the lateral load is not applied at the level of the base of the pile cap, thus imposing a moment that must be balanced by the appearance of compressive forces in the leading piles and tensile forces in the trailing piles. The terms "leading" and "trailing" refer here to the direction of application of the load; the leading piles deflect in the direction of the load and thus away from the trailing piles, which deflect also in the direction of the load and toward the leading piles. As the pile heads turn with the pile cap, perfectly fixed pile head conditions are clearly not in existence.

It is not easy to accurately quantify how much the pile head will rotate and what the value of the moment will be at the pile head. One of the factors this clearly depends on is the uplift capacity of the trailing piles and the compressive capacity of the leading piles. Piles with small axial bearing capacity will allow more rotation of the pile cap and thus lead to conditions of less fixity at the pile head. The degree of fixity can be quantified using the rotation stiffness K_M for the pile cap, defined as

$$K_M = \frac{M}{\theta} \tag{15.11}$$

where M is the moment on the pile cap and θ is the resulting rotation. Mokwa and Duncan (2003) obtained values as low as 55 MN·m for short floating piles and as high as 365 MN·m for long floating piles for pile groups with four piles. The degree of fixity increases with the number of piles in the group, just as the resisting moment M would increase substantially not only because there are more piles, but also because their leverage arms will be larger, given the larger plan dimensions of the pile group.

15.4.2.2 Simplified approach using the p–y method

Pile group design using a simplified method based on p–y curves follows the following general steps:

1. Selection of a suitable p–y relationship for a single pile.
2. Reduction of the p values for each pile to reflect its interaction with other piles.
3. Performance of a set of calculations aiming to obtain a lateral load versus lateral deflection relationship for the whole group.

The modification of the p–y curves of the individual piles to account for pile interaction effects is necessary because the leading piles (the piles in the front row of the group) move in the direction of the load, away from the trailing piles. This in effect reduces the net resistance p of the trailing piles at a set value of lateral deflection y. The load that these piles cannot carry is, in turn, transferred forward to the leading piles, so that they move more because of the added load and, in effect, the p–y curves of all piles must be corrected down.

The reduction in resistance caused in one pile by another pile depends on their relative positions in the pile group (whether they are side-by-side, in line, or offset by some angle θ). In-line piles can lead or trail the pile under consideration. A pile leads another if it is further out in the direction of load application. The reduction in p values can in theory be taken into account by using a p *multiplier* f_i for each pile i in the group. The values of the p multipliers for a given pile depend on its position with respect to all other piles in the group. Referring to Figure 15.13, f_i is given by the product of interaction coefficients β_{ij} between pile i and pile j; j=1, 2,…,n_p, where n_p is the number of piles in the group:

$$f_i = \prod_{j-1}^{n_p} \beta_{ij} \tag{15.12}$$

where $\beta_{ij} = 1$ if $i = j$ (because the pile does not modify its own p–y curve).

In practice, finding theoretically sound β_{ij} relationships is a problem that has not yet been solved entirely satisfactorily, but the literature does contain recommendations based on data from load tests on pile groups (in most cases, model pile groups, which are not directly applicable to real-scale pile groups) that allow us to design laterally loaded pile groups, so long as we understand there is still uncertainty in what the values of the β_{ij} should be. The following equations were proposed by Reese et al. (2006) based on the work of Brown et al. (1987):

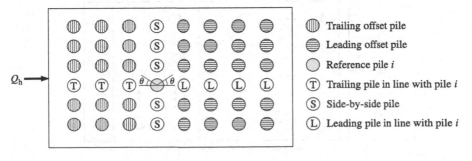

Figure 15.13 Schematic diagram of pile group acted upon by a lateral load.

$$\beta_{iL} = 0.48 \left(\frac{s_{piL}}{B} \right)^{0.38} \leq 1 \qquad (15.13)$$

for a leading pile that is in line with pile i;

$$\beta_{iT} = 0.7 \left(\frac{s_{piT}}{B} \right)^{0.26} \leq 1 \qquad (15.14)$$

for a trailing pile that is in line with pile i (this can also be used if the pile is not offset by more than $B/2$ to pile i, where B is the pile diameter);

$$\beta_{iS} = 0.64 \left(\frac{s_{piS}}{B} \right)^{0.34} \leq 1 \qquad (15.15)$$

for a pile that is side-by-side with pile i;

$$\beta_{i\theta} = \sqrt{\beta_{iL}^2 \cos^2 \theta + \beta_{iS}^2 \sin^2 \theta} \leq 1 \qquad (15.16)$$

for a leading offset pile; and

$$\beta_{i\theta} = \sqrt{\beta_{iT}^2 \cos^2 \theta + \beta_{iS}^2 \sin^2 \theta} \leq 1 \qquad (15.17)$$

for a trailing offset pile, where θ is the angle (always positive) between the direction of load application and the line joining the centers of the two offset piles.

Note that s_{piL}, s_{piT}, and s_{piS} are not the spacing s_p between adjacent piles in the group, but the center-to-center spacings or distances between the two interacting piles (pile i and a leading, trailing, or side pile).

Simpler equations may be used for the p multiplier f directly. In these equations, the same p multiplier is used for all the piles in a given row. Referring to Figure 15.14 and taking pile rows perpendicular to the line of action of the applied lateral load, the leading row is the farthest row from the point of application of the lateral load. Trailing rows are the rows following the leading row. For the leading row:

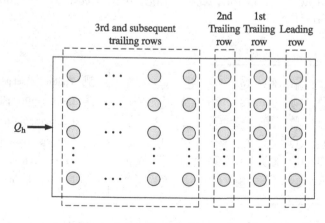

Figure 15.14 Simplified approach to determining pile p multipliers based only on the row in which the pile is located.

$$f_L = 0.64 + 0.06 \left(\frac{s_p}{B} \right) \leq 1 \tag{15.18}$$

For the first trailing row:

$$f_{T1} = 0.34 + 0.11 \left(\frac{s_p}{B} \right) \leq 1 \tag{15.19}$$

For the second trailing row:

$$f_{T2} = 0.16 + 0.14 \left(\frac{s_p}{B} \right) \leq 1 \tag{15.20}$$

For the third and subsequent trailing rows:

$$f_{T3} = 0.04 + 0.16 \left(\frac{s_p}{B} \right) \leq 1 \tag{15.21}$$

These equations are based on the design chart of Mokwa (1999), which was developed based on the values of p multipliers obtained from lateral pile load test data reported in the literature. Mokwa (1999) recommends the use of the chart only for projects where lateral loads are not critical. Note that the p multiplier depends only on the row (leading, first trailing, second trailing, or third trailing row) to which the pile belongs.

After suitable p–y relationships are obtained, the following steps are followed:

1. Impose a sequence of lateral deflections from small all the way to a value larger than what is likely to be the tolerable deflection for the pile group.
2. Calculate the lateral loads at the top of all the piles for each imposed deflection and then add them together.
3. Steps 1 and 2 produce a lateral load versus lateral deflection curve for the pile group; we now read the deflection corresponding to the lateral load applied on the cap from this curve.
4. Repeat the whole procedure for different pile diameters and arrangements until the lateral deflection resulting from Step 3 is below the tolerable lateral deflection.

When the lateral load is obtained for a single pile in Step 2 using a computer program (as discussed in Chapter 13), it is necessary to specify the boundary conditions at the pile head; since it is difficult to obtain an accurate assessment of the pile cap rotation stiffness K_M, it is usually necessary to assume perfectly fixed conditions.

This procedure also generates the maximum bending moment for each pile, which allows the sizing, choice, and structural design of the piles. It is usually more economical and pragmatic to size every pile the same, as opposed to ending up with piles with different dimensions and structural design within the same group; consequently, the maximum bending moment of all piles is usually used in the design of the piles and pile group.

Example 15.4

Consider the case of Example 13.19. Assume that a group of four drilled shafts identical to that analyzed in the example, spaced 2 diameters center-to-center, are to be used to carry a lateral load of 150 kN applied parallel to one of its sides without deflecting more than 25 mm. Using a suitable laterally loaded pile software to which you may have access, determine whether the four-shaft arrangement can satisfactorily absorb that load.

Solution:

This solution uses the program PYGMY. Readers will benefit most from solving this example with a suitable *p–y* analysis software to which they may have access. Consider a plan view of the four piles. To keep track of the piles, we will number them, proceeding from left to right and top to bottom. Now, to calculate the f_i value for each pile, we need to determine the interaction factors β_{ij} between piles.

<u>Pile 1 (top-left pile)</u>

Using Equation (15.13), the interaction factor between pile 1 and pile 2 is given by

$$\beta_{12} = 0.48\left(\frac{s_{p12}}{B}\right)^{0.38} = 0.48\left(\frac{1}{0.5}\right)^{0.38} = 0.625$$

The interaction factor between pile 1 and pile 3 is obtained from Equation (15.15):

$$\beta_{13} = 0.64\left(\frac{s_{p13}}{B}\right)^{0.34} = 0.64\left(\frac{1}{0.5}\right)^{0.34} = 0.810$$

The interaction factor between pile 1 and pile 4 is obtained from Equation (15.16):

$$\beta_{14} = \sqrt{\beta_{1L}^2\cos^2\theta + \beta_{1S}^2\sin^2\theta}$$

where $\beta_{1L} = 0.48\left(\frac{\sqrt{2}}{0.5}\right)^{0.38} = 0.713$ and $\beta_{1S} = 0.64\left(\frac{\sqrt{2}}{0.5}\right)^{0.34} = 0.911$

So,

$$\beta_{14} = \sqrt{0.713^2 \times \cos^2 45° + 0.911^2 \times \sin^2 45°} = 0.818$$

The *p* multiplier can now be obtained from Equation (15.12):

$$f_i = \prod_{j=1}^{n_p} \beta_{ij}$$

$$f_1 = \prod_{j=1}^{4} \beta_{ij} = \beta_{11}\beta_{12}\beta_{13}\beta_{14} = (1)(0.625)(0.810)(0.818) = 0.414$$

<u>Pile 2 (top-right pile)</u>

Using Equation (15.14), the interaction factor between pile 2 and pile 1 is given by

$$\beta_{21} = 0.7\left(\frac{s_{p21}}{B}\right)^{0.26} = 0.7\left(\frac{1}{0.5}\right)^{0.26} = 0.838$$

The interaction factor between pile 2 and pile 3 is obtained from Equation (15.17):

$$\beta_{23} = \sqrt{\beta_{2T}^2\cos^2\theta + \beta_{2S}^2\sin^2\theta}$$

where $\beta_{2T} = 0.7\left(\frac{\sqrt{2}}{0.5}\right)^{0.26} = 0.917$ and $\beta_{2S} = 0.64\left(\frac{\sqrt{2}}{0.5}\right)^{0.34} = 0.911$

So,

$$\beta_{23} = \sqrt{0.917^2 \times \cos^2 45° + 0.911^2 \times \sin^2 45°} = 0.914$$

The interaction factor between pile 2 and pile 4 is obtained from Equation (15.15):

$$\beta_{24} = 0.64\left(\frac{s_{p24}}{B}\right)^{0.34} = 0.64\left(\frac{1}{0.5}\right)^{0.34} = 0.810$$

The p multiplier is now obtained from Equation (15.12):

$$f_i = \prod_{j=1}^{n_p} \beta_{ij}$$

$$f_2 = \prod_{j=1}^{4} \beta_{ij} = \beta_{21}\beta_{22}\beta_{23}\beta_{24} = (0.838)(1)(0.914)(0.810) = 0.620$$

Pile 3 (bottom-left pile) and Pile 4 (bottom-right pile)

Now we can use the symmetry of the pile group geometry to calculate the p multipliers for pile 3 and pile 4. Observing that $\beta_{31} = \beta_{13}$, $\beta_{34} = \beta_{12}$, and $\beta_{32} = \beta_{14}$:

$$f_3 = f_1 = 0.414$$

Similarly, observing that $\beta_{41} = \beta_{23}$, $\beta_{42} = \beta_{24}$, and $\beta_{43} = \beta_{21}$:

$$f_4 = f_2 = 0.620$$

Now, we have p multipliers for all piles. Inputting these values in our laterally loaded pile software, we obtain reduced lateral capacities of each pile for 25 mm tolerable deflection. After summation of these capacities of each pile, we obtain the lateral capacity of the pile group. Finally, we obtain 121 kN for piles 1 and 3, and 132 kN for piles 2 and 4 under the 25 mm tolerable deflection. Therefore, the total lateral capacity of this pile group for 25 mm deflection is 506 kN, which is greater than the applied lateral load of 150 kN. So, the pile group will successfully absorb the applied load.

Now, following Mokwa's approach, we have a single p multiplier for each row. For the leading row (that is, for piles 2 and 4) for a spacing equal to 2B, the p multiplier is 0.76 using Equation (15.18), whereas, for the first trailing row (that is, for piles 1 and 3), this value is 0.56 using Equation (15.19). Performing an analysis with our laterally loaded pile analysis software with this different set of p multipliers, we obtain a pile lateral capacity equal to 129 kN for piles 1 and 3, and 139 kN for piles 2 and 4 for the 25 mm tolerable deflection. Therefore, the total lateral capacity of this pile group for 25 mm deflection is 536 kN, a slightly greater resistance than the first method we used to obtain the p multipliers.

15.5 CHAPTER SUMMARY

Loads that are too large or too closely spaced to be carried by single foundation elements are transferred to the ground by relying on groups of piles. If design relies on the bearing resistance of the structural element placed on top of the piles, the structural element is usually a **piled mat**; otherwise, it is a **pile cap**.

Pile caps are stiff (with little to no differential settlement between piles of the same group), while mats are flexible.

The distribution of loads among the piles supporting a pile cap may be approximated by

$$Q_i = \frac{Q}{n_\mathrm{p}} + \frac{M_y x_i}{\displaystyle\sum_{j=1}^{n_\mathrm{p}} x_j^2} + \frac{M_x y_i}{\displaystyle\sum_{j=1}^{n_\mathrm{p}} y_j^2} \tag{15.1}$$

where Q_i = load at the top of pile i due to both the vertical load Q and the moments M_y with respect to the y axis and M_x with respect to the x axis applied at the pile cap; n_p = total number of piles in the pile group; and x_i, y_i = coordinates of the center of pile i with respect to the y and x axes, respectively (these can be either positive or negative).

Comparing the loads calculated using Equation (15.1) with the individual pile capacities, we can ascertain that no individual pile will be subjected to a load approaching its ultimate load capacity.

It is also important, more so in the case of clays (and soft clays, in particular), to check for the possibility of a limit bearing capacity failure of the pile group as a block. This can be done using:

$$q_\mathrm{bL} = 5s_\mathrm{u}\left(1 + 0.2\frac{B_\mathrm{g}}{L_\mathrm{g}}\right)\left(1 + \frac{L}{12B_\mathrm{g}}\right) \tag{15.2}$$

where B_g, L_g = plan dimensions of the group, L = embedded pile length, and s_u is a representative undrained shear strength.

In pile groups, piles interact with one another, so that a given pile will settle more under the same load than if it were alone supporting that load. It is possible to calculate a **pile group efficiency** or, alternatively, a **pile group stiffness** based on considerations of the interactions between the piles. The best way to do these calculations is by using purpose-written computer programs, but charts such as those in Figure 15.9 are useful as well.

Piled mats are used mostly in connection with large bridge foundations and high-rise buildings, that is, to support large loads. Piles are used to reduce either differential settlements in the case of flexible mats, or bending moments in the case of stiff mats.

The design of a piled mat requires that we specify the following:

1. the raft (mat) thickness and reinforcement;
2. the number of piles to be used;
3. the location of each pile; and
4. the pile diameter and length.

There is evidence that piles should be concentrated more toward the center of the mat to achieve the greatest reduction in differential settlements or bending moments. Design of these foundations involves a greater degree of sophistication in calculations, and there are no simplified design procedures that can be referred to.

Piles subjected to lateral loads are at present usually designed using the p–y method. This method gives both the relationship between the lateral load at the pile head and the lateral deflection, and the location and magnitude of the maximum bending moment in the pile. The method can be used also to obtain the relationship between the lateral load and the deflection of a pile group by accounting for the interaction between the piles. This is accomplished through the use of p **multipliers**. The p multipliers reduce the values of p for a given deflection y from those that a single, isolated pile would have.

15.5.1 Symbols and notations

Symbol	Quantity represented	US units	SI units
A_p	Cross-sectional area of the pile	ft^2	m^2
B	Pile diameter	ft	m
B_g	Width of pile group	ft	m
E_p	Young's modulus of the pile	psf	kPa
G_b	Soil shear modulus at the pile base	psf	kPa
G_t	Soil shear modulus at the pile top	psf	kPa
K_g	Overall pile group stiffness	lb/ft	kN/m
K_M	Rotation stiffness for the pile cap	lb·m/deg	kN·m/deg
K_t	Individual pile stiffness	lb/ft	kN/m
K_{tj}	Stiffness of pile j	lb/ft	kN/m
L	Embedded pile length	ft	m
L_g	Depth of pile group	ft	m
M	Moment on the pile cap	lb·ft	kN·m
M_x	Moment with respect to the x axis	lb·ft	kN·m
M_y	Moment with respect to the y axis	lb·ft	kN·m
n_p	Number of piles	Unitless	Unitless
Q	Vertical load	lb or ton	kN
Q_b	Pile base load	lb or ton	kN
q_{bL}	Limit unit base capacity	psf	kPa
Q_j	Load acting on the pile j	lb or ton	kN
Q_L	Total limit capacity of the pile group	lb or ton	kN
q_{sL}	Limit unit shaft resistance	psf	kPa
Q_{total}	Total load applied on the pile group	lb or ton	kN
r_m	"Magical" radius at which w becomes zero	ft	m
r_s	Equivalent pile radius	ft	m
s_p	Pile spacing	ft	m
s_{piL}, s_{piT}, and s_{piS}	Spacings between interacting piles in a pile group; used for calculating p multipliers	ft	m
s_u	Undrained shear strength	psf	kPa
w_{avg}	Average settlement	in.	mm
w_b	Pile base settlement	in.	mm
w_i	Settlement of pile i	ft	m
x, y, z	Cartesian coordinates	ft	m
x_i	x-coordinate of the center of pile i with respect to the centroid of the pile group	ft	m
y_i	y-coordinate of the center of pile i with respect to the centroid of the pile group	ft	m
α_{ij}	Influence factor between piles i and j	Unitless	Unitless
α_{pr}	Coefficient of pile raft performance	Unitless	Unitless
β_{ij}	Pile interaction coefficient	Unitless	Unitless
β_{iL}	Interaction coefficient for a leading pile that is in line with pile i	Unitless	Unitless

(Continued)

Symbol	Quantity represented	US units	SI units
β_{iS}	Interaction coefficient for a pile that is side-by-side with pile i	Unitless	Unitless
β_{iT}	Interaction coefficient for a trailing pile that is in line with pile i	Unitless	Unitless
$\beta_{i\theta}$	Interaction coefficient for a trailing offset pile	Unitless	Unitless
η_g	Group efficiency	Unitless	Unitless
θ	Degree of rotation	deg	deg
Λ	Interaction adjustment factor	Unitless	Unitless
λ	Winkler factor	1/ft	1/m
ν	Poisson's ratio	Unitless	Unitless
Ω	Dimensionless base stiffness	Unitless	Unitless

15.6 PROBLEMS

15.6.1 Conceptual problems

Problem 15.1 Define all the terms in bold contained in the chapter summary.

Problem 15.2 If two piles are sufficiently close to one another, both piles deflect even if only one of the piles is loaded. Discuss the physical processes involved.

Problem 15.3 In Problem 15.2, you explained how a pile moves due to the load applied on a separate pile. Consider now the case of two piles rigidly connected at the top by a pile cap. Assume that a vertical load is applied at the pile cap and, because the soil, piles, and load are ideally symmetric, the load is distributed equally among the two piles. How can your previous reasoning be used to explain the observation that, for realistic pile spacings, the two piles settle more together than each would settle alone under the same load per pile.

15.6.2 Quantitative problems

Problem 15.4 Due to space restrictions, an eccentric load equal to 2000 kN is to be supported by a pile group as detailed in P-Figure 15.1. Calculate the loads to be supported by each pile.

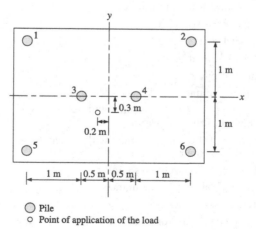

P-Figure 15.1 Dispositions of piles and load for pile group of Problem 15.4.

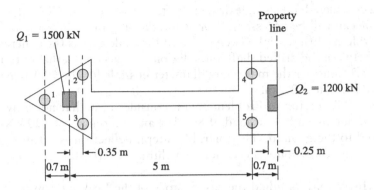

P-Figure 15.2 Sketch of pile and pile cap arrangements for Problem 15.5.

Problem 15.5 Two pile caps supporting a column each (with loads $Q_1 = 1500$ kN and $Q_2 = 1200$ kN) are connected by a beam as shown in P-Figure 15.2. This solution was adopted due to the proximity of column 2 to the property line. Compute the loads to be supported by each pile.

Problem 15.6 Redo Example 15.1 for an axial compressive load of 10,000 kN and a maximum moment of 20,000 kN m due to wind action acting in only one of the directions at any given time.

Problem 15.7 Redo Example 15.2 for 300 mm diameter piles that are 12 m long. Assume $s_u/\sigma'_v = 0.25$ for your calculations.

15.6.3 Design problems

Problem 15.8 Study pile group arrangements to absorb a vertical load of 20,000 kN with a moment of 6500 kN·m due to wind action acting in only one of the directions at any given time. Assume that the lateral loads are easily absorbed by the piles and can thus be neglected in calculations. The piles are 0.5 m square precast concrete piles that can take 500 kN in compression and 250 kN in tension.

Problem 15.9 Calculations show that individual pile stiffness for square precast concrete piles with 400 mm side is 500 kN/mm for total settlement of the order of 10 mm for a given soil profile. The piles are 27 m long and have a 25 GPa elastic modulus. The soil shear modulus is equal to $12 + 1.25z$ with z in meters and G in MPa. Poisson's ratio is 0.2. Using a spacing of 3 pile diameters, what is the settlement of a pile group with a square nine-pile arrangement under a vertical load of 20,000 kN.

Problem 15.10* Given that the tolerable ground displacement can be no more than 15 mm, calculate the number of piles required if the vertical load in Problem 15.9 is 25,000 kN.

Problem 15.11 For the same soil conditions as in Example 15.3 and for piles with the same diameter and length, calculate the efficiency of a group of nine piles arranged as a square.

Problem 15.12* Consider the 400 mm drilled shafts of Problems 13.23 and 13.26 to be part of a group of four piles in a square arrangement with a spacing of 2.5 pile diameters. Find the total axial and lateral load capacity of this group. The tolerable deflections for this group are 40 mm (vertical) and 20 mm (lateral). Make assumptions as needed.

Problem 15.13 Redo Problem 15.12 for nine piles in a square arrangement.

Problem 15.14* An axial load of 4000 kN is to be supported by a foundation system installed in soil with the profile as indicated in P-Figure 13.3. The following options are being considered: (a) a single drilled shaft; (b) two or more CFA piles. In the second case, a pile cap will be necessary. Proportion the piles in both cases, and calculate the concrete volume differential. Given the need for a pile cap, is the concrete volume differential between the drilled shaft and CFA piles large enough that option (b) might be economical? Consider the maximum diameter possible for the CFA pile to be 1 m. For the drilled shaft, it is 2.2 m.

Problem 15.15* Referring to Problem 15.14, consider now that not only the 4000 kN axial load, but also a lateral load of 800 kN and a moment of 300 kN·m need to be transferred to the ground. The tolerable lateral deflection is 40 mm. Reproportion the piles and recalculate the concrete volume differential. Has your conclusion changed?

Problem 15.16 We have studied the case history of the Leaning Tower of Pisa in Parts I (Chapter 2), II (Chapter 6), III (Chapter 9), and IV (Chapter 11). Provide a design for the foundations of the Tower of Pisa using current piling technology and current design methods. Make whatever assumptions may be necessary.

REFERENCES

References cited

AASHTO. (2020). AASHTO LRFD bridge design specifications. 9th Edition, American Association of State Highway and Transportation Officials, Washington, D.C., USA.

Brown, D. A., Reese, L. C., and O'Neill, M. W. (1987). "Cyclic lateral loading of a large-scale pile group." Journal of Geotechnical Engineering, 113(11), 1326–1343.

CSRI. (1992). CRSI handbook. Concrete Reinforcing Steel Institute, Schaumburg, Illinois, USA.

Duan, L. and McBride, S. B. (1995). "The effects of cap stiffness on pile reactions." Concrete International, 17(1), 42–44.

Han, F., Salgado, R., Prezzi, M., and Lim, J. (2019). "Axial resistance of nondisplacement pile groups in sand." Journal of Geotechnical and Geoenvironmental Engineering, 145(7), 04019027.

Loukidis, D. and Salgado, R. (2009). "Modeling sand response using two-surface plasticity." Computers and Geotechnics, 36(1–2), 166–186.

Mokwa, R. L. (1999). "Investigation of the resistance of pile caps to lateral loading." Ph.D. Thesis, Virginia Polytechnic Institute and State University, Virginia, USA.

Mokwa, R. L. and Duncan, J. M. (2003). "Rotational restraint of pile caps during lateral loading." Journal of Geotechincal and Geoenvironmental Engineering, 129(9), 829–837.

Mylonakis, G. and Gazetas, G. (1998). "Settlement and additional internal forces of grouped piles in layered soil." Géotechnique, 48(1), 55–72.

Randolph, M. F. (2003a). "Science and empiricism in pile foundation design." Géotechnique, 53(10), 847–875.

Randolph, M. F. (2003b). PIGLET User Manual.

Randolph, M. F. and Wroth, C. P. (1978). "Analysis of deformation of vertically loaded piles." Journal of the Geotechnical Engineering Division, 104(12), 1465–1488.

Reese, L. C., Isenhower, W. M., and Wang, S.-T. (2006). Analysis and Design of Shallow and Deep Foundations. Wiley, New York.

Reul, O. and Randolph, M. F. (2004). "Design strategies for piled rafts subjected to nonuniform vertical loading." Journal of Geotechnical and Geoenvironmental Engineering, 130(1), 1–13.

Salgado, R. (2020). "Forks in the road: rethinking modeling decisions that defined the teaching and practice of geotechnical engineering." Geotechincal Keynote Lecture in International Conference on Geotechnical Engineering Education, Athens, Greece.

Skempton, A. W. (1951). "The bearing capacity of clays." Proceedings of Building Research Congress, London, UK, 1, 180–189.

Additional references

Chow, Y. K. (1987). "Axial and lateral response of pile groups embedded in nonhomogeneous soils." *International Journal for Numerical and Analytical Methods in Geomechanics*, 11(6), 621–638.

McVay, M., Casper, R., and Shang, T. I. (1995). "Lateral response of three-row groups in loose to dense sands at 3D and 5D pile spacing." *Journal of Geotechnical Engineering*, 121(5), 436–441.

McVay, M., Bloomquist, D., Vanderlinde, D., and Clausen, J. (1994). "Centrifuge modeling of laterally loaded pile groups in sands." *Geotechnical Testing Journal*, 17(2), 129–137.

McVay, M., Zhang, L., Molnit, T., and Lai, P. (1998). "Centrifuge testing of large laterally loaded pile groups in sands." *Journal of Geotechnical and Geoenvironmental Engineering*, 124(10), 1016–1026.

McVay, M. C., Shang, T. I., and Casper, R. (1996). "Centrifuge testing of fixed-head laterally loaded battered and plumb pile groups in sand." *Geotechnical Testing Journal*, 19(1), 41–50.

Ng, C. W., Zhang, L., and Nip, D. C. (2001). "Response of laterally loaded large-diameter bored pile groups." *Journal of Geotechnical and Geoenvironmental Engineering*, 127(8), 658–669.

Rollins, K. M., Peterson, K. T., and Weaver, T. J. (1998). "Lateral load behavior of full-scale pile group in clay." *Journal of Geotechnical and Geoenvironmental Engineering*, 124(6), 468–478.

Stewart, D. P. (2000). *User manual for program PYGMY*. The University of Western Australia.

Sun, K. (1993). "Static analysis of laterally loaded piles." *Proceedings of 11th Southeast Asian Geotechnical Conference*, Singapore, 589–594.

Sun, K. (1994a). "Laterally loaded piles in elastic media." *Journal of Geotechnical Engineering*, 120(8), 1324–1344.

Sun, K. (1994b). "A numerical method for laterally loaded piles." *Computers and Geotechnics*, 16(4), 263–289.

Winkler, E. (1867). *Die lehre von elastizitat und festigkeit (on elasticity and fixity)*. H. Dominicus, Prague.

Chapter 16

Retaining structures

The sciences do not try to explain, they hardly even try to interpret, they mainly make models. By a model is meant a mathematical construct which, with the addition of certain verbal interpretations, describes observed phenomena. The justification of such a mathematical construct is solely and precisely that it is expected to work.

John Von Neumann

Theory attracts practice as the magnet attracts iron.

Johann Carl Friedrich Gauss

The two main purposes of retaining structures are to create space where it is needed and to provide the support required to build up the ground when it would not be stable otherwise. The first use is illustrated by the creation of underground space for parking garages, retail stores, or subway stations. This space is often created with the aid of retaining structures. An example of the second purpose is seen in roadways or railways, when retained soils are used to construct bridge abutments. Another possible use of retaining structures is to either permanently or temporarily stabilize slopes or excavations. In this chapter, we examine the most common types of retaining structures and cover the design methods for each.

16.1 PURPOSE AND TYPES OF RETAINING STRUCTURES

16.1.1 The function of retaining structures

Retaining structures are often needed to create temporary or permanent space. Temporary space is needed to create the conditions for construction and foundation installation activities at construction sites. Permanent space is needed for underground parking, subway stations, and various other purposes. Retaining structures retain soil in place when the natural tendency of the soil would be to slide due to self-weight or existing loading. When the ground needs to be built up, the space behind a retaining wall is backfilled with soil. A clear example of this use of retaining walls is the construction of bridge abutments, nowadays most frequently done using mechanically stabilized earth (MSE) walls.

As foundations and slopes, retaining structures must also be designed in a way such that serviceability and ultimate limit states are not reached. Serviceability limit states are mostly associated with excessive deflections of the structures toward the open space and away from the ground they support. This leads to a corresponding loss of ground behind the retaining structure; if a structure is present there, it may be damaged. Ultimate limit states include bearing capacity failure, sliding, overturning, and general instability of the retaining structure and the soil it supports (Figure 16.1). Loukidis and Salgado (2012) showed that these

DOI: 10.1201/b22079-16

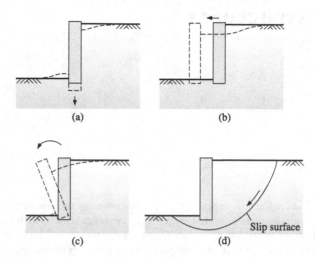

Figure 16.1 Ultimate limit states for retaining structures: (a) limit bearing capacity failure, (b) sliding, (c) overturning, and (d) general loss of stability.

ultimate limit states do not, under realistic conditions, happen completely independently from each other. In other words, for the wall to overturn, a bearing capacity failure will tend to happen at the same time (and there may be significant sliding of the wall as well). Only sophisticated numerical simulations of wall response can capture such complex response, so routine design is based on analyzing each of these limit states separately, as if they happened that way. That approach is typically safe. In addition to checking for all of these geotechnical limit states, engineers must also ascertain that the retaining structure will retain structural integrity throughout its useful life.

Serviceability and ultimate limit states become particularly important for buildings with underground parking space or similar structures, with *tieback* walls[1] (see Figure 16.3c) used as retaining structures. It occasionally happens that tiebacks move after installation. This leads in turn to removal of support from adjacent building foundations, which, if severe, can lead to serious damage. Figure 16.2a and b shows a four- to five-story-high tieback *slurry wall*[2] that moved significantly, causing structural damage to a neighboring building. The steel bracing used to control the severe cracking observed in structural elements of the building is shown in Figure 16.2c. Although the retaining wall was ultimately shown to be stable, the movement was large enough to cause the damage shown in the figure, which required costly repair.

To summarize, it is essential in the design of retaining structures to either explicitly or implicitly check that the following limit states are not attained:

- sliding (horizontal translation of the wall away from the retained soil);
- overturning;
- bearing capacity (see Chapters 10 and 13); and
- general instability (see Chapter 17).

---

[1] A tieback is an anchor that uses soil resistance to restrain the wall from moving away from the retained soil and toward the open space in front of it. We will discuss tiebacks and tieback walls in detail later in this chapter.

[2] Slurry walls are cast-in-place concrete walls; the name comes from the use of a slurry to prevent soil cave-ins during the construction of the walls. Slurry walls are discussed in detail later in this chapter.

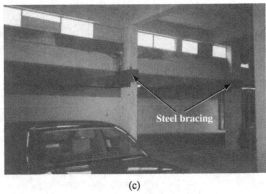

(a)

(c)

(b)

Figure 16.2 Consequences of excessive movements of a tieback wall: (a) aerial view of the retaining wall, (b) the slurry wall viewed from the side of the excavation, and (c) damage to neighboring building, contained with the use of steel bracing of the structure.

16.1.2 Types of retaining structures

Retaining structures can initially be classified in terms of the mechanism by which they develop the resistance necessary to overcome the earth pressures from the retained ground and any other surcharge applied on top of the retained soil. The mechanisms can be broadly classified as external or internal. External sources include:

1. gravity (self-weight of the retaining structure, which applies a counter-moment that balances the moment applied by the earth pressures) and
2. passive resistance (resulting from the embedment of the wall into the soil, forming a cantilevered structure).

Tiebacks (which pull on the wall, opposing the applied earth pressures) and bracing (which pushes on the wall when the wall tends to move into the excavation) are also considered external sources of stability.

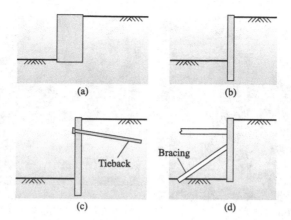

Figure 16.3 Externally stabilized retaining walls: (a) gravity structure, (b) cantilevered structure, (c) tieback structure, and (d) braced excavation.

Retaining structures whose stability derives from external sources are said to be *externally stabilized*. Figure 16.3 shows four types of externally stabilized retaining walls.

Figure 16.4 shows a variety of gravity walls. Not shown in the figure are reinforced concrete walls, which are relatively expensive and have been used less frequently in recent years due to the rise in the use of MSE walls. Figures 16.2b and 16.5 show a tieback wall. Tieback walls are particularly suited to situations where a tall wall is needed, as multiple layers of tiebacks can be used to ensure stability and keep wall movement in check. Figure 16.6 shows a secant drilled shaft cantilever wall. Unpropped, embedded cantilever walls with heights of as much as 5–6 m (in the case of slurry or secant pile walls) and 3–4 m (in the case of steel sheet pile walls) are economically feasible in situations when a structure does not exist very close to the wall crest.

Sheet piles are often used to construct retaining structures. Sheet piles may be made of steel, concrete, or wood, but the vast majority of work is done with steel sheet piles. The pile cross sections (see Figure 16.7) are such that the piles interlock when driven side-by-side, leaving no space between the piles for the soil to move through; the interlocking also offers some watertightness.

Hot-rolled sheet piles are manufactured from liquid steel in a process that produces a homogeneous metallurgical fabric and uniform mechanical properties. It is also possible to vary sheet pile thickness with the hot-rolled process. The other method, in which steel coil is cold-rolled into a sheet pile shape, is subject to several limitations, including a maximum section thicknesses of 0.5 in. and looser connections at the interlocks. In most sheet pile applications in the United States, the hot-rolled Z-sections shown in Figure 16.7 are used. Sheet piles are manufactured in a range of sizes. They are specified according to the dimensions shown in Figure 16.7b.

A common application of sheet piling is waterfront structures, such as piers and docks. Figure 16.8a shows a retaining structure along the Chicago River designed to support the banks on which residential buildings are located. Figure 16.8b also shows another common application of sheet piling: temporary excavation support during the construction of foundations or other construction activities. Additional applications include the stabilization of slopes and the construction of *cofferdams*.[3] Ordinary sheet pile walls are, like slurry walls or secant pile walls, either cantilever (embedded) walls or tieback walls.

---

[3] A cofferdam is a temporary or permanent structure built to exclude water from an area that is normally submerged as in, for example, a river. The water exclusion is necessary for construction activity, as in building the foundations of a bridge pier.

Figure 16.4 Various types of gravity walls: (a) masonry wall, (b) gabion wall, and (c)–(e) crib walls.

Sheet piles are driven, pushed, or vibrated into the ground. In sands, particularly if the water table is high, vibratory hammers produce the best results, as sand liquefaction due to the vibration aids the rate of advance. For short sheet piles, jacking can be quite efficient, as agile, hydraulic machines now exist that can quickly push sheet piles into the ground.

Internally stabilized (also referred to as *mechanically stabilized*) retaining structures are of the reinforced soil type, where inserts of different types provide the necessary stability. The most common types of reinforcing elements are the so-called *soil nails*, geotextiles/ geogrids (Figure 16.9), and metallic strips (Figure 16.10). Soil nails are installed in natural soil (Figure 16.11), whereas geogrids, geotextiles, or metallic strips are placed between lifts of soil when a fill is built. Either way, soils reinforced with nails, geogrids, geotextiles, or

Figure 16.5 Tieback wall.

(a)   (b)

Figure 16.6 Cantilever secant drilled shaft wall: (a) immediately after construction (with a temporary slope to aid stability) and (b) used as final wall of an underground parking garage.

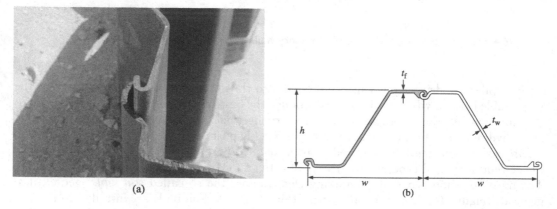

Figure 16.7 Typical sheet pile cross sections: (a) a view of a Z-section sheet pile and (b) schematic of Z-section sheet pile.

(a) (b)

Figure 16.8 Typical applications of sheet piles: (a) a permanent retaining structure along the Chicago River; (b) temporary excavation support. ((a) Courtesy of Ioannis Zevgolis.)

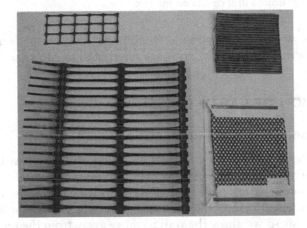

Figure 16.9 Geogrids (left) and geotextiles (right): biaxial geogrid on upper left, uniaxial geogrid on lower left, and woven geotextiles on right.

Figure 16.10 Metallic strips: the photo shows them attached to the panels that make up the facing of the wall. (Courtesy of the Indiana Department of Transportation.)

Figure 16.11 Installation of a soil nail: the photo shows a hose used for grouting being inserted in the pre-drilled soil nail hole; a drill rig sits on the left.

metallic strips are known as reinforced soils. Retaining structures made up of soil reinforced with geotextiles, geogrids, or metallic strips are referred to as MSE walls.

16.2 CALCULATION OF EARTH PRESSURES

16.2.1 Mobilization of active and passive pressures

Figure 16.12 shows how the pressure behind a retaining wall varies with lateral deflection. Initially, with the lateral deflection equal to zero, the pressure is the at-rest pressure, and the lateral earth pressure coefficient K is equal to K_0. If we push on the wall, increasing the pressures behind it, the soil behind the wall will eventually reach a passive Rankine state, for which $K = K_p$. If instead we allow the wall to move away from the retained soil mass, an active state is established, and $K = K_A$. Chapter 4 discusses in detail the attainment of active and passive states in soil.

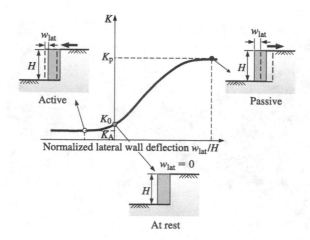

Figure 16.12 Coefficient of lateral earth pressure behind retaining wall versus lateral deflection w_{lat} normalized with respect to wall height H.

Table 16.1 Movement required to mobilize active and passive pressures

	Horizontal displacement/wall height	
Soil	Active state	Passive state
Dense sand	0.001	0.020
Loose sand	0.004	0.060
Stiff clay	0.010	0.020
Soft clay	0.020	0.040

Based on values proposed in CGS (1992).

Much more movement is required to mobilize the passive than the active state, as shown in Table 16.1. For a 10-meter-high wall retaining dense sand, a movement at the top of the wall of as little as 10 mm away from the soil is sufficient to establish an active state in the retained soil. This means that, in most gravity or cantilever (embedded) retaining walls, the pressure acting on (behind) the wall is active. On the other hand, the total earth force acting behind a tieback wall is on the order of 20% greater than the active force (for one line of tiebacks), and even greater for multiple lines of tiebacks or for supported (braced) excavations. This is so because the tieback, by restricting wall movement, prevents or at the least disrupts the formation of an active Rankine state in the soil.

16.2.2 Calculation of active earth pressures using the formulation of Rankine

16.2.2.1 Active pressures for level soil masses

In Chapter 4, we introduced the concept of the active and passive Rankine states for a soil mass with a horizontal ground surface. A soil mass (or part of a soil mass) is in a Rankine state when the stress state at every point of it satisfies the failure criterion, which means that the soil is in a plastic state, and the principal stress directions are the same at every point of the soil mass. In a Mohr's diagram, the Mohr circle representing every one of these states of stress touches the strength envelope. Let us see now how that knowledge can be used to calculate earth pressures acting on retaining structures.

Figure 16.13 shows how an active Rankine state develops behind retaining structures that move in translation (Figure 16.13a) and rotation (Figure 16.13b) when acted upon by the

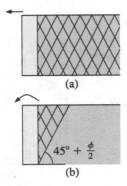

(a)

$45° + \dfrac{\phi}{2}$

(b)

Figure 16.13 Mobilization of active Rankine state in the soil retained by (a) a retaining structure that moves away from the soil horizontally in translation mode and (b) a retaining structure that rotates away from the soil around its toe.

pressure from the soil. Note that, in the case of rotation, an active state is mobilized only within a wedge of soil because the extent of the Rankine zone at a given depth is directly proportional to the amount of movement of the wall at that depth. In the case of translation, a more extensive active zone develops.

To calculate active pressures using the Rankine formulation covered in Chapter 4, we must assume that the shear stress along the back face of the retaining structure is zero (which means the pressures are normal to the wall). This is because, in Rankine theory, the vertical plane is a principal plane, as is the horizontal plane. This also means σ'_v and σ'_h are principal effective stresses.

It is convenient at this point to recall the equations for the at-rest lateral pressure coefficient K_0 and for the Rankine lateral earth pressure coefficients K_A and K_P introduced in Chapter 4:

$$K_0 = K_{0,NC} \sqrt{OCR} \tag{4.57}$$

where $K_{0,NC}$ is a value usually in the 0.4–0.5 range for sands and 0.5–0.75 range for clays, and

$$K_A = \frac{1}{N} = \frac{1 - \sin\phi}{1 + \sin\phi} \tag{4.59}$$

$$K_P = N = \frac{1 + \sin\phi}{1 - \sin\phi} \tag{4.60}$$

In Chapter 4, we did not consider either the existence of the water table, or a soil profile containing more than one soil type. We will introduce these two cases with two examples.

Example 16.1

Calculate the active earth pressure distribution and the resultant force on the retaining wall in E-Figure 16.1a.

Solution:

The active earth pressure coefficient for the sand retained by the wall is given by Equation (4.59):

$$K_A = \frac{1}{N} = \frac{1 - \sin 40°}{1 + \sin 40°} = 0.217$$

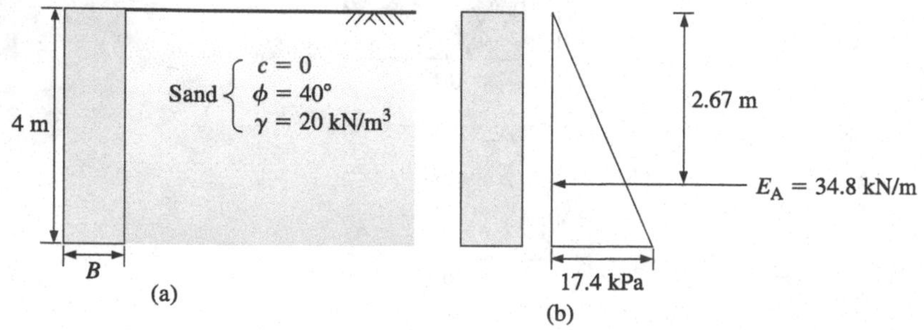

E-Figure 16.1 Retaining wall of Example 16.1: (a) geometry of the wall and soil properties and (b) pressure diagram and resultant active force on the wall.

The lateral effective stress for any depth z is given by

$$\sigma'_{hA} = K_A \sigma'_v = K_A \gamma z = 0.217 \times 20 \times z = 4.34z$$

Specifically, at the base of the wall, at a depth of 4 m:

$$\sigma'_{hA} = 4.34z = 4.34 \times 4 = 17.4 \text{ kPa}$$

as seen in E-Figure 16.1b.
The total force on the wall is obtained as either

$$E_A = \frac{1}{2} \times 17.4 \times 4 = 34.8 \approx 35 \text{ kN/m}$$

or, if the retained soil mass is homogeneous with a deep water table (as in the present case), directly from:

$$E_A = \frac{1}{2} K_A \gamma H^2 = \frac{1}{2} \times 0.217 \times 20 \times 16 = 34.8 \approx 35 \text{ kN/m} \text{ answer}$$

Because of the triangular pressure distribution, E_A is applied at one-third of the wall height from the base of the wall, or at a depth of 2.67 m, as shown in E-Figure 16.1b.

Example 16.2

For the retaining wall and soil profile in E-Figure 16.1, consider now the water table to be at a depth of 1 m. Recalculate the pressures acting on the back of the wall, considering now also that due to the water. Assume hydrostatic conditions (that is, assume no water flow anywhere).

Solution:

The difference between this and the previous example is the water table at 1 m depth below the ground surface. The presence of the water changes the analysis in two ways: (1) the vertical effective stresses decrease and (2) the force on the wall due to the water pressures must be added to the active force to find the total force on the wall.
The active stress is calculated as follows:

$$\sigma'_{hA} = K_A \sigma'_v$$

$$= K_A \gamma_m z = 0.217 \times 20 \times z = 4.34z \text{ for } z \le 1\text{m}$$

$$= K_A \left[\gamma_m z - \gamma_w (z - z_w) \right] = 0.217 \left[20z - 9.81(z-1) \right] = 2.21z + 2.13 \text{ for } z > 1\text{m}$$

E-Figure 16.2 shows the resulting active stress distribution, which can be decomposed in two triangular and one rectangular stress distribution to yield the total active force E_A:

$$E_A = \frac{1}{2} \times 1 \times 4.34 + 3 \times 4.34 + \frac{1}{2} \times 3 \times 6.66 = 2.2 + 13 + 10 = 25.2 \approx 25 \text{kN/m} \text{ answer}$$

The height of the point of application of E_A measured from the base of the retaining wall can be determined from the points of application of the resultants of the three component stress distributions (refer to E-Figure 16.2b):

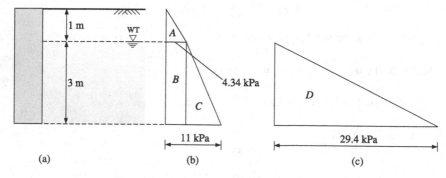

E-Figure 16.2 Solution of Example 16.2: (a) retaining wall with water table at a depth of 1 m, (b) active stress distribution and its decomposition in three separate stress distributions, and (c) distribution of water pressure on the back of wall.

$$E_A h_E = 25.2 h_E = \left[\left(3+\frac{1}{3}\times 1\right)\times 2.2 + 1.5\times 13 + \frac{1}{3}\times 3\times 10\right]$$

where h_E is the distance from the base of the wall to the point of application of E_A. From this equation, we get

$$h_E = 1.46\,\mathrm{m} \quad \text{answer}$$

$$z_E = 4-1.46 = 2.54\,\mathrm{m}$$

The water pressure distribution is simply a triangular distribution starting at $z=1$ m with slope 9.81 (refer to E-Figure 16.2c). So the total water force on the wall is

$$E_w = \frac{1}{2}\gamma_w h_w^2 = \frac{1}{2}\times 9.81\times 3^2 = 44.1 \approx 44\,\mathrm{kN/m} \quad \text{answer}$$

Note how the water force can quickly exceed the active force, hence the desirability or even necessity of providing sufficient drainage behind retaining walls to prevent water from accumulating behind them.[4]

16.2.3 Calculation of earth pressures using the formulation of Coulomb

Coulomb (1776) also studied the problem of earth pressures on retaining walls. In particular, Coulomb allowed the soil–wall interface friction angle to take any value δ. Coulomb also allowed for a sloping retained soil mass (with angle α_g to the horizontal) and an inclined wall (with angle β_w to the horizontal), as shown in Figure 16.14. According to Coulomb's theory, if the active force E_A on the wall shown in the figure reaches a threshold value, a soil wedge forms, with a slip surface developing in the soil mass extending all the way from

---

[4] It is not at all clear that a designer should rely on the availability of effective drainage. Although designers commonly do assume effective drainage, and this common practice would be evidence in support of a standard of care based on the assumption of functioning drains and weep holes, that would not necessarily be dispositive. It can be argued that it is foreseeable that drainage might be impeded at some point in the future, in which case water forces would be brought to bear on the wall. Accordingly, it is possible that a court would find a designer negligent for not anticipating that.

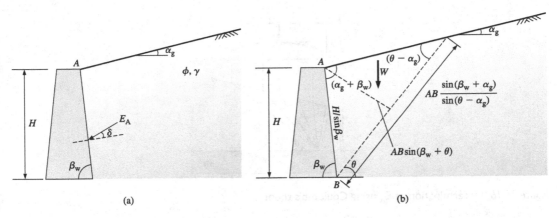

Figure 16.14 Coulomb (1776) formulation for the retaining wall problem: (a) wall geometry and applied earth force and (b) wall and wedge geometry.

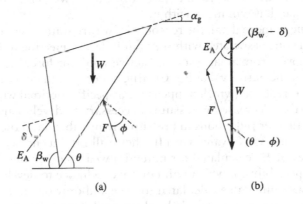

Figure 16.15 Potential sliding soil wedge: (a) free body diagram and (b) force polygon.

point B at the base of the wall to the ground surface. A question that arises immediately is: of all the potential active soil wedges behind the wall, how do we determine which one is the one that actually develops and slides together with the wall?

To answer this question, consider first the free body diagram of a soil wedge, shown in Figure 16.15a. Note that the forces acting on the wedge are its own self-weight W (passing through its center of mass), the resisting force F (a combination of a normal force F_N and a tangential force F_T) exerted by the stationary soil mass, and the active force E_A exerted by the wall on the soil wedge. If the wedge is in equilibrium under the action of these three forces, the polygon of forces must close, as shown in Figure 16.15b. This analysis ignores the possibility that the three forces would not pass through the same point, which would require a check on moment equilibrium as well. In practice, the error introduced by ignoring moment equilibrium is small, and the analysis produces satisfactory results.

Now, referring to Figure 16.16, consider that a very thin wedge, with a steep slip surface through the soil mass and a small self-weight W, would get most of its weight supported by the soil resistance F on the potential slip surface (and mostly by the tangential component F_T of F), with E_A, as a result, being very small. E_A is also very small for a long, large wedge, with very flat slip surface. In this case, the projection of W in the direction of the slip surface is small, and the soil wedge is easily supported by the soil resistance F (this time, mostly by

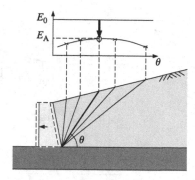

Figure 16.16 Determination of E_A using Coulomb's theory.

the normal component F_N of F), so that the balance of force left for the wall to hold back is small. It is clear that, for some wedge, one with a base that is not too steep, but not too flat either, E_A is maximum. This is shown in Figure 16.16. If the wall fails in sliding or overturning, it is this soil wedge that will move with it.

Now let us examine in more detail the reason why we are stating that the wedge for which E_A is maximum is the one associated with the wall failure. Referring to Figure 16.12, as the wall is loaded and moves from an at-rest to an active state, the force on the wall drops from the value E_0 at rest to the active value E_A. Clearly, as soon as the maximum value of E_A of the Coulomb theory is reached (which happens for a specific, optimal wedge), an active state is established: no further decrease in pressure is necessary to develop an active state behind the wall. Indeed, no further reductions in pressure are possible,[5] and the pressures stabilize, as shown by the constant $K=K_A$ value with further wall movement (see Figure 16.12). This means that the values of E_A calculated for potential wedges other than the optimal wedge are only theoretical possibilities, which will not have a chance to develop, and that the real value of E_A is the maximum value calculated for one and only one potential sliding wedge, which will be reached first as pressures behind the wall drop from the at-rest to the active values.

For the case in Figure 16.14, in which there are no surcharges on top of the retained soil mass, maximization of E_A leads to the following expression for the Coulomb active force E_A:

$$E_A = \frac{1}{2} K_A \gamma H^2 \tag{16.1}$$

where

$$K_A = \frac{\sin^2(\phi + \beta_w)}{\sin^2 \beta_w \sin(\beta_w - \delta)\left[1 + \sqrt{\dfrac{\sin(\phi + \delta)\sin(\phi - \alpha_g)}{\sin(\beta_w - \delta)\sin(\alpha_g + \beta_w)}}\right]^2} \tag{16.2}$$

When $\delta = \alpha_g = 0$ and $\beta_w = 90°$, Equation (16.2) reduces to the Rankine solution, Equation (4.59). In Problem 16.10, you will be asked to show that this is indeed the case.

---

[5] This is strictly true for a material with constant ϕ. We will discuss later in this chapter how to select a value of ϕ for use in design when there is a peak friction angle at relatively small values of shear strain followed by values that drop with continuing shearing until the critical state is reached, as is the case with frictional–dilative soils sheared under drained conditions.

16.2.4 Calculation of earth pressures using the formulation of Lancellotta*

Lancellotta (2002) proposed a solution for the active and passive earth pressure coefficients for a wall with interface friction angle δ based on a lower bound analysis.[6] The resulting expression for the passive and active lateral earth pressure coefficients for level ground is

$$K_{P,A} = \left[\frac{\cos\delta}{1\mp\sin\phi}\left(\cos\delta \pm \sqrt{\sin^2\phi - \sin^2\delta}\right) \right]\exp\left(\pm 2\theta\tan\phi\right) \qquad (16.3)$$

where

$$\theta = \frac{1}{2}\left[\arcsin\left(\frac{\sin\delta}{\sin\phi}\right) \pm \delta \right] \qquad (16.4)$$

In these equations, when a choice has to be made between a plus and a minus, the upper symbol is used for the passive case and the lower symbol for the active case. Note that, when $\delta=0$, Equation (16.3) reduces to the Rankine solution. Additionally, given that the solution is based on lower bound analysis (in effect, finding a soil unit weight that is no greater than the unit weight that would lead to an active state), it overestimates the ratio of lateral to vertical effective stresses and is thus conservative for the calculation of active pressures.

16.2.5 Calculation of earth pressures accounting for soil arching effects using the formulation of Paik and Salgado (2003)*

The use of any of the solutions developed thus far for obtaining the distribution of earth pressures behind a retaining wall would lead to a linear variation with depth of active pressures on the wall. In reality, for a wall with nonzero interface friction angle δ, the lateral stress distribution behind the wall is nonlinear. This is due to arching effects that develop as the wall moves away from the retained soil mass and the soil wedge attempts to slide down behind the retreating wall.

The wall provides vertical support to the soil wedge through friction between the soil wedge and the wall. This reduces the vertical effective stress in the soil at every depth, thus the lateral effective stress that must be transferred to the wall. This means that the increase in lateral earth pressure is less than that predicted by the Rankine theory, which assumes interface friction angle $\delta=0$. Additionally, as the width of the soil wedge reduces with depth, approaching zero at the toe of the wall, there is initially a nonlinear increase and after some point a decrease in the active lateral earth pressure coefficient with depth until it becomes zero at the toe of the wall (Figure 16.17).

According to Paik and Salgado (2003), the lateral effective stress σ'_{ahw} acting on the retaining wall at a given depth z is given by

$$\sigma'_{ahw} = K_{awn}\bar{\sigma}'_v \qquad (16.5)$$

where the lateral earth pressure coefficient K_{awn} is given by

---

[6] Lower bound analysis is a rigorous analysis based on a theorem of plasticity theory that allows us to determine with 100% certainty the lower bound on the load causing collapse of a given structure. It is discussed and used in other parts of the text together with its frequent companion: the upper bound theorem. Together, the lower and upper bound theorems allow us to establish the range within which a collapse load falls with 100% certainty, so long as the shear strength of the soil is known deterministically.

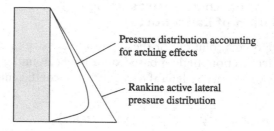

Pressure distribution accounting for arching effects

Rankine active lateral pressure distribution

Figure 16.17 Earth pressure distribution behind a retaining wall with nonzero δ.

$$K_{awn} = \frac{\sigma'_{ahw}}{\overline{\sigma}'_v} = \frac{3\left(N\cos^2\theta + \sin^2\theta\right)}{3N - (N-1)\cos^2\theta} \tag{16.6}$$

the average vertical stress $\overline{\sigma}_v$ within the soil wedge is given by

$$\overline{\sigma}'_v = \frac{\gamma H}{1 - K_{awn}\tan\delta\tan\left(45° + \frac{\phi}{2}\right)}\left[\left(1 - \frac{z}{H}\right)^{K_{awn}\tan\delta\tan\left(45° + \frac{\phi}{2}\right)} - \left(1 - \frac{z}{H}\right)\right] \tag{16.7}$$

the angle θ that σ'_1 makes with the horizontal is given by

$$\theta = \tan^{-1}\left[\frac{(N-1) + \sqrt{(N-1)^2 - 4N\tan^2\delta}}{2\tan\delta}\right] \tag{16.8}$$

and N is the flow number, from Equation (4.45):

$$N = \frac{1 + \sin\phi}{1 - \sin\phi}$$

Integration of the lateral stress along the height of the wall leads to a resultant lateral force E_{Ah} given by

$$E_{Ah} = \frac{1}{2}K_{AW}\gamma H^2 \tag{16.9}$$

applied at a distance h from the base of the wall given by

$$h = \frac{2\left(1 + K_{awn}\sqrt{N}\tan\delta\right)}{3\left(2 + K_{awn}\sqrt{N}\tan\delta\right)}H \tag{16.10}$$

where

$$K_{AW} = \frac{K_{awn}}{1 - K_{awn}\sqrt{N}\tan\delta}\left(\frac{2}{1 + K_{awn}\sqrt{N}\tan\delta} - 1\right) \tag{16.11}$$

As the method applies only to walls with vertical backs, the total active force E_A can be calculated from the vector addition of the lateral active force E_{Ah} normal to the wall and the shearing force T acting tangentially to the wall. E_A, which makes an angle δ with the normal to the back of wall, can be obtained as

$$E_A = \sqrt{E_{Ah}^2 + T^2} = \sqrt{E_{Ah}^2 + E_{Ah}^2 \tan^2 \delta} = \frac{E_{Ah}}{\cos \delta} \qquad (16.12)$$

Example 16.3

Recalculate the active force acting on the back of the retaining wall in E-Figure 16.1 and Example 16.1 using the Coulomb (1776), Lancellotta (2002), and Paik and Salgado (2003) analyses. The wall is made of a material such that $\delta=20°$.

Solution:

For this problem:

$$\alpha_g = 0°$$

$$\beta_w = 90°$$

$$\delta = 20°$$

$$\phi = 40°$$

Solution using Coulomb's method

We use Equation (16.2) directly to compute K_A:

$$K_A = \frac{\sin^2(\phi + \beta_w)}{\sin^2 \beta_w \sin(\beta_w - \delta) \left[1 + \sqrt{\dfrac{\sin(\phi + \delta)\sin(\phi - \alpha_g)}{\sin(\beta_w - \delta)\sin(\alpha_g + \beta_w)}} \right]^2}$$

$$= \frac{\sin^2(40° + 90°)}{\sin^2 90° \sin(90° - 20°) \left[1 + \sqrt{\dfrac{\sin(40° + 20°)\sin(40° - 0°)}{\sin(90° - 20°)\sin(0° + 90°)}} \right]^2}$$

$$= 0.199$$

The force on the wall is given by Equation (16.1):

$$E_A = \frac{1}{2} K_A \gamma H^2 = \frac{1}{2} \times 0.199 \times 20 \times 4^2 = 31.8 \approx 32 \, \text{kN/m} \quad \text{answer}$$

The Coulomb equation for K_A reduces to the Rankine equation when $\delta=0°$. On this basis, it is tacitly assumed that the Coulomb method, like Rankine, implies a linear stress distribution behind the wall. Thus, the force E_A is applied at one-third the height of the wall, corresponding to:

$$h_E = \frac{1}{3} \times 4 = 1.33 \, \text{m}$$

$$z_E = 2.67 \, \text{m}$$

just as in the Rankine method.

Solution using Lancellotta's method

The angle θ required to calculate K_A is given by Equation (16.4):

$$\theta = \frac{1}{2}\arcsin\left(\frac{\sin\delta}{\sin\phi}\right) - \delta = \frac{1}{2}\arcsin\left(\frac{\sin 20°}{\sin 40°}\right) - 20° = -3.93° = -0.0686\,\text{radians}$$

Taking that and the other input values into Equation (16.3) gives

$$K_A = \left[\frac{\cos\delta}{1+\sin\phi}\left(\cos\delta - \sqrt{\sin^2\phi - \sin^2\delta}\right)\right]\exp(-2\theta\tan\phi)$$

$$= \left[\frac{\cos 20°}{1+\sin 40°}\left(\cos 20° - \sqrt{\sin^2 40° - \sin^2 20°}\right)\right]\exp(0.137\tan 40°)$$

$$= 0.254$$

The active force on the wall can now be calculated from Equation (16.1):

$$E_A = \frac{1}{2}K_A\gamma H^2 = \frac{1}{2}\times 0.254 \times 20 \times 4^2 = 40.6 \approx 41\,\text{kN/m} \quad^{\text{answer}}$$

with the same point of application as obtained using Coulomb's method, given that a linear stress distribution is implied. As indicated previously, this value is conservative.

Solution using Paik and Salgado's Method

We first calculate the flow number N:

$$N = \frac{1+\sin\phi}{1-\sin\phi} = \frac{1+\sin 40°}{1-\sin 40°} = 4.6$$

and then the angle θ from Equation (16.8):

$$\theta = \tan^{-1}\left[\frac{(N-1)+\sqrt{(N-1)^2 - 4N\tan^2\delta}}{2\tan\delta}\right]$$

$$= \tan^{-1}\left[\frac{(4.6-1)+\sqrt{(4.6-1)^2 - 4\times 4.6\tan^2 20°}}{2\tan 20°}\right]$$

$$= 83.9°$$

The ratio of σ'_{ahw} acting on the wall to the average vertical effective stress is given by Equation (16.6):

$$K_{\text{awn}} = \frac{3(N\cos^2\theta + \sin^2\theta)}{3N - (N-1)\cos^2\theta} = \frac{3(4.6\cos^2 83.9° + \sin^2 83.9°)}{3\times 4.6 - (4.6-1)\cos^2 83.9°} = 0.227$$

The coefficient K_{AW} that allows the calculation of the total earth force on the wall is given by Equation (16.11):

$$K_{AW} = \frac{K_{awn}}{1 - K_{awn}\sqrt{N}\tan\delta}\left(\frac{2}{1 + K_{awn}\sqrt{N}\tan\delta} - 1\right)$$

$$= \frac{0.227}{1 - 0.227\sqrt{4.6}\tan 20°}\left(\frac{2}{1 + 0.227\sqrt{4.6}\tan 20°} - 1\right)$$

$$= 0.193$$

The horizontal active force on the wall can now be calculated from Equation (16.9):

$$E_{Ah} = \frac{1}{2}K_{AW}\gamma H^2 = \frac{1}{2}\times 0.193 \times 20 \times 4^2 = 30.9 \approx 31\,\text{kN/m}\ \text{answer}$$

applied at

$$h_E = \frac{2\left(1 + K_{awn}\sqrt{N}\tan\delta\right)}{3\left(2 + K_{awn}\sqrt{N}\tan\delta\right)}H$$

$$= \frac{2\left(1 + 0.227\sqrt{4.6}\tan 20°\right)}{3\left(2 + 0.227\sqrt{4.6}\tan 20°\right)}\times 4$$

$$= 1.44\,\text{m}$$

from the base of the wall or

$$z_E = H - h_E = 4 - 1.44 = 2.56\,\text{m}$$

from the top.

16.2.6 Choice of friction angle for use in calculations of active and passive pressures*

In contrast to foundations, which cause substantial increases in normal stresses in the soil below them when loaded, retaining structures move away from the soil they retain, doing the opposite: unloading the soil. This implies that the range of normal stresses observed in the soil backfill or retained soil mass is of the order of the initial stress range. This means that an applicable strength envelope can be obtained by fitting to the results of strength tests, but we should resist the temptation to use a straight-line fit that is defined by a cohesive intercept c and a "friction angle" ϕ for reasons discussed earlier in the text, Chapter 4 in particular.

Given that sandy and gravelly backfills are purely frictional soils, the question of what friction angle ϕ to use in calculations reduces to whether any contribution from dilatancy should be considered. If the answer were in the affirmative, then the resulting friction angle ϕ would be greater than the critical-state friction angle ϕ_c and the pressures on the wall, lower. However, if a contribution to strength from dilatancy is not available when a limit state is reached, then ϕ_c should be used in design.

As we have seen earlier, the wall does not need to move very significantly for an active state to develop behind the wall. When that happens, shear bands form, the most notable being the one that forms the base of the potential sliding wedge. Along that shear band, significant shear deformation takes place, with the soil reaching or approaching critical state. Loukidis and Salgado (2012) performed finite element analyses to study what happens in the soil when walls reach an active state and approach the most critical limit state. Based on the results of that analysis, it is reasonable to use ϕ_c in design.

Clays are not typically used as backfill in constructed walls, but can be part of a retained soil mass when excavations are done. In the case of clays, short-term stability, except in soft clays, is not usually critical because only soft clays will develop positive excess pore pressures in the short term. So a drained, effective stress analysis is an appropriate basis for design, and again ϕ_c should be used in design. There could be situations in which even the residual friction angle ϕ_r might be an appropriate choice.

Some designers might be tempted to use a Mohr–Coulomb straight-line fit to perform stability analysis. This would typically yield curve-fitting parameters c and ϕ. These cannot be assumed to be sustainable, whether the soil is a clay or a sand. Take et al. (2004) discuss (for slopes, but the reasoning applies here as well) how, even if all or some of the dilative component of shear strength were to be present in the active state, wetting and drying associated with seasonal patterns may lead to gradual degradation of the dilativeness of the soil and thus of its "cohesive" intercept. Thus, it is necessary to model retained clay as a purely frictional soil, with ϕ equal to the critical-state friction angle ϕ_c, to neglect the contribution to stability from a dilative component of shear strength that will likely degrade as shear bands form or that will, in any case, degrade over time.

Soft clays can be unstable even in the short term. It would be incorrect, however, to carry out a total stress analysis with $\phi = 0$ and $K_A = 1$, which would be excessively conservative.

We will later see that passive pressures are often relied on by retaining structures to resist active pressure loadings. Passive pressures result from compressive loading of the soil, a process that is similar to what happens in the loading of foundations. The stress range in the soil where the passive pressures develop is thus greater than observed in active zones. Additionally, passive pressures require much more movement for full mobilization, so that to achieve compatibility between the movements required to mobilize active and passive pressures, we can only account for a fraction of the total, limit passive pressure. It is possible that a sustainable friction angle in excess of the critical-state friction angle is operative for this specific calculation, but there is no simple way to account for that.

Chapters 5 and 6 discuss in some detail the mobilization of shear strength in sands and clays, the occurrence of peaks in shear strength in drained loading of dilative soils (relatively dense sands and clays with large OCR) and in undrained loading of contractive soils (extremely loose sands and normally consolidated clays). These features in soil response should be considered in establishing appropriate design shear strengths.

16.3 DESIGN OF EXTERNALLY STABILIZED WALLS

16.3.1 Gravity walls

Gravity walls derive their resistance to both sliding and overturning almost exclusively from their own self-weight. Take the wall with the simple rectangular cross section in Figure 16.18 supporting an equally simple, uniform backfill. The problem is to find the width B of the wall such that it will be able to support the height H of soil with unit weight γ_m and friction angle ϕ without either overturning or sliding. Let us start by determining B such that the wall will not overturn.

For checking stability against overturning, we must first identify the forces acting on the wall. Clearly, the earth force E_A is trying to overturn the wall (as is any water force E_w behind the wall due to a high water table), which the weight W_{wall} of the wall is resisting. The question that arises next is: what is the role of the soil pressure at the base of the wall?

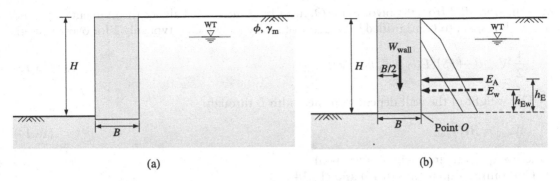

Figure 16.18 Gravity wall with rectangular cross section holding back (a) a soil mass with height H, unit weight γ_m, and friction angle ϕ plus water table behind the wall or (b) the equivalent pressures applied by the soil (with resultant E_A) and water (with resultant E_w).

As we saw when we studied footings subjected to moment loadings in Chapter 10, the pressure distribution at the base of the footing is skewed toward the side opposite the direction from which the moment is applied. The pressure distribution becomes increasingly skewed if either (1) the magnitude of the moment increases for a constant wall width, or (2) the wall width decreases for a constant moment. In this second case, the equivalent vertical force starts getting outside the middle third of the base as the width decreases, overloading areas increasingly closer to the edge of the foundation of the wall. Figure 16.19 shows three walls with decreasing base width from (a) to (c). Note that for the same supported soil mass, the pressure distribution becomes increasingly concentrated until, on the verge of overturning, the wall is supported only by its heel (point O). It is clear from Figure 16.19 that, on the verge of overturning, the moment of the soil force acting on the base of the wall at point O with respect to point O is zero. Our overturning check is done for the wall on the verge of overturning, and thus, the only forces to account for are the weight W_{wall} of the wall, the earth pressure force E_A, and any water force E_w that may be present.

For the simple case we are now considering, the back face of the wall is vertical and the vertical component of the active force E_A on the wall is zero. This means that E_A is a horizontal force applied at a distance h_E from the base of the wall. In the case of a water table located above the base of the wall, there is also a water force E_w applied at a distance h_{Ew} from the base of the wall. This force is always normal to the wall, given that there is no friction between water and the back face of the wall. The weight W_{wall} of the wall, which is applied

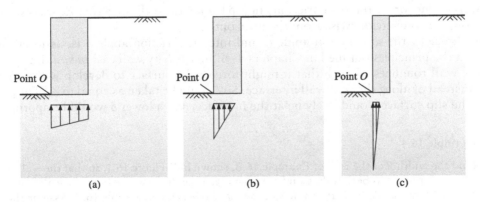

Figure 16.19 Soil pressure at base of a retaining wall for decreasing wall width from (a) to (c).

at a distance $B/2$ from the pivot point O, must be sufficient to balance the moment of E_A and E_w with respect to O magnified by a factor of safety FS (which is typically 2 for overturning):

$$\frac{1}{2} W_{wall}B - (FS)(E_A h_E + E_w h_{Ew}) = 0 \tag{16.13}$$

The weight of the wall depends on the width B through:

$$W_{wall} = BH\gamma_{wall} \tag{16.14}$$

where γ_{wall} is the unit weight of the wall.
   Combining Equations (16.13) and (16.14):

$$B = \sqrt{\frac{2(FS)(E_A h_E + E_w h_{Ew})}{H\gamma_{wall}}} \tag{16.15}$$

The source of the resistance to sliding is the shear strength that develops between the base of the wall and the supporting soil. This resistance must be greater than or equal to the active force on the wall plus any water force in case of a high water table. This can be written, with due account of the safety factor, as

$$HB\gamma_{wall}\mu = (FS)(E_A + E_w)$$

Rewriting this equation in terms of the wall width:

$$B = (FS)\frac{(E_A + E_w)}{H\gamma_{wall}\mu} \tag{16.16}$$

where μ is the coefficient of friction between the soil and the base of the wall, and the factor of safety against sliding is taken to be at least 1.5.
   The coefficient of friction μ is the same as that discussed in Chapter 11 for the sliding resistance of footings. It can be taken as

$$\mu = \tan\delta \tag{11.2}$$

where, for the rough surfaces typical of gravity walls, δ may be conservatively taken as 0.85–$0.95\ \phi_c$.
   In Problem 16.8, the reader is asked to rederive Equations (16.15) and (16.16) considering a nonzero value for the interface friction angle δ behind the wall, such that E_A is now rotated by an angle δ counterclockwise from the horizontal.
   The choice of the soil friction angle ϕ and interface friction angle δ is, as usual, made based on the principles outlined in Chapters 4–6. In a gravity wall cast *in situ*, it is possible that the wall roughness is such that it might force a slip surface to develop within the soil mass, instead of along the soil–wall interface. So δ could be taken as equal to ϕ_c. For precast walls, the slip surface would likely be at the interface, and a lower δ would be appropriate.

Example 16.4

Find the width B of the wall of Example 16.3, shown in E-Figure 16.1, so that the wall is safe with respect to both sliding and overturning. Take the unit weight of the wall material as 22 kN/m³. Assume the same values of ϕ and δ used in Example 16.3. Assume the same δ can be used for the base of the wall.

Solution:

The width of the wall will obviously vary depending on the method used to calculate the active force on the wall. Let us use the values calculated using the Mohr–Coulomb method, namely:

$$E_A = 32 \, \text{kN/m}$$

$$h_E = 1.33 \, \text{m}$$

There is a stabilizing effect of the vertical component of E_A, which is small and will thus be neglected.

The coefficient of friction μ is calculated using Equation (11.2):

$$\mu = \tan\delta = \tan 20° = 0.36$$

Using this value of μ and a factor of safety of 1.5 for sliding in Equation (16.16) gives

$$B = (FS)\frac{(E_A + E_w)}{H\gamma_{\text{wall}}\mu} = 1.5\frac{32}{4 \times 22 \times 0.36} = 1.52 \, \text{m}$$

Using now Equation (16.15) to determine the value of B required for overturning stability with $FS = 2$, we get

$$B = \sqrt{\frac{2(FS)(E_A h_E + F_w h_{Ew})}{H\gamma_{\text{wall}}}} = \sqrt{\frac{2 \times (2)(32 \times 1.33)}{4 \times 22}} = 1.39 \, \text{m}$$

It is obvious that, in this case, sliding controls the design because of the relatively low coefficient of friction. The design value of B, assuming the wall passes the bearing capacity and general stability checks, is

$$B = 1.52 \, \text{m}$$

which would typically be specified as 1.55 m.[answer]

16.3.2 Cantilever (embedded) walls

Figure 16.20a shows a cantilever (embedded) wall (which could be a drilled shaft, slurry, or sheet pile wall) supporting uniform soil with a deep water table. It also shows the movement of the wall under the action of the soil pressures. The wall has an above-ground height H and is embedded d into the ground. When acted upon by the soil pressures, the wall rotates around a point O located at a distance f from the ground surface. To determine where active and passive pressures act, we need to locate the parts of the wall that move against and those that move away from the soil. Figure 16.20b shows the pressure distributions behind and in front of the wall. Figure 16.20c shows a simplified form of this diagram, in which the active and passive pressures acting below point O are combined into a single force E_{P2} acting at O. This replacement is conservative given that there would be a net resisting moment with respect to O had we not made this replacement.

Referring to the free body diagram in Figure 16.20c, taking moments with respect to O gives

$$E_{P1}\frac{f}{3} = E_A\frac{H+f}{3}$$

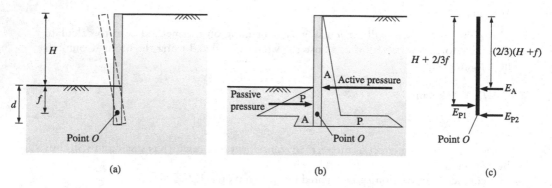

Figure 16.20 Cantilever wall: (a) wall before and after the application of active earth pressures; (b) free body diagram with earth pressure distributions on both sides of the wall; and (c) simplified free body diagram.

or simply

$$E_A(H+f) - E_{P1}f = 0 \tag{16.17}$$

Assuming the same unit weight for the soil behind and in front of the wall, we may write the Rankine expressions for the active and passive forces:

$$E_A = \frac{1}{2}K_A\gamma(H+f)^2$$

$$E_{P1} = \frac{1}{2}K_P\gamma f^2$$

Taking these into Equation (16.17), we get

$$K_A(H+f)^3 - K_P f^3 = 0 \tag{16.18}$$

Equation (16.18) is an equation of third degree on f. It can be easily solved with a calculator. Of the three roots of the equation, we discard the two that make no sense (e.g., complex or negative roots) and retain the third, which will be our f value. The actual embedment of the wall is d, which may be taken as

$$d = mf \tag{16.19}$$

where m is generally taken as 1.2. A factor of safety is implicit in this value of m. According to Powrie (1996), this value of m leads to slightly conservative designs.

The only other task remaining is to find the maximum bending moment and shear force acting on the wall, so that a structural engineer can properly size and reinforce the drilled shafts or slurry wall, or that suitable sheet piles may be selected.

Example 16.5

A slurry wall will be used to retain 4 m of a sandy soil. The friction angle of the soil is 30°. The water table behind the wall is deep below the ground surface. Find the depth of embedment.

Solution:

E-Figure 16.3 shows a schematic representation of the wall. Using Equation (16.18), we get

$$K_P f^3 - K_A \left(H + f \right)^3 = 0$$

So,

$$\frac{1 + \sin 30°}{1 - \sin 30°} \times f^3 - \frac{1 - \sin 30°}{1 + \sin 30°} \times \left(4 + f \right)^3 = 0$$

$$3f^3 - \frac{1}{3} \left(4 + f \right)^3 = 0$$

$$9f^3 - \left(4 + f \right)^3 = 0$$

$$2f^3 - 3f^2 - 12f - 16 = 0$$

Using a calculator, we find the three roots of the equation, two of which are complex numbers, which we accordingly neglect. The third one is

$$f = 3.7 \, \text{m}$$

To find the depth of embedment, we use Equation (16.19):

$$d = mf$$

where $m = 1.2$. So,

$$d = 1.2 \times 3.7 = 4.4 \, \text{m} \quad ^{\text{answer}}$$

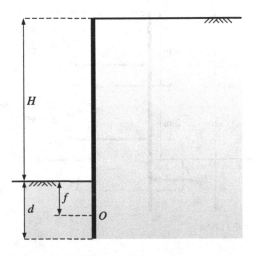

E-Figure 16.3 Schematic of retaining wall of Example 16.5.

In addition to using an embedment depth d that is 20% greater than the depth f to the pivot O to guarantee stability against collapse and (implicitly) keep deflections to within serviceable limits, Powrie (1996) lists a few other possible approaches for this type of wall:

1. The shear strength $\tan \phi$ is reduced by division by a safety factor FS (in other words, we may use a reduced value of ϕ to calculate the resistance, that is, the passive pressures).
2. The passive (resisting) pressures are reduced by division by a factor F_P.
3. The net resisting force, calculated by considering the net earth pressure coefficient $(K_P - K_A)$ below the formation level (that is, the level of the ground surface on the front of the wall), is divided by a factor F_N (Burland et al. 1981).

Given the nonlinear relationship between passive pressures and ϕ, implicit in method (2) is a variable safety factor on strength (lower safety factors for higher values of ϕ), which is not consistent with normal geotechnical practice in retaining structure design. Method (3) does not have this shortcoming, but offers no advantage with respect to method (1), in which the factor of safety takes its usual conceptual form and is transparent. So the best alternative to the approach of calculating f with unfactored pressures and then stipulating $d = 1.2f$ appears to be method (1).

The calculations for a cantilever wall would of course be more complicated if we had a multilayered soil and a high water table. Let us consider this through an example.

Example 16.6

Assume that the same wall of Example 16.5 will retain two layers of sandy soil, each 2 m thick, with $\gamma = 18$ kN/m³, $\phi = 30°$ (top layer), and $\gamma = 22$ kN/m³, $\phi = 35°$ (bottom layer). Find the depth of embedment in this case.

Solution:

The problem can now be represented as shown in E-Figure 16.4.
In this case, we cannot use Equation (16.18) because the properties of the sand change below the depth of 2 m. Referring to the free body diagram in E-Figure 16.4, we take the moments with respect to the potential point of rotation O:

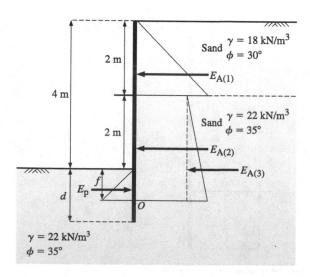

E-Figure 16.4 Schematic illustration of the wall of Example 16.6.

$$E_P \frac{f}{3} - E_{A(1)}\left(\frac{1}{3}\times 2 + 2 + f\right) - E_{A(2)}\left(\frac{1}{2}(2+f)\right) - E_{A(3)}\left(\frac{1}{3}(2+f)\right) = 0$$

To proceed, we need to calculate the active and passive earth pressures. The vertical effective stress at depth $z=2$ m is

$$\sigma'_v = 18 \times 2 = 36\,\text{kPa}$$

So the horizontal effective stress at the bottom of the top layer is

$$\sigma'_{h1} = K_{A1}\sigma'_v = \frac{1 - \sin 30°}{1 + \sin 30°} \times 36 = 12\,\text{kPa}$$

At the same depth, at the bottom of the first layer, the horizontal effective stress is

$$\sigma'_{h2} = K_{A2}\sigma'_v = \frac{1 - \sin 35°}{1 + \sin 35°} \times 36 = 0.27 \times 36 = 9.72\ \text{kPa}$$

The discontinuity at $z=2$ m exists because of the different friction angles of the two layers.
At depth $z=4+f$ meters,

$$\sigma'_v = 36 + 22(2+f) = 80 + 22f$$

and

$$\sigma'_{h3} = K_{A2}\sigma'_v = \frac{1 - \sin 35°}{1 + \sin 35°}(80 + 22f) = 0.27(80 + 22f) = 21.6 + 5.94f$$

So the active and passive earth pressures are

$$E_{A(1)} = \frac{1}{2}\times 12 \times 2 = 12\ \text{kN/m}$$

$$E_{A(2)} = 9.72(2+f) = 19.44 + 9.72f$$

$$E_{A(3)} = \frac{1}{2}(2+f)(21.6 + 5.94f - 9.72) = (1 + 0.5f)(5.94f + 11.88)$$

$$= 2.97f^2 + 11.88f + 11.88$$

$$E_P = \frac{1}{2}K_{P2}\gamma_2 f^2 = \frac{1}{2}\times \frac{1 + \sin 35°}{1 - \sin 35°} \times 22f^2 = 3.69 \times 11f^2 = 40.59f^2$$

Now we can go back to the moment equilibrium equation and solve it for f:

$$13.53f^3 - 12(2.67 + f) - (19.44 + 9.72f)(1 + 0.5f)$$

$$- (2.97f^2 + 11.88f + 11.88)(0.67 + 0.33f) = 0$$

$$\Rightarrow 13.53f^3 - 32.04 - 12f - 19.44 - 9.72f - 9.72f - 4.86f^2 - 1.99f^2$$

$$- 0.98f^3 - 7.96 - 3.92f - 7.96f - 3.92f^2 = 0$$

$$\Rightarrow 12.55f^3 - 10.77f^2 - 43.32f - 59.44 = 0$$

The positive solution of this equation is:

$$f = 2.74\,\text{m}$$

So using Equation (16.19) with $m = 1.2$:

$$d = mf = 1.2 \times 2.74 = 3.3\,\text{m} \quad ^{\text{answer}}$$

We notice that the depth of embedment is approximately 1 m less than in Example 16.5. This happens because below the depth of 2 m, the properties of the sand are better than they were for that example; that is, its shear strength is greater.

16.3.3 Tieback walls

16.3.3.1 The basic design problem

If the height H of retained soil is too large (typically greater than 5 m, approximately 15 ft), the embedment depth d of cantilever walls will be excessive. Tieback walls become then an economical alternative. A tieback is an anchoring element that is fixed at some point behind the wall and thus pulls on the wall with a certain force T. It is typically inclined at a small angle to the horizontal, but, for simplicity, is often assumed to be horizontal in calculations. We will discuss the details of how tiebacks are installed later in this section. In many cases, a single line of tiebacks is sufficient to achieve stability economically; this is the case we examine in this chapter. When the height of the retained soil mass is very large, multiple lines of tiebacks are used. In practice, it is not feasible to do a proper analysis of these walls by hand. Simplified analyses might rely on pre-assumed simplified earth pressure distributions, the same approach discussed later for braced excavations, but specialized computer programs are required if higher-quality analysis is sought.

Figure 16.21a shows a tieback wall supporting a soil mass with height H, embedded d into the ground, with a tieback located at a distance H_T from the surface of the retained soil mass. Under the action of the retained soil mass, the wall will rotate out around a point O below the soil surface. The depth of embedment to the point of rotation O is f. Figure 16.21b and c shows the pressure diagrams that develop in front of and behind the wall if we assume an active state develops behind the wall. The diagram in Figure 16.21c is a simplification of the diagram in Figure 16.21b, where the active and passive pressure distributions below point O are neglected. This means that the stabilizing action of the resultant of the pressures

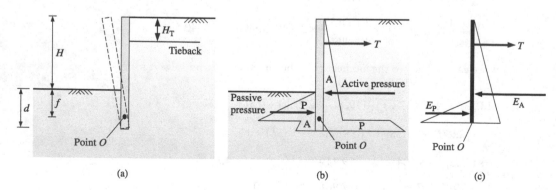

Figure 16.21 Tieback wall: (a) wall before and after the application of earth pressures; (b) pressure diagrams in front and back of the wall; and (c) free body diagram for free-earth support method.

below O is neglected. The assumption – that we can neglect the contribution of resistances below the point of rotation O – is known as the free-earth support assumption.

The tieback wall design problem clearly is to find the value of the tieback force T per unit length into the plane and the embedment depth d such that the wall will perform adequately. A second part of the problem is to find the load T_d acting on each tieback, which will differ from T because tieback spacing is not necessarily equal to one unit of length in the direction normal to the plane of the figure. In the next section, we will detail how to do these calculations.

16.3.3.2 Analysis based on free-earth support assumption

Figure 16.21c shows the free body diagram for a tieback wall. The first design decision is to set the value of the distance H_T from the tieback line to the top of the wall at a value that is effective based on experience. This makes it possible to determine the embedment length f and the tieback loads. The value of H_T can always be changed later by repeating calculations, if necessary. We can write the equations of equilibrium for rotation with respect to the point of application of the tieback force and for translation in the horizontal direction as

$$E_P \left[\frac{2}{3} f + (H - H_T) \right] - E_A \left[\frac{2}{3}(H + f) - H_T \right]$$

(16.20)

$$T + E_P - E_A = 0$$

(16.21)

The unknowns in these equations are the embedment f, which is obtained directly from Equation (16.20), and the anchor force T per unit length of wall, which is obtained from Equation (16.21). The design values of the anchor force and the embedment d are then calculated as

$$d = mf$$

(16.22)

$$T_d = n S_T T$$

(16.23)

where S_T is the tieback spacing, m is in the 1.2–1.4 range, and $n > 1$ (it has been traditionally taken as 1.2) reflects the fact that a fully active Rankine state does not develop in the retained soil due to the restraint imposed by the tiebacks. Equations (16.22) and (16.23) already account for uncertainties and have an implicit factor of safety. The same alternative methods available for embedded cantilever walls are available for tieback walls, namely the use of a factor of safety on shear strength, the use of a factor of safety on E_P, or the use of a factor of safety on net passive resistance (Powrie 1996).

For a tieback wall in a single, homogeneous soil layer, the active and passive forces are given by

$$E_A = \frac{1}{2} K_A \gamma (H + f)^2$$

(16.24)

$$E_P = \frac{1}{2} K_P \gamma f^2$$

(16.25)

If these equations are substituted into Equation (16.20), we can obtain an equation for f directly in terms of the height H to be supported and the soil properties (γ and ϕ); however, it is much simpler to obtain compact expressions for the active and passive forces separately, by using the numerical values of ϕ and γ, and then to plug the resulting expressions into Equation (16.20).

Example 16.7

A slurry wall together with tiebacks will be used to retain 7 m of a sandy soil. The tie-backs will be put 2 m below the ground surface. The friction angle of the soil is 30°, and the unit weight is 20 kN/m³. The water table behind the wall is deep below the ground surface. Find the depth of embedment and the design force on the tiebacks so that the wall will perform adequately. Assume the tiebacks to be inclined 15° with respect to the horizontal.

Solution:

We will use the free-earth support method to determine the embedment of the wall and the force on the tiebacks. The tieback wall is shown in E-Figure 16.5. Although we show the tieback inclined in the figure, the tieback is often represented as horizontal for the wall analysis, the reasoning being that the vertical component of the tieback force is small and thus may be neglected. For a complete treatment of the problem, we will consider the inclination and thus first calculate the horizontal component of the tieback force by requiring that equilibrium holds and then use it to obtain the design force by dividing it by cos 15°.
We will use Equation (16.20):

$$E_P\left[\frac{2}{3}f + (H - H_T)\right] - E_A\left[\frac{2}{3}(H + f) - H_T\right]$$

We have

$$E_A = \frac{1}{2}K_A\gamma(H + f)^2 = \frac{1}{2} \times \frac{1 - \sin 30°}{1 + \sin 30°} \times 20(7 + f)^2 = 3.33(7 + f)^2$$

$$E_P = \frac{1}{2}K_P\gamma f^2 = \frac{1}{2} \times \frac{1 + \sin 30°}{1 - \sin 30°} \times 20f^2 = 30f^2$$

It follows that

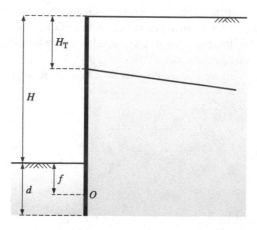

E-Figure 16.5 Schematic of wall of Example 16.7.

$$30f^2\left[\frac{2}{3}f+(7-2)\right]-3.33(7+f)^2\left[\frac{2}{3}(7+f)-2\right]=0$$

$$\Rightarrow 20f^3+150f^2-3.33\left(49+14f+f^2\right)(2.67+0.67f)=0$$

$$\Rightarrow 17.77f^3+109.87f^2-233.80f-435.66=0$$

Using a calculator, we find three roots, two of which are negative and can be neglected. The third one is

$$f=2.6\,\text{m}$$

To find the depth of embedment, we use Equation (16.22). We know that m is in the 1.2–1.4 range. Assuming a value of 1.3, the depth of embedment will be:

$$d=mf=1.3\times2.6=3.38\approx3.4\,\text{m}\ ^{\text{answer}}$$

To find the design tieback force, we will use Equation (16.21) with E_P and E_A now known because we determined the value of f:

$$T+E_P-E_A=0$$

We have

$$E_A=3.33(7+2.6)^2=306.9\approx307\,\text{kN/m}$$

$$E_P=30\times2.6^2=202.8\approx203\,\text{kN/m}$$

So Equation (16.19) gives

$$T_A=307-203=104\,\text{kN/m}$$

The design value of the tieback force is given by Equation (16.23):

$$T_d=nS_T T$$

where S_1 is the spacing between tiebacks, which we assume to be 1.5 m. So, dividing it by $\cos 15°$ to correct for the slight inclination of the tieback:

$$T_d=\frac{1.2\times1.5\times104}{\cos15°}=\frac{187.2}{\cos15°}=193.8\approx194\,\text{kN}\ ^{\text{answer}}$$

It can be seen that the inclination of 15° is inconsequential to the value of T_d. So designers often use Equation (16.23) directly, without the division by the cosine of the angle the tieback makes with the horizontal.

16.3.3.3 Design of tieback

A tieback is most commonly a drilled hole in which a steel tendon is inserted, followed by grouting of the hole. The grouting is usually done in such a way as to leave part of the length of the tieback without any significant bond with the soil, and the other part with as much bonding as possible. There are several techniques for debonding, such as using grease on the tendons and/or the use of a very weak grout.

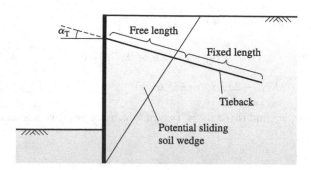

Figure 16.22 The concepts of free and fixed lengths of a tieback.

The length of the tieback without bonding with the soil is known as the free length; the length with significant bonding is known as the fixed or effective length. Figure 16.22 shows that the free length should extend beyond the potential sliding soil wedge, while the fixed length depends on the magnitude of the design load T_d and the unit resistance along the interface of the tieback fixed length and the soil. The figure also shows that the tieback is installed at an inclination (of usually 15°–25°) with the horizontal. This is usually done so that the grout can flow down aided by gravity.

The head of the tieback is connected with the wall; it is also suited with a jack or other systems allowing preloading of the tieback. This means that we can apply the design load of the tieback with any desired safety factor at the time of construction of the wall. This is advantageous from the point of view of quality control, as it is immediately known if any of the tiebacks will not perform as required.

The design of the tiebacks requires specification of their location, diameter, and free and fixed lengths. A suitable steel tendon must also be specified, as well as any special installation requirements. We will focus here on the geotechnical aspects of the design, namely the determination of the diameter and length of the tieback. A single line of tiebacks is usually placed at $0.2H$ to $0.3H$ below the top of the wall. By performing the analysis for various possible positions of the head of the tieback, an optimal design can be found in terms of the tieback length, wall embedment, and bending moments and shear forces on the wall.

The free length is determined with reference to the potential slip surface, which is assumed inclined at an angle $45° + \phi/2$ to the horizontal. The free length must extend beyond the slip surface by a margin of 1–5 m (3–15 ft). This becomes a simple geometry problem, as the inclination of the tieback and the slip surface, together with the location of the tieback head on the wall, are used to determine the free length of the tieback. This procedure will be illustrated in Example 16.8.

The fixed length starts where the free length ends. The tieback develops its resistance by the bonding of its fixed length with the soil. The available resistance T_{res} is given as

$$T_{res} = \pi B \sum_{i=1}^{n} q_{sLi} L_{fi} \qquad (16.26)$$

where B is the diameter of the tieback, L_{fi} is the length of the tieback crossing soil layer i, q_{sLi} is the limit unit shaft resistance along the tieback–soil interface for layer i, and n is the number of soil layers. The number of layers crossed by the tieback in typical, horizontally layered soil is usually very small because the angle of the tieback with the horizontal is small. Figure 16.23 shows a tieback crossing two layers.

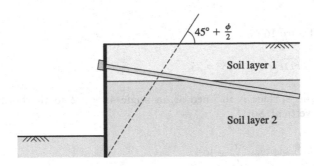

Figure 16.23 Tieback installed in a two-layer soil deposit; the potential slip surface is also shown.

The normal range for tieback diameters is from 75 (3 in.) to 200 mm (8 in.). Large diameters may be necessary for keeping the length of the tieback within a practical range.

The unit shaft resistance q_{sLi} can be very conservatively calculated, for sands, as

$$q_{sLi} = \left(\frac{1+K_{0i}}{2}\right)\sigma'_{vi}\tan\delta_i \tag{16.27}$$

where σ'_{vi} is the vertical effective stress acting at the midpoint of the segment of the tieback that crosses layer i, K_{0i} is the coefficient of earth pressure at rest for layer i, and δ can be taken as 95% of the critical-state friction angle ϕ_c (= 28°–36°, as seen in Chapter 5). This approach is conservative because, in using K_0, it does not consider the contribution of dilatancy upon pulling of the tieback. A less conservative approach would resemble the calculation of q_{sL} for nondisplacement piles done using the Purdue Pile Design Method (see Chapter 13), but recognizing that the tieback is not vertical. Such an approach would produce a higher effective normal stress between the soil nail and soil.

For clays, q_{sLi} is calculated as

$$q_{sLi} = \alpha_i s_{ui} \tag{16.28}$$

where α may be assumed to be in the 0.45–0.6 range. Analyses similar to those done for nondisplacement piles in clay (see Chapter 13) could be done to obtain more realistic expressions.

The goal of design is to make sure that the available resistance T_{res} is a sufficient multiple of the tieback design load T_d:

$$T_{res} \geq (FS)T_d \tag{16.29}$$

where the factor of safety FS can be taken as 1.5 – a value that is relatively low in the context of geotechnical design – in light of the fact that tiebacks are preloaded usually to 1.5 times the design load, thus reducing the uncertainty in the estimation of resistance quite significantly.

Example 16.8

For the slurry wall and the soil profile of Example 16.7, find the free and the fixed lengths of the tiebacks. The tiebacks cross one soil layer and are installed at an inclination of 15°. The critical-state friction angle of the retained soil is 30°. The soil is normally consolidated.

Solution:

Referring to E-Figure 16.6:

$$\widehat{EDA} = 90° - \widehat{FDE} = 90° - 15° = 75°$$

The potential slip surface is inclined at an angle $45° + \phi/2$ to the horizontal, that is $45° - \phi/2$ to the vertical. So

$$\widehat{DAE} = 45° - \frac{\phi}{2} = 45° - \frac{30°}{2} = 30°$$

It follows that

$$\widehat{DEA} = 180° - 75° - 30° = 75°$$

Using the law of sines:

$$\frac{(DA)}{\sin \widehat{DEA}} = \frac{(DE)}{\sin \widehat{DAE}}$$

$$(DE) = \frac{(DA)}{\sin \widehat{DEA}} \sin \widehat{DAE} = \frac{(10.4 - 2)}{\sin 75°} \sin 30° = 4.35\,\text{m}$$

The free length of the tieback must extend beyond the slip surface by a margin of 1–5 m. We will use 2 m. The total free length is then

$$L_{\text{free}} = 4.35 + 2 = 6.35 \text{ m} \,^{\text{answer}}$$

The necessary available resistance of the tiebacks is given by Equation (16.29):

$$T_{\text{res}} \geq (FS)T_{\text{d}}$$

Taking $FS = 1.5$ and $T_{\text{d}} = 194$ kN, the preceding inequality becomes:

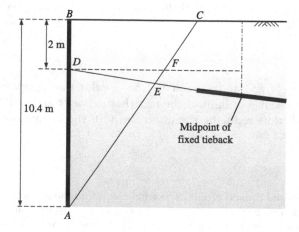

E-Figure 16.6 Wall and tieback of Example 16.8.

$$T_{res} \geq (1.5)194$$

$$T_{res} \geq 291 \text{ kN}$$

So we wish to have at least

$$T_{res} = 291 \text{ kN}$$

To determine the fixed length of the tiebacks, we will use Equation (16.26):

$$T_{res} = \pi B \sum_{i=1}^{n} q_{sLi} L_{fi}$$

Tieback diameters are normally in the 75–200 mm range. We will try a design with a tieback with a diameter of 150 mm. We have only one soil layer, so $n=1$. Since we have a sandy soil, we will use Equation (16.27) to calculate the unit shaft resistance:

$$q_{sL} = \left(\frac{1+K_0}{2}\right)\sigma_v' \tan\delta$$

where we can take $K_0=0.5$ because the soil is normally consolidated.
The midpoint of the fixed length of the tieback is at the depth calculated next:

$$z = 2 + \left(6.35 + \frac{L_f}{2}\right) \times \sin 15° = 2 + 1.64 + 0.13 l_f = 3.64 + 0.13 l_f$$

So

$$\sigma_v' = \gamma z = 20 \times (3.64 + 0.13 L_f) = 72.8 + 2.6 L_f$$

We take δ as 95% of the critical-state friction angle; so

$$\delta = 0.95\phi_c = 0.95 \times 30° = 28.5°$$

We can now write Equation (16.27) as

$$q_{sL} = \left(\frac{1+K_0}{2}\right)\sigma_v' \tan\delta = \left(\frac{1+0.5}{2}\right)(72.8 + 2.6 L_f)\tan 28.5°$$

$$q_{sL} = 29.65 + 1.06 L_f$$

Finally, Equation (16.26) can be used to calculate the fixed length of the tieback:

$$L_f = \frac{T_{res}}{\pi B q_{sL}} = \frac{291}{3.14 \times 0.15 \times (29.65 + 1.06 L_f)}$$

$$L_f = \frac{617.83}{29.65 + 1.06 L_f}$$

$$1.06 L_f^2 + 29.65 L_f - 617.83 = 0$$

$$L_f = 13.92 \text{ m}^{\text{answer}}$$

The total length of the tiebacks is obtained by the simple addition of the free and fixed lengths:

$$L = L_{\text{free}} + L_f = 6.35 + 13.92 = 20.27 \approx 20 \text{ m}$$

If this length is not acceptable, then the two variables that we would adjust would be tieback spacing and tieback diameter.

16.3.4 Braced excavations

If, instead of using tiebacks to support essentially vertical excavations, we used struts or rakers, we would have what is referred to as braced excavations (Figure 16.24a has an illustration). The movement of the retaining systems (tieback walls and braced excavations) is very similar, but, with struts or rakers, a cluttered workspace may result. Thus, braced excavations are not used as often today for foundation works. They can be indicated as an earth-retaining method in cases where the use of tiebacks is not possible due to the inability to secure permission for tieback installation from adjacent property owners, for example.

The classical way of building braced excavations, using "soldier beams with lagging," makes use of I-section steel piles (the so-called soldier beams) that are either driven into the ground or inserted into pre-drilled holes down to a depth below the bottom of the excavation. The spacing between adjacent soldier beams is in the 2–4 m range. Soil is then excavated between each pair of "soldier beams." As the excavation proceeds, wooden planks (the "lagging") are inserted between the soldier beams to provide the retention. One or more wales (usually large wooden planks, but a steel beam can also be used) are placed in front of the wall. The struts or rakers push against the wale. Tiebacks can be used in place of bracing.

The design of braced excavations requires the assumption of a reasonable pressure distribution. Figure 16.24 and Table 16.2 summarize the Peck (1969) and Tschebotarioff (1973) lateral earth pressure distributions for different soils. These distributions are intended to represent a state between the at-rest and active states.

Once the pressure distribution is established, the engineer can design the structural elements of the wall itself (the lagging, wale, soldier piles, struts, or rakers) or tiebacks (if used in place of struts or rakers).

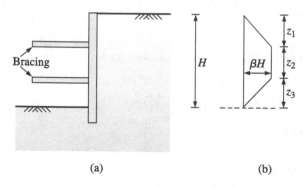

Figure 16.24 (a) Braced cut and (b) lateral earth pressure distribution on a braced cut.

Table 16.2 Peck (1969) and Tschebotarioff (1973) idealized pressure distributions in braced excavations (refer also to Figure 16.24)

Soil	References	z_1/H	z_2/H	z_3/H	β
Sand	Peck (1969)	0	1.0	0	$0.65\gamma K_A$
Stiff, fissured clay	Peck (1969)	0.25	0.50	0.25	0.2γ to 0.4γ
Soft-to-medium stiff clay[a]	Peck (1969)	0.25	0.75	0	$\gamma\left(1-m\dfrac{4s_u}{\gamma H}\right), m \le 1$
Sand	Tschebotarioff (1973)	0.1	0.7	0.2	0.25γ
Medium-stiff clay (permanent support)	Tschebotarioff (1973)	0.6	0	0.4	0.3γ
Medium-stiff clay (temporary support)	Tschebotarioff (1973)	0.75	0	0.25	0.375γ

[a] The value of m, according to Terzaghi and Peck (1967), is dependent on the stress–strain response of the clay, but should be determined empirically from experience with a given clay.

Example 16.9

A 7-meter-deep vertical cut will be done in a soft-to-medium stiff clay. Determine a design lateral pressure distribution on the braced wall using both the Peck (1969) and the Tschebotarioff (1973) distributions.

Solution:

According to the Peck diagrams for pressure distribution (E-Figure 16.7a and Table 16.2), we have the following for soft-to-medium stiff clays:

$$z_1 = 0.25H, z_2 = 0.75H, z_3 = 0, \quad \text{and} \quad \beta = \gamma\left(1 - m\frac{4s_u}{\gamma H}\right)$$

$$z_1 = 1.75\,\text{m}, z_2 = 5.25\,\text{m}, z_3 = 0$$

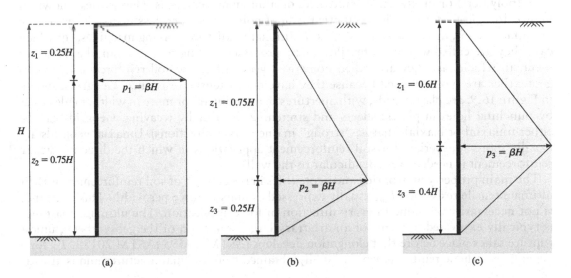

E-Figure 16.7 Pressure distributions: (a) according to Peck, (b) according to Tschebotarioff for a temporary excavation, and (c) according to Tschebotarioff for a permanent excavation.

We assume that the clay unit weight is 16 kN/m³, the undrained shear strength is 15 kPa, and m equals 0.8. So the maximum pressure p_1 in the pressure diagram will be

$$p_1 = \beta H = \gamma \left(1 - m\frac{4s_u}{\gamma H}\right) H = 16\left(1 - 0.8\frac{4 \times 15}{16 \times 7}\right)7 = 64 \text{ kPa}$$

Tschebotarioff's approach takes into account whether the cut is permanent or temporary. According to him, if the cut is temporary (E-Figure 16.7b and Table 16.2), we have

$$z_1 = 0.75H, z_2 = 0, z_3 = 0.25H, \text{ and } \beta = 0.375\gamma$$

$$z_1 = 5.25\,\text{m}, z_2 = 0, z_3 = 1.75\,\text{m}, \text{ and } \beta = 0.375\gamma$$

So the maximum pressure p_2 in the diagram is given by

$$p_2 = \beta H = 0.375\gamma H = 0.375 \times 16 \times 7 = 42 \text{ kPa}$$

If the excavation is permanent (refer to E-Figure 16.7c and Table 16.2), then

$$z_1 = 0.6H, z_2 = 0, z_3 = 0.4H, \text{ and } \beta = 0.3\gamma$$

$$z_1 = 4.2 \text{ m}, z_2 = 0, z_3 = 2.8 \text{ m}, \text{ and } \beta = 0.3\gamma$$

and the pressure, p_3, will be:

$$p_3 = \beta H = 0.3\gamma H = 0.3 \times 16 \times 7 = 33.6 \approx 34 \text{ kPa}.$$

16.4 DESIGN OF MECHANICALLY STABILIZED EARTH (MSE) WALLS

16.4.1 Materials

Mechanically stabilized walls are commonly built of a suitable soil (traditionally sandy soil with less than 15% fines) reinforced typically using strips, sheets, or grids made of extensible (polymers) or inextensible (mild steel or aluminum) materials. Geotextiles, shown in Figure 16.9, are a form of geosynthetic made of polymer or fiberglass fibers. They may be manufactured in much the same way as traditional textiles in weaving machines,[7] in which case they are called woven geotextiles, or by processes leading to a random fiber arrangement, in which case they are called nonwoven geotextiles. For soil reinforcement, woven geotextiles are typically used because they have higher tensile resistance. Geogrids, shown in Figure 16.9, are plastic grids, with apertures of a centimeter or more in width, made either by punching holes in plastic sheets and stretching them or by weaving. Geogrids can be either uniaxial or biaxial (that is, "strong" in one or two directions). Uniaxial geogrids are usually more appropriate for soil reinforcement applications in which the direction where reinforcement is needed is perpendicular to the wall.

The main property of geogrids and geotextiles in the context of soil reinforcement is their ultimate tensile resistance T_{ult}, usually expressed as units of force per width. This resistance is not necessarily the same in every direction of force application. The ultimate resistance is typically expressed in terms of an arbitrary large elongation of the geosynthetic, unless rupture takes place before that elongation develops [ASTM D4595 (ASTM 2017)]. T_{ult} for a given geosynthetic reinforcement is usually obtained from the manufacturer and is already

---

[7] In the case of traditional textiles, natural fibers are used.

Table 16.3 Creep reduction factor values (FHWA 2009)

Polymer type	$(RF)_{cr}$
Polyester	2.5–1.6
Polypropylene	5.0–4.0
High-density polyethylene	5.0–2.6

reduced to account for installation damage, creep, chemical damage, and biological degradation. In case the value provided does not contain these reductions, we must reduce it by dividing it by the following reduction factors (FHWA 2009):

$(RF)_{id}$ to account for possible installation damage (1.1 to 3);
$(RF)_{cr}$ to account for possible creep (see Table 16.3); and
$(RF)_d$ to account for chemical damage and biological degradation (1.1–2).

In design, we must further reduce the ultimate resistance T_{ult} (already reduced as discussed earlier) by dividing it by a factor of safety to account for the consequences of a stability failure of the MSE wall. This factor of safety *FS* is typically 1.5 or greater. The maximum tensile force in the reinforcement must not then exceed the reduced tensile resistance divided by *FS* (which is an allowable tensile force T_{all} expressed per unit width of reinforcement):

$$T_{all} = \frac{T_{ult}}{FS} \tag{16.30}$$

The T_{all} of Equation (16.30) is a force per unit width of reinforcement. It must be compared with the force T_{max} pulling on the reinforcement per unit length of wall. To make that comparison possible, the coverage ratio R_c, which converts a load per unit length of wall T_{max} into a load per unit width of reinforcement, is introduced. The design inequality is

$$T_{all} \geq \frac{T_{max}}{R_c} \tag{16.31}$$

The force T_{max}/R_c is the maximum tensile force experienced by the reinforcement, right at the point where the reinforcement and the slip surface cross (see Figure 16.25).

There is an equivalent calculation for metallic strips. The allowable tensile force T_{all} for steel strip reinforcement can be calculated in terms of the yield stress f_y of steel as

$$T_{all} = \frac{0.55 f_y A_c}{b} = \frac{f_y A_c}{1.82b} \tag{16.32}$$

where *b* is the width of the steel strips and A_c is the cross-sectional area of the reinforcement, given as

$$A_c = bt_{LT} \tag{16.33}$$

in which t_{LT} is the long-term thickness of the reinforcement element, accounting for any likely corrosion. In the case of steel grids, we use 0.48 in place of 0.55 in Equation (16.32). Expanding Equation (16.33), the design cross-sectional area of the strip can be written as

$$A_c = bt_{LT} = b(t_n - t_R) \tag{16.34}$$

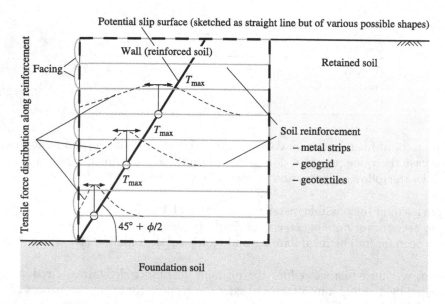

Figure 16.25 Overview of an MSE wall showing slip surface and distribution of tension along the reinforcements.

where t_{LT} is the thickness of the reinforcement at the end of the design life, t_n is the nominal thickness at the time of construction, and t_R is the sacrificial thickness of metal expected to be lost by corrosion during the design life of the wall.

The nominal thickness of a metal strip is usually 4 mm. Thus, to calculate t_{LT}, we need to be able to estimate how much metal is corroded over time. Typically, the metal strips are made of galvanized steel with a minimum galvanization coating of 86 μm thickness. The zinc is oxidized during the first years of design life. According to FHWA (2009),

1. the zinc corrosion rate for the first 2 years is 15 μm/year/side,
2. the zinc corrosion takes place at a rate of 4 μm/year/side for $t > 2$ years, and
3. the carbon steel corrosion rate is 12 μm/year/side.

The service life of zinc is accordingly:

$$\text{Service Life} = 2\,\text{years} + \frac{86\,\mu\text{m} - 2\,\text{years} \times 15\,\mu\text{m/year}}{4\,\mu\text{m/year}} = 16\,\text{years}$$

This means that, in 16 years, the zinc is completely corroded. At that point, consumption of the base metal starts. According to FHWA (2009), the steel loss is 12 μm/year. If we know what the design life of our structure is, we can calculate how much thickness will be left at the end of the design life (our t_{LT}).

16.4.2 General design considerations

The general idea behind a reinforced soil wall is to alternate layers of reinforcement and soil (Figure 16.25). The reinforcement should extend beyond the expected slip surface to develop enough resistance against pullout, as shown schematically in Figure 16.26a for inextensible reinforcement and in Figure 16.26b for extensible reinforcement. Since extensible reinforcement tends to deform to comparable levels of magnitude as soil, potential slip

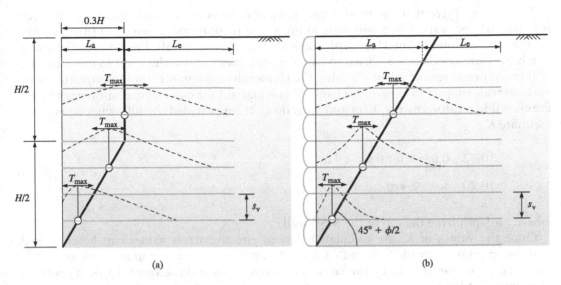

Figure 16.26 Idealized slip surfaces for reinforced soil walls: (a) for the case of inextensible reinforcement and (b) for the case of extensible reinforcement. (After FHWA 2009.)

surfaces resemble those that we would find in unreinforced soil. Figure 16.26b shows the critical slip surface that would be associated with the Rankine analysis. For inextensible reinforcements, more restraint is imposed on the wall by the reinforcement. The slip surface in Figure 16.26a is often assumed in design in this case. When the potential failure wedge attempts to slide down the slip surface, it is held up by the resistance that develops between the parts of the reinforcing layers that extend beyond the slip surface, assuming there is enough of this resistance. Friction (with some contribution from passive resistance) is the predominant load transfer mechanism: the soil wedge transfers load to the reinforcement by friction, as it attempts to slide out, and the stationary soil mass restrains movement of the reinforcement, and thus of the soil wedge, also by friction. In this process, the reinforcement is subjected to tensile forces. These tensile forces reach a peak at the point at which the slip surface crosses the reinforcement, as illustrated in Figure 16.25.

The vertical spacing between reinforcement layers is denoted by s_v. This is usually in the 300 to 600 mm range (1 to 2 ft range). If reinforcement is not continuous across the width of the wall, the reinforcement width b and the horizontal spacing, center-to-center, denoted by s_h, are also design variables that must be specified.

For the reinforced soil wall concept to work well, it is necessary that two limit states not be reached:

1. The reinforcement must not elongate excessively or break under the tensile forces (this is accomplished by using the inequality in Equation (16.31) to check that this will not happen).
2. The reinforcement must not be pulled out from within the stationary soil mass (this is accomplished by making the reinforcement layers sufficiently long, so that there is sufficient length of reinforcement beyond the slip surface).

Additionally, even in the absence of complete pullouts, the front panels of the MSE wall may become misaligned, and that may lead to serviceability or ultimate limit states (see Lin et al. (2019) for a discussion of design criteria regarding this).

The pullout force that the reinforcement must resist is the summation of lateral forces imposed on the wall, which will be held up by the reinforcement, which in turn develops its resistance along its interface with the soil. For a vertical wall, the ratio of the lateral earth pressure coefficient K_r (used to calculate the pressures on the wall) to the active lateral earth pressure coefficient K_A using the Coulomb equation (or Rankine equation if the vertical wall supports a horizontal backfill) is equal to 1 for extensible reinforcements. For inextensible reinforcements, Kim and Salgado (2012a) provided the following equation to estimate K_r:

$$K_r = \begin{cases} 0.32 - 0.03z & \text{for } 0 \leq z < 4\,\text{m} \\ 0.20 & \text{for } z \geq 4\,\text{m} \end{cases} \tag{16.35}$$

where z=depth from the top of the MSE wall.

Once the values of K_r are calculated, lateral effective stress values can be computed, and the maximum tensile force on each level of reinforcement, per unit length of wall, is obtained by multiplying K_r by the vertical effective stress and the vertical spacing between reinforcement layers:

$$T_{\max} = K_r \left(\gamma z + q_0 \right) s_v \tag{16.36}$$

where q_0=uniform surcharge applied on top of the MSE wall.

The interaction between the reinforcing element and the backfill is key to the design of MSE walls. In particular, we are interested in the resistance to slipping of the reinforcement with respect to the soil (known as the pullout resistance R_{PO}), which, except at shallow depths, is mostly, but not exclusively, due to friction. Passive resistance to relative slip of the reinforcement with respect to the soil is also present to some extent, particularly in the case of grids, as soil particles can actually occupy positions across the plane of the geogrid, as shown in Figure 16.27.

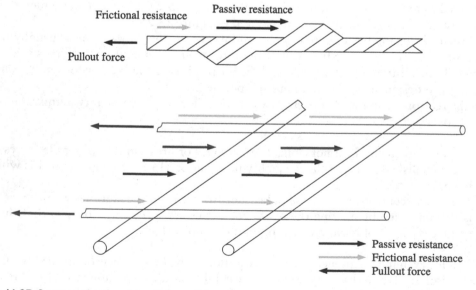

Figure 16.27 Sources of soil resistance for reinforcements.

The passive resistance is usually not calculated separately, but rather lumped in with friction. The total pullout resistance R_{PO} calculated according to FHWA (2009) is

$$R_{PO} = C_S C_P C_R \sigma'_v L_e R_c \tag{16.37}$$

where C_S is a scale effect coefficient used to correct for the nonlinear stress reduction over the embedded length of highly extensible reinforcements (it is equal to 1.0 for metallic reinforcements; in the absence of laboratory tests, it may be taken as 0.6 for geotextiles and 0.8 for geogrids); C_P is the reinforcement effective perimeter coefficient (typically taken as equal to 2 because friction develops on both sides of the reinforcement); C_R is the pullout resistance coefficient; and L_e is the effective length of reinforcement, along which the resistance develops.

The pullout resistance coefficient C_R is given by (Kim and Salgado 2012a; FHWA 2009)

$$C_R = \begin{cases} \tan \delta_c & \text{for metallic reinforcement} \\ C_i \tan \phi & \text{for geosynthetic reinforcement} \end{cases} \tag{16.38}$$

where δ_c=critical-state interface friction angle between the reinforced soil and steel reinforcement (which can be estimated using Figure 13.19) and C_i=coefficient of interaction that attempts to capture resistance to pullout due not only to friction, but also to the passive resistance that is found in geogrids, for example. In the absence of test data, the value of ϕ is limited to a maximum of 34° for reinforced soil, while the value of C_i may conservatively be taken as 0.67 (FHWA 2009).

The effective length L_e is the length of reinforcement that extends beyond the slip surface. L_e should be no less than 1 m. For simple wall geometries and inextensible reinforcement (metallic reinforcement), the slip surface is defined by Figure 16.26a. The distance L_a from the wall facing to the slip surface is (FHWA 2009)

$$L_a = \begin{cases} 0.3H & \text{for } 0 \leq z \leq H/2 \\ 0.6(H-z) & \text{for } H/2 \leq z \leq H \end{cases} \tag{16.39}$$

where z is the depth from the top of the MSE wall and H is the MSE wall height. For extensible reinforcements (geosynthetics), an active Rankine slip surface is assumed (see Figure 16.26b) so that

$$L_a = (H-z)\tan\left(45 - \frac{\phi}{2}\right) \tag{16.40}$$

Reinforced soil walls typically have a facing, but this facing is not assumed to contribute to wall stability. The facing can be made of precast concrete panels, cast-in-place concrete, shotcrete, metal, gabion, or fabric (in the case of geotextiles used as reinforcing elements). The reinforcing elements are typically linked to the facing by suitable connections. The one exception is walls built with geotextiles, where geotextile layers may be wrapped around the soil to create the "facing" of the wall, as shown in Figure 16.28b. In this case, it is often necessary to add a protective layer of gunite or some other similar material to avoid degradation of the fabric by ultraviolet light.

(a)

(b)

Figure 16.28 Examples of MSE wall facings: (a) precast concrete panels and (b) facing obtained by wrapping of geotextiles around soil lifts.

16.4.3 External stability design checks using WSD and LRFD

16.4.3.1 Sliding limit state

The working stress design (WSD) check for the sliding limit state for an MSE wall is:

$$\frac{1}{(FS)_{\text{sliding}}} W \tan\delta \geq E_{A1} + E_{A2} \tag{16.41}$$

where $(FS)_{\text{sliding}}$ is the factor of safety against sliding (usually taken as 1.5), W is the self-weight of the reinforced soil volume, δ is the interface friction angle at the base of the MSE wall, and E_{A1} and E_{A2} are the horizontal forces due to the active earth pressures caused by the self-weight of the retained soil and the uniform surcharge applied on top of the MSE wall, respectively.

The corresponding load and resistance factor design (LRFD) check is:

$$RF(W \tan\delta) \geq LF_{EA1}E_{A1} + LF_{EA2}E_{A2} \tag{16.42}$$

where RF is a resistance factor, and LF_{EA1} and LF_{EA2} are load factors corresponding to the horizontal forces E_{A1} and E_{A2}, respectively. Kim and Salgado (2012b) proposed the following RF values for the sliding limit state: 0.80 for 5 m $\leq H <$ 7.5 m, 0.90 for 7.5 m $\leq H <$ 10 m, and 1.00 for $H \geq$ 10 m. These RF values correspond to the AASHTO (2020) load factors, $LF_{EA1} = 1.50$ and $LF_{EA2} = 1.75$, and a target reliability index β_T of 3.5.

16.4.3.2 Overturning limit state

The WSD check for the overturning limit state for an MSE wall is:

$$\frac{1}{(FS)_{overturning}} W\left(\frac{L_r}{2}\right) \geq E_{A1}\left(\frac{H}{3}\right) + E_{A2}\left(\frac{H}{2}\right) \tag{16.43}$$

where $(FS)_{overturning}$ is the factor of safety against overturning (usually taken as 2.0), L_r is the reinforcement length, W is the self-weight of the reinforced soil volume, H is the MSE wall height, and E_{A1} and E_{A2} are the horizontal forces due to the active earth pressures caused by the self-weight of the retained soil and the uniform surcharge applied on top of the MSE wall, respectively.

The corresponding LRFD check is:

$$RF\left[W\left(\frac{L_r}{2}\right)\right] \geq LF_{EA1}E_{A1}\left(\frac{H}{3}\right) + LF_{EA2}E_{A2}\left(\frac{H}{2}\right) \tag{16.44}$$

where RF is the resistance factor, and LF_{EA1} and LF_{EA2} are the load factors corresponding to horizontal forces E_{A1} and E_{A2}, respectively. Kim and Salgado (2012b) proposed the following RF values for the overturning limit state: 0.75 for 5 m $\leq H <$ 7.5 m, 0.80 for 7.5 m $\leq H <$ 12.5 m, and 1.00 for $H \geq$ 12.5 m. These RF values correspond to the AASHTO (2020) load factors, $LF_{EA1} = 1.50$ and $LF_{EA2} = 1.75$, and a target reliability index β_T of 3.5.

16.4.3.3 Bearing capacity limit state

The WSD check for the bearing capacity limit state for an MSE wall is:

$$\frac{1}{(FS)_{BC}} q_{bL} \geq q_b \tag{16.45}$$

where $(FS)_{BC}$ is the factor of safety against bearing capacity failure, q_{bL} is the limit unit bearing capacity of the MSE wall, and q_b is the unit load at the base of the MSE wall.

The corresponding LRFD check is:

$$RF(R_n) \geq LF_{DL}DL_n + LF_{LL}LL_n \tag{16.46}$$

where RF is the resistance factor, R_n is the nominal resistance of the MSE wall foundation, DL_n is the nominal dead load (=W, the self-weight of the reinforced soil volume), LL_n is the nominal live load (=$q_0 L_r$), q_0 is the uniform surcharge applied on top of the MSE wall, L_r is the reinforcement length, and LF_{DL} and LF_{LL} are the load factors for dead load and live load, respectively.

16.4.4 Internal stability design checks using WSD and LRFD

16.4.4.1 Reinforcement rupture limit state

As we saw earlier, the WSD check for the reinforcement rupture (breakage) limit state for an MSE wall is:

$$T_{\text{all}} \geq \frac{T_{\text{max}}}{R_c} \tag{16.31}$$

The coverage ratio R_c is equal to the gross width b of each reinforcement strip divided by the center-to-center horizontal spacing s_h of the reinforcements. For continuous reinforcement, the coverage ratio would obviously be 1. Note that, for geosynthetic reinforcing elements, $R_c = 1$.

The corresponding LRFD check is:

$$RF\left(T_{\text{all}}R_c\right) \geq LF_{\text{DL}}T_{\text{max,DL}} + LF_{\text{LL}}T_{\text{max,LL}} \tag{16.47}$$

where RF is the resistance factor; $T_{\text{max,DL}}$ and $T_{\text{max,LL}}$ are the maximum tensile forces in the reinforcement due to the lateral earth pressures caused by the self-weight of the reinforced soil and the uniform surcharge applied on top of the MSE wall, respectively; and LF_{DL} and LF_{LL} are the load factors corresponding to forces $T_{\text{max,DL}}$ and $T_{\text{max,LL}}$, respectively. Kim and Salgado (2012a) proposed RF values of 0.70, 0.80, and 0.86 for the rupture limit state for target reliability indices β_T of 3.0, 2.33, and 2.0, respectively. These RF values are valid for steel strip reinforcement and correspond to load factors $LF_{\text{DL}} = 1.50$ and $LF_{\text{LL}} = 1.75$ (AASHTO 2020).

16.4.4.2 Reinforcement pullout limit state

The WSD check for the reinforcement pullout limit state for an MSE wall is:

$$\frac{1}{\left(FS\right)_{\text{PO}}} R_{\text{PO}} \geq T_{\text{max}} \tag{16.48}$$

where $(FS)_{\text{PO}}$ is the factor of safety against pullout, usually taken as 1.5.

The corresponding LRFD check is:

$$RF\left(R_{\text{PO}}\right) \geq LF_{\text{DL}}T_{\text{max,DL}} + LF_{\text{LL}}T_{\text{max,LL}} \tag{16.49}$$

Table 16.4 summarizes the RF values proposed by Kim and Salgado (2012a) for the reinforcement pullout ultimate limit state as a function of the target reliability index β_T for load factors $LF_{\text{DL}} = 1.50$ and $LF_{\text{LL}} = 1.75$. Kim and Salgado (2012a) recommended that, for $\beta_T = 3.0$, a conservative value of L_a that is 20% greater than that obtained from Equation (16.39) be used.

Now that we know the limit states to check for and how to do these checks, the design of a reinforced soil wall is best illustrated with an example. The example shows design for an MSE wall with metal strips, an inextensible form of reinforcement, but the general approach is similar for inextensible and extensible reinforcements.

Table 16.4 RF values for reinforcement pullout limit state (Kim and Salgado 2012a)

Target reliability index β_T	Resistance factor RF	Length of reinforcement in active zone L_a
2.0	0.72	Equation (16.39)
2.33	0.65	Equation (16.39)
3.0	0.53	1.2 times the value obtained from Equation (16.39)

Note: The RF values are valid for steel strips and correspond to load factors $LF_{\text{DL}} = 1.50$ and $LF_{\text{LL}} = 1.75$.

Example 16.10

An MSE wall with ribbed steel strip reinforcement (with width=50 mm and thickness=4 mm) and a design life of 75 years is to be constructed. The total (design) height H of the wall is 7 m, including 0.5 m embedment into the natural ground. The reinforced, retained, and foundation soils are all sandy materials with the following properties: $\gamma = 19$ kN/m³ and $\phi_c = 36°$ for the reinforced soil, $\gamma = 19$ kN/m³ and $\phi_c = 32°$ for the retained soil, and $\gamma = 16$ kN/m³ and $\phi_c = 30°$ for the foundation soil. The relative density D_R of the foundation soil is 70%, and a uniform surcharge q_0 of 20 kPa is applied on top of the wall. Assuming that the overall (global) stability requirement is met, design this wall against both internal and external stabilities. Specifically, determine reinforcement length and vertical and horizontal spacing.

Solution:

We will start by examining the MSE wall with respect to external stability. The coefficient of active earth pressure of the retained soil is given by

$$K_A = \frac{1 - \sin\phi_c}{1 + \sin\phi_c} = \frac{1 - \sin 32°}{1 + \sin 32°} = 0.307$$

The horizontal, active earth pressure is given by

$$E_{A1} = \frac{1}{2} K_A \gamma H^2 = \frac{1}{2} \times 0.307 \times 19 \times 7^2 = 143 \, \text{kN/m}$$

and the horizontal force due to the uniform surcharge q_0 is simply the lateral stress due to the surcharge times the wall height:

$$E_{A2} = K_A q_0 H = 0.307 \times 20 \times 7 = 43 \, \text{kN/m}$$

To calculate the self-weight W of the reinforced soil volume, we need to know the length L_r of the reinforcements. A reasonable first estimate for this length is obtained by making $L_r = 0.7H$. So L_r will be

$$L_r = 0.7H = 0.7 \times 7 = 4.9 \, \text{m}$$

The corresponding MSE wall weight is:

$$W = \gamma H L_r = 19 \times 7 \times 4.9 = 652 \, \text{kN/m}$$

We will use both WSD and LRFD for external and internal stability design checks in this example.

Sliding limit state:

The WSD check for sliding of the MSE wall is:

$$\frac{1}{(FS)_{\text{sliding}}} W \tan\delta \geq E_{A1} + E_{A2}$$

As sliding can occur either along the foundation soil or along the reinforced soil near the bottom of the MSE wall, the interface friction angle δ is taken as the minimum of the ϕ_c values of the reinforced and foundation soils, which in this case is 30°. Therefore, the resistance to sliding of the MSE wall, with due account for the safety factor, is:

$$\frac{1}{(FS)_{sliding}} W \tan\delta = \frac{1}{1.5} \times 652 \times \tan 30° = 251\,kN/m > 186\,kN/m = E_{A1} + E_{A2}$$

According to the LRFD method, the factored load for the sliding limit state check is:

$$LF_{EA1}E_{A1} + LF_{EA2}E_{A2} = 1.5(143) + 1.75(43) = 290\ kN/m$$

and the factored resistance is:

$$RF(W\tan\delta) = 0.80 \times 652 \times \tan 30° = 301\ kN/m > 290\ kN/m = LF_{EA1}E_{A1} + LF_{EA2}E_{A2}$$

Thus, the wall passes the sliding limit state check for both WSD and LRFD.

Overturning limit state:

According to the WSD method, the driving moment with respect to the wall toe is given by

$$E_{A1}\left(\frac{H}{3}\right) + E_{A2}\left(\frac{H}{2}\right) = 143\left(\frac{7}{3}\right) + 43\left(\frac{7}{2}\right) = 484\,kNm/m$$

while the resisting moment, with due account for the safety factor, is:

$$\frac{1}{(FS)_{overturning}} W\left(\frac{L_r}{2}\right) = \frac{1}{2} \times 652 \times \left(\frac{4.9}{2}\right) = 799\ kNm/m > 484\ kNm/m = E_{A1}\left(\frac{H}{3}\right) + E_{A2}\left(\frac{H}{2}\right)$$

According to the LRFD method, the factored driving moment with respect to the wall toe is given by

$$LF_{EA1}E_{A1}\left(\frac{H}{3}\right) + LF_{EA2}E_{A2}\left(\frac{H}{2}\right) = \left[1.5 \times 143 \times \left(\frac{7}{3}\right)\right] + \left[1.75 \times 43 \times \left(\frac{7}{2}\right)\right] = 764\ kNm/m$$

while the factored resisting moment with respect to the wall toe is:

$$RF\left[W\left(\frac{L_r}{2}\right)\right] = 0.75 \times 652 \times \left(\frac{4.9}{2}\right)$$

$$= 1198\ kNm/m > 764\ kNm/m = LF_{EA1}E_{A1}\left(\frac{H}{3}\right) + LF_{EA2}E_{A2}\left(\frac{H}{2}\right)$$

Thus, the wall passes the overturning limit state check for both WSD and LRFD. Note that the load and resistance factors for the sliding and overturning limit state checks correspond to a target reliability index of 3.5, which corresponds to a target probability of failure of 0.00023 (or about 1 in 5000). In the WSD and LRFD checks for the sliding and overturning limit states, the uniform surcharge q_0 acting on top of the reinforced soil is conservatively neglected because it contributes to the resistance, while that acting on top of the retained soil is considered because it contributes to the load (AASHTO 2020; FHWA 2009).

Bearing capacity limit state:

To ensure stability against bearing capacity failure, we must start by requiring that the eccentricity e be less than one-sixth of the width of the wall (that is, the length L_r of the reinforcements). The eccentricity is given by

$$e = \frac{\sum M}{R_v} = \frac{E_{A1}\left(\dfrac{H}{3}\right) + E_{A2}\left(\dfrac{H}{2}\right)}{W + q_0 L_r} = \frac{143\left(\dfrac{7}{3}\right) + 43\left(\dfrac{7}{2}\right)}{652 + 20(4.9)} = 0.65\,\text{m}$$

where R_v=vertical component of the resultant force acting on the base of the MSE wall. Now we check whether this value is acceptable:

$$e = 0.65\,\text{m} < \frac{L_r}{6} = 0.82\,\text{m}$$

So the wall satisfies our maximum eccentricity constraint.
The effective width of the MSE wall foundation is calculated using Equation (10.63):

$$B_{\text{eff}} = B\left(1 - 2.273\frac{e}{B}\right)^{0.8} = 4.9 \times \left(1 - 2.273 \times \frac{0.65}{4.9}\right)^{0.8} = 3.7\,\text{m}$$

The unit load at the base of the wall is given by

$$q_b = \frac{W + q_0 L_r}{B_{\text{eff}}} = \frac{652 + (20 \times 4.9)}{3.7} = 203\,\text{kPa}$$

Although the MSE wall is typically embedded into the natural ground to a depth of the order of 0.5 m, the beneficial effect of the embedment is usually neglected in the calculation of the bearing capacity of the MSE wall. This means that the only term of the bearing capacity equation that is left is the γ term. Given our idealization of the wall as long enough for plane-strain conditions to apply, the shape factor s_γ is equal to 1. The depth factor d_γ, we recall from Chapter 10, is also equal to 1. Thus, the limit unit bearing capacity of the wall is given by

$$q_{bL} = \frac{1}{2}\gamma B_{\text{eff}} N_\gamma i_\gamma \tag{16.50}$$

where N_γ=bearing capacity factor and i_γ=inclination factor, both of which depend on the representative peak friction angle ϕ_p of the foundation soil. We must start by calculating the representative mean effective stress for this MSE wall by using Equation (10.39):

$$\sigma'_{mp} = 20 p_A \left(\frac{\gamma B_{\text{eff}}}{p_A}\right)^{0.7}\left(1 - 0.32\frac{B_{\text{eff}}}{L}\right) = 20 \times 100 \times \left(\frac{16 \times 3.7}{100}\right)^{0.7} \times (1 - 0) = 1386\,\text{kPa}$$

The corresponding relative dilatancy index I_R is calculated using Equation (5.12)

$$I_R = \frac{D_R}{100}\left[Q - \ln\left(\frac{100\sigma'_{mp}}{p_A}\right)\right] - R_Q = \frac{70}{100}\left[10 - \ln\left(\frac{100 \times 1386}{100}\right)\right] - 1 = 0.94$$

The representative peak friction angle ϕ_p and peak dilatancy angle ψ_p of the foundation soil are calculated using Equations (5.20) and (5.19):

$$\phi_p = \phi_c + A_\psi I_R = 30° + (5 \times 0.94) = 34.7°$$

$$\psi_p = \sin^{-1}\left(\frac{I_R}{6.7 + I_R}\right) = \sin^{-1}\left(\frac{0.94}{6.7 + 0.94}\right) = 7.1°$$

The nonassociativity factor $F(\phi_p, \psi_p)$ and the bearing capacity factors N_q and N_γ follow:

$$F(\phi_p, \psi_p) = 1 - \tan\phi_p\left[\tan\left(0.8(\phi_p - \psi_p)\right)\right]^{2.5}$$

$$= 1 - \tan 34.7°\left[\tan\left(0.8 \times (34.7° - 7.1°)\right)\right]^{2.5} = 0.927$$

$$N_q = \frac{1 + \sin\phi_p}{1 - \sin\phi_p}\exp\left[F(\phi_p, \psi_p)\pi\tan\phi_p\right] = \frac{1 + \sin 34.7°}{1 - \sin 34.7°}\exp[0.927 \times \pi\tan 34.7°] = 27.4$$

$$N_\gamma = (N_q - 0.6)\tan(1.33\phi_p) = (27.4 - 0.6)\tan(1.33 \times 34.7°) = 27.9$$

The inclination angle α of the resultant force acting at the base of the MSE wall with respect to the vertical is:

$$\alpha = \tan^{-1}\left(\frac{R_h}{R_v}\right) = \tan^{-1}\left(\frac{E_{A1} + E_{A2}}{W + q_0 L_r}\right) = \tan^{-1}\left(\frac{143 + 43}{652 + 20 \times 4.9}\right) = 13.9°$$

where R_h is the horizontal component of the resultant force acting on the base of the MSE wall.

The inclination factor i_γ is calculated using Equation (10.64):

$$i_\gamma = \left(1 - 0.94\frac{\tan\alpha}{\tan\phi_p}\right)^{(1.5\tan\phi_p + 0.4)^2} = \left(1 - 0.94 \times \frac{\tan 13.9°}{\tan 34.7°}\right)^{(1.5\tan 34.7° + 0.4)^2} = 0.43$$

We are now ready to compute the limit unit bearing capacity of the MSE wall:

$$q_{bL} = \frac{1}{2}\gamma B_{eff} N_\gamma i_\gamma = \frac{1}{2} \times 16 \times 3.7 \times 27.9 \times 0.43 = 355\,\text{kPa}$$

The factor of safety against bearing capacity failure is

$$(FS)_{BC} = \frac{q_{bL}}{q_b} = \frac{355}{203} = 1.75$$

which is reasonable. Note that due to the much greater flexibility of MSE walls when compared with footings, lower values of $(FS)_{BC}$ can be accepted than for typical foundations (for foundations, we saw in Chapters 11 and 13, typical values of factors of safety, when proper design equations are used and proper site characterization is performed, would be in the 2–3 range).

We can now check the internal stability of the MSE wall. The internal stability design includes a check for the rupture of the reinforcements and another for pullout of the reinforcements. Considering the relatively large height of the wall, we will work with ten reinforcement layers. The first one will be placed 0.3 m above the base of the wall. The next nine layers will be placed at a vertical spacing of 0.7 m. The last layer will be placed 0.4 m below the top of the wall. For each of the ten reinforcement layers, we must calculate the lateral earth pressure coefficient K_r and the maximum tensile force T_{max} in the reinforcement.

For the first reinforcement layer, we have:

$z_1 = 6.7$ m (because the layer is located 0.3 m above the base of the wall).
Since $z > 4$ m, $K_r = 0.2$ [according to Equation (16.35)].

We now calculate the maximum tensile force T_{max} in the reinforcement per unit width of wall, based on the vertical spacing s_v. Note that for the first reinforcement layer, the vertical spacing $s_v = 0.65$ m; for the second through ninth layers, $s_v = 0.70$ m; and for the tenth and last layer, $s_v = 0.75$ m. So for the first layer,

$$T_{max(1)} = K_r \left(\gamma z + q_0 \right) s_v = 0.2 \times \left[19(6.7) + 20 \right] \times 0.65 = 19 \, \text{kN/m}$$

Rupture limit state:

The first check we must perform is against rupture (breakage) of the reinforcements. Stability with respect to rupture requires that

$$T_{all} \geq \frac{T_{max}}{R_c}$$

where T_{all} is the allowable tensile force per unit width of reinforcement and R_c is the coverage ratio. The coverage ratio is equal to the gross width b of each reinforcement strip divided by the center-to-center horizontal spacing s_h of the reinforcements. For continuous reinforcement, the coverage ratio would obviously be 1. For this case, we will take a typical value for s_h of 0.70 m, which gives:

$$R_c = \frac{b}{s_h} = \frac{0.05 \, \text{m}}{0.7 \, \text{m}} = 0.071$$

The allowable tensile force per unit width of reinforcement is given by

$$T_{all} = \frac{f_y}{FS} \left(\frac{A_c}{b} \right)$$

where f_y is the yield strength of steel, FS is a factor of safety applied to f_y, A_c is the design cross-sectional area of the strip, and b is the gross width of the reinforcing strip.

The factor of safety here is a number that should be reasonably consistent with general steel design approaches. The AASHTO (1996) specifications for highway bridge design required a factor of safety $FS = 1.82$. If we use 65-grade steel, $f_y = 65$ ksi $= 448$ MPa. The gross width of the reinforcement is typically $b = 50$ mm $= 0.05$ m. The design cross-sectional area of the strip is given by Equation (16.34):

$$A_c = bt_{LT} = b \left(t_n - t_R \right)$$

Considering that the design life of the structure is 75 years, that the service life of the zinc coating is 16 years (as seen earlier), and that the corrosion rate of steel is 12 μm per year for each side of the reinforcement, t_R is given by

$$t_R = 12 \, \mu\text{m /year /side} \times (75 - 16) \, \text{years} \times 2 = 1416 \, \mu\text{m} = 1.416 \, \text{mm}$$

The cross-sectional area of the reinforcement and the allowable tensile force are then calculated as

$$A_c = 50 \times (4 - 1.416) = 129.2 \text{ mm}^2$$

$$T_{all} = 0.55 \times 448 \text{ MPa} \times \frac{129.2 \text{ mm}^2}{50 \text{ mm}} = 636.7 \times 10^{-3} \text{ MN/m} = 637 \text{ kN/m}$$

$$T_{all} \times R_c = 637 \times 0.071 = 45 \text{ kN/m} > 19 \text{ kN/m} = T_{max}$$

Therefore, the first reinforcement layer passes the breakage check according to the WSD approach. We can do the same check using the LRFD approach. In this case, we use this inequality:

$$RF(T_{all}R_c) \geq LF_{DL}T_{max,DL} + LF_{LL}T_{max,LL}$$

For the first reinforcement layer, the maximum tensile forces $T_{max,DL}$ and $T_{max,LL}$ due to the lateral earth pressures caused by the self-weight of the reinforced soil and the uniform surcharge q_0 acting on top of the MSE wall, respectively, are given by

$$T_{max,DL} = K_r \gamma z s_v = 0.2 \times 19 \times 6.7 \times 0.65 = 16.5 \text{ kN/m}$$

$$T_{max,LL} = K_r q_0 s_v = 0.2 \times 20 \times 0.65 = 2.6 \text{ kN/m}$$

The maximum factored tensile force in the first reinforcement layer is given by

$$LF_{DL}T_{max,DL} + LF_{LL}T_{max,LL} = 1.5(16.5) + 1.75(2.6) = 29 \text{ kN/m}$$

and

$$RF \times (T_{all}R_c) = 0.80 \times 637 \times 0.071 = 36 \text{ kN/m} > 29 \text{ kN/m} = LF_{DL}T_{max,DL} + LF_{LL}T_{max,LL}$$

where $RF = 0.80$ for a target reliability index of 2.33, which corresponds to a target probability of failure of 0.01 (or 1 in 100). Assuming that this probability of failure for a reinforcement element is acceptable, the first reinforcement layer passes the breakage check according to the LRFD approach. The reason higher probabilities of failure may be acceptable for this check is the considerable redundancy that exists in the internal stability performance of MSE walls, with there being many other reinforcing elements to carry any load that the individual ruptured element may fail to carry.

Pullout limit state:

The second check we must perform concerns safety against pullout of the reinforcements. Combining Equations (16.37) and (16.48), stability with respect to pullout requires that:

$$\frac{1}{(FS)_{PO}} R_{PO} = \frac{1}{(FS)_{PO}} C_S C_P C_R \sigma'_v L_e R_c \geq T_{max}$$

where $C_S = 1.0$ for metallic reinforcements, C_P = reinforcement effective perimeter coefficient (= 2 for strips), and C_R = pullout resistance coefficient, which can be determined from Equation (16.38):

$$C_R = \tan \delta_c$$

Considering the surface of the steel strips to be rough, the value of δ_c may be taken as $0.85\phi_c$. This leads to a pullout resistance coefficient equal to

$$C_R = \tan(0.85 \times 36°) = 0.59$$

The effective length L_e of the reinforcement is the length that extends beyond the slip surface. If we add L_e to the length L_a that is embedded in the active wedge, we obtain the total length L_r of the reinforcement:

$$L_r = L_a + L_e$$

For simple wall geometries, as in our case, we use Figure 16.26a to determine L_a:

$$L_a = \begin{cases} 0.3H & \text{for} \quad 0 \le z \le H/2 \\ 0.6(H - z) & \text{for} \quad H/2 \le z \le H \end{cases}$$

So for $z=6.7$ m and $H=7$ m, we have $L_a=0.18$ m. We can now compute L_e as

$$L_e = 4.9 - 0.18 = 4.72 \text{ m}$$

So we can now compute $R_{PO}/(FS)_{PO}$ as

$$\frac{R_{PO}}{(FS)_{PO}} = \frac{1}{1.5} \times 1 \times 2 \times 0.59 \times (19 \times 6.7) \times 4.72 \times 0.071 = 34 \text{ kN/m} > 19 \text{ kN/m} = T_{max}$$

which means the first reinforcement layer passes the pullout check according to the WSD approach. The minimum effective length of the first reinforcement layer, for which the WSD-based pullout criterion is met, is

$$L_{e,min} = \frac{T_{max}(FS)_{PO}}{C_S C_P C_R \upsilon_v' R_c} = \frac{19 \times 1.5}{1 \times 2 \times 0.59 \times (19 \times 6.7) \times 0.071} = 2.67 \text{ m}$$

We can do the same check using the LRFD approach. In this case, we use this inequality:

$$RF(R_{PO}) \ge LF_{DL} T_{max,DL} + LF_{LL} T_{max,LL}$$

The pullout resistance of the reinforcement is

$$R_{PO} = C_S C_P C_R \sigma_v' L_e R_c = 1 \times 2 \times 0.59 \times (19 \times 6.7) \times 4.72 \times 0.071 = 50 \text{ kN/m}$$

So we can now compute $RF(R_{PO})$ as

$$RF(R_{PO}) = 0.65 \times 50 = 32.5 \text{ kN/m} > 29 \text{ kN/m} = LF_{DL} T_{max,DL} + LF_{LL} T_{max,LL}$$

where $RF=0.65$ for a target reliability index of 2.33, which corresponds to a target probability of failure of 0.01 (or 1 in 100). Thus, the first reinforcement layer passes the pullout check according to the LRFD approach. The minimum effective length of the first reinforcement layer, for which the LRFD-based pullout criterion is met, is

$$L_{e,min} = \frac{LF_{DL} T_{max,DL} + LF_{LL} T_{max,LL}}{RF(C_S C_P C_R \sigma_v' R_c)} = \frac{29}{0.65 \times 10.7} = 4.17 \text{ m}$$

We have completed the checks for the first reinforcement layer (counting from the base of the wall). The same process must be followed for all other reinforcement layers.

E-*Table 16.1* Results for all reinforcement layers using the WSD approach

Layer	z (m)	$\gamma z + q_0$ (kPa)	K_r	T_{max} (kN/m)	$T_{all}R_c$ (kN/m)	L_a (m)	L_e (m)	$R_{PO}/(FS)_{PO}$ (kN/m)	$L_{e,min}$ (m)	$L_{r,min}$ (m)
10	0.4	28	0.31	6.4	45	2.1	2.8	1	14.9	17.0
9	1.1	41	0.29	8.2	45	2.1	2.8	3	7.0	9.1
8	1.8	54	0.27	10.1	45	2.1	2.8	5	5.2	7.3
7	2.5	68	0.25	11.6	45	2.1	2.8	7	4.3	6.4
6	3.2	81	0.22	12.7	45	2.1	2.8	10	3.7	5.8
5	3.9	94	0.20	13.4	45	1.9	3.0	13	3.2	5.1
4	4.6	107	0.20	15.0	45	1.4	3.5	17	3.1	4.5
3	5.3	121	0.20	16.9	45	1.0	3.9	22	3.0	4.0
2	6.0	134	0.20	18.8	45	0.6	4.3	28	2.9	3.5
1	6.7	147	0.20	19.1	45	0.2	4.7	34	2.7	2.9

E-*Table 16.2* Factors of safety for the as-designed MSE wall of Example 16.10

Layer	z (m)	Actual FS against rupture	Actual FS against pullout
10	0.4	13.0	0.3
9	1.1	10.1	0.6
8	1.8	8.2	0.8
7	2.5	7.1	1.0
6	3.2	6.5	1.1
5	3.9	6.2	1.4
4	4.6	5.5	1.7
3	5.3	4.9	2.0
2	6.0	4.4	2.2
1	6.7	4.3	2.7

Note that, for the uppermost layer, $s_v = 0.75$ m. The results for all ten layers using the WSD approach are shown in E-Table 16.1. We see that all reinforcement layers pass the rupture check ($T_{all}R_c \geq T_{max}$), but only layers 1–4 pass the pullout check [$R_{PO}/(FS)_{PO} \geq T_{max}$]. The values of T_{all} and R_{PO} incorporate FS values of 1.82 for breakage and 1.5 for pullout. The actual FS values for each mode of failure are shown in E-Table 16.2. To satisfy the pullout limit state check for reinforcement layers 5–10 using the WSD approach, we would need to increase the effective length L_e of the reinforcement. The minimum effective and total lengths, $L_{e,min}$ and $L_{r,min}$, respectively, of each reinforcement layer needed to satisfy the WSD-based pullout check are summarized in E-Table 16.1.

The results for all ten reinforcement layers using the LRFD approach for $\beta_T = 2.33$ are summarized in E-Table 16.3. Recall that a target reliability index of 2.33 corresponds to a target probability of failure of 0.01 (or 1 in 100). We see that all reinforcement layers pass the rupture check ($RFT_{all}R_c \geq LF_{DL}T_{max,DL} + LF_{LL}T_{max,LL}$), but only layer 1 passes the pullout check ($RFR_{PO} \geq LF_{DL}T_{max,DL} + LF_{LL}T_{max,LL}$). To satisfy the pullout limit state check for reinforcement layers 2–10 using the LRFD approach, we would need to (a) decrease the reinforcement spacing and/or (b) increase the effective length L_e of the reinforcement. Following option (b) for example, the minimum effective and total lengths, $L_{e,min}$ and $L_{r,min}$, respectively, of each reinforcement layer needed to satisfy the LRFD-based pullout check are listed in E-Table 16.3.

Problem 16.20 asks the reader to redo this design, adjusting reinforcement spacing, to obtain an optimal design for this wall.

E-Table 16.3 Results for all reinforcement layers using the LRFD approach for $\beta_T=2.33$

Layer	z (m)	K_r	$T_{max,DL}$ (kN/m)	$T_{max,LL}$ (kN/m)	$LF_{DL}T_{max,DL}+LF_{LL}T_{max,LL}$ (kN/m)	$(RF)T_{all}R_c$ (kN/m)	L_a (m)	L_e (m)	$(RF)R_{PO}$ (kN/m)	$L_{e,min}$ (m)	$L_{r,min}$ (m)
10	0.4	0.31	1.8	4.6	10.7	36	2.1	2.8	1	25.7	27.8
9	1.1	0.29	4.2	4.0	13.3	36	2.1	2.8	3	11.6	13.7
8	1.8	0.27	6.4	3.7	16.1	36	2.1	2.8	5	8.6	10.7
7	2.5	0.25	8.1	3.4	18.2	36	2.1	2.8	7	7.0	9.1
6	3.2	0.22	9.5	3.1	19.8	36	2.1	2.8	9	5.9	8.0
5	3.9	0.20	10.5	2.8	20.8	36	1.9	3.0	12	5.1	7.0
4	4.6	0.20	12.2	2.8	23.3	36	1.4	3.5	17	4.8	6.3
3	5.3	0.20	14.1	2.8	26.0	36	1.0	3.9	21	4.7	5.7
2	6.0	0.20	16.0	2.8	28.8	36	0.6	4.3	27	4.6	5.2
1	6.7	0.20	16.5	2.6	29.4	36	0.2	4.7	33	4.2	4.4

16.5 SOIL NAILING

There are two basic types of soil nails: (1) small-diameter steel rods with 20 to 30 mm diameter (rebars are sometimes used) jacked, driven, or launched at high velocity into place and (2) drilled grouted holes with a steel bar at the center. This second type of soil nail is installed in a way that is similar to how tiebacks are installed, with the same type of equipment (see, for example, Figure 16.29). The hole (75–200 mm in diameter) is drilled, a 20–30 mm rebar is inserted, and grout is poured or pumped in around the rebar. If the nails

Figure 16.29 Typical rig used for soil nail installation.

are installed to support an excavation, the excavation proceeds in steps: a small height that can stand unsupported is excavated, a line of nails is installed, the excavation is continued, and an additional line of nails is installed.

The heads of the soil nails are often connected with a wall or facing (see, for example, Figure 16.30). Connection to the wall is usually accomplished by means of a plate and a nut. The head of a soil nail is said to be fixed if it moves with the wall or facing, and free if the retained ground can slide past the nail head without any impediment. An intermediate possibility is that the soil nail head is fixed so long as the force in the nail is less than some limiting value, but free otherwise. In the case of a nailed wall, when a head plate and nut are used, the nail head can usually be assumed to be fixed. When the nail heads are not connected with the wall or facing by any means, interacting directly with the ground surface (as is sometimes the case with nailed slopes), then the head is clearly a free head.

Despite the apparent similarity, there are important differences between soil nails and tiebacks. In contrast to tiebacks, soil nails are not prestressed. Tiebacks have a free length that is designed to have as little contact with the surrounding soil as possible and that extends through the potential sliding soil wedge. Beyond the potential sliding soil wedge, the tieback develops its resistance along the fixed length, where it develops a sufficiently large resistance along its interface with the soil. Prestressing fully mobilizes the tieback resistance before the wall moves in any significant way. Soil nails, on the other hand, develop their resistance only upon movement of the wall and the potentially critical soil wedge behind it, which attempts to drag the nails with it.

In the case of soil nails with a fixed head, as the sliding soil wedge moves down and out, it attempts to drag the nails with it (which is made possible by the shear stress that develops along its interface with the nails), as illustrated in Figure 16.31b. The resistance between a given nail and the soil beyond the sliding wedge is only then mobilized. The wedge pulls on the nail, the stable soil mass pulls on the nail from the other direction, and the nail pulls on the soil wedge. If the resistance of the soil nail is sufficient, the wall–nail–soil system will be stable; otherwise, the soil wedge will slide, carrying the nails with it. It is clear from this discussion that more movement is required for the mobilization of nail resistance than of tieback resistance. This makes nailed excavations suitable for temporary excavations or for

Figure 16.30 Soil-nailed cut: note the shotcrete on the back of the reinforcement mesh for a final concrete facing under construction. (Courtesy of the Indiana Department of Transportation.)

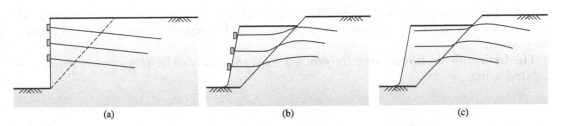

(a)   (b)   (c)

Figure 16.31 Soil-nailed wall: (a) undeformed configuration, (b) failure mechanism in which the nails are pulled out of the stationary soil mass (the only possible failure mechanism for fixed-head nails), and (c) failure mechanism in which the nails stay behind as sliding develops.

permanent excavations with supported heights that are not large and are distant from any structure that might be adversely affected by relatively large displacements.

If the soil nails have free heads (meaning the heads are not attached to the facing of the excavation or slope), then an additional failure mechanism (illustrated in Figure 16.31c) is possible. The wall or excavation face can slide, together with the failure wedge, while the soil nails stay in place. This would happen if the resistance developed by the soil nails beyond the sliding soil wedge is larger than the total resistance between the nail and the sliding wedge. This means that, in effect, the resistance required for stabilizing the wall and the soil wedge, although available, cannot be put to use because of insufficient coupling of the soil nails with the soil wedge.

Based on our discussion of the possible failure mechanisms and the way in which soil nails work, it is clear that the shape of the sliding soil wedge is a key design factor. The best way to solve this problem is to use a slope stability program with arbitrarily shaped slip surfaces that can handle soil nails, as discussed in the next chapter. An approximate design method relies on an assumed shape for the slip surface and on an assumed lateral earth pressure distribution. The earth pressure distributions used for braced excavations (Figure 16.24) are often used for this purpose (FHWA 1990). Start by setting a first estimate for the lateral and vertical spacings s_h and s_v of the soil nails based on previous experience. This will define an influence area for each nail, which may be multiplied by the corresponding lateral stress and a factor of safety FS to compute the nail resistance T_{req} required of the nail:

$$T_{req} = (FS)\sigma'_h s_v s_h \tag{16.51}$$

where s_h and s_v are the lateral and vertical spacings of the soil nails and FS is usually taken as at least 2. Equation (16.51) does not account for any unbalanced water forces, whose contribution, also magnified by a factor of safety, must be added to T_{req}.

The next step in the design process is to determine the nail diameters and lengths required for mobilizing T_{req}. The range of possible nail diameters is usually small, as seen earlier. For steel rods, a 20–30 mm (1–1.375 in.) range is typical. For drilled, grouted nails, diameters typically range from 75 to 200 mm (3–8 in.). The nail cross section must be structurally strong to resist the tension on the nail. Once the diameter B is determined, then the "fixed" (effective) length L_f (the length beyond the slip surface) must be determined from:

$$T_{req} = \pi B \sum_{i=1}^{n} q_{sLi} L_{fi} \tag{16.52}$$

where B is the diameter of the nail, L_{fi} is the length of the nail beyond the slip surface crossing soil layer i, q_{sLi} is the limit unit shaft resistance along the nail–soil interface for layer

i, and n is the number of soil layers. The number of layers crossed by the nail in a typical, horizontally layered soil is usually very small because the angle of the nail with the horizontal is small.

The limit unit shaft resistance for grouted nails in sands can be very conservatively calculated using:

$$q_{sLi} = \left(\frac{1 + K_{0i}}{2}\right)\sigma'_{vi} \tan\delta_i \tag{16.53}$$

where σ'_{vi} is the vertical effective stress acting at the midpoint of the segment of the nail that crosses layer i, K_{0i} is the coefficient of earth pressure at rest for layer i, and δ_i can be taken as 95% of the critical-state friction angle ϕ_{ci} ($= 28°-36°$, as seen in Chapter 5) for layer i. For jacked, driven, or launched nails, a higher K value (comparable to what it would be for displacement piles in sand) may be used; the interface friction angle for nails of this type is 85% of the critical-state friction angle ϕ_{ci}. The equations for unit shaft resistance of nondisplacement piles in the Purdue Pile Design Method (see Chapter 13) can be used for greater accuracy, with some caution given that nails are subhorizontal, not vertical elements.

For clays, the limit unit shaft resistance q_{sLi} of nails is calculated as

$$q_{sLi} = \alpha_i s_{ui} \tag{16.54}$$

where α_i is in the 0.4–0.6 range for grouted nails and in the 0.6–1 range for jacked, driven, or launched nails; s_{ui} is the undrained shear strength of layer i.

Once the fixed length L_{fi} is determined, the length from the nail head to the slip surface is added to L_{fi} to obtain the total length L to complete the design.

16.6 CHAPTER SUMMARY

Retaining structures are used often in foundation engineering projects as a way to create or preserve open space. This is done by retaining in place soil that would tend to slide into the open space as a result of the action of gravity, of applied surface loads or of inertial forces. Retaining structures must support the retained soil without reaching any of the following limit states:

1. limit bearing capacity failure;
2. sliding;
3. overturning; and
4. general loss of stability.

In most cases, the loading on the retaining structure is due to the active stresses behind it caused by the retained soil mass. The two most common methods of calculating the active pressures and forces on the retaining structure are the **Rankine method** and the **Coulomb method**. The calculation of the Rankine stresses was covered in Chapter 4. The active force calculated by the Coulomb method is given by

$$E_A = \frac{1}{2}K_A\gamma H^2 \tag{16.1}$$

where

$$K_A = \frac{\sin^2(\phi + \beta_w)}{\sin^2 \beta_w \sin(\beta_w - \delta)\left[1 + \sqrt{\dfrac{\sin(\phi + \delta)\sin(\phi - \alpha_g)}{\sin(\beta_w - \delta)\sin(\alpha_g + \beta_w)}}\right]^2} \quad (16.2)$$

Depending on the source of stabilizing action, retaining structures can be **externally stabilized** or **internally stabilized**. Externally stabilized retaining walls include **gravity walls** (the weight of the wall is the source of stability), **cantilever embedded walls** (the passive pressures in front of the embedded part of the wall are the source of stability), and **tieback walls** (both the passive pressures in front of the embedded part of the wall and the tieback force pulling back on it are the source of stability). Design of these walls consists in checking that the four limit states listed above are not attained. The starting point for that is to draw a free body diagram of the wall, with all the forces acting on it, and checking for sliding and overturning. In embedded cantilever walls and in tieback walls, a simplifying assumption must be made regarding the forces acting below the assumed center of rotation of the wall. Bearing capacity checks were extensively discussed in Chapters 10 and 11, and general stability is discussed in the next chapter.

A **tieback** is an anchor that exerts a restraining force on the retaining wall. It is usually a cylindrical hole with a steel cable separated from the soil by grout. The cable is firmly connected with the wall. The tieback develops its resistance by side friction with the soil beyond the potential sliding wedge. Little to no side resistance exists between the tieback and the potential soil wedge. Tiebacks are usually preloaded to 150% of their design load, and that load is locked in. They differ from **soil nails** mostly in this regard: soil nails develop side resistance against the surrounding soil throughout, including along the contact with the potential sliding wedge, and soil nails are not preloaded. This makes for a much less stiff response from soil nails, which will only mobilize their resistance after significant movement of the wall has taken place.

MSE walls are walls constructed of soil reinforced using strips, sheets, or grids made of steel or a geosynthetic material. In addition to the limit states for other types of walls, we must also check the reinforcement rupture and pullout limit states for MSE walls.

16.6.1 Symbols and notations

Symbol	Quantity represented	US units	SI units
A_c	Cross-sectional area of the reinforcement	in.^{2}	mm^2
B	Diameter of the tieback	ft	m
B	Width of the wall	ft	m
b	Width of the reinforcement	in.	mm
c	Cohesion	psf	kPa
C_p	Reinforcement effective perimeter coefficient	Unitless	Unitless
C_R	Pullout resistance coefficient	Unitless	Unitless
C_s	Scale effect coefficient	Unitless	Unitless
C_U	Coefficient of uniformity	Unitless	Unitless
d	Embedded length of the wall	ft	m

(Continued)

Symbol	Quantity represented	US units	SI units
E_A	Active force per unit length of wall	lb/ft	kN/m
E_P	Passive force per unit length of wall	lb/ft	kN/m
E_w	Water force per unit length of wall	lb/ft	kN/m
F	Soil resisting force on the slip surface	lb/ft	kN/m
f	Distance from the pivot point to the ground surface	ft	m
F_N	Normal force per unit length of wall	lb/ft	kN/m
F_T	Tangential force per unit length of wall	lb/ft	kN/m
f_y	Yield stress of steel	psf	kPa
H	Height of the wall	ft	m
h_E	Distance from the base of the wall to the point of application of E_A	ft	m
h_{Ew}	Distance from the base of the wall to the point of application of E_W	ft	m
H_T	Distance from the surface of backfill to the location of tieback	ft	m
K_0	Coefficient of lateral earth pressure at rest	Unitless	Unitless
K_A	Coefficient of active lateral earth pressure	Unitless	Unitless
K_{AW}	Modified active earth pressure coefficient accounting for soil arching effects	Unitless	Unitless
K_{awn}	Active lateral earth pressure coefficient accounting for soil arching effects	Unitless	Unitless
K_P	Coefficient of passive lateral earth pressure	Unitless	Unitless
K_r	Coefficient of lateral earth pressure in reinforced soil wall	Unitless	Unitless
L_a	Distance from the wall facing to the slip surface	ft	m
L_e	Effective length of reinforcement	ft	m
L_{fi}	Fixed length of tieback or nail crossing soil layer i	ft	m
N	Flow number	Unitless	Unitless
q_{sLi}	Limit unit shaft resistance along the tieback–soil interface for layer i	psf	kPa
R_c	Coverage ratio, used to convert a load per unit length of wall (T_{max}) into a load per unit length of reinforcement (T_{all})	Unitless	Unitless
R_{PO}	Pullout resistance	lb/ft	kN/m
s_h	Horizontal spacing of reinforcements	ft	m
s_T	Tieback spacing	ft	m
s_v	Vertical spacing of reinforcements	ft	m
T	Tieback force per unit length of wall	lb/ft	kN/m
T_{all}	Allowable tensile force of reinforcement layer	lb/ft	kN/m
T_d	Design load of tieback	lb	kN
t_{LT}	Long-term thickness of the reinforcement	in.	mm
T_{max}	Maximum tensile force on reinforcement layer	lb/ft	kN/m
t_n	Normal thickness of metal	in.	mm
t_R	Sacrificial thickness of metal	in.	mm
T_{req}	Required nail resistance	lb	kN
T_{res}	Resistance of tieback	lb	kN
T_{ult}	Ultimate tensile resistance of reinforcement layer	lb/ft	kN/m

(Continued)

Symbol	Quantity represented	US units	SI units
W	Weight of the soil wedge per unit length of wall	lb/ft	kN/m
α_g	Backfill slope angle with horizontal	deg	deg
β_w	Wall inclination angle with horizontal	deg	deg
γ_m	Unit weight of soil	pcf	kN/m³
γ_{wall}	Unit weight of the wall	pcf	kN/m³
δ	Soil–wall interface friction angle	deg	deg
σ_{ahw}	Active lateral stress accounting for soil arching effects	psf	kPa
σ_h'	Horizontal effective stress	psf	kPa
$\bar{\sigma}_v$	Average vertical stress	psf	kPa
σ_v'	Vertical effective stress	psf	kPa
σ_{vi}'	Vertical effective stress acting at the midpoint of the segment of the tieback crossing soil layer i	psf	kPa
ϕ	Friction angle	deg	deg
ϕ_c	Critical-state friction angle	deg	deg
ϕ_p	Peak friction angle	deg	deg

16.7 PROBLEMS

16.7.1 Conceptual problems

Problem 16.1 Define all the terms in bold contained in the chapter summary.

Problem 16.2 When is a retaining structure typically used?

Problem 16.3 What is the difference between internally and externally stabilized retaining structures? Can you give examples of each? How could you further classify retaining structures? Give examples.

Problem 16.4 Define the active and passive lateral earth pressure coefficients in sloping ground conditions?

Problem 16.5 Is it always true that the angle of wall friction is zero when we use Rankine's earth pressure theory? Explain.

Problem 16.6 Is it unconservative or conservative to assume too large a value of cohesion under (a) active and (b) passive conditions? Justify your answer.

Problem 16.7 List the ultimate limit states against which we must ensure that a retaining wall has an adequate margin of safety. What additional limit states must we guard against when designing an MSE wall?

Problem 16.8 Rederive the equations for the base width B of a gravity wall with a rectangular cross section based on overturning, Equation (16.15), and sliding, Equation (16.16), considering now that there is a nonzero interface friction angle δ between the soil and the back of the wall.

Problem 16.9 Why, in using Coulomb's method, do we search for the wedge that gives a maximum value for the active pressure on the wall?

Problem 16.10 Show that (16.2) reduces to (4.59) when $\alpha_g=0°$, $\beta_w=90°$, and $\delta=0°$.

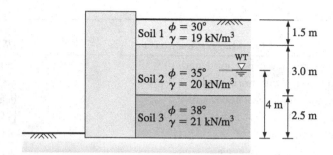

P-Figure 16.1 Wall of Problem 16.15.

16.7.2 Quantitative problems

Problem 16.11 A retaining wall 7 m high, with a vertical face, supports a sandy soil of dry unit weight $\gamma=16$ kN/m³ and friction angle$=35°$. The surface of the soil is horizontal and aligned with the top of the wall. If an active state is established behind the wall, find the distribution of active pressures on the wall and the resultant active force. The water table is located below the level of the base of the wall.

Problem 16.12 Redo Problem 16.11 for the passive case.

Problem 16.13 Consider a wall and retained soil mass with the following properties: smooth back ($\delta=0$), $H=4$ m, $\gamma=19$ kN/m³, and $\phi=33°$; deep water table. Determine the following: (a) the pressure diagram behind the wall if it is prevented from yielding and (b) the pressure diagram behind the wall if it yields far enough for the soil to enter the active state.

Problem 16.14 The water level behind the wall of Problem 16.13 rises to an elevation 1 m below the crest. Determine the magnitude and point of application of the resultant of the total force exerted on the wall, including the active earth force and the water force.

Problem 16.15 Plot the vertical effective stress, the active Rankine lateral effective stresses, the water pressures, and the total pressures for the wall in P-Figure 16.1. In addition, calculate the total resistance and the distance from its point of application to the base of the wall. Soils 1, 2, and 3 have friction angles equal to 30°, 35°, and 38° and unit weights of 19, 20, and 21 kN/m³.

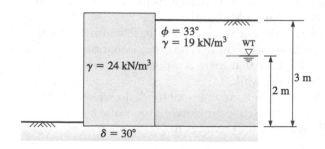

P-Figure 16.2 Wall of Problem 16.17.

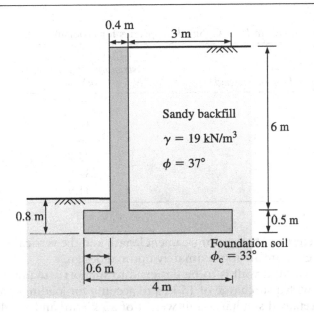

0.4 m

3 m

Sandy backfill

$\gamma = 19$ kN/m³

$\phi = 37°$

6 m

0.8 m

0.5 m

Foundation soil
$\phi_c = 33°$

0.6 m

4 m

P-Figure 16.3 Wall whose stability is to be checked as part of Problem 16.18.

16.7.3 Design problems

Problem 16.16 Find the width B of the wall in Problem 16.14 so that it is safe with respect to both overturning ($FS=2$) and sliding ($FS=1.5$). The unit weight of the wall is 24 kN/m³, and the friction angle at the base–foundation soil interface is 30°.

Problem 16.17 Consider the wall and retained soil mass in P-Figure 16.2 with the following properties: smooth back ($\delta=0$), $H=3$ m, $\gamma=19$ kN/m³, and $\phi=33°$. The water level behind the wall is at an elevation 1 m below the crest.

a. Determine the vectorial resultant of the active earth force exerted on the wall.
b. Find the width B of the wall so that it is safe with respect to both overturning ($FS=2$) and sliding ($FS=1.5$). The unit weight of the wall is 24 kN/m³, and the friction angle at the wall base–foundation soil interface is 30°.

Problem 16.18 A reinforced concrete ($\gamma=24$ kN/m³) wall constructed several decades ago, shown in P-Figure 16.3, has had its stability questioned. Check the horizontal and overturning stability of this wall. What are the factors of safety with respect to each of these limit states? Are they satisfactory? Neglect passive resistance.

Problem 16.19* A slurry wall 40 cm thick will be used to retain 7 m of a sandy soil with friction angle equal to 32° and unit weight equal to 20 kN/m³. The water table is deep. Find the depth of embedment for two conditions: (a) without and (b) with tiebacks. Assume that one line of tiebacks is installed 1.5 m from the top of the wall. For the tieback wall, find the spacing, length, and diameter of the tiebacks. For determining spacing, consider that the maximum tensile force the tiebacks can withstand is 400 kN and that the concrete panel width is a design decision (but that this width should be between 2.5 and 5 m). Make any remaining assumptions you find necessary, but justify them. Discuss the two designs from the point of view of likely costs. Is the unpropped wall likely to be economically competitive with the tieback wall?

Problem 16.20 Revise the MSE wall design in Example 16.10 to ensure that both external and internal stability design checks are satisfied using the WSD and LRFD approaches.

P-Table 16.1 Geogrid properties for Problem 16.22

Type of geogrid	Long-term tensile strength after 30 years (kN/m)
1	7.8
2	10.4
3	16.8
4	25.7
5	32.1
6	41.9

Specifically, determine both reinforcement length and the vertical and horizontal spacings that will generate an approximately optimal design.

Problem 16.21* An MSE wall is to be designed to support a roadway. The MSE wall is to be 8 m tall with a surcharge of 15 kPa to account for loadings applied on top of the backfill. The retained soil has a unit weight of 23 kN/m³ and $\phi = 40°$. The reinforced soil has $\gamma = 22$ kN/m³ and $\phi = 38°$. The foundation soil has a unit weight of 20 kN/m³ and $\phi = 34°$. The design life of the wall is 30 years. Design this wall for external stability (bearing capacity, overturning, and sliding) and internal stability (reinforcement rupture and pullout). Use 65-grade galvanized steel strips ($f_y = 448$ MPa) with width = 50 mm and thickness = 4 mm. Assume that the galvanization corrodes fully in 16 years and that the corrosion rate of the steel strip is 12 μm/year.

Problem 16.22* Redo Problem 16.21 using geosynthetics. As required by the imposed forces, use geogrids of types 1–6, with the properties shown in P-Table 16.1.
Use these coefficients to calculate pullout resistance:
$(FS)_{PO} = 1.5$
$C_i = 0.8$ (coefficient of interaction)
$C_P = 2$ (reinforcement effective perimeter coefficient)
$C_S = 0.8$ (scale effect coefficient)

Problem 16.23* An excavation of height 8 m is to be done. A vertical facing is desired. Three lines of soil nails will be used. The natural soil is a clayey sand with unit weight equal to 19 kN/m³. The soil is nearly all saturated due to capillary rise from the water table, located at around the level of the base of the excavation. The critical-state friction angle of this soil is 31°. Design a soil nailing scheme for this excavation.

REFERENCES

References cited

AASHTO. (1996). *AASHTO standard specifications for highway bridges.* 16th Edition, American Association of State Highway and Transportation Officials, Washington, D.C., USA.

AASHTO. (2020). *AASHTO LRFD bridge design specifications.* 9th Edition, American Association of State Highway and Transportation Officials, Washington, D.C., USA.

Burland, J. B., Potts, D. M., and Walsh, N. M. (1981). "The overall stability of free and propped embedded cantilever retaining walls." *Ground Engineering*, 14(5), 28–38.

CGS. (1992). *Foundation Engineering Manual*, 3rd Edition, Canadian Geotechnical Society, Bitech, Vancouver, Canada.

Coulomb, C. (1776). "Essai sur une application des règles de maximis & minimis à quelques problèmes de statique, relatifs à l'architecture." *Memoires de Mathematique de l'Academie Royale de Science*, Paris, France, 7.

FHWA. (1990). *Reinforced soil structures–Volume I.* Design and Construction Guidelines, Report No. FHWA-RD-89-043, Federal Highway Administration, U.S. Department of Transportation, Washington, D.C., USA.

FHWA. (2009). *Design and construction of mechanically stabilized earth walls and reinforced soil slopes–Volume I.* Report No. FHWA-NHI-10-024, Federal Highway Administration, U.S. Department of Transportation, Washington, D.C., USA.

Kim, D., and Salgado, R. (2012a). "Load and resistance factors for internal stability checks of mechanically stabilized earth walls." *Journal of Geotechnical and Geoenvironmental Engineering*, 138(8), 910–921.

Kim, D., and Salgado, R. (2012b). "Load and resistance factors for external stability checks of mechanically stabilized earth walls." *Journal of Geotechnical and Geoenvironmental Engineering*, 138(3), 241–251.

Lancellotta, R. (2002). "Analytical solution of passive earth pressure." *Géotechnique*, 52(8), 617–619.

Lin, Y.-J., Habib, A., Bullock, D., and Prezzi, M. (2019). Application of high-resolution terrestrial laser scanning to monitor the performance of mechanically stabilized earth walls with precast concrete panels." *Journal of Performance of Constructed Facilities*, 33(5), 04019054.

Loukidis, D., and Salgado, R. (2012). "Active pressure on gravity walls supporting purely frictional soils." *Canadian Geotechnical Journal*, 49(1), 78–97.

Paik, K. H., and Salgado, R. (2003). "Estimation of active earth pressure against rigid retaining walls considering arching effects." *Géotechnique*, 53(7), 643–653.

Peck, R. B. (1969). "Deep excavations and tunneling in soft ground." *Proceedings of 7th International Conference on Soil Mechanics and Foundation Engineering*, Mexico City, Mexico, 225–290.

Powrie, W. (1996). "Limit equilibrium analysis of embedded retaining walls." *Géotechnique*, 46(4), 709–723.

Take, W. A., Bolton, M. D., Wong, P. C. P., and Yeung, F. J. (2004). "Evaluation of landslide triggering mechanisms in model fill slopes." *Landslides*, 1, 173–184.

Terzaghi, K., and Peck, R. B. (1967). *Soil Mechanics in Engineering Practice*. 2nd Edition, Wiley, New York.

Tschebotarioff, G. P. (1973). *Foundations, Retaining and Earth Structures*. 2nd Edition, McGraw-Hill, New York.

Additional references

Daniel, D. E., and Olson, R. E. (1982). "Failure of an anchored bulkhead." *Journal of the Geotechnical Engineering Division*, 108(10), 1318–1327.

Golder, H. Q. (1948). "Coulomb and earth pressure." *Géotechnique*, 1(1), 66–71.

Fang, Y. S., and Ishibashi, I. (1986). "Static earth pressures with various wall movements." *Journal of Geotechnical Engineering*, 112(3), 317–333.

Finno, R. J., and Roboski, J. F. (2005). "Three-dimensional responses of a tied-back excavation through clay." *Journal of Geotechnical and Geoenvironmental Engineering*, 131(3), 273–282.

Handy, R. L. (1985). "The arch in soil arching." *Journal of Geotechnical Engineering*, 111(3), 302–318.

Harrop-Williams, K. O. (1989). "Geostatic wall pressures." *Journal of Geotechnical Engineering*, 115(9), 1321–1325.

Hashash, Y. M., and Whittle, A. J. (2002). "Mechanisms of load transfer and arching for braced excavations in clay." *Journal of Geotechnical and Geoenvironmental Engineering*, 128(3), 187–197.

Matsuo, M., Kenmochi, S., and Yagi, H. (1978). "Experimental study on earth pressure of retaining wall by field tests." *Soils and Foundations*, 18(3), 27–41.

Sadek, S., and El-Khoury, M. (2005). "The ABCs of soil nailing: an integrated tutorial and knowledge-based approach to teaching design." *International Journal of Engineering Education*, 21(5), 993–1002.

Terzaghi, K. V. (1920). "Old earth pressure theories and new test results." *Engineering News Record*, 85(14), 632–637.

Terzaghi, K. (1943). *Theoretical Soil Mechanics*. Wiley, New York.

Terzaghi, K. (1934). "Large retaining wall test." *Engineering News Record*, 112, 136–140.

Tsagareli, Z. V. (1965). "Experimental investigation of the pressure of a loose medium on retaining walls with a vertical back face and horizontal backfill surface." *Soil Mechanics and Foundation Engineering*, 2(4), 197–200.

Wang, Y. Z. (2000). "Distribution of earth pressure on a retaining wall." *Géotechnique*, 50(1), 83–88.

Relevant ASTM standards

ASTM. (2017). *Standard test method for tensile properties of geotextiles by the wide-width strip method*. ASTM D4595, American Society for Testing and Materials, West Conshohocken, Pennsylvania, USA.

Chapter 17

Soil slopes

> Earthwork gives way by the slipping or sliding of its parts on each other; and its stability arises from resistance to the tendency to slip.
>
> William John Macquorn Rankine

Geotechnical and foundation engineers must deal routinely with slopes in their work. This is because foundation works are often carried out in sloping ground and also because excavations or fills may create slopes where none previously existed. In addition, there is a range of applications not directly related to foundations of which slopes are an integral part. In this chapter, we are concerned with soil slopes. The emphasis of this chapter is on the analysis that enables engineers to make an assessment of the stability of a slope against sliding. A slope stability failure, a landslide, occurs for the reason described by Rankine: the shear strength available at the interface between the moving soil mass and a stable soil mass is not sufficient to prevent sliding.

17.1 THE ROLE OF SLOPE STABILITY ANALYSIS IN FOUNDATION ENGINEERING PROJECTS

17.1.1 Engineering analysis of soil slopes

In soils, slides are the most serious type of slope movement. The stability of a soil slope is dependent on the existence of enough shear strength within the soil mass to balance the gravity loads of the soil mass itself, any existing surface loads, and the forces of water moving through the slope. In the most common type of slope stability analysis, we consider *limit equilibrium* conditions between these forces and the soil shear strength, suitably minimized by a factor of safety. By limit equilibrium, we mean equilibrium on the verge of sliding along a given slip surface (Figure 17.1), such that either a minimal reduction in strength, or a minimal increase in one of the loads would cause sliding or collapse of the slope.

There are other methods that may also be used for slope stability calculations. The first decision engineers must make is the choice of the method of calculation. Two important additional decisions are whether to use effective or total stress analysis and which values of shear strength to use in the analysis. Finally, based on the analysis, engineers obtain an estimate of the factor of safety of the slope and must then decide whether the factor of safety is adequate. If the factor of safety is not adequate, redesign of an engineered slope or remediation of a natural slope is required. The decision can also be made based on an LRFD check of the stability of the slope. Some of the engineering decisions required for slope stability analysis are discussed in this section. The methods of analysis are discussed in subsequent sections.

DOI: 10.1201/b22079-17

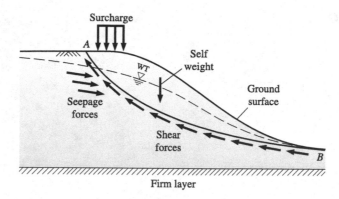

Figure 17.1 The limit equilibrium method: the weight of the potential sliding mass extending from A to B, the applied surcharge on top of the slope, and the seepage forces due to groundwater flow are destabilizing forces that must be matched by shear strength along the slip surface AB to prevent sliding.

17.1.2 Stability and deformation analyses

Soil response to load is a continuous process. Soil first undergoes small deformations, then larger deformations, and finally deformations so large as to characterize loss of stability; however, stability and deformation analyses are usually done using different methods. Even though the finite element method (FEM) has been developed to a stage at which it could be used alone to describe the whole range of deformations all the way to stability loss, this application still requires considerable analytical expertise and a significant amount of time to compile, enter, and process information. Accurate results also require automatic or manual remeshing to guarantee sufficient fineness of the mesh along the eventual slip surface (truly, a shear band, as discussed in Chapters 4 and 6) if stability is the focus of the finite element analysis. Another method that can be used is the material point method (MPM), but sophistication of the analyst is again required, for different reasons. This makes these analyses considerably more expensive than the limit equilibrium analyses we will discuss in this chapter, which is an important reason why limit equilibrium analyses continue to be used in most projects. Additionally, slope stability analyses based on limit equilibrium have been shown to be reliable and are reasonably simple to use. Simplified deformation analyses or empirically based deformation estimation methods, however, are not very reliable. If FEM or MPM are used, a realistic soil constitutive model is recommended if deformation prediction with sufficient accuracy is required. Table 17.1 summarizes the main methods used in the engineering of soil slopes and their applicability.

In connection with slopes, much focus is placed on the ultimate limit state defined by stability loss and very large, practically instantaneous deformations. But both ultimate and serviceability limit states exist associated with slope displacements. Sufficiently large deformations of a slope may distort the foundations of a structure so much that structural damage or other serious limit states may result. Less severe slope movements may lead to

Table 17.1 Main types of analyses for assessing slope load response

Type of analysis	Applicability
Limit equilibrium (LE)	Slope stability
Limit analysis (LA)	Slope stability
Finite element method (FEM)	Slope deformation and stability
Material point method (MPM)	Slope deformation and stability

Figure 17.2 Large slope movements can lead to severe limit states when structures rest on the sliding mass, as illustrated by this bridge across the Missouri River (Oahe Reservoir), in Forest City, South Dakota, which was built on a landslide. (Courtesy of Vern Bump, SDDOT.)

architectural damage. In cases where it is shown that classical instability does not develop, but very large deformations may develop over time, a deformation analysis must be carried out. As an illustration of what may happen when we build on landslides, Figure 17.2 shows a bridge across the Missouri River (Oahe Reservoir), in Forest City, South Dakota, which was built on a landslide. The bridge was completed in 1958. The first signs of distress were noted in 1962, followed by a realization in 1968 that geotechnical problems lay at the heart of the problem (Vernon Schaefer, Personal communication). The landslide presented problems regarding the road, the approach embankment, and the bridge itself well into the 1990s, during which a comprehensive intervention was done using excavations, stone columns, and shear pins.

Figure 17.3 Collapse of slope at a construction site.

In contrast, not every sudden slope collapse leads to an ultimate limit state. The collapse of the small temporary slope in Figure 17.3 at a construction site is an inconvenience and has a small cost associated with the removal of the material to free up space for construction operations to proceed. Naturally, if there had been a person or expensive equipment standing on top of the slope and that was to be expected, it would arguably have been an ultimate limit state.

17.1.3 Effective versus total stress analysis

As discussed in Chapter 3 and expanded in Chapters 5 and 6, effective stresses – not total stresses – are operative in the mechanical response of soil. It follows that effective stress analysis is, in principle, preferable to total stress analysis. However, it is typically not possible to perform effective stress analysis because of difficulties in the determination of excess pore pressures, whose knowledge is required if we wish to perform effective stress analysis. A soil model that can capture the effective stress response of the soil at every point in the soil mass would be required for this purpose. Instead, slope stability analyses are typically done based solely on the knowledge of a strength envelope, which is not sufficient to perform an effective stress analysis.

Consider, for example, an excavation in clay. Immediately after the excavation, there is a total stress reduction along any potential slip surface within the slope and, consequently, corresponding drops in the pore pressures (since the soil is slow-draining, there is initially no change in effective stresses). Shear stresses induced in the slope by the excavation would lead to the development of excess pore pressures. An effective stress analysis must capture all these effects.

In total stress analysis, total stresses are used together with shear strength determined and expressed in terms of total stresses. In contrast, in an effective stress analysis, effective stresses are used together with shear strength expressed in terms of effective stresses.[1] Total stress analyses are typically used for clayey soils for short-term stability analyses. In these cases, the total stress strength criterion (the Tresca yield criterion) discussed in Chapter 6 can be used:

$$\tau = s = s_u \tag{17.1}$$

where τ is the shear stress, s is the general notation used for shear strength, and s_u is the undrained shear strength of the clay. The strength criterion in essence states that, for sliding to occur, τ must equal s_u.

In effective stress analysis, the failure criterion is expressed as

$$\tau = s = c + \sigma' \tan\phi = c + (\sigma - u)\tan\phi \tag{17.2}$$

In certain derivations in this chapter, we use the generic equation:

$$\tau = s = c + \sigma \tan\phi \tag{17.3}$$

to describe soil shear strength. In calculations, the appropriate values of c and ϕ, corresponding to either Equation (17.2) or (17.3), should be used. In some other derivations, the formulations are developed directly in terms of effective stresses because effective stress formulations are used exclusively in those cases.

---

[1] We have seen these analyses before. For example, in Chapter 10, we studied the analysis of bearing capacity of footings in clays, a total stress analysis, and in sands, an effective stress analysis.

It is important to note that slope stability problems, like retaining wall stability problems, but unlike foundation bearing capacity problems, are problems in which the stress range is either close to or less than initial stress conditions. This means that c and ϕ values from linear fits on the results of laboratory strength testing can be determined so that they are representative of stress conditions observed in the real problem and thus may be used directly in analysis. The only qualification to this statement relates to the possibility of degradation of any component of strength that is due to dilatancy (typically, this shows up as a cohesive intercept c if curve fitting is used, as discussed in Chapters 5 and 6). This degradation is typically caused by creep (slipping of the slope at a very slow pace) or by seasonal oscillations of water table, which alternate periods of stability with periods during which some limited shearing of the material happens (as described in Take et al. 2004).

17.1.4 Typical slope problems

17.1.4.1 Sandy/silty/gravelly fills built on firm soil or rock

This case, shown in Figure 17.4a, encompasses fills made of gravels, sands, nonplastic silts, or mixtures thereof built on top of firm ground. The stability of such slopes is a function of the friction angle ϕ of the sandy soil, the slope angle, and pore pressures, including any excess pore pressures that may develop within the sandy soil as a result of fast loading. The failure mechanism is usually surface raveling or shallow sliding. One of the important questions in connection with this type of problem is how to get the friction angle ϕ of the soil to use in analyses. In the laboratory, drained triaxial tests or direct shear tests on samples prepared in a way that simulates fill construction as closely as possible are the best options. Correlations in terms of particle size distribution, particle shape, relative density, or *in situ* test results may also be used to obtain ϕ. Sustainability of any degradable component of shear strength needs to be ascertained.

Slopes made of fine sands, silty sands, or silts may be subjected to erosion and surface runoff. Saturated slopes may also undergo flow liquefaction[2] during an earthquake if the soil is not sufficiently dense. Erosion and surface runoff can be mitigated by use of suitable cover. Liquefaction susceptibility can be mitigated by using a variety of soil improvement techniques.

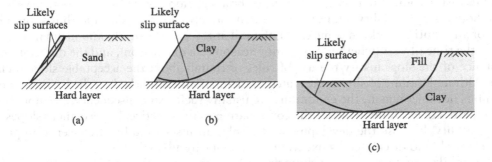

Figure 17.4 Some typical slope problems and likely slip surface locations: (a) sandy soil on top of a firm soil or rock layer, (b) clayey soil on top of a firm soil or rock layer, and (c) fill on top of soft clay.

---

[2] Flow liquefaction is the failure of a slope, often associated with large soil mass movements, caused by a sharp drop in the soil shear strength (below the existing shear stresses) due to shearing accompanied by a rise in pore pressures.

17.1.4.2 Clayey fills built on firm soil or rock

This case, shown in Figure 17.4b, applies to fills built of clays or soils containing relatively large percentages of clay on firm ground. The stability of such slopes depends on soil shear strength, unit weight, slope height, slope angle, and pore pressures. The failure mechanism is usually a deep failure surface tangent to the top of the firm soil or rock foundation. If drainage is slow, an investigation of the stability should be done for short-term (end-of-construction) conditions as well as long-term conditions.

As discussed previously, the short-term stability analysis is usually done in terms of total stresses, with shear strengths from UU triaxial tests on samples compacted to the same density and water content as in the field. Pore pressures are accounted for implicitly in the tests. External water, as when the slope is in contact with a water body (for example, a lake, river, or reservoir), if present, provides a stabilizing effect by applying a normal stress against the part of the slope with which it is in contact. The long-term stability analysis is done in terms of effective stresses, with shear strengths from CU triaxial tests (with pore pressure measurements) or CD triaxial tests on samples compacted to the field density and water content.

17.1.4.3 Fills built on soft subsoil

The stability of this type of slope, shown in Figure 17.4c, depends on the strength of the fill material, the unit weight of the fill material, the height of the fill, the slope angle, the shear strength of the foundation soils, and any pore pressures – including excess pore pressures – that may develop within the slope or in the foundation soil. The failure mechanism is usually a deep slip surface tangent to a firmer layer within the foundation soil. Short-term conditions are usually the most critical because the foundation consolidates, thereby gaining strength, over time.

17.1.4.4 Excavation slopes

Excavations are common in foundation engineering work because underground space is often needed for parking, for example. Temporary or shallow excavations are sometimes unsupported, although soil nails and other techniques we examined in Chapter 16 are sometimes used for stabilizing them. Permanent excavations are usually carried out with simultaneous installation of tiebacks, soil nails, or bracing. Slopes vary from vertical to very flat.

Although slope stability analysis can be done simply enough today and thus should be a part of any routine work, in some cases, the soil at the site, the height and inclination of the slope, the time of year (whether it is a rainy season or a dry season), and the duration of the existence of the slope justify the use of rules of thumb about the acceptable slope inclination. Although failure of slopes excavated to inclinations established on the basis of rules of thumb is not uncommon, the conditions are usually such that consequences are not serious and repairs are easily done, hence the continuation of the practice. In permanent slopes, not only stability, but also the development of displacements should be of concern; no permanent slope should be created by excavation without careful analysis.

The stability of an excavation slope depends on the strength of the natural soil, its unit weight, the slope height, the slope angle, and pore pressures generated by the excavation. The slip mechanism depends on soil uniformity. For a uniform sandy soil (or soil made up predominantly of gravel or nonplastic silt), the mechanism is likely to be a shallow slip surface because shear strength increases substantially with depth for these materials, being very low near the surface of the slope. For a uniform clayey soil (or soil made up predominantly of plastic silt), the critical slip surface is likely to be deep because strength does not increase significantly with depth, but the weight of a potential sliding mass does. The exception to that is the case of steep slopes in clay (with an angle with the horizontal in excess of 53°, as shown by Taylor (1948)), for which the critical slip surface would be through the toe of the slope. The slip mechanism in nonuniform soil slopes is controlled by weakness zones.

17.1.4.5 Natural slopes

Natural slopes are often characterized by heterogeneous soil profiles and groundwater regimes that depend strongly on hydrological patterns. The engineer has limited control over drainage conditions and can acquire knowledge of the soil profile only through site investigation. This contrasts with the situation for fill slopes, for which the material is much better known because it is formed in place according to specifications.

In the absence of any grading, the slope has existed for many years in equilibrium with prevailing groundwater conditions. Failure of such slopes usually results from construction activities or unusually heavy rainstorms, during which pore pressure changes will occur due to seepage.[3] Stability analyses in slopes known to be of problematic or in areas where intense rainstorms are possible during part of the year should take account of the increased pore pressures during these events. Some degree of monitoring of pore pressure generation and movement is often indicated, not only for providing a basis for issuing warnings of possible slides, but also for providing the data for improved analysis.

17.1.4.6 "Special" cases

Some soils require special attention. Slopes in stiff, fissured clays, and shales, for example, are likely to develop slip surfaces along preexisting fissures. Additionally, as discussed in Chapter 6, any previous sliding along existing fissures usually aligns clay particles in the direction of the sliding surface, reducing the shear strength there to a residual value, potentially much lower than that related to the critical-state friction angle. Finally, such materials are usually dilative, which means that long-term conditions are usually substantially more critical than short-term conditions.

Slopes in loess, a lightly cemented wind-deposited silty material (see Chapter 3), present their own problems. Upon inundation or shock waves from such activities as pile driving or the use of explosives, much of the cementation present in loess may break down, leading to considerable loss of strength. Other collapsible soils, including many residual soils found in tropical and subtropical regions, are subject to the same problems.

A last example of unusual dangers to slope stability comes from the rather unusual behavior of marine clays. As discussed in Chapter 3, such materials, present notably in Canada and Scandinavia, have an inherently unstable structure that, upon even small disturbances, breaks down, leading to almost complete loss of strength. Less extreme cases of clay sensitivity still require some care because local reduction in shear strength may lead to different values of shear strength in different parts of an assumed slip mechanism, contrary to the assumption often made of constant shear strength along the slip surface. This means that the factor of safety can reach 1 along parts of the slip surface while it remains above 1 elsewhere along the slip surface. This is referred to as progressive failure. If it is believed that progressive failure is possible, appropriate reduction in design strength (by as much as 30%) may be needed.

---

[3] In unsaturated soils, the increase in degree of saturation will also eliminate the stability provided by matric suction.

17.1.4.7 Need for computations

Manual calculations of slope stability are time-consuming, and not even possible for the more modern methods of analysis. Thanks to the wide availability of powerful computers, slope stability analyses are now routinely done in engineering work. In the past, engineers often used slope stability charts, prepared in many cases by running cases for simple slopes (with uniform soil properties or properties varying in some simple way with depth) in mainframe computers (which were less powerful than the personal computers of today). These charts today should be used at most as a general check on results from computer analyses for the engineer's own assurance that the answers are in general agreement as to order of magnitude and direction with what would be expected based on the charts.

Limit equilibrium analysis of slopes in an inexpensive personal computer in the year 2021 can be performed in seconds. Data input is also done quite easily. We do not cover slope stability charts in this text because the low cost of access to satisfactory slope stability analysis software practically makes it a requirement of good engineering to use them.

Stability analyses should be done using the expected values of soil properties and conditions as well as using credibly low and high values of these parameters to assess the sensitivity of stability to possible variations in assumed parameters. The rest of this chapter focuses on stability analyses of slopes and the related decision on whether a given slope is "sufficiently safe" from a design perspective.

17.1.5 The basics of limit equilibrium analysis

Limit equilibrium analysis relies on the concept of the slip surface. As seen in Chapter 4, a slip surface[4] separates two blocks of soil, one typically being stationary, that behave as rigid blocks. Loss of stability is therefore associated with large movements of one block of soil with respect to another stationary mass.

For an arbitrary slip surface idealized to pass through the slope (Figure 17.5), if the destabilizing shear stresses exceed the shear strengths along the slip surface, sliding occurs. If sliding does not come to a stop before large displacements take place, an ultimate limit state is almost assured to be reached regardless of the details of the project (that is, anything constructed on top of the slope or near the slope is likely to be damaged). If sliding starts, but then stabilizes after relatively small displacements, a serviceability limit state and even an ultimate limit state may still be reached. Limit equilibrium is concerned only with the start of sliding from some initial slope configuration; therefore, it is not a very suitable method to determine the magnitude of displacements after sliding has started and thus to assess whether ultimate or serviceability limit states will develop after slide initiation. However, if it can be shown using limit equilibrium that sliding will not start for a given slope, then neither an ultimate nor a serviceability limit state will develop, and this may be (and often is) all that is required in design.

Of the infinite number of potential slip surfaces through a slope, there is one for which the ratio of driving (destabilizing) to resisting (stabilizing) forces is maximum. This is referred to as the critical slip surface, for, if sliding were to occur, it would naturally develop along this surface. It is customary in limit equilibrium slope stability analysis to define the factor of safety as the number by which to divide the shear strength of soil along the critical slip surface to bring the soil to a state of imminent sliding (or marginal stability). So, a factor of safety of 1.5, for example, would suggest that the shear strength along the slip surface

---

[4] The slip surface idealization is adequate for virtually all analyses of interest in practice, but, in reality, this "surface" is often a shear zone with a nonzero thickness and therefore with some volume.

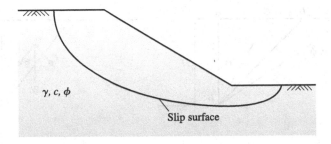

Figure 17.5 Slip surface through soil slope.

is 50% greater than would be required to balance the driving forces. A tacit assumption is made that the shear strength is fully, uniformly mobilized along the slip surface so that progressive failure does not take place. The solution of a stability problem using limit equilibrium usually involves searching for the critical slip surface and then calculating the ratio of resisting to driving forces for it.

The discussion so far has not discerned between two-dimensional (2D) and three-dimensional (3D) slope stability problems. Although virtually all slope stability problems are in truth 3D, most can be adequately modeled using 2D analysis. The assumption is made of plane-strain conditions, which means that no deformation takes place in the direction of the axis of the slope, normal to the cross section we usually show in figures and sketches to represent slopes. This assumption is acceptable so long as the slope is much longer than it is tall or wide; otherwise, 3D analysis must be used. In this text, we focus on 2D analysis, which is used in at least 90% of the problems encountered in practice.

Example 17.1 illustrates, for the simple case of a slope in a homogeneous soil, the search for the critical slip plane and how the factor of safety would be calculated.

Example 17.1

Consider E-Figure 17.1, which shows a simple vertical cut in a homogeneous soil. The soil has properties $\phi=0$, γ, and c. The height of the cut is H. Assuming only planar slip surfaces, what is the factor of safety of this cut? What value of H would lead to its collapse?

Solution:

E-Figure 17.1 shows potential slip planes. The steepest of the three planes shown is also the shortest; this means the resisting forces along it are the smallest. On the other hand, the destabilizing forces due to gravitational effects are also smallest because the volume of the sliding mass is smallest, leading to a comparatively large factor of safety. For the leftmost plane shown in the figure, the weight of the soil wedge is large, but its projection in the direction of the slip surface is not; additionally, the resisting forces are very large because of the great length of the slip surface. So, the factor of safety is also large. The critical slip plane is the one that balances the length of the slip plane (which would tend to increase the resisting shear forces) with the projection of the weight of the potential sliding mass in the direction of the slip surface (the driving force) in such a way as to minimize the factor of safety.

To mathematically locate the critical slip plane, we must consider the following quantities, all expressed per unit length normal to the cross section of the slope:

$$W = \frac{1}{2}\gamma H^2 \tan\beta \quad \text{weight of potential sliding wedge}$$

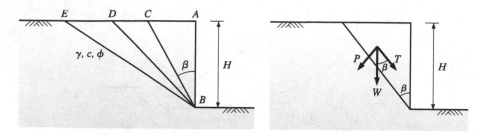

E-Figure 17.1 Search for critical slip surface in a vertical cut.

$$T = W \cos \beta \quad \text{component of weight in the direction of potential slip plane}$$

$$A = \frac{H}{\cos \beta} \quad \text{area of slip plane}$$

The factor of safety is the ratio of resisting forces to driving forces:

$$FS = FS(\beta) = \frac{\dfrac{cH}{\cos \beta}}{W \cos \beta} = \frac{4c}{\gamma H \sin 2\beta}$$

The critical slip plane corresponds to the minimum factor of safety. To minimize the factor of safety, we differentiate the expression above with respect to β, equate the resulting expression to zero, and solve for β. The resulting β value is 45°. Substitution of this value of β back into the equation for the factor of safety gives:

$$\left(FS \right)_{\text{crit}} = \frac{4c}{\gamma H}$$

This implies that, according to limit equilibrium analysis, a vertical cut in clay is vulnerable to collapse in the short term when $H = \dfrac{4 s_u}{\gamma}$, for that makes $(FS)_{\text{crit}} = 1$.

Most slope stability computations are not as simple as that of Example 17.1. Slopes are usually irregular, composed of soils with properties varying across the slope, with or without the presence of a water table. Additionally, planar slip surfaces develop in real situations only under special conditions. In slopes of homogeneous soils, circles or log-spirals are more adequate representations of real sliding mechanisms. In other cases, arbitrarily shaped surfaces should sometimes be assumed for realistic results. In a subsequent section, we will examine some limit equilibrium methods that are well suited to dealing with complex soil slopes. We examine next some analyses applicable to relatively simple slope problems.

17.2 SOME BASIC LIMIT EQUILIBRIUM METHODS

17.2.1 Wedge analysis

In some cases, the slip surface through a slope can be approximated by a few straight-line segments. Such cases are distinguished by the presence of sharp discontinuities in strength, due to the presence of rock layers or weak soil layers. The sliding mass can be idealized as being composed of blocks. These blocks must be in equilibrium with each other and with the shear resistance along the slip surface when the factor of safety is just large enough to prevent sliding from initiating. Graphically, a polygon of forces is drawn for each sliding block, and it must close for the block to be in equilibrium. The factor of safety of the slope for a specific block mechanism will be that for which the polygons all close. As always, several mechanisms must be investigated in order for the minimum factor of safety to be found.

An assumption on the side forces is required to eliminate indeterminacy. Sultan and Seed (1967) assumed the resultant force acting on the interface of two blocks to be inclined at an angle arctan(tanϕ/(FS)) with respect to the normal to the interface. That is conceptually correct for $c=0$, but, if $c>0$, the assumption must reflect that. Lowe and Karafiath (1960) made a different assumption. They assumed that the force acting at an interface between two blocks to make an angle, with respect to the horizontal, will be equal to the average of the angles of the top and base of each of the two blocks separated by the interface. The calculation procedure is best understood through an example.

Example 17.2

Use the wedge method to analyze the stability of the slope shown in E-Figure 17.2a. Use wedges A and B drawn in the figure.

Solution:

The geometry of the dam is shown in E-Figure 17.2a, which also shows two wedges (wedges A and B, with areas A_A and A_B) forming a potential slip mechanism. E-Figure 17.2b and c shows both wedges and the forces acting on each. E-Figure 17.2d shows the force polygon that, for practical purposes, closes (that is, is in equilibrium) for the "right" factor of safety. Note that the force Z is common to both slices: it is the contact force between the wedges. Its angle θ with the interface of the two wedges is given by

$$\theta = \tan^{-1}\left(\frac{\tan\phi}{FS}\right)$$

Now on to the details of the calculations.

Wedge A
Vertical force equilibrium

$$W_1 - P_1\cos\alpha - T_1\sin\alpha + Z\sin(\theta - 23°) = 0$$

Horizontal force equilibrium

$$-P_1\sin\alpha + T_1\cos\alpha - Z\cos(\theta - 23°) = 0$$

where

$$T_1 = \frac{P_1\tan\phi}{FS}$$

$$W_1 = \gamma_m \times A_A = 21 \times 1031 = 21651 \text{ kN/m}$$

$$\alpha = \tan^{-1}\left(\frac{17.6}{100}\right) = 9.98°$$

Wedge B

Vertical force equilibrium

$$W_2 - P_2 \cos 30° - T_2 \sin 30° - Z \sin(\theta - 23°) = 0$$

Horizontal force equilibrium

$$-P_2 \sin 30° + T_2 \cos 30° + Z \cos(\theta - 23°) = 0$$

where

$$T_2 = \frac{s_u L_2}{FS}$$

$$W_2 = \gamma_m \times A_B = 21 \times 701 = 14,721 \text{ kN/m}$$

Horizontal and vertical force equilibrium equations from both wedge A and wedge B enable us to calculate P_1, P_2, Z, and FS. The forces in E-Table 17.1 cause the force polygon to close, as shown in E-Figure 17.2d, and are thus the values of the forces at equilibrium. The factor of safety is equal to $2.38 \approx 2.4$.

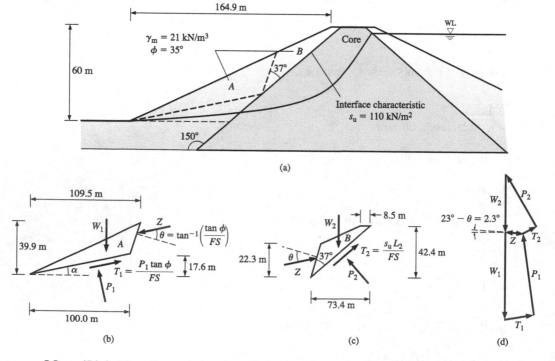

E-Figure 17.2 Stability of an earth dam slope: (a) slope, (b) free body diagram of wedge A; (c) free body diagram of wedge B, and (d) force polygon corresponding to static equilibrium.

E-*Table 17.1* Values of forces at limit equilibrium

Force	Value (kN/m)
W_1	21,651
W_2	14,721
T_1	7733
T_2	3914
P_1	20,465
P_2	14,918
Z	4072

17.2.2 The infinite slope method

This method is suitable when the ground surface may be idealized as an infinite plane with potential slip surfaces parallel to it (Skempton and Delory 1957). Long natural slopes are the best candidates to be analyzed by this method. Submarine slopes have often been analyzed with it.

The analysis is simple. Consider a dry slope with the ground surface making an angle β with the horizontal (Figure 17.6a). Taking a slip surface at a depth z below the ground surface, it follows that the weight W of a slice of soil with width $b=l\cos\beta$, where l is the length of the corresponding slip surface segment, is $\gamma_m bz$, where γ_m is the total unit weight of the soil. So the tangential and normal forces on the base of the slice are:

$$T = \gamma_m zl \sin\beta \cos\beta \qquad (17.4)$$

and

$$P = \gamma_m zl \cos^2\beta \qquad (17.5)$$

The slope is infinite, so the forces Q_R and Q_L on either side of the slice of soil shown in Figure 17.6 are the same due to symmetry. Considering, additionally, that the soil has strength parameters c and ϕ at limit equilibrium, we have:

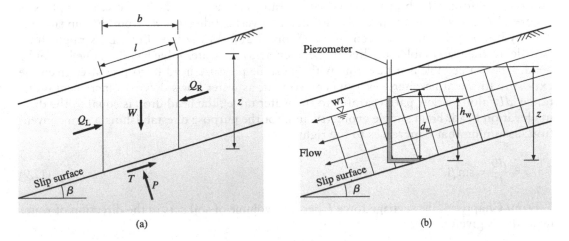

Figure 17.6 Infinite slope method: (a) dry slope and (b) slope with water table parallel to slope surface making an angle β with horizontal.

$$T = \frac{cl + P\tan\phi}{FS} \tag{17.6}$$

so that the factor of safety of the slope is obtained from substitution of Equations (17.4) and (17.5) into Equation (17.6):

$$FS = \frac{c}{\gamma_m z \sin\beta \cos\beta} + \frac{\tan\phi}{\tan\beta} \tag{17.7}$$

In Equation (17.7), it is apparent that failure will take place as deep as possible because higher values of z lead to lower values of FS. If rock or a firmer soil layer is present, sliding will take place tangent to it. If $c=0$, Equation (17.7) reduces to:

$$FS = \frac{\tan\phi}{\tan\beta} \tag{17.8}$$

Solely on the basis of Equation (17.8), it is often stated that sands will stand at an angle of repose equal to its friction angle. Since FS cannot be less than one, any attempt to deposit sand at a steeper angle will cause sliding, which will be arrested only when the slope angle β reaches its equilibrium value – ϕ – again. This concept is somewhat intuitive because we are quite used to making piles of soils or frictional materials on construction sites, industrial sites, or the beach. We know that if we try to make the pile of soil too steep, there will be sliding until a slope that restores equilibrium forms. This angle of repose should not be assumed to exceed the critical-state friction angle ϕ_c because the dilatancy component of shear strength can degrade over time due to cycles in temperature and weather and other causes of slope movement.

A variation of Equation (17.7) can be obtained if the slope can be considered to be submerged, but without any groundwater flow. Such a condition can be assumed to exist for certain submarine slopes, for example. In this case, the buoyant unit weight γ_b should be used in place of γ_m in Equation (17.7). For $c=0$, Equation (17.8) still applies.

The case of an infinite slope with a groundwater table parallel to the ground surface, a condition that would be found often in natural slopes in areas of the world with substantial rainfall, offers an interesting illustration of two different ways to analyze the same stability problem. One uses total unit weights and pore pressures; the other, buoyant unit weights and seepage forces. Both produce the same result. Figure 17.6b shows the same slope as in Figure 17.6a, now with a water table located at a vertical distance d_w from the slip surface. The hydraulic gradient i is given by Equation (3.26). As the flow lines are straight lines parallel to the water table, any line perpendicular to the water table has the same hydraulic head. The water table is a flow line with the same pressure head (zero pressure, given the exposure to the atmosphere) everywhere, so that, as water flows down an incremental distance $|dL|$ along a flow path (parallel to the water table), the head drop is equal to the drop in elevation head. From simple geometry and, for the purpose of establishing a sign convention, assuming that L increases to the right:

$$i = \frac{dh}{dL} = \sin\beta \tag{17.9}$$

From Chapter 3, the seepage force f_s per unit volume of soil acts in the direction of water flow and is given by

$$f_s = i\gamma_w \tag{17.10}$$

The soil volume subjected to the seepage forces is bd_w, so the total force acting on the volume of the soil slice is

$$F_s = bd_w \gamma_w \sin\beta \tag{17.11}$$

pointing down the slope.

The factor of safety can now be expressed as the ratio of the magnitude of the forces resisting a downward move of the soil slice to the magnitude of those pushing it down:

$$FS = \frac{\dfrac{cb}{\cos\beta} + P'\tan\phi}{T + F_s} \tag{17.12}$$

The driving force T is:

$$T = \left[\gamma_b d_w + \gamma_m (z - d_w)\right] b\sin\beta$$

The normal effective force P' on the slip surface is:

$$P' = \left[\gamma_b d_w + \gamma_m (z - d_w)\right] b\cos\beta$$

Taking the expressions for P' and T into Equation (17.12), we get:

$$FS = \frac{\dfrac{c}{\left[\gamma_b d_w + \gamma_m (z - d_w)\right]\sin\beta\cos\beta} + \dfrac{\tan\phi}{\tan\beta}}{1 + \dfrac{\gamma_w d_w}{\left[\gamma_b d_w + \gamma_m (z - d_w)\right]}} \tag{17.13}$$

or

$$FS = \frac{c\sec^2\beta + \left[\gamma_b d_w + \gamma_m (z - d_w)\right]\tan\phi}{\left[\gamma_m (z - d_w) + \gamma_{sat} d_w\right]\tan\beta}$$

Equation (17.13) is quite general. For example, if there is no water within the slope, $d_w = 0$, and Equation (17.13) reduces to Equation (17.7). If the water table is flush with the slope surface, $d_w = z$, and Equation (17.13) reduces to:

$$FS = \frac{\dfrac{c}{\gamma_b z\sin\beta\cos\beta} + \dfrac{\tan\phi}{\tan\beta}}{1 + \dfrac{\gamma_w}{\gamma_b}} \tag{17.14}$$

For a sand, gravel, or nonplastic silt slope, $c = 0$, and Equation (17.14) reduces to:

$$FS = \frac{\gamma_b \tan\phi}{\gamma_{sat} \tan\beta} \tag{17.15}$$

It is clear from Equation (17.15) that the factor of safety for a sand slope with seepage down the slope and the water table at the surface of the slope is about one half of what it would be if the slope were dry (compare Equations (17.15) and (17.8)). We conclude this from the typical values of the unit weights, which are roughly $\gamma_m = 20$ kN/m³ and $\gamma_b = \gamma_m - \gamma_w \approx 20 - 10 = 10$ kN/m³.

The same equations (Equations 17.11–17.15) would have been obtained if we had used total unit weights and pore pressures instead of buoyant unit weights and seepage forces. This derivation is left as an exercise (see Problem 17.3). The key to these derivations is the expression of the pore pressure on the slip surface as a function of d_w.

Finally, this method only works if no excess pore pressures develop as the soil is sheared; in other words, the analysis is drained with pore pressures being exclusively due to the hydrodynamic state of the slope.

Example 17.3

Consider an infinite soil slope making an angle of 30° with the horizontal with a firm soil layer parallel to the ground surface at a depth of 5 m. The soil properties are: $\gamma_m = 20$ kN/m³, $c = 10$ kPa, and $\phi = 28°$. Determine the factor of safety for two cases: (a) dry conditions and (b) groundwater table at the surface.

Solution:

Equation (17.13) can be used for both cases. The interface between the weaker and firmer soil is the natural candidate for the slip plane, so $z = 5$ m. In case (a), there is no water, so $d_w = 0$ and $\gamma_m = 20$ kN/m³. Plugging these values, the slope angle, and the soil parameters into Equation (17.13), we get

$$FS = \frac{\dfrac{c}{\left[\gamma_b d_w + \gamma_m \left(z - d_w\right)\right] \sin\beta \cos\beta} + \dfrac{\tan\phi}{\tan\beta}}{1 + \dfrac{\gamma_w d_w}{\left[\gamma_b d_w + \gamma_m \left(z - d_w\right)\right]}}$$

$$= \frac{c}{\gamma_m z \sin\beta \cos\beta} + \frac{\tan\phi}{\tan\beta}$$

$$= \frac{10}{20 \times 5 \sin 30° \cos 30°} + \cot 30° \tan 28° = 1.15 \quad \text{answer}$$

In case (b), taking $\gamma_b = 20 - 9.81 = 10.2$ kN/m³ and $d_w = 5$ m into Equation (17.13) gives

$$FS = \frac{\dfrac{c}{\left[\gamma_b d_w + \gamma_m \left(z - d_w\right)\right] \sin\beta \cos\beta} + \dfrac{\tan\phi}{\tan\beta}}{1 + \dfrac{\gamma_w d_w}{\left[\gamma_b d_w + \gamma_m \left(z - d_w\right)\right]}}$$

$$= \frac{\dfrac{10}{10.2 \times 5 \sin 30° \cos 30°} + \dfrac{\tan 28°}{\tan 30°}}{1 + \dfrac{9.81 \times 5}{10.2 \times 5}} = 0.70 \quad \text{answer}$$

The influence of the water table is therefore significant, reducing the factor of safety from 1.15 to 0.7, which illustrates why landslides are so often associated with intense rainfall or other processes that lead to soil saturation.

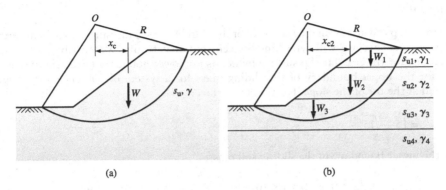

Figure 17.7 Swedish circle method: (a) slope with single soil layer and (b) multilayered soil slope.

17.2.3 The Swedish circle method

The Swedish circle method, also known as the $\phi=0$ method, is the simplest of the circular methods. It is a total stress analysis applicable to end-of-construction, short-term conditions. The undrained shear strength does not necessarily have to be constant along the circle; if it is, the resulting expression for the factor of safety is

$$FS = \frac{s_u LR}{Wx_c} \tag{17.16}$$

where L is the length of the slip surface, R is the radius of the circular surface, W is the weight of the sliding mass, and x_c is the horizontal distance from the centroid of the sliding mass to the center of the circle defining the slip surface. Figure 17.7a illustrates the meaning of each of these quantities. The most time-consuming part of this procedure is determining the centroid of the sliding mass. Note that calculations must be made for many circles so that the one yielding the lowest FS is found.

Example 17.4

A 45° cut is made in a homogeneous clay with $s_u = 50$ kPa and $\gamma_m = 17$ kN/m³ (see E-Figure 17.3). The height of the cut is 10 m. The soil gets considerably stronger below the base of the cut, meaning that any failure would take place along a slip surface through the toe of the excavation. Calculate the FS for the slip surface shown.

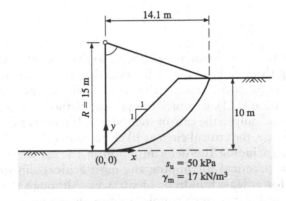

E-Figure 17.3 Cut slope and slip surface for which factor of safety is desired.

Solution:

The first step is determining the centroid of the sliding mass. This can be done in different ways. In this case, we will determine the centroid analytically. We start by writing the equations for the circular slip surface (which is the lower boundary of the sliding mass) and for the upper boundary of the sliding mass for a system of reference with origin located at the toe of the slope. For the slip surface:

$$y_{ss} = 15 - \sqrt{15^2 - x^2}$$

For the upper boundary of the sliding soil mass:

$$y_{\text{upper boundary}} = \begin{cases} x & \text{for} \quad 0 \le x \le 10 \\ 10 & \text{for} \quad 10 \le x \le 14.1 \end{cases}$$

Therefore, the product of the cross-sectional area of the sliding mass by the abscissa x of the centroid of the sliding mass can be calculated as

$$Ax_c = \int_0^{10} \left[x - \left(15 - \sqrt{15^2 - x^2} \right) \right] x \, dx + \int_{10}^{14.1} \left[10 - \left(15 - \sqrt{15^2 - x^2} \right) \right] x \, dx$$

$$= 242.5 + 174.2 = 416.7$$

The length of the slip surface can be determined from the calculation of the corresponding central angle. Based on E-Figure 17.3, the angle can be written as

$$\theta = 90° - \arccos\left(\frac{14.1}{15} \right) = 70.1° = 1.22 \, \text{radians}$$

We can now calculate the factor of safety *FS*:

$$FS = \frac{s_u LR}{Wx_c} = \frac{s_u (R\theta) R}{\gamma_m Ax_c} = \frac{50(15 \times 1.22)15}{17 \times 416.7} = 1.94 \quad \text{answer}$$

For a stratified soil mass (Figure 17.7b), the factor of safety is expressed as

$$FS = \frac{\sum s_{ui} L_i R}{\sum W_i x_{ci}} \tag{17.17}$$

where the subscript i refers to individual soil layers, so that L_i, for example, is the length of the slip surface passing through layer i, which has undrained shear strength s_{ui}.

The critical surface is found by trial and error. As will be discussed in a later section, for soils following Equation (17.1), to which the present method is applicable, circles passing through the toe are critical only in the case of steep slopes. If there is a firm soil layer at a depth that is not excessively large, the critical circle is likely to be tangent to the top of this layer.

Unless the method, including the determination of the centroid of the sliding mass, is automated, it is not practical for finding the most critical slip surface, a search that requires calculations for at least hundreds of surfaces. Although this method was useful in pre-computer days, today it is much simpler to directly use one of the so-called *method of slices*.

17.3 THE SLICE METHODS OF LIMIT EQUILIBRIUM ANALYSIS OF SLOPES

17.3.1 General formulation

To perform limit equilibrium computations in an effective and efficient way, it is usual to break the sliding mass into slices. In most methods, these slices are vertical. The base of each slice is part of the slip surface. So the stationary part of the soil mass, located underneath the slip surface, exerts normal and shear forces on the slice bases. The shear forces exerted by the stationary mass on the slice bases are stabilizing forces (resistances) that depend on the shear strength of the soil. Additionally, each slice is acted upon by any surface forces directly applied to it, its own self-weight, and forces exerted by the two slices on either side of it. These forces can be either resistances or driving forces, depending on the geometry of the slice and its position within the sliding mass.

Figure 17.8 shows the free body diagram of a typical slice. The slice has width b. The following forces can be identified as acting on a given slice: P'=normal effective force on the slice base; U=water force on the slice base (known from the prevailing groundwater conditions in the slope); T=mobilized resisting shear force (on the slice base); X_L and X_R=horizontal forces exerted on the slice by the slice on its left and the slice on its right, respectively; Y_L and Y_R=vertical shear forces exerted on the slice by the slice on its left and the slice on its right, respectively; and W=slice's self-weight. The slip surface shown in the figure is of a general shape, not necessarily circular. Moment equilibrium for a slip surface with arbitrary shape is considered with respect to an arbitrary point O. The distance from the center of the slice base to O has then two components: a distance R normal to the slice base and a distance r parallel to the slice base. The distance from the center of the slice base to point O is $(R^2+r^2)^{1/2}$. Obviously, for a circular surface and O coinciding with the center of the circle, $r=0$ and R is itself the distance from the center of the slice base to O (that is, R is the radius of the slip surface).

To assess whether a slope is in equilibrium for a given factor of safety FS, we must consider the three equations of equilibrium for each slice. The forces acting on a slice, shown in Figure 17.8, would need to be shown to be in equilibrium with each other in the x and y directions (force equilibrium), as well as with respect to rotation (moment equilibrium). Besides the equilibrium equations, a sliding criterion, establishing that sliding is imminent

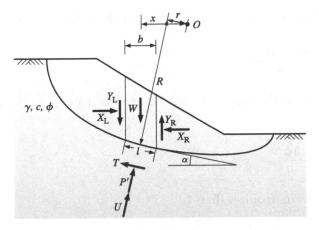

Figure 17.8 General slip surface with center of rotation O and free body diagram of an arbitrary slice of the potential sliding mass.

once the driving shear force on each slice base is equal to the resisting force there, must also be considered. For each slice, this sliding criterion, expressed by either Equation (17.1) or (17.2), is used to obtain the available resisting shear force T from the normal force acting on the base of the slice. For effective stress analysis, the appropriate normal force is $P'=P-U$. So, for N slices, a total of $3N$ (equilibrium)$+N$ (the sliding criterion applied to all the slices) equations are available for determining the remaining unknowns. With respect to the unknowns, the factor of safety is one, but many of the slice forces are themselves unknowns. Specifically, the total number of unknowns for a problem where the sliding mass is subdivided into N slices is $6N-2$, as detailed next:

1 factor of safety FS
N total forces P acting on slice bases
N points of application of forces P
N forces T acting on the slice bases
$N-1$ interslice forces
$N-1$ inclinations (Y/X) of interslice forces
$N-1$ points of application of interslice forces
_____
$6N-2$ total number of unknowns

The number of unknowns exceeding the $4N$ equations available is $2N-2$. It is customary to assume the P (or P') forces to act through the center of the base of each slice. This is a very good assumption for computer calculations, in which slice widths are very small. It eliminates N unknowns (points of application of P). This still leaves us with $N-2$ unknowns. This deficit is made up for by making $N-2$ assumptions. Some methods don't use all equations of equilibrium, which means they require more than $N-2$ assumptions. The selection of the assumptions to do away with excess unknowns is what differentiates most of the important limit equilibrium methods.

For an arbitrarily shaped slip surface (Figure 17.8), consideration of vertical equilibrium for a slice gives:

$$P' = \frac{1}{m_\alpha}\left[W - (Y_R - Y_L) - ul\cos\alpha - \frac{cl\sin\alpha}{FS}\right] \tag{17.18}$$

where u=pore pressure at center of slice base, l=length of slice base, and

$$m_\alpha = \cos\alpha\left(1 + \frac{\tan\alpha\tan\phi}{FS}\right) \tag{17.19}$$

The shear force T on the slice base can be expressed in terms of c, ϕ, and FS using Equation (17.2):

$$T = \frac{cl + (P - ul)\tan\phi}{FS} \tag{17.20}$$

Horizontal equilibrium requires that:

$$X_R - X_L = P\sin\alpha - \frac{\left[cl + (P - ul)\tan\phi\right]\cos\alpha}{FS} \tag{17.21}$$

To assure overall moment equilibrium, we take moments with respect to the arbitrary point O discussed earlier. The vector from the center of the base of the slice to point O has a component R normal to the slice base and another r parallel to it, so the moment of P with respect to O is Pr and that of T is TR. If the horizontal distance from the center of the base of the slice to O is x, the moment of W with respect to O is Wx. The interslice forces are internal forces, which cancel out and produce no net moment. Referring again to Figure 17.8, overall moment equilibrium is thus expressed as

$$\sum_{i=1}^{N} W_i x_i = \sum_{i=1}^{N} T_i R_i + \sum_{i=1}^{N} P_i R_i \tag{17.22}$$

Substitution of Equation (17.20) into (17.22) and some rearrangement leads to:

$$(FS)_M = \frac{\sum R\left[cl + (P - ul)\tan\phi\right]}{\sum(Wx - Pr)} \tag{17.23}$$

where the summations are always over all the slices (from 1 to N) and the subscript M indicates that the factor of safety was obtained from overall moment equilibrium considerations. For a circular slip surface, $r=0$ and $x=R\sin\alpha$, so Equation (17.23) becomes:

$$(FS)_M = \frac{\sum\left[cl + (P - ul)\tan\phi\right]}{\sum(W\sin\alpha)} \tag{17.24}$$

A factor of safety can also be determined from consideration of force equilibrium. Vertical equilibrium is assured slice by slice. Horizontal equilibrium can be obtained by summation of Equation (17.21) for all slices:

$$\sum(X_R - X_L) = \sum P\sin\alpha - \sum\left[cl + (P - ul)\tan\phi\right]\cos\alpha/FS \tag{17.25}$$

Because the horizontal components of the interslice forces are internal to the sliding mass, they cancel out when added together. It follows that:

$$(FS)_F = \frac{\sum\left[cl + (P - ul)\tan\phi\right]\cos\alpha}{\sum P\sin\alpha} \tag{17.26}$$

To compute both $(FS)_F$ and $(FS)_M$, it is necessary to have values for the P forces, which in turn depend on the values of the Y forces. An assumption is needed to make the problem a determinate one. The literature shows that different researchers made different assumptions; some of the more well-known assumptions and their authors are the following:

1. The resultant of the interslice forces is parallel to the slice base (Fellenius 1936).
2. The resultant of the interslice forces is horizontal, $Y_R - Y_L = 0$ (Bishop 1955).
3. The point of application of the interslice forces is some fraction of the slice height, usually of the order of one-third of the height (Janbu 1954a,b, 1973).
4. The direction of the interslice forces is the same for every slice, with the slope of the interslice force given by $Y/X = $ constant (Spencer 1967).
5. The direction of the interslice forces is a function of the abscissa x of the center of the slice, with the slope of the interslice force given by $Y/X = f(x)$, with $f(x)$ a function of x (Morgenstern and Price 1965).

Each of these has become known as a separate "method" of slope stability analysis, although we see that they are in fact the same method with a different assumption on interslice forces. We will not discuss all methods, covering instead the simpler ones – which will illustrate the calculations involved – and the best ones – those that produce factors of safety closest to what has been called the "exact" or "true" factor of safety of a given slope, but is better understood as the best factor of safety that can be obtained using the limit equilibrium method.

17.3.2 Ordinary method of slices

In the ordinary method of slices (OMS), the assumption is made for every slice that the resultant of the two side forces is parallel to the slice base. This is convenient because the normal force P acting on the slice base is then independent of the side forces and given by

$$P = W \cos \alpha \tag{17.27}$$

where α = angle the slice base makes with the horizontal. Furthermore, individual slice moment equilibrium is not considered. These assumptions also eliminate the $3N-3$ unknowns related to the interslice forces, leaving $3N+1$ to be determined. Assuming the P forces to be located at the center of the base of each slice reduces the unknowns further to $2N+1$. These can now be fully determined from consideration of equilibrium perpendicular to each slice base (N equations), from relating T to N using the sliding criterion (N equations) and from overall rotational equilibrium (one equation). This means we have not used all equilibrium equations available to us. So, this method is not based on full satisfaction of equilibrium requirements.[5]

The factor of safety $(FS)_{OMS}$ for a circular slip surface with center O and radius R according to the OMS is given by

$$(FS)_{OMS} = \frac{\text{Moment of resisting forces w/respect to } O}{\text{Moment of driving forces w/respect to } O}$$

which can be expressed in terms of c and ϕ, the angle α, the length l of the base of each slice, the weight W of each slice, the radius R of the slip surface, and the horizontal distance $x = R \sin \alpha$ from the center of each slice to the center of rotation O:

$$(FS)_{OMS} = \frac{R \sum [cl + W \cos \alpha \tan \phi]}{\sum Wx} \tag{17.28}$$

or simply

$$(FS)_{OMS} = \frac{\sum [cl + W \cos \alpha \tan \phi]}{\sum W \sin \alpha} \tag{17.29}$$

The formulation of this method so far is absolutely general: c and ϕ may be strength envelope fit parameters, or they may be physically meaningful shear strength parameters. Additionally, c and ϕ may be effective stress or total stress parameters.[6] If the slip surface were going through clay and we wanted to perform a total stress analysis, we would make

---

[5] It is tacitly assumed that equilibrium is satisfied; in most problems, the assumption is quite reasonable and computed factors of safety are not far from values produced by other methods.

[6] It is advisable to use total stress analysis only for saturated clays, in which case the only total stress shear strength parameter is $c = s_u$.

$\phi=0$ and $c=s_u$ in Equation (17.29). Taking c and ϕ to be effective stress parameters and introducing pore pressures in the formulation, we have

$$(FS)_{OMS} = \frac{\Sigma\left[cl+\left(W\cos\alpha-ul\right)\tan\phi\right]}{\Sigma W\sin\alpha} \tag{17.30}$$

In the absence of seepage, there are two ways to perform the analysis as far as the properties of soil below the water table is concerned: (1) with γ, u, ϕ, and c or (2) with γ_b, ϕ, and c. So, the choice here is between using total unit weight together with pore pressures and using buoyant (effective) unit weight without the need to consider pore pressures. If there is seepage, seepage forces would need to be considered explicitly in the calculation using option (2). Seepage forces do not need to be considered explicitly in (1) because the pore pressures should already reflect their effects. Notice that these are the same two approaches discussed when we presented the infinite slope method. They are applicable to any of the methods of slope stability analysis discussed in this chapter. As noted before, effective stress analysis of undrained stability of a slope for a slow-draining material – clay and clayey soils – is not feasible without a soil constitutive model that can correctly produce the excess pore pressures generated by shearing of the material. In this chapter, effective stress analyses involve always drained analyses, with any pore pressures due only to the presence of water, whether stationary or in flow.

The OMS has limitations due to its over-simplistic assumption about the interslice forces. An example of a serious distortion of the result caused by this assumption is the case in which a slice with large pore pressure at its base and a small value of $\cos\alpha$ could lead to $(P-ul)<0$, which would in turn lead to unrealistically small and even negative shear strength for that slice and consequently excessively conservative results. This is especially problematic in deep slip surfaces, which are nearly vertical as they dip into the ground at the top of the slope and rise up beyond the toe. In such cases, where certain slices have nearly vertical bases, the use of this method is discouraged, or, at the very least, $(P-ul)$ must be restrained numerically from taking a negative value.

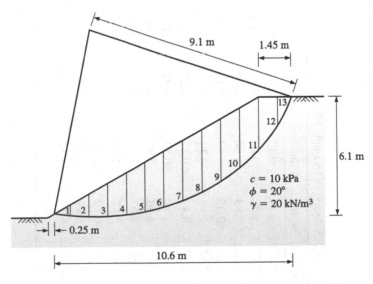

E-Figure 17.4 Slope and slip surface for which factor of safety is desired in Example 17.5.

E-Table 17.2 Solution of Example 17.5

Slice	y_1 (top)	y_2 (top)	y_1 (bottom)	y_2 (bottom)	x_1	x_2	W_i (kN)	α_i (rad)	$W_i \sin \alpha_i$ (kN/m)	$P_{il}(= W_i \cos \alpha_i)$ (kN/m)	l_i (m)	$cl_i + P_i \tan \phi$ (kN/m)
1	0.17	0.73	0.17	0.04	0.25	1.10	5.87	−0.15	−0.88	5.80	0.86	10.71
2	0.73	1.32	0.04	0.00	1.10	1.99	17.89	−0.04	−0.72	17.88	0.89	15.41
3	1.32	1.83	0.00	0.03	1.99	2.75	23.71	0.04	0.95	23.69	0.76	16.22
4	1.83	2.37	0.03	0.14	2.75	3.55	32.24	0.14	4.50	31.92	0.81	19.72
5	2.37	2.90	0.14	0.31	3.55	4.35	38.56	0.21	8.04	37.71	0.82	21.93
6	2.90	3.43	0.31	0.57	4.35	5.15	43.60	0.31	13.3	41.52	0.84	23.51
7	3.43	3.97	0.57	0.91	5.15	5.94	46.77	0.41	18.64	42.89	0.86	24.21
8	3.97	4.50	0.91	1.35	5.94	6.75	50.30	0.50	24.12	44.14	0.92	25.27
9	4.50	5.03	1.35	1.90	6.75	7.55	50.24	0.60	28.37	41.46	0.97	24.79
10	5.03	5.57	1.90	2.60	7.55	8.35	48.80	0.72	32.18	36.69	1.06	23.95
11	5.57	6.10	2.60	3.49	8.35	9.15	44.64	0.84	33.24	29.80	1.20	22.85
12	6.10	6.10	3.49	4.52	9.15	9.85	29.33	0.97	24.19	16.58	1.24	18.43
13	6.10	6.10	4.52	6.10	9.85	10.6	11.85	1.13	10.72	5.06	1.76	19.44
Σ									196.65			266.44

Example 17.5

Find the factor of safety for the circular slip surface shown in E-Figure 17.4 using the OMS. The soil, which has a unit weight of 20 kN/m³, is modeled as cohesive-frictional with $c=10$ kPa and $\phi=20°$. The water table is very deep. To calculate the weight of the slices, you may approximate them using trapezoids.

Solution:

The solution consists of finding the terms appearing in Equation (17.30) for each slice and adding them up to determine the numerator and denominator of that equation. It is best presented in the form of a table (E-Table 17.2). The factor of safety follows directly from the two summations in the last row of the table:

$$(FS)_{OMS} = \frac{\sum_{i=1}^{N} \left[cl_i + (W_i \cos\alpha_i - u_i l_i)\tan\phi \right]}{\sum_{i=1}^{N} W_i \sin\alpha_i} = \frac{266.44}{196.65} = 1.35 \text{ answer}$$

17.3.3 Bishop's simplified method

Bishop's simplified method (BSM), proposed by Bishop (1955), is also known as the modified or routine Bishop's method. The only assumption that needs to be made to obtain Bishop's formulation is that the difference between the vertical components of each pair of interslice forces is zero. With this assumption, by projecting slice forces in the vertical direction, it is possible to express P in terms of quantities not including the interslice forces. Once the boundary conditions on the first and last slices (for which one of the two side forces is zero) are considered, the assumption implies also that the interslice forces themselves are horizontal, not only their difference. As in the OMS, $N+1$ equations of equilibrium are used. Besides vertical equilibrium, overall rotational (moment) equilibrium is also satisfied.

The assumption that interslice forces are horizontal implies Y_R and Y_L in Equation (17.18) are equal to zero, so vertical equilibrium of a slice yields:

$$P' = \frac{1}{m_\alpha} \left[W - ul\cos\alpha - \frac{cl\sin\alpha}{FS} \right] \tag{17.31}$$

where m_α is given by Equation (17.19).

Overall rotational equilibrium for circular slip surfaces is satisfied by Equation (17.24), repeated here as

$$FS = \frac{\sum(cl + P'\tan\phi)}{\sum(W\sin\alpha)} \tag{17.32}$$

Substituting P' from Equation (17.31) gives

$$(FS)_{BSM} = \frac{\sum \dfrac{cb + ((W-ub)\tan\phi)}{m_\alpha}}{\sum W\sin\alpha} \tag{17.33}$$

Taking $W=\gamma bh$ into Equation (17.33), we get

$$(FS)_{\text{BSM}} = \frac{\sum \dfrac{b\left[c+\left((\gamma h - u)\tan\phi\right)\right]}{m_\alpha}}{\sum b\gamma h \sin\alpha} \tag{17.34}$$

The factor of safety appears on both sides of Equation (17.34) because m_α is a function of the factor of safety, so an iterative procedure is required. The procedure usually consists of assuming a value for FS, calculating m_α, calculating FS using (17.34), and comparing the calculated FS with the assumed FS. This is repeated until convergence of the assumed and calculated values of FS is obtained. The BSM can be extended to noncircular slip surfaces by using Equation (17.23) instead of Equation (17.24) to define the factor of safety.

Bishop's method has some numerical shortcomings. In the case of an embankment on top of a clay foundation, the critical slip surface is often a deep surface that is nearly vertical (α in the $70°–90°$ range) through the embankment. If c is large (as it might be if we are modeling a compacted sandy soil embankment using curve-fitting c and ϕ parameters), then, according to Equation (17.31), P' would be excessively small or even negative, leading to unrealistically low values of FS. Another case where numerical problems could arise would be that of a very deep slip surface rising practically vertically beyond the toe, for which the values of α there would be in the $-70°$ to $-90°$ range. From Equation (17.19), large negative values of α might lead to large negative values of m_α, and consequently to unreasonably high values of FS.

17.3.4 Janbu's method

Janbu's method (JM), applicable to both circular and general slip surfaces, was first presented by Janbu (1954a,b, 1973). Janbu also developed a simplified version of his method. We focus here on the complete or general method, sometimes called Janbu's "rigorous" method.[7] Janbu considered force equilibrium, obtaining in the process an expression for $(FS)_F$ equivalent to Equation (17.26). He then stipulated that an assumption be made on the locations of the points of application of the interslice forces. The line connecting these points for all slices is known as the thrust line; therefore, the distance from the point of application of the side forces (that is, from the thrust line) to the base of the slice will be denoted by h_t. The inclination of the thrust line (the inclination of the resultant of the side forces) will be denoted by α_t so that $\tan\alpha_t = Y/X$. In this method, the position of the line of thrust is assumed (that is, chosen by the engineer). The values assigned to h_t are usually in the vicinity of one-third (1/3) of the slice height. Knowledge of the thrust line position allows determination of the interslice forces themselves, by taking the sum of moments with respect to the center of the base of the slice. The values of the interslice forces can then be taken into the equation for $(FS)_F$ to compute the factor of safety satisfying both force equilibrium and moment equilibrium (moment equilibrium having been satisfied when moments were taken with respect to the center of the slice base).

The algorithm calls for initially assuming that $Y_R - Y_L = 0$. Then the factor of safety is estimated from

$$(FS)_F = \frac{\sum\left[cl+(P-ul)\tan\phi\right]\sec\alpha}{\sum\left[W-(Y_R-Y_L)\right]\tan\alpha} \tag{17.35}$$

---

[7] A misnomer, as discussed earlier, since limit equilibrium methods are not strictly rigorous.

Because $P = P' + ul$ is also a function of the factor safety FS (see Equation (17.18)), Equation (17.35) is solved iteratively by assuming an initial value $(FS)_{F0}$ for $(FS)_F$ (say, a value of 1.5), calculating a new value of $(FS)_F$ and using the value to calculate P and then once again $(FS)_F$ using Equation (17.35).

With $FS = (FS)_F$, the horizontal and vertical side forces follow from the horizontal force equilibrium for each slice:

$$X_R = X_L + \left[W - (Y_R - Y_L) \right] \tan \alpha - \left[cl + (P - ul) \tan \phi \right] \sec \alpha / FS \tag{17.36}$$

For the first slice, X_L is typically equal to 0, providing a convenient starting point for the calculations. New estimates of the vertical interslice forces are then obtained from moment equilibrium for each slice (moments taken about the center of the slice base):

$$Y_R = X_R \tan \alpha_t - (X_R - X_L) h_t / b \tag{17.37}$$

Having new estimates of Y_R, and thus of $Y_R - Y_L$, we return to Equation (17.35) and repeat the process [starting with Equation (17.35) and ending with Equation (17.37)] until convergence is reached on the values of interslice forces and factor of safety FS.

17.3.5 Spencer's method

Spencer (1967) proposed an analysis in which the inclination θ of the interslice forces is constant throughout the slope ($Y/X = \tan \theta$). Intuitively, we know that this assumption is not completely realistic, but this is generally the case for all variants of the limit equilibrium method. The methods still produce in almost all cases sufficiently accurate values of factor of safety. See, for example, Yu et al. (1998), who used rigorous *limit analysis* to test the Bishop method as calculated using the limit equilibrium computer program STABL, concluding that STABL produced very accurate values of factor of safety for the benchmark problems considered. Taking $Y/X = \tan \theta$ into Equation (17.18) gives

$$P' = \frac{1}{m_\alpha} \left[W - (X_R - X_L) \tan \theta - ul \cos \alpha - \frac{cl \sin \alpha}{FS} \right] \tag{17.38}$$

and, repeating Equation (17.19),

$$m_\alpha = \cos \alpha \left(1 + \frac{\tan \alpha \tan \phi}{FS} \right)$$

The factors of safety considering force equilibrium and moment equilibrium are obtained from Equations (17.24) and (17.26) for a circular slip surface [but the method can be easily extended for the case of irregular surfaces using Equation (17.23)]:

$$(FS)_M = \frac{\Sigma \left[cl + (P - ul) \tan \phi \right]}{\Sigma (W \sin \alpha)} \tag{17.39}$$

$$(FS)_F = \frac{\Sigma \left[cl + (P - ul) \tan \phi \right] \sec \alpha}{\Sigma \left[W - (Y_R - Y_L) \right] \tan \alpha} \tag{17.40}$$

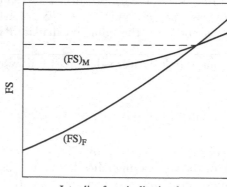

Figure 17.9 Factor of safety from force and moment equilibria as a function of side force inclination angle θ.

Substituting Equation (17.38) into Equations (17.39) and (17.40), it is possible to compute $(FS)_M$ and $(FS)_F$. Because P is also a function of the factor of safety FS, FS appears on both the left-hand and the right-hand sides of Equations (17.39) and (17.40). Therefore, Equations (17.39) and (17.40) are solved iteratively by assuming an initial value of FS (say, equal to 1.5).

The solution algorithm for Spencer's method according to Nash (1987), which differs from the one proposed originally by Spencer (1967), calls for initially setting $Y_R - Y_L = 0$. A first set of $(FS)_M$ and $(FS)_F$ values are calculated based on Equation (17.39) and (17.40). The X forces may then be calculated as in Janbu's method using Equation (17.36). For the first slice, X_L is normally equal to 0. The new estimates for the Y forces are obtained as $Y = X \tan \theta$, using an assumed (trial) value for θ. Having new estimates of the interslice forces, we return to Equations (17.39) and (17.40) to obtain new estimates for the safety factors. The computational procedure is repeated until the values of the interslice forces converge.

This procedure, repeated calculations using Equations (17.36), (17.39), and (17.40), produces two values for the factor of safety, $(FS)_M$ and $(FS)_F$, which are not necessarily equal. Having $(FS)_M \neq (FS)_F$ means that moment equilibrium and force equilibrium are not simultaneously satisfied. So this computational procedure must be repeated several times with different trial values of θ until $(FS)_F$ matches $(FS)_M$. When $(FS)_F$ and $(FS)_M$ have converged to the same value, this value is taken as the factor of safety $(FS)_{SM}$ of the slope.

Spencer (1967) studied the variation of $(FS)_F$ and $(FS)_M$ with the angle θ. A schematic representation of his results is shown in Figure 17.9. It can be seen that $(FS)_F$ is substantially more sensitive to θ than $(FS)_M$. Spencer's studies also indicate that the value of $(FS)_M$ for $\theta = 0$ is not very different, in most cases, from the value obtained using his method. The method that assumes $\theta = 0$ and determines FS based on moment equilibrium is BSM. So Spencer (1967) showed, in essence, that BSM is quite satisfactory in engineering applications under most conditions.

17.3.6 Other methods

There are other limit equilibrium methods or modifications to limit equilibrium methods available in the literature. For reviews of other available methods and attempts to put them within the same framework, see, for example, Nash (1987) and Leshchinsky and Huang (1992). There is not much to be gained from exploring limit equilibrium methods beyond the methods we have covered. All methods satisfying all the equations of equilibrium, such as Spencer's and Janbu's, produce comparable results. It is much more critical to concentrate on other computational aspects, such as how to model the ground water regime in the slope and how to search for critical slip surfaces. This discussion is left for the section on limit equilibrium slope stability software.

17.3.7 Comparison of different methods of stability analysis

It is interesting now to review the methods we studied and point out the main differences between them. The most important factor of differentiation is the assumption on the side forces. Table 17.2 summarizes what this assumption is for the different methods.

Other important distinctions between the different methods of analysis are the possible shapes of the slip surface and the number and type of equilibrium equations satisfied along the slip surface. Table 17.3 summarizes these differences for the methods discussed here. It is of note that some methods were originally developed for circular slip surfaces, but were later extended to slip surfaces of general geometry. Depending on the number of equilibrium

Table 17.2 Assumptions in limit equilibrium analyses

Method	Assumption
Force equilibrium methods	Inclinations of side force (value of θ) at each interslice boundary
Ordinary method of slices (OMS)	Resultant of side forces is parallel to the base of each slice
Bishop's simplified method (BSM)	Resultant of side forces on every slice is horizontal (corresponds to $\theta=0$ in Spencer's method)
Janbu's method (JM)	Point of application of side force resultants on sides of slices
Spencer's method (SM)	Side forces are parallel ($\theta=$constant); corresponds to $f(x)=$constant in Morgenstern and Price's method
Morgenstern and Price's method	Side force inclination (θ) varies according to $\theta=\lambda f(x)$, where x is the position of the center of the slice. Each slice has its own value of $f(x)$; the value of λ is an unknown determined in the course of the analysis

Source: Modified after Duncan and Wright (1980).

Table 17.3 Characteristics of equilibrium methods

Method	Equilibrium conditions satisfied				No. of equations[a]	Shape of slip surface	Calculations by	
	Overall moment	Slice moment	Vertical force	Horizontal force			Hand	Computer
Ordinary method of slices	Yes	No	No[b]	No[b]	$N+1$	Circular	Yes	Yes
Bishop's simplified method	Yes	No	Yes	No	$N+1$	Circular[c]	Yes	Yes
Janbu's general method	Yes	Yes	Yes	Yes	$3N$	Any	Yes	Yes
Morgenstern and Price's method	Yes	Yes	Yes	Yes	$3N$	Any	No	Yes
Spencer's method	Yes	Yes	Yes	Yes	$3N$	Any	No	Yes
Force equilibrium	No	No	Yes	Yes	$2N$	Any	Yes	Yes

Source: Modified after Duncan and Wright (1980).

[a]N, the number of slices.

[b]The OMS satisfies force equilibrium for each slice in a direction normal to the base of the slice.

[c]The original method can be extended to noncircular slip surfaces.

equations satisfied and the shape of the slip surface, it may be possible to use the method in hand calculations; Table 17.3 indicates which methods are amenable to hand calculations.

17.3.8 Acceptable values of factor of safety and resistance factors for stability analysis

When we perform a slope stability analysis and obtain our best estimate of the minimum factor of safety $(FS)_{crit}$ of a slope, how do we judge whether it is acceptable? Clearly, $(FS)_{crit}$ must be greater than 1, but by how much?

It is important to recognize that a first major difference exists between slopes that are considered temporary and those that are permanent. Permanent slopes require greater factors of safety. Second, the consequences of a slope failure must be taken into account. It is not uncommon for deaths to result from slope failures (landslides). All we need to do is to turn on the news when intense rainfall develops somewhere in the world that is hilly (in the United States, California comes to mind) for us to learn that slides have happened and that people have perished. An extreme example of a risky failure would be the failure of the slope of the face of an earth dam, which could, as a result, let out the enormous volume of water in its reservoir, submerging entire towns and killing thousands of people. Such a design requires a large factor of safety because failure of a dam represents a risk of economic loss and loss of life that is "unacceptable."[8] In comparison, the slope of a low-volume-road embankment or a temporary slope at a construction site is near the other end of the spectrum of severity of the consequences of a failure. A situation involving a slope on top of which a large residential building is located would fall somewhere in the middle.

Although recommendations exist for factors of safety as low as 1.1 for favorable conditions and noncritical structures (structures whose failure would be unlikely to lead to loss of life or large economic loss and that are easily repairable), it is wiser to work with a higher minimum FS (a factor of safety of 1.3 may be used). For permanent constructions, a factor of safety of at least 1.5 is advisable, with numbers as large as 1.8 or more for highly risky situations.

In LRFD checks, load and resistance factors replace the factor of safety. Salgado and Kim (2014) proposed load and resistance factors for various typical slope configurations for different probabilities of failure. The probability of failure was calculated as discussed in Chapter 2 using random field descriptions of individual soil layers. For each slope configuration considered, a large number of slope realizations were generated from the expected values and coefficients of variation of the strength parameters and unit weights and the scale of fluctuation of each soil layer. The scale of fluctuation is a length scale that expresses how far apart two points in the soil must be for the properties of the two points to be essentially uncorrelated.

The advantage of this approach is that the values of resistance factors are tied to the target probability of failure. The probability of failure is the ratio of random realizations of the slope for which sliding would occur divided by the total number of random realizations of the slope. Knowing the mean values of soil properties and any loading on the slope and knowing the coefficient of variation for these properties, probability distributions can be generated for each of these variables. With these probability distributions available, it is possible to randomly generate a large number of combinations of values of these properties. Dividing the number of combinations leading to instability by the total number of combinations generated

---

[8] There is really no such thing as an "unacceptable" risk. We simply do not have the resources to have infinite factors of safety in any project and force determinism on a probabilistic world. But we do wish to be very comfortable that the risk of failure is very low when consequences are of this magnitude.

Table 17.4 Resistance factors and equivalent factors of safety for a range of coefficient of variation values and scale of fluctuation of c, ϕ, and γ for drained analysis and s_u and γ for undrained analysis for use in slope stability checks with load factors of 1.0 (for dead loads) and 1.2 (for live loads)

Analysis	Probability of failure	Surcharge (kPa)	RF	FS
Undrained	0.01	12	0.55–0.76	1.840–1.339
		—	0.58–0.83	1.702–1.203
	0.001	12	0.48–0.73	2.119–1.399
		—	0.49–0.79	2.009–1.264
	0.0001	12	0.44–0.71	2.327–1.442
		—	0.44–0.76	2.269–1.307
Drained	0.01	12	0.80–0.93	1.267–1.092
		—	0.83–0.95	1.191–1.053
	0.001	12	0.75–0.90	1.300–1.132
		—	0.78–0.92	1.275–1.076
	0.0001	12	0.70–0.87	1.447–1.170
		—	0.72–0.90	1.377–1.100

Source: Modified after Salgado and Kim (2014).

Note: The ranges of RF and FS are defined by coefficients of variation of c or s_u (0.4–0.2), ϕ (0.1–0.05), and γ (0.1–0.05) and scale of fluctuation of 20–10 m.

leads to the value of probability of failure. We can use lower values of probability of failure for a dam, for example, than for a low-consequence slope. Table 17.4 provides a range for the resistance factors and corresponding factors of safety for proposed load factors equal to 1 for dead loads and 1.2 for live loads. The range results from consideration of two extreme sets of the coefficient of variation of the various properties and the scale of fluctuation.

17.3.9 Computational issues associated with limit equilibrium slope stability analysis

17.3.9.1 Groundwater modeling

When there is groundwater flow within a slope, it is necessary to consider its effects on slope stability. The two ways to do that, as discussed earlier, are to consider buoyant unit weights with explicitly considered seepage forces, or total unit weights with pore pressures. The use of pore pressures is most common. Pore pressures in the past were obtained from flow nets.[9] Flow nets can be drawn by hand, but computer programs are available that generate not only the flow net, but also the pore pressures throughout the slope.

An alternative, approximate method to estimate pore pressures within a slope is illustrated in Figure 17.10, which shows a slope with steady-state groundwater flow through it. The water table is denoted as AB. The actual pressure head at point C (the center of the slice base) can be found by drawing a flow net through the slope and reading the hydraulic head from the equipotential line going through point C. In the simplified method, the pressure head is estimated by taking an average of two approximate pressure heads, referred to as the vertical and perpendicular pressure heads. The vertical pressure head h_{pv} is computed based on an assumption that it is a hydrostatic pressure; that is, the pressure head is simply

---

[9] Refer back to Chapter 3 for a discussion of flow nets and an example of how they are used to determine pore pressures.

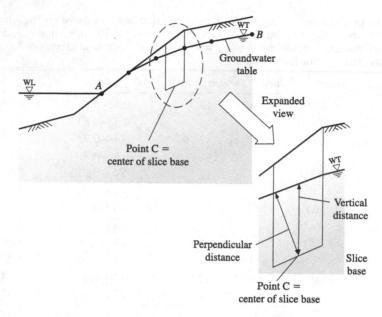

Figure 17.10 Approximate method to estimate pore pressure from given water table.

the vertical distance from point C to the water table immediately above point C. The perpendicular pressure head h_{pp}, in turn, is calculated by approximating the equipotential line as a straight line from point C to the water table in a direction normal to the water table. The vertical pressure head is a conservative estimate (higher than the actual value), while the perpendicular pressure head is an unconservative estimate (lower than the actual value). When the angle of the water table with the horizontal is 35°, the vertical pressure head can be as much as 30% higher than the actual pressure head and the perpendicular pressure head can be as much as 10% lower than the actual pressure head. Therefore, the degree of conservatism can be controlled by taking the average of the vertical and the perpendicular pressure heads (Salgado and Bandini 1999). The average value is slightly conservative. In fact, the pressure head calculated in this manner is about 9% higher than the actual pressure head when the angle of the water table with the horizontal is 35°. Errors are less than that for a flatter water table.

Yet another method exists to compute the pore pressure. It relies on the use of the pore pressure ratio defined by

$$r_u = \frac{u}{\gamma z} \tag{17.41}$$

where u=pore pressure, γ=total unit weight of soil, and z=vertical distance from the ground surface to the point of interest. The pore pressure ratio has traditionally been used in slope stability analysis, especially in stability charts. In reality, the actual value of the pore pressure ratio is not constant throughout the slope, except when the water table coincides with the ground surface and the flow lines are straight. Therefore, the slope is often divided into several subregions so that, for each subregion, the average value, calculated from contours of pore pressure ratio, is assigned. Although the pore pressure ratio concept is easily implemented in numerical slope stability calculation procedures, great care is needed in the estimation and assignment of its value throughout the slope for accurate solutions to be obtained.

17.3.10 Search for the critical slip surface

A key requirement for successful slope stability analysis using limit equilibrium is that the critical slip surface be reasonably well located. This requires a good search algorithm. It also requires that the user of the program specify search parameters that will focus the search on the zone where the critical slip surface is to be found. Several searches with different parameters are often needed, whether done manually or automatically, until the critical surface can be located with a high degree of confidence.

When circular slip surfaces are used, searches must be sufficiently broad to cover two types of circular surfaces: circles passing through some fixed point (such as the toe of the slope) and circles tangent to some elevation. Toe circles are often critical when most of the shear strength is frictional because, in such cases, strength increases with depth, and deep circles lead to increasingly higher factors of safety. This can be expressed quantitatively in terms of the parameter $\lambda_{c\phi}$ (Janbu 1954b):

$$\lambda_{c\phi} = \frac{\gamma H \tan\phi}{c} \tag{17.42}$$

The higher the $\lambda_{c\phi}$ is, the more frictional in nature the shear strength is. As an approximate rule, the critical circle is likely to pass through the toe whenever

$$\lambda_{c\phi} = \begin{cases} \geq 0.80 & \text{for } \beta = 15° \\ \geq 0.35 & \text{for } \beta = 30° \\ \geq 0.08 & \text{for } \beta = 45° \\ \geq 0 & \text{for } \beta \geq 53° \end{cases} \tag{17.43}$$

where β is the slope angle.

There are a variety of ways to search for the critical slip surface. The search scheme adopted in the computer program STABL consists of generating random surfaces within a zone defined by the user (Siegel et al. 1981). Usually, the user will specify two points on the ground surface between which slip surfaces initiate and two others between which slip surfaces terminate. It is necessary to specify how many surfaces should be generated. It is also possible to specify the angle of initiation for the slip surfaces and to define boundaries within the slope that cannot be crossed by the slip surfaces. Of all the surfaces generated, the program picks the one leading to the lowest factor of safety as the critical slip surface. Refinements in the search parameters can then be made until one can be reasonably sure that the lowest factor of safety or one very close to it was indeed determined.

Other techniques are available to search for the critical slip surface (e.g., Baker 1980; Siegel et al. 1981). Greco (1996) presented an efficient way to determine the critical slip surface of a slope by means of an iterative procedure based on the generation of random numbers using the Monte Carlo technique.

17.4 SLOPE STABILITY ANALYSIS PROGRAMS: AN EXAMPLE

The main steps in the analysis of a slope stability problem using a computer program are the following:

1. Enter the geometry of the slope.

2. Enter the shear strength parameters for each soil layer in the soil profile.
3. Enter data for the groundwater regime (for example, enter one or more water tables[10]).
4. Select a method (e.g., OMS, Bishop, or Spencer) to use.
5. Specify the number of slip surfaces for which calculations are to be done and how these slip surfaces are to be generated.
6. Select the size of the linear segment constituting the slip surface (that is, select the width of the slices).

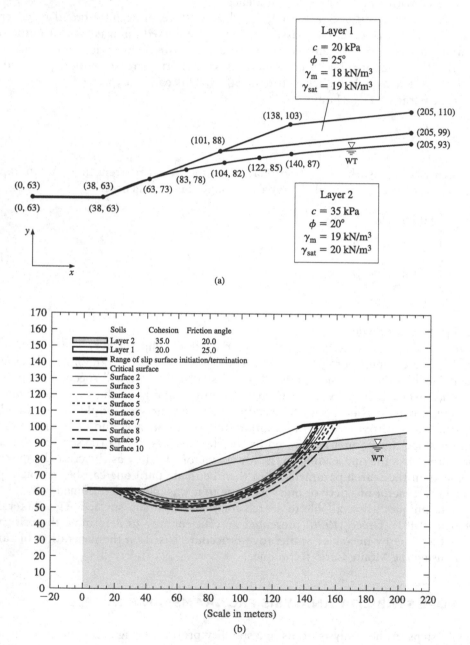

(a)

(b)

7. Analyze and interpret the results (which consist of slip surfaces and the corresponding factors of safety).

8. Rerun the analysis until there is reasonable certainty that the factor of safety found is the minimum representative factor of safety for the slope.

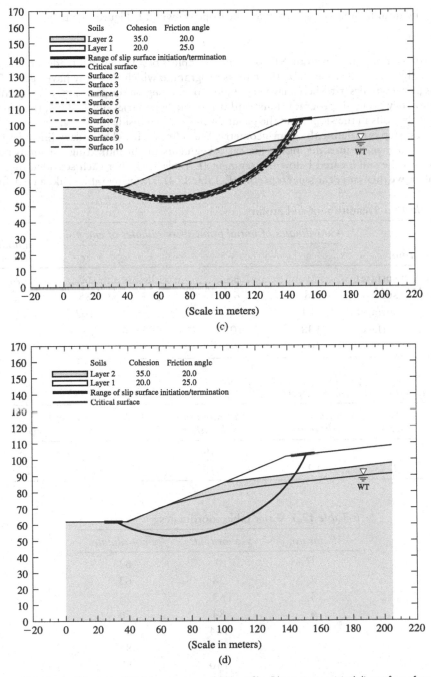

E-Figure 17.5 Slope of Example 17.6: (a) geometry and soil profile, (b) ten most critical slip surfaces for a wide search zone, (c) ten most critical slip surfaces for a more focused search, and (d) most critical slip surface.

Running an analysis in a program is simply part of the more comprehensive engineering of the problem, which requires satisfactory characterization of the slope, a decision concerning short- and long-term stability, and finally a decision as to what the minimum factor of safety must be for the slope.

Example 17.6

Find the minimum factor of safety for the slope shown in E-Figure 17.5a.

Solution:

The solution uses the program STABL. The example is best followed if readers reproduce the analysis using a slope stability analysis program to which they may have access. The steps as we will describe here may vary slightly depending on the program, but regardless of how they do it, all programs must still define the boundaries of the slope, the properties of the soils in the slope, and the groundwater pattern inside the slope.

The first step is to define the slope geometry. We define each soil boundary (including the ground surface) by introducing a sequence of segments and by indicating the number that identifies the soil located below the segment (E-Table 17.3). For each soil identification number, we then enter the soil properties (E-Table 17.4). The water table is likewise defined

E-Table 17.3 Definition of soil profile

	Coordinates of initial point		Coordinates of end point		Soil under segment
Segment no.	x	y	x	y	
1 (ground surface)	0	63	38	63	2
2 (ground surface)	38	63	101	88	2
3 (ground surface)	101	88	138	103	1
4 (ground surface)	138	103	205	110	1
5	101	88	205	99	2

E-Table 17.4 Soil properties

Layer	Wet unit weight (kN/m³)	Saturated unit weight (kN/m³)	Cohesive intercept (kPa)	ϕ (°)
1	18	19	20	25
2	19	20	35	20

E-Table 17.5 Water table coordinates

Point no.	x-coordinate	y-coordinate
1	0	63
2	38	63
3	63	73
4	83	78
5	104	82
6	122	85
7	140	87
8	205	93

as a sequence of points (E-Table 17.5). When there is a single water table,[11] we number it water table No. 1. This water table is used to calculate the pore pressures for all the soils. With the slope properly defined, we can now analyze its stability. Let us say we would like to use BSM for that. We must specify the parameters for the analysis, which include where to generate slip surfaces and how many surfaces to generate. We must also indicate how to discretize the slip surfaces (that is, specify the length of the linear segments that make it up, which is the same as specifying the width of the slices in which the sliding mass is divided), and we may impose certain geometric constraints on the slip surfaces, such as the angles of the initial segments with the horizontal and the lowest elevation that any point of the slip surface may reach. Initially, we may not be very certain as to where the critical slip surface (the one associated with the lowest factor of safety) will be located, so we are very liberal in specifying where to generate slip surfaces, and we may obtain a result such as that shown in E-Figure 17.5b, where the ten most critical slip surfaces are widely spaced, suggesting we have not narrowed down the search sufficiently. So now we can perform more targeted searches: E-Figure 17.5c shows an improved one, and E-Figure 17.5d shows one we may consider final, as all ten slip surfaces fall essentially on top of each other. For this slope, according to the BSM, $FS = 1.15$, and the slope is probably safe (but the FS is not sufficiently large for most applications).

17.5 ADVANCED METHODS OF ANALYSIS: LIMIT ANALYSIS*

17.5.1 Basic concepts of limit analysis

The limit equilibrium method does not produce rigorous solutions in accordance with the principles of mechanics. A rigorous solution of a boundary-value problem requires that the equilibrium equations, the compatibility condition, and the constitutive response of the material, along with prescribed boundary conditions, be satisfied. The limit equilibrium method considers equilibrium conditions only in part of the soil mass (along a specified slip surface). An alternative to limit equilibrium analysis is provided by two theorems from plasticity theory, which are the basis for what is known as *limit analysis*.

Limit analysis solutions are of two types: lower bound and upper bound solutions. In lower bound solutions, a stress field that (1) satisfies the equations of equilibrium everywhere, (2) violates the yield (strength) criterion nowhere, and (3) is consistent with stress boundary conditions is determined. The stress field is in equilibrium with imposed loads at the boundaries of the soil mass, which means collapse is not underway, although it could be imminent. In upper bound solutions, a velocity field is found that (1) is consistent with the yield criterion, (2) satisfies compatibility of velocities (which basically assures that no voids are created or overlapping of material occurs within the soil mass), and (3) satisfies the velocity boundary conditions. The velocity field associated with an upper bound solution is compatible with imposed velocities, which means collapse is either underway or imminent. Lower bounds imply equilibrium; upper bounds imply collapse. If they match, that implies that both are also equal to the exact collapse load. Otherwise, the lower and upper bound solutions define the range in which the solution lies under given soil and slope conditions.

It is interesting to express the two theorems in words in the way they are usually used in computations. The upper bound theorem would read as follows: given an assumed kinematically admissible slip mechanism, the total work done per unit of time by both body forces and surface tractions on the assumed slip mechanism, which is equal to the power dissipated by that mechanism, is not less than the power dissipated in the real or optimum

---

[11] When there are perched water tables, more than one water table exists, and we than need to specify which water table to use for each soil.

mechanism. In computations, we would assume a slip mechanism (which might be a single slip surface or a network of slip surfaces separating adjacent wedges or blocks) that is kinematically admissible (that is, does not violate the displacement boundary conditions or the flow rule[12] for the soil in question) and compute from considerations involving work and energy dissipation an upper bound to the collapse load. Ideally, this upper bound would be as low as possible to closely cap the true collapse load, but that depends on how closely the assumed slip mechanism matches the true slip mechanism. Power dissipates along the slip mechanism because the shear stresses that develop along slip surfaces do work against the incremental movement between two masses of soils, and this work is irrecoverable. In other words, plastic (irrecoverable) deformations are required for energy dissipation to be present.

The lower bound theorem would state that if a stress field can be found that is compatible with the load boundary conditions and for which the slip criterion ($\tau = s_u$ or $\tau = c + \sigma' \tan \phi$) is not violated anywhere, that stress field is in equilibrium with a lower bound to the collapse load.

Despite the great rigor of limit analysis over the limit equilibrium methods, it has been underused. This is mainly due to the difficulties encountered in constructing proper stress fields and velocity fields, and the even larger difficulties in determining optimal solutions (the highest lower bound and the lowest upper bound). When, in addition to such difficulties, the effects of pore water, a heterogeneous soil profile, and irregular slope geometry are considered, computational difficulties are considerably increased, and the manual construction of stress fields and velocity fields becomes impossible. Yu et al. (1998), Kim et al. (1999, 2002), and Loukidis et al. (2003) presented a way to perform finite element limit analysis of complex slopes. Sloan (2013) discussed the use of finite element limit analysis (FELA), which can be used to find the solution to problems with arbitrarily complex geometry. This is discussed later.

For a simple example showing how the two principles of limit analysis are used in computations, let's revisit the vertical cut in the clay of Example 17.1.

Example 17.7

Find the lower and upper bounds on the critical height (the height leading to collapse) for the vertical cut in the clay of Example 17.1.

Solution:

Upper bound

We postulate a slip mechanism consisting of a slip plane making an angle $\beta = 45°$ with the horizontal and going through the heel of the cut (E-Figure 17.6a). We first verify that it indeed does satisfy displacement boundary conditions. The work done by the self-weight of the sliding wedge per unit length of the cut per unit of time is equal to the component of weight parallel to the slip plane multiplied by the velocity v of the wedge parallel to the slip plane:

$$\dot{W}_w = \frac{1}{2\sqrt{2}} \gamma H^2 v$$

The power $\dot{W}_d$ dissipated along the slip plane is equal to the product of the area of the slip plane (per unit length of the cut), the shear stress there, and the velocity along the slip plane:

$$\dot{W}_d = \sqrt{2} H c v$$

---

[12]The flow rule specifies the relationship between plastic strain increments in the principal directions.

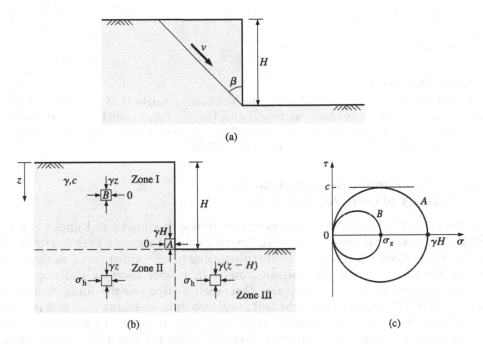

E-Figure 17.6 Calculation of vertical cut stability using limit analysis: (a) upper bound slip mechanism, (b) lower bound stress field, and (c) Mohr's circles for the stress field.

By the upper bound theorem, the power (work per unit time) of external forces is equal to the power dissipated on the mechanism. H follows from combining the two preceding expressions:

$$H = \frac{4c}{\gamma}$$

as our best (least) upper bound for this mechanism.

Lower bound

We divide the soil mass in three zones: zone I is the area of soil with depth z less than H, above the base of the cut. Zone III is the mass of soil on the right of the cut. Zone II is the transition zone between zones I and III, as shown in E-Figure 17.6b. The assumed states of stress in these zones are

$$\sigma_v = \gamma z; \qquad \sigma_h = 0 \text{ in zone I}$$

$$\sigma_v = \gamma z; \qquad \sigma_h \text{ in zone II}$$

$$\sigma_v = \gamma(z - H); \quad \sigma_h \text{ in zone III}$$

We then verify that the vertical stress is continuous between zones I and II. The continuity of lateral stress between zones II and III is also satisfied, but the value of this lateral stress is unknown. Using the failure criterion $\sigma_1 - \sigma_3 = 2c$ allows us to determine this unknown lateral stress and the height H by requiring that it not be violated anywhere in the soil mass. Point A in zone I is the point with maximum shear stress (E-Figure 17.6c), where the failure criterion would be satisfied first. Applying the failure criterion there gives

$$\gamma H - 0 = 2c$$

So, $H = 2c/\gamma$

True collapse height

According to these simple calculations, the true collapse height H of the cut would fall between $2c/\gamma$ (the lower bound we found) and $4c/\gamma$ (the upper bound we found). More sophisticated statically admissible stress fields and kinematically admissible velocity fields show that $3.76 \leq \gamma H/c \leq 3.78$ (Pastor et al. 2000).

17.5.2 Finite element modeling for limit analysis of complex soil slopes

In Example 17.7, we calculated in a very simple way the lower and upper bounds on the height of a vertical cut in a material with $c>0$ and zero ϕ. Even for that simple slope, a precise answer could only be obtained using more sophisticated analysis, which approximates the exact slip mechanism closely. In a more complex soil slope, sophisticated analysis is absolutely necessary. One such analysis is the finite element limit analysis (FELA). Discretization of the soil mass into finite elements is an efficient way to construct various statically admissible stress fields and kinematically admissible velocity fields. Using finite elements, the proper stress field and velocity field can be efficiently found. Another benefit of using finite elements is that both rigid body rotation and continuous deformation of the soil mass are admitted as possible failure mechanisms in upper bound limit analysis.

Triangular elements with three nodes may be used for the construction of stress and velocity fields for a given slope under plane-strain conditions. For the construction of statically admissible stress fields, triangular and rectangular extension elements are used to simulate a semi-infinite soil mass in both the vertical and the lateral directions (Kim et al. 1999). Each node of a finite element mesh belongs uniquely to a particular element; that is, more than one node may have identical coordinates; thus, statically admissible stress discontinuities and kinematically admissible velocity jumps are allowed to occur along the edges shared by adjacent elements.

To model a heterogeneous soil profile, the mesh should be formed in a way that the edges shared by two adjacent elements with different soil types lie along the interface of those two soil types, so that shear strength parameters c and ϕ and total unit weight of soil γ can be element constants. Figure 17.11 shows a finite element mesh for an irregular soil slope with a heterogeneous soil profile.

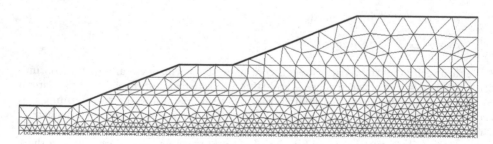

Figure 17.11 Typical finite element mesh for irregular soil slope.

17.5.3 Optimization of lower and upper bound solutions

In lower and upper bound analyses, it is essential to maximize the lower bound and minimize the upper bound solutions. To achieve this goal, a mathematical optimization technique (either linear or nonlinear programming) is required. The first work of this kind was done by Lysmer (1970), who found lower bound solutions to plane-strain soil mechanics problems using three-noded triangular elements and linear programming. The first formulation of the upper bound limit analysis using constant-strain triangular finite elements and linear programming was proposed by Anderheggen and Knöpfel (1972) for plate problems. Since Lysmer's and Anderheggen and Knöpfel's pioneering work, much work has been done on FELA with the aim of (1) reducing the computation time (Sloan 1988a,b, 1989), (2) simulating the yield criterion more realistically using nonlinear programming (Casciaro and Cascini 1982; Munro 1982; Jiang and Magnan 1997; Lyamin 1999), and (3) reducing the restrictions imposed on the number of possible velocity discontinuities in the optimization process (Sloan and Kleeman 1995). Kim et al. (2002) developed the complete set of techniques for performing limit analysis of slope stability problems with soil profile complexities of any kind, the presence of water, and the existence of boundary loadings.

If collapse of soil slopes caused by self-weight is of interest, the quantity to optimize (the objective function, in optimization terminology) is the total unit weight γ of the soil. However, for a heterogeneous soil profile, the total unit weight of soil is different from element to element, depending on the location of the element. To overcome this difficulty, an arbitrary reference value γ is selected such that each element's unit weight γ_e can be set as a multiple of γ; $\gamma_e = \alpha_e \gamma$. The problem of finding the optimum (maximum) lower bound solution and the problem of finding the optimum (minimum) upper bound solution are reduced to optimizing an objective function (reference unit weight of soil) under constraints describing the static admissibility conditions (for the lower bound) or the kinematic admissibility (for the upper bound). A detailed presentation of the limit analysis formulation may be found in Sloan (2013), Sloan (1988a), and Kim et al. (1999, 2002).

17.5.4 Factor of safety and other results

In lower and upper bound analyses, the factor of safety for a heterogeneous slope with given soil properties (γ, c, and ϕ for each soil type), groundwater conditions, and geometry can be calculated from iterative computation of the optimized reference unit weight. For a given iteration, if the optimized reference unit weight from computations is greater (less) than the actual reference unit weight, then the shear strength parameters should be reduced (increased) by increasing (reducing) the factor of safety. This process is repeated until the optimized reference unit weight is equal to the actual reference unit weight for the slope. The corresponding factor of safety is the factor of safety for the slope. Both lower bound and upper bound factors of safety are determined by this process. Throughout the calculation of lower and upper bounds and their corresponding factors of safety, the relative proportion of unit weights of different soils remains unchanged; the reference unit weight is optimized by starting with a factor of safety equal to 1 and then adjusting it until the optimized reference unit weight matches the actual reference unit weight. Since this process is based on the same definition of factor of safety as is used in the limit equilibrium method, results from limit equilibrium and limit analysis can be compared directly.

The factor of safety is not the only result of limit analysis computations. The output of upper bound computations includes, for example, the optimized velocity field and zones where significant plastic energy dissipation has taken place. The slip surface from a limit equilibrium method, if it is an accurate portrayal of how the slope would fail, should be consistent with both the velocity field and the zones where high plastic power dissipation is observed in the optimum upper bound solution.

Figure 17.12 Photos and illustrations of toppling of building due to landslide in Kuala Lumpur, Malaysia. (a) Plan view of the locations of the collapse building, the neighboring buildings, the landslide, and some effects of the landslide on the surroundings of the collapsed building (the letters A through D are used to indicate these effects in the photos); (b) initiation of collapse; (c) collapse in progress; (d) collapsed building and the other two buildings in the condominium complex; (e) compressed ground just behind the collapsed building, and (f) leaning pole, debris, and crack caused by slope failure. (Reprinted from International Newsletter Landslide News, No. 8, 1994.)

17.6 CASE STUDY: BUILDING COLLAPSE CAUSED BY LANDSLIDE

A shocking example of limit states of type III, discussed in Chapter 11 and illustrated in Figure 11.3, and one of the most vivid illustrations of why slope stability analysis should be an integral part of design work related to building a new structure, is the collapse of a building due to a landslide in Kuala Lumpur, Malaysia. The building was only 12-story high, but 50 people died in this tragic event, which happened on Saturday, December 11, 1993, during the rainy season. The building, founded on steel piles, was one of the three buildings that made up the Highland Towers condominium complex. The slope next to it was a natural slope made of residual soils of granitic rock.[13]

There had been an intense rainfall in the evening preceding the accident. A resident of one of the other two buildings, Bruce Mitchell, relates the failure: "I heard the crack of the retaining wall giving way. The whole block started to move. The building first slid forward in a vertical position for about 10 ft, making little noise. When it started to topple, it went over very quickly." It is clear that what is being described is a landslide, which carried the building with it as it developed. Mr. Mitchell also took photographs, shown in Figure 17.12. There had been some minor mud slides a year earlier, some rock falls, and other indications that not all was well. The night before, a nearby resident reportedly observed some earth movement that might have been the start of the formation of the landslide. So there were some warning signs, but perhaps not obvious ones.

This case history leaves us with a lesson. Hills and mountains offer great views whether we are looking at them or from them. People will continue to construct on or near slopes. Foundation engineering in a strict sense is the selection of a type of foundations for build ings or structures, their configuration, and their proportioning. However, it serves no one to design the foundations of a building with perfection just to see that building carried away by a landslide, as in this case. So we must be very mindful of limit state III and, in particular, the possibility of collapse of a structure caused by slope failure.

17.7 CHAPTER SUMMARY

Slope stability analysis is most often done using the limit equilibrium method. In this method, sliding is assumed to develop along a slip surface, which separates a moving mass of soil from a stationary soil mass. Of the infinite number of **potential slip surfaces** through a slope, there is one for which the ratio of driving to resisting forces is maximum. This is referred to as the **critical slip surface**. If sliding were to occur, it would naturally develop along this surface. It is customary in limit equilibrium slope stability analysis to define the factor of safety as the number by which to divide the shear strength of soil along the critical slip surface to bring the soil to a state of imminent sliding.

The simplest limit equilibrium method is the **infinite slope method**, which is applicable to very extensive slopes in which a plane slip surface develops parallel to the slope. According to this method, the factor of safety of the slope is given as

$$FS = \frac{\dfrac{c}{\left[\gamma_b d_w + \gamma_m \left(z - d_w\right)\right]\sin\beta\cos\beta} + \dfrac{\tan\phi}{\tan\beta}}{1 + \dfrac{\gamma_w d_w}{\left[\gamma_b d_w + \gamma_m \left(z - d_w\right)\right]}} \qquad (17.13)$$

[13]The material in this case history comes entirely from T. Nonaka's description of the event, published in Landslide News, No. 8, August 1994, 2–4. This newsletter is available online at http://www.landslide-soc.org/publications/l-news/.

For more general conditions, circular slip surfaces (or slip surfaces that are general, adapting to the geometry of the slope and the shear strength of the soil layers) are required. The part of the slope located above the slip surface is divided into **vertical slices,** the equilibrium equations are written for the slices and for the moving mass as a whole, and the factor of safety is calculated. Without simplifying assumptions, the problem is indeterminate (there are more unknowns than there are equations to find the values of these unknowns). A variety of assumptions can be made to remove excess unknowns. Each set of assumptions leads to a different **limit equilibrium method.** The most widely used slice method is **Bishop's simplified method,** according to which the factor of safety is given by

$$
(FS)_{BSM} = \frac{\sum \dfrac{b\left[c+(\gamma h - u)\tan\phi\right]}{m_\alpha}}{\sum b \gamma h \sin\alpha}
\tag{17.34}
$$

where:

$$
m_\alpha = \cos\alpha\left(1 + \frac{\tan\alpha\tan\phi}{FS}\right)
\tag{17.19}
$$

b, h = slice base width and height;
α = angle of slice base with the horizontal;
γ = unit weight of soil within slice; and
c, ϕ = shear strength parameters of soil at base of slice.

For a given slip surface, an iterative procedure is required. First, the potential sliding mass is divided into a number of slices. Then a value for FS is assumed, m_α is calculated using Equation (17.19), and FS is calculated using Equation (17.34). This value of FS is compared with the previously assumed value of FS. This is repeated until convergence of the assumed and calculated values of FS is obtained.

The **critical factor of safety** is the smallest value of FS that can be found for the slope with a comprehensive slip surface search. This type of calculation is done using computer programs because they would be prohibitive by hand. Even in computer programs, the user has control over how and where to generate slip surfaces, and over the shape of these surfaces. Although circular slip surfaces are suitable in many cases, in others the slip surface must be allowed to be of whatever shape it "wants" to be, so the flexibility of not having to assume a specific shape for the slip surface in the method of analysis is desirable. It is important to span the regions of the slope where the critical slip surface is likely to be located.

17.7.1 Symbols and notations

Symbol	Quantity represented	US units	SI units
A	Area of slip plane	ft^2	m^2
b	Width of slip surface segment (or of slice base)	ft	m
c	Cohesive intercept	psf	kPa
f_s	Seepage force per unit volume of soil	pcf	kN/m^3
h_{pp}	Perpendicular pressure head	ft	m
			(continued)

Symbol	Quantity represented	US units	SI units
h_{pv}	Vertical pressure head	ft	m
i	Hydraulic gradient	Unitless	Unitless
L	Length of slip surface	ft	m
l	Length of slip surface segment (or of slice base)	ft	m
N	Number of slices	Unitless	Unitless
P	Normal force per unit length of slope on the base of vertical slice	lb/ft	kN/m
Q_L	Force per unit length of slope acting on the left side of slice in infinite slope analysis	lb/ft	kN/m
Q_R	Force per unit length acting on the right side of slice in infinite slope analysis	lb/ft	kN/m
R	Radius of circular slip surface	ft	m
r_u	Pore pressure ratio	Unitless	Unitless
s	Shear strength	psf	kPa
s_u	Undrained shear strength	psf	kPa
T	Resisting shear force on the slip plane per unit length of slope	lb/ft	kN/m
U	Water force on the slip surface per unit length of slope	lb/ft	kN/m
W	Weight of soil mass, wedge, or slice per unit length of slope	lb/ft	kN/m
$\dot{W}_d$	Power dissipated along the slip plane	lb.ft/s	W
$\dot{W}_w$	Power of self-weight of sliding wedge	lb.ft/s	W
x_c	Horizontal distance from the centroid of soil mass to the center of the circle	ft	m
X_L, X_R	Horizontal forces exerted on the left and right sides of the slice per unit length of slope	lb/ft	kN/m
Y_L, Y_R	Vertical forces exerted on the left and right sides of the slice per unit length of slope	lb/ft	kN/m
α	Angle the slice base makes with the horizontal	deg	deg
β	Slope angle	deg	deg
γ	Total unit weight of soil	pcf	kN/m^3
γ_b	Buoyant unit weight of soil	pcf	kN/m^3
γ_e	Elemental unit weight in finite element limit analysis	pcf	kN/m^3
γ_w	Unit weight of water	pcf	kN/m^3
ϕ	Friction angle	deg	deg
σ'	Normal effective stress	psf	kPa
τ	Shear stress	psf	kPa

17.8 PROBLEMS

17.8.1 Conceptual problems

Problem 17.1 Define all the terms in bold contained in the chapter summary.

Problem 17.2 The critical height of a cut in soil is the height at which a vertical cut will stand with a factor of safety equal to 1. Develop expressions for this height for (a) a soil with nonzero c and $\phi=0$ and (b) a cohesive-frictional soil with both c and ϕ greater than zero.

Problem 17.3 Rederive Equations (17.11)–(17.15) using total stresses and pore pressures instead of effective stresses and seepage forces, as done in the chapter.

Problem 17.4 Consider a simple vertical cut in a homogeneous soil with properties ϕ, γ, and c. The height of the cut is H. Assuming only planar slip surfaces, what is the factor of safety of this cut? What value of H would lead to its collapse? Use limit equilibrium in your solution.

Problem 17.5 Redo Problem 17.4 using lower and upper bound analyses.

17.8.2 Quantitative problems

Problem 17.6 A vertical bank was formed during the excavation of a plastic clay having a unit weight of 16 kN/m³. When the depth of excavation reached 5.5 m, the bank failed. Assuming a planar slip surface, what is the value of undrained shear strength s_u that we can estimate for this clay from this failure?

Problem 17.7 Use the wedge method to analyze the stability of the slope shown in P-Figure 17.1. The pore water forces acting on wedge A and wedge B are 252 and 15 kN/m, respectively. The pore water forces have already been calculated by multiplying the average pore pressure at the base of each block by the length of the base of the block. The pore pressures were calculated from a flow net (not shown).

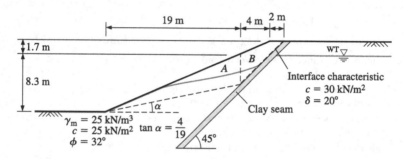

P-Figure 17.1 Slope of Problem 17.7.

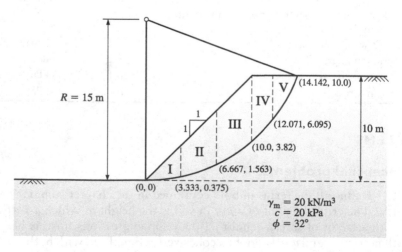

P-Figure 17.2 Slope of Problem 17.8.

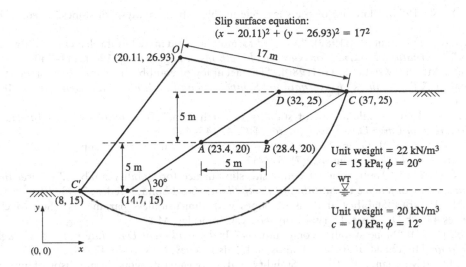

Slip surface equation:
$(x - 20.11)^2 + (y - 26.93)^2 = 17^2$

P-Figure 17.3 Slope of Problem 17.11.

Problem 17.8 Find the factor of safety for the circular slip surface shown in P-Figure 17.2 using the ordinary method of slices. The slope height is 10 m, and the slope inclination is 45°. The soil, which has a unit weight of 20 kN/m³, is cohesive-frictional with $c=20$ kPa and $\phi = 32°$. The water table is very deep. Consider five slices with base coordinates as shown in P-Figure 17.2. To calculate the weight of the slices, you may approximate them using trapezoids.

Problem 17.9 Redo Problem 17.8 using Bishop's simplified method.

Problem 17.10 Solve Example 17.5 using Bishop's simplified method. Solve it using both a spreadsheet (note that an iteration is required to determine the factor of safety) and a computer program of your choice.

17.8.3 Design problems

Problem 17.11 Computations show the slope with cross section as in P-Figure 17.3 to be unsafe. As an attempt to enhance the safety of the slope, the material occupying the volume $ABCD$ in the figure was excavated out. For the slip surface provided, calculate the factor of safety before and after excavation.

Problem 17.12 Use a computer program to find the minimum (critical) factor of safety for the slope of Problem 17.11 both before and after the excavation. Do your computations confirm the original conclusion that the slope was unsafe? Is the proposed measure sufficient to make it safe?

REFERENCES

References cited

Anderheggen, E. and Knöpfel, H. (1972). "Finite element limit analysis using linear programming." *International Journal of Solids and Structures*, 8(12), 1413–1431.

Baker, R. (1980). "Determination of the critical slip surface in slope stability computations." *International Journal for Numerical and Analytical Methods in Geomechanics*, 4(4), 333–359.

Bishop, A. W. (1955). "The use of the slip circle in the stability analysis of slopes." *Géotechnique*, 5(1), 7–17.

Casciaro, R. and Cascini, L. (1982). "A mixed formulation and mixed finite elements for limit analysis." *International Journal for Numerical Methods in Engineering*, 18(2), 211–243.

Duncan, J. M. and Wright, S. G. (1980). "The accuracy of equilibrium methods of slope stability analysis." *Proceedings of International Symposium on Landslides*, New Delhi, India, 16(1–2), 5–17.

Fellenius, W. (1936). "Calculation of stability of earth dam." *Transactions of 2nd International Congress on Large Dams*, Washington, DC, 4, 445–462.

Greco, V. R. (1996). "Efficient Monte Carlo technique for locating critical slip surface." *Journal of Geotechnical Engineering*, 122(7), 517–525.

Janbu, N. (1954a). "Application of composite slip surface for stability analysis." *Proceedings of European Conference on Stability of Earth Slopes*, Stockholm, Sweden, 3, 43–49.

Janbu, N. (1954b). "Stability analysis of slopes with dimensionless parameters." Soil Mechanics Series 46, Ph.D. Thesis, Harvard University, Cambridge, MA.

Janbu, N. (1973). "Slope stability computations." In *Embankment Dam Engineering: Casagrande Volume*. Hirschfeld, R. C. and Poulos, S. J. (eds.), Wiley, New York, 47–86.

Jiang, G. J. and Magnan, J. P. (1997). "Stability analysis of embankments: Comparison of limit analysis with method of slices." *Géotechnique*, 47(4), 857–872.

Kim, J., Salgado, R., and Yu, H. S. (1999). "Limit analysis of soil slopes subjected to pore-water pressures." *Journal of Geotechnical and Geoenvironmental Engineering*, 125(1), 49–58.

Kim, J., Salgado, R., and Lee, J. (2002). "Stability analysis of complex soil slopes using limit analysis." *Journal of Geotechnical and Geoenvironmental Engineering*, 128(7), 546–557.

Leshchinsky, D. and Huang, C.-C. (1992). "Generalized slope stability analysis: Interpretation, modification, and comparison." *Journal of Geotechnical Engineering*, 118(10), 1559–1576.

Loukidis, D., Bandini, P., and Salgado, R. (2003). "Stability of seismically loaded slopes using limit analysis." *Géotechnique*, 53(5), 463–479.

Lowe, J. and Karafiath, L. (1960). "Stability of earth dams upon drawdown." *Proceedings of 1st Pan American Conference on Soil Mechanics and Foundation Engineering*, Mexico City, Mexico, 537–552.

Lyamin, A. V. (1999). "Three-dimensional lower bound limit analysis using non-linear programming." Ph.D. Thesis, Department of Civil, Surveying and Environmental Engineering, University of New Castle, New South Wales, Australia.

Lysmer, J. (1970). "Limit analysis of plane problems in soil mechanics." *Journal of the Soil Mechanics and Foundations Division*, 96(4), 1311–1334.

Morgenstern, N. R. and Price, V. E. (1965). "The analysis of the stability of general slip surfaces." *Géotechnique*, 15(1), 79–93.

Munro, J. (1982). "Plastic analysis in geomechanics by mathematical programming." *Proceedings of the NATO Advanced Study Institute: Numerical Methods in Geomechanics*, Vimeiro, Portugal, 4, 247–272.

Nash, D. (1987). "Chapter 2: A comparative review of limit equilibrium methods of stability analysis." In *Slope Stability: Geotechnical Engineering and Geomorphology*. Anderson, M. G. and Richards, K. S. (eds.), Wiley, New York, 11–75.

Pastor, J., Thai, T. H., and Francescato, P. (2000). "New bounds for the height limit of a vertical slope." *International Journal for Numerical and Analytical Methods in Geomechanics*, 24(2), 165–182.

Salgado, R. and Bandini, P. (1999). PCSTABL User Manual.

Salgado, R. and Kim, D. (2014). "Reliability analysis of load and resistance factor design of slopes." *Journal of Geotechnical and Geoenvironmental Engineering*, 140(1), 57–73.

Siegel, R. A., Kovacs, W. D., and Lovell, C. W. (1981). "Random surface generation in stability analysis." *Journal of the Geotechnical Engineering Division*, 107(7), 996–1002.

Skempton, A. W. and Delory, F. A. (1957). "Stability of natural slopes in London clay." *4th International Conference on Soil Mechanics and Foundation Engineering*, London, UK, 378–381.

Sloan, S. W. (1988a). "Lower bound limit analysis using finite elements and linear programming." *International Journal for Numerical and Analytical Methods in Geomechanics*, 12(1), 61–77.

Sloan, S. W. (1988b). "A steepest edge active set algorithm for solving sparse linear programming problems." *International Journal for Numerical Methods in Engineering*, 26(12), 2671–2685.

Sloan, S. W. (1989). "Upper bound limit analysis using finite elements and linear programming." *International Journal for Numerical and Analytical Methods in Geomechanics*, 13(3), 263–282.

Sloan, S. W. (2013). "Geotechnical stability analysis." *Géotechnique*, 63(7), 531–572.

Sloan, S. W. and Kleeman, P. W. (1995). "Upper bound limit analysis using discontinuous velocity fields." *Computer Methods in Applied Mechanics and Engineering*, 127(1–4), 293–314.

Spencer, E. (1967). "A method of analysis of the stability of embankments assuming parallel inter-slice forces." *Géotechnique*, 17(1), 11–26.

Sultan, H. A. and Seed, H. B. (1967). "Stability of sloping core earth dams." *Journal of the Soil Mechanics and Foundations Division*, 93(4), 45–67.

Take, W. A., Bolton, M. D., Wong, P. C. P., and Yeung, F. J. (2004). "Evaluation of landslide trigger-ing mechanisms in model fill slopes." *Landslides*, 1(3), 173–184.

Taylor, D. W. (1948). *Fundamentals of Soil Mechanics*. Wiley, New York.

Yu, H. S., Salgado, R., Sloan, S. W., and Kim, J. M. (1998). "Limit analysis versus limit equilibrium for slope stability." *Journal Geotechnical and Geoenvironmental Engineering*, 124(1), 1–11.

Additional references

Arai, K. and Tagyo, K. (1985). "Determination of noncircular slip surface giving the minimum factor of safety in slope stability analysis." *Soils and Foundation*, 25(1), 43–51.

Bjerrum, L. (1967). "Progressive failure in slopes of overconsolidated plastic clay and clay shales." *Journal of the Soil Mechanics and Foundations Division*, 93(SM5), 1–49.

Bjerrum, L. and Kenney, T. C. (1967). "Effect of structure on the shear behaviour of normal consoli-dated quick clays." *Proceedings of Geotechnical Conference on Shear Strength Properties of Natural Soils and Rocks*, Oslo, Norway, 2, 19–27.

Cedergren, H. R. (1967). *Seepage, Drainage and Flow Nets*. 1st Edition. Wiley, New York.

Chen, W. F. (1975). *Limit Analysis and Soil Plasticity*. 1st Edition, Elsevier, Amsterdam, Netherland.

Cousins, B. F. (1978). "Stability charts for simple earth slopes." *Journal of the Geotechnical Engineering Division*, 104(2), 267–279.

Freeze, R. A. and Cherry, J. A. (1979). *Groundwater*, Prentice Hall, Englewood Cliffs, NJ.

Gibson, R. E. and Morgenstern, N. R. (1962). "A note on the stability of cuttings in normally consoli-dated clays." *Géotechnique*, 12(3), 212–216.

Janbu, N. (1957). "Earth pressures and bearing capacity calculations by generalized procedure of slices." *Proceedings of 4th International Conference on Soil Mechanics and Foundation Engineering*, London, UK, 2, 207–212.

Manzari, M. T. and Nour, M. A. (2000). "Significance of soil dilatancy in slope stability analysis." *Journal of Geotechnical and Geoenvironmental Engineering*, 126(1), 75–80.

Michalowski, R. L. (1995). "Slope stability analysis: A kinematic approach." *Géotechnique*, 45(2), 283–293.

Michalowski, R. L. (2002). "Stability charts for uniform slopes." *Journal of Geotechnical and Geoenvironmental Engineering*, 128(4), 351–355.

Ramalho-Ortigao, J. A., Werneck, M. L., and Lacerda, W. A. (1983). "Embankment failure on clay near Rio de Janeiro." *Journal of Geotechnical Engineering*, 109(11), 1460–1479.

Skempton, A. W. (1977). "Slope stability of cuttings in brown London clay." *Proceedings of 9th International Conference on Soil Mechanics and Foundation Engineering*, Tokyo, Japan, 3, 261–270.

Sridevi, B. and Deep, K. (1992). "Application of global-optimization technique to slope-stability analysis." *Proceedings of 6th International Symposium on Landslides*, Christchurch, New Zealand, 573–578.

Appendix A: Unit conversions

Multiply	By	To obtain
Length		
Centimeter	0.01	Meter
Yard	0.9144	Meter
Meter	3.281	Foot
Meter	39.37	Inch
Meter	1.094	Yard
Meter	6.214×10^{-4}	Mile
Meter	100	Centimeter
Mile	1609.3	Meter
Foot	0.3048	Meter
Foot	12	Inch
Inch	0.0254	Meter
Area		
Acre	4046.856	Square meter
Square centimeter	1×10^{-4}	Square meter
Square meter	10.764	Square foot
Square meter	1×10^{4}	Square centimeter
Square meter	1550	Square inch
Square meter	3.86×10^{-7}	Square mile
Square mile	2.59×10^{6}	Square meter
Square foot	0.0929	Square meter
Square inch	6.452×10^{-4}	Square meter
Volume		
Barrel	0.159	Cubic meter
Barrel	42	Gallon
Barrel	5.614	Cubic foot
Cubic centimeter	1×10^{-6}	Cubic meter
Gallon	2.38×10^{-2}	Barrel
Gallon	3.785	Liter
Gallon	3.785×10^{-3}	Cubic meter
Liter	6.29×10^{-3}	Barrel
Liter	1×10^{3}	Cubic centimeter
Liter	3.531×10^{-2}	Cubic foot

(Continued)

Multiply	By	To obtain
Liter	0.2642	Gallon
Cubic meter	6.29	Barrel
Cubic meter	1×10^6	Cubic centimeter
Cubic meter	264.2	Gallon
Cubic meter	1000	Liter
Cubic meter	61,023	Cubic inch
Cubic meter	35.31	Cubic foot
Cubic foot	0.1781	Barrel
Cubic foot	1728	Cubic inch
Cubic foot	2.832×10^{-2}	Cubic meter
Cubic foot	7.481	Gallon
Cubic foot	28.32	Liter
Cubic inch	1.639×10^{-5}	Cubic meter
Mass		
Gram	1×10^{-3}	Kilogram
Pound	453.6	Gram
Kilogram	1000	Gram
Kilogram	2.205	Pound
Kilogram	0.102	MTU
Kilogram	0.0685	Slug
Slug	14.59	Kilogram
Slug	32.174	Pound
Metric ton	1000	Kilogram
Ton (US)	2000	Pound
MTU	9.81	Kilogram
Mass Density		
Gram/cubic centimeter	1000	Kilogram/cubic meter
Gram/cubic centimeter	3.618×10^{-2}	Pound/cubic inch
Gram/cubic centimeter	62.43	Pound/cubic foot
Gram/cubic centimeter	8.34	Pound/gallon
Pound/cubic foot	1.602×10^{-2}	Gram/cubic centimeter
Pound/cubic foot	16.02	Kilogram/cubic meter
Pound/cubic inch	27,680	Kilogram/cubic meter
Pound/gallon	119.9	Kilogram/cubic meter
Kilogram/cubic meter	1×10^{-3}	Gram/cubic centimeter
Kilogram/cubic meter	6.243×10^{-2}	Pound/cubic foot
Kilogram/cubic meter	3.613×10^{-5}	Pound/cubic inch
Kilogram/cubic meter	1×10^{-3}	Kilogram/liter
Kilogram/cubic meter	8.34×10^{-3}	Pound/gallon
Force		
Dyne	1×10^{-5}	Newton
Dyne	1.0197×10^{-6}	Kilogram
Pound	0.4536	Kilogram
Pound	4.45	Newton
Pound	32.174	Poundal
Newton	1×10^5	Dyne

(Continued)

Multiply	By	To obtain
Newton	0.225	Pound
Newton	0.102	Kilogram
Newton	7.23	Poundal
Poundal	0.1383	Newton
Kilogram	980,665	Dyne
Kilogram	2.205	Pound
Kilogram	9.81	Newton
Pressure		
Atmosphere	14.7	Pound/square inch
Atmosphere	2116.8	Pound/square foot
Atmosphere	760	Millimeter of Hg
Atmosphere	10,333	Kilogram/square meter
Atmosphere	101,325	Pascal
Atmosphere	1.0333	Kilogram/square meter
Dyne/square centimeter	9.87×10^{-7}	Atmosphere
Dyne/square centimeter	1.02×10^{-2}	Kilogram/square meter
Dyne/square centimeter	1.45×10^{-5}	Pound/square inch
Dyne/square centimeter	0.1	Pascal
Pound/square inch	144	Pound/square foot
Pound/square inch	6.804×10^{-2}	Atmosphere
Pound/square inch	51.7	Millimeter of Hg
Pound/square inch	703.1	Kilogram/square meter
Pound/square inch	0.0703	Kilogram/square centimeter
Pound/square inch	6895	Pascal (Pa)
Pound/square foot	0.0479	Kilopascal (kPa)
Millimeter of Hg	0.0193	Pound/square inch
Millimeter of Hg	1.36×10^{-3}	Kilogram/square centimeter
Millimeter of Hg	133.32	Pascal
Pascal	9.869×10^{-6}	Atmosphere
Pascal	10	Dyne/square centimeter
Pascal	1.45×10^{-4}	Pound/square inch
Pascal	7.5×10^{-3}	Millimeter of Hg
Pascal	1.02×10^{-5}	Kilogram/square centimeter
Kilopascal (kPa)	20.9	Pound/square foot
Kilogram/square centimeter	0.968	Atmosphere
Kilogram/square centimeter	14.22	Pound/square inch
Kilogram/square centimeter	980,665	Dyne/square centimeter
Kilogram/square centimeter	735	Millimeter of Hg
Kilogram/square centimeter	98,066	Pascal
Work		
BTU	777.5	Pound-foot
BTU	107.5	Kilogram-meter
BTU	1054.6	Joule
BTU	0.252	Kilocalorie
BTU	3.927×10^{-4}	Horsepower-hour
BTU	2.928×10^{-4}	Kilowatt-hour

(Continued)

Multiply	By	To obtain
Calorie	3.968×10^{-3}	BTU
Calorie	4.186	Joule
Horsepower-hour	2547	BTU
Horsepower-hour	2.684×10^{-6}	Joule
Horsepower-hour	2.737×10^{5}	Kilogram-meter
Horsepower-hour	0.7457	Kilowatt-hour
Joule	1×10^{7}	Erg
Joule	2.39×10^{-4}	Kilocalorie
Joule	0.102	Kilogram-meter
Joule	2.778×10^{-4}	Watt-hour
Joule	0.7376	Pound-foot
Joule	9.486×10^{-4}	BTU
Kilogram-meter	9.302×10^{-3}	BTU
Kilogram-meter	2.724×10^{-6}	Kilowatt-hour
Kilogram-meter	9.81	Joule
Kilowatt-hour	3415	BTU
Kilowatt-hour	2.655×10^{-6}	Pound-foot
Kilowatt-hour	1.341	Horsepower-hour
Kilowatt-hour	3.6×10^{-6}	Joule
Kilowatt-hour	860.5	Kilocalorie
Watt-hour	3.6×10^{-9}	Joule
Power		
BTU/hour	3.927×10^{-4}	Horsepower
BTU/hour	0.293	Watt
Horsepower	42.44	BTU/minute
Horsepower	33,000	Pound-foot/minute
Horsepower	10.7	Kilocalorie/minute
Horsepower	745.7	Watt
Kilocalorie/minute	69.72	Watt
Kilowatt	1000	Watt
Kilowatt	1.341	Horsepower
Watt	59.62	BTU-minute
Watt	44.26	Pound-foot/minute
Watt	1.341×10^{-3}	Horsepower
Watt	1.434×10^{-2}	Kilocalorie/minute
Viscosity		
Centipoise	0.01	Poise
Centipoise	2.42	Pound/foot-hour
Centipoise	6.72×10^{-4}	Pound/foot-second
Centipoise	3.60	Kilogram/meter-hour
Centipoise	0.001	Pascal-second
Centipoise	0.01	Dyne-second/square centimeter
Pound/foot-second	1488	Centipoise
Pound/foot-second	1.488	Pascal-second
Pascal-second	1000	Centipoise
Pascal-second	0.672	Pound/foot-second

Appendix B: Useful relationships and typical values of various quantities

TEMPERATURE

$$\frac{°C}{100} = \frac{°F - 32}{180} = \frac{K - 273}{100} = \frac{°R - 492}{180}$$

WATER PRESSURE GRADIENT

$0.1\,\dfrac{kgf/cm^2}{m}$	$0.433\,psi/ft$
$0.42\,psi/m$	$9.81\,kPa/m$

ACCELERATION OF GRAVITY AT SEA LEVEL

$9.806\,m/s^2$	$32.174\,ft/s^2$

MASS DENSITY OF WATER

$1000\,kg/m^3$	$1.0\,kg/dm$
$1.0\,g/cm^3$	$62.4\,pcf$
$8.34\,lb/gal$	$350\,lb/bbl$

TYPICAL MASS DENSITY/UNIT WEIGHT OF SOILS

Sands:

115–135 pcf
18–21 kN/m³
Up to 22 kN/m³ with some gravel content

Clays:

90–120 pcf
14–19 kN/m³

TYPICAL VOID RATIOS OF SANDS

$e_{min} = 0.38 - 0.5$

$e_{max} = 0.75 - 1$

DEGREE OF COMPACTNESS VERSUS RELATIVE DENSITY OF SANDS

D_R (%)	Qualitative assessment of degree of compactness
0–15	Very loose
15–35	Loose
35–65	Medium dense
65–85	Dense
85–100	Very dense

COEFFICIENT OF LATERAL EARTH PRESSURE AT REST

$K_{0,NC} = 0.38-0.5$ in sands
$K_{0,NC} = 0.5-0.75$ in clays
$K_0 \approx K_{0,NC}\sqrt{OCR}$

TYPICAL VOID RATIOS OF CLAYS

From $e = 2$ (extremely soft, weak clays) to $e = 0.7$ (very stiff clays).
There are extreme examples of softer clays (with e as high as 5) and stiffer clays.

CRITICAL-STATE FRICTION ANGLE ϕ_c

Silica sands: 28°–36°
Clays: 15°–30°

RESIDUAL FRICTION ANGLE ϕ_r IN CLAYS

As low as 5°–7° for smectites. For low confining stress levels and/or large sand content, ϕ_r can be as high as ϕ_c.

PEAK FRICTION ANGLE ϕ_p IN SAND

$$\phi_p = \phi_c + A_\psi I_R = \phi_c + \left[5 - 2(k-1)\right]I_R$$

$$I_R = I_D\left[Q - \ln\left(\frac{100\sigma'_{mp}}{p_A}\right)\right] - R_Q$$

$$\sigma'_{mp} = \frac{1}{k+1}(\sigma'_1 + k\sigma'_3)$$

$Q = 10$, $R_Q = 1$ for silica sands

$$k = \begin{cases} 1 & \text{for plane-strain compression} \\ 2 & \text{for triaxial compression} \end{cases}$$

UNDRAINED SHEAR STRENGTH INCREASE WITH DEPTH FOR CLAY

$$\left(\frac{s_u}{\sigma'_v}\right)_{NC} = 0.15 - 0.30$$

$$\left(\frac{s_u}{\sigma'_v}\right) = \left(\frac{s_u}{\sigma'_v}\right)_{NC} OCR^{0.8}$$

SMALL-STRAIN (MAXIMUM) SHEAR MODULUS AND POISSON'S RATIO

$$\left(\frac{G_0}{p_A}\right) = C_g \frac{(e_g - e)^2}{1+e}\left(\frac{\sigma'_m}{p_A}\right)^{n_g} OCR^{m_g}$$

with typical values $C_g = 650$, $e_g = 2.17$, $n_g = 0.45$, and $m_g = 0$ for sands and up to 0.5 for clays
$\nu = 0.1 - 0.3$ (drained)
$\nu = 0.5$ (undrained)

ATTERBERG INDICES

LL = liquid limit (water content at which soil transitions to a "liquid") = 30%–160% for clays
PL = plastic limit (water content at which soil transitions from a "brittle" to a "plastic solid") = 20%–50%
PI = Plasticity index = $LL - PL$

COMPRESSION INDEX C_c AND RECOMPRESSION INDEX C_s OF CLAY

$$C_c = \frac{1}{200}G_s PI(\%)$$

$$C_s = 0.1C_c - 0.2C_c$$

COEFFICIENT OF CONSOLIDATION FOR CLAY

$c_v = 10^{-8}$ to 5×10^{-7} m^2/s

COEFFICIENT OF SECONDARY COMPRESSION
FOR SHALE, MUDSTONE, CLAY, AND PEAT

$$\frac{C_\alpha}{C_c} = 0.02 - 0.07$$

(lower values for shale and mudstone; higher values for peat)

APPROXIMATE VALUES OF HYDRAULIC CONDUCTIVITY

$$K = \begin{cases} 10^{-3} - 1 \text{ m/s for gravel} \\ 10^{-7} - 10^{-2} \text{ m/s for sand} \\ 10^{-9} - 10^{-5} \text{ m/s for silt} \\ 10^{-13} - 10^{-9} \text{ m/s for clay} \end{cases}$$

Appendix C: Measurement of hydraulic conductivity in the laboratory using the falling-head permeameter

In the falling-head permeameter, a pipe is filled with water and the water is allowed to flow through a soil sample, as depicted in Figure C.1. If the area of the pipe is A_p and the water level in the pipe (and thus the hydraulic head) drops dh in a period of time dt, then the rate of water flow through the sample is equal to:

$$Q = -A_p \frac{dh}{dt} \qquad (C.1)$$

where the minus sign accounts for the fact that a negative dh (a drop in water level) generates positive flow through the soil sample.

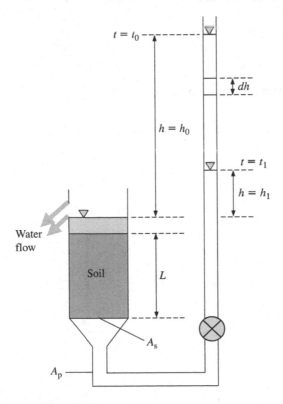

Figure C.I Schematic diagram of falling-head permeameter.

The flow rate may also be determined using Darcy's law:

$$Q = -A_s K i = -A_s K \frac{\Delta h}{L} \tag{C.2}$$

Conservation of mass requires that both flow rates be the same. Combining Equations (C.1) and (C.2):

$$-A_p \frac{dh}{dt} = -A_s K \frac{\Delta h}{L}$$

Reorganizing and observing that we can arbitrarily set the head at the soil sample as zero, so that $\Delta h = -h$, we get the following differential equation:

$$-\frac{dh}{h} = \frac{A_s K}{A_p L} dt \tag{C.3}$$

which we can integrate to give:

$$\ln h = -\frac{A_s K}{A_p L} t + C \tag{C.4}$$

where C is an integration constant we can determine from the initial condition that $h = h_0$ at time $t = 0$. Taking this into Equation (C.4), we can determine C and then rewrite Equation (C.4) as:

$$\ln \frac{h}{h_0} = -\frac{A_s K}{A_p L} t \tag{C.5}$$

and, finally, we can write an equation for K:

$$K = \frac{A_p L}{A_s t} \ln \frac{h_0}{h} \tag{C.6}$$

Appendix D: Determination of preconsolidation pressure, compression and recompression indices, and coefficient of consolidation from consolidation test data

CASAGRANDE CONSTRUCTION

The Casagrande construction (Casagrande 1936) is used to estimate the preconsolidation pressure σ'_{vp} *in situ* from the results of a consolidation test. To use it, we need to plot void ratio versus vertical effective stress for the consolidation test, as shown in Figure D.1. The estimate of σ'_{vp} is then determined by following these steps:

1. Locate the point P on the e–log σ'_v curve where the curvature is maximum.
2. Draw a tangent T to the e–log σ'_v curve through point P.
3. Draw a horizontal line H through P.
4. Bisect the angle between H and P with line B.
5. Project the laboratory normal consolidation (straight) portion of the e–log σ'_v curve upward until it intersects line B at point Q.
6. The abscissa of point Q is log σ'_{vp}.

SCHMERTMANN'S GRAPHICAL CONSTRUCTION TO CORRECT FOR DISTURBANCE

Once σ'_{vp} has been determined, the *in situ* e–log σ'_v curve may be estimated using Schmertmann's graphical construction (Schmertmann 1953), illustrated in Figure D.2 for normally consolidated clay and in Figure D.3 for overconsolidated clay. The steps are as follows:

1. Determine the in situ void ratio e_0 from the water content $(wc)_0$ using:

$$e_0 = (wc)_0 G_s$$

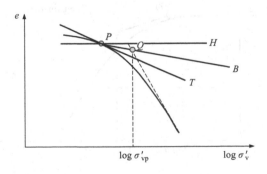

Figure D.1 Casagrande graphical construction to determine the preconsolidation stress σ'_{vp}.

2. Compute the *in situ* vertical effective stress σ'_{v0} from the soil profile (including the position of the water table). This may be equal to σ'_{vp} (in which case the soil is normally consolidated and Figure D.2 applies), less than σ'_{vp} (in which case the soil is overconsolidated and Figure D.3 applies), or greater than σ'_{vp} (in which case the soil is undergoing consolidation, which requires special considerations, as discussed in Chapter 6).

3. Plot the point O ($\log \sigma'_{v0}, e_0$).

4. Plot the point R, which is the intersection of the horizontal line $e = 0.42e_0$ with the extension of the laboratory normal consolidation (straight) portion of the e–$\log \sigma'_v$ curve.

5. Plot the compression line from point Q to point R (point Q is the same point determined using the Casagrande construction).

6. The compression line is the *in situ* normal consolidation line, with slope C_c, where C_c is the compression index discussed in Chapter 6.

7. The line from point O to point Q is a recompression line, with slope equal to C_s, where C_s is the recompression index discussed in Chapter 6.

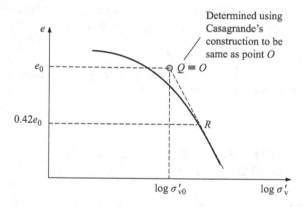

Figure D.2 Schmertmann's graphical construction for normally consolidated clay.

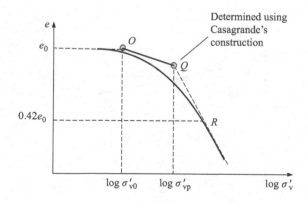

Figure D.3 Schmertmann's graphical construction for overconsolidated clay.

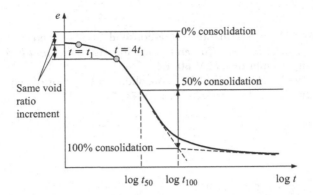

Figure D.4 Variation of void ratio with the logarithm of time for a given stress increment $\Delta\sigma_v$ in a consolidation test.

ESTIMATION OF c_v USING THE LOGARITHM-OF-TIME METHOD

Figure D.4 shows how the void ratio of a sample varies with time after the application of a stress increment $\Delta\sigma_v$ in a consolidation test. The initial portion of the plot is in most cases approximately parabolic in shape. This fact can be used to determine the void ratio at the moment the stress increment is applied ($t = 0$), or, equivalently, at zero consolidation. For that, we take any two values of void ratio at times t_1 and $4t_1$ in the range where the curve is parabolic. Say the difference between these two void ratios is Δe. The void ratio at zero consolidation is then simply the void ratio at time t_1 plus Δe.

After the determination of the void ratio at zero consolidation, we must find the void ratio at 100% consolidation. The procedure is illustrated in Figure D.4: we extend forward the portion of the plot that is linear with the log of time that immediately follows the parabolic portion of the plot and also, back in time, the linear portion that is found at the end of the test. The point where they cross marks 100% consolidation. The continuing drop in void ratio after 100% consolidation is secondary compression.

The void ratio at 50% consolidation is simply the average of the void ratios at 0% and 100% consolidation. The time corresponding to that on the e–log t plot is t_{50}. Now we can use our knowledge from Chapter 6 that T_v corresponding to 50% consolidation is 0.196 to determine c_v using the definition of the time factor:

$$T_v = \frac{c_v t}{H^2}$$

from which:

$$c_v = \frac{H^2 \left(T_v\right)_{50}}{t_{50}} = 0.196 \frac{H^2}{t_{50}}$$

where H is the height of the sample.

The values of c_v are determined for all the stress increments so that a relationship between c_v and the stress level can be established and used in deciding which value of c_v to use in solving a given consolidation problem.

REFERENCES

Casagrande, A. (1936). "The determination of pre-consolidation load and it's practical significance." *Proceedings of 1st International Conference on Soil Mechanics and Foundation Engineering,* Harvard University, Cambridge, MA, 60–64.

Schmertmann, J. H. (1953). "The undisturbed consolidation behavior of clay." *Transactions of the American Society of Civil Engineers,* 120, 1208–1216.

Appendix E: Stress rotation analysis

In this appendix, we use stress rotation analysis to develop an expression that relates the major principal stresses in two zones A and B between which the stresses rotate (Figure E.1). The formulation is general in that, if the friction angle is positive, it is applicable to effective stresses, and, if the friction angle is equal to zero, it applies to total stresses. Accordingly, no primes are used to denote effective stresses as the context defines when total stresses or effective stresses apply.

The zones A and B, in plane-strain conditions, usually possess a line in common, which shows as a projected point O in Figure E.1, and are separated by a transition zone T. Stress rotation takes place continuously across zone T. That means that (1) there can only be an infinitesimal stress rotation equal to $d\Psi$ across a plane X' inside zone T, such as the one illustrated in Figure E.2, and (2) the apex angle of the transition zone, measured at O, must be exactly equal to the angle that the major principal stress in zone A makes with its counterpart in zone B. If either of these is not true, then the stresses must have had a finite rotation across one or more planes, and that is contrary to our assumption of continuous stress rotation.

In the Mohr diagram corresponding to the infinitesimal stress rotation across plane X' there are two infinitely close circles, each associated with one of the zones separated by plane X', both tangent to the failure envelope at point X' with coordinates (σ_p, τ_p) in Figure E.3. This situation can be pictured as the limit when the Mohr circles in Figure E.4 are drawn increasingly close to each other so that their point of intersection X tends to point X' in Figure E.3. The angle of major principal stress rotation in the situation described by Figure E.1 is half the difference between the central angles XBD and XAC in Figure E.4,

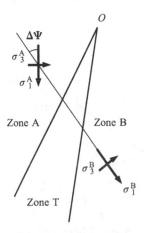

Figure E.1 Stress rotation between zones A and B takes place in the transition zone T.

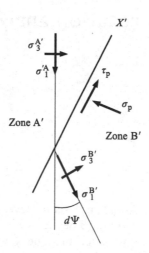

Figure E.2 Infinitesimal stress rotation across a plane X'.

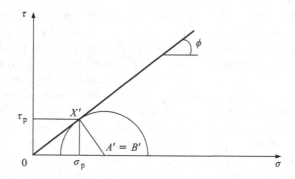

Figure E.3 Mohr's circle of stresses for two zones, A' and B', separated by plane X' across which an infinitesimal stress rotation occurs. Point X' in the diagram corresponds to the stresses acting on plane X'.

which is in agreement with the property of Mohr circles that angles that define stress directions appear in Mohr circles as central angles twice as large as the original angles. The major principal stress rotation $\Delta\Psi$ is accordingly indicated in Figure E.4.

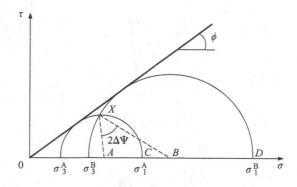

Figure E.4 Mohr's circles for two zones A and B between which a stress rotation takes place across a transition zone T.

Let the plane-strain stress variable s be given by

$$s = \frac{\sigma_1 + \sigma_3}{2}$$

(E.1)

so that it is also the abscissa of the center of a generic Mohr's circle defined by σ_1 and σ_3.
   Applying the law of sines to triangle ABX (Bolton 1979):

$$\frac{\sin 2\Delta\Psi}{AB} = \frac{\sin A\hat{B}X}{AX}$$

(E.2)

In the limit, as the circles are drawn infinitesimally close to each other,

$$\sin 2\Delta\Psi = \sin 2d\Psi = 2d\Psi$$

(E.3)

$$\sin A\hat{B}X = \cos\phi$$

(E.4)

$$AX = OA\sin\phi = s\sin\phi$$

(E.5)

$$AB = ds$$

(E.6)

Expression (E.2) becomes

$$\frac{2d\Psi}{ds} = \frac{\cos\phi}{s\sin\phi}$$

which is reorganized into

$$\frac{ds}{s} = 2\tan\phi\, d\Psi$$

(E.7)

As the transition zone T in Figure E.1 is a succession of transition planes, Equation (E.7) may be integrated to yield the value of the plane-strain stress variable s in zone B in Figure E.1 in terms of its value (s_A) in zone A and the stress rotation $\Delta\Psi$ between zones A and B:

$$\ln\frac{s}{s_A} = 2\Delta\Psi\tan\phi$$

(E.8)

or, alternatively,

$$s = s_A e^{2\Delta\Psi\tan\phi}$$

(E.9)

and so s, given by

$$s = \frac{1}{2}(\sigma_3 + \sigma_1) = \frac{1}{2}\sigma_1\left(\frac{\sigma_3}{\sigma_1} + 1\right) = \frac{1}{2}\sigma_1\left(\frac{1}{N} + 1\right)$$

relates to σ_1 through

$$\frac{s}{\sigma_1} = \frac{1}{2}\left(\frac{1}{N}+1\right) = \text{constant}$$

and we may write $\sigma_1/\sigma_1^A = s/s_A$. It follows that Equation (E.9) may be rewritten as

$$\sigma_1 = \sigma_1^A e^{2\Delta\Psi\tan\phi} \tag{E.10}$$

REFERENCE

Bolton, M. D. (1979). *A Guide to Soil Mechanics*. 1st Edition, MacMillan Education, London, UK.

Index